Mechanics of Solids: An Introduction

Made in the USA
Middletown, DE
18 March 2023

26950789R00119

Mechanics of Solids

AN INTRODUCTION

T. J. LARDNER R. R. ARCHER

Structural Engineering and Mechanics, Department of Civil Engineering
University of Massachusetts at Amherst

McGRAW-HILL, INC.

New York St. Louis San Francisco Auckland Bogotá Caracas Lisbon
London Madrid Mexico City Milan Montreal New Delhi
San Juan Singapore Sydney Tokyo Toronto

DISCLAIMER: This book includes a diskette containing eleven programs useful for the analysis of problems in the mechanics of solids. Neither the authors nor the publisher shall be held liable for any damages whatsoever resulting in connection with or arising from the use of any programs in this book. Questions about the programs can be addressed to the authors at FAX: 413-545-2840.

MECHANICS OF SOLIDS: AN INTRODUCTION

This book is printed on acid-free paper.

2 3 4 5 6 7 8 9 0 VNH VNH 9 0 9 8 7 6 5 4

P/N 036403-6
PART OF
ISBN 0-07-833358-X

This book was set in Century Oldstyle by Monotype Composition Company.
The editors were John J. Corrigan and Jack Maisel;
the designer was Joan Greenfield;
the production supervisor was Elizabeth J. Strange.
Von Hoffmann Press, Inc., was printer and binder.

Library of Congress Cataloging-in-Publication Data

Lardner, Thomas J.
Mechanics of solids: an introduction / T. J. Lardner, R. R. Archer.
p. cm.
Includes bibliographical references and index.
ISBN 0-07-833358-X (set)
1. Mechanics, Applied. 2. Solids. I. Archer, Robert R.
II. Title.
TA350.L32 1994
620.1'05—dc20 93-2372

About the Authors

T. J. LARDNER received his B. Aero. Eng. in 1958, his M.S. in 1959, and his Ph.D. in 1961 from the Polytechnic Institute of New York. After serving two years in the United States Army, he joined the faculty of the Massachusetts Institute of Technology in 1963 as an instructor in mathematics, becoming assistant professor of applied mathematics and then associate professor of mechanical engineering. In 1973 he joined the faculty at the University of Illinois as professor of theoretical and applied mechanics. Since 1978 he has been on the faculty of the University of Massachusetts at Amherst.

He has published over 80 papers in the general areas of solid and structural mechanics, mechanical behavior of materials, and applied mathematics, among other subjects.

He was an editor and contributing author for *An Introduction to the Mechanics of Solids* (McGraw-Hill, 1978).

He is a fellow of the American Society of Mechanical Engineers.

R. R. ARCHER received his S.B. in 1952 and his Ph.D. in 1956 from the Massachusetts Institute of Technology. He joined the faculty of MIT in 1956 as assistant professor of mechanical engineering. From 1959 to 1961 he was assistant and then associate professor of mathematics at the University of Massachusetts at Amherst. In 1961 he joined the faculty of Case Institute of Technology as associate professor of civil engineering. Finally, in 1966 he returned to the University of Massachusetts at Amherst as professor of civil engineering.

He has published over 60 papers in areas of structural mechanics, applied mathematics, and the analysis of growth mechanics of trees, among other subjects.

He was a contributing author of *An Introduction to the Mechanics of Solids* (McGraw-Hill, 1978) and author of *Growth Stresses and Strains in Trees* (Springer-Verlag, 1987).

He is a fellow of the American Society of Mechanical Engineers.

TO ANNE JEANNE, AND NAN

Contents

CHAPTER 3 Torsion of Circular Shafts 115

CHAPTER 4 Shear Forces and Bending Moments in Beams 176

CHAPTER 5 Stresses due to Bending 255

Preface

This is a textbook that provides an introduction to the subject of the mechanics of solids. As a course in the engineering curriculum, mechanics of solids comes after a course in statics in which the concepts of force, moment, and equilibrium equations are introduced for the analysis of *rigid* bodies. These concepts are developed in statics by using free-body diagrams to master techniques for the solution of equilibrium problems for simple structures.

Mechanics of solids extends the analysis of statics to include the study of the deformation of the materials making up the structures. The response of the materials treated in this textbook will be "solidlike," so that it is appropriate to think of this textbook as covering the application of the principles of mechanics to solids. An excellent overview of the mechanics of solids by James R. Rice, one of the outstanding practitioners in mechanics, can be found in the article *Mechanics* in the 1993 edition of the Encyclopedia Britannica.

It is the purpose of this textbook to build on the concepts of statics with three general goals in mind: (1) to develop the major concepts of the deformation of solid elastic materials under load, (2) to develop a systematic approach to solving problems of deformable bodies that lays out a general, effective method in place of ad hoc approaches for different problems, and (3) to present methods which lend themselves to computer-assisted problem solving.

We present a methodology for problem solving that draws on the three-step approach introduced in an earlier textbook, *An Introduction to the Mechanics of Solids,* 2d ed., edited by S. H. Crandall, N. C. Dahl, and T. J. Lardner (McGraw-Hill, New York, 1978). Our combined almost fifty years of involvement with this textbook as authors and as an editor has greatly influenced our approach to a number of topics in the study of Mechanics of Solids.

The three-step approach introduced in Chap. 2 of using equilibrium equations, force-deformation relations, and geometry arguments to formulate and set up governing equations is a powerful approach. We carry this approach throughout the text. In so doing we develop the governing equations for a solution to a problem in a systematic and organized fashion. We emphasize the use of coordinate axes and sign conventions associated with the axes as an important part of the formulation of the solution to an engineering problem.

Once we have developed a systematic approach to problem solving, we turn to the use of computer techniques for certain classes of problems. The use of computer methods is a unique feature of this textbook and the accompanying diskette. The class of problems selected for the use of computer methods includes those for which—*once the fundamental concepts are understood*—the details of obtaining the solution can be tedious. We advocate the use of computer computation as an adjunct to obtaining a fundamental *understanding* of the mechanics of solids. As we demonstrate, a systematic formulation of a problem often leads naturally to computer implementation of the solution.

It might be argued that there is no time for the use of computers in a course in the mechanics of solids. However, we argue that personal computers are readily available for use, that engineers make considerable use of computers in practice, and that the solutions to many problems in mechanics of solids are more interesting when the tedium of excessive hand numerical computations is avoided. In addition, with the use of the computer, more thought can be given to the significance of the solution and to the effect on the solution of changing input parameters in the problem. Greater insight into the solution is also provided by making use of graphics to plot results in the solution. In our programs we use computer graphics to exhibit clearly the solution.

In developing the computer programs on the diskette, we had two objectives:

1. We wanted to demonstrate that certain classes of problems encountered in the mechanics of solids course could be solved interactively with the aid of a personal computer. Considerable effort has gone into discussions of the underlying methods and derivations upon which the computer programs are based. Detailed examples are worked out which make the reader familiar with the systematic notation and standardized procedures which are then translated to a computer program on the diskette.
2. We wanted to enrich the present course in the mechanics of solids by providing an interactive computer experience that would be a *natural* part of the course. It is obvious that engineering students are going to make considerable use of computer-assisted analysis and design in their remaining years of schooling and for their careers in engineering. We felt that it is important to show students early in their career how certain methods of stress and displacement analysis are readily translated to computer programs. But equally important in using com-

puter-assisted analysis is the opportunity to show how easily solutions can be obtained corresponding to changes in the values of loads or parameters of the structure. As a consequence, *design changes* can be explored more fully, and the programs can be used effectively with the text to introduce notions of a *design approach.*

We feel that the discussions of the programs enrich the teaching of a course in the mechanics of solids. Special topics that might have been displaced by including computer applications in a textbook at this level can be covered in the next level course in machine design or in structural analysis. We have emphasized material in this textbook that is usually covered in a one-semester course in the mechanics of solids, and as such, this textbook is focused on these topics. Note that the computer-related topics are included after the basic topics are covered in certain of the chapters, so that an instructor who does not wish to cover these subjects can simply skip over them.

Acknowledgments

The authors wish to thank their students and colleagues, especially Karl Jakus who contributed advice and criticism toward improving this work when parts of it were used in the classroom. Special mention should be made of the help which Tom Service gave us in the early stages of the work. The following reviewers provided detailed comments and suggestions for improvement: L. Bucciarelli, Massachusetts Institute of Technology; Daniel Haines, Manhattan College; Dewey H. Hodges, Georgia Institute of Technology; Robert E. Miller, University of Illinois at Urbana-Champaign; Michael E. Plesha, University of Wisconsin at Madison; Michael Santare, University of Delaware; Robert Sennett, California Polytechnic State University; Carl Vilmann, Michigan Technological University; and George Voyiadjiis, Louisiana State University.

Borliang Chen, Hsiaocheng Chen, Weigun Gu, Hsin-Hsi Lu, Pam Stephan, and Wei-Jong Sun provided expert help in word processing and line drawings. Sao-Jeng Chao and Tsung-Ju Gwo worked with us in coding many of the computer programs included on the MECHMAT diskette. We also express our gratitude to William Highter for encouragement during the initial phase of this project. Some of the input/output formatting and menu structures were influenced by each of us having taught for several years from the textbook by Mario Paz, *Structural Dynamics,* 2d ed. (Van Nostrand Reinhold Co., New York, 1985).

T. J. Lardner
R. R. Archer

Mechanics of Solids: An Introduction

CHAPTER 1

Introduction to Stress and Strain

1.1 Introduction to the Mechanics of Solids

We are familiar from a course in Physics with the concepts of force and moment vectors and with the force and moment equilibrium equations used to analyze simple structures at rest. These concepts were developed further in a course in Statics in which the notions of *free-body diagrams* applied to simple structures were studied in detail. The free-body diagram of a structure or a portion of a structure is the pictograph that allows us to extract and write down easily in a systematic way the governing equations of equilibrium of the structure. In analyzing equilibrium of bodies at rest, we assumed that the bodies or portions thereof were made up of *rigid* materials in which no deformations or movements within the bodies occurred. Naturally, in real structural elements we expect materials to deform and change shape. As a consequence, we need to investigate the application of the concepts of force and moment equilibrium to deformable solid bodies; this is the main goal of this textbook.

Before we turn to this study of deformable bodies, it is worthwhile to note what we should know or at least remember from the study of Statics. The following list of topics covers the material we need to have mastered (or be prepared to review as appropriate):

- The use of significant figures and appropriate units in the solution of problems
- Force and moment vectors
- Reactions at supports in simple structures
- Free-body diagrams
- The use of equilibrium equations
- The idea of statically determinate and statically indeterminate prob-
- lems
- Centroids of composite plane areas; moments of areas
- Area moments of inertia of composite plane areas

The above list of topics is usually covered in a course in Statics. We will develop all the topics in this list further as we progress in this book. Appendix A presents a brief review of the calculation of centroids and moments of inertia for plane areas. However, we want to emphasize the importance of *free-body diagrams* and the use of force and moment equilibrium equations as the basic building blocks for subsequent work.

Historically, a course in the Mechanics of Solids was often referred to as *Strength of Materials.* This title no doubt arose from the need to know for simple structures if the structure had sufficient "strength" to support or carry the applied loads. S. P. Timoshenko's book on the *History of Strength of Materials* (McGraw-Hill, New York, 1953) provides an interesting account of the approaches taken to analyze structures early in the study of the subject. Timoshenko (1878–1972) himself was a major contributor to the subject of mechanics of solids and has written a number of classic books on the subject.

In a very early text on Mechanics, *An Introduction to Natural Philosophy* (by Denison Olmstead, 3d ed., New Haven, Conn., 1838), we find an attempt to address the issue of the Strength of Materials:

> The importance to the architect and the engineer of ascertaining the form and position of the materials which he employs, in order to secure the greatest degree of strength and stability at the least expense, has led mathematicians and writers on mechanics, to devote much attention to this subject. How is the strength of a beam affected by giving to it different shapes and different positions? How must a given quantity of matter be disposed in order that it may have the greatest possible strength? And upon what principles depends the stability of columns, roofs and arches? These, and many similar inquiries, have been the objects of profound investigation. . . . Strength is the power to *resist* fracture.

The questions posed above still exist today, and fortunately since that time (1838!) our understanding of the applications of mechanics has improved substantially. In addition, our understanding of materials and of *materials science* has progressed rapidly since about 1960, when newer materials began to be fabricated and newer techniques and instruments were developed to probe the internal structural detail of materials. New computerized laboratory equipment has given us the ability to determine the quantitative constants associated with material behavior in a convenient manner. Engineering and physics courses in materials and materials science are now common in the engineering curriculum and are often taken concurrently with a Mechanics of Solids course. Engineering courses in materials science emphasize the properties of materials,[1] while

[1]M. F. Ashby and D. R. H. Jones, *Engineering Materials,* vols. 1 and 2, Pergamon Press, New York, 1980.

T. H. Courtney, *Mechanical Behavior of Materials,* McGraw-Hill, New York, 1990.

W. F. Smith, *Principles of Materials Science and Engineering,* 2d ed., McGraw-Hill, New York, 1990.

courses in the mechanics of materials, such as the one using this textbook, emphasize the formulation and solution of problems, assuming the material properties are known. The line of demarcation between these two disciplines can become fuzzy, especially at a more advanced level.

In this textbook we emphasize the behavior of elastic solids under load by analyzing a number of structural components of increasing complexity. We begin with what we call one-dimensional problems and then proceed to two- and three-dimensional problems. Our approach is to build on intuition developed from the solution of simple problems as we go along. An alternate approach, which tends to build on solutions obtained from a more general formulation of the governing equations together with a more complete discussion of material behavior and failure modes, can be found, e.g., in *An Introduction to the Mechanics of Solids, with SI units,* 2d ed. (by S. H. Crandall, N. C. Dahl, and T. J. Lardner, McGraw-Hill, New York, 1978). The authors of this textbook are contributing authors and an editor of Crandall, Dahl, and Lardner.

Three books that enrich the appreciation of the mechanics of solids and materials and of problem solving are by J. E. Gordon:

The New Science of Strong Materials or Why You Don't Fall Through the Floor, 2d ed., Princeton University Press, 1984

Structures, or Why Things Don't Fall Down, Plenum Press, New York, 1978

The Science of Structures and Materials, Scientific American Library, New York, 1988

These three books by Gordon are recommended for supplementary reading to this textbook to provide a deeper understanding of materials and mechanics.

A more recent book—*Why Buildings Fall Down,* M. Levy and M. Salvadori, Norton, New York, 1992—provides engaging discussions of why structures in general fall down (!); much of the discussion relates to topics to be covered in this textbook. On the other hand, why buildings do not fall down is discussed in an earlier book by M. Salvadori: *Why Buildings Stand Up, The Strength of Architecture,* Norton, New York, 1980.

The book by M. Salvadori with R. Heller, *Structure in Architecture, The Building of Buildings,* 3d ed. (Prentice-Hall, Englewood Cliffs, N.J., 1986), also provides interesting reading on the nature of structures.

From these books and others—and we hope this textbook—we can draw the conclusion that the study of the mechanics of materials can be an exciting and rewarding experience.

As we solve problems in this textbook, we will emphasize the proper use of coordinate axes, sign conventions, and units. It is easy to assume, e.g., that the oft repeated concern over checking units in problems is a contrived college or university concern without application in the world beyond academia. However, at times we find in this world that even the apparently simple task of converting between units cannot be done without a need for attention to detail. One newspaper article describes how two airline pilots, in converting their fuel volume readings to total fuel weight, used the wrong conversion factors. As a consequence they thought that the figure for fuel weight on board their aircraft was in kilograms when it was really in pounds! Since 1 kg equals about 2.2 lb, the plane took off with about half the fuel required for the flight.[2] Examples can be found also in which plus and minus signs have been interchanged in computer instructions because of different assumed coordinate systems in the analyses. Axes, sign conventions, and units are important! And the need for free-body diagrams was even noted in a recent popular book on building a steel skyscraper.[3]

1.2 Uniaxial Normal Stress and Uniaxial Deformation

Since we are concerned with deformation in the study of the mechanics of solids, we need to ask how materials deform under load. In Sec. 1.5, we discuss in greater detail experimental techniques used to quantify the description of the behavior of materials. However, at this point we describe in general terms the determination of the constants used to characterize the behavior of elastic materials under axial loading.

Consider, as shown in Fig. 1.1*a*, a solid bar of elastic material of original length L_1 and cross-sectional area A_1 attached at its upper end through the fixtures to a rigid support. We assume that the weight of the bar and fixtures is negligible compared to the loads that we will apply to the lower end of the bar. If we apply a load P to the lower end, then we expect intuitively that the lower end of the bar will move downward in the direction of the load P, as in Fig. 1.1*b*. The amount of movement or the displacement of the lower end will be designated by the Greek letter delta: δ. As we increase the load, the value of δ will increase. When we remove the load P, the value of δ will return to zero; i.e., the bar will return to its original undeformed length L_1. We say that the material of the bar is *elastic,* or behaves in an elastic manner, if, after the load is removed, the bar returns to its original length without any permanent deformation.

---

[2] *New York Times,* "Airliner Ran Out of Fuel after Two Metric Errors," July 30, 1983, p. 7.

[3] K. Sabbagh, *Skyscraper, The Making of a Building,* Viking Press, New York, 1990, p. 111.

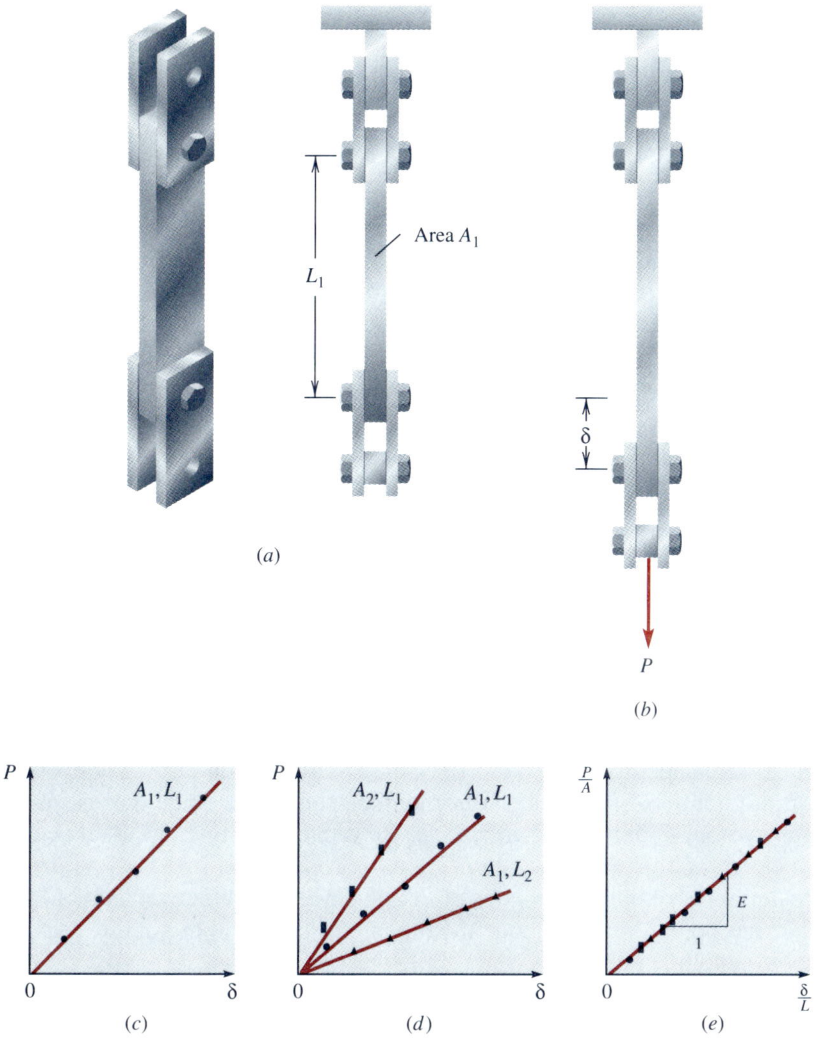

Figure 1.1 Tensile test: (*a*) specimen, (*b*) load P with displacement δ, (*c*) plot of load P versus displacement δ, (*d*) plot of P versus δ for different areas A and lengths L, (*e*) plot of P/A versus δ/L giving the elastic modulus E.

If we measure the value of the displacement δ for each value of the load P that we apply to the bar, we obtain a series of data points through which we can sketch a curve of load P versus displacement δ, as shown in Fig. 1.1*c*. When the load-displacement curve is linear, we say the material is *linear elastic*. Of course, as we will see later, we cannot continue to increase the load on the bar without causing large displacements leading to nonelastic behavior or even fracture of the bar. As long as the load is less than some critical value, the material will behave in a linear elastic fashion. We should emphasize at this point that in our sketches and drawings throughout the book we often exaggerate the displacement of points. For example, in Fig. 1.1*b* we show for clarity in the drawing a displacement at the lower end of the bar that is almost 25 percent of the original length L_1 in the undeformed state; such a large value for the elastic displacement is unrealistic in most engineering applications.

The load-displacement curve of Fig. 1.1*c* is for a bar of linear elastic material of length L_1 and cross-sectional area A_1. If we load another bar whose cross-sectional area is equal to a value greater than A_1 while the original undeformed length is still equal to L_1, we get the load-displacement curve marked A_2, L_1 shown in Fig. 1.1*d*. It appears intuitively obvious that increasing the area of the bar while keeping the undeformed length equal to L_1 will lead to a smaller displacement δ for the same load P. If we now load a bar of length L_2 greater than L_1 with an area equal to the original A_1, we get the load-displacement curve marked A_1, L_2 in Fig. 1.1*d*. Since we have a longer bar, there is more material to deform and the load-displacement curve is therefore below the curve for the area equal to A_1 and length equal to L_1. For the same load P the displacement is greater. It is clear that plotting load-displacement curves for different values of area A and length L might become tedious and bring us no closer to a simple means to characterize the material from which the bar is made.

However, if we now take the data from many different load-displacement curves, such as those in Fig. 1.1*d*, and plot the *load intensity* P/A on the vertical axis and the change in length δ divided by the original length L on the horizontal axis, we find that the different load-displacement curves fall approximately on the same straight line, as shown in Fig. 1.1*e*, where A is the original cross-sectional area of the bar and L is the original length of the bar. The slope E of the line in Fig. 1.1*e* depends on the nature of the material from which the bar is made and is given the symbol E, for elastic modulus. We often refer to E as the *Young's modulus* of the material, after Thomas Young (1773–1829). Different materials have different values of E, and since the horizontal axis of Fig. 1.1*e* has no units associated with it, the units of E are those of the vertical axis, namely, units of force per unit area.

Table 1.1 Typical Values for E

Material	E, psi	E, GPa
Tungsten	58×10^6	400
Aluminum oxide	47×10^6	325
Steel and iron	$28\text{–}30 \times 10^6$	194–205
Brass	15×10^6	103
Aluminum	10×10^6	69
Glass	10×10^6	69
Cast iron	$10\text{–}20 \times 10^6$	69–138
Wood	$1\text{–}2 \times 10^6$	6.9–13.8

Using the units of newton for force and meter for length, we find E has the units of newtons per square meter (N/m^2), which we call a *Pascal* (Pa). Using the units of pound for force and inch for length, we see that E has units of pounds per square inch (lb/in^2 or psi). For steel, E is approximately 200 GPa (gigapascals, or 10^9 N/m^2). Typical values for E for some materials are given in Table 1.1, and the range of values for different materials is shown in Fig. 1.2. Appendix F gives additional values for E.

The equation of the straight line in Fig. 1.1*e* is given by

$$\frac{P}{A} = E\frac{\delta}{L} \tag{1.1}$$

which can be solved for the displacement δ at the end of the bar in the form

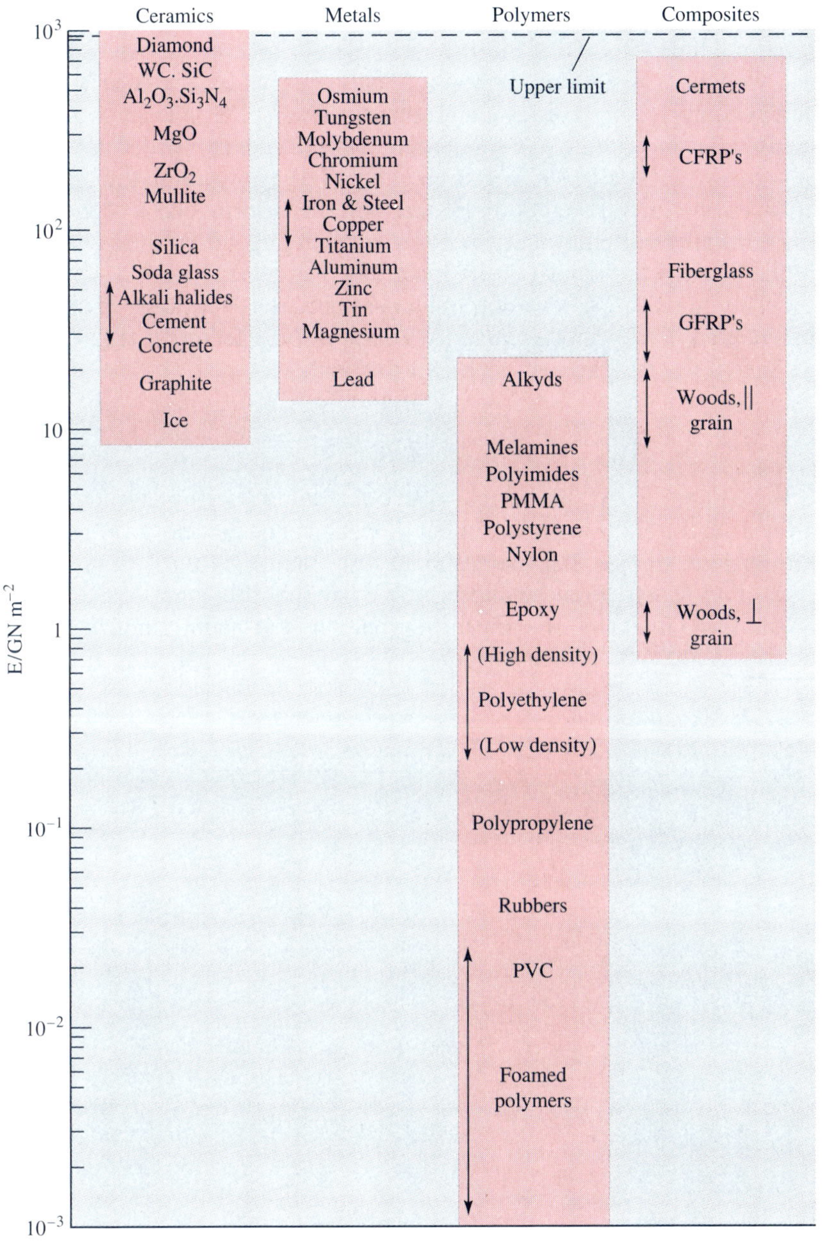

Figure 1.2 Bar chart of data for elastic modulus E (from M. F. Ashby and D. R. H. Jones, *Engineering Materials,* vol. 1, 1980). GFRPs and CFRPs are glass fiber and carbon fiber reinforced polymers. (*Courtesy of Pergamon Press.*)

$$\delta = \frac{PL}{AE} \tag{1.2}$$

Equation (1.2) is referred to often as *Hooke's law,* named after Robert Hooke (1635–1703), who was the first to record that many materials have a linear relation between load and displacement. For many materials the same relation will hold if the bar is shortened by a compressive force, and we will assume that the same relation, Eq. (1.2), holds in tension and in compression. We should emphasize that Eq. (1.1) is an empirical result following from experiments on materials; it was *not* derived from any theoretical principles. We think of Eq. (1.1) as providing the relation between the force P acting on the bar and the deformation δ of the bar; the constants in this *force-deformation* relation depend on the geometry of the bar and the nature of the material.

The quantity P/A is a force intensity; i.e., it is the force P divided by the area A. We call this quantity a *normal stress* acting on area A, and we give it the symbol σ (Greek letter sigma):

$$\sigma = \frac{P}{A} \tag{1.3}$$

If we section or cut the bar at some point along its length far enough away from the ends, as shown in Fig. 1.3*a*, and consider the lower portion in Fig. 1.3*b*, we see that the resultant force acting on the section must equal P to satisfy force equilibrium. We say that this force P gives rise to the force intensity or uniform normal stress σ acting on area A, as shown in Fig. 1.3*c*. When the force is stretching the bar as in Fig. 1.3,

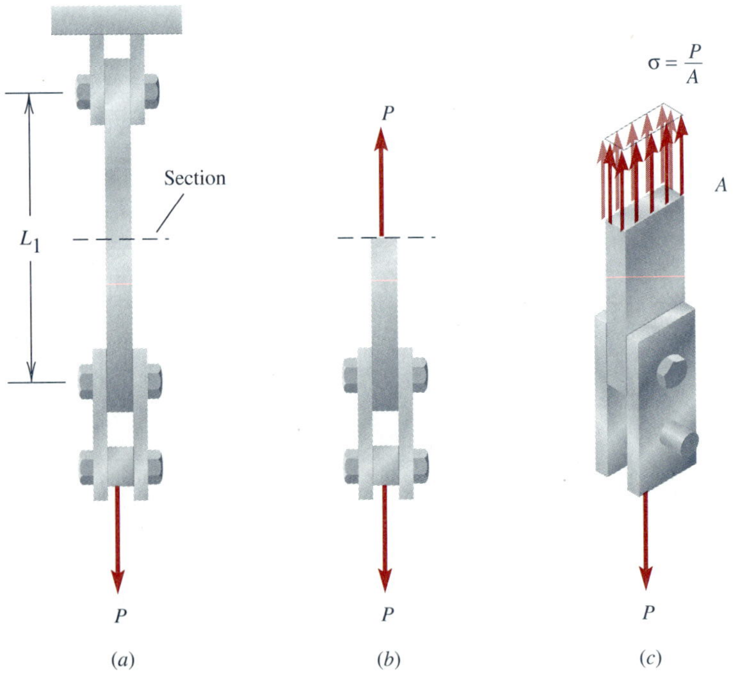

Figure 1.3 (*a*) Section of a loaded bar, (*b*) free-body diagram of lower segment, (*c*) normal stress σ acting on cross section of area A.

we say the corresponding stress σ is a *tensile stress*; if the force is acting to compress or shorten the bar, we say the corresponding stress is a *compressive stress*.

We define the ratio of the deformation δ of the bar, i.e., its change in length, to its original length L in Eq. (1.1) as the *normal strain* ϵ (Greek letter epsilon) of a bar of length L

$$\epsilon = \frac{\delta}{L} \tag{1.4}$$

Normal strain has no units, and as we will see, for most engineering applications the normal strain is small, of order of magnitude 10^{-3}.

Equation (1.1) can be written now in the form of a one-dimensional *stress-strain relation* for a material:

$$\sigma = E\epsilon \tag{1.5}$$

Equation (1.5) is called often a one-dimensional Hooke's law for stress and strain or a one-dimensional stress-strain relation for the material. The normal stress σ is given by Eq. (1.3) while the normal strain ϵ is given by Eq. (1.4). We discuss stress-strain relations in more detail in Chaps. 2 and 8.

In loading the bar of Fig. 1.1, we found that the bar elongated an amount δ under load P. Our experience with stretching elastic bands or plastic strips leads to the observation that as we stretch a material in one direction, it will contract in the transverse or perpendicular directions. We show in Fig. 1.4 a transverse contraction away from the end of the bar. The amount of contraction in the transverse directions will depend on the amount of elongation in the loaded direction. Experiments have shown for a given material that the change in length per unit length of line elements in the perpendicular or transverse directions, i.e., the normal strains in the transverse directions, are a fixed fraction of the normal strain in the loaded direction. Therefore, for a given material, the ratio

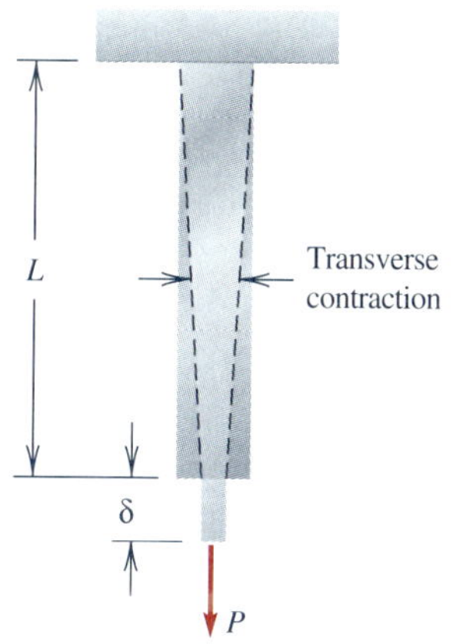

Figure 1.4 Transverse contraction of a longitudinally loaded bar.

$$\nu = -\frac{\text{normal strain in transverse direction}}{\text{normal strain in loaded direction}}$$

is a constant. The minus sign is inserted before the ratio of strains so that the constant ν (Greek letter nu) will be positive; the normal strain in the transverse direction is negative because of the contraction.

The constant ν is called *Poisson's ratio* [after S. D. Poisson (1781–1840)] of the material, and we refer to it as an elastic constant of the material. For most metals ν is approximately 0.33. The value of ν for cork is approximately zero, which makes cork useful for stoppers on bottles.

The types of problems we will consider in this book are those for which loads are placed on a structure and we determine the stresses and displacements in components of the structure. A typical problem follows.

In Chap. 2, however, we will turn to a more careful approach to problems involving uniaxial deformation, and we will introduce our general approach to problem solving.

EXAMPLE 1.1

A portion of a model hydraulic force system is shown in Fig. 1.5*a*. When pressurized, the hydraulic cylinder exerts a downward force P at point B on the rigid lever BCD. Member DF is a steel member with $E = 30 \times 10^6$ psi, and it will stretch under load. If point C is fixed, we wish to find the elongation of member DF when $P = 900$ lb; the cross-sectional area of DF is 0.125 in^2. We neglect friction at all pins, and we neglect the weight of each component.

To find the elongation of DF, we need to find first the force in DF. A free-body diagram of the rigid bar BCD is shown in Fig. 1.5*b*. The reaction at pin C is vertical and equal to R_C. Summation of the moments about point C gives the force F_{DF}

$$F_{DF} = \frac{14}{12} P = 1050 \text{ lb} \tag{a}$$

The normal stress in DF is then

$$\sigma_{DF} = \frac{F_{DF}}{0.125 \text{ in}^2} = 8400 \text{ psi} \tag{b}$$

Finally the displacement of point D (Fig. 1.5*c*) is given by Eq. (1.2):

$$\delta = \frac{F_{DF} L}{AE} = \frac{1.050 \times 10^3 \times 12}{0.125 \times 30 \times 10^6} = 3.36 \times 10^{-3} \text{ in} \tag{c}$$

Member BCD will rotate as a consequence of this displacement from the horizontal by an angle of 1.60×10^{-2} degrees.

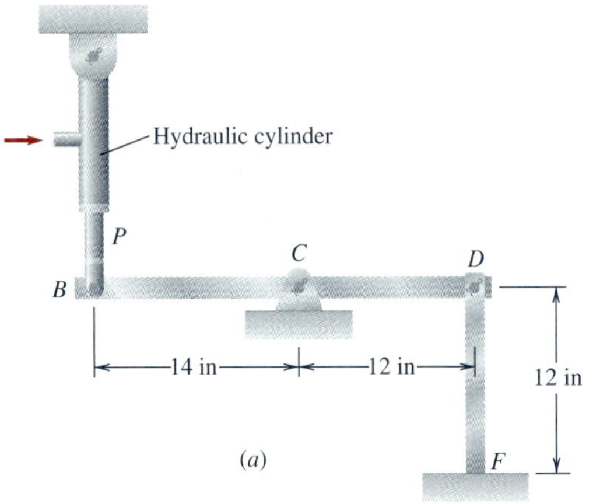

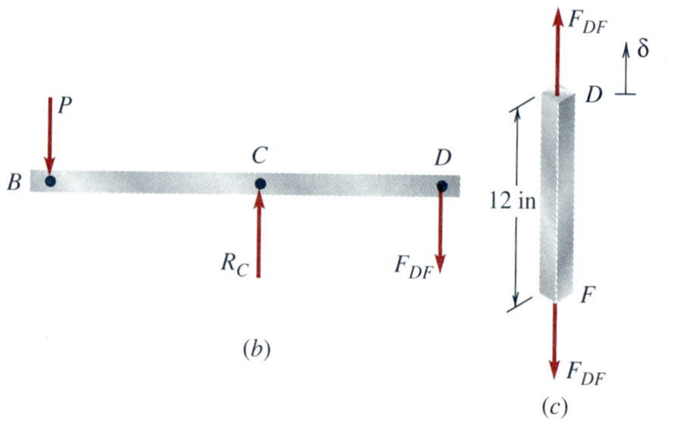

Figure 1.5 Example 1.1

As we will discuss in greater detail in Chap. 2, normal tensile and compressive stresses occur in components of many common engineering problems. As a consequence, the first step for the calculation of the normal stresses is the visualization of the existence of the normal *force* in the component of interest. If we can determine the value of the axial tensile or compressive force in the component, then we can determine the value of the normal stress acting at different sections along the component.

We now turn to a discussion of a stress that acts in the plane of the area on which the force acts.

1.3 Shear Stress and Shear Strain

To characterize the behavior of a bar of material in tension and compression, we discussed in general terms in Sec. 1.2 an experiment that related the normal stress σ to the normal strain ϵ (Fig. 1.1). We now wish to turn to a method to characterize a sample of material when it is loaded in *shear.*

Consider a block of material of cross-sectional area A and height h, as shown in Fig. 1.6*a*. The block of material is attached firmly on its lower surface to a rigid table; at the top of the block of the material a rigid plate is firmly attached. In Fig. 1.6*b* we show a force F applied to the rigid plate *in the plane of the plate.* We call a force acting in the plane of an area a *shear force* acting on the area. A shear force causes or tends to cause different parts of a body to slide relative to one another in the direction of the plane of the shear force. As a consequence of the shear force F, planes parallel to area A slide relative to one another, and an element $ABCD$ located away from the ends distorts through a small angle γ, in radians, as shown in Fig. 1.6*b*. This angle γ (Greek letter gamma) in Fig. 1.6*c* measures the *change in right angle* of a line element originally

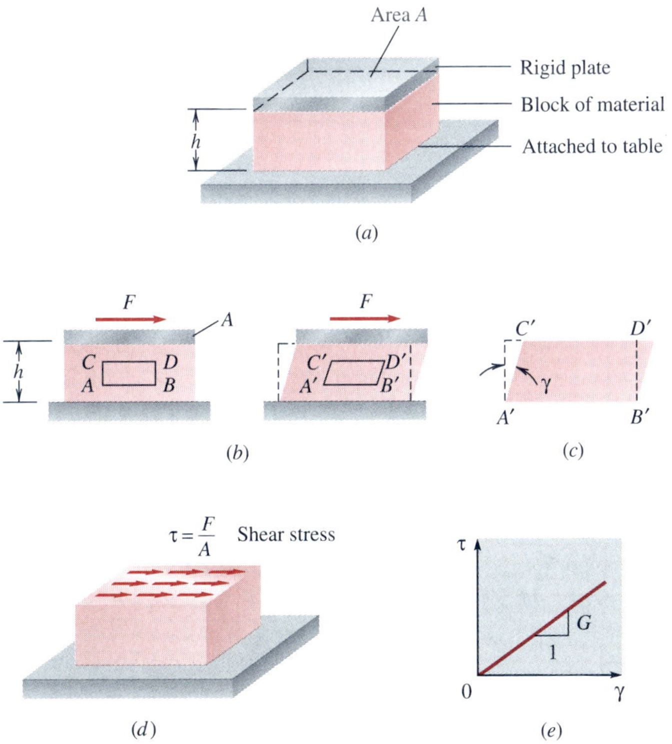

Figure 1.6 Shear stress: (*a*) block of elastic material, (*b*) under load F, (*c*) shear strain of element $ABCD$, (*d*) shear stress, (*e*) shear stress versus shear strain.

perpendicular to the table, and we refer to this angle γ as the *shear strain* in the material.

If we *assume* that the shear force F applied to the plate is spread uniformly by the action of the rigid plate *onto* the top surface of the material (Fig. 1.6*d*), we can think of the force intensity F/A as giving rise to a uniform average *shear stress* in the plane of area A. We usually identify shear stress with the Greek letter τ (tau); see Fig. 1.6*d*. Even with the rigid plate distributing the force F onto the top surface, the shear stress will not be uniform near the edges of the material. However, it is convenient to think of the shear stress τ as uniform on the top surface, as shown in Fig. 1.6*d*.

As we increase the value of the shear force F, the value of the shear strain will increase (Fig. 1.6*e*). If we limit ourselves to small values of the angle γ and to a material that behaves in a linear elastic fashion, we find that there is a linear relation between the shear stress τ and the shear strain γ [completely analogous to the linear relation between the normal stress σ and the normal strain ϵ, given by Eq. (1.5)] of the form

$$\tau = G\gamma \tag{1.6}$$

where G is defined as the *shear modulus* of the material. Since γ is nondimensional, in radians, the units of G are those of τ, namely, the units of stress. For many metals the value of G is about $(3/8)E$. In fact, as we will derive in Chap. 8, there exists a relation between Young's modulus E, the shear modulus G, and Poisson's ratio ν of the form

$$G = \frac{E}{2(1 + \nu)} \tag{1.7}$$

If we know the values of E and ν for a material, we can use Eq. (1.7) to calculate the value of G.

We have discussed shear stress and shear strain in the context of the experimental arrangement depicted in Fig. 1.6. However, in contrast to the experimental arrangements depicted in Fig. 1.1, shear stress–shear strain experiments on a specimen in the form of Fig. 1.6 are seldom carried out except to obtain approximate values of shear modulus for very flexible materials. It is usually difficult to attach the material uniformly to the table and to attach the rigid plate uniformly to the material in an attempt to approximate a uniform average shear stress across the top surface. In addition, the distribution of stresses in the material along the surface of attachment to the table will involve both shear stresses and normal stresses in order to keep the specimen in equilibrium. Further, the shear stress distribution in the material will not be uniform across planes parallel to the rigid plate on the top. In view of these difficulties, it is usually easier to test a circular cylindrical specimen of material with a twisting moment along the axis to obtain information on the behavior of the material under shear stress and shear strain; we consider this in Chap. 3.

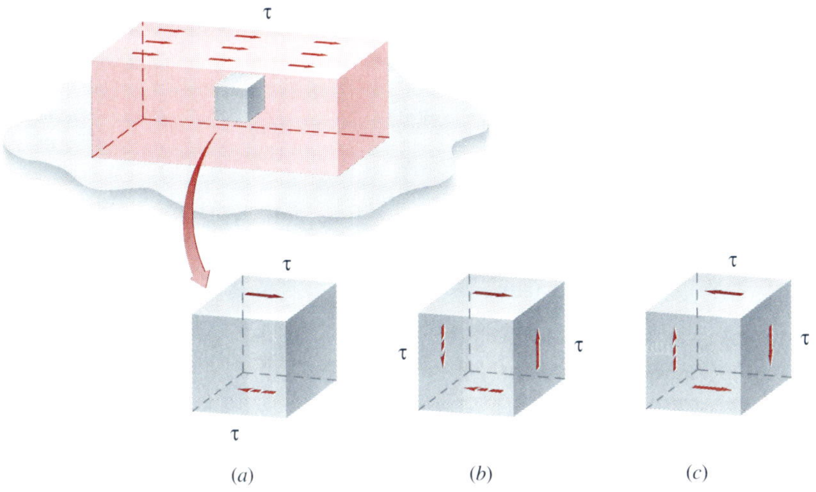

Figure 1.7 (*a*) Average stress τ acting on a block of material and on an infinitesimal element. (*b*), (*c*) Equal shear stresses acting on perpendicular faces of an infinitesimal element.

However, away from the edges of the material, the shear stress on a plane parallel to the top surface will be approximately equal to τ. Therefore, if we consider an ***infinitesimal*** cubic element cut out from the middle of the specimen away from the bounding surfaces, such as in Fig. 1.7*a*, the top and bottom surfaces of the element from equilibrium arguments will experience a shear stress τ as shown.

We immediately observe that this infinitesimal element in Fig. 1.7*a* is not in moment equilibrium! We must have additional shear stresses equal to τ acting on the vertical faces of the element (Fig. 1.7*b*) to maintain the force and moment equilibrium of the element. This infinitesimal element is experiencing what we call ***pure shear*** under the action of the shear stresses τ. In general, in the interior of a body under stress, it is necessary for moment equilibrium of any infinitesimal element at a point to have the shear stresses on perpendicular faces equal, as shown in Fig. 1.7*b* and *c*; see Sec. 8.4.

The concept of average shear stresses often arises in engineering problems when loads are applied to components so as to tend to move the components relative to one another across a surface. We show a number of typical situations in Fig. 1.8.

In Fig. 1.8*a*, a load P is applied to the center block through a rigid plate. This block tends to shear between the two side blocks across the areas common to the blocks. We assume that an average uniform shear stress acts on these areas of the blocks to maintain equilibrium. From Fig. 1.8*a* and from force equilibrium of the center block, we see that this average shear stress is given by

$$\tau = \frac{P}{2aL} \tag{1.8}$$

where aL is the area over which the average shear stress τ is acting and the factor of 2 arises from the presence of an area on each side.

In Fig. 1.8*b*, we show a rigid cylindrical die punch used to punch out a circular plug of radius r from a plate of thickness t. When the punch under a load P is applied to the plate, average shear stresses are set up on the area around the plug to balance force P. If the maximum average shear stress that the plate can carry is τ, then as the plug is punched out, we have from force equilibriumn

$$P = \tau(2\pi rt) \tag{1.9}$$

where $2\pi rt$ is the area over which the shear stress τ is acting (Fig. 1.8*b*).

A section of an angle member bolted to a support is shown in Fig. 1.8*c*. When a load W is applied to the angle member, the angle member under the load tends to shear the bolt connection at the support. As a consequence, an average shear stress τ is set up over the cross-sectional area of the bolt

$$\tau = \frac{W}{A} \tag{1.10}$$

where A is the cross-sectional area of the bolt. We often refer to this case as *single shear* in the bolt.

Finally we show three steel straps riveted together at a joint, as shown in Fig. 1.8*d*. The force transmitted across the joint is P. As a consequence of the joint configuration and the loading, the rivet is exposed to what we call *double shear*. An average shear stress τ acts in the rivet

$$\tau = \frac{P}{2A} \tag{1.11}$$

where A is the cross-sectional area of the rivet.

Keep in mind that in the illustrations in Fig. 1.8 we have made a number of implicit assumptions about the nature of the loading and how it was transmitted to the section or member experiencing the average shear stress. In any engineering application, the nature of the loading on the structural component must be considered carefully before any subsequent calculations are made.

EXAMPLE 1.2

The shear strength of wood is often determined by testing small samples in a holding device such as that shown in Fig. 1.9*a*. A force P on a rigid block shears out the central portion of the wood in the sample along the two planes AB. A number of oak samples have been tested, and the average value of load P to shear out the central portion was found to be 8.8 kN. We wish to find the average shear strength of the wood samples.

In Fig. 1.9*b* we show the central portion of the sample as it shears out of the holding device. The average shear strength of the wood sample is obtained from vertical force equilibrium

$$2\tau(40 \times 18 \times 10^{-6}\ \text{m}^2) = P \tag{a}$$

and with $P = 8.8$ kN, we find

$$\tau = 6.11\ \text{MPa} \tag{b}$$

as the average shear strength for the oak samples.

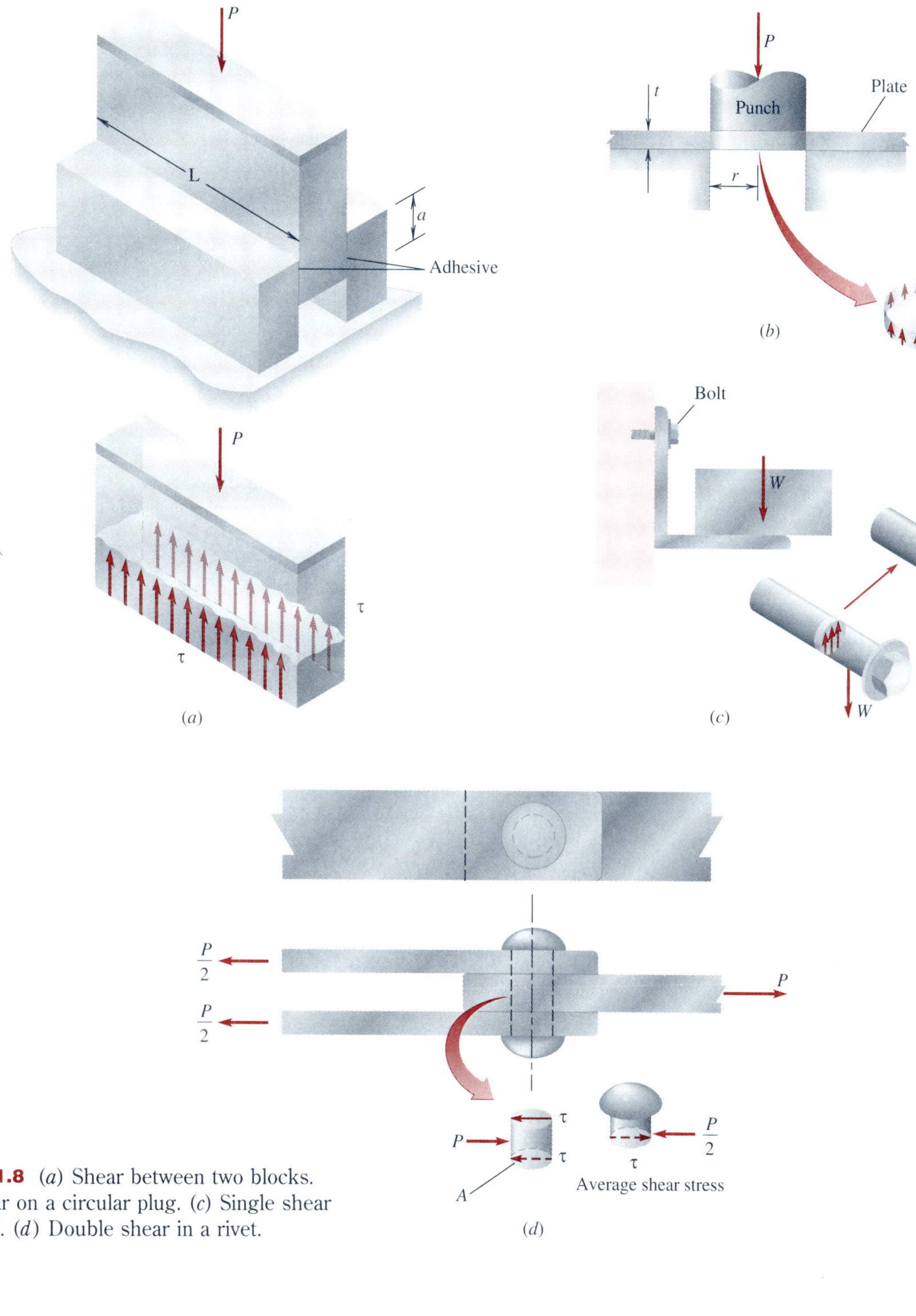

Figure 1.8 (*a*) Shear between two blocks. (*b*) Shear on a circular plug. (*c*) Single shear in a bolt. (*d*) Double shear in a rivet.

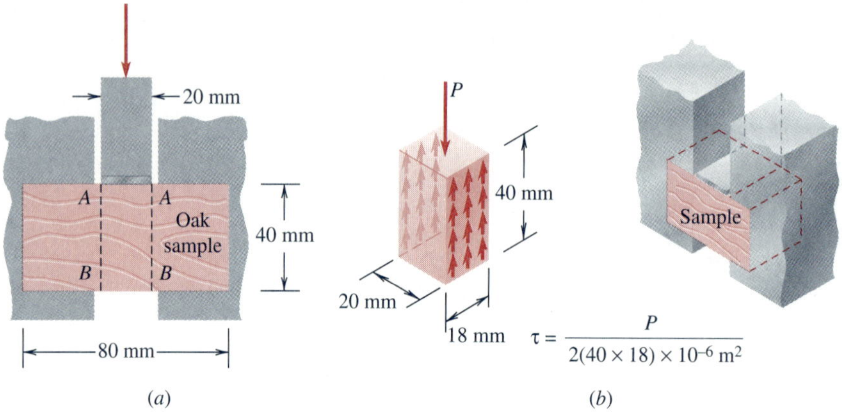

Figure 1.9 Example 1.2

EXAMPLE 1.3

A proposed design of a multistory university library uses a brick facade on the exterior surfaces; see Fig. 1.10. Each section of the brick facade is carried on a 6-in by 6-in angle section that is in turn connected by two ⅝-in-diameter bolts to the concrete exterior wall, as shown in Fig. 1.10. Each section of the brick facade supported by the angle section is 5 ft wide and 30 ft high and has a weight of approximately 6000 lb. We wish to calculate the shear stress in each bolt due to the weight of the brick facade carried by the angle section.

Since the brick facade sections are on top of one another, the loading onto the angle brackets is not simple. However, a reasonable estimate for the shear stress is to assume that each of the two bolts carries in shear a load of 3000 lb. This case is then one of single shear in the bolt, as in Fig. 1.8c, and the average shear stress in the bolts is

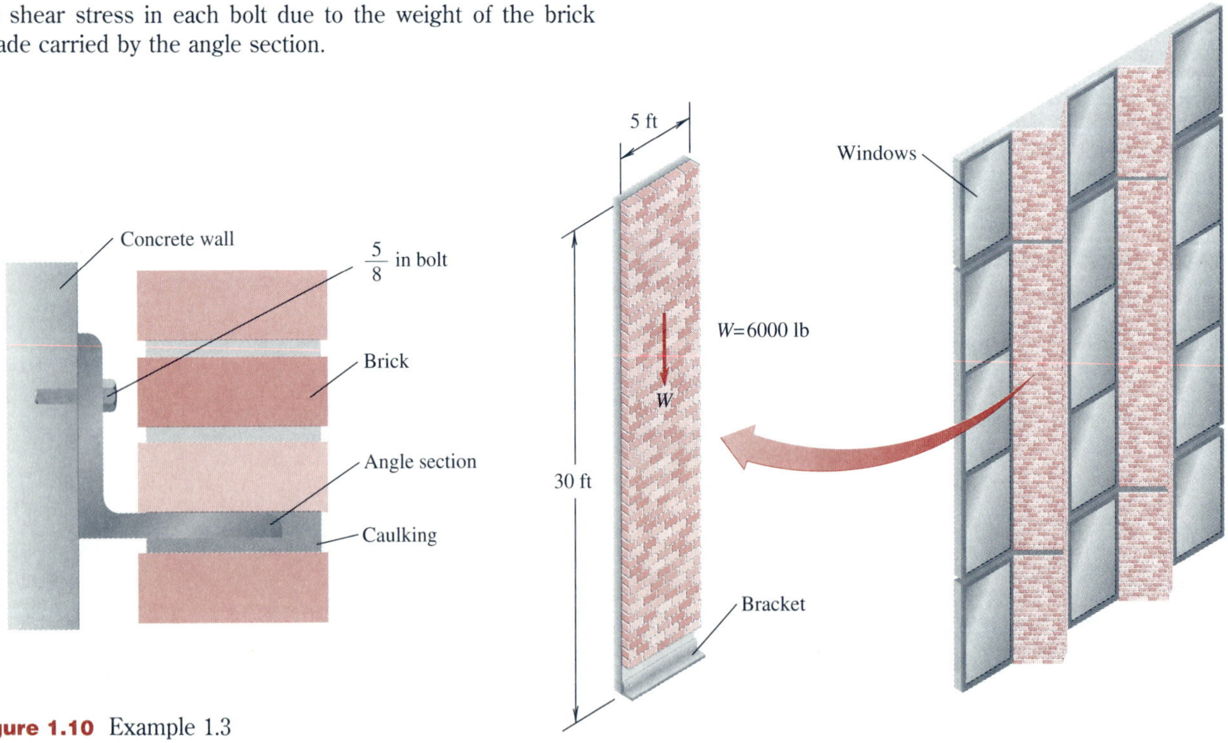

Figure 1.10 Example 1.3

$$\tau = \frac{3000}{A_{\text{bolt}}} = \frac{3000}{(\pi/4)(5/8)^2} = 9780 \text{ psi} \qquad (a)$$

It turns out that in a design of this nature there are specifications—allowable loads and stresses, as we will discuss in the next section—that set maximum values for the shear stresses. These specifications allow for uncertainty in the loading and tend to be conservative. In this case if the maximum shear stress permitted on the bolts is specified as 7000 psi, the proposed design would not be acceptable.

1.4 Allowable Loads

Thus far we have discussed the determination of normal stresses and shear stresses in simple structural components. In many engineering applications, we need to know the load or loads that a structure can support before a specified *maximum* value of normal stress or shear stress in one of the components is reached. The values specified for the maximum stress allowed in a component are often obtained from experiments or experience with the component. At other times, the values are obtained from knowledge of the stress to cause permanent deformation or fracture and are then reduced by some factor to ensure that the component will not fail. In many applications, especially in structures that may affect the safety of people, the maximum values are specified in building codes or in product codes, e.g., pressure vessel codes. Designers in these cases have a legal requirement to ensure that any component under a code specification has a stress or load value in it that is less than that allowed by the code. The values for the stresses permitted in a structural component are called the *allowable stresses.*

A *factor of safety n* is a factor by which we reduce a maximum stress or even an allowable stress to obtain a new allowable stress, e.g.,

$$\text{Working allowable stress} = \frac{\text{allowable stress}}{n}$$

where n is the factor of safety. Values for the factor of safety depend on the choice of the allowable stress and can be as high as 3 or as low as 1. The specific application in engineering practice will set the factor of safety either by code or by experience.

EXAMPLE 1.4

Figure 1.11*a* shows a triangular truss carrying a load P at position D. Member BD is a steel rod of circular cross section while member CD is a steel I beam as shown. Members BD and CD are connected to the support fixtures at B and C and to each other at D by high-strength steel pins 10 mm in diameter. The allowable tensile stress in member BD and the allowable compressive stress in member CD are 100 MPa, and the allowable shear stress in each pin is 150 MPa. We wish to find the maximum allowable load P that the truss is capable of carrying without exceeding the allowable stresses in the members or in the pins. We neglect friction in the pins and the weight of the members, and we assume that the support fixtures at the wall are themselves adequate to carry the maximum load. Based on intuition we might try to guess the maximum load before doing any calculations.

To find the maximum allowable load, we obtain first expressions in terms of P for the normal stresses in BD and CD and then expressions for the shear stresses in the pins. Figure 1.11*b* shows a simplified sketch of the truss. Bar BD is a two-force member with a force F_{BD} assumed as tension along its axis as shown in Fig. 1.11*c*. Similarly, the force in member CD is along the bar and is compressive (Fig. 1.11*d*). Force equilibrium at point D (Fig. 1.11*e*) shows that

$$F_{CD} = P \qquad F_{BD} = 1.414P \tag{a}$$

Therefore, the tensile stress in member BD and the compressive stress in member CD are

$$\sigma_{BD} = \frac{F_{BD}}{A_{BD}} = \frac{1.414P}{491 \times 10^{-6}\ \mathrm{m}^2} \tag{b}$$

$$\sigma_{CD} = \frac{F_{CD}}{A_{CD}} = \frac{P}{3.2 \times 10^{-3}\ \mathrm{m}^2} \tag{c}$$

If the stress in each member is equal to the maximum allowable stress

$$\sigma_{BD} = \sigma_{CD} = 100\ \mathrm{MPa} = 100 \times 10^6\ \mathrm{N/m^2}$$

then we find from Eq. (*b*) that

$$P = 34.7\ \mathrm{kN} \tag{d}$$

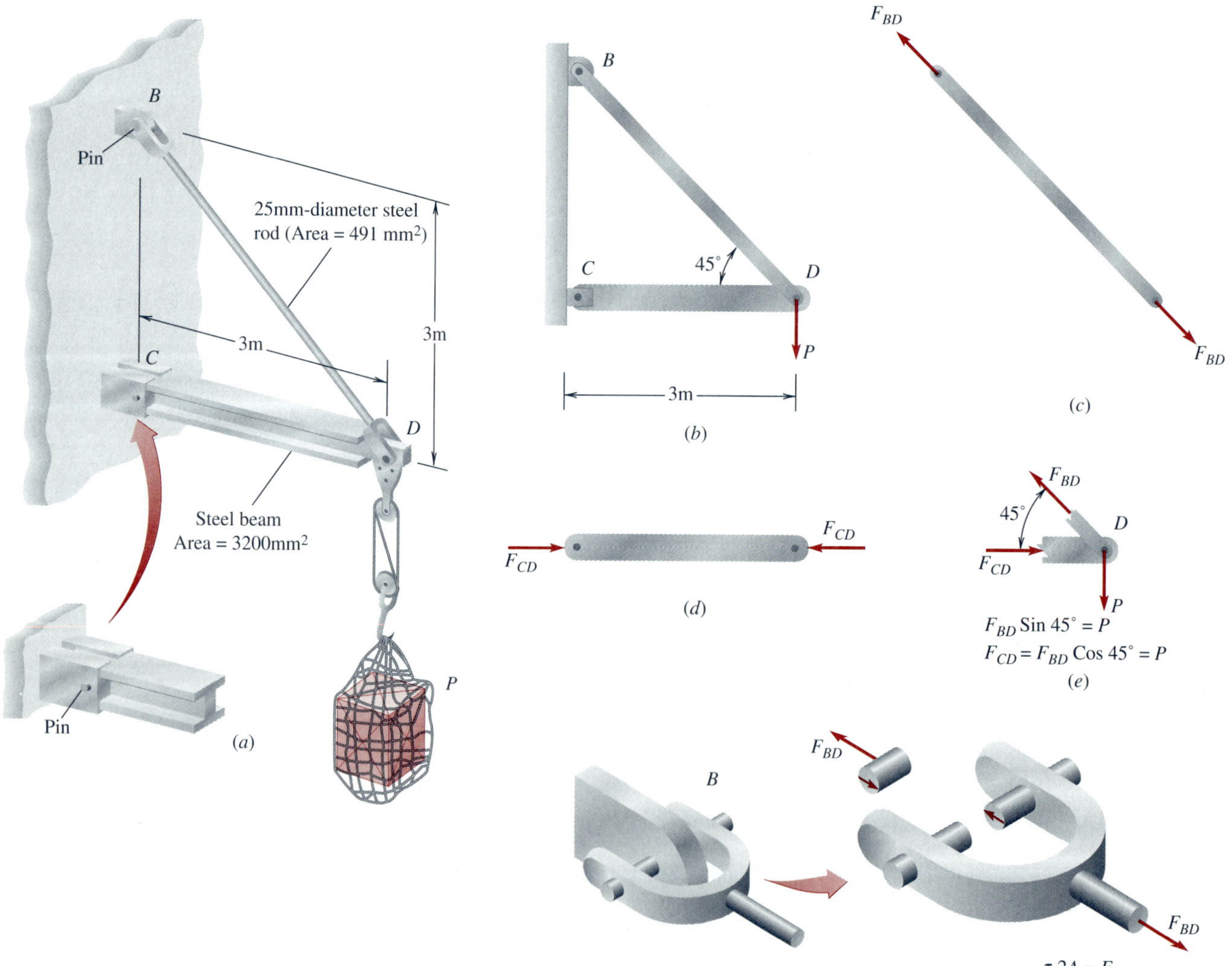

Figure 1.11 Example 1.4

and we find from Eq. (c) that

$$P = 320 \text{ kN} \tag{e}$$

If we were limited to the maximum load controlled by the normal stress in each member, the maximum load would be $P = 34.7$ kN. At this load, the maximum tensile stress in BD is 100 MPa, and the maximum compressive stress in CD is 10.84 MPa. It remains, however, to determine the shear stress in the pins.

Figure 1.11f shows the fixture at point B and the pin in double shear. A similar situation occurs at locations D and C. The pins at locations B and D are exposed to the maximum load $F_{BD} = 1.414P$. It follows from Fig. 1.11f that

$$\tau = \frac{F_{BD}}{2A} = \frac{1.414P}{2(\pi \times 10^2 \times 10^{-6}/4)} \tag{f}$$

The value of P needed to give rise to an allowable shear stress in the pin of $\tau = 150$ MPa is, from Eq. (f),

$$P = 16.7 \text{ kN} \tag{g}$$

This value of P is less than the 34.7 kN from Eq. (d), and so it is the maximum allowable load to be carried by the truss. At this load, the shear stresses in the pins at B and D are 150 MPa, the shear stress in the pin at C is 106 MPa, the tensile stress in member BD is 48.1 MPa, and the compressive stress in member CD is 5.22 MPa. Of course, in a practical situation we would also check the strength at the maximum load of the fixtures supporting the lifting tackle and hooks; we would also investigate buckling (Chap. 10) of member CD.

EXAMPLE 1.5

An aluminum-alloy bracket in an aircraft component is shown in Fig. 1.12a. A load P is to be transferred to a rigid support from member AB across the 0.30-in-diameter aluminum bolt. If the allowable tensile stress in the aluminum alloy is 30 kilopounds (kips) per square inch (ksi) and the allowable shear stress in the bolt and the material is 20 ksi, we wish to find the maximum allowable load P that can be carried across the bracket.

To find the maximum allowable load, we need to consider how the bracket might fail; we carry out what we call a *mode-of-failure analysis.* First, the bolt might fail in double shear, for which we have

$$\tau = \frac{P}{2A_{\text{bolt}}} \tag{a}$$

If $\tau = 20$ ksi, then the load P to cause shear failure of the bolt is 2830 lb.

Strap AB has a tensile stress that will be a maximum value across the minimum cross section at the bolt hole (Fig. 1.12b);

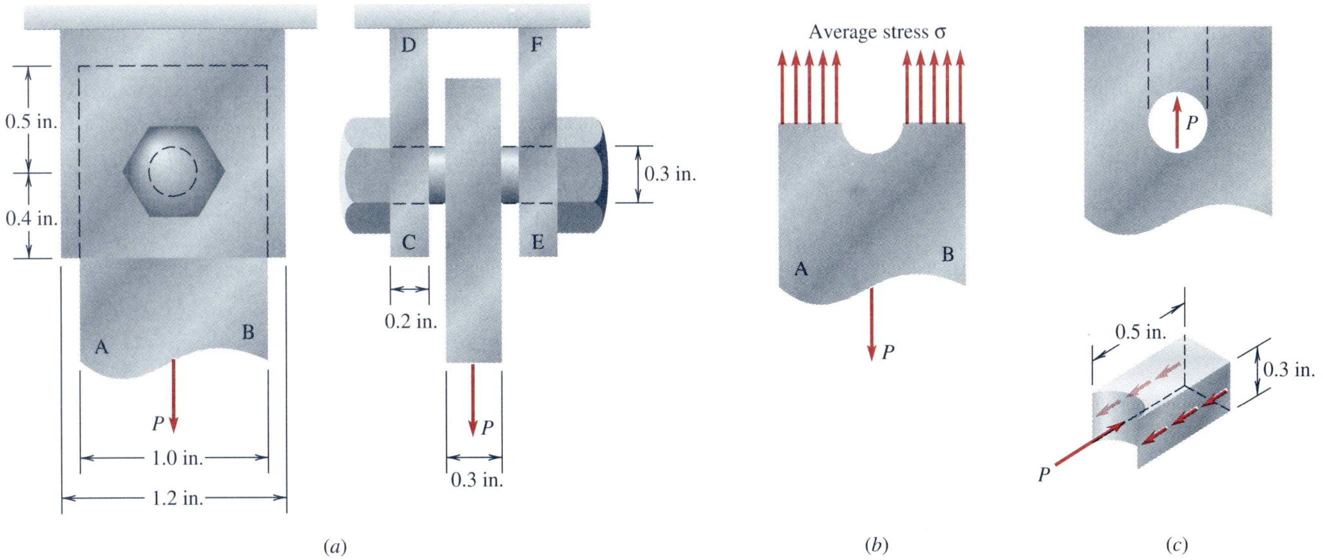

Figure 1.12 Example 1.5

therefore, the average normal stress at this section is

$$\sigma = \frac{P}{0.7(0.3)} \qquad (b)$$

If $\sigma = 30$ ksi, then the load to cause tensile stress failure in the strap is $P = 6300$ lb. Brackets *CD* and *EF* support a load of $P/2$ so that the tensile stress in each of these brackets at the section with the smallest area is

$$\sigma = \frac{P/2}{0.9(0.2)} \qquad (c)$$

from which the load to cause tensile stress failure in the brackets is $P = 10{,}800$ lb.

We also need to consider the possibility that load *P* can "shear out" the material above the bolt hole, as shown in Fig. 1.12*c*. In this case the load *P* is related to the average shear stress in the material by

$$P = 2\tau A = 2\tau(0.3 \times 0.5) = 6000 \text{ lb} \qquad (d)$$

A similar calculation for bracket *CD* gives

$$\frac{P}{2} = 2\tau A \qquad P = 4\tau A = 4\tau(0.2 \times 0.4) = 6400 \text{ lb} \qquad (e)$$

From our calculations, it would appear that the maximum allowable load in the bracket is governed by the failure of the bolt in double shear at a load of 2830 lb. At this load the average tensile stress in strap *AB* at the minimum cross section is

$$\sigma = \frac{2830}{0.7(0.3)} = 13{,}480 \text{ psi} \qquad (f)$$

This is less than one-half of the allowable tensile stress in the material, and it will allow for some ***stress concentration*** effect due to the presence of the hole. It appears, then, that the bracket can carry a load of $P = 2830$ lb. Of course, we would want to check that brackets *CD* and *EF* are appropriately attached to the rigid support and that the bolt is securely fastened so as to not fall out due to vibration under a no-load condition.

Here is a summary of the calculations we carried out:

Mode of Failure	Load *P*, lb
Double shear in bolt	2,830
Tensile in *AB*	6,300
Tensile in *CD*	10,800
Shear out of pin in *AB*	6,000
Shear out of pin in *CD*	6,400
$\sigma_{\text{allow}} = 30$ ksi	$\tau_{\text{allow}} = 20$ ksi

1.5 Tensile Stress-Strain Test

We introduced the idea of the normal stress–normal strain test in Sec. 1.2, Fig. 1.1. In the earlier discussion our interest was to characterize the behavior of the material when the material was loaded in tension. We now wish to obtain a deeper understanding of the stress-strain behavior in a tensile test. In Sec. 1.1 we pointed out that many books provide an excellent introduction to the behavior of materials, and so our discussion here will provide only a brief overview of material behavior, emphasizing linear elastic behavior for one-dimensional loading. Our aim is to use experimental information from a tensile test to formulate relations between stress and strain that can be used in practical engineering situations; we will return to the discussion of stress-strain relations for three-dimensional bodies in Chap. 8.

To perform a tensile test, a piece of material is usually machined into the form of a cylindrical test specimen, such as that shown in Fig. 1.13. Either mechanical or electronic gages are used to measure the elongation and the lateral contraction. In Fig. 1.13 we show mechanical dial gage devices to measure the elongation between two marked loca-

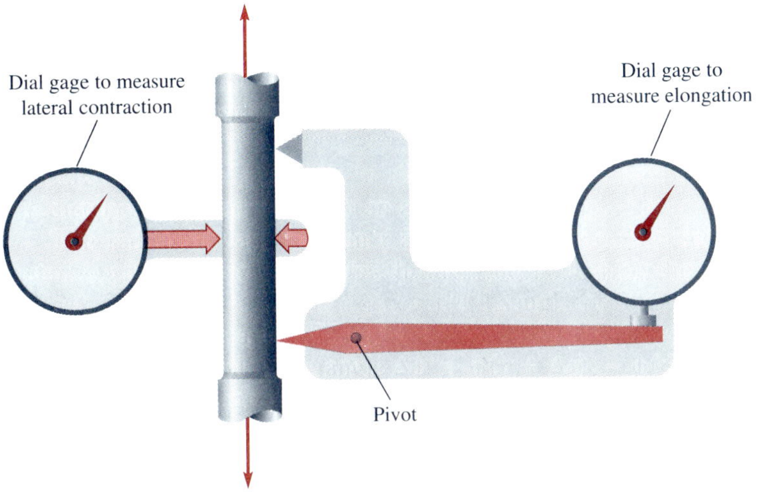

Figure 1.13 A tensile test specimen and mechanical gages for measuring deformation.

tions, or the gage length on the specimen, and to measure the change in diameter of the specimen; a number of different devices are available for these measurements. The specimen is placed in a tensile testing machine, and the ends of the specimen are moved apart by the machine by increasing the load. The original cross-sectional area and the gage length of the specimen are known, from which the relation between the stress and the strain of the specimen can be recorded. The stress is found by dividing the load by the original cross-sectional area. The strain is found by noting the change in length ΔL of the gage length L (Fig. 1.14) to obtain the strain as $\epsilon = \Delta L/L$. Typical results[4] of room-temperature tests on steel and aluminum alloys are shown in Fig. 1.15. These tests were carried out to strains of 0.020 (2 percent). This value of strain is significantly less than the strain necessary to cause fracture of the specimen. For example, steels will have strains at fracture over a 2-in gage length as high as 10 to 40 percent depending on the composition of the steel.

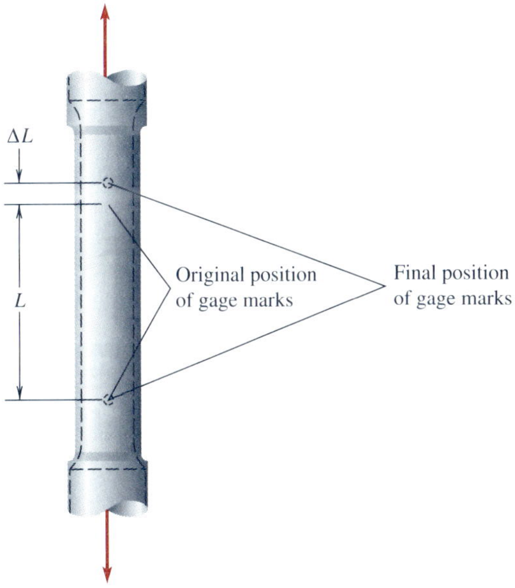

Figure 1.14 Displacements in a tensile test.

For the stress-strain curves in Fig. 1.15, we note that there is an initial region where the stress is very nearly proportional to the strain ϵ, that is,

$$\sigma = E\epsilon$$

where E is the elastic modulus of the material. This is the linear response of materials that we discussed in the context of Fig. 1.1 to obtain the elastic modulus E.

[4]S. H. Crandall, N. C. Dahl, and T. J. Lardner, *An Introduction to the Mechanics of Solids,* 2d ed., McGraw-Hill, New York, 1978, Chap. 4; see discussion of this book in Sec. 1.1.

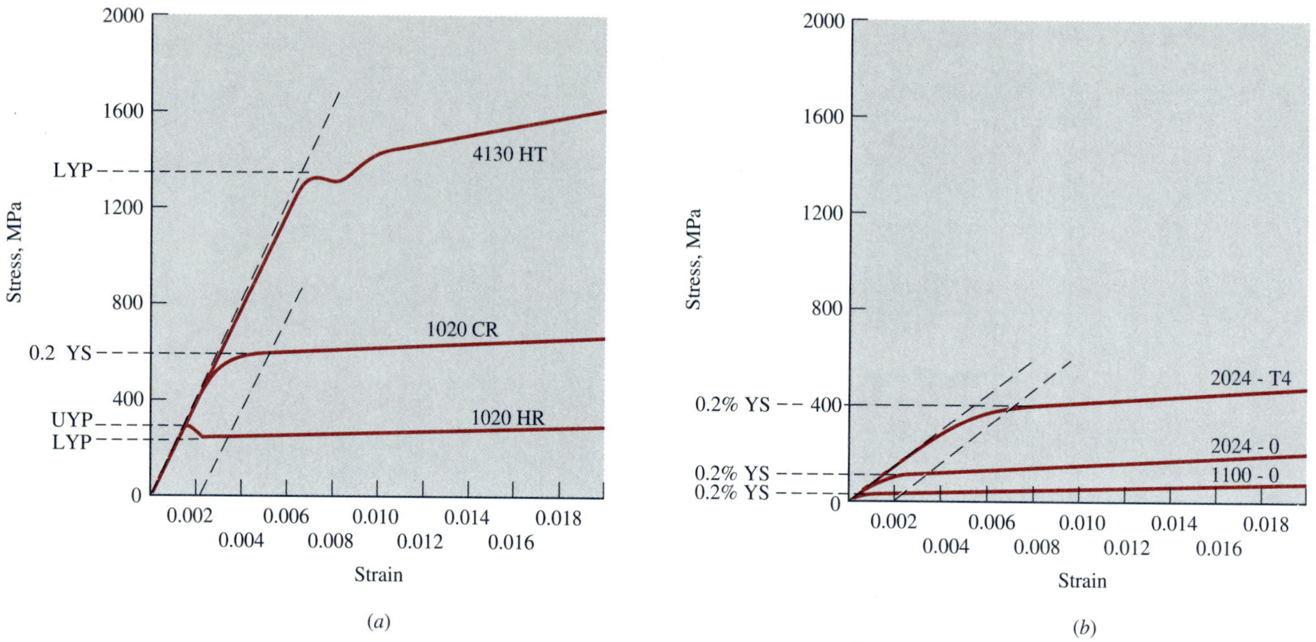

Figure 1.15 Typical stress-strain curves (*a*) for three steels and (*b*) for aluminum (1100-0) and two aluminum alloys (from Crandall, Dahl, and Lardner, chap. 4).

The *proportional limit* of a material is defined as the largest value of stress for which the stress is still proportional to the strain. The *elastic limit* of a material is defined as the largest value of stress that can be applied without causing any permanent strain upon removal of the stress. For the materials shown in Fig. 1.15, the proportional and elastic limits coincide. If the material is loaded beyond the elastic limit and the stress is removed, the stress-strain curve has the shape shown in Fig. 1.16. The unloading portion *BC* of the curve is approximately parallel to the initial loading portion *OA*, and we often say that the material has *unloaded elastically*. The strain remaining after unloading is the *permanent* or *plastic strain OC*; that is, the specimen is longer than its original length. If the material is loaded again from point *C* (Fig. 1.16), the stress-strain curve appears as in Fig. 1.16. The total strain corresponding to a stress at *D* can be considered as made up of an elastic part *FH* and a plastic part *OF*; the elastic part *FH* is recovered upon unloading from *D*, and the plastic part *OF* is the permanent strain remaining upon unloading.

Figure 1.16 Specimen loaded beyond the elastic limit.

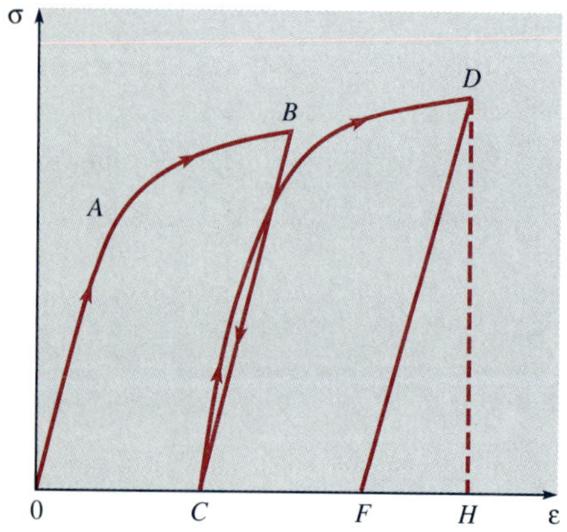

It is difficult to determine precisely either the proportional or the elastic limits. Instead it is usual to obtain the *yield strength* of the material, which is the stress required to produce a certain arbitrary plastic strain. The yield strength of a material is determined by drawing through the point on the horizontal strain axis corresponding to an arbitrary plastic strain, usually 0.002 (0.2 percent), a line that is parallel to the initial tangent to the stress-strain curve. The intersection of this line with the stress-strain curve defines the yield strength. This construction is illus-

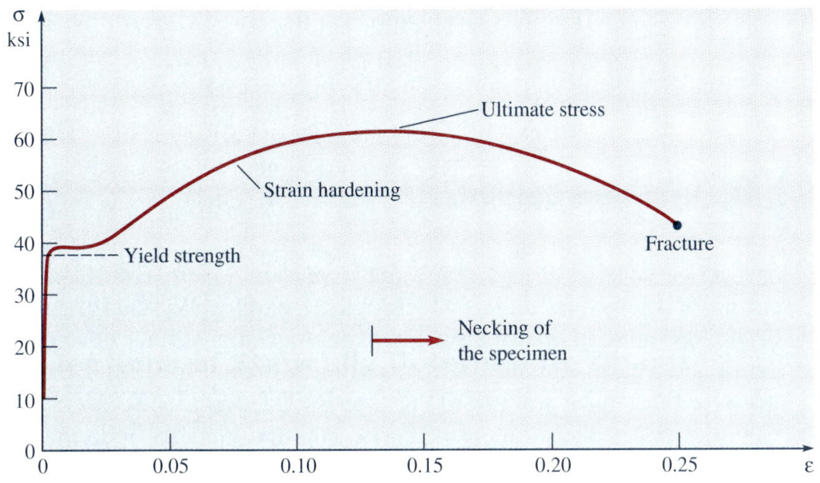

Figure 1.17 Typical stress-strain curve for structural steel.

trated in several cases in Fig. 1.15. Yield strengths are more sharply defined than are the proportional limits.

For many of the common steels we find that plastic deformation begins abruptly, resulting in an increase of strain with no increase, or perhaps even a decrease, in stress. For such materials a *yield point* is defined as the stress level, less than the maximum attainable stress, at which an increase in strain occurs without an increase in stress. The stress at which such plastic deformation first begins is called the *upper yield point*; subsequent plastic deformation may occur at a lower stress, called the *lower yield point*, as shown for 1020 HR steel in Fig. 1.15*a*.

Once the stress exceeds the yield strength, the stress required for further plastic deformation rises. This characteristic of the material in which further strain after the yield point requires an increase in the stress is referred to as *strain hardening* of the material. If the stress continues to rise, *necking* of the specimen starts at a stress level called the *ultimate stress*, and the specimen subsequently fails by fracture. A typical stress-strain curve for a structural steel is shown in Fig. 1.17.

Figure 1.18 shows a bar chart giving the range of values for the yield strength of a number of materials; see also App. F.

The discussion of stress-strain curves presented here is only a brief introduction to the richness of material behavior. In engineering practice it is essential in the analysis and design of engineering components that the material response of the components be fully understood. No design or analysis can be expected to be correct if the wrong material behavior or an incorrect material constant is used.

In this book we emphasize linear elastic material behavior (Fig. 1.19*a*), and occasionally we will consider elastic–perfectly plastic materials (Fig. 1.19*b*) for which no strain hardening occurs. In Fig. 1.19*b*, Y

denotes the yield strength of the material, and ϵ_Y is the value of the strain at yield. We also assume that materials behave in compression in a manner similar to their behavior in tension.

1.6 Problem Solving

As we proceed, we will find that a major portion of our efforts will be spent on problem formulation and problem solution. We found this to be true in a course in Statics also, and by now we should have developed an approach to problem solving that works. However, it is important to emphasize that in the practice of engineering for design and analysis, it is vital that our calculations, procedures, and assumptions be in such a form that they can be checked by someone else. Checking calculations is an important part of engineering practice; everyone makes mistakes. For these reasons we think it is of value to outline an approach to problem formulation and solution that we have found useful in engineering practice:

1. Review what is required for the solution. That is, ask yourself, What is the problem? It is useful to write down a brief statement of the objectives of the solution when we begin.
2. Draw reasonably neat diagrams to an approximate scale. In many problems if you have an approximate to-scale diagram, you can see geometric and physical relationships that might not be obvious at first. Also develop the habit of drawing sketches of a problem beyond those you might be given initially. In working on sketches of the problem, ask yourself, e.g., How big is this structure? Is it larger than I am, or is it a small microelectronics part?
3. Consider and draw neatly the appropriate *free-body diagrams* for the system. This is a critical step because it will allow you to see the interrelations between the forces and moments acting on parts of the system. Are these parts in equilibrium? Neat and clear free-body diagrams are the most important part of the solution. Can you write down enough equations to determine the solution?
4. As we discuss in detail in Chap. 2, the next step is to apply the *three-step* approach to the solution of solid mechanics problems. The three-step approach allows you to formulate an approach to the solution in an efficient and organized manner.
5. Express the final results in a form that allows for checking. If the result is in symbolic form, can you check the limiting cases? Are the equations dimensionally correct? Does the combination of variables in the solution look right and have the expected form? Do the units check?

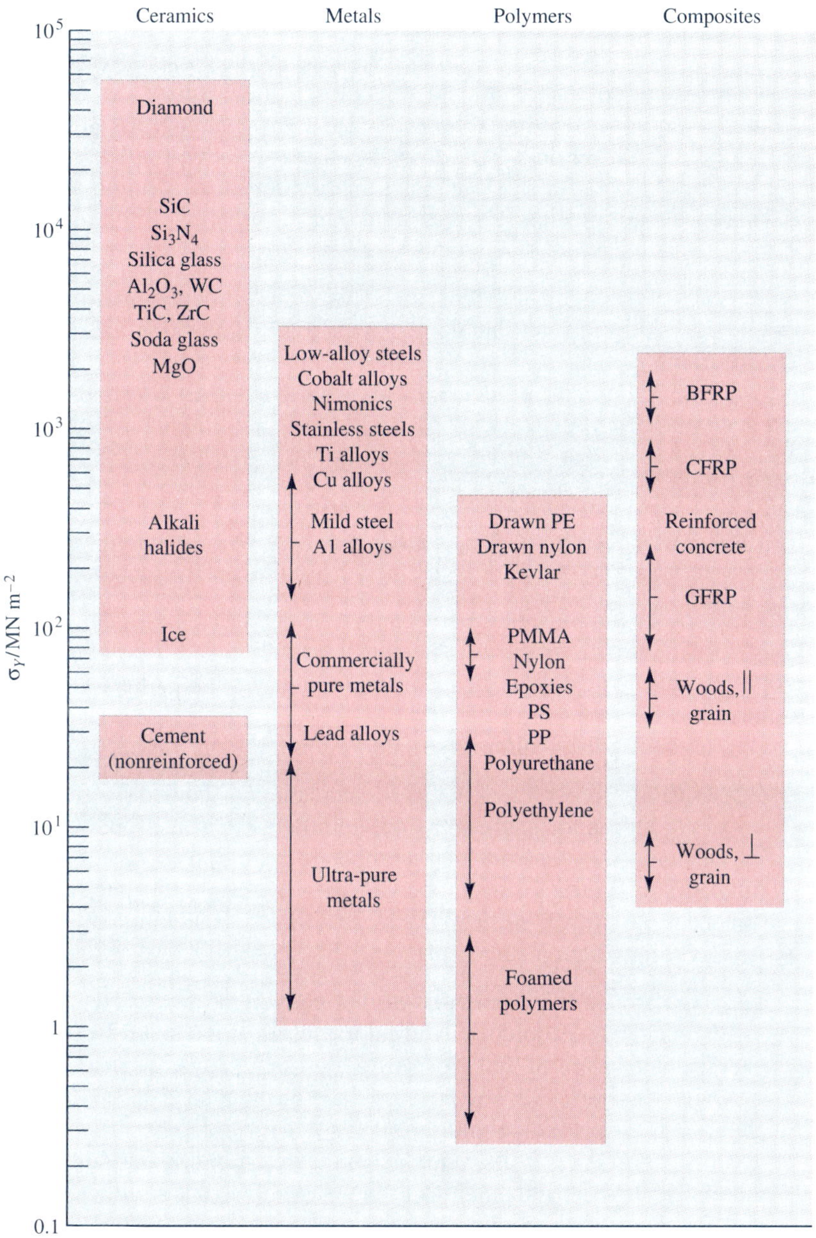

Figure 1.18 Bar chart of data for yield strength (from M. F. Ashby and D. R. H. Jones, *Engineering Materials,* vol. 1, 1980, see Fig. 1.2).

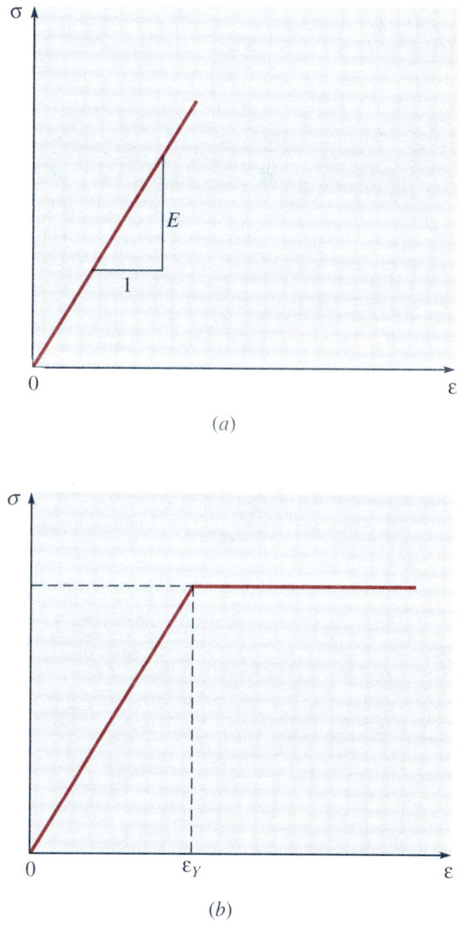

Figure 1.19 (*a*) Linear elastic stress-strain curve; (*b*) linear elastic–perfectly plastic stress-strain curve.

If the result is obtained numerically, does it have the right magnitude? Is it too great or too small? In view of your grasp of the size of the problem, does the value for the answer make sense? Develop the habit of scrutinizing your answers to see if the result agrees with your intuition. After a time, some answers will begin to look right, and this will be the start of a "feeling" for the correctness of solutions.

6. Finally, again question the statement and formulation of the problem to see if indeed you found what was required.

Problem solving is a skill that can be learned by practice; an important part of this skill is to continually ask yourselves questions about what you are doing as you work through a solution. And, of course, carry out the work so that it can be checked by others; in other words, be neat and systematic in your work.

1.7 Summary of Appendices

This textbook contains a number of appendices that we think are useful. They contain review material, specific information for the solution of problems, and information on the computer programs to be used with the text.

We present a brief summary of and comments on the appendices:

Appendix A Procedures for the calculation of centroids and moments of inertia for plane areas are summarized. You should be able to calculate the centroid of a composite area and to calculate the moment of inertia of a composite area by using the parallel-axis theorem. We review some of the techniques for calculating moments of inertia in Chap. 5.

Appendix B In this appendix we list conversion factors useful for solid mechanics problems. We find useful the conversion factors for stress values from SI units (MPa) to the English system (psi). Often we need to check that the stress values in megapascals (MPa) in a solution to a problem are not unreasonable for the materials in question. Keep this appendix handy as you check numerical results. You should write in specific values as needed.

Appendix C Properties of Selected Structural-Steel Shapes. We will use the properties of standard structural shapes when we analyze stresses and deflections of beams in Chaps. 5, 6, and 7.

Appendix D Section Properties of Sawn Lumber and Timber. Again we will find these tabular values useful for the calculation of stresses and deflections. We note from this table that a 2 × 4 piece of lumber is actually 1.5 in by 3.5 in!

Appendix E Section Properties of Common Piping Sections.

Appendix F Typical Mechanical Properties of Selected Materials. These tables give representative data values for a number of common materials. For example, it is useful as we work the solutions to problems to check that the stress values obtained are less than the yield stress or ultimate stress of the material in the problem. For some problems we will need to look up the material properties to proceed with the solution.

Appendix G Deflections and Slopes of Beams. We will find this table useful when we need to add simple solutions of beam problems using superposition (covered in Chaps. 6 and 7).

Appendix H Instructions are given for running the programs on the diskette that comes with this textbook.

Appendix I This is a summary of the computer programs on the diskette that comes with this textbook. Here we show the menus from the programs that are on the diskette.

Appendix J Instructions are given for running specific programs on the diskette with reference to text Examples.

PROBLEMS

1.2-1 A 2-in-diameter steel rod is loaded in tension by an axial load of 50,000 lb. Estimate the values of the longitudinal and transverse strains in the rod. For steel, $E = 30 \times 10^6$ psi and $\nu = 0.3$.

1.2-2 A steel wire and an aluminum wire, each 2 m long, are stretched separately by an amount equal to 2 mm. The cross-sectional area of each wire is 5×10^{-4} m^2, and $E_{st} = 205$ GPa and $E_{al} = 69$ GPa. Find the force and normal stress in each wire.

1.2-3 Tie rods are used to support the walls in a utility shed which stores sand. The sand acts against the wall at a given location with an approximate pressure of 15 kPa, as shown in Fig. P1.2-3. Estimate the tensile stress in the tie rod if we assume that the total pressure of the sand at this location acting on the wall is carried by the washer plate. This will give a higher value than would be expected in practice.

1.2-4 A sign of weight W is supported by two elastic rods AB and BC, as shown in Fig. P1.2-4. The diameter of rod AB is d_1, and the diameter of rod BC is d_2. Determine the normal stress in each rod. Take $W = 300$ lb, $d_1 = \frac{3}{8}$ in, $d_2 = \frac{1}{2}$ in, $\theta_1 = 60°$, and $\theta_2 = 45°$. Neglect the weight of the rods.

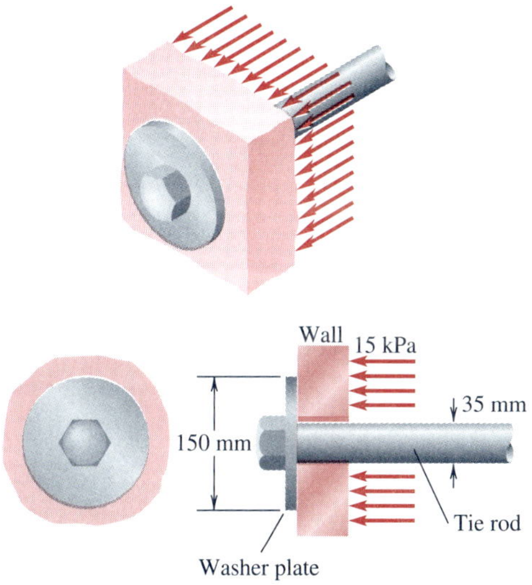

Fig. P1.2-3

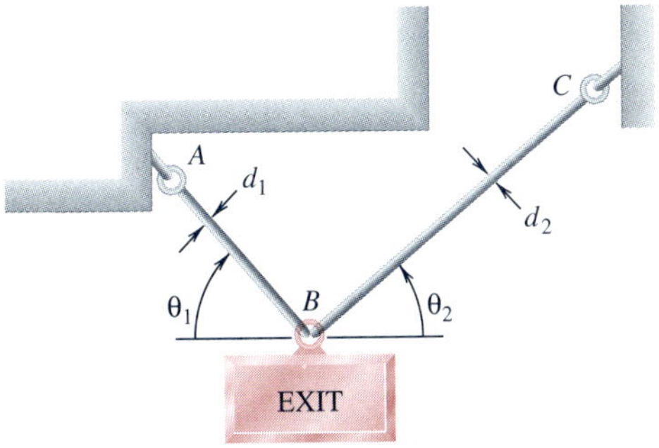

Fig. P1.2-4

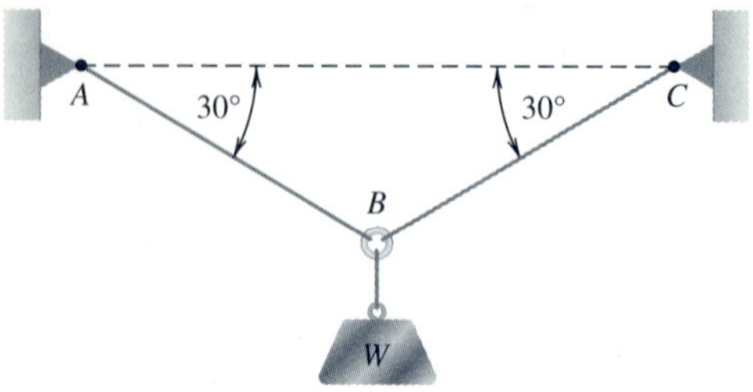

Fig. P1.2-6

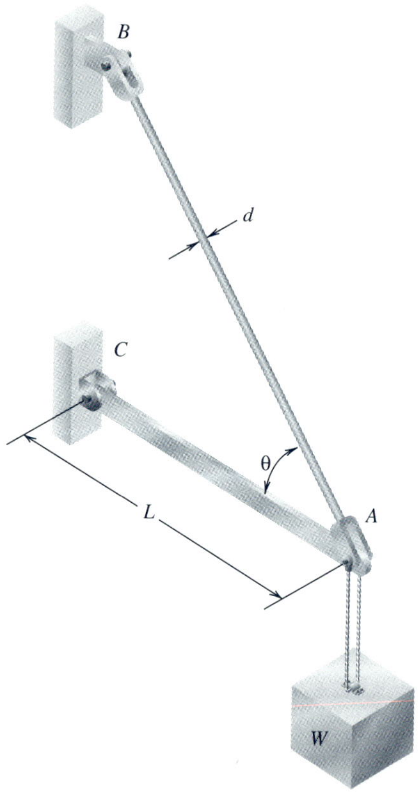

Fig. P1.2-7

1.2-5 Solve Prob. 1.2-4 if $W = 800$ kN, $d_1 = 20$ mm, $d_2 = 10$ mm, $\theta_1 = 45°$, and $\theta_2 = 45°$.

1.2-6 Two wires AB and BC are attached to walls at A and C and support a box of weight $W = 500$ lb, as shown in Fig. P1.2-6. If the diameter of the wire is 0.5 in, find the stress in the wire.

1.2-7 A light rod AB attached to a wall at B supports a rigid bar AC which is pinned to the wall at C, as shown in Fig. P1.2-7. Find an expression for the stress in the rod in terms of W, θ, L, and the diameter d of the rod.

1.2-8 Rods AB and BC are pinned at A, B, and C and carry a load W at B, as shown in Fig. P1.2-8. Find an expression for the maximum stress in the rods if the cross-sectional area of the rods is A_R.

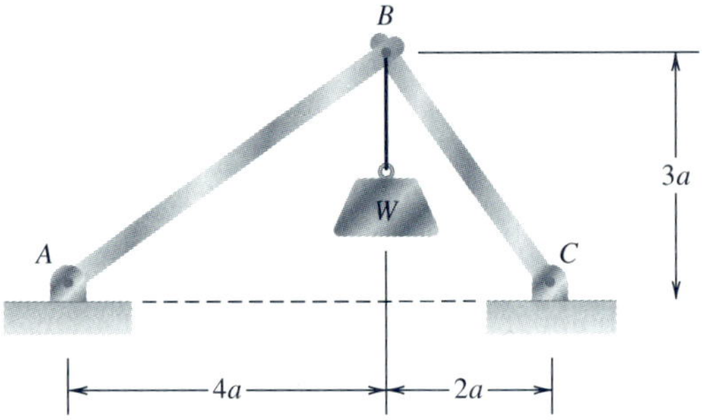

Fig. P1.2-8

1.2-9 Suppose for a given material that testing indicates that

$$\sigma = E_Q \epsilon^2$$

where E_Q is an experimentally determined elastic constant in pounds per square inch and the relation is only valid for $\sigma > 0$. For two bars of length L and cross-sectional area A, one made up of a material obeying Hooke's law and the other obeying the quadratic relation given above, compare the predicted elongations of the two bars for a given load P.

1.2-10 Two solid-steel circular rods are attached to rigid walls at A and C and fixed together at B, as shown in Fig. P1.2-10. In a test of this structural component, a displacement of $\delta = 0.425$ mm is measured for an applied load of $P = 500$ kN. Find the force carried by each of the rods and the stresses in each. Take $A_{AB} = 1.96 \times 10^{-3}$ m^2, $A_{BC} = 7.85 \times 10^{-3}$ m^2, and $E = 200$ GPa.

1.3-1 A rod and yoke, as shown in Fig. P1.3-1, are subjected to a load $P = 5000$ lb. If the pin diameter is 0.5 in and the rod diameter is 1 in, find the average shear stress in the pin and the average normal stress in the rod.

1.3-2 What force is required to punch a 2-in-diameter hole through a $\frac{5}{8}$-in-thick aluminum-alloy plate for the configuration shown in Fig. 1.8b? The average shear strength of the alloy is 30,000 psi.

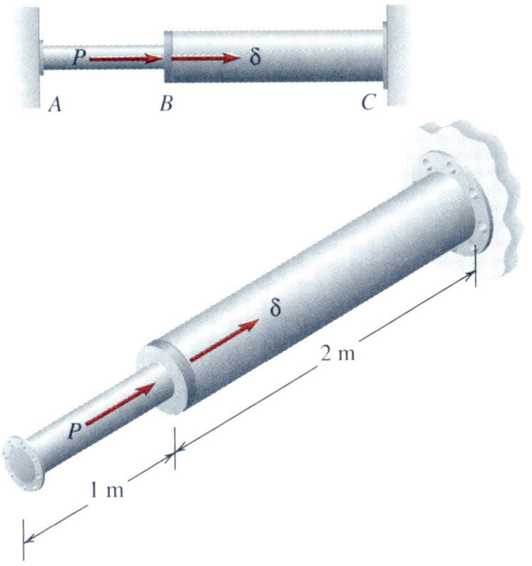

Fig. P1.2-10

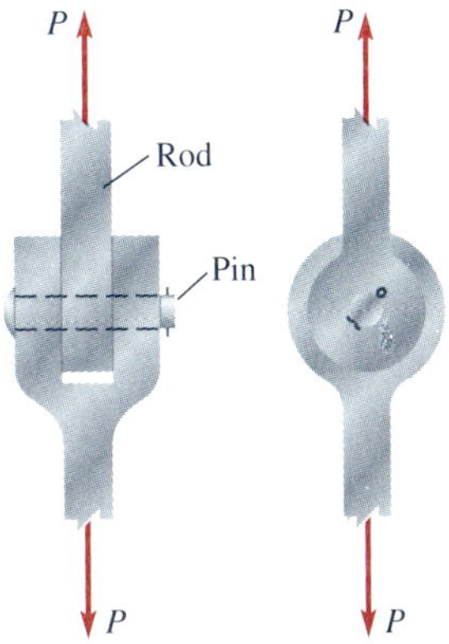

Fig. P1.3-1

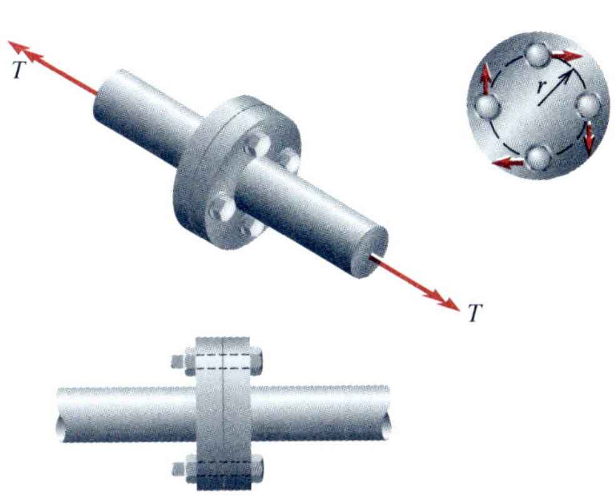

Fig. P1.3-3

1.3-3 A bolted flange connection for a shaft is shown in Fig. P1.3-3. Four $\frac{3}{8}$-in-diameter bolts whose allowable average shear stress is 4000 psi are used. If $r = 3.5$ in, what is the maximum twisting moment that can be transmitted across the flange connection? Neglect friction in the flange connection, and assume that the twisting moment T is carried totally by the bolts.

1.3-4 A load P is carried by the rigid bar ABF and a steel supporting link DB as shown in Fig. P1.3-4. The link BD has cross-sectional area of 0.75 in^2. Find the deflection under the load P and the normal stress in link DB. Take $P =$ 10,000 lb and $E = 30 \times 10^6$ psi.

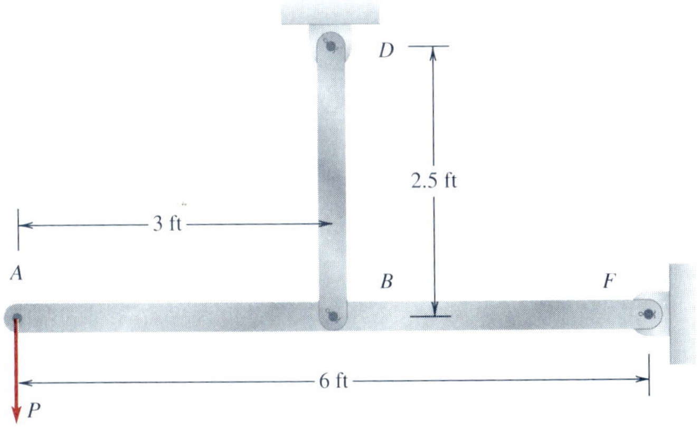

Fig. P1.3-4

1.3-5 The snowblower blade components are connected to the 1-in-diameter drive shaft by a shear pin, as shown in Fig. P1.3-5. If an object is jammed in the blades, the shear pin will fail, preventing damage to the drive shaft. If the shear strength in the pin is 3500 psi, estimate the force on the blades that will cause the pin to fail.

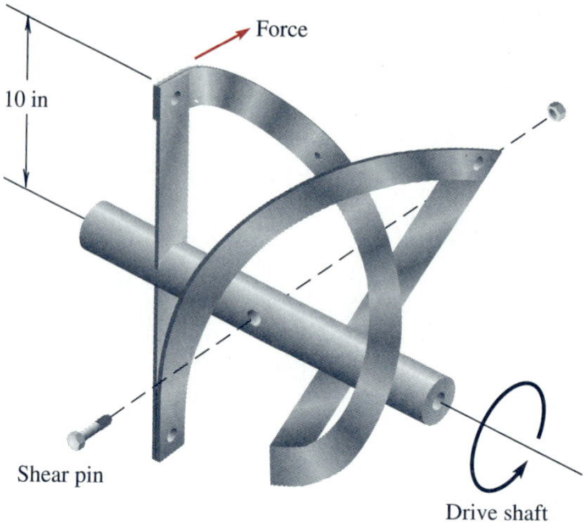

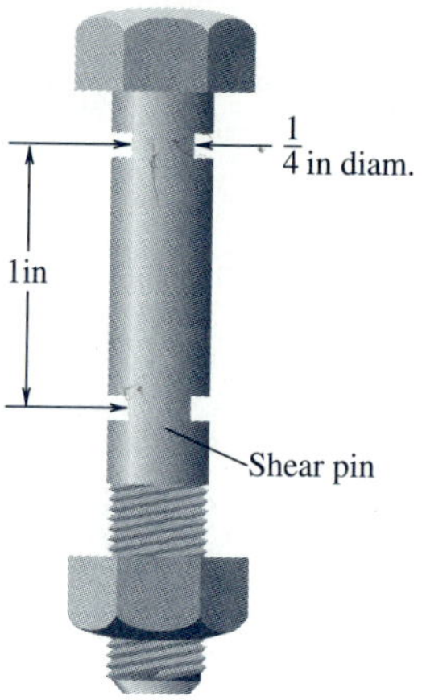

Fig. P1.3-5

1.3-6 To isolate the vibration modes of adjacent buildings during an earthquake, seismic joints are sometimes employed to allow independent movement of each building, as shown in Fig. P1.3-6. If $\frac{3}{4}$-in bolts are used in the links, estimate the force needed to cause failure of the bolts if the maximum allowable average shear strength in the bolts is 6500 psi.

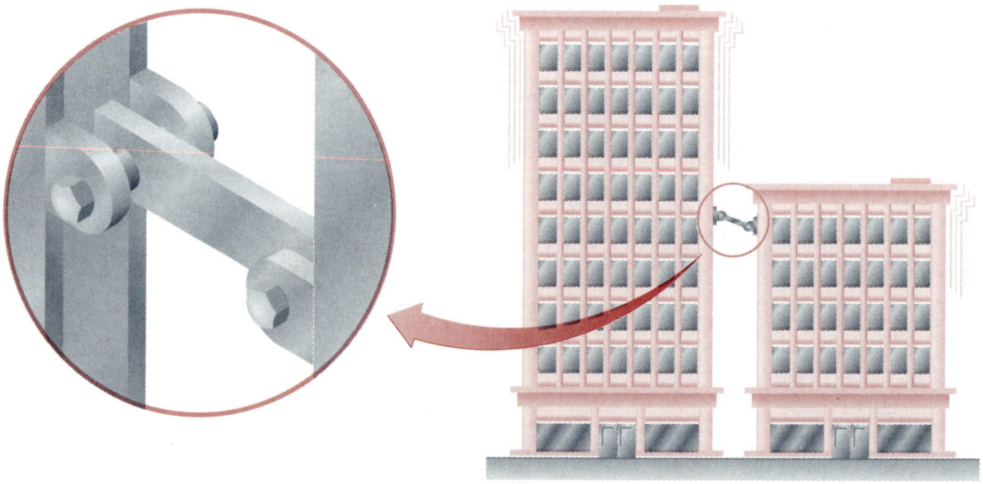

Fig. P1.3-6

1.3-7 An aircraft engine pylon is supported by a thrust link and a $\frac{1}{2}$-in bolt connection to a support fitting in the wing, as sketched in Fig. P1.3-7. If the bolt failed in shear, estimate the force that acted across the bolt. Take the maximum allowable shear strength in the bolt as 10,000 psi.

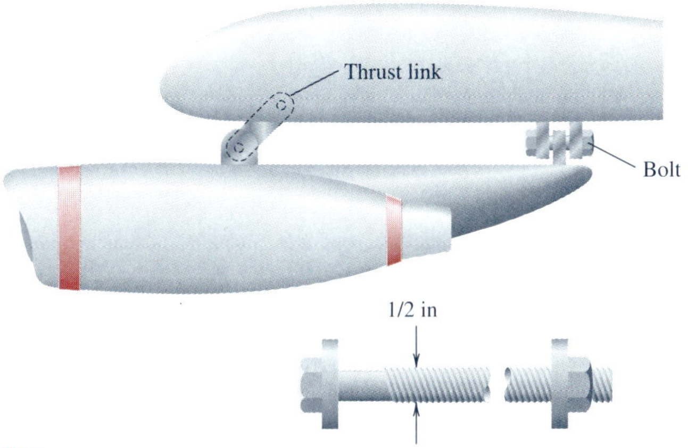

Fig. P1.3-7

1.3-8 Estimate the average shear stress on the cross section of the 15-mm bolt at location *A* connecting the boat davit to the side of the ship, as shown in Fig. P1.3-8. The boat weighs 7 kN, and the weight of the davit structure itself can be neglected.

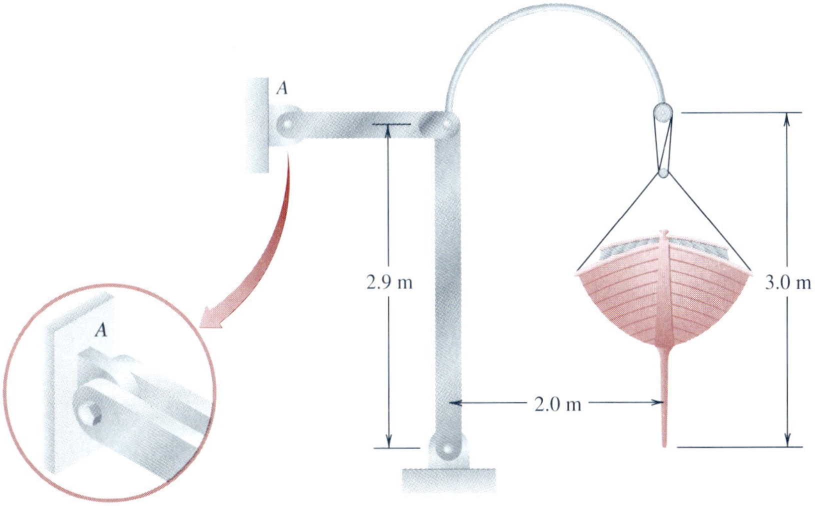

Fig. P1.3-8

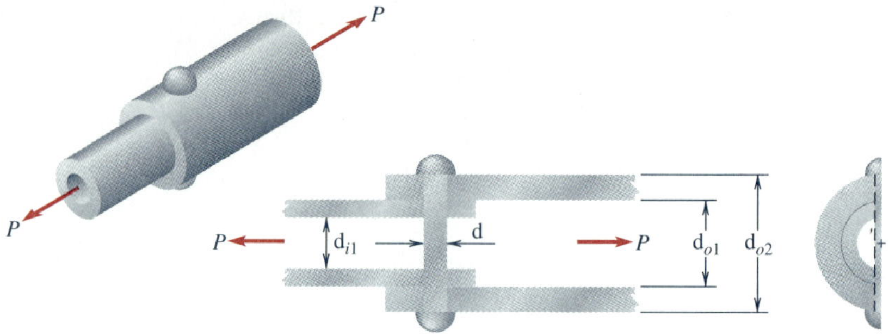

Fig. P1.3-9

1.3-9 Two tubes are connected as shown in Fig. P1.3-9 by a pin of diameter d. If a load $P = 40$ kN is carried across the pin, calculate the average shear stress in the pin. Take $d_{o1} = 50$ mm, $d_{i1} = 30$ mm, $d_{o2} = 80$ mm, and $d =$ 10 mm.

1.3-10 Consider the assembly in Fig. P1.3-9. If the maximum average shear stress which can be allowed in the pin is 5000 psi, find the maximum allowable load P that can be carried across the tubes. Take $d_{o1} = 3.00$ in, $d_{i1} = 2.25$ in, $d_{o2} = 3.50$ in, and $d = 0.375$ in.

1.3-11 A 100-mm-diameter drive shaft shown in Fig. P1.3-11 transmits a twisting moment of $T = 15$ kN · m to the outer hollow shaft through a shear key as shown. Find the average shear stress in the key.

1.3-12 A copper disk 100 mm in diameter and 0.125 mm thick is fitted into the casing of a low-pressure test air chamber, as shown in Fig. P1.3-12. The disk

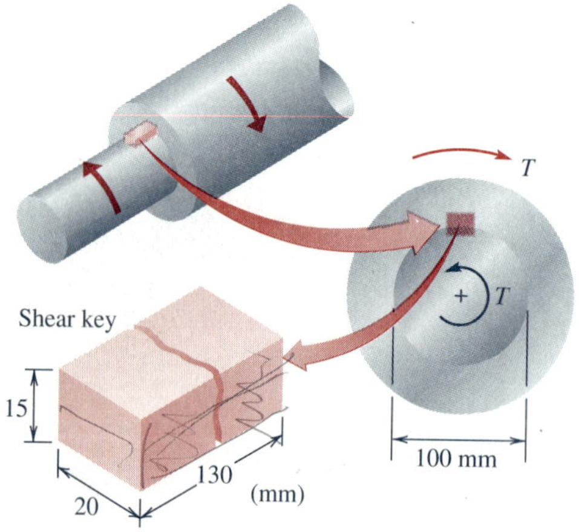

Fig. P1.3-11

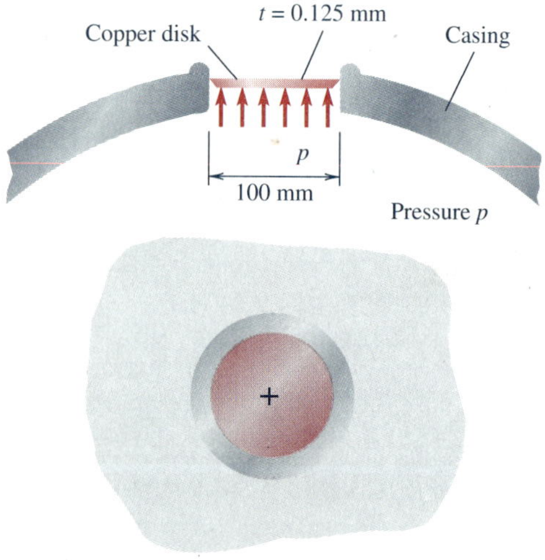

Fig. P1.3-12

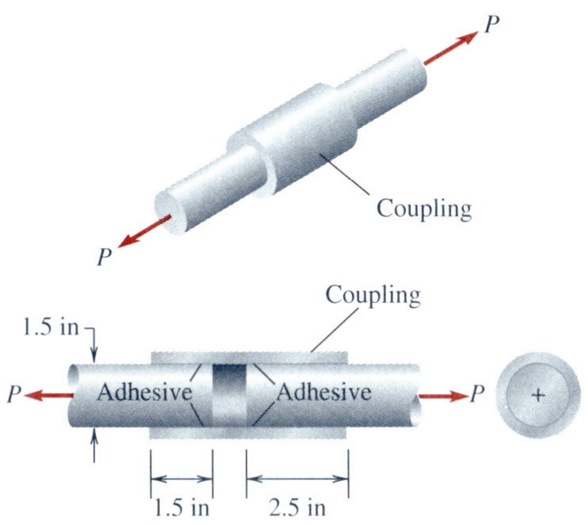

Fig. P1.3-13

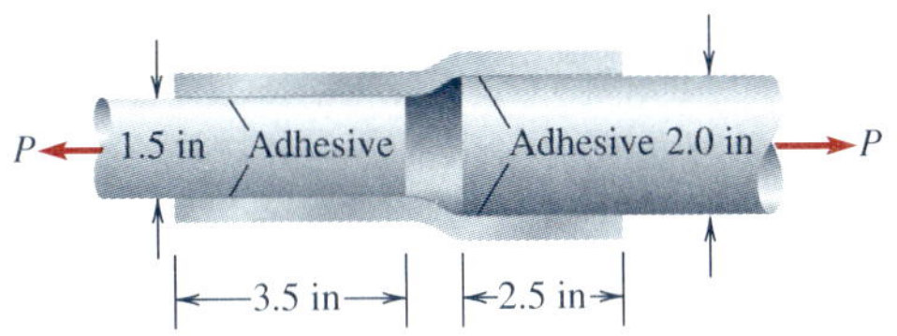

Fig. P1.3-14

is designed to blow out if the pressure within the chamber reaches a critical value. If we assume that disk failure occurs by shear failure around the edges of the disk, estimate the required pressure p to cause failure of the disk. Assume that shear failure of the material of the disk occurs at about 1.4 MPa.

1.3-13 A coupling tube connects two 1.5-in-diameter plastic bars as shown in Fig. P1.3-13. The coupling tube is bonded to each of the bars by a quick-acting adhesive. If a load $P = 8000$ lb is applied, estimate the average uniform shear stress acting on the bonded surfaces.

1.3-14 A coupling used to connect two different-diameter lightweight plastic truss members is shown in Fig. P1.3-14. If the maximum allowable uniform shear stress in the adhesive is 500 psi, find the maximum load that can be transmitted across the coupling.

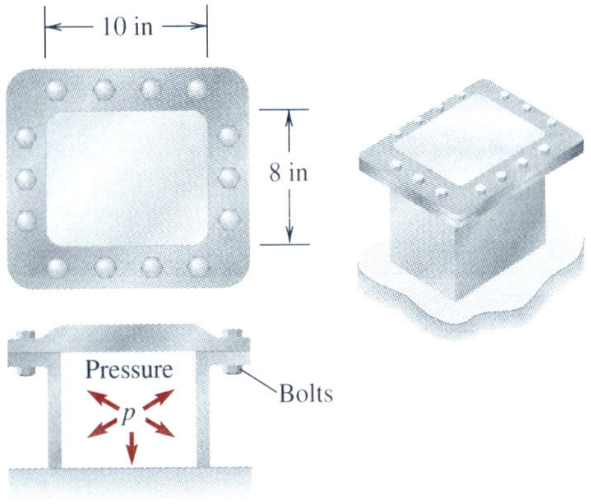

Fig. P1.4-1

1.4-1 A pressurized tank is closed by a cover plate connected to the tank with steel bolts, as shown in Fig. P1.4-1. If the pressure in the tank is 200 psi, the diameter of the bolts is 0.5 in, and the allowable tensile stress in the bolts is 8000 psi, find the minimum number of bolts required to keep the cover plate fastened under pressure.

1.4-2 Two steel members are connected by a bolt CD as shown in Fig. P1.4-2. If the load is $P = 40$ kN and the allowable shear stress in the bolt is 100 MPa, find the minimum required diameter of the bolt.

1.4-3 Consider Example 1.5, Fig. 1.12. If the thickness of component AB is now 0.2 in instead of 0.3 in, what is the maximum load P that can be carried by the bracket?

1.4-4 Consider Example 1.3, Fig. 1.10. If the original $\frac{5}{8}$-in bolts are inadequate with respect to the allowable shear stress of 7000 psi in the bolts, what diameter bolt will meet this requirement?

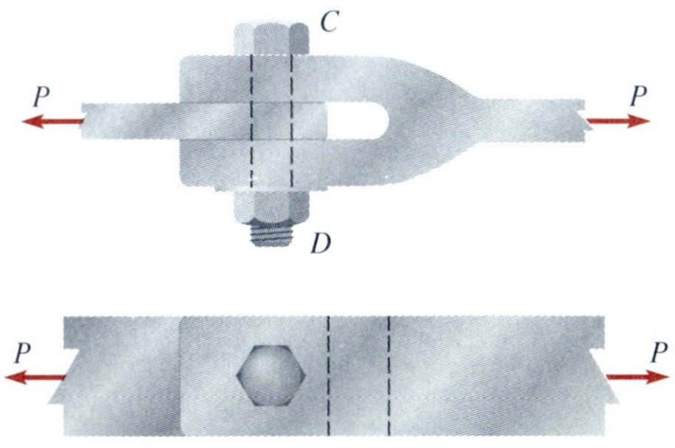

Fig. P1.4-2

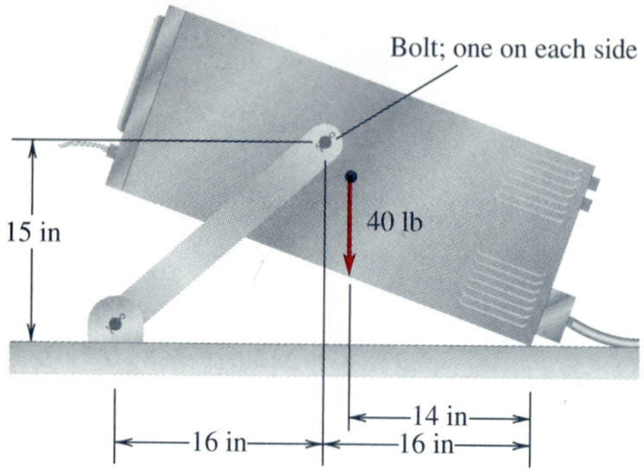

Fig. P1.4-6

1.4-5 The human femur is estimated to have an elastic modulus of approximately 17 GPa and a breaking stress in tension of approximately 200 MPa. Find the strain at breakage and the change in length of a bone 0.30 m in length.

1.4-6 An instrument holder sketched in Fig. P1.4-6 is connected to the instrument by two plastic bolts as shown. If the instrument weighs 40 lb and the bolts are $\frac{1}{4}$ in in diameter, what is the minimum shear strength required of the bolts to support the instrument? Neglect friction.

1.4-7 A short, hollow, cylindrical steel member is to support a load of 30,000 lb as shown in Fig. P1.4-7. If the maximum allowable normal compressive stress is 12,000 psi, find the values of the inside and outside diameters if $d_o = 1.5d_i$ and $E = 30 \times 10^6$ psi.

1.4-8 The maximum gas pressure in an 8-in-diameter engine cylinder is 700 psi. Eight ⅞-in cap screws fasten the rigid head to the cylinder as shown in Fig. P1.4-8, and the screws are tightened in the absence of pressure so that each screw has an axial load of 8000 lb. Estimate the average normal stress in each screw when the cylinder is pressurized.

1.4-9 A steel link from a mine bucket loader is shown in Fig. P1.4-9. If the maximum normal stress in the material is not to exceed 10,000 psi (a value that includes a factor of safety), find the maximum value of the force P that can be carried by the link. Investigate the stresses in both sections AA and BB.

1.4-10 A small device for crushing cans for recycling is shown in Fig. P1.4-10. The main components are steel tubes, and strap AB is connected by $\frac{1}{4}$-in-diameter pins at points A and B. If a load P is applied when a rigid rock is wedged into the crushing space, what load P will cause the pins to shear if the maximum allowable average shear strength in the pins is $\tau = 10,000$ psi?

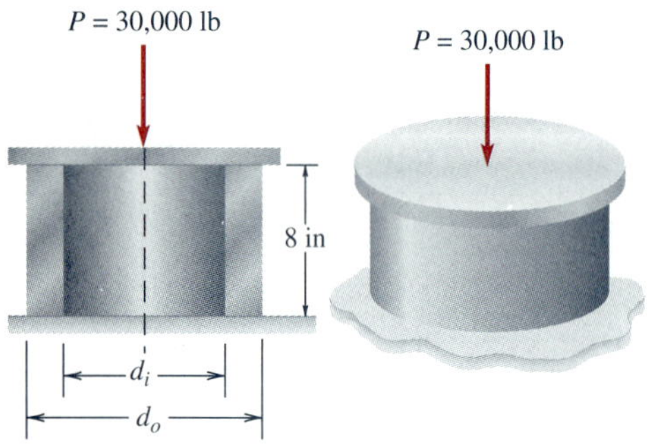

Fig. P1.4-7

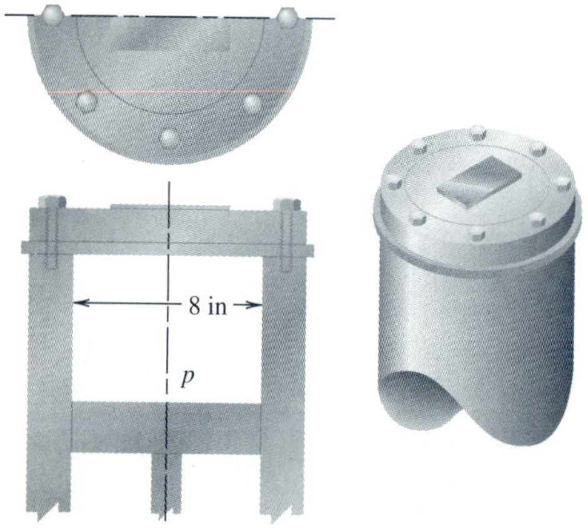

Fig. P1.4-8

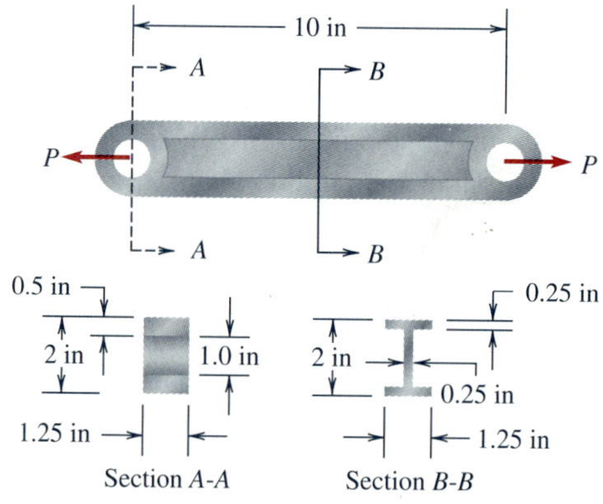

Fig. P1.4-9

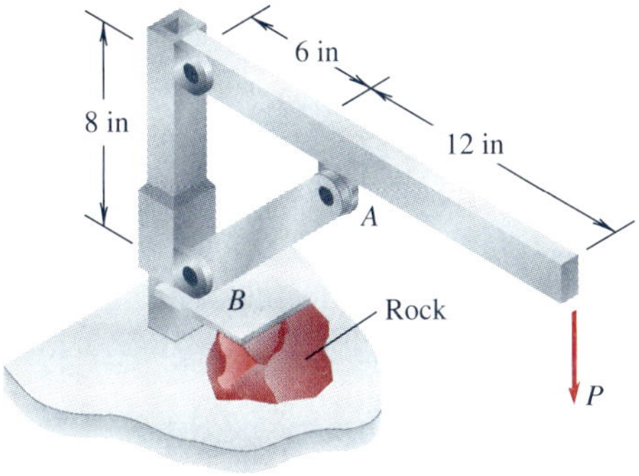

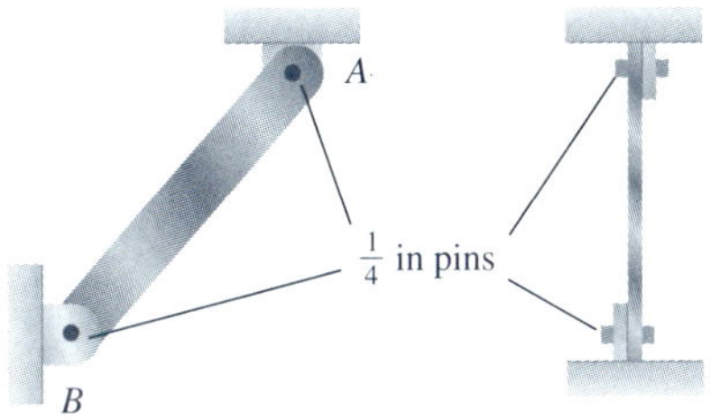

Fig. P1.4-10

1.4-11 A simple hoist is constructed from a steel rod AB and a steel tube BC, as shown in Fig. P1.4-11. If the maximum allowable shear stress in the bolt at point B is 48 MPa (the bolt is in double shear), estimate the maximum allowable load P that the hoist can support. The diameter of the bolt at B is 25 mm. Neglect the weight of the components.

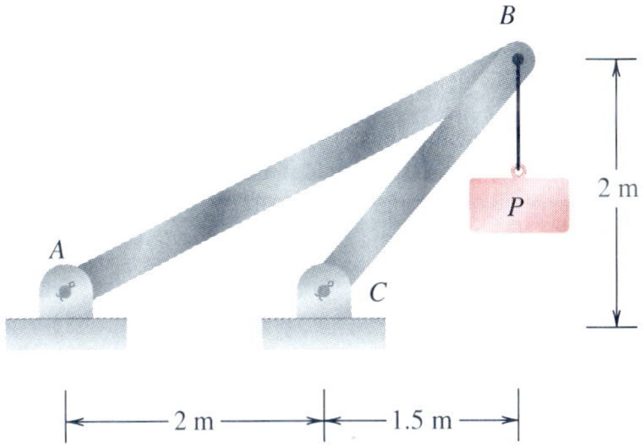

Fig. P1.4-11

1.4-12 A rigid member DAC as shown in Fig. P1.4-12 carries the load P. Bar AB is an aluminum bar whose cross-sectional area is 0.5 in^2, and it is pinned at A and B by 0.5-in-diameter pins in a single-shear configuration. If the maximum allowable normal stress in AB is 7000 psi and the maximum allowable shear stress in the pins is 15,000 psi, find the maximum allowable load P. For aluminum $E_{al} = 10 \times 10^6$ psi.

1.4-13 If in addition to the requirements specified in Prob. 1.4-12 the maximum elongation of member AB is not to exceed 0.10 in, find the maximum allowable load P.

1.4-14 A rigid member AB as shown in Fig. P1.4-14 is supported by a high-strength steel rod BC 0.75 in in diameter through pins of 0.625-in diameter loaded in double shear at locations B and C. We wish to estimate the maximum load P that can be applied to the member if the maximum allowable normal stress in the rod is 60 ksi and the maximum allowable shear stress in each pin is 50 ksi. Neglect the weight of the rigid member AB and any friction at the pins.

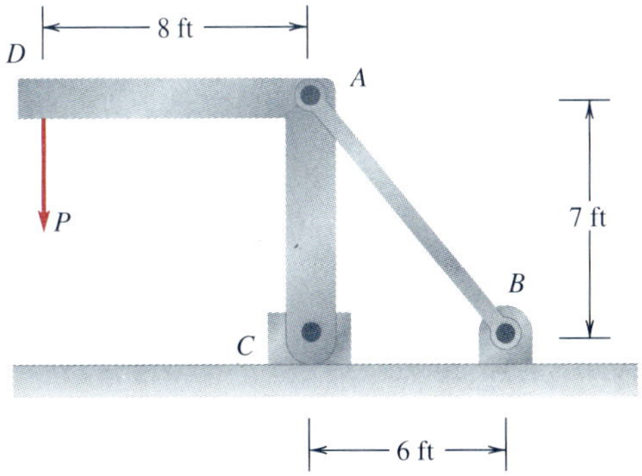

Fig. P1.4-12

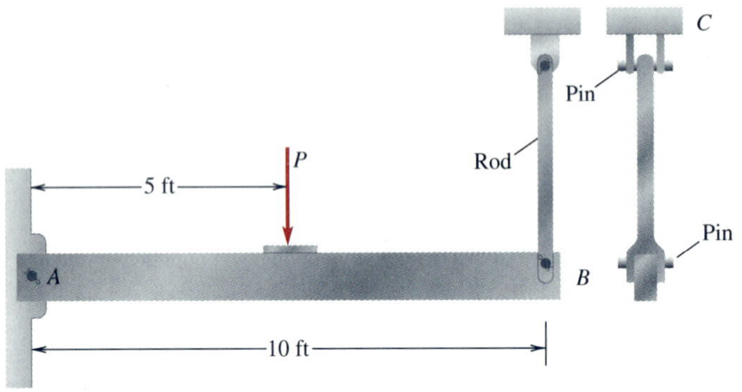

Fig. P1.4-14

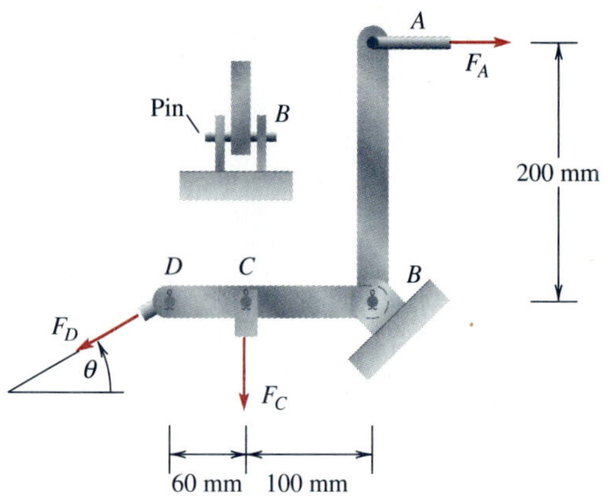

Fig. P1.4-15

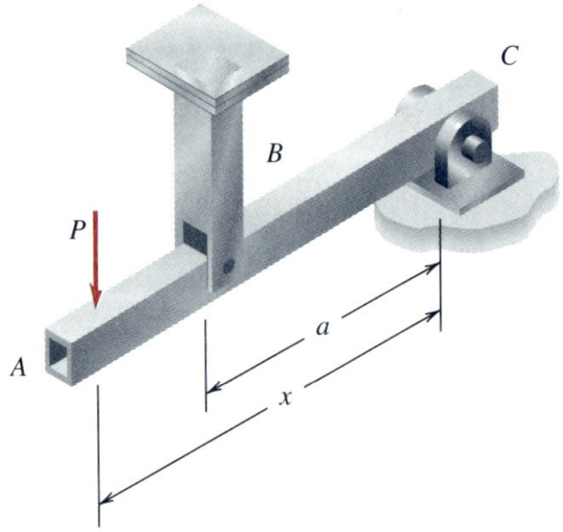

Fig. P1.4-17

1.4-15 A rigid control arm *ABCD* is loaded as shown in Fig. P1.4-15. We wish to find the required diameter of the pin in the bracket at *B* if the allowable shear stress of the pin material is 60 MPa. Take $F_A = 12$ kN, $F_C = 14$ kN, and $\theta = 30°$. The force F_D is found from equilibrium requirements.

1.4-16 A rigid control arm *ABCD* is loaded as shown in Fig. P1.4-15. We wish to find the required diameter of the pin in the bracket at *B* if the allowable shear stress for the pin material is 60 MPa. Take $F_C = 10$ kN, $\theta = 40°$, and $F_D = 20$ kN. The force F_A is found from equilibrium requirements.

1.4-17 A rigid bar *ABC* as shown in Fig. P1.4-17 is to carry a load *P*. The pins at supports *B* and *C* are 0.25 in in diameter with an allowable shear stress of 20 ksi. Estimate the range of possible locations *x* over which a load of $P = 1100$ lb can be placed on the bar without exceeding the shear stress in the pins at either *B* or *C*. Take $a = 1$ ft.

1.4-18 A 2-in-diameter turbine shaft is mounted vertically as shown in Fig. P1.4-18. A collar connected to the shaft is supported on the lower bearing surface. If the maximum allowable shear stress along the cylindrical surface *AC-BD* is 20 ksi, estimate the maximum thrust *P* that can be put on the shaft.

1.4-19 The technology of paper and paperboard uses notation for stress and failure stress that attempts to avoid the uncertainty associated with the thickness of paper as it is processed. For example, newsprint is approximately 0.003 in thick and weighs about 25.6 lb/3000 ft^2. The tensile strength is approximately 5.9×10^5 lb/ft^2. Find the breaking length in kilometers for newsprint if the breaking length is the length of a strip of newsprint whose weight is equivalent to the force that would break it. Carry out a similar calculation for linerboard that is 93.7 lb/3000 ft^2 and 0.011 in thick and has a tensile strength of 5500 psi.

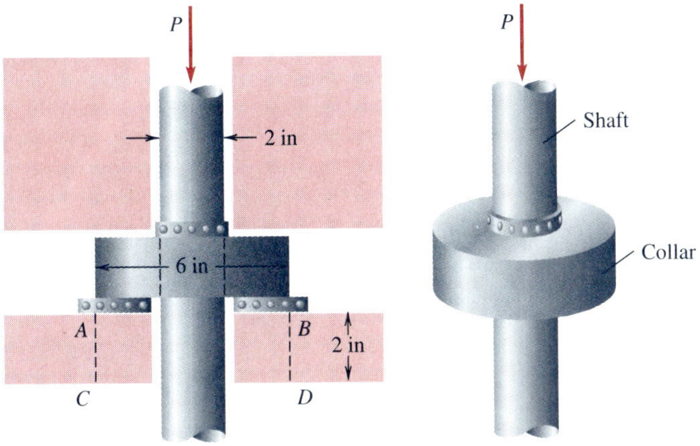

Fig. P1.4-18

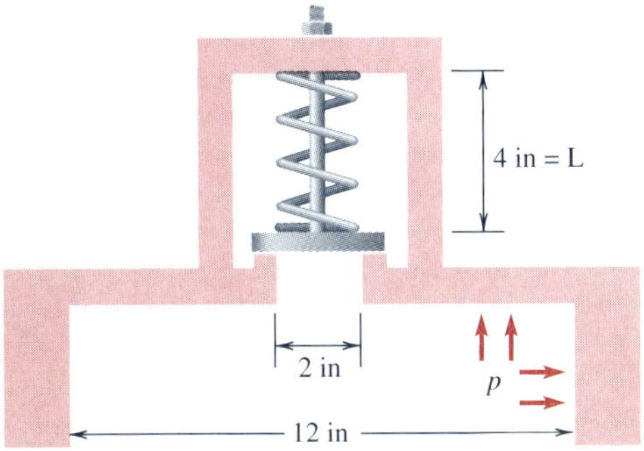

Fig. P1.4-20

1.4-20 A spring whose free length is 5 in is compressed to a length $L = 4$ in and used in a safety valve for a pressure system, as shown in Fig. P1.4-20. The diameter of the discharge hole is 2 in, and the spring has a spring constant of 700 lb/in. Estimate the pressure at which the valve will open. How does the opening pressure vary with the length L?

1.4-21 For the truss shown in Fig. P1.4-21, find the minimum cross-sectional area for members 2-4, 1-4, 6-8, 5-7 if the allowable normal stress is 15 ksi.

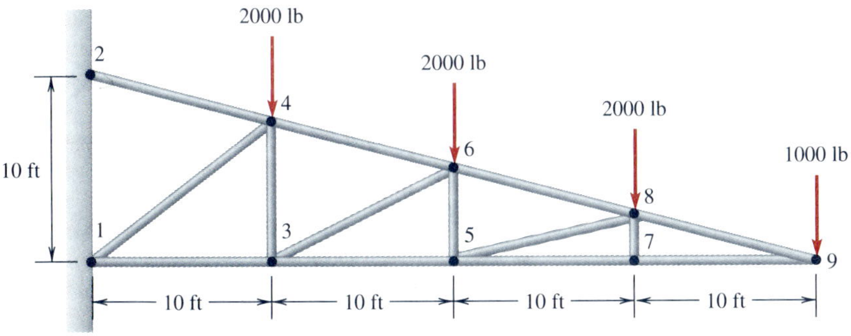

Fig. P1.4-21

1.4-22 A person resting in a hammock as shown in Fig. P1.4-22 weighs 240 lb. If the single nylon ropes AE and CB supporting the hammock have an allowable stress of 150 psi, find the minimum diameter of these ropes to support the person.

Fig. P1.4-22

1.4-23 The concrete bucket shown in Fig. P1.4-23 weighs 25 kN. If the allowable stress in the cable is 35 MPa and the coefficient of friction μ between the bucket and the slide is 0.3, find the minimum diameter of the cable. Check both operations of the bucket when it is being lowered as well as being raised.

1.4-24 If the second deepest well in the world is approximately 31,440 ft deep, what is the normal stress in the steel drill pipe near the top supporting the weight of the pipe? The pipe has an outside diameter of 6 in and a wall thickness of 0.5 in. What is the depth of the deepest well in the world?

1.4-25 For the truss shown in Fig. P1.4-25, find the minimum allowable areas A_1 and A_2 in terms of the allowable stress σ_a in the members and the load P.

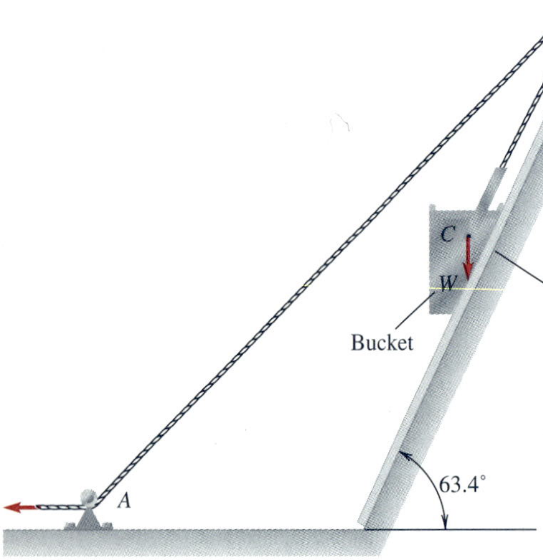

Fig. P1.4-23

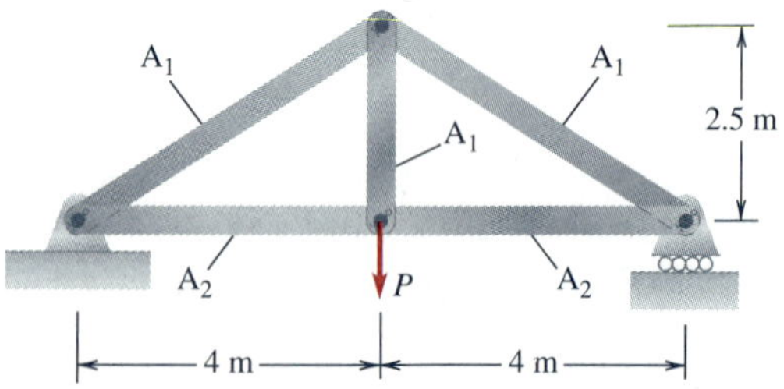

Fig. P1.4-25

CHAPTER 2

Uniaxial Loading and Deformation

2.1 Introduction

In Chap. 1, we reviewed equilibrium of forces in bars and introduced the concept of a normal stress acting in a bar. We also noted that most engineering materials obey Hooke's law, which states that the change in length per unit length of a bar, the strain, is proportional to the force intensity in the bar, the stress, i.e.,

$$\frac{\delta}{L} = \epsilon = \frac{F/A}{E} = \frac{\sigma}{E} \tag{2.1}$$

where δ is the change in length of the bar of original length L.

ϵ is the strain in the bar.

F is the force in the bar.

A is the cross-sectional area of the bar.

σ is the stress in the bar.

E is the elastic modulus or Young's modulus of the material.

We usually write this stress-strain relation for a bar in the form

$$\sigma = E\epsilon \tag{2.2}$$

Equation (2.2) can be considered as the relation that gives the stress in the bar arising from stretching or compression of the bar. Many important technological problems involve long, slender members or components modeled as bars interconnected to one another and subjected to applied loads and temperature changes. For this reason, it is important to have a clear notion of the concepts of stress and strain in a bar so as to be able to solve problems. The important first step is to define carefully all force and displacement quantities relative to a coordinate system; we do this next in Sec. 2.2.

2.2 Axial Deformation of a Bar

We will consider forces applied to bars in such a way that the resulting force in the bar acts along the centroidal axis of the bar; since the bar is deformable, this loading gives rise to what we call axial deformation. Figure 2.1*a* shows a uniform bar of rectangular cross section subjected to an axial force F. In Fig. 2.1*a* we show the coordinate axis x passing through the centroid of the cross section of the bar, and the force acts along this centroidal axis.

It is convenient in the solution of most problems in mechanics of materials to refer force and deformation variables to a set of coordinate axes. In this way we can keep track of directions and movement in a consistent manner. We start introducing coordinate axes here for the simple case of one-dimensional or uniaxial deformation. If a force or displacement at a section acts in the positive direction of the axis, it is positive; if a force or displacement at a section acts in the negative direction of the axis, it is negative. We can locate the origin of the coordinate system in Fig. 2.1*a* at any convenient position along the bar, and the location or position of any cross section along the bar will be specified by a numerical value of x in appropriate units, e.g., meters or feet measured from this origin. Since the force in the bar acts along the centroidal axis, all points on a given cross section will experience the same axial deformation and stress depending only on the value of x at the section.

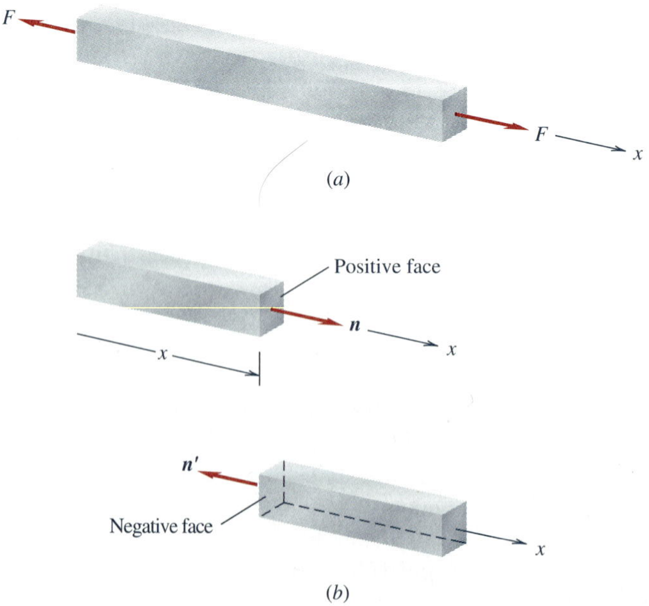

Figure 2.1 (*a*) Axial force in a bar; coordinate axis passes through the centroid of the cross section. (*b*) Positive and negative faces at section x.

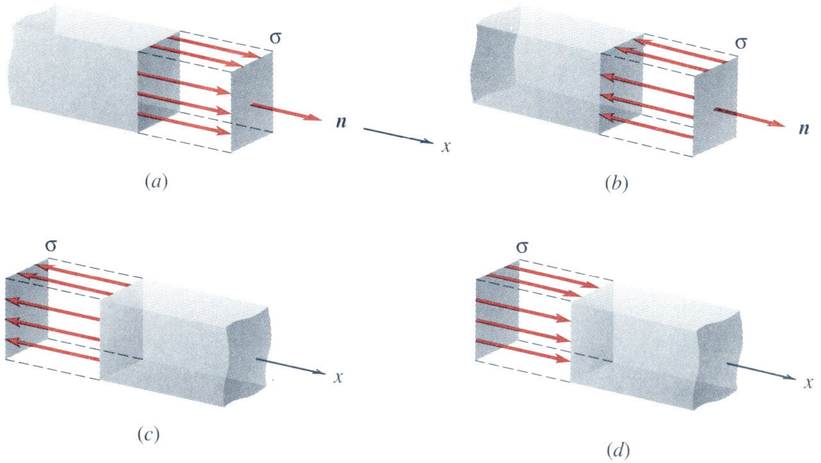

Figure 2.2 Stresses acting on cross-sectional area A. (a) Positive stress on a positive face; (b) negative stress on a positive face; (c) positive stress on a negative face; (d) negative stress on a negative face.

A cross section or face of the bar at a location x is defined as a *positive* face if the outward-pointing normal vector **n** to the face is in the same direction as the positive x coordinate. A face is defined as a *negative* face if the outward-pointing normal vector $\mathbf{n}'$ is in the direction of the negative coordinate axis; see Fig. 2.1b.

If an axial force F acts on a face with area A, it will give rise to a normal stress given by

$$\sigma = \frac{F}{A} \tag{2.3}$$

If the normal stress on a *positive* face is in the positive direction of the axis, we say this is a positive stress (Fig. 2.2a). If the normal stress on a *positive* face is in the *negative* direction of the axis, we say this is a negative stress (Fig. 2.2b). In the same way, if the normal stress on a *negative* face is in the negative direction of the axis, we say this is a positive stress (Fig. 2.2c); if the normal stress on a negative face is in the positive direction, we say this is a negative stress (Fig. 2.2d).

A bar in *tension* has a positive stress acting on each end of the bar; a bar in *compression* has a negative stress acting on each end of the bar. See Fig. 2.3. Each of the bars shown in Fig. 2.3 is in equilibrium, and the *internal* stress at any section in each bar has a constant value equal to σ.

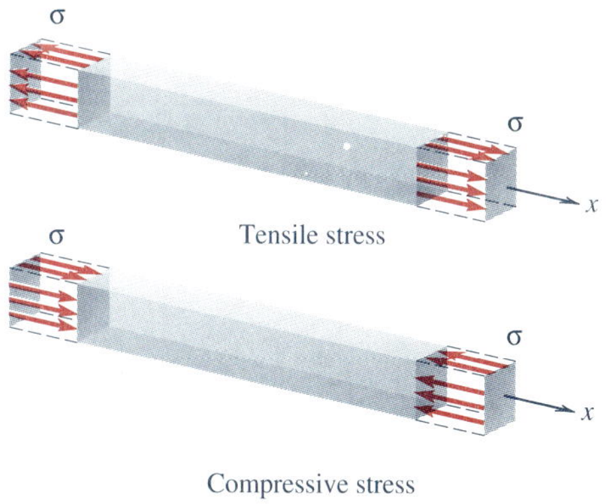

Figure 2.3 Tensile and compressive stress in a bar.

An axial bar AB of initial length L (see Fig. 2.4) under load will experience an axial deformation or change in length which we define as δ. To find an expression for the deformation of the bar, we consider the *displacements* of the endpoints A and B of the bar. Point A located initially, relative to the coordinate axis, at x_A in general *moves* due to the loading to a new position $x_A + u_A$, where u_A is called the *displacement* of point A. Similarly for point B; see Fig. 2.4. The displacements shown in Fig. 2.4 are highly exaggerated since the relative axial displacements of most

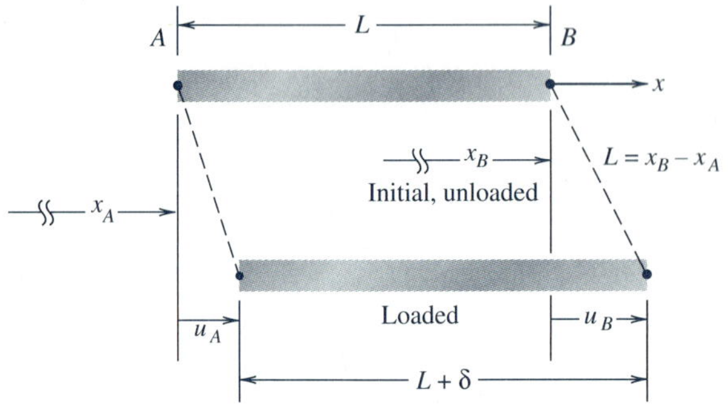

Figure 2.4 Displacements u_A and u_B at the ends of a bar and the deformation δ of the bar of length L.

engineering structures are small (compared to the dimensions of the structure). The displacements denoted by u_A and u_B are positive in the positive x direction as shown. The new length of the bar under the load is $x_B + u_B - (x_A + u_A)$ or $x_B - x_A + u_B - u_A = L + u_B - u_A$. Since the initial length of the member is $L = x_B - x_A$, the *deformation* δ or *change in length* of the bar is $u_B - u_A$. The displacements at the ends of the bar and the deformation of the bar are thus related by

$$\delta = u_B - u_A \tag{2.4}$$

When δ is positive, we say the bar has elongated or stretched; when δ is negative, we say the bar has shortened or contracted. For example, if $u_A = 0$, that is, point A is fixed, then the bar elongates when $u_B > 0$ and the bar contracts when $u_B < 0$. The deformation in Eq. (2.4) depends on the *relative* displacements of the ends of the bar. That is, if point B has a positive displacement relative to end A, then δ is positive and the bar elongates. If point B has a negative displacement relative to end A, then the bar contracts.

If an axial bar of length L as shown in Fig. 2.4 undergoes a deformation δ, the normal (or axial) strain in the bar is given by

$$\epsilon = \frac{\delta}{L} = \frac{\text{change in length of bar}}{\text{original length of bar}} = \frac{u_B - u_A}{L} \tag{2.5}$$

where the strain ϵ is positive if the bar elongates and is negative if the bar shortens.

A convenient formula for the deformation follows by substituting for the stress and strain in Hooke's law, Eq. (2.2), using Eqs. (2.3) and (2.5) to obtain

$$\delta = \frac{FL}{AE} \tag{2.6}$$

where $\delta = u_B - u_A$. Alternatively, we can express the force in the bar in terms of the displacements at the ends of the bar

$$F = \frac{EA}{L}(u_B - u_A) = k(u_B - u_A) \tag{2.7}$$

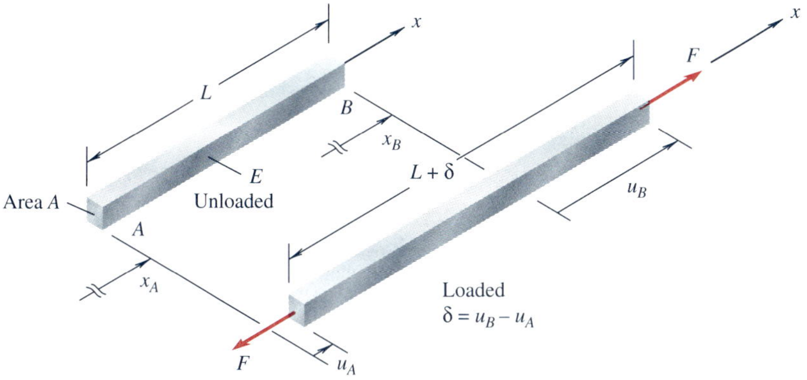

Figure 2.5 Forces and displacements in a bar under axial load. The locations of points A and B are given by x_A and x_B; the displacements of points A and B are u_A and u_B.

where $k = EA/L$ is called the *axial stiffness* of the bar with units of force/length, e.g., newtons per meter, N/m. We refer to Eq. (2.7) as the *force-deformation relation* of the bar; see Fig. 2.5. In obtaining Eq. (2.7) we have assumed that the bar is in tension, as shown in Fig. 2.5.

EXAMPLE 2.1

A steel link shown in Fig. 2.6 carries a load F. If the maximum allowable stress in the link is 15,000 psi, we wish to find the maximum allowable load in the link and the corresponding elongation of the link. We neglect end effects at the end fixtures and take $E = 30 \times 10^6$ psi. The approximate weight of the link is 5 lb, so we neglect the weight of the link in the stress calculations. The tensile stress in the link is

$$\sigma = \frac{F}{A} \tag{a}$$

If $\sigma = 15{,}000$ psi, and $A = 1\ \text{in}^2$, the force in the link is equal to 15,000 lb. The corresponding elongation or deformation of the link is, from Eq. (2.6),

$$\delta = \frac{FL}{AE} = \frac{(15 \times 10^3)(10)}{(1)(30 \times 10^6)} = 5 \times 10^{-3}\ \text{in} \tag{b}$$

The result of this calculation indicates, as we have noted previously, that the deformation of typical engineering components is small. The strain in the link is

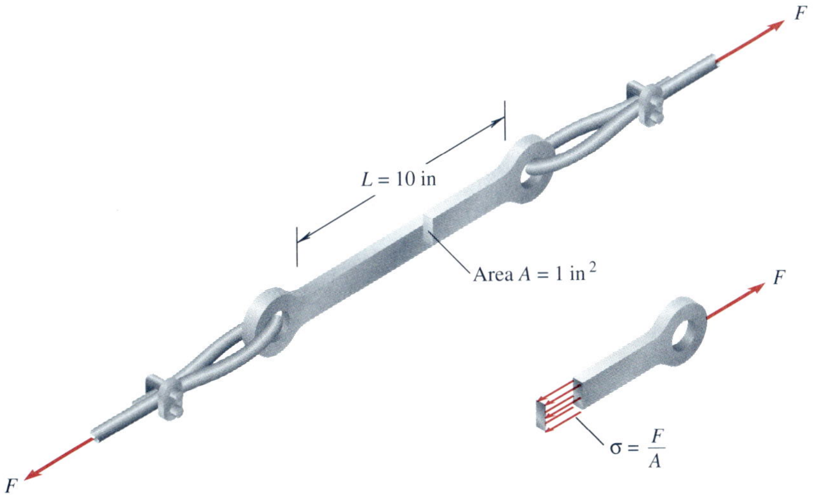

Figure 2.6 Example 2.1

$$\epsilon = \frac{\delta}{L} = 5 \times 10^{-4}$$
$$= 500 \times 10^{-6} \text{ in/in} = 500 \ \mu\text{in/in} \qquad (c)$$

We see from the results of this example that a steel member is capable of supporting a substantial load in tension relative to its weight and that the corresponding deformation and strain of the member are small.

2.3 Analysis of Deformable Bodies

In this section we begin developing the procedures for the analysis of mechanical and structural problems involving deformable bodies. It will be seen that the main steps to be followed for the analysis will remain the same in spite of a wide range of problem types encountered. As we proceed through the chapters of this book, more complicated models relating the forces and moments acting on a deformable structural element to the geometric deformation of the element will be developed. Our objective will be to develop a general approach for solving a wide class of practical problems.

We assume, as we discussed in Chap. 1, that the reader is familiar with the procedures and techniques for equilibrium analysis taught in an introductory course on the statics of rigid bodies. The process of identifying a so-called free-body diagram where attention is focused on the force equilibrium aspects of the problem of interest is the important first step in the solution of problems in the statics of rigid bodies. In the analysis of *deformable* bodies, we add two additional steps to obtain:

1. Statics
2. Force-deformation relations (2.8)
3. Geometry of deformation

We refer to these three steps as Eq. (2.8) to be used for the solution of problems.

If the free-body diagram for a problem of interest involves only a simple bar, as in Example 2.1, then stress and deformation information follows immediately from using Eqs. (2.2) to (2.5). However, many problems of practical interest consist of interconnected structural or machine elements which have different sizes and material properties with different loads applied at different points. For the solution to such problems it is necessary to use the three steps of Eq. (2.8), as illustrated in the following examples.

EXAMPLE 2.2

A model of a steel column *ABC* shown in Fig. 2.7*a* is part of a two-story building and carries a roof load of $P_A = 450$ kN and a second-floor load of $P_B = 900$ kN. The length L of each segment is 4 m, and the cross-sectional areas of *AB* and *BC* are 40 and 100 cm^2, respectively. We wish to find the normal stress in each segment and the displacements of sections *A* and *B*. We neglect

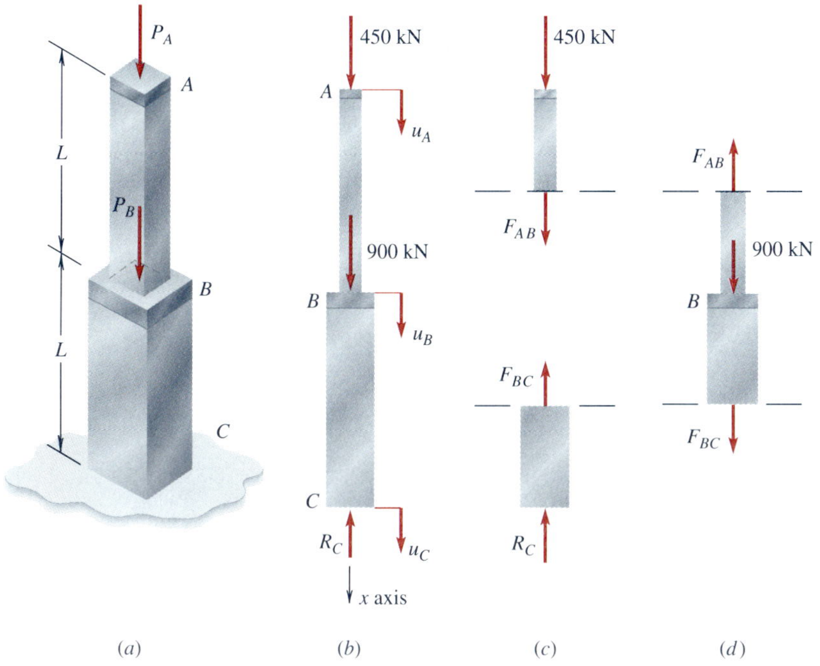

Figure 2.7 Example 2.2

the weight of the column and take $E = 200$ GPa. The applied loads P_A and P_B are assumed to act along the centroidal axis of the column.

Proceeding in a systematic fashion which leads to the use of the three steps of Eq. (2.8), we first set up a centroidal coordinate axis with origin at point A, as shown in Fig. 2.7*b*. The possible displacements at sections A, B, and C in the direction of the axis are indicated as shown in Fig. 2.7*b* by u_A, u_B, and u_C, respectively. These displacements are positive as shown in the positive direction of the axis; e.g., the displacement u_A is the distance that section A moves from its original unloaded position at $x = 0$. Note that the origin of the axis *and* the corresponding point on the column coincide before the loads are applied. The origin does not move; the point on the column, however, will experience a displacement.

The first step in applying the three steps of Eq. (2.8) is to use the principles of statics to determine forces in different parts of the system by using the appropriate equilibrium equations.

In Fig. 2.7*b*, we show the reaction R_C at the bottom support in the free-body diagram. Force equilibrium in the x direction gives

$$\Sigma F_x = 0 \qquad 450\text{ kN} + 900\text{ kN} - R_C = 0$$
$$R_C = 1350\text{ kN} \qquad (a)$$

We next turn to the internal forces in each of the segments, as shown in Fig. 2.7*c*, where F_{AB} and F_{BC} are the internal forces as shown. It is convenient to assume that the internal forces in each segment are tensile (as, e.g., we assumed for members when analyzing trusses in a statics course); if an internal force in a segment turns out to be negative, then the force in that segment is compressive.

Equilibrium of a portion of segment AB gives (Fig. 2.7*c*)

$$F_{AB} = -450\text{ kN} \qquad (b)$$

indicating that the force in segment AB is compressive, which we would expect. The force in bar BC is

$$F_{BC} = -R_C = -1350\text{ kN} \qquad (c)$$

Thus the normal stresses in AB and BC are compressive and given by

$$\sigma_{AB} = \left(\frac{F}{A}\right)_{AB} = \frac{-450\text{ kN}}{40 \times 10^{-4}\text{ m}^2} = -112\text{ MPa}$$
$$\sigma_{BC} = \left(\frac{F}{A}\right)_{BC} = \frac{-1350\text{ kN}}{100 \times 10^{-4}\text{ m}^2} = -135\text{ MPa} \qquad (d)$$

To calculate the displacements of sections A and B, first we find the change in length of each segment, using the second step of Eq. (2.8). The *force-deformation relation* of each segment gives the deformation of each segment:

$$\delta_{AB} = \left(\frac{FL}{EA}\right)_{AB} = \frac{(-450 \times 10^3\ \text{N})(4\ \text{m})}{(200 \times 10^9\ \text{N/m}^2)(40 \times 10^{-4}\ \text{m}^2)}$$

$$= -2.25 \times 10^{-3}\ \text{m} = -2.25\ \text{mm}$$

$$\delta_{BC} = \left(\frac{FL}{EA}\right)_{BC} = \frac{(-1350 \times 10^3\ \text{N})(4\ \text{m})}{(200 \times 10^9\ \text{N/m}^2)(100 \times 10^{-4}\ \text{m}^2)} \qquad (e)$$

$$= -2.70 \times 10^{-3}\ \text{m} = -2.70\ \text{mm}$$

Remember that the deformations calculated in Eqs. (*e*) are the changes in length of each segment. The displacements follow from geometry, step 3 of Eq. (2.8), consistent with our definition of the relation between deformation and displacements of the end sections

$$\delta_{BC} = u_C - u_B = -2.70\ \text{mm} \qquad (f)$$

Since $u_C = 0$, that is, section C is fixed, it follows from Eq. (*f*) that

$$u_B = 2.70\ \text{mm} \qquad (g)$$

Section B moves downward in the positive direction of the x axis. Finally,

$$\delta_{AB} = u_B - u_A \qquad (h)$$

from which

$$u_A = u_B - \delta_{AB} = 2.70 - (-2.25) = 4.95\ \text{mm}$$

Again as expected, u_A moves downward. The displacement at A relative to the total length of the column is about 0.06 percent.

EXAMPLE 2.3

A structural bar in service is fixed at A and carries axial loads applied at sections B, C, and D, as shown in Fig. 2.8*a*. If $P_1 = 1000$ lb, $P_2 = 700$ lb, and $P_3 = 500$ lb, let us find the stress in each segment of the bar and the displacement of sections B, C, and D. Take $A_{AB} = A_{BC} = 1\ \text{in}^2$, $A_{CD} = 0.3\ \text{in}^2$, and $E = 30{,}000$ ksi. Neglect the weight of the bar segments.

The first step in the solution is to introduce the centroidal coordinate axis x whose origin is located at section A, as shown in Fig. 2.8*b*. We sketch the free-body diagram of the bar where R_A is taken as the wall reaction, as shown in Fig. 2.8*b*. We show the assumed displacements at sections A, B, C, and D in Fig. 2.8*d*. We apply the three steps of Eq. (2.8).

Statics: Based on the free-body diagram for the entire bar, Fig. 2.8*b*, and summation of forces in the x direction, we have

$$R_A = 200\ \text{lb} \qquad (a)$$

The value of the internal axial force in each segment (we assume that the internal unknown forces in the segments are tensile forces) follows from an axial force balance using the free bodies in Fig. 2.8*c*

$$F_{CD} = -500\ \text{lb}$$

$$F_{BC} = -1200\ \text{lb} \qquad (b)$$

$$F_{AB} = -200\ \text{lb}$$

where the minus signs indicate that the force in each segment is compressive. Normal stresses are found by dividing the force in each segment by the appropriate cross-sectional area to get

$$\sigma_{CD} = \left(\frac{F}{A}\right)_{CD} = \frac{-0.5}{0.3} = -1.67\ \text{ksi}$$

$$\sigma_{BC} = \left(\frac{F}{A}\right)_{BC} = \frac{-1.2}{1.0} = -1.2\ \text{ksi} \qquad (c)$$

$$\sigma_{AB} = \left(\frac{F}{A}\right)_{AB} = \frac{-0.2}{1.0} = -0.2\ \text{ksi}$$

Force-Deformation Relations (Hooke's Law): Repeated application of Hooke's law in the form of Eq. (2.6) gives the axial deformation of each segment

$$\delta_{AB} = \left(\frac{FL}{EA}\right)_{AB} = \frac{(-0.2)(12)}{(30 \times 10^3)(1)} = -0.8 \times 10^{-4}\ \text{in}$$

$$\delta_{BC} = \left(\frac{FL}{EA}\right)_{BC} = \frac{(-1.2)(12)}{(30 \times 10^3)(1)} = -4.8 \times 10^{-4}\ \text{in} \qquad (d)$$

$$\delta_{CD} = \left(\frac{FL}{EA}\right)_{CD} = \frac{(-0.5)(16)}{(30 \times 10^3)(0.3)} = -8.89 \times 10^{-4}\ \text{in}$$

Geometry of Deformation: Since δ_{AB} is the axial deformation of segment AB, it follows from Fig. 2.8*d* that

$$\delta_{AB} = u_B - u_A \qquad (e)$$

Since section A is fixed, $u_A = 0$, it follows that $u_B = -0.8 \times 10^{-4}$ in. In a similar fashion

$$\delta_{BC} = u_C - u_B$$

$$\delta_{CD} = u_D - u_C$$

or

$$u_C = u_B + \delta_{BC} \qquad u_C = (-0.8 - 4.8) \times 10^{-4} = -5.6 \times 10^{-4}\ \text{in} \qquad (f)$$

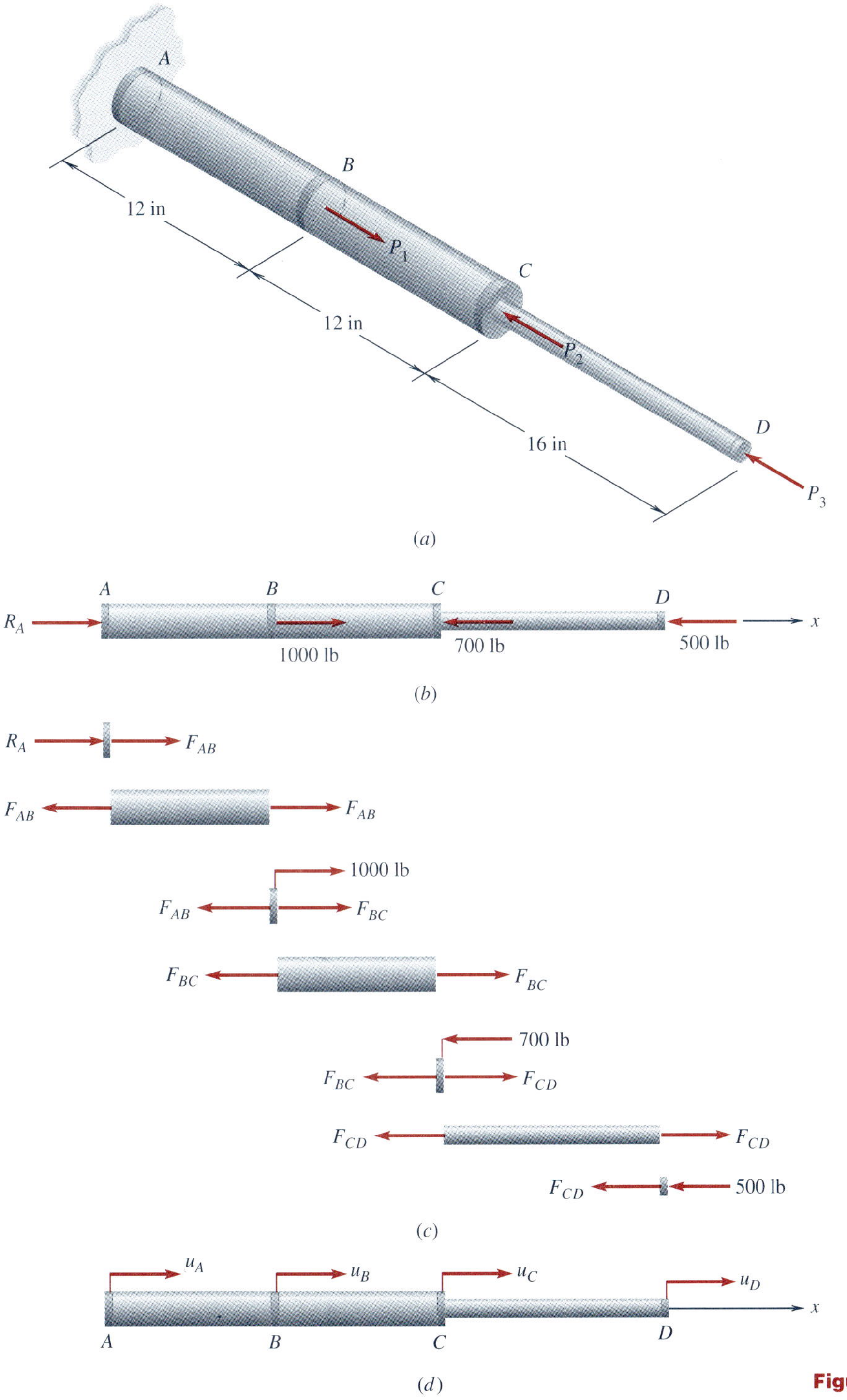

Figure 2.8 Example 2.3

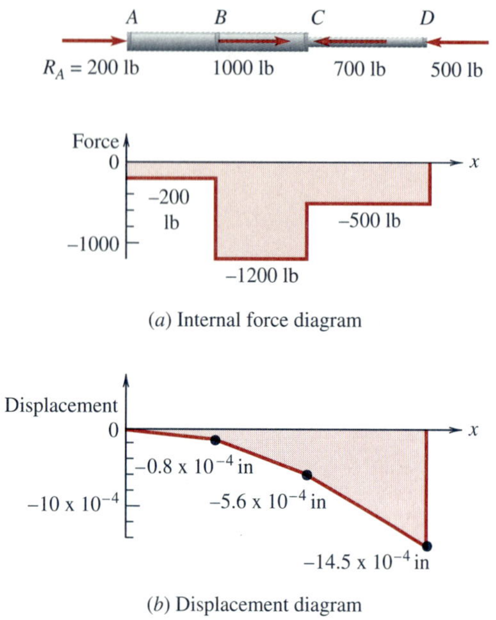

Figure 2.9 Example 2.3: (*a*) internal force diagram; (*b*) displacement diagram.

and

$$u_D = u_C + \delta_{CD} \qquad \begin{aligned} u_D &= (-5.6 - 8.89) \times 10^{-4} \\ &= -14.5 \times 10^{-4} \text{ in} \end{aligned} \tag{g}$$

Note that u_B, u_C, and u_D are the geometric movements or displacements of sections *B, C,* and *D* relative to the fixed support at *A*. The difference of the end displacements of a segment contributes to its deformation; for example, $u_C - u_B$ contributes to the deformation δ_{BC} of bar *BC*.

We can summarize the results for the force in each segment by sketching an internal force diagram, as shown in Fig. 2.9*a*. The constant force in each segment is plotted as a function of *x* by a series of horizontal lines. At each point where a load is applied to the bar, a jump in the internal force occurs, as shown in Fig. 2.9*a*. The negative values for all segments in the internal force diagram indicate that the forces in all segments are compressive.

The displacements u_A, u_B, and u_C are plotted in the displacement diagram shown in Fig. 2.9*b*. The negative values of the axial displacements at all points in the displacement diagram indicate that *all* sections along the bar move to the left. The displacement is linear between the endpoints of each segment as shown since the force is constant in each segment.

2.4 Statically Indeterminate Problems

In Examples 2.2 and 2.3 we systematically used the three steps of Eq. (2.8) to determine the unknowns. In both examples, only equilibrium considerations were used to find the external reactions and the internal forces from which the deformations due to these forces were found.

However, there are many problems of practical interest for which it is not possible to determine all the reactions and internal forces based on equilibrium analysis alone. It is necessary in solving such problems to carry out all three of the steps outlined in Eq. (2.8) simultaneously before a complete set of forces and displacements can be determined. Such problems are referred to as *statically indeterminate* problems. The following examples will indicate the method to solve such problems, again using the three steps of Eq. (2.8).

EXAMPLE 2.4

A short composite column is constructed by filling a round steel pipe with concrete as shown in Fig. 2.10*a*. If the column is placed between two rigid plates and loaded in compression, as shown in Fig. 2.10*b*, we would like to find the load-deflection relation for the column. Also, we wish to find the maximum allowable load if the allowable stresses for the steel and concrete are 100 MPa and 8 MPa. For a particular problem of interest $L = 2$ m, $d = 500$ mm, $t = 13$ mm, $E_S = 200$ GPa, and $E_C = 14$ GPa. Neglect the weight of the materials.

We will use the three steps of Eq. (2.8) to determine the

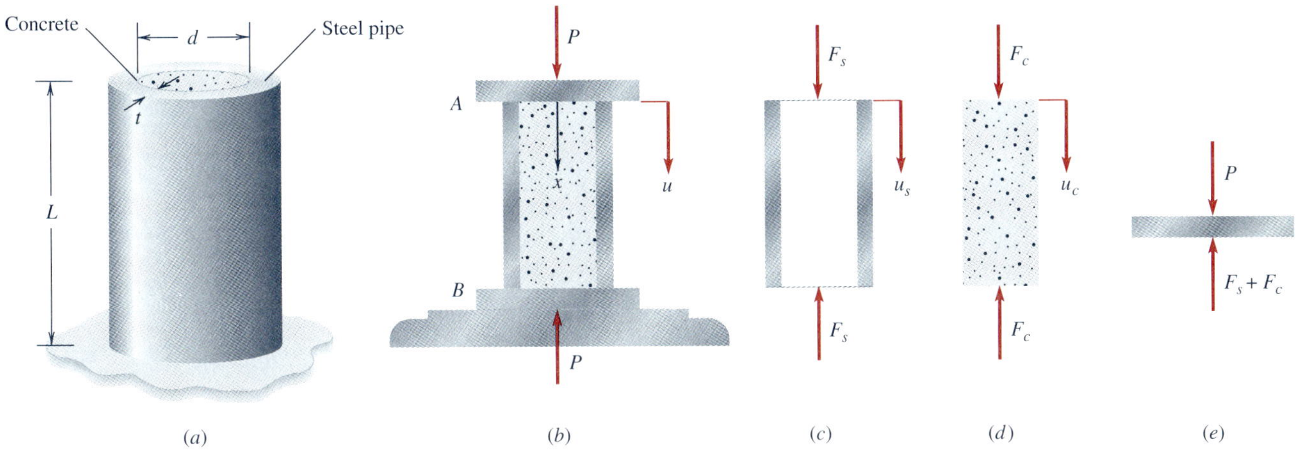

Figure 2.10 Example 2.4

relation between the force in the steel pipe and the force in the concrete. We assume that the resultant forces in the steel and concrete act along the axis of the column. We show the force F_S in the steel pipe in Fig. 2.10*c* and the force F_C in the concrete in Fig. 2.10*d*. The forces F_S and F_C are assumed positive as shown. The *x* axis is along the axis of the column with the origin at the top plate, as shown in Fig. 2.10. The displacements of the top plate, the top of the steel pipe, and the top of the concrete are shown as u, u_S, and u_C in Fig. 2.10*b*, *c*, and *d*, respectively.

Statics: From the free-body diagram of the top plate, as shown in Fig. 2.10*e*, we find from force equilibrium that

$$F_S + F_C = P \qquad (a)$$

We cannot determine the force in the steel and in the concrete without additional information on the deformation of the pipe and concrete; the problem is *statically indeterminate.* We continue with the steps of Eq. (2.8).

Force-Deformation Relations: Recall that the force-deformation relation, Eq. (2.7), was derived under the assumption that the force in the bar element was tensile. In this case, F_S and F_C are shown as compressive forces in Fig. 2.10*c* and *d*. Therefore in the force-deformation relations they appear with minus signs

$$\begin{aligned} -F_S &= \left(\frac{EA}{L}\right)_S (u_B - u_A) = -\left(\frac{EA}{L}\right)_S u_S \\ -F_C &= \left(\frac{EA}{L}\right)_C (u_B - u_A) = -\left(\frac{EA}{L}\right)_C u_C \end{aligned} \qquad (b)$$

where u_B, the displacement at the bottom, is zero. It follows, e.g., from Eqs. (*b*), that F_S as shown in Fig. 2.10*c* will give rise to the positive displacement u_S.

Geometry: The displacement of the pipe and concrete must conform to the displacement of the rigid end plate, as shown in Fig. 2.10*b*. Thus we have

$$u_S = u_C = u \qquad (c)$$

We have used in Eqs. (*b*) the geometric information that u_B is zero. From Eqs. (*b*) and (*c*) we have

$$\begin{aligned} F_S &= (EA)_S\left(\frac{u}{L}\right) \\ F_C &= (EA)_C\left(\frac{u}{L}\right) \end{aligned} \qquad (d)$$

Upon substitution of Eqs. (*d*) into Eq. (*a*), we can find an expression for the load-deformation relation of the pipe

$$\begin{aligned} \frac{u}{L}\,[(EA)_S + (EA)_C] &= P \\ \frac{u}{L} &= \frac{P}{E_S A_S + E_C A_C} \end{aligned} \qquad (e)$$

Inserting numerical values, we find

$$\begin{aligned} A_S &= \frac{\pi}{4}(526^2 - 500^2) = 20.95 \times 10^3\ \text{mm}^2 = 20.95 \times 10^{-3}\ \text{m}^2 \\ A_C &= \frac{\pi}{4}(500^2) = 196.3 \times 10^3\ \text{mm}^2 = 196.3 \times 10^{-3}\ \text{m}^2 \end{aligned} \qquad (f)$$

and

$$u = \frac{(P)(2)}{(200 \times 10^9)(20.95 \times 10^{-3}) + (14 \times 10^9)(196.3 \times 10^{-3})}$$

$$= 2.88 \times 10^{-10} P \text{ m} \qquad P \text{ in newtons} \qquad (g)$$

Equation (*g*) gives the displacement of the end plate in terms of the load *P*.

To find the maximum allowable load that we can put on the column, we have from Eq. (*d*)

$$F_S = F_C \frac{E_S A_S}{E_C A_C} = 1.525 F_C \qquad (h)$$

Since it is not clear a priori whether the stress in the concrete will reach its allowable value before the stress in the steel does, it is necessary to assume that one or the other reaches its allowable stress and then to check the validity of the assumption. If we assume that the allowable stress in the concrete is reached first, then

$$F_C = A_C(\sigma_C)_a = (196.3 \times 10^{-3})(8 \times 10^6) = 1570 \text{ kN}$$

and from Eq. (*h*)

$$F_S = 1.525 F_C = 2394 \text{ kN}$$

The stress corresponding to this load in the steel pipe is

$$\sigma_S = \frac{F_S}{A_S} = \frac{2394}{20.95 \times 10^{-3}} = 114 \text{ MPa}$$

which is greater than the allowable stress of 100 MPa. Therefore, since our original assumption leads to an unacceptable stress in the steel pipe, we consider the other possibility, i.e., that the steel reaches the allowable value of stress

$$F_S = A_S(\sigma_S)_a = 2095 \text{ kN}$$

and from Eq. (*h*)

$$F_C = \frac{F_S}{1.525} = 1374 \text{ kN}$$

and finally from Eq. (*a*)

$$P_a = F_S + F_C = 3469 \text{ kN} \qquad (i)$$

At a load $P = 3469$ kN, the maximum stress in the steel pipe is at its allowable value of 100 MPa, and the stress in the concrete is 7 MPa, which is less than its allowable value of 8 MPa. Equation (*i*) therefore gives the value of the maximum load. The displacement of the top of the column at the allowable load is, from Eqs. (*g*) and (*i*),

$$u = (2.88 \times 10^{-10})(3.469 \times 10^6) = 9.99 \times 10^{-4} \text{ m} = 1.0 \text{ mm}$$

EXAMPLE 2.5

In Fig. 2.11*a*, we show a rigid bar *DHC* attached to two elastic wires *AD* and *BC* of diameters *d* and 2*d*, modulus of elasticity *E*, and length L_1. A load *P* is applied at point *H*. For an arbitrary location of the load *P*, the bar *DHC* will move downward and rotate from the horizontal. We wish to find the specific distance *x* from *D* at which the load *P* should be applied so that the rigid bar remains horizontal after the load is applied. We neglect the weight of the rigid bar and the wires in the calculations.

We apply the three steps of Eq. (2.8) to find *x*.

Statics: From the free-body diagram shown in Fig. 2.11*b*, the equilibrium equations give

$$F_{AD} + F_{BC} = P \qquad (a)$$

$$\Sigma M_H = 0 \qquad F_{BC}(L - x) = xF_{AD} \qquad (b)$$

The unknowns in the problem are the forces F_{AD} and F_{BC} and the location *x*. The problem is statically indeterminate.

Force-Deformation Relations: The displacements of points *D* and *C*, u_D and u_C, respectively, are shown in Fig. 2.11*b*. Since points *A* and *B* are fixed, we have

$$u_D = \left(\frac{FL}{EA}\right)_{AD} \qquad u_C = \left(\frac{FL}{EA}\right)_{BC} \qquad (c)$$

where the force in each of the wires is tensile. The displacements are positive downward as shown.

Geometry: The requirement that the bar *DHC* remain horizontal is

$$u_D = u_C \qquad (d)$$

Solving Eqs. (*c*) and (*d*), with $A_{BC} = 4A_{AD}$, we have

$$\frac{F_{AD}L_1}{EA_{AD}} = \frac{F_{BC}L_1}{E(4A_{AD})}$$

or

$$F_{BC} = 4F_{AD} \qquad (e)$$

From Eqs. (*b*) and (*e*), we find

$$\frac{x}{L - x} = \frac{F_{BC}}{F_{AD}} = 4 \qquad (f)$$

which gives

$$x = \tfrac{4}{5} L \qquad (g)$$

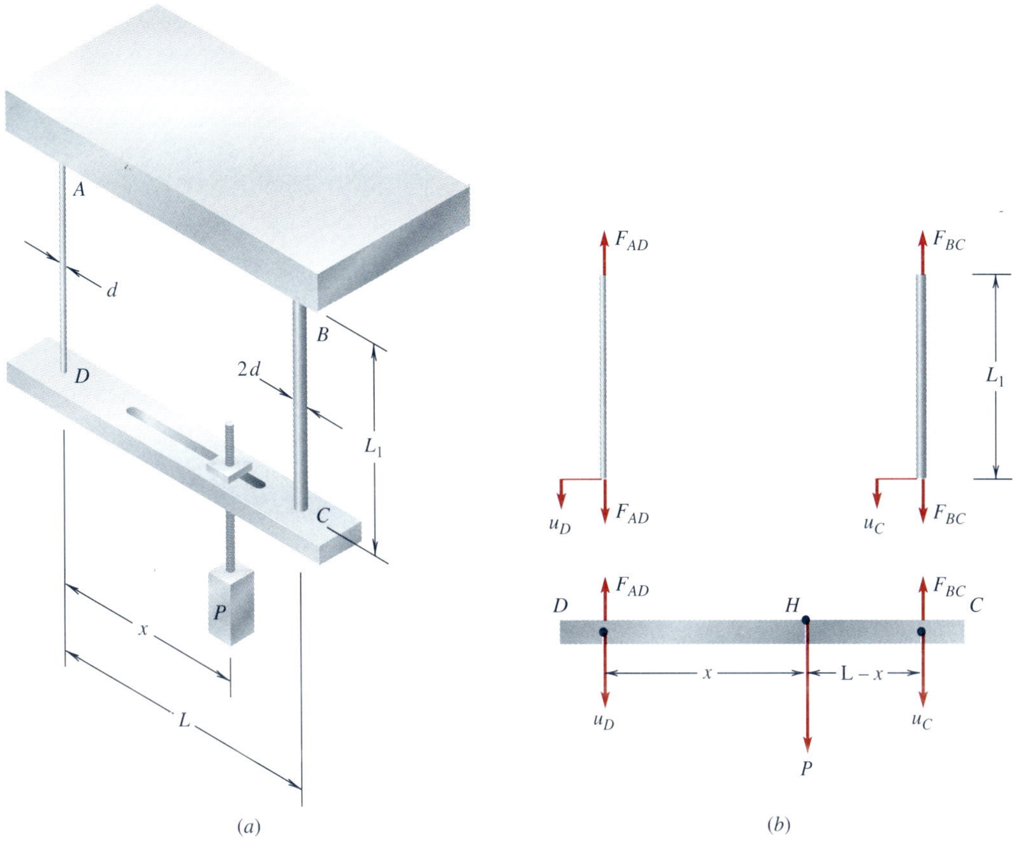

Figure 2.11 Example 2.5

as the location of the load for which the bar will remain horizontal.

For a value of x different from that given by Eq. (g), the bar will move downward and rotate by an angle θ; see Probs. 2.4-7 and 2.4-8.

EXAMPLE 2.6

An idealized two-dimensional model of a vertical force platform is shown in Fig. 2.12*a*. A load of $P = 150$ lb is placed at position A on the rigid platform, as shown in the figure. We wish to determine the vertical displacement under the load. The supporting members have an axial stiffness of $k_1 = k_2 = 100$ lb/in and $k_3 = 130$ lb/in. Neglect the weight of the force platform.

Statics: We first use equilibrium of the platform as shown in Fig. 2.12*b* to determine expressions connecting the forces in the support members and the applied load P. Summation of the forces and moment equilibrium about the left end give

$$F_1 + F_2 + F_3 = P$$
$$15F_2 + 30F_3 = 22P \qquad (a)$$

As expected, equilibrium considerations only give us two independent equations for the three unknown forces F_1, F_2, and F_3; the problem is statically indeterminate.

Geometry: We assume the displacement of the rigid platform is as shown in Fig. 2.12*c*. The displacements u_1, u_2, and u_3 are the displacements at the top of the support members, assumed positive in the direction downward. The angle θ in radians is

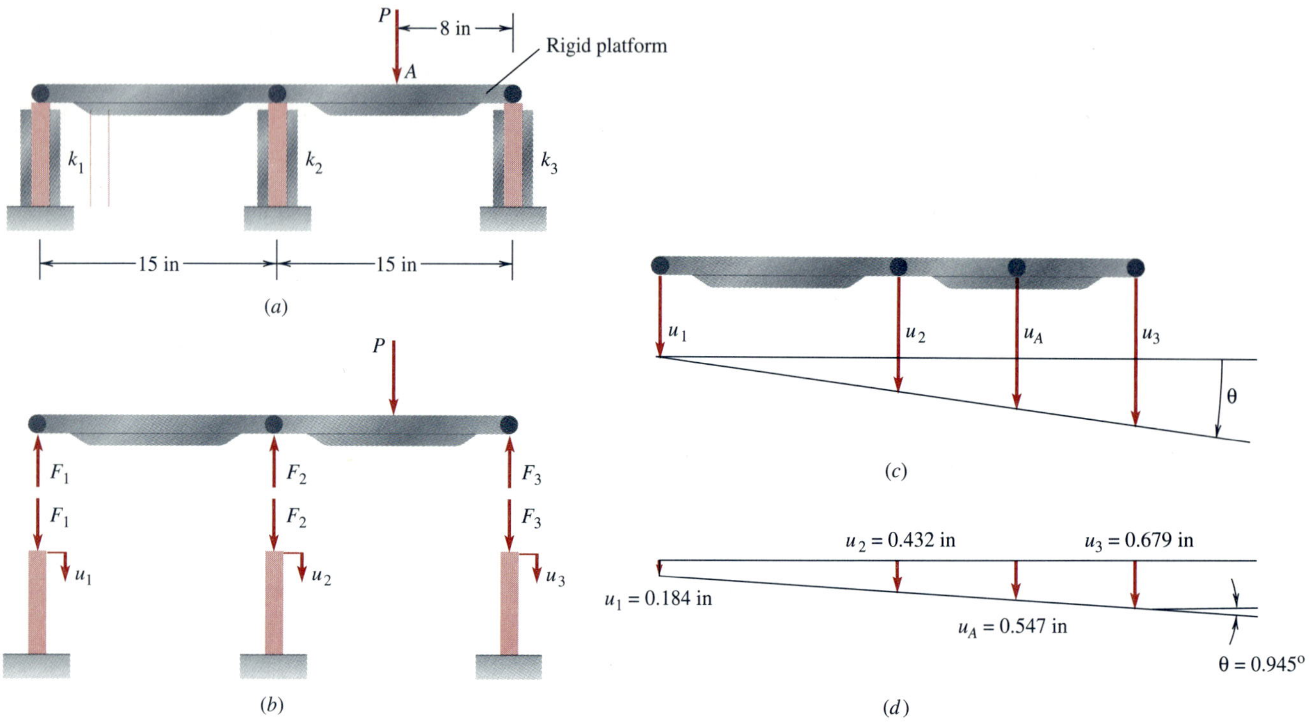

Figure 2.12 Example 2.6

small; the condition that the platform is rigid requires

$$
\begin{aligned}
u_2 &= u_1 + (15)(\theta) \\
u_3 &= u_1 + (30)(\theta) \qquad (b) \\
u_A &= u_1 + (22)(\theta)
\end{aligned}
$$

Force-Deformation Relations: The force in each support is related to the displacement at the top of the support in the form

$$
\begin{aligned}
F_1 &= k_1 u_1 \\
F_2 &= k_2 u_2 = k_2(u_1 + 15\theta) \qquad (c) \\
F_3 &= k_3 u_3 = k_3(u_1 + 30\theta)
\end{aligned}
$$

If u_1 and θ are determined, the force in each of the supports and the displacement u_A can be found. If we substitute Eqs. (*c*) into Eqs. (*a*), we have two equations for u_1 and θ in the form

$$
\begin{aligned}
k_1 u_1 + k_2(u_1 + 15\theta) + k_3(u_1 + 30\theta) &= P \\
15k_2(u_1 + 15\theta) + 30k_3(u_1 + 30\theta) &= 22P
\end{aligned} \qquad (d)
$$

or upon substituting numerical values $k_1 = k_2 = 100$ lb/in and $k_3 = 130$ lb/in,

$$
\begin{aligned}
u_1(330) + \theta(5400) &= 150 \\
u_1(5400) + \theta(139{,}500) &= 3300
\end{aligned} \qquad (e)
$$

Solving Eqs. (*e*), we find

$$
\begin{aligned}
u_1 &= 0.184 \text{ in} \\
\theta &= 0.0165 \text{ rad}
\end{aligned} \qquad (f)
$$

We note that the top of the left support moves downward and that each of the supports will be in compression. The remaining displacements are as shown in Fig. 2.12*d*, with

$$
\begin{aligned}
u_1 &= 0.184 \text{ in} \\
u_2 &= 0.432 \text{ in} \\
u_3 &= 0.679 \text{ in} \\
u_A &= 0.547 \text{ in}
\end{aligned} \qquad (g)
$$

EXAMPLE 2.7

A steel pipe AC is attached at its ends to rigid supports, as shown in Fig. 2.13*a*. If a load P is applied at section B, we wish to determine the stress in each of segments AB and BC and the displacement of section B. The cross-sectional area of the pipe $A = 2.5 \text{ in}^2$, $E = 30 \times 10^6$ psi, $L_1 = 24$ in, $L_2 = 36$ in, and $P = 10{,}000$ lb. We neglect the weight of the pipe.

Again we think of the three steps of Eq. (2.8) as the guide to the solution of this problem. We first introduce a coordinate axis along the pipe and denote the support reactions by R_A and R_C, as shown in Fig. 2.13*b*. The origin of the coordinate system is at A.

Statics: The internal forces F_{AB} and F_{BC} in segments AB and BC are shown in Fig. 2.13*c*. We assume that each segment is in tension as shown. From equilibrium of each segment, we have

$$F_{AB} = R_A \qquad F_{BC} = -R_C \tag{a}$$

From equilibrium of the small, rigid segment at B at which the load is applied, Fig. 2.13*d*, or from equilibrium of the entire pipe and using Eqs. (*a*), we have

$$F_{AB} = P + F_{BC} \tag{b}$$

Equilibrium considerations alone will not give the forces in each segment, so the problem is statically indeterminate. We need to consider the deformation of each segment.

Force-Deformation Relations: The deformation in each segment is given by

$$\delta_{AB} = \left(\frac{FL}{AE}\right)_{AB} = \frac{F_{AB}L_1}{AE}$$
$$\delta_{BC} = \left(\frac{FL}{AE}\right)_{BC} = \frac{F_{BC}L_2}{AE} \tag{c}$$

Geometry: Finally, the deformation in each segment can be expressed in terms of the displacements at the ends of each segment (Fig. 2.13*e*),

$$\delta_{AB} = u_B - u_A = u_B$$
$$\delta_{BC} = u_C - u_B = -u_B \tag{d}$$

where we have used $u_A = u_C = 0$, corresponding to the fixed end sections.

From Eqs. (*c*) and (*d*), we have

$$F_{AB} = \left(\frac{AE}{L_1}\right)u_B \qquad F_{BC} = \left(\frac{AE}{L_2}\right)(-u_B) \tag{e}$$

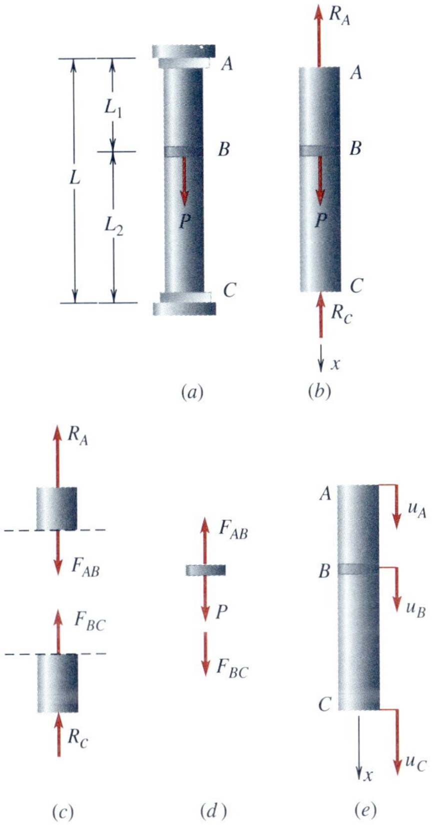

Figure 2.13 Example 2.7

Once the displacement of section B is determined, the internal force in each segment is found from Eqs. (*e*). To find u_B, we substitute the results from Eqs. (*e*) into Eq. (*b*):

$$u_B\left(\frac{AE}{L_1}\right) = P + \left(\frac{AE}{L_2}\right)(-u_B)$$

or

$$u_B\left(\frac{AE}{L_1} + \frac{AE}{L_2}\right) = P \tag{f}$$

Upon substitution of numerical values, we have

$$u_B = \frac{10 \times 10^3}{(2.5)(30 \times 10^6)(\frac{1}{24} + \frac{1}{36})}$$
$$= 1.92 \times 10^{-3} \text{ in} \tag{g}$$

Finally the stress in each segment follows from Eqs. (*e*) and (*g*),

$$\sigma_{AB} = \frac{F_{AB}}{A} = \frac{E}{L_1} u_B = 2400 \text{ psi}$$
$$\sigma_{BC} = \frac{F_{BC}}{A} = \frac{E}{L_2}(-u_B) = -1600 \text{ psi} \qquad (h)$$

We see that segment *AB* is in tension and the wall reaction $R_A = F_{AB} = A\sigma_{AB} = 6000$ lb. Segment *BC* is in compression as expected, and the wall reaction $R_C = -F_{BC} = 4000$ lb. The sum of the wall reactions is equal to *P*.

Finally, it is of interest to find the expressions for the reactions in the general case by using Eqs. (*a*), (*e*), and (*f*). We have

$$R_A = F_{AB} = \frac{AE}{L_1} u_B = \frac{AE}{L_1} \frac{P}{AE/L_1 + AE/L_2} = \frac{L_2}{L} P \qquad (i)$$

$$R_C = -F_{BC} = \frac{AE}{L_2} u_B = \frac{AE}{L_2} \frac{P}{AE/L_1 + AE/L_2} = \frac{L_1}{L} P \qquad (j)$$

where $L = L_1 + L_2$. We see from Eqs. (*i*) and (*j*) that the wall reactions can be expressed in terms of the ratios of segment length to total length (in this case where *AE* is constant in each segment) in a rather simple form.

EXAMPLE 2.8

A portion of a wire-controlled mechanical linkage system is shown in Fig. 2.14*a*. A load *P* is applied at point *A* to the rigid member *AE* which is free to rotate about point *B*. Two elastic links are connected to the rigid member at points *C* and *D*. If $P = 200$ lb, we need to find the displacement point *E*. We neglect any friction in the system, and the cross-sectional areas of the links are as given in Fig. 2.14*a*. We take $E_{st} = 30 \times 10^6$ psi and $E_{al} = 10 \times 10^6$ psi.

The model problem is shown again in Fig. 2.14*b*. In Fig. 2.14*c*, we show the forces acting on member *AE* where F_{al} is the force in the aluminum link and F_{st} is the force in the steel link; R_B is the reaction at the frictionless support located at *B*.

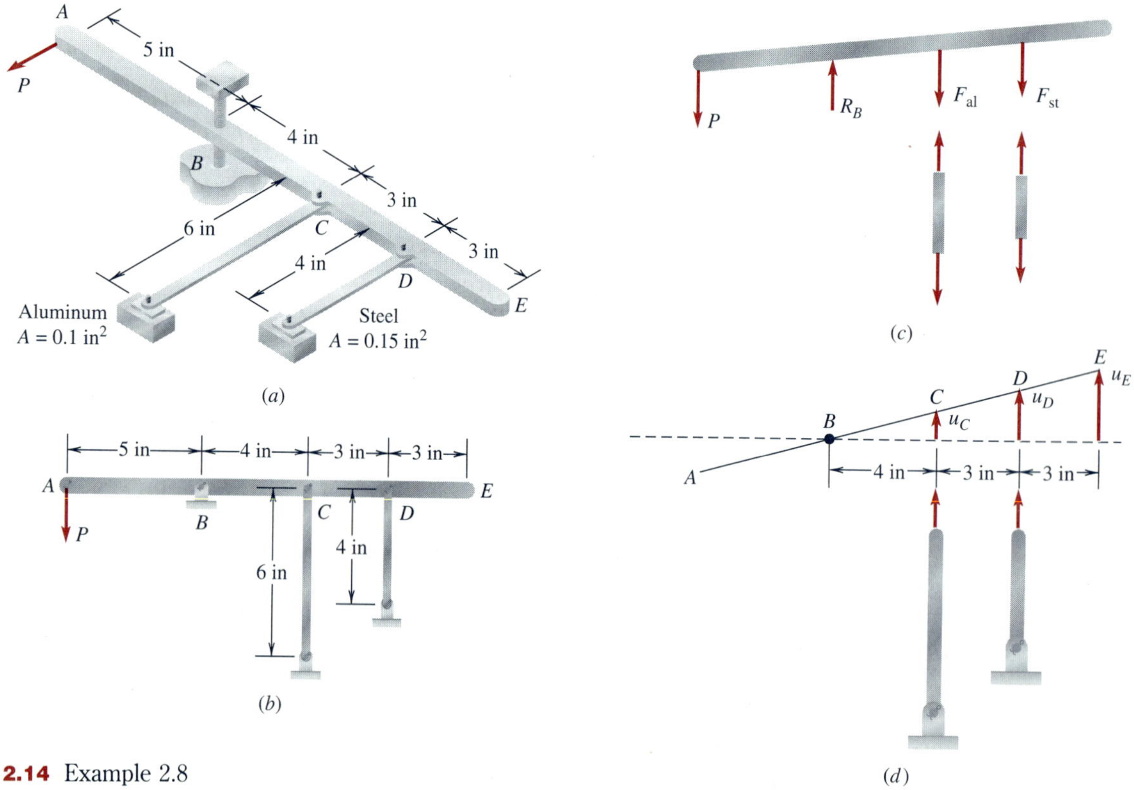

Figure 2.14 Example 2.8

To find the displacement at E, we will follow the three steps.

Statics: Moment equilibrium of the forces about point B so as to eliminate consideration of the unknown reactive force at B gives

$$4F_{al} + 7F_{st} = 5P \tag{a}$$

We assume that the unknown force in each link is tensile.

Geometry: If we exaggerate the displacements of points on the rigid member, we have the displacements as shown in Fig. 2.14*d*. It follows that

$$\frac{u_C}{4} = \frac{u_D}{7} = \frac{u_E}{10} \tag{b}$$

or

$$u_D = \frac{7}{4}u_C \qquad u_E = \frac{10}{4}u_C \tag{c}$$

Force-Deformation Relations: Finally we use force-deformation relations for each of the links in the form

$$F_{al} = \left(\frac{EA}{L}\right)_{al} u_C \qquad F_{st} = \left(\frac{EA}{L}\right)_{st} u_D \tag{d}$$

If we use Eqs. (c) and (d) in Eq. (a), we can obtain a single equation for the determination of u_C

$$\left[4\left(\frac{EA}{L}\right)_{al} + 7\left(\frac{EA}{L}\right)_{st}\left(\frac{7}{4}\right)\right]u_C = 5P \tag{e}$$

Upon substituting numerical values, we have

$$\left[4\left(\frac{10 \times 10^6 \times 0.1}{6}\right) + \frac{49}{4}\left(\frac{30 \times 10^6 \times 0.15}{4}\right)\right]u_C = 1000$$

or

$$1.445 \times 10^7\, u_C = 1000 \qquad u_C = 6.92 \times 10^{-5}\text{ in} \tag{f}$$

Finally, the displacement at point E follows from Eqs. (c)

$$u_E = 2.5u_C = 1.73 \times 10^{-4}\text{ in} \tag{g}$$

The force F_{al} in the aluminum link can be calculated from the known value of u_C by using Eqs. (d) to get

$$F_{al} = 11.53\text{ lb} \tag{h}$$

Now that F_{al} is known, we can use equilibrium, Eq. (a), to get

$$F_{st} = \frac{1000 - 4(11.53)}{7} = 136.3\text{ lb} \tag{i}$$

which, as expected, indicates that both links are in tension.

EXAMPLE 2.9

A steel pipe with a different cross-sectional area in each of its three segments is loaded by the axial force F at section B, as shown in Fig. 2.15*a*. The load F is considered to be applied along the axis to a rigid plate shown as shaded at section B.

Let $F = 3$ kips, A_{AB}, A_{BC}, and A_{CD} have areas 1, 3, and 2 in^2, respectively, and $E = 30 \times 10^6$ psi. We wish to find (a) the stress in each of the three segments and (b) the axial displacement of section B.

As a first step we introduce an x coordinate axis along the pipe with origin at A as shown, and we introduce the wall reactions R_A and R_D as shown in Fig. 2.15*b*. We find it convenient to assume that all pipe segments are in tension, as shown in the free-body diagrams of the segments in Fig. 2.15*b*. We also show in Fig. 2.15*b* the displacements of sections A, B, C, and D on the pipe.

Statics: Overall axial force equilibrium of the pipe, Fig. 2.15*b*, gives

$$R_D - F - R_A = 0 \tag{a}$$

This represents one equation for the two unknowns R_D and R_A. Once R_D and R_A are known, the forces in each segment of the pipe can be found as shown by the free-body diagrams in Fig. 2.15*b*, from which we have

$$F_{AB} = R_A \qquad F_{AB} = -F + F_{BC} \qquad F_{BC} = F_{CD} \qquad F_{CD} = R_D \tag{b}$$

In what follows we express the forces in each of the segments in terms of R_A and R_D.

Force-Deformation Relations and Geometry: For AB since $u_A = 0$, using Eqs. (b) for the force gives

$$\delta_{AB} = u_B - u_A = u_B = R_A\left(\frac{L}{EA}\right)_{AB} \tag{c}$$

For BC,

$$\delta_{BC} = u_C - u_B = R_D\left(\frac{L}{EA}\right)_{BC} \tag{d}$$

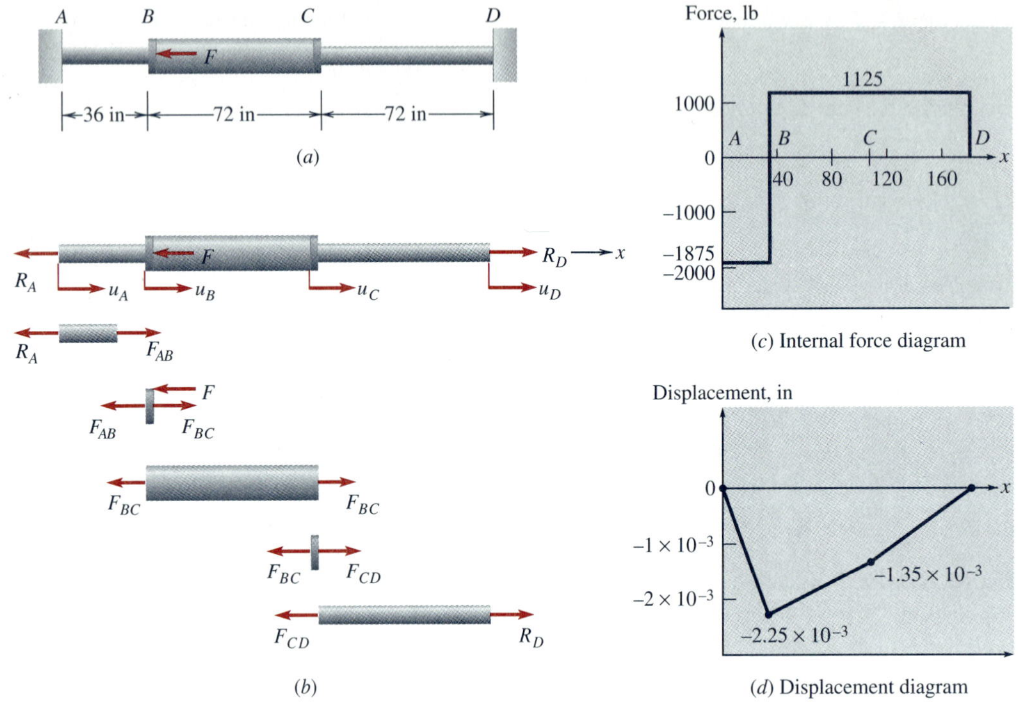

Figure 2.15 Example 2.9

For CD since $u_D = 0$,

$$\delta_{CD} = u_D - u_C = -u_C = R_D\left(\frac{L}{EA}\right)_{CD} \tag{e}$$

The unknown displacements are u_B and u_C. Adding Eqs. (d) and (e) gives u_B in terms of R_D:

$$-u_B = R_D\left[\left(\frac{L}{EA}\right)_{BC} + \left(\frac{L}{EA}\right)_{CD}\right]$$

or

$$u_B = -R_D\left[\frac{72}{(30)(3)} + \frac{72}{(30)(2)}\right] \times 10^{-6}$$

or

$$R_D = -\frac{10^6}{2}u_B \tag{f}$$

Using Eq. (c), we have

$$R_A = \frac{(30 \times 10^6)(1)}{36}u_B = \frac{5}{6} \times 10^6 u_B \tag{g}$$

Substituting Eqs. (f) and (g) into Eq. (a) gives

$$-(10^6)(\tfrac{5}{6} + \tfrac{1}{2})u_B = 3 \times 10^3$$

$$u_B = -2.25 \times 10^{-3}\ \text{in} \tag{h}$$

Returning to Eqs. (f) and (g), we can now insert the result, Eq. (h), for u_B to get the reactions

$$R_D = \frac{2.25 \times 10^{-3}}{2} \times 10^6 = 1125\ \text{lb} \tag{i}$$

$$R_A = \tfrac{5}{6}(-2.25 \times 10^{-3}) \times 10^6 = -1875\ \text{lb} \tag{j}$$

and it is easily checked again that equilibrium, Eq. (a), is satisfied.

The forces in each segment follow from Eqs. (b)

$$\begin{aligned} F_{AB} &= -1875\ \text{lb} \\ F_{BC} &= 1125\ \text{lb} \\ F_{CD} &= 1125\ \text{lb} \end{aligned} \tag{k}$$

The stress in each segment is found by dividing each segment force by the appropriate area to get

$$\sigma_{AB} = \frac{R_A}{A_{AB}} = \frac{-1875}{1} = -1875 \text{ psi}$$
$$\sigma_{BC} = \frac{R_D}{A_{BC}} = \frac{1125}{3} = 375 \text{ psi} \qquad (l)$$
$$\sigma_{CD} = \frac{R_D}{A_{CD}} = \frac{1125}{2} = 562 \text{ psi}$$

We can summarize the forces in each segment by plotting a diagram of the internal force in each segment as a function of distance x from the origin, as shown in Fig. 2.15*c*. In Fig. 2.15*c* a positive force is tensile and a negative force is compressive. The displacement of each point along the rod is shown in the displacement diagram in Fig. 2.15*d*. A negative value of displacement indicates that the displacement is in the negative direction of the axis.

EXAMPLE 2.10

A steel bolt is placed in an aluminum tubular sleeve of length L, and a nut is tightened onto the bolt by hand so that it just makes contact with the sleeve, as shown in Fig. 2.16*a*. If the bolt has 16 threads per inch and the nut is further tightened by wrenches an extra quarter-turn, estimate the tensile stress in the bolt and the compressive stress in the tube. Assume that the resultant forces in the bolt and tube are along the axis of the bolt; the bolt cross-sectional area is $A_B = 1.00 \text{ in}^2$, the tube cross-sectional area is $A_T = 0.60 \text{ in}^2$, and $L = 6$ in. Take $E_T = 10 \times 10^6$ psi and $E_B = 30 \times 10^6$ psi.

The first step here is to convince ourselves that as we further tighten the nut, the portion of the bolt inside the tube will go into tension and the tube will go into compression. Limiting cases often help with problems of this nature. For example, if the bolt is rigid and the tube is elastic, then turning the nut clearly compresses the tube. Similarly, if the tube is rigid and the bolt is elastic, then turning the nut "brings up" the bolt and puts the bolt into tension. Therefore, in the case when both the tube and the bolt are elastic, we can expect the tube to be in compression and the bolt to be in tension. We now follow our three-step method, starting this time with the geometry. The x axis with origin at the head of the bolt is along the axis of the bolt.

Geometry: It is convenient to focus attention on a point on the nut in contact with the tube and to note that its distance on the x axis is initially L, which we show in Fig. 2.16*b*. When we tighten the nut, the nut moves to a new position.

$$L - d + u_B \qquad (a)$$

where d is the displacement of the nut *on* the bolt and u_B is the displacement of the bolt at the position of the nut (Fig. 2.16*b*). At the same time, the top of the tube at the nut moves to a position $L - u_T$, where u_T is the displacement of the top of the

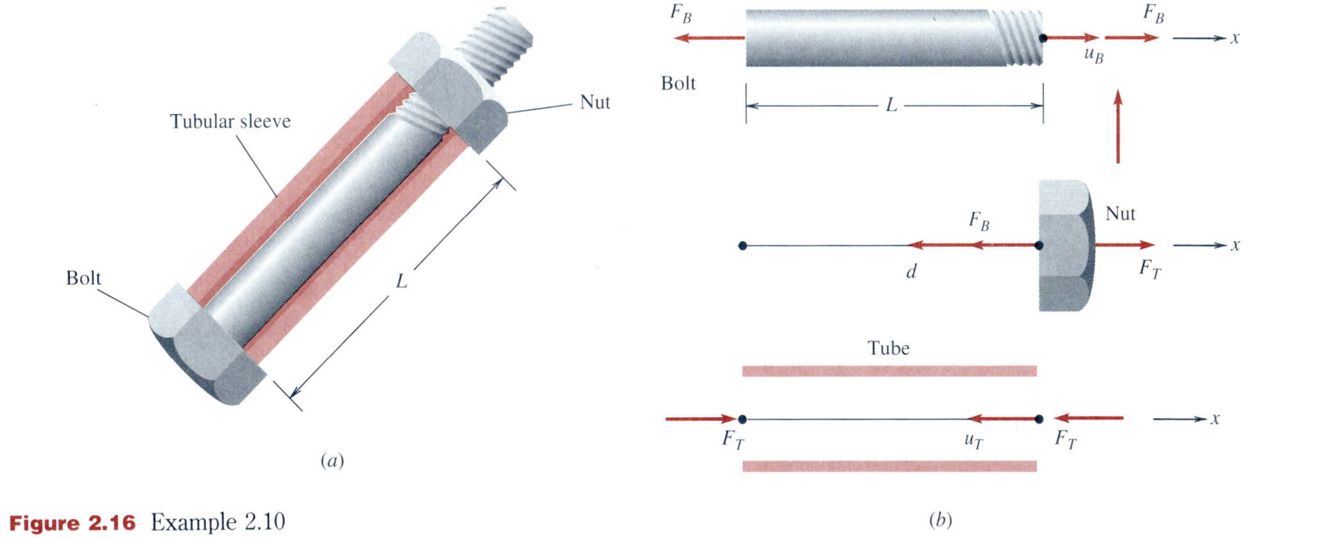

Figure 2.16 Example 2.10

tube. The nut remains in contact with the tube so that these altered lengths must be the same, and we have

$$L - d + u_B = L - u_T$$

or

$$u_T + u_B = d \qquad (b)$$

The displacement quantities are positive, as shown in Fig. 2.16*b*. In this case $d = (\frac{1}{4})(\frac{1}{16})$ in since we turned the nut an extra quarter-turn.

Statics: Since the nut is in equilibrium,

$$F_B = F_T \qquad (c)$$

where F_B is the axial tensile force in the bolt and F_T is the axial compressive force in the tube.

Force-Deformation Relations: We are assuming that the forces in the bolt and tube are axial so that the force-displacement relations are

$$F_B = \frac{E_B A_B}{L} u_B$$
$$F_T = \frac{E_T A_T}{L} u_T \qquad (d)$$

where the forces are as shown in Fig. 2.16*b*. Note that the force-displacement relation for the bolt assumes that both the force and displacement are positive; for the tube the force is compressive, and the displacement as shown in Fig. 2.16*b* shortens the tube. Therefore, combining Eqs. (*b*), (*c*), and (*d*), we have

$$F_B\left(\frac{L}{E_T A_T} + \frac{L}{E_B A_B}\right) = \frac{1}{64}$$

or

$$F_B = \frac{(\frac{1}{64})}{\dfrac{6}{(10 \times 10^6)(0.60)} + \dfrac{6}{(30 \times 10^6)(1)}} = 13{,}020 \text{ lb} \qquad (e)$$

for the tensile force in the bolt. From Eq. (*c*), we have

$$F_T = 13{,}020 \text{ lb} \qquad (f)$$

for the compressive force in the tube.

Finally the stresses from Eqs. (*e*) and (*f*) are

$$\sigma_B = \frac{F_B}{1 \text{ in}^2} = 13{,}020 \text{ psi}$$
$$\sigma_T = \frac{13{,}020 \text{ lb}}{0.6 \text{ in}^2} = 21{,}700 \text{ psi} \qquad (g)$$

The elongation u_B from Eqs. (*d*) and (*e*) is 2.604×10^{-3} in. The shortening of the tube can be obtained from Eq. (*b*) to get

$$u_T = \frac{1}{64} - 2.604 \times 10^{-3} = 1.302 \times 10^{-2} \text{ in} \qquad (h)$$

EXAMPLE 2.11

A small three-bar truss structure carries a load $P = 50$ kN at joint B, as shown in Fig. 2.17*a*. If $L = 1$ m, $E = 200$ GPa, and the cross-sectional area $A = 450$ mm^2 for all three bars, we wish to find the displacement of joint B and the stresses in each of the three bars.

This problem differs from those considered previously because we need to consider the geometric compatibility of several axially loaded bars pinned together at a joint. In previous cases we had either bars placed end to end or bars attached to rigid structures with imposed displacements.

Geometry: Let us begin by considering the geometry which relates the elongation of each of the bars to the downward displacement of point B. By symmetry, point B will displace downward an amount u_B to point B'. Symmetry requires that bars AB and CB carry the same forces, and thus they will elongate by the same amount δ so that

$$\delta_{AB} = \delta_{CB} = \delta \qquad (a)$$

For small u_B and δ, we can interpret Fig. 2.17*b* as representing the elongation of bars AB and CB by the amount δ followed by rotations of the bars about their pinned ends so that B ends up at B'. For small displacements, we can approximate the location of point B' by the location of point B'', using the tangent to the arc of rotation, so that from Fig. 2.17*b* it follows that δ and u_B are related by

$$\delta = u_B \sin\phi \qquad (b)$$

Statics: Force equilibrium at the joint B gives, Fig. 2.17*c*

$$F_{BD} + 2F\sin\phi = P \qquad (c)$$

Force-Displacement Relations: For each of the inclined bars we have

$$F = \frac{EA\delta}{2L/\cos\phi} \qquad (d)$$

Figure 2.17 Example 2.11

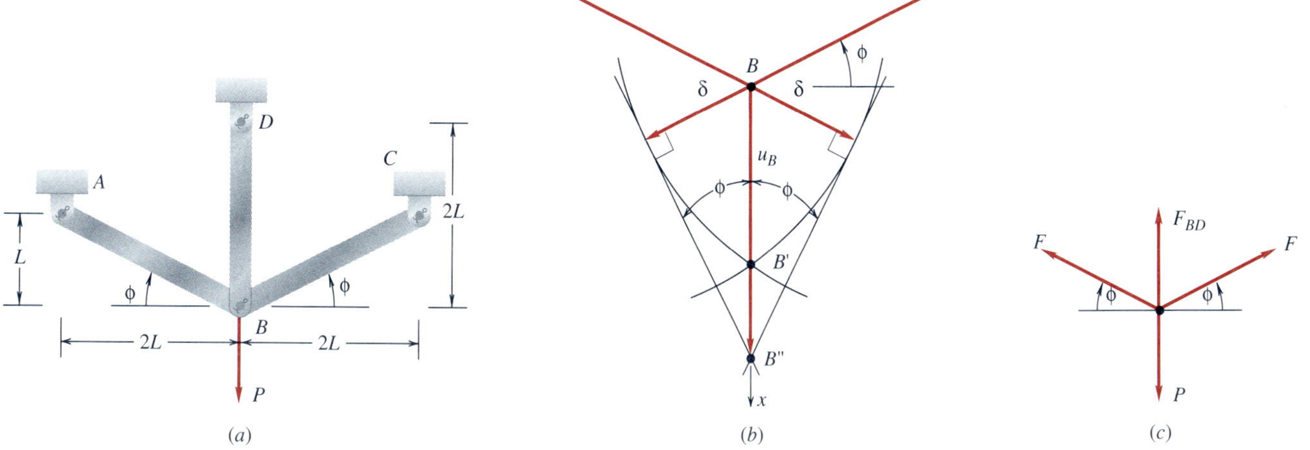

and for bar BD we have

$$F_{BD} = \frac{EAu_B}{2L} \tag{e}$$

We can find a single equation for the unknown displacement u_B by using Eq. (*b*) to write F as

$$F = \frac{EA}{2L} u_B \cos\phi \sin\phi \tag{f}$$

and then using Eqs. (*e*) and (*c*) to write

$$\frac{EA}{2L}(1 + 2\cos\phi \sin^2\phi)u_B = P \tag{g}$$

Inserting the numerical values gives

$$\frac{(200 \times 10^9)(450 \times 10^{-6})}{2(1)}[1 + 2(0.8944)(0.4472)^2]u_B = 50 \times 10^3$$

or

$$u_B = 0.818 \times 10^{-3} \text{ m} = 0.818 \text{ mm} \tag{h}$$

Returning to the force-displacement relation Eq. (*e*), we use the displacement u_B to find F_{BD}

$$F_{BD} = \frac{(200 \times 10^9)(450 \times 10^{-6})}{2(1)}(0.818 \times 10^{-3}) = 36.8 \text{ kN} \tag{i}$$

Using Eqs. (*e*) and (*f*), we can write F in the form

$$F = F_{BD} \cos\phi \sin\phi$$

or

$$F = 36.8(0.8944)(0.4472) = 14.7 \text{ kN} \tag{j}$$

We note in this problem that we applied the requirements of force equilibrium to the original undeformed geometry as in Fig. 2.17*c* and used the force-deformation relations with geometric compatibility of the displacements to obtain the displacements. The effect of the displacements on the calculated values of the forces will be small in problems of this type.

2.5 Temperature Effects

A temperature change will produce a change in length or a deformation of an *unconstrained* axial bar. Experiments show that a nearly linear relation exists between the change in length divided by the original length L or strain of the bar and the change in temperature. Therefore, the strain due to the temperature change can be written in the form

$$\epsilon_T = \alpha(\Delta T) \tag{2.9}$$

where ϵ_T is the thermal strain, ΔT is the temperature change from a reference value (usually room temperature), and α is the *coefficient of thermal expansion* of the material of the bar. The coefficient α has the units of 1/°F or 1/°C and it is assumed to be independent of the temperature change; it is a material constant. A brief tabulation of values of α for different materials is given below, and a more extensive listing can be found in Table F.1 of App. F. For example, for steel at room temperature, $\alpha = 6.5 \times 10^{-6}/°F$. In addition, for moderate temperature changes of the order of 100°C, the elastic constants for many engineering materials remain essentially at their "room-temperature" values.

Coefficient of Thermal Expansion α

	$10^{-6}/°F$	$10^{-6}/°C$
Aluminum and aluminum alloys	12	23
Bronze	10	19
Copper	9.5	17
Structural steel	6.5	12
Tungsten	2.4	4.3

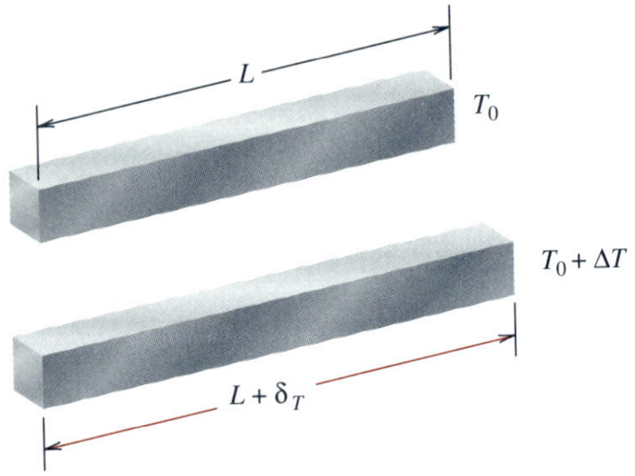

Figure 2.18 Change in length of a bar of length L due to a temperature rise ΔT.

Therefore, the thermal change in length or deformation of a bar of initial length L is given by

$$\delta_T = \epsilon_T L = \alpha(\Delta T)L \tag{2.10}$$

where L is the initial length of the bar at some reference temperature T_0 before the change in temperature to $T_0 + \Delta T$; see Fig. 2.18. If α is positive, as it is for most materials, and if ΔT is positive for a temperature increase, then δ_T is positive, indicating that the bar has increased in length.

Since the thermal strain in the bar is assumed to take place independently of any axial loads, we can modify Hooke's law, Eq. (2.2), as originally stated to account for the addition of the thermal strain and write

$$\epsilon = \frac{\sigma}{E} + \alpha\Delta T \tag{2.11}$$

where now the *total* strain ϵ consists of the sum of the "mechanical" strain σ/E or the strain due to the internal mechanical tensile stress in the bar and the thermal strain $\alpha(\Delta T)$. Equation (2.11) is the basic force-deformation equation for analyzing the response of a bar with a temperature change. It is convenient to write it in the form

$$\sigma = E\epsilon - E\alpha\Delta T \tag{2.12}$$

or in terms of the force F in the bar

$$F = EA\epsilon - EA\alpha\,\Delta T \tag{2.13}$$

Finally the strain ϵ in the bar can be expressed in terms of the displacements at the ends of the bar and the length L, Eq. (2.5), to obtain

$$\begin{aligned} F &= \frac{EA}{L}(u_B - u_A) - EA\alpha\,\Delta T \\ &= k(u_B - u_A) - EA\alpha\,\Delta T \end{aligned} \tag{2.14}$$

where $k = EA/L$ is the axial stiffness of the bar. Equation (2.14) is the force-deformation relation for a bar in the presence of a temperature change. Again, in Eq. (2.14) we assume that the force in the bar is tensile.

The inclusion of temperature effects in our analysis of problems does not change the basic steps as stated in Eq. (2.8), except for step 2 which now should make use of the modified force-deformation relation given by Eq. (2.14).

EXAMPLE 2.12

A long steel pipe is mounted between two fixed supports, as shown in Fig. 2.19*a*. When it was mounted, the temperature of the pipe was 20°C. In use, however, cold fluid moves through the pipe, causing it to cool to −15°C. Estimate the force on the supports due to this cooling. Neglect the weight of the pipe; assume $E = 205$ GPa and $\alpha = 12 \times 10^{-6}/°\text{C}$.

We assume that the pipe is a one-dimensional axial force member, and we use three steps in the solution. The x axis is along the pipe centerline with origin at the left end. The reactions R at the supports, as shown in Fig. 2.19*b*, are equal and oppositely directed and equal to the internal axial force in the pipe F, $R = F$. As the pipe cools, it will tend to shorten and thus tend to pull away from the walls.

The force-deformation relation from Eq. (2.14) is

$$F = \frac{EA}{L}(u_B - u_A) - EA\alpha\,\Delta T \tag{a}$$

and u_A and u_B are the displacements at the ends. However, the fact that the ends of the pipe are fixed gives $u_B = u_A = 0$, so that

$$F = -EA\alpha\,\Delta T \tag{b}$$

Substituting numerical values into Eq. (*b*), we find

$$A = \frac{\pi}{4}(d_o^2 - d_i^2) = 6.38 \times 10^{-3}\ \text{m}^2$$

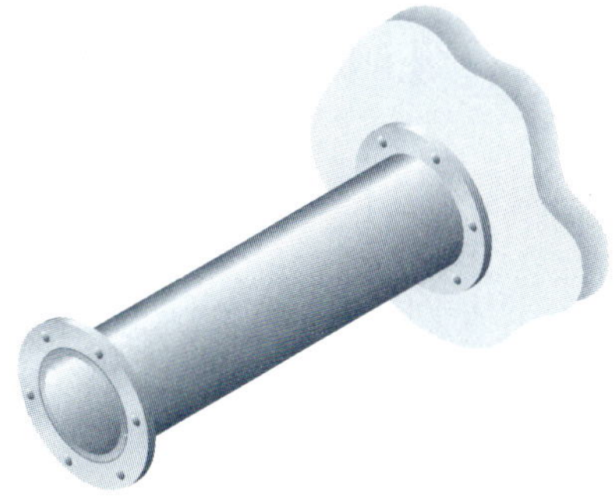

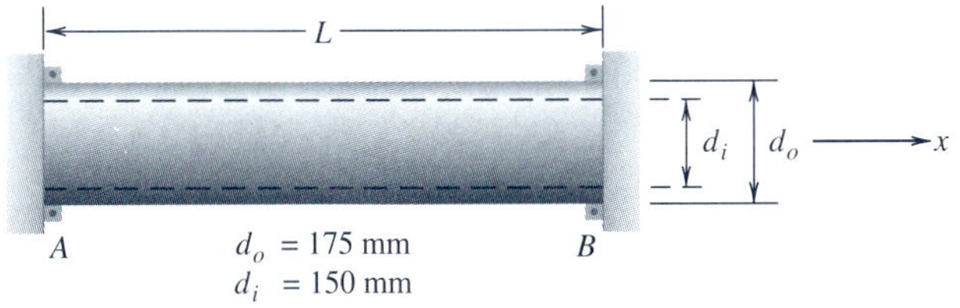

(*a*)

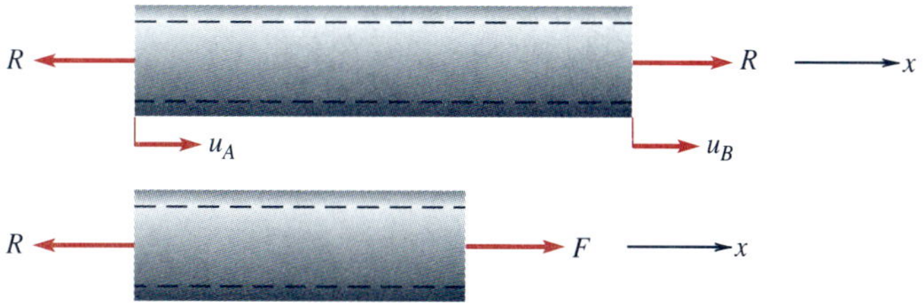

(*b*)

Figure 2.19 Example 2.12

and

$$F = -(205 \times 10^9 \text{ N/m}^2)(6.38 \times 10^{-3} \text{ m}^2) \cdot (12 \times 10^{-6}/^\circ\text{C})(-35^\circ\text{C}) = 5.49 \times 10^5 \text{ N} = 549 \text{ kN}$$

The force in the pipe is tensile since ΔT is negative. If the temperature of the pipe increased, $\Delta T > 0$, the force in the pipe would be compressive and the pipe would be pushing against the supports. We see in this example that the effect of constraining the pipe at the supports can give rise to large reactions at the supports. In most piping systems, allowance is made for thermal changes by including thermal expansion/contraction joints in the system.

EXAMPLE 2.13

Two standard-weight steel pipes of nominal diameters 3 in (segment AB) and 2 in (segment BC) (see App. E) are joined end to end at B and constrained between rigid walls, as shown in Fig. 2.20*a*. Find the stress in each pipe and the displacement of joint B due to a temperature increase of $\Delta T = 100^\circ$F. Use the values $E = 30{,}000$ ksi and $\alpha = 6.5 \times 10^{-6}/^\circ$F; $A_{AB} = 2.23 \text{ in}^2$, $A_{BC} = 1.07 \text{ in}^2$.

The problem is statically indeterminate. As a first step, we introduce an x coordinate axis and draw the free-body diagram of each segment as shown in Fig. 2.20*b*. Since no external forces act on the system, the left and right wall reactions are equal in magnitude and are taken as R. We proceed with our three-step approach to formulate the basic equations.

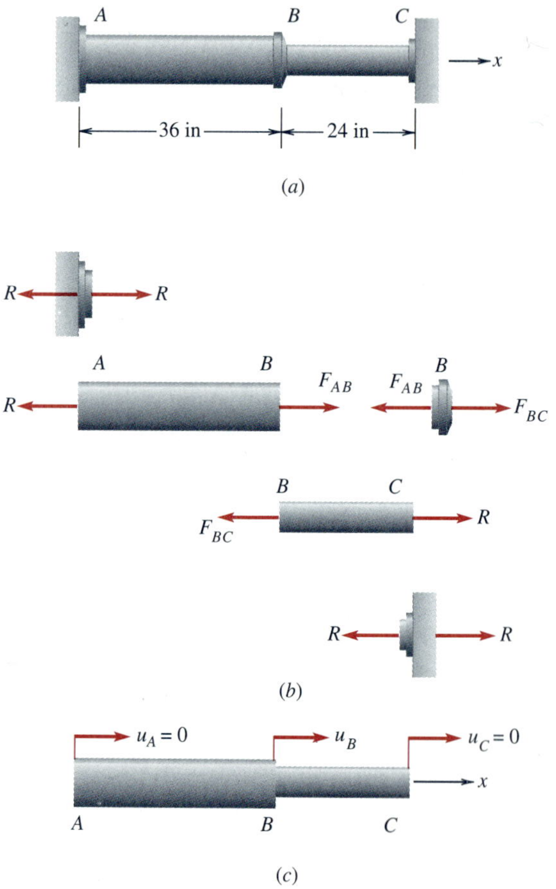

Figure 2.20 Example 2.13

Statics: Since the wall reactions are equal and opposite, we have from the free-body diagrams in Fig. 2.20*b*

$$F_{AB} = R \qquad F_{AB} = F_{BC} \qquad F_{BC} = R \tag{a}$$

where we assume for convenience that the force in each segment is tensile and that the reactions are tensile at the walls. If a force or reaction is found to be negative, then that force acts in the opposite direction.

Force-Deformation Relations: Using Eq. (2.11) for each pipe, we obtain for pipe AB

$$\epsilon_{AB} = \frac{F_{AB}}{EA_{AB}} + \alpha \Delta T$$

or

$$F_{AB} = EA_{AB}\epsilon_{AB} - EA_{AB}\alpha\Delta T \tag{b}$$

and for pipe BC,

$$\epsilon_{BC} = \frac{F_{BC}}{EA_{BC}} + \alpha \Delta T$$

or

$$F_{BC} = EA_{BC}\epsilon_{BC} - EA_{BC}\alpha\Delta T \tag{c}$$

Geometry: The displacements of sections A, B, and C are shown in Fig. 2.20*c*, and the wall constraints require $u_A = u_C = 0$. Therefore,

$$\epsilon_{AB} = \frac{u_B - u_A}{L_{AB}} = \frac{u_B}{L_{AB}} \tag{d}$$

$$\epsilon_{BC} = \frac{u_C - u_B}{L_{BC}} = \frac{-u_B}{L_{BC}} \tag{e}$$

It follows from Eqs. (*a*) to (*e*) with $F_{AB} = F_{BC}$ that

$$EA_{AB}\frac{u_B}{L_{AB}} - EA_{AB}\alpha\,\Delta T = -EA_{BC}\frac{u_B}{L_{BC}} - EA_{BC}\alpha\,\Delta T$$

If we solve for u_B, we have

$$u_B\left(\frac{A_{AB}}{L_{AB}} + \frac{A_{BC}}{L_{BC}}\right) = \alpha\,\Delta T(A_{AB} - A_{BC})$$

or

$$u_B\left(\frac{2.23}{36} + \frac{1.07}{24}\right) = 6.5 \times 10^{-6} \times 100 \times 1.16$$

$$u_B = 7.08 \times 10^{-3}\ \text{in}$$

Now that u_B is known, we can determine R from Eqs. (*a*) and (*b*). We find

$$R = -30.33\ \text{kips}$$

Therefore the stresses in the segments are compressive

$$\sigma_{AB} = \frac{R}{A_{AB}} = -13.6\ \text{ksi}$$

$$\sigma_{BC} = \frac{R}{A_{BC}} = -28.3\ \text{ksi}$$

and the wall reactions are compressive. The pipe segments try to expand and push against the supports.

EXAMPLE 2.14

Consider the wire-controlled mechanical linkage system that was investigated in Example 2.8; we show the model system of Fig. 2.14*b* again in Fig. 2.21*a*. Bar *AE* is rigid. If the system experiences a temperature increase of 100°F in the absence of any load *P* at point *A*, find the forces in links *CF* and *DG* connected at *C* and *D*. Link *CF* is aluminum with $E_{CF} = 10 \times 10^6$ psi, $A_{CF} = 0.1\ \text{in}^2$, and $\alpha_{CF} = 12 \times 10^{-6}/°\text{F}$; link *DG* is steel with $E_{DG} = 30 \times 10^6$ psi, $A_{DG} = 0.15\ \text{in}^2$, and $\alpha_{DG} = 6.5 \times 10^{-6}/°\text{F}$.

We assume that the force in each link is tensile, as shown in Fig. 2.21*b*, and the displacements u_C and u_D at points *C* and *D* are as shown in Fig. 2.21*c*. We neglect friction and the weight of all members.

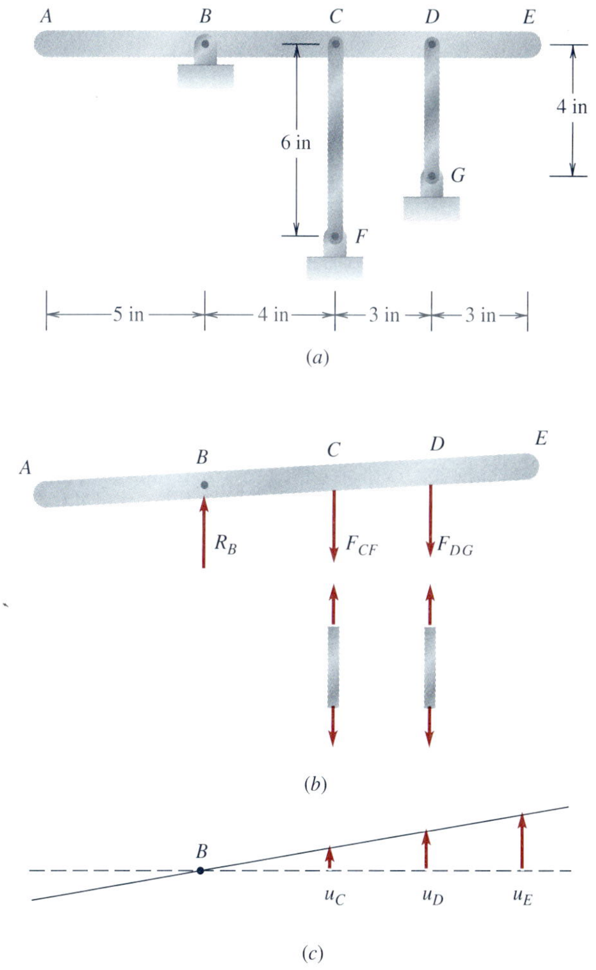

Figure 2.21 Example 2.14

Statics: We take moments about point *B* to obtain (Fig. 2.21*b*)

$$4F_{CF} + 7F_{DG} = 0 \qquad (a)$$

Geometry: From the displacement diagram shown in Fig. 2.21*c*, we have

$$\frac{u_C}{4} = \frac{u_D}{7} = \frac{u_E}{10} \qquad (b)$$

Force-Deformation Relations: Up to this point the method of solution is identical to that of Example 2.8 with the load $P = 0$ [see Eqs. (*a*) and (*b*) of Example 2.8]. In the presence of a temperature change, however, the force-deformation relation for each link becomes

$$F_{CF} = \left(\frac{EA}{L}\right)_{CF} u_C - (EA\alpha)_{CF}\,\Delta T \qquad (c)$$

$$F_{DG} = \left(\frac{EA}{L}\right)_{DG} u_D - (EA\alpha)_{DG}\,\Delta T \qquad (d)$$

where the forces and displacements are assumed positive in each link as shown in Fig. 2.21*b* and *c*. If Eqs. (*c*), (*d*), and (*b*) are substituted into Eq. (*a*), we find

$$u_C\left[4\left(\frac{EA}{L}\right)_{CF}+7\left(\frac{EA}{L}\right)_{DG}\left(\frac{7}{4}\right)\right] = \Delta T[4(EA\alpha)_{CF}+7(EA\alpha)_{DG}] \qquad (e)$$

Upon substitution of numerical values, we have

$$u_C\left[\frac{(4)(10\times10^6)(0.1)}{6}+\frac{(49)(30\times10^6)(0.15)}{(4)(4)}\right] = \Delta T[(4)(10\times10^6)(0.1)(12\times10^{-6}) + (7)(30\times10^6)(0.15)(6.5\times10^{-6})]$$

from which, with $\Delta T = 100°$F, we find

$$u_C = 1.75\times10^{-3}\text{ in} \qquad (f)$$

Upon substitution of the value for u_C into Eqs. (*c*) and (*d*) we find

$$F_{CF} = -908\text{ lb}$$

and

$$F_{DG} = 520\text{ lb} \qquad (g)$$

Link *CF* is in compression while link *DG* is in tension. The values for F_{CF} and F_{DG} satisfy the equilibrium equation, Eq. (*a*).

EXAMPLE 2.15

The steel bolt and aluminum tube assembly of Example 2.10 (Fig. 2.16*a*) is to be investigated for the case of a temperature increase. The assembly is redrawn in Fig. 2.22*a*. Two cases are to be considered. In the first case, the nut is tightened by hand, and the temperature is increased by $\Delta T = 50°$F. In the second case, after being tightened by hand, the nut is further tightened by one-eighth turn, and the temperature is increased by 50°F. We wish to find the force in the bolt and tube in each case.

The bolt has 16 threads per inch, the cross-sectional area of the bolt is $A_B = 1.00\text{ in}^2$, the cross-sectional area of the tube is $A_T = 0.60\text{ in}^2$, and $L = 6$ in. Take $E_T = 10\times10^6$ psi, $\alpha_T = 12\times10^{-6}/°\text{F}$, $E_B = 30\times10^6$ psi, and $\alpha_B = 6.5\times10^{-6}/°\text{F}$.

As we have seen so far in the solution of problems with a temperature change, the use of the three steps for the solution still applies when we account for the presence of the temperature in the force-deformation relations. It is important to realize that temperature changes in general do not effect the formulation of equilibrium equations or geometry arguments. We will emphasize this again in the solution of this example.

Statics: In the case of a temperature rise, it is convenient in the solution procedure to *assume* that both the bolt and the tube are in tension. Equilibrium of the nut gives (Fig. 2.22*b*)

$$F_T + F_B = 0 \qquad (a)$$

where F_T and F_B are the forces in the tube and bolt. Of course, we expect the bolt to be in tension and the tube to be in compression.

Force-Deformation Relations: If we assume that the tube and the bolt are in tension and that the left end of the bolt and tube are held fixed, then the force-deformation relations, Eqs. (2.14), are

$$F_B = \frac{E_BA_B}{L}u_B - E_BA_B\alpha_B\Delta T$$

$$F_T = \frac{E_TA_T}{L}u_T - E_TA_T\alpha_T\Delta T \qquad (b)$$

The displacements u_B and u_T are positive in the positive *x* direction, as shown in Fig. 2.22*b*. The convenience of assuming all members in tension arises in the force-deformation relations, Eqs. (*b*), when the temperature term is present.

Geometry: We now need to consider two cases. In the first case, the nut is tightened by hand, and the temperature increases by $\Delta T = 50°$F. In this case, the nut moves with the bolt and the end of the tube, so that

$$u_B = u_T \qquad (c)$$

In the second case, the nut is first tightened by hand and further tightened one-eighth turn while the temperature increases by $\Delta T = 50°$F. In this case the location of the nut is then

$$L - d + u_B \qquad (d)$$

where *d* is the movement of the nut on the bolt and is positive, as shown in Fig. 2.22*b*, with $d = (\frac{1}{8})(\frac{1}{16})$ in. The location of the end of the tube is

$$L + u_T \qquad (e)$$

Since the tube and nut must remain in contact, we have from Eqs. (*d*) and (*e*)

$$L - d + u_B = L + u_T$$

or

$$u_B - d = u_T \qquad (f)$$

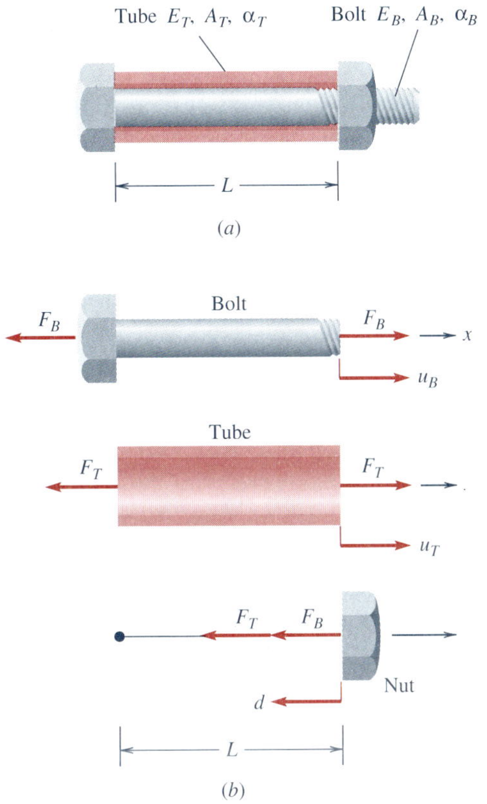

Figure 2.22 Example 2.15

Equation (f), of course, reduces to Eq. (c) in the absence of the additional tightening of d. Note that the temperature-increase term only arises in Eqs. (b). To determine the forces in both cases, we add Eqs. (b) and use Eqs. (a) and (f). For the solution with no tightening we can set $d = 0$.

Adding Eqs. (b) with the use of (a), we find

$$\frac{E_B A_B}{L} u_B + \frac{E_T A_T}{L} u_T = \Delta T (E_B A_B \alpha_B + E_T A_T \alpha_T) \qquad (g)$$

With $u_T = u_B - d$ from Eq. (f), we can solve Eq. (g) for u_B,

$$u_B = d \frac{E_T A_T}{E_B A_B + E_T A_T} + L \Delta T \frac{E_B A_B \alpha_B + E_T A_T \alpha_T}{E_B A_B + E_T A_T} \qquad (h)$$

We now have solved for the bolt displacement for each of the required cases. First with $d = 0$, we have

$$u_B = \frac{(30 \times 10^6)(1)(6.5 \times 10^{-6}) + (10 \times 10^6)(0.6)(12 \times 10^{-6})}{(30 \times 10^6)(1) + (10 \times 10^6)(0.6)} \cdot (6)(50) \qquad (i)$$
$$= 2.225 \times 10^{-3} \text{ in}$$

The force in the bolt is given by the first of Eqs. (b) with the value of u_B from Eq. (i)

$$F_B = (30 \times 10^6)(1) \left[\frac{2.225 \times 10^{-3}}{6} - (6.5 \times 10^{-6})(50) \right]$$

or

$$F_B = 1375 \text{ lb} \qquad (j)$$

The force in the tube is then $F_T = -F_B = -1375$ lb.

In the second case with $d = \frac{1}{8}(\frac{1}{16}) = \frac{1}{128}$ in and $\Delta T = 50°$F, we have from Eq. (h)

$$u_B = \frac{1}{128} \frac{(10 \times 10^6)(0.6)}{(30 \times 10^6)(1) + (10 \times 10^6)(0.6)} + 2.225 \times 10^{-3}$$
$$= 1.302 \times 10^{-3} + 2.225 \times 10^{-3} \qquad (k)$$
$$= 3.527 \times 10^{-3} \text{ in}$$

The force in the bolt follows from Eqs. (b) as before, and we find

$$F_B = 7885 \text{ lb} \qquad (l)$$

We note that the effect of tightening the nut in the present case is significant.

Finally, it is of value to obtain algebraically the expression for the force in the bolt by combining Eq. (h) for u_B with Eqs. (b) to find

$$F_B = \frac{E_B A_B}{1 + (E_B A_B / E_T A_T)} \left[\frac{d}{L} + (\alpha_T - \alpha_B) \Delta T \right] \qquad (m)$$

We then have $F_T = -F_B$ and $u_T = u_B - d$.

2.6 Displacement Method for Axially Loaded Members

Frequently problems of practical interest involve a large number of interconnected structural components or machine parts. In some cases, as we have seen, the analysis of such systems consists of the repeated use of fairly simple free-body diagrams and associated equilibrium and force displacement equations. For such problems with the thought of computer-assisted analysis in mind, it is worthwhile to make the effort to set up a systematic notation that lends itself to an organized derivation of the equations which govern the problem.

A class of problems that can be solved by a systematic derivation of the basic equations includes those involving axially loaded bars or members placed in series (similar to those treated in Examples 2.2, 2.3, 2.7, 2.9, 2.12, and 2.13). The bars in the system can have different areas and lengths (see Fig. 2.23), can be made of different materials, can undergo temperature changes, and can be subjected to a variety of applied load and support conditions. In Fig. 2.23 we show the x axis along the centroidal axis of the members.

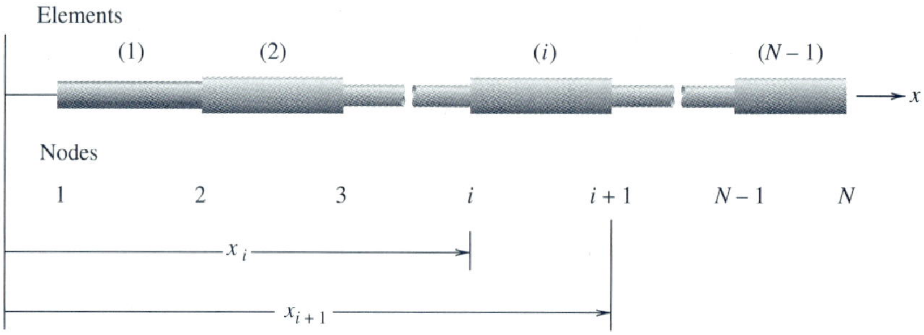

Figure 2.23 Axially loaded bars in series.

If we wish to analyze a system of bars as shown in Fig. 2.23, we subdivide the total system into bar *elements* labeled (1) to (N – 1) with the *endpoints* of the bar elements called *nodes* labeled 1 to N. We assume that the system has the following properties:

1. Applied axial forces act only at nodes (i.e., at the junction of two bar elements or at the ends of the first or last bar element) and are denoted by P (lb or N). *Applied* forces are positive in the positive x direction.
2. The stiffness of each bar element of length L is constant for each element and is denoted by $k_i = (EA/L)_i$. The coefficient of thermal expansion α is constant for each element.
3. The axial displacement of each node is denoted by u_i.

Again, all applied axial forces and displacements are considered *positive* when they are directed in the *positive x direction* of the bar. It is also convenient to assume initially that all the bar elements are in tension.

In using this notation it is clearly necessary to introduce new nodes for each change in geometry (i.e., change in cross-sectional area), each change in material, each location where an applied load is specified, or each location where an axial displacement is specified (e.g., at points where supports require that no displacement occur). When a temperature change occurs, we assume it is the same for all bar elements in the system.

Our approach to solving problems involving bar systems like the one shown in Fig. 2.23 will be to set up a procedure which will generate a system of simultaneous linear equations for the unknown node displacements of the problem. After we solve these linear equations and obtain the unknown displacements, it will be possible to calculate internal forces in each bar element as well as the unknown external reactions. The steps in the analysis are very systematic and lend themselves to being organized into a computer program. The displacement method presented in this section is a simple version of methods presented in advanced courses.

Derivation of Equations for Unknown Node Displacements

Our objective is to relate the force in each element to the displacements at the ends or nodes of the element and to the temperature change of the element. We use the three steps in our formulation.

Statics. Consider a free body as shown in Fig. 2.24 of a typical *i*th element from the bar system shown in Fig. 2.23. The force in the *i*th element is f_i and is assumed to be tensile, u_i and u_{i+1} are the end or node displacements, and x_i and $x_{i+1} = x_i + L_i$ are the initial coordinates of the nodes i and $i + 1$; L_i is the length of the element. Again, the force f_i in the element is constant and is assumed to be tensile.

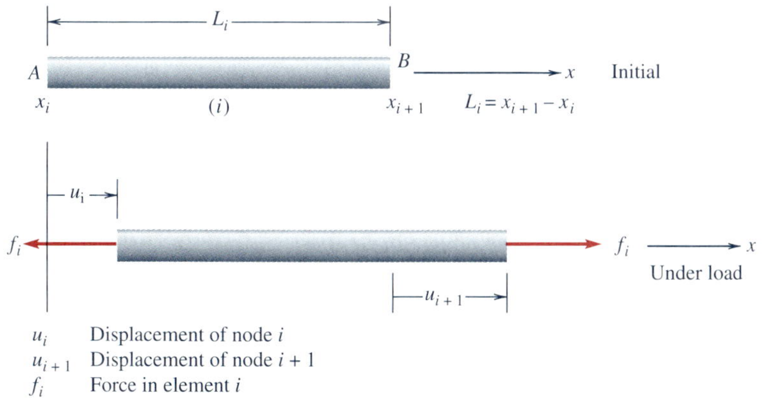

Figure 2.24 Free-body diagram of the *i*th element showing end displacements u_i and u_{i+1} and internal element force f_i.

Force-Deformation Relations. Using Hooke's law as modified to account for temperature changes, Eq. (2.11), we have

$$\epsilon = \frac{\sigma}{E} + \alpha \Delta T$$

or

$$\sigma = E(\epsilon - \alpha \Delta T) \qquad (2.15)$$

Now the axial strain $\epsilon = (u_{i+1} - u_i)/L_i$ and the axial stress $\sigma = f_i/A_i$ can be inserted into Eq. (2.15) to obtain

$$f_i = \frac{E_i A_i}{L_i}(u_{i+1} - u_i) - E_i A_i \alpha_i (\Delta T)_i \tag{2.16}$$

or in terms of the stiffness $k_i = (EA/L)_i$ for the ith element,

$$f_i = k_i(u_{i+1} - u_i) - k_i L_i \alpha_i (\Delta T)_i \tag{2.17}$$

Equation (2.17) is the fundamental force-displacement relation for the ith element. It is easy to interpret the terms in Eq. (2.17). If there is no temperature change, that is, $\Delta T = 0$ in the element, then the first term gives the force in the bar required to cause the deformation $u_{i+1} - u_i$. If the deformation of the element is zero, that is, $u_{i+1} = u_i$, then the second term represents the compressive force in the bar which results when the thermal expansion of the bar given by $L_i \alpha_i (\Delta T)_i$ is prevented.

Geometry. Note that by our choice of notation the node displacements are the end displacements of the bar elements. That is, the end displacement u_{i+1} at the right end of bar element (i) is the same as the end displacement at the left end of element $(i + 1)$.

In general, for a given problem there will be one or more specified values of the node displacements. The remaining unknown displacements will be found by solving appropriate equilibrium equations expressed in terms of these unknown displacements. First we illustrate this solution procedure for the problem in Example 2.9, and then we briefly discuss how a more general procedure can be used to solve this class of problems involving bar systems. Finally we discuss the use of a computer program to solve this class of problems.

EXAMPLE 2.16

Consider again the steel pipe from Example 2.9 with different cross-sectional areas in each of its three segments, as shown in Fig. 2.25*a*. A load $F = 3$ kips is applied at section B. We wish to find the force in each segment and the displacements of sections B and C, using the displacement method as outlined above. Given: $A_{AB} = 1$ in^2, $A_{BC} = 3$ in^2, $A_{CD} = 2$ in^2, and $E = 30 \times 10^6$ psi. There is no temperature change.

The procedure for solving this problem will involve the use of node equilibrium equations to derive a set of equations for the unknown node displacements. Element forces, stresses, and external reactive forces will then be found from the node displacements.

Statics: It is convenient to introduce four nodes and three elements and to orient the x axis with origin at A as shown in Fig. 2.25*b*. Also shown in Fig. 2.25*b* are the node displacements. Free-body diagrams for the three segments of the pipe and for infinitesimal node elements at the ends of the elements where the external forces are applied or where two elements join are shown in Fig. 2.25*c*. Axial equilibrium of the node elements at nodes 2 and 3 requires that

$$\begin{aligned} f_2 - f_1 - F &= 0 \\ f_3 - f_2 &= 0 \end{aligned} \tag{a}$$

Force-Deformation Relations: The bar forces can be expressed in terms of the node displacements by Eq. (2.17), with $\Delta T = 0$, to obtain

$$\begin{aligned} f_1 &= k_1(u_2 - u_1) \\ f_2 &= k_2(u_3 - u_2) \\ f_3 &= k_3(u_4 - u_3) \end{aligned} \tag{b}$$

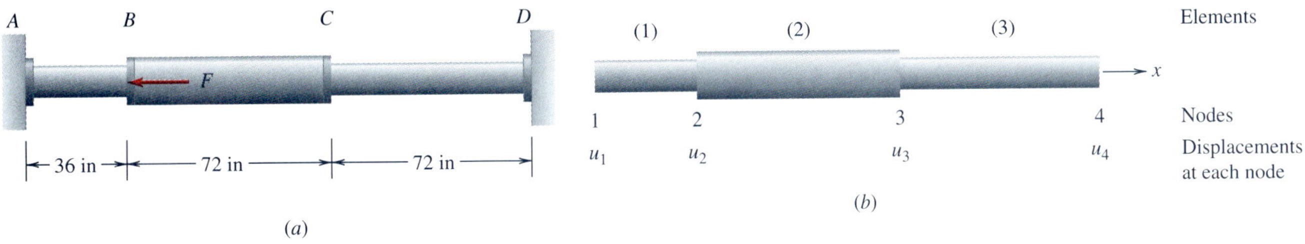

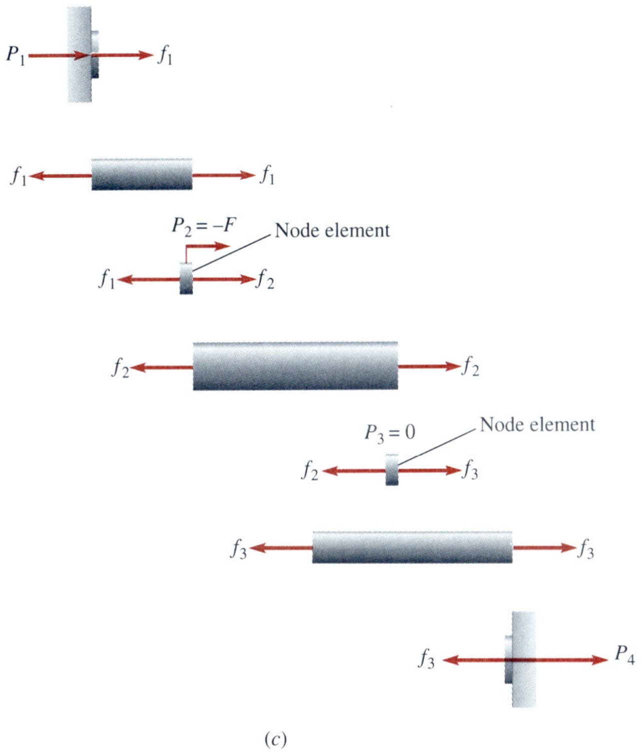

Figure 2.25 Example 2.16

where the k_i are the stiffnesses in each element, $k_i = (EA/L)_i$.

Geometry: From Fig. 2.25*a* we note that nodes 1 and 4 are fixed so that

$$u_1 = 0 \qquad u_4 = 0 \tag{c}$$

and thus u_2 and u_3 are the unknown node displacements for the problem.

The element forces can be eliminated from the two node equilibrium equations, Eqs. (*a*), by using the force-deformation relations, Eqs. (*b*), to get

$$\begin{aligned} k_2(u_3 - u_2) - k_1(u_2 - u_1) - F &= 0 \\ k_3(u_4 - u_3) - k_2(u_3 - u_2) &= 0 \end{aligned} \tag{d}$$

which can be written as

$$\begin{aligned} -k_1u_1 + (k_1 + k_2)u_2 - k_2u_3 &= -F \\ -k_2u_2 + (k_2 + k_3)u_3 - k_3u_4 &= 0 \end{aligned} \tag{e}$$

If we use the conditions $u_1 = u_4 = 0$, $F = 3$ kips and

$$k_1 = \left(\frac{EA}{L}\right)_1 = \frac{(30 \times 10^3)(1)}{36} = 8.33 \times 10^2 \text{ kips/in}$$

$$k_2 = \left(\frac{EA}{L}\right)_2 = \frac{(30 \times 10^3)(3)}{72} = 1.25 \times 10^3 \text{ kips/in}$$

$$k_3 = \left(\frac{EA}{L}\right)_3 = \frac{(30 \times 10^3)(2)}{72} = 8.33 \times 10^2 \text{ kips/in}$$

then Eqs. (*e*) take the form

$$\begin{aligned} 2.083 \times 10^3(u_2) - 1.25 \times 10^3(u_3) &= -3 \\ -1.25 \times 10^3(u_2) + 2.083 \times 10^3(u_3) &= 0 \end{aligned} \tag{f}$$

To illustrate the general approach, we used the node equilibrium equations at nodes 2 and 3 corresponding to the two unknown node displacements. This gave the correct number of equations for the determination of these displacements. The solution of the two equations in Eqs. (*f*) for the two unknown displacements yields

$$\begin{aligned} u_2 &= -2.25 \times 10^{-3} \text{ in} \\ u_3 &= -1.35 \times 10^{-3} \text{ in} \end{aligned} \tag{g}$$

which agrees with the result found for the displacement at section B in Example 2.9, Eq. (*h*).

Now that u_2 and u_3 are known, we can use Eqs. (*b*) to find the bar forces f_1, f_2, and f_3:

$$f_1 = k_1 u_2 = (8.33 \times 10^2)(-2.25 \times 10^{-3}) = -1875 \text{ lb}$$
$$f_2 = k_2(u_3 - u_2) = (1.25 \times 10^3)(-1.35 + 2.25) \times 10^{-3} = 1125 \text{ lb} \qquad (h)$$
$$f_3 = k_3(-u_3) = (8.33 \times 10^2)(1.35 \times 10^{-3}) = 1125 \text{ lb}$$

The node reactive forces at the walls follow from the free-body diagrams in Fig. 2.25*c* and from the values of f_1 and f_3 in the form

$$P_1 = -f_1 = 1875 \text{ lb}$$
$$P_4 = f_3 = 1125 \text{ lb} \qquad (i)$$

General Solution Procedure

The solution procedure for the displacement method for a general bar system with N node points (see Fig. 2.23) is similar to that used in Example 2.16. We outline the steps for the general solution procedure in what follows.

Statics. At each internal node element of the N-node case, there will be end element forces caused by the elements on the left and right sides of the node element, and these two forces will enter in the equilibrium equation for the node element along with the external force (if any) acting at that node.

A typical node equilibrium equation at the ith node will then be of the form

$$f_{i-1} - f_i = P_i \qquad (2.18)$$

where P_i is the applied force or the reactive force at the node i.

Force-Deformation Relations. For the ith element, we use Eq. (2.17)

$$f_i = k_i(u_{i+1} - u_i) - [k\alpha(\Delta T)L]_i \qquad (2.19)$$

to relate the bar force to the node displacements.

A system of N equations for the N node displacements can be derived by substituting the expression for the element forces from Eqs. (2.19) into the equilibrium equations, Eqs. (2.18), to get

$$-k_{i-1}u_{i-1} + (k_{i-1} + k_i)u_i - k_i u_{i+1} = H_i \qquad (2.20)$$

where

$$H_i = [k\alpha(\Delta T)L]_{i-1} - [k\alpha(\Delta T)L]_i + P_i \qquad (2.21)$$

where the P_i, $i = 1$ to N, are the external forces at the nodes, either an applied force or a reaction.

Geometry. In general, for each problem there will be nodes where the node displacements are specified. Once these node displacements are specified, it is possible to reduce the system of equations to a smaller number of equations. Once this reduction procedure is carried out, the system of remaining simultaneous linear equations can be solved and the unknown node displacements found.

The organization of the steps in the above analysis and the systematic notation introduced greatly facilitate the programming of the displacement method for a computer solution. A computer program called BARMECH (see the MECHMAT program on the diskette which includes the BARMECH program) has been written which allows the user to introduce a description of an axially loaded bar system, and the program produces printed and graphics output describing the bar forces and node displacements.

In the next section details on the use of the computer program will be given.

2.7 Use of Computer Program BARMECH

The procedure for using the program BARMECH to solve problems with axially loaded bars is explained in the following examples. The displacement method as described in Sec. 2.6 forms the basis for this program. Both statically determinate and indeterminate problems can be handled, with and without temperature changes. We treat first a statically determinate system, then a statically indeterminate system, and finally a statically indeterminate system with a temperature change.

EXAMPLE 2.17

A steel pipe shown in Fig. 2.26*a* is loaded by three concentrated axial loads as shown. If the cross-sectional area of the pipe is 1800 mm^2 and E = 200 GPa, (*a*) find the displacement at the right end, (*b*) find the stresses in each segment of the pipe, and (*c*) find the distance from the left end support to that section along the pipe which does not displace under the action of the applied loads.

As the first step in the use of the program, we introduce the x coordinate axis as shown in Fig. 2.26*b* with the origin at the left end of the pipe. We will measure all locations and

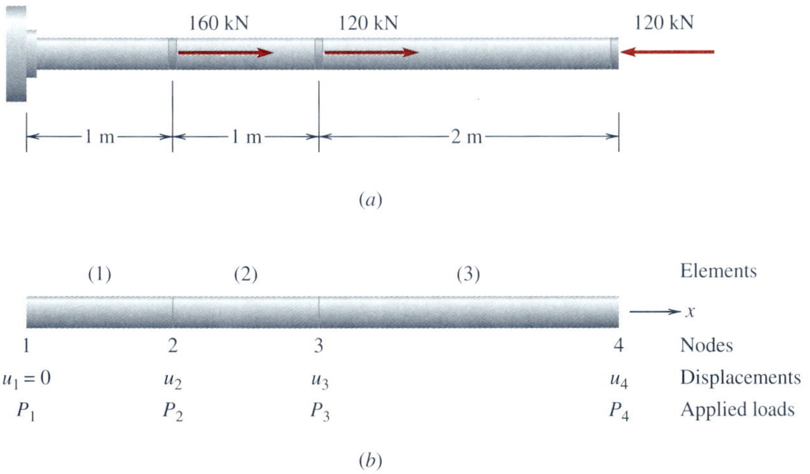

Figure 2.26 Example 2.17

displacements on the pipe with respect to this axis.

The program BARMECH is included in the MECHMAT program on the computer diskette. In preparation for using BARMECH, it is important to recast the original problem into the notation of the displacement method. Separate elements must be defined so that each element is made up of the same material and has the same constant cross-sectional area. Also, applied loads can be applied only at end nodes of the system or at the nodes between elements. A choice for the present problem is shown in Fig. 2.26*b* where we have introduced four nodes and three elements. Three elements are chosen so that the specified loads are all applied at the ends of elements as shown. At nodes 2, 3, and 4, the node forces are identified with the specified *external* loads $P_2 = 160$ kN, $P_3 = 120$ kN, and $P_4 = -120$ kN. At the support (node 1), the node force P_1 represents the reactive force exerted by the support on the node element at node 1. Since the support at the left end of the bar is fixed, $u_1 = 0$, the displacements u_2, u_3, and u_4 become the three unknown displacements for the problem.

To Use BARMECH

The first step is to follow the instructions in App. H to load MECHMAT. From the main menu of the MECHMAT program, Fig. 2.27*a,* we select 1: Axially loaded members, as shown in Fig. 2.27*a,* to get the BARMECH menu shown in Fig. 2.27*b.* Since we are creating a new data file, we proceed as shown in Fig. 2.27*b* and select 1: Input new data file. Usually we will save the data file in drive *A*; we then give a name to the file, and in this case we use EX217. If we wish, later we can modify the input data in the file or run again; these are choices 2 and 3 in the BARMECH menu.

The program then prompts the input data required, as shown in Fig. 2.28. Note that the multiple entries giving element properties *must be separated by commas.* The units of E are 10^6 psi or MPa; in this problem $E = 200$ GPa $= 200{,}000$ MPa, as shown in Fig. 2.28. There are three applied forces specified, and the location and value of each specified force are given by typing the node number followed by a comma and the appropriate force with the correct sign; see Fig. 2.28. The specified displacements are handled in a similar fashion. There are options to print a table giving the input data for the problem and the solution as shown in Figs. 2.29 and 2.30.

The program BARMECH derives three node equilibrium equations (at nodes 2, 3, 4) for the three unknown displacements u_2, u_3, and u_4 and solves these equations and prints the results; see Fig. 2.30. Thus the solution to part (*a*) of this problem, the displacement of the free end (node 4), is given by $u_4 = -0.222$ mm. Since there is no temperature change, the element forces are found from Eq. (2.17) with $\Delta T = 0$, that is,

$$f_i = k_i(u_{i+1} - u_i) \qquad i = 1, 2, 3$$

These are given in Fig. 2.30 along with the stresses in each element. Units of force are newtons and units of stress are pascals. Finally, a graph of the axial displacement is shown in Fig. 2.31. The displacement curve shows the displacement of each section along the pipe. The circled points correspond to the nodes. The displacement of any section is measured from the undeformed location of that section on the bar. From the graph

```
*********************************************************

*** *MECHANICS OF MATERIALS* ****
PROGRAM: MECHMAT                    mod 5.0

*********************************************************

     MAIN MENU

           1. AXIALLY LOADED MEMBERS
           2. TORSION
           3. V + M DIAGRAMS OF BEAMS
           4. DEFLECTIONS OF BEAMS
           5. MOHR'S CIRCLE
           6. CENTROIDS + MOMENTS OF INERTIA
           7. EXIT

     SELECT NUMBER = ? 1
```

(*a*)

```
*********************************************************

*** AXIALLY LOADED MEMBERS ***
        ***BARMECH***

*********************************************************

*** DATA INPUT CHOICE ***

   1. INPUT NEW DATA FILE
   2. MODIFY EXISTING DATA FILE
   3. USE EXISTING DATA FILE
OR 4. RETURN TO MAIN MENU
      SELECT NUMBER? 1


          DRIVE USED (A:, B:, or C:) ? a:
        FILE NAME (OMIT DRIVE LETTER)? ex217
```

(*b*)

Figure 2.27 (*a*) Main menu of MECHMAT; (*b*) menu for BARMECH, Example 2.17.

we see there is one section other than the fixed left end that has zero displacement. By straight linear interpolation, if x_0 is the distance from the left wall support to the section of zero displacement, then

$$\frac{x_0 - 2}{4.44} = \frac{4 - x_0}{2.22}$$

and

$$x_0 = \tfrac{10}{3} = 3.33 \text{ m}$$

This gives us the section on the bar where the zero displacement occurs.

```
   INPUT THE FOLLOWING DATA:

NUMBER OF ELEMENTS                 NM=? 3

CHANGE IN THE TEMPERATURE        Y/N: ? n

INPUT THE LENGTH, AREA, AND MODULUS OF ELASTICITY:
            L        A             E
           (in)     (inxin)     (10**6; psi)
           (m)      (mxm)          (MPa)

ELEMENT(1)---? 1,1.8e-03,200000
ELEMENT(2)---? 1,1.8e-03,200000
ELEMENT(3)---? 2,1.8e-03,200000

   ARE THE ABOVE VALUES CORRECT Y/N? y

NUMBER OF APPLIED FORCES:  ? 3

  NODE?, FORCE (lb,N)?

? 2,160000
? 3,120000
? 4,-120000

NUMBER OF SPECIFIED DISPLACEMENTS:  ? 1

  NODE?, VALUE (in,m)?

? 1,0

   ARE THE ABOVE VALUES CORRECT  Y/N? y
```

Figure 2.28 Input format for Example 2.17.

```
***MECHANICS OF MATERIALS***  DATA FILE: ex217
                    ***BAR***

   INPUT DATA:

NUMBER OF ELEMENTS      NM = 3

CHANGE IN THE TEMPERATURE?     n

 ELEMENT        L (in)     A (inxin)        E (10**6; psi)
                  (m)        (mxm)              (MPa)

     1          1.00       1.80E-03         2.00E+05
     2          1.00       1.80E-03         2.00E+05
     3          2.00       1.80E-03         2.00E+05

                CONDITION
***********************************

     NODE            FORCE
       2           1.60E+05
       3           1.20E+05
       4          -1.20E+05

     NODE       SPECIFIED DISP.
       1           0.00E+00
```

Figure 2.29 Input summary of input data for Example 2.17.

Figure 2.30 Output of solution for Example 2.17.

```
********************************************
       NODE          DISPLACEMENT (in,m)
********************************************
          1                0.00E+00
          2                4.44E-04
          3                4.44E-04
          4               -2.22E-04


*******************************************************
 ELEMENT       FORCE            STRESS              E
*******************************************************
     1       1.6000E+05       8.8889E+07       2.00E+11
     2       0.0000E+00       0.0000E+00       2.00E+11
     3      -1.2000E+05      -6.6667E+07       2.00E+11
```

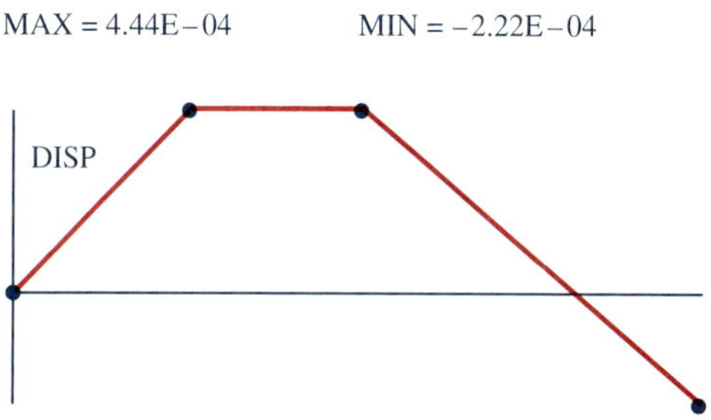

Figure 2.31 Displacement diagram for Example 2.17.

EXAMPLE 2.18

To contrast the hand solution procedure with the computer solution procedure, we reconsider the statically indeterminate Example 2.16, which is shown again in Fig. 2.32. A load $F = 3$ kips is applied at section B. We wish to find the force in each segment and the displacements of sections B and C, using the BARMECH program. Take

$$A_{AB} = 1 \text{ in}^2$$
$$A_{BC} = 3 \text{ in}^2$$
$$A_{CD} = 2 \text{ in}^2$$

and $\quad E = 30 \times 10^6$ psi

For this problem we have three elements and four nodes. The coordinate axis is along the pipe with the origin at section A. The applied load $F = 3$ kips is applied at node 2. Note that the applied force at node 2 is then $P_2 = -3$ kips. The displacement is zero at nodes 1 and 4. Figure 2.33 shows the input data and the solution while Fig. 2.34 shows the displacement curve along the pipe. The results are in complete agreement with those found in Example 2.16.

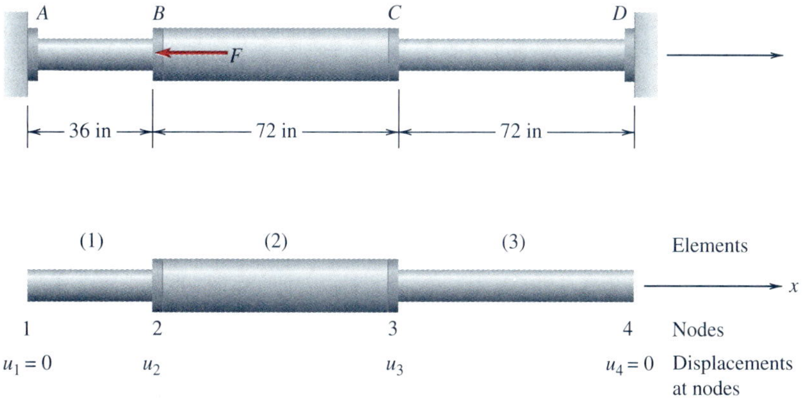

Figure 2.32 Example 2.18

```
        ***MECHANICS OF MATERIALS***  DATA FILE: ex218
                          ***BAR***

   INPUT DATA:

NUMBER OF ELEMENTS     NM = 3

CHANGE IN THE TEMPERATURE?     n

 ELEMENT       L (in)     A (inxin)    E (10**6; psi)
                (m)         (mxm)          (MPa)

    1          36.00      1.00E+00        3.00E+01
    2          72.00      3.00E+00        3.00E+01
    3          72.00      2.00E+00        3.00E+01


             CONDITION
***********************************

     NODE            FORCE
       2           -3.00E+03

     NODE      SPECIFIED DISP.
       1           0.00E+00
       4           0.00E+00


SOLUTION:

********************************************

       NODE         DISPLACEMENT (in,m)

********************************************

         1                0.00E+00
         2               -2.25E-03
         3               -1.35E-03
         4                0.00E+00


*******************************************************

 ELEMENT       FORCE           STRESS             E

*******************************************************

    1      -1.8750E+03     -1.8750E+03       3.00E+07
    2       1.1250E+03      3.7500E+02       3.00E+07
    3       1.1250E+03      5.6250E+02       3.00E+07
```

Figure 2.33 Summary of input data and solution for Example 2.18.

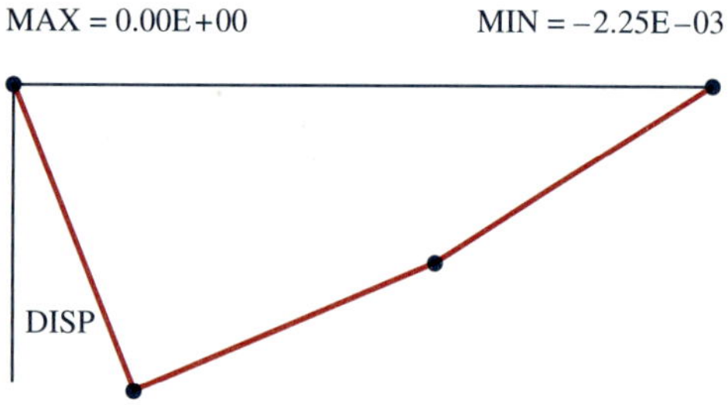

Figure 2.34 Displacement diagram for Example 2.18.

EXAMPLE 2.19

A model of a linkage system with three steel rod segments as shown in Fig. 2.35 is loaded at section C by a force $F = 2$ kips. The model is to be used to investigate the effect of a 7°F reduction in temperature. In particular we wish to find (*a*) the force in each segment and the displacement of sections B and C using the BARMECH program ($A_{AB} = 1$ in^2, $A_{BC} = 3$ in^2, $A_{CD} = 2$ in^2, $\alpha = 6.5 \times 10^{-6}$/°F, and $E = 30 \times 10^6$ psi), and (*b*) the temperature change ΔT such that section B does not move due to the combined action of the load and the temperature change.

We divide the bar into three elements and four nodes as shown in Fig. 2.35. Nodes 1 and 4 are fixed. There is initially a temperature change of -7°F. The summary of the input data is shown in Fig. 2.36. The solution is given in Figs. 2.37 and 2.38.

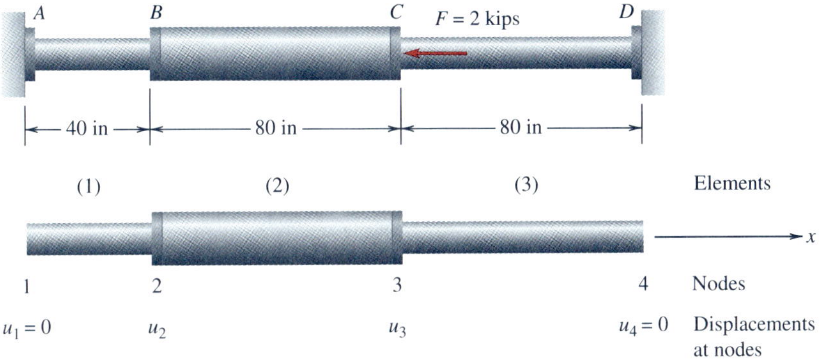

Figure 2.35 Example 2.19

```
     ***MECHANICS OF MATERIALS***   DATA FILE: ex219
                        ***BAR***

  INPUT DATA:

NUMBER OF ELEMENTS      NM = 3

CHANGE IN THE TEMPERATURE?       y

CHANGE OF TEMPERATURE VALUE = -7

 ELEMENT   L (in)   A (inxin)   E (10**6; psi)   AL (10**-6/F)
             (m)      (mxm)        (MPa)            (10**-6/C)

    1      40.00    1.00E+00      3.00E+01          6.500
    2      80.00    3.00E+00      3.00E+01          6.500
    3      80.00    2.00E+00      3.00E+01          6.500


              CONDITION
***********************************

    NODE          FORCE
     3          -2.00E+03

    NODE      SPECIFIED DISP.
     1          0.00E+00
     4          0.00E+00
```

Figure 2.36 Summary of input data for Example 2.19 with $\Delta T = -7$°F.

```
SOLUTION:

********************************************

        NODE          DISPLACEMENT (in,m)

********************************************

          1                  0.00E+00
          2                  5.92E-04
          3                 -1.44E-03
          4                  0.00E+00


*******************************************************

  ELEMENT       FORCE           STRESS             E

*******************************************************

     1        1.8094E+03      1.8094E+03      3.00E+07
     2        1.8094E+03      6.0313E+02      3.00E+07
     3        3.8094E+03      1.9047E+03      3.00E+07
```

Figure 2.37 Output of solution for Example 2.19 with $\Delta T = -7°F$.

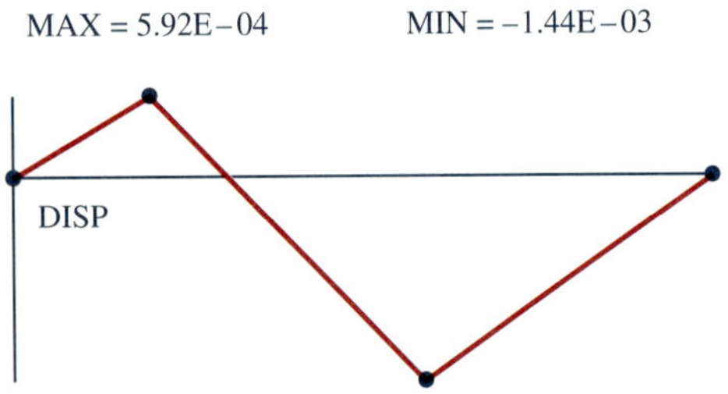

Figure 2.38 Displacement diagram for Example 2.19 with $\Delta T = -7°F$.

From Fig. 2.38 we note that there is a section between sections B and C that does not move due to the applied load and temperature change. In part (*b*) we would like to locate this section of zero displacement at section B by changing the value of ΔT. To carry out an iteration to find this unknown ΔT, we make another BARMECH run, using option 2 in the BARMECH menu to modify the existing data file and using a lower value of the temperature change, say, $\Delta T = -3°F$ (keeping F at 2 kips); the displacement in this case at B is -3.18×10^{-4} in.

Thus from these two runs we have the results:

$$\Delta T = -7.0°F \qquad u_B = 5.92 \times 10^{-4} \text{ in}$$

$$\Delta T = -3.0°F \qquad u_B = -3.18 \times 10^{-4} \text{ in}$$

By linear interpolation of these results we can obtain the temperature change corresponding to zero displacement at section B. We have

$$\frac{\Delta T + 3.0}{3.18} = \frac{-7.0 - \Delta T}{5.92}$$

or

$$\Delta T = -4.40°F$$

The results for a run with $\Delta T = -4.40°F$ are given in Fig. 2.39, and the graph of the displacement given in Fig. 2.40 does indeed show a zero displacement at section B. Note that the force in element (1) is not zero.

```
        ***MECHANICS OF MATERIALS***  DATA FILE: ex219
                          ***BAR***

    INPUT DATA:

NUMBER OF ELEMENTS      NM = 3

CHANGE IN THE TEMPERATURE?       y

CHANGE OF TEMPERATURE VALUE = -4.4

 ELEMENT   L (in)    A (inxin)    E (10**6; psi)    AL (10**-6/F)
             (m)       (mxm)          (MPa)             (10**-6/C)

    1       40.00    1.00E+00       3.00E+01          6.500
    2       80.00    3.00E+00       3.00E+01          6.500
    3       80.00    2.00E+00       3.00E+01          6.500

              CONDITION
***********************************

     NODE            FORCE
      3            -2.00E+03

     NODE       SPECIFIED DISP.
      1            0.00E+00
      4            0.00E+00

SOLUTION:
********************************************

        NODE         DISPLACEMENT (in,m)

********************************************

          1                0.00E+00
          2                1.00E-06
          3               -1.52E-03
          4                0.00E+00

*******************************************************

 ELEMENT       FORCE             STRESS             E
*******************************************************

    1       8.5875E+02      8.5875E+02        3.00E+07
    2       8.5875E+02      2.8625E+02        3.00E+07
    3       2.8588E+03      1.4294E+03        3.00E+07
```

Figure 2.39 Summary of input data and solution for Example 2.19 with $\Delta T = -4.4°F$.

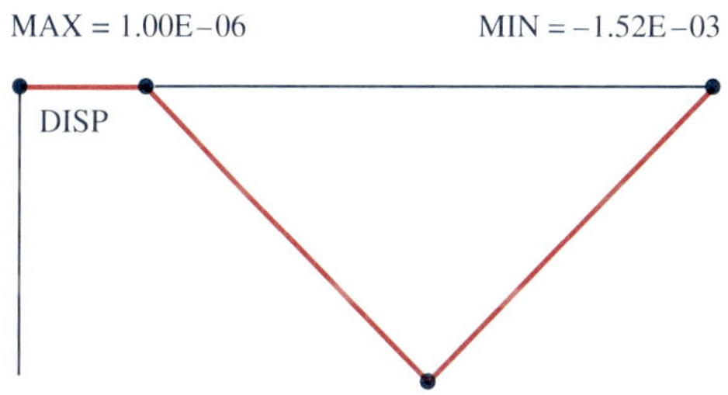

Figure 2.40 Displacement diagram for Example 2.19 with $\Delta T = -4.4°F$.

2.8 Differential Equations for Axial Force and Deformation

In our analysis of the deformation of axial members, we have assumed that the loads were applied at the ends of the members. In every case the members were assumed to have constant cross-sectional areas. If, however, loads are distributed along the length of the member and if the area varies along the length, then we need to derive equations for the force and displacement as a function of position along the member. To do this, we consider an axial member that carries an axial load per unit length $q(x)$ in the direction of the centroidal axis, as shown in Fig. 2.41*a*. In this case $q(x)$ is the load intensity per unit length distributed along the length of the member. Applied loads or reactive forces may also be present at the ends to keep the member in equilibrium, but are not shown.

We propose to investigate the conditions imposed by the requirements of equilibrium on a small element of length Δx cut out from the member as the length Δx shrinks to zero. Any portion or element of an axial member in equilibrium is also in equilibrium. The use of *differential* elements to derive differential equations *at a point* in a material is a common practice in the study of solid mechanics.

In Fig. 2.41*a*, we show a bar of variable cross-sectional area loaded

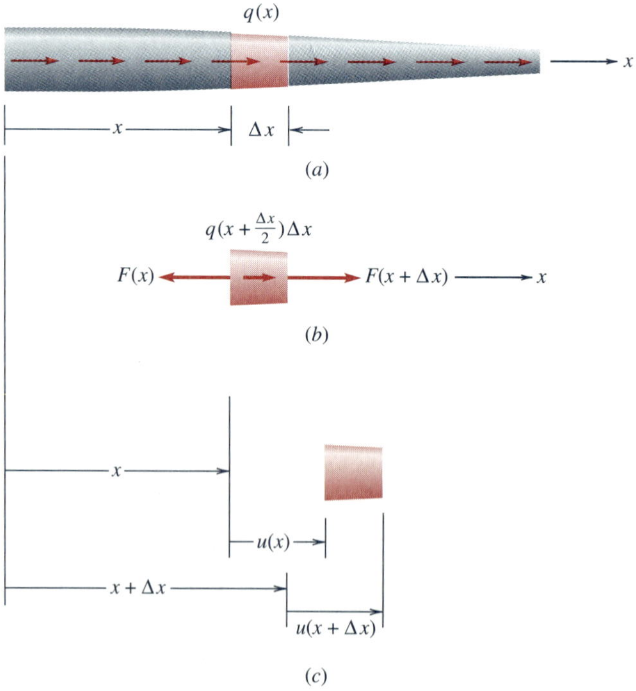

Figure 2.41 (*a*) Bar of variable cross section under distributed load per unit length $q(x)$. (*b*) An element of the bar showing internal forces acting on the ends. (*c*) Displacements of the ends of the two elements.

by a distributed force $q(x)$ applied in the positive x direction, e.g., to the surface of the bar. The distributed force $q(x)$ is applied to the bar such that it acts along the axis of the bar; the units of q are force/length. We also show an element of length Δx at location x along the bar; the left end of the element is at coordinate x, and the right end is at coordinate $x + \Delta x$. We wish to investigate this element, using the three steps of Eq. (2.8); moment equilibrium of the element is identically satisfied because the forces acting on the element are along the axis.

Statics. In Fig. 2.41*b*, we show the internal force $F(x)$ at position x and the internal force $F(x + \Delta x)$ at position $x + \Delta x$. In general, F will be a function of x along the member. All these internal forces are positive on their respective faces with our convention of positive forces acting on an element of area. The distributed load acting on the element has an approximate resultant force

$$q\left(x + \frac{\Delta x}{2}\right)\Delta x$$

where we evaluate $q(x)$ at the midpoint of the element at $(x + \Delta x/2)$.

Force equilibrium of the element in the positive x direction gives

$$-F(x) + F(x + \Delta x) + q\left(x + \frac{\Delta x}{2}\right)\Delta x = 0 \tag{2.22}$$

or upon rewriting,

$$\frac{F(x + \Delta x) - F(x)}{\Delta x} + q\left(x + \frac{\Delta x}{2}\right) = 0 \tag{2.23}$$

In the limit as $\Delta x \to 0$, that is, as the element shrinks to zero, we have an equation that must be satisfied by the internal force $F(x)$ in the bar

$$\frac{dF}{dx} + q(x) = 0 \tag{2.24}$$

Geometry. In Fig. 2.41*c*, we again show the element and the corresponding displacements at the ends of the element $u(x)$ and $u(x + \Delta x)$. The change in length per unit length of the element or the strain of the element of length Δx is

$$\epsilon = \frac{[x + \Delta x + u(x + \Delta x)] - [x + u(x)] - \Delta x}{\Delta x} = \frac{u(x + \Delta x) - u(x)}{\Delta x} \tag{2.25}$$

In the limit as $\Delta x \to 0$, the strain of element Δx is the strain at position x, and it follows from Eq. (2.25)

$$\epsilon(x) = \frac{du}{dx} \tag{2.26}$$

Therefore, the strain at any point x is the derivative of the displacement function $u(x)$ at that value of x.

Force-Deformation Relations. If the bar is linear elastic, then according to Eq. (2.11) the relation between the strain and the force at any point x along the bar is

$$\epsilon = \frac{F}{AE} + \alpha \Delta T$$

or

$$F = AE\frac{du}{dx} - AE\alpha \Delta T \tag{2.27}$$

If Eq. (2.27) is substituted into Eq. (2.24), we have

$$\frac{d}{dx}\left(AE\frac{du}{dx}\right) - \frac{d}{dx}(AE\alpha \Delta T) + q(x) = 0 \tag{2.28}$$

Equation (2.28) relates the distributed load $q(x)$ and the temperature change ΔT in the bar to the displacement $u(x)$ along the bar. The solution of this differential equation will contain two constants of integration that can be evaluated from the specified conditions on the displacements of the problem.

In the special case when $\Delta T = 0$, A and E are constant, and q is zero, we have from Eq. (2.24)

$$\frac{dF}{dx} = 0 \tag{2.29}$$

which gives the result that the force in the bar is independent of x and thus is equal to a constant. Equation (2.28) reduces to

$$\frac{d}{dx}\left(\frac{du}{dx}\right) = 0$$

so that upon integration we have

$$\frac{du}{dx} = C_1$$

and upon integration again

$$u(x) = C_1 x + C_2 \tag{2.30}$$

where C_1 and C_2 are constants of integration. If the displacement for bar AB shown in Fig. 2.42a and b at $x = x_A$ is u_A, and if the displacement at $x = x_B$ is u_B, then constants C_1 and C_2 can be found from the conditions

$$u_A = C_1 x_A + C_2$$

$$u_B = C_1 x_B + C_2$$

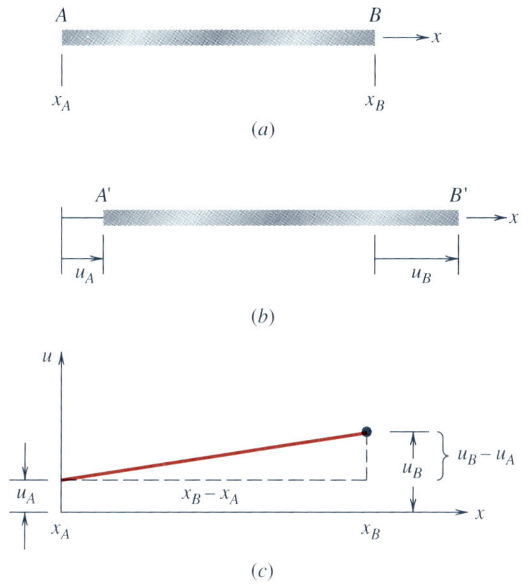

Figure 2.42 Displacement of an element. The displacements of the ends are u_A and u_B. The displacement diagram of the element in (c) shows the linear variation of displacement u from u_A at $x = x_A$ to u_B at $x = x_B$.

or

$$C_1 = \frac{u_B - u_A}{x_B - x_A} \qquad C_2 = -C_1 x_B + u_B$$

Therefore, the displacement in the bar in this case is linear with

$$u(x) = \frac{u_B - u_A}{x_B - x_A}(x - x_B) + u_B \tag{2.31}$$

We have used this result that the displacement is a linear function of x in a bar with constant end loads a number of times in earlier discussions. A graph of $u(x)$ is given in Fig. 2.42*c* for the case when $u_B > u_A > 0$.

If $q(x)$ is not zero, the internal force $F(x)$ will be a function of x along the bar.

EXAMPLE 2.20

A heavy, uniform prismatic pipe of length L, cross-sectional area A, weight density ρ, and *total* weight W is fixed at the top end to a support, as shown in Fig. 2.43*a*. The pipe hangs under its own weight. We wish to find expressions for the axial stress and displacement along the pipe arising from the weight of the pipe. We assume that the stress in the pipe is uniform across each section; the x axis is along the pipe as shown, with the origin at the upper fixed support.

Statics: To use the differential relation for the variation of the axial internal force in the pipe, Eq. (2.24),

$$\frac{dF}{dx} + q(x) = 0 \tag{a}$$

we need to determine an expression for the external distributed force $q(x)$ acting on the pipe. To do this, we consider an element of the pipe of length Δx, and we determine the force acting on this element due to its weight.

From Fig. 2.43*b*, we have the weight ΔW of the element given by $\Delta W = \rho A\, \Delta x$, so that

$$q(x) = \rho A = \frac{\rho A L}{L} = \frac{W}{L} \tag{b}$$

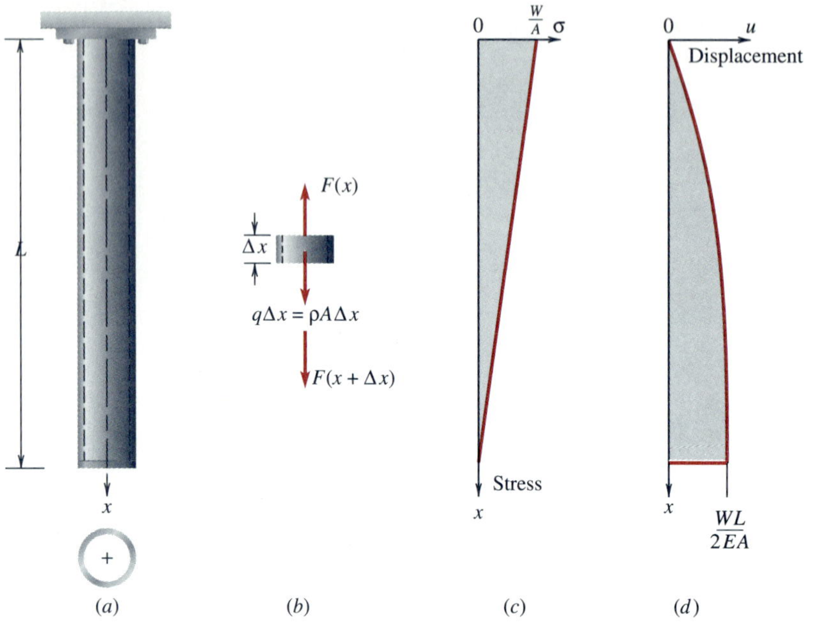

Figure 2.43 Example 2.20

where W is the total weight of the pipe of length L. Upon substitution of Eq. (*b*) into Eq. (*a*), we have

$$\frac{dF}{dx} = -q(x) = -\frac{W}{L} \tag{c}$$

Integrating leads to

$$F(x) = -\frac{W}{L}x + C_1 \tag{d}$$

where C_1 is a constant of integration. The boundary condition of zero force in the pipe at $x = L$ requires that

$$F(L) = -W + C_1 = 0$$

and thus $C_1 = W$. We have therefore

$$F = W\left(1 - \frac{x}{L}\right) \tag{e}$$

and

$$\sigma_x = \frac{F}{A} = \frac{W}{A}\left(1 - \frac{x}{L}\right) \tag{f}$$

which shows that the axial stress varies linearly with axial position along the pipe; we plot the stress distribution in Fig. 2.43*c*.

Force-Deformation Relations: Using Eq. (2.27) with $\Delta T = 0$, we have

$$\frac{du}{dx} = \frac{F}{EA} = \frac{W}{EA}\left(1 - \frac{x}{L}\right) \tag{g}$$

and integrating, we obtain

$$u(x) = \frac{W}{EA}\left(x - \frac{x^2}{2L}\right) + C_2 \tag{h}$$

where C_2 is the constant of integration.

Geometry: Since the top of the pipe, $x = 0$, is fixed, the choice $C_2 = 0$ in Eq. (*h*) forces $u(0)$ to be zero at the top. Therefore the displacement of the pipe is

$$u(x) = \frac{W}{EA}\left(x - \frac{x^2}{2L}\right) \tag{i}$$

That is, the axial displacement varies quadratically with the distance from the top of the pipe, as shown in Fig. 2.43*d*. The axial displacement at the free end of the pipe is

$$u_{max} = \frac{WL}{2EA} \tag{j}$$

which is half of the end deflection of a weightless pipe loaded at its free end by a force equal to W.

EXAMPLE 2.21

A steel pipe AB of length L is constrained between two rigid supports, as shown in Fig. 2.44*a*. If the steady-state increase in temperature of the pipe above room temperature is linear along the pipe in the form

$$\Delta T = \Delta T_B \frac{x}{L} \qquad (a)$$

where ΔT_B is the increase in temperature at end B as shown in Fig. 2.44*b*, we wish to determine the wall reactions at A and B and the displacement of the midpoint of the pipe. The origin of the coordinate system is at section A.

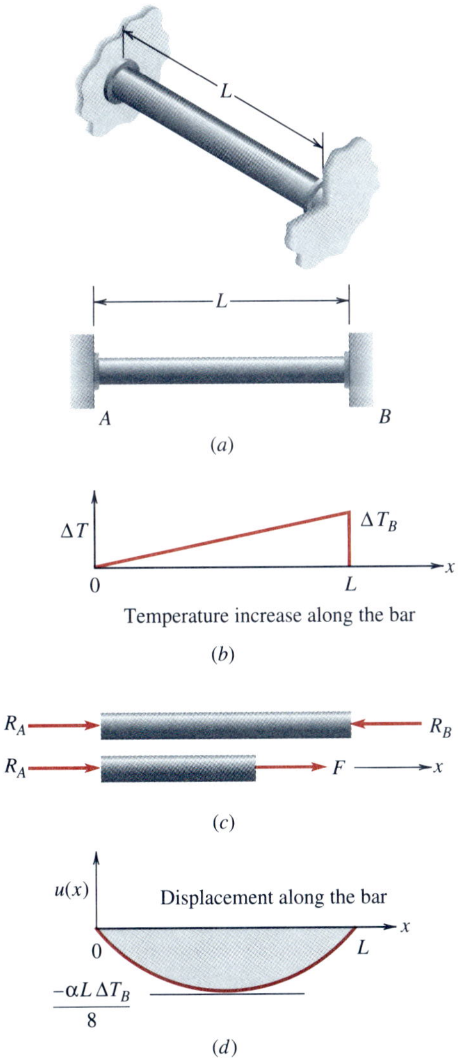

Figure 2.44 Example 2.21

The wall reactions R_A and R_B are equal, as shown in Fig. 2.44*c*. Since no external forces act on the pipe, the force F in Fig. 2.44*c* is constant along the length of the pipe and is taken as tensile according to our usual sign convention for internal forces.

In the case of the linear temperature increase, we have from Eq. (2.28) and Eq. (*a*) with $q(x) = 0$

$$AE\frac{d}{dx}\left(\frac{du}{dx}\right) - AE\alpha\frac{d}{dx}\left(\Delta T_B \frac{x}{L}\right) = 0$$

or

$$\frac{d}{dx}\left(\frac{du}{dx}\right) = \frac{\alpha}{L}\Delta T_B \qquad (b)$$

Integrating Eq. (*b*) twice gives

$$u(x) = \frac{\alpha \Delta T_B}{2L}x^2 + C_1 x + C_2 \qquad (c)$$

where C_1 and C_2 are constants of integration. To determine C_1 and C_2, we have the conditions at the ends of the pipe, $u(0) = u(L) = 0$. Therefore,

$$C_2 = 0 \qquad 0 = \frac{\alpha \Delta T_B}{2L}L^2 + C_1 L \qquad (d)$$

from which, we find

$$u(x) = \frac{\alpha \Delta T_B}{2}\left(\frac{x}{L}\right)(x - L) \qquad (e)$$

The displacement distribution along the pipe is shown in Fig. 2.44*d*, and the displacement u at the midpoint $x = L/2$ is

$$u\left(\frac{L}{2}\right) = -\frac{\alpha L \Delta T_B}{8} \qquad (f)$$

Finally, from Eq. (2.27), we have

$$F = AE\frac{du}{dx} - AE\alpha\Delta T$$

$$= AE\left(\frac{\alpha \Delta T_B}{L}\right)\left(x - \frac{L}{2}\right) - AE\frac{\alpha \Delta T_B x}{L} \qquad (g)$$

$$= -AE\frac{\alpha \Delta T_B}{2}$$

As expected, the force in the pipe is constant independent of x. Therefore, from Fig. 2.44*c*, the wall reactions are

$$R_A = R_B = -F = \frac{AE\alpha\,\Delta T_B}{2} \qquad (h)$$

The value of the displacement at the center of the pipe, Eq. (f), and the wall reactions, Eq. (h), should be compared to the case of a pipe in which there is a constant temperature increase ΔT along the pipe. In that case, the reactions at the wall are equal to

$$AE\alpha\,\Delta T \qquad (i)$$

2.9 Concluding Remarks

In this chapter we have introduced an approach which makes use of the three steps of Eq. (2.8) for the solution of one-dimensional axial loading and deformation problems. This approach is not limited to one-dimensional problems, as will be seen as we progress through this book. This approach provides an orderly procedure for the solution of problems that allows us to consider the requirements of equilibrium, to ensure geometric compatibility, and to make use of the force-deformation behavior of each member.

We have also seen how the consideration of temperature changes leads to a thermal-strain term in the force-deformation relation. The three-step approach, however, remains the same in any problem with a temperature change. We have found in a number of examples that in the presence of a temperature change, it is possible to have a stress in a member in the absence of deformation, e.g., in a constrained bar.

In the case of interconnected bars in series, we found that it is possible to approach the solution of such problems in a systematic way that can be implemented on a computer. The BARMECH program allows us to analyze problems easily without the necessity of extensive algebra and numerical work. The graphics output allow us to visualize the nature of the displacement variation. Also we can investigate easily changes to a given problem in terms of applied load, temperature changes, and properties of elements, such as in Example 2.19. However, to emphasize a theme to which we will return often in this book, it is important to study carefully the output of the programs to ensure that the results make sense. Of importance in all problems is the checking of magnitudes of answers against your own intuition, e.g., Is it too large? Too small? As you do more problems, you will develop a deeper intuition and gain the experience needed to spot wrong results.

Finally, note that while many of the examples in this chapter look simple and certainly the real world is not made up of one-dimensional bars, the class of problems we have treated is important often as specific problems but more importantly as models for more complicated engineering problems. As we might expect, if a reasonably complicated problem can be reduced to a simple one-dimensional model, then we have made major progress in the solution.

PROBLEMS

2.3-1 Consider Example 2.2, Fig. 2.7. If the cross-sectional area of segment AB is increased to 100 cm^2, find the displacements of sections A and B. All other values are to remain the same.

2.3-2 Consider Example 2.2, Fig. 2.7. If the loads P_A and P_B are doubled and the cross-sectional areas of each segment are doubled, what are the displacements of sections A and B? All other values remain the same. Can we draw any conclusions about the effect of scaling the areas and the loads?

2.3-3 Consider Example 2.2, Fig. 2.7. If the loads P_A and P_B, the areas of each segment, and the length L are doubled, find the forces and normal stresses in each segment as well as the displacements of sections A and B.

2.3-4 Consider a schematic representation of a steel column which is part of a two-story building, as shown in Fig. P2.3-4. The loads are $P_A = 100$ kips and $P_B = 300$ kips. The lengths are $L_1 = 12$ ft and $L_2 = 18$ ft, and the cross-sectional areas of AB and BC are 8 and 20 in^2. We wish to find the normal stress in each segment and the displacements at sections A and B. Neglect the weight of the column, and take $E = 30 \times 10^6$ psi.

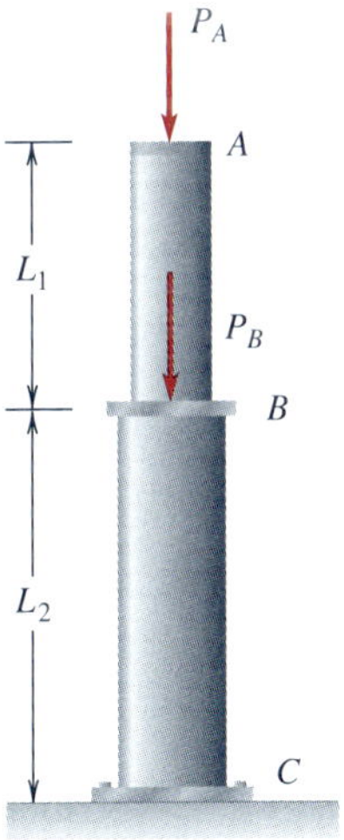

Fig. P2.3-4

2.3-5 Consider the column shown in Fig. P2.3-4. Derive a general expression for the displacements of sections A and B in terms of P_A, P_B, L_1, L_2, A_{AB}, A_{BC}, and E. Check the formula derived against the numerical results of Prob. 2.3-4 and Example 2.2.

2.3-6 Consider Example 2.3, Fig. 2.8. If P_1 is set equal to zero, find the displacements of sections B, C, and D and the internal force in each segment. All other values remain the same. Also sketch the internal force diagram and the displacement diagram similar to that shown in Fig. 2.9.

2.3-7 Consider Example 2.3, Fig. 2.8. If P_2 is set equal to zero, find the displacements of sections B, C, and D and the internal force in each segment. All other values are to remain the same. Also sketch the internal force diagram and the displacement diagram similar to that shown in Fig. 2.9.

2.3-8 Consider Example 2.3, Fig. 2.8. If P_3 is set equal to zero, find the displacements of sections B, C, and D and the internal force in each segment. All other values remain the same. Also sketch the internal force diagram and the displacement diagram similar to that shown in Fig. 2.9.

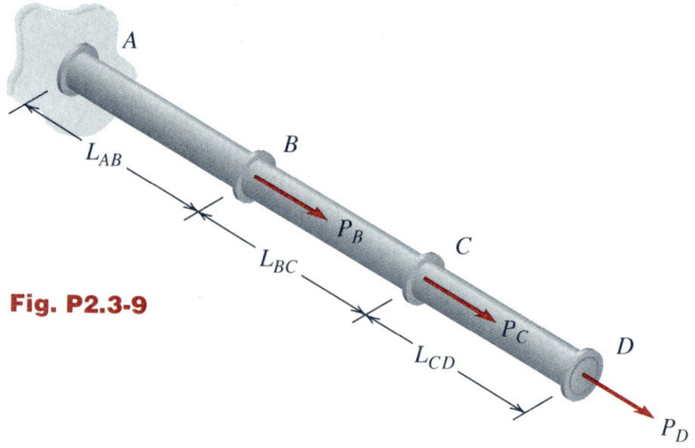

Fig. P2.3-9

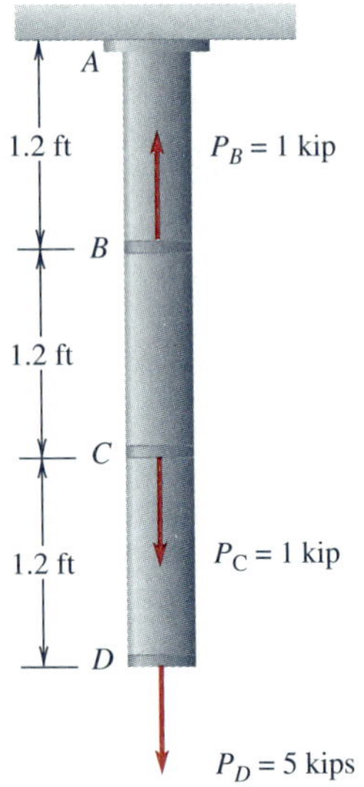

Fig. P2.3-11

2.3-9 A model steel component shown in Fig. P2.3-9 is subjected to the three loads P_B, P_C, and P_D as shown. If $P_B = 30$ kN, $P_C = 20$ kN, and $P_D = 2$ kN, we wish to find the stresses in each segment of the component and the displacements of sections B, C, and D. The cross-sectional areas of the segments are $A_{AB} = A_{BC} = 1000\ \text{mm}^2$ and $A_{CD} = 400\ \text{mm}^2$; and $L_{AB} = L_{BC} = 0.3$ m and $L_{CD} = 0.2$ m; and the elastic modulus is $E = 200$ GPa. Neglect the weight of the component.

2.3-10 Consider the model steel component shown in Fig. P2.3-9 with $A_{AB} = A_{BC} = 1000\ \text{mm}^2$, $A_{CD} = 400\ \text{mm}^2$, $L_{AB} = L_{BC} = 0.3$ m, $L_{CD} = 0.2$ m, and $E = 200$ GPa. If the loads are $P_C = 20$ kN and $P_D = 2$ kN, what value of P_B will give a zero displacement at section B?

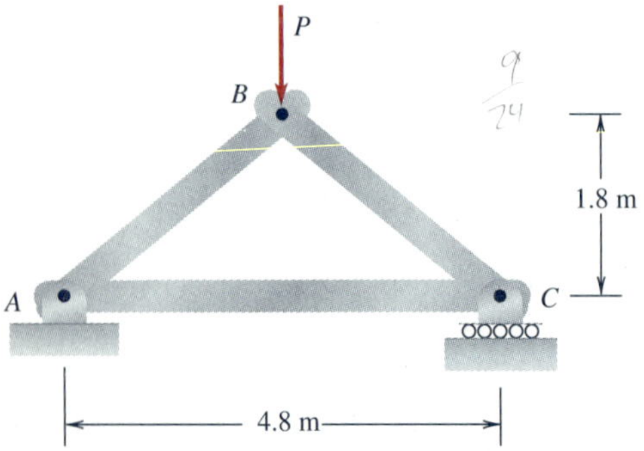

Fig. P2.3-12

2.3-11 A steel bar $ABCD$ of cross-sectional area 0.67 in^2, as shown in Fig. P2.3-11, is loaded by three axial loads. Find the axial displacements at sections B, C and D, and plot the curve of the internal force distribution and the displacement along the bar. Neglect the weight of the bar, and take $E = 30 \times 10^6$ psi.

2.3-12 A simple truss made up of three members as shown in Fig. P2.3-12 is loaded at joint B by a load $P = 20$ kN. If the member AC is steel with a cross-sectional area of $2 \times 10^{-4}\ \text{m}^2$, estimate the horizontal motion of the roller at joint C. Take $E = 205$ GPa.

2.3-13 A steel rod as shown in Fig. P2.3-13 is loaded by two loads P_1 and P_2. Estimate the normal stress in each segment of the rod and the displacements at sections B and C. The cross-sectional area of the rod is 0.30 in^2, and take $E = 30 \times 10^6$ psi.

Fig. P2.3-13

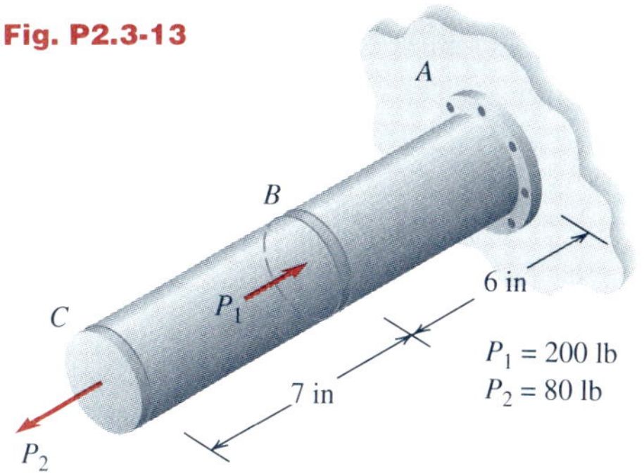

2.3-14 A model steel rod made up of three circular rods of different diameters joined together as shown in Fig. P2.3-14 is loaded by four loads. Estimate the force in each segment and the displacement at sections B, C, D, and E. Also sketch the internal force diagram and the displacement diagram for the rod. Take $E = 200$ GPa.

(*a*) Take $P_B = 20$ kN, $P_C = 30$ kN, $P_D = 0$, and $P_E = 100$ kN.
(*b*) Take $P_B = 0$, $P_C = 40$ kN, $P_D = -10$ kN, and $P_E = 30$ kN.
(*c*) Take $P_B = 0$, $P_C = 50$ kN, $P_D = -50$ kN, and $P_E = 0$ kN.
(*d*) Take $P_B = 0$, $P_C = -30$ kN, $P_D = 30$ kN, and $P_E = 50$ kN.
(*e*) Take $P_B = P_C = P_D = 0$ and $P_E = 100$ kN.

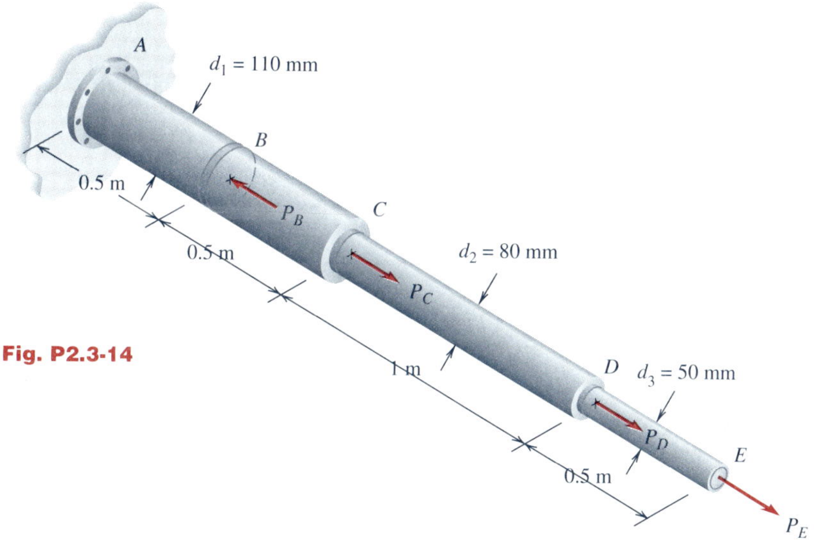

Fig. P2.3-14

2.3-15 A rigid member BD is supported by two elastic rods as shown in Fig. P2.3-15. Estimate the normal stress in each rod and the deflection under the load P. Take $E_1 = E_2 = 205$ GPa, $L_1 = 1.5$ m, $L_2 = 1.65$ m, $A_1 = 90$ mm^2, $A_2 = 60$ mm^2, $L = 3$ m, $x = 2$ m, and $P = 5$ kN.

2.3-16 A rigid member BD is supported by two elastic rods as shown in Fig. P2.3-15. Determine the location x of the load along the rigid member BD such that the normal stress in each rod is the same. Take $E_1 = E_2 = 205$ GPa, $L_1 = 1.5$ m, $L_2 = 1.8$ m, $A_1 = 12$ mm^2, $A_2 = 10$ mm^2, $L = 3$ m, and $P = 10$ kN.

2.3-17 Estimate the deflection under the load for Prob. 2.3-16.

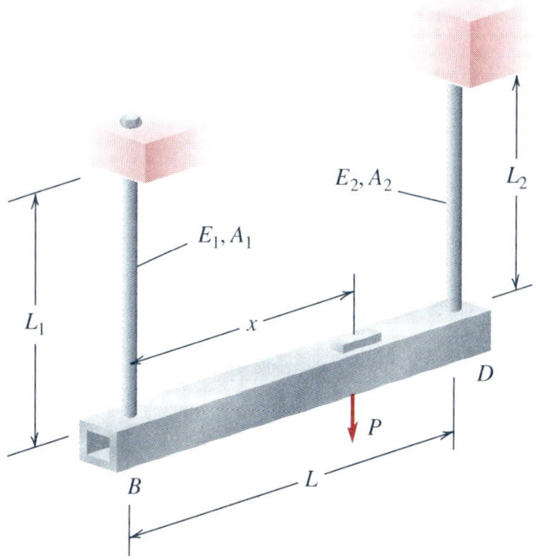

Fig. P2.3-15

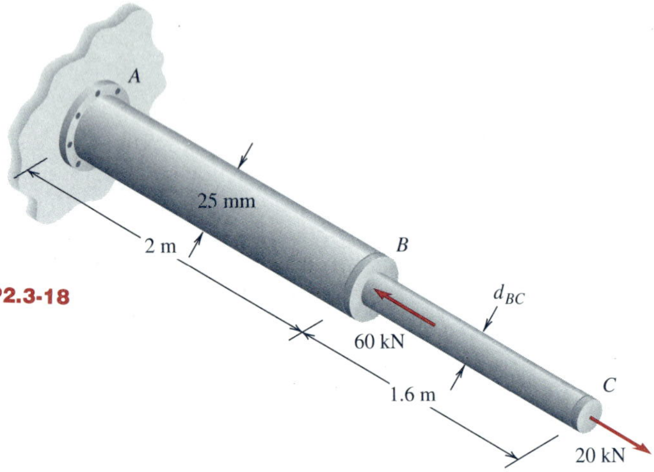

Fig. P2.3-18

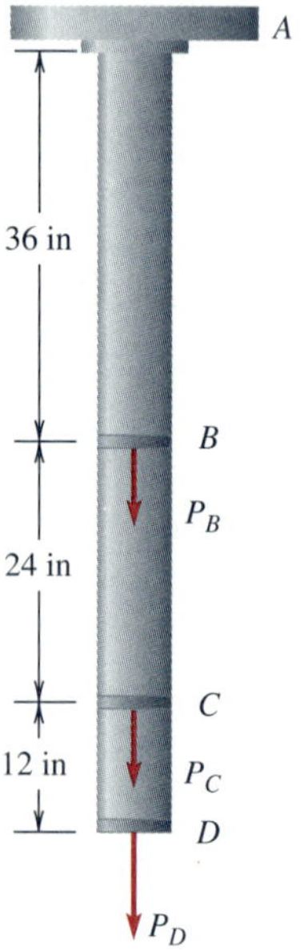

Fig. P2.3-19

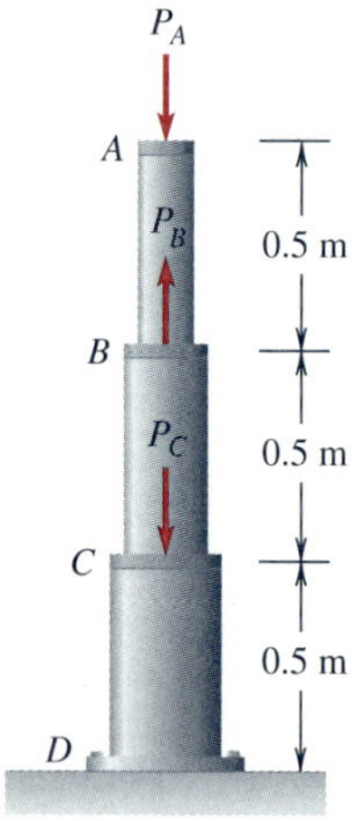

Fig. P2.3-20

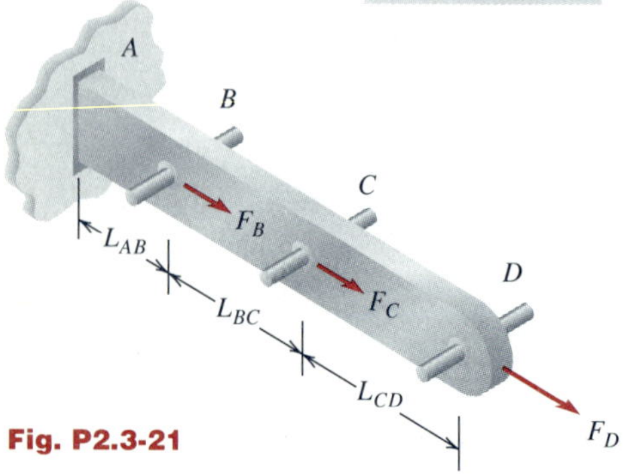

Fig. P2.3-21

2.3-18 Two steel rods are attached at B and fixed to a wall at A and are loaded at B and C as shown in Fig. P2.3-18. If $E = 200$ GPa for both rods, $d_{AB} = 25$ mm, the maximum allowable stress is 100 MPa, and the maximum allowable displacement in either direction of section C is 1 mm, find the minimum allowable diameter d_{BC}.

2.3-19 A steel bar of constant cross section carries axial loads at sections B, C, and D and is fixed to a support at A as shown in Fig. P2.3-19. Given $P_B = 2000$ lb, $P_C = 3000$ lb, and $P_D = 5000$ lb:
(*a*) Find the minimum cross-sectional area A_{min} of the bar if the maximum allowable normal stress in the bar is taken to be 40,000 psi.
(*b*) For the value of A_{min} found in part (*a*) find the displacements of sections B, C, and D. Take $E = 30{,}000$ ksi.

2.3-20 Axial loads are applied to the three-segment system made up of circular members as shown in Fig. P2.3-20. If the maximum allowable normal stresses in tension and compression for the segments are $\sigma_{AB} = 120$ MPa, $\sigma_{BC} = 80$ MPa, and $\sigma_{CD} = 85$ MPa, find the minimum allowable diameter for each segment. Take $P_A = 50$kN, $P_B = 60$ kN, and $P_C = 100$ kN.

2.3-21 A steel bar has three loads applied along its length as shown in Fig. P2.3-21. Sketch the plots of the internal force versus distance along the bar and the displacement versus distance along the bar.
(*a*) Take $F_B = 250$ kN, $F_C = -100$ kN, $F_D = 200$ kN, $L_{AB} = 0.8$ m, $L_{BC} = 1.0$ m, $L_{CD} = 1.1$ m, area $= 2000$ mm^2, $E = 200$ GPa.
(*b*) Take $F_B = 50$ kips, $F_C = 30$ kips, $F_D = -40$ kips, $L_{AB} = 2.5$ ft, $L_{BC} = 2.0$ ft, $L_{CD} = 2.0$ ft, area $= 3.0$ in^2, $E = 30{,}000$ ksi.

2.3-22 A rigid beam AB of length 10 m and weight 2 MN is temporarily supported by a composite column as shown in Fig. P2.3-22. The beam rests on frictionless supports at each end. Calculate the stress in each segment of the composite column and the displacement at the top of the column. Take E_A = 200 GPa, A_A = 100 cm^2, E_B = 240 GPa, and A_B = 225 cm^2.

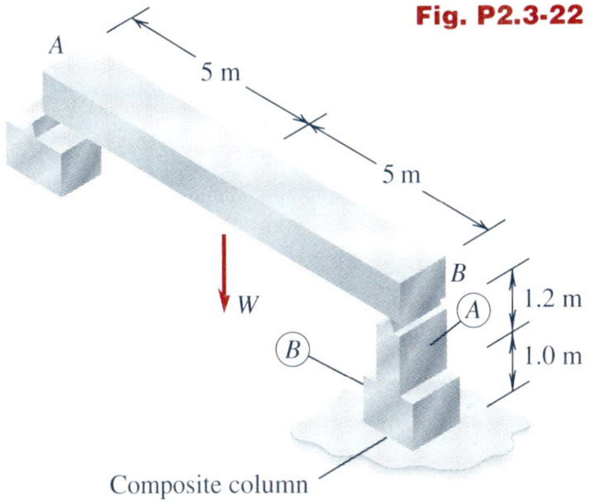

Fig. P2.3-22

2.3-23 A simple truss shown in Fig. P2.3-23 is loaded by a load P at B. Show that the expression for the small vertical deflection at B is given by

$$\delta_v = \frac{PL}{AE}\left(\frac{1}{\sin^2\theta\,\cos\theta} + \frac{1}{\tan^2\theta}\right)$$

Find the value of θ such that the deflection is a minimum, and show that the minimum value is

$$\delta_{v,\,\min} = \frac{3PL}{AE}$$

Fig. P2.3-23

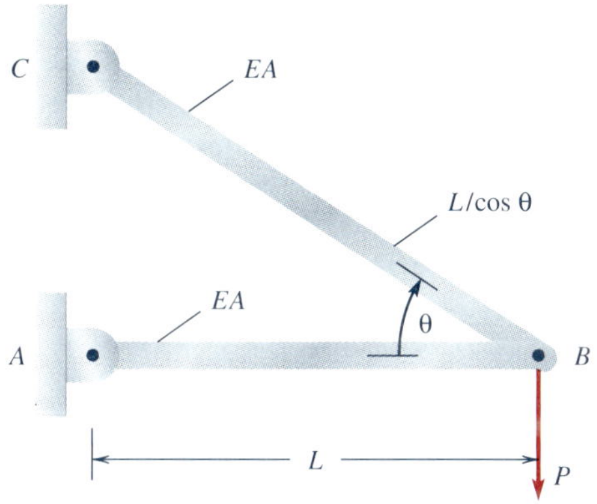

2.3-24 A rigid bar ABC is supported by an aluminum (E_{al} = 10 × 10^6 psi) and a steel link (E_{st} = 30 × 10^6 psi) as shown in Fig. P2.3-24. If a load P = 2500 lb is applied at point C, find the value of the deflection under the load and the angle the bar ABC makes with the horizontal. Take the area of the aluminum link as 0.50 in^2 and the area of the steel link as 0.75 in^2. Neglect the weight of all components.

Fig. P2.3-24

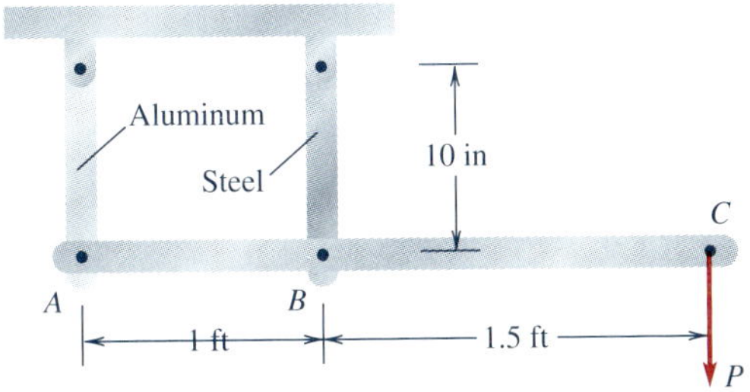

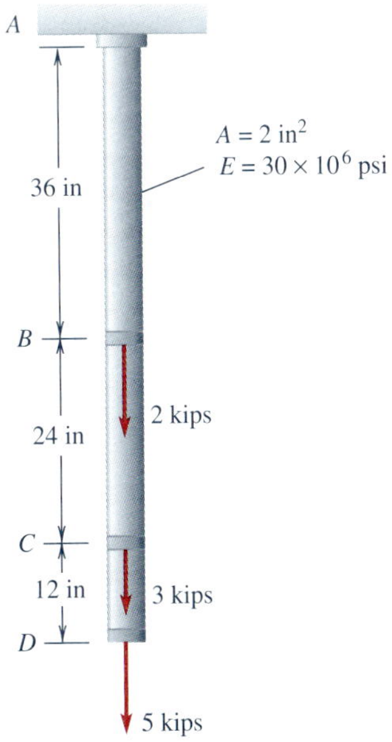

Fig. P2.3-25

2.3-25 The bar structure $ABCD$ as shown in Fig. P2.3-25 is loaded by three loads. Neglect the weight of the bar and find the axial stress in each section and the displacements of sections B, C, and D.

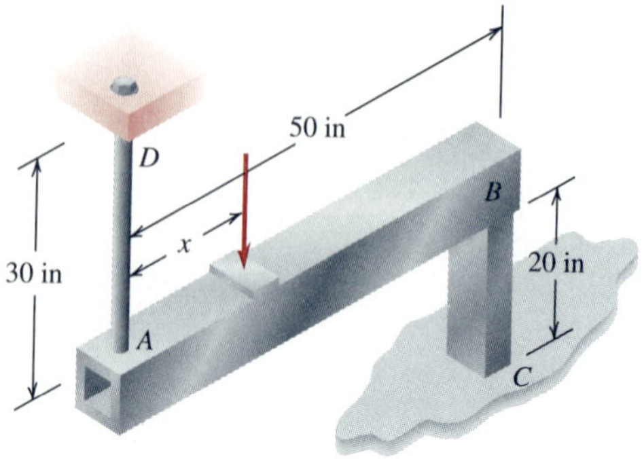

Fig. P2.3-26

2.3-26 A rigid bar AB is supported by a steel bar DA (cross-sectional area = 6 in^2 and $E = 30 \times 10^6$ psi) and a concrete pile BC (cross-sectional area = 120 in^2 and $E = 3 \times 10^6$ psi), as shown in Fig. P2.3-26. At what location x should the P = 200-kip load be placed so that the bar AB remains horizontal after loading? For this condition, what is the stress in the steel bar DA?

2.4-1 Consider Example 2.4, Fig. 2.10. Equation (g) gives the load-displacement relation for the composite column. Sketch a graph of the straight-line load-displacement relation, and compare the result with the case for which the steel pipe does not contain any concrete.

2.4-2 A composite material is fabricated by aligning stiff fibers in a matrix of a less stiff material and bonding them together in a layer, as shown in Fig. P2.4-2. If a load P is applied, the fiber and the matrix each carry a portion of the load. We wish to derive an expression for the load-deflection curve of a layer of length L when the area ratio of the fibers to the matrix is 0.2. If the fibers are boron with $E = 5 \times 10^7$ psi and the matrix is an epoxy resin with $E = 5 \times 10^5$ psi, compare the equivalent stiffness of the layer to that of steel for which $E = 30 \times 10^6$ psi.

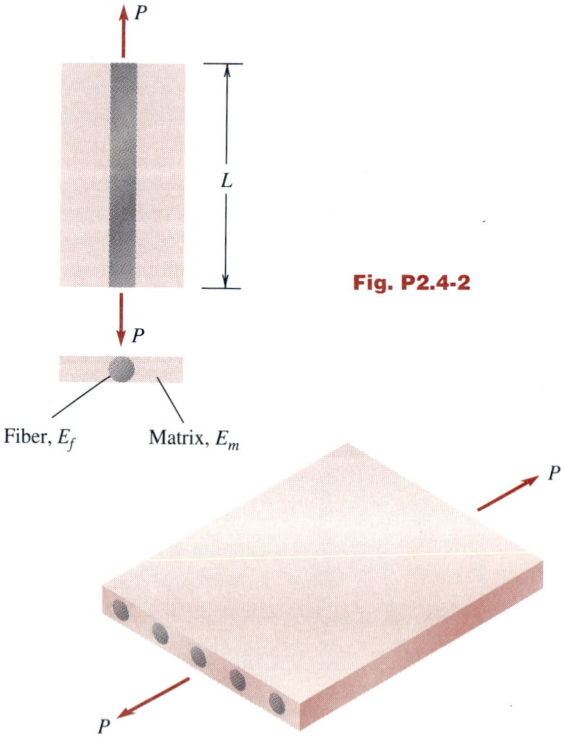

Fig. P2.4-2

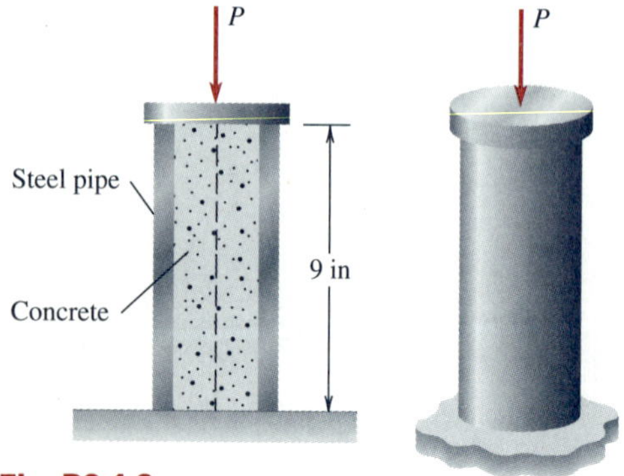

Fig. P2.4-3

2.4-3 A circular steel pipe of 4-in outside diameter and $\frac{1}{2}$-in wall thickness, as shown in Fig. P2.4-3, is filled with concrete and used as a pier for a dock. Find the displacement at the top of the pipe and the normal stresses in the pipe and concrete if $P = 2000$ lb. Take $E_C = 2 \times 10^6$ psi, $E_S = 30 \times 10^6$ psi, and $L = 9$ in.

2.4-4 Consider the model composite structure made up of a single central bar labeled 1 and two identical bars on either side labeled 2, as shown in Fig. P2.4-4. All three bars have the same length L, and the cross-sectional areas and elastic moduli are as indicated in the figure. A load P is applied to the structure. Derive a general formula for the displacement of the top plate as a function of load P. If the areas are all doubled in value with all other values remaining the same, what is the load-displacement relation?

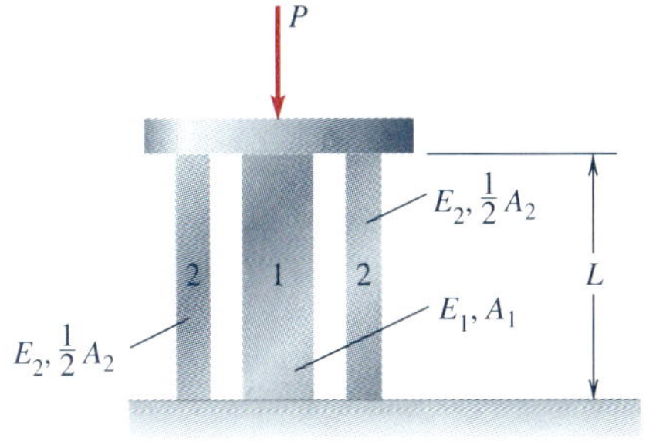

Fig. P2.4-4

2.4-5 Consider the following dimensions and properties for the model composite structure shown in Fig. P2.4-4: $A_2 = 40$ in^2, $A_1 = 100$ in^2, $L = 10$ in, $E_2 = 30 \times 10^6$ psi, and $E_1 = 3 \times 10^6$ psi. If $P = 100$ kips, find the displacement at the top of the structure and the normal stress in components 1 and 2.

2.4-6 Consider the model composite structure shown in Fig. P2.4-6. A gap Δ exists in the unloaded state between the top of the A_2 component and the rigid load plate as shown. If a load P is applied to the rigid load plate, find expressions for the downward displacement of the load plate in terms of E_1, E_2, A_1, A_2, L, and Δ. Assume all displacements are small and neglect the weight of all components.

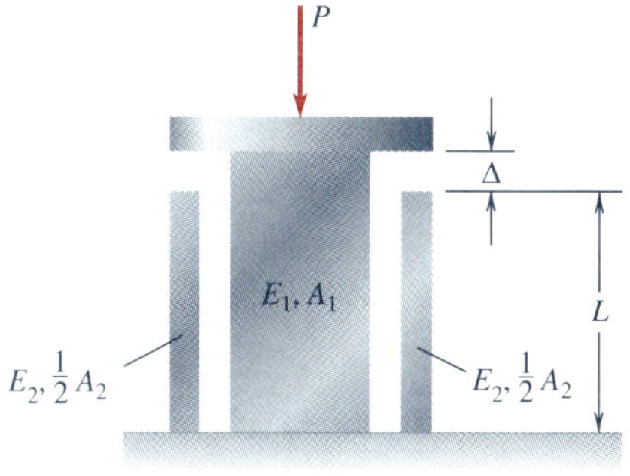

Fig. P2.4-6

2.4-7 Consider Example 2.5, Fig. 2.11. If the load P is placed at position $x = L/2$, find the downward displacements of points C and D.

2.4-8 Consider Example 2.5, Fig. 2.11. The load P can be located at any point x. Sketch the curve of the displacements at points D and C as a function of the value of x. The value of $x = 4L/5$ as derived in Eq. (g) of Example 2.5 should give equal displacements for points C and D.

2.4-9 Consider Example 2.6, Fig. 2.12. If the load P is now located a distance of 10 in from the right end of the platform, find the displacement u_A under the load, the displacements at the supports, and the angle that the loaded platform makes with the horizontal. All other values remain the same.

2.4-10 Consider Example 2.6, Fig. 2.12. If the load P is now located a distance 8 in from the left end of the platform, find the displacement u_A under the load, the displacements at the supports, and the angle that the loaded platform makes with the horizontal. All other values remain the same.

2.4-11 Consider Example 2.6, Fig. 2.12. If the load P is located a distance x from the right end of the platform, derive general expressions in terms of x for the displacements at the supports, the displacement under the load P, and the angle that the loaded platform makes with the horizontal. The values of k_1, k_2 and k_3 are to be taken now as equal.

2.4-12 Consider Example 2.7, Fig. 2.13. If the cross-sectional area of the pipe is 1000 mm^2, $E = 205$ GPa, $L_1 = 0.6$ m, $L_2 = 1$ m, and $P = 30$ kN, find the displacement of section B. Also find the values of the wall reactions at A and C and the axial stresses in segments AB and BC. Neglect the weight of the pipe.

Fig. P2.4-13

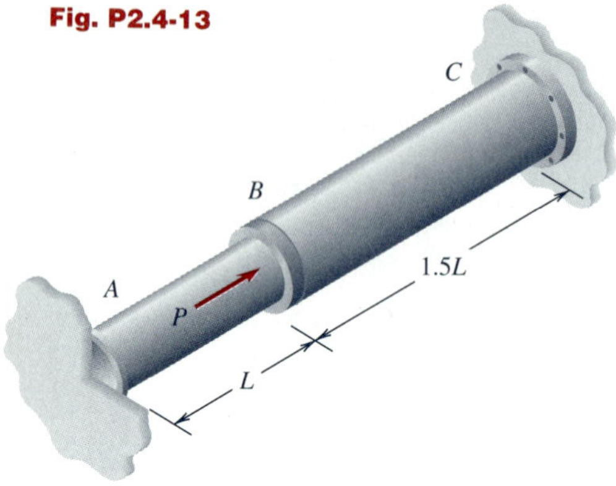

2.4-13 A segment of a piping system is loaded as shown in Fig. P2.4-13. Find the axial stress in each segment of the piping and the displacement of section B. Take $P = 50$ kN, $A_{AB} = 400$ mm^2, $E_{AB} = 205$ GPa, $A_{BC} = 900$ mm^2, $E_{BC} = 70$ GPa, and $L = 0.3$ m.

2.4-14 Consider Example 2.7, Fig. 2.13. Verify the expressions for the reactions given by Eqs. (i) and (j) in the discussion of the problem.

2.4-15 A pipe of length L made up of two components, as shown in Fig. P2.4-15, is loaded by a force P at section B as shown. Derive an expression for the displacement at section B under load P. Also determine the internal force in each segment and the values of the reactions. Take the cross-sectional area of the pipe as A and the elastic modulus as E. Neglect the weight of the components.

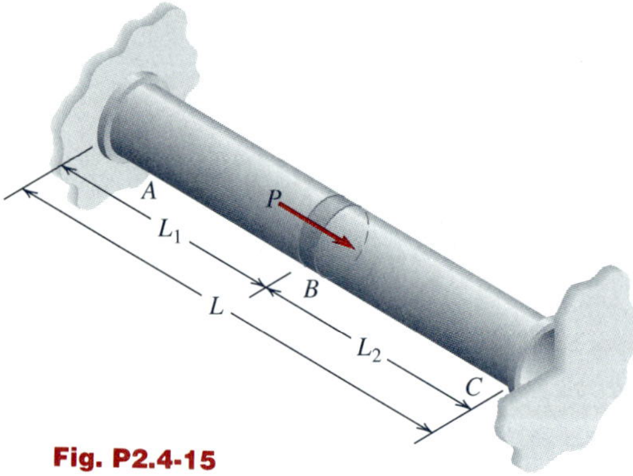

Fig. P2.4-15

2.4-16 A portion of a steel piping system supports a load $P = 10$ kips at section B, as shown in Fig. P2.4-15. Find the axial stress in each segment and the displacement of section B. Take the outside diameter of the pipe equal to 3.25 in and the inside diameter equal to 3.00 in. Also $L_1 = L_2 = 15$ in and $E = 30 \times 10^6$ psi.

2.4-17 The steel piping made up of two segments shown in Fig. P2.4-15 has a load of 800 kN applied at section B. Find the axial stress in each segment and the displacement of section B. Take the area of the pipe equal to 6000 mm^2, $L_1 = 2$ m, $L_2 = 1.0$ m, and $E = 200$ GPa.

2.4-18 A pipe ABC of modulus E, length L, and cross-sectional area A is to be connected between two walls, as shown in Fig. P2.4-18. However a small gap Δ exists between the end of the pipe and the wall. To close the gap, a force F is applied at the midpoint of the pipe, and the pipe is connected to the flange. What is the force in each segment of the pipe and what are the values of the wall reactions when the gap is just closed and the pipe is connected with the load F still applied? What are the values of the wall reactions given in terms of the gap once the pipe is connected firmly to the flange and the force is removed?

Fig. P2.4-18

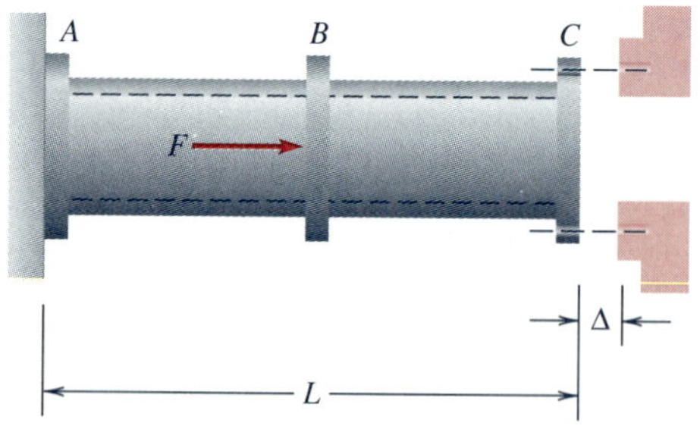

2.4-19 A pipe ABC of total length L and a known value of the axial modulus EA is to be connected between two rigid supports as shown in Fig. P2.4-19. A gap s exists between the end of the pipe at C and the wall. A load P is applied to the pipe at section B, and we wish to find

(*a*) the value of the load P_C required to move the end C of the pipe into contact with the lower wall support,
(*b*) the reactions at the walls if the load is increased by the amount ΔP beyond P_C, $P = P_C + \Delta P$, and
(*c*) the value of the load in terms of s such that reactions at the supports are of equal magnitude. Also sketch the curves of the support reactions as a function of the load P starting from $P = 0$.

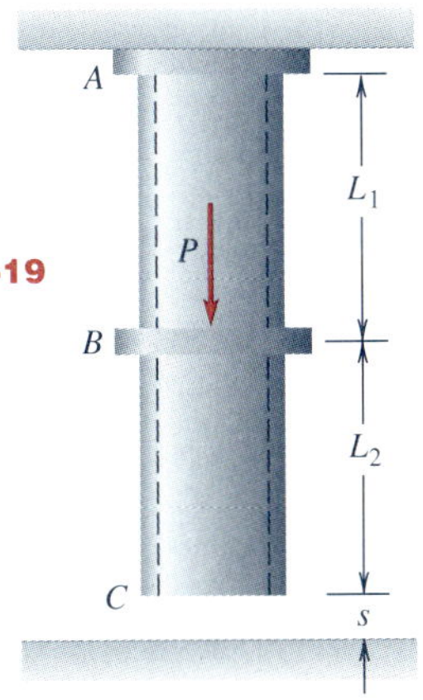

Fig. P2.4-19

2.4-20 Consider Example 2.8, Fig. 2.14. In the design of the linkage system, if the area of the aluminum link is increased to 0.20 in^2, find the displacement at point E and the angle of rotation of the rigid member AE. All other values remain the same.

2.4-21 Consider Example 2.8, Fig. 2.14. In the design of the linkage system, if the aluminum and steel links are interchanged, find the displacement at point E and the angle of rotation of the rigid member AE. All other values remain the same.

2.4-22 Consider Example 2.9, Fig. 2.15. If the area of segment AB is increased to 2 in^2, that is, $A_{AB} = 2$ in^2, and all other values remain the same, find the stress in each segment and the displacements of sections B and C. Also plot the internal force diagram and the displacement diagram, similar to Fig. 2.15*c* and *d*.

2.4-23 Consider Example 2.9, Fig. 2.15. If the maximum allowable normal stress in pipe AD is 450 psi, find the maximum allowable value of the applied load F. All the values of the parameters are the same as those given in Example 2.9.

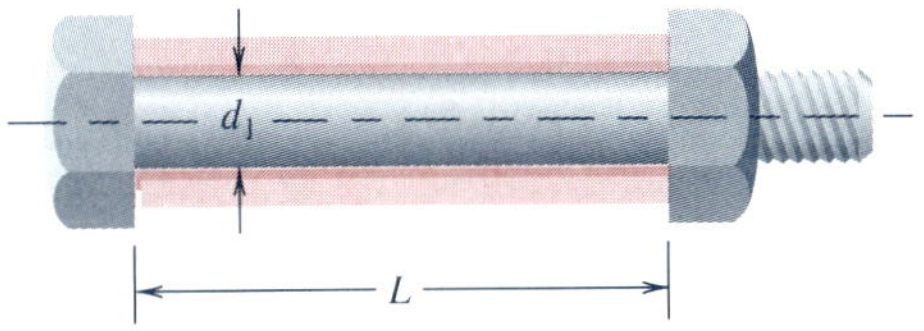

Fig. P2.4-26

2.4-24 Consider Example 2.10, Fig. 2.16. If the area of the tube is increased to $A_T = 0.75$ in^2, determine the tensile stress in the bolt and the compressive stress in the tube. All other values remain the same.

2.4-25 Consider Example 2.10, Fig. 2.16. If the maximum allowable stress in the tube is 23 ksi and the maximum allowable stress in the bolt is 15 ksi, find the maximum allowable number of turns of the nut. All geometric and material properties are as in Example 2.10.

2.4-26 A brass bolt of diameter $d_1 = \frac{3}{4}$ in is inserted into a cast-iron tube of inside diameter $\frac{13}{16}$ in and outside diameter of $1\frac{1}{2}$ in, as shown in Fig. P2.4-26. The pitch of the threads on the bolt is $\frac{1}{8}$ in. Determine the stresses in the bolt and tube if the nut is tightened by one-fourth of a turn. Take $E_{\text{bolt}} = 12 \times 10^6$ psi, $E_{\text{tube}} = 15 \times 10^6$ psi, and $L = 9$ in.

2.4-27 A steel bolt is inserted into a tube as shown in Fig. P2.4-27*a*. The diameter of the bolt and the inner diameter of the tube are 20 mm, and the outer diameter of the tube is 40 mm. The pitch of the threads on the bolt is 3 mm. If the nut is further tightened by one-fourth of a turn after it was tightened by hand, what is the stress in the tube and bolt? If an additional load of $P = 10$ kN is applied to the steel bolt as shown in Fig. P2.4-27*b*, what are the resultant stresses in the tube and bolt? We take $E_{\text{bolt}} = 200$ GPa and $E_{\text{tube}} = 100$ GPa.

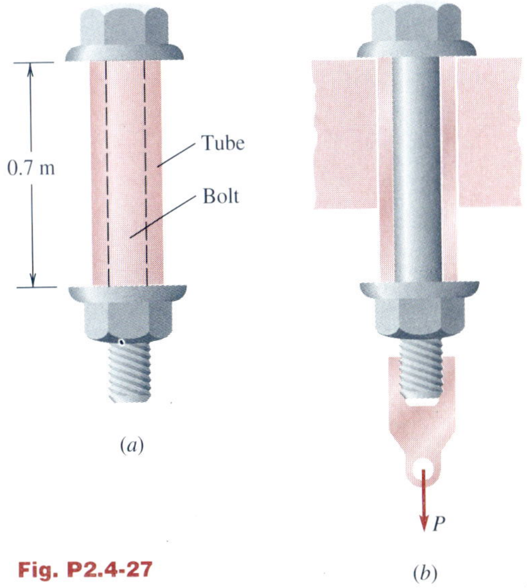

Fig. P2.4-27

2.4-28 Consider Example 2.10, Fig. 2.16. If the elastic modulus of the tube is increased while all other values stay the same as in the example, does the stress in the tube increase or decrease?

2.4-29 Consider Example 2.11, Fig. 2.17. If the cross-sectional area of members AB and CB is 500 mm^2 and the cross-sectional area of member DB is 400 mm^2, find the deflection of point B. All other values are the same.

2.4-30 Consider Example 2.11, Fig. 2.17. What values of ϕ will lead to maximum and minimum values of the deflection of B under load P?

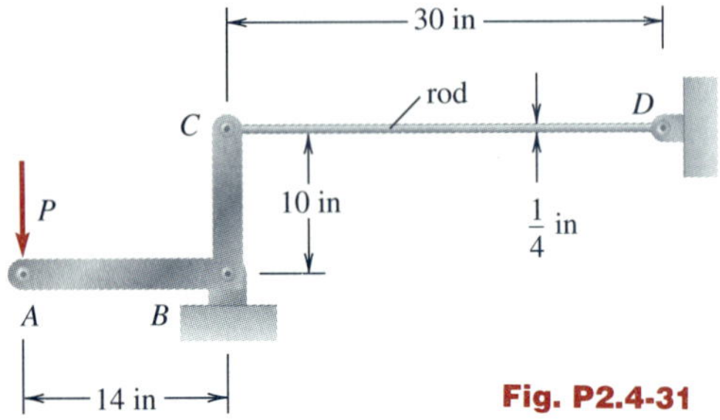

Fig. P2.4-31

2.4-31 A steel rod of diameter $\frac{1}{4}$ in connects points C and D as shown in Fig. P2.4-31. Fixture ABC is a rigid body. If a load $P = 800$ lb is applied at A, find the normal stress in the rod and the deflection under the load P. Neglect the weight of the members and friction in all joints, and take $E = 30 \times 10^6$ psi.

Fig. P2.4-32
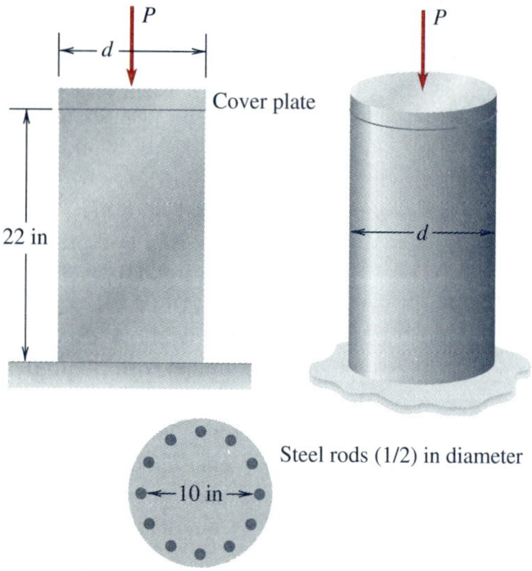

2.4-32 A concrete cylinder is reinforced by 12 steel rods of $\frac{1}{2}$-in diameter, as shown in Fig. P2.4-32. If a load $P = 70$ kips is applied as shown, find the average normal stress in the concrete and in each rod, as well as the downward deflection of the coverplate. Take $E_S = 30 \times 10^6$ psi, $E_C = 2 \times 10^6$ psi, and $d = 12$ in.

2.4-33 Consider Prob. 2.4-32, but now take $d = 13$ in for the diameter of the concrete cylinder. Comment on the effect of d on the stress and displacement.

2.4-34 A concrete cylinder of diameter $d = 12.5$ in contains 12 steel rods of diameter $\frac{1}{2}$ in, as shown in Fig. P2.4-32. If the maximum allowable stress in the concrete is 750 psi and the maximum allowable normal stress in the steel rods is 11,000 psi, estimate the maximum load P that can be put on the composite cylinder. Take $E_S = 30 \times 10^6$ psi and $E_C = 2 \times 10^6$ psi.

2.4-35 A lightweight rigid plank of length $2L = 16$ ft is attached to two identical springs at C and D, as shown in Fig. P2.4-35. The value of the spring constant of each spring is $k = 1000$ lb/ft. The plank is initially a distance $h = 1$ ft off the floor. If a person of weight $W = 160$ lb steps onto the center of the plank and begins to walk slowly toward one end, estimate the distance b at which one end just makes contact with the floor. Neglect the weight of the plank. Take $a = 1.5$ ft.

2.4-36 A lightweight rigid plank of length $2L$ is attached to two identical springs at C and D, as shown in Fig. P2.4-35. If a person of weight W steps onto the center of the plank and begins to walk slowly toward one end, we would like to find the distance b at which one end just makes contact with the floor. The plank is initially a distance h off the floor, and the weight of the plank can be neglected. The values of the linear spring constant are k. Show that the value of b is given by

$$b = \frac{a^2}{L}\left(\frac{2kh}{W} - 1\right)$$

or

$$\frac{b}{L} = \left(\frac{a}{L}\right)^2\left[\frac{h}{W/(2k)} - 1\right]$$

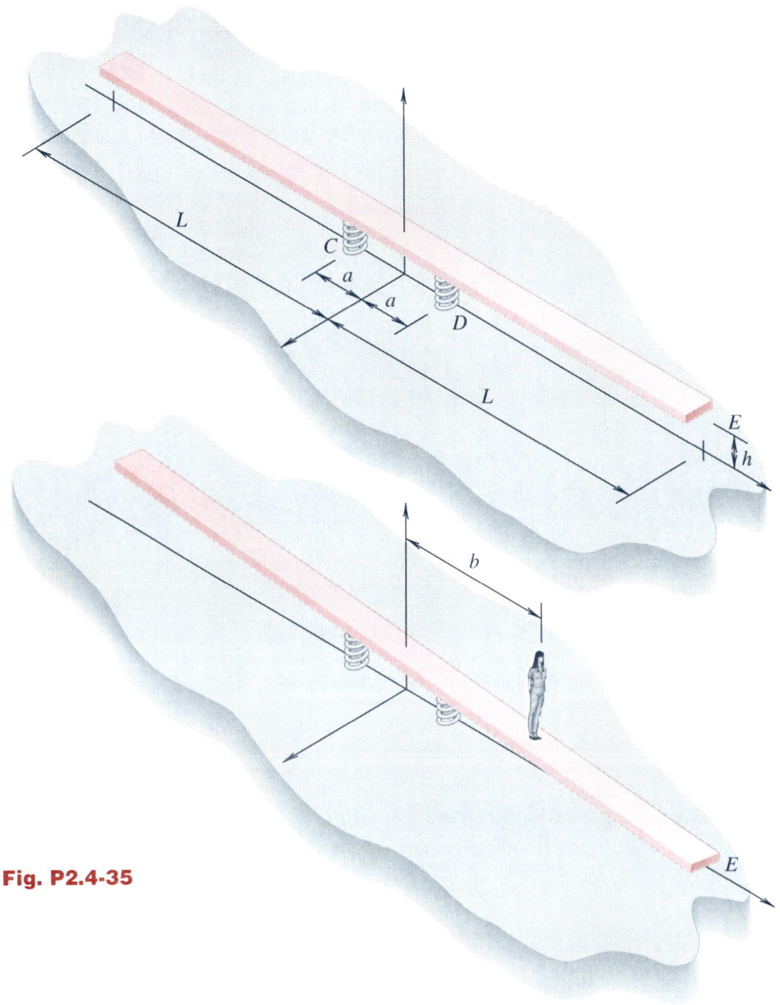

Fig. P2.4-35

2.4-37 Consider Prob. 2.4-36, but this time the spring constants of the springs at C and D are k_C and k_D. If we define $\gamma = k_C/k_D$, show that the expression for b/L is given by

$$\frac{b}{L} = \frac{(a/L)^2}{1 - \frac{1}{2}(1 - \gamma)(1 + a/L)}\left[\frac{2k_C h}{W} - 1 + \frac{1}{2}(1 - \gamma)\left(1 + \frac{L}{a}\right)\right]$$

Verify that the above expression reduces to the result in Prob. 2.4-36 when $\gamma = 1$.

2.4-38 If the values of the spring constants in Prob. 2.4-35 are now $k_C = 900$ lb/ft and $k_D = 1100$ lb/ft, estimate the value of b. What is the effect of a plus or minus 10 percent change in the value of the spring constants on the value of b?

2.4-39 A light, rigid bar *ABC* is supported by three linear springs, as shown in Fig. P2.4-39. The bar is horizontal before the load *P* is applied. The distance from the center of the bar to the point of application of the load *P* is λa, where λ is a dimensionless parameter that can vary between $\lambda = -1$ and $\lambda = 1$. Find the deflection of each spring when $k_A = 0.5k$, $k_B = k$, and $k_C = 1.5k$, and find the orientation of the bar under load *P*. Neglect the weight of bar *ABC* and assume small displacements.

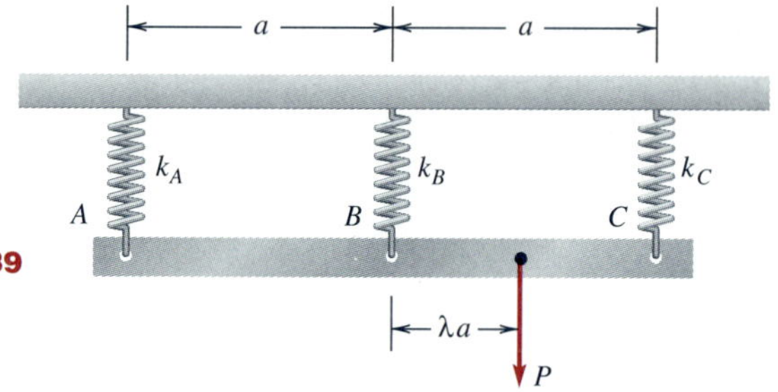

Fig. P2.4-39

2.4-40 Determine the deflections of each spring in Prob. 2.4-39 if the spring constants are in general k_A, k_B, and k_C. Show that the expressions for the deflections are given by

$$\delta_A = P[2k_C - \lambda(k_B + 2k_C)]/K$$

$$\delta_B = P[k_A + k_C + \lambda(k_A - k_C)]/K$$

$$\delta_C = P[2k_A + \lambda(k_B + 2k_A)]/K$$

$$K = k_A k_B + 4k_A k_C + k_B k_C$$

2.4-41 From the results in Prob. 2.4-40, find the location of the load, λ_0, such that all the spring deflections are equal. Show that

$$\lambda_0 = (k_C - k_A)/(k_A + k_B + k_C)$$

and that the deflection at this location of the load is

$$\delta_0 = \frac{P}{k_A + k_B + k_C}$$

2.4-42 A rigid member *ABC* is supported by three steel bars of length 0.5 m, as shown in Fig. P2.4-42. Bar *BE* is midway between *AD* and *CF*.

(*a*) Find the stresses in the bars and the displacement of *ABC* if $P = 100$ kN. Take $E_1 = E_2 = 200$ GPa, $A_1 = 1300$ mm^2, and $A_2 = 2600$ mm^2.

(*b*) Suppose bar 2 is initially 0.5 mm shorter than the other two bars, then solve part (*a*) again.

2.4-43 A rigid bar of weight *W* hangs on three equally spaced vertical wires of equal length, two of steel and one of aluminum, as shown in Fig. P2.4-43. The wires also support a load *P* acting at the center of the bar. What load *P* can be supported if the allowable stresses for the steel and aluminim wires are 20 and 12 ksi, respectively? Take $W = 80$ lb, $E_{st} = 30{,}000$ ksi, $E_{al} = 10{,}000$ ksi, $d_{st} = \frac{1}{8}$ in, and $d_{al} = \frac{3}{16}$ in.

Fig. P2.4-42

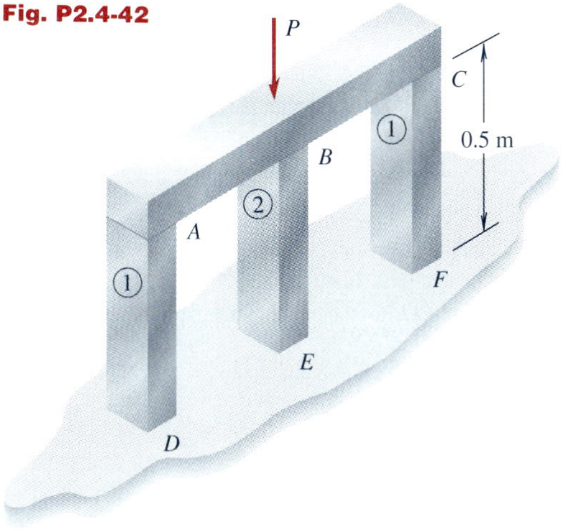

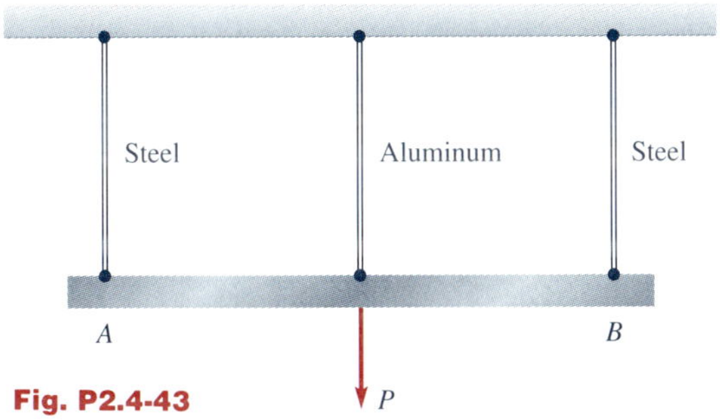

Fig. P2.4-43

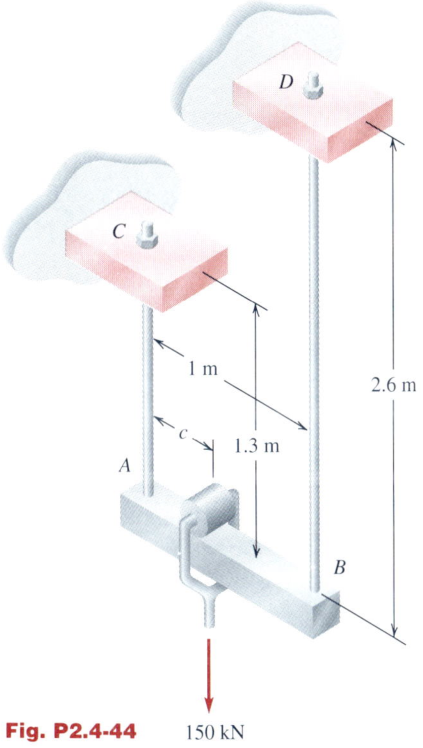

Fig. P2.4-44

2.4-44 A stiff horizontal beam as shown in Fig. P2.4-44 is supported by two soft copper rods AC and BD of cross-sectional area $A = 1300\ \text{mm}^2$ and lengths as shown. A vertical load of 150 kN is to be suspended from a roller that rides on the horizontal beam. We do not want the roller to move after the load is put on. Estimate the location c of the roller so that the beam will still be horizontal in the deflected position; take $E = 100$ GPa.

2.4-45 Consider Prob. 2.4-44, this time with $E = 90$ GPa. How sensitive is the value of c to the value of E?

2.4-46 A clock mechanism is shown in Fig. P2.4-46. The 2.7-lb weight is supported by three circular rods, two of which are brass and one is steel. Find the amount of load carried by each rod; take $E_S = 30{,}000$ and $E_B = 15{,}000$ ksi.

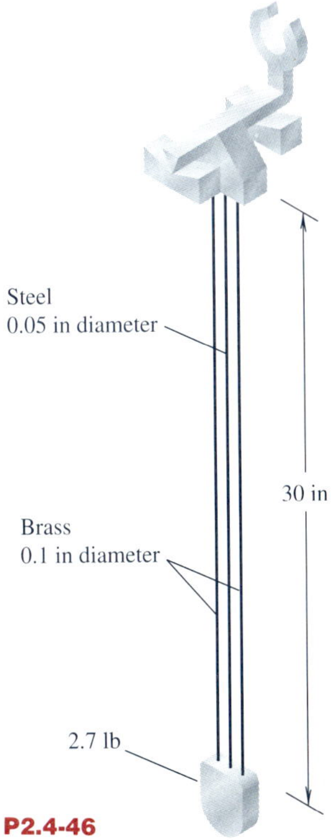

Fig. P2.4-46

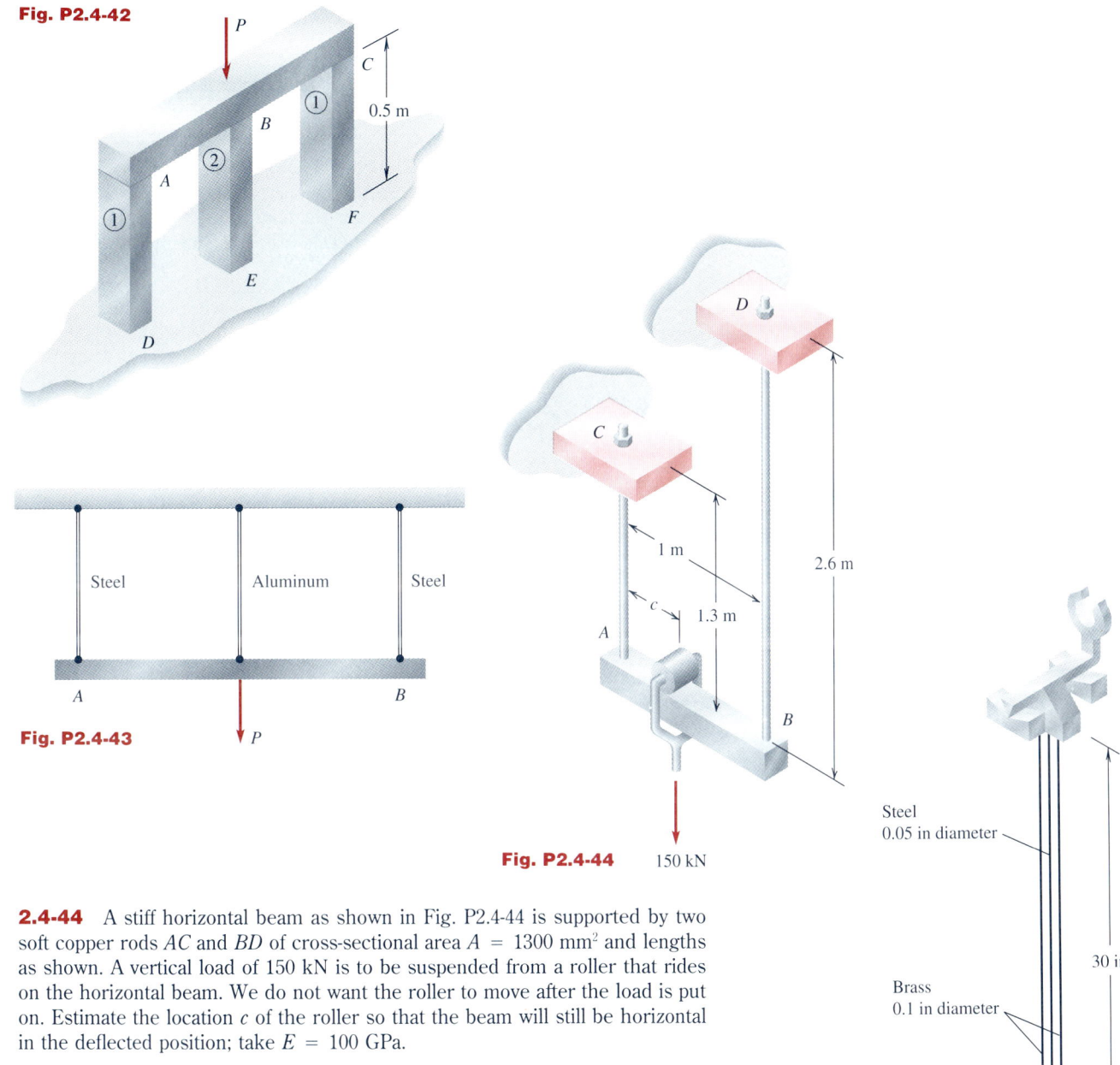

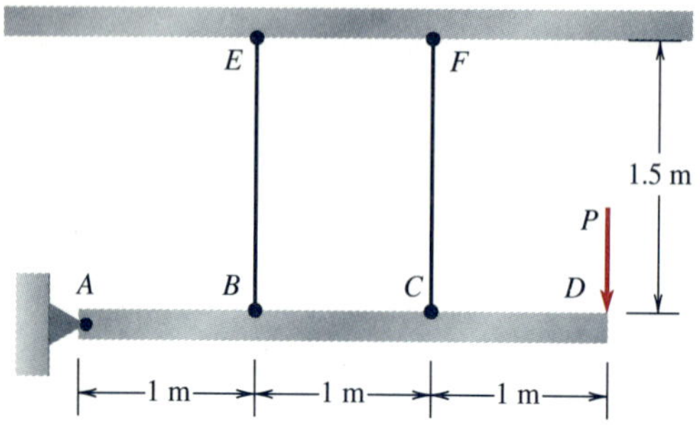

Fig. P2.4-47

2.4-47 A rigid bar *ABCD* is supported by a hinge at *A* and is held horizontal (when $P = 0$) by two identical wires *BE* and *CF*, as shown in Fig. P2.4-47. Find the stress in each wire and the displacement of point *D* due to load $P = 1300$ N. Neglect the weight of the bar compared to the load. For wires *BE* and *CF*, the area is 64 mm^2 and $E = 200$ GPa.

2.4-48 A load of 200 lb is applied to a rigid lever *AD* as shown in Fig. P2.4-48. Wires *BE* and *CF* are of 0.01-in^2 cross-sectional area and are unstressed before the 200-lb load is applied. Neglecting any bending of the lever or sag effect in the wires, what will be the final load in each wire and how much will point *A* move to the right? ($E = 30 \times 10^6$ psi.)

Fig. P2.4-48

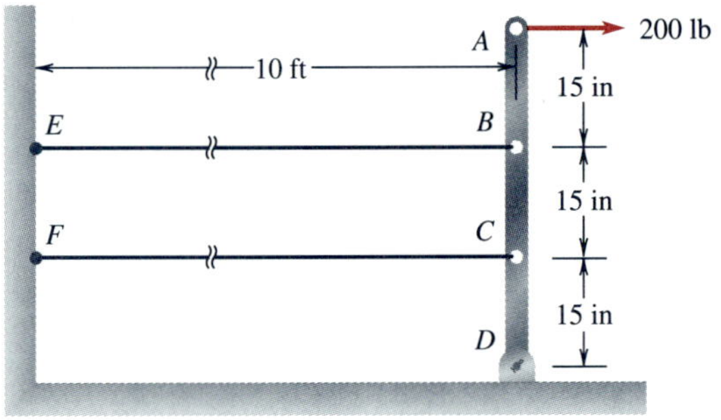

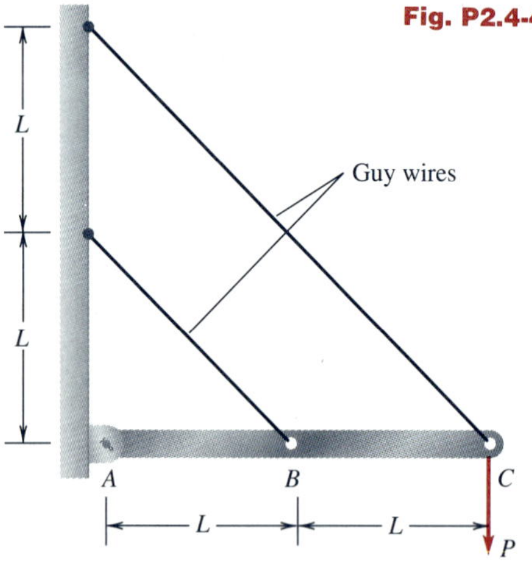

Fig. P2.4-49

2.4-49 The rigid member *ABC* is initially horizontal as shown in Fig. P2.4-49, and the guy wires, each with modulus of elasticity *E* and cross-sectional area *A*, are initially unstressed. Determine the small vertical displacement at point *C* when a load *P* is applied at *C*.

2.4-50 A reinforced-concrete column is constructed with 12 steel reinforcing bars, each of diameter 1 in, as shown in Fig. P2.4-50. Find the maximum allowable load *P* that can be applied to the column if the allowable stresses are $\sigma_S = 12{,}000$ psi (steel) and $\sigma_C = 1200$ psi (concrete); take $E_S = 29 \times 10^6$ psi and $E_C = 3.6 \times 10^6$ psi. Neglect the weight of the steel and the concrete.

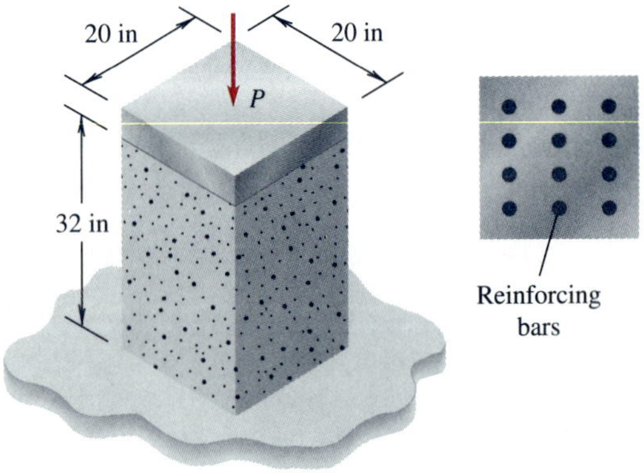

Fig. P2.4-50

2.4-51 A section of heavy rigid piping AB weighs 10,000 lb and is shown supported by three wire cables as in Fig. P2.4-51. The end cables are steel, and the inner cable is aluminum. Find the vertical displacement of the pipe due to its weight and the stress in each of the cables. Take $E_A = 10 \times 10^6$ psi, $E_S = 30 \times 10^6$ psi, $A_A = 0.5$ in^2, and $A_S = 0.3$ in^2.

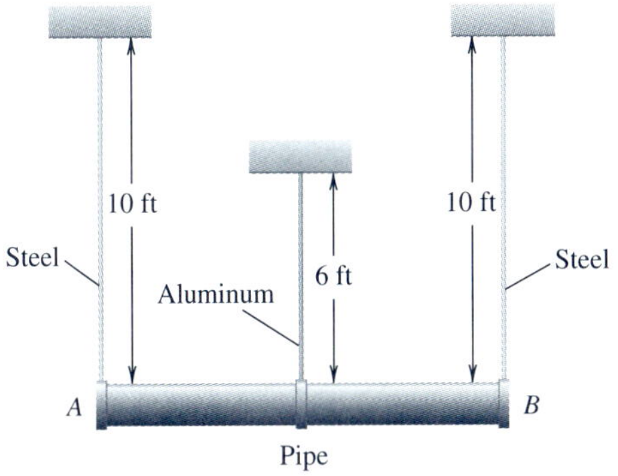

Fig. P2.4-51

2.4-52 A composite column is made up of a steel sleeve and an aluminum and copper core, as shown in Fig. P2.4-52. The column is loaded by a load P through a rigid plate. Find the fraction of the load P carried by the steel sleeve.

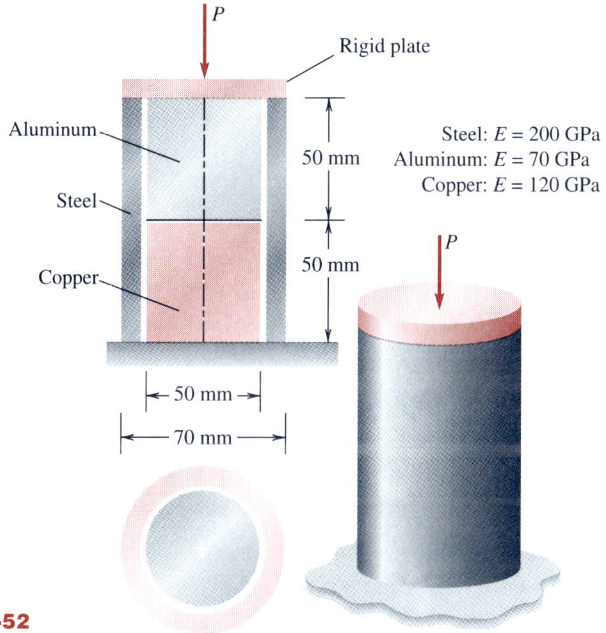

Fig. P2.4-52

2.4-53 Consider the structure shown in Fig. P2.4-53. Bar $ABCD$ is rigid; a linear spring with spring constant k is connected at B, and an aluminum bar is connected at C. A load $P = 5000$ lb is applied at D. Find the deflection at point D. Assume that the pin at A is frictionless.

Fig. P2.4-53

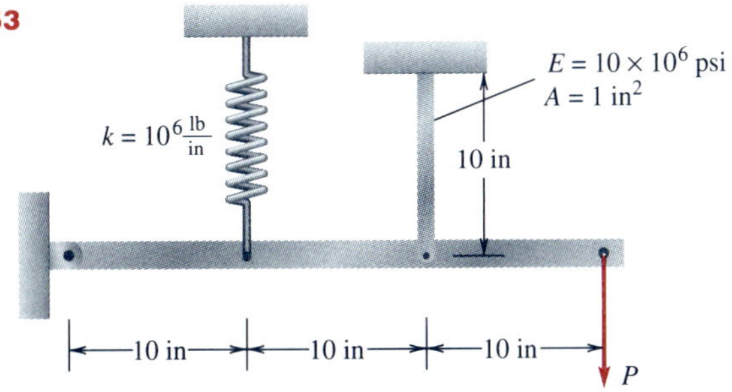

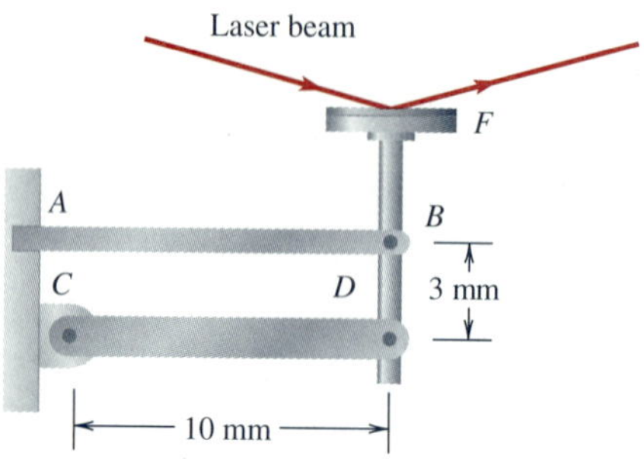

Fig. P2.5-2

2.5-1 A shaft connecting two units in a power plant is 20 m long. The shaft is supported by bearings along its length and is installed at 25°C. The operating range of temperature for the shaft is −20 to 65°C. Estimate the amount of expansion and contraction that must be designed into the support system to allow for operation over the full temperature range. Take $\alpha = 14.0 \times 10^{-6}/°C$.

2.5-2 A small lightweight temperature sensor is fabricated with a brass rod *AB* and an aluminum-alloy rod *CD* as shown in Fig. P2.5-2. A small mirror surface at *F* reflects a laser beam to a sensor. If the temperature of the system is raised 5°C, find the new orientation of the mirror at *F*. All pins are assumed to be frictionless, and member *FBD* is assumed to be rigid. Neglect the weight of all members; take $\alpha_{AB} = 17 \times 10^{-6}/°C$ and $\alpha_{CD} = 24 \times 10^{-6}/°C$.

2.5-3 Steel railroad rails 60 ft long are laid with a small gap between the end of one 60-ft section and the beginning of the next. What gap should be built in between the sections at 50°F so that the ends of the sections will just touch at 130°F? Once the gap size is found, what is its value if the temperature of the rails drops to 0°F? Take $\alpha = 6.5 \times 10^{-6}/°F$.

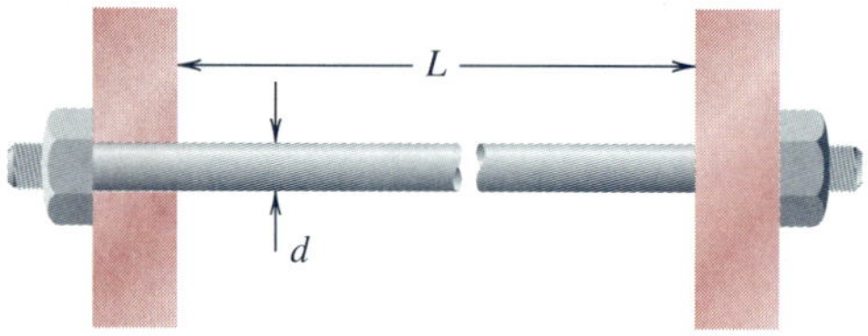

Fig. P2.5-5

2.5-4 What gap should be introduced between 50-ft-long paving slabs of concrete on a road section laid at a temperature of 60°F if the slabs are to just touch one another at 100°F? If the temperature increases to 120°F, estimate the compressive stress in the slabs. Take $\alpha = 5.5 \times 10^{-6}/°F$ and $E = 3 \times 10^3$ ksi.

2.5-5 A steel tie rod of length L and diameter d ties two walls of a building together as shown in Fig. P2.5-5. The rod is tightened so that it has an initial tension of F. What is the normal stress in the rod if the temperature changes by ΔT? Take $L = 0.6$ m, $d = 25$ mm, $F = 30$ kN, and $\Delta T = 15°C$. Take $E =$ 200 GPa and $\alpha = 12 \times 10^{-6}/°C$.

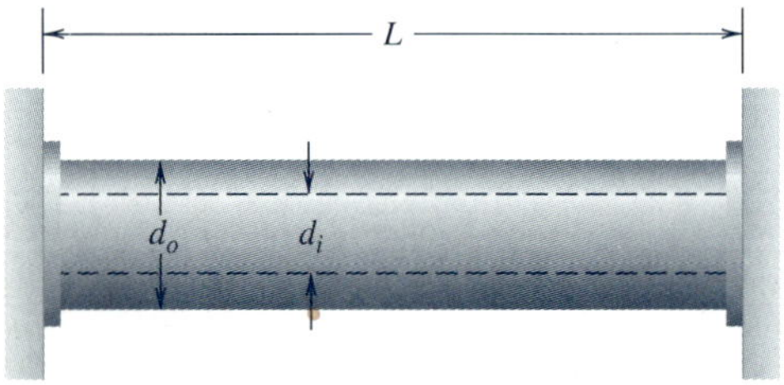

Fig. P2.5-8

2.5-6 Consider Prob. 2.5-5. Derive a general expression for the normal stress in the rod.

2.5-7 Consider Example 2.12, Fig. 2.19. If the piping system is now exposed to hot fluid so that the temperature of the pipe rises to 50°C, estimate the forces on the supports.

2.5-8 A steel pipe as shown in Fig. P2.5-8 is heated during operation by $\Delta T = 120°F$. If during this heating the right-end support moves a distance 12×10^{-3} in to the right, estimate the forces at the supports. Take $E = 30 \times 10^6$ psi, $\alpha = 6.5 \times 10^{-6}/°F$, $L = 30$ in, $d_o = 3.0$ in, $d_i = 2.75$ in, and neglect the weight of the pipe.

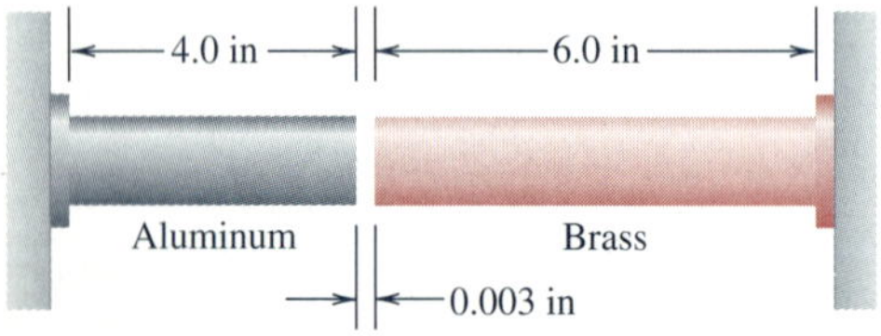

Fig. P2.5-9

2.5-9 An aluminum rod and a brass rod are connected to rigid supports as shown in Fig. P2.5-9. The area of each rod is 1.5 in^2. The temperature of the system is to increase.

(*a*) Find the temperature increase to close the gap of 0.003 in.

(*b*) Find the normal stress in each rod if the temperature increase is 120°F above the temperature found in part (*a*).

Take, for aluminum, $E_A = 10 \times 10^6$ psi, $\alpha_A = 12.5 \times 10^{-6}/°F$; for brass, $E_B = 10.4 \times 10^6$ psi, $\alpha_B = 10.5 \times 10^{-6}/°F$.

2.5-10 A model of a hydraulic system is shown in Fig. P2.5-10. If a load $P = 1.70$ kN is applied at B and the temperature increases by 10°C, find the displacement at section B and the reactions at A and C. Use the following properties:

Bar	A	E	α
1	2000 mm²	200 GPa	$12 \times 10^{-6}/°C$
2	1300 mm²	70 GPa	$17 \times 10^{-6}/°C$

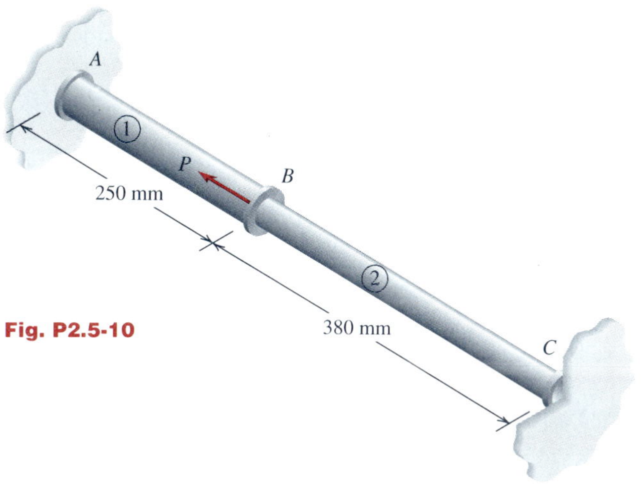

Fig. P2.5-10

2.5-11 A section of heavy rigid piping of weight W is supported by three elastic hangers during a construction operation, as shown in Fig. P2.5-11. The hangers have the same elastic modulus E and cross-sectional area A.

(*a*) Obtain an expression for the vertical displacement of the pipe in terms of W, E, A, and a.

(*b*) If the temperature decreases by an amount ΔT, obtain an expression for the temperature change which results in a combined deflection of the piping equal to zero due to the load W and the temperature change. Take α_T as the coefficient of thermal expansion.

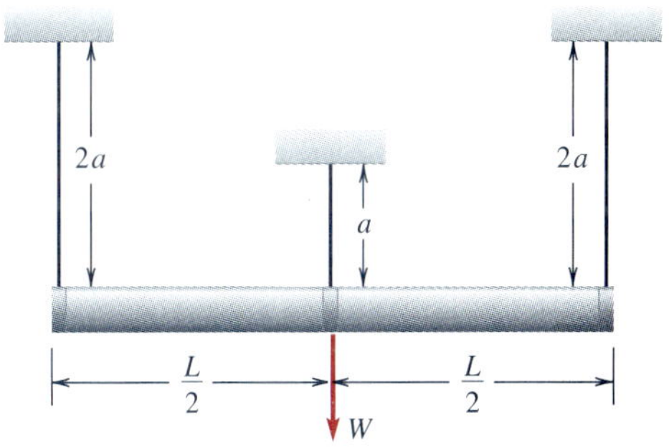

Fig. P2.5-11

2.5-12 A steel bar AB of length 1 m is placed between rigid walls at room temperature with a gap of 0.5 mm, as shown in Fig. P2.5-12. If $\alpha_T = 12 \times 10^{-6}/°C$ and $E = 200$ GPa:

(*a*) What increase in temperature ΔT (°C) would close the gap?

(*b*) What additional temperature change ΔT_a would be allowed if the allowable axial stress is 180 MPa?

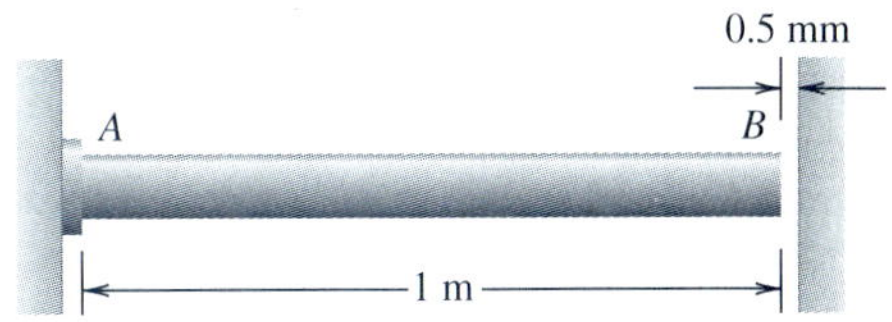

Fig. P2.5-12

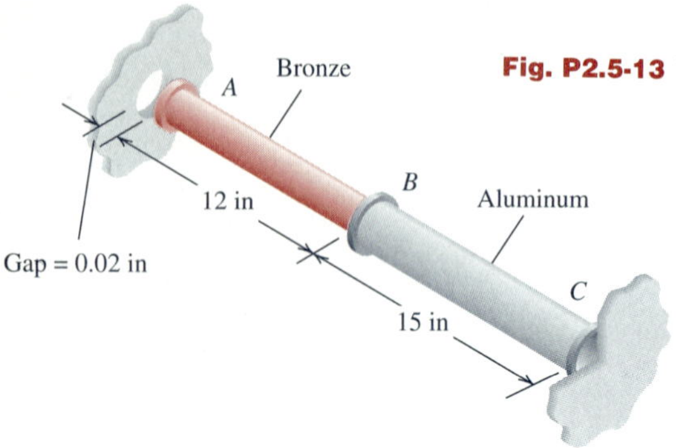

Fig. P2.5-13

2.5-13 An aluminum pipe is attached to a bronze pipe at B and to a rigid support at C, as shown in Fig. P2.5-13. (*a*) If the areas are $A_B = 2.5\ \text{in}^2$ and $A_A = 3\ \text{in}^2$, the elastic moduli are $E_B = 15{,}000$ ksi and $E_A = 10{,}000$ ksi, and the coefficients of thermal expansion $\alpha_B = 10.1 \times 10^{-6}/°\text{F}$ and $\alpha_A = 12.8 \times 10^{-6}/°\text{F}$, what temperature increase will just close the gap of 0.02 in at A? (*b*) What additional temperature increase could be allowed if the allowable stresses in bronze and aluminum are taken as 10 and 20 ksi, respectively?

2.5-14 A proposed design of a machine component connects a bronze rod of cross-sectional area A_B to a hollow copper tube of cross-sectional area A_C by a small ceramic pin of area A_p, as shown in Fig. P2.5-14. If the temperature of the system increases by ΔT, obtain an expression for the average shear stress in the pin. The rod and tube are connected to the base which is assumed rigid.

Fig. P2.5-14

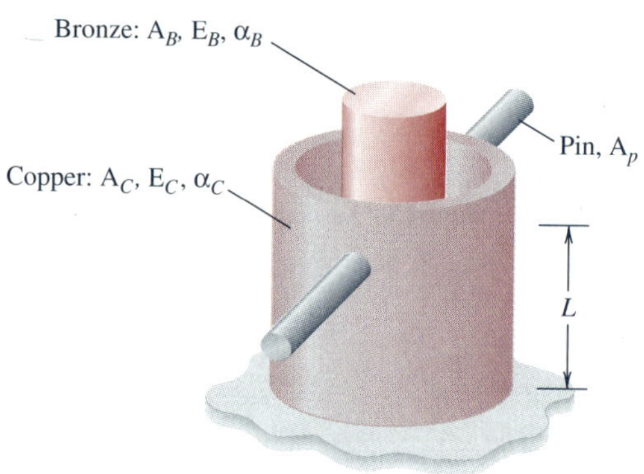

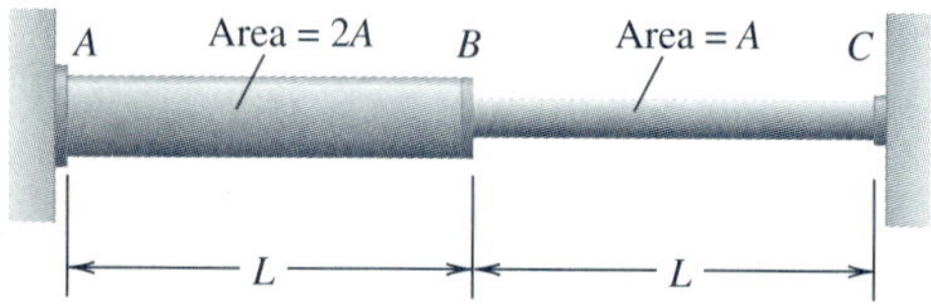

Fig. P2.5-15

2.5-15 Two solid-steel bars of length L with different cross-sectional areas are placed end to end between rigid walls at A and C, as shown in Fig. P2.5-15. Find the stress in each bar due to a temperature increase of $\Delta T = 100°\text{F}$ and the displacement of section B. Take $E = 30 \times 10^6$ psi, $\alpha = 6.5 \times 10^{-6}/°\text{F}$, $L = 40$ in, and $A = 1\ \text{in}^2$.

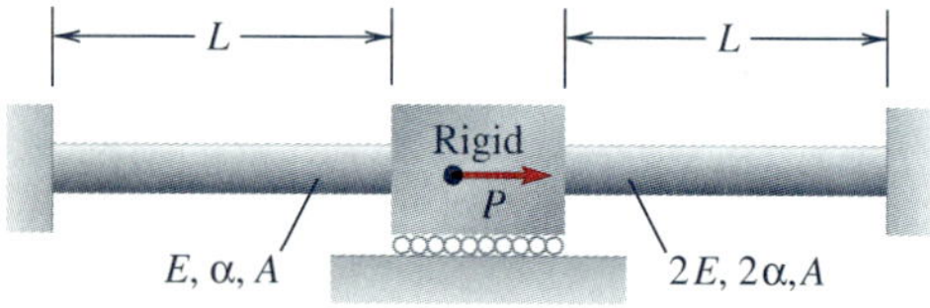

Fig. P2.5-16

2.5-16 A model of a structural system is shown in Fig. P2.5-16. Initially there are no forces in the bars of length L and cross-sectional area A. The temperature of the system is raised by an amount ΔT, and a load P is applied to the center of the rigid block. Find the load P such that the block moves a prescribed amount δ to the right.

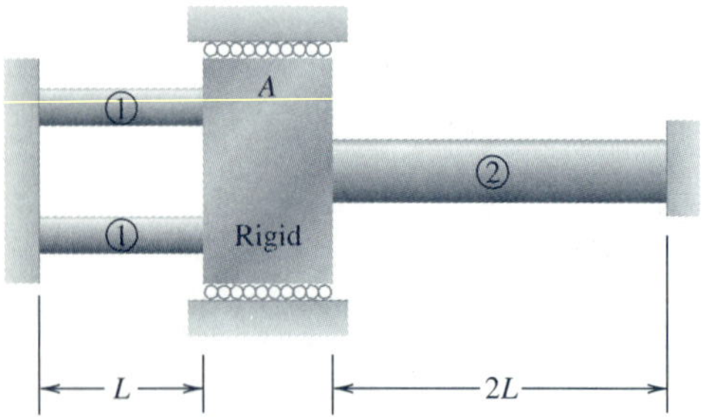

Fig. P2.5-17

2.5-17 In a model system of a structure as shown in Fig. P2.5-17, the rigid block A is constrained by the three elastic rods. The system undergoes a temperature increase of $\Delta T = 100°\text{F}$. Find the forces in each bar and the displacement of block A. The properties of bar 1 are $A_1 = 1\ \text{in}^2$, $\alpha_1 = 6 \times 10^{-6}/°\text{F}$, $L = 10$ in, and $E_1 = 10 \times 10^6$ psi; the properties of bar 2 are $A_2 = 2\ \text{in}^2$, $\alpha_2 = 12 \times 10^{-6}/°\text{F}$, $L_2 = 2L$, and $E_2 = 40 \times 10^6$ psi.

2.5-18 A 10,000-N rigid weight is supported by a concentric aluminum bar and a steel tube assembly, as shown in Fig. P2.5-18.

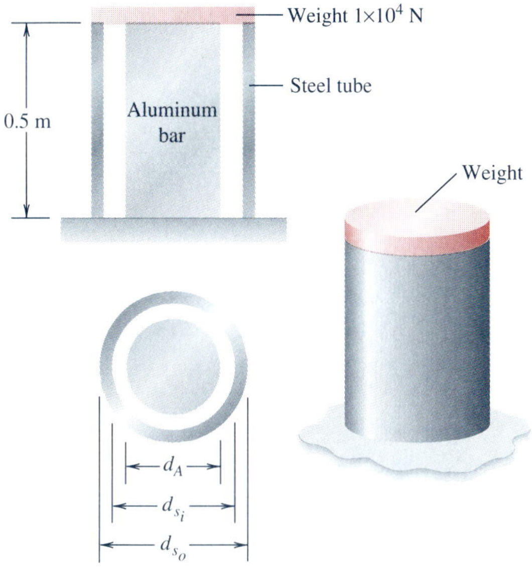

Fig. P2.5-18

(*a*) Find the displacement of the weight when there is no temperature change.
(*b*) If the temperature is lowered, what ΔT will cause all the weight to be supported by the steel tube?

Take for the aluminum bar $d_A = 0.02$ m, $E_A = 73$ GPa, and $\alpha_A = 23 \times 10^{-6}/°\text{C}$; and for the steel tube $d_{s_i} = 0.03$ m, $d_{s_o} = 0.035$ m, $E_S = 200$ GPa, and $\alpha_S = 12 \times 10^{-6}/°\text{C}$.

2.5-19 A model of a structural system is shown in Fig. P2.5-19. Each bar has area A and modulus E, but $\alpha_2 = 2\alpha_1$. If a load P is applied to the rigid block and the temperature drops by an amount ΔT, find an expression for the displacement u of the rigid block.

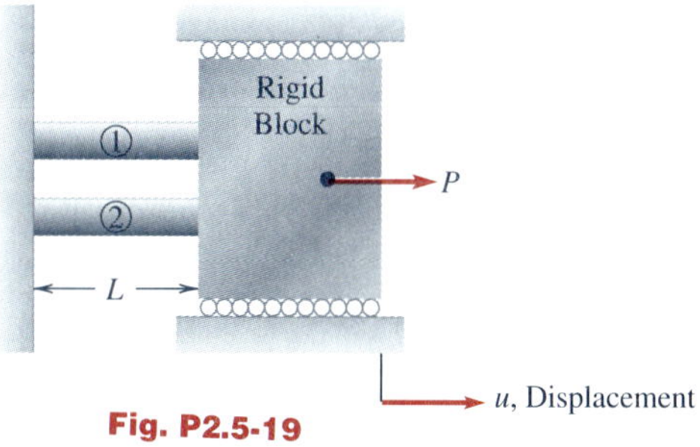

Fig. P2.5-19

2.5-20 An elastic component AB of length 6.0005 in, as shown in Fig. P2.5-20, is to be inserted into a 6.0000-in gap by lowering the temperature of the component from a room temperature of 68°F and then sliding it into the gap. The gap is set

Fig. P2.5-20

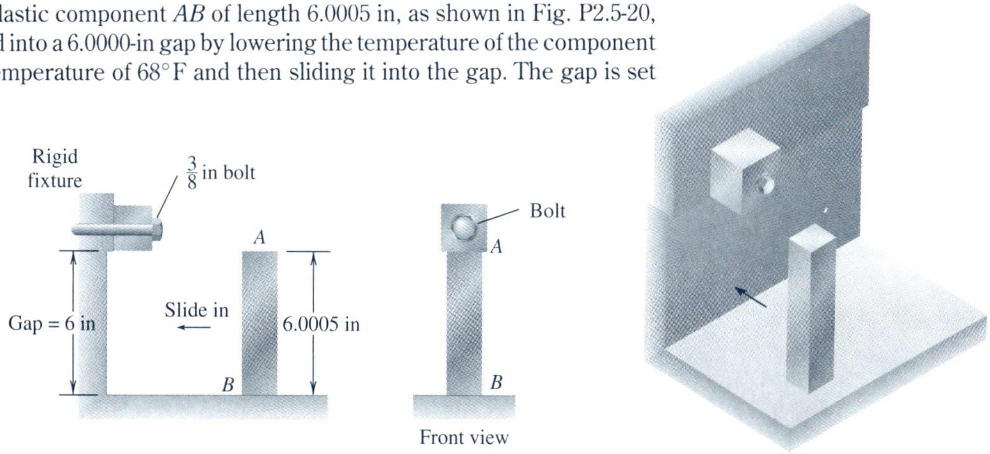

by attaching a rigid piece with a $\frac{3}{8}$-in bolt to the rigid fixture. The maximum allowable average shear stress across the bolt is $\tau_A = 7000$ psi.

(*a*) What minimum temperature drop will allow AB to be inserted?

(*b*) Once the component is inserted, what temperature increase will cause the shear stress across the bolt to equal τ_A?

Take area of $AB = 0.25$ in^2, $E = 10 \times 10^6$ psi, and $\alpha_{AB} = 10 \times 10^{-6}$/°F.

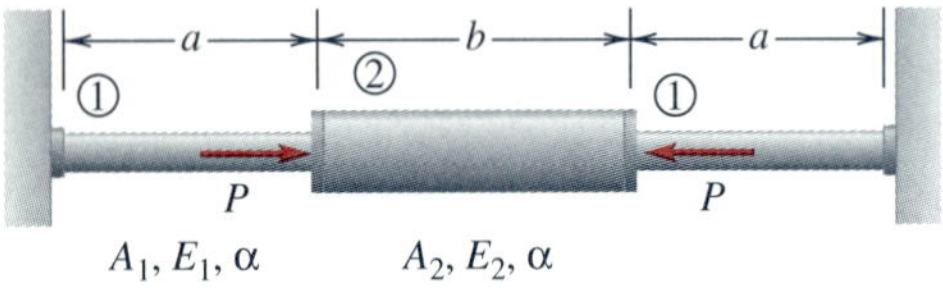

Fig. P2.5-21

2.5-21 Three bars are placed between rigid supports as shown in Fig. P2.5-21. Two loads P are applied, and in addition the system experiences a temperature change ΔT. Show that the expression for the stress in the middle bar is

$$\sigma_2 = -\frac{\left[\dfrac{2a}{b}\dfrac{A_2}{A_1}\dfrac{P}{A_2} + E\alpha\,\Delta T\left(1 + \dfrac{2a}{b}\right)\right]}{1 + \dfrac{2a}{b}\dfrac{A_2}{A_1}}$$

2.5-22 Compare the thermal stresses generated in pipes of different materials that are fixed at both ends to rigid walls and subjected to a temperature increase of 100°F. Consider the following materials:

	E, ksi	α, /°F
(*a*) Brass	15,000	11×10^{-6}
(*b*) Copper	17,000	9.5×10^{-6}
(*c*) Steel	29,000	8×10^{-6}
(*d*) Titanium	15,000	5×10^{-6}

2.6-1 Consider Example 2.16, Fig. 2.25. Verify that Eqs. (*f*) are correct and that the solutions for u_2 and u_3 are as given. Plot the internal force versus distance along the pipe and the displacement versus distance along the pipe.

2.6-2 Verify that the equilibrium equation, Eq. (2.18), upon use of force deformation relations, Eq. (2.19), reduces to the form given in Eq. (2.20). Write out Eqs. (2.20) if $N = 5$. Can you draw any conclusions about the structure of the equations?

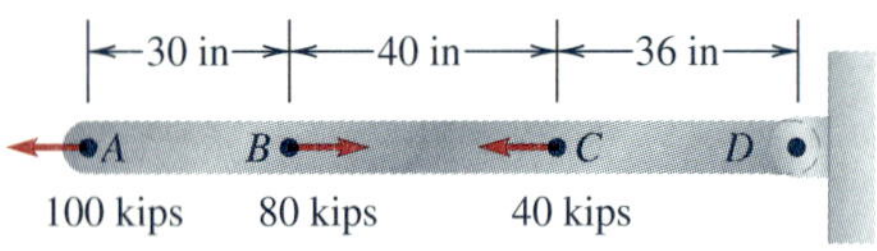

Fig. P2.7-1

2.7-1 A flat steel bar carries three axial loads at sections A, B, and C and is pinned at D, as shown in Fig. P2.7-1. The cross-sectional area is 4 in^2.

(*a*) Find the axial stress in each of the segments of the bar.

(*b*) Find the maximum axial displacement and its location along the bar. (Use $E = 30 \times 10^3$ ksi.)

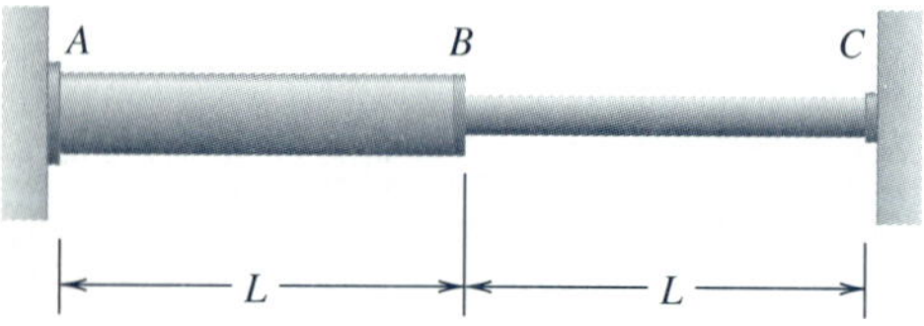

Fig. P2.7-2

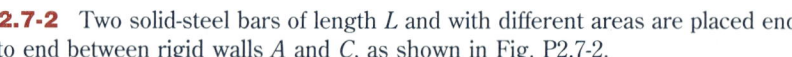

2.7-2 Two solid-steel bars of length L and with different areas are placed end to end between rigid walls A and C, as shown in Fig. P2.7-2.

(*a*) Find the stress in each bar due to a temperature increase of 100°F.

(*b*) Find the displacement of section B, and determine the direction of the displacement (left or right).

Take $A_{AB} = 2$ in^2, $A_{BC} = 1$ in^2, $L = 40$ in, $E = 30 \times 10^6$ psi, $\alpha = 6.5 \times 10^{-6}$/°F.

2.7-3 In the assembly of a structure shown in Fig. P2.7-3, the steel bar AB is fixed to a rigid wall at A at room temperature 20°C, and it is separated by a gap of 0.8 mm from an aluminum bar $B'C$ which is fixed to a rigid wall at C.

(*a*) Find the temperature increase ΔT such that the two bars just touch.

(*b*) If the temperature is further increased by 50°C above the temperature for contact found in part (*a*), find the stresses in each bar and the total displacements of sections B and B' from their initial locations at room temperature.

Take $E_S = 190$ GPa, $E_A = 70$ GPa, $\alpha_S = 18 \times 10^{-6}/°C$, and $\alpha_A = 23 \times 10^{-6}/°C$.

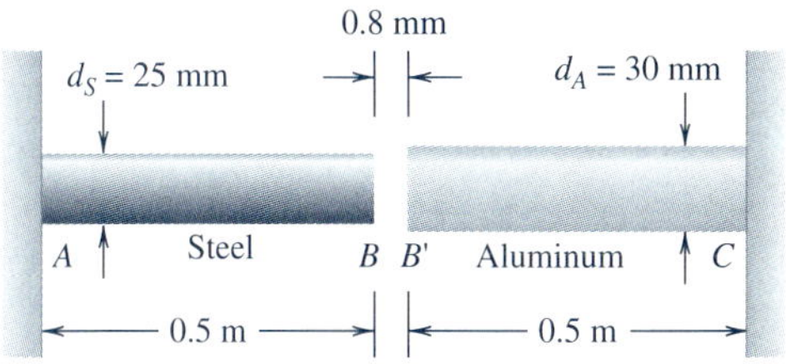

Fig. P2.7-3

2.7-4 A 1.5-in-diameter solid-steel rod ABC is joined to a brass rod CD of the same diameter to form a 14-ft-long rod and is loaded as shown in Fig. P2.7-4. The weight of the rod is neglected.

(*a*) Find the displacement of sections B and D.

(*b*) Find the axial stress in segments AB, BC, and CD.

Take $E_S = 30 \times 10^3$ ksi and $E_B = 15 \times 10^3$ ksi.

Fig. P2.7-4

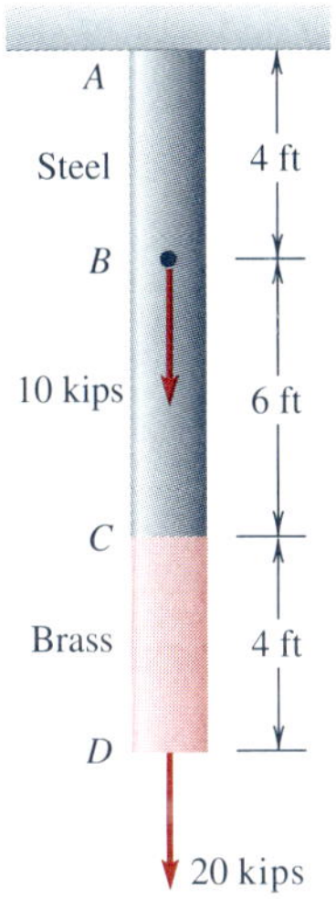

2.7-5 Two circular steel bars of different diameters and different lengths are joined at B, as shown in Fig. P2.7-5, and fixed to a rigid wall at A and loaded by the two loads as shown. A dial gage measurement at C indicated a displacement toward A of 0.02 in due to this load system. It is given that $E = 30{,}000$ ksi.

(*a*) Find the value of P which causes the given displacement at C.

(*b*) Find the displacement of section B.

(*c*) Find the axial stress in each bar.

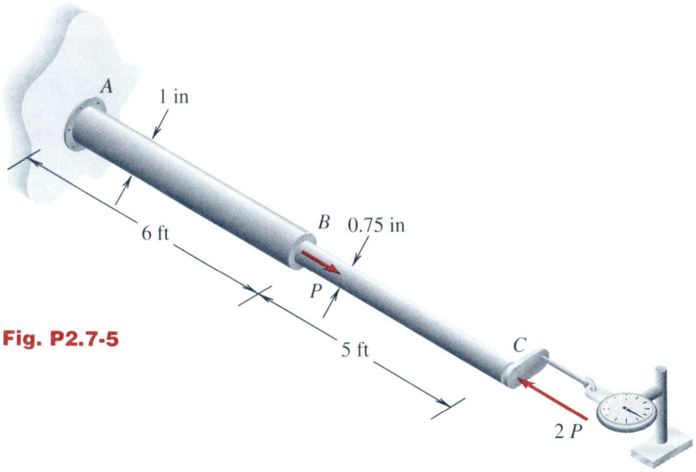

Fig. P2.7-5

2.7-6 Two aluminum bars AB and CD and a steel bar BC are placed end to end and attached to rigid walls at A and D, as shown in Fig. P2.7-6. Suppose that the cross-sectional area of each of the aluminum bars is 4 in² and the cross-sectional area of the steel bar is 2 in² and $E_{al} = 10{,}000$ ksi and $E_{st} = 30{,}000$ ksi.

(*a*) Find the maximum load P if the allowable stress in the steel is 24 ksi and in the aluminum is 16 ksi.

(*b*) If the temperature increases by 100°F, find the new maximum load P. Take $\alpha_{al} = 12.5 \times 10^{-6}/°F$ and $\alpha_{st} = 6.5 \times 10^{-6}/°F$.

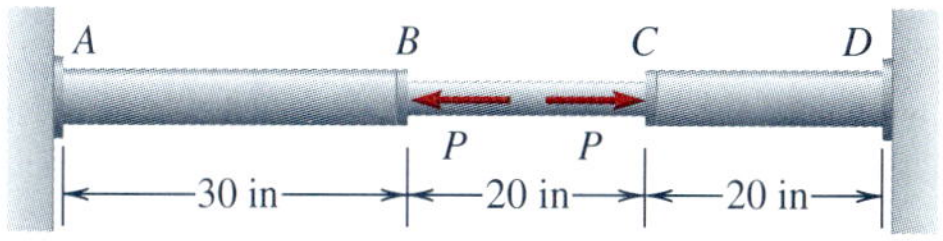

Fig. P2.7-6

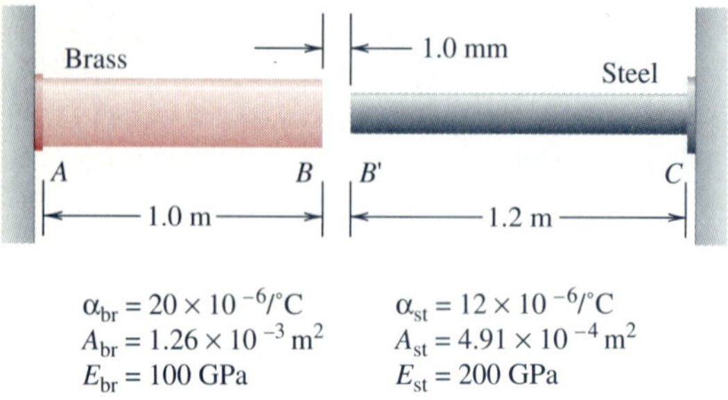

Fig. P2.7-7

2.7-7 The brass and steel rods shown in Fig. P2.7-7 are firmly attached to rigid walls at A and C. They are separated by a gap of 1 mm at B at room temperature.

(*a*) If the temperature increases by 60°C, find the stress in each rod. (Note: Find the temperature increase required to close the gap, and use the additional temperature increment required to reach 60°C for the stress calculation.)

(*b*) If the allowable stresses for brass and for steel are 50 and 125 MPa, find the maximum temperature increase so as not to exceed the allowable stresses.

2.7-8 A circular steel bar is loaded and supported as shown in Fig. P2.7-8. Suppose that the allowable axial stress in the bar is 20 ksi.

(*a*) Find the minimum diameter d for which the stresses in all sections of the bar will not exceed the allowable stress in tension or compression. Obtain a plot of axial displacement versus distance along the bar.

(*b*) Using the results for the diameter d from part (*a*), find the temperature change ΔT which will cause section D to remain fixed due to the combined effect of temperature change and the loading as shown. Obtain a plot of axial displacement versus distance from the support, and comment on the changes in the displacement plot from part (*a*).

Take $E = 30 \times 10^3$ ksi and $\alpha = 6.5 \times 10^{-6}/°F$; neglect the weight of the bar.

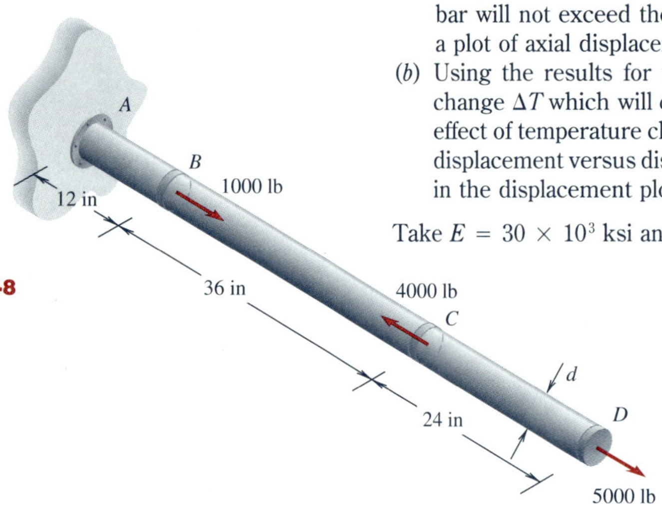

Fig. P2.7-8

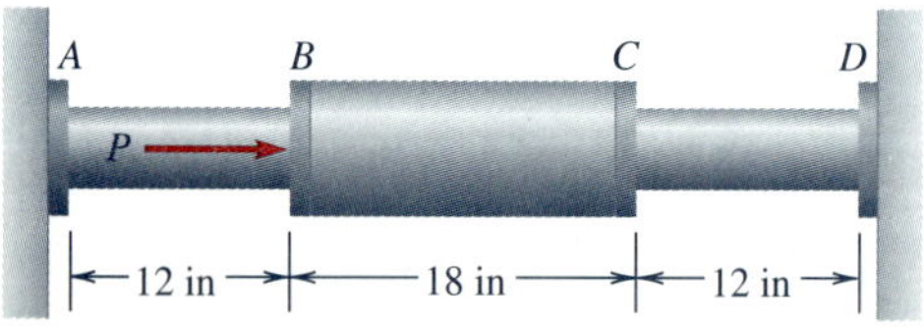

Fig. P2.7-9

2.7-9 Three steel bars are firmly attached and placed between rigid walls as shown in Fig. P2.7-9. The cross-sectional areas of AB, BC, and CD are 1, 2, and 1 in^2, respectively. The allowable stress is 20 ksi in tension and compression, and $E = 30 \times 10^3$ ksi.

(*a*) Find the maximum allowable load P.

(*b*) Find the temperature change ΔT such that the tensile stress in AB due to the maximum load P of part (*a*) is reduced to zero at the new temperature, and plot the displacement along the bar. Take $\alpha = 6.5 \times 10^{-6}/°F$.

2.7-10 A composite bar ABC as shown in Fig. P2.7-10 is made up of two circular bars to be fitted into the 0.1-m-wide rigid space.

(*a*) If the bar is to be fitted by compressing it with a force P, how large should this force be?

(*b*) What is the stress from this force in the brass and in the aluminum bars?

(*c*) What is the displacement of section B?

(*d*) What minimum temperature decrease ΔT for the bar in the absence of load P would allow it to fit in the space?

(*e*) If only half of the force P found in part (*a*) is applied, then what temperature decrease ΔT is needed to fit the bar into the space?

Fig. P2.7-10

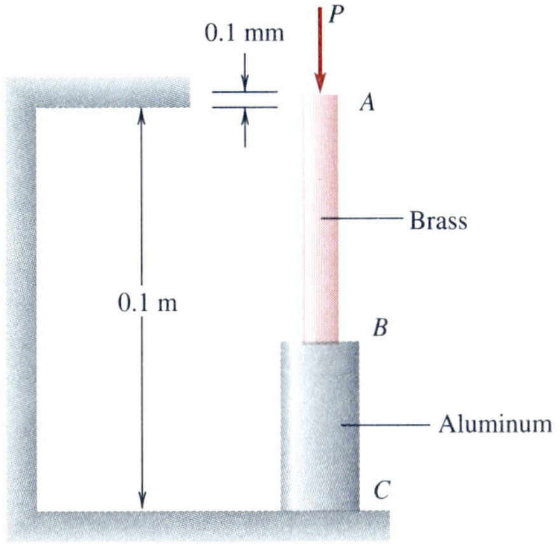

Brass: $E_B = 100$ GPa $\alpha_B = 21 \times 10^{-6}/°C$ $L_B = 0.06$ m $d_B = 10$ mm
Aluminum: $E_A = 70$ GPa $\alpha_A = 23 \times 10^{-6}/°C$ $L_A = 0.04$ m $d_A = 20$ mm

2.8-1 Verify the form of Eq. (2.28) arising from the substitution of the force-deformation relation, Eq. (2.27), into the equilibrium equation, Eq. (2.24). Verify that Eq. (2.31) follows from Eq. (2.30).

2.8-2 A 1-in-diameter steel cable weighs 1.60 lb/ft. The equivalent EA of the steel cable is 1.5×10^6 lb. How much will a 6000-ft segment of the cable elongate under its own weight?

2.8-3 During construction a 40-ft section of steel pipe with 10-in outside diameter and 9.0-in inner diameter is supported vertically as shown in Fig. P2.8-3. How much does the pipe elongate under its own weight, and what is the maximum normal stress in the pipe? Steel weighs 490 lb/ft³. Take $E = 30{,}000$ ksi.

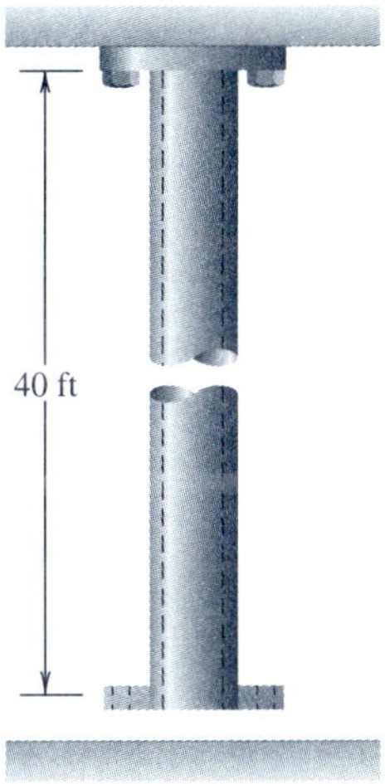

Fig. P2.8-3

2.8-4 Consider Example 2.21, Fig. 2.44. If the steel pipe has an outside diameter of 1 in, wall thickness of $\frac{1}{8}$ in, $E = 30{,}000$ ksi, and $\alpha = 6.5 \times 10^{-6}/°F$, and if $\Delta T_B = 100°F$, find the stress in the pipe and the axial displacements of sections at distances of $L/4$ and $L/2$ from the left support. Compare the results of this calculation with the results for the same pipe but with a uniform temperature increase along the pipe. Take $L = 8$ ft.

2.8-5 Consider Example 2.21, Fig. 2.44, but remove the support at end B so that the pipe is free to expand. Find an expression for the displacement of the end section B, and sketch the displacement curve $u(x)$ versus position along the pipe (see Fig. 2.44d).

2.8-6 A uniformly tapered column of square cross section carries a load P at the top, as shown in Fig. P2.8-6. If the cross-sectional areas at the top and bottom are A and $2A$ and the height of the column is L, find the overall shortening of the column if we neglect the self-weight of the column.

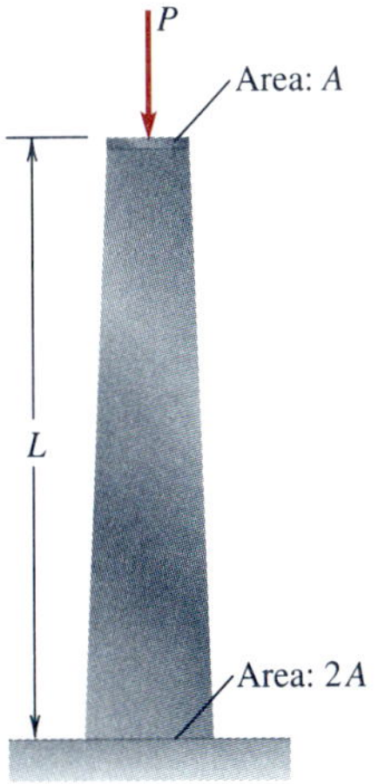

Fig. P2.8-6

2.8-7 A flat bar has a constant thickness t, but the width $w(x)$ varies uniformly from the value w at the left end to the value $w/2$ at the other end, as shown in Fig. P2.8-7. If ends A and B are fixed, find the maximum stress in the bar and the displacement of the midsection C due to a uniform increase of the temperature ΔT of the bar. Neglect the self-weight of the bar.

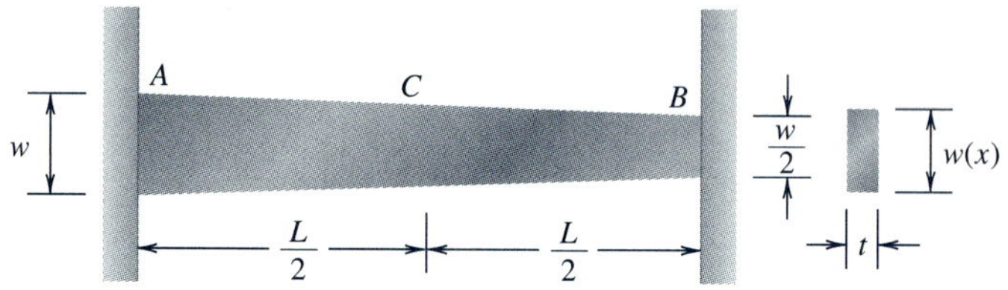

Fig. P2.8-7

2.8-8 For the flat bar in Prob. 2.8-7, find the maximum stress in the bar and the displacement at the midsection when a load P is applied at the midpoint C of the bar and acting to the right and ΔT is equal to zero.

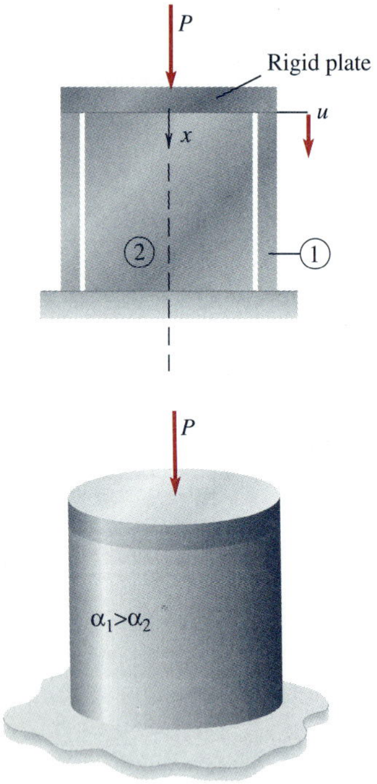

Fig. P2.9-1

2.9-1 Material 2 is contained within a cylindrical tube of material 1, as shown in Fig. P2.9-1. A rigid plate is attached to the top of the system to which a load P is applied. The properties of the materials are given by E_i, A_i, α_i where $i = 1$ and 2 and the length is L. If the temperature of the system is changed by ΔT and the load P is applied, verify that the expression for the downward displacement in the direction of the load is given by

$$\frac{u}{L} = \frac{P - \Delta T\,(A_2E_2\alpha_2 + A_1E_1\alpha_1)}{A_2E_2 + A_1E_1}$$

The force in each material, assuming that positive forces are compressive, is given by

$$F_1 = \frac{(E_1A_1)P + (E_1A_1)(E_2A_2)\,\Delta T\,(\alpha_1 - \alpha_2)}{A_1E_1 + A_2E_2}$$

$$F_2 = \frac{(E_2A_2)P + (E_2A_2)(E_1A_1)\,\Delta T\,(\alpha_2 - \alpha_1)}{A_1E_1 + A_2E_2}$$

In the absence of the load P, the expression for the force in each member can be written in the form

$$F_1 = -F_2 = \frac{\Delta T\,(\alpha_1 - \alpha_2)}{1/(E_1A_1) + 1/(E_2A_2)}$$

2.9-2 A thin ring of internal radius r, thickness t, and width b is subjected to uniform pressure p over the entire internal surface, as shown in Fig. P2.9-2a and b. We wish to determine the forces and deformation of the ring.

(a) Show that the hoop force in the ring is given by $F_T = prb$ obtained upon balancing the pressure forces in the y direction, Fig. P2.9-2c.

(b) If we assume that the hoop is a flat plate of thickness t, width b, and length $2\pi r$, show that the increase in length of the circumference is

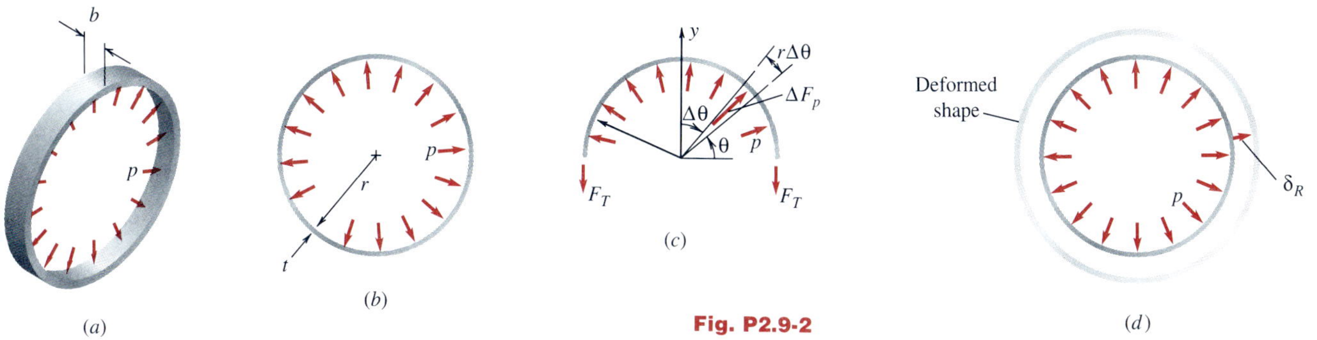

Fig. P2.9-2

$$\delta_T = \frac{F_T(2\pi r)}{btE} = \frac{2\pi p r^2}{tE}$$

We assume $t/(2r) \ll 1$.

(c) Show that the radial expansion of the ring is given by (Fig. P2.9-2d)

$$\delta_R = \frac{\delta_T}{2\pi} = \frac{pr^2}{tE}$$

2.9-3 A brass ring of 300-mm internal radius and 3-mm thickness is subjected to an internal pressure of 0.6 MPa, as shown in Fig. P2.9-3. Estimate the hoop force in the ring and the radial expansion of the ring. Take E = 100 GPa.

Fig. P2.9-3

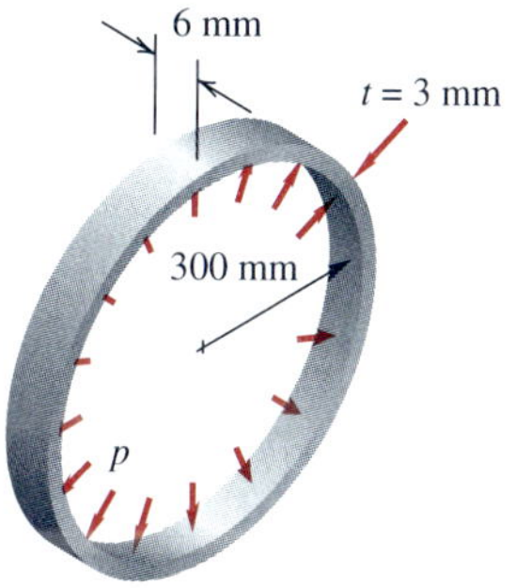

2.9-4 A composite hoop consists of a brass hoop of 300-mm internal radius and 3-mm thickness and a steel hoop of 303-mm internal radius and 6-mm thickness, as shown in Fig. P2.9-4. Both hoops are 6 mm thick normal to the plane of the hoop. If a radial pressure of p = 1.4 Pa is applied to the brass hoop, estimate the forces in the brass and steel hoops and the total radial expansion of the brass hoop. Take E_B = 103 GPa and E_S = 205 GPa.

Fig. P2.9-4

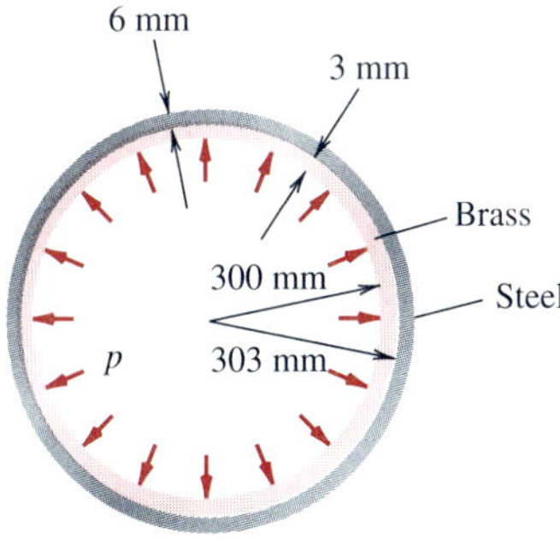

2.9-5 A thin-walled aluminum ring is placed inside a thin-walled steel ring, as shown in Fig. P2.9-5, with a gap of 10^{-3} in (shown exaggerated in the figure). If a pressure p = 30 psi is applied as shown, find the value of the radial expansion of the aluminum ring.

Fig. P2.9-5

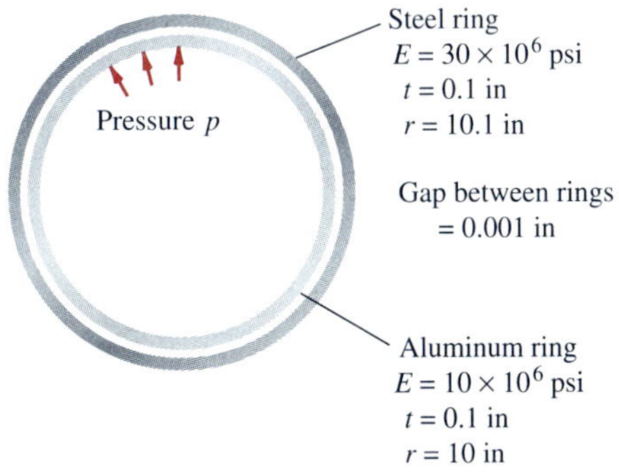

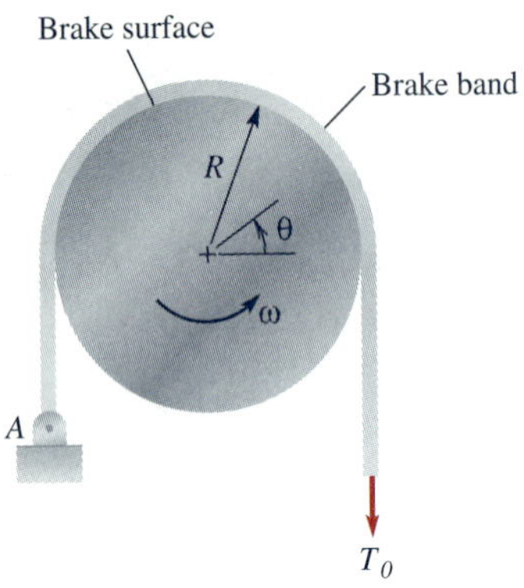

Fig. P2.9-6

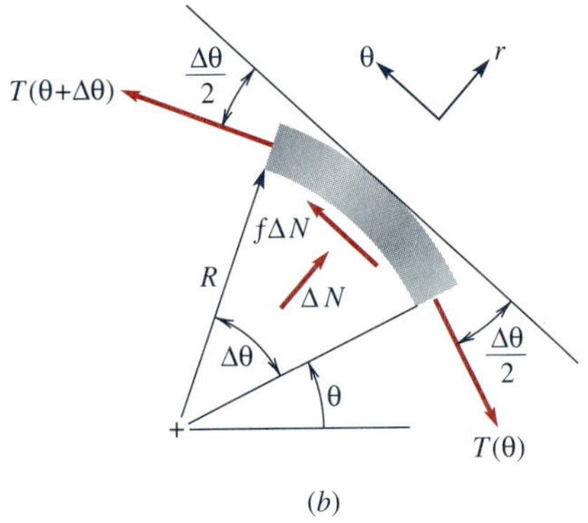

2.9-6 If a brake band is subjected to frictional forces arising from the relative motion between the brake surface and the band, as shown in Fig. P2.9-6*a*, the tension force in the band will vary with distance along the band. We wish to obtain an equation for the tension force in the band as a function of distance along the brake surface.

(*a*) We show a small element of the band with forces acting upon it in Fig. P2.9-6*b*. The force T is the tension force in the band, ΔN is the normal force between the element and the brake surface, and $f\,\Delta N$ is the frictional force acting on the element. Show that force equilibrium in the radial and circumferential directions gives (Fig. P2.9-6*b*)

$$\Delta N - T(\theta)\frac{\Delta\theta}{2} - T(\theta + \Delta\theta)\frac{\Delta\theta}{2} = 0$$
$$T(\theta + \Delta\theta) - T(\theta) + f\Delta N = 0 \qquad (a)$$

(*b*) Show that in the limit as $\Delta\theta \to 0$, we obtain

$$\frac{dT}{d\theta} = -fT \qquad (b)$$

(*c*) Show that $T = T_0 e^{-f\theta}$, where T_0 is the constant of integration. The tension force in the band varies exponentially. We note that the direction of rotation of the brake surface and the direction of θ coincide.

(*d*) Show that the tension in the band at A (Fig. P2.9-6*a*) if $f = 0.4$ is given by $T_A = T_0 e^{-0.4\pi}$ or $T_A = 0.285T_0$.

2.9-7 A rope from a small sailboat is wrapped once around a rotating capstan, as shown in Fig. P2.9-7, and pulled with a force of 200 N. What is the maximum force that can be exerted on the sailboat if the coefficient of friction between the capstan and the rope is 0.3?

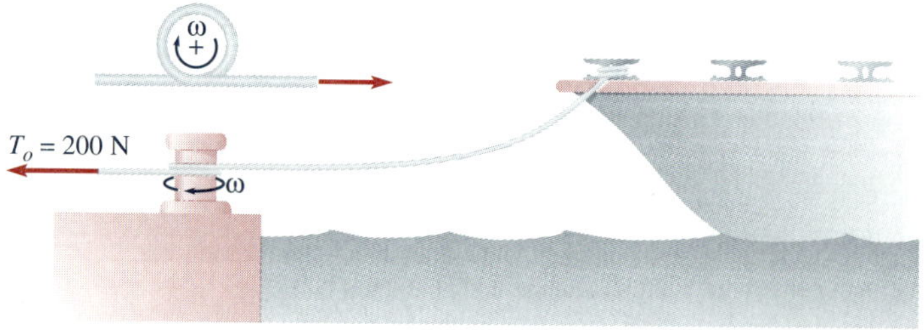

Fig. P2.9-7

2.9-8 A stiff bar ABC carries the load F at its midpoint and is supported by three steel wires as shown in Fig. P2.9-8. The stress-strain behavior of the high-strength alloy steel used in wires AD and CF and of the stainless steel used in BE is given in Fig. P2.9-8*b* and *c* where at the yield strength the stress remains constant with increasing strain (see Fig. 1.19*b*).

(*a*) Find the load F_1 which corresponds to the initiation of yielding in one of the wires.

(*b*) Find the load F_2 which is the maximum load that can be carried.

(*c*) Sketch a force-versus-displacement curve for bar *ABC*. Take $E = 30{,}000$ ksi for all wires, $L = 4$ ft, and the areas for all wires equal 0.05 in^2.

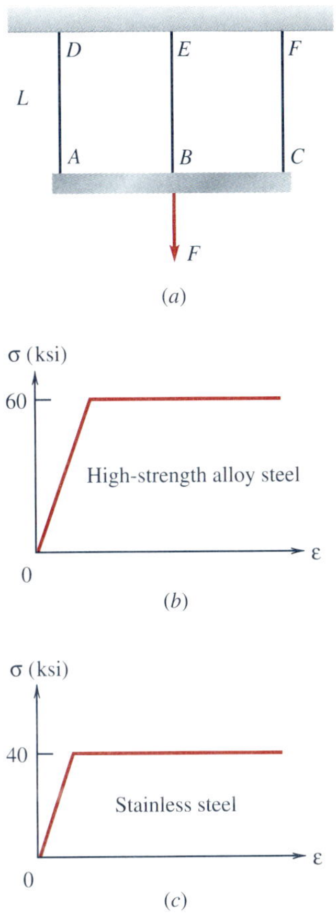

Fig. P2.9-8

2.9-9 A cylindrical tube *A* and a circular rod *B* are concentrically placed in a test fixture and loaded to failure by an axial load *P*, as shown in Fig. P2.9-9.

(*a*) Find the load P_1 which corresponds to the initiation of yielding in either the tube or the rod.

(*b*) Find the load P_2 which will cause both the tube and the rod to yield.

(*c*) Sketch a graph of the force-versus-axial displacement of the stiff test plate.

Assume that the tube is made of high-strength alloy steel and the rod of stainless steel, and make use of the stress-strain curves given in Fig. P2.9-8*b* and *c*. Take $E = 30{,}000$ ksi, $L = 10$ in, $d_o = 4$ in, $d_i = 3.8$ in for the tube, and $d = 0.75$ in for the rod.

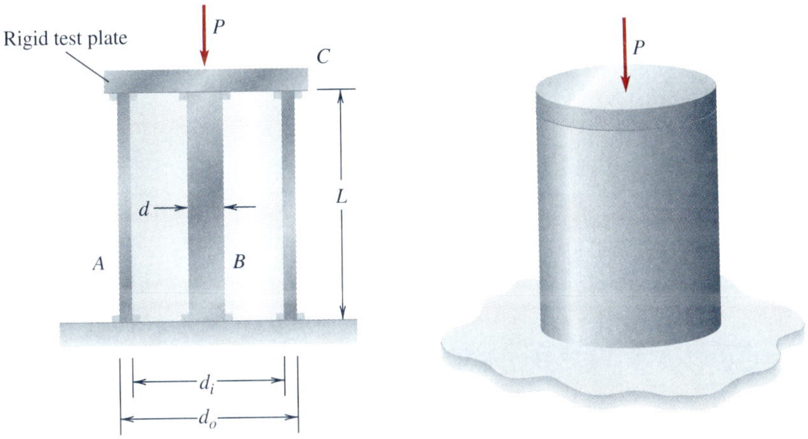

Fig. P2.9-9

2.9-10 In the case when *AE* is a constant for a bar, show that Eq. (2.28) can be integrated once to give

$$AE\frac{du}{dx} - AE\alpha\,\Delta T + \int_0^x q(\xi)\,d\xi = C_1 \qquad (a)$$

where C_1 is the constant of integration. Finally show that the expression for the displacement along the bar can be written in the form

$$u(x) = \alpha\int_0^x \Delta T(\xi)\,d\xi - \frac{1}{AE}\int_0^x\left[\int_0^y q(\xi)\,d\xi\right]dy + \frac{C_1 x}{AE} + C_2 \qquad (b)$$

or in the form

$$u(x) = \int_0^x [\alpha\,\Delta T(\xi) - (1/AE)(x-\xi)q(\xi)]\,d\xi + (C_1/AE)\,x + C_2 \qquad (c)$$

If the bar is fixed at $x = 0$, then $C_2 = 0$.

2.9-11 Consider the results from Prob. 2.9-10. If $q(x) = 0$, the bar is fixed at $x = 0$ and L, that is, $u(0) = u(L) = 0$, and $\Delta T(x) = \Delta T_0(x/L)$, use Eq. (*c*) of Prob. 2.9-10 to evaluate the expression for the displacement of the bar. Show that the result for the displacement agrees with Eq. (*e*) in Example 2.21.

2.9-12 Consider the results of Prob. 2.9-10. If $\Delta T(x) = 0$, the bar is fixed at $x = 0$ and free at $x = L$, and $q(x) = q_0$, a constant, show that the expression for the displacement becomes

$$\begin{aligned} u(x) &= -\frac{q_0}{AE}\int_0^x (x - \xi)\, d\xi + \frac{C_1 x}{AE} \\ &= -\frac{q_0 x^2}{2AE} + \frac{C_1 x}{AE} \end{aligned} \tag{a}$$

Therefore, with $du/dx = 0$ at $x = L$, $C_1 = q_0 L$, or

$$u(x) = -\frac{q_0}{AE}\left(\frac{x^2}{2} - Lx\right) = \frac{q_0 L}{AE}\left(x - \frac{x^2}{2L}\right) \tag{b}$$

Compare this result to Eq. (*i*) of Example 2.20.

Torsion of Circular Shafts

3.1 Introduction

Many technological applications involve the twisting of circular elastic shafts. A very common application is the drive shaft of an automobile in which the power from the engine is transmitted to the wheels; see Fig. 3.1*a*. Another automobile example is the torsion bar used in the front suspension of some models (Fig. 3.1*b*). We see in Fig. 3.1*b* that as the wheel moves up and down relative to the frame, the torsion bar twists and untwists, absorbing in a manner similar to a spring some of the forces transmitted from bumps in the road to the frame.

Drill shafts are another common example of circular shafts exposed to twisting. The most common cause of a broken bone in skiing accidents—as might be expected—arises from twisting the tibia bone in the lower leg beyond a critical angle of twist. Many other examples can be considered to convince us that torsion or twisting of slender circular shafts arises in a number of practical situations.

In this chapter we develop the basic equations that determine the angle of twist and stress distribution in a circular shaft under the action of a twisting moment or a torque.

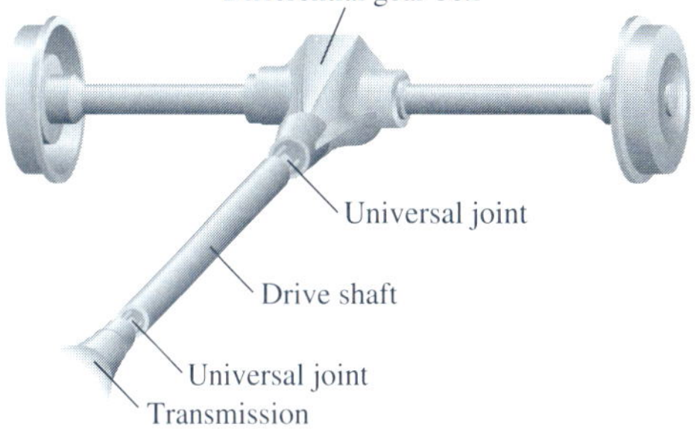

(*a*)

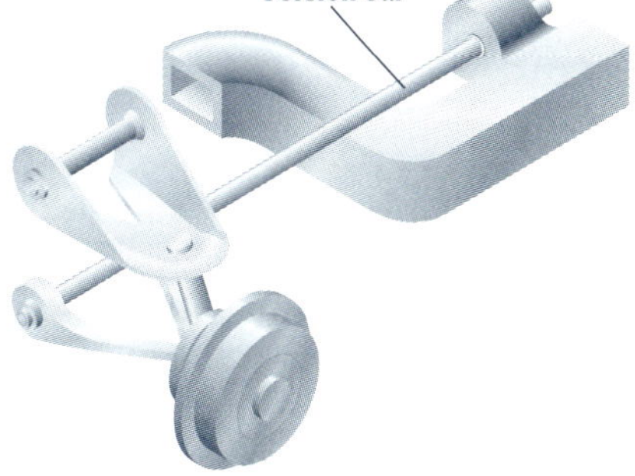

(*b*)

Figure 3.1 (*a*) Example of torsion of a circular drive shaft in an automotive drive train. (*b*) Torsion bar in an automotive front-end suspension.

3.2 Geometry of Deformation

We start first with a study of the deformation that takes place in a circular shaft when a twisting moment is applied. Consider a long, slender circular shaft of length L and radius a fixed at one end as shown in Fig. 3.2*a*. We assume that the shaft is slender enough that a/L is much less than 1, say, less than $\frac{1}{5}$. It is necessary that the shaft be long enough so that

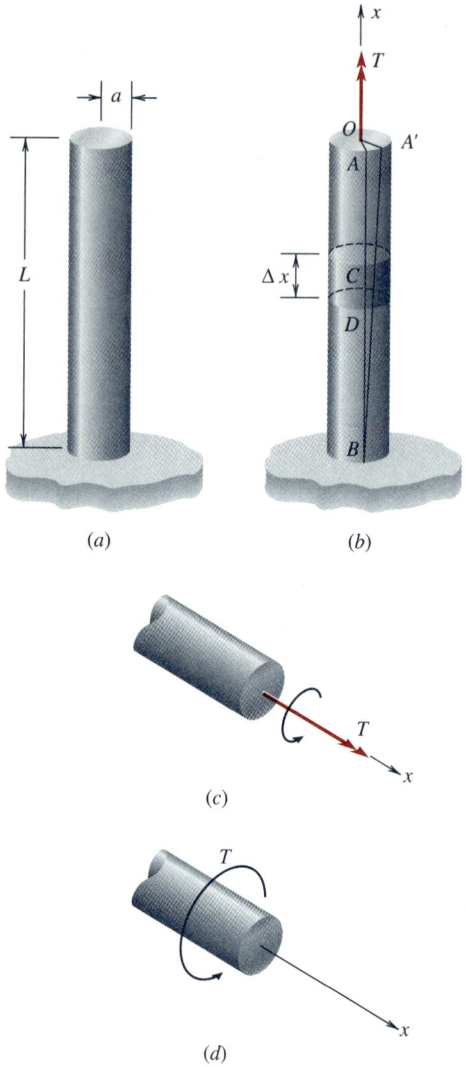

Figure 3.2 (*a*) A circular shaft of radius a and length L built-in at one end. (*b*) A circular shaft twisted by a twisting moment T. (*c*) and (*d*) Positive twisting moment T acting on a positive x face.

end effects arising from the nature of the support or from the details of how the loading is applied are not significant. To investigate the nature of the deformation in the shaft, we will appeal to a number of symmetry arguments, invoking the circular symmetry of the shaft.

First we introduce an x axis along the axis of the shaft, as shown in Fig. 3.2*b*. The origin of the x axis will be set at the fixed end at section B. As was also the case for the axial deformation of a bar considered in Chap. 2, we will introduce the notion of a positive cross-sectional face for the shaft as one for which the outward-pointing normal is in the positive x direction. Twisting moments or torques will be denoted by double-headed vectors with their length indicating magnitude and their direction indicating the line of action and sense of the moment according to the right-hand rule (Fig. 3.2*c*). Also at times we will use a curved arrow at the section where the twisting moment is applied to the shaft to show the twisting moment (Fig. 3.2*d*). A positive twisting moment T occurs on a positive face when the direction of T is in the direction of the positive x axis. A negative twisting moment T occurs on a positive face when the direction of T is in the direction of the negative x axis. The twisting moments in Fig. 3.2*b, c,* and *d* are positive twisting moments. We now turn to the deformation of the shaft when a twisting moment T is applied to the top end.

Consider, as shown in Fig. 3.2*b,* a line AB drawn on the outside surface of the cylinder and the radial line OA on the top section before the twisting moment T is applied. Once the twisting moment T is applied, radius OA will rotate together with line AB. We will exaggerate all angles of rotation or twist for clarity; in engineering materials, the actual angles of rotation will be very small. We will further investigate the nature of the deformation by considering a cylindrical segment of length Δx cut out from the deformed shaft (Fig. 3.2*b*). The portion of the original line AB on the outside surface of this segment is CD.

We show this segment in Fig. 3.3*a*. We note from moment equilibrium about the axis that the internal twisting moment acting at any section of the shaft is equal to the applied twisting moment T at the end. We show the positive twisting moment T on each end of the segment of length Δx in Fig. 3.3*a*. It follows from circular symmetry that the deformation of any segment of length Δx cut from the shaft will be similar to the deformation of any other segment of length Δx, because the twisting moment acting on each end of any segment will be the same. From symmetry arguments we conclude that the deformation in any plane circular section perpendicular to the axis will be such that the plane will remain plane and perpendicular to the axis. That is, no deformation out of the plane will occur under the action of the twisting moment.

The deformation must consist of a rotation of radial lines about the axis of the shaft. In particular, the plane $OCDO$ shown in Fig. 3.3*a* will deform such that radial line OD at position x rotates to OD' through an angle $\phi(x)$, where $\phi(x)$ is called the *angle of twist* at section x, and radial line OC at position $x + \Delta x$ rotates to OC' through an angle $\phi(x + \Delta x)$. Line $C''D'$ is constructed parallel to line CD. We started by considering line CD on the outside surface of the shaft; however, the same type of deformation is expected again from symmetry arguments for any cylindrical plug of radius r, as shown in Fig. 3.3*b*. We can therefore associate the distances OC and OC' with an arbitrary radius r.

The radius OD at a section x along the axis, i.e., the section at a distance x from the origin, rotates through the angle $\phi(x)$; the radius OC at a section $x + \Delta x$ along the axis rotates through the angle $\phi(x + \Delta x)$. Angle COC'' equals angle DOD', which is angle $\phi(x)$ by construction. Angle $C''OC'$ is equal to the angle $\Delta\phi = \phi(x + \Delta x) - \phi(x)$. Therefore, the *change* in the angle of rotation of radial lines OD and OC in two sections separated by a distance Δx is equal to $\Delta\phi$. This change in angle causes a shear strain or change in right angle given by the angle γ between line elements $D'C''$ and $D'C'$, as shown in Fig. 3.3*a*.

The value of the angle γ is given by

$$\tan\gamma \approx \gamma = \frac{C''C'}{\Delta x} \tag{3.1}$$

where for a small angle in radians the tangent of γ is approximately equal to γ. Further, the *distance* $C''C'$ is equal to $r\,\Delta\phi$, so that the shear strain can be written in the form

$$\gamma = r\,\frac{\Delta\phi}{\Delta x} \tag{3.2}$$

If we shrink the segment so that in the limit $\Delta x \to 0$, the shear strain becomes

$$\gamma = r\,\frac{d\phi}{dx} = r\theta \tag{3.3}$$

where θ is called the *rate of change of the angle of rotation,* or *twist,* along the x axis. The rate of change θ is a measure of how much the shaft twists in radians as we move along the shaft; the units of θ are, e.g., radians/meter. Since all segments of length Δx along the shaft behave in the same way, the quantity θ is constant along the shaft. We can think of θ as a measure of the *relative* angle of twist per unit length between any two sections along the shaft.

Since γ is the shear strain, we need to relate it to the shear stress in the material. It is important to emphasize that thus far in our derivation we used only arguments for the geometry of deformation and arguments invoking circular symmetry to deduce the nature of the shear strain distribution in the shaft. At this point we have not used any information concerning the response of the material making up the shaft; we do this next.

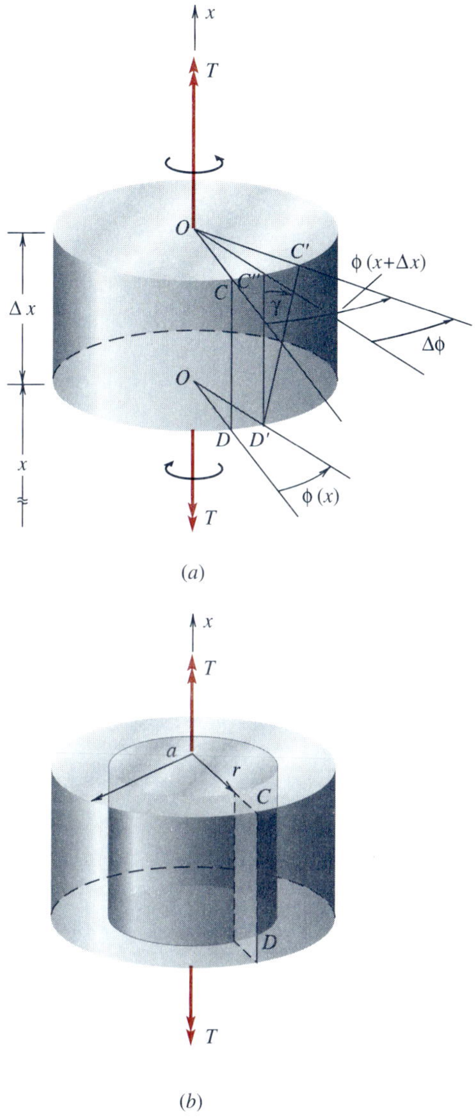

Figure 3.3 (*a*) Geometry of deformation of an element of length Δx. (*b*) An element of length Δx and of radius r.

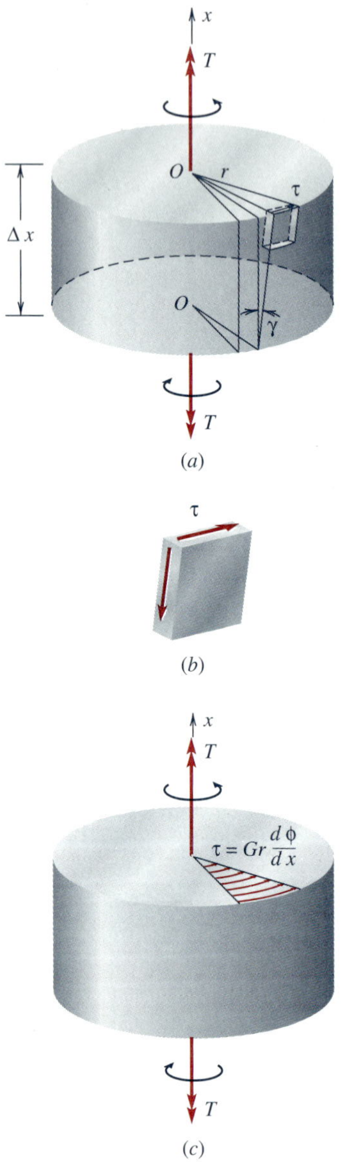

Figure 3.4 (*a*) Geometry of deformation showing the shear strain γ and shear stress τ at radius r. (*b*) An element showing the shear stresses τ acting on perpendicular faces. (*c*) Linear distribution of shear stress τ over the section.

3.3 Stress Distribution and Equilibrium Requirements

We assume that the material of the shaft is linearly elastic, so that the shear stress is related to the shear strain, Eq. (1.6), by

$$\tau = G\gamma \tag{3.4}$$

where G is the shear modulus of the material. A twisted segment of the shaft is shown in Fig. 3.4*a* with the shear stress τ acting on the top surface of the segment on a small curvilinear rectangular element.

Since the rectangular element is experiencing shear stress on the top and bottom surfaces, equal and opposite shear stresses are acting on the perpendicular sides, as shown in Fig. 3.4*b*. We often say that a circular shaft with an applied twisting moment is undergoing a state of *pure shear*. We discuss this state of stress again in Chap. 8, where we investigate the general states of stress in a deformed body. The shear strain is given by Eq. (3.3), so that the value of the shear stress at a distance r from the axis is given by

$$\tau = Gr\frac{d\phi}{dx} \tag{3.5}$$

We see from Eq. (3.5) that the shear stress acting on the circular cross section is linear in the radius r. We show this shear stress distribution on the section in Fig. 3.4*c*.

This shear stress distribution gives rise to a resultant twisting moment which is statically equivalent to the twisting moment acting on the section. We see from Fig. 3.5 that

$$\int_A r\tau\, dA = T \tag{3.6}$$

where $\tau\, dA$ is the *force* acting on an element of area dA which when multiplied by r, produces the moment of this force about the x axis. To obtain the resultant moment, we integrate the moment contributions over the total area A. Upon substitution of Eq. (3.5) into Eq. (3.6), we find

$$G\frac{d\phi}{dx}\int_A r^2\, dA = T \tag{3.7}$$

The integration is over the total area A, and the quantity

$$J = \int_A r^2\, dA \tag{3.8}$$

is called the *polar moment of inertia* of the cross section about the axis.

The integration can be carried out for the present case of a circular shaft to obtain (Fig. 3.5)

$$J = \int_A r^2\,dA = \int_0^a \int_0^{2\pi} r^3\,dr\,d\theta = \frac{\pi a^4}{2} = \frac{\pi d^4}{32} \qquad (3.9)$$

where $d = 2a$ is the diameter of the shaft and a is the radius of the shaft.

Therefore from Eq. (3.7), the rate of change of the angle of twist along the shaft can be expressed in terms of the twisting moment T in the shaft in the form

$$\frac{d\phi}{dx} = \frac{T}{GJ} \qquad (3.10)$$

Finally, to obtain the angle of twist at the end of the shaft where the twisting moment T is applied (Fig. 3.6), we need to integrate Eq. (3.10) with respect to x, the distance along the axis of the shaft. If we rewrite Eq. (3.10), we have

$$d\phi = \frac{T}{GJ}\,dx \qquad (3.11)$$

where now the right and left sides of Eq. (3.11) can be integrated between the fixed end at $x = 0$ where $\phi(0) = 0$ to the end at $x = L$ where $\phi = \phi(L)$. Since $T/(GJ)$ is constant along the shaft, integration gives

$$\int_{x=0}^{x=L} d\phi = \phi(L) - \phi(0) = \int_{x=0}^{x=L} \frac{T}{GJ}\,dx = \frac{TL}{GJ}$$

or

$$\phi(L) = \frac{TL}{GJ} \qquad (3.12)$$

Equation (3.12) expresses the angle of twist at the end of the shaft relative to the fixed end in terms of the applied twisting moment T, the material property G, and the geometric cross-sectional property J (Fig. 3.6). The units of ϕ are radians; care should be taken with the terms on the right side of Eq. (3.12) so that the units cancel.

The shear stress distribution on a section can be expressed now in terms of the applied twisting moment T from Eqs. (3.5) and (3.10):

$$\tau = \frac{Tr}{J} \qquad (3.13)$$

The shear stress distribution is linear in r, and the maximum value of the shear stress occurs at the outside radius a

$$\tau_{\max} = \frac{Ta}{J} \qquad (3.14)$$

where, from Eq. (3.9), $J = \pi a^4/2$.

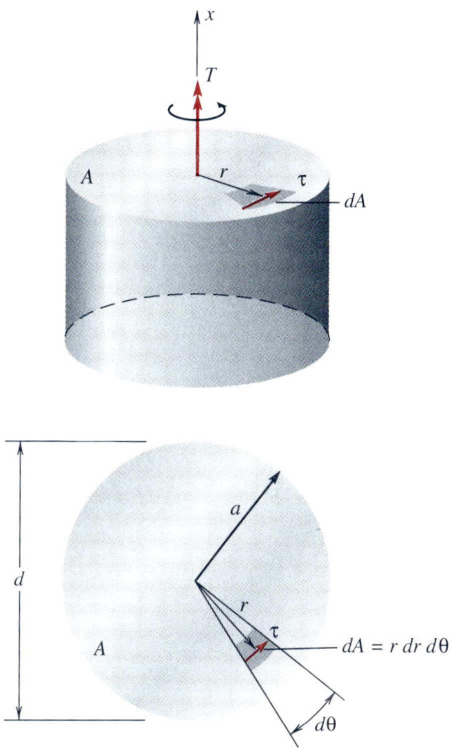

Figure 3.5 Shear stress τ acting on an element of area dA at radius r.

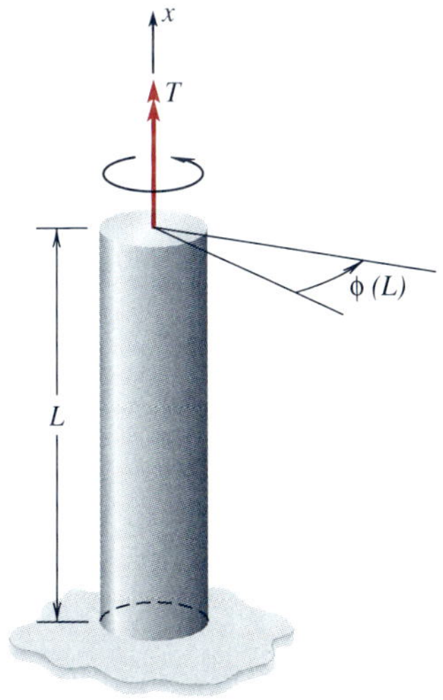

Figure 3.6 Angle of twist $\phi(L)$ of a circular shaft under a twisting moment T.

3.4 Equations for Torsion of Circular Shafts

Equations for the rate of change of the angle of twist along the shaft, Eq. (3.10); the relative angle of twist at the end loaded by T, Fig. 3.6 and Eq. (3.12); the shear stress distribution, Eq. (3.13); and the maximum shear stress, Eq. (3.14), are the equations we will use to analyze the twisting of circular shafts. In solving problems it turns out to be important

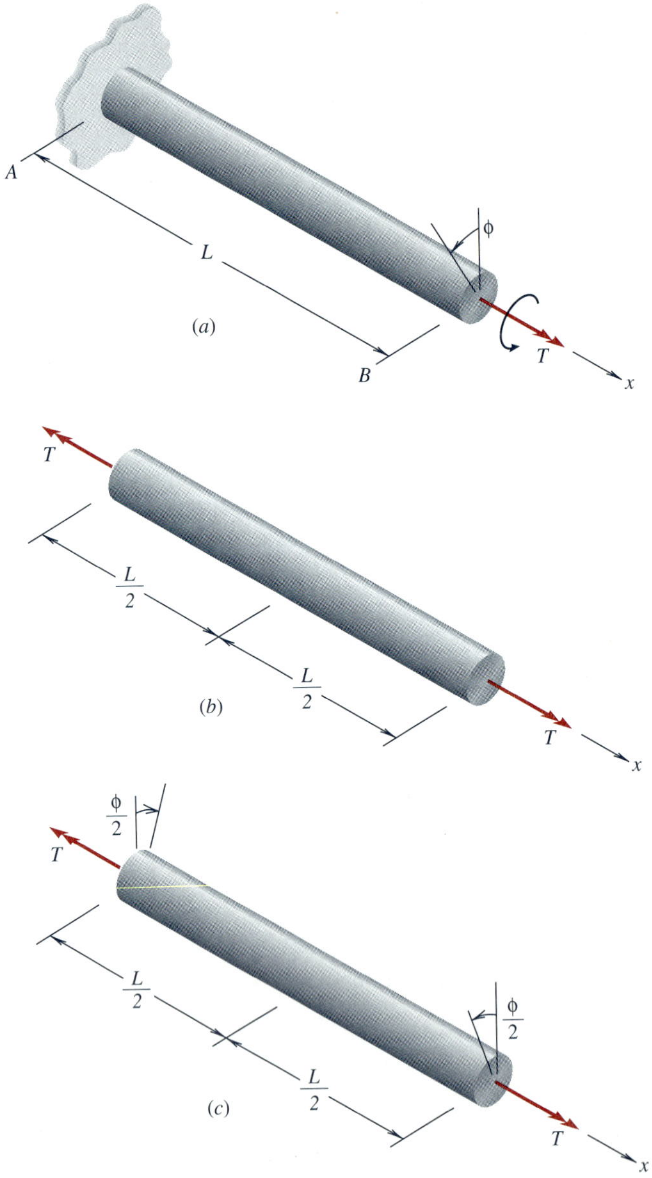

Figure 3.7 (*a*) Angle of twist ϕ of a circular shaft built-in at A under a twisting moment T. (*b*) Free-body diagram of shaft in (*a*). (*c*) Angle of twist $\phi/2$ at each end of a free shaft under twisting moment T. The relative angle of rotation between the ends is ϕ.

to make clear the difference between the absolute angle of twist at a section and the relative angle of the twist between two sections (an *angular deformation* analogous to the axial deformation in uniaxial bars, discussed in Chap. 2).

Figure 3.7*a* shows a shaft of length L fixed at one end A and twisted by a twisting moment T at the other end B. The angle of twist at end B relative to end A is given by Eq. (3.12) and is shown as positive in Fig. 3.7*a* according to the right-hand rule. A free-body diagram of the shaft of Fig. 3.7*a* shown in Fig. 3.7*b* shows that the twisting moment at the fixed end A is equal to T to satisfy equilibrium. If we were to consider now the shaft shown in Fig. 3.7*b*, which has the same dimensions as the shaft in Fig. 3.7*a*, with applied twisting moments T at each end as a *free* shaft, we see that the twisting moments at each end are positive. The angle of twist at each end in the positive direction by the right-hand rule along T equals $\phi/2$, as shown in Fig. 3.7*c*, because by symmetry the middle section of the shaft does not twist. The *relative* angle of twist is equal to ϕ. We see from this illustration that it is the *relative* angle of twist between the ends of the shaft that is determined by T in the shaft.

Recall that a similar situation arises for the case of uniaxial deformation of a bar in that the *deformation* δ of a bar, Eq. (2.6), is determined by the force in the bar. The deformation in Chap. 2 was expressed in terms of the relative displacement of the ends of the bar.

To keep track of the angle of rotation at each section along a shaft, we can adopt the convention that the angle of rotation at a section is positive according to the right-hand rule, as shown in Fig. 3.8. It is convenient at times to associate the rotation angle with a vector, shown with a double-headed arrow, and to use the right-hand rule to identify the direction of rotation, as in Fig. 3.8. All vector quantities are positive in the positive direction of the axis.

For example, we assume that the shaft AB shown in Fig. 3.9 is part of a larger system of interconnected shafts and that the twisting moment T in shaft AB is constant. The locations of the ends of the shaft along

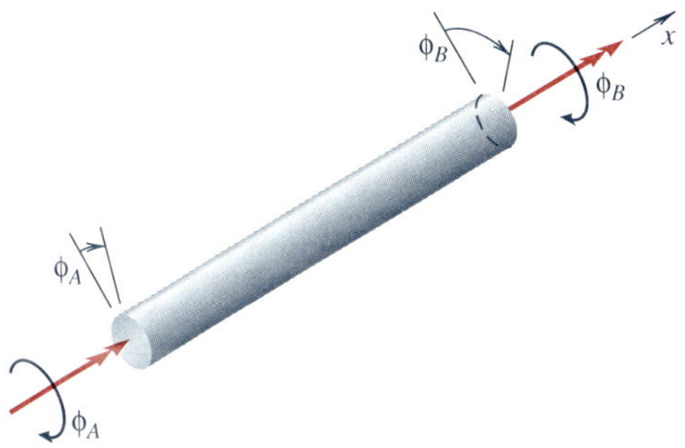

Figure 3.8 Positive angles of twist at the ends of a shaft.

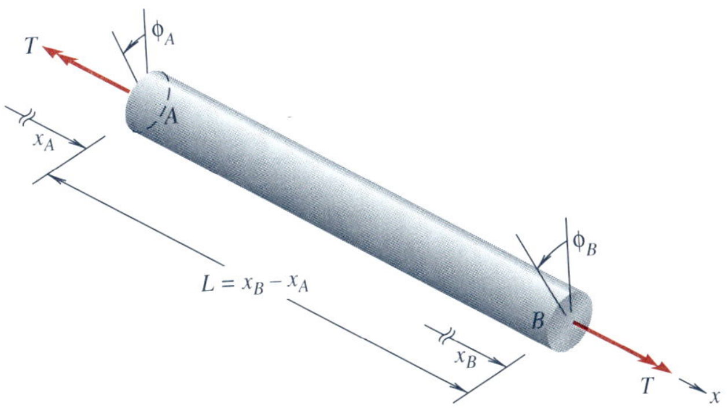

Figure 3.9 A shaft of length L under a twisting moment T.

the axis are given by x_B and x_A, and the length of the shaft is $L = x_B - x_A$. To determine the relation between the angles of twist at sections A and B, we integrate Eq. (3.10) between $x = x_A$, where $\phi = \phi_A$, and $x = x_B$, where $\phi = \phi_B$, to get

$$\int_{\phi_A}^{\phi_B} d\phi = \frac{T}{GJ}\int_{x_A}^{x_B} dx$$

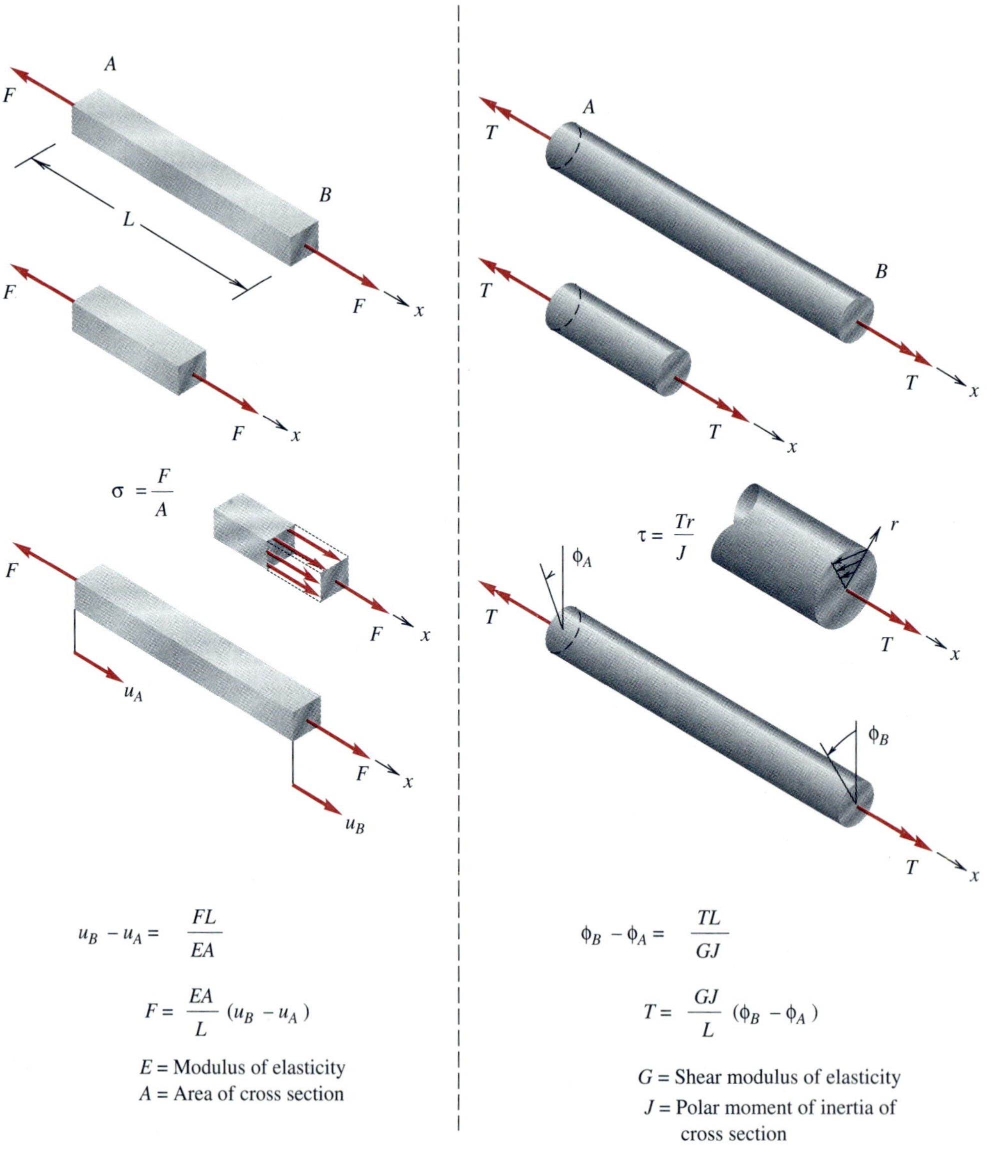

Figure 3.10 Analogy between uniaxial loading of a bar and torsional loading of a circular shaft.

since $T/(GJ)$ is constant along the shaft. Therefore

$$\phi_B - \phi_A = \frac{TL}{GJ} \tag{3.15}$$

where $L = x_B - x_A$ [again recall Eq. (2.6)].

Finally we can rewrite Eq. (3.15) as a twisting moment–twisting angle relation in the form

$$T = \frac{GJ}{L}(\phi_B - \phi_A) \tag{3.16}$$

where ϕ_B is the angle of twist at section B, measured positive according to the right-hand rule *along the axis,* and ϕ_A is the angle of twist at section A, again measured positive according to the right-hand rule *along the axis*; see Fig. 3.9. Equation (3.16) is the generalized force-deformation relation for a circular shaft of length L.

In view of the sign conventions we have adopted for the twisting moment and the angle of rotation or twist at a section, we see that a parallel exists between the equations for uniaxial deformation discussed in Chap. 2 and the equations for twisting of a circular shaft. In Fig. 3.10 we summarize the equations for each case. To carry out the solution of a problem for torsion of a circular shaft, we proceed as we did in Chap. 2, using the three steps of Eq. (2.8). The force-deformation relation for a circular shaft with a twisting moment is given by Eq. (3.16).

3.5 Torsion of Hollow Circular Shafts

The steps in the derivation of the equation relating the angle of twist to the applied twisting moment, Eq. (3.12), carry over immediately to the case of a hollow circular cylindrical shaft. In particular, the integration in Eqs. (3.7) and (3.8) leading to the polar moment of inertia for a hollow shaft should be carried out over that part of the cross-sectional area where the shear stress τ is acting

$$J = \int_A r^2 \, dA \tag{3.8}$$

where the integration is to be carried out over the annular region between the inside radius r_i and the outside radius r_o (Fig. 3.11). This integration gives

$$J = \int_A r^2 \, dA = \int_0^{2\pi} \int_{r_i}^{r_o} r^3 \, dr \, d\theta = \frac{\pi}{2}(r_o^4 - r_i^4) \tag{3.17}$$

or

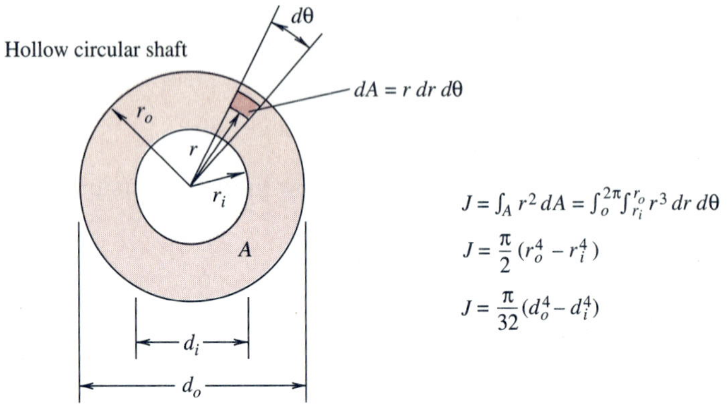

Figure 3.11 Integration over an annular region for calculation of the polar moment of inertia J.

$$J = \frac{\pi}{32}(d_o^4 - d_i^4) \tag{3.18}$$

For the case when $d_i = 0$, a solid shaft, Eq. (3.18) reduces the result given by Eq. (3.9). If the hollow circular shaft is such that its wall thickness t is small, i.e.,

$$r_o = r_i + t \qquad \frac{t}{r_i} \ll 1 \tag{3.19}$$

then the expression for J can be approximated as follows:

$$J = \int_A r^2\,dA = r_i^2\,(2\pi r_i t) = 2\pi r_i^3 t \tag{3.20}$$

Equation (3.20) is useful for analysis of torsion of thin-walled tubes; however, when torsion of thin-walled tubes is considered, the thickness t of the tube must be great enough to prevent torsional buckling of the tube.

Equations (3.10), (3.12), and (3.16) apply to hollow shafts where J is now given by Eq. (3.18). The maximum shear stress in a hollow cylindrical shaft occurs on the outside surface of the shaft and is given by Eq. (3.14)

$$\tau_{\max} = \frac{T(d_o/2)}{J} \tag{3.21}$$

where J is given by Eq. (3.18).

As an example of the application of the relations we have derived, we note that if a solid shaft and a hollow shaft of the same outside diameter are transmitting the same twisting moment T, then the ratio of the maximum shear stress from Eqs. (3.14) and (3.21) is given by

$$\frac{(\tau_{\max})_{\text{hollow}}}{(\tau_{\max})_{\text{solid}}} = \frac{J_{\text{solid}}}{J_{\text{hollow}}} = \frac{1}{1 - (d_i/d_o)^4} \tag{3.22}$$

If for a particular case the inside diameter is half of the outside diameter, that is, $d_i/d_o = 0.5$, the ratio of the maximum stresses from Eq. (3.22) is

$$\frac{(\tau_{max})_{hollow}}{(\tau_{max})_{solid}} = \frac{16}{15} = 1.067 \tag{3.23}$$

We see from Eq. (3.23) that the maximum stress has increased by 6.7 percent. However, the ratio of the weight of the hollow shaft to that of the solid shaft which is determined by the ratio of the cross-sectional areas is given by

$$\frac{(\text{Weight})_{hollow}}{(\text{Weight})_{solid}} = 1 - \left(\frac{d_i}{d_o}\right)^2 = 0.75 \tag{3.24}$$

or there is a weight decrease of 25 percent. In those cases in which weight is important, hollow shafts are often more efficient.

3.6 Torsion of Statically Determinate Systems

As in the case of the uniaxial deformation of bars studied in Chap. 2, it is possible in many torsion problems involving circular shafts to determine the twisting moments or torques without consideration of the angular deformations. In such statically determinate problems, the three steps of Eq. (2.8) are used with the step involving statics giving us the required twisting moments or torques. We treat four statically determinate problems in the following examples.

EXAMPLE 3.1

A steel shaft AB of length 1.5 m has a twisting moment of $T = 1100$ N · m applied to end B, as shown in Fig. 3.12*a*. If the diameter d of the shaft is 50 mm, we wish to find the maximum shear stress developed in the shaft and the angle of twist at section B. The end at A is built into the wall, and $G = 80$ GPa; we neglect the weight of the shaft.

The approach to the solution of torsion problems, as we have noted already, is very similar to the approach used for the solution of uniaxial deformation problems of Chap. 2. We will use the three steps of Eq. (2.8) to organize the solution. Of course, the first step is the location of a coordinate x axis along the axis of the shaft with the origin at A.

Statics: A free-body diagram of the shaft is shown in Fig. 3.12*b* with the applied twisting moment T; the twisting moment reaction at A follows immediately from moment equilibrium about the x axis and is equal to T. A cut at any value of x along the shaft (Fig. 3.12*b*) shows that the internal twisting moment is constant along the shaft and equal to T.

Force-Deformation Relations and Geometry: Since the shaft is fixed at section A, $\phi_A = 0$, the angle of twist at section B follows immediately from Eq. (3.16)

$$\phi_B = \frac{TL}{GJ} \tag{a}$$

where ϕ_B is the angle of twist at B. The direction of the angle of twist is positive; in this case ϕ_B is in the same direction as T.

Upon substitution of numerical values, we find

$$J = \frac{\pi d^4}{32} = \frac{\pi(50 \times 10^{-3})^4}{32} = 6.136 \times 10^{-7}\ \text{m}^4 \tag{b}$$

and

$$\phi_B = \frac{(1100\ \text{N}\cdot\text{m})(1.5\ \text{m})}{(80\times 10^9\ \text{N/m}^2)(6.136\times 10^{-7}\ \text{m}^4)} = 3.36\times 10^{-2}\ \text{rad}$$

or

$$\phi_B = 1.93° \qquad (c)$$

The maximum shear stress in the shaft occurs on the outside surface where the radius $a = d/2$ so that, Eq. (3.14)

$$\tau_{max} = \left(\frac{d}{2}\right)\left(\frac{T}{J}\right) = \frac{(25\times 10^{-3}\ \text{m})(1100\ \text{N}\cdot\text{m})}{6.136\times 10^{-7}\ \text{m}^4} = 44.8\ \text{MPa} \qquad (d)$$

This value of τ_{max} is reasonable since for most steels the maximum shear stress to cause yielding of the material is greater than 100 MPa.

Figure 3.12*c* shows the internal twisting moment along the shaft and the twisting angle from $\phi = 0$ at the fixed end, $x = 0$, to $\phi = \phi_B$ at the loaded end, $x = L$.

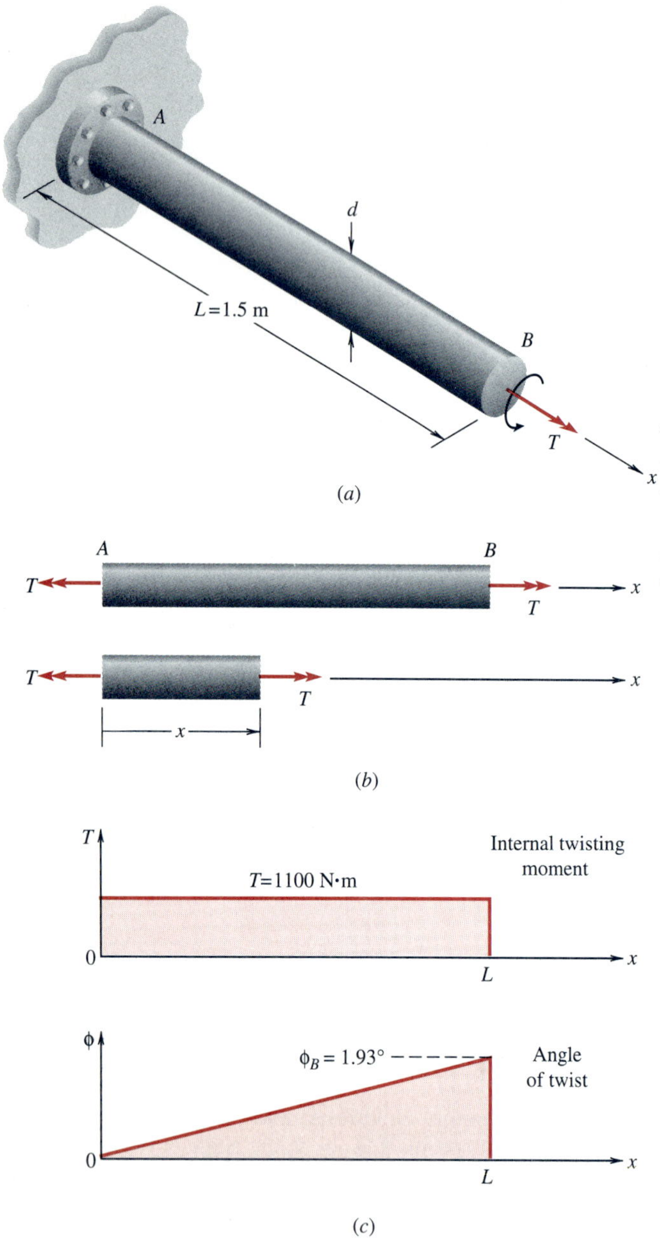

Figure 3.12 Example 3.1

EXAMPLE 3.2

An aluminum tube AB is attached at B to a brass tube BC and is fixed to a rigid wall at section A, as shown in Fig. 3.13*a*. The inside and outside diameters for AB are 1 and 2 in, and the inside and outside diameters for BC are 0.5 and 1 in, respectively. Twisting moments $T_B = 4$ kip · in and $T_C = 2.5$ kip · in are applied at sections B and C, as shown in Fig. 3.13*a*.

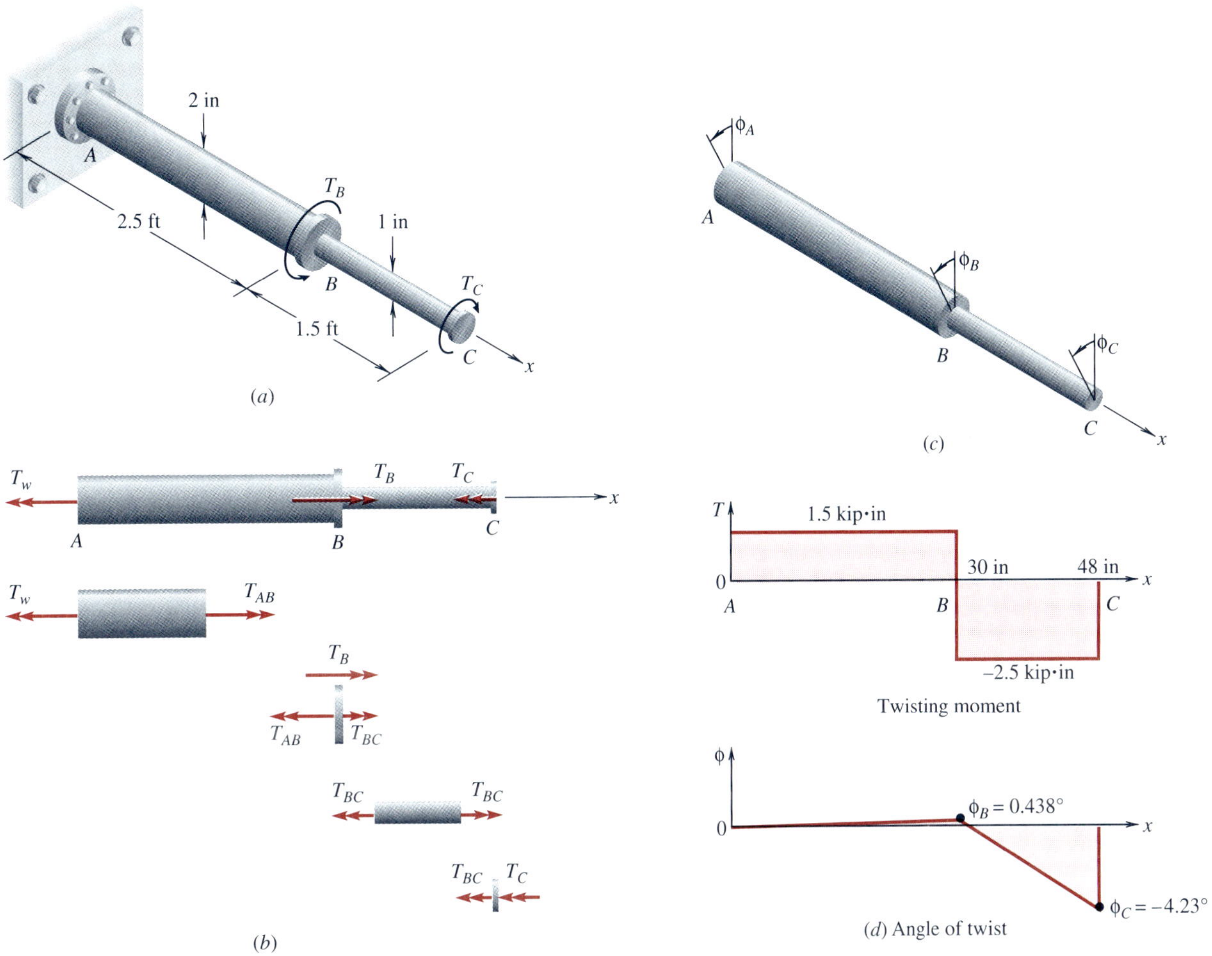

Figure 3.13 Example 3.2

We wish to find the maximum shear stress in each tube, the variation of the angle of twist along tubes AB and BC, and, in particular, the angle of twist at sections B and C. We will take $G_{AB} = 4 \times 10^6$ psi and $G_{BC} = 6 \times 10^6$ psi. Note that in Fig. 3.13*a*, as is the case with many drawings of torsion problems, the scale of the diameter of the shafts is different from the axial length scale. We neglect the weight of the shafts. First we locate the x axis along the axis of the two tubes with the origin at section A, as shown in Fig. 3.13*a*.

Statics: In Fig. 3.13*b* we show a free-body diagram of the system with T_w as the wall twisting-moment reaction. All twisting moments act along the axis, although for clarity we will sometimes show the applied twisting moment off the axis. The internal twisting moments in the segments are T_{AB} and T_{BC} shown as positive on each segment.

From equilibrium of the rigid elements at sections B and C where the twisting moments are applied and of segment AB we have

$$T_{BC} = -T_C = -2.5 \text{ kip} \cdot \text{in}$$

$$T_{AB} = T_{BC} + T_B = -2.5 + 4 = 1.5 \text{ kip} \cdot \text{in} \qquad (a)$$

$$T_w = T_{AB} = 1.5 \text{ kip} \cdot \text{in}$$

We can use overall moment equilibrium to check the above results; i.e., from Fig. 3.13*b*, we have

$$T_w = -T_C + T_B = -2.5 + 4 = 1.5\ \text{kip}\cdot\text{in} \qquad (b)$$

which checks the result in Eqs. (*a*).

Force-Deformation Relations and Geometry: If we associate positive angles of twist with sections *A*, *B*, and *C*, as shown in Fig. 3.13*c*, the force-deformation relation of each segment takes the form, from Eq. (3.16),

$$T_{AB} = \left(\frac{GJ}{L}\right)_{AB}(\phi_B - \phi_A)$$
$$T_{BC} = \left(\frac{GJ}{L}\right)_{BC}(\phi_C - \phi_B) \qquad (c)$$

The polar moment of inertia of each tube is given by [Eq. (3.18)]

$$(J)_{AB} = \frac{\pi}{32}(2^4 - 1^4) = 1.47\ \text{in}^4$$
$$(J)_{BC} = \frac{\pi}{32}(1^4 - 0.5^4) = 9.20 \times 10^{-2}\ \text{in}^4 \qquad (d)$$

The units in the formula for ϕ must be consistent; ϕ is in radians.

Further, since section *A* is fixed at the wall, $\phi_A = 0$; therefore, from Eqs. (*a*), (*c*), and (*d*) we have

$$\phi_B = \left(\frac{TL}{GJ}\right)_{AB} = \frac{(1.5 \times 10^3\ \text{lb}\cdot\text{in})(30\ \text{in})}{(4 \times 10^6\ \text{lb/in}^2)(1.47\ \text{in}^4)}$$
$$= 7.65 \times 10^{-3}\ \text{rad} \qquad (e)$$
$$= 0.438°$$

and

$$\phi_C = \left(\frac{TL}{GJ}\right)_{BC} + \phi_B$$
$$= \frac{(-2.5 \times 10^3\ \text{lb}\cdot\text{in})(18\ \text{in})}{(6 \times 10^6\ \text{lb/in}^2)(9.20 \times 10^{-2}\ \text{in}^4)} + \phi_B \qquad (f)$$
$$= -8.15 \times 10^{-2}\ \text{rad} + 7.65 \times 10^{-3}\ \text{rad}$$
$$= -7.385 \times 10^{-2}\ \text{rad} = -4.23°$$

The negative sign for ϕ_C indicates that the angle is opposite to that shown in Fig. 3.13*c*.

A plot of the internal twisting moment versus distance along the tubes and a plot of the angle of twist along the tubes are shown in Fig. 3.13*d*. The twisting moment in each segment is constant, and at section *B* where the applied twisting moment $T_B = 4$ kip · in occurs, there is a jump of magnitude 4 kip · in in the graph. The variation in the angle of twist along each segment is linear, as shown in Fig. 3.13*d*. Note that the slope $d\phi/dx$ of the linear segments in Fig. 3.13*d* for ϕ versus x undergoes a jump from a positive slope in the *AB* portion of the shaft to a negative slope in *BC*. Referring to Eq. (3.10), we see that $d\phi/dx$ is proportional to *T*. The discontinuity and change in sign of $d\phi/dx$ can be anticipated from the graph of *T* in Fig. 3.13*d*. Finally, we calculate the maximum magnitude of the shear stress in each tube. From Eq. (3.14) we have for *AB*

$$\tau_{max} = \frac{(1\ \text{in})(1.5 \times 10^3\ \text{lb}\cdot\text{in})}{1.47\ \text{in}^4} = 1.02\ \text{ksi} \qquad (g)$$

and for *BC*

$$\tau_{max} = \frac{(0.5\ \text{in})(2.5 \times 10^3\ \text{lb}\cdot\text{in})}{9.20 \times 10^{-2}\ \text{in}^4} = 13.6\ \text{ksi} \qquad (h)$$

EXAMPLE 3.3

A small motor drives two gear systems as shown in Fig. 3.14*a*. If the gear system at *B* "takes off" a twisting moment of $T_B = 10$ N · m from the main shaft and the system at *C* takes off a twisting moment of $T_C = 10$ N · m, we want to find the maximum shear stress in the shaft due to twisting and the angle of twist between the ends of the shaft. The motor is delivering a twisting moment of 20 N · m. We take $G = 80$ GPa and neglect friction in the bearings and weight of the components.

We approach the solution using the three steps. First we show a simple model of the problem in Fig. 3.14*b*. The *x* axis is along the shaft, and the origin is at *A*. The angle of twist at section *A* is taken as equal to zero.

Statics: The free-body diagram of each segment of the shaft is shown in Fig. 3.14*c*. The twisting moment in each segment follows immediately from equilibrium

$$T_{AB} = -T_A = -20\ \text{N}\cdot\text{m}$$
$$T_{BC} = -T_C = -10\ \text{N}\cdot\text{m} \qquad (a)$$

where the minus signs indicate that the directions of the twisting moments acting on the elements are opposite to those shown in Fig. 3.14*c*. As we have discussed, it is often convenient to initially assume twisting moments as positive in each segment.

Force-Deformation Relations: For each segment we have the force-deformation relation

$$T_{AB} = \frac{GJ}{L_{AB}}(\phi_B - \phi_A) \qquad T_{BC} = \frac{GJ}{L_{BC}}(\phi_C - \phi_B) \qquad (b)$$

where the angles of twist at each section are taken as positive according to the right-hand rule along the axis (Fig. 3.14*b*).

Geometry: The only geometric constraint on the system is

$$\phi_A = 0 \qquad (c)$$

It follows therefore from Eqs. (*b*) that

$$\phi_B = \frac{T_{AB}L_{AB}}{GJ} = \frac{(-20\ \text{N}\cdot\text{m})(0.5\ \text{m})}{(80 \times 10^9\ \text{N/m}^2)(9.82 \times 10^{-10}\ \text{m}^4)} = -0.127\ \text{rad} = -7.28° \qquad (d)$$

In a similar fashion we find

$$\phi_C = \phi_B + \frac{T_{BC}L_{BC}}{GJ} = -14.6° \qquad (e)$$

The maximum magnitude of the shear stress occurs at the outside of the shaft in segment AB since $|T_{AB}| > |T_{BC}|$; from Eq. (3.14) we find

$$\tau_{max} = \frac{aT_{AB}}{J} = \frac{(5 \times 10^{-3}\ \text{m})(20\ \text{N}\cdot\text{m})}{9.82 \times 10^{-10}\ \text{m}^4} = 102\ \text{MPa}$$

A plot of the twisting moment and the angle of twist along the shaft is shown in Fig. 3.14*d*.

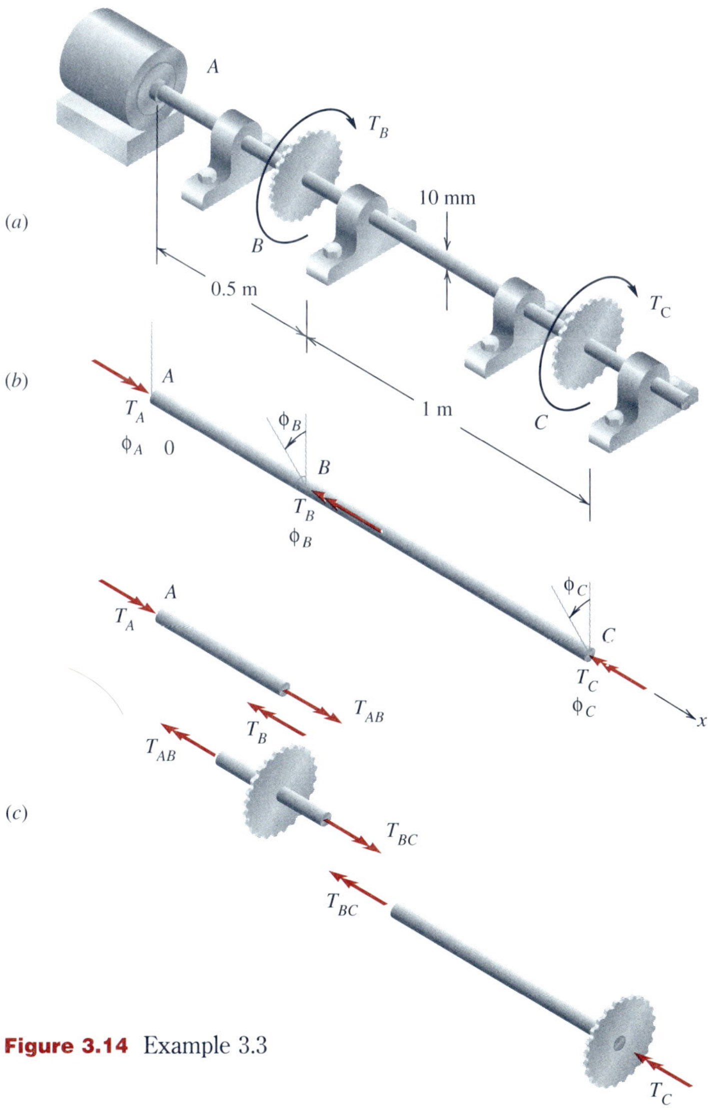

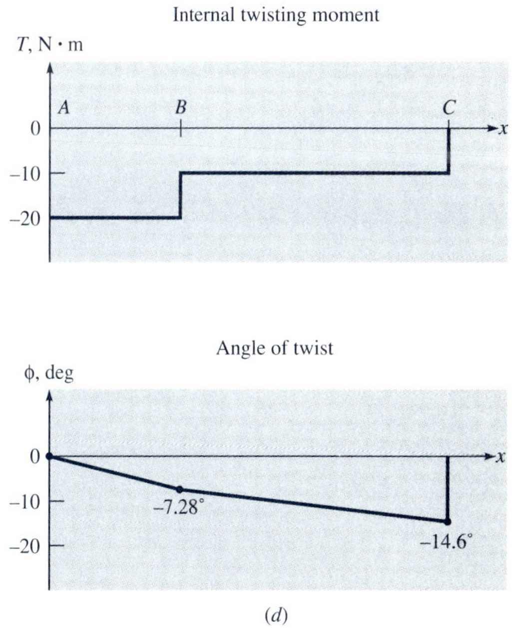

Figure 3.14 Example 3.3

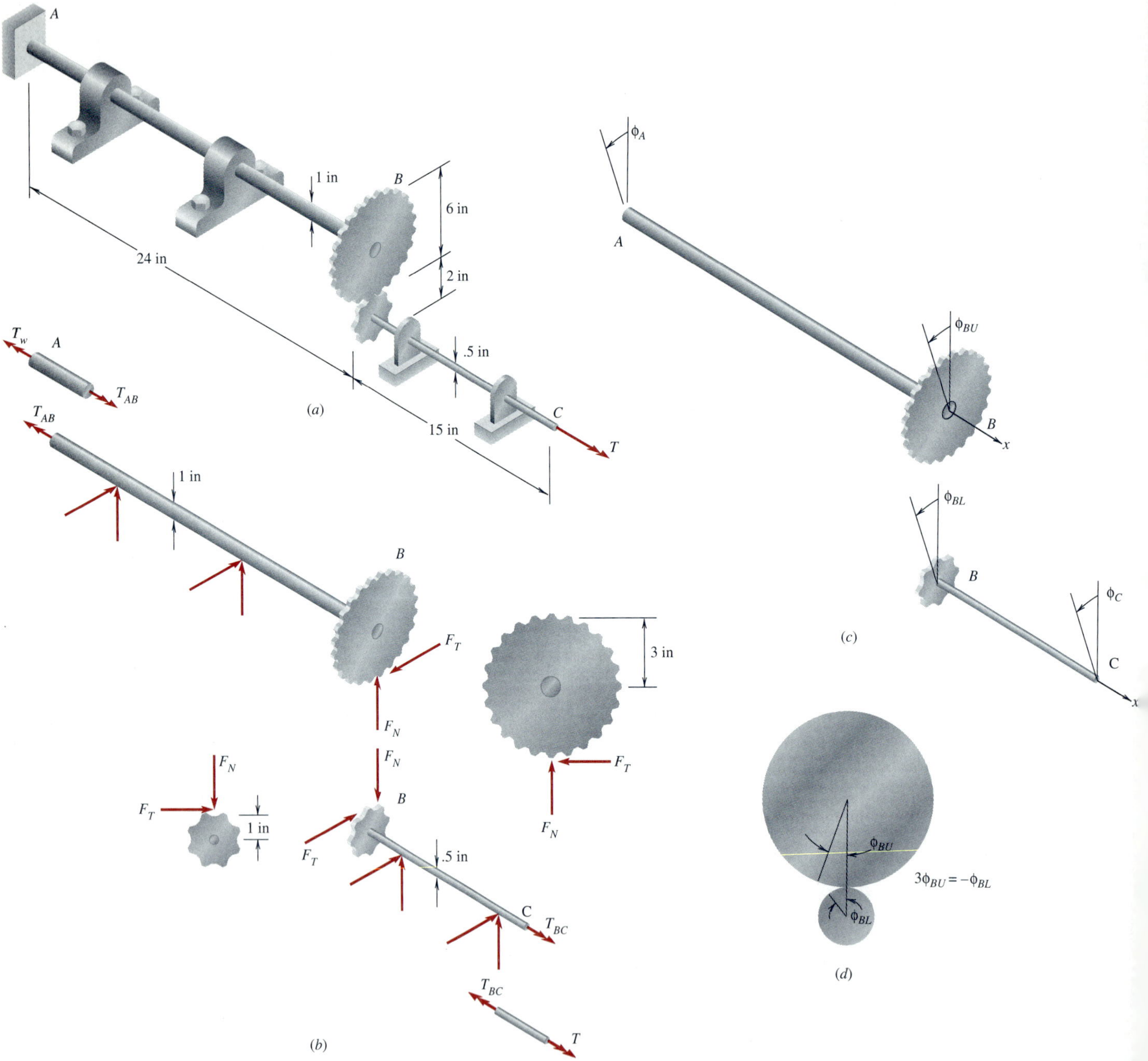

Figure 3.15 Example 3.4

EXAMPLE 3.4

A twisting moment of $T = 7.5 \text{ lb} \cdot \text{ft}$ is applied at section C to the model system consisting of two shafts geared together as shown in Fig. 3.15*a*. Section A is fixed. We would like to determine the total angle of twist at C where the twisting moment T is applied and to find the maximum shear stress in the system. Assume that the gears are rigid compared to the shafts. We neglect the weight of the shafts, assume friction is negligible at the bearings, and take $G = 8.0 \times 10^6$ psi.

To determine the angle of twist at C, we need to ask what twisting deformations contribute to the total angle of twist at C. Consideration of limiting cases might help. For example, if shaft AB is rigid and shaft BC is elastic, then the angle of twist at C arises only from the twisting of the lower shaft BC. If shaft BC is rigid and shaft AB is elastic, and if shaft AB undergoes an angle of twist at section B, then the two gears at B will cause shaft BC to rotate through an angle determined by the gear ratio of the two gears. We therefore need to investigate the deformation of each shaft and the geometry associated with the gears; we use the three steps.

Statics: In Fig. 3.15*b*, free-body diagrams of shafts AB and BC and sketches of the gears at section B are given. The bearing reaction forces are shown at each bearing support location but unlabeled (since they do not concern us in this problem), and we show the twisting moment along the top shaft as T_{AB} and the twisting moment along the bottom shaft as T_{BC}. The reactive force between the gears is broken up into a tangential-force component F_T and a normal-force component F_N, as shown in Fig. 3.15*b*. The wall moment reaction is T_w. Moment equilibrium about the axis of the lower shaft BC gives

$$\begin{aligned} T_{BC} &= T = 7.5 \text{ lb} \cdot \text{ft} = 90 \text{ lb} \cdot \text{in} \\ &= (F_T)(1 \text{ in}) \end{aligned} \tag{a}$$

while moment equilibrium about the axis of the upper shaft gives

$$T_{AB} + 3F_T = 0 \tag{b}$$

From Eqs. (*a*) and (*b*) we have

$$\begin{aligned} T_{BC} &= 7.5 \text{ lb} \cdot \text{ft} \\ T_{AB} &= -3F_T = -3T_{BC} = -22.5 \text{ lb} \cdot \text{ft} \\ &= -270 \text{ lb} \cdot \text{in} \end{aligned} \tag{c}$$

The minus sign for T_{AB} indicates that the twisting moment in shaft AB is in the direction opposite to that shown in Fig. 3.15*b*.

Force-Deformation Relations and Geometry: The force-deformation relation for the lower shaft is from Fig. 3.15*b* and *c*

$$T_{BC} = \left(\frac{GJ}{L}\right)_{BC} (\phi_C - \phi_{BL}) \tag{d}$$

where ϕ_{BL} is the rotation of the 2-in gear at section B on the *lower* shaft. For the upper shaft we have

$$T_{AB} = \left(\frac{GJ}{L}\right)_{AB} (\phi_{BU} - \phi_A) \tag{e}$$

where ϕ_{BU} is the rotation of the 6-in gear at section B on the *upper* shaft. The angles of twist in Fig. 3.15*c* are positive as shown. Section A is fixed, so that

$$\phi_A = 0 \tag{f}$$

Further since the gears must stay in contact without slipping, we have from equality of the distances moved on the gear surfaces (Fig. 3.15*d*)

$$-(3)\phi_{BU} = (1)\phi_{BL} \tag{g}$$

The minus sign in Eq. (*g*) arises because if one gear has a positive rotation, the other gear has a negative rotation, consistent with the convention of positive angles of twist shown in Fig. 3.15*c*. The expression for the angle of twist at section C then follows from Eqs. (*d*) to (*g*):

$$\begin{aligned} \phi_C &= \left(\frac{TL}{GJ}\right)_{BC} + \phi_{BL} \\ &= \left(\frac{TL}{GJ}\right)_{BC} - 3\left(\frac{TL}{GJ}\right)_{AB} \\ &= \frac{(90)(15)}{(8.0 \times 10^6)(\pi/32)(0.5)^4} \\ &\quad - (3)\frac{(-270)(24)}{(8.0 \times 10^6)(\pi/32)} \\ &= 2.75 \times 10^{-2} + 2.48 \times 10^{-2} \\ &= 5.23 \times 10^{-2} \text{ rad} \\ &= 3.0^\circ \end{aligned} \tag{h}$$

Note that each shaft contributes a positive term to the angle of twist at C.

The maximum magnitude of the shear stress in each shaft follows from Eq. (3.14); for shaft AB,

$$\begin{aligned} \tau_{max} &= \left(\frac{T}{J}\frac{d}{2}\right)_{AB} = \frac{(270)(1)}{(\pi/32)(2)} \\ &= 1375 \text{ psi} \end{aligned} \tag{i}$$

And for shaft BC,

$$\tau_{max} = \left(\frac{T}{J}\frac{d}{2}\right)_{BC} = \frac{(90)(0.5)}{(\pi/32)(0.5)^4(2)} \qquad (j)$$
$$= 3667 \text{ psi}$$

The simple model of this example is useful in showing the effect of the gear ratio on the angle of twist and on the value of the twisting moment in the top shaft.

3.7 Torsion of Statically Indeterminate Systems

In the previous four examples we were able to solve for the twisting moments in each part of the system by a direct appeal to the equations of equilibrium. However, as we know from our analysis of systems of bars under axial deformation in Chap. 2, many problems require the *simultaneous* determination of the forces and displacements, i.e., the problems are statically indeterminate. Similar situations arise in torsion problems, as the following two examples illustrate.

EXAMPLE 3.5

A twisting moment of $T = 50$ lb · ft is applied to a 1-in-diameter aluminum-alloy shaft ABC at section B, as shown in Fig. 3.16*a*. Ends A and C are built in to prevent rotation. We wish to determine the twisting moment in each segment of the shaft, the angle of twist variation along the shaft, and τ_{max} in each of the two segments. We will take $G = 4 \times 10^6$ psi. The x axis is taken along the shaft with the origin at section A; we will solve this problem by using the three steps.

Statics: In Fig. 3.16*b* we show the twisting moment in each segment. The wall reaction at A is T_A; the wall reaction at C is T_C. Moment equilibrium of the segment at section B about the x axis gives

$$T_{AB} = T_{BC} + T = T_{BC} + 50 \qquad (a)$$

In a like manner at sections A and C, we have

$$T_A = T_{AB} \qquad T_C = T_{BC} \qquad (b)$$

where T_A and T_C are the wall reactions at A and C.

The two unknown twisting moments in the segments are T_{AB} and T_{BC} for which we have only one equilibrium equation, Eq. (*a*); the problem is statically indeterminate.

Force-Deformation Relations: For each segment we have the relation between the twisting moment and the difference of the angle of twist at the ends (Fig. 3.16*c*); the angles shown in Fig. 3.16*c* are positive. For AB, we have

$$T_{AB} = \frac{GJ}{L_{AB}}(\phi_B - \phi_A) \qquad (c)$$

and for BC

$$T_{BC} = \frac{GJ}{L_{BC}}(\phi_C - \phi_B) \qquad (d)$$

Geometry: Since sections A and C are fixed,

$$\phi_A = \phi_C = 0 \qquad (e)$$

Therefore upon substitution of Eqs. (*c*), (*d*), and (*e*) into Eq. (*a*), we find a single equation to determine ϕ_B

$$\frac{GJ}{L_{AB}}\phi_B = -\frac{GJ}{L_{BC}}\phi_B + T$$

or

$$\phi_B\, GJ\left(\frac{1}{L_{AB}} + \frac{1}{L_{BC}}\right) = T$$

Upon substitution of numerical values, we obtain

$$\phi_B = \frac{T}{GJ\left(\frac{1}{L_{AB}} + \frac{1}{L_{BC}}\right)} = \frac{(50)(12)}{(4 \times 10^6)(\pi/32)(\frac{1}{12} + \frac{1}{36})} \qquad (f)$$
$$= 1.375 \times 10^{-2} \text{ rad}$$
$$= 0.788°$$

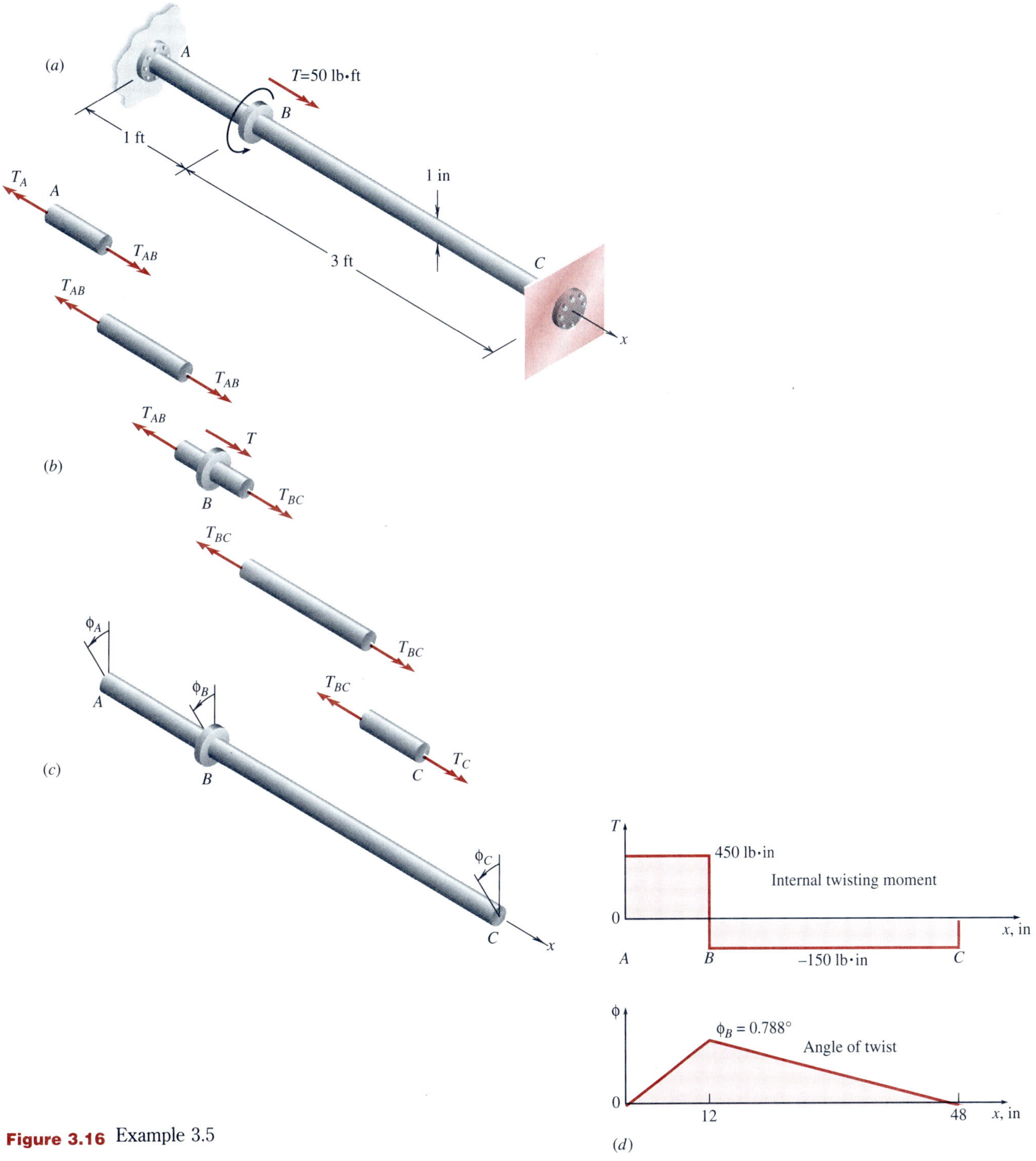

Figure 3.16 Example 3.5

Therefore the twisting moment in each segment can be found from Eqs. (*c*) and (*d*); for segment *AB*, we have

$$T_{AB} = \frac{GJ}{L_{AB}}\phi_B = \frac{(4 \times 10^6)(\pi/32)}{12}(1.375 \times 10^{-2}) \\ = 450 \text{ lb} \cdot \text{in} \tag{g}$$

and for segment *BC*

$$T_{BC} = -\frac{GJ}{L_{BC}}\phi_B = -150 \text{ lb} \cdot \text{in} \tag{h}$$

The minus sign on T_{BC} shows that the actual directions for T_{BC} in Fig. 3.16*b* are opposite to those shown.

A plot of the twisting moment versus the distance along the shaft is shown in Fig. 3.16*d* together with the variation of the angle of twist along the shaft. A jump of magnitude 600 lb · in occurs in the twisting moment diagram at point *B* where the twisting moment is applied. The angle of twist is linear in each segment of the shaft.

Once the values of T_{AB} and T_{BC} are known, the maximum magnitudes of the shear stresses follow by using Eq. (3.14). For *AB*

$$\tau_{max}\big|_{AB} = \left(\frac{Ta}{J}\right)_{AB} = \frac{(450)(0.5)}{\pi/32} \\ = 2292 \text{ psi} \tag{i}$$

Since both segments of the shaft have the same diameter, the ratio of the maximum shear stresses in the segments equals the ratio of the twisting moments. Thus the magnitude of $\tau_{max}|_{BC}$ is only one-third of $\tau_{max}|_{AB}$, or 764 psi.

EXAMPLE 3.6

In Fig. 3.17*a*, we consider a problem similar to Example 3.5 except we allow for differences in material and in the polar moments of inertia in segments *AB* and *BC*. The twisting moment *T* is applied at section *B*, and we wish to determine the angle of twist at section *B* and the twisting moments in segments *AB* and *BC*. We parallel the arguments given in Example 3.5.

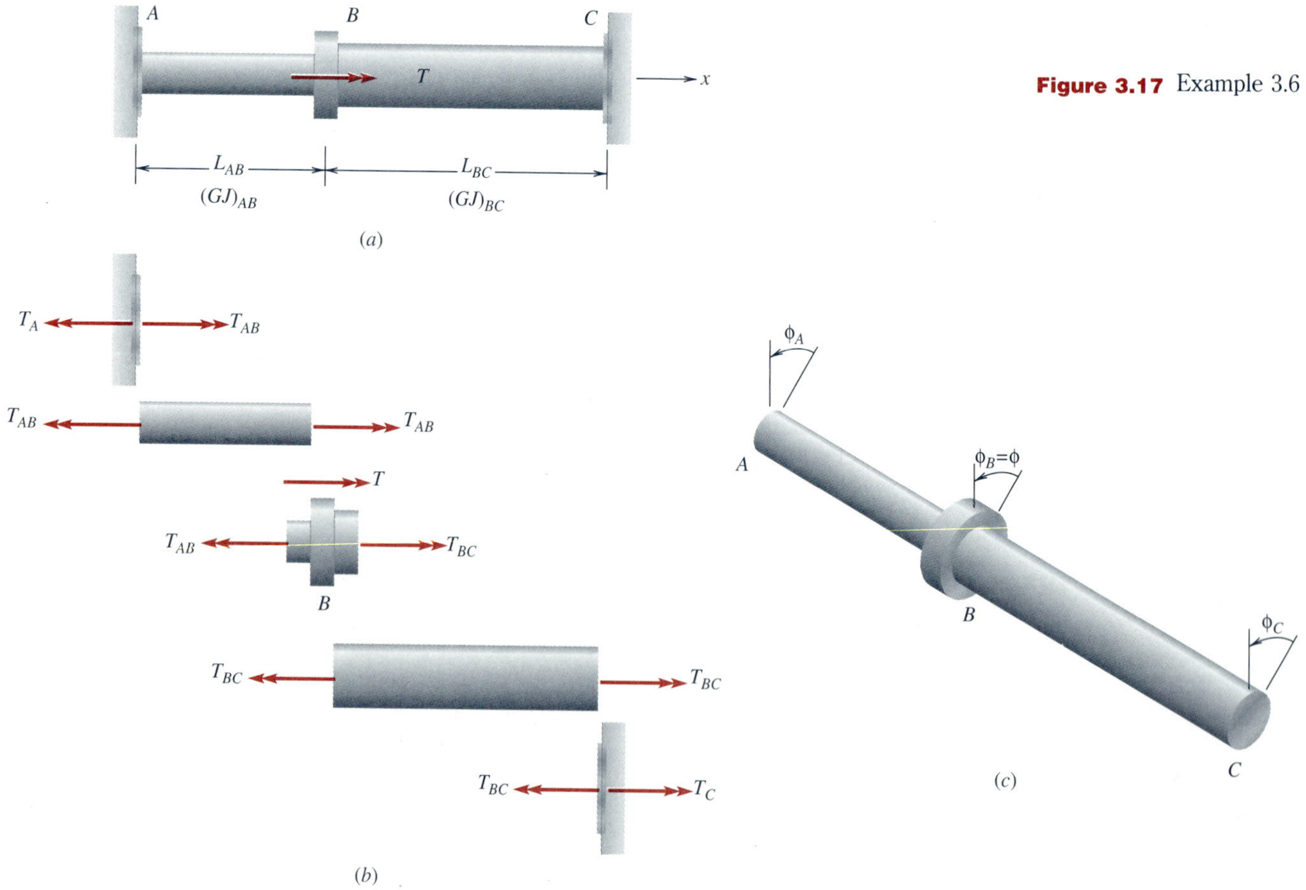

Figure 3.17 Example 3.6

Statics: In Fig. 3.17*b* we show the free-body diagrams of the segments along the shaft with the internal twisting moments T_{AB} and T_{BC}. The wall reactions at A and C are T_A and T_C. At section B, we have

$$T_{AB} = T_{BC} + T \tag{a}$$

Force-Deformation Relations: For each segment we have (Fig. 3.17*c*)

$$T_{AB} = \left(\frac{GJ}{L}\right)_{AB} (\phi_B - \phi_A) \tag{b}$$

$$T_{BC} = \left(\frac{GJ}{L}\right)_{BC} (\phi_C - \phi_B) \tag{c}$$

Geometry: Since the shaft is built in at A and C, we have

$$\phi_A = \phi_C = 0 \tag{d}$$

Substitution of Eqs. (*b*) and (*c*) into Eq. (*a*) with the zero values of the angles ϕ_A and ϕ_C gives us an equation for angle ϕ_B in the form

$$\phi_B = \frac{T}{(GJ/L)_{AB} + (GJ/L)_{BC}} \tag{e}$$

The twisting moment in each segment follows from Eqs. (*b*), (*c*), and (*e*):

$$T_{AB} = \frac{T(GJ/L)_{AB}}{(GJ/L)_{AB} + (GJ/L)_{BC}} \tag{f}$$

$$T_{BC} = \frac{-T(GJ/L)_{BC}}{(GJ/L)_{AB} + (GJ/L)_{BC}} \tag{g}$$

The minus sign for T_{BC} indicates that the twisting moment is in the opposite direction to that shown in Fig. 3.17*b*.

In the special case when $(GJ)_{AB} = (GJ)_{BC} = GJ$, as in Example 3.5, we have from Eqs. (*e*), (*f*), and (*g*)

$$\phi_B = \frac{T}{GJ\left(\frac{1}{L_{AB}} + \frac{1}{L_{BC}}\right)} = \frac{T}{GJ}\frac{L_{AB}L_{BC}}{L_{AB} + L_{BC}}$$

$$T_{AB} = T\frac{L_{BC}}{L_{AB} + L_{BC}} \qquad T_{BC} = -T\frac{L_{AB}}{L_{AB} + L_{BC}} \tag{h}$$

The wall reactions T_A and T_C then follow from Fig. 3.17*b*

$$T_A = T_{AB} = T\frac{L_{BC}}{L_{AB} + L_{BC}} \qquad T_C = T_{BC} = -T\frac{L_{AB}}{L_{AB} + L_{BC}} \tag{i}$$

Again we note that the results in Eqs. (*h*) and (*i*) hold only for the case of uniform GJ. Finally we should note that this problem is the torsion analog to the axial load problem in Example 2.7.

3.8 Displacement Method for Torsion of Circular Shafts

The displacement method was used in Sec. 2.6 to solve problems of axially loaded bars, and the same method can be applied with only slight changes to problems of shafts of circular cross section loaded by axial twisting moments.

Suppose that a shaft is made up of segments with different lengths and with different circular cross sections, as shown in Fig. 3.18. Let the shaft be subdivided into shaft *elements* labeled (1) to $(N - 1)$ with *nodes* defined by the endpoints of the elements and labeled 1 to N with the following properties:

1. Applied axial twisting moments act only at node points (i.e., at the junction of two shaft elements or at the end of the first or last shaft element) and are denoted by Q_i (lb · in or N · m), where i can take on values from $i = 1$ to $i = N$.
2. The shear modulus G_i and the polar moment of inertia J_i are constant

Figure 3.18 Interconnected shaft elements showing elements and nodes.

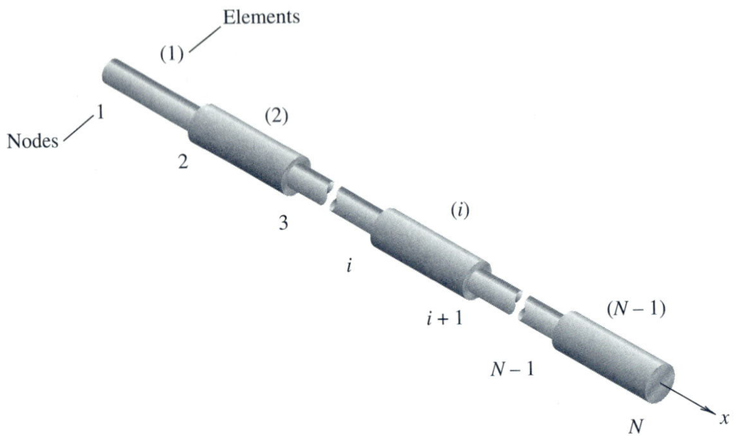

in each shaft element of length L_i, where i can take on values from $i = 1$ to $i = N - 1$.

3. The angular displacement (the angle of twist in radians) of each node point is denoted by ϕ_i, and all applied twisting moments or torques and angular displacements are *positive* when they are directed in the positive x direction of the shaft. The vector twisting moments and vector angular displacements follow the right-hand rule.

The method of solution for axially twisted shafts is completely analogous to that followed for axially loaded bars. First, we need to relate the twisting moment T_i in the ith shaft element to the angles of twist or angular displacements ϕ_i and ϕ_{i+1} at each end or node of the element. The twisting moment–twisting angle relation in terms of the twisting moment T in a shaft element of length L is, according to Eq. (3.16), of the form,

$$T = \frac{GJ}{L}(\phi_B - \phi_A) \tag{3.25}$$

where G is the shear modulus and J is the polar moment of inertia.

In the notation of Fig. 3.19, the relative angle of twist can be written in terms of the angle of twist at each end of the element

$$\phi_B - \phi_A = \phi_{i+1} - \phi_i \tag{3.26}$$

so that the twisting moment in each element is

$$T_i = \left(\frac{GJ}{L}\right)_i (\phi_{i+1} - \phi_i) \tag{3.27}$$

where $L_i = x_{i+1} - x_i$ is the length of the ith element. The quantity

$$\hat{k}_i = \left(\frac{GJ}{L}\right)_i \tag{3.28}$$

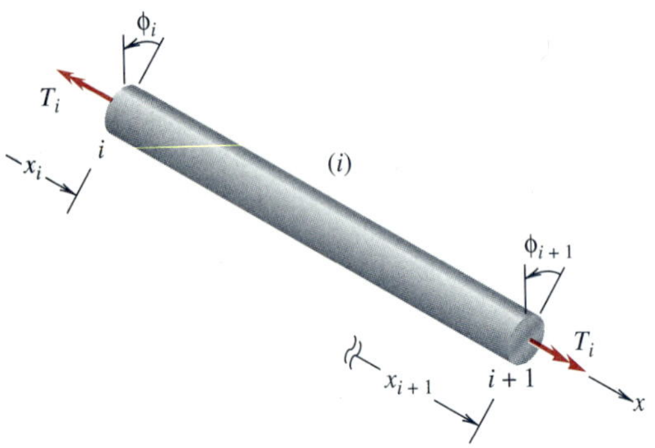

Figure 3.19 Shaft element (i) with nodes i and $i + 1$ and twisting moment T_i.

is called the *torsional stiffness* of the ith element, shown in Fig. 3.19. Thus we can write

$$T_i = \hat{k}_i(\phi_{i+1} - \phi_i) \tag{3.29}$$

We see that Eq. (3.29) is completely analogous to Eq. (2.17) (without the temperature-change term) for the axially loaded bar element. The procedure for the solution of problems with a number of segments subjected to twisting moments is exactly analogous to the procedure followed for bar systems in Chap. 2.

Moment equilibrium about the x axis at the nodes yields N equations which can be written in terms of the N angular displacements of the nodes. Specified (known) angular displacements are inserted, and a system of reduced linear equations for the remaining unknown displacements is solved to produce values for these unknowns. Once the angular displacements at the nodes are known, we can find all shaft element twisting moments from Eq. (3.29) and thus the maximum shear stress in each element from Eq. (3.14).

In Example 3.7 a simple two-element problem is solved to demonstrate the procedure. Then some further remarks are included to indicate how the displacement method can be applied to a shaft with N node points and $(N - 1)$ elements.

EXAMPLE 3.7

For the stepped circular shaft shown in Fig. 3.20*a*, shaft AB has a diameter equal to 4 in with a shear modulus $G_{AB} = 10 \times 10^6$ psi, while the diameter of shaft BC is 2 in and $G_{BC} = 15 \times 10^6$ psi. The applied torques are $T_B = 100{,}000$ in · lb and $T_C = 20{,}000$ in · lb and are directed as shown.

(*a*) Find the maximum shear stress in each segment of the shaft.
(*b*) Calculate the angle of twist of sections B and C relative to the fixed section A.

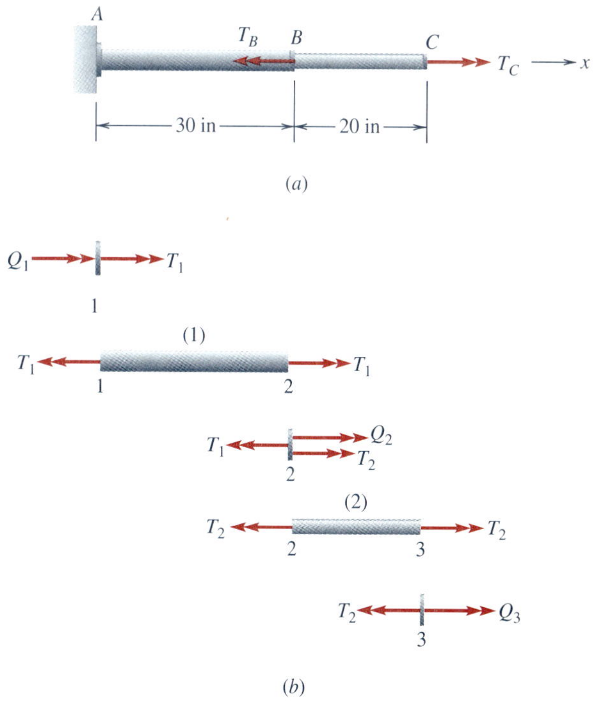

Figure 3.20 Example 3.7

We introduce the coordinate x axis along the shaft with origin at A, and we divide the shaft into two elements and three nodes, as shown in Fig. 3.20*b*. To solve this problem, we use twisting moment equilibrium at nodes 2 and 3 to derive two equations for the unknown node angular displacements ϕ_2 and ϕ_3. Element twisting moments or torques, stresses, and reaction torques can then be found from the node angular displacements after this pair of equations is solved.

Statics: Free-body diagrams for the two shaft elements and for the infinitesimal node elements at nodes 1, 2, and 3 with the applied twisting moments and the reactive moment acting are shown in Fig. 3.20*b*. Twisting moments, of course, are along the axis of the node elements, but we often show them off the axis, as in Fig. 3.20*b*, for clarity. Moment equilibrium about the x axis of the two node elements at nodes 2 and 3 gives

$$T_2 - T_1 + Q_2 = 0$$
$$-T_2 + Q_3 = 0 \qquad (a)$$

where Q_2 and Q_3 are the twisting moments applied at nodes 2 and 3, respectively, assumed positive.

Force-Deformation Relations: The twisting moments in each element are related to the node angular displacements from Eq. (3.29) by

$$T_1 = \hat{k}_1(\phi_2 - \phi_1)$$
$$T_2 = \hat{k}_2(\phi_3 - \phi_2) \qquad (b)$$

where $\hat{k}_1$ and $\hat{k}_2$ are the torsional stiffnesses of the segments, Eq. (3.28).

Geometry: From Fig. 3.20*a*, we see that node 1 is fixed so that

$$\phi_1 = 0 \qquad (c)$$

and thus ϕ_2 and ϕ_3 are the unknown angular displacements for this problem.

The element torques T_1 and T_2 given by Eqs. (*b*) are substituted into Eq. (*a*) to find

$$\hat{k}_2(\phi_3 - \phi_2) - \hat{k}_1(\phi_2 - \phi_1) + Q_2 = 0$$
$$-\hat{k}_2(\phi_3 - \phi_2) + Q_3 = 0 \qquad (d)$$

which can also be written in the form

$$-\hat{k}_1\phi_1 + (\hat{k}_1 + \hat{k}_2)\phi_2 - \hat{k}_2\phi_3 = Q_2$$
$$-\hat{k}_2\phi_2 + \hat{k}_2\phi_3 = Q_3 \qquad (e)$$

Since for the current problem $Q_2 = -100{,}000$ in · lb, $Q_3 = 20{,}000$ in · lb, $J_1 = \pi(4^4/32) = 25.13$ in^4, and $J_2 = \pi(2^4/32) = 1.571$ in^4, we have

$$\hat{k}_1 = \left(\frac{GJ}{L}\right)_1 = \frac{(10 \times 10^6)(25.13)}{30} = 8.377 \times 10^6 \text{ lb} \cdot \text{in}$$

$$\hat{k}_2 = \left(\frac{GJ}{L}\right)_2 = \frac{(15 \times 10^6)(1.571)}{20} = 1.178 \times 10^6 \text{ lb} \cdot \text{in}$$

Then, since $\phi_1 = 0$, Eqs. (*e*) become

$$9.555\phi_2 - 1.178\phi_3 = -100 \times 10^{-3}$$
$$-1.178\phi_2 + 1.178\phi_3 = 20 \times 10^{-3} \qquad (f)$$

We solve for the angular displacements ϕ_2 and ϕ_3 to get

$$\phi_2 = -9.55 \times 10^{-3} \text{ rad}$$
$$\phi_3 = 7.43 \times 10^{-3} \text{ rad} \qquad (g)$$

Since ϕ_2 is now known, we can use the first equation in Eqs. (*b*) to obtain

$$T_1 = (8.377)(-9.55) \times 10^3$$
$$= -80{,}000 \text{ in} \cdot \text{lb} \qquad (h)$$

The wall twisting moment reaction at node 1 follows from Fig. 3.20*b*, by using moment equilibrium of the wall element,

$$Q_1 = -T_1 = 80{,}000 \text{ in} \cdot \text{lb} \qquad (i)$$

The maximum shear stress in shaft element 1 is

$$\tau_{1,\max} = \frac{T_1 r_1}{J_1} = \frac{(80{,}000)(2)}{25.13} = 6370 \text{ psi} \qquad (j)$$

where r_1 is the radius of shaft element 1. A similar calculation for shaft element 2 yields

$$\tau_{2,\max} = \frac{T_2 r_2}{J_2} = 12{,}730 \text{ psi} \qquad (k)$$

General Case

In Example 3.7 the shaft consisted of two shaft elements and three node points. Since one node displacement was specified ($\phi_1 = 0$), only two node displacements remained unknown. In the general case shown in Fig. 3.18, there are N node points and ($N - 1$) elements. We briefly outline the method of solution for this case. As always, we use three steps.

Statics. For each internal *node element* in the N-node case, there will be an element twisting moment acting on each side of the node element together with the applied twisting moment, if any. At a typical internal node element, the twisting moment equilibrium equation will therefore be of the general form

$$T_{i-1} - T_i = Q_i \tag{3.30}$$

where Q_i is the applied twisting moment or reaction at the node. At the first node element, using equilibrium, we have $-T_1 = Q_1$ and at the last node element we have $T_{N-1} = Q_N$.

Force-Deformation Relations. For the ith shaft element we have

$$T_i = \hat{k}_i(\phi_{i+1} - \phi_i) \tag{3.31}$$

We can express the equilibrium equations in terms of the node displacements $\phi_1, \ldots, \phi_N$ by substituting Eq. (3.31) into Eq. (3.30) to obtain

$$-\hat{k}_{i-1}\phi_{i-1} + (\hat{k}_{i-1} + \hat{k}_i)\phi_i - \hat{k}_i\phi_{i+1} = Q_i \tag{3.32}$$

For the first and last node, making use of Eq. (3.31) in the node equilibrium equation will give one equation with two unknowns ϕ_1 and ϕ_2 and one equation with two unknowns ϕ_{N-1} and ϕ_N.

Geometry. In general, for each specified value of a node displacement, there will be a corresponding unknown twisting moment reaction Q_i. The procedure for handling these known node displacements and unknown reactions is the same procedure that was used for axial bar systems.

As was also the case for systems of axially loaded bars, the organization of the basic steps in the analysis of twisted shafts and the systematic notation introduced facilitate the programming of the displacement method for computer solutions.

A computer program TORMECH on the diskette has been written which allows the user to give a description of a torsion problem (interactively), and then the program produces the solution with graphics output for the angle of twist and tables for the twisting moments, the angles of twist, and the maximum shear stresses.

3.9 Computer Program TORMECH

A computer program to analyze torsion problems of circular shafts attached end to end as in Fig. 3.18 is on the diskette with this book. The procedure for running the program TORMECH is similar to the procedure for running the program BARMECH discussed in Sec. 2.7; also see Apps. H, I, and J. From the main menu, we select program number 2 for torsion, which brings us to the TORMECH menu. We discuss the use of this program in the examples that follow. As in the case of axial deformation, the program will treat both statically determinate and statically indeterminate problems.

EXAMPLE 3.8

As the first computer example, we will show how the problem previously solved in Example 3.7 can be solved by using TORMECH. For the stepped circular shaft shown in Fig. 3.21*a*, the diameter of shaft *AB* is 4 in and its shear modulus $G_{AB} = 10 \times 10^6$ psi, while the diameter of shaft *BC* is 2 in and $G_{BC} = 15 \times 10^6$ psi. The applied twisting moments are $T_B = 100{,}000$ in · lb and $T_C = 20{,}000$ in · lb and are directed as shown.

(*a*) Find the maximum shear stress in each segment of the shaft.

(*b*) Calculate the angle of twist of sections *B* and *C* relative to the fixed section *A*.

In Fig. 3.21*b*, we have located the *x* axis along the shaft, and we have divided the shaft into two elements and three nodes. At node 1, the angular displacement is zero. At nodes 2 and 3 the applied twisting moments are $Q_2 = -100{,}000$ in · lb and $Q_3 = 20{,}000$ in · lb. The unknowns in the displacement method are ϕ_2 and ϕ_3. Since the shaft is solid, the inside diameter of each segment is DI = 0. We use DI to denote the inside diameter and DO to denote the outside diameter.

In the main menu of MECHMAT we first select item 2 for the torsion option. From the TORMECH menu, since this is a new problem, we make the data input choice as 1: Input new data file. Usually we store the data file on *A*: and in this case we give the name EX38 to this data file.

Figure 3.22 gives the summary of the input data, while Fig. 3.23 gives a table of the solution for the node rotation angles, the torque or twisting moment in each element, and the maximum shear stress in each element. In the input data, the units of *G* are 10^6 psi or MPa; the format of the input is similar to that for the BARMECH problem. The results in Fig. 3.23 agree with those found in Example 3.7.

It should be emphasized that applied twisting moments Q_i and node angles ϕ_i follow the sign convention of vector notation. The element twisting moments or torques given in the solution table of Fig. 3.23 refer to torques acting on a free body of the given element, where a positive vector torque on a positive *x* face coincides with the direction of the outward normal, as shown in Fig. 3.24.

Finally, a plot of the twisting angle at each section along the shaft is given in Fig. 3.25 together with the maximum and minimum values of ϕ in *radians*.

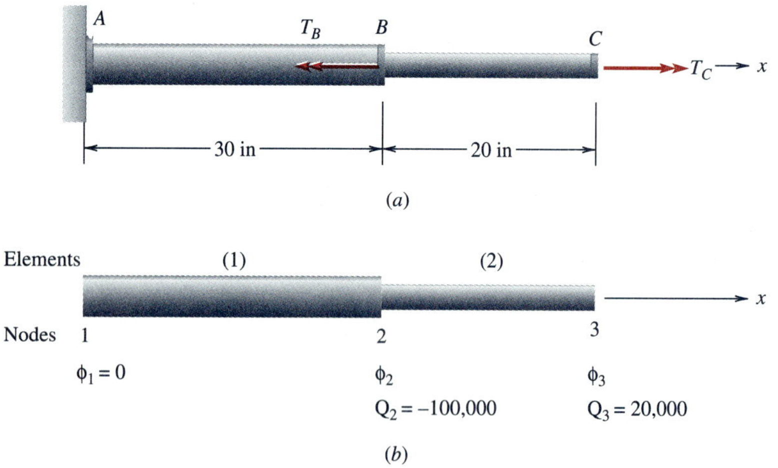

Figure 3.21 Example 3.8

```
        ***MECHANICS OF MATERIALS***  DATA FILE: ex38
                        ***TORSION***

   INPUT DATA:

NUMBER OF ELEMENTS     NM = 2

 ELEMENT     L (in)       DI (in)        DO (in)       G (10**6;psi)
              (m)          (m)            (m)              (MPa)

    1        30.00       0.0000         4.0000          1.00E+01
    2        20.00       0.0000         2.0000          1.50E+01


             CONDITION
***********************************

     NODE         TORQUE (lb-in;N-m)
      2                -1.00E+05
      3                 2.00E+04

     NODE       SPECIFIED ANGLE (rad.)
      1                0.00E+00
```

Figure 3.22 Example 3.8: Summary of input data.

```
SOLUTION:

*********************************************************

       NODE        ANGLE (radians)       ANGLE (degrees)

*********************************************************

         1            0.0000E+00            0.0000E+00
         2           -9.5493E-03           -5.4713E-01
         3            7.4272E-03            4.2555E-01


*********************************************************

                              MAX SHEAR
 ELEMENT      TORQUE                         J            G
                                STRESS
*********************************************************

    1      -8.0000E+04   -6.3662E+03   2.5133E+01   1.00E+07
    2       2.0000E+04    1.2732E+04   1.5708E+00   1.50E+07
```

Figure 3.23 Example 3.8: Solution.

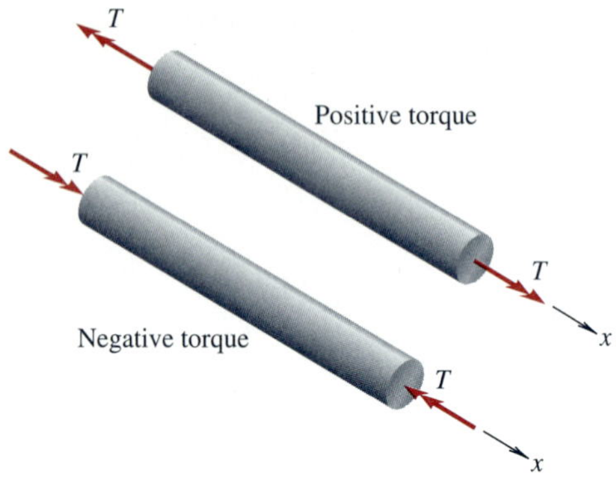

Figure 3.24 Positive and negative twisting moments or torques acting on an element.

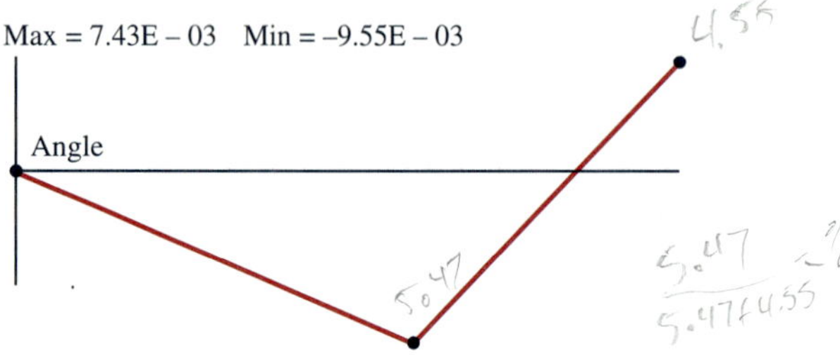

Figure 3.25 Example 3.8: Angle-of-twist diagram.

EXAMPLE 3.9

A twisting moment of 70 N · m is applied to a 20-mm-diameter aluminum shaft, as shown in Fig. 3.26*a*. Ends *A* and *C* are prevented from rotating. Using TORMECH, we wish to find the angle of twist as a function of position along the shaft and the maximum shear stresses in *AB* and *BC*. Take $G = 25$ GPa.

The x axis is taken along the shaft, as shown in Fig. 3.26*b*. We have three nodes and two elements with $\phi_1 = \phi_3 = 0$, and the applied twisting moment at node 2 is 70 N · m.

Figure 3.27 shows the input data, the solution, and the maximum shear stresses while Fig. 3.28 shows a plot of the angle of twist versus distance along the shaft from the left end.

Figure 3.26 Example 3.9

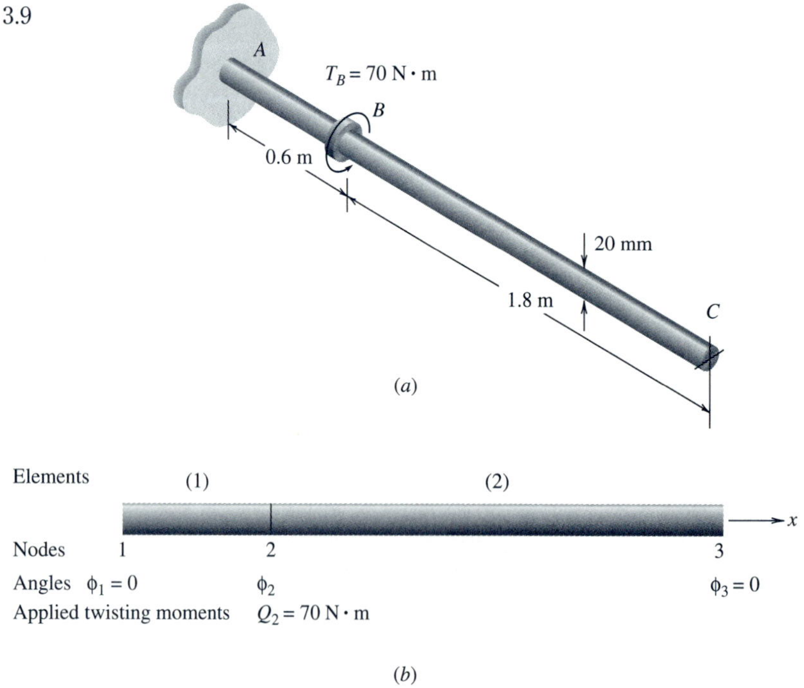

```
         ***MECHANICS OF MATERIALS***   DATA FILE: ex39
                          ***TORSION***

    INPUT DATA:

NUMBER OF ELEMENTS     NM = 2

 ELEMENT   L (in)     DI (in)        DO (in)       G (10**6;psi)
            (m)         (m)            (m)             (MPa)

    1       0.60      0.0000         0.0200          2.50E+04
    2       1.80      0.0000         0.0200          2.50E+04


             CONDITION
****************************************
      NODE         TORQUE (lb-in;N-m)
       2                7.00E+01

      NODE       SPECIFIED ANGLE (rad.)
       1                0.00E+00
       3                0.00E+00


SOLUTION:

**************************************************
        NODE      ANGLE (radians)     ANGLE (degrees)
**************************************************
         1          0.0000E+00          0.0000E+00
         2          8.0214E-02          4.5959E+00
         3          0.0000E+00          0.0000E+00


**********************************************************
                             MAX SHEAR
 ELEMENT        TORQUE         STRESS          J             G
**********************************************************
    1        5.2500E+01     3.3423E+07    1.5708E-08    2.50E+10
    2       -1.7500E+01    -1.1141E+07    1.5708E-08    2.50E+10
```

Figure 3.27 Example 3.9: Input data and solution.

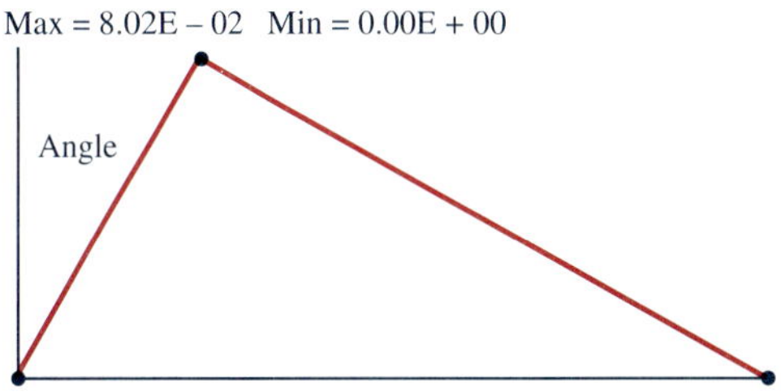

Figure 3.28 Example 3.9: Angle-of-twist diagram.

EXAMPLE 3.10

The stepped circular shaft shown in Fig. 3.29*a* is to be installed as a machine component in which sections *A*, *C*, and *D* have torques applied to give the specified angular rotations, as indicated in Fig. 3.29*a*: $\phi_A = -1°$, $\phi_C = \frac{1}{2}°$, and $\phi_D = 1°$. Use the program TORMECH to:

(*a*) Find the angle of twist at section *B*.

(*b*) Find the maximum shear stress in each of the three shaft segments.

(*c*) Find the values of the torques at sections *A*, *C*, and *D* which cause the specified angular rotations.

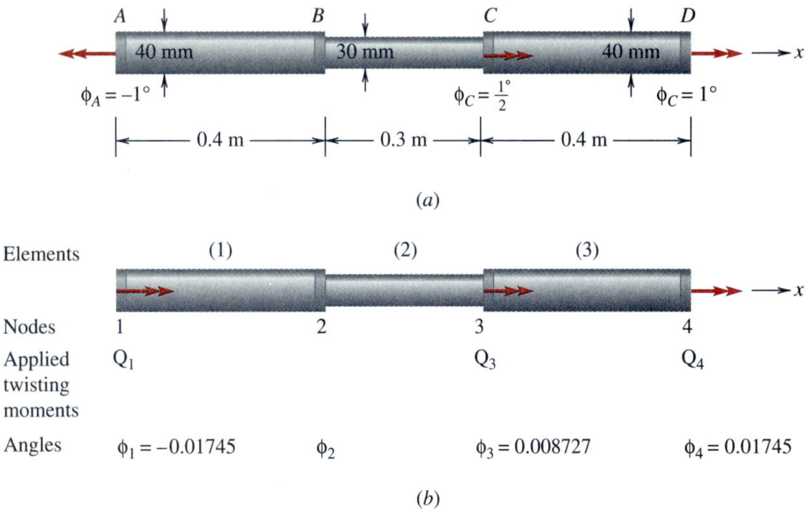

Figure 3.29 Example 3.10

```
***MECHANICS OF MATERIALS***   DATA FILE: ex310
                ***TORSION***

INPUT DATA:

NUMBER OF ELEMENTS        NM = 3

ELEMENT   L (in)     DI (in)          DO (in)        G (10**6;psi)
            (m)        (m)              (m)              (MPa)

   1      0.40      0.0000           0.0400           7.50E+04
   2      0.30      0.0000           0.0300           7.50E+04
   3      0.40      0.0000           0.0400           7.50E+04


                 CONDITION
****************************************

     NODE          TORQUE (lb-in;N-m)

     NODE        SPECIFIED ANGLE (rad.)
       1               -1.74E-02
       3                8.73E-03
       4                1.74E-02
```

Figure 3.30 Example 3.10: Input data.

The value of $G = 75$ GPa.

We first recast the problem into the notation of the displacement method (Fig. 3.29*b*). We divide the shaft into four nodes and three elements. At the nodes where the angular rotations are specified, we show the unknown twisting moments that give rise to the specified rotations as Q_1, Q_3, and Q_4. The specified angles have to be converted to radians. Note that section D twists a total of 2° *relative* to section A. Figure 3.30 gives a summary of the input data which describes the problem. Note G is in units of megapascals (MPa) for the input data table.

TORMECH solves the twisting-moment equilibrium equation at node 2 in terms of the unknown angle ϕ_2. The solution for ϕ_2 along with the specified angles is given in Fig. 3.31, and a plot of ϕ is given in Fig. 3.32.

The twisting moments in each element are calculated from the stiffness for each element and the known angles of twist at each end of the element. The twisting moments given in the solution table of Fig. 3.31 are used in TORMECH to calculate the maximum shear stresses by using the relation

$$\tau_{max} = \frac{Tr_0}{J}$$

These stresses are given in Fig. 3.31, and we find that the maximum shear stress occurs in element 2 and has the value of 69 MPa.

Finally, we can calculate the external twisting moments which cause the specified angular rotations by considering the moment equilibrium of node elements at nodes 1, 3, and 4. Equilibrium at node 1 of Fig. 3.33*a* gives

$$Q_1 + T_1 = 0$$

or

$$Q_1 = -T_1 = -366 \text{ N} \cdot \text{m}$$

Since no moment is applied at node 2, we have $T_2 = T_1 = 366$ N · m. At node 3 we have (Fig. 3.33*b*)

$$Q_3 + T_3 = T_2$$

or

$$Q_3 = 366 - 411 = -45 \text{ N} \cdot \text{m}$$

and at node 4, we have (Fig. 3.33*c*)

$$Q_4 = T_3 = 411 \text{ N} \cdot \text{m}$$

The three torques which must be applied to the shaft at nodes 1, 3, and 4 in order to deform it into the required shape are shown in Fig. 3.33*d*.

SOLUTION:

NODE	ANGLE (radians)	ANGLE (degrees)
1	-1.7450E-02	-9.9981E-01
2	-9.6832E-03	-5.5481E-01
3	8.7270E-03	5.0002E-01
4	1.7450E-02	9.9981E-01

ELEMENT	TORQUE	MAX SHEAR STRESS	J	G
1	3.6600E+02	2.9126E+07	2.5133E-07	7.50E+10
2	3.6600E+02	6.9038E+07	7.9522E-08	7.50E+10
3	4.1106E+02	3.2711E+07	2.5133E-07	7.50E+10

Figure 3.31 Example 3.10: Solution.

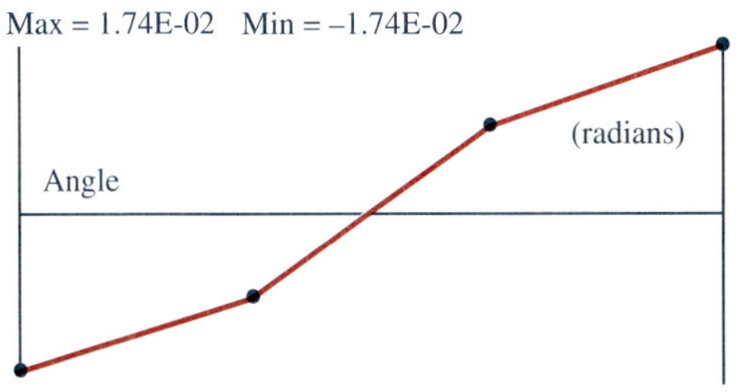

Figure 3.32 Example 3.10: Angle-of-twist diagram.

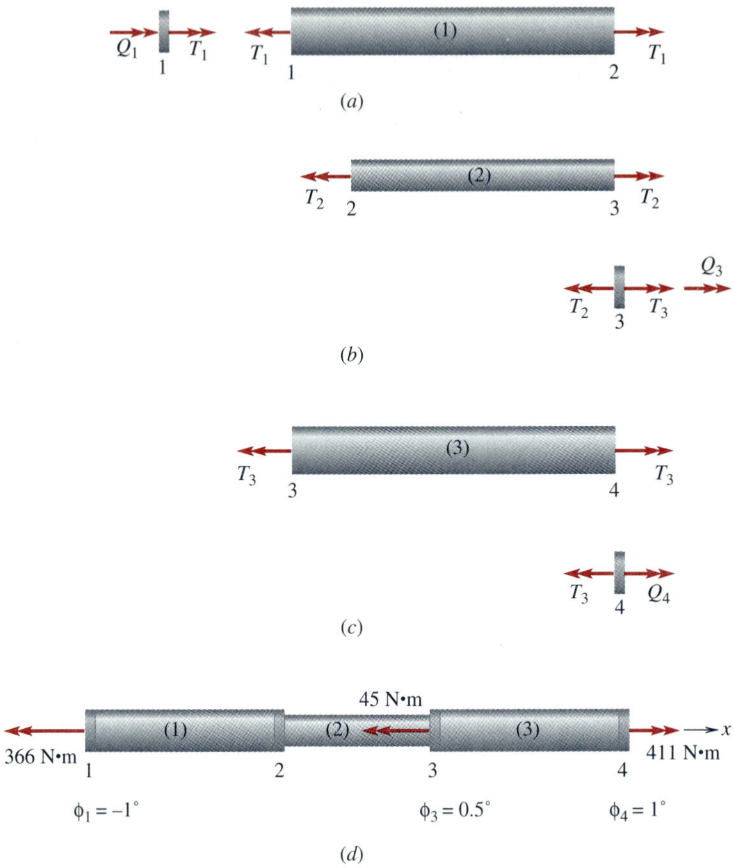

Figure 3.33 Example 3.10

3.10 Design of Circular Shafts for Power Transmission

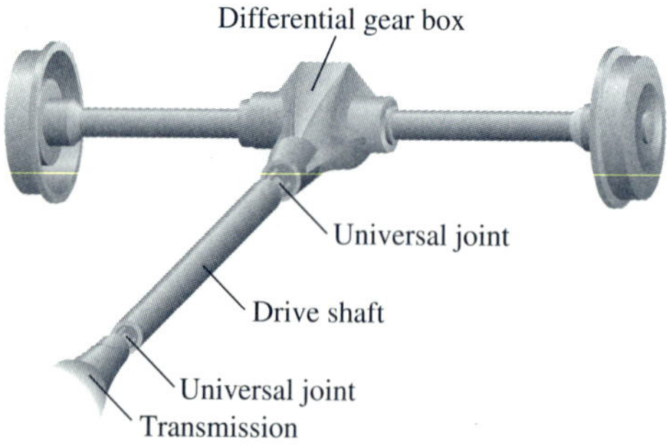

Figure 3.34 Drive shaft for power transmission.

An important use of circular shafts is to transfer mechanical power from one component to another. Of all the uses of circular shafts, probably the most common is the drive shaft in an automobile (Fig. 3.34); the twisting moment from the automobile engine-transmission system is transmitted by the drive shaft to the rear gearbox (in a rear-wheel drive) which in turn exerts twisting moments on the rear axles and drives the wheels.

We would like to relate the power transmitted to a shaft to the maximum shear stress in the shaft and to the dimensions of the shaft. In many situations we would like to *design* the shaft; i.e., we would like to specify the diameter of the shaft, or the inside and outside diameter of a hollow shaft, to transmit a specific amount of power so as not to exceed an allowable value of shear stress and/or a critical value of the angle of twist in the shaft.

Consider a schematic diagram of a shaft driven by a twisting moment T at section A, shown in Fig. 3.35. When the shaft comes up to a constant rotational speed ω rad/s, section A will have undergone a *relative* angle of twist ϕ to section B, as shown. At the constant operating speed ω of the shaft, the angle of twist of section A relative to section B is given by Eq. (3.12)

$$\phi = \frac{TL}{GJ} \tag{3.12}$$

where L is the length of the shaft. The maximum shear stress in the shaft is given by Eq. (3.14)

$$\tau_{max} = \frac{T(d/2)}{J} \tag{3.14}$$

where d is the outside diameter of the shaft.

It remains to express T in terms of the power transmitted by the shaft; recall that power is the rate of doing work. The power P transmitted by the shaft is given by

$$P = T\omega \tag{3.33}$$

where T is the twisting moment and ω is the rotational velocity in radians per second. The units of power P are usually given in watts (W) or foot-pounds per second. The following table shows the primary units:

	$P = T\omega$	Power
T	$N \cdot m$	$ft \cdot lb$
ω	$\frac{rad}{s}$	$\frac{rad}{s}$
P	$\frac{N \cdot m}{s} = W$	$\frac{ft \cdot lb}{s}$

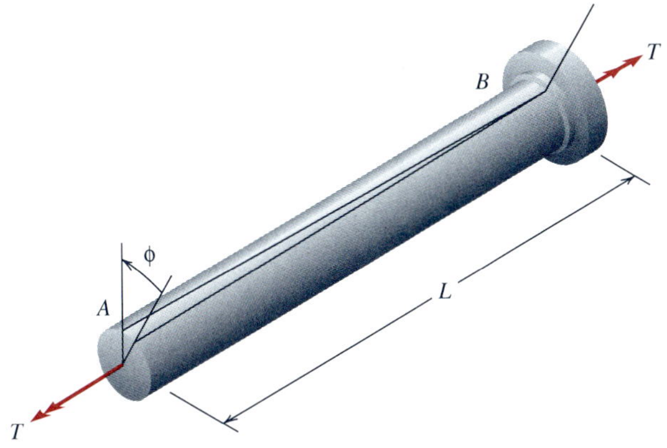

Figure 3.35 Schematic diagram of a circular shaft driven by a twisting moment T at section A.

Other units occur frequently in shaft design especially for the angular velocity ω, for example,

$$\omega = 2\pi f = 2\pi\left(\frac{n}{60}\right) \qquad \text{rad/s} \tag{3.34}$$

where f is the frequency in units of 1/s or hertz (Hz) and n is the frequency in units of revolutions per minute (rpm), that is, n represents the number of times the shaft rotates per minute.

Finally, an additional unit, generally familiar to us, is that of horsepower (hp), given by

$$1 \text{ hp} = 550 \text{ ft} \cdot \text{lb/s} = 6600 \text{ in} \cdot \text{lb/s} \tag{3.35}$$

For example, in the case of a 10-hp motor rotating at 60 rpm, we can find the torque acting from

$$10 \text{ hp} = T\omega$$

or

$$(10)(6600 \text{ in} \cdot \text{lb/s}) = T(2\pi)\left(\frac{60}{60}\right)$$

or

$$T = 1.05 \times 10^4 \text{ in} \cdot \text{lb}$$

Care must be taken with the units.

EXAMPLE 3.11

A 28-ft power boat has a 260-hp engine that delivers full power to the propeller at about 3800 rpm. If the shaft is a high-strength steel alloy with an allowable shear strength of 30 ksi, we wish to find the minimum allowable diameter of a solid shaft connecting the propeller to the final drive connection, as shown schematically in Fig. 3.36.

The power is given by Eq. (3.33)

$$P = T\omega \tag{a}$$

where in this case ω is in radians per second and T is in inch-pounds. We have for the angular velocity

$$\omega = 2\pi\left(\frac{n}{60}\right) = 2\pi\left(\frac{3800}{60}\right) = 398 \text{ rad/s} \tag{b}$$

where n = 3800 rpm.

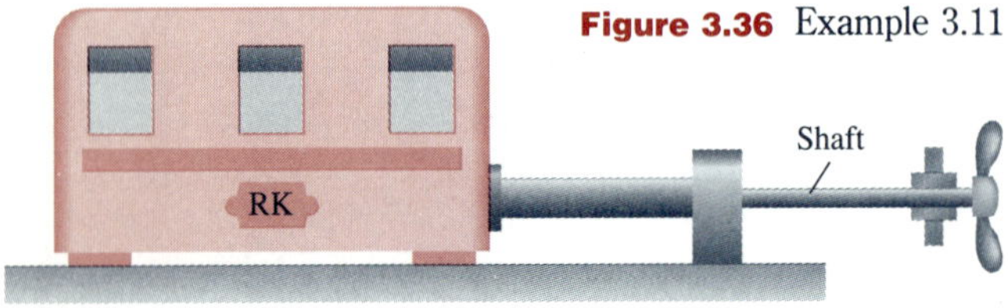

Figure 3.36 Example 3.11

From Eq. (3.35), the horsepower rating can be written as

$$P = 260 \text{ hp} = 260(6600 \text{ in} \cdot \text{lb/s}) = 1.716 \times 10^6 \text{ in} \cdot \text{lb/s} \tag{c}$$

From Eqs. (a) to (c), we have

$$T = \frac{P}{\omega} = \frac{1.716 \times 10^6 \text{ in} \cdot \text{lb/s}}{398 \text{ rad/s}} = 4.31 \times 10^3 \text{ in} \cdot \text{lb} \tag{d}$$

To determine the diameter of the shaft, we use Eq. (3.14) for the maximum shear stress in a shaft

$$\tau_{max} = \frac{T(d/2)}{J} \tag{e}$$

where $J = \pi d^4/32$.

Therefore, we have

$$\tau_{max} = \frac{16T}{\pi d^3} \tag{f}$$

or upon solving for the diameter and by use of Eq. (d), we have

$$d^3 = \frac{16T}{\pi \tau_{max}} = \frac{(16)(4.31 \times 10^3 \text{ in} \cdot \text{lb})}{(\pi)(30 \times 10^3 \text{ lb/in}^2)} = 0.732 \text{ in}^3 \tag{g}$$

Finally, the required diameter of the shaft is given by

$$d = 0.90 \text{ in} \qquad (h)$$

A standard 1-in shaft would be appropriate since weight is not a major consideration; the maximum shear stress in the shaft would then equal about 22 ksi, which is less than the specified allowable stress of 30 ksi.

EXAMPLE 3.12

We wish to determine the minimum diameter of a steel shaft required to transmit 8 kW at a rotational speed of 15 Hz if the allowable maximum shear stress in the shaft is not to exceed 30 MPa.

Again we use Eq. (3.33) to determine the torque T from which the maximum shear stress using Eq. (3.14) can be calculated

$$P = T\omega \qquad (a)$$

In this case, $\omega = 2\pi f = 2\pi(15)$ rad/s, and

$$P = 8 \text{ kW} = 8 \times 10^3 \text{ N} \cdot \text{m/s} \qquad (b)$$

Therefore,

$$T = \frac{P}{\omega} = \frac{8 \times 10^3 \text{ N} \cdot \text{m/s}}{30\pi \text{ rad/s}} = 84.88 \text{ N} \cdot \text{m} \qquad (c)$$

The diameter of the shaft is determined from Eq. (3.14) and from Eq. (c):

$$d^3 = \frac{16T}{\pi\tau_{max}} = \frac{(16)(84.88)}{(\pi)(30 \times 10^6)} = 1.44 \times 10^{-5} \text{ m}^3$$

or

$$d = 24.3 \text{ mm} \qquad (d)$$

The required diameter of the shaft is 24.3 mm; a 1-in shaft would do the job.

3.11 Differential Equations for the Twisting Moment and Angle of Twist

In developing the twisting moment–deformation relation for a circular shaft, Eq. (3.16) and Fig. 3.9, we assumed that the radius of the shaft was constant and that an equal and opposite twisting moment was applied to each end of the shaft. If, however, a distributed twisting moment is applied *along* the shaft, then the internal twisting moment in the shaft will vary along the axis of the shaft. In this case, we wish to derive an equation to determine the change in the twisting moment and the angle of twist along the shaft when an applied distributed twisting moment acts along the shaft. The formulation will follow an approach similar to that used for forces in bars in Sec. 2.8.

In Fig. 3.37*a* we show a circular shaft whose radius changes along the axis and upon which an applied distributed twisting moment $q(x)$ is acting. The x axis is along the axis of the shaft. The units of $q(x)$ are force × length/length, or force. The shaft is of length $L = x_B - x_A$, and in addition to $q(x)$ that is distributed along the axis of the shaft, twisting moments T_A and T_B are applied at the end sections A and B, as shown in Fig. 3.37*a*. All twisting moments are shown as positive.

We will investigate the equilibrium of a small element of length Δx cut from the shaft, as shown in Fig. 3.37*b*. The internal twisting moment at section x is $T(x)$, and the internal twisting moment at section $x + \Delta x$

is $T(x + \Delta x)$; both are shown as positive. The twisting moment contribution of $q(x)$ to the element is $q(x + \Delta x/2)\,\Delta x$. Moment equilibrium about the axis of the element of length Δx gives

$$-T(x) + T(x + \Delta x) + q\left(x + \frac{\Delta x}{2}\right)\Delta x = 0 \tag{3.36}$$

In the limit as $\Delta x \to 0$, we have

$$\frac{dT}{dx} + q(x) = 0 \tag{3.37}$$

Equation (3.37) relates the rate of change in T at x to the value of q at that value of x.

The relative angle of twist along the shaft satisfies, from Eq. (3.10),

$$\frac{d\phi}{dx} = \frac{T}{GJ} \tag{3.10}$$

or

$$T = GJ\frac{d\phi}{dx} \tag{3.38}$$

Therefore upon combining Eqs. (3.37) and (3.38), we have

$$\frac{d}{dx}\left(GJ\frac{d\phi}{dx}\right) + q(x) = 0 \tag{3.39}$$

Equation (3.39) is completely analogous to Eq. (2.28) in the absence of temperature change. If $q = 0$ and GJ is constant, then from Eq. (3.37)

$$\frac{dT}{dx} = 0 \tag{3.40}$$

which gives the result that the twisting moment along the shaft is independent of x; that is, upon integration of Eq. (3.40), we find that T is a constant.

In the general case in which $q = q(x)$ is known, Eq. (3.37) can be integrated once to find $T = T(x)$. With $T(x)$ known, we can integrate Eq. (3.38) once to find $\phi(x)$. The two constants of integration arising from the two integrations can be evaluated from two constraints on the system. Alternatively, with $q = q(x)$ known, we can directly integrate Eq. (3.39) twice to find $\phi(x)$; again two conditions imposed on the solution are used to evaluate the two constants of integration.

For example, with $q(x) = 0$ and GJ equal to a constant, the angle of twist satisfies

$$\frac{d^2\phi}{dx^2} = 0 \tag{3.41}$$

Upon integration we find

$$\phi(x) = C_1 x + C_2 \tag{3.42}$$

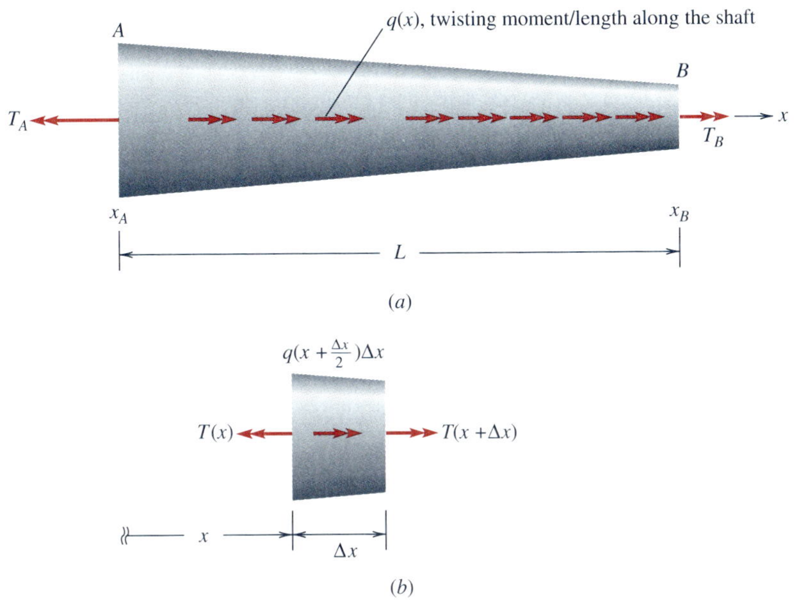

Figure 3.37 (*a*) A circular shaft of variable radius under a twisting moment per unit length $q(x)$ along the shaft. (*b*) An element of length Δx from the shaft of (*a*).

where C_1 and C_2 are constants of integration. If the angle of twist at $x = x_A$ is ϕ_A and the angle of twist at $x = x_B$ is ϕ_B, the constants are found from the conditions

$$\begin{aligned} \phi_A &= C_1 x_A + C_2 \\ \phi_B &= C_1 x_B + C_2 \end{aligned} \tag{3.43}$$

or

$$C_1 = \frac{\phi_B - \phi_A}{x_B - x_A} \qquad C_2 = -C_1 x_B + \phi_B \tag{3.44}$$

Therefore the angle of twist in the shaft in this case is linear with

$$\phi(x) = \frac{\phi_B - \phi_A}{x_B - x_A}(x - x_B) + \phi_B \tag{3.45}$$

We have used the result that ϕ is linear with x in our earlier discussion. Further, still with $q(x) = 0$ and constant GJ, we have from Eqs. (3.45) and (3.38)

$$\frac{d\phi}{dx} = \frac{\phi_B - \phi_A}{x_B - x_A} = \frac{T}{GJ}$$

or

$$T = \frac{GJ}{L}(\phi_B - \phi_A) \tag{3.46}$$

where $T = \text{constant} = T_A = T_B$. Equation (3.46) is the usual twisting moment–deformation relation for a uniform shaft, Eq. (3.16).

EXAMPLE 3.13

A steel shaft AB of diameter $d = 50$ mm is loaded by a distributed twisting moment $q_0 = 200\ \text{N}\cdot\text{m/m}$ uniform along its length $L = 1.25$ m, as shown in Fig. 3.38*a*. We wish to determine the twisting moment and angle-of-twist distribution along the shaft. We take $G = 80$ GPa, and we neglect the weight of the shaft.

The x axis is taken along the axis of the shaft with the origin at section A. Section A is fixed at the wall so that $\phi_A = 0$. In addition, the *internal* twisting moment $T(x)$ must be equal to zero at the free end where $x = L$. We can see this by considering the moment equilibrium of an element of length Δx at the end of the shaft; see Fig. 3.38*b*. In Fig. 3.38*b*, $T(L - \Delta x)$ is the twisting moment, assumed positive, acting on the section $L - \Delta x$. Moment equilibrium about the axis of the small element Δx gives

$$-T(L - \Delta x) + q_0\,\Delta x = 0$$

In the limit as $\Delta x \to 0$, we have $T(L - \Delta x) \to T(L)$ and $q_0\,\Delta x \to 0$; therefore

$$T(L) = 0 \qquad (a)$$

Equation (*a*) tells us that since there is no applied concentrated moment at $x = L$, the internal moment at $x = L$ is zero. Equation (*a*) together with $\phi_A = 0$ provides two conditions to

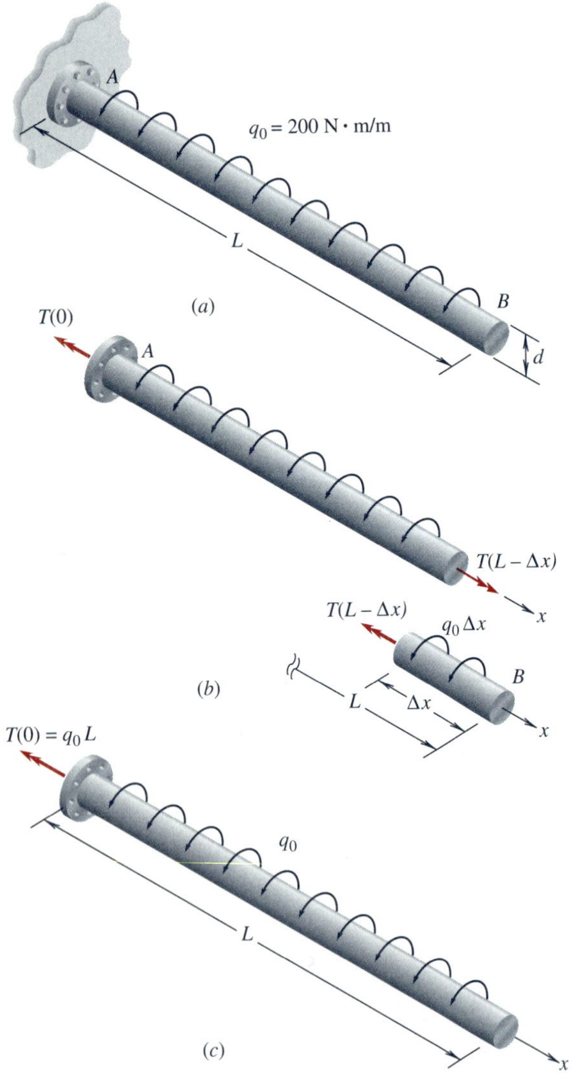

Figure 3.38 Example 3.13

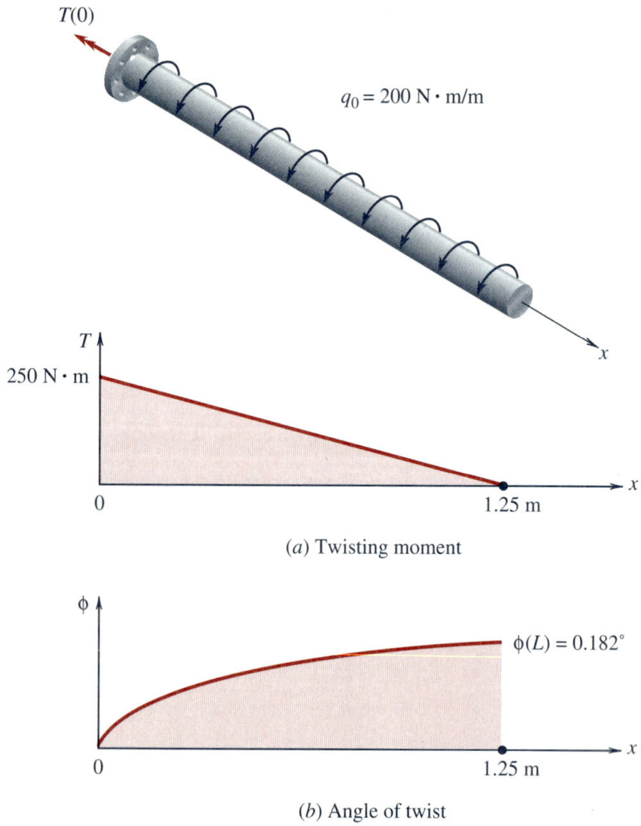

Figure 3.39 Example 3.13: Twisting moment and angle-of-twist diagrams.

be imposed on the solution. Since q_0 is a constant, we have from Eq. (3.37)

$$\frac{dT}{dx} = -q = -q_0$$

or upon integration

$$T(x) = -q_0 x + C_1 \qquad (b)$$

where C_1 is the constant of integration. If we use the condition that $T(L) = 0$, we have $C_1 = q_0 L$ and

$$T(x) = q_0(L - x) \qquad (c)$$

The value of the twisting moment from Eq. (*c*) at $x = 0$ is $T(0) = q_0 L$, which also follows from the overall twisting moment equilibrium of the shaft; see Fig. 3.38*c*.

From Eq. (3.10) and Eq. (*c*) we have

$$\frac{d\phi}{dx} = \frac{T}{GJ} = \frac{q_0(L - x)}{GJ} \qquad (d)$$

from which upon integration,

$$\phi(x) = \frac{q_0}{GJ}\left(Lx - \frac{x^2}{2}\right) + C_2 \qquad (e)$$

The constant $C_2 = 0$ since $\phi(0) = \phi_A = 0$. Therefore

$$\phi(x) = \frac{q_0 L}{GJ} x\left(1 - \frac{x}{2L}\right) \qquad (f)$$

so that the angle of twist varies quadratically with the distance x along the axis. At $x = L = 1.25$ m, we have

$$\begin{aligned} \phi(L) = \phi_B &= \frac{(200\ \text{N}\cdot\text{m/m})(1.25\ \text{m})(1.25\ \text{m})(0.5)}{(80 \times 10^9\ \text{N/m}^2)[(\pi)(0.05^4/32)\ \text{m}^4]} \\ &= 3.183 \times 10^{-3}\ \text{rad} \\ &= 0.182^\circ \end{aligned} \qquad (g)$$

The twisting moment at $x = 0$ is

$$T(0) = T_A = q_0 L = (200\ \text{N}\cdot\text{m/m})(1.25\ \text{m}) = 250\ \text{N}\cdot\text{m}$$

Plots of the distribution of the twisting moment and the angle of twist along the shaft are shown in Fig. 3.39*a* and *b*.

EXAMPLE 3.14

A circular shaft of length L, shown in Fig. 3.40*a*, is fixed at each end and loaded by a constant distributed twisting moment q_0 (with units of force × length/length). The material of the shaft has a shear modulus G, and the polar moment of inertia of the shaft is J. We wish to determine an expression for the twisting moment reactions at each support and the distribution of twisting angle along the shaft. In addition, if the diameter of the shaft is d, we want to find the maximum shear stress in the shaft. The x axis is along the shaft with the origin at section A.

The shaft is shown in Fig. 3.40*b* with the wall twisting reactions T_A and T_B. Overall moment equilibrium about the x axis gives

$$T_A = q_0 L + T_B \qquad (a)$$

Equation (3.39) with GJ equal to a constant and $q = q_0$ gives

$$\frac{d^2\phi}{dx^2} = -\frac{q_0}{GJ} \qquad (b)$$

Integration of Eq. (*b*) twice leads to an expression for ϕ in the form

$$\phi(x) = -\frac{q_0 x^2}{2GJ} + C_1 x + C_2 \qquad (c)$$

where C_1 and C_2 are the constants of integration. The geometric constraints on the shaft are

$$\phi(0) = \phi_A = 0 \qquad \phi(L) = \phi_B = 0 \qquad (d)$$

Therefore $C_2 = 0$ and $C_1 = q_0 L/(2GJ)$, and the expression for the angle distribution $\phi(x)$ along the shaft becomes

$$\phi(x) = \frac{q_0 L}{2GJ} x\left(1 - \frac{x}{L}\right) \qquad (e)$$

To determine the distribution of twisting moment, we have from Eq. (3.38)

$$T = GJ\frac{d\phi}{dx} = \frac{q_0 L}{2}\left(1 - \frac{2x}{L}\right) \qquad (f)$$

Therefore the twisting moment distribution along the shaft is linear, and

$$T_A = T(0) = \frac{q_0 L}{2} \qquad T_B = T(L) = -\frac{q_0 L}{2} \qquad (g)$$

where the minus sign for T_B indicates that the direction of T_B is to be reversed. We show the shaft with the wall reactions in Fig. 3.40*c*. Plots of the distribution of the internal twisting moment and the angle of twist along the shaft are shown in Fig. 3.40*d*.

The maximum and minimum values of the twisting moment are equal in magnitude at the supports, $|T_{max}| = |T_{min}| =$

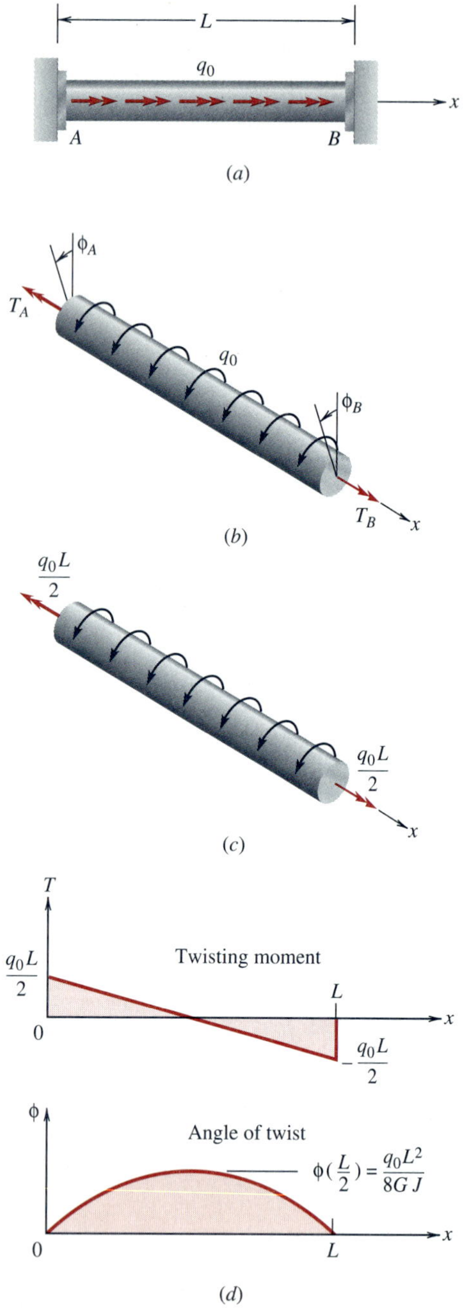

Figure 3.40 Example 3.14

$q_0 L/2$, so that the maximum shear stress occurs at each support, and from Eq. (3.14) the magnitude of the maximum shear stress is given by

$$|\tau_{max}| = \frac{d}{2}\frac{T_{max}}{J} = \frac{d}{4}\frac{q_0 L}{J} \tag{h}$$

3.12 Concluding Remarks

In this chapter we have investigated the response of circular elastic shafts to applied twisting moments. In each case the shaft was either a single shaft or made up of different interconnected shafts. In all cases we found that we could determine the twisting moments and angles of twist along the shafts by using the three-step solution method of Eq. (2.8). The method works well for both statically determinate and statically indeterminate shafts. When we used the equilibrium equations, force-deformation relations, and geometric constraints in a systematic fashion, we found that interconnected shafts could be analyzed by a simple computer program. The computer program TORMECH allows us to analyze and solve such problems in a simplified manner. Again we emphasize that attention must be paid to the coordinate axis and to the convention for positive twisting moments and for positive angles of twist with respect to the axis that we use in the program.

PROBLEMS

3.2-1 Use the discussion in Sec. 3.2 to present an argument based on symmetry that no deformation out of the plane of the cross section will occur for a circular shaft acted upon by a twisting moment.

3.3-1 Verify the integration in Eq. (3.9) to obtain the expression for the polar moment of inertia J.

3.3-2 Show that the expression for the maximum shear stress in a solid circular shaft, Eq. (3.14), also can be written in the form

$$\tau_{\max} = \frac{2T}{\pi a^3} = \frac{16T}{\pi d^3}$$

where a is the radius and d the diameter of the shaft.

3.4-1 The torsional stiffness is given by GJ/L in Eq. (3.16). What is the value of the torsional stiffness for a steel rod 1 ft long and 1 in in diameter? If the diameter doubles, what is the change in stiffness? $G = 11.5 \times 10^6$ psi.

3.5-1 Verify that the integration in Eq. (3.17) to obtain the polar moment of inertia J for a hollow tube is correct.

3.5-2 Show that Eq. (3.18) will reduce to Eq. (3.20) when the thickness $t = r_o - r_i$ is small compared to r_i. If $t = 0.1r_i$, what is the percentage difference between the two formulas for J?

Fig. P3.5-3

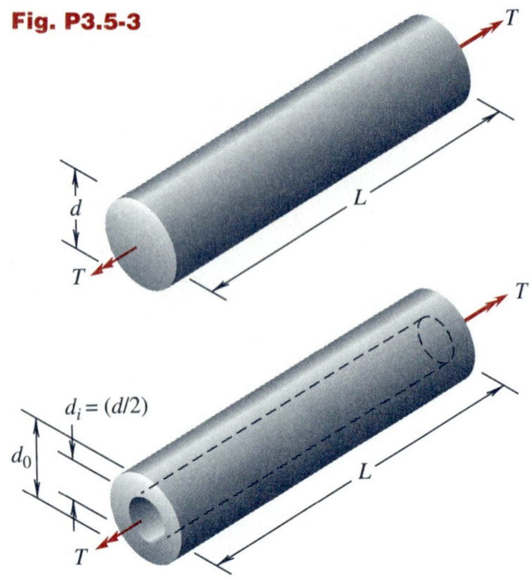

3.5-3 A solid shaft of diameter d and a hollow shaft of inside diameter $d_i = d/2$ of the same length and weight transmit the same twisting moment T, as shown in Fig. P3.5-3. Find the maximum shear stress in each shaft and the relative angle of twist between the ends of each shaft.

3.6-1 Consider Example 3.1, Fig. 3.12. If the maximum allowable shear stress of the shaft shown in Fig. 3.12*a* is 70 MPa, what is the maximum value of the twisting moment that can be applied to the shaft? If the diameter of the shaft is doubled, what is the maximum value of the twisting moment?

3.6-2 Consider Example 3.1, Fig. 3.12. If the maximum allowable angle of twist at section B is limited to 2°, what is the maximum value of the twisting moment that can be applied to the shaft?

3.6-3 A hollow steel shaft 2.0 m long is to transmit a torque of 15 kN · m. The inside radius of the shaft is 50 mm, and the outside radius of the shaft is 75 mm. Find the relative angle of twist between the ends of the shaft and the maximum shear stress. Take $G = 80$ GPa.

3.6-4 A high-strength steel shaft of 1.25-in diameter and 2-ft length transmits a twisting moment T just below that which causes permanent deformation of the shaft. Permanent deformation of the shaft will occur at a maximum shear stress of 15,000 psi; $G = 11.5 \times 10^6$ psi. Estimate the angle of twist between the ends of the shaft under the twisting moment T.

3.6-5 A hollow steel shaft 2.5 m long must transmit a twisting moment of 25 kN · m. The total angle of twist between the ends of the shaft is not to exceed 2.0°, and the maximum allowable shearing stress is 82 MPa. Find the values of the inside and outside diameters of the shaft. $G = 80$ GPa.

3.6-6 A hollow steel shaft of 2-in outside diameter is to transmit a twisting moment without exceeding a maximum allowable shear stress of 40,000 psi. What is the maximum internal diameter of the shaft which will allow it to transmit a twisting moment without exceeding the allowable shear stress of (*a*) 2000 ft · lb and (*b*) 4000 ft · lb? Comment on the nature of the nonlinearity of the dependence of the inside diameter on the value of the twisting moment.

3.6-7 A series of circular rods of fresh compact bone from a human femur was tested in torsion. If the samples had a cross-sectional diameter of approximately 1.9 mm and four tests yielded breaking twisting moments of 0.100, 0.125, 0.110, and 0.120 N · m, find the average torsional shear stress at breaking. Could you fracture a human femur in torsion in your hands?

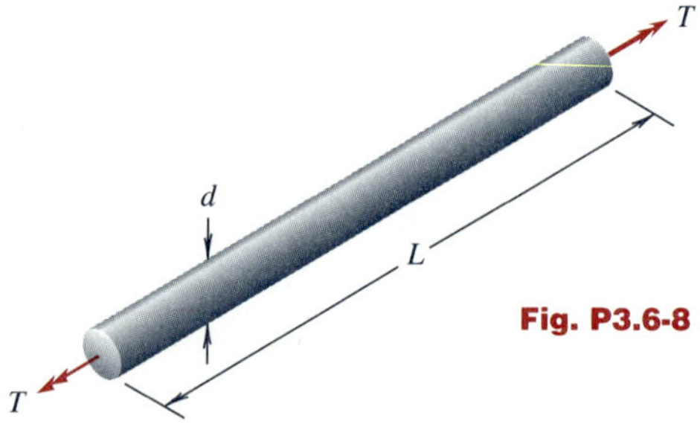

Fig. P3.6-8

3.6-8 An aluminum-alloy shaft is twisted by twisting moments T at each end, as shown in Fig. P3.6-8. If the angle of rotation of one end of the bar with respect to the other end is 4°, find the maximum shear stress in the shaft; $L = 1.1$ m, $d = 25$ mm, and $G = 25$ GPa.

3.6-9 Find the diameter d (Fig. P3.6-8) of a circular shaft subjected to a twisting moment $T = 3$ kN · m if the allowable shear stress in the shaft is 60 MPa and the allowable angle of twist per unit length is 0.3°/m. Take $G = 70$ GPa.

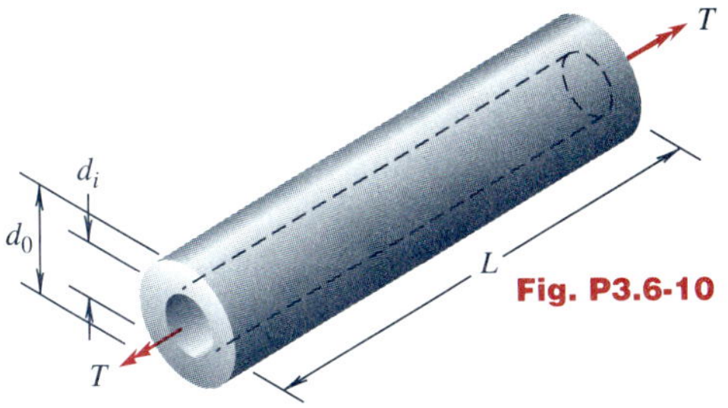

Fig. P3.6-10

3.6-10 A hollow steel shaft (Fig. P3.6-10) whose outside diameter $d_o = 4$ in and whose inside diameter $d_i = 3$ in is twisted by a twisting moment T. If $L = 10$ ft and $G = 11.5 \times 10^6$ psi, find the relative angle of twist of the ends when the maximum shear stress is 8000 psi.

3.6-11 A stepped shaft ABC is subjected to a twisting moment T at C and is fixed at A, as shown in Fig. P3.6-11. We wish to apply a countermoment directed opposite to T and of magnitude s times T, so that when both twisting moments are acting either (*a*) section B does not rotate or (*b*) section C does not rotate. Find values of s for each of these cases, using the values of the polar moments of inertia and lengths as indicated.

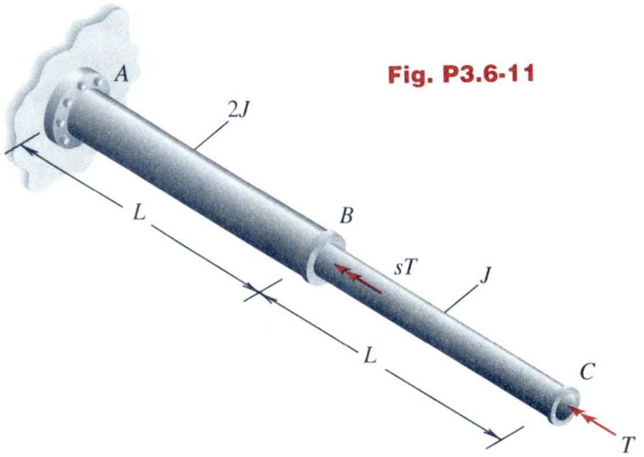

Fig. P3.6-11

3.6-12 A torsional spring device consists of a hollow cylindrical tube AB welded to a very stiff plate E and a solid cylindrical shaft which is welded to the plate at D. The solid shaft carries a twisting moment T at end C, as shown in Fig. P3.6-12. The hollow tube has an outside diameter of 2.5 in, a wall thickness of 0.25 in, and a length of 12 in. $G = 12{,}000$ ksi for the tube and shaft.

(*a*) Find the torsional stiffness of the device; i.e., find k where $T = k\phi_C$.

(*b*) Find the maximum shear stresses in the two parts of the device in terms of the moment T.

Fig. P3.6-12

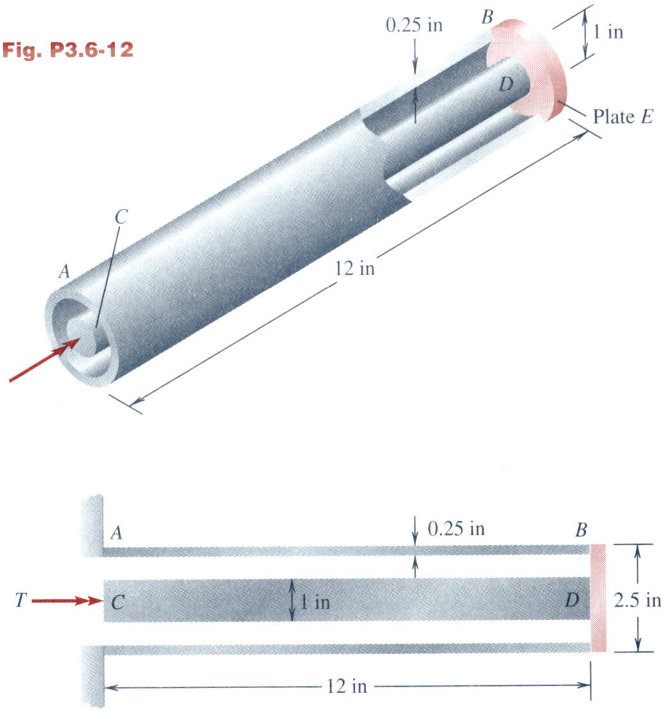

3.6-13 Consider Example 3.2, Fig. 3.13. If $T_B = 0$, find the angle of twist at sections B and C and the maximum shear stress in each tube. All other values are to remain the same. Plot the twisting moment distribution and the angle of twist along the tubes.

3.6-14 Consider Example 3.2, Fig. 3.13. Find the value of T_B such that the angle of twist at section B is zero. Also find the maximum shear stress in each tube. All other values remain the same. Plot the twisting-moment distribution and the angle of twist along the tubes.

3.6-15 Consider Example 3.2, Fig. 3.13. Find the value of T_B such that the angle of twist at section B is the negative of the value at section C. Plot the twisting-moment distribution and the angle of twist along the tubes. Assume all other values remain the same.

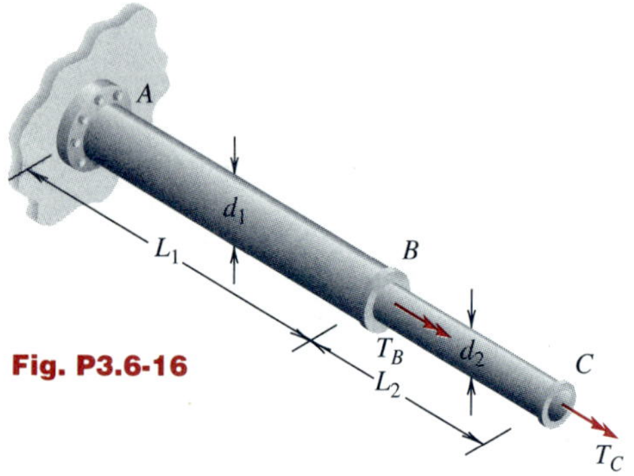

Fig. P3.6-16

3.6-16 A model of a steel shaft in a drive train consists of two segments, as shown in Fig. P3.6-16. Two twisting moments are applied as shown at sections B and C. The values are $L_1 = 1$ m, $d_1 = 75$ mm, $L_2 = 0.8$ m, $d_2, = 50$ mm, and $G = 75$ GPa. If $T_B = 1$ kN · m and $T_C = 0.65$ kN · m, plot the twisting moment and the angle of twist distribution along the shaft, and determine the maximum shear stress in each segment.

3.6-17 Consider Prob. 3.6-16. If the sign of T_C is now changed so that $T_C = -0.65$ kN · m, solve the problem as stated.

3.6-18 A hollow aluminum shaft is made up of two segments, as shown in Fig. P3.6-18. The values are $L_1 = 3$ ft, $d_{o1} = 2$ in, $d_{i1} = 1$ in, $L_2 = 2$ ft, $d_{o2} = 1.75$ in, $d_{i2} = 1.0$ in, and $G = 3.8 \times 10^6$ psi. What is the maximum twisting moment T that can be applied to the shaft if the allowable angle of twist between the ends of the shaft is 1.2° and the maximum allowable shear stress is 5000 psi?

Fig. P3.6-18

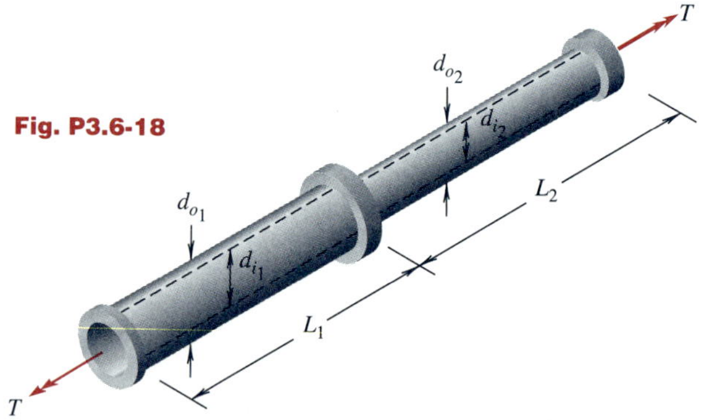

3.6-19 Consider Example 3.3, Fig. 3.14. If the system is changed so that now $T_B = 5$ N · m and $T_C = 15$ N · m with all other values the same, plot the twisting moment and angle of twist distribution along the shaft. Also determine the maximum shear stress in each segment of the shaft.

3.6-20 Consider Example 3.3, Fig. 3.14. If the diameter of the shaft is increased to 17 mm with all other values remaining the same, determine the twisting moment in each segment of the shaft and the angles of twist at sections B and C. How does the twisting moment in each segment vary with the diameter of the shaft?

3.6-21 Consider Example 3.3, Fig. 3.14. If the maximum angle of twist at section C relative to section A is to be limited to 7°, find the appropriate diameter of the shaft. All other values remain the same.

3.6-22 Consider Example 3.3, Fig. 3.14. If the maximum shear stress in the shaft is to be limited to 60 MPa, find the minimum diameter of the shaft. Also find the angle of twist of section C relative to section A. All other values remain the same.

3.6-23 Consider Example 3.4, Fig. 3.15. If the gears at B are replaced with two 4-in-diameter gears, calculate the angle of twist at section C relative to section A. All other values remain the same. What is the effect of changing the gear ratio?

3.6-24 Consider Example 3.4, Fig. 3.15. If the gears are replaced with two gears such that the radius of the upper gear is r_U and the radius of the lower gear is r_L, with $r_U + r_L = 4$ in, obtain a general expression for the angle of twist at section C relative to section A. All other values remain the same.

3.6-25 A solid shaft whose diameter is 225 mm is to be replaced by a hollow shaft whose ratio of outside to inside diameter is 2. We wish to find the diameters of the new shaft if the maximum twisting moment and the shear stress experienced by both shafts are the same. Also determine the savings in weight in changing from the old solid shaft to the new hollow shaft.

3.6-26 A propeller shaft has an outside diameter of 400 mm and an inside diameter of 200 mm and is subjected to a twisting moment of 400 kN · m. If $G = 75$ GPa, estimate the maximum shear stress and the angle of twist in a length of the shaft equal to 20 diameters. What diameter would be required for a solid shaft with the same maximum stress and twisting moment?

3.6-27 A steel tube as shown in Fig. P3.6-27 is subjected to a twisting moment of 600 lb · in. If the outside diameter d of the tube is 1.5 in, the thickness t is 0.025 in, and the length L is 10 ft, find the maximum shear stress and the relative angle of twist between the ends of the tube. Take $G = 12.5 \times 10^6$ psi.

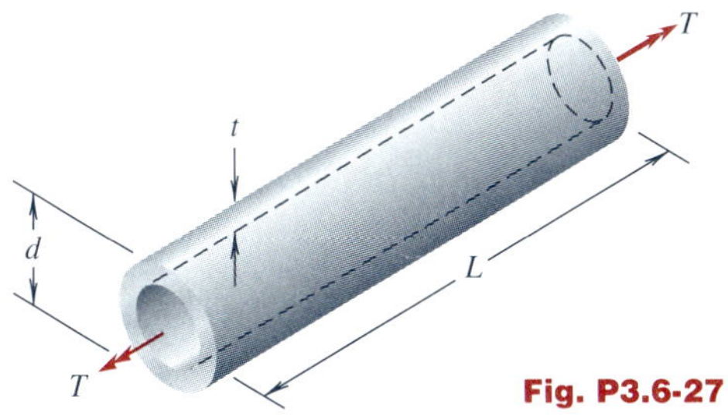

Fig. P3.6-27

3.6-28 A hollow steel tube as shown in Fig. P3.6-27 has an outside diameter $d = 10$ in and a wall thickness of $t = 1$ in. What is the value of the twisting moment T to cause a maximum shearing stress of 11,000 psi in the tube?

3.6-29 If a solid-steel shaft transmitting power has a maximum allowable shear stress of 6000 psi, find the angle of twist for a length equal to 20 times the diameter of the shaft. Take $G = 12.5 \times 10^6$ psi.

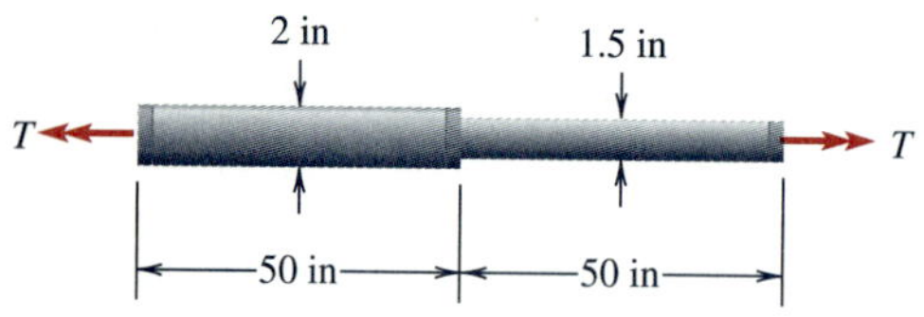

Fig. P3.6-30

3.6-30 A shaft subjected to a twisting moment T consists of two segments of equal length but with different diameters, as shown in Fig. P3.6-30. What is the allowable torque T_a if the relative angle of twist between the ends is not to exceed 1°? Take G = 12,000 ksi.

3.6-31 A compact torsional spring design (see Prob. 3.6-12) is shown in Fig. P3.6-31. A twisting moment T_0 is applied at A to the elastic shaft AB with shear modulus G_1. Member AB is connected to the thin-walled tube CD of radius a and shear modulus G_2; the thin-walled tube in turn is fixed to the wall at D. Derive an expression for the total angle of twist at A arising from T_0. Express the result in terms of G_1, G_2, L, a, d, t, and T_0.

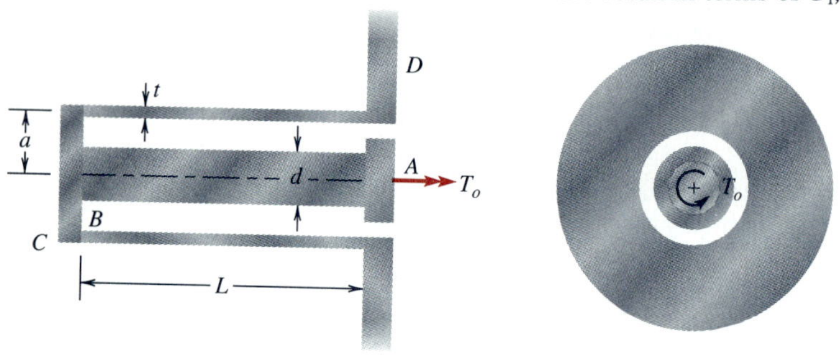

Fig. P3.6-31

3.6-32 A hollow aluminum shaft BC is attached at B to a solid-brass shaft AB and at C to a rigid support, as shown in Fig. P3.6-32. Twisting moments are applied at sections A and B as shown. Find the maximum shear stress in each shaft, and find the angle of rotation at section A and at section B. Take G_{al} = 4000 ksi, G_{br} = 6000 ksi, and $(d_i/d_o)_{BC}$ = 0.8 with d_o = 2 in.

3.6-33 A schematic diagram of a torsion bar for an automobile front suspension is shown in Fig. P3.6-33. The maximum load P from the wheels is 900 lb, d = 1.5 in, a = 20 in, and L = 30 in. Determine the spring constant of the torsion

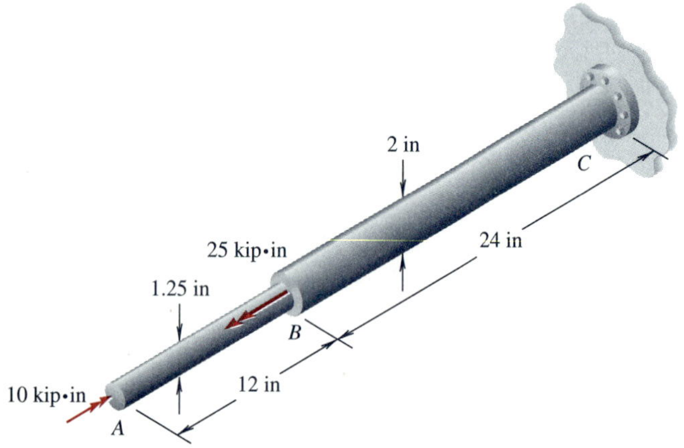

Fig. P3.6-32

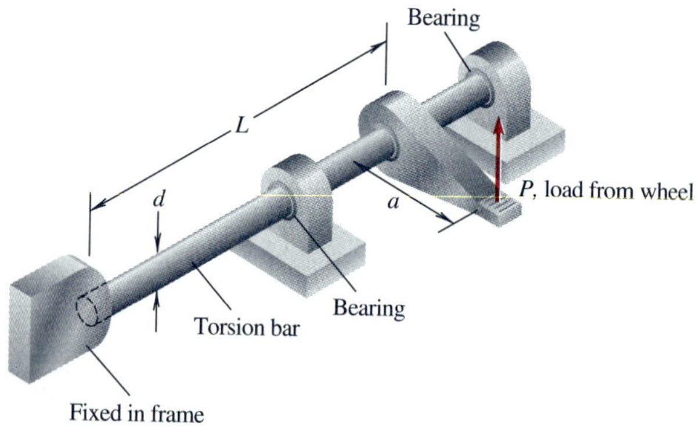

Fig. P3.6-33

bar, i.e., the ratio of the load P to the deflection at the load, in (lb/in) and the maximum shear stress in the bar. Neglect the weight of all components, and take $G = 12 \times 10^6$ psi.

3.6-34 Carry out the solution to Prob. 3.6-33 if now $d = 1.75$ in. How does the spring constant vary with the diameter d?

3.6-35 A model of a microbalance component shown in Fig. P3.6-35 consists of a quartz fiber $d = 10$ μm in diameter loaded by a twisting moment at B of $T_B = 2 \times 10^{-11}$ N · m. Find the angle of twist at A if the twisting moment T at A is sufficient to prevent rotation of the section at B. Neglect the weight of the components, and take $G = 30$ GPa.

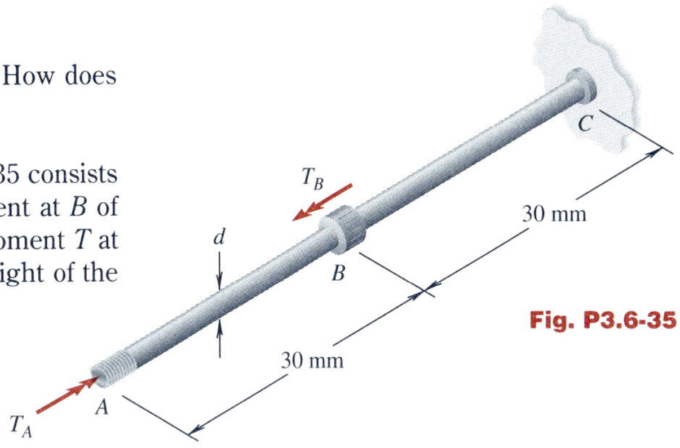

Fig. P3.6-35

3.6-36 Estimate the maximum shear stress in the shank of a screw driver if the diameter d is 6 mm, as shown in Fig. P3.6-36. The twisting torque on the shaft from a hand is about 1 N · m. If the length L of the shaft is about 130 mm, determine the angle of twist between the ends. Take $G = 70$ GPa.

Fig. P3.6-36

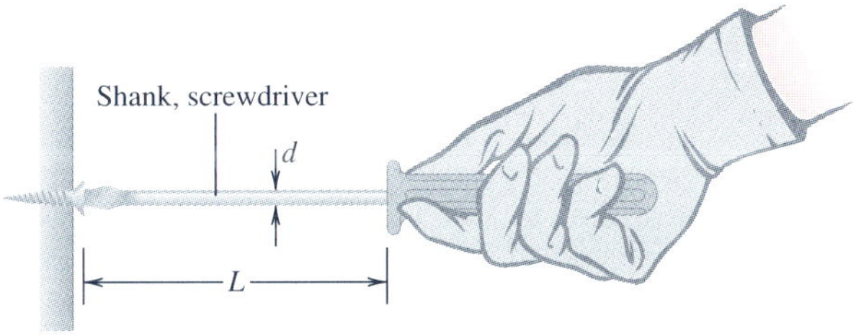

3.6-37 A shaft supported by two bearings is subjected to three twisting moments at the gears, as shown in Fig. P3.6-37. Determine the maximum shear stress in each segment between the gears and the relative angle of twist between the end gears. Take $d = 1.5$ in and $G = 12 \times 10^6$ psi.

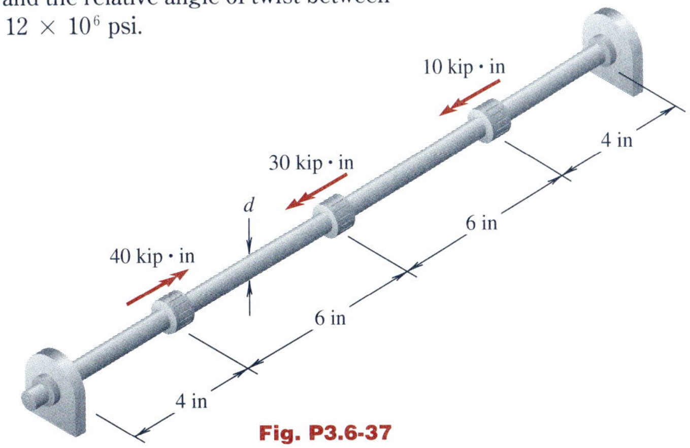

Fig. P3.6-37

3.6-38 A circular cylindrical shaft AB, with 2-in outside diameter, has a concentric hole of diameter $d = 1.45$ in drilled in from end B, as shown in Fig. P3.6-38. If a twisting moment $T = 15{,}000$ lb · in is applied at each end, find the angle of twist of end B relative to end A. Take $G = 20 \times 10^6$ psi.

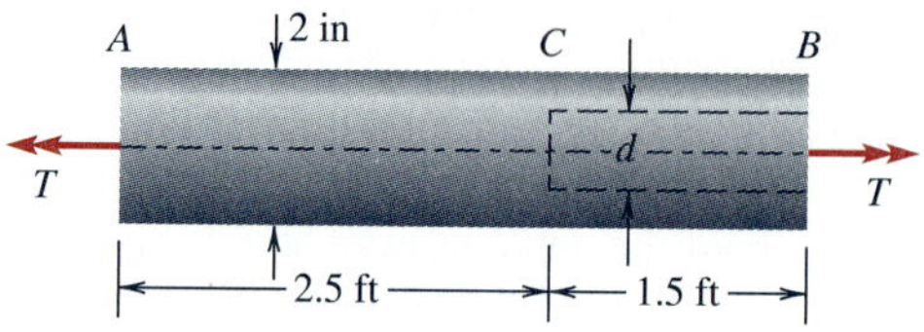

Fig. P3.6-38

3.6-39 An elastic shaft AB of diameter d and length L has a concentric hole of diameter $d/2$ bored halfway along the axis, as shown in Fig. P3.6-39. A twisting moment T is applied at end B as shown. Determine an expression for the angle of twist at end B and for the maximum shear stress in the shaft. The shear modulus is G.

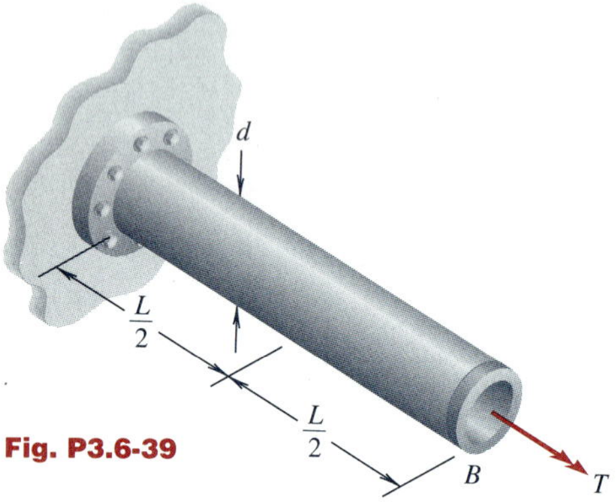

Fig. P3.6-39

3.6-40 The steel tube and shaft shown in Fig. P3.6-40 are welded together and subjected to two twisting moments as indicated. Determine T_1 and T_2 so that the maximum shear stress in the tube and the maximum shear stress in the shaft are the same and each equal to 10,000 psi. Also determine the angle of twist of the free end of the tube. Take $G = 12 \times 10^6$ psi.

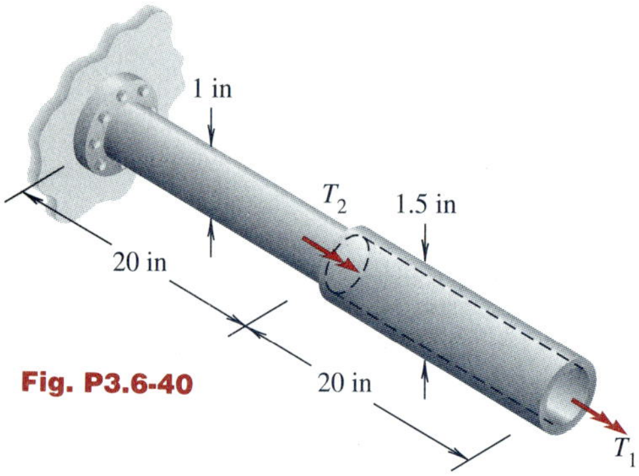

Fig. P3.6-40

3.6-41 The stepped circular shaft shown in Fig. P3.6-41 carries a torque of T. What is the allowable value of T if the angle of twist between the ends of the shaft is not to exceed 0.02 rad and the shear stress is not to exceed 4000 psi? Take $G = 12 \times 10^6$ psi.

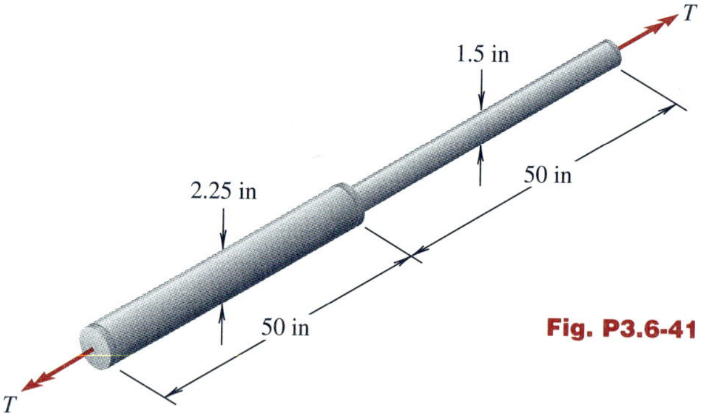

Fig. P3.6-41

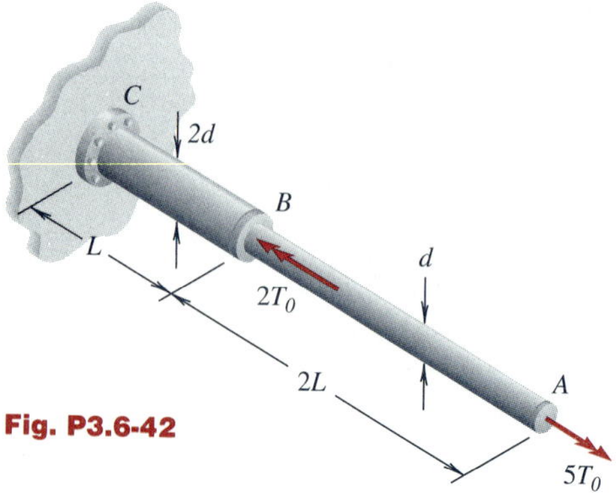

Fig. P3.6-42

3.6-42 Two shafts are joined at B and subjected to the two twisting moments as shown in Fig. P3.6-42. Obtain an expression for the total angle of twist at section A and the maximum shear stress in each segment of the shaft. The shear modulus is G.

3.6-43 The model circular shaft shown in Fig. P3.6-43 is subjected to a twisting moment $T_C = 2000$ in · lb at section C and a second twisting moment at section B, T_B, and is fixed at section A. It is required that the total angle of twist at section C not exceed 0.5°. What must the twisting moment at section B equal to meet this condition? Take $G = 15 \times 10^6$ psi.

Fig. P3.6-43

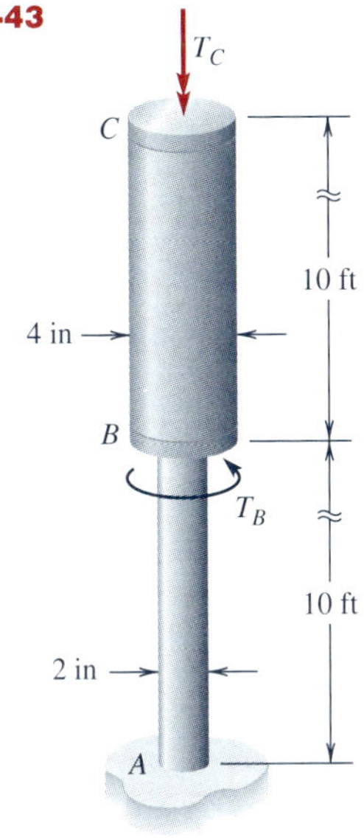

3.7-1 Consider Example 3.5, Fig. 3.16. If the maximum allowable shear stress in the shaft is 5000 psi, what is the maximum value of T that can be applied to the shaft? All other values remain the same.

3.7-2 Consider Example 3.5, Fig. 3.16. If the maximum allowable shear stress in the shaft is 4000 psi and $T = 80$ lb · ft, what is the minimum allowable diameter of the shaft so as not to exceed the maximum allowable shear stress? All other values remain the same.

3.7-3 Consider Example 3.5, Fig. 3.16. If the diameter of the shaft is increased to 1.25 in, what is the maximum shear stress in the shaft? All other values remain the same. How does the maximum shear stress vary with the diameter of the shaft?

3.7-4 Consider Example 3.5, Fig. 3.16. If all geometric dimensions are doubled, what is the maximum shear stress in the shaft? The values of T and G remain the same. Sketch the twisting moment and angle of twist distribution along the shaft.

3.7-5 Consider Example 3.5, Fig. 3.16. For a twisting moment of $T = 50$ lb · ft applied at section B, it was found that the resulting rotation of section B was $\phi_B = 1.375 \times 10^{-2}$ rad. Thus the stiffness $k_T = T/\phi_B$ is given by $50/(1.375 \times 10^{-2}) = 3636$ lb · ft/rad. For a given application, segment BC is to be redesigned by replacing it with a solid shaft of diameter d so as to increase k_T to twice its previous value. Find this new diameter d.

3.7-6 Consider Example 3.6, Fig. 3.17. Verify the results in Eqs. (h) and (i). If $L = L_{AB} + L_{BC}$, we have

$$T_{AB} = T\frac{L_{BC}}{L} \qquad T_{BC} = -T\frac{L_{AB}}{L}$$

3.7-7 A composite shaft is made up by bonding an inner circular cylindrical elastic shaft with shear modulus G_1 to an outer circular elastic tube with shear modulus G_2, as shown in Fig. P3.7-7. On the basis of symmetry arguments, we can show that the shear strain distribution is still given by Eq. (3.2), i.e.,

$$\gamma = r\frac{d\phi}{dx} \qquad (a)$$

Show that the shear stress distribution across the section is given by

$$\tau = G_1 r\frac{d\phi}{dx} \quad 0 < r < r_i \qquad \tau = G_2 r\frac{d\phi}{dx} \quad r_i < r < r_o \qquad (b)$$

Therefore the twisting moment in terms of the shear stress can be written

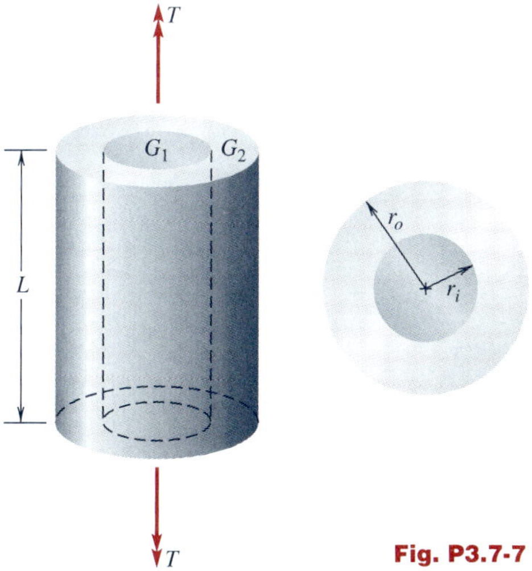

Fig. P3.7-7

$$T = \int_A r\tau \, dA = \frac{d\phi}{dx}(G_1 J_1 + G_2 J_2) \qquad (c)$$

where

$$J_1 = \frac{\pi}{2} r_i^4 \qquad J_2 = \frac{\pi}{2}(r_o^4 - r_i^4)$$

Show that the relative angle of twist between the ends of the shaft of length L, as shown in Fig. P3.7-7, is given by

$$\phi = \frac{TL}{G_1 J_1 + G_2 J_2} \qquad (d)$$

and the shear stress distribution is given by

$$\tau = \begin{cases} \dfrac{G_1 T r}{G_1 J_1 + G_2 J_2} & 0 < r < r_i \\ \dfrac{G_2 T r}{G_1 J_1 + G_2 J_2} & r_i < r < r_o \end{cases} \qquad (e)$$

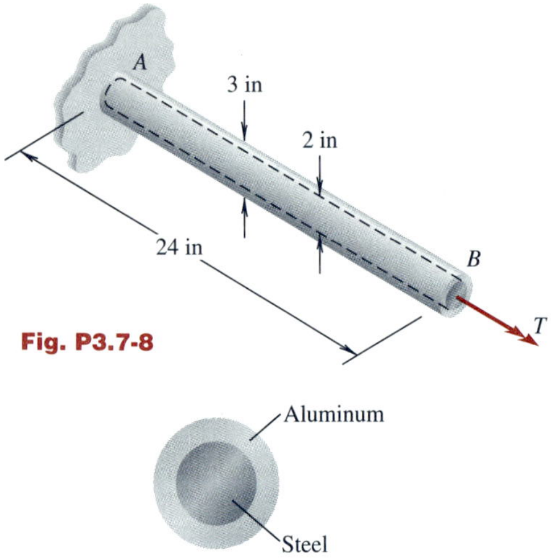

Fig. P3.7-8

3.7-8 For corrosion protection a 2-in-diameter steel shaft is encased in a hollow aluminum shaft with outside diameter 3 in, as shown in Fig. P3.7-8. The maximum allowable shear stress in the aluminum is 10 ksi and in the steel is 16 ksi. Find the maximum allowable value of T. Take G for steel and aluminum as 12,000 and 5000 ksi, respectively.

3.7-9 Consider the composite shaft of Prob. 3.7-8. In another application the same shaft is fixed to rigid walls at *both* ends A and B, and the twisting moment T is applied at the midpoint, as shown in Fig. P3.7-9. Find the maximum value of T such that the maximum allowable stresses in both the steel and aluminum are not exceeded. Also find the angle of twist at C.

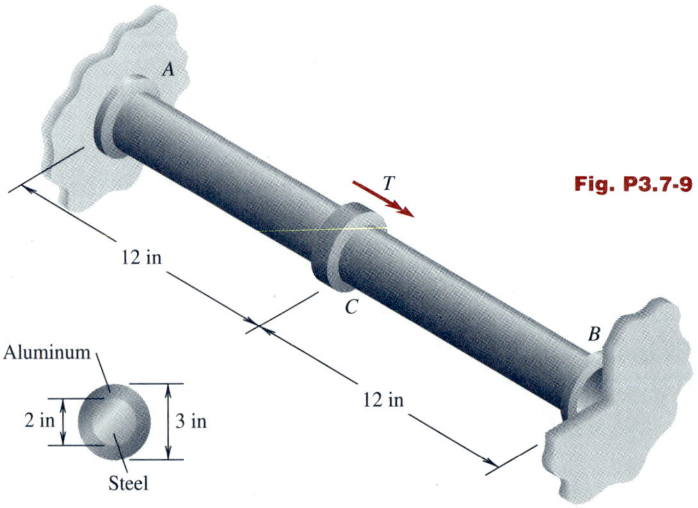

Fig. P3.7-9

3.7-10 A steel shaft with diameter $d = 20$ mm is loaded as shown in Fig. P3.7-10. Find the wall reactions, the maximum shear stress in the shaft, and the angle of twist at the section at which T is applied. If the direction of T is reversed, what happens to the values just calculated? Take $G = 75$ GPa.

Fig. P3.7-10

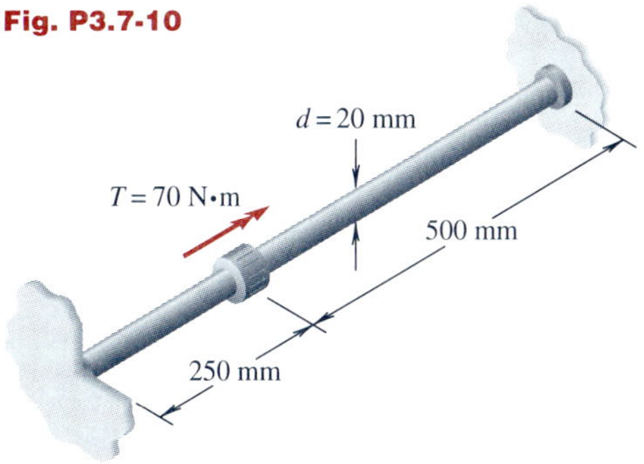

3.7-11 Solve Prob. 3.7-10 if $d = 30$ mm. What is the effect of increasing d on the values of the maximum shear stress, the angle of twist, and the wall reactions?

3.7-12 A circular steel shaft of diameter d_1 is provided at sections A and B with an enlarged diameter portion as shown in Fig. P3.7-12 for the purpose of attaching a tube of thickness t over the shaft. The shaft is first twisted by a twisting moment T and the tube is placed on the shaft and glued to the enlarged portions of the twisted shaft. The twisting moment remains until the glue is set, and the tube is firmly set on the shaft. The twisting moment T is then removed. This process sets up a "built-in" angle of twist in the shaft and tube. Find the twisting moment in the tube and the shaft after T is removed. Take $G_1 = G_2 = 70$ GPa, $d_2 = 50$ mm, $d_1 = 25$ mm, $t = 1.5$ mm, and $T = 80$ N · m.

Fig. P3.7-12

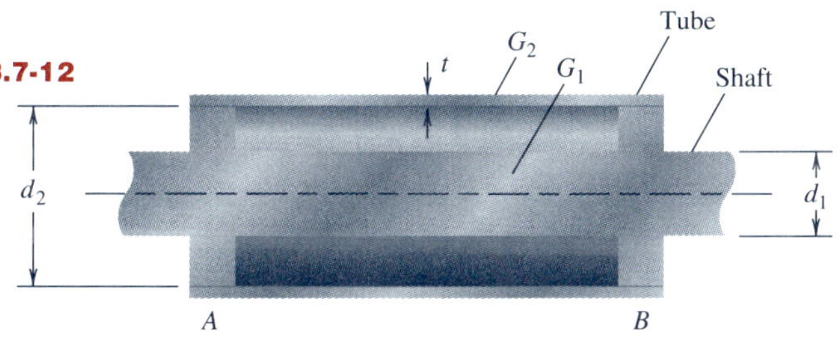

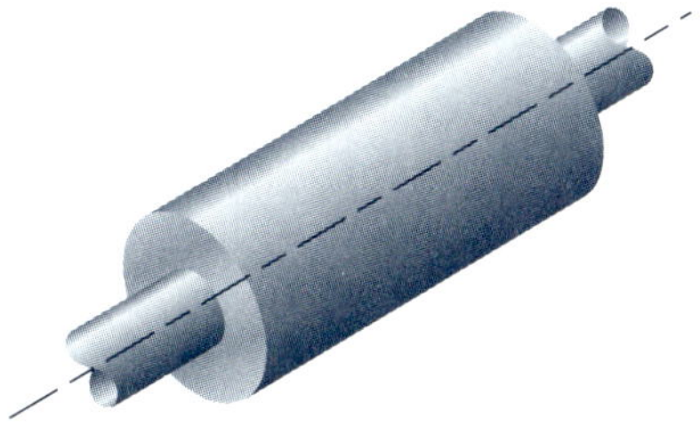

3.7-13 Solve Prob. 3.7-12 with $d_1 = 1.5$ in, $d_2 = 2.5$ in, $t = 0.08$ in, and $G_1 = G_2 = 12 \times 10^6$ psi with $T = 800$ lb · in.

3.7-14 Solve Prob. 3.7-12 with $t = 3.0$ mm. Comment on the effect of the tube thickness on the twisting moments remaining in the shaft and the tube.

3.7-15 Solve Prob. 3.7-12 with $G_1 = 12.5$ GPa and $G_2 = 18.7$ GPa. Comment on the effect of the difference of the shear modulus on the twisting moments in the tube and the shaft.

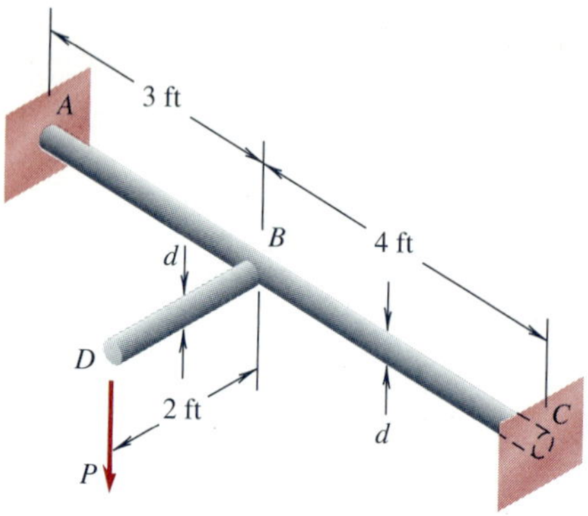

Fig. P3.7-16

3.7-16 A T bar with diameter $d = 2$ in is fixed at both ends A and C and is subjected to a load P at D, as shown in Fig. P3.7-16. If we restrict attention only to the torsional stresses in bar ABC due to the load P and the maximum allowable shear stress is taken as 20,000 psi, find the maximum allowable load P. $G = 11 \times 10^6$ psi.

3.7-17 A solid circular shaft is attached to rigid supports at A and B and carries a torque T at C, as shown in Fig. P3.7-17. The allowable shear stress is 10,000 psi. (*a*) For what value of the diameter d will the maximum stress in each segment be the same? (*b*) For the value of the diameter found in part (*a*), find the maximum allowable value for the torque. $G = 11.5 \times 10^6$ psi.

Fig. P3.7-17
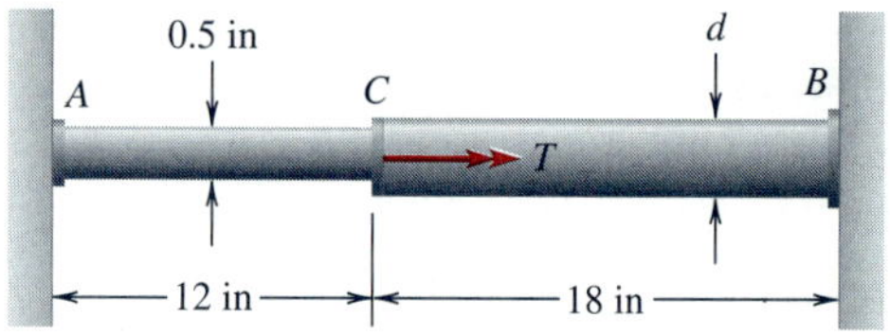

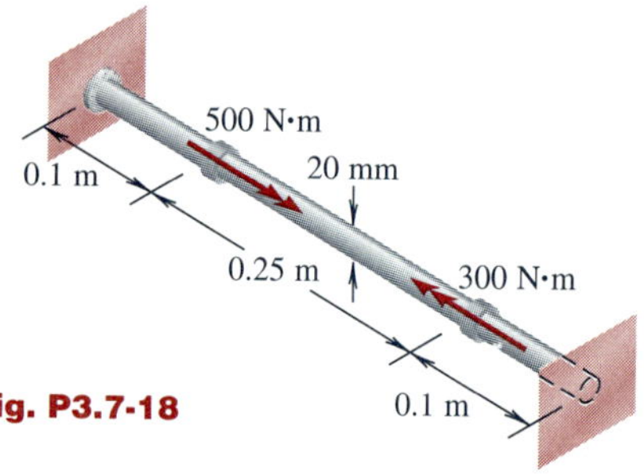

Fig. P3.7-18

3.7-18 A model of a shaft system is shown in Fig. P3.7-18. The ends of the 20-mm-diameter shaft are fixed, and the two twisting moments are applied as shown. Determine the wall reactions, and plot the twisting moment distribution and angle-of-twist distribution along the shaft. Neglect the weight of all components, and take $G = 75$ GPa.

3.7-19 A composite shaft (see Prob. 3.7-7) is shown in Fig. P3.7-19. The shaft is made from a steel tube and a brass core. Determine the maximum shear stress in the steel and in the brass core and the value of the angle of twist at the end of the shaft. Neglect the weight of the shaft, and take $G_{st} = 70$ GPa and $G_{br} = 36$ GPa.

Fig. P3.7-19
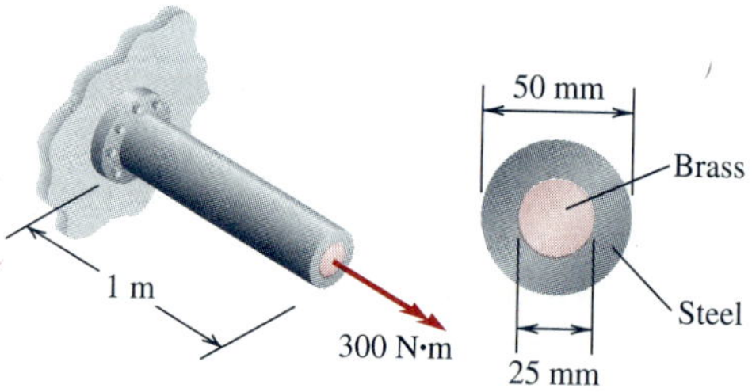

Fig. P3.7-20
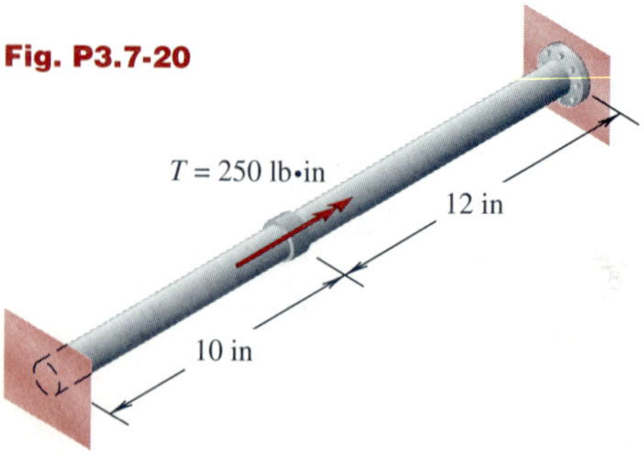

3.7-20 A 1.5-in nominal diameter standard-weight pipe (see App. E) is attached at its ends as shown in Fig. P3.7-20. Determine the angle of twist at the section at which T is applied, and plot the distribution of the twisting moment and the angle of twist along the pipe. Also determine the maximum shear stress in each segment. Neglect the weight of the pipe, and take $G = 6.2 \times 10^6$ psi.

3.7-21 Solve Prob. 3.7-20, but use an extra-strong pipe with a 1.5-in nominal diameter.

3.7-22 Three segments of a shaft are joined together and fixed at the ends as shown in Fig. P3.7-22. A twisting moment of 700 N · m is applied at the section shown. Plot the twisting moment distribution and the angle-of-twist distribution along the shaft. Take $G = 75$ GPa, $d_1 = 75$ mm, and $d_2 = d_3 = 125$ mm. Neglect the weight of the shaft.

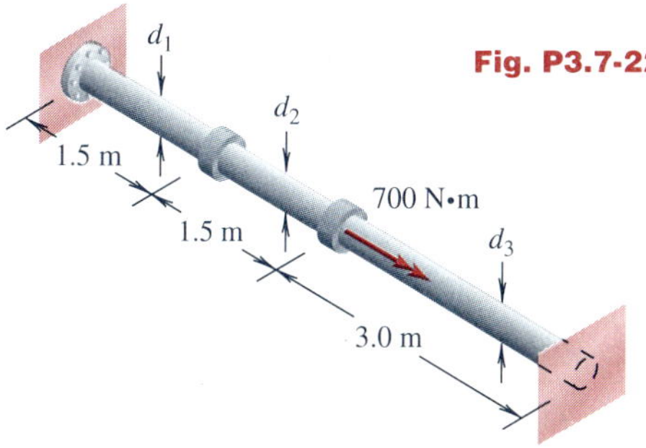

Fig. P3.7-22

3.7-23 Solve Prob. 3.7-22 with $d_1 = d_2 = 75$ mm and $d_3 = 125$ mm.

3.7-24 A 3-in-diameter shaft *ABCD* as shown in Fig. P3.7-24 is made up of two different materials joined at section *C*. Sections *A* and *D* are fixed. Determine the maximum shear stress in each segment, and plot the twisting-moment distribution and the angle-of-twist distribution along the shaft. Take:
(*a*) $G_{AC} = 10.5 \times 10^6$ psi, $G_{CD} = 15 \times 10^6$ psi, $T_B = 700$ lb · ft, $T_C = 0$
(*b*) $G_{AC} = 10.5 \times 10^6$ psi, $G_{CD} = 15 \times 10^6$ psi, $T_B = 0$, $T_C = 700$ lb · ft
(*c*) $G_{AC} = 10.5 \times 10^6$ psi, $G_{CD} = 10.5 \times 10^6$ psi, $T_B = 700$ lb · ft, $T_C = 700$ lb · ft

Fig. P3.7-24

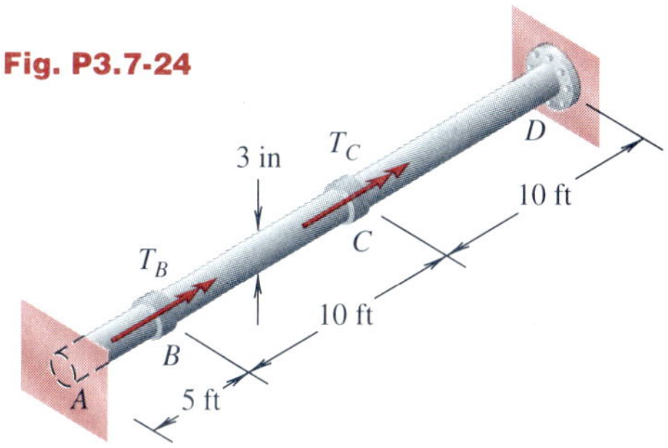

3.7-25 A small shaft component *ABC* shown in Fig. P3.7-25 is connected to a rigid plate at section *B* so that part of the applied twisting moment at *A* is picked up by the cylindrical tube of radius *a* and thickness *t*. Find the spring constant of the system, that is, $k = T/\phi_A$, and compare it to the spring constant in the absence of the tube, that is, $k = GJ/(L_1 + L_3)$. Take $d = 2$ mm, $L_1 = 25$ mm, $a = 10$ mm, $t = 1$ mm, $L_2 = 25$ mm, $L_3 = 35$ mm, $G_{\text{shaft}} = 50$ GPa, and $G_{\text{tube}} = 70$ GPa, with $T = 0.1$ N · m.

Fig. P3.7-25

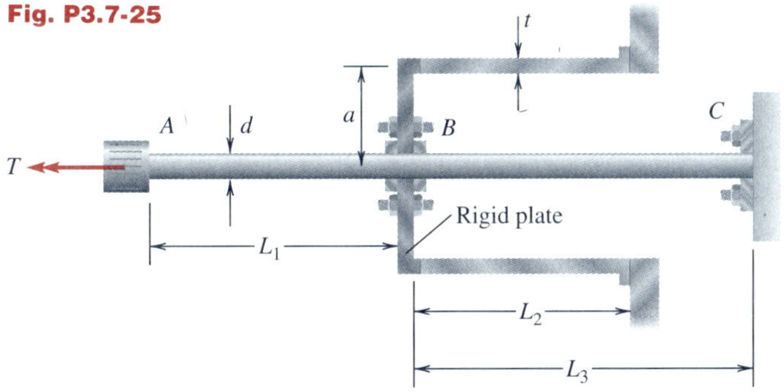

3.7-26 Consider Prob. 3.7-25. Determine the maximum shear stress in the shaft, and compare it to the value of the maximum shear stress without the tube.

3.7-27 Derive a general expression for the spring constant of the system shown in Fig. P3.7-25; see Prob. 3.7-25.

3.7-28 A composite shaft consisting of an inner steel shaft connected by rigid plates at the ends to an aluminum tube is shown in Fig. P3.7-28. The dimensions of the cross section in millimeters for the shaft and tube are as shown. Find the relative angle of twist and the maximum shear stress in the steel and aluminum components. Take $G_{st} = 90$ GPa and $G_{al} = 60$ GPa.

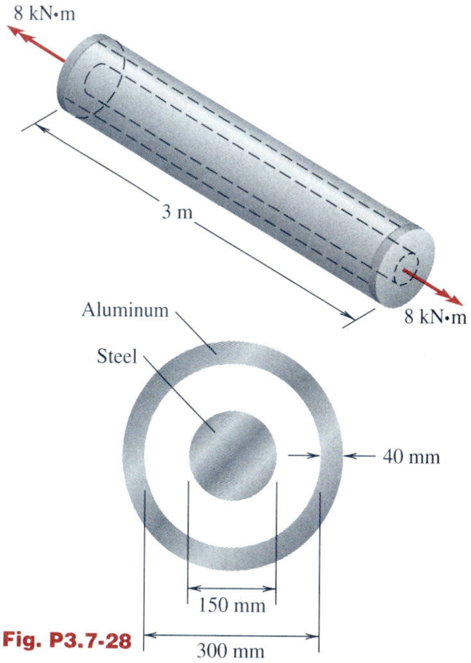

Fig. P3.7-28

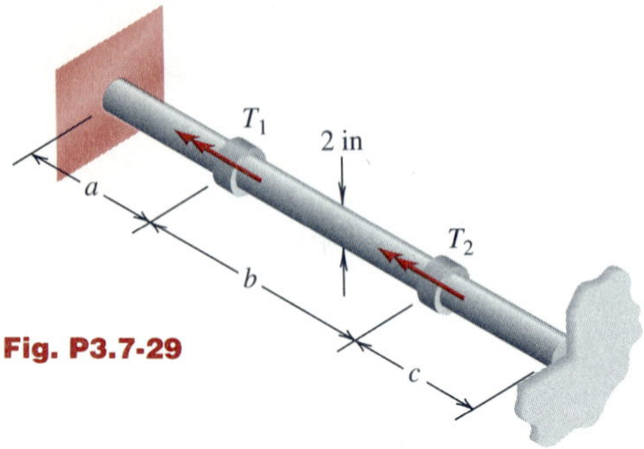

Fig. P3.7-29

3.7-29 A circular shaft of diameter $d = 2$ in with built-in ends is subjected to the action of the two twisting moments T_1 and T_2, as shown in Fig. P3.7-29. What are the internal twisting moments in the segments of the shaft if $a = 30$ in, $b = 50$ in, $c = 40$ in, $T_1 = 12{,}000$ in · lb, $T_2 = 24{,}000$ in · lb, and $G = 12 \times 10^6$ psi? Plot a graph of the internal twisting moment versus distance along the shaft.

3.7-30 Determine the maximum shear stress in the shaft of Prob. 3.7-29.

3.7-31 Given is the torsional system shown in Fig. P3.7-31 with a twisting moment T applied at section B. Find the maximum allowable value of the diameter d if the maximum allowable shear stress is 50 MPa. Take $T = 1715$ N · m and $G = 85$ GPa.

Fig. P3.7-31

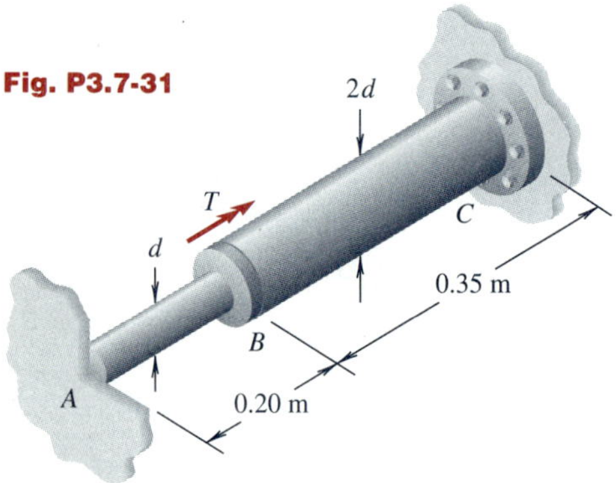

3.7-32 Given is a circular shaft under the two twisting moments T, as shown in Fig. P3.7-32. Each end is built into the wall. Find an expression for the angle of twist at section A. Take G as the shear modulus.

Fig. P3.7-32

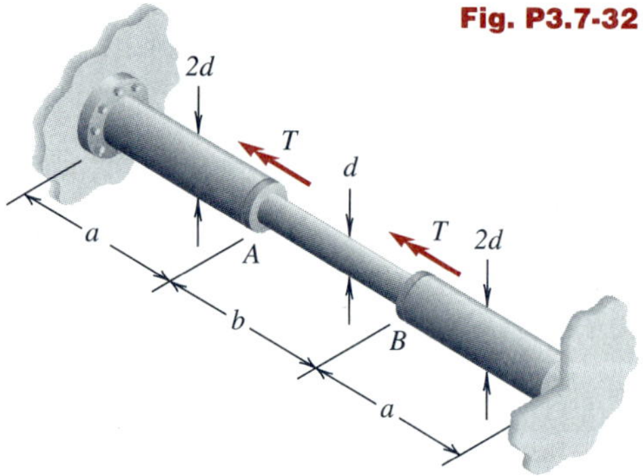

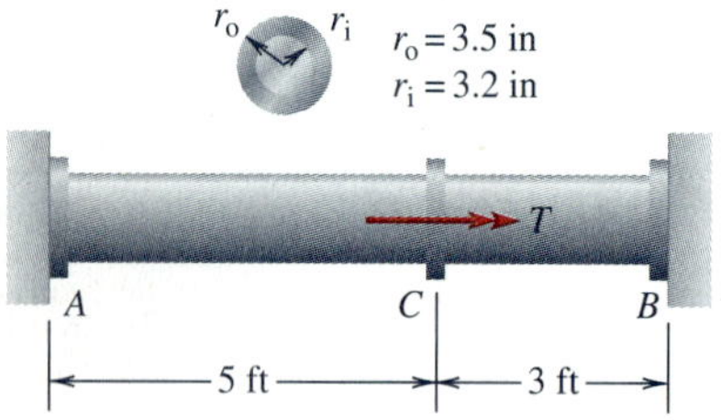

Fig. P3.7-33

3.7-33 A pipe as shown in Fig. P3.7-33 is used as a torsion member. Because of faulty fabrication, section B can twist from the unloaded position through an angle of 0.25° before encountering a rigid stop. If a twisting moment $T = 26{,}000$ lb · in is applied at section C as shown, find the maximum shear stress in segment AC and segment CB. $G = 11 \times 10^6$ psi.

3.8-1 Consider Example 3.7, Fig. 3.20. Verify Eqs. (*f*) and the solutions, Eqs. (*g*), and plot the twisting moment distribution and the angle-of-twist distribution along the shaft, similar to Fig. 3.16*d*.

3.8-2 Consider Example 3.7, Fig. 3.20. Determine the location of the section other than that at section A where the angle of twist is zero.

3.9-1 Consider Example 3.8, Fig. 3.21. Plot the twisting moment distribution along the shaft.

3.9-2 Consider Example 3.9, Fig. 3.26. Plot the twisting moment distribution along the shaft.

3.9-3 Consider Example 3.10, Fig. 3.29. Plot the twisting moment distribution along the shaft.

3.9-4 An aluminum tube AB is attached at B to a brass tube BC and is fixed to a rigid wall at A, as shown in Fig. P3.9-4. The inside and outside diameters for AB are 1 and 2 in and for BC, 0.5 and 1 in; and $T_B = 4$ kip · in and $T_C = 2.5$ kip · in.

(*a*) Find the maximum shear stress in each tube due to the applied torques.

(*b*) Find the angle of rotation at sections B and C. Take $G_{AB} = 4 \times 10^6$ psi and $G_{BC} = 6 \times 10^6$ psi.

(*c*) Obtain a graph of the angle of twist versus position along the tubes measured from A.

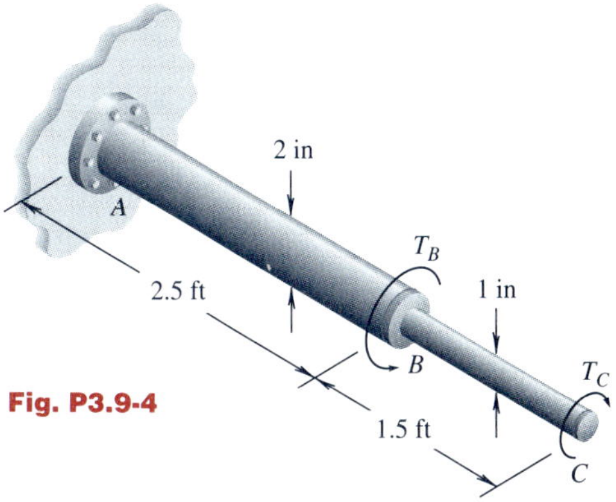

Fig. P3.9-4

3.9-5 A solid-steel stepped shaft is fixed at the left end and subjected to a torque T at the right end, as shown in Fig. P3.9-5.

(*a*) For the diameters shown, find the maximum allowable torque T_a if the allowable shear stress is 12 ksi and the allowable total twisting angle over the full length ABC is 0.12 rad.

(*b*) Obtain a graph of angle of twist versus position along the shaft for $T = T_a$. Take $G = 12 \times 10^6$ psi.

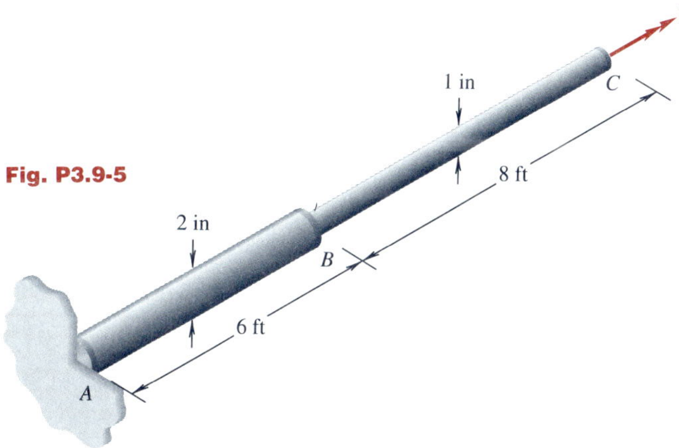

Fig. P3.9-5

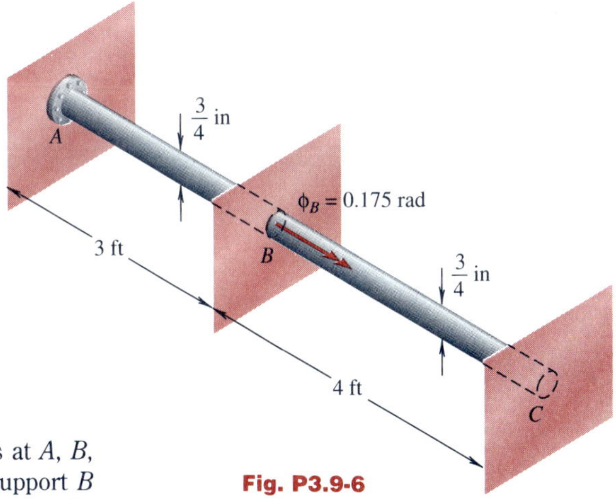

Fig. P3.9-6

3.9-6 A solid circular steel shaft ABC is firmly attached to supports at A, B, and C as shown in Fig. P3.9-6. Supports A and C remain fixed, and support B undergoes a rotation of 0.175 rad in the direction as shown by the vector at B.

(*a*) Find the maximum shear stress in each segment of the shaft.

(*b*) Find the twisting moment which must be applied at B in order to cause the prescribed twisting angle. $G = 12 \times 10^3$ ksi.

3.9-7 A solid circular shaft of 1-in diameter is fixed at its ends A and C and loaded at B by a torque T, as shown in Fig. P3.9-7.

(*a*) If the allowable shear stress is 12 ksi and the allowable angle of twist is 0.05 rad, find the maximum allowable torque T.

(*b*) Find the minimum allowable diameter d so that this shaft can carry an applied torque of $T = 6000$ lb · in. $G = 12 \times 10^6$ psi.

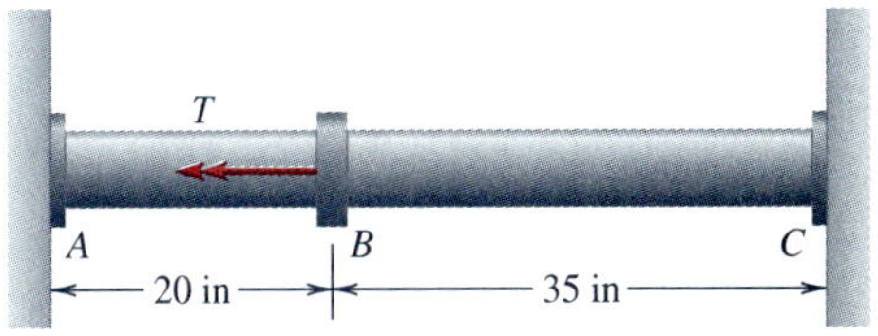

Fig. P3.9-7

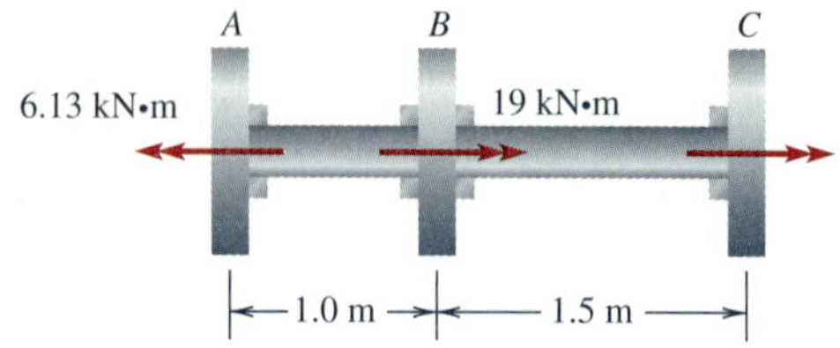

Fig. P3.9-8

3.9-8 A torque of 19 kN · m is applied to the shaft system at *B*, as shown in Fig. P3.9-8. A torque of 6.13 kN · m is to be taken off at *A*, and the balance at *C*. The allowable shear stress is 70 MPa.

(*a*) Find the minimum allowable diameter for each of the two shafts *AB* and *BC*.

(*b*) If a given design requires that the two shafts have a common diameter of 80 mm, what are the angles of twist at *A* and *C* relative to *B* for the applied torques? Obtain a graph of the angle of twist versus distance. $G = 80$ GPa.

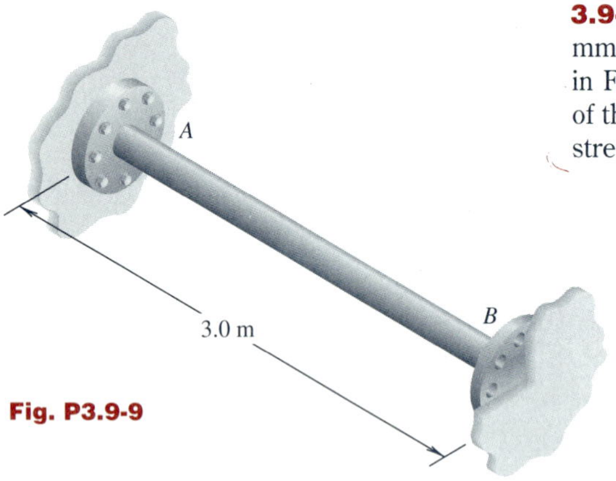

Fig. P3.9-9

3.9-9 A hollow steel pipe of inside diameter 70 mm and outside diameter 80 mm is to be installed in brackets between two rigid walls at *A* and *B*, as shown in Fig. P3.9-9. To line up the bolt holes at *A*, it was necessary to twist one end of the pipe by an angle of 1° relative to the other end. Find the maximum shear stress due to this installation process. $G = 80$ GPa.

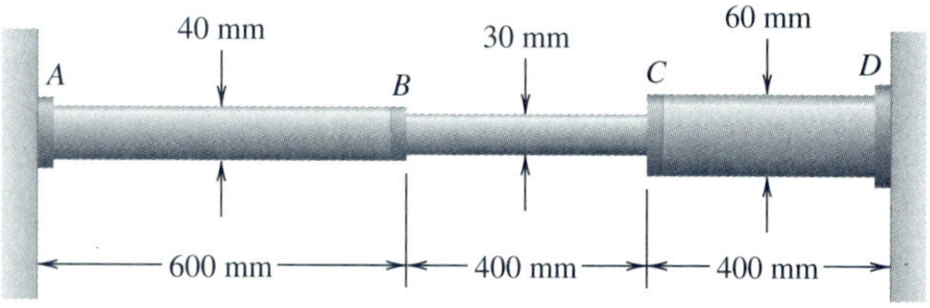

Fig. P3.9-10

3.9-10 When the circular steel shafts *AB*, *BC*, and *CD* shown in Fig. P3.9-10 are about to be attached to the walls at *A* and *D*, it is found that a 2° mismatch exists at *D* in the alignment of the flanges for bolting the shaft at *D*. An external torque is applied at *C* so that the attachment can be made and the external torque is then removed.

(*a*) Find the maximum shear stress in each shaft after the shaft is bolted at *D*.

(*b*) Obtain a graph of the twisting angle versus distance from the left end of the shaft system. $G = 80$ GPa.

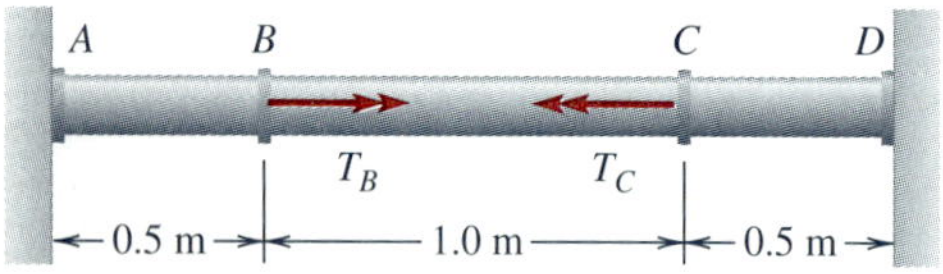

Fig. P3.9-11

3.9-11 A solid-steel shaft of diameter 30 mm is attached to rigid walls at *A* and *D* and loaded by torques T_B and T_C, as shown in Fig. P3.9-11. The allowable shear stress in the shaft is 70 MPa.

(*a*) Find the maximum allowable torque T_B when T_C is taken as zero.

(*b*) If T_B is fixed at the value of 495 N · m, find the range of positive and negative allowable values that T_C may have.

(*c*) If $T_C = T_B = T$ (same magnitude but oppositely directed as shown), find the allowable value for *T*. It is given that $G = 80$ GPa.

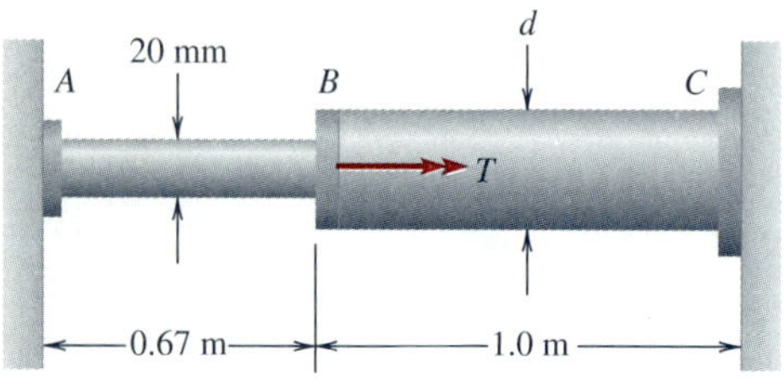

Fig. P3.9-12

3.9-12 A solid stepped circular shaft is attached to rigid walls at *A* and *C* as shown in Fig. P3.9-12. The allowable shear stress is 70 MPa.

(*a*) Find the diameter *d* such that the maximum shear stress (in magnitude) in each segment will be the same.

(*b*) For that diameter, what is the maximum allowable torque *T*? Given: $G = 80$ GPa.

3.9-13 A stepped steel shaft is loaded by twisting moments as shown in Fig. P3.9-13. The diameters of segments *AB*, *BC*, and *CD* are 60, 40, and 20 mm; $G = 75$ GPa.
(*a*) Find the maximum shear stress in each segment of the shaft.
(*b*) Obtain a graph of the angle of twist as a function of distance from the support at *A*.

3.9-14 Use the TORMECH program to solve the following problems:
(*a*) Prob. 3.6-16
(*b*) Prob. 3.6-17
(*c*) Prob. 3.6-18
(*d*) Prob. 3.6-27
(*e*) Prob. 3.6-30
(*f*) Prob. 3.6-32
(*g*) Prob. 3.6-37
(*h*) Prob. 3.6-38
(*i*) Prob. 3.6-40
(*j*) Prob. 3.6-41
(*k*) Prob. 3.6-43
(*l*) Prob. 3.7-10
(*m*) Prob. 3.7-17
(*n*) Prob. 3.7-18
(*o*) Prob. 3.7-20
(*p*) Prob. 3.7-22
(*q*) Prob. 3.7-23
(*r*) Prob. 3.7-31

Fig. P3.9-13

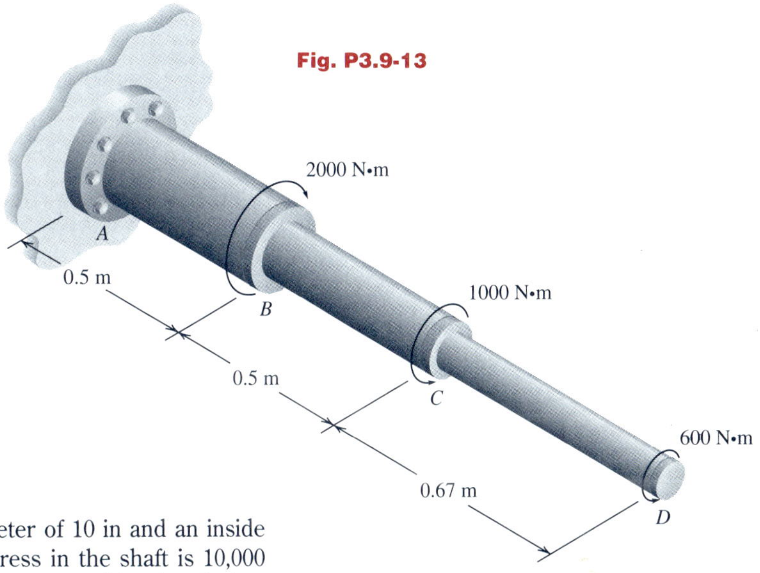

3.10-1 A ship propeller shaft has an outside diameter of 10 in and an inside diameter of 7 in. If the maximum allowable shear stress in the shaft is 10,000 psi, estimate the maximum horsepower that can be transmitted by the shaft at 105 rpm. Also estimate the angle of twist in a 25-ft length of the shaft. Take $G = 12.5 \times 10^6$ psi.

3.10-2 A large hollow propeller shaft is to transmit 10,000 hp at 200 rpm. The inside diameter is equal to 6 in, and the outside diameter is 9 in. Estimate the maximum shear stress in the shaft. $G = 12.5 \times 10^6$ psi.

3.10-3 If the shaft of Prob. 3.10-2 is to experience a maximum shear stress of 16 ksi and the inside diameter is retained at 6 in, find the minimum allowable external diameter.

3.10-4 A steel turbine shaft 80 ft long transmits 12,000 hp at 60 rpm. The outside diameter of the shaft is 24 in, and the inside diameter is 10 in. Find the maximum shear stress in the shaft and the relative angle of twist between the ends of the shaft. Take $G = 12.5 \times 10^6$ psi.

3.10-5 A hollow steel shaft transmits 9000 hp at 120 rpm. If the maximum shear stress in the material is 10,000 psi and the outside diameter of the shaft is 1.6 times the inner diameter, find the dimensions of the shaft. Take $G = 11 \times 10^6$ psi.

3.10-6 A solid-steel shaft of length 10 ft is required to transmit 80 hp at 60 rpm. The maximum allowable shear stress in the material is 8000 psi. Find the minimum allowable diameter of the shaft and the relative angle of twist between the ends of the shaft. Take $G = 12.5 \times 10^6$ psi.

3.10-7 Compare the weight of a solid shaft with that of a hollow shaft whose inside diameter is two-thirds of its outside diameter if both shafts are to transmit the same horsepower at the same rpm with a given allowable shear stress.

3.10-8 A hollow propeller shaft has an internal diameter of d_i and an outside diameter of d_o. It is designed to transmit 8000 hp at 150 rpm. If $d_i = 5$ in and the maximum shear stress in the material is 12 ksi, find the outside diameter d_o. Take $G = 12.5 \times 10^6$ psi.

3.10-9 Solve Prob. 3.10-8 if $d_i = 6$ in.

3.11-1 Verify Eqs. (3.45) and (3.46).

3.11-2 Consider Example 3.13, Fig. 3.38. However, now in addition to the constant distributed twisting moment of $q_0 = 200$ N · m/m, we have a twisting moment of $T_B = 1000$ N · m applied at section B. Plot the twisting moment distribution and angle-of-twist distribution for the shaft in this case.

3.11-3 Consider Example 3.14, Fig. 3.40. Verify Eq. (*e*) in the solution.

3.11-4 A self-tapping screw can be modeled as a circular shaft of length L and radius r under constant torsional frictional resistance per unit length q_0, as shown in Fig. P3.11-4. If the maximum allowable shear stress in the screw is τ_{max}, obtain expressions for the maximum possible depth L that the screw can be sunk and for the total angle of twist at the head of the screw in terms of τ_{max}, q_0, and r.

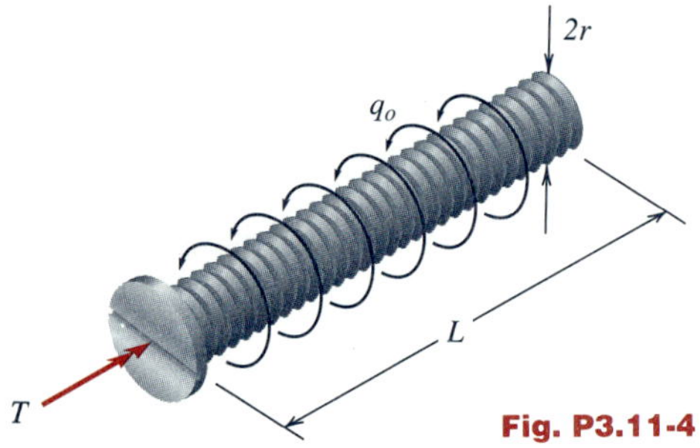

Fig. P3.11-4

3.11-5 A twisting moment T_A is applied to a steel shaft of diameter $d = 4$ mm that is carried in a hollow sheathing and connected to a switch at B as shown in Fig. P3.11-5. If a twisting moment of 0.5 N · m is needed to activate the switch

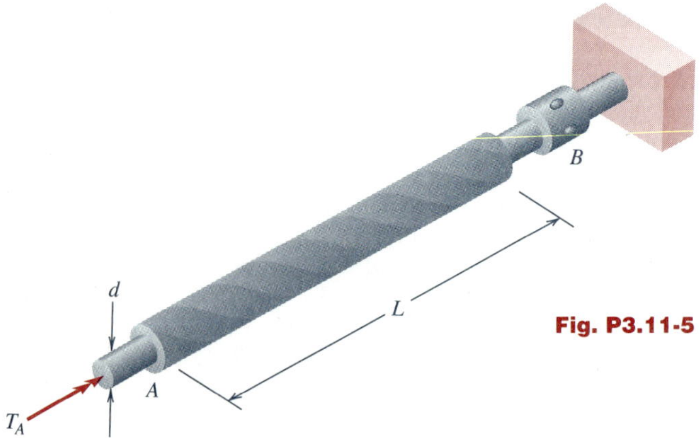

Fig. P3.11-5

at B, what value of applied twisting moment T_A is needed if the sheath imposes a frictional twisting moment per unit length of 0.05 N · m/m on the shaft with $L = 1$ m? What is the angle of twist at section A relative to B when T_A is applied? Take $G = 84$ GPa.

3.11-6 A tapered solid shaft of circular cross section and length L is fixed at both ends and loaded by a constant distributed twisting moment q_0 (with units of force × length/length), as shown in Fig. P3.11-6. The material of the shaft has a shear modulus of G, and the polar moment of inertia J of the shaft varies along the shaft according to the relation $J = J_0/(1 + x/L)$, where x is the distance from the left end A.

(*a*) Show that using Eq. (3.37) results in

$$T(x) = T_A - q_0 x \qquad (a)$$

where T_A is the twisting moment reaction at A.

(*b*) Show that Eq. (3.38) becomes

$$\frac{GJ_0}{1 + x/L}\frac{d\phi}{dx} = T_A - q_0 x \qquad (b)$$

and integrate to obtain

$$\phi = \frac{1}{GJ_0}\left[T_A x + \frac{x^2}{2}\left(\frac{T_A}{L} - q_0\right) - \frac{q_0}{L}\frac{x^3}{3}\right] + C_2 \qquad (c)$$

(*c*) Show that application of the geometric boundary conditions at $x = 0$ and $x = L$ gives

$$T_A = \tfrac{5}{9} q_0 L \qquad T_B = -\tfrac{4}{9} q_0 L \qquad (d)$$

(*d*) Plot the expression $\phi/[q_0 L^2/(GJ_0)]$ versus x/L.

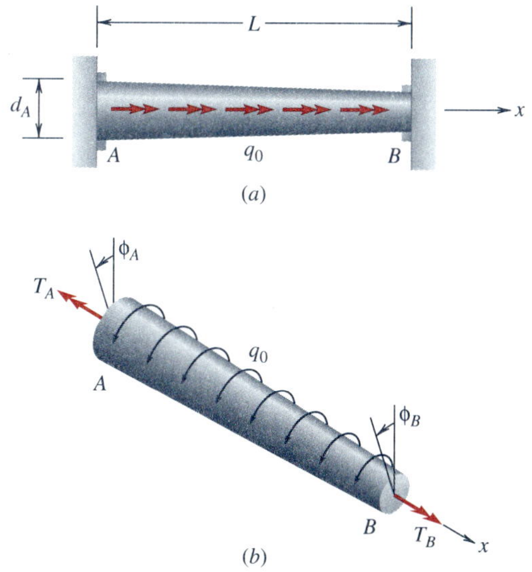

Fig. P3.11-6

Fig. P3.12-2

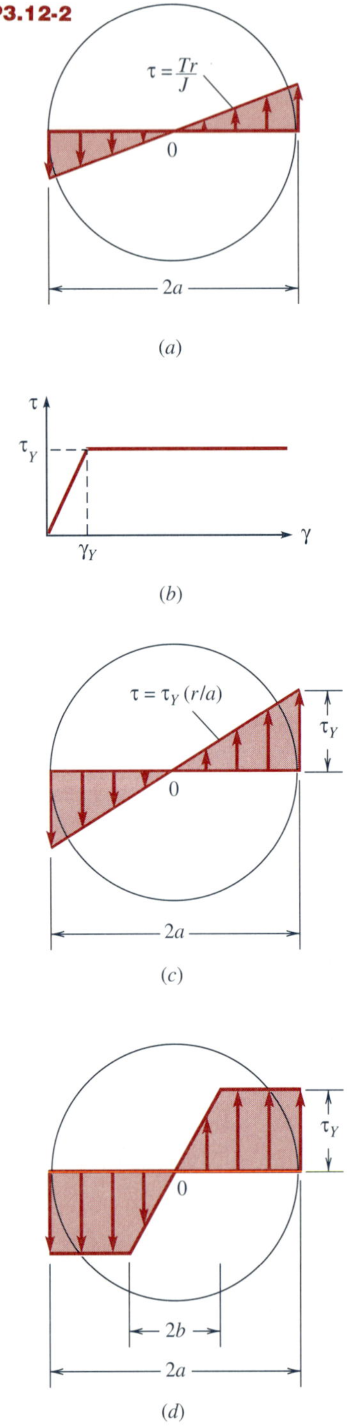

3.12-1 If an elastic material has a nonlinear relation between shear strain γ and shear stress τ in the form $\gamma = k\tau^2$, obtain a relation for the relative angle of twist between the ends of a circular shaft of length L of this material. Also obtain an expression for the maximum shear stress in the shaft, and compare it to the expression for the shear stress in the case of a linear elastic material.

3.12-2 A solid shaft of circular cross section subjected to a twisting moment T has a linear distribution of shear stress as shown in Fig. P3.12-2*a*. As T increases, according to the maximum shear stress criterion yielding will be initiated (Fig. P3.12-2*c*) when $\tau = \tau_Y$, where τ_Y is the shear stress at yielding.

(*a*) Show that at yielding the values of T and the twisting angle ϕ are

$$T_Y = \frac{\tau_Y J}{a} \quad \text{and} \quad \phi_Y = \frac{\tau_Y L}{G\,a} \qquad (a)$$

where L is the length of the shaft and G is the shear modulus.

(*b*) If we assume that the stress-strain curve for this material is of the elastic–perfectly plastic type (Fig. P3.12-2*b*), that is, $\tau = \tau_Y$ for $\gamma \geq \gamma_Y$, it follows that for $T > T_Y$ there will be an annular region $b \leq r \leq a$ where $\tau = \tau_Y$ (Fig. P3.12-2*d*). Show that if we write $\tau = \tau_Y r/b$ for $0 \leq r \leq b$ and $\tau = \tau_Y$ for $b \leq r \leq a$, it follows by calculating the twisting moment T, using $T = \int r\tau\, dA$, that

$$T = \frac{\pi}{2} b^3 \tau_Y + \frac{2\pi}{3}(a^3 - b^3)\,\tau_Y = \frac{2\pi}{3}\tau_Y a^3\left(1 - \frac{1}{4}\frac{b^3}{a^3}\right) \qquad (b)$$

3.12-3 For the elastic-plastic torsion problem in Prob. 3.12-2, show that

(*a*) By using the results given there with $J/a = \pi a^3/2$, we get

$$T_Y = \tau_Y \frac{\pi a^3}{2} \qquad (a)$$

(*b*) An increasing twisting moment T will cause the size of the "elastic core," $0 \leq r \leq b$, to approach zero (Fig. P3.12-3), and the section will become "fully plastic." Show that using Eq. (*a*) in this problem and Eq. (*b*) in the previous problem gives

$$T = \frac{4}{3} T_Y \left(1 - \frac{1}{4}\frac{b^3}{a^3}\right) \qquad (b)$$

Fig. P3.12-3

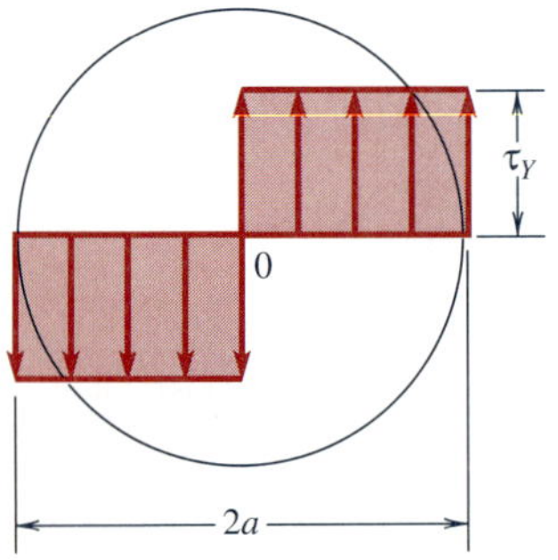

and the fully plastic moment T_{FP} can be found by setting $b = 0$ in Eq. (*b*) to find

$$T_{FP} = \frac{4}{3} T_Y \qquad (c)$$

According to this analysis, the twisting moment may be increased by 33 percent above the value required to initiate yielding before the fully plastic state is reached.

3.12-4 Experimental evidence indicates that a simple model to account for the unloading of elastic–perfectly plastic materials can be obtained by taking the unloading portion *BC* in Fig. P3.12-4*a* as parallel to segment *OA*. Suppose that the shaft in Prob. 3.12-2 is loaded by a twisting moment $T_0 > T_Y$ such that $b = a/2$, as shown in Fig. P3.12-4*b*.

(*a*) If T_0 is then removed by analytically superposing a linearly distributed stress over the entire section, $\tau = -T_0 r/J$, Fig. P3.12-4*c*, show that by combining the stresses in Figs. P3.12-4*b* and P3.12-4*c* we obtain the residual-stress distribution as given in Fig. P3.12-4*d*. These residual stresses remain in the shaft which carries no external twisting moment.

(*b*) Find the minimum and maximum values of the residual-stress distribution in Fig. P3.12-4*d* by finding the value of T_0 corresponding to $b = a/2$ from Eq. (*b*) of Prob. 3.12-2 and adding the stress states in Figs. P3.12-4*b* and P3.12-4*c*.

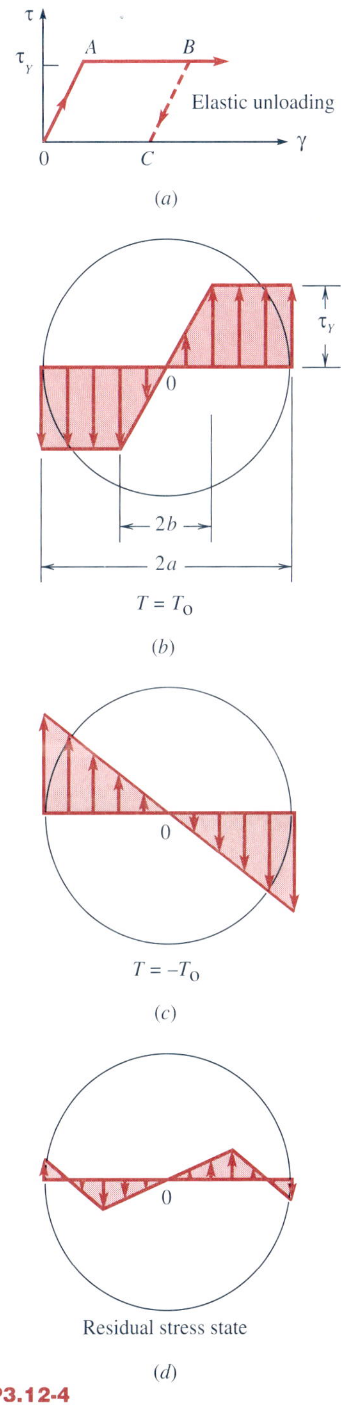

Fig. P3.12-4

CHAPTER 4

Shear Forces and Bending Moments in Beams

4.1 Introduction

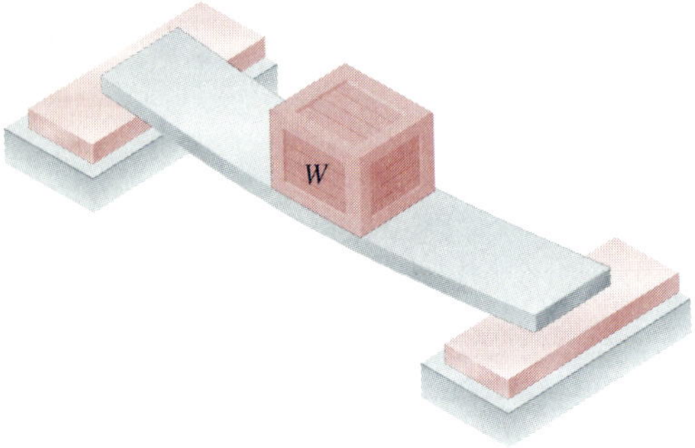

Figure 4.1 A beam, supported at its ends, carrying a crate of weight W.

Figure 4.2 Beams resting on support beams.

When we look closely at structures and machines designed by engineers such as bridges, automobiles, electric power line transmission towers, or the framing in wooden houses, we see that a great many of the load-carrying members can be classified as *slender* members.[1] As we recall from Chaps. 2 and 3, a slender member is a member whose length is much greater (say, at least 5 times greater) than either of its cross-sectional dimensions. This classification includes the bars under uniaxial load discussed in Chap. 2, the circular shafts under twisting moments discussed in Chap. 3, and beams and columns. In this chapter we study the behavior of beams.

A beam is a slender structural member that resists forces or loads acting transverse or perpendicular to its longitudinal axis. A simple illustration of loading on a beam is a wooden plank placed between two supports on which a crate of weight W is placed, as shown in Fig. 4.1. The plank shown in Fig. 4.1 is resting on the end supports and carries the weight of the crate by *internal* forces and moments arising from bending of the plank. Another illustration is the case of beams under self-weight and resting on supports (see Fig. 4.2); the beams bend due to the weight distributed along their length and so experience internal forces and moments. In this chapter, we develop a general method for determining the internal forces and moments in beams under transverse loads. This is the first step for finding the load-deformation behavior of beams.

[1]See, for example, *The Builders, Marvels of Engineering,* National Geographical Society, Washington, D.C., 1992

4.2 General Method

The general method for determining the internal forces and moments acting across any section of a slender member that is in equilibrium is to imagine a *hypothetical cut* or *section* across the member at the point of interest. Recall that we used cuts or sections for determining internal forces and twisting moments in Chaps. 2 and 3. We consider either part of the slender member so divided as an isolated free body, and the force and moment required at the section to keep that part of the member in equilibrium are found by applying the conditions for equilibrium. In general, there will be both a resultant force *vector* and a resultant moment *vector* acting across the section.

For convenience, we usually resolve the force vector and the moment vector on the section into components normal and parallel to the axis of the member; see Fig. 4.3. In Fig. 4.3*a* we show a section cut across a slender member; the *x* axis is oriented so as to coincide with the longitudinal axis of the member. The *y* and *z* axes lie in the plane of the cross section. The orientation of the *y* and *z* axes is usually dictated by the shape of the cross section, as in Fig. 4.3*b*, or by the direction of the transverse loading applied to the member.

The notation F_{xx}, . . . of the components shown in Fig. 4.3*a* is used to indicate both the orientation of the cross section and the direction of the particular force or moment component. The first subscript indicates the direction of the *outwardly* directed normal vector to the *face* of the cross section. As we discussed in Chap. 2, the cross-sectional *face* will be called *positive* when the outward-directed normal vector points in the positive coordinate direction and *negative* when the outward-directed normal vector points in the negative coordinate direction. The cross-sectional faces in Fig. 4.3*a* and *b* are positive *x* faces. The second subscript indicates the coordinate direction of the force or moment component. Thus, F_{xy} is the force component acting on the *x* section in the *y* direction, and M_{xz} is the moment component in the *z* direction. Force F_{xx} is the axial force on the member that we treated in Chap. 2, and M_{xx} is the twisting moment that we treated in Chap. 3. The different components have different effects on the member, and hence they have been given special names:

Axial force F_{xx}: This component tends to elongate or shorten the member and is often given the symbol F or F_x. We discussed such forces in Chaps. 1 and 2.

Shear force F_{xy}, F_{xz}: These components tend to shear one part of the member relative to the adjacent part and are often given the symbols V, or V_y and V_z. Again, we discussed such shear forces in Chap. 1.

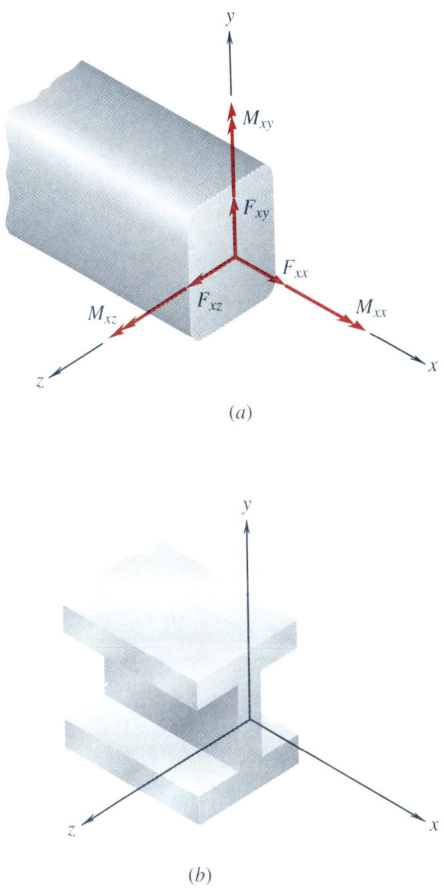

Figure 4.3 (*a*) Components of force and moment vectors acting on a section of a member. (*b*) Orientation of axes on an I-beam section.

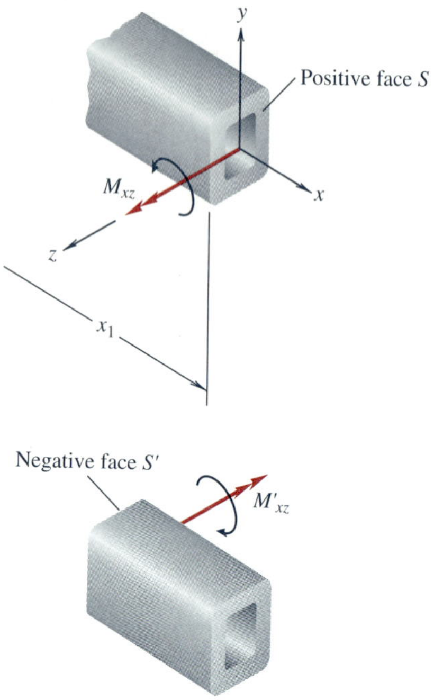

Figure 4.4 Equal and opposite bending moments $M_{xz} = M'_{xz}$ acting on a positive and negative face.

Twisting moment M_{xx}: This component is responsible for the twisting of the member about its longitudinal axis and is often given the symbol M_t or T. We discussed twisting moments and torques for torsion of circular shafts in Chap. 3.

Bending moments M_{xy}, M_{xz}: These components cause the member to bend and are often given the symbols M_b and M, or M_{by} and M_{bz}.

To ensure consistency and reproducibility of analyses of beams, it will be convenient to define sign conventions for the axial force, shear force, twisting moment, and bending moment. We define these quantities to be positive when the force or moment component acts on a *positive face* in a *positive coordinate direction;* the force and moment components shown in Fig. 4.3 all are positive according to this convention. We used this definition in Chaps. 2 and 3.

When a positive force or moment acts on a positive face, then on the opposite negative face, the positive force or moment acts in the negative coordinate direction. For example, in Fig. 4.4 in which we show a slender member cut at position x_1 along its longitudinal axis, a positive bending moment M_{xz} on the (positive) face S will give rise by Newton's law of action and reaction to a bending moment M'_{xz} in the negative z direction on the negative face S'; $M'_{xz} = M_{xz}$. The bending moment M'_{xz} is a positive bending moment because a negatively directed component acting on a negative face is positive.

Figure 4.3*a* illustrates the general case where three components of force and three components of moment act across a section. However, in many instances the applied transverse forces act in one plane, and the problems as a consequence are considerably simpler.

For example, if the plane of loading on the beam is the xy plane and the beam is symmetric about the plane $z = 0$, we have only two force components and one moment component: an axial force F_x (F) and a shear force F_{xy} (V) and the bending moment M_{xz} (M_b), as shown in Fig. 4.5. In Fig. 4.5, it is important to emphasize that positive *components* of force and moment act on a positive face in the positive *coordinate* direction. In previous analyses we have seen that a clear sketch of the orienta-

Figure 4.5 Two-dimensional loading in the xy plane on a beam.

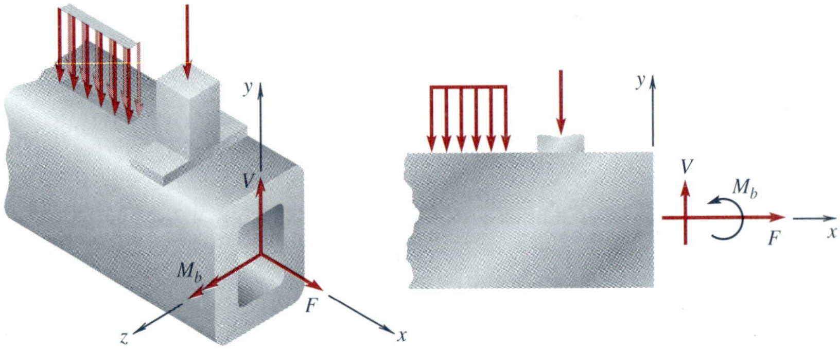

tion of the coordinate axes is an important first step in formulating and solving problems. In many two-dimensional problems, such as shown in Fig. 4.5, we will use the symbol M without a subscript to indicate the bending moment M_b at a particular section. Often in two-dimensional problems the transverse loading will be such that the axial force acting on a section will be zero.

While we will emphasize the solution of two-dimensional problems, the general approach to be taken for the determination of the internal forces and moments in a slender member is organized as follows:

1. Idealize the actual problem; i.e., create a model of the system showing all forces acting on the member. This is a step which follows from our knowledge of statics. Clearly show the axes.
2. Use the equations of equilibrium to calculate any unknown reaction or support forces.
3. Cut the member at a section of interest, and insert the internal forces and moments acting at that section. Isolate one or more of the segments, and use the equations of equilibrium to determine the values of the internal forces and moments acting on the section.

In the case of a straight slender member of symmetric cross section for which the transverse loading is in a plane, e.g., the loading shown in the xy plane in Fig. 4.5, we call the slender member a *straight beam* or simply a beam. Under the action of the transverse loading, the beam will deform or bend in the plane of loading. The determination of the internal forces and moments is the important first step for structural design and analysis of beams and for machine design of beamlike components.

We usually refer to straight beams that have a fixed frictionless pin support at one end and a roller support at the other end as *simply supported beams.* Beams can have supports located anywhere along them. Beams that are built into a wall for support at one end and are unsupported at the other end are referred to as *cantilever beams.* Figure 4.6*a* shows a typical simply supported beam under three concentrated loads and a constant distributed load, and Fig. 4.6*b* shows a cantilever beam under the action of a concentrated load and a constant distributed load.

Distributed loads are loads spread out or distributed along the beam, and they are specified by giving values for the load intensity or *force per unit length* at points along the beam, e.g., in units of newtons per meter (N/m). Common examples of distributed loads arise from water pressure acting on submerged surfaces or the weight per unit length of a beam.

While we have shown the sketches of beams in Fig. 4.6 with the loads applied to the beams, it is important to emphasize that we have not shown a coordinate system with the beams. Whenever we wish to find internal forces and moments in a beam, it is important to use a coordinate system since internal quantities are defined relative to a set of coordinate axes; see Fig. 4.5.

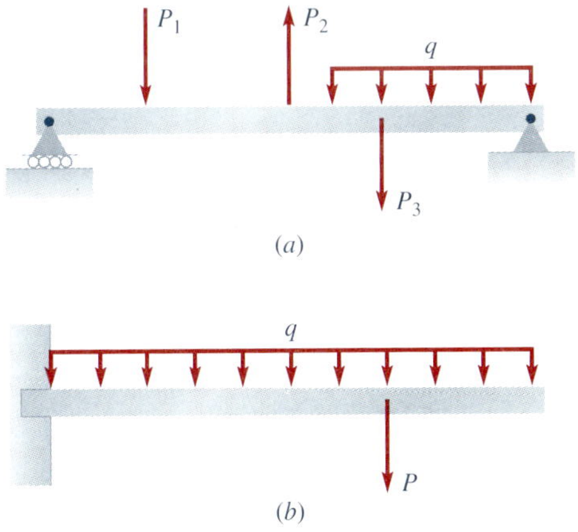

Figure 4.6 (*a*) Simply supported beam under the action of three concentrated loads P_1, P_2, and P_3 and a constant distributed load of intensity q. (*b*) A cantilever beam under the action of a constant distributed load of intensity q and a concentrated load P.

EXAMPLE 4.1

Let us consider a simply supported beam carrying a load P near its center, as shown in Fig. 4.7*a*. We wish to find the internal forces and moments acting at section C located at a distance x along the beam. The total length of the beam is L.

In the absence of friction, the roller support at B will support only a vertical reaction R_B; the only load on the beam is in the vertical direction so that the reaction R_A at A is vertical, as shown in Fig. 4.7*b*. The load is represented by a concentrated load P acting at a point given by $x = a$. We assume that the weight of the beam itself is small *compared* to the load P applied to the beam. This assumption is often valid for many structural engineering problems, but of course it should be investigated in each problem.

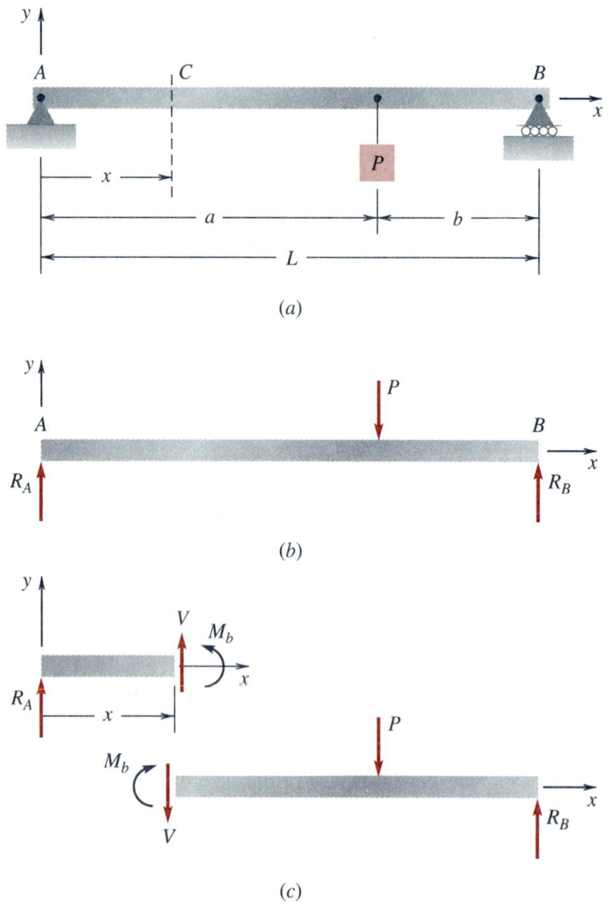

Figure 4.7 Example 4.1

We have completed the first step for the determination of the internal forces and moments in that we have idealized the physical problem to the problem shown in Fig. 4.7*b*. To calculate the values of the support forces or reactions at A and B, we apply the equations of equilibrium.

The requirement of $\Sigma M_A = 0$ gives

$$R_B L - Pa = 0$$

therefore

$$R_B = P\frac{a}{L} \qquad (a)$$

The requirement of $\Sigma M_B = 0$ gives

$$Pb - R_A L = 0$$

therefore

$$R_A = P\frac{b}{L} \qquad (b)$$

We can check our results by noting that the summation of forces in the y or vertical direction is identically satisfied:

$$R_A + R_B - P = 0 \qquad (c)$$

We have now determined the forces external to our idealized system.

We now wish to determine the internal forces at section C. To do this, we imagine the beam to be cut at C, and we show the force and moment acting on the cut section external to the segments resulting from the cut, as shown in Fig. 4.7*c*. The shear force V and bending moment M_b are the internal force and bending moment acting on the cut section.

The segment from A to C shows at cut C a positive shear force V (positive in the positive direction of the y axis since it acts on a positive face) and a positive bending moment M_b (positive in the positive direction of the z axis normal to the page since it acts on a positive face). The segment from C to B shows the equal and opposite V and M_b (on the negative face) at cut C. In general, we would have forces in the x, y, and z directions and moments about these three axes. Our idealized model is two-dimensional with no horizontal forces so that no longitudinal force is required and only the shear force V and bending moment M_b act at section C.

If we apply the equations of equilibrium to the left-hand segment, we find

$$\begin{aligned} \Sigma F_y = 0: &\quad R_A + V = 0 \quad V = -R_A \\ \Sigma M_z = 0: &\quad M_b - R_A x = 0 \quad M_b = R_A x \end{aligned} \qquad (d)$$

for the shear force and bending moment at location x.

We note that since R_A is positive, equal to Pb/L from Eq. (*b*), the shear force V is negative. The bending moment M_b is positive since R_A is positive.

EXAMPLE 4.2

We wish to determine the values of the shear force and the bending moment at sections C_1 and C_2 of the simply supported beam *AB* under the action of two concentrated loads shown in Fig. 4.8*a*. Neglect the weight of the beam. After inserting the coordinate axes as shown in Fig. 4.8*b*, we idealize the beam, introducing the support reactions in Fig. 4.8*b*.

Equilibrium gives the values of the reactions:

$$\Sigma F_y = 0: \quad R_A + R_B - 100 = 0$$

$$R_A = R_B \quad \text{symmetry} \qquad (a)$$

$$R_A = 50 \text{ kN}$$

Equilibrium based on the cut at C_1, $x = 2$ m, gives (Fig. 4.8*c*)

$$\Sigma F_y = 0: \quad R_A + V = 0 \qquad V = -50 \text{ kN} \qquad (b)$$

$$\Sigma M_{C_1} = 0: \quad M - 2R_A = 0 \qquad M = 100 \text{ kN} \cdot \text{m} \qquad (c)$$

Equilibrium based on the cut at C_2, $x = 7$ m, gives

$$\Sigma F_y = 0: \quad R_A - 50 + V = 0 \qquad V = 0 \qquad (d)$$

$$\Sigma M_{C_2} = 0: \quad M + 4 \cdot 50 - 7 \cdot 50 = 0 \qquad M = 150 \text{ kN} \cdot \text{m} \qquad (e)$$

Note the change in the value of V from C_1 to C_2; the value of V at C_2 is equal to the value of V at C_1 plus an amount equal to the value of the concentrated load located between C_1 and C_2. Note also from symmetry that V must equal zero at the midpoint of the beam.

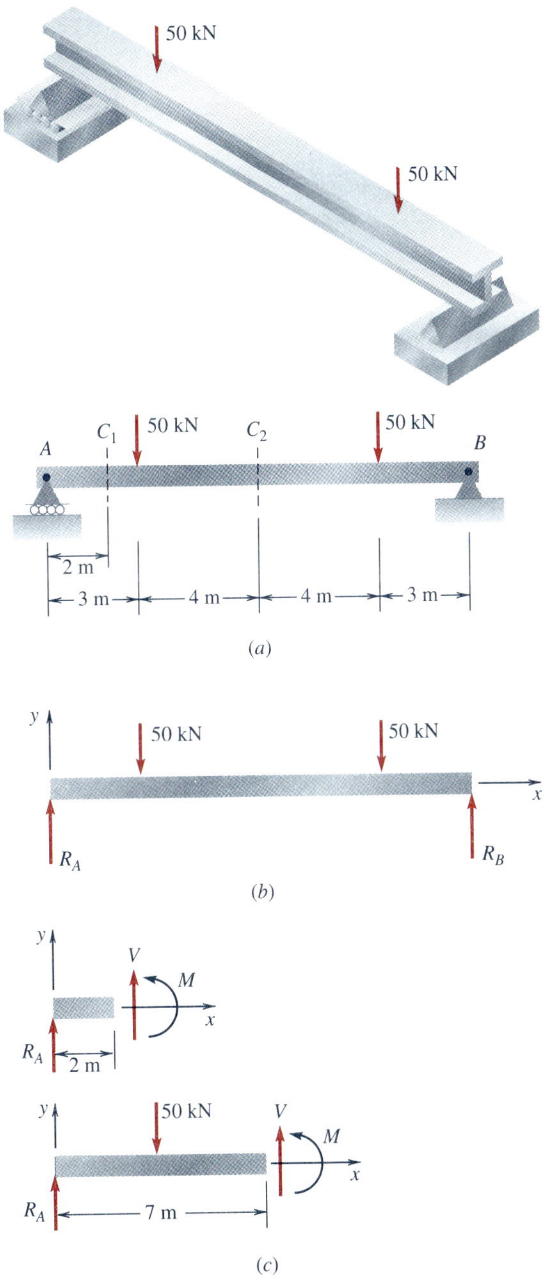

Figure 4.8 Example 4.2

EXAMPLE 4.3

Determine the shear force, axial force, and bending moment at section C, $x = b$, for the cantilever box beam AB shown in Fig. 4.9*a*. The horizontal load P is applied to the vertical bracket welded to the beam. A free-body diagram of the bracket in Fig. 4.9*b* shows that the bracket transmits a positive horizontal force P (positive in the positive x direction) and concentrated moment $M_0 = Pd$ to the axis of the beam.

The reactions at the wall include a horizontal force obtained from $\Sigma F_x = 0$. The other reactions at the wall follow by using the free body for the beam in Fig. 4.9*b* for summation of forces in the y direction and from moment equilibrium,

$$\Sigma F_y = 0: \quad R_A = 0$$
$$\Sigma M = 0: \quad M_A = M_0 = Pd \qquad (a)$$

To obtain the shear force, axial force, and bending moment at section C, we cut the beam at section C and use equilibrium (Fig. 4.9*c*) to find

$$\Sigma F_x = 0: \quad N - P = 0 \qquad N = P$$
$$\Sigma F_y = 0: \quad R_A + V = 0 \qquad V = 0 \qquad (b)$$
$$\Sigma M_C = 0: \quad M + M_A - R_A b = 0 \qquad M = -M_A$$

where N is the axial force in the beam.

Note that the above results also follow directly from applying equilibrium conditions to the segment to the right of the cut segment (Fig. 4.9*d*); in this case, the wall reactions are not needed. However, it is usually good practice to determine values of the support reactions as part of the solution of a beam problem.

Figure 4.9 Example 4.3

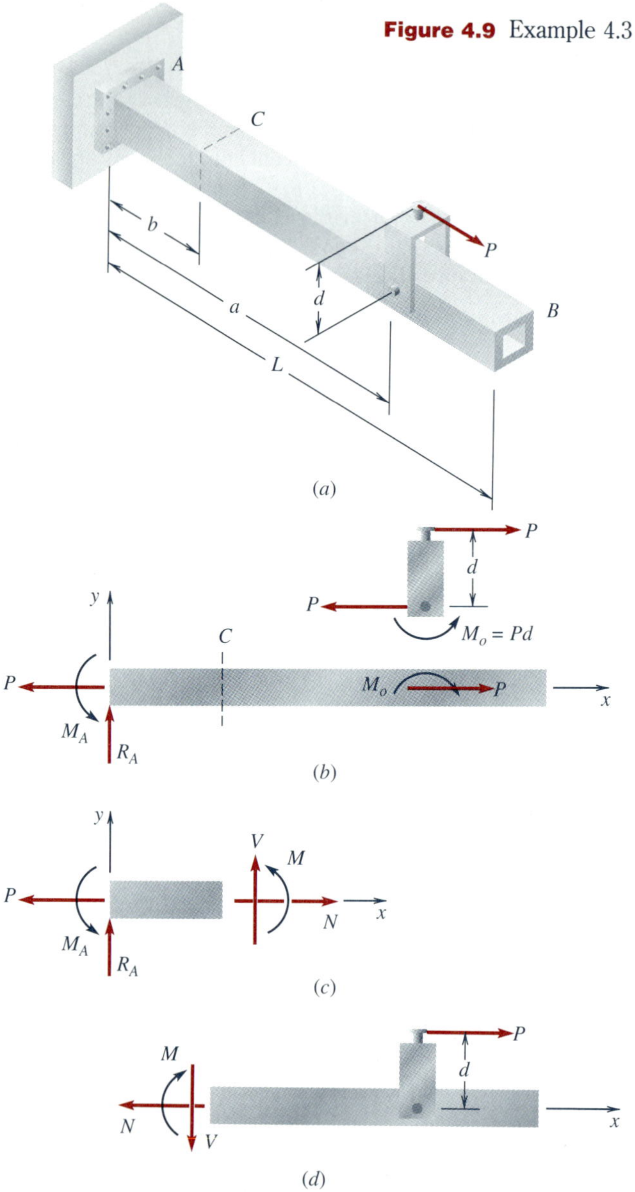

EXAMPLE 4.4

Determine the value of the normal force, the shear force and the bending moment at section C defined by angle θ for the circular beam loaded as shown in Fig. 4.10*a*. Since the centerline of the beam is circular, it is convenient to locate cuts along the beam in terms of polar coordinates. A cut *normal* to the axis of the beam at angle θ is shown in Fig. 4.10*b*. The normal force is specified by N; overall equilibrium of the entire beam (Fig. 4.10*a*) results in the reactions at A, as shown in Fig. 4.10*b*.

To determine N, V, and M at the section, we use force equilibrium equations in the normal and tangential directions at section C and moment equilibrium about section C for segment CB (Fig. 4.10*c* and *d*):

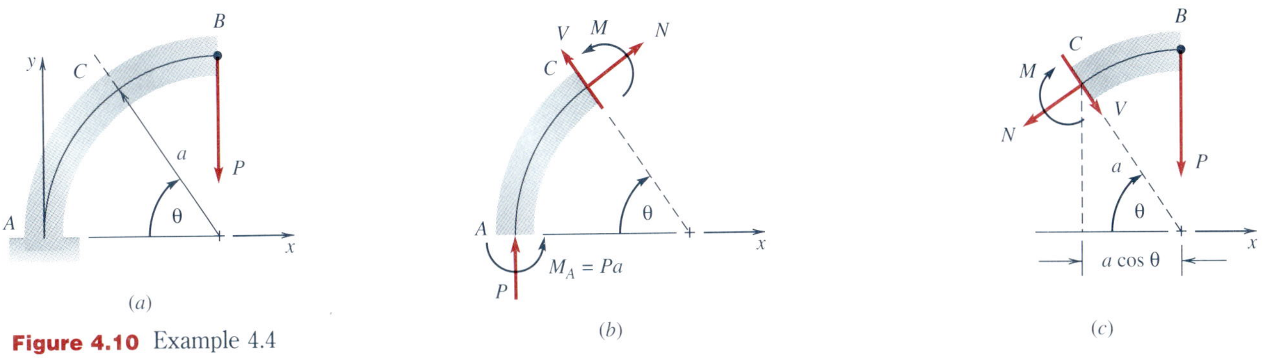

Figure 4.10 Example 4.4

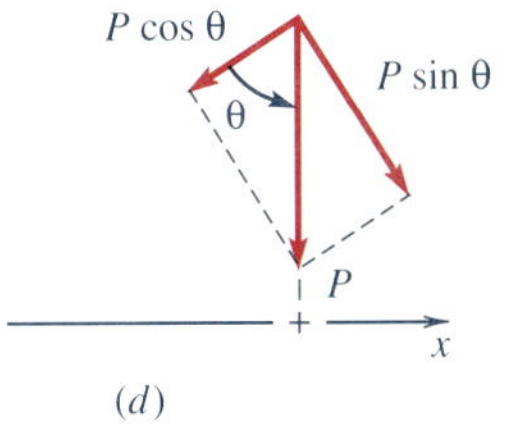

$$\Sigma M_C = 0: \quad -M - Pa\cos\theta = 0 \qquad M = -Pa\cos\theta$$

$$\Sigma F_n = 0: \quad -V - P\sin\theta = 0 \qquad V = -P\sin\theta$$

$$\Sigma F_t = 0: \quad -N - P\cos\theta = 0 \qquad N = -P\cos\theta$$

Note that in this case of a curved slender member, a load transverse to the curved axis of the beam gives rise to an axial force N which will vary as angle θ changes. Also note that the above results could have been derived from segment AC in Fig. 4.10*b*, but with the use of more algebra.

In many situations, the determination of the shear force and bending moment at a particular section will be sufficient. However, it is often necessary to determine how these internal forces and moments vary along the length of the beam so as to find the maximum and minimum values. These values are then used to find maximum stresses in the beam.

4.3 Shear Force and Bending Moment Diagrams

A graph which shows the values of the shear force plotted against distance along the beam is called a *shear force diagram.* A similar graph showing the values of the bending moment plotted against distance along the beam is called a *bending moment diagram.* We have already used axial force and twisting moment diagrams in Chaps. 2 and 3; see, e.g., Figs. 2.9, 3.12, 3.13, 3.14, and 3.16.

To construct a shear force and a bending moment diagram, it is only necessary to consider the location x of the cut as a variable along the beam and to determine the values of V and M at different values of x. The values of V and M as a function of x can then be plotted. To illustrate, we return to the idealized models of Examples 4.1 and 4.2.

EXAMPLE 4.5

We wish to construct the shear force and bending moment diagrams for the idealized beam of Example 4.1; see Fig. 4.7. In Fig. 4.11*a* and *b* we show the idealized beam and the cut segment for $x < a$. For $x < a$, from Eqs. (*d*) of Example 4.1, we have

$$V = -R_A = -\frac{Pb}{L}$$
$$M = R_A x = \frac{Pbx}{L} \qquad (a)$$

These values are correct for any value of x to the left of the point of application of the concentrated load P at $x = a$. To determine the shear force and the bending moment to the right of the concentrated load, we cut the beam again, as shown in Fig. 4.11*c*. In this case $x > a$, and the equations of equilibrium give

$$R_A - P + V = 0 \qquad V = \frac{Pa}{L}$$
$$M + P(x - a) - R_A x = 0 \qquad M = \frac{Pa}{L}(L - x) \qquad (b)$$

The expressions for V and M in Eqs. (*b*) follow more simply from equilibrium conditions applied to the right-hand segment shown in Fig. 4.11*d*.

The shear force and bending moment diagrams now can be plotted by using the expressions in Eqs. (*a*) and (*b*) as shown in Fig. 4.11*e* and *f*. There is a discontinuity in the shear force diagram and a discontinuity in the *slope* of the bending moment diagram at the point of application of the concentrated load. Note in the special case when the load is located at the center of the beam, the maximum value of the shear force is equal to $P/2$ and the maximum value of the bending moment is equal to $PL/4$.

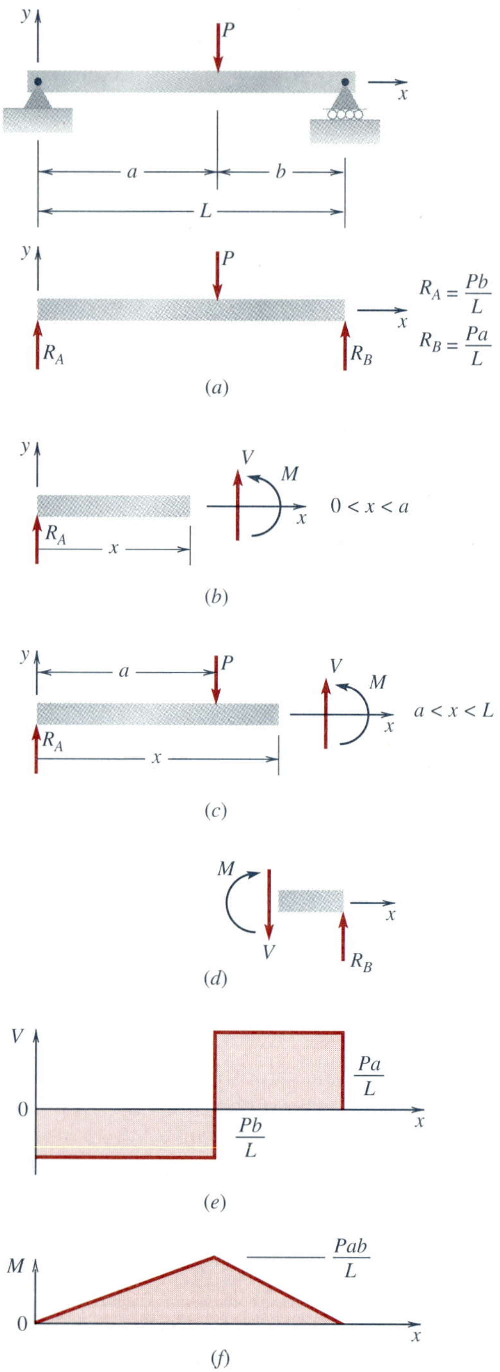

Figure 4.11 Example 4.5

EXAMPLE 4.6

Construct the shear force and bending moment diagrams for the beam *AB* of Example 4.2, shown again in Fig. 4.12.

We draw the beam in Fig. 4.12*a*, showing the values of the reactions at the supports. We then take different sections along the beam to find the values of the shear force and bending moments. For $0 < x < 3$, we show a typical segment of the beam in Fig. 4.12*b* and sum the forces in the *y* direction and take moments about the cut section to get

$$V = -50 \text{ kN} \qquad M = 50x \quad \text{kN} \cdot \text{m} \tag{a}$$

For $3 < x < 11$, shown in Fig. 4.12*c*, we again take a section at a distance *x* from the left end, showing the positive *V* and *M* acting. From conditions of force and moment equilibrium, we have

$$50 - 50 + V = 0 \qquad V = 0$$
$$M + 50(x - 3) - 50x = 0 \tag{b}$$
$$M = 150 \text{ kN} \cdot \text{m}$$

In a similar way, for the segment corresponding to $11 < x < 14$, shown in Fig. 4.12*d*, we have

$$V = 50 \text{ kN}$$
$$M + 50(x - 11) + 50(x - 3) - 50x = 0 \tag{c}$$
$$M = 50(14 - x) \quad \text{kN} \cdot \text{m}$$

The shear force and bending moment diagrams are constructed from the expressions for *V* and *M* in each segment, as shown in Fig. 4.12*e* and *f*. Again we find a discontinuity in *V* and in *dM*/*dx* at the points corresponding to the points of application of the concentrated loads. We note from the loading on this beam that the bending moment is equal to a constant value of 150 kN · m in the center 8-m segment.

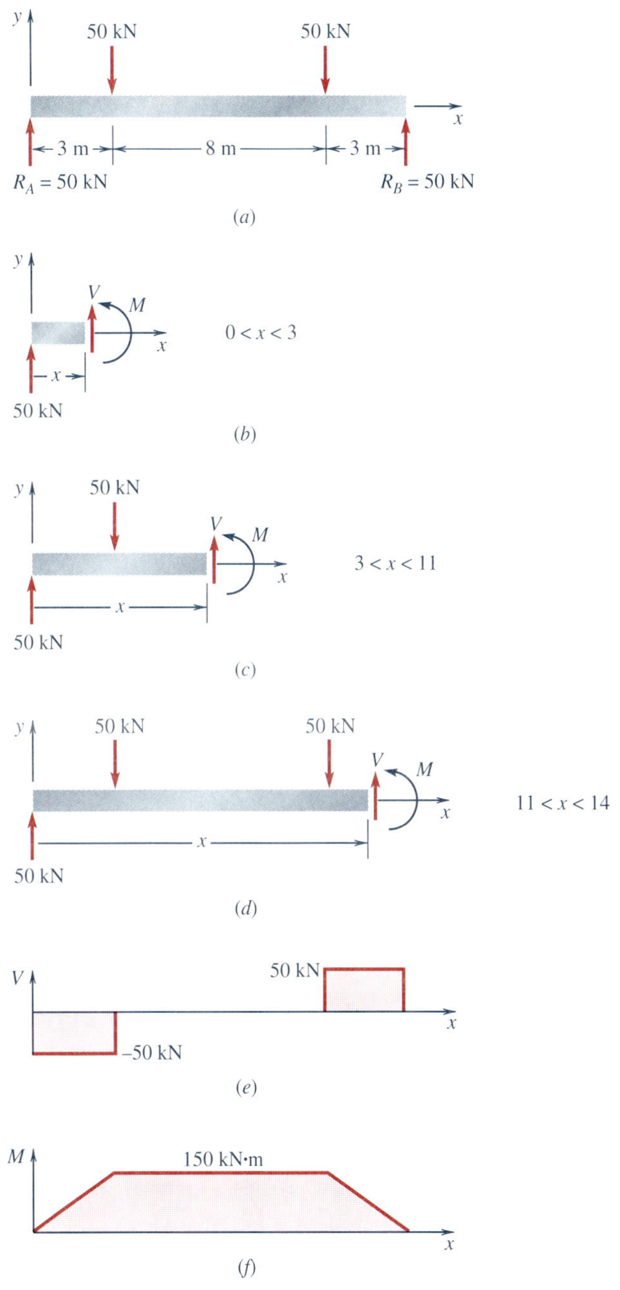

Figure 4.12 Example 4.6

So far in this chapter it has been assumed that the transverse loads acting on a beam and the support forces are concentrated or point forces. In the case of a cantilever beam, the support forces at the wall include a concentrated bending moment. However, in many cases loads are often spread out or continuously distributed along a beam, giving rise to an *intensity* of loading q, defined as the limit

$$q = \lim_{\Delta x \to 0} \frac{\Delta F}{\Delta x} \tag{4.1}$$

where ΔF is the amount of force applied over a given distance increment Δx of the beam; q has the dimensions of force per unit length, e.g., newtons per meter. For example, often material of a given total weight is distributed uniformly along a beam over a certain total length, and for this case we say that the beam is loaded by so many pounds per foot.

In general, the intensity of loading will vary with position along the beam so that q is a function of x, $q = q(x)$; see Fig. 4.13*a*. The two most common distributions of loading on a beam are a uniform or constant distribution, where $q(x)$ is constant, and a linearly varying distribution, where $q(x)$ has the form $Ax + B$. These two cases are shown in Fig. 4.13*b* and *c*.

To be consistent with our use of force and moment components defined with respect to the coordinate axes, we will adopt the following sign convention for the intensity of loading on the beam: Loading is positive when it acts in the direction of the positive coordinate axis. The distributed loads shown in Fig. 4.13 are all positive loads since they are acting in the positive y direction.

It is convenient in many cases to think of the loading on the beam in terms of a *loading diagram* in which the value of q is plotted as a function of distance along the beam. The loading diagram shows all the loads on the beam, and it is convenient to line up vertically the shear force diagram and the bending moment diagram with the loading diagram.

We recall from our study of statics that the *resultant* force of a distributed load is equal to the area of the loading diagram and that this resultant of the distributed load passes through the *centroid of the loading diagram*. We can use these results to help us obtain shear force and bending moment diagrams for beams loaded by distributed loads.

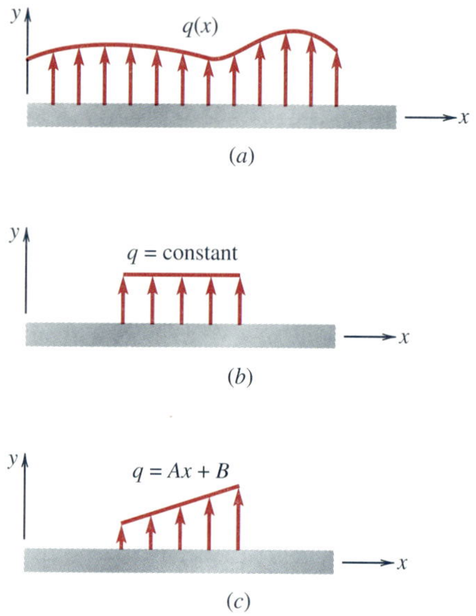

Figure 4.13 (*a*) Distributed load of intensity $q(x)$; (*b*) constant distributed load; (*c*) linear distributed load.

EXAMPLE 4.7

Determine the shear force and bending moment diagrams for a simply supported beam of length L loaded by a uniformly distributed load, as shown in Fig. 4.14*a*. We first idealize the loading and support reactions, as shown in Fig. 4.14*a*. The load intensity $q(x)$ is negative—i.e., it acts in the negative y direction—thus $q = -w_0$, where w_0 is positive. We will neglect the weight of the beam; or since it is so easy to do in this case, we can assume that the weight of the beam per unit length is included in the term w_0.

Figure 4.14*b* gives a graph of the loading diagram showing $q(x)$ versus x for the beam. In addition to the applied load w_0, we show the values of the reactions at each end in the loading diagram.

The values of the support reactions follow from equilibrium and symmetry:

$$R_A = R_B = \frac{w_0 L}{2}$$

A cut at a distance x from the left end gives the segment shown in Fig. 4.14c. Equilibrium of forces in the vertical direction gives

$$R_A + V - w_0 x = 0 \tag{a}$$

where the term $w_0 x$ arises from the resultant of the uniform load w_0 acting on the segment of length x. It follows that

$$V = w_0\left(x - \frac{L}{2}\right) \tag{b}$$

Moment equilibrium about the cut section x gives

$$M + w_0 x\left(\frac{x}{2}\right) - R_A x = 0$$

where the term $w_0 x(x/2)$ arises from the moment contribution of the resultant force $w_0 x$ with the moment arm $x/2$ about the cut section x. The bending moment is then given by

$$M = \frac{w_0 L x}{2} - \frac{w_0 x^2}{2} \tag{c}$$

The shear force and the bending moment expressions in Eqs. (b) and (c) are valid for all values of x; the shear force and bending moment diagrams can be constructed by using Eqs. (b) and (c) and are given in Fig. 4.14d.

The maximum absolute value of the shear force is $w_0 L/2$, and this value occurs at the supports. The maximum value of the bending moment occurs at the center of the beam with the value

$$M_{\max} = \frac{w_0 L^2}{8} \tag{d}$$

This maximum value of the bending moment should be compared to the value of the maximum bending moment $PL/4$ for the case of a simply supported beam with a concentrated load P at the center; see Example 4.5. If the total distributed load $w_0 L$ were concentrated at the center of the beam, then with $P = w_0 L$, the maximum bending moment would be $PL/4 = w_0 L^2/4$, which is double the value given by Eq. (d) for the distributed load. We also note from Fig. 4.14d and from Eqs. (b) and (c) that the slope of the bending moment diagram, that is, dM/dx, is equal to $-V$ at all points $0 \le x \le L$; we will derive this general result in Sec. 4.4.

Figure 4.14 Example 4.7

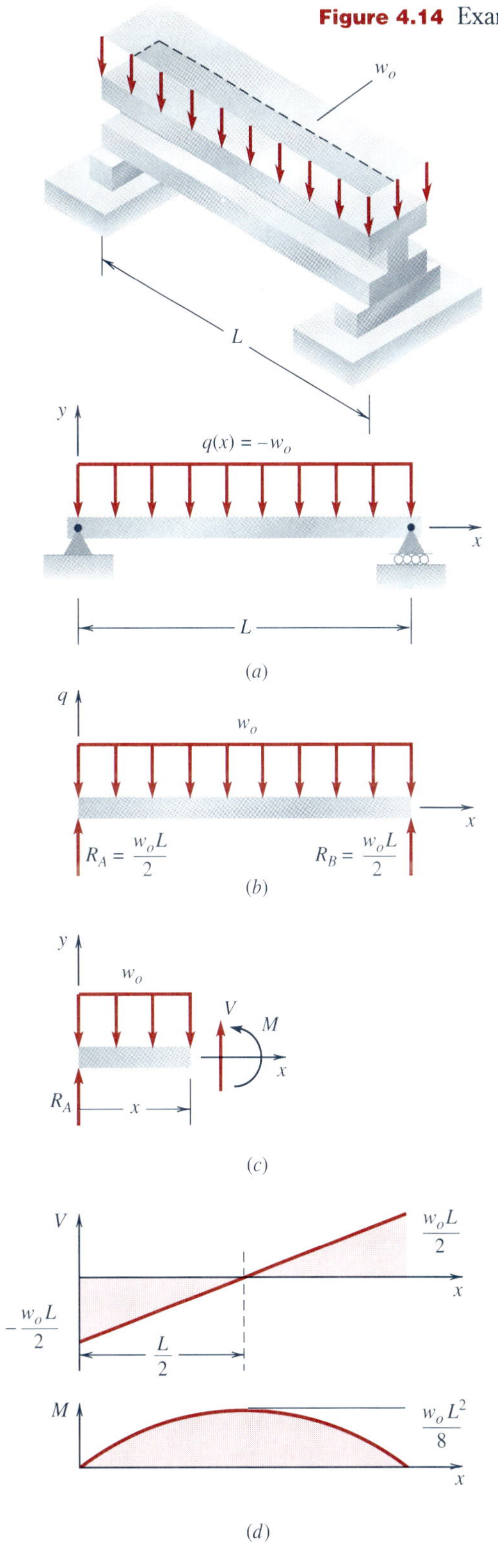

EXAMPLE 4.8

We wish to determine the shear force and bending moment diagrams for a cantilever beam AB loaded with a linearly distributed load, as shown in Fig. 4.15*a*. We neglect the weight of the beam. In Fig. 4.15*b*, we show the loading, with the distributed loading increasing from zero at the tip A of the cantilever to its maximum value w_0 at wall B, and the wall reactions R_B and M_B.

To determine the wall reactions at B, we replace the triangular load distribution by its resultant R equal to the area of the loading diagram. Since the loading diagram is a triangle, its area is one-half the product of the base and the altitude, and its centroid is two-thirds the distance from the vertex to the opposite side. Therefore, as shown in Fig. 4.15*b*,

$$R = \frac{w_0 L}{2} \qquad \bar{x} = \frac{2L}{3} \tag{a}$$

where $\bar{x}$ is the distance to the centroid, measured from the left end. The *external* support reactions R_B and M_B at the wall shown in Fig. 4.15*b* are now easily obtained by applying the conditions of equilibrium:

$$\Sigma F_y = R_B - R = 0 \qquad \text{or} \qquad R_B = \frac{w_0 L}{2} \tag{b}$$

$$\Sigma M_z = R(L - \bar{x}) + M_B = 0 \qquad \text{or} \qquad M_B = -\frac{w_0 L^2}{6} \tag{c}$$

We cannot use the resultant R which was calculated for the whole beam in the calculation of the shear force and bending moment at any section x along the beam. To find the shear force and bending moment acting at any section x, we cut the beam at x, and the shear force and bending moment on that section become external forces for the isolated beam segment, as shown in Fig. 4.15*c*. We can replace the distributed load acting on this portion of the beam by its resultant R', as shown in Fig. 4.15*c*. Now using the same method as was used above to find the wall

Figure 4.15 Example 4.8

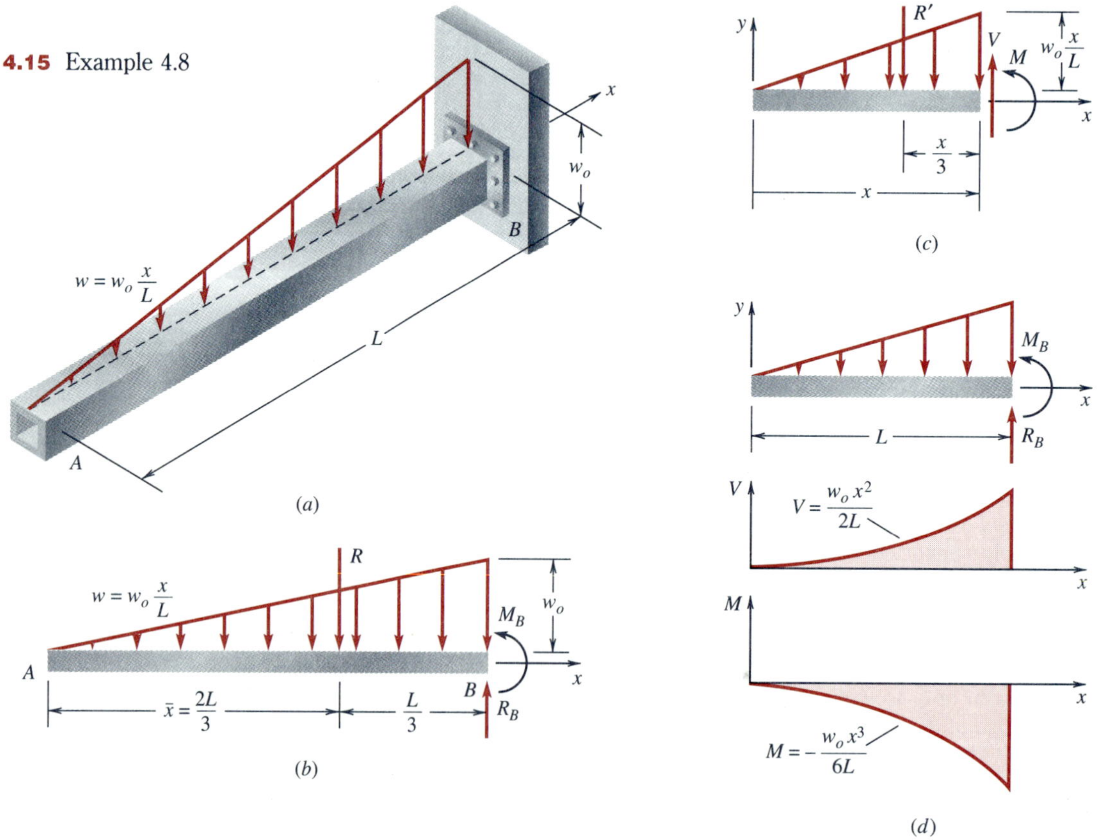

reactions in Eqs. (*b*) and (*c*), we obtain from the equilibrium conditions

$$V = R' = \left(\frac{w_0 x}{L}\right)\left(\frac{x}{2}\right) = \frac{w_0 x^2}{2L}$$
$$M = -R'\frac{x}{3} = -\frac{w_0 x^3}{6L} \qquad (d)$$

These values are now used to sketch the shear force and bending moment diagrams, as shown in Fig. 4.15*d*. In interpreting these diagrams, we note that the shear force remains positive over the entire beam and increases in magnitude from zero at the tip to a maximum at the base. The free body of an arbitrary segment of the cantilever of length x (Fig. 4.15*c*) clearly shows that the shear force V supplied by the rest of the beam situated to the right of the section to maintain equilibrium must be positive and must increase with distance x to a magnitude of $R_B = w_0 L/2$ at $x = L$. The same free-body diagram shows that the bending moment M must remain negative and increase in magnitude to a peak of $M_B = -w_0 L^2/6$ at $x = L$.

EXAMPLE 4.9

Determine the shear force and bending moment diagrams for the beam loaded with a uniform load intensity w_0 over one-half of its length, as shown in Fig. 4.16*a*. We wish also to determine the value of the maximum bending moment and its location along the length of the beam. Neglect the weight of the beam.

In Fig. 4.16*a* we show the idealized model of the beam and the values of the reactions at the supports. Equilibrium gives

$$R_A = \frac{3w_0 L}{8} \qquad R_B = \frac{w_0 L}{8} \qquad (a)$$

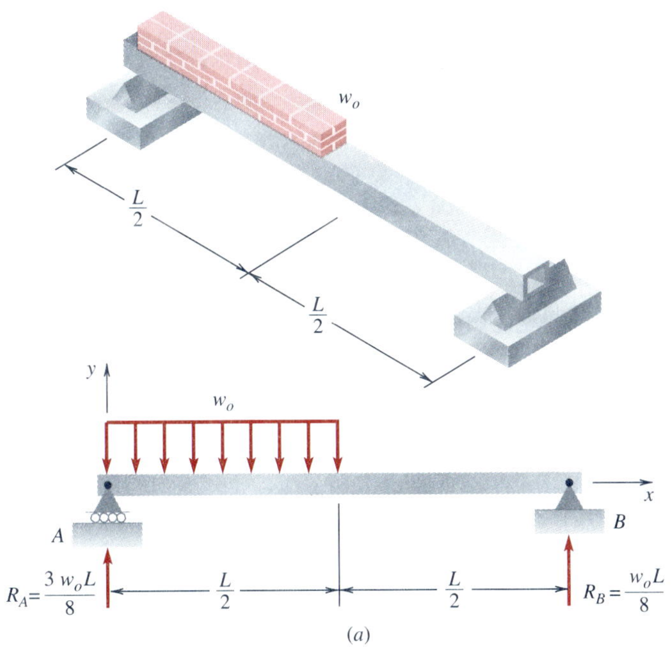

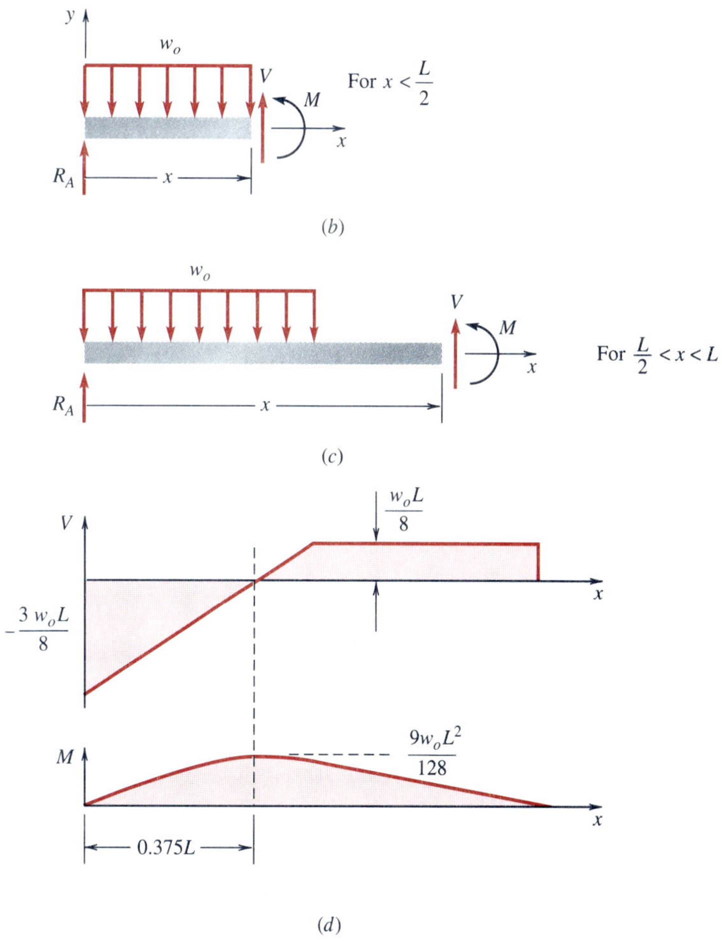

Figure 4.16 Example 4.9

Since the load is discontinuous, we need to consider two different segments along the beam, one for $0 < x < L/2$ and the other for $L/2 < x < L$. For each segment, we use the equilibrium equations to find the expressions for the shear force and bending moment.

For $0 < x < L/2$, we have from Fig. 4.16*b*

$$R_A + V - w_0 x = 0$$

$$V = -w_0\left(\frac{3L}{8} - x\right) \qquad (b)$$

$$M + w_0\frac{x^2}{2} - R_A x = 0$$

$$M = w_0\left(\frac{3Lx}{8} - \frac{x^2}{2}\right)$$

For $L/2 < x < L$, we have from Fig. 4.16*c*

$$V = \frac{w_0 L}{8}$$

$$M + \frac{w_0 L}{2}\left(x - \frac{L}{4}\right) - R_A x = 0 \qquad (c)$$

$$M = -\frac{w_0 L}{8}(x - L)$$

From Eqs. (*b*) and (*c*), the shear force and bending moment diagrams can be drawn as shown in Fig. 4.16*d*. The maximum bending moment at $x = 0.375L$ is equal to $9w_0L^2/128$. We note that the maximum bending moment occurs when $V = 0$; we also note from Eqs. (*b*) and (*c*) that the derivative of the moment with respect to x, dM/dx, is equal to $-V$.

4.4 Differential Equilibrium Relations

In each of the above examples in which we determined the shear force and bending moment diagrams, we followed the general procedure outlined in Sec. 4.2 of cutting the beam and using equilibrium equations for one of the segments. Clearly by considering the equilibrium of a small beam element located at a general point along the beam, we should be able to derive relations for the changes in the values of V and M in terms of the applied loading q. In Chap. 2, an equation for the rate of change of the internal axial force in terms of the applied axial loading on a bar was derived by using force equilibrium; see Sec. 2.8 and Eq. (2.24). Equation (2.24) involving the derivative of the internal axial force can be used to determine by integration an expression for the variation of the internal axial force along a bar from the knowledge of the applied loading. In a similar way, Eq. (3.37) can be used to find an expression for the variation of the internal twisting moment in a circular shaft.

In this section, we will determine the equations (we now need to consider force *and* moment equilibrium so that two equations will be necessary) relating the applied loading on a beam to the internal shear force and bending moment. Instead of cutting a beam into two segments and considering the equilibrium of one of the segments, we consider a small element cut from the beam as a free body with shear forces and bending moments acting on it. We then let the length of this element go to zero.

Figure 4.17*a* shows a portion of a beam with a distributed load $q(x)$ and a beam element of length Δx. The small element of length Δx is redrawn in Fig. 4.17*b*. The external actions on this element are the distributed load of intensity q acting over the length Δx and the shear forces and bending moments on the two faces, as shown in Fig. 4.17*b*. The shear force and bending moment at section x are $V(x)$ and $M(x)$; the shear force and bending moment at section $x + \Delta x$ are $V(x + \Delta x)$ and $M(x + \Delta x)$. Note the positive directions of the shear force and bending moment applied to the positive (right) face and to the negative (left) face. In Fig. 4.17*c* we have replaced the distributed loading acting on the element by its resultant R. It is clear that if the variation of $q(x)$ is smooth and if Δx is very small, then R is very nearly given by $q(x)\,\Delta x$, and the line of action of R will pass through close to the midpoint O of the element. We shall assume in writing the equilibrium conditions that Δx is already so small that we can approximate R by $q\,\Delta x$ which passes through O. The conditions of equilibrium applied to Fig. 4.17*c* are then

$$\Sigma F_y = V(x + \Delta x) + q\,\Delta x - V(x) = 0$$
$$\Sigma M_O = M(x + \Delta x) + V(x + \Delta x)\frac{\Delta x}{2} + V(x)\frac{\Delta x}{2} - M(x) = 0 \tag{4.2}$$

Before completing the limiting process of letting the length Δx of the element shrink to zero, we rearrange Eqs. (4.2) as follows:

$$\frac{V(x + \Delta x) - V(x)}{\Delta x} + q(x) = 0$$
$$\frac{M(x + \Delta x) - M(x)}{\Delta x} + \frac{V(x + \Delta x) + V(x)}{2} = 0 \tag{4.3}$$

Now as Δx approaches zero, the ratios in Eqs. (4.3) with Δx in the denominator tend to differential quotients or derivatives, and $V(x + \Delta x)$ approaches $V(x)$. Thus the limiting forms of Eqs. (4.3) when Δx goes to zero are

$$\frac{dV}{dx} + q(x) = 0 \tag{4.4}$$

$$\frac{dM}{dx} + V(x) = 0 \tag{4.5}$$

These are the two basic differential equations relating the load intensity $q(x)$ to the shear force $V(x)$ and bending moment $M(x)$ along the beam. We refer to these equations as differential equations because they involve the derivatives of V and M.

If we think of the loading diagram as given by a plot of the curve $q(x)$ versus x, then upon rewriting Eq. (4.4) in the form

$$\frac{dV}{dx} = -q(x) \tag{4.6}$$

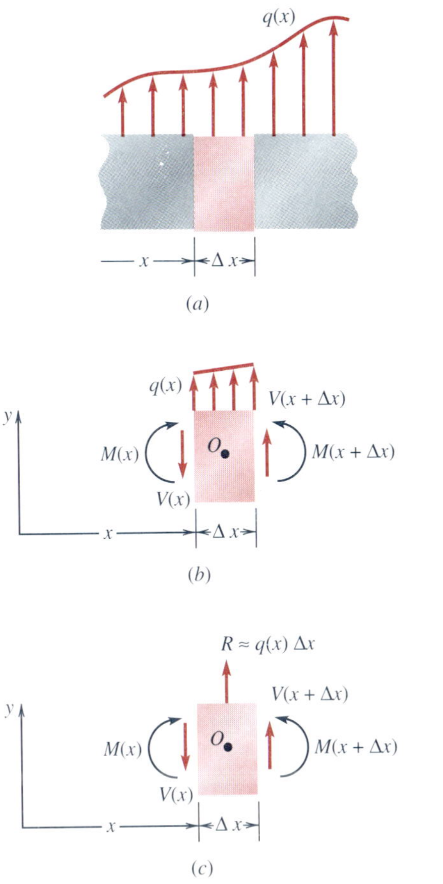

Figure 4.17 (*a*) A segment of a beam loaded by a distributed load $q(x)$. (*b*) A small element of length Δx. (*c*) A small element with resultant R of the distributed load $q(x)$ acting through the center O of the element.

we see that the *slope* of the curve of the shear force $V(x)$ in a shear force diagram at a point x is equal to the negative of the *value* of $q(x)$ in the loading diagram at that point. In a similar way, we conclude from Eq. (4.5)

$$\frac{dM}{dx} = -V(x) \tag{4.7}$$

so that the *slope* of the curve of the bending moment $M(x)$ in a bending moment diagram at a point x is equal to the negative of the *value* of $V(x)$ in the shear force diagram at that point. Figure 4.18*a* shows these results graphically, assuming that V is positive. These relations, Eqs. (4.4) and (4.5), hold at a point x located along the beam where $q(x)$, $V(x)$, and $M(x)$ are continuous functions of x. These results are very useful for visualizing the relations between the curves in the loading, shear force, and bending moment diagrams.

If instead of considering Eqs. (4.6) and (4.7) at a *point* x, we integrate each equation with respect to x between point x_1 and point x_2, we find

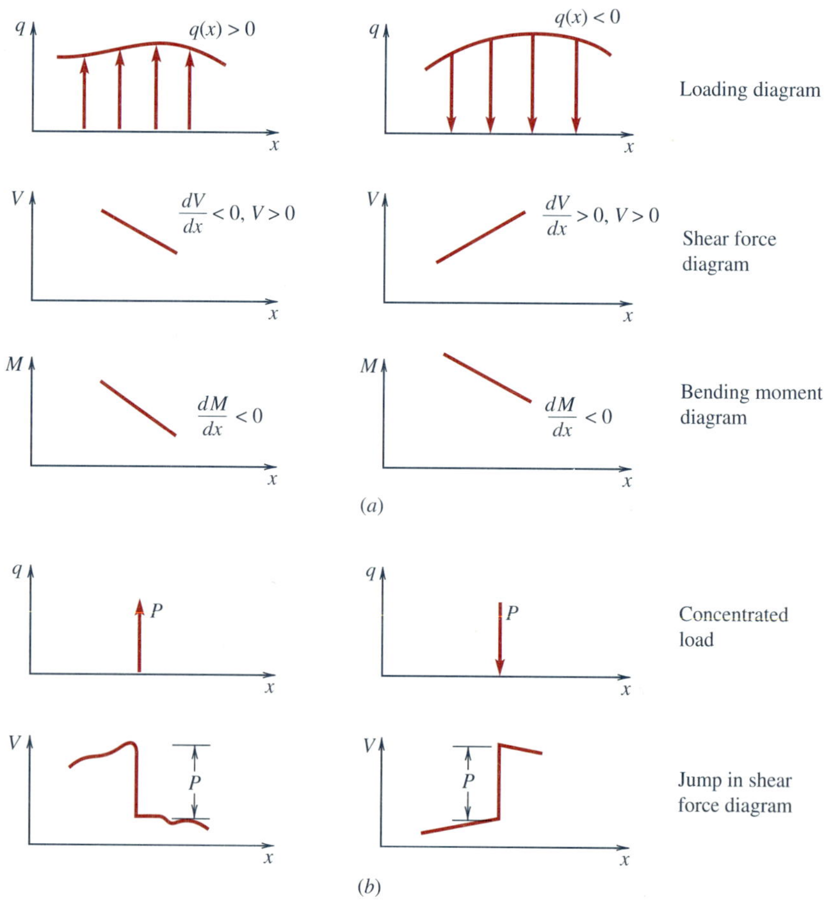

Figure 4.18 (*a*) Geometric relations between the loading diagram, shear force diagram, and bending moment diagram. (*b*) A positive or negative concentrated load in the loading diagram causes a jump or discontinuity in the shear force diagram.

$$V(x_2) - V(x_1) = -\int_{x_1}^{x_2} q(x)\,dx \tag{4.8}$$

$$M(x_2) - M(x_1) = -\int_{x_1}^{x_2} V(x)\,dx \tag{4.9}$$

where $V(x_1)$ and $M(x_1)$ are the values of the shear force and bending moment at point x_1 and $V(x_2)$ and $M(x_2)$ are the values at point x_2.

If we recall that the *integral* of a continuous curve between two values of x can be interpreted as the *area* under the curve between the two values of x, then Eqs. (4.8) and (4.9) can be interpreted as follows. First we rewrite Eqs. (4.8) and (4.9) in the form

$$V(x_2) = V(x_1) - \int_{x_1}^{x_2} q(x)\,dx \tag{4.10}$$

$$M(x_2) = M(x_1) - \int_{x_1}^{x_2} V(x)\,dx \tag{4.11}$$

Equation (4.10) states that the value of the shear force at a point x_2, $V(x_2)$, is equal to the value of the shear force at point x_1, $V(x_1)$, *minus* the *area* under the loading diagram between point x_1 and point x_2. Equation (4.11) states that the value of the bending moment at a point x_2, $M(x_2)$, is equal to the value of the bending moment at point x_1, $M(x_1)$, *minus* the *area* under the shear force diagram between point x_1 and point x_2.

The use of Eqs. (4.10) and (4.11), together with Eqs. (4.6) and (4.7), allows us to construct—*with a little experience*—shear force and bending moment diagrams with ease. We will return to some of the previous examples to see how this is done; the experience follows from doing a number of problems. In working out examples, we will line up the loading diagram, shear force diagram, and bending moment diagram so that the relations between slopes and areas in each diagram are clear. The integrals in the following examples should be related to the areas in the diagrams as we verify the steps.

EXAMPLE 4.10

Consider the beam shown in Fig. 4.19*a*. We wish to use the equilibrium Eqs. (4.4) to (4.11) to construct the shear force and bending moment diagrams. Figure 4.19*a* shows a coordinate system with the origin at the left end.

In Fig. 4.19*b*, we show a graph of the *loading* diagram, $q(x)$ versus x, in which the applied concentrated load P and the reaction concentrated loads R_A and R_B are shown. The loading $q(x)$ is zero except at the reactions and at $x = a$. The values of the reactions are found from equilibrium

$$R_A = \frac{Pb}{L} \qquad R_B = \frac{Pa}{L} \tag{a}$$

At the left support $x = 0$, we consider a small element of length Δx. The shear force and bending moment acting on the positive face of this element are $V(\Delta x)$ and $M(\Delta x)$. Equilibrium of this small element Δx as shown in Fig. 4.19*c* gives

$$\begin{aligned} V(\Delta x) + R_A &= 0 \\ M(\Delta x) - R_A\,\Delta x &= 0 \end{aligned} \tag{b}$$

where $V(\Delta x)$ and $M(\Delta x)$ are the values of the shear force and bending moment at section $x = \Delta x$. In the limit as Δx goes to zero, we find

$$\begin{aligned} V(0) &= -R_A = -\frac{Pb}{L} \\ M(0) &= 0 \end{aligned} \tag{c}$$

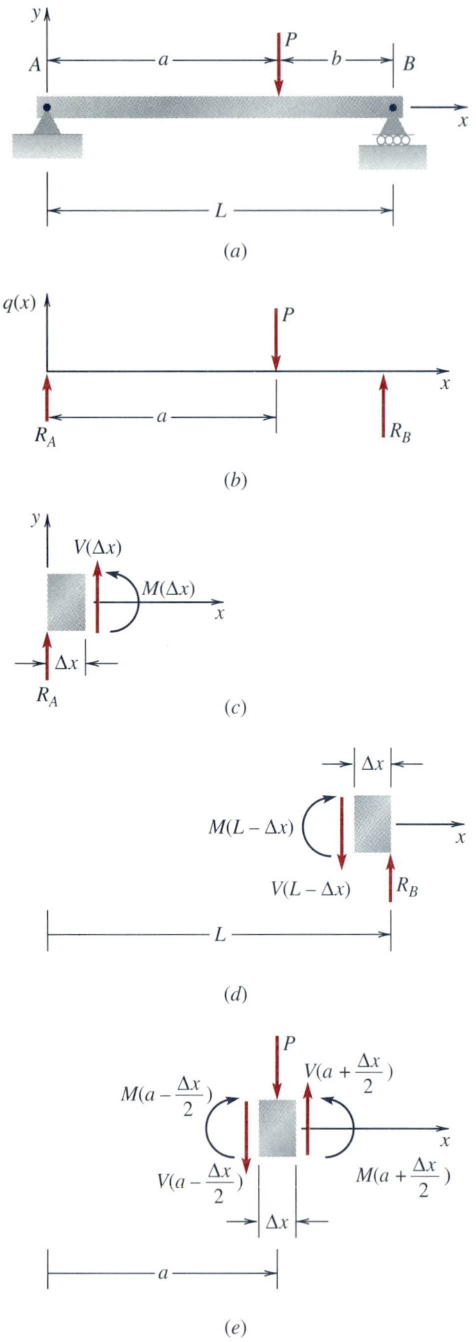

Figure 4.19 Example 4.10

This analysis of equilibrium of an infinitesimal element at the left end confirms the fact that at a pinned support without an applied bending moment the bending moment is zero. The value of the shear force at the left support for this case is the negative of the value of the reaction force.

At the right support we again consider a vanishing small element of length Δx, as shown in Fig. 4.19*d*. Equilibrium equations applied to this element give

$$-V(L - \Delta x) + R_B = 0$$
$$-M(L - \Delta x) + R_B \Delta x = 0 \qquad (d)$$

where $V(L - \Delta x)$ and $M(L - \Delta x)$ are the values of the shear force and bending moment at section $x = L - \Delta x$. They are positive on the negative face of the element.

In the limit as Δx goes to zero, we find

$$V(L) = R_B = \frac{Pa}{L}$$
$$M(L) = 0 \qquad (e)$$

Again, we find the bending moment is zero, and the value of the shear force at $x = L$ in the beam to the left of the support is the reaction force. Note the signs associated with the values of the shear forces in the beam at $x = 0$ and $x = L$ in terms of the values of the reactions.

With the value of the shear force at $x = 0$ known, we may use Eq. (4.10) to find the values of the shear force for $0 < x < a$

$$V(x) = V(0) - \int_0^x q(x)\,dx$$
$$= V(0) = -\frac{Pb}{L} \qquad (f)$$

since $q(x)$ is equal to zero for $0 < x < a$.

At the section $x = a$, we have a concentrated load P applied in the negative y direction. In Fig. 4.19*e*, we show a small element of length Δx centered on the section $x = a$. The shear force and bending moment acting on each face of this element at $x = a \pm \Delta x/2$ are shown. Equilibrium of this element gives

$$-V\left(a - \frac{\Delta x}{2}\right) - P + V\left(a + \frac{\Delta x}{2}\right) = 0$$

$$-M\left(a - \frac{\Delta x}{2}\right) + M\left(a + \frac{\Delta x}{2}\right)$$
$$+ V\left(a + \frac{\Delta x}{2}\right)\frac{\Delta x}{2} + V\left(a - \frac{\Delta x}{2}\right)\frac{\Delta x}{2} = 0 \qquad (g)$$

In the limit as Δx goes to zero, we find

$$V(a^+) = V(a^-) + P \tag{h}$$

$$M(a^+) = M(a^-) \tag{i}$$

where a^+ is the value of x just to the right of $x = a$ as Δx goes to zero, i.e., just beyond the location of the concentrated load, and a^- is the value of x just to the left of $x = a$ as Δx goes to zero. Equation (*h*) shows that the shear force will undergo a jump or a discontinuity at $x = a$ of positive magnitude P, since P was a negative concentrated load. Equation (*i*) shows that the bending moment is continuous at $x = a$.

From Eq. (*f*) the value of $V(a^-)$ is known. Therefore from Eq. (*h*) we have

$$V(a^+) = V(a^-) + P = -\frac{Pb}{L} + P = \frac{Pa}{L} \tag{j}$$

Then from Eq. (4.10), for $a < x < L$, with $q(x) = 0$,

$$V(x) = V(a^+) - \int_a^x q(x)\,dx = \frac{Pa}{L} \tag{k}$$

which is the value of the shear force at the right end support. Also since $q(x)$ is equal to zero in each of the segments $0 < x < a$, $a < x < L$, the *slope* of the shear force curve is zero, that is, $V(x)$ is constant in each segment. The shear force diagram is placed in Fig. 4.20*b* directly under the loading diagram in Fig. 4.20*a*.

To determine the bending moment along the beam, we first note that according to Eq. (4.7) since the *value* of V in each segment is equal to a constant, the slope of the bending moment diagram in each segment is equal to a constant.

For $0 < x < a$,

$$\frac{dM}{dx} = -V = R_A = \frac{Pb}{L} > 0$$

For $a < x < L$,

$$\frac{dM}{dx} = -V = -R_B = -\frac{Pa}{L} < 0 \tag{l}$$

Since the slopes are constant in each segment, the bending moment is a linear function of x in each segment.

For $0 < x < a$, we have from Eq. (4.11)

$$M(x) = M(0) - \int_0^x V(x)\,dx = \int_0^x R_A\,dx = R_A x \tag{m}$$

so that $M(x)$ is equal to the negative of the area of the shear force diagram between 0 and x because for this case $M(0) = 0$.

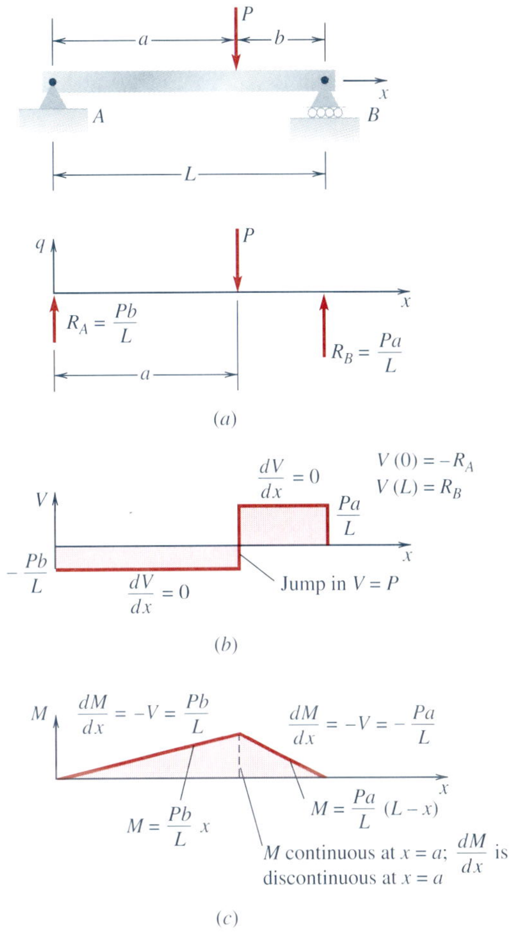

Figure 4.20 Example 4.10: Loading diagram, shear force diagram, and bending moment diagram.

For $a < x < L$, since we already know from Eq. (*m*) that the value of $M(a) = R_A a$, we can use Eq. (4.11) to write

$$\begin{aligned} M(x) &= M(a) - \int_a^x V(x)\,dx = R_A a - \int_a^x R_B\,dx \\ &= R_A a - R_B(x - a) \\ &= \frac{Pa}{L}(L - x) \end{aligned} \tag{n}$$

The bending moment diagram is shown in Fig. 4.20*c*.

Again, it is important to emphasize that jumps or discontinuities occur in the value of the shear force at points at which concentrated loads are applied (Fig. 4.18*b*), and at these points there will be a sudden change or discontinuity in the slope of the bending moment curve in the bending moment diagram. We see these discontinuities in the shear force and bending moment diagrams in Fig. 4.20. If a concentrated moment is applied to a beam at a point, there will be a jump in the moment diagram at that point; see Example 4.15.

EXAMPLE 4.11

Construct the shear force and bending moment diagrams for the uniformly loaded beam of Example 4.7, which is shown in Fig. 4.21*a*. In this case we have a constant distributed load acting over the entire length of the beam so that the loading diagram consists of the uniform load plus the two reactions, as shown in Fig. 4.21*b*. The loading $q(x)$ acts downward so it is negative, $q = -w_0$. Based on equilibrium of a small element of length Δx located at the left end of the beam, we have in the limit as $\Delta x \to 0$

$$V(0) = -\frac{w_0 L}{2} \qquad M(0) = 0 \tag{a}$$

and for $0 < x < L$,

$$V(x) = V(0) - \int_0^x q(x)\,dx$$

or

$$V(x) = -\frac{w_0 L}{2} + w_0 x \tag{b}$$

Also we have

$$\frac{dV}{dx} = -q = w_0 \qquad \text{a constant} \tag{c}$$

Therefore, we can easily sketch the shear force diagram, making sure that $V(L)$ agrees with the value of the reaction at $x = L$ (Fig. 4.21*c*).

To obtain the bending moment diagram, we first note that since V changes sign at $x = L/2$, the slope of the bending moment curve changes from positive to negative at $x = L/2$. Further, since $V(x)$ is linear in x, $M(x)$ must be quadratic in x. Starting at $x = 0$, with $M(0) = 0$, and a positive slope $dM/dx = -V > 0$, the M curve will be positive and increasing, and the slope dM/dx will be positive and decreasing until at $x = L/2$ the slope will reach zero. In the right half of the beam $dM/dx =$

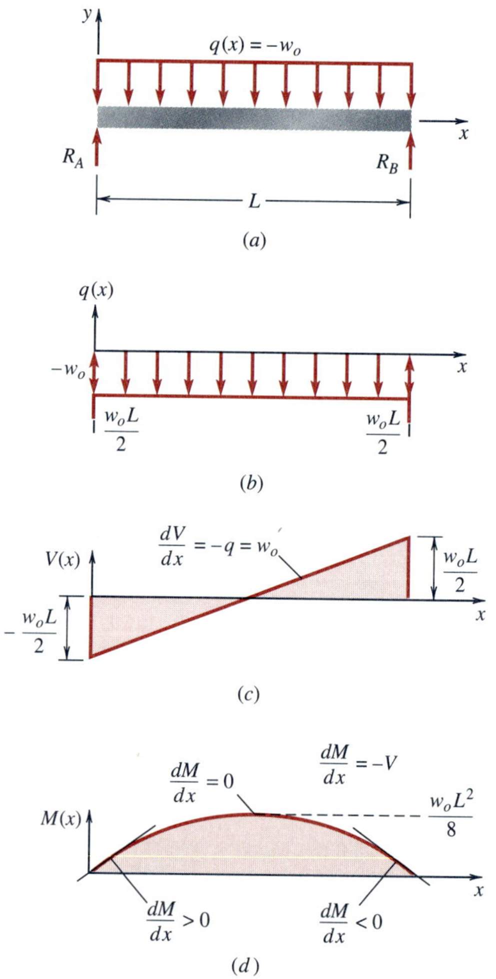

Figure 4.21 Example 4.11

$-V < 0$; therefore M will remain positive and decrease to the value zero at the right end support. With this knowledge we can then sketch $M(x)$ as shown in Fig. 4.21*d*.

An analytic expression for the $M(x)$ curve is easily derived by making use of Eq. (4.11) to write

$$M(x) = M(0) - \int_0^x V(x)\,dx \qquad (d)$$

or by using Eqs. (*a*) and (*b*)

$$M(x) = w_0 \int_0^x \left(\frac{L}{2} - x\right) dx = \frac{w_0 x}{2}(L - x) \qquad (e)$$

The maximum bending moment occurs at the center of the beam and is given by

$$M_{\text{max}} = \frac{w_0 L^2}{8} \qquad (f)$$

EXAMPLE 4.12

We wish to use the methods of this section to construct the shear force and bending moment diagrams for the beam shown in Fig. 4.22*a*; see Example 4.9. We show the beam with the applied loads and support forces in Fig. 4.22*a*, and directly under this figure in Fig. 4.22*b* we show the corresponding loading diagram. By repeated use of the Eqs. (4.6) to (4.11) we propose to construct the shear force and bending moment diagrams for this beam.

The support reactions were obtained earlier in Example 4.9 and are shown in Fig. 4.22*a*. Starting at the left support, we have

$$V(0) = -\frac{3w_0 L}{8} \qquad (a)$$

Over segment AB we have

$$q(x) = -w_0$$

Therefore, according to Eq. (4.6)

$$\frac{dV}{dx} = -q(x) = w_0 > 0$$

and we conclude that over this segment V is linear with a positive slope. From Eq. (4.10) we have

$$V\left(\frac{L}{2}\right) = V(0) - \int_0^{L/2} q(x)\,dx$$

or

$$V\left(\frac{L}{2}\right) = -\frac{3w_0 L}{8} + \frac{w_0 L}{2} = \frac{w_0 L}{8} \qquad (b)$$

The linear plot in segment AB is shown in Fig. 4.22*c*.

Over segment BC, $q(x) = 0$, so that from Eq. (4.6)

$$\frac{dV}{dx} = 0$$

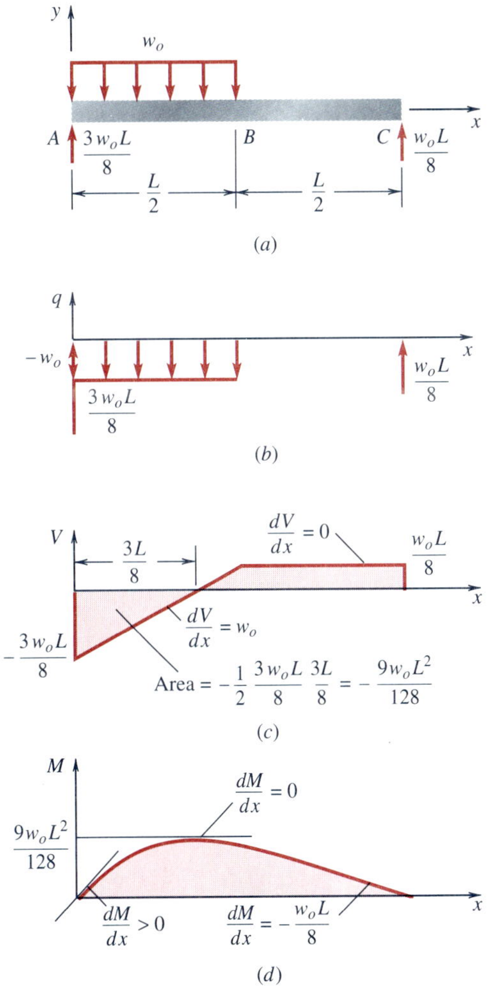

Figure 4.22 Example 4.12

That is, the slope of V is zero, so that V remains constant at the value

$$V = \frac{w_0 L}{8}$$

To construct the M curve, we again start at the left support where $M(0) = 0$. From Eq. (4.7) we have $dM/dx = -V$, and from Fig. 4.22*c* we observe that over segment AB as x increases, $-V$ starts out positive, decreases to zero at $x = 3L/8$, and then takes on negative values. Since M is the integral of V which is linear over segment AB, as described above, M is a quadratic curve. This quadratic curve starts with a positive slope which decreases to zero at $x = 3L/8$. At this point, M takes on a maximum value, and Eq. (4.11) can be used to find this value since

$$M\left(\frac{3L}{8}\right) = M(0) - \int_0^{3L/8} V(x)\,dx$$

$$= 0 + \frac{1}{2}\left(\frac{3w_0 L}{8}\right)\left(\frac{3L}{8}\right) \qquad (c)$$

$$= \frac{9}{128} w_0 L^2$$

The quadratic curve over the segment $x = 0$ to $x = 3L/8$ is shown in Fig. 4.22*d*. The quadratic curve ends at midpoint B with the value $M(L/2)$, which by Eq. (4.11) is given by

$$M\left(\frac{L}{2}\right) = M\left(\frac{3L}{8}\right) - \int_{3L/8}^{L/2} V(x)\,dx$$

or

$$M\left(\frac{L}{2}\right) = \frac{9}{128} w_0 L^2 - \frac{1}{2}\left(\frac{L}{8}\right)\left(\frac{w_0 L}{8}\right)$$

so that

$$M\left(\frac{L}{2}\right) = \frac{8 w_0 L^2}{128} = \frac{w_0 L^2}{16} \qquad (d)$$

Since $V = w_0 L/8$ over segment BC of the beam, we have

$$\frac{dM}{dx} = -\frac{w_0 L}{8}$$

and M is linear with a constant negative slope. As an overall check on the bending moment diagram, we can use Eq. (4.11) to write

$$M(L) = M\left(\frac{L}{2}\right) - \int_{L/2}^{L} V(x)\,dx$$

or

$$M(L) = \frac{w_0 L^2}{16} - \frac{L}{2}\frac{w_0 L}{8} = 0 \qquad (e)$$

We see that the zero end moment at $x = L$ predicted by the integration method, which makes use of Eqs. (4.10) and (4.11), checks with the result found by equilibrium considerations.

EXAMPLE 4.13

Consider the beam shown in Fig. 4.23*a* loaded by a constant distributed load over a segment AB and a concentrated load at its right end. We wish to sketch the shear force and bending moment diagrams for this beam. We neglect the weight of the beam, and we use Eqs. (4.10) and (4.11).

As the first step, we draw in the coordinate axes, as shown in Fig. 4.23*a*, and use equilibrium to determine the values of the reactive forces at A and C. We find

$$R_A = 17 \text{ kN} \qquad R_C = 23 \text{ kN} \qquad (a)$$

In Fig. 4.23*b* we sketch the loading diagram, including the concentrated reactive forces at A and C. The loading diagram is a depiction of the loads acting on the beam.

We first note that the bending moment at each end of the beam is zero and that the shear force at $x = 0$ is

$$V(0) = -R_A = -17 \text{ kN} \qquad (b)$$

For $0 < x < 6$, from the loading diagram $q(x) = -5$ kN/m, so that dV/dx is positive and V is linear. Further, the value of V at $x = 6$ can be obtained from

$$V(6) = V(0)$$
$$- \text{(area of loading diagram between 0 and 6)}$$
$$= -17 - (-5 \cdot 6) = 13 \text{ kN} \qquad (c)$$

We can therefore draw a straight line connecting the V values at $x = 0$ and $x = 6$.

To find the section where $V = 0$, we need to find the value of x such that the corresponding area under the loading curve cancels the value of V at $x = 0$, that is,

$$V(x) = 0 = V(0) - (-5x)$$
$$= -17 + 5x \qquad (d)$$

or

$$x = 3.4 \text{ m}$$

For $6 < x < 10$, the loading $q(x) = 0$ so that $dV/dx = 0$ and V remains constant at 13 kN. At point C, where the concentrated reactive force is present, the shear force will undergo a jump discontinuity of -23 kN since the reactive force is positive. Therefore the value of V on the right of C is -10 kN; since $q(x) = 0$ for $10 < x < 14$, the value of V remains at -10 kN. The completed shear force diagram is shown in Fig. 4.23*c*.

To determine the bending moment diagram, we will use the areas from the shear force diagram. The shear force is linear for $0 < x < 6$; therefore M will be quadratic in this interval. For $0 < x < 3.4$, the *slope* of the M curve will be positive and decreasing in value, and at $x = 3.4$ the slope of the M curve will be zero. To find the value of M at $x = 3.4$, we have from the area of the shear force diagram

$$M(3.4) = M(0) - \tfrac{1}{2}(-17)(3.4) = 28.9 \text{ kN} \cdot \text{m} \qquad (e)$$

At $x = 6$, we have

$$M(6) = M(3.4) - \tfrac{1}{2}(13)(2.6) = 12 \text{ kN} \cdot \text{m} \qquad (f)$$

We can sketch therefore the quadratic curve for $0 < x < 6$ as shown in Fig. 4.23*d*.

For $6 < x < 10$, V is a constant positive value so that dM/dx is negative. The value at $x = 10$ for M follows from the area of the V diagram

$$M(10) = M(6) - (13)(4) = -40 \text{ kN} \cdot \text{m} \qquad (g)$$

Finally for $10 < x < 14$, again M is linear this time with a positive slope since V is negative in this interval, and we have, as expected,

$$M(14) = M(10) - (-10)(4) = 0 \qquad (h)$$

The complete bending moment diagram is shown in Fig. 4.23*d*. We see that using the areas under the loading diagram provides a convenient way to sketch the shear force diagram; similarly, using the areas under the shear force diagram gives the bending moment diagram.

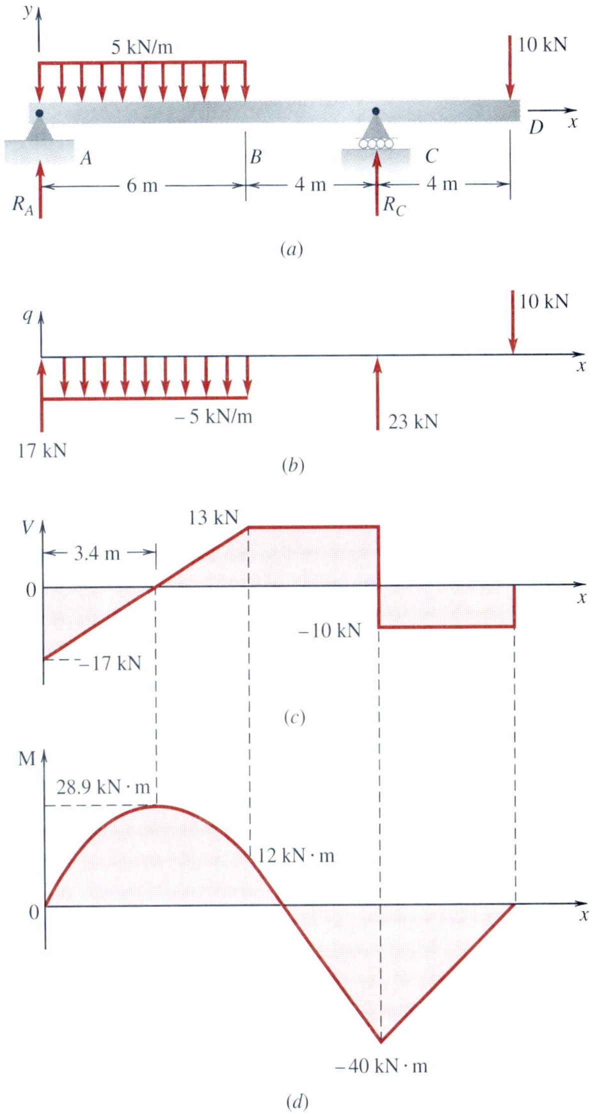

Figure 4.23 Example 4.13

Thus far in plotting the shear force and bending moment diagrams, we have used the integrated form of Eqs. (4.6) and (4.7), either directly in the form of Eqs. (4.10) and (4.11) or through the use of areas, together with the relations between the slopes of the shear force and bending moment diagrams and values of q and V. We have found that using the equations in this way provides a systematic means of plotting the diagrams. However, we can also integrate Eqs. (4.6) and (4.7) directly as differential equations. That is, if $q(x)$ is known, we can integrate once to find $V(x)$, and then knowing $V(x)$, we can integrate again to find $M(x)$. Each integration will introduce a *constant of integration,* which is to be determined from a knowledge of values of V or M at points along or at the ends of the beam. The relative simplicity of this direct method of integration depends on the nature of the applied loading. If the applied loading is made up of a number of different portions of distributed loads, concentrated loads, and concentrated moments, then a systematic way of integrating the equations is necessary, which we discuss in the next section. We first consider two illustrations of this integration method to demonstrate how the constants of integration are evaluated.

EXAMPLE 4.14

Consider the beam shown in Fig. 4.24*a* carrying a constant distributed load with the values of the reactions shown; see Example 4.11. We wish to sketch the shear force and bending moment diagrams by using Eqs. (4.6) and (4.7) directly.

The applied loading is $q(x) = -w_0$, a constant. Equation (4.6) becomes

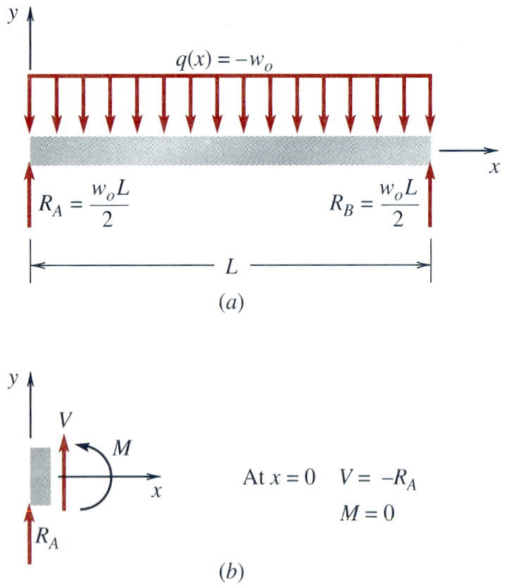

Figure 4.24 Example 4.14

$$\frac{dV}{dx} = -q = w_0 \qquad 0 < x < L \tag{a}$$

Upon integration, we find

$$V(x) = w_0x + C_1 \tag{b}$$

where C_1 is the constant of integration. This constant is to be evaluated from information regarding the support of the beam. Since we know the value of the reaction at $x = 0$, we can express the shear force at $x = 0$ in terms of the reaction; we have from Fig. 4.24*b*

$$V(0) = -R_A = C_1 \tag{c}$$

Therefore, $C_1 = -w_0L/2$ and from Eq. (*b*), $V(x) = w_0(x - L/2)$. Now that $V(x)$ has been found, we have, from Eq. (4.7),

$$\frac{dM}{dx} = -V = -w_0\left(x - \frac{L}{2}\right) \tag{d}$$

Upon integration, we have

$$M(x) = -w_0\left(\frac{x^2}{2} - \frac{Lx}{2}\right) + C_2 \tag{e}$$

where C_2 is the constant of integration. Since the beam is simply supported at $x = 0$, we have $M(0) = 0$ and so $C_2 = 0$. Therefore

$$M(x) = \tfrac{1}{2}\, w_0x(L - x) \tag{f}$$

The loading diagram, shear force diagram, and bending moment diagram are shown in Fig. 4.21.

If we wish to use the integration of Eqs. (4.6) and (4.7) directly when concentrated forces or moments or discontinuous loading acts on the beam, we have to divide the beam into different segments and integrate over each segment, making certain that we evaluate the constants of integration from information in each segment. We illustrate this method in the next example.

EXAMPLE 4.15

Consider a simply supported beam subjected to a concentrated bending moment M_B acting at B, as shown in Fig. 4.25*a*. We wish to obtain the shear force and bending moment diagrams for this beam by direct integration of Eqs. (4.6) and (4.7).

We first determine the reactions at the supports. By using the free-body diagram in Fig. 4.25*b*, force equilibrium gives

$$R_A + R_C = 0 \qquad (a)$$

and moment equilibrium about A yields

$$LR_C + M_B = 0 \qquad (b)$$

Thus

$$R_C = -\frac{M_B}{L} \qquad (c)$$

and by Eq. (*a*)

$$R_A = \frac{M_B}{L} \qquad (d)$$

Over segment AB, $q(x) = 0$; if $V_1(x)$ denotes the shear force expression for segment AB, then according to Eq. (4.6),

$$\frac{dV_1}{dx} = 0$$

and upon integration

$$V_1(x) = C_1$$

where C_1 is the arbitrary constant of integration. But at $x = 0$ we have

$$V_1(0) = C_1 = -R_A = -\frac{M_B}{L}$$

so we obtain

$$V_1(x) = -\frac{M_B}{L} \qquad (e)$$

If $M_1(x)$ denotes the moment expression over segment AB, then by Eq. (4.7)

$$\frac{dM_1}{dx} = -V_1 = \frac{M_B}{L}$$

and integrating produces

$$M_1(x) = \frac{M_B x}{L} + C_2$$

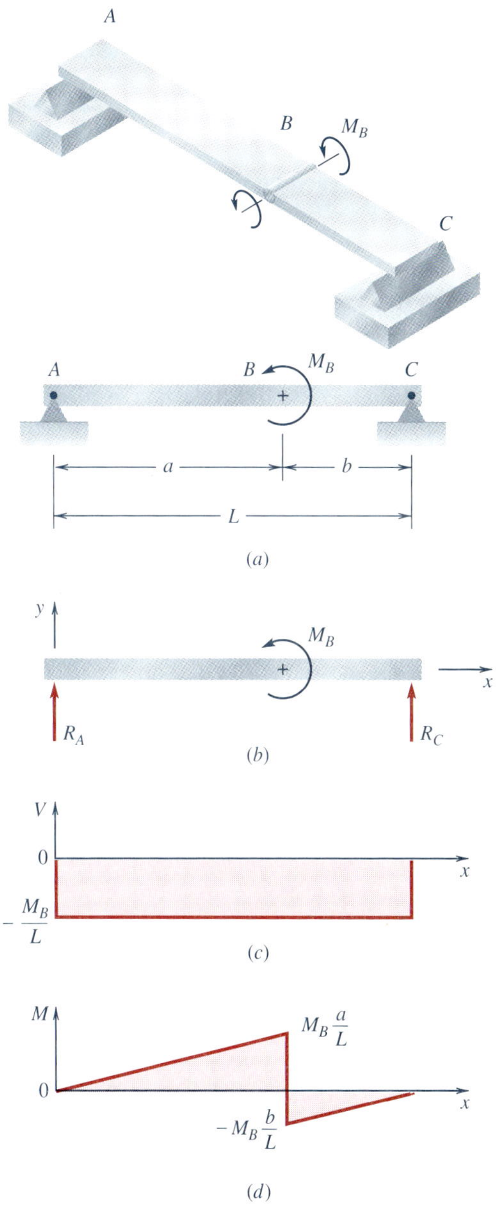

Figure 4.25 Example 4.15

where C_2 is the arbitrary constant of integration. Since at $x = 0$

$$M_1(0) = C_2 = 0$$

it follows that

$$M_1(x) = \frac{M_B}{L}x \qquad (f)$$

The shear force and bending moment curves for segment AB are plotted in Fig. 4.25*c* and *d*.

Expressions for the shear force and bending moment in segment BC can be found by noting that over $a < x < L$ we have $q(x) = 0$, so that from Eq. (4.6)

$$\frac{dV_2}{dx} = 0$$

Following the procedure used above for segment AB, we integrate to obtain

$$V_2(x) = C_3$$

where C_3 is the constant of integration. At $x = L$

$$V_2(L) = C_3 = R_C = -\frac{M_B}{L}$$

so that

$$V_2(x) = -\frac{M_B}{L} \qquad (g)$$

Using Eq. (4.7), we have

$$\frac{dM_2}{dx} = -V_2 = \frac{M_B}{L}$$

and integrating leads to

$$M_2(x) = \frac{M_B x}{L} + C_4$$

Finally, at $x = L$

$$M_2(L) = M_B + C_4 = 0$$

so that upon evaluation of C_4 we have

$$M_2(x) = \frac{M_B}{L}(x - L) \qquad (h)$$

The V and M diagrams can now be completed by using expressions for V_2 and M_2 from Eqs. (*g*) and (*h*), as shown in Fig. 4.25*c* and *d*. The shear force is constant along the beam. The bending moment diagram shows a discontinuity of magnitude M_B at the point $x = a$ at which M_B is applied, that is, $M(a^+) = M(a^-) - M_B$. The discontinuity is equal to $-M_B$ because the moment applied at $x = a$ is positive.

4.5 Singularity Functions

In Example 4.14 we demonstrated how the relatively routine procedure of integration of Eqs. (4.6) and (4.7) could be used to obtain the shear force and bending moment diagrams for a beam with a constant distributed load. However, whenever there are concentrated force and concentrated moment loadings or when the distributed load suddenly changes its magnitude, the procedure shown in Examples 4.14 and 4.15 becomes cumbersome unless a special mathematical apparatus is available to handle concentrated and discontinuous loadings. In this section we introduce a family of singularity functions which allows us to analyze in a systematic way beams under concentrated and discontinuous loadings.

In introducing this set of functions, we seek to account for distributed loads which are applied over a specified segment of the beam or for concentrated loads or moments acting at a specified location on the beam. Concentrated loads and moments must be treated differently from distributed loads.

We define first the functions

$$f_n(x) = \langle x - a \rangle^n \qquad n > 0 \tag{4.12}$$

where n is positive and the ***pointed*** brackets in Eq. (4.12) have special significance. For our purposes, n is usually an integer. If the value of the expression inside the pointed brackets is ***positive,*** i.e., if $x > a$, then the value of $f_n(x)$ is equal to $(x - a)^n$, that is, it is equal to the nth power of $(x - a)$. If the value of the expression inside the pointed brackets is ***negative,*** i.e., if $x < a$, then the value of $f_n(x)$ is equal to zero. We often say that $f_n(x)$ is zero until $x = a$, at which point $f_n(x)$ is "turned on" to be the value of $(x - a)^n$. When $n > 0$, $f_n(x)$ at $x = a$ is equal to zero. The *pointed* brackets are like ordinary parentheses except that they have the value zero for negative arguments inside of the pointed brackets.

When $n = 0$, the function

$$f_0(x) = \langle x - a \rangle^0$$

is called a ***unit step function*** starting at $x = a$. It is zero until $x = a$, at which point it is turned on to the value 1. The function $f_0(x)$ jumps to the value 1 at $x = a$. The functions $f_0(x)$, $f_1(x)$, and $f_2(x)$ are shown in Fig. 4.26*a;* the function $f_1(x)$ is often called, for obvious reasons, a ***unit ramp function*** starting at $x = a$: $f_1 = 0, x < a; f_1 = (x - a), x \geq a$. These functions will be used to represent distributed loads applied along segments of a beam.

If we are to use these functions with Eqs. (4.6) and (4.7), we need

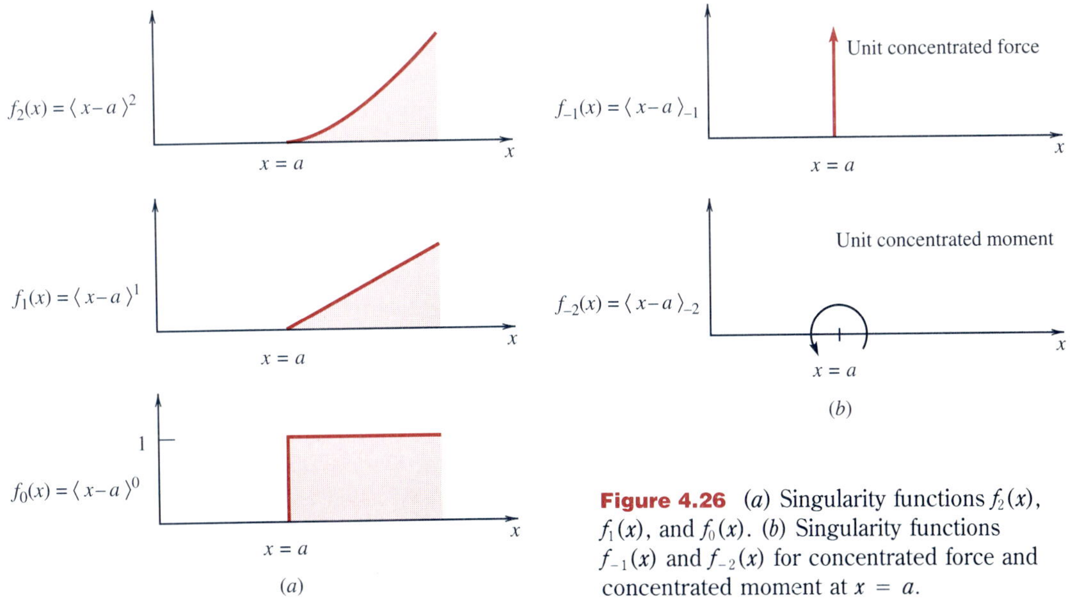

Figure 4.26 (*a*) Singularity functions $f_2(x)$, $f_1(x)$, and $f_0(x)$. (*b*) Singularity functions $f_{-1}(x)$ and $f_{-2}(x)$ for concentrated force and concentrated moment at $x = a$.

to consider integration. We can integrate these singularity functions similarly ($a \geqslant 0$) to ordinary functions, and the integration law is

$$\int_0^x \langle x - a \rangle^n \, dx = \frac{\langle x - a \rangle^{n+1}}{n + 1} \qquad n \geq 0 \tag{4.13}$$

We also need to *represent* concentrated loads and concentrated moments acting at $x = a$ on a beam in such a way that upon integration these representations are physically consistent with the expressions for the shear force and bending moment along the beam. In particular, we expect a jump at the point of application of a concentrated load in the expression for the shear force and a jump at the point of application of a concentrated moment in the expression for the bending moment; see Fig. 4.18*b* and Fig. 4.25*d*.

We introduce a representation for a unit concentrated moment acting at the point $x = a$ in the form

$$f_{-2}(x) = \langle x - a \rangle_{-2} \tag{4.14}$$

and for a unit concentrated force acting at the point $x = a$ in the form

$$f_{-1}(x) = \langle x - a \rangle_{-1} \tag{4.15}$$

These functions are truly singular, and *to emphasize this,* we *write the exponent below the pointed brackets instead of above;* these functions are zero everywhere except at $x = a$, where they are infinite! However, they are infinite in such a way that they satisfy the integration laws (see Probs. 4.5-39 and 4.5-41):

$$\begin{aligned} \int_0^x \langle x - a \rangle_{-2} \, dx &= -\langle x - a \rangle_{-1} \\ \int_0^x \langle x - a \rangle_{-1} \, dx &= \langle x - a \rangle^0 \end{aligned} \tag{4.16}$$

Note carefully the change in sign in the integration of $\langle x - a \rangle_{-2}$. We define the integration of these functions in this way for consistency in the definition of positive loads and moments on a beam.

Figure 4.26*b* shows the representation of these functions as a positive concentrated force and a positive concentrated moment acting at $x = a$. We will use these functions to *represent* concentrated loads in the loading diagrams.

The function $\langle x - a \rangle_{-1}$ is called the *unit concentrated force,* or the *delta function.* The function $\langle x - a \rangle_{-2}$ is called the *unit concentrated moment,* or the *unit doublet function.* Additional discussions of these functions can be found in Probs. 4.5-37 and 4.5-38.

The singularity functions introduced above can be used to represent concentrated loads, concentrated moments, and distributed loads on a beam. That is, as we will show, it is possible to write an algebraic expression for $q(x)$, the loading on the beam, in terms of these singularity functions. Once $q(x)$ is known, integration of Eqs. (4.6) and (4.7)

$$\frac{dV}{dx} = +q(x)$$

$$\frac{dM}{dx} = +V(x)$$

will give $V(x)$ and $M(x)$. We can therefore obtain expressions for the shear force and the bending moment from any loading distribution $q(x)$ that we are able to represent in terms of the singularity functions. Figure 4.27 illustrates examples of load distributions and how they are repre-

Figure 4.27 Singularity function representation of loadings.

Concentrated force P at $x = a$

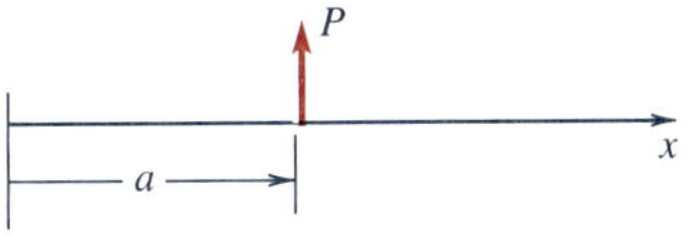

$$q(x) = P\langle x-a\rangle_{-1}$$

Concentrated moment M at $x = a$

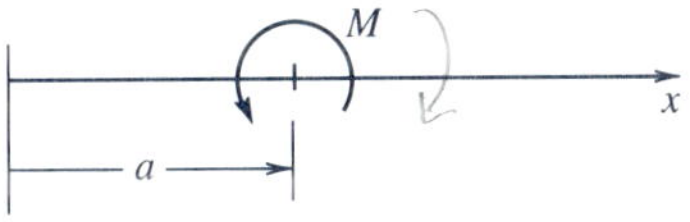

$$q(x) = M\langle x-a\rangle_{-2}$$

Constant distributed load q_0 from $x = a_1$ to $x = a_2$

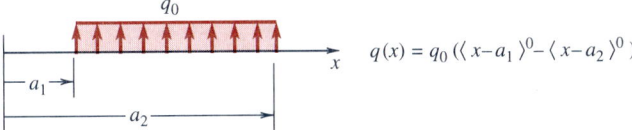

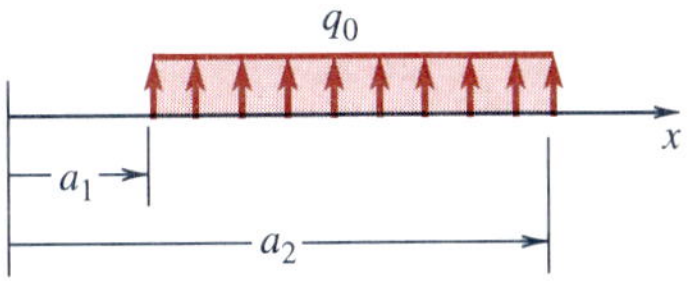

$$q(x) = q_0\left(\langle x-a_1\rangle^0 - \langle x-a_2\rangle^0\right)$$

Linearly varying load from $x = a_1$ to $x = a_2$

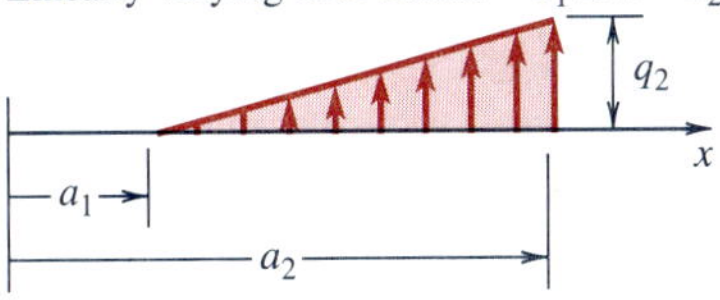

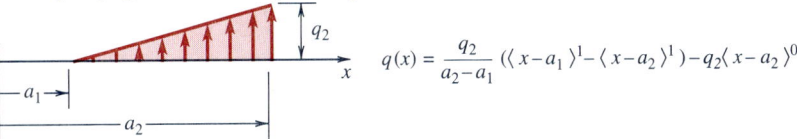

$$q(x) = \frac{q_2}{a_2-a_1}\left(\langle x-a_1\rangle^1 - \langle x-a_2\rangle^1\right) - q_2\langle x-a_2\rangle^0$$

Linearly varying load from $x = a_1$ to $x = a_2$

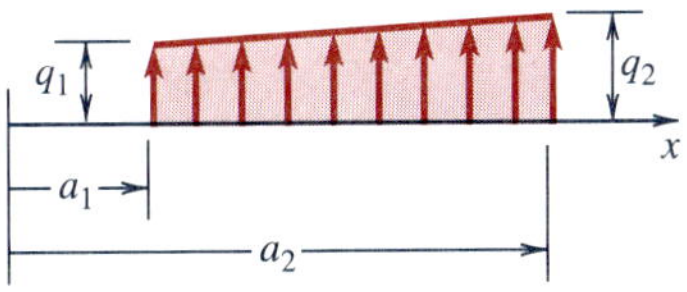

$$q(x) = q_1\left(\langle x-a_1\rangle^0 - \langle x-a_2\rangle^0\right)$$

$$+\frac{q_2-q_1}{a_2-a_1}\left(\langle x-a_1\rangle^1 - \langle x-a_2\rangle^1\right)$$

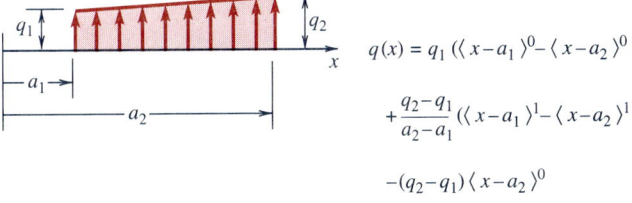

$$-(q_2-q_1)\langle x-a_2\rangle^0$$

sented by singularity functions. Many practical cases of beam loading can be built up by combinations of the cases shown in the loading diagrams of Fig. 4.27. Note, e.g., that in the case of a constant distributed load q_0 acting *over a segment* from $x = a_1$ to $x = a_2$, the loading expression for $q(x)$ is made up of two terms—one term that turns *on* the load at a_1 and the other term that turns *off* the load at a_2 by canceling the contribution of the first load:

$$q(x) = q_0\langle x - a_1\rangle^0 - q_0\langle x - a_2\rangle^0 = q_0(\langle x - a_1\rangle^0 - \langle x - a_2\rangle^0)$$

The expressions for the linearly varying loads are obtained in a similar fashion. All loads shown in Fig. 4.27 are positive.

Once $q(x)$ is known from the loading on the beam, then from Eqs. (4.6) and (4.7) we can find the shear force and bending moment

$$V(x) = -\int q(x)\,dx$$

$$M(x) = -\int V(x)\,dx$$

The use of $P\langle x - a\rangle_{-1}$ in the expression for $q(x)$ to represent a positive concentrated load P at $x = a$ becomes clear when we realize that upon integration once, using Eqs. (4.6) and (4.16), we obtain $P\langle x - a\rangle^0$ in the shear force expression. This term will result in a jump at the point of application of the concentrated load at $x = a$; that is, if

$$q(x) = \cdots + P\langle x - a\rangle_{-1} + \cdots$$

then

$$V(x) = \cdots - P\langle x - a\rangle^0 - \cdots + C_1$$

where the ellipses indicate other loading terms in the $q(x)$ expression and C_1 is the constant of integration (Fig. 4.18*b*).

The fact that $M_a\langle x - a\rangle_{-2}$ represents a positive concentrated moment M_a at $x = a$ follows upon integrating this expression twice, using Eqs. (4.6) and (4.7) and Eqs. (4.16) (with attention paid to the minus sign in the first integration) to give $-M_a\langle x - a\rangle^0$ in the moment expression, which corresponds to a jump in the expression for the moment at $x = a$; that is, if

$$q(x) = \cdots + M_a\langle x - a\rangle_{-2} + \cdots$$

then

$$V(x) = \cdots + M_a\langle x - a\rangle_{-1} - \cdots + C_1$$

and

$$M(x) = \cdots - M_a\langle x - a\rangle^0 + \cdots - C_1x + C_2$$

The procedure to generate the shear force and bending moment diagrams by using singularity functions in view of the above results is a systematic one.

First we determine the beam reactions and construct the expression for the loading $q(x)$ from the loading diagram for the problem; we include *all* loads acting on the beam *including* the reactions. Upon integration of Eqs. (4.6) and (4.7) and evaluation of the constants of integration, we obtain $V(x)$ and $M(x)$. The expressions for $V(x)$ and $M(x)$ are then *evaluated* at points along the beam from the definitions of the singularity functions. In the following examples we will illustrate this procedure; our major emphasis will be on constructing the expression for $q(x)$ from the loading diagrams. We will limit ourselves to the loadings shown in Fig. 4.27 since most engineering loads are of that nature. We will treat only statically determinate beams for which we can determine the beam reactions from equilibrium conditions alone. Consideration of statically indeterminate problems where in general not all the reactions can be found from equilibrium arguments alone will be discussed in Chap. 7.

EXAMPLE 4.16

Consider a simply supported beam with the two concentrated loads, as shown in Fig. 4.28*a*. We wish to obtain the expressions for the shear force and bending moment by using singularity functions. The advantage of the singularity-function representation of the loads is that a single expression for the loading function $q(x)$ valid along the entire length of the beam can be written down immediately, using the results shown in Fig. 4.27. In Fig. 4.28*a* we show the x axis along the beam with the origin at the left end.

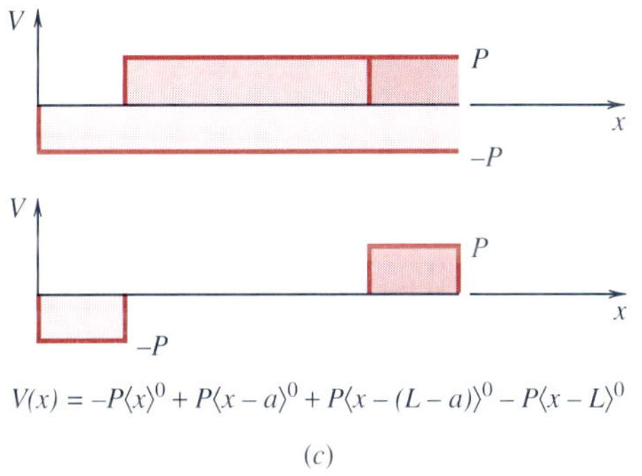

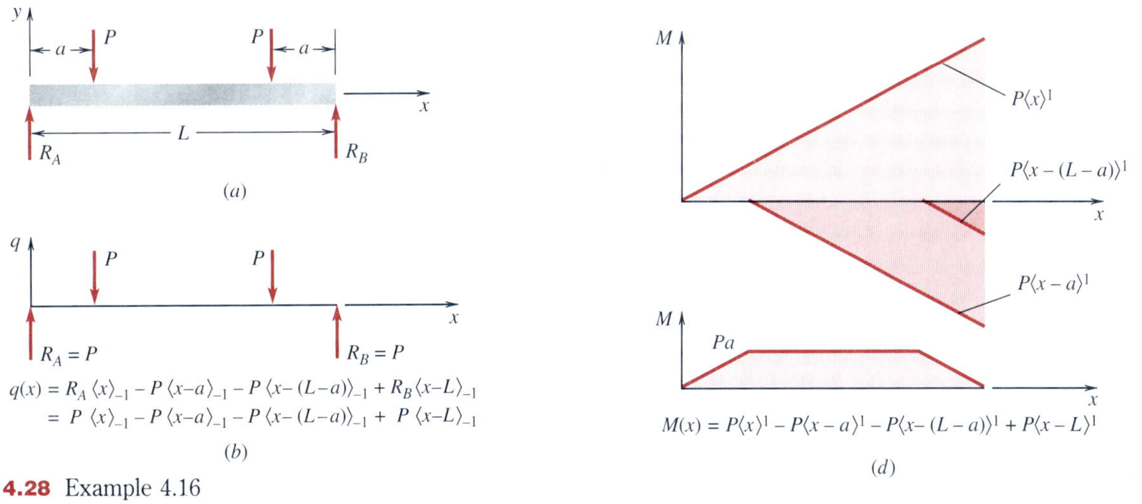

Figure 4.28 Example 4.16

Under the free-body diagram showing the beam we sketch the loading diagram of the loads acting on the beam. In this case, we have the two concentrated loads *and* the reactions at the ends of the beam. Therefore the expression for $q(x)$—pay close attention to the sign convention consistent with the coordinate axes—is

$$q(x) = R_A\langle x\rangle_{-1} - P\langle x - a\rangle_{-1} - P\langle x - (L - a)\rangle_{-1} + R_B\langle x - L\rangle_{-1} \qquad (a)$$

where we have included all applied loads and reactions. This is a representation of the loading on the beam; reactions R_A and R_B are known, having been determined initially from overall equilibrium, $R_A = R_B = P$. Therefore, from Fig. 4.28*b*

$$q(x) = P\langle x\rangle_{-1} - P\langle x - a\rangle_{-1} - P\langle x - (L - a)\rangle_{-1} + P\langle x - L\rangle_{-1} \qquad (b)$$

Upon use of Eq. (4.6), together with the integration rule, Eqs. (4.16), for the singularity functions, we obtain the expression for the shear force

$$V(x) = -P\langle x\rangle^0 + P\langle x - a\rangle^0 + P\langle x - (L - a)\rangle^0 - P\langle x - L\rangle^0 + C_1 \qquad (c)$$

The constant of integration C_1 is identically zero since $V(0^+) = -P$. The constant of integration will usually be zero if the loading is expressed in terms of singularity functions and the values of the reaction forces are included in the expression for the loading function. Upon integration again, we determine the expression for the bending moment along the beam

$$M(x) = P\langle x\rangle^1 - P\langle x - a\rangle^1 - P\langle x - (L - a)\rangle^1 + P\langle x - L\rangle^1 + C_2 \qquad (d)$$

where again the constant of integration C_2 is identically zero since $M(0) = 0$. Equations (*c*) and (*d*) give the expressions for the shear force and bending moment along the beam. For $V(x)$, we can evaluate each term in Eq. (*c*) separately as sketched in Fig. 4.28*c* and combine the terms to get the final result as shown, or we can evaluate the expression at different values of x along the beam.

For $M(x)$ in Eq. (*d*) we can proceed in a similar way to obtain the bending moment diagram shown in Fig. 4.28*d*.

EXAMPLE 4.17

Consider the beam shown in Fig. 4.29*a*. We wish to construct the shear force and bending moment diagrams by using singularity functions. Neglect the weight of the beam. As our first step we calculate the reactions on the beam and construct the loading diagram as shown in Fig. 4.29*b*. The units of load are kips, and the units of length are feet. From the loading diagram, we can write down the expression for $q(x)$:

$$q(x) = 7.5\langle x\rangle_{-1} - 20\langle x - 10\rangle_{-1} + 22.5\langle x - 20\rangle_{-1} - 1\langle x - 20\rangle^0 \qquad (a)$$

We have three concentrated loads plus a uniform load. Note the ease in constructing the expressions for the loading intensity

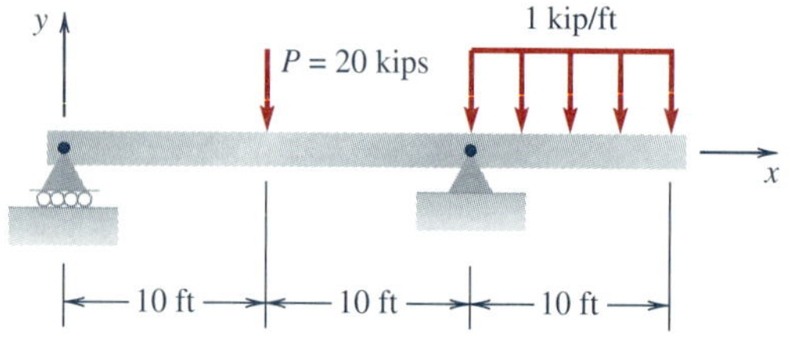

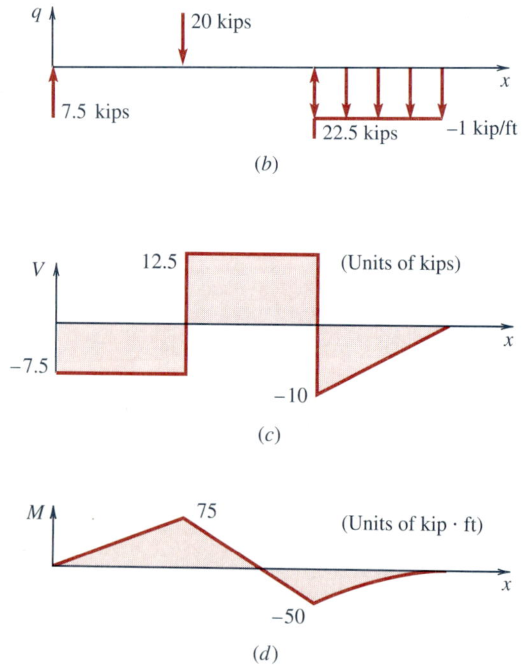

Figure 4.29 Example 4.17

$q(x)$. The shear force and bending moment expressions follow upon integration by using Eqs. (4.6) and (4.7) and integration laws Eqs. (4.13) and (4.16)

$$V(x) = -7.5\langle x\rangle^0 + 20\langle x - 10\rangle^0 - 22.5\langle x - 20\rangle^0 + \langle x - 20\rangle^1$$

$$M(x) = 7.5\langle x\rangle^1 - 20\langle x - 10\rangle^1 + 22.5\langle x - 20\rangle^1 - \tfrac{1}{2}\langle x - 20\rangle^2$$

where the constants of integration turn out to be zero. The shear force and bending moment diagrams in Fig. 4.29*c* and *d* follow upon evaluation of the expressions for $V(x)$ and $M(x)$ when numerical values of x are substituted. For example,

$$V(15) = -7.5 + 20 + 0 + 0 = 12.5 \text{ kips}$$

$$M(15) = 7.5(15) - 20(5) + 0 - 0 = 12.5 \text{ kip} \cdot \text{ft}$$

The sketching of the shear force and bending moment diagrams is carried out by plotting the values of $V(x)$ and $M(x)$ computed for different values of x along the beam. Of course, we are guided in sketching the diagrams (Fig. 4.29*c* and *d*) by our knowledge of the relationships between the slopes and areas of the different diagrams. In fact, it is good practice to check the areas between the diagrams as a check on the correctness of the diagrams.

EXAMPLE 4.18

Consider the simply supported beam shown in Fig. 4.30*a* with two intervals of uniform loads applied to the beam. Neglecting the weight of the beam, we wish to obtain the shear force and bending moment diagrams for the beam. The axes are as shown.

We first determine reactions R_A and R_B and construct the loading diagram as shown in Fig. 4.30*b*. We show the negative distributed loading below the axis in the loading diagram. Using the loading diagram, we can write

$$\begin{aligned} q(x) &= -15(\langle x\rangle^0 - \langle x - 4\rangle^0) + R_A\langle x - 2\rangle_{-1} \\ &\quad - 10\langle x - 6\rangle^0 + R_B\langle x - 9\rangle_{-1} \end{aligned} \qquad (a)$$

where the values of the reactions are found from equilibrium considerations to be

$$R_A = 66.43 \text{ kN} \qquad R_B = 23.57 \text{ kN} \qquad (b)$$

The units of force are kilonewtons (kN) and the units of distance are meters. Upon integration of Eq. (*a*), using Eq. (4.6), we have

$$\begin{aligned} V(x) &= 15(\langle x\rangle^1 - \langle x - 4\rangle^1) - R_A\langle x - 2\rangle^0 \\ &\quad + 10\langle x - 6\rangle^1 - R_B\langle x - 9\rangle^0 \end{aligned} \qquad (c)$$

where the constant of integration is zero. And from Eq. (4.7) we have

$$\begin{aligned} M(x) &= -\tfrac{15}{2}(\langle x\rangle^2 - \langle x - 4\rangle^2) + R_A\langle x - 2\rangle^1 \\ &\quad - \tfrac{10}{2}\langle x - 6\rangle^2 + R_B\langle x - 9\rangle^1 \end{aligned} \qquad (d)$$

with the constant of integration again equal to zero.

From Eqs. (*c*) and (*d*) we can evaluate the expressions for $V(x)$ and $M(x)$ to construct the shear force and bending moment diagrams as shown in Fig. 4.30*c* and *d*. For example,

$$\begin{aligned} M(7) &= -7.5(7^2 - 3^2) + R_A(5) - 5 + 0 \\ &= 27.15 \text{ kN} \cdot \text{m} \end{aligned} \qquad (e)$$

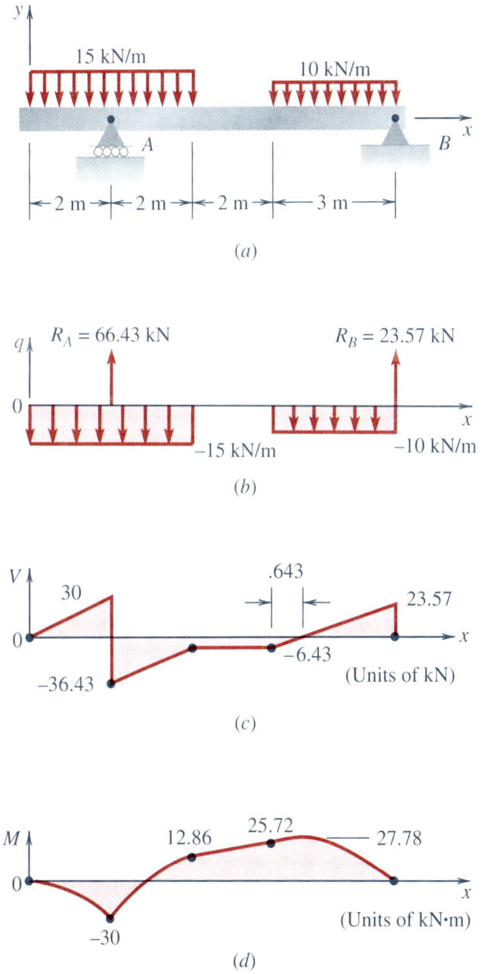

Figure 4.30 Example 4.18

Again we should keep in mind that the expression for $q(x)$ is to include all loads acting on the beam including the reactions. When we do this, we will find that the constants of integration usually turn out to be zero; however, they should be checked (e.g., see Prob. 4.5-13).

EXAMPLE 4.19

Consider the cantilever beam loaded as shown in Fig. 4.31*a*. We wish to sketch the shear force and bending moment diagrams for the beam, neglecting the weight of the beam. The axes are shown with the origin at the wall at section *A* in Fig. 4.31*a*. We also show the values of the wall reaction force and the wall concentrated moment acting on the beam, as determined from equilibrium considerations.

From the loading diagram shown in Fig. 4.31*b* we have

$$q(x) = M_w\langle x\rangle_{-2} + R_w\langle x\rangle_{-1} - w_0\left(\langle x\rangle^0 - \left\langle x - \frac{L}{2}\right\rangle^0\right) \qquad (a)$$

where

$$M_w = \frac{w_0 L^2}{8} \qquad R_w = \frac{w_0 L}{2} \qquad (b)$$

The expressions for the shear force and bending moment follow from Eq. (*a*) upon integration and can be written as

$$V(x) = M_w\langle x\rangle_{-1} - R_w\langle x\rangle^0 + w_0\left(\langle x\rangle^1 - \left\langle x - \frac{L}{2}\right\rangle^1\right) \qquad (c)$$

$$M(x) = -M_w\langle x\rangle^0 + R_w\langle x\rangle^1 - \frac{w_0}{2}\left(\langle x\rangle^2 - \left\langle x - \frac{L}{2}\right\rangle^2\right) \qquad (d)$$

where the constants of integration are identically zero.

Using the expressions in Eqs. (*c*) and (*d*), we can construct the shear force and bending moment diagrams as shown in Fig. 4.31*c* and *d*. Note that the expressions for the shear force, Eq. (*c*), and bending moment, Eq. (*d*), result in zero values over segment *BC*.

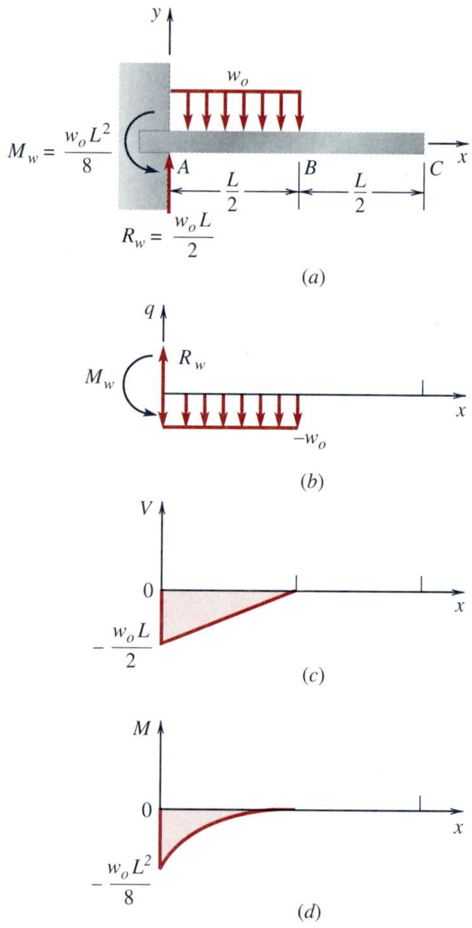

Figure 4.31 Example 4.19

EXAMPLE 4.20

Consider the beam shown in the Fig. 4.32*a* with six concentrated loads applied along its length. We wish to construct the shear force and bending moment diagrams, using singularity functions and neglecting the weight of the beam. We first determine the reactions and construct the loading diagram as shown in Fig. 4.32*b*, from which the expression for $q(x)$ is obtained. Once $q(x)$ is known, we can find upon integration the expressions for $V(x)$ and $M(x)$ and evaluate them by substituting numerical values for *x*. The expression for $q(x)$ is

$$q(x) = -2\langle x\rangle_{-1} + 4.625\langle x - 6\rangle_{-1} - \langle x - 9\rangle_{-1} - 4\langle x - 15\rangle_{-1} - 2\langle x - 20\rangle_{-1} + 7.375\langle x - 22\rangle_{-1} - \langle x - 25\rangle_{-1} - 2\langle x - 28\rangle_{-1}$$

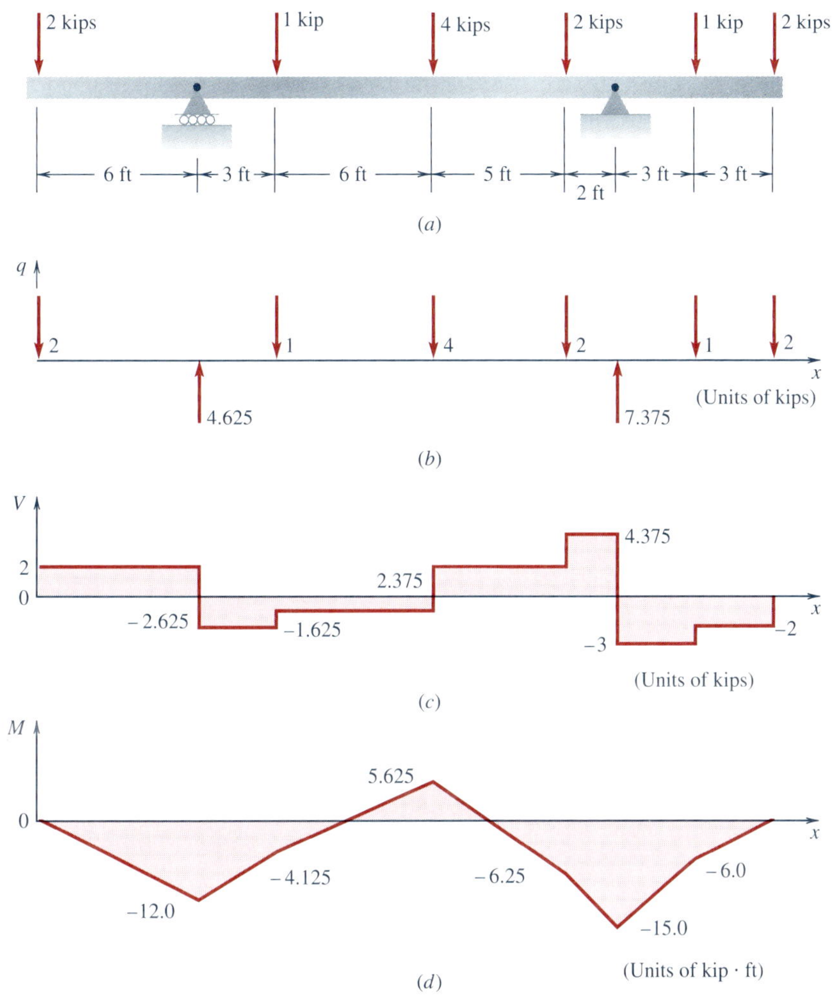

Figure 4.32 Example 4.20

and the shear force and bending moment diagrams are shown in Fig. 4.32*c* and *d*. In this case when the beam is only loaded by concentrated loads, the shear force diagram consists of steps of constant value between the loads, and the bending moment diagram is made up of straight-line segments between the loads.

4.6 Computer Method for Shear Force and Bending Moment Diagrams

The method that uses singularity functions to obtain the shear force and bending moment expressions from the expression for the loading function can be programmed for a computer to obtain the shear force and bending moment diagrams. We will discuss briefly how such a computer program might be written for a simple beam problem. Once the programming approach to be taken is clear, it will be obvious that a general program can be written to plot the shear force and bending moment diagrams for any reasonable transverse loading on a statically determinate beam.

Consider a simply supported beam of length L loaded by a negative concentrated load P at $x = b$, as shown in Fig. 4.33. First, we introduce a coordinate system with the origin at the left end of the beam, as shown. Second, we need the information listed under step 1 in Fig. 4.33. Once these values are specified, the equilibrium equations will give the values of the reactions at supports A and B

$$R_A = \frac{P(c - b)}{c - a}$$

$$R_B = \frac{P(b - a)}{c - a}$$

This is step 2 from Fig. 4.33.

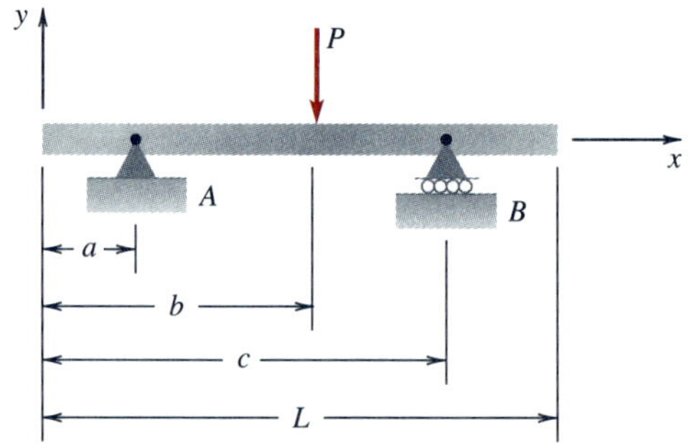


```
1. INPUT: LENGTH OF BEAM, L
          LOCATION OF SUPPORTS, a, c
          LOCATION OF CONCENTRATED LOAD, b
          VALUE OF CONCENTRATED LOAD, -P
1A. NUMBER OF POINTS ALONG THE BEAM, N
2. USE OF EQUILIBRIUM EQUATIONS TO DETERMINE THE
   REACTIONS R_A, R_B
3. LOADING EXPRESSION, q(x)
4. SHEAR FORCE EXPRESSION, V(x)
5. BENDING MOMENT EXPRESSION, M(x)
6. EVALUATE q(x), V(x), M(x) NUMERICALLY AT POINTS ALONG
   THE BEAM.
7. PLOT q(x), V(x), M(x)
```

Figure 4.33 Outline of steps for a computer implementation of shear force and bending moment diagrams.

After the reactions are found, the expression for the loading function on the beam can be written down in terms of singularity functions in the form

$$q(x) = R_A\langle x - a\rangle_{-1} + R_B\langle x - c\rangle_{-1} - P\langle x - b\rangle_{-1}$$

This is step 3.

The shear force and bending moment expressions follow upon integration to give steps 4 and 5

$$V(x) = -R_A\langle x - a\rangle^0 - R_B\langle x - c\rangle^0 + P\langle x - b\rangle^0$$

$$M(x) = R_A\langle x - a\rangle^1 + R_B\langle x - c\rangle^1 - P\langle x - b\rangle^1$$

where the constants of integration are zero.

To explicitly evaluate the expressions for $V(x)$ and $M(x)$ numerically at points along the beam, step 6, we divide the beam uniformly so as to obtain N points, step 1A, and numerically evaluate the expressions at each of these points. A graphics program can then be used to graph the values of q, V, and M, step 7.

It may happen that the exact maximum or minimum values of the shear force or bending moment occur at a point not corresponding to a point arising from the division of the beam into N points. In that case, the maximum or minimum values may not be accurate; accuracy can be improved, however, by dividing the beam into a greater number of points.

We give a simple illustration of the steps in Fig. 4.33 in the following example.

EXAMPLE 4.21

Given is the simply supported beam with a concentrated load P applied as shown in Fig. 4.34. We wish to write a portion of a BASIC program to obtain the expressions for the shear force and bending moment along the beam and to evaluate them for a given set of x values. Neglect the weight of the beam. We start with the origin of the coordinate system at the left end of the beam.

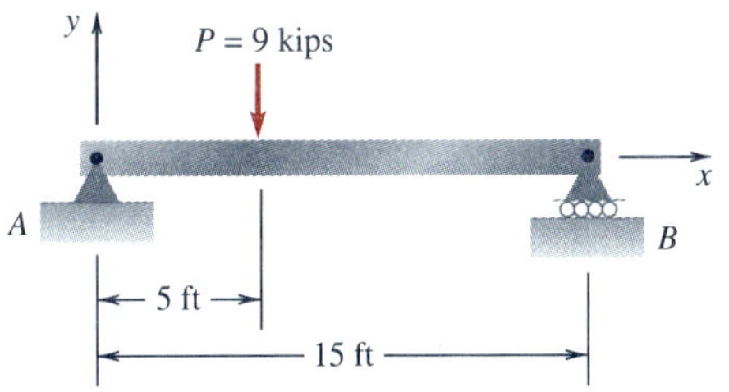

Figure 4.34 Example 4.21

First we input the geometric information:

```
INPUT ''LENGTH OF BEAM = '' ; 15
INPUT ''LOCATION OF LEFT SUPPORT'' ; 0
INPUT ''LOCATION OF RIGHT SUPPORT'' ; 15
INPUT ''LOCATION OF CONCENTRATED LOAD'' ; 5
INPUT ''VALUE OF CONCENTRATED LOAD'' ; -9
INPUT ''NUMBER OF POINTS ALONG BEAM'' ; 300
```

Usually, sufficient accuracy will be obtained if the number of points along the beam is taken to be equal to 20 times the length of the beam. We set $N = 300$. The next step is to calculate the reactions from equilibrium, to find

$$R_A = \frac{P(10)}{15} = 6 \text{ kips}$$

$$R_B = \frac{P(5)}{15} = 3 \text{ kips} \qquad (a)$$

The loading function for the beam is given in the form

$$q(x) = 6\langle x\rangle_{-1} - 9\langle x - 5\rangle_{-1} + 3\langle x - 15\rangle_{-1} \qquad (b)$$

The shear force and bending moment expressions are obtained upon integration of Eq. (*b*) in the form

$$\begin{aligned} V(x) &= -6\langle x\rangle^0 + 9\langle x - 5\rangle^0 - 3\langle x - 15\rangle^0 \\ M(x) &= 6\langle x\rangle^1 - 9\langle x - 5\rangle^1 + 3\langle x - 15\rangle^1 \end{aligned} \qquad (c)$$

where the constants of integration are zero. To evaluate the shear force and bending moment expressions, Eqs. (*c*), at points along the beam by using the definitions of the singularity functions, we can write a portion of a BASIC program as follows:

```
FOR I = 0 TO 300
X = 15*I/300
F1 = 1 : F2 = 1
IF X < 5, F1 = 0
IF X < 5, F2 = 0
V(X) = -6 + 9*F1
M(X) = 6*X - 9*(X-5)*F2
NEXT I
```

We see that in this case the numerical values of $V(x)$ and $M(x)$ can be obtained easily and subsequently graphed.

The program on the diskette V + M DIAGRAMS OF BEAMS generalizes the approach used in Example 4.21. For each loading on the beam of the type shown in Fig. 4.27, an expression for the shear force and bending moment along the beam can be obtained and evaluated at the specified points along the beam. The program is simple to use and provides a convenient means by which the shear force and bending moment diagrams can be obtained.

The number of points along the beam at which the shear force and bending moment are to be evaluated is usually taken as 20 times the length of the beam. If this number of points is not adequate, then we should rescale the length and/or use more points.

The program prints a summary of the solution, giving the values of the reactions and the maximum and minimum values of the shear force and bending moment. Thereafter, plots of the loading, shear force, and bending moment diagrams are given. These diagrams should be reviewed in light of our knowledge of the relations between slopes and areas. Finally, if required, the values of the shear force and bending moments at points along the beam can be printed. Usually, however, all that is needed are the maximum and minimum values of the shear force and bending moment.

4.7 Computer Program to Plot Shear Force and Bending Moment Diagrams

Program 3, from the main menu of the MECHMAT program on the diskette supplied with this book, V + M DIAGRAMS OF BEAMS, can be used to plot shear force and bending moment diagrams. The beams must be statically determinate.

From the main menu given in Fig. 4.35 we select no. 3 to obtain the menu for the V and M diagrams. In the program 3 menu for bending of

```
*********************************************************

*** *MECHANICS OF MATERIALS* ****

PROGRAM: MECHMAT                 mod 5.0

*********************************************************

     MAIN MENU

           1. AXIALLY LOADED MEMBERS
           2. TORSION
           3. V + M DIAGRAMS OF BEAMS
           4. DEFLECTIONS OF BEAMS
           5. MOHR'S CIRCLE
           6. CENTROIDS + MOMENTS OF INERTIA
           7. EXIT

     SELECT NUMBER = ? 3


*********************************************************

     BENDING OF BEAMS                   PROGRAM

           1. BEAM                       BEAM
           2. CANT BEAM                  CANB
       OR  3. RETURN TO MAIN MENU

     SELECT NUMBER = ? 1
```

Figure 4.35 Menus from the MECHMAT program for bending of beams.

beams (Fig. 4.35), program 1 (BEAM) is for simply supported beams and program 2 (CANT BEAM) is for cantilever beams.

To obtain the shear force and bending moment diagrams for a given beam problem, it is convenient to have a drawing which is approximately to scale for the problem from which all relevant dimensions and loadings can be read easily. *The origin of the coordinate system must be set at the left end of the beam.* For cantilever beams, the built-in end of the beam at the wall is to be located at the left end at the origin of the coordinate system. If the cantilever beam is built in at the right end, we redraw the beam and loadings with the built-in end on the left, taking care to observe the correct direction of any concentrated moments. The sign conventions for the positive shear force and the positive bending moment on a positive face are the same as those we have consistently used previously; i.e., quantities are positive in the direction of the positive coordinate axes. Loadings are positive in the positive coordinate directions.

The program allows for linear distributed loads, as shown in Fig. 4.36, and includes the cases of a uniform load and a triangular load. Distributed load segments must not overlap in the program input.

The following examples demonstrate how the program can be used.

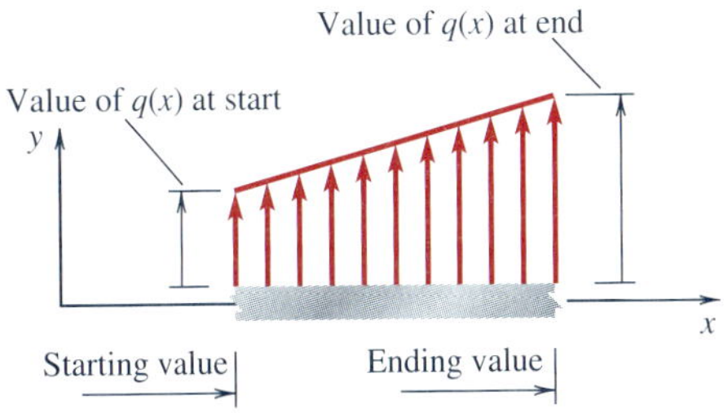

Figure 4.36 Load distribution format for program for bending of beams.

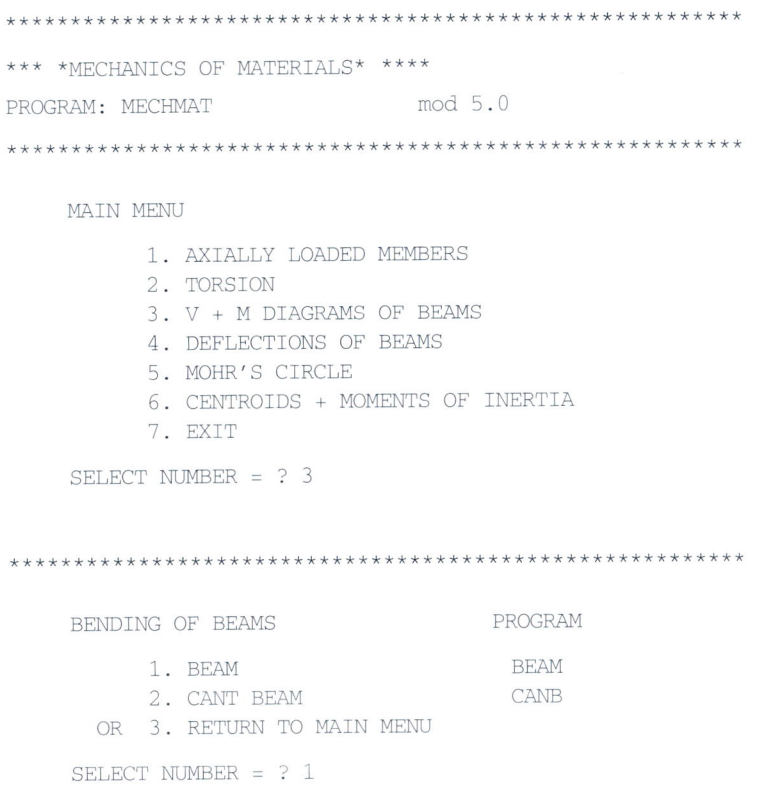

EXAMPLE 4.22

Obtain the shear force and bending moment diagrams for the beam shown in the Fig. 4.37. Neglect the weight of the beam.

From the program 3 menu, we select no. 1 since the beam is simply supported. We then obtain the simple beam menu shown in Fig. 4.38 and prepare a new data file. The input data file is stored on the diskette in drive *A*, and it is given a file name, in this case ex422. Figure 4.39 gives the input data and a summary of the solution. Finally, the shear force and bending moment diagrams are obtained, and by using the *print screen key* on the computer keyboard a copy of the diagrams shown in Fig. 4.40 can be produced.

In the accompanying table we show the effect of *N*, the number of points along the beam, on the accuracy of the maximum and minimum values.

We see that when $N = 20$ times the length of the beam, the error in the value of the minimum shear is less than 2 percent. The value $N = 20L$ therefore is a reasonable value to select to shorten the running time of the program. The program as it is now written is limited to $N \leq 700$ in the dimension statement.

N	Maximum V	Minimum V	Maximum M	Minimum M
180	32.14	−44.06	51.43	−32.00
200	31.68	−44.46	51.35	−31.36
400	32.01	−44.82	51.41	−31.89
600	31.98	−44.70	51.42	−31.84
Exact	32.00	−44.86	51.43	−32.00

An option also exists in the program to obtain a table of *V* and *M* with *x*.

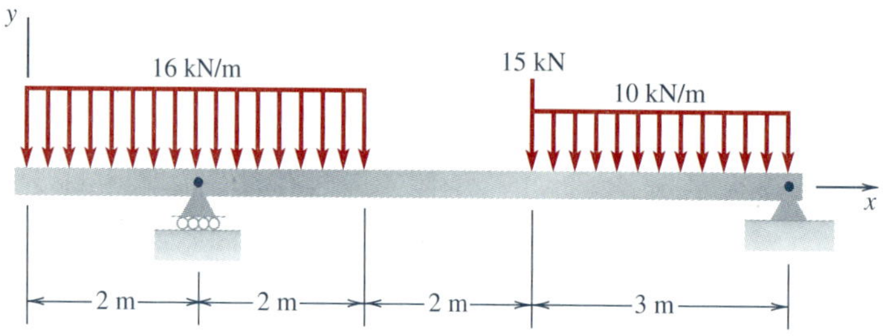

Figure 4.37 Example 4.22

```
*********************************************************

          ***SIMPLE BEAM***

*********************************************************

   ***DATA FILE INFORMATION***

 1. PREPARE NEW DATA FILE
 2. MODIFY EXISTING DATA FILE
 3. USE EXISTING DATA FILE
       SELECT NUMBER? 1

            DRIVE USED FOR DATA FILES (A:, B:, or C:) ? a:
            FILE NAME (OMIT DRIVE LETTER)? ex422
```

Figure 4.38 Menu for simple beam program.

```
 ***SIMPLE BEAM***            DATA FILE: ex422

        INPUT DATA:
             THE LENGTH OF BEAM:                          9
THE NUMBER OF POINTS ALONG THE BEAM:                    180
THE LOCATION OF LEFT SIDE SUPPORT:                        2
THE LOCATION OF RIGHT SIDE SUPPORT:                       9
THE NUMBER OF CONCENTRATED LOADS:                         1
THE LOCATION AND MAGNITUDE OF CONCENTRATED LOADS:
       LOCATION = 6             LOAD = -15
THE NUMBER OF CONCENTRATED MOMENTS:                       0
THE NUMBER OF SEGMENTS OF UNIFORM OR LINEAR LOADS:        2
THE START AND END LOCATION AND VALUE OF UNIFORM OR
   LINEAR LOADS:
       START LOCATION = 0             LOAD = -16
       END LOCATION   = 4             LOAD = -16
       START LOCATION = 6             LOAD = -10
       END LOCATION   = 9             LOAD = -10

              ****** SUMMARY SOLUTION ******

*********************************************************
 LEFT SUPPORT REACTION               RIGHT SUPPORT REACTION
        76.857                              32.143

MAXIMUM SHEAR FORCE AND MOMENT
     MAXIMUM SHEAR FORCE =  32.14
     MINIMUM SHEAR FORCE = -44.06
     MAXIMUM MOMENT      =  51.43
     MINIMUM MOMENT      = -32.00
```

Figure 4.39 Example 4.22: Input data and solution.

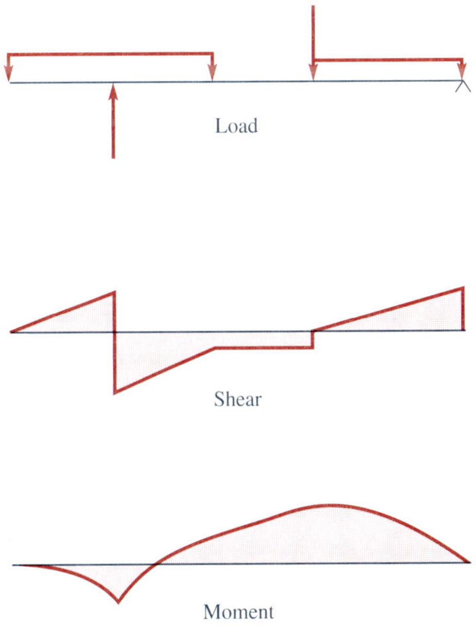

Figure 4.40 Example 4.22: Loading diagram, shear force diagram, and bending moment diagram.

EXAMPLE 4.23

Obtain the shear force and bending moment diagrams for the beam shown in Fig. 4.41, using the program BEAM. Neglect the weight of the beam. For this problem, first we need to replace the effect of member *BEF* by a statically equivalent force and moment acting on the beam at section *B*. Once we do this, we have the necessary data to run the program. The input and output of the program are given in Fig. 4.42. The exact *V* and *M* diagrams are given in Fig. 4.41; the computer plots in Fig. 4.43 can be compared to the exact diagrams in Fig. 4.41. We note the jump in the bending moment diagram corresponding to the concentrated negative moment of 5 kN · m at section *B*.

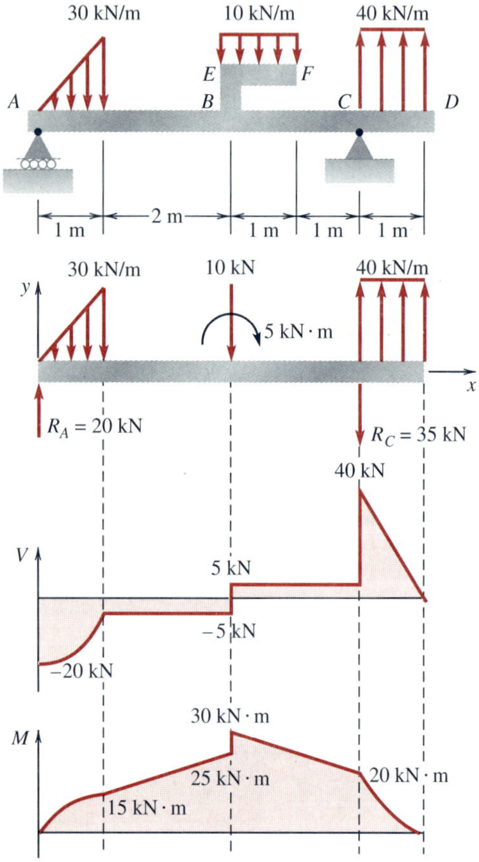

Figure 4.41 Example 4.23

```
***SIMPLE BEAM***                    DATA FILE: ex423

              INPUT DATA:
                   THE LENGTH OF BEAM:                             6
THE NUMBER OF POINTS ALONG THE BEAM:                             120
THE LOCATION OF LEFT SIDE SUPPORT:                                 0
THE LOCATION OF RIGHT SIDE SUPPORT:                                5
THE NUMBER OF CONCENTRATED LOADS:                                  1
THE LOCATION AND MAGNITUDE OF CONCENTRATED LOADS:
        LOCATION = 3              LOAD = -10
THE NUMBER OF CONCENTRATED MOMENTS:                                1
THE LOCATION AND MAGNITUDE OF CONCENTRATED MOMENTS:
        LOCATION = 3              MOMENT = -5
THE NUMBER OF SEGMENTS OF UNIFORM OR LINEAR LOADS:                 2
THE START AND END LOCATION AND VALUE OF UNIFORM OR
   LINEAR LOADS:
        START LOCATION = 0              LOAD =   0
        END LOCATION   = 1              LOAD = -30
        START LOCATION = 5              LOAD =  40
        END LOCATION   = 6              LOAD =  40

                ****** SUMMARY SOLUTION ******
***********************************************************
 LEFT SUPPORT REACTION                 RIGHT SUPPORT REACTION
        20.000                                 -35.000

MAXIMUM SHEAR FORCE AND MOMENT
     MAXIMUM SHEAR FORCE =  38.00
     MINIMUM SHEAR FORCE = -20.00
     MAXIMUM MOMENT      =  29.75
     MINIMUM MOMENT      =   0.00
```

Figure 4.42 Example 4.23: Input data and solution.

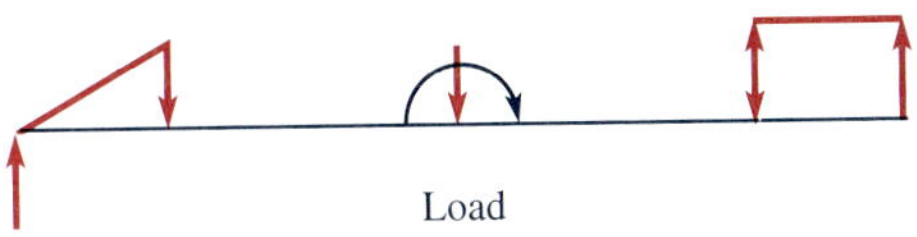

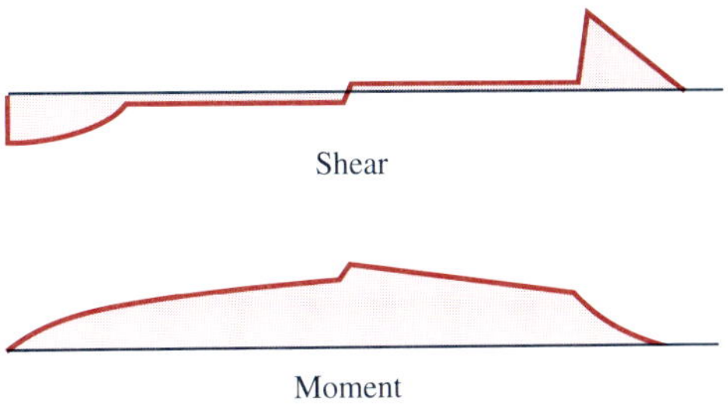

Figure 4.43 Example 4.23: Loading diagram, shear force diagram, and bending moment diagram.

EXAMPLE 4.24

Obtain the shear force and bending moment diagrams for the cantilever beam shown in Fig. 4.44. Neglect the weight of the beam. Since this is a cantilever beam built in at the left end, we are immediately able to use the program CANT BEAM. If the cantilever beam were built in at the right end, then the beam would have to be looked at from the other side so that the built-in end was at the left end; this is necessary because of the structure of the computer program. The solutions are given in Figs. 4.45 and 4.46.

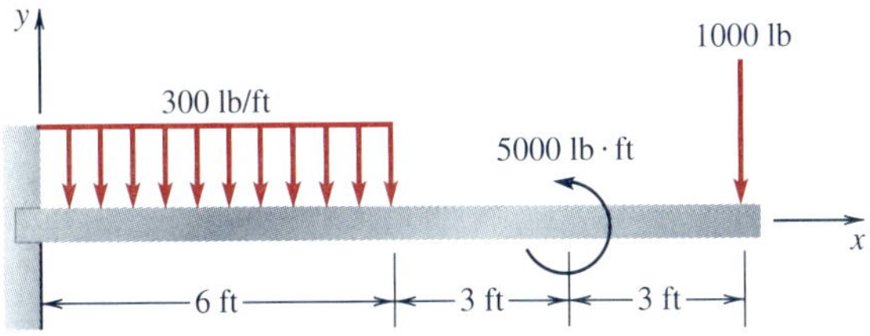

Figure 4.44 Example 4.24

```
***CANTILEVER BEAM***          DATA FILE: ex424

       INPUT DATA:
            THE LENGTH OF BEAM:                             12
THE NUMBER OF INTEGRATION STEPS:                           240
THE NUMBER OF CONCENTRATED LOADS:                            1
THE LOCATION AND VALUE OF CONCENTRATED LOADS:
       LOCATION = 12              LOAD = -1000
THE NUMBER OF CONCENTRATED MOMENTS:                          1
THE LOCATION AND VALUE OF CONCENTRATED MOMENTS:
       LOCATION = 9               MOMENT = 5000
THE NUMBER OF SEGMENTS OF UNIFORM (OR LINEAR) LOADS:         1
THE START, END LOCATION AND VALUE OF UNIFORM (OR
   LINEAR) LOADS:
       START LOCATION = 0              LOAD = -300
       END LOCATION   = 6              LOAD = -300

                 ****** SUMMARY SOLUTION ******

*********************************************************
 LEFT SUPPORT REACTION                RIGHT SUPPORT REACTION
       2800.000                              12400.000

MAXIMUM SHEAR FORCE AND MOMENT
     MAXIMUM SHEAR FORCE =       0.00
     MINIMUM SHEAR FORCE =   -2800.00
     MAXIMUM MOMENT      =    2000.02
     MINIMUM MOMENT      =  -12400.00
```

Figure 4.45 Example 4.24: Input data and solution.

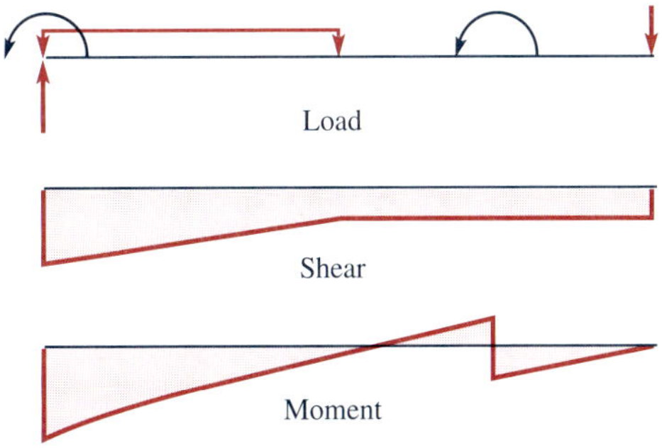

Figure 4.46 Example 4.24: Loading diagram, shear force diagram, and bending moment diagram.

4.8 Three-Dimensional Problems

Thus far we have determined internal forces and internal moments for loading in the symmetric plane of a beam. However, the general procedure outlined in Sec. 4.2 still applies if the beam is three-dimensional and the loading is three-dimensional. As we did in the study of statics, we can analyze three-dimensional problems in terms of components of forces and moments or directly in terms of the vector forces and moments. We present two examples of these approaches.

EXAMPLE 4.25

A section of ducting during fabrication is exposed to a load P, as shown in Fig. 4.47*a*. Determine the internal force and moment at the section C indicated. Neglect the weight of the ducting; dimensions are to the centerline of the ducting. We show a coordinate system at section A, and all internal forces and moments are defined relative to this coordinate system. In Fig. 4.47*b*, we show a free-body diagram of the total structure with the wall reactive force vector $\mathbf{R}_A$ and wall reactive moment vector $\mathbf{M}_A$ as shown.

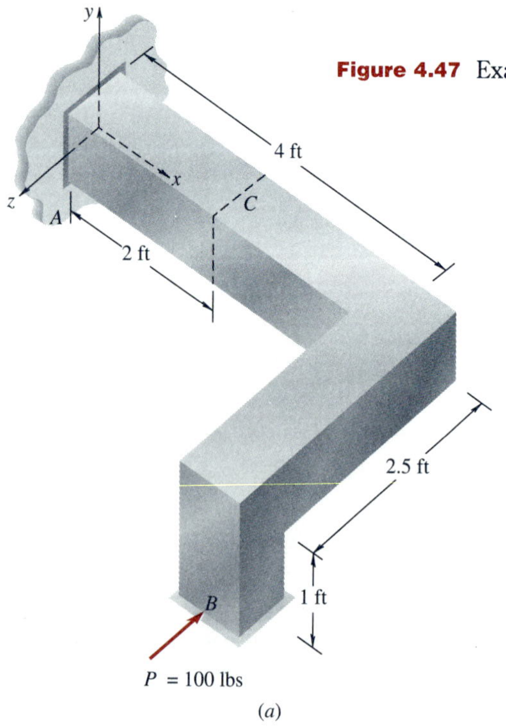

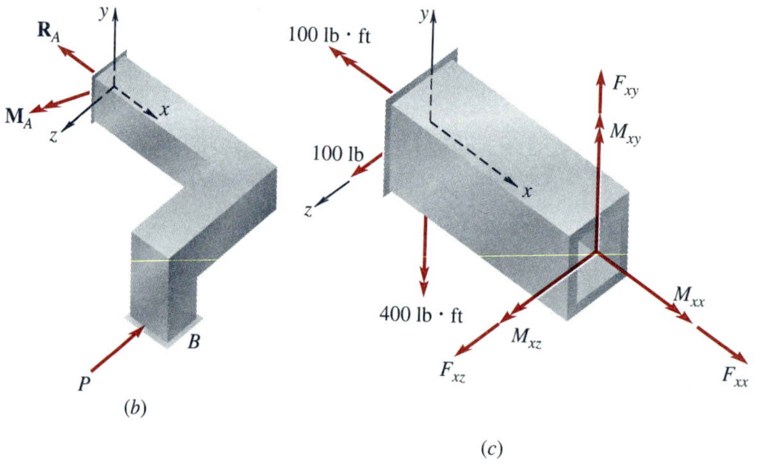

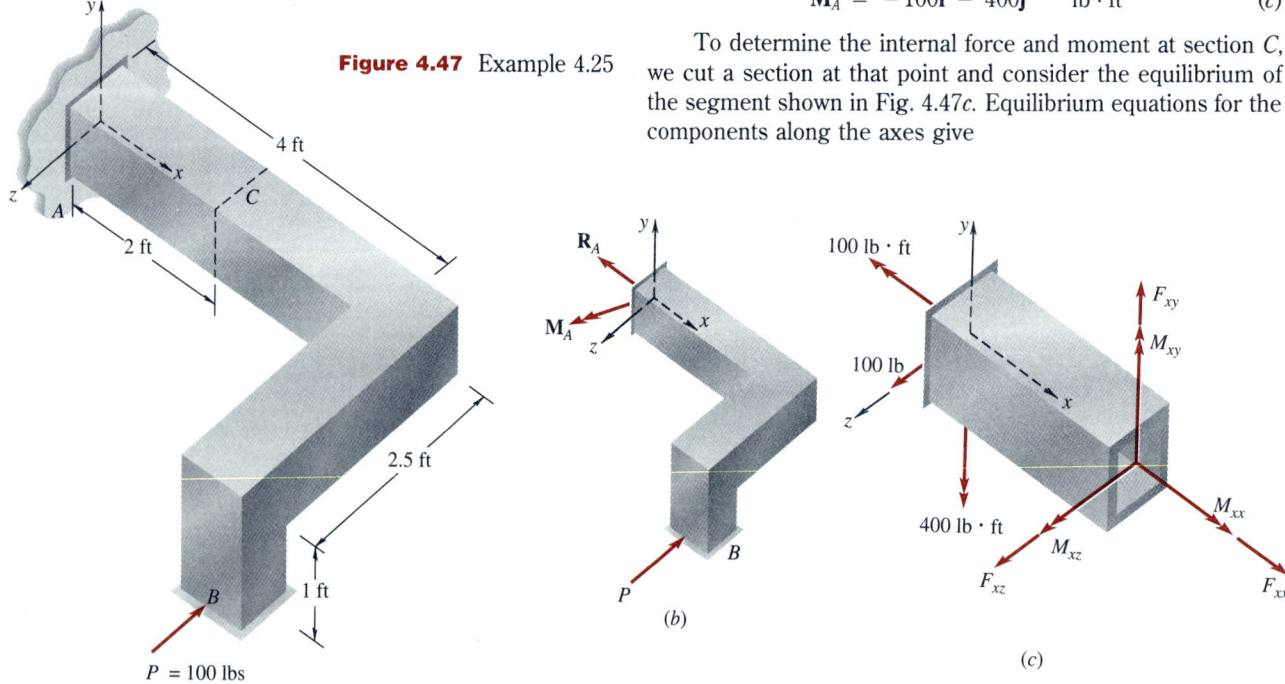

Figure 4.47 Example 4.25

Equilibrium conditions for the free-body diagram in Fig. 4.47*b* give

$$\mathbf{R}_A - 100\mathbf{k} = 0 \qquad \mathbf{R}_A = 100\mathbf{k} \quad \text{lb} \qquad (a)$$
$$\mathbf{M}_A + \mathbf{r}_{AB} \times (-100\mathbf{k}) = 0$$

We have

$$\mathbf{r}_{AB} = 4\mathbf{i} - \mathbf{j} + 2.5\mathbf{k} \qquad (b)$$

From which

$$\mathbf{M}_A = -100\mathbf{i} - 400\mathbf{j} \quad \text{lb} \cdot \text{ft} \qquad (c)$$

To determine the internal force and moment at section C, we cut a section at that point and consider the equilibrium of the segment shown in Fig. 4.47*c*. Equilibrium equations for the components along the axes give

$$F_{xx} = F_{xy} = 0 \qquad F_{xz} = -100 \text{ lb}$$
$$M_{xx} = 100 \text{ lb} \cdot \text{ft} \qquad M_{xz} = 0 \qquad (d)$$
$$M_{xy} = 400 - 2(100) = 200 \text{ lb} \cdot \text{ft}$$

We note that the bending moment about the y axis is a function of the distance—in this case 2 ft—of section C from section A and is a maximum at section A.

EXAMPLE 4.26

Consider a section of piping in a refinery in the form of a quarter-circle subjected to a load P transverse to its plane, as shown in Fig. 4.48*a*. We wish to determine the components of the internal force and moment at section C defined by angle θ. We carried out an analysis of a beam of this shape when the loading was in the plane of the beam in Example 4.4. In this case, the transverse loading will give rise to out-of-plane internal forces and moments. The origin of the coordinate system is located at section A.

In Fig. 4.48*b* we show the section cut at C defined by angle θ. At this section we introduce a local right-handed coordinate system defined by the vectors ($\mathbf{u}_\theta$, $\mathbf{u}_r$, $\mathbf{k}$). The vector $\mathbf{u}_\theta$ is in the θ direction of the normal to the section at C, vector $\mathbf{u}_r$ is in the outward radial direction at section C, and vector $\mathbf{k}$ is in the positive z direction perpendicular to the plane defined in Fig. 4.48*d*. We define internal forces and moments at section C in terms of this local coordinate system. In Fig. 4.48*b* we also show the internal force vector $\mathbf{F}_C$ and moment vector $\mathbf{M}_C$ acting on section C. In terms of the local coordinate system,

$$\mathbf{F}_C = F_\theta \mathbf{u}_\theta + F_r \mathbf{u}_r + F_k \mathbf{k} \qquad (a)$$
$$\mathbf{M}_C = M_\theta \mathbf{u}_\theta + M_r \mathbf{u}_r + M_k \mathbf{k}$$

Figure 4.48 Example 4.26

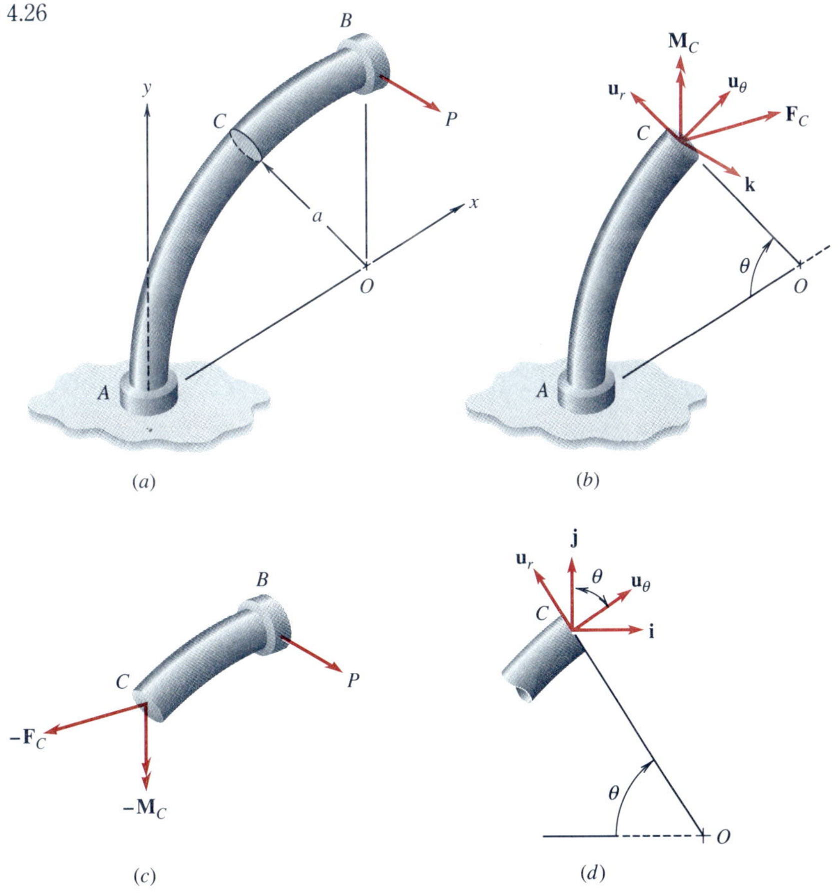

so that F_θ is the axial force, F_r and F_k are shear forces, M_θ is the twisting moment, and M_r and M_k are bending moments at the section. In Fig. 4.48*c*, we show the equal and opposite vectors acting at *C* on segment *CB*; the minus signs arise since the force and moment are equal and opposite to the quantities given as positive in Eqs. (*a*).

To determine the values of $\mathbf{F}_C$ and $\mathbf{M}_C$, we use the equilibrium conditions for the segment *CB* shown in Fig. 4.48*c*.

$$-\mathbf{F}_C + P\mathbf{k} = 0 \qquad (b)$$

$$-\mathbf{M}_C + \mathbf{r}_{CB} \times (P\mathbf{k}) = 0 \qquad (c)$$

From Eq. (*b*), we find for the force components

$$F_\theta = F_r = 0 \qquad F_k = P \qquad (d)$$

To find the moment components, we first note that

$$\mathbf{r}_{CB} = \mathbf{r}_{OB} - \mathbf{r}_{OC} = a\mathbf{j} - a\mathbf{u}_r$$

where $\mathbf{j}$ is the unit vector in the *y* direction. It follows from the geometry of Fig. 4.48*d* that

$$\mathbf{j} = \mathbf{u}_\theta \cos\theta + \mathbf{u}_r \sin\theta$$

so that

$$\mathbf{r}_{CB} = a\cos\theta\,\mathbf{u}_\theta - a(1 - \sin\theta)\mathbf{u}_r \qquad (e)$$

Upon substitution of Eq. (*e*) into Eq. (*c*) and solving for the components of the moment vector, we find

$$M_\theta = -Pa(1 - \sin\theta) \qquad M_r = -Pa\cos\theta \qquad M_k = 0 \quad (f)$$

The internal forces and moments are thus given by Eqs. (*d*) and (*f*). We note that there is a twisting moment and bending moment variation along the pipe. These results can be combined with the results in Example 4.4 and with the results in Prob. 4.2-9 to give the internal forces and moments for a pipe in this shape loaded at the end by a force in an arbitrary direction.

4.9 Concluding Remarks

In this chapter we have focused on the determination of the shear forces and bending moments in beams. Of major interest is the construction of the shear force and bending moment diagrams for such beams. Figure 4.49 shows four sets of diagrams for cases that arise often in engineering applications, namely, either a simply supported beam or a cantilever beam under the action of a concentrated load or a constant distributed load. By now you should be able to sketch quickly the shear force and bending moment diagrams for these cases, and it is of value to remember these results.

The use of a computer program to obtain the shear force and bending moment diagrams does simplify the effort. In fact, most distributed loading functions can be approximated by piecewise linear trapezoidal loading of the form shown in Fig. 4.36. Therefore, the computer program can be used to obtain the shear force and bending moment diagrams. However, the output from a program always needs to be reviewed and checked, and this can be done only if the theory underlying the construction of the diagrams is fully understood. Remember also that the accuracy of the computer's numerical results must always be considered in reviewing the summary of results given in the output.

Finally, it is important to realize that constructing shear force and bending moment diagrams is not an end in itself; they are the first steps in the design or analysis of a beam or a machine component.

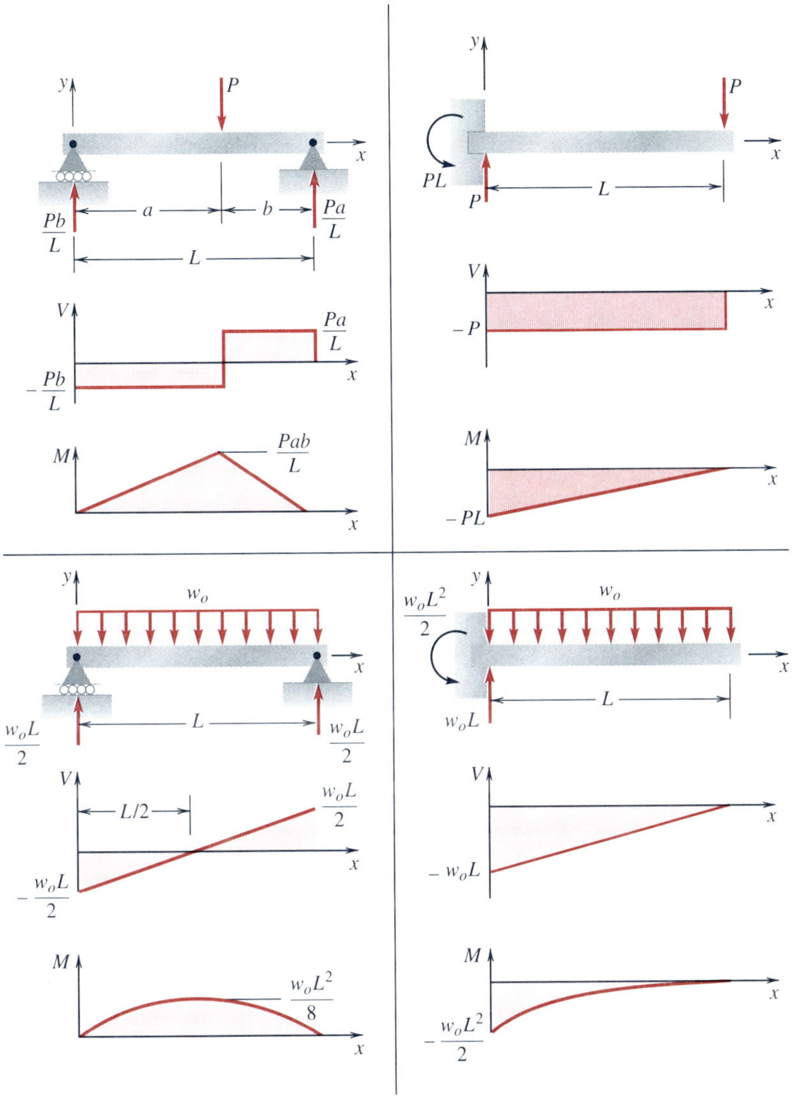

Figure 4.49 Shear force and bending moment diagrams for simply supported and cantilever beams under a concentrated load or a constant distributed load.

PROBLEMS

4.2-1 Consider Example 4.2, Fig. 4.8. Give an argument to show from symmetry that the shear force at the midpoint of the beam must equal zero.

4.2-2 Consider Example 4.2, Fig. 4.8. If all geometric lengths are doubled, what changes occur in the values of the shear force and bending moment at sections C_1 and C_2?

4.2-3 Determine the shear force V and bending moment M at section C for the beams shown in
(*a*) Fig. P4.2-3*a*
(*b*) Fig. P4.2-3*b*
(*c*) Fig. P4.2-3*c*
(*d*) Fig. P4.2-3*d*
(*e*) Fig. P4.2-3*e*
(*f*) Fig. P4.2-3*f*
(*g*) Fig. P4.2-3*g*
(*h*) Fig. P4.2-3*h*
Neglect the weight of the beams.

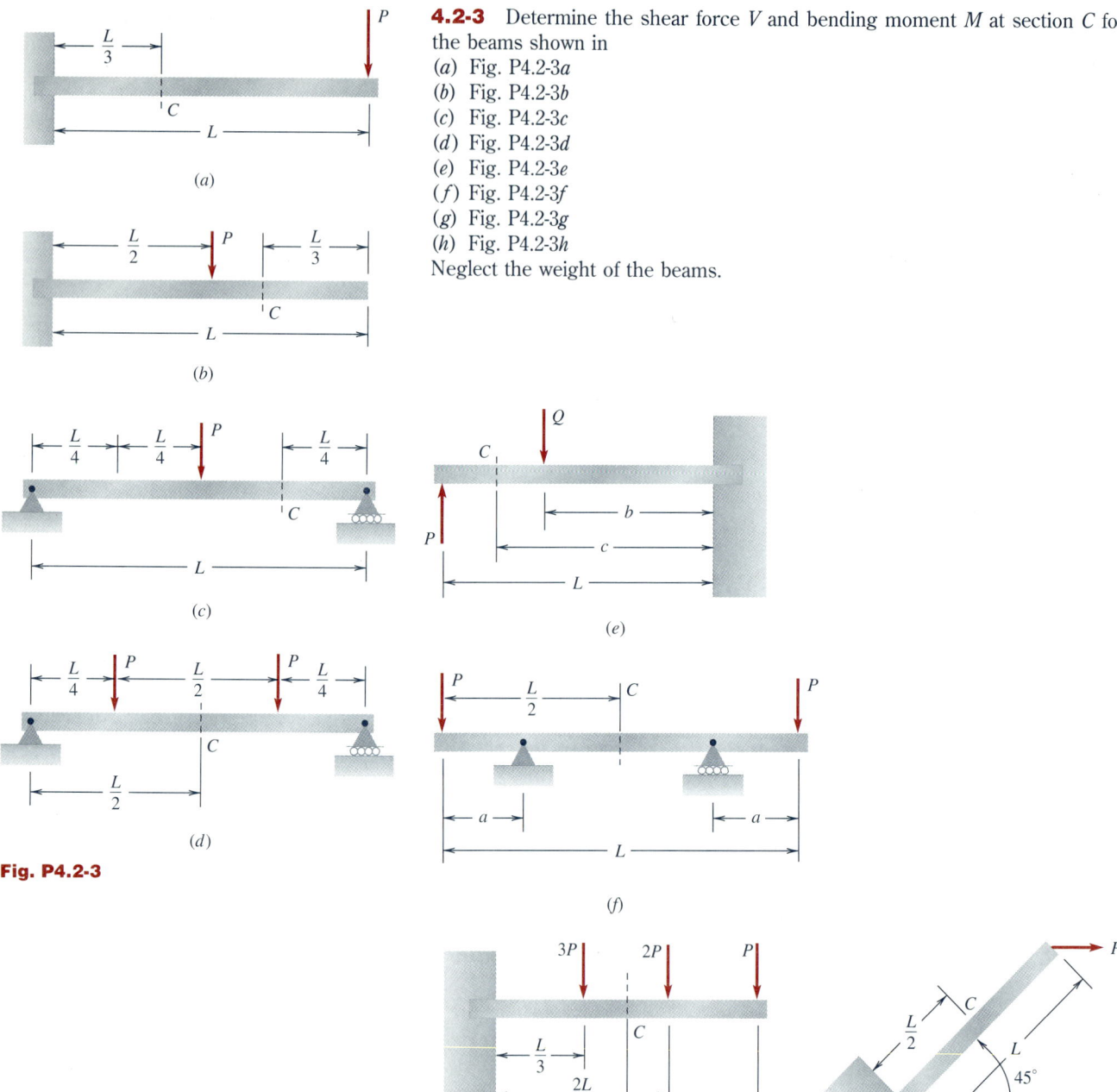

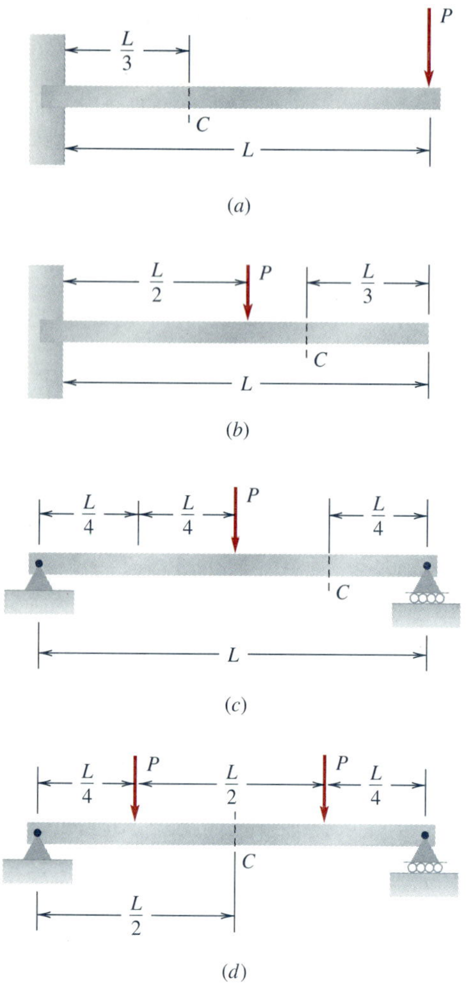

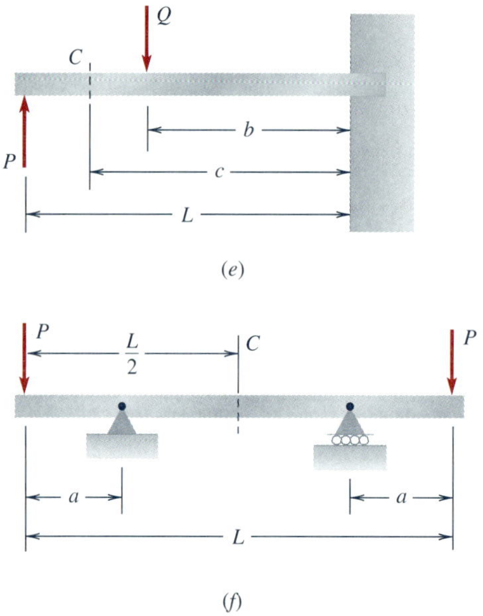

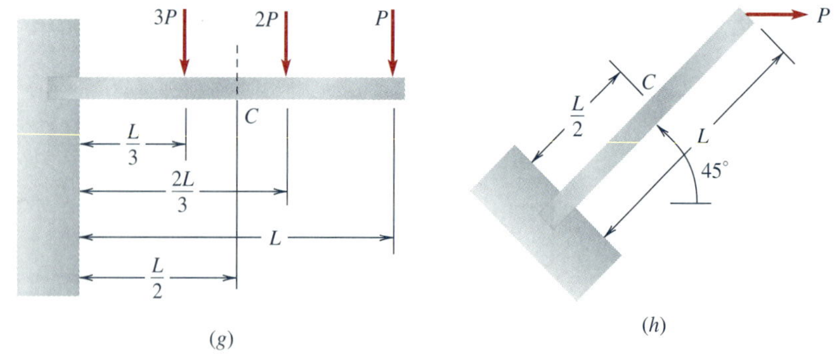

Fig. P4.2-3

4.2-4 Determine the value of the shear force and the bending moment at section C for the beams shown in
(*a*) Fig. P4.2-4*a*
(*b*) Fig. P4.2-4*b*
(*c*) Fig. P4.2-4*c*
(*d*) Fig. P4.2-4*d*
(*e*) Fig. P4.2-4*e*
(*f*) Fig. P4.2-4*f*
(*g*) Fig. P4.2-4*g*
(*h*) Fig. P4.2-4*h*
Neglect the weight of the beams.

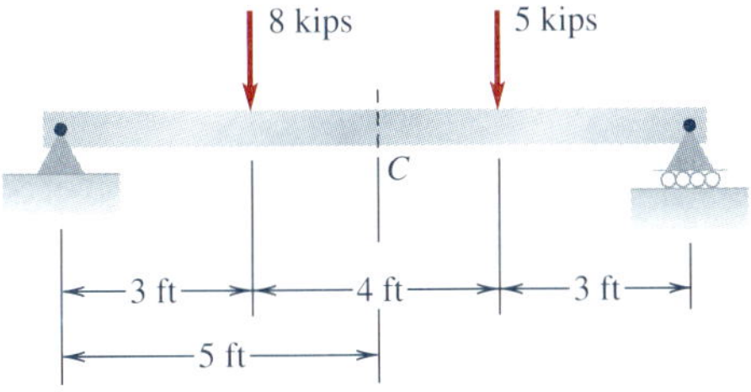

(*a*)

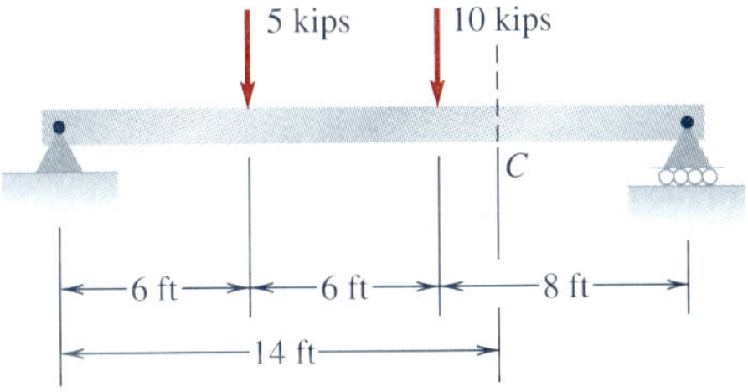

(*b*)

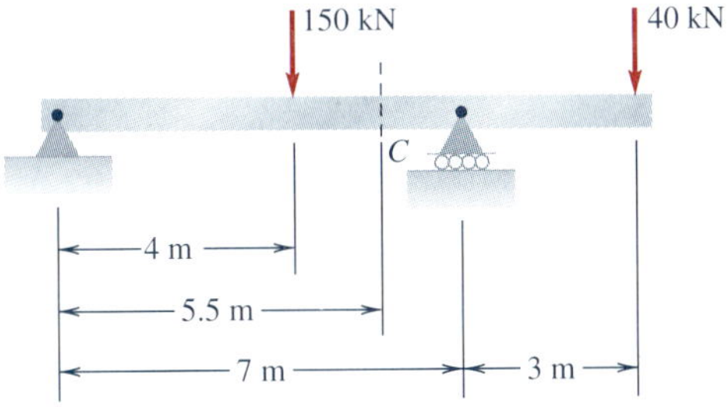

(*c*)

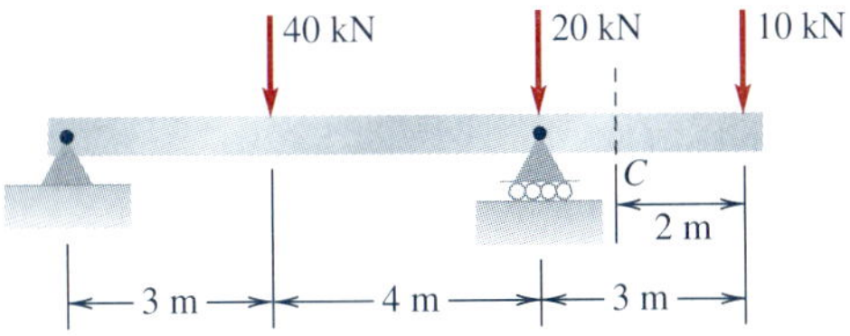

(*d*)

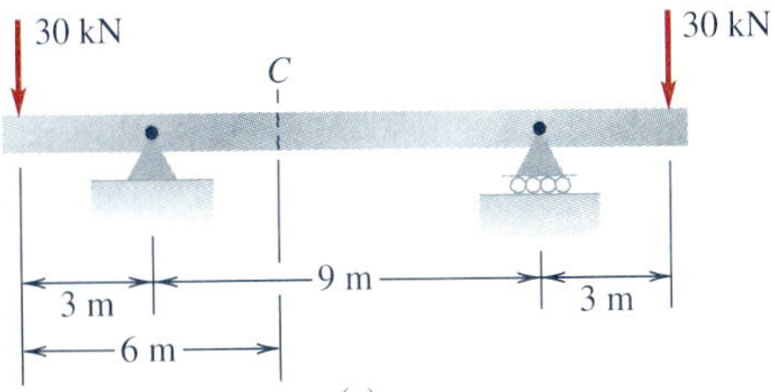

(*e*)

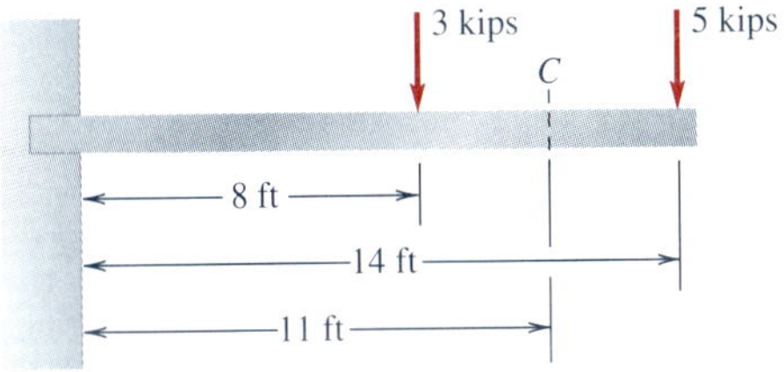

(*f*)

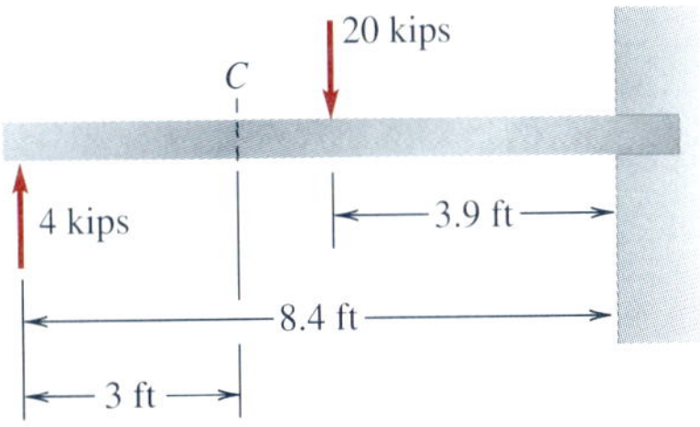

(*g*)

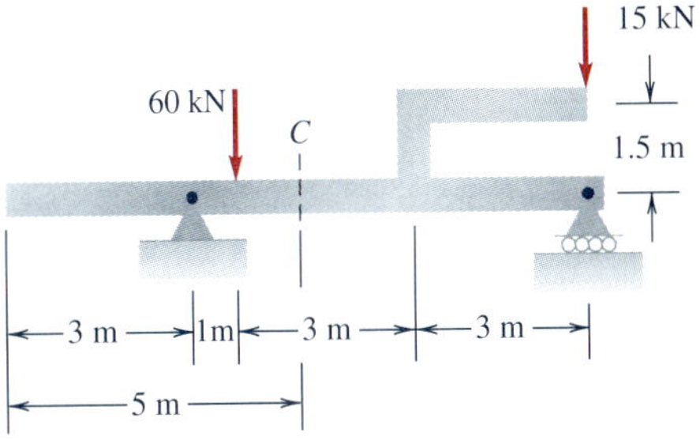

(*h*)

Fig. P4.2-4

Fig. P4.2-5

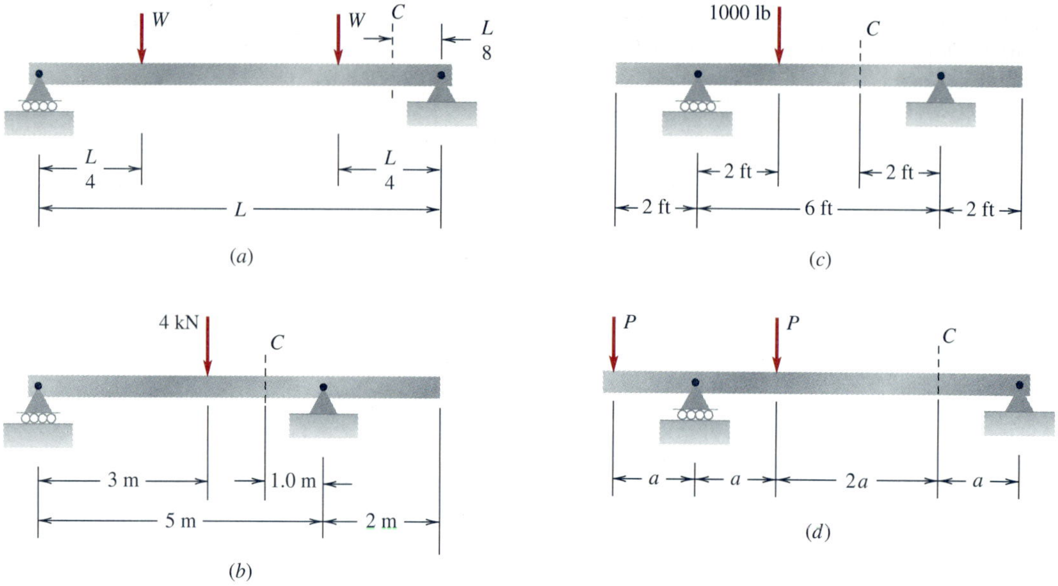

4.2-5 Determine the value of the shear force and the bending moment at section C for the beams shown in

(*a*) Fig. P4.2-5*a*
(*b*) Fig. P4.2-5*b*
(*c*) Fig. P4.2-5*c*
(*d*) Fig. P4.2-5*d*

Neglect the weight of the beams.

4.2-6 The 1-kip weight is supported by a flexible cable passing over a (frictionless) roller at position G and connected at point H, as shown in Fig. P4.2-6. Neglect the weight of the beam. Determine the value of the shear force and bending moment at (*a*) section D and (*b*) section C. Clearly indicate the coordinate axes.

Fig. P4.2-6

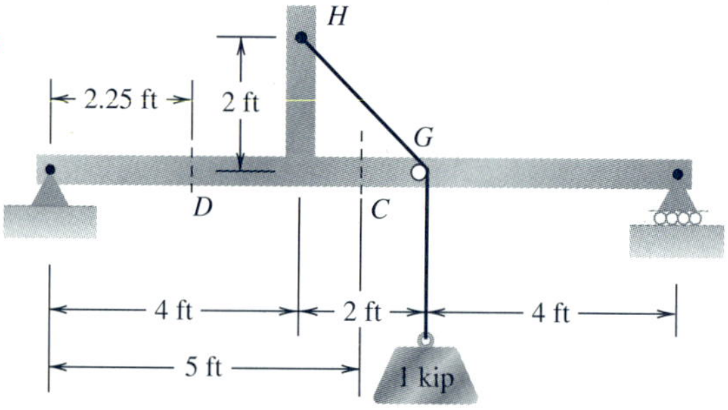

4.2-7 Determine the value of the shear force, axial force, and bending moment at section C defined by angle θ for the circular beam loaded as shown in Fig. P4.2-7. Compare the results to those in Example 4.4, Fig. 4.10.

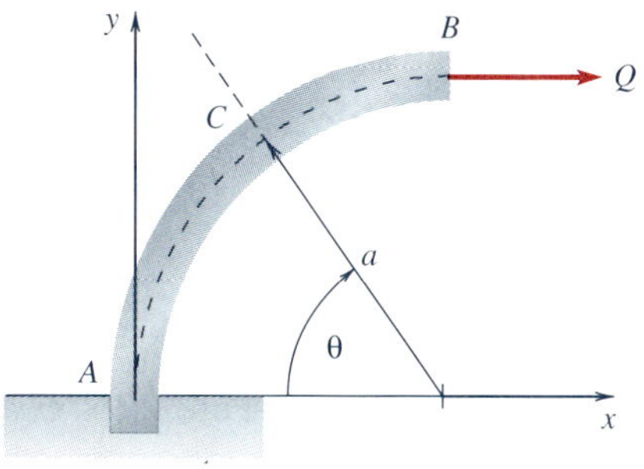

Fig. P4.2-7

4.2-8 A beam AB with a welded offset is loaded as shown in Fig. P4.2-8 (also see Fig. 4.9) by load P at angle θ to the horizontal. Determine the value of the shear force, bending moment, and axial force acting on section C. Neglect the weight of the beam.

Fig. P4.2-8

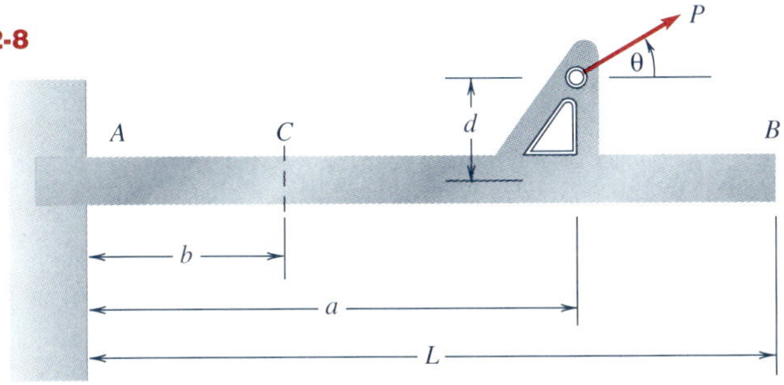

4.2-9 Determine the value of the shear force, axial force, and bending moment at section C defined by angle θ for the circular beam loaded as shown in Fig. P4.2-9. Compare the results to those in Example 4.4 and Prob. 4.2-7 when $\phi = 0$ and $\phi = \pi/2$.

Fig. P4.2-9

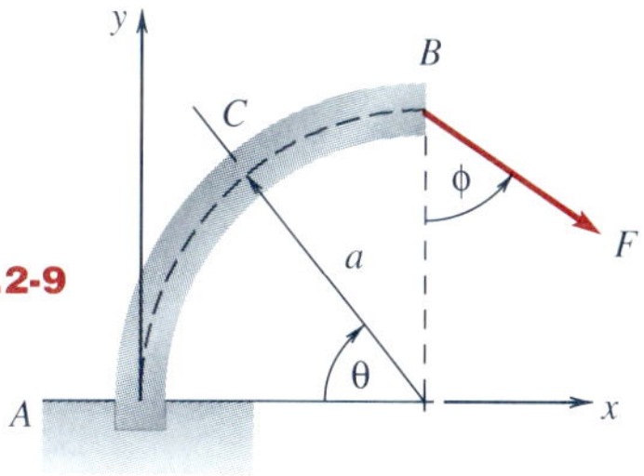

4.2-10 Determine the value of the shear force, axial force, and bending moment at section C defined by angle θ for the semicircular pipe AB of radius a loaded as shown in Fig. P4.2-10 when

(*a*) $\phi = 0°$

(*b*) $\phi = 90°$

(*c*) ϕ is arbitrary, $0° < \phi < 360°$

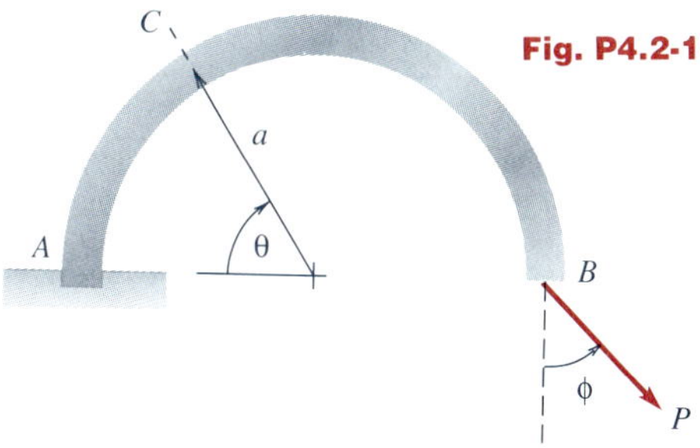

Fig. P4.2-10

4.3-1 Consider Example 4.5, Fig. 4.11. Draw to approximate scale the shear force and bending moment diagrams on the same plot for the three cases $a = L/4$, $a = L/2$, and $a = 3L/4$. For which value of a is the bending moment in the beam the largest? Plot the maximum bending moment as a function of a.

4.3-2 Sketch the shear force and bending moment diagrams to approximate scale for the three beams shown in Fig. P4.3-2. Can you draw any conclusions about approximating a uniform load with a number of concentrated loads? Neglect the weight of the beams.

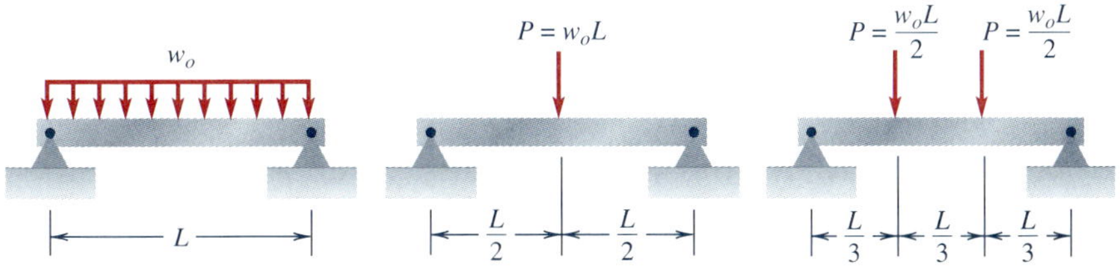

Fig. P4.3-2

4.3-3 A cantilever beam of length L is loaded by a linear load, as shown in Fig. P4.3-3. Sketch the shear force and bending moment diagrams for the beam; note Example 4.8, Fig. 4.15. Neglect the weight of beam.

Fig. P4.3-3

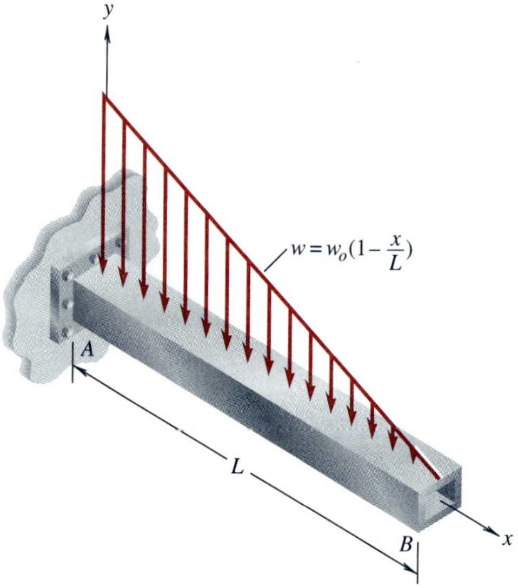

4.3-4 Consider Example 4.9, Fig. 4.16. Verify upon differentiation of the expression for M in Eqs. (*b*) that the maximum value of the bending moment occurs at $x = 0.375L$ and that the maximum value is $9w_0L^2/128$. Note that the condition for the maximum value of the bending moment is $V = 0$.

4.3-5 Draw to approximate scale, with the coordinate axes clearly identified, the shear force and bending moment diagrams for the beams shown in Fig. P4.3-5. Identify all critical values, and neglect the weight of the beam.

(*a*) Fig. P4.3-5*a*
(*b*) Fig. P4.3-5*b*
(*c*) Fig. P4.3-5*c*
(*d*) Fig. P4.3-5*d*
(*e*) Fig. P4.3-5*e*
(*f*) Fig. P4.3-5*f*
(*g*) Fig. P4.3-5*g*
(*h*) Fig. P4.3-5*h*
(*i*) Fig. P4.3-5*i*

Fig. P4.3-5*a* to *d*

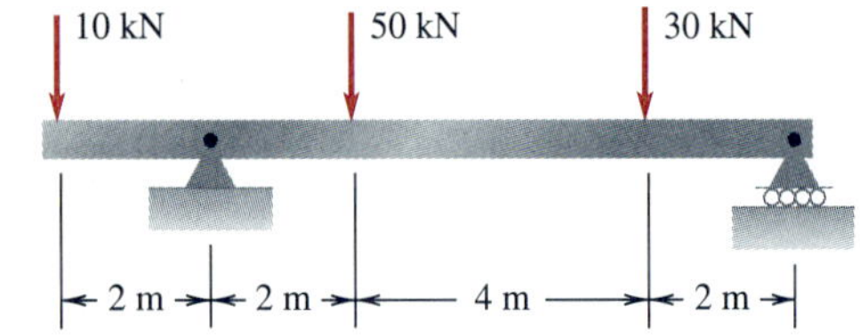

(*a*)

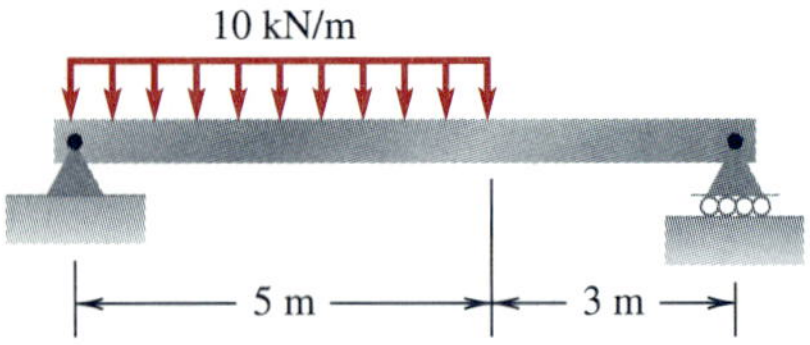

(*b*)

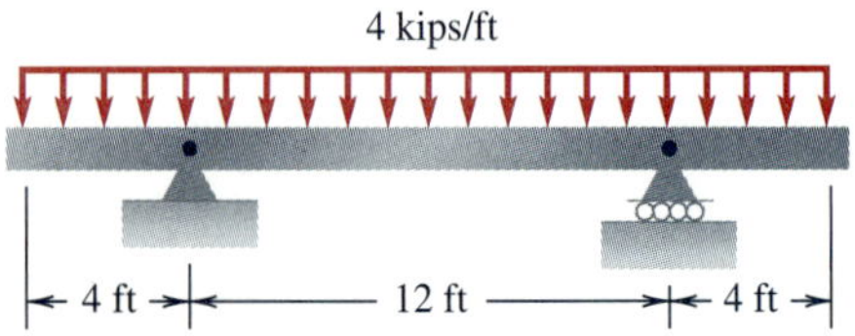

(*c*)

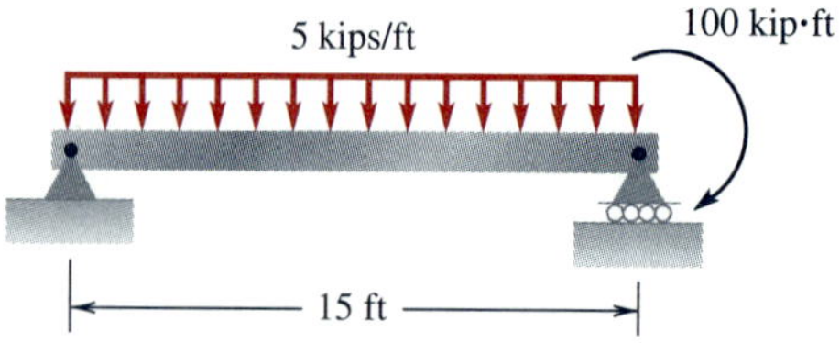

(*d*)

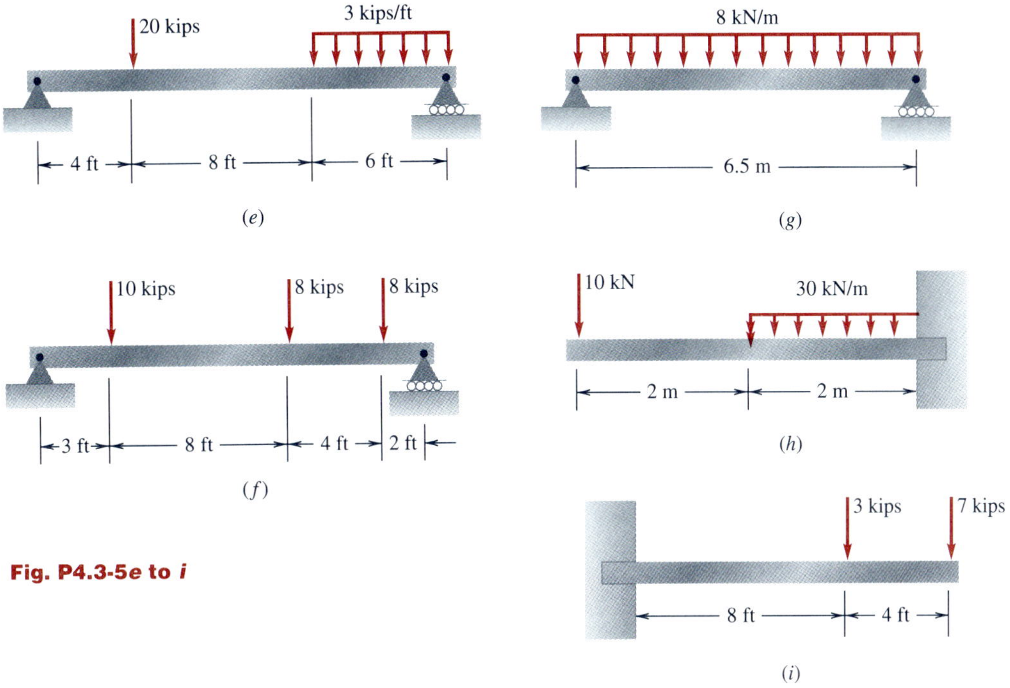

Fig. P4.3-5*e* to *i*

4.3-6 A beam of length L is loaded as shown in Fig. P4.3-6 by a concentrated load $P = w_0 L/2$. Sketch the shear force and the bending moment diagrams for this beam, and determine the maximum value of the bending moment. Compare the value of the maximum bending moment to that obtained in Example 4.9, Fig. 4.16. Neglect the weight of the beam.

Fig. P4.3-6

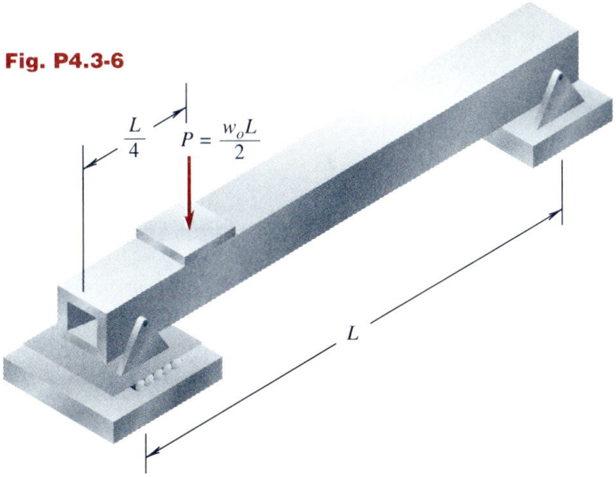

4.3-7 A simply supported beam carries twenty-seven 400-N barrels, as shown in Fig. P4.3-7. Draw the shear force and bending moment diagrams for the beam. Neglect the weight of the beam.

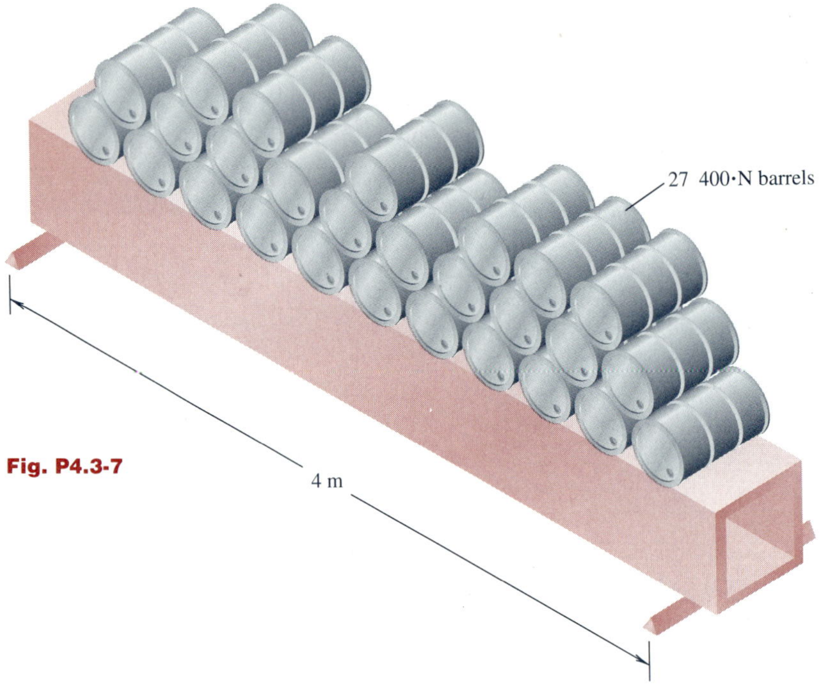

Fig. P4.3-7

4.3-8 Consider the semicircular beam shown in Fig. P4.3-8. Neglect the weight of the beam, and sketch the shear force, axial force, and bending moment as a function of angle θ when

(*a*) $b = a/4$
(*b*) $b = a/2$
(*c*) $b = 3a/4$

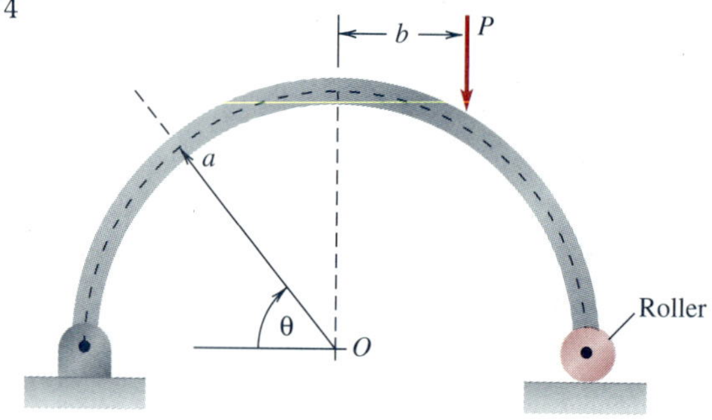

Fig. P4.3-8

4.4-1 Verify that the results shown in Fig. 4.18*a* are consistent with Eqs. (4.6) and (4.7).

4.4-2 Consider Example 4.10, Fig. 4.19. Verify the correctness of Eqs. (*c*), (*e*), (*h*), and (*i*). If an additional constant distributed load is applied along the total length of the beam, are these equations still valid?

4.4-3 Consider Example 4.11, Fig. 4.21. If the length of the beam is increased by a factor k such that the new length is now kL and the total *load* on the lengthened beam is still equal to $w_0 L$, what are the values of the maximum shear force and bending moment in the lengthened beam?

4.4-4 Consider Example 4.12, Fig. 4.22. If $w_0 = 2$ kN/m and $L = 6$ m, sketch the shear force and bending moment diagrams. Locate all critical values on the diagrams. Neglect the weight of the beam.

4.4-5 Consider Example 4.12, Fig. 4.22. If $w_0 = 15$ lb/in and $L = 15$ ft, sketch the shear force and bending moment diagrams. Locate all critical values on the diagrams. Neglect the weight of the beam.

4.4-6 Draw the shear force and bending moment diagrams for the beams shown in Fig. P4.4-6. Identify all critical values, and determine the maximum values of the shear force and bending moment; neglect the weight of the beam.
(*a*) Fig. P4.4-6*a*
(*b*) Fig. P4.4-6*b*
(*c*) Fig. P4.4-6*b*. Take $a = b = c = L/3$.
(*d*) Fig. P4.4-6*b*. Take $a = 0$ and $b = L/3$.
(*e*) Fig. P4.4-6*b*. Take $L = 1$ m, $P = 0$, $a = b = 250$ mm, and $M_0 = 60$ N · m.
(*f*) Fig. P4.4-6*c*
(*g*) Fig. P4.4-6*d*

4.4-7 Draw the shear force and bending moment diagrams for the beam shown in Fig. P4.4-7. Neglect the weight of the beam.

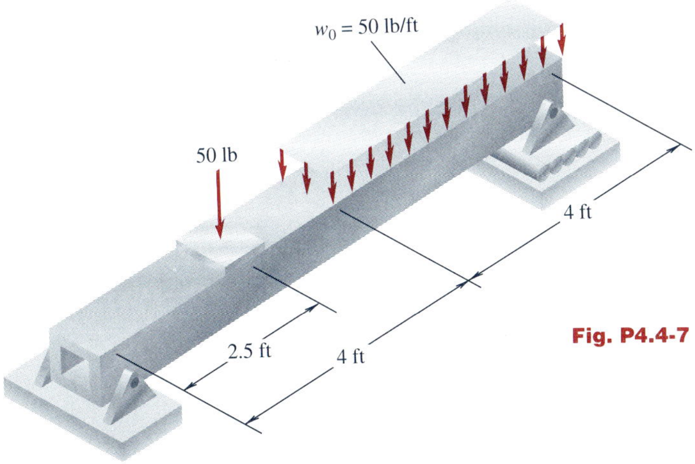

Fig. P4.4-7

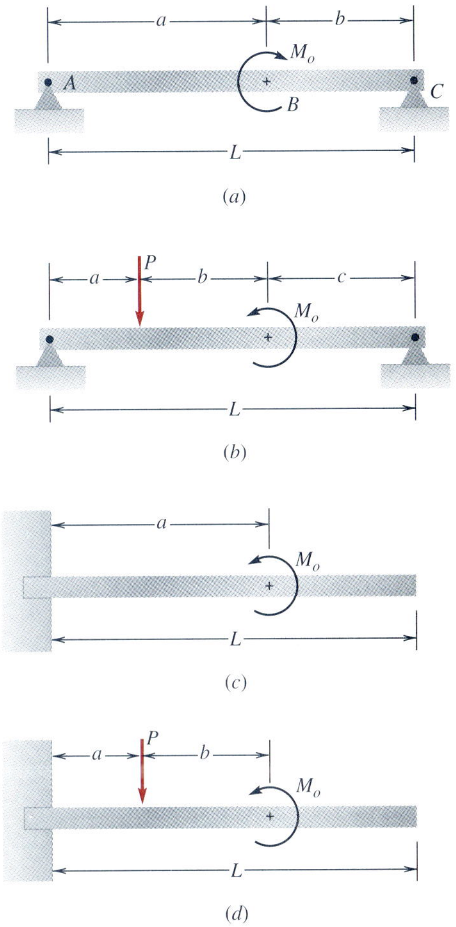

Fig. P4.4-6

4.4-8 Consider Fig. P4.4-8. Sketch the shear force and bending moment diagrams for the beam. Neglect the weight of the beam.

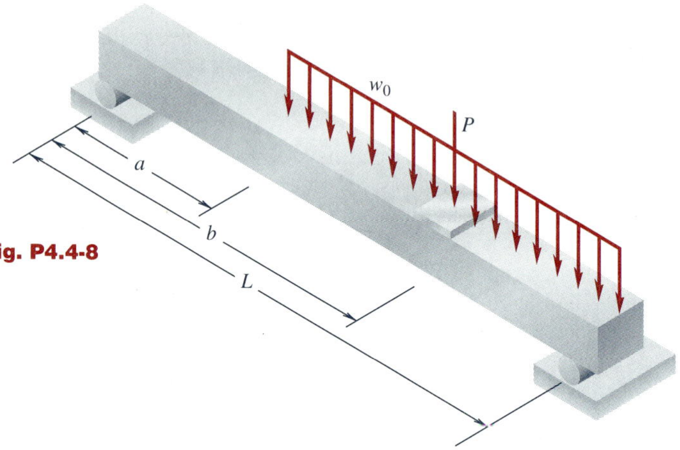

Fig. P4.4-8

4.4-9 Consider Fig. P4.4-8. Sketch the shear force and bending moment diagrams for the beam shown with the values below. Neglect the weight of the beams.

(*a*) $w_0 = 1000$ lb/ft, $P = 0$, $a = 8$ ft, $L = 32$ ft
(*b*) $w_0 = 800$ lb/ft, $P = 8$ kips, $b = 14$ ft, $a = 6$ ft, $L = 24$ ft
(*c*) $w_0 = 900$ lb/ft, $P = 10$ kips, $b = 9$ ft, $a = 0$, $L = 18$ ft
(*d*) $w_0 = 12$ kN/m, $P = 2$ kN, $b = 3$ m, $a = 0$, $L = 6$ m
(*e*) $w_0 = 3$ kN/m, $P = 10$ kN, $b = 2$ m, $a = 4$ m, $L = 8$ m

4.4-10 Sketch the shear force and bending moment diagrams for the beams shown in Fig. P4.4-10. Neglect the weight of the beams.

(*a*) Fig. P4.4-10*a*
(*b*) Fig. P4.4-10*b*
(*c*) Fig. P4.4-10*c*
(*d*) Fig. P4.4-10*d*
(*e*) Fig. P4.4-10*e*
(*f*) Fig. P4.4-10*f*
(*g*) Fig. P4.4-10*g*
(*h*) Fig. P4.4-10*h*
(*i*) Fig. P4.4-10*i*
(*j*) Fig. P4.4-10*j*
(*k*) Fig. P4.4-10*k*
(*l*) Fig. P4.4-10*l*
(*m*) Fig. P4.4-10*m*

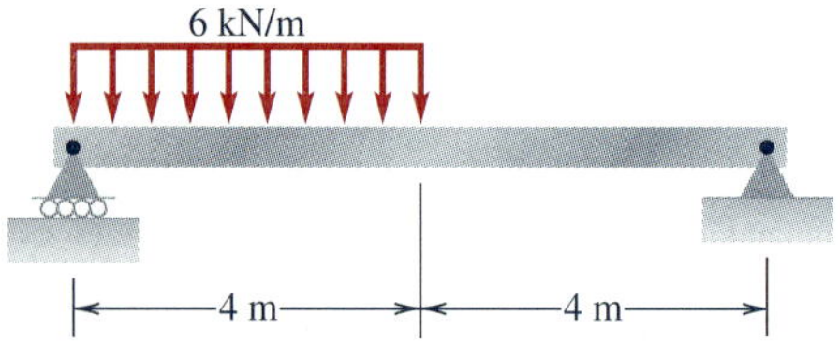

(*a*)

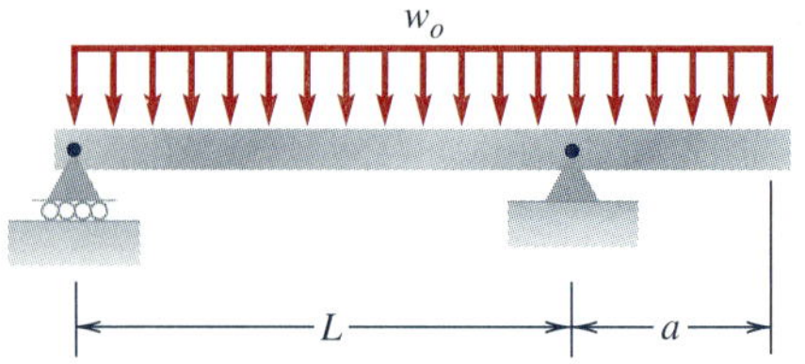

(*b*)

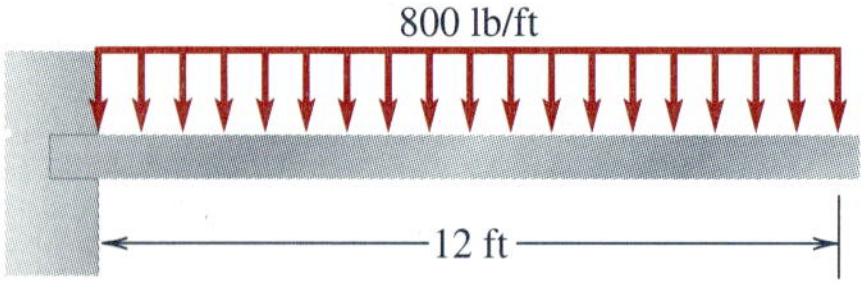

(*c*)

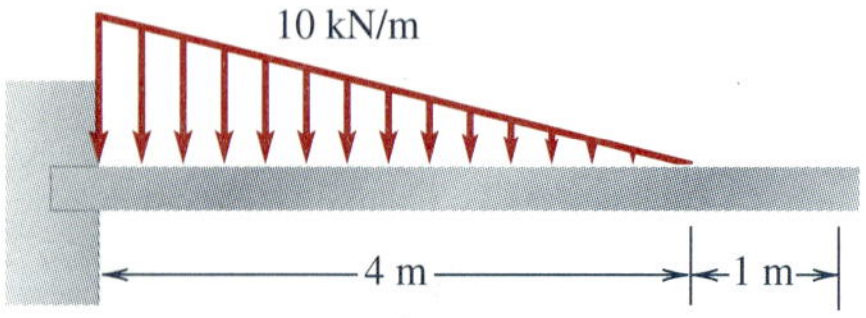

(*d*)

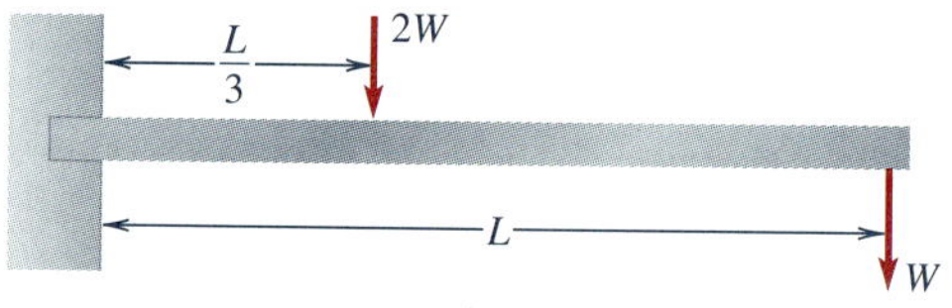

(*e*)

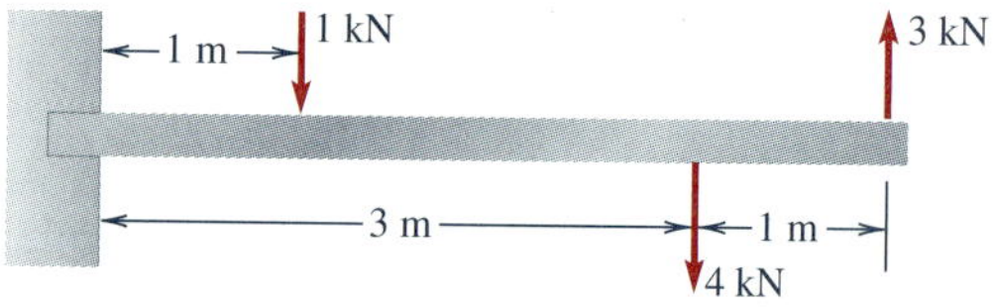

(*f*)

Fig. P4.4-10*a* to *f*

Fig. P4.4-10*g* to *m*

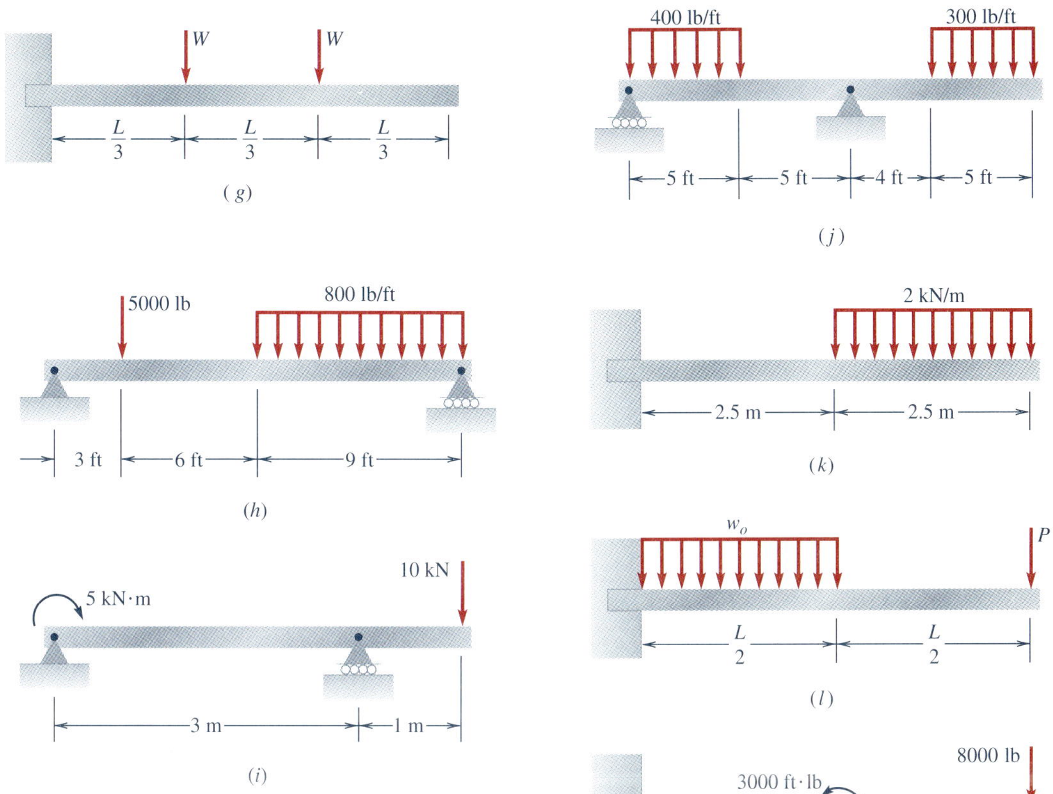

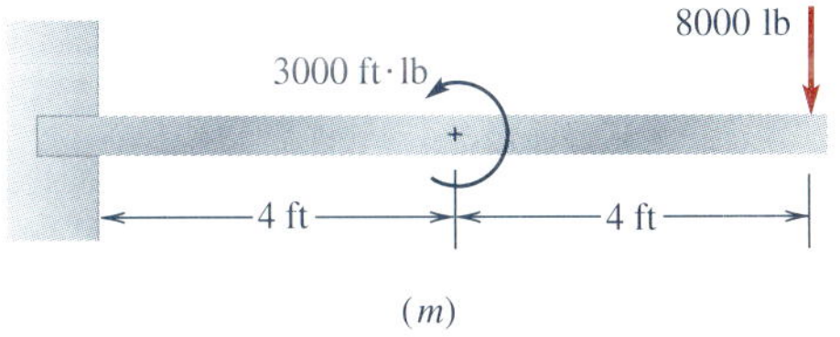

4.4-11 Sketch the shear force and bending moment diagrams for the beams shown in the indicated figure. Neglect the weight of the beams.

(*a*) Fig. P4.2-3*a*
(*b*) Fig. P4.2-3*b*
(*c*) Fig. P4.2-3*c*
(*d*) Fig. P4.2-3*d*
(*e*) Fig. P4.2-3*e*
(*f*) Fig. P4.2-3*f*
(*g*) Fig. P4.2-3*g*

4.4-12 Sketch the shear force and bending moment diagrams for the beams shown in the indicated figure. Neglect the weight of the beams.

(*a*) Fig. P4.2-4*a*
(*b*) Fig. P4.2-4*b*
(*c*) Fig. P4.2-4*c*
(*d*) Fig. P4.2-4*d*
(*e*) Fig. P4.2-4*e*
(*f*) Fig. P4.2-4*f*
(*g*) Fig. P4.2-4*g*
(*h*) Fig. P4.2-4*h*

4.4-13 Sketch the shear force and bending moment diagrams for the beams shown in the indicated figure. Neglect the weight of the beams.
(*a*) Fig. P4.2-5*a*
(*b*) Fig. P4.2-5*b*
(*c*) Fig. P4.2-5*c*
(*d*) Fig. P4.2-5*d*

4.4-14 Two people of identical weight W start at the midpoint of a simply supported beam and walk in opposite directions so as to remain symmetrically located with respect to the midpoint of the beam, as shown in Fig. P4.4-14. Show a graph of $|M|_{max}$ versus s, the separation distance. Take at least the four cases for s equal to 0, a, $2a$, and $4a$ in preparing the graph.

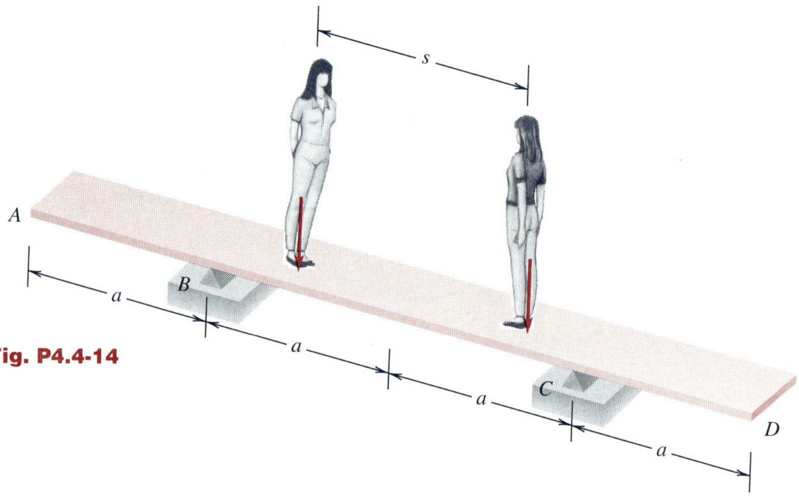

Fig. P4.4-14

4.4-15 Sketch the shear force and bending moment diagrams for the 22-ft-long beam shown in Fig. P4.4-15. Neglect the weight of the beam.

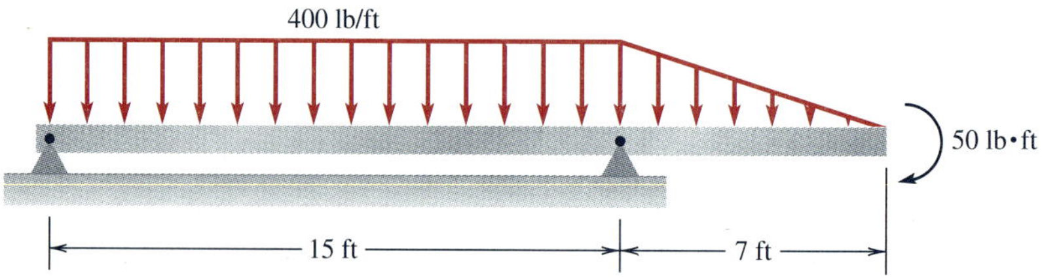

Fig. P4.4-15

4.5-1 Sketch the shear force and bending moment diagrams for the beams shown in Fig. P4.5-1. As the first step in sketching the diagrams, write out the expression for the loading $q(x)$ on the beam in terms of singularity functions;

then integrate to obtain $V(x)$ and $M(x)$. Clearly indicate all critical values and coordinate axes. Neglect the weight of the beams.

(*a*) Fig. P4.5-1*a*
(*b*) Fig. P4.5-1*b*
(*c*) Fig. P4.5-1*c*
(*d*) Fig. P4.5-1*d*
(*e*) Fig. P4.5-1*e*
(*f*) Fig. P4.5-1*f*
(*g*) Fig. P4.5-1*g*
(*h*) Fig. P4.5-1*h*
(*i*) Fig. P4.5-1*i*
(*j*) Fig. P4.5-1*j*

Fig. P4.5-1

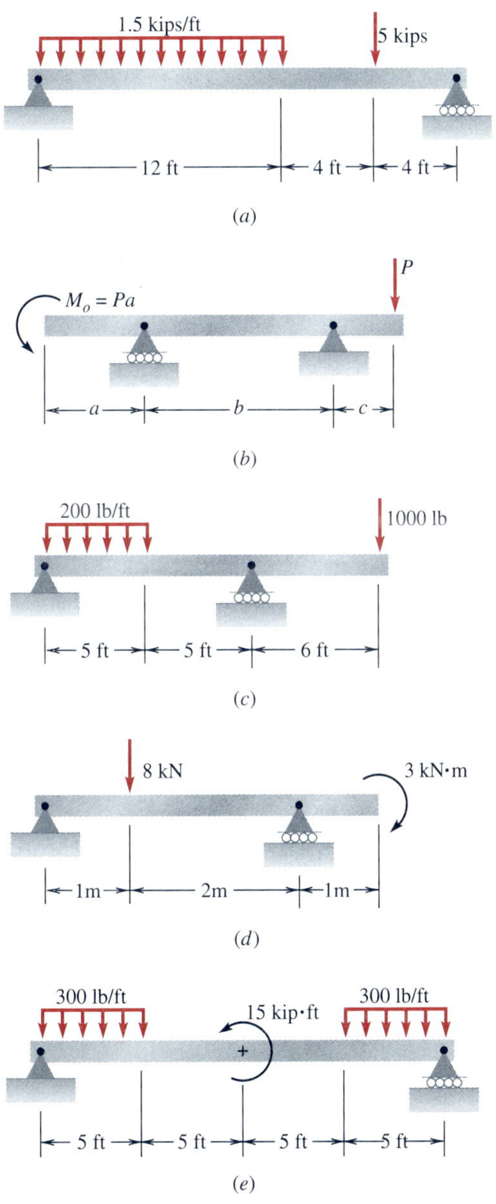

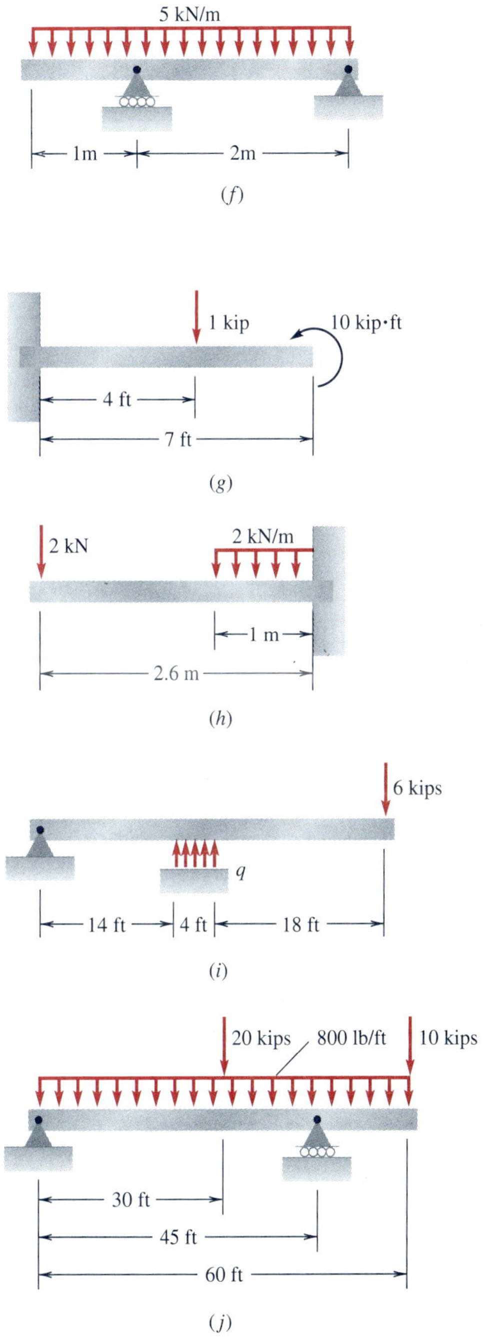

4.5-2 Sketch the shear force and bending moment diagrams for the beams shown in Fig. P4.2-3*a* to *g*. As the first step in sketching the diagrams, write out the expression for the loading $q(x)$ on the beam in terms of singularity functions, and then integrate to obtain $V(x)$ and $M(x)$. Clearly indicate all critical values and coordinate axes. Neglect the weight of the beams.
(*a*) Fig. P4.2-3*a*
(*b*) Fig. P4.2-3*b*
(*c*) Fig. P4.2-3*c*
(*d*) Fig. P4.2-3*d*
(*e*) Fig. P4.2-3*e*
(*f*) Fig. P4.2-3*f*
(*g*) Fig. P4.2-3*g*

4.5-3 Sketch the shear force and bending moment diagrams for the beams shown in Fig. P4.2-4*a* to *h*. Write out the expression for the loading $q(x)$ in terms of singularity functions, and integrate to obtain $V(x)$ and $M(x)$. Clearly indicate all critical values and coordinate axes. Neglect the weight of the beams.
(*a*) Fig. P4.2-4*a*
(*b*) Fig. P4.2-4*b*
(*c*) Fig. P4.2-4*c*
(*d*) Fig. P4.2-4*d*
(*e*) Fig. P4.2-4*e*
(*f*) Fig. P4.2-4*f*
(*g*) Fig. P4.2-4*g*
(*h*) Fig. P4.2-4*h*

4.5-4 Sketch the shear force and bending moment diagrams for the beam shown in Fig. P4.5-4. Consider the special cases when $P = 0$ or $M = 0$. Neglect the weight of the beam.

Fig. P4.5-4

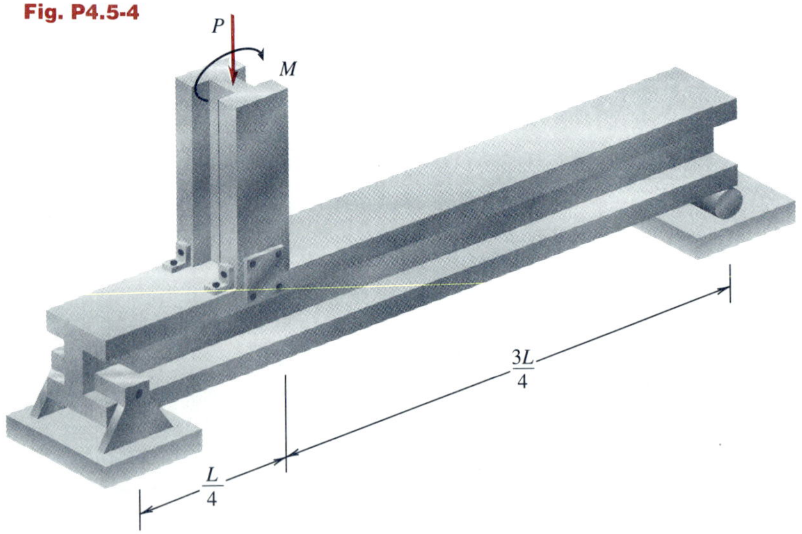

4.5-5 Consider the beam shown in Fig. P4.5-5. Sketch the shear force and bending moment diagrams for the beam. Neglect the weight of the beam.

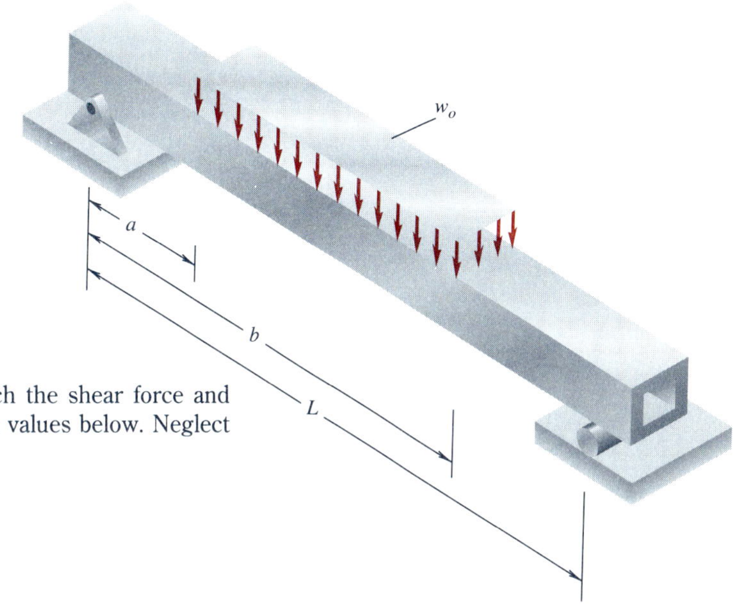

Fig. P4.5-5

4.5-6 Consider the beam shown in Fig. P4.5-5. Sketch the shear force and bending moment diagrams for the beam for the specific values below. Neglect the weight of the beam.
(*a*) $a = L/2$, $b = L$
(*b*) $a = 0$, $b = \frac{3}{4}L$
(*c*) $w_0 = 2$ kN/m, $a = 4$ m, $b = 10$ m, $L = 12$ m
(*d*) $w_0 = 150$ lb/ft, $a = 3$ ft, $b = 9$ ft, $L = 12$ ft

4.5-7 A beam is loaded by a uniform load w_0 as shown in Fig. P4.5-7. Determine the value of a in terms of L such that the magnitudes of the maximum and minimum values of the bending moment are equal. Compare the maximum bending moment in this case to the cases when $a = L$ and $a = L/2$. Sketch the bending moment diagrams for each of the three values of a.

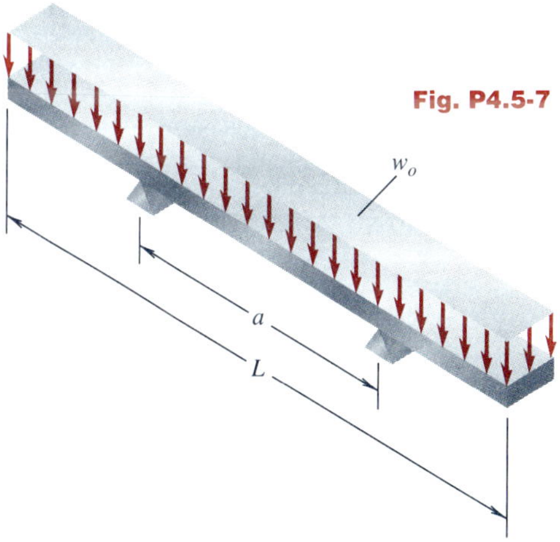

Fig. P4.5-7

4.5-8 For the cantilever beam shown in Fig. P4.5-8, determine an expression for the shear force and bending moment along the length of the beam. Neglect the weight of the beam.

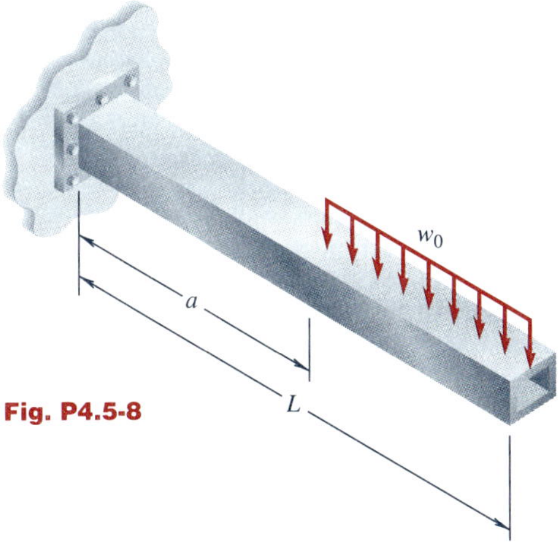

Fig. P4.5-8

4.5-9 Oil flows through a 3.5-in-inner-diameter pipe *ABC* from a tank at *A* to a holding tank at *C*, as shown in Fig. P4.5-9. If the oil weighs 45 lb/ft^3, sketch the shear force and bending moment diagrams for the pipe. The pipe itself weighs 9.1 lb/ft, and the supports at *A* and *B* are simple supports.

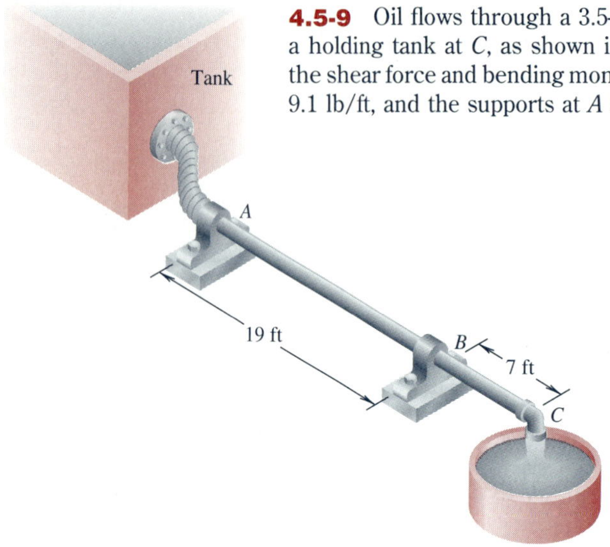

Fig. P4.5-9

4.5-10 Consider the beam shown in Fig. P4.5-10. Sketch the shear force and bending moment diagrams for the beam for the special cases given below. Neglect the weight of the beam.

(*a*) $a = L/4,\ b = L/2,\ M_a = 2M_b$

(*b*) $a = 0,\ b = L,\ M_a = -M_b = -M_0$

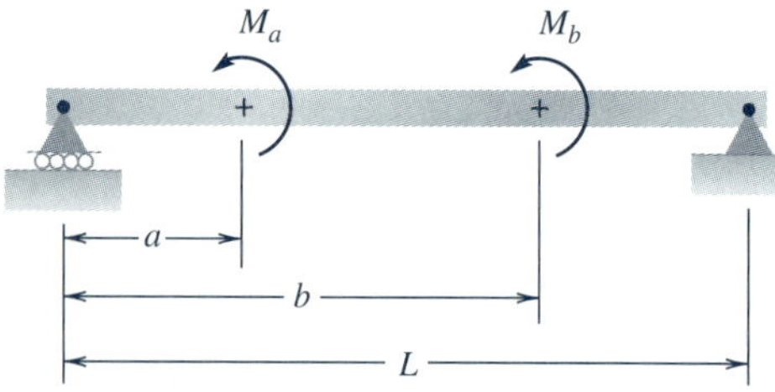

Fig. P4.5-10

4.5-11 Consider the beam shown in Fig. P4.5-11. Sketch the shear force and bending moment diagrams for the beam. Neglect the weight of the beam.

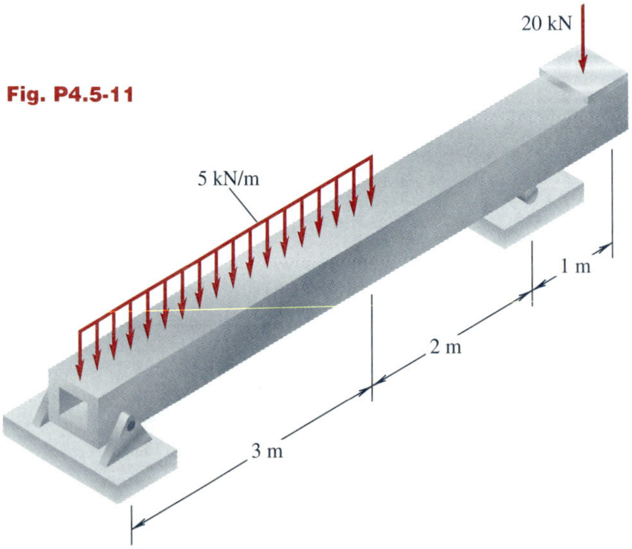

Fig. P4.5-11

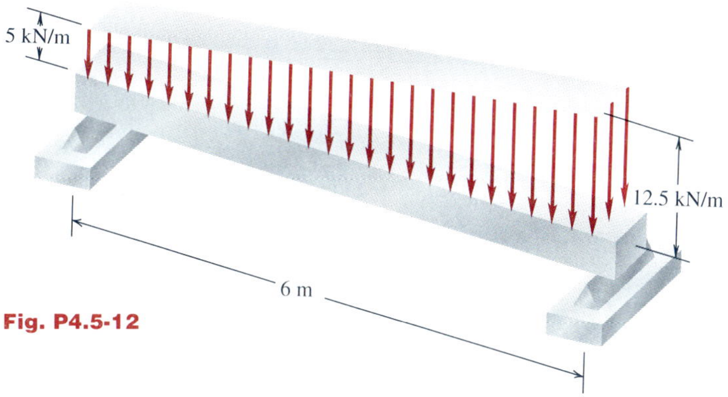

Fig. P4.5-12

4.5-12 A simply supported beam carries a distributed loading which increases from left to right along the beam, as shown in Fig. P4.5-12. Sketch the shear force and bending moment diagrams for the beam. Neglect the weight of the beam.

4.5-13 A simply supported beam is subjected to a distributed loading whose amplitude varies sinusoidally with position along the beam, as shown in Fig. P4.5-13. Find expressions for the shear force V and the bending moment M by integrating a suitable loading function $q(x)$ for this problem. In this problem, special care must be taken to make use of the constants of integration to satisfy the conditions of force and moment equilibrium for the beam. The reaction forces are each equal to w_0L/π. Also find the value of a such that M vanishes at $x = L/2$. Neglect the weight of the beam.

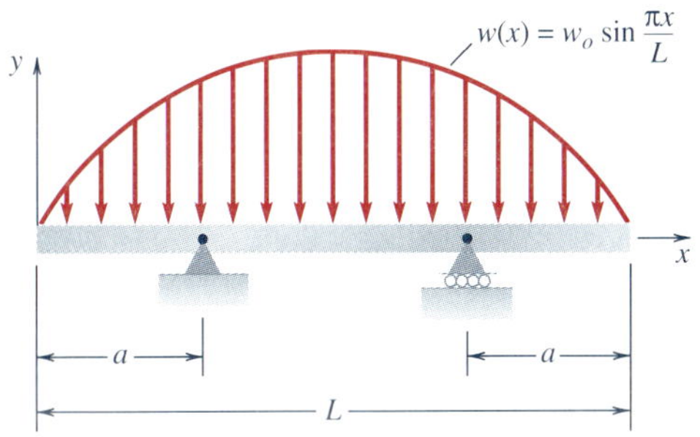

Fig. P4.5-13

4.5-14 Verify that the expressions are correct for the loading functions for a constant distributed loading and for a linearly varying loading in Fig. 4.27 in terms of the singularity functions.

4.5-15 Consider Example 4.16, Fig. 4.28. Show by sketching each term to approximate scale that the expressions for $V(x)$ in Eq. (*c*) and $M(x)$ in Eq. (*d*) will give the shear force and bending moment diagrams shown in Fig. 4.28*c* and *d*.

4.5-16 Consider Example 4.17, Fig. 4.29. Find new locations of the load P (originally at $x = 10$ ft) such that the maximum and the minimum values of the bending moments along the beam are equal.

4.5-17 Sketch the shear force and bending moment diagrams for the beam loaded by two concentrated bending moments as shown in Fig. P4.5-17. Neglect the weight of the beam, and assume that the supports are such that they do not lift off the ground. Consider the special cases:

(*a*) $M_1 = M_2 = M$
(*b*) $M_2 = 0$
(*c*) $M_2 = -M_1$
(*d*) Arbitrary M_1 and M_2

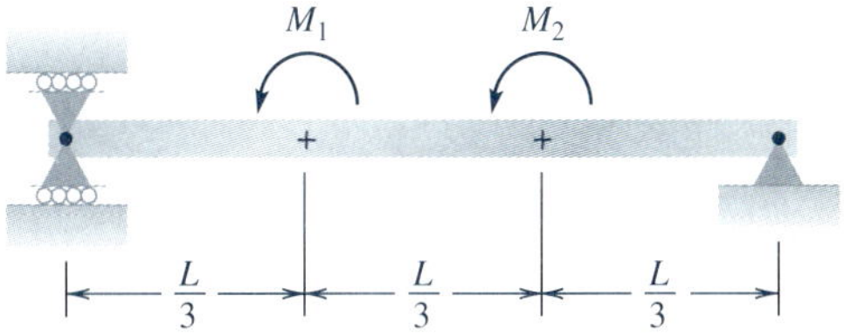

Fig. P4.5-17

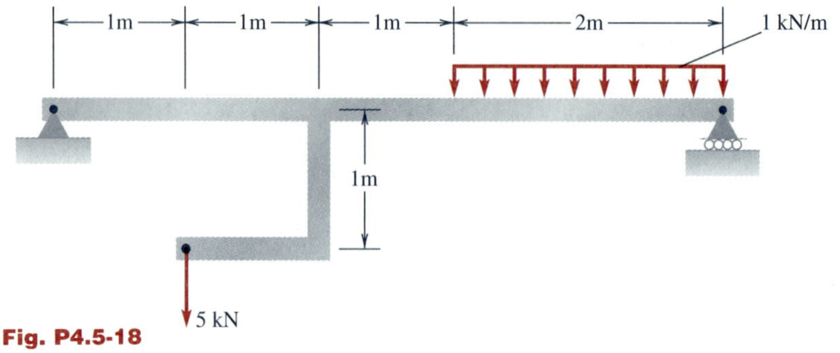

Fig. P4.5-18

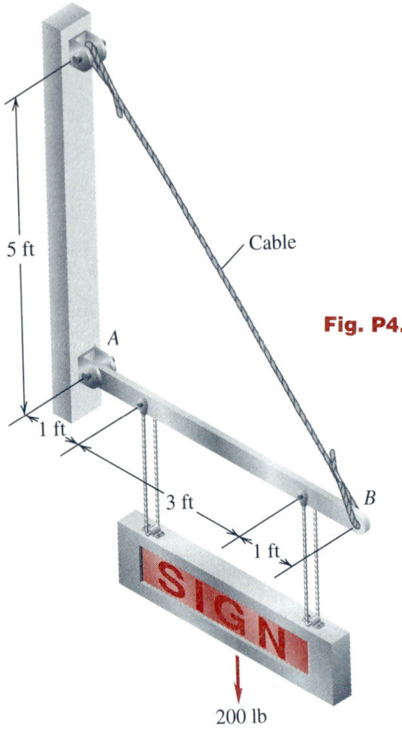

Fig. P4.5-19

4.5-18 Sketch the shear force and bending moment diagrams for the beam shown in Fig. P4.5-18. Neglect the weight of the beam.

4.5-19 Sketch the shear force and bending moment diagrams for the horizontal pinned member AB carrying the sign of weight 200 lb as shown in Fig. P4.5-19.

4.5-20 A heavy 10-in-outside-diameter steel pipe is lifted by a crane, as shown in Fig. P4.5-20. Estimate the value of the distance a so that the maximum bending moment in the pipe is as small as possible. Check the limiting cases of $a = 0$ and $a = 25$ ft. The pipe weighs 100 lb/ft.

Fig. P4.5-20

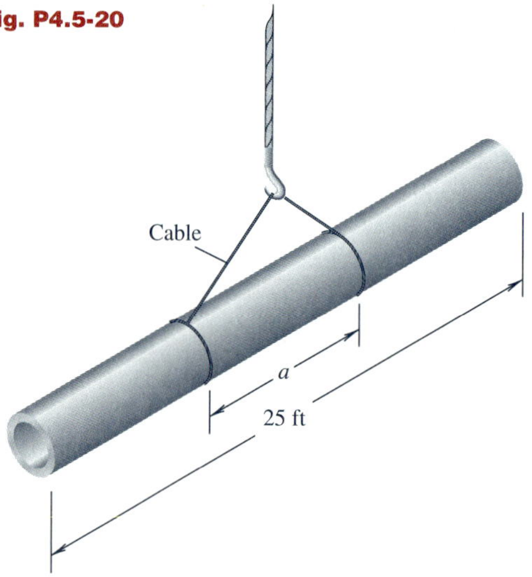

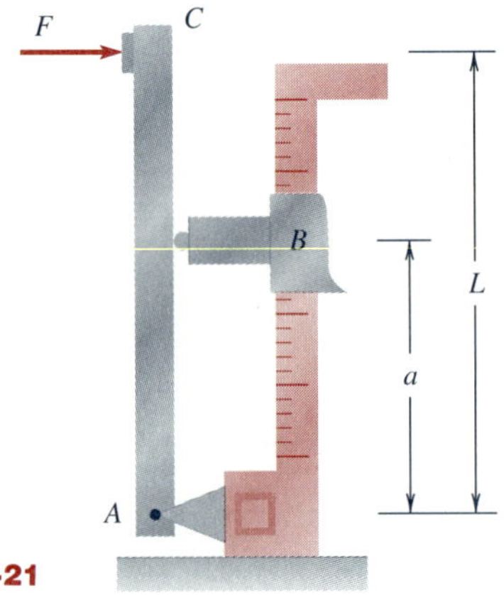

Fig. P4.5-21

4.5-21 A portion of a load-sensing device sketched in Fig. P4.5-21 is designed such that the vertical beam member AC is to have the same value of the maximum bending moment independent of the value of the force F applied at C. How should the distance a be varied as a function of F such that the maximum bending moment is the same for all values of F? Neglect the weight of the beam.

4.5-22 A simply supported beam is loaded by a triangular loading as shown in Fig. P4.5-22. Neglect the weight of the beam, and show that the maximum bending moment occurs at a point $x = L/\sqrt{3}$. Compare the value of the maximum bending moment to the value of the bending moment at $x = 2L/3$, and comment on the difference.

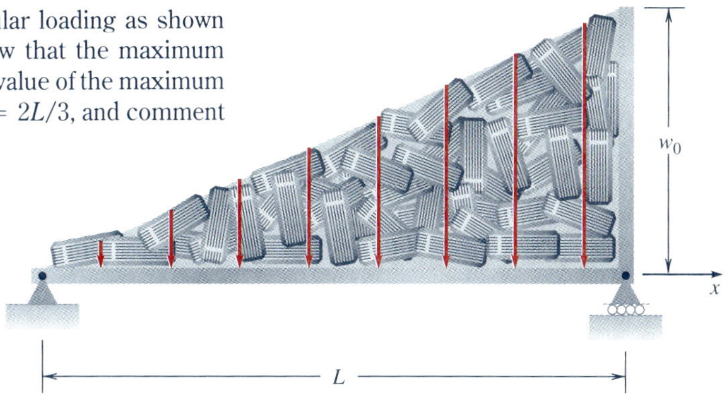

Fig. P4.5-22

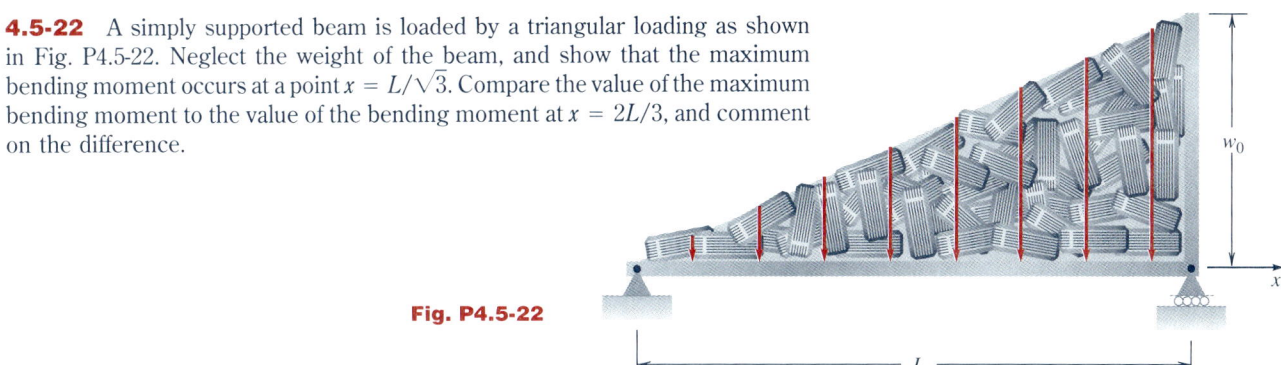

4.5-23 An 80-ft-long steel culvert is to be transported by a railroad flatcar, as shown in Fig. P4.5-23. The culvert weighs 300 lb/ft. Sketch the shear force and bending moment diagrams for the culvert.

Fig. P4.5-23

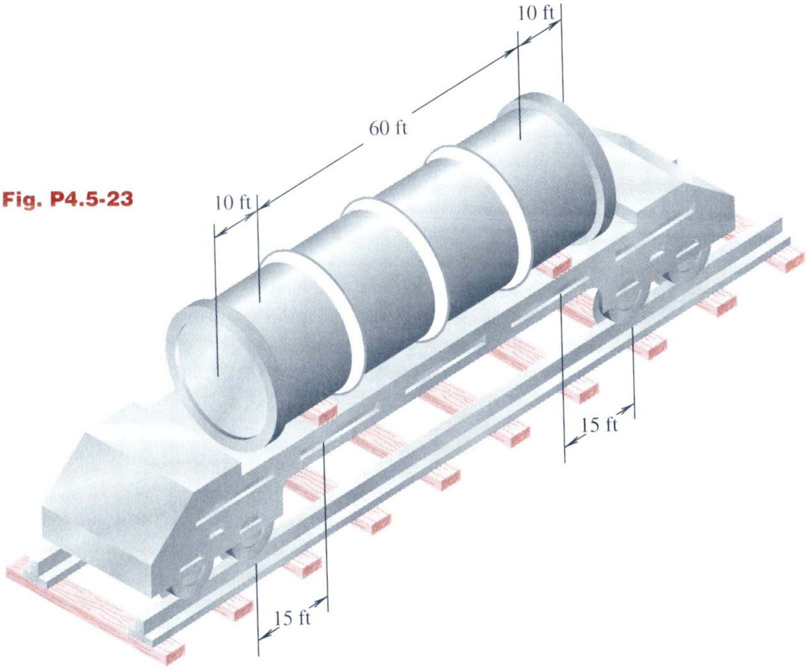

4.5-24 Consider Prob. 4.5-23 again, but this time sketch the shear force and bending moment diagrams for the railroad flatcar. Neglect the weight of the flatcar.

4.5-25 A beam of length L carries two concentrated loads at its ends and is supported by a uniform distributed load, as shown in Fig. P4.5-25. Sketch the shear force and bending moment diagrams for the beam.

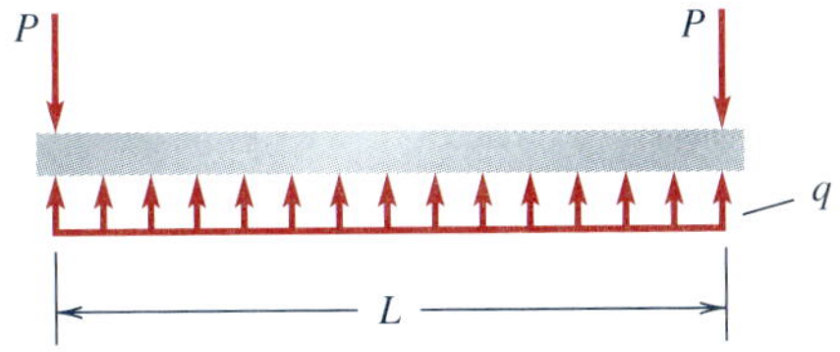

Fig. P4.5-25

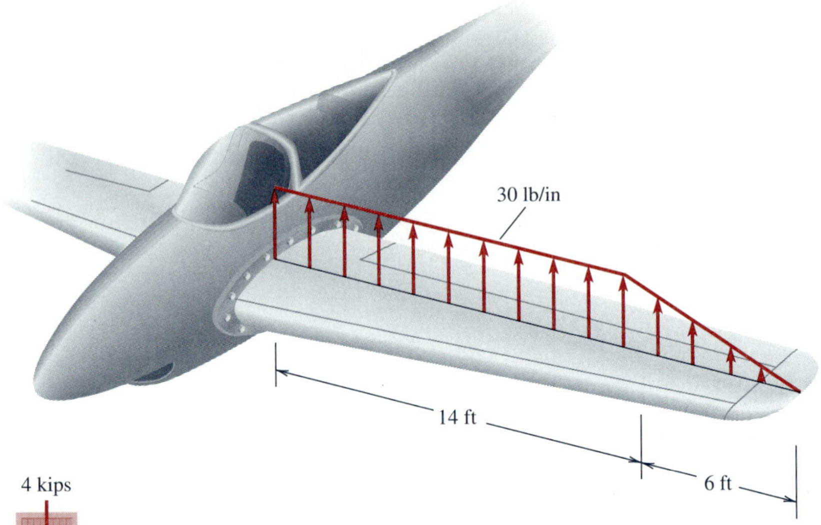

Fig. P4.5-26

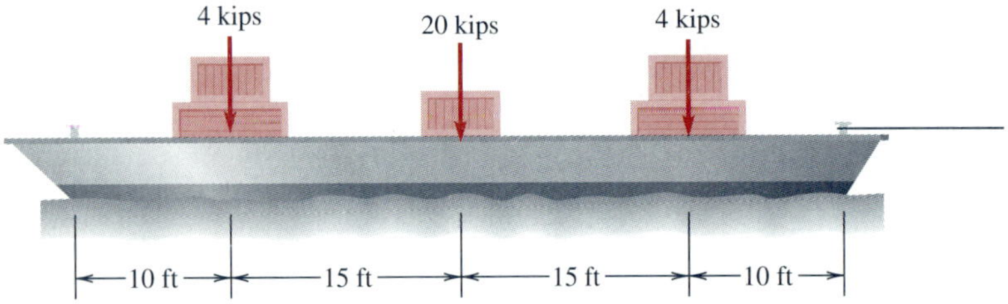

Fig. P4.5-27

4.5-26 The loading on the wing of a small sports airplane is shown in Fig. P4.5-26. Sketch the shear force and bending moment diagrams for the wing.

4.5-27 A small barge is carrying loads as shown in Fig. P4.5-27. The barge weighs 10 kips/ft. We can assume that the intensity of the buoyant force supporting the barge is constant along the length of the barge. Determine the magnitude of the buoyant force intensity, and then sketch the shear force and bending moment diagrams for the barge.

4.5-28 Preliminary design of small ships requires the knowledge of the shear force and bending moment distributions along the ship under the loading to be expected and the upward buoyant thrust. Figure P4.5-28 shows the loading and

Fig. P4.5-28

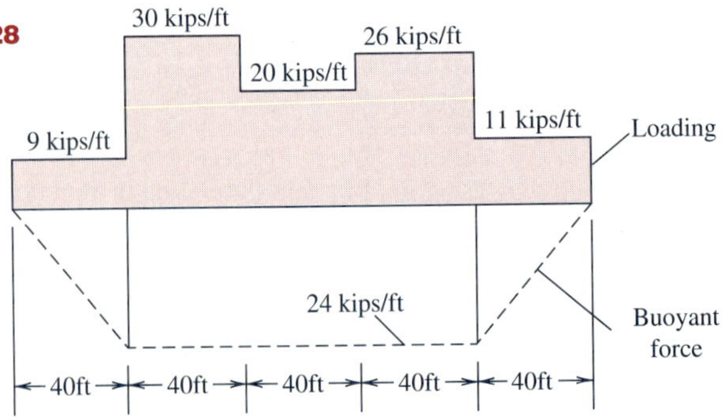

the buoyant force for a small ship. Sketch the shear force and bending moment diagrams for the ship.

4.5-29 In a fluid at rest, the pressure at a point in the fluid is the same in all directions. In addition, the pressure on a submerged surface in a fluid is normal to the surface and equal to the weight density of the fluid times the distance z to the surface, as shown in Fig. P4.5-29*a*. We usually neglect the atmospheric pressure p_0 acting on the free surface of the fluid.

Sketch the shear force and bending moment diagrams for gate AB shown in Fig. P4.5-29*b* if the gate is 0.75 m wide perpendicular to the plane of the paper. Take $\gamma = 9.8\ \text{kN/m}^3$.

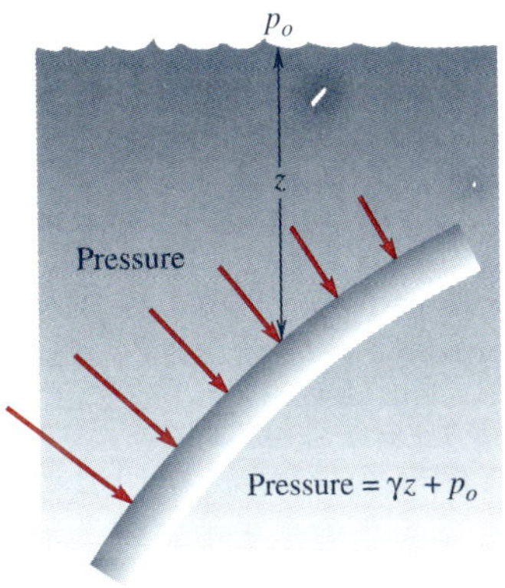

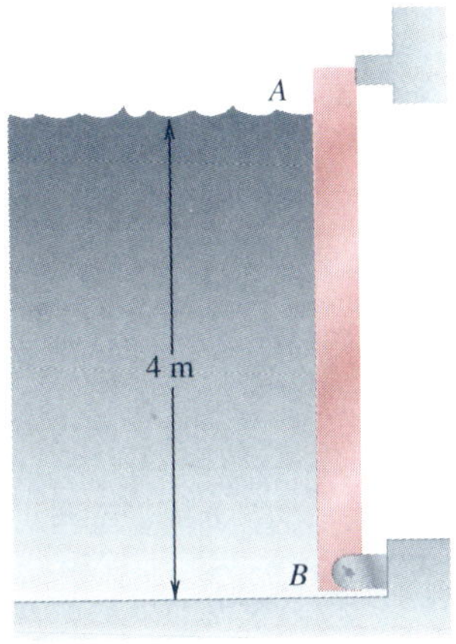

Fig. P4.5-29

4.5-30 A small wooden dam in an irrigation system is made of planks fastened to uprights driven into the canal bed. The uprights are 8.0 ft apart, and the water is 5.0 ft deep, as shown in Fig. P4.5-30. Sketch the shear force and bending moment diagrams for the uprights. Take $\gamma = 62.4\ \text{lb/ft}^3$.

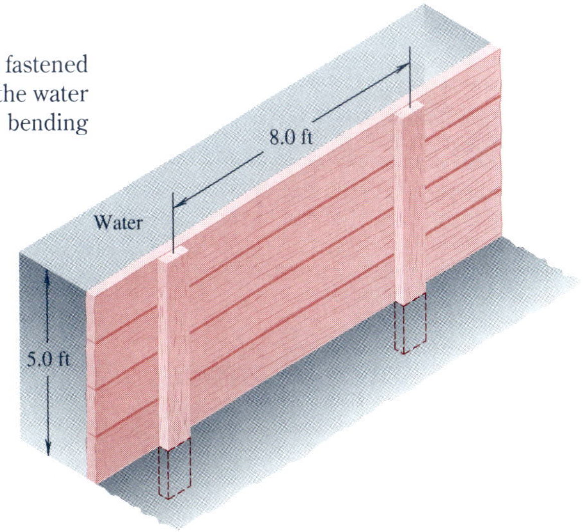

Fig. P4.5-30

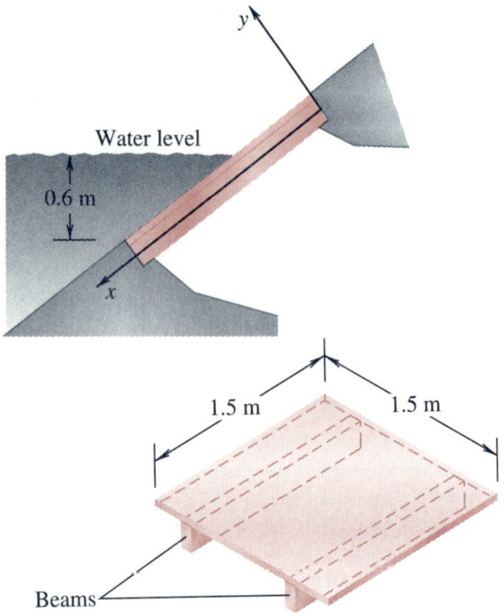

Fig. P4.5-31

4.5-31 A 1.5-m square gate is retaining water at half the length of the gate, as shown in Fig. P4.5-31. The pressure load on the gate is transmitted to the supports by means of two symmetrically located simply supported beams, as shown. We wish to find the maximum bending moment in the beams. The weight density of the water is $\gamma = 9.8\ \text{kN/m}^3$.

(*a*) Obtain the loading at the bottom of each beam as $w_0 = (1.5/2)(0.6)(9.8)$ kN/m.

(*b*) Obtain the reaction at the top as $\frac{1}{16}w_0$.

(*c*) Finally obtain the shear force and bending moment expressions to get $M_{max} = 263\ \text{N} \cdot \text{m}$.

4.5-32 A temporary dam is constructed to hold back an overflow of sludge, as shown in Fig. P4.5-32. Sketch the shear force and bending moment diagrams for the 3-in by 8-in planks (actual dimensions), and determine the maximum bending moment in the planks.

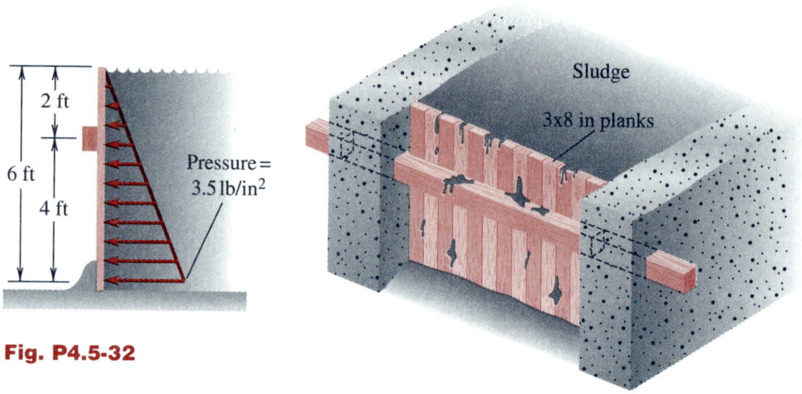

Fig. P4.5-32

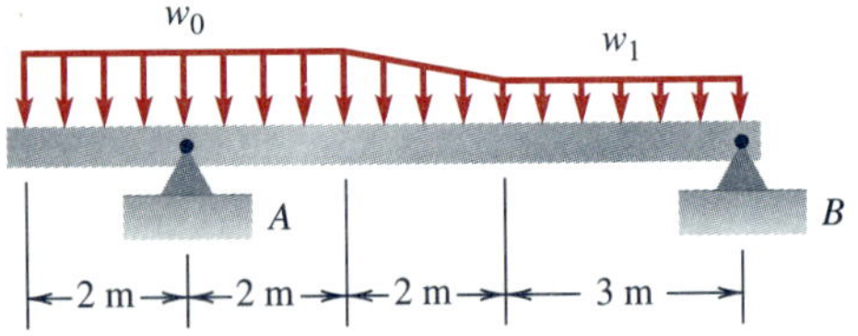

Fig. P4.5-33

4.5-33 A beam is loaded by a distributed load, as shown in Fig. P4.5-33. Sketch the shear force and bending moment diagrams for the following special cases:

(*a*) $w_0 = 15\ \text{kN/m}$, $w_1 = 10\ \text{kN/m}$. Compare this case to the results in Example 4.18, Fig. 4.30.

(*b*) $w_0 = w_1 = 15\ \text{kN/m}$

(*c*) $w_0 = w_1 = 10\ \text{kN/m}$

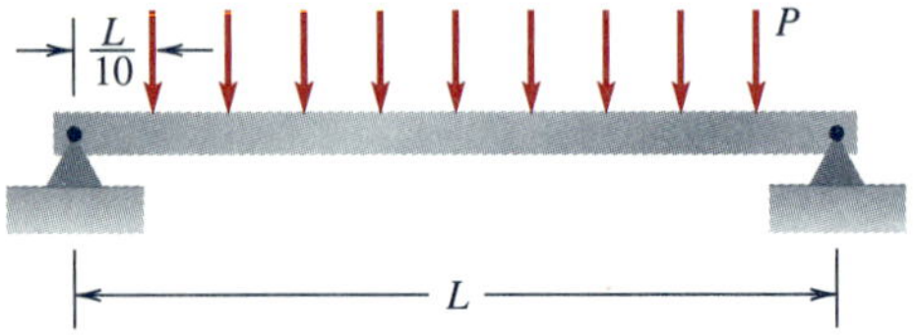

Fig. P4.5-34

4.5-34 Consider a beam of length L under nine equal loads of magnitude P, as shown in Fig. P4.5-34. The loads are spaced a distance $L/10$ apart, as shown. If we neglect the weight of the beam and use the observations about the structure of the shear force and bending moment diagrams given in Example 4.20, Fig. 4.32, we can develop a simple scheme to sketch the shear force and bending moment diagrams for this case. Sketch the shear force and bending moment diagrams, and compare them to the diagrams in Fig. 4.21, Example 4.11.

4.5-35 The sketch in Fig. P4.5-35 shows a possible set of muscle forces acting on the femur of a man who is running upstairs. Find the unknown reactions R_A and R_D in terms of P, and show how the transverse force varies along the femoral

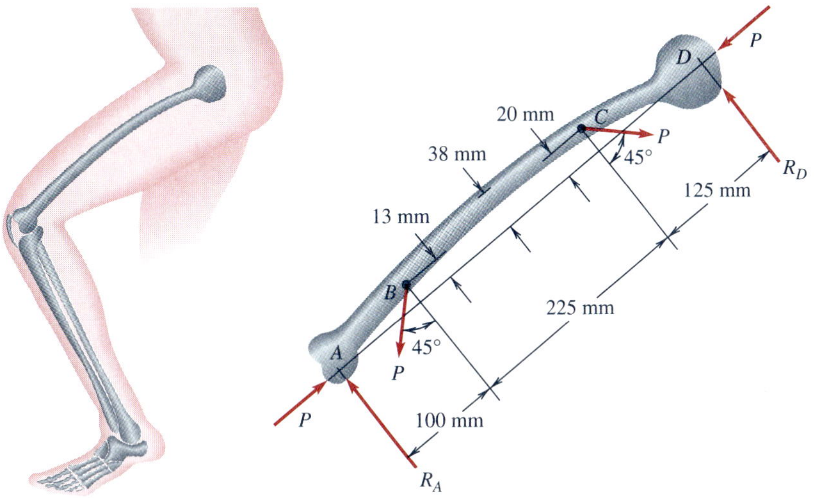

Fig. P4.5-35

shaft. Show how the bending moment varies along the shaft, and comment on the compensating effect of the muscles attached at B and C in terms of reducing the bending moments in the shaft.

4.5-36 Cilia are motile hairlike appendages on the free surfaces of certain cells, as shown in Fig. P4.5-36. They are present in the trachea and in the reproductive tracts of humans as well as lower animals. Their motion can be considered as made up of an effective stroke with a pendular motion with constant angular velocity through an angle of approximately 140° and a return or recovery stroke, as shown in Fig. P4.5-36*a* and *b*.

(*a*) For the configuration in Fig. P4.5-36*c* in which the cilium is arrested by a force $F = 2.2 \times 10^{-9}$ N, calculate the bending moment at the cell boundary.

(*b*) If a cilium moving in a viscous fluid rotates through its effective stroke, estimate the driving bending moment at the cell boundary. The viscous force on an element of length of the cilium may be taken to be proportional to the length of the element, to the angular velocity, to the viscosity, and to a function which depends upon the position along the cilium.

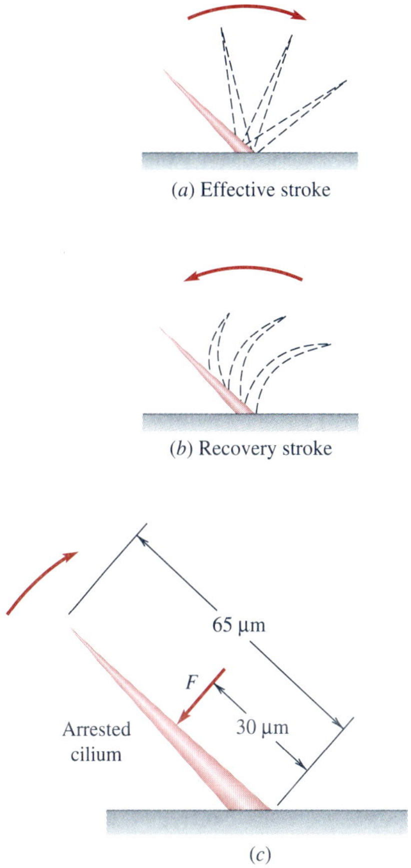

Fig. P4.5-36

4.5-37 Consider the loading function $f_{-1}(x, u)$ defined by

$$f_{-1}(x, u) = \begin{cases} 0 & \text{for } -\infty < x < 0 \\ \dfrac{x}{u^2} & \text{for } 0 < x < u \\ \dfrac{2}{u} - \dfrac{x}{u^2} & \text{for } u < x < 2u \\ 0 & \text{for } 2u < x < \infty \end{cases}$$

as shown in Fig. P4.5-37. Show that in the limit as $u \to 0$, $f_{-1}(x, u)$ approaches

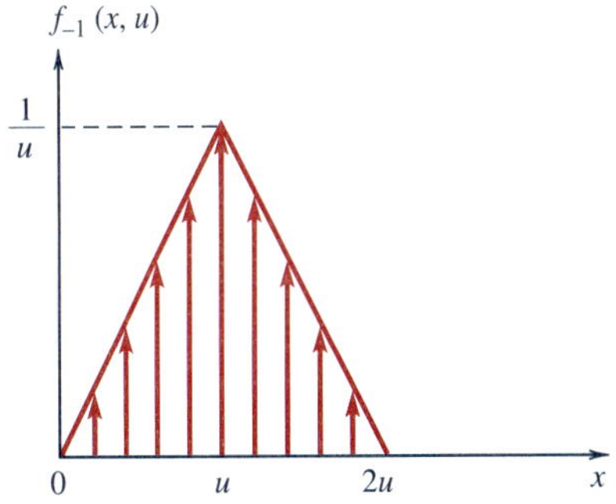

Fig. P4.5-37

a unit "concentrated force" $\langle x\rangle_{-1}$ located at the point $x = 0$. The limit of the function $f_{-1}(x - a, u)$ is denoted by $\langle x - a\rangle_{-1}$.

Consider a cantilever beam, free at $x = 0$ and built in at $x = L$, subjected to the loading $f_{-1}(x, u)$. Sketch the shear force and bending moment diagrams, and discuss their limiting forms when $u \to 0$.

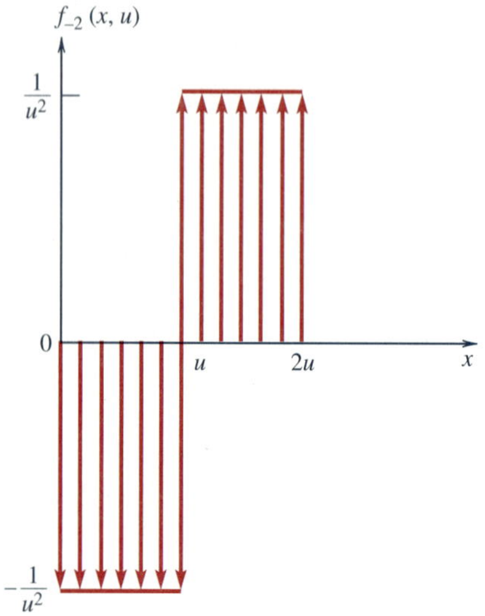

Fig. P4.5-38

4.5-38 Consider the loading function $f_{-2}(x, u)$ defined by

$$f_{-2}(x, u) = \begin{cases} 0 & \text{for } -\infty < x < 0 \\ \dfrac{-1}{u^2} & \text{for } 0 < x < u \\ \dfrac{1}{u^2} & \text{for } u < x < 2u \\ 0 & \text{for } 2u < x < \infty \end{cases}$$

as shown in Fig. P4.5-38. Show that in the limit as $u \to 0$, $f_{-2}(x, u)$ approaches a unit "concentrated moment" $\langle x\rangle_{-2}$ located at the point $x = 0$. The limit of the function $f_{-2}(x - a, u)$ is denoted by $\langle x - a\rangle_{-2}$.

Consider a cantilever beam, free at $x = 0$ and built in at $x = L$, subjected to the loading $f_{-2}(x, u)$. Sketch the shear force and bending moment diagrams, and discuss their limiting forms when $u \to 0$.

4.5-39 Show that for any fixed u, Probs. 4.5-37 and 4.5-38,

$$\int_0^x f_{-2}(x, u)\, dx = -f_{-1}(x, u)$$

and, therefore, assuming that the limit as $u \to 0$ and the integration operation can be interchanged, we have

$$\int_0^x \langle x - a\rangle_{-2}\, dx = -\langle x - a\rangle_{-1} \qquad a \geq 0$$

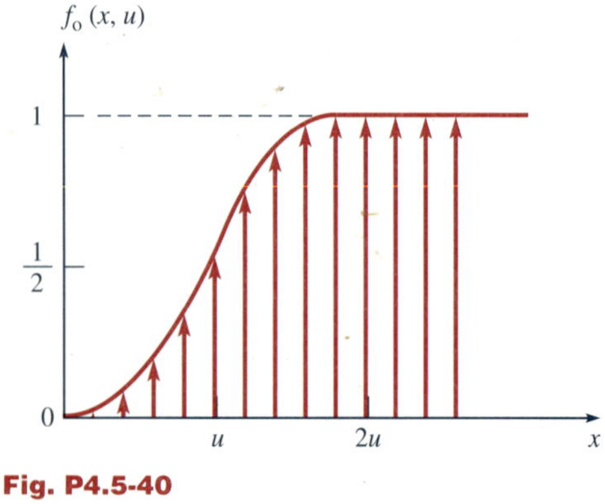

Fig. P4.5-40

4.5-40 Consider the loading function $f_0(x, u)$ defined by

$$f_0(x, u) = \begin{cases} 0 & -\infty < x < 0 \\ \dfrac{x^2}{2u^2} & 0 < x < u \\ \dfrac{1}{2} + \dfrac{2}{u}(x - u) - \dfrac{1}{2u^2}(x^2 - u^2) & u < x < 2u \\ 1 & 2u < x < \infty \end{cases}$$

as shown in Fig. P4.5-40. Show that in the limit as $u \to 0$, $f_0(x, u)$ approaches a unit step distributed loading $\langle x\rangle^0$ which has the value zero for $x < 0$ and takes on a unit value for $x > 0$. The limit of the function $f_0(x - a, u)$ is denoted by $\langle x - a\rangle^0$.

Consider a cantilever beam, free at $x = 0$ and built in at $x = L$, subjected to the loading $f_0(x, u)$. Sketch the shear force and bending moment diagrams, and discuss their limiting forms when $u \to 0$.

4.5-41 Show that for any fixed u, Probs. 4.5-37 and 4.5-40,

$$\int_0^x f_{-1}(x, u)\, dx = f_0(x, u)$$

and, therefore, assuming that the limit as $u \to 0$ and the integration operation can be interchanged, we have

$$\int_0^x \langle x - a \rangle_{-1}\, dx = \langle x - a \rangle^0 \qquad a \geq 0$$

4.7-1 Consider the beams shown in Fig. P4.7-1. Use the computer program V + M DIAGRAMS OF BEAMS to obtain the shear force and bending moment diagrams for the beams. Clearly indicate critical values and locations, and neglect the weight of the beams.

(*a*) Fig. P4.7-1*a*
(*b*) Fig. P4.7-1*b*
(*c*) Fig. P4.7-1*c*
(*d*) Fig. P4.7-1*d*
(*e*) Fig. P4.7-1*e*
(*f*) Fig. P4.7-1*f*
(*g*) Fig. P4.7-1*g*
(*h*) Fig. P4.7-1*h*
(*i*) Fig. P4.7-1*i*
(*j*) Fig. P4.7-1*j*

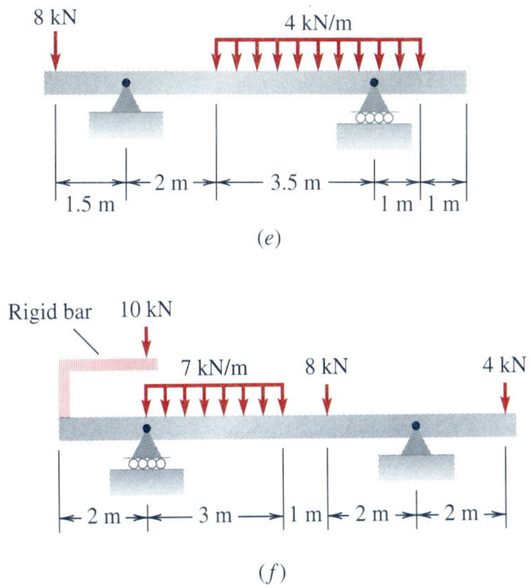

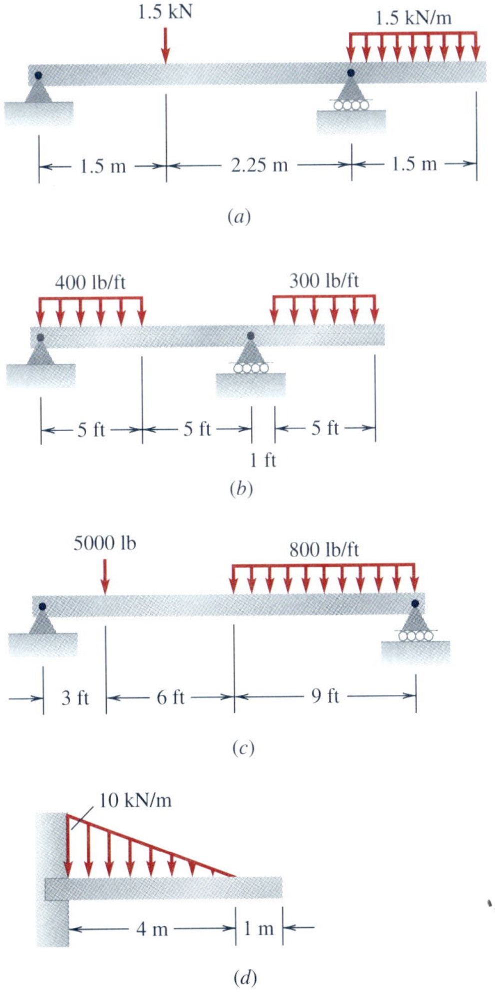

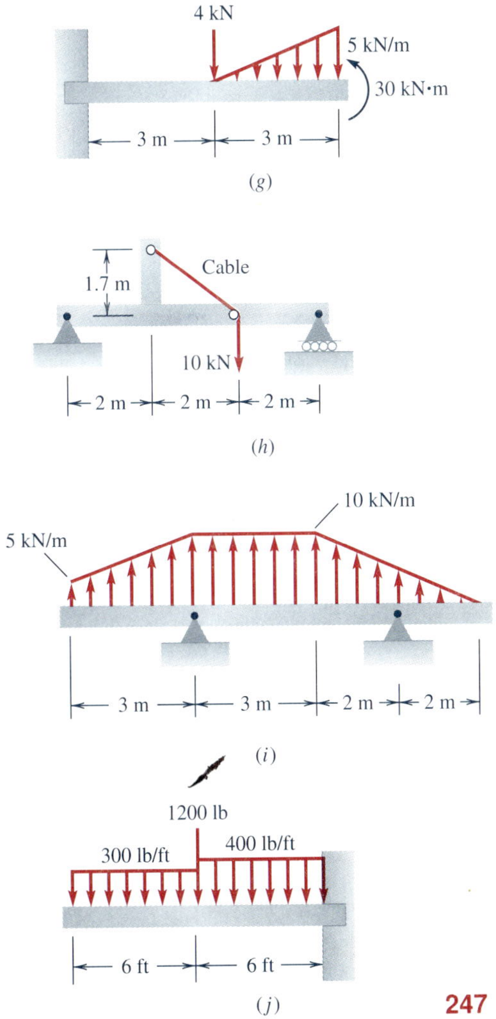

Fig. P4.7-1

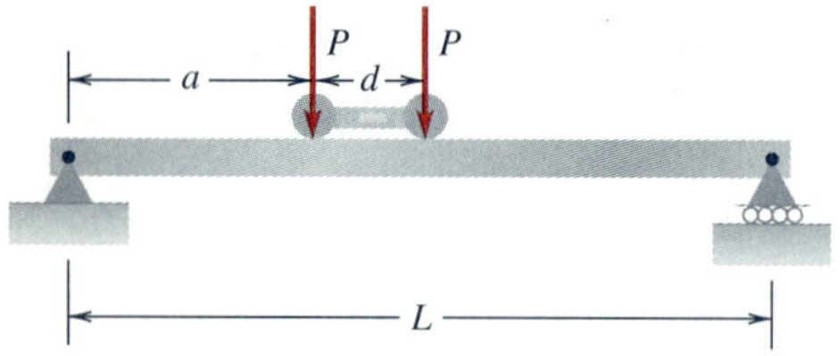

Fig. P4.7-2

4.7-2 For the beam shown in Fig. P4.7-2, use the computer program V + M DIAGRAMS OF BEAMS to obtain the shear force and bending moment diagrams for the beam with $L = 30$ ft, $P = 10$ kips, and $d = 5$ ft. Take a equal to 5, 10, and 15 ft, and use the computer output to sketch all three V and M diagrams on the same graph.

4.7-3 For the beam shown in Fig. P4.7-3, use the computer program V + M DIAGRAMS OF BEAMS to obtain the shear force and bending moment diagrams with $P_3 = 10$ kN, $P_1 = P_2 = 55$ kN when $a = 0, 3, 6, 9, 11, 14$ m. Draw the resulting diagrams on the same graph.

Fig. P4.7-3

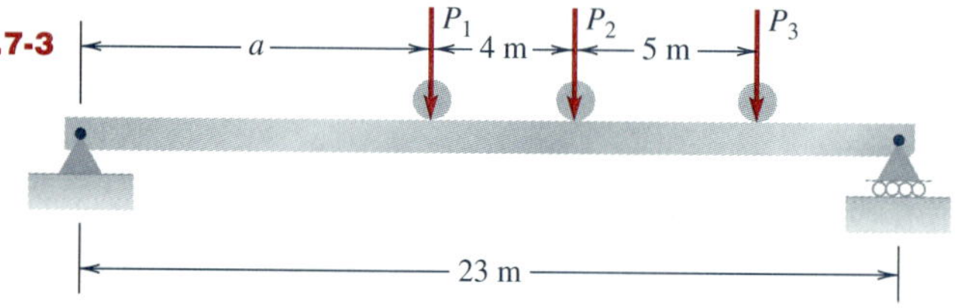

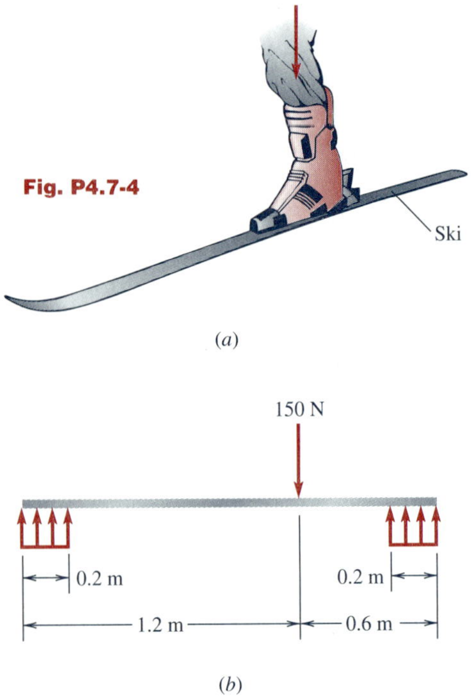

Fig. P4.7-4

4.7-4 A skier is standing on her skis so that half of her weight of 300 N is supported by each ski, as shown in Fig. P4.7-4*a*. Only the toe and heel regions of the ski are in contact with the snow, and the reaction in these regions from the snow is assumed to be a constant loading. This situation can be modeled by considering each ski as a beam loaded as shown in Fig. P4.7-4*b*. Sketch the shear force and bending moment diagrams for the model beam, showing all critical values.

4.7-5 Sketch the shear force and bending moment diagrams for the beam shown in Fig. P4.7-5. Clearly indicate all important values, and neglect the weight of the beam.

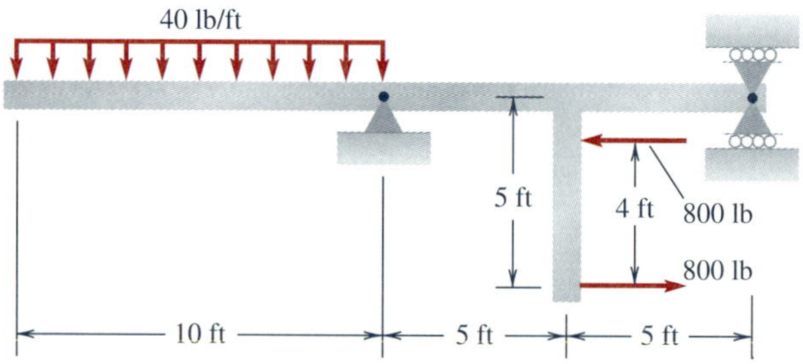

Fig. P4.7-5

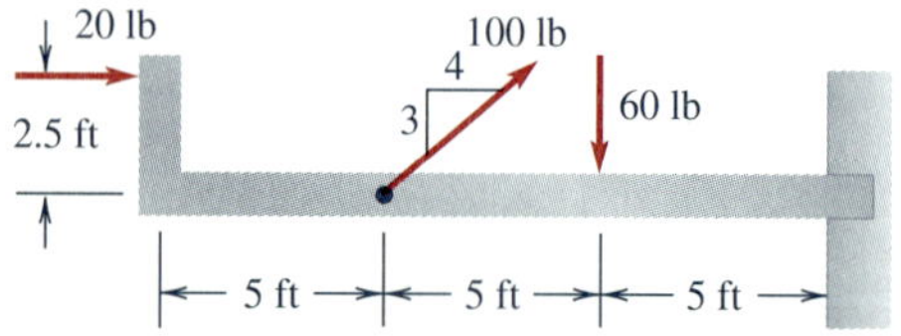

Fig. P4.7-6

4.7-6 Sketch the shear force and bending moment diagrams for the beam shown in Fig. P4.7-6. Clearly indicate all important values, and neglect the weight of the beam.

4.7-7 Sketch the shear force and bending moment diagrams for the beams indicated in Fig. P4.7-7. Clearly indicate all important values, and neglect the weight of the beam.
(*a*) Fig. P4.7-7*a*
(*b*) Fig. P4.7-7*b*
(*c*) Fig. P4.7-7*c*

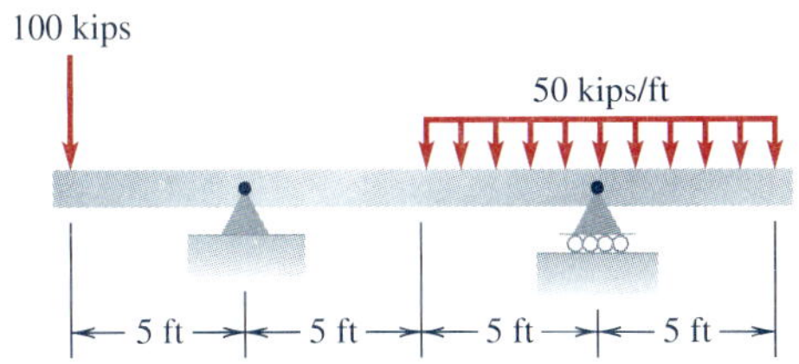

(*a*)

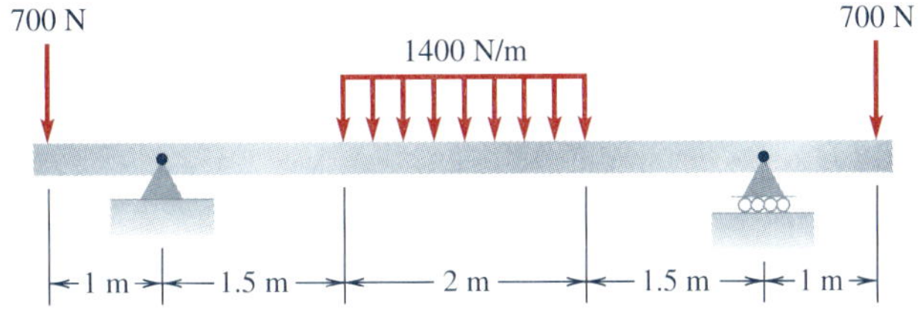

(*b*)

Fig. P4.7-7

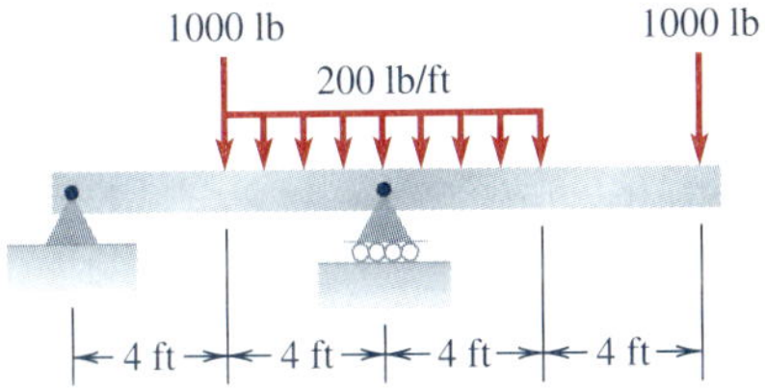

(*c*)

4.7-8 Use the computer program V + M DIAGRAMS OF BEAMS to obtain the shear force and bending moment diagrams for the beams shown in the figure indicated. Neglect the weight of the beams, and clearly indicate all important points on the diagrams.
(*a*) Fig. P4.2-4*a*
(*b*) Fig. P4.2-4*b*
(*c*) Fig. P4.2-4*c*
(*d*) Fig. P4.2-4*d*
(*e*) Fig. P4.2-4*e*
(*f*) Fig. P4.2-4*f*
(*g*) Fig. P4.2-4*g*
(*h*) Fig. P4.2-4*h*
(*i*) Fig. P4.3-5*a*
(*j*) Fig. P4.3-5*b*
(*k*) Fig. P4.3-5*c*
(*l*) Fig. P4.3-5*d*
(*m*) Fig. P4.3-5*e*
(*n*) Fig. P4.3-5*f*
(*o*) Fig. P4.3-5*g*
(*p*) Fig. P4.3-5*h*
(*q*) Fig. P4.3-5*i*
(*r*) Fig. P4.4-10*c*
(*s*) Fig. P4.4-10*d*
(*t*) Fig. P4.4-10*f*
(*u*) Fig. P4.4-10*k*
(*v*) Fig. P4.4-10*m*
(*w*) Fig. P4.4-15
(*x*) Fig. P4.5-1*d*
(*y*) Fig. P4.5-1*e*
(*z*) Fig. P4.5-1*h*

Fig. P4.8-1

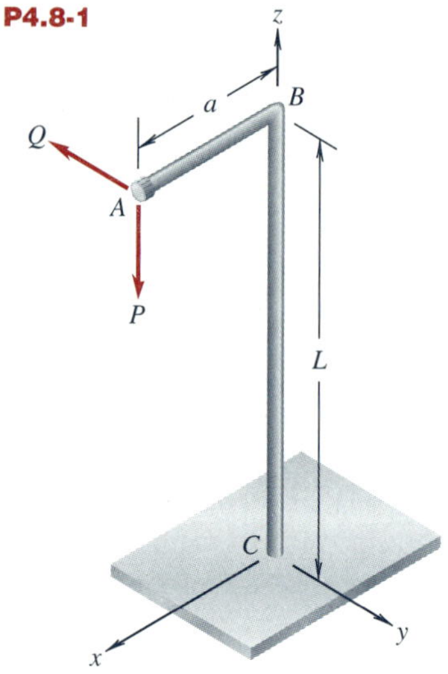

4.8-1 A lamppost consists of a vertical member BC and a horizontal arm AB attached at B, as shown in Fig. P4.8-1.

(*a*) For a vertical load P applied at A, sketch the force and moment diagrams for ABC.

(*b*) Repeat part (*a*) except add the horizontal load Q applied at A.

4.8-2 A crankshaft AF is supported in bearings at A and F, and the connecting rod force of 2000 lb is in equilibrium with the torque M_0, as shown in Fig. P4.8-2. Sketch the shear force, bending moment, and twisting moment diagrams for segments AB and EF.

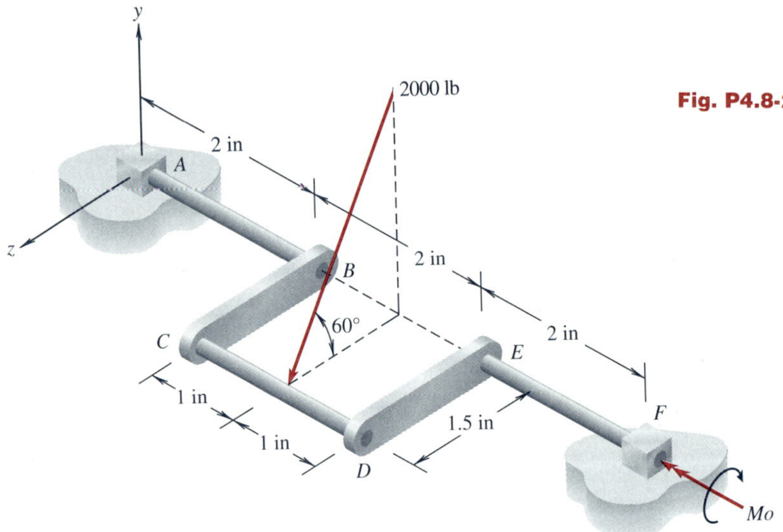

Fig. P4.8-2

4.8-3 A semicircular segment of piping OA is built in at end O and subjected to a force P acting perpendicular to the plane of OA as shown in Fig. P4.8-3. Sketch the shear force, bending moment, and twisting moment diagrams for the piping segment. Neglect the weight of the pipe.

Fig. P4.8-3

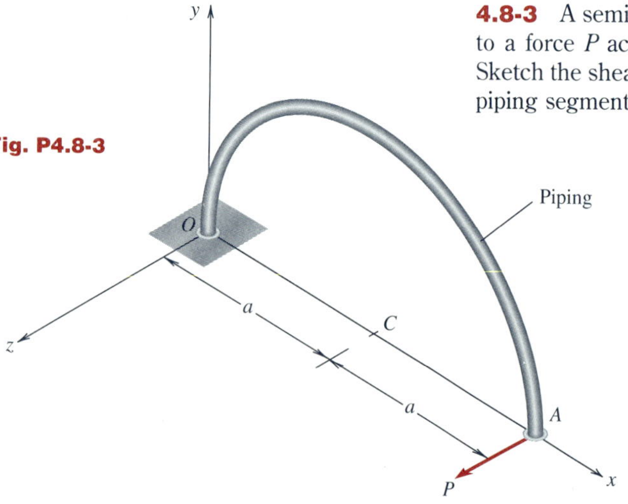

4.8-4 If the weight per unit length of the pipe shown in Fig. P4.8-3 is q and the pipe is oriented so that the y axis in the figure points upward in the vertical direction, sketch the shear force, bending moment, and twisting moment diagrams for the pipe under the load q.

4.8-5 A curtain is supported by rings attached to a horizontal bar OA, and the bar is supported by two wires AB and DC and a ball-and-socket joint at O, as shown in Fig. P4.8-5. If the curtain weighs 1 kN, sketch the shear force and bending moment diagrams for the horizontal member OA, and determine the magnitude of the maximum bending moment.

4.8-6 Gear B drives pulley C, and both the pulley and the gear are mounted on shaft AD which is supported by frictionless bearings at A and D, as shown in Fig. P4.8-6. Find the gear tooth force F (in the negative z direction), and sketch the shear force, bending moment, and twisting moment diagrams for the shaft AD.

4.8-7 A winch is used to raise a 200-lb weight as shown in Fig. P4.8-7. Find the force F required to maintain equilibrium. Sketch the shear force, bending moment, and twisting moment diagrams for the shaft $EOAB$. Assume that the bearings at O and A do not carry any axial forces.

Fig. P4.8-5

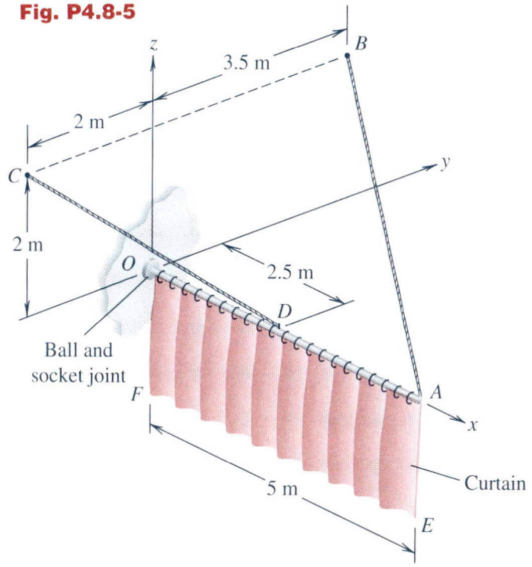

Fig. P4.8-6

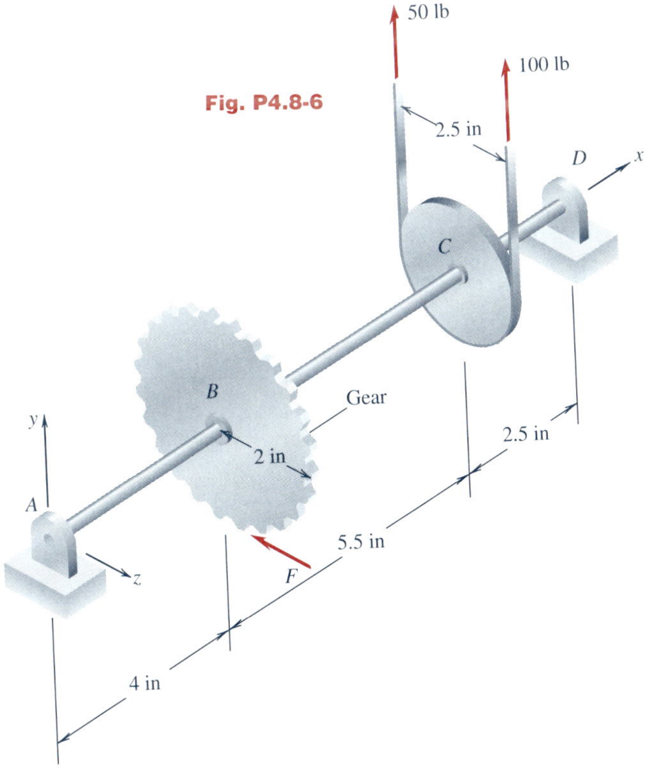

Fig. P4.8-7

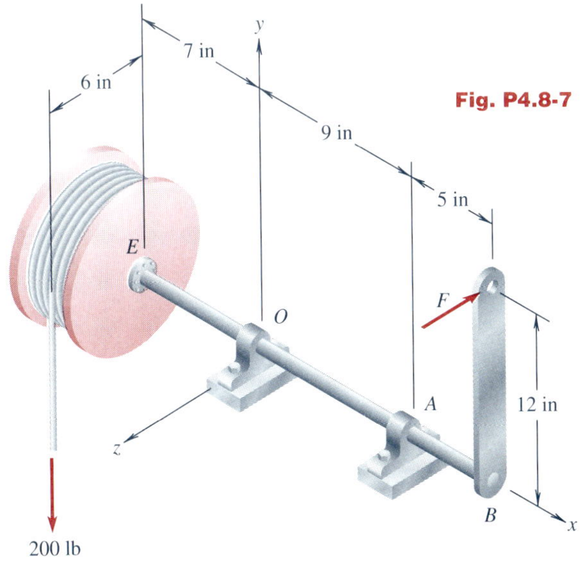

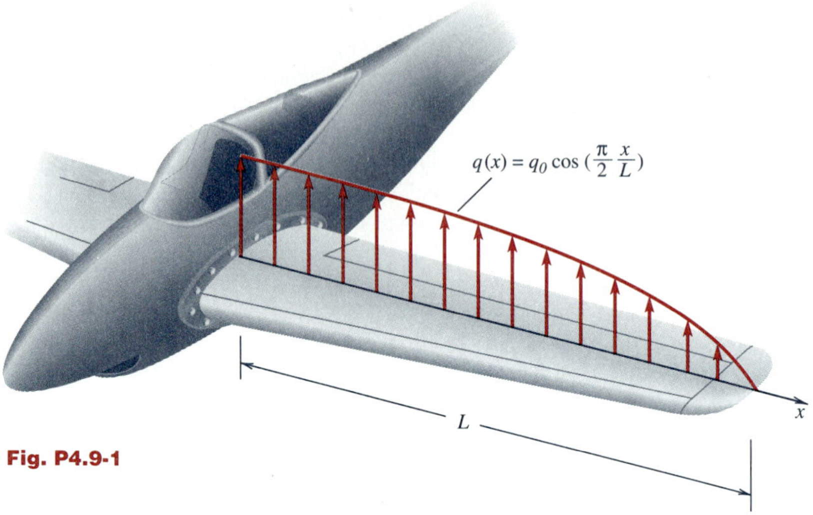

Fig. P4.9-1

4.9-1 A cantilever beam is subjected to a distributed loading which varies cosinusoidally as shown in Fig. P4.9-1.

(*a*) Find expressions for the shear force $V(x)$ and bending moment $M(x)$ by integrating a suitable loading function $q(x)$. In this problem special care must be taken to make use of the constants of integration to satisfy the conditions of force and moment equilibrium.

(*b*) Sketch the V and M diagrams, showing values at all important points.

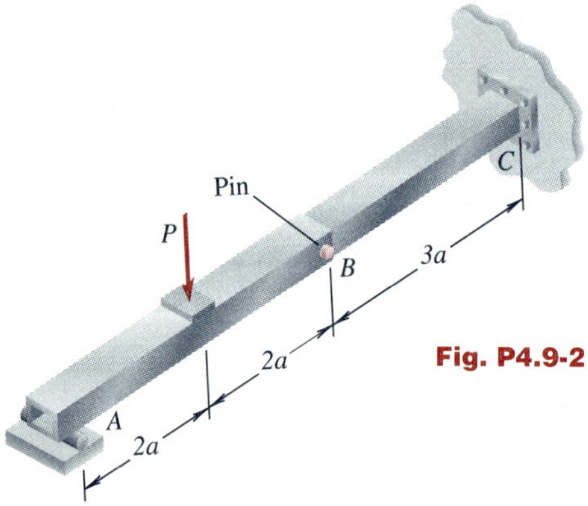

Fig. P4.9-2

4.9-2 A beam consists of a segment AB simply supported at A and hinged at B to a second segment BC, which is cantilevered from a wall at C, as shown in Fig. P4.9-2. If a load P is applied at the midpoint of AB, find the reactions and sketch separate shear force and bending moment diagrams for each segment. The hinge at B cannot carry a moment.

4.9-3 A simply supported beam is subjected to a distributed loading with an amplitude that varies sinusoidally, as shown in Fig. P4.9-3. Find expressions for the shear force $V(x)$ and bending moment $M(x)$ by integrating a suitable loading function $q(x)$. In this problem special care must be taken to make use of the constants of integration to satisfy the conditions of force and moment equilibrium for the beam.

Fig. P4.9-3

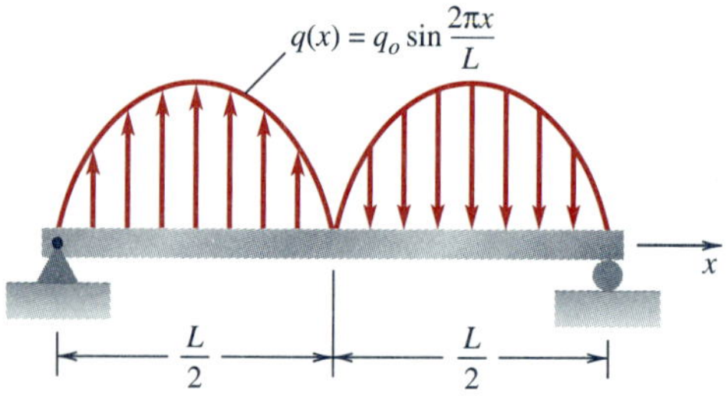

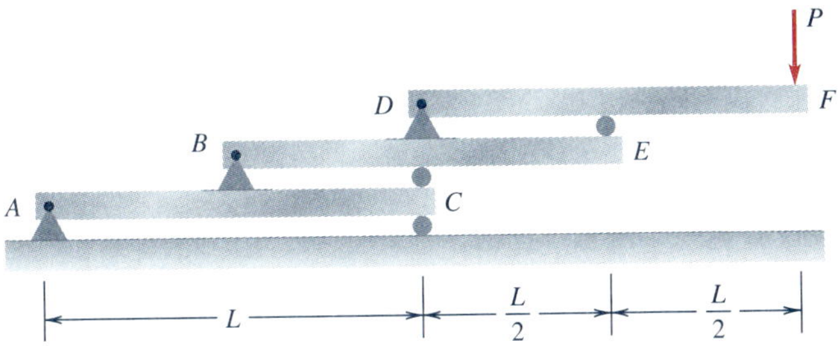

Fig. P4.9-4

4.9-4 A special diving board is made up of an array of three identical boards, as shown in Fig. P4.9-4. If a load P is applied at F, sketch separate shear force and bending moment diagrams for each of beams AC, BE, and DF. The boards are of length L.

4.9-5 A uniform concrete pole AC is hoisted into place by cables BD and CD, as shown in Fig. P4.9-5. If the pole is homogeneous and weighs 21 kN, sketch the shear force, bending moment, and axial force diagrams for the pole for the configuration shown.

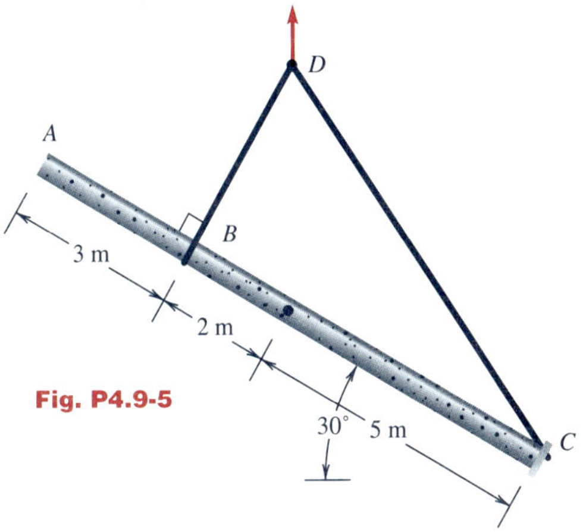

Fig. P4.9-5

4.9-6 Two frames ABC and CEF are hinged at C, as shown in Fig. P4.9-6. If a load P is applied at the midpoint D of CE, find the values of the shear force, bending moment, and axial force at the midpoints G, H, and I of the segments of the frame. Show the results on appropriate free-body diagrams of the segments. Take $AB = EF = 2L$ and $BC = CE = L$. The pin at C cannot carry a moment.

4.9-7 Beams AB and BC are pinned together at B and pinned to supports at A and C, as shown in Fig. P4.9-7. If beam AB carries a constant distributed loading of 30 kN/m, sketch the shear force and bending moment diagrams for AB.

Fig. P4.9-6

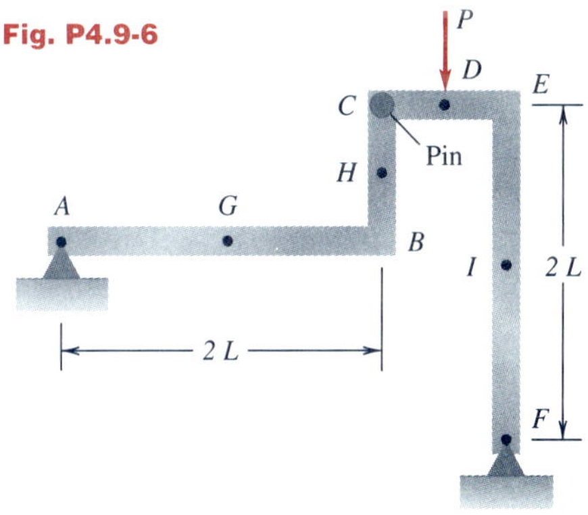

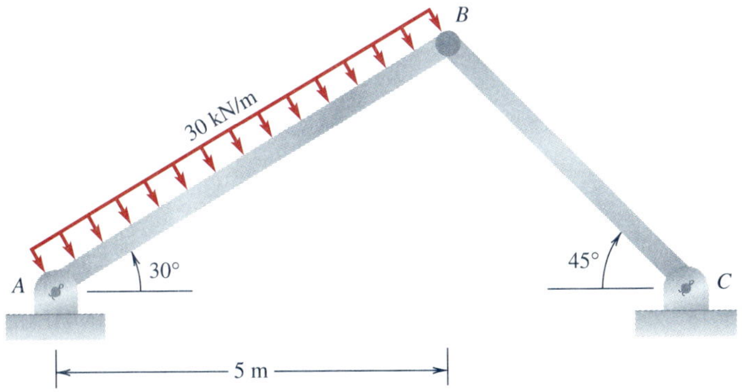

Fig. P4.9-7

4.9-8 For the bent beam shown in Fig. P4.9-8, $P = 1000$ lb. Sketch the shear force and bending moment diagrams for the beam segment *ABCD*.

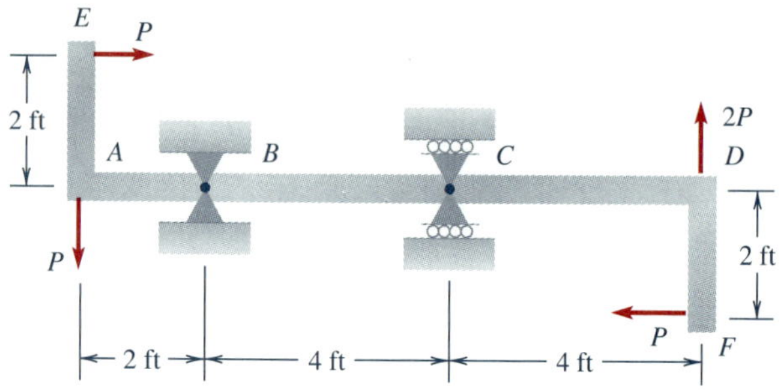

Fig. P4.9-8

CHAPTER 5

Stresses Due to Bending

5.1 Introduction

In Chap. 2 we found the normal stress distribution σ that has a resultant equal to an axial force F acting on a bar (see Fig. 5.1*a*). In Chap. 3 we investigated the shear stress distribution τ that has a resultant equal to a twisting moment T acting on a circular shaft (see Fig. 5.1*b*). In the case of the axial force acting on a bar, the normal stress distribution was constant across the section, while in the case of the twisting moment acting on a circular shaft, the shear stress varied linearly with the radial distance from the axis.

In Chap. 4 equilibrium arguments were used to determine the shear force V and the bending moment M at a section x in a beam under transverse loading $q(x)$ (see Fig. 5.1*c*). In this chapter we wish to find the distribution of stresses acting at a section of a beam that has a resultant force equal to the shear force V and a resultant moment equal to the bending moment M.

Our approach will follow the general method we used for finding the shear stress distribution in a circular shaft in Chap. 3. We first investigate the geometry of the deformation, using symmetry arguments to obtain the strain distribution in a beam. The stress-strain relations are then used to obtain stresses from the strains. Finally, the stress distribution must satisfy overall equilibrium. The equation we will derive for the normal stress in a beam due to bending has wide-ranging application in analysis and design.

However, before we discuss beam deformation, let us review the important concepts of curvature and radius of curvature of a plane curve.

Consider a plane curve $ACDB$ in the xy plane as shown in Fig. 5.2; we can think of this curve as specified by $y = y(x)$. The slope angle or angle of the line tangent to the curve at point C measured from a line parallel to the x axis is θ; the slope angle at a nearby point D is $\theta + \Delta\theta$.

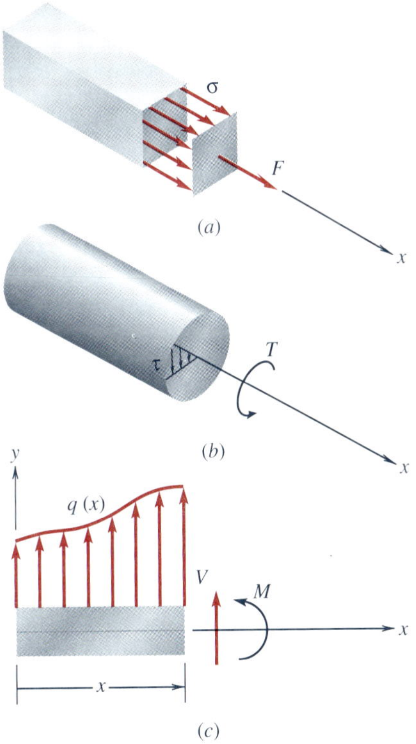

Figure 5.1 (*a*) Uniform normal stress σ statically equivalent to an axial force F. (*b*) Linear shear stress τ statically equivalent to a twisting moment T. (*c*) Shear force V and bending moment M acting on a section of a beam.

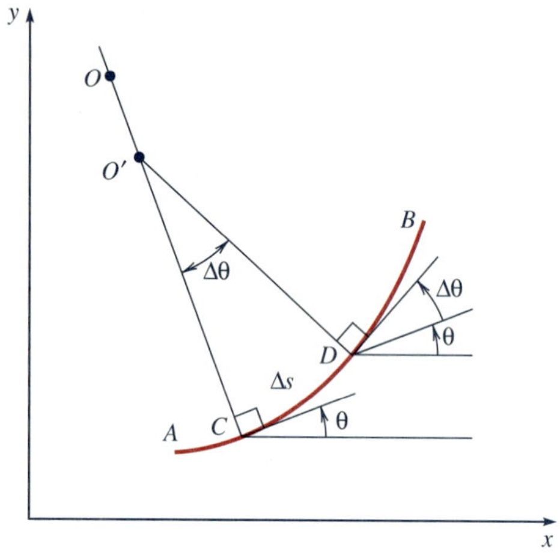

Figure 5.2 Construction for radius of curvature of a plane curve at point C.

We wish to investigate the rate of change in the value of the slope angle as we move along the curve from point C to point D. A line normal to the curve at point C and a line normal to the curve at point D intersect at O', and the angle $CO'D$ is $\Delta\theta$. The arc length distance between points C and D is written as Δs; the Δs symbol is used to signify that C and D are nearby points on the curve. When $\Delta\theta$ is small, the arc length distance Δs along the curve between points C and D is approximately the distance $O'D$ times $\Delta\theta$.

The *curvature* of a plane curve at a point on the curve is defined as the rate of change at that point of the slope angle with respect to the distance along the curve. The curvature, which we usually write as the Greek letter κ (kappa), at point C follows from the definition as

$$\kappa = \frac{d\theta}{ds} = \lim_{\Delta s \to 0} \frac{\Delta\theta}{\Delta s} = \lim_{\Delta s \to 0} \frac{1}{O'D} = \frac{1}{OC} \tag{5.1}$$

Curvature has the dimension of 1/length, and so we define the *radius of curvature* at point C as the reciprocal of the curvature; $\rho = 1/\kappa$. The radius of curvature ρ (Greek letter rho) has the dimension of length.

We note from Fig. 5.2 that the normals to the curve at points C and D intersect at O'. In the limit as D approaches C, the point O' approaches O, and the distance OC equals the radius of curvature. A circle can be constructed which is centered at point O in Fig. 5.2 with a radius OC. This so-called circle of curvature is tangent to the curve at C and has a radius equal to the value of the radius of curvature of the curve at C. If the plane curve is itself a circle of radius R, then the radius of curvature is constant at every point on the circle, $\rho = R$, and the curvature $\kappa = 1/R$. In general, for a plane curve the curvature will be a function of position along the curve. Later in Chap. 6 we will review how to find the curvature κ from the equation $y = y(x)$ for the curve.

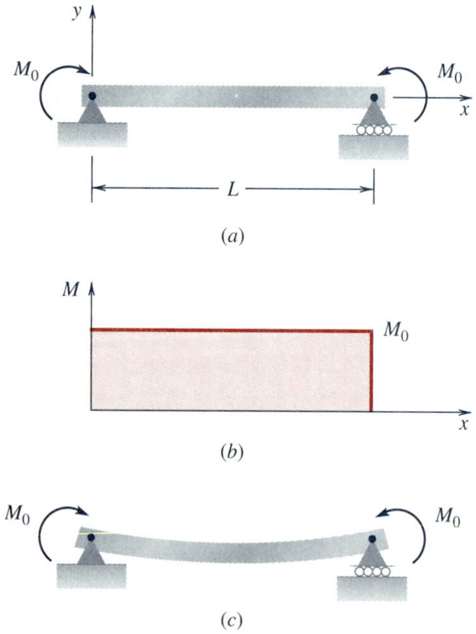

Figure 5.3 (*a*) A simply supported beam under applied bending moments M_0 at the ends. (*b*) The bending moment diagram. (*c*) The approximate deflected shape.

5.2 Geometry of Deformation

We start our discussion of the stress distribution in beams by considering a straight beam whose cross section is symmetric about the plane of transverse loading of the beam. The material properties of the beam are assumed constant along the length of the beam and are symmetric about the plane of loading. We consider first the case of a beam with a constant bending moment along its length, which we will refer to as a beam in *pure bending*, e.g., as shown in Fig. 5.3. Beams with a constant bending moment over a portion of their length are used often for establishing the failure strength of materials in a test device called a *four-point bending test rig*, shown schematically in Fig. 5.4. In Fig. 5.4*a*, a pair of loads P is applied to the ends of the beam, resulting in a state of pure bending over a central portion of length L as shown in Fig. 5.4*b*; the approximate deflected shape is shown in Fig. 5.4*c*.

Consider the geometry of the deformation of a beam under pure bending in its plane of symmetry, as shown in Fig. 5.5, where we exaggerate the deformation; in most engineering applications the deflections of beams are small. Element $ABED$ and element $BCFE$ are identical and are subjected to the same constant value of bending moment on each of their faces. Because of the symmetry of the elements, we conclude that each face of an element must deform as a plane; i.e., a plane surface such as CF perpendicular to the longitudinal axis of the beam before bending will remain plane and perpendicular to the axis of the beam after bending. Therefore, initially parallel planes AD, BE, and CF will have a common point of intersection O after deformation, as shown in Fig. 5.5*b*. Line elements initially straight and perpendicular to planes AD, BE, and CF will bend into arcs of circles centered at point O. In particular, arcs $A_1B_1C_1$, and $D_1E_1F_1$ are arcs of concentric circles centered at O. We also assume that planes AD, BE, and CF do not deform in the y direction so that line elements AD and A_1D_1 are the same length.

Based on these observations, we can calculate the normal strain due to bending of line elements *in the plane of symmetry* lying parallel to the axis of the beam. First, we see from the nature of the deformation in Fig. 5.5*b* that line elements on the top of the beam will shorten while line elements on the bottom of the beam will elongate. It follows, therefore, that there is one line element in the plane of symmetry that has not changed its length. We call this line element the *neutral axis* of the beam, and we set up our coordinate system in the undeformed beam so that the x axis coincides with or is along the direction of the neutral axis.

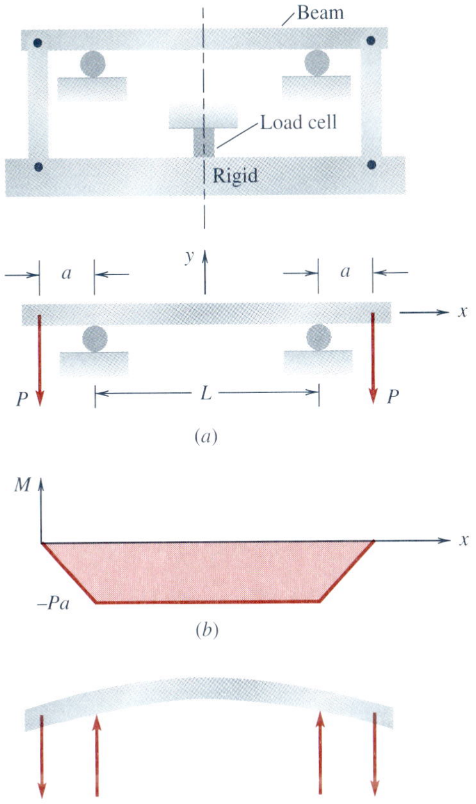

Figure 5.4 A four-point bending test of an elastic beam.

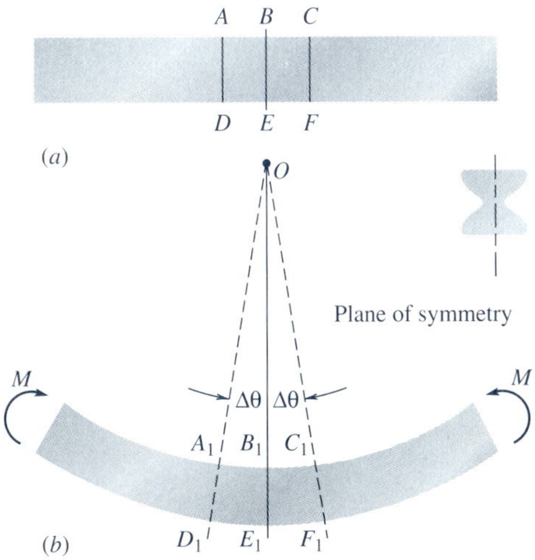

Figure 5.5 Deformation of a symmetric beam subjected to pure bending in its plane of symmetry.

The exact location of the x axis in the plane of symmetry of the cross section has not been specified yet. The plane perpendicular to the plane of symmetry and passing through the neutral axis is called the ***neutral surface***.

Figure 5.6*a* again shows a segment of the beam under pure bending and a line element PQ that is located a positive distance y above the line element RS which lies along the neutral axis. Also shown are the set of coordinate axes and the distance Δs between the sections containing PR and QS. After deformation caused by a constant bending moment, the line element RS on the neutral axis deforms into R_1S_1 on an arc with radius of curvature ρ, as shown in Fig. 5.6*b*. The line element R_1S_1 because it is on the neutral axis does not change its length; thus

$$R_1S_1 = \rho\,\Delta\theta = RS = \Delta s \tag{5.2}$$

where $\Delta\theta$ is the angle between the normals to the neutral surface passing

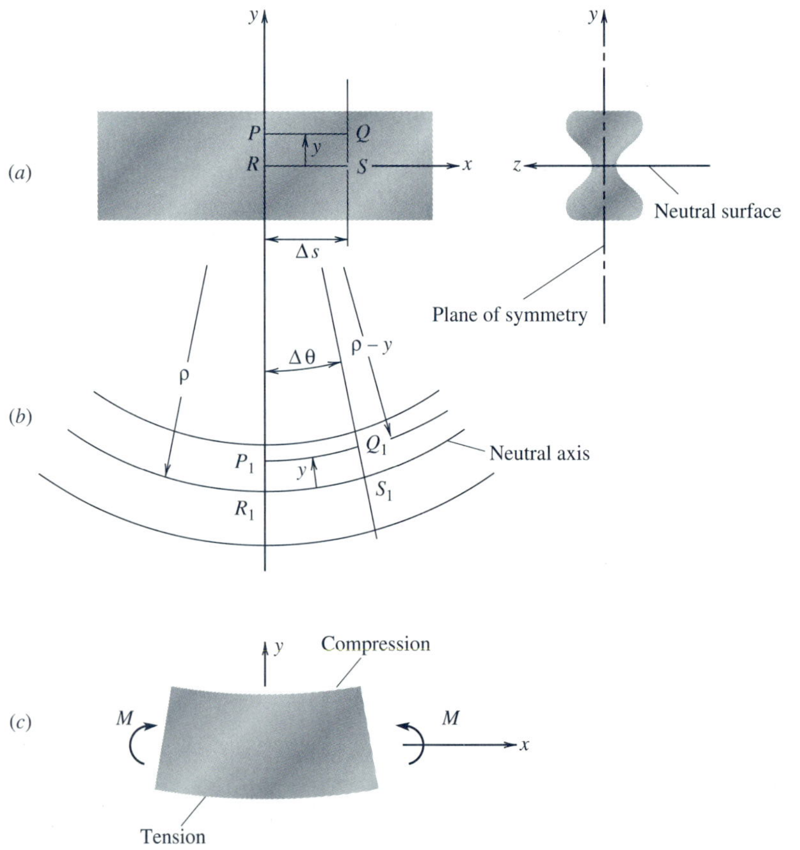

Figure 5.6 Geometry of deformation.

through points P_1R_1 and points Q_1S_1 and where Δs is the original undeformed length of RS.

The line element PQ, however, does experience a change in length. The length of P_1Q_1 in the deformed state is given by

$$P_1Q_1 = (\rho - y)\,\Delta\theta \tag{5.3}$$

where we have assumed that no deformation takes place in line elements oriented along the y direction so that the distance y of PQ from RS remains constant during bending.

The axial strain ϵ_x of the line element PQ, that is, the change in length divided by the original length, is given by

$$\epsilon_x = \frac{P_1Q_1 - PQ}{PQ} \tag{5.4}$$

where the subscript on the strain ϵ_x is used to denote strain of a line element originally lying parallel to the x axis. Upon substituting the value for P_1Q_1 from Eq. (5.3), using Eq. (5.2) with $PQ = \Delta s$, we find

$$\epsilon_x = \frac{(\rho - y)\,\Delta\theta - \Delta s}{\Delta s} = -y\frac{\Delta\theta}{\Delta s} \tag{5.5}$$

From Eqs. (5.1) and (5.5), upon shrinking Δs to zero, we find

$$\epsilon_x = -y\frac{d\theta}{ds} = -\frac{y}{\rho} \tag{5.6}$$

where ρ is the radius of curvature of the neutral axis of the bent beam.

Equation (5.6) gives the result that the axial strain of a line element parallel to the neutral axis in the plane of symmetry of the cross section varies linearly with its distance y measured from the neutral axis. For a positive bending moment as shown in Fig. 5.6*c* and with y positive, the strain will be negative, indicating that line elements above the neutral axis have shortened, i.e., that the line elements have undergone compression. If y is negative, the strain will be positive, indicating that line elements below the neutral axis have elongated or undergone tension.

Our derivation of Eq. (5.6) holds for points lying in the plane of symmetry; we will assume, however, that the normal strain at all points across the section is given by Eq. (5.6). We note that the derivation of Eq. (5.6) follows from symmetry arguments on the nature of the deformation under pure bending and does not depend on the nature of the beam material. Problems 5.7-1 and 5.7-2 show how Eq. (5.6) can be used in the case of nonhomogeneous beams.

In Sec. 1.5 we noted that a normal longitudinal strain is accompanied by normal strains in the transverse directions. The strains in the transverse directions are equal to minus Poisson's ratio times the longitudinal

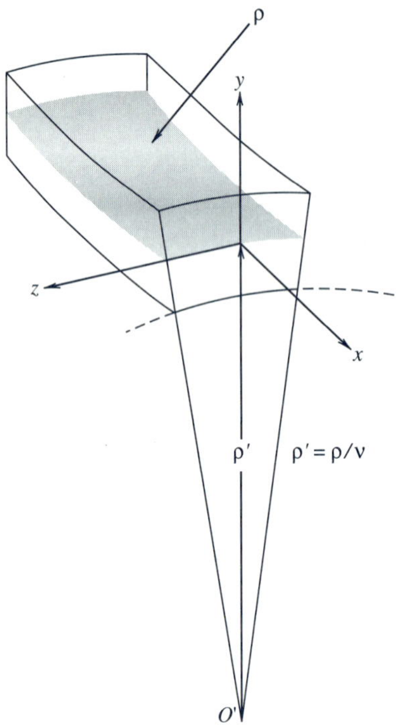

Figure 5.7 Transverse deformation of a beam of rectangular cross section.

strain. For example, if ϵ_x is the longitudinal strain, the strains in the y and z directions are

$$\epsilon_y = -\nu\epsilon_x \qquad \epsilon_z = -\nu\epsilon_x \tag{5.7}$$

where ν is Poisson's ratio.

For the case of a beam in pure bending, Eq. (5.6), the longitudinal strain is given by

$$\epsilon_x = -\frac{y}{\rho}$$

and therefore from Eqs. (5.7) we have the associated transverse strains

$$\epsilon_y = \frac{\nu y}{\rho} \qquad \epsilon_z = \frac{\nu y}{\rho} \tag{5.8}$$

Therefore, line elements in the plane of the cross section undergo a linear variation of normal strain. For positive values of y, the distance from the neutral axis to a given point, the normal strain ϵ_x will be negative and both of the strains ϵ_y and ϵ_z will be positive, i.e., tensile. As a result, the section will undergo transverse expansion at the top, and by a similar argument the section will undergo transverse contraction at the bottom of the section, as shown in Fig. 5.7 for a deformed beam that was originally of rectangular section before loading. Those line elements originally oriented parallel to the z axis will deform into arcs of circles, and the curvature of the z axis after loading will be opposite to that for the deformed x axis. This transverse curvature $1/\rho'$ is called the ***anticlastic curvature***. The geometry of a deformed cross section and the anticlastic curvature can be made quite visible if a large rubber eraser is bent between the thumb and forefinger.

EXAMPLE 5.1

A steel bar of rectangular section carries a bending moment M, as shown in Fig. 5.8. If the maximum tensile strain which can be sustained before yielding of the material occurs is known to be 1×10^{-3}, find the radius of curvature and the angle change between the ends of the deformed bar.

We treat the bar as a beam under pure bending. We set the origin of the coordinate system at the centroid of section A which (as we will see later) corresponds to the location of the neutral axis.

The maximum tensile strain will occur on the bottom surface of the beam at $y = -37.5$ mm; therefore, from Eq. (5.6),

$$\epsilon_{x_{\max}} = 1 \times 10^{-3} = -\frac{y_{\max}}{\rho} = -\frac{-37.5\text{ mm}}{\rho}$$

or

$$\rho = 37.5 \times 10^3\text{ mm} = 37.5\text{ m}$$

The angle change $\Delta\theta$ follows if we note from Eq. (5.1) that $d\theta/ds = 1/\rho =$ constant for the present case and therefore

$$\frac{\Delta\theta}{\Delta s} = \frac{1}{\rho}$$

and thus

$$\Delta\theta = \frac{1}{\rho}\Delta s = \left(\frac{1}{37.5\text{ m}}\right)(1.5\text{ m}) = 0.04\text{ rad} = 2.29^\circ$$

The angle change is small, and for most engineering structures we will find that deflections and angle changes will be small.

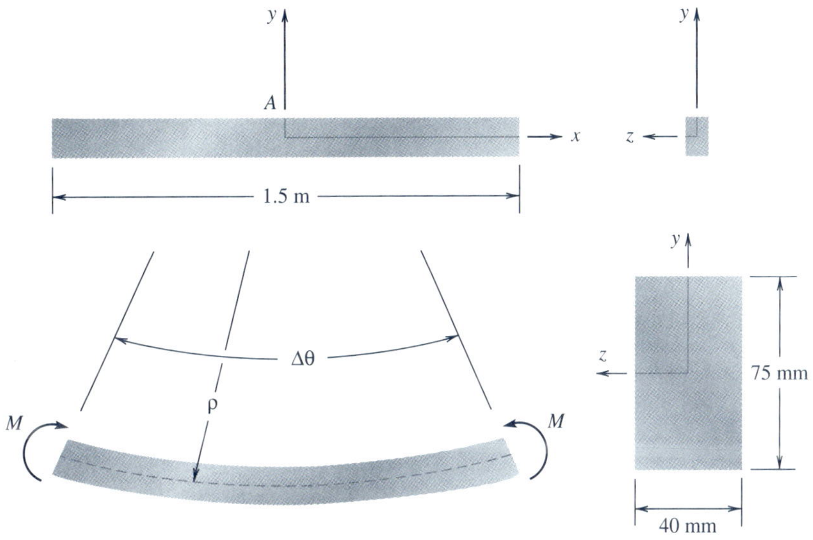

Figure 5.8 Example 5.1

5.3 Stress Distribution and Equilibrium Requirements

In the previous section, in which we discussed the geometry of the deformation in pure bending, we obtained the distribution of the normal strain ϵ_x over the section of the beam. To obtain the associated stress distribution, we need to use some kind of force-deformation relation. If we assume that the beam can be modeled as a collection of linear elastic material elements that obey Hooke's law, then using Eq. (5.6), we find that the stress at points on the cross section is given by

$$\sigma_x = E\epsilon_x = -E\frac{y}{\rho} \tag{5.9}$$

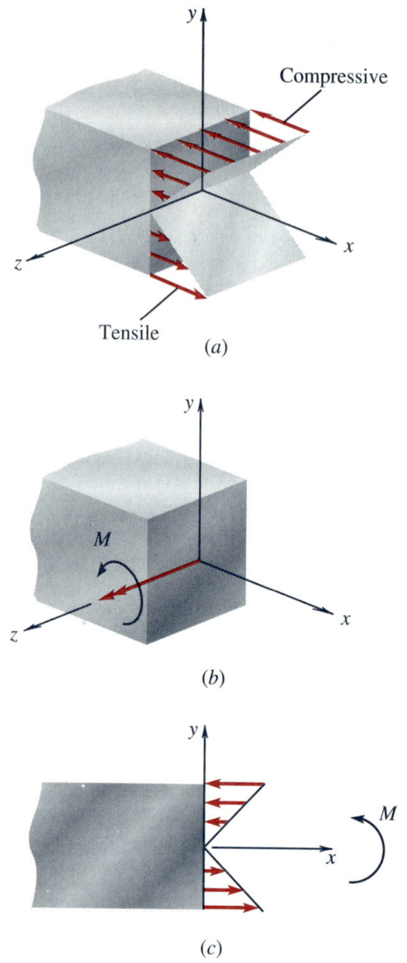

Figure 5.9 (*a*) Distribution of normal stress across the section. (*b*) Bending moment acting on the cross section. (*c*) Two-dimensional representation of stress distribution and bending moment.

where σ_x is the normal stress. The distribution of stress on a rectangular cross section is shown in Fig. 5.9*a*; above the neutral axis the stress is compressive while below the neutral axis the stress is tensile. This stress distribution must have a resultant equal to the bending moment M acting on the section, as shown in Fig. 5.9*b* and *c*. That is, the stress distribution must give rise to a zero axial force resultant and to a bending moment resultant M. Figure 5.10 shows the force $\sigma_x\, dA$ acting on an element of area dA of the section.

The requirement that the stress distribution be statically equivalent to the bending moment M is given by the equations

$$\sum F_x = \int_A \sigma_x\, dA = 0$$

$$\sum M_y = \int_A z\,\sigma_x\, dA = 0 \qquad (5.10)$$

$$\sum M_z = -\int_A y\,\sigma_x\, dA = M$$

where $\sigma_x\, dA$ is an element of force acting on the infinitesimal element of area dA and where the integrals are to be taken over the total area A of the cross section (Fig. 5.10). The minus sign in the moment equation about the z axis in Eqs. (5.10) arises from our convention that in general a normal stress on a positive face is positive, as shown in Fig. 5.10.

Our goal now is to relate the bending moment M to the deformation of the beam as given by the radius of curvature ρ. Upon substituting Eq. (5.9) into the first of Eqs. (5.10), we find

$$\sum F_x = \int_A \sigma_x\, dA = -\int_A \frac{Ey}{\rho}\, dA = -\frac{E}{\rho}\int_A y\, dA = 0 \qquad (5.11)$$

If the material properties of the beam are uniform across the section, then Eq. (5.11) tells us that the first moment of the total cross-sectional area A about the z axis $\int_A y\, dA$ must vanish. *Therefore, the x axis must pass through the centroid of the cross section.* (See App. A for a discussion of centroids of plane areas.) This result locates the position of the neutral axis; we used this result in Example 5.1.

If the material of the beam is not uniform across the section, Eq. (5.11) can still be used to determine the location of the neutral axis which will not in general be located at the centroid of the cross-sectional area; see Probs. 5.7-1 and 5.7-2.

Substituting Eq. (5.9) into the second equation of Eqs. (5.10), we find

$$\sum M_y = \int_A z\,\sigma_x\,dA = -\int_A E\frac{y}{\rho} z\,dA = -\frac{E}{\rho}\int_A yz\,dA = 0 \qquad (5.12)$$

Since the cross section is assumed to be symmetric with respect to the xy plane, the last integral in Eq. (5.12) is identically zero.

Substituting Eq. (5.9) into the third equation of Eqs. (5.10), we have

$$\sum M_z = -\int_A y\,\sigma_x\,dA = \int_A yE\frac{y}{\rho}\,dA = \frac{E}{\rho}\int_A y^2\,dA = M \qquad (5.13)$$

The integral

$$\int_A y^2\,dA = I_z \qquad (5.14)$$

is known as the ***moment of inertia*** about the z axis of the cross-sectional area; i.e., it is the moment of inertia of the area about the z axis that passes through the neutral axis. The value of I_z can be calculated and is therefore a known quantity once the specific shape of the cross section is given. We will review useful techniques for the calculation of the moment of inertia below; also see App. A.

Substituting Eq. (5.14) into Eq. (5.13) with $1/\rho = d\theta/ds$ from Eq. (5.6), we obtain the following important relation connecting the bending moment and the curvature:

$$M = \frac{EI_z}{\rho} = EI_z\frac{d\theta}{ds} \qquad (5.15)$$

Finally, upon solving for the curvature $1/\rho$ in terms of the moment, we can go back to obtain the distribution of the strain and the stress across the section in the form

$$\epsilon_x = -\frac{y}{\rho} = -\frac{My}{EI_z} \qquad (5.16)$$

$$\sigma_x = -\frac{My}{I_z} \qquad (5.17)$$

Equation (5.17) is one of the more important and useful equations we will derive in this book; we will find that its applicability is far-reaching. Again we note from Eq. (5.17) that when M is positive, the stress distribution across the section is linear with compression on the top surface and tension on the bottom surface, as shown in Fig. 5.9*a*.

To calculate the stress distribution at a given section along a beam, we need to know the value of the bending moment at that section and the value of the moment of inertia I_z of the cross section about the z axis passing through the centroid. Often we wish to calculate the maximum bending stress in the beam.

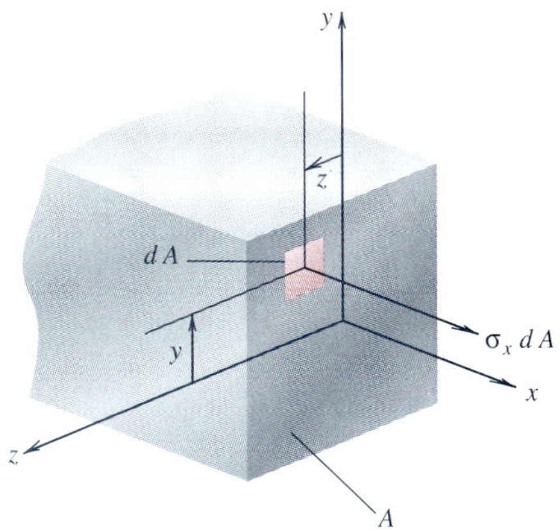

Figure 5.10 Resultant force $\sigma_x\,dA$ acting on an element of area dA on the beam cross section. The element is located so that the values of y and z are positive.

EXAMPLE 5.2

A small, 99.5 percent pure polycrystalline alumina beam specimen is to be tested in a small four-point bending test rig, as shown schematically in Fig. 5.11*a*. From an independent set of tests it is known that alumina fractures at a tensile stress of approximately 240 MPa. Estimate the load P to cause fracture of the specimen in this test setup.

The cross section of the specimen is a rectangle, as shown in Fig. 5.11*c*. The coordinate axes are shown in Fig. 5.11*a* and *c*. The x axis passes through the neutral axis of the beam which is located at the centroid of the cross section with the origin at the left end, and the y and z axes are oriented as shown. The loading on the beam gives rise to the bending moment curve, as shown in Fig. 5.11*b*. The center portion of the beam experiences pure bending of magnitude $-P \times 10^{-2}$ N · m when the load P is expressed in newtons. The maximum stress occurs in this region of the beam according to Eq. (5.17)

$$\sigma_x = -\frac{My}{I_z} \tag{a}$$

where I_z is the moment of inertia about the z axis which passes through the neutral axis. The moment of inertia of the rectangular cross section can be calculated as follows:

$$I = \int_A y^2\,dA = \int_{-h/2}^{h/2} y^2 b\,dy = \frac{1}{12}bh^3 \tag{b}$$

We will use this result often in what follows.

The expression for the tensile stress on the top surface of the specimen becomes

$$\begin{aligned}\sigma_x &= -\frac{(-P \times 10^{-2}\ \text{N}\cdot\text{m})(0.9 \times 10^{-3}\ \text{m})}{\frac{1}{12}(1.2 \times 10^{-3}\ \text{m})(1.8 \times 10^{-3}\ \text{m})^3} \\ &= 1.543 \times 10^7 P \qquad \text{N/m}^2\end{aligned} \tag{c}$$

If the maximum tensile stress in the specimen is

$$240\ \text{MPa} = 240 \times 10^6\ \text{N/m}^2$$

then the load P to cause fracture is, from Eq. (*c*),

$$\begin{aligned}P &= \frac{240 \times 10^6}{1.543 \times 10^7} \\ &= 15.55\ \text{N}\end{aligned} \tag{d}$$

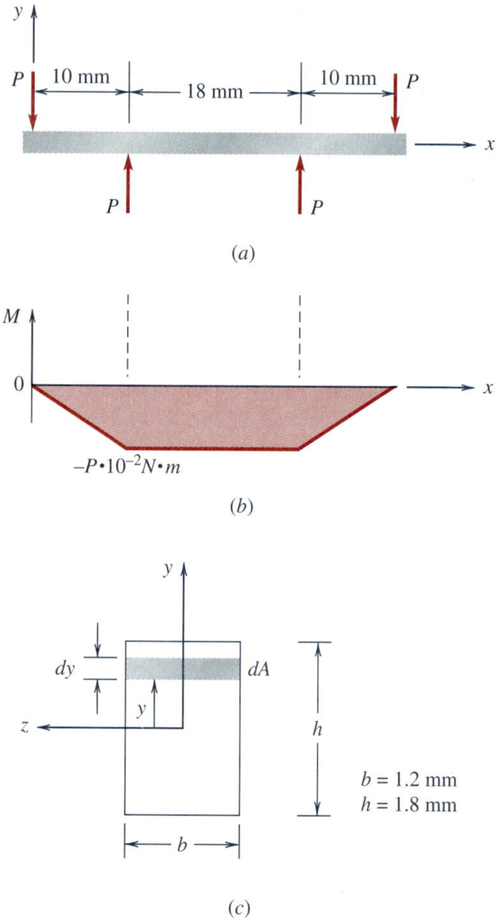

Figure 5.11 Example 5.2

EXAMPLE 5.3

In the previous example we obtained the moment of inertia of a rectangular section about its neutral axis, Eq. (*b*) above. In this example let us review the use of composite areas to determine the location of the neutral axis at the centroid and the use of the parallel-axis theorem to determine the moment of inertia of the cross-sectional area about the neutral axis. Further discussion of these topics can be found in App. A. We wish to determine the expressions for the centroid and the moment of inertia of a channel section beam and a T section beam.

Consider first a symmetric channel section of uniform thickness t, as shown in Fig. 5.12*a*. The origin of the yz axes must lie on the line of symmetry of the cross section at the centroid location. The exact location of the origin on this line of symmetry is to be determined. In general, we divide the total area of a section into a number of sub-areas whose centroid locations and moments of inertia are known; in this case, we divide the channel section into three areas shown as rectangles I and II in Fig. 5.12*b*. The total area of the section is the sum of the composite areas I and II

$$\text{Area} = wt + 2t(d - t) \tag{a}$$

The location of the centroid of the channel section measured from the bottom of the section, c, as shown in Fig. 5.12*b* along the line of symmetry is given in terms of the location of the centroids of the *composite areas*, each measured from the bottom:

$$\begin{aligned} c(\text{Area}) &= A_I c_I + 2A_{II} c_{II} \\ &= wt\left(d - \frac{t}{2}\right) + 2t(d - t)\left(\frac{1}{2}\right)(d - t) \end{aligned} \tag{b}$$

where A_I and A_{II} are the areas of rectangles I and II and c_I and c_{II} are the centroid locations of the rectangles I and II. From Eq. (*b*) we can determine the distance c that locates the centroid of the cross section, as shown in Fig. 5.12*b*.

The moment of inertia of the channel cross section about the neutral axis, i.e., about the z axis passing through the centroid, will be determined by using the *parallel-axis theorem*.

As a brief review of the parallel-axis theorem, consider a rectangle of area A, shown in Fig. 5.12*c*; the mn axes pass through the centroid of the rectangular area. We wish to determine the moment of inertia of the rectangle about an axis *parallel* to the n axis; in particular, we wish to determine the moment of inertia of the rectangle about the n' axis, a distance s from the n axis (Fig. 5.12*c*). The parallel-axis theorem gives the result in the form

$$I_{n'} = I_n + As^2 \tag{c}$$

where I_n is the moment of inertia about the centroidal axis of the rectangle, A is the area, and s is the distance between the axes. The moment of inertia about the centroidal axis is (Example 5.2)

$$I_n = \frac{1}{12} bh^3 \tag{d}$$

For the case of the channel section in Fig. 5.12*b*, the parallel-axis theorem can be applied to each of the composite rectangular areas and the results added to find the total moment of inertia about the centroidal axis of the channel section. The result can be written as follows

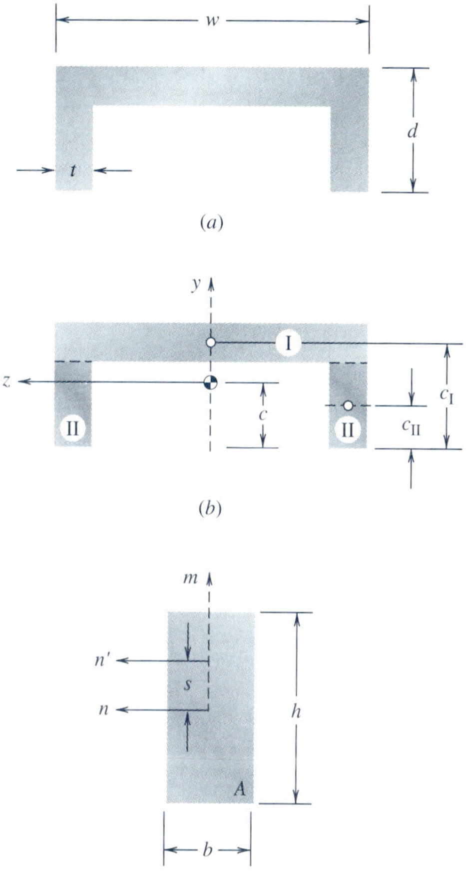

Figure 5.12 Example 5.3

$$I_z = (I_z)_I + 2(I_z)_{II}$$

where

$$\begin{aligned} (I_z)_I &= (I_c)_I + A_I(c_I - c)^2 \\ (I_z)_{II} &= (I_c)_{II} + A_{II}(c - c_{II})^2 \end{aligned} \tag{e}$$

The terms $(I_c)_I$ and $(I_c)_{II}$ represent the moments of inertia of rectangles I and II about their respective centroidal axis. The final expression for the moment of inertia of the channel section is given in the form

$$\begin{aligned} I_z &= \frac{1}{12} wt^3 + wt\left(d - \frac{t}{2} - c\right)^2 \\ &+ 2\left\{\frac{1}{12} t(d - t)^3 + t(d - t)\left[c - \frac{1}{2}(d - t)\right]^2\right\} \end{aligned} \tag{f}$$

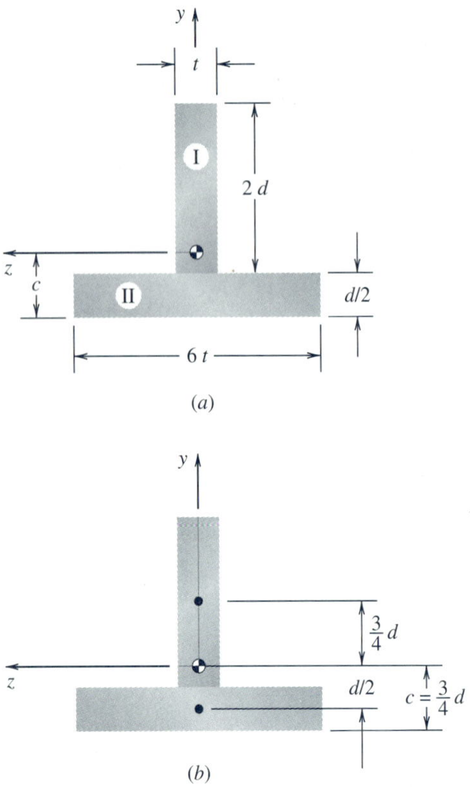

Figure 5.13 Example 5.3: T-beam section.

Once the dimensions of the channel section are given, the numerical values of c and I_z can be found from Eqs. (*b*) and (*f*).

As a second illustration, consider the specific T-beam section shown in Fig. 5.13*a*; the values of t and d are specified. We consider the section to be made up of rectangle I and rectangle II. The centroid location on the line of symmetry, c, is then given by

$$c = \frac{\frac{3}{2}d(2dt) + (d/4)(3td)}{2dt + 3td} = \frac{3}{4}d \qquad (g)$$

The moment of inertia about the centroidal z axis follows from using the parallel-axis theorem for each rectangle (Fig. 5.13*b*)

$$I_z = (I_z)_{\mathrm{I}} + (I_z)_{\mathrm{II}} \qquad (h)$$

where

$$(I_z)_{\mathrm{I}} = \frac{t(2d)^3}{12} + 2dt\left(\frac{3}{4}d\right)^2 = \frac{43}{24}td^3$$
$$(I_z)_{\mathrm{II}} = \frac{6t(d/2)^3}{12} + 3dt\left(\frac{d}{2}\right)^2 = \frac{13}{16}td^3 \qquad (i)$$

We obtain for I_z

$$I_z = \frac{125}{48}td^3 \qquad (j)$$

The use of composite areas to determine the location of the centroid and the value of I_z for sections made up of rectangles is straightforward, as the above illustrations demonstrate. For this reason it is amenable to simple computer programming to obtain results. The program MOMENTS OF INERTIA on the diskette with this book gives the location of the neutral axis and the moment of inertia about the neutral axis for the sections shown in Fig. 5.14, i.e., for a

Channel section

T beam

I beam

Unsymmetric I beam

The program is listed on the main menu and is simple to use. We use this program in some of the examples to follow.

While the computer is convenient for finding the location of the centroid and the value of I_z about the centroidal axes for the sections shown in Fig. 5.14, it is recommended that, in the initial phase of learning how to calculate the values of I_z, the values be worked out directly, and then, if appropriate, the computer program can be used to verify the results.

The cross sections given in the MOMENTS OF INERTIA program are only representative of different beam cross sections used in engineering practice. For many design problems, standard tables of properties of common sections can be used, and Apps. C, D, and E give properties for different standard sections; we will discuss these later. It should be obvious that an I-shape cross section will give a smaller value of the maximum bending stress in comparison to a symmetric section of the same area in which the area is not moved as far away from the neutral axis. However, the thickness of the web of an I beam is controlled by stability and shear stress considerations and cannot be fabricated less than a specified thickness if instability and excessive shear stresses in the web are to be avoided when the beam is loaded.

Figure 5.14 Cross-sectional shapes in MOMENTS OF INERTIA program.

5.4 Stresses in Symmetric Elastic Beams with Variable Bending Moment

Our development of Eq. (5.17) for the normal stress distribution across the section of a symmetric elastic beam assumed that the bending moment was constant along the length of the beam.

In practice, this is not a common situation. For most of the problems considered in Chap. 4, we found that the transverse loading on a beam resulted in nonzero shear forces and as a consequence in bending moments that were not constant along the length of the beam. Many of the symmetry arguments that were used in the derivation of Eq. (5.17) are valid only for the case of pure bending. However, solutions to beam problems obtained by more rigorous methods beyond the scope of this book indicate that the normal stress distribution given by Eq. (5.17) can be used as a good approximation for long, slender beams *even when shear forces are present* and *when bending moments vary along the beam.* The approximate theory based upon Eq. (5.17) is often called the *engineering theory of beam stresses* and is widely used. It is the basis for many design "codes," e.g., for commercial structures and homes. We will use this engineering theory in this book to determine the normal stresses in beams under transverse loading. Accurate solutions for beam stresses for certain special cases of loading and support have been obtained by methods beyond the scope of this book and by the engineering theory of beam stresses, and the results agree closely.

In many situations we wish to determine the *maximum* tensile or compressive bending stress in a beam subjected to transverse loads. In other situations we wish to select a particular beam to carry prescribed transverse loads without exceeding a specified value of stress in the beam or without exceeding a specified maximum value for the deflection of the beam. In both of these situations, the analysis of the normal stresses arising from bending requires the determination of the maximum or minimum bending moment in the beam, the determination of the location of the centroid, and the determination of the moment of inertia about the axis through the centroid of the beam cross section. Each step can be done separately, and the results combined in Eq. (5.17) to determine the normal stress distribution.

The largest positive or negative normal stresses for a beam with a symmetric cross section about the z axis occur where M is a maximum or a minimum and occur at either the top or the bottom of the section. The signs associated with the value of the bending moment and the value of y must be included in Eq. (5.17) to determine if the normal stress is tensile or compressive. Care must be taken in the determination of the maximum compressive or tensile stress for a beam whose cross section is not symmetric about the z axis, as e.g., in a channel section shown in Fig. 5.12, since the numerical value of y in Eq. (5.17) will be different for compression and tension. We turn to a number of examples demonstrating the steps to determine the largest value of the bending moment, the value of the moment of inertia, and the maximum normal bending stresses.

EXAMPLE 5.4

A simply supported rectangular wooden beam of length L supports a load $P = 1000$ lb at the midpoint, as shown in Fig. 5.15*a*. The cross section of the beam is rectangular, as shown. We wish to determine the expressions for the maximum tensile and compressive stresses due to bending in the beam and the longest allowable span L_a in the case of a square beam such that $|\sigma_{max}| < 1000$ psi with $b = h = 6$ in. We will neglect the weight of the beam in this example; see Prob. 5.4-5 to include the weight of the beam.

The bending moment diagram shown in Fig. 5.15*b* (recall Fig. 4.49) gives the maximum bending moment at midspan as

$$M_{max} = \frac{PL}{4} \tag{a}$$

Therefore, the normal stress distribution at the section at which the bending moment is a maximum is given by

$$\sigma_x = -\frac{PL}{4}\frac{y}{I_z} = -\frac{3PLy}{bh^3} \tag{b}$$

where the moment of inertia of the rectangular section is given by $I_z = \frac{1}{12}bh^3$. The maximum tensile stress occurs when $y = -h/2$, at the bottom of the beam,

$$\sigma_x = \frac{3}{2}\frac{PL}{bh^2} \tag{c}$$

and the maximum compressive stress occurs when $y = h/2$, at the top of the beam,

$$\sigma_x = -\frac{3}{2}\frac{PL}{bh^2} \tag{d}$$

In this case where the beam is symmetric about the z axis, the maximum tensile and compressive stresses differ only in sign. The stress distribution across the section is shown in Fig. 5.15*c*.

To find the allowable span, we use Eq. (*d*) to write

$$1000 = |\sigma_x| = \frac{3}{2}\frac{PL_a}{bh^2} = \frac{3}{2}\frac{1000L_a}{6 \cdot 6^2} \tag{e}$$

or

$$L_a = 144 \text{ in} = 12 \text{ ft} \tag{f}$$

If the span is greater than 12 ft, the maximum normal bending stress in the beam will be greater than 1000 psi.

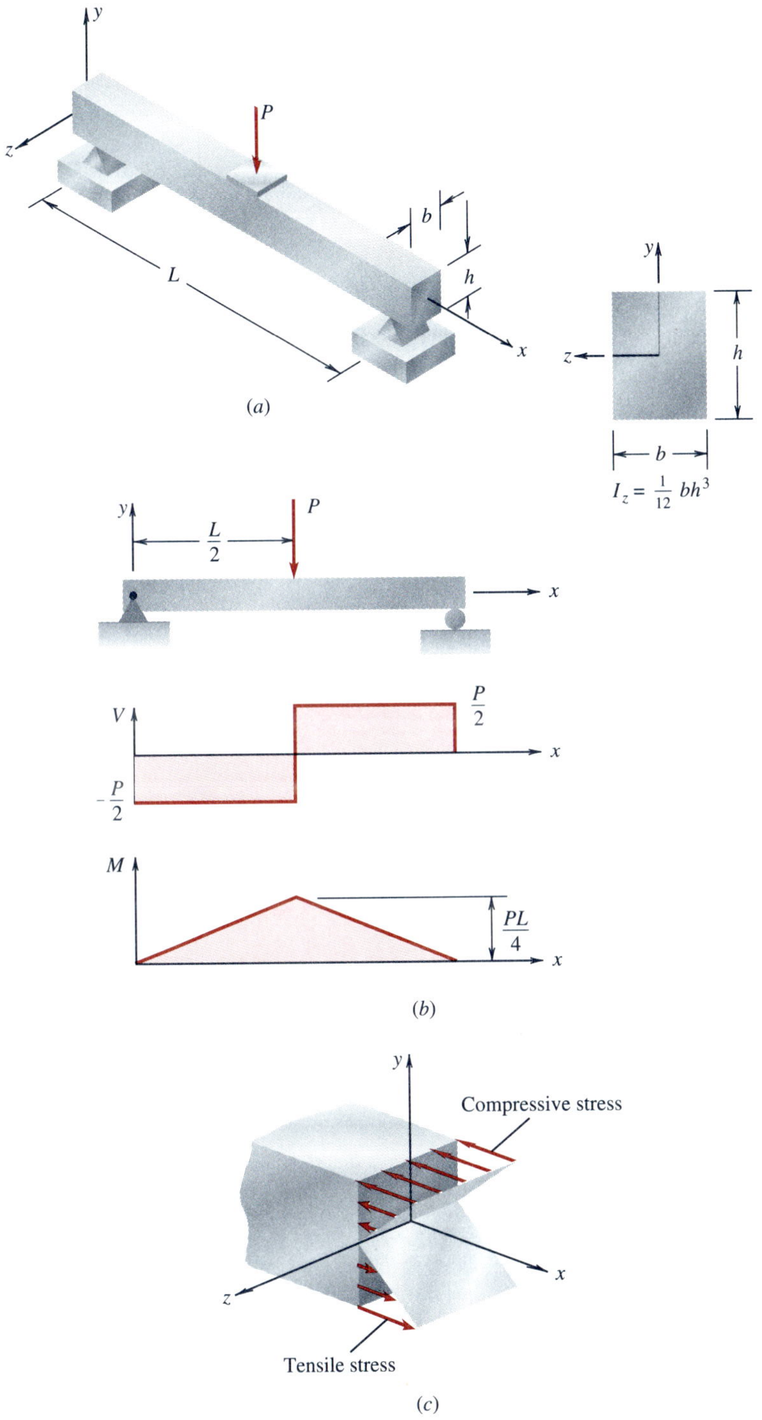

Figure 5.15 Example 5.4

EXAMPLE 5.5

A steel cantilever beam 6 m long is loaded at its end by a load of $P = 7$ kN, as shown in Fig. 5.16*a*. The cross section of the beam is a symmetric I-beam shape, as shown in Fig. 5.16*b*. We wish to determine the maximum tensile bending stress in the beam. First we will neglect the weight of the beam and then estimate the effect of the weight on the maximum stress. The specific weight of steel used in this beam is 77.0 kN/m³.

The maximum bending moment due to the load P occurs at the wall support (see Fig. 4.49 and Fig. 5.16*c*),

$$M_{max} = -PL = -42 \text{ kN} \cdot \text{m} \tag{a}$$

The neutral axis passes through the centroid of the section. The moment of inertia of the cross section (Fig. 5.16*b*) about the z axis is given by

$$\begin{aligned} I_z &= \tfrac{1}{12}(5.75)(184.5)^3 + 2[\tfrac{1}{12}(130)(7.75)^3 \\ &\quad + 130(7.75)(96.125)^2] \qquad (b) \\ &= 2.164 \times 10^7 \text{ mm}^4 \end{aligned}$$

The maximum bending stresses occur when $y = \pm 100$ mm; the maximum tensile stress occurs at the top of the beam and is given by

$$\begin{aligned} \sigma_x &= -\frac{My}{I_z} = -\frac{(-42 \text{ kN} \cdot \text{m})(10^{-1} \text{ m})}{2.164 \times 10^{-5} \text{ m}^4} \\ &= 194 \text{ MN/m}^2 = 194 \text{ MPa} \end{aligned} \tag{c}$$

This maximum tensile stress occurs at the wall support at the top of the beam. The maximum compressive stress occurs at the bottom of the beam at the wall support. The yield stress of mild steel is approximately 250 MPa or greater depending on the quality of the steel. Depending upon the factor of safety against yielding, the maximum stress in Eq. (*c*) may or may not be acceptable.

If the weight of the beam is included, we need to determine the stress in a cantilever beam under a constant distributed load given by

$$\begin{aligned} w &= 77.0 \text{ kN/m}^3 \times \text{area} \\ &= 77.0 \times 3.076 \times 10^{-3} \text{ kN/m} = 0.237 \text{ kN/m} \end{aligned} \tag{d}$$

The maximum bending moment in this case occurs at the wall support (Fig. 4.49) and is given by

$$M_{max} = -\frac{wL^2}{2} = -4.27 \text{ kN} \cdot \text{m} \tag{e}$$

The corresponding tensile stress is

$$\sigma_x = 19.7 \text{ MPa}$$

In this case the additional stress due to the weight of the beam is approximately 10 percent of the stress due to the applied load. In general, the effect of the weight of the beam on the value of the stresses in beams is usually less than 10 percent and is often ignored. However, in many cases the effect of the weight should be checked, especially in cases of design where exact design code specifications are to be met.

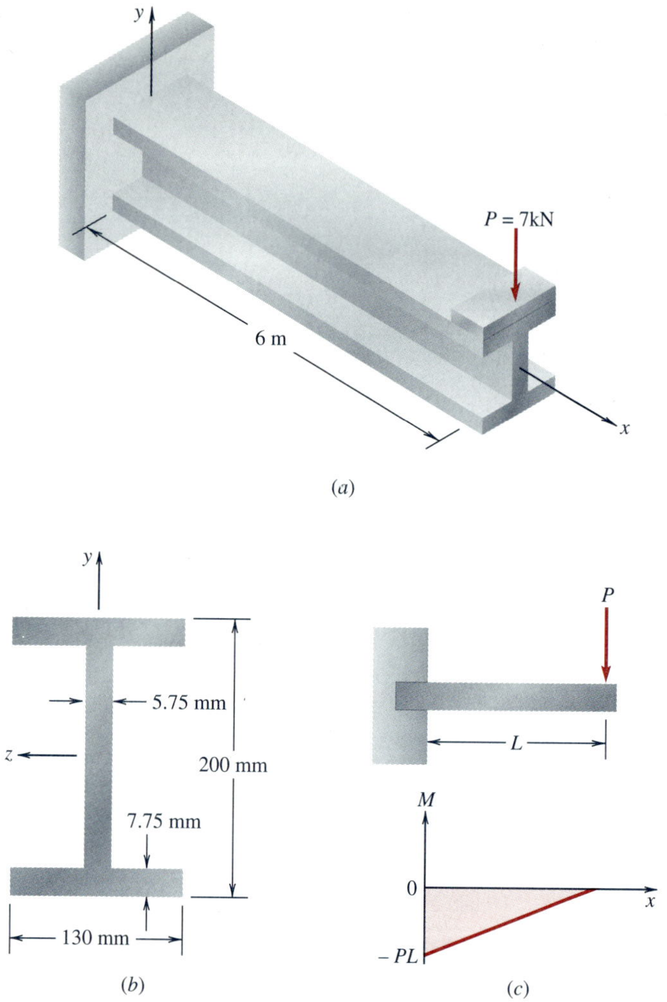

Figure 5.16 Example 5.5

EXAMPLE 5.6

A T beam shown in Fig. 5.17*a* is loaded by a set of three loads where $P = 2$ kips. Determine the maximum value of the tensile and compressive normal stresses in the beam due to bending. Neglect the weight of the beam. The shear force and bending moment diagrams are shown in Fig. 5.17*b*; the bending moment with the largest magnitude occurs at the supports and has a value equal to $-4P$.

Since the T section is not symmetric about the z axis, we need to determine the location of the centroid and the moment of inertia of the cross section about the z axis passing through the centroid. We can divide the area into two rectangular parts, I and II, as shown Fig. 5.17*c* and use composite areas together with the parallel-axis theorem to determine the centroid and the moment of inertia. First, we have

$$c = \frac{(2 \times 5 \times 4.5) + (8 \times 2 \times 1)}{2 \times 5 + 8 \times 2} = 2.346 \text{ in} \qquad (a)$$

$$I_z = \tfrac{1}{12}(2)(5)^3 + 2(5)(2.15)^2 + \tfrac{1}{12}(8)(2)^3 + 8(2)(1.35)^2 = 101.6 \text{ in}^4 \qquad (b)$$

Alternatively, we can use the program MOMENTS OF INERTIA, and the output is given in Fig. 5.18. The maximum compressive stress occurs at the bottom of the beam at the support and is given by

$$\sigma_x = -\frac{My}{I_z} = -\frac{(-4 \cdot 12 \cdot 2000)(-2.346)}{101.6} = -2220 \text{ psi} \quad (c)$$

The maximum tensile stress occurs at the top of the beam at the support and is given by

$$\sigma_x = -\frac{My}{I_z} = -\frac{(-4 \cdot 12 \cdot 2000)(4.654)}{101.6} = 4400 \text{ psi} \qquad (d)$$

In this case we note that the tensile stress is about 2 times greater than the compressive stress.

At the midpoint of the beam $M = P$, so that the stresses at the top and bottom of the beam at this section are opposite in sign from the stresses at the section over the support and are reduced by a factor of 4; therefore at the midpoint the maximum tensile stress is 550 psi, and the maximum compressive stress is 1100 psi. Thus the maximum of the absolute value of the stress occurs over the supports, as given in Eqs. (*c*) and (*d*).

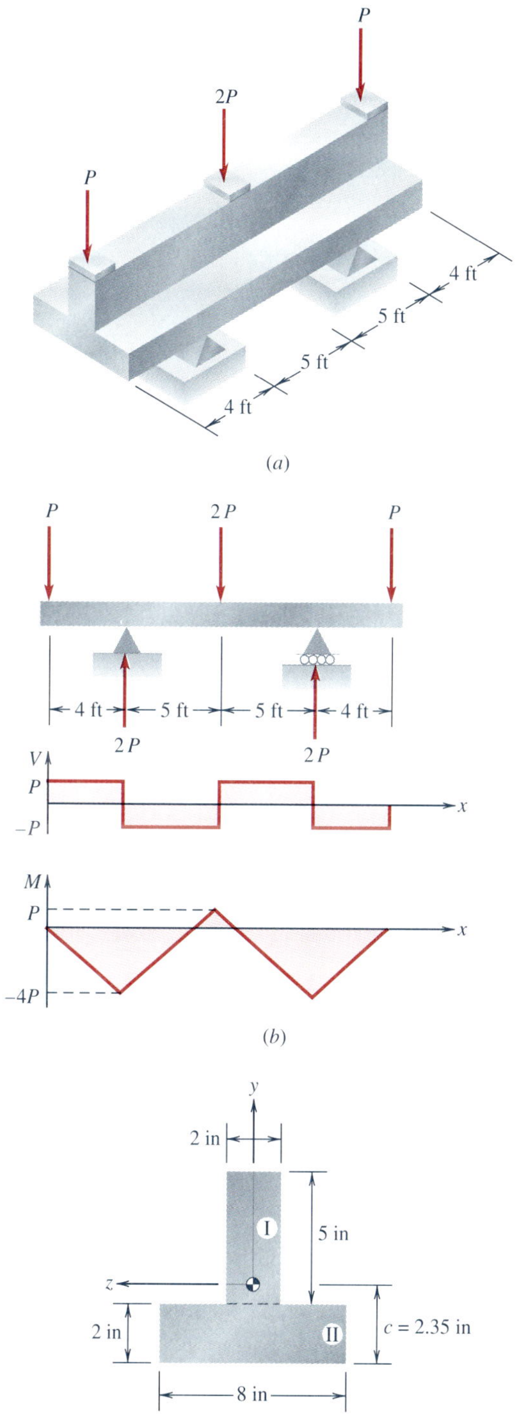

Figure 5.17 Example 5.6 (*c*)

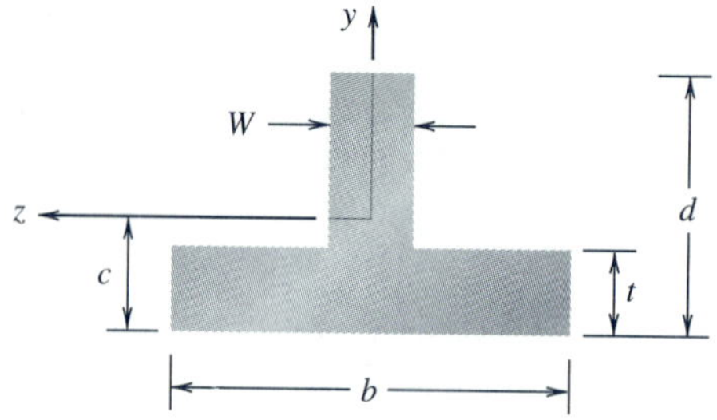

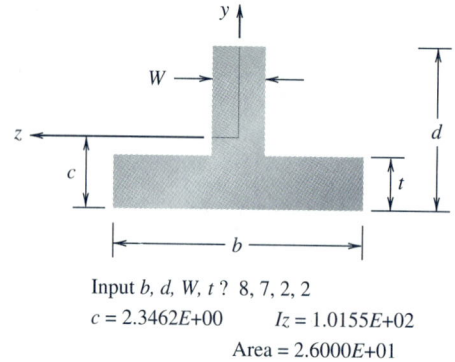

Input b, d, W, t ? 8, 7, 2, 2

c = 2.3462E+00 Iz = 1.0155E+02

Area = 2.6000E+01

Figure 5.18 Example 5.6: Cross-sectional properties.

EXAMPLE 5.7

A steel box beam supports two segments of a constant distributed load, as shown in Fig. 5.19. We wish to determine the maximum tensile bending stress in the beam; we neglect the weight of the beam. It is convenient in this case to use the computer program BENDING OF BEAMS, program 3 on the diskette, to determine the shear force and bending moment diagrams from which the maximum bending moment along the beam can be found; see Sec. 4.7. If we use 20 × 21 = 420 points on the beam, we obtain the results given in Fig. 5.20*a* and *b*. The accuracy of the results, as we discussed in Sec. 4.7, depends on the number of points selected for the evaluation of the shear force and bending moment. The maximum bending moment from Fig. 5.20*a* is

$$M_{\text{max}} = 1.333 \times 10^5 \text{ ft} \cdot \text{lb} \qquad (a)$$

The minimum moment, of course, is zero, but the computer program gives a small negative value which can be taken as zero when compared to the value of the maximum moment. The exact value of the maximum moment is 1.334×10^5 ft · lb.

From the shear force and bending moment diagrams in Fig. 5.20*b*, we see (e.g., from scaling the bending moment diagram) that the maximum value of the moment occurs at a section

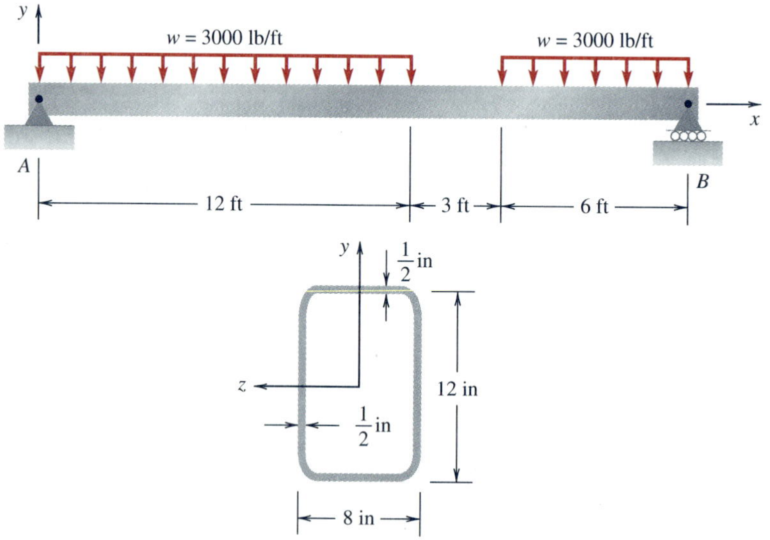

Figure 5.19 Example 5.7

```
***SIMPLE BEAM***                DATA FILE: ex57

       INPUT DATA:
            THE LENGTH OF BEAM:                                21
THE NUMBER OF POINTS ALONG THE BEAM:                          420
THE LOCATION OF LEFT SIDE SUPPORT:                              0
THE LOCATION OF RIGHT SIDE SUPPORT:                            21
THE NUMBER OF CONCENTRATED LOADS:                               0
THE NUMBER OF CONCENTRATED MOMENTS:                             0
THE NUMBER OF SEGMENTS OF UNIFORM OR LINEAR LOADS:              2
THE START AND END LOCATION AND VALUE OF UNIFORM OR
   LINEAR LOADS:
       START LOCATION = 0            LOAD = -3000
       END LOCATION   = 12           LOAD = -3000
       START LOCATION = 15           LOAD = -3000
       END LOCATION   = 21           LOAD = -3000

                ****** SUMMARY SOLUTION ******

*********************************************************
 LEFT SUPPORT REACTION                  RIGHT SUPPORT REACTION
      28285.720                               25714.290

MAXIMUM SHEAR FORCE AND MOMENT
     MAXIMUM SHEAR FORCE =   25714.29
     MINIMUM SHEAR FORCE =  -28135.72
     MAXIMUM MOMENT      =  133346.50
     MINIMUM MOMENT      =      -3.94
```

(*a*)

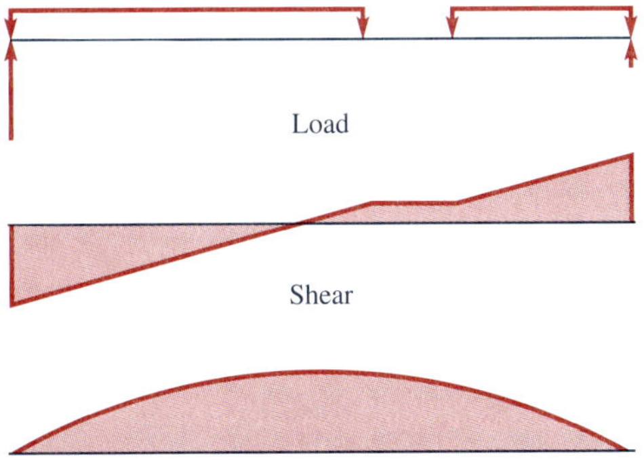

(*b*)

x	$V(x)$	$M(x)$
9.200	–685.715	133268.800
9.250	–535.715	133299.400
9.300	–385.715	133322.400
9.350	–235.715	133337.900
9.400	–85.715	133346.000
9.450	64.285	133346.500
9.500	214.285	133339.600
9.550	364.285	133325.100
9.600	514.285	133303.100
9.650	664.285	133273.700
9.700	814.285	133236.700

(*c*)

Figure 5.20 Example 5.7. (*a*) Input data and solution. (*b*) Loading, shear force, and bending moment diagrams. (*c*) Partial table of shear force $V(x)$ and bending moment $M(x)$ at different values of x.

between $x = 9$ ft and $x = 10$ ft. A review of a portion of the output of the shear force and bending moment table shown in Fig. 5.20*c* leads to the conclusion that the bending moment is a maximum near $x = 9.43$ ft. (The exact value is given by $x = R/3000 = 9.429$ ft, where R is the value of the reaction at the left support.) Therefore, we conclude from the results in Fig. 5.20 that the maximum value of the bending moment is given by Eq. (*a*) and occurs at $x = 9.43$ ft. The moment of inertia of the cross section given in Fig. 5.19 is

$$I_z = \tfrac{1}{12}[8(12)^3 - 7(11)^3] = 375.6 \text{ in}^4 \qquad (b)$$

The maximum tensile stress therefore is given by

$$\sigma_x = -\frac{M_{\max}(-6)}{I_z} = \frac{1.333 \times 10^5 \times 12 \times 6}{375.6} = 25{,}550 \text{ psi} \qquad (c)$$

and occurs on the bottom of the beam at $x = 9.43$ ft.

If we had wished to obtain a quick approximation to the value of the maximum stress, we could have assumed that the beam was a simply supported beam with a constant load of 3000 lb/ft × 18 ft/21 ft = 2570 lb/ft applied over the entire 21-ft length of the beam.

The maximum bending moment in this case then occurs at the center $x = 10.5$ ft and is

$$M_{\max} = \frac{wL^2}{8} = \frac{2570(21)^2}{8} = 1.417 \times 10^5 \text{ ft} \cdot \text{lb} \qquad (d)$$

which gives a stress of

$$\sigma_x = 27{,}160 \text{ psi} \qquad (e)$$

or 6.3 percent greater than the exact value given in Eq. (*c*).

EXAMPLE 5.8

A heavy steel channel section is to be used as a simply supported beam, as shown in Fig. 5.21. The beam is loaded by a uniform load of 10 kN/m. Determine the maximum tensile and compressive bending stresses in the beam. Investigate the effect of the weight of the beam on the stress values as the thickness t of the section is increased from 50 to 70 mm. The specific weight of steel in this beam is 77 kN/m³.

Since the channel section is not symmetric about a horizontal axis, we need first to determine the location of the neutral axis and the moment of inertia about the neutral axis. A direct appeal to the program MOMENTS OF INERTIA (see Fig. 5.22) with $t = 50$ mm gives

$$c = 175 \text{ mm}$$
$$I_z = 2.604 \times 10^8 \text{ mm}^4 \qquad (a)$$
$$A = 5.0 \times 10^4 \text{ mm}^2$$

The weight per unit length of the beam is therefore

$$\frac{\text{Weight}}{\text{m}} = (77 \text{ kN/m}^3)(5 \times 10^{-2} \text{ m}^2) = 3.85 \text{ kN/m}$$

Therefore, the total distributed load acting along the beam is 13.85 kN/m.

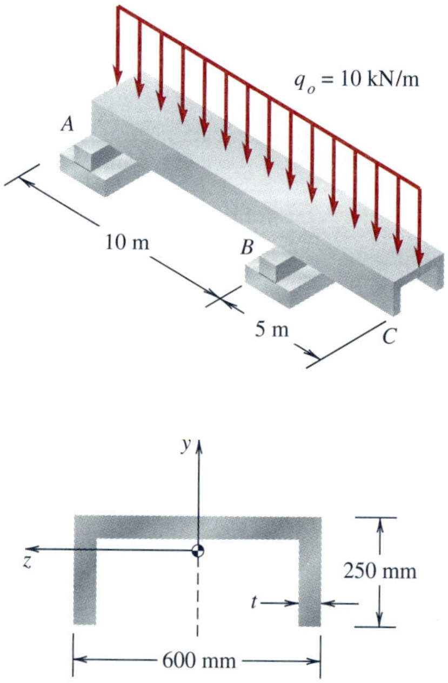

Figure 5.21 Example 5.8

If we use program 4 for the shear force and bending moment diagrams (Sec. 4.7), we obtain the values of the bending moments from Figs. 5.23 and 5.24 as

$$M = -173.1 \text{ kN} \cdot \text{m} \tag{b}$$

at support B, $x = 10$ m, and a value of

$$M = 97.38 \text{ kN} \cdot \text{m} \tag{c}$$

at a point $x = 3.75$ m from support A. Because the section is not symmetric, both locations should be checked for the maximum bending stress. At support B, the maximum tensile stress is given by

$$\sigma_x = -\frac{(-173.1 \text{ kN} \cdot \text{m})(75 \times 10^{-3} \text{ m})}{2.604 \times 10^{-4} \text{ m}^4} = 49.86 \text{ MPa} \tag{d}$$

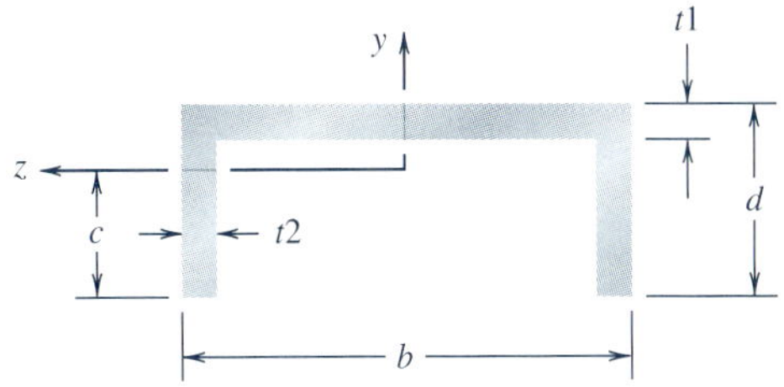

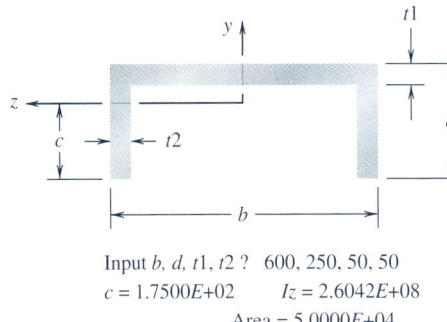

Input b, d, $t1$, $t2$? 600, 250, 50, 50
c = 1.7500E+02 Iz = 2.6042E+08
Area = 5.0000E+04

Figure 5.22 Example 5.8: Cross-sectional properties.

```
***SIMPLE BEAM***                DATA FILE: ex58

       INPUT DATA:
            THE LENGTH OF BEAM:                              15
THE NUMBER OF POINTS ALONG THE BEAM:                        300
THE LOCATION OF LEFT SIDE SUPPORT:                            0
THE LOCATION OF RIGHT SIDE SUPPORT:                          10
THE NUMBER OF CONCENTRATED LOADS:                             0
THE NUMBER OF CONCENTRATED MOMENTS:                           0
THE NUMBER OF SEGMENTS OF UNIFORM OR LINEAR LOADS:            1
THE START AND END LOCATION AND VALUE OF UNIFORM OR
   LINEAR LOADS:
       START LOCATION = 0               LOAD = -13.85
       END LOCATION   = 15              LOAD = -13.85

                ****** SUMMARY SOLUTION ******

*********************************************************
 LEFT SUPPORT REACTION               RIGHT SUPPORT REACTION
        51.938                              155.813

MAXIMUM SHEAR FORCE AND MOMENT
     MAXIMUM SHEAR FORCE =     86.56
     MINIMUM SHEAR FORCE =    -68.56
     MAXIMUM MOMENT      =     97.38
     MINIMUM MOMENT      =   -173.13
```

Figure 5.23 Example 5.8: Input data and solution.

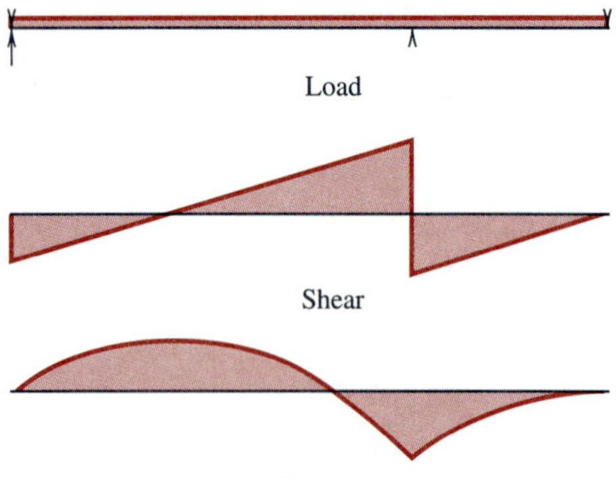

Figure 5.24 Example 5.8: Loading, shear force, and bending moment diagrams.

and the maximum compressive stress is given by

$$\sigma_x = -\frac{(-173.1\ \text{kN}\cdot\text{m})(-175\times10^{-3}\ \text{m})}{2.604\times10^{-4}\ \text{m}^4} = -116.3\ \text{MPa} \qquad (e)$$

At the location $x = 3.75$ m from A, the maximum tensile stress is given by

$$\sigma_x = -\frac{(97.38\ \text{kN}\cdot\text{m})(-175\times10^{-3}\ \text{m})}{2.604\times10^{-4}\ \text{m}^4} = 65.44\ \text{MPa} \qquad (f)$$

and the maximum compressive stress is given by

$$\sigma_x = -\frac{(97.38\ \text{kN}\cdot\text{m})(75\times10^{-3}\ \text{m})}{2.604\times10^{-4}\ \text{m}^4} = -28.05\ \text{MPa} \qquad (g)$$

For this case we see that the maximum compressive stress occurs at $x = 10$ m while the maximum tensile stress occurs at $x = 3.75$ m, as summarized below:

Location	Compressive	Tensile
$x = 3.75$ m	−28.05 MPa	65.44 MPa
$x = 10.00$ m	−116.3 MPa	49.86 MPa
$t = 50$ mm		

If the thickness of the section is now increased to 70 mm, the section properties will change (Fig. 5.25) to $t = 70$ mm, $c = 168.1$ mm, $I_z = 3.313\times10^8$ mm^4, and area $= 6.72\times10^4$ mm^2, and the weight per length will increase by the ratio of the areas to a value 3.85 (6.72/5.00) = 5.17 kN/m.

The bending moment at $x = 10$ m will change, therefore, in proportion to the applied loading to give

$$M = -173.1\left(\frac{15.17}{13.85}\right) = -189.6\ \text{kN}\cdot\text{m} \qquad (h)$$

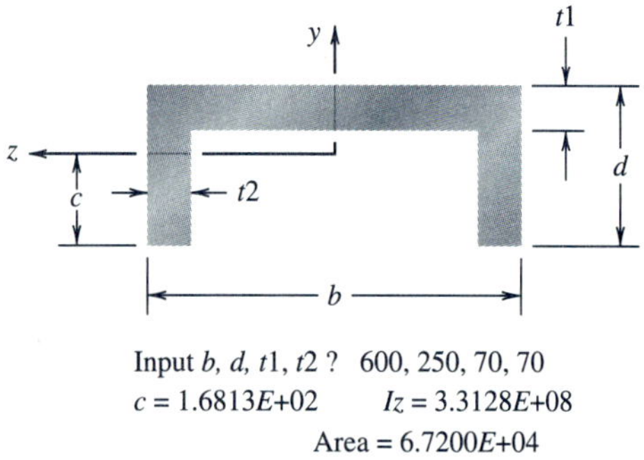

Figure 5.25 Example 5.8: Cross-sectional properties for $t = 70$ mm.

The maximum compressive stress at the support $x = 10$ m, however, will decrease because of the change in the section properties

$$\sigma_x = -\frac{(-189.6\ \text{kN}\cdot\text{m})(-168.1\times10^{-3}\ \text{m})}{3.313\times10^{-4}\ \text{m}^4} = -96.20\ \text{MPa} \qquad (i)$$

This value is about 17 percent less than the value of the maximum compressive stress with $t = 50$ mm, Eq. (e).

The maximum tensile stress at the support becomes

$$\sigma_x = -\frac{(-189.6\ \text{kN}\cdot\text{m})(81.9\times10^{-3}\ \text{m})}{3.313\times10^{-4}\ \text{m}^4} = 46.87\ \text{MPa} \tag{j}$$

We see, therefore, that while the weight of the beam has increased, the maximum stresses at support B have decreased.

At $x = 3.75$ m, we have for the maximum tensile stress, using for the bending moment $(15.17/13.85) \times 97.38 = 106.7$ kN · m,

$$\sigma_x = -\frac{(106.7\ \text{kN}\cdot\text{m})(-168.1\times10^{-3}\ \text{m})}{3.313\times10^{-4}\ \text{m}^4} = 54.14\ \text{MPa} \tag{k}$$

and for the maximum compressive stress

$$\sigma_x = -\frac{(106.7\ \text{kN}\cdot\text{m})(81.9\times10^{-3}\ \text{m})}{3.313\times10^{-4}\ \text{m}^4} = -26.38\ \text{MPa} \tag{l}$$

Therefore, for the case of $t = 70$ mm, the stresses are

Location	Compressive	Tensile
$x = 3.75$ m	−26.38 MPa	54.14 MPa
$x = 10.00$ m	−96.20 MPa	46.87 MPa
$t = 70$ mm		

Again we find that the maximum compressive stress occurs at $x = 10$ m while the maximum tensile stress occurs at $x = 3.75$ m. We also note that by increasing the weight of the beam, we have decreased the maximum values of the stresses. However, we have increased the cost of the beam by about 35 percent.

EXAMPLE 5.9

A hollow steel pipe is clamped to a wall at A and subjected to concentrated loads P acting in opposite directions at sections B and C, as shown in Fig. 5.26a. If the outside and inside radii of the pipe are 3 and 2.75 in, respectively, we wish to find the maximum allowable magnitude P_a of the loads that can be applied to the pipe so that the normal bending stresses in the pipe do not exceed 24 ksi.

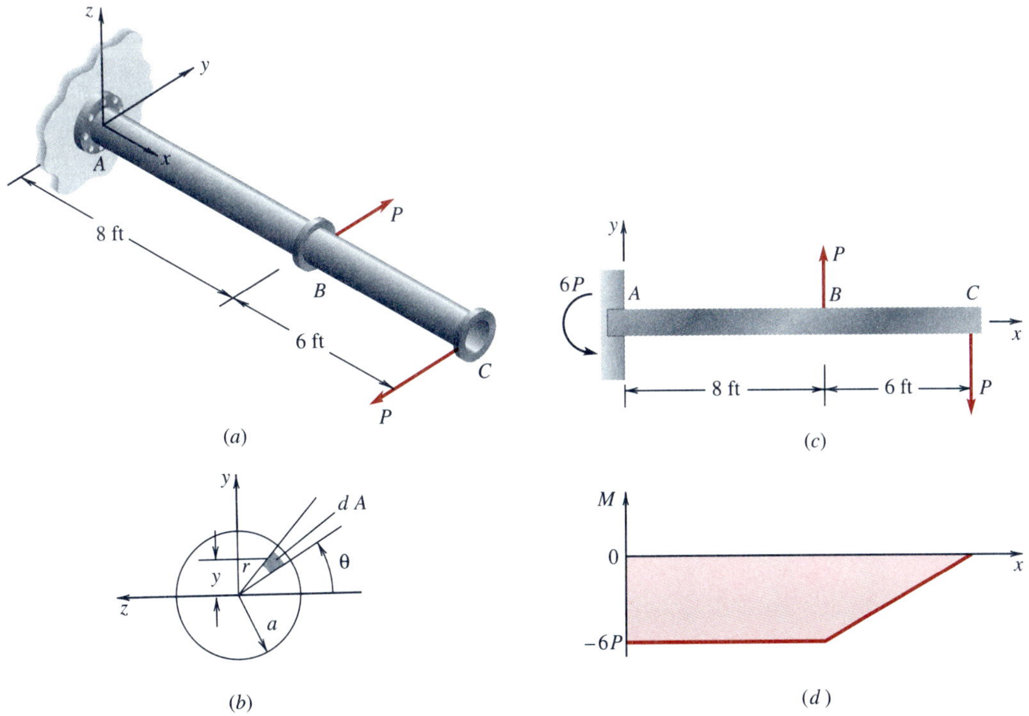

Figure 5.26 Example 5.9

The coordinate axes are located as shown in Fig. 5.26*a*. We neglect the weight of the pipe.

The centroid of the annular section is at the center. The moment of inertia of the section about the *z* axis can be calculated by subtracting the moment of inertia for a solid circular section of radius r_i from that for a solid circular section of radius r_o. In general, for a circular area of radius *a*, we have (Fig. 5.26*b*)

$$I_z = \int_A y^2\, dA$$

$$= \int_0^{2\pi}\int_0^{a} r^2 \sin^2\theta\, r\, dr\, d\theta \qquad (a)$$

$$= \frac{\pi a^4}{4}$$

Therefore for the hollow section in the present problem,

$$I_z = \frac{\pi}{4}(r_o^4 - r_i^4)$$

$$= \frac{\pi}{4}(3^4 - 2.75^4) \qquad (b)$$

$$= 18.70 \text{ in}^4$$

Using the free-body diagram shown in Fig. 5.26*c*, we find the unknown wall moment to be equal to 6*P* lb · ft. A sketch of the bending moment variation along the beam (Fig. 5.26*d*) shows that the maximum bending moment equals −6*P* and occurs over the portion *AB* of the beam. Over this portion

$$\sigma_{max} = -\frac{Mr_o}{I_z} \qquad (c)$$

or

$$24 = \frac{6 \times 12 \times P \times 3}{18.70}$$

and

$$P_a = 2.08 \text{ kips} \qquad (d)$$

Because the section is symmetric about the *z* axis, the allowable load P_a = 2.08 kips would give equal maximum tensile and compressive stresses of magnitude 24 ksi over the portion *AB*.

For each of the problems above we *analyzed* the given beam to determine the normal stress due to bending. However, in engineering practice we often need to select a beam for a particular application. This means, in general, that we know the loading to be expected across a span of length *L* and we wish to select or design a beam cross section such that the maximum stress in the beam is less than an allowable value. Other considerations such as cost or availability of specific beams also enter into the design of a beam. The selection of a beam cross section for engineering applications is often facilitated by tables of beam properties such as those given in the App. C. Appendix C gives the properties for beams of different cross sections, for example, W, S, C, and angle shapes, listing the areas and geometric properties including the moment of inertia about each axis and the section modulus about each axis.

The *section modulus* *S* about the *z* axis of the cross section of a beam is *defined* as the moment of inertia I_z about the *z* axis divided by the distance y_{max} from the centroid to the most distant element in the section along the *y* axis

$$S = \frac{I_z}{y_{max}} \qquad (5.18)$$

A corresponding formula holds for the section modulus about the *y* axis.

The maximum normal stress in a beam symmetric about the *z* axis without regard to whether it is a tensile or a compressive stress can therefore be expressed in terms of the section modulus in the form

$$\sigma_{max} = \frac{M_{max}\, y_{max}}{I_z} = \frac{M_{max}}{S}$$

If the numerically largest bending moment M_{max} and the allowable stress σ_{max} in the beam are known, the required section modulus S for the beam is given by

$$S = \frac{M_{max}}{\sigma_{max}} \tag{5.19}$$

For the design of such sections, Eq. (5.19) is used to find a value of S when the maximum bending moment and allowable bending stress are given. Once a beam section is selected, the weight of the beam should then be considered to ensure that the combination of the applied loads and the weight of the beam does not give a stress larger than the allowable stress. We show how to proceed in the following two examples.

EXAMPLE 5.10

A steel cantilever beam is loaded at the end, as shown in Fig. 5.27. If the maximum allowable stress in the beam is 20 ksi, select a wide-flange beam (W beam) to support the load when $P = 1200$ lb.

The maximum bending moment in the beam is, of course,

$$M_{max} = 1200 \times 18 \times 12 = 2.592 \times 10^5 \text{ lb} \cdot \text{in} \tag{a}$$

The value of S required for the cross section is, therefore, from Eq. (5.19)

$$S = \frac{2.592 \times 10^5 \text{ lb} \cdot \text{in}}{20 \times 10^3 \text{ lb/in}^2} = 12.96 \text{ in}^3 \tag{b}$$

A review of Table C.1 shows that a number of beams could be used. We select the section

$$\text{W10} \times 15$$

which has an S value of 13.8 in^3 and a weight per foot of 15 lb. The weight of the beam gives rise to an additional maximum bending moment $wL^2/2$. Therefore, the maximum stress including the weight of this beam is

$$\sigma = \frac{M}{S} = \frac{2.592 \times 10^5 + 0.292 \times 10^5}{13.8} = 20{,}900 \text{ psi} \tag{c}$$

The stress in Eq. (c) is greater than the allowable stress, so that the section W10 × 15 is not adequate. It was not a good choice! Instead we select the section

$$\text{W12} \times 14$$

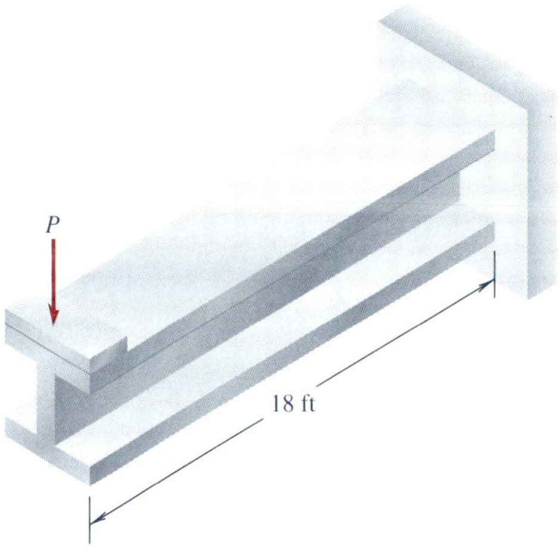

Figure 5.27 Example 5.10

which has an S value of 14.9 in^3 and a weight per foot of 14 lb. If we use this section, the maximum stress including the weight is

$$\sigma = \frac{M}{S} = \frac{2.592 \times 10^5 + 0.272 \times 10^5}{14.9} = 19{,}200 \text{ psi} \qquad (d)$$

which is below the maximum allowable stress of 20,000 psi. Therefore the beam selected is adequate.

EXAMPLE 5.11

A simply supported beam shown in Fig. 5.28 is to carry the uniform load of 300 lb/ft. We wish to determine an I-beam section (S beam) to carry the applied load if the maximum allowable stress in the beam is 20,000 psi.

The first step in determining the required section modulus is to find the maximum bending moment in the beam. To do this, we can use the computer program for the bending of beams discussed in Sec. 4.7. The summary of the input and output and the shear force and bending moment diagrams are given in Figs. 5.29 and 5.30.

The maximum bending moment from the program is

$$M = 13{,}184 \text{ lb} \cdot \text{ft} = 1.582 \times 10^5 \text{ lb} \cdot \text{in} \qquad (a)$$

Therefore the section modulus required is

$$S = \frac{M_{\max}}{\sigma_{\max}} = \frac{1.582 \times 10^5 \text{ lb} \cdot \text{in}}{20 \times 10^3 \text{ lb/in}^2} = 7.91 \text{ in}^3 \qquad (b)$$

A review of Table C.2 suggests a section S 6 × 17.25 with S = 8.77 in^3 and 17.25 lb/ft as the weight per length.

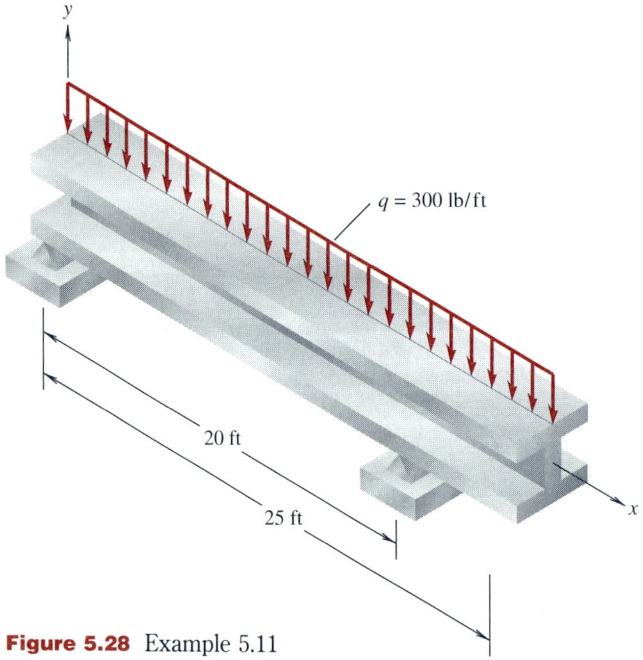

Figure 5.28 Example 5.11

If we include the weight of the beam, the loading on the beam is 317.25 lb/ft, and the required S is given by

$$S = 7.91 \left(\frac{317.25}{300.00}\right) = 8.36 \text{ in}^3 \qquad (c)$$

This value of S is less than the section modulus of 8.77 in^3 of the beam that we selected, so that an S 6 × 17.25 is adequate to carry the load.

```
***SIMPLE BEAM***             DATA FILE: ex511

       INPUT DATA:
            THE LENGTH OF BEAM:                            25
THE NUMBER OF POINTS ALONG THE BEAM:                      500
THE LOCATION OF LEFT SIDE SUPPORT:                          0
THE LOCATION OF RIGHT SIDE SUPPORT:                        20
THE NUMBER OF CONCENTRATED LOADS:                           0
THE NUMBER OF CONCENTRATED MOMENTS:                         0
THE NUMBER OF SEGMENTS OF UNIFORM OR LINEAR LOADS:          1
THE START AND END LOCATION AND VALUE OF UNIFORM OR
   LINEAR LOADS:
       START LOCATION = 0              LOAD = -300
       END LOCATION   = 25             LOAD = -300

                ****** SUMMARY SOLUTION ******

*********************************************************
 LEFT SUPPORT REACTION                RIGHT SUPPORT REACTION
       2812.500                              4687.500

MAXIMUM SHEAR FORCE AND MOMENT
     MAXIMUM SHEAR FORCE =   3187.50
     MINIMUM SHEAR FORCE =  -2797.50
     MAXIMUM MOMENT      =  13183.53
     MINIMUM MOMENT      =  -3750.14
```

Figure 5.29 Example 5.11: Input data and solution.

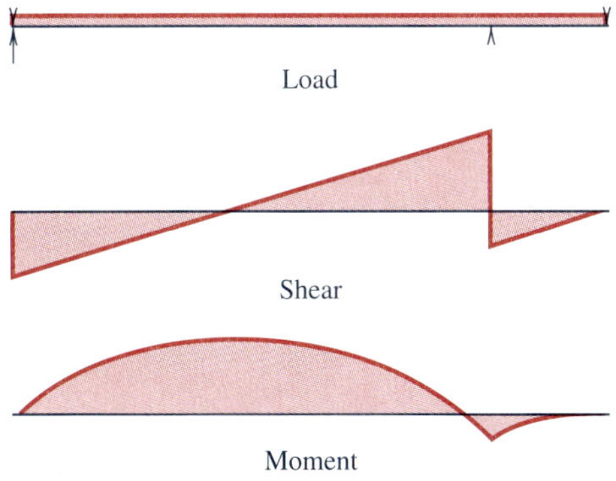

Figure 5.30 Example 5.11: Loading, shear force, and bending moment diagrams.

5.5 Shear Stress Distribution in Symmetric Beams with Variable Bending Moment

In the last section, we derived Eq. (5.17) for the normal stress on a cross section of a beam due to bending. The normal stress distribution was derived under the assumption that it gives rise to no net axial resultant force on the section and that it was statically equivalent to the bending moment M acting on the section. This formula was obtained initially for beams that had a constant bending moment along their length. By assuming that the normal stress formula is applicable also to beams with bending moments that vary along the length of the beam, i.e., to beams in which a shear force is also present, we obtained the *engineering theory of beams.* Recall from the equilibrium equation, Eq. (4.7), that $dM/dx = -V$, so that if dM/dx is not zero, a shear force V will act on the section along with the bending moment M.

We now wish to find the approximate distribution of shear stresses acting on the section that is statically equivalent to the shear force V acting on the section. We begin by considering the shear stress distribution in a beam with a rectangular section; see Figs. 5.31 and 5.32. We should also recall the result derived in Sec. 1.3, Fig. 1.7, concerning the equivalence of shear stresses acting on perpendicular planes at a point. In Fig. 5.31*a*

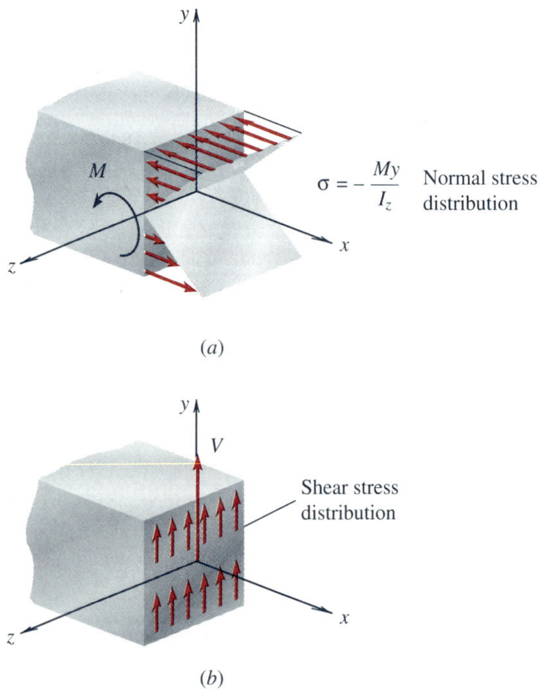

Figure 5.31 (*a*) Normal stress distribution. (*b*) Shear stress distribution.

we show the normal stress distribution acting on the section and the moment M, and in Fig. 5.31*b* we show the shear stresses acting on the section and the shear force V. By introducing appropriate free bodies of segments of the beam, we will be able to deduce the particular distribution of these shear stresses over the section.

Shear stresses acting on a section of the beam give rise to shear stresses acting on planes perpendicular to the section, as shown in Fig. 5.32. We know from the moment equilibrium of the infinitesimal element shown in Fig. 5.32 that the magnitude of the shear stress acting on a horizontal plane perpendicular to the section is equal to the magnitude of the shear stress acting on the section. It is this observation that allows us to calculate the distribution of the shear stress τ_{xy} across the section.

We will focus our attention on a horizontal plane $y = y_1$ in Fig. 5.32 and use equilibrium arguments to determine an expression for the average shear stress acting on the horizontal plane. Once the shear stress on the horizontal plane $y = y_1$ is found, we have from equality of the shear stress on the perpendicular plane (Fig. 5.32) the value of the vertical shear stress acting in the y direction at location y_1 on the section. These arguments will become clear as we proceed.

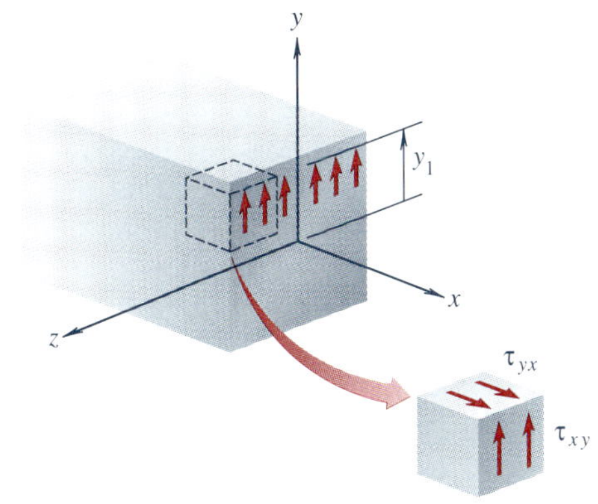

Figure 5.32 Shear stresses on perpendicular faces are equal.

Consider, as shown in Fig. 5.33*a*, a segment of length Δx cut from a beam that has a section symmetric about the y axis but otherwise of general shape and that is subjected to both a bending moment and a shear force. We show the shear forces $V(x)$ and $V(x + \Delta x)$ and bending moments $M(x)$ and $M(x + \Delta x)$ acting on sections x and $x + \Delta x$. The normal stress distributions that are statically equivalent to $M(x)$ and $M(x + \Delta x)$ at each section are given by Eq. (5.17). In Fig. 5.33*b* we have taken the case for which $dM/dx > 0$ so that the bending moment at $x + \Delta x$, $M(x + \Delta x)$, is larger than the bending moment at x, $M(x)$; thus the bending stresses acting on section $x + \Delta x$ will be larger than the bending stresses acting on section x for the same value of y (Fig. 5.33*b*). The discussion of the alternate case would follow similar lines. We now isolate part of the beam segment above the *horizontal* plane $y = y_1$, as shown in Fig. 5.33*c*. We wish to determine the shear stress acting on *this* plane.

We see that there is an unbalance of the forces arising from the force resultant of the (different) bending stresses acting on the ends of this part, and so there must be a force ΔF_{yx} acting on the bottom negative y face to maintain force balance in the x direction, as shown in Fig. 5.33*c*. Recall from Sec. 4.2 that the subscripts on force components indicate the face and direction of the force. We show the force as ΔF_{yx} to emphasize that it is acting on an element of area of length Δx and of width b whose normal points in the negative y direction. The direction of ΔF_{yx} is in the negative x direction in Fig. 5.33*c* which is consistent with our convention. The value of b is the width of the horizontal cut. We do not show the vertical loads in Fig. 5.33*c* in the interest of clarity.

Force equilibrium of the segment in Fig. 5.33*c* in the *x* direction gives

$$\left[\int_{A_1} \sigma_x \, dA\right]_{x+\Delta x} - \Delta F_{yx} - \left[\int_{A_1} \sigma_x \, dA\right]_x = 0 \tag{5.20}$$

where A_1 is the area over which the normal bending stresses are acting, i.e., over the area between $y = y_1$ and $y = c$ (Fig. 5.33*d*). Upon substitution of Eq. (5.17) into Eq. (5.20), we find

$$\begin{aligned} \Delta F_{yx} &= -\int_{A_1} \frac{M(x+\Delta x)}{I_z} y \, dA + \int_{A_1} \frac{M(x)}{I_z} y \, dA \\ &= -\frac{M(x+\Delta x) - M(x)}{I_z} \int_{A_1} y \, dA \end{aligned} \tag{5.21}$$

If we divide by Δx and take the limit as $\Delta x \to 0$, we have

$$\frac{dF_{yx}}{dx} = \lim_{\Delta x \to 0} \frac{\Delta F_{yx}}{\Delta x} = -\frac{dM}{dx} \frac{1}{I_z} \int_{A_1} y \, dA \tag{5.22}$$

However, from moment equilibrium, Eq. (4.7),

$$\frac{dM}{dx} = -V$$

and we rewrite Eq. (5.22) in the form

$$\frac{dF_{yx}}{dx} = \frac{V}{I_z} \int_{A_1} y \, dA \tag{5.23}$$

The integral in Eq. (5.23) represents the first moment of area A_1 about the *z* axis, and its value depends on the geometry of the cross section and the value of y_1. We find it convenient to introduce the symbol $Q(y_1)$ to represent it so that

$$Q(y_1) = \int_{A_1} y \, dA \tag{5.24}$$

We therefore rewrite Eq. (5.23) in the form

$$f = \frac{dF_{yx}}{dx} = \frac{V(x)Q(y_1)}{I_z} \tag{5.25}$$

where f is called the ***shear flow,*** and it is defined as the rate of change of the shear force acting on the horizontal plane $y = y_1$ per unit length along the beam. Shear flow has units of force per length. If V is constant, the shear flow along the beam is constant.

The moment of inertia I_z in Eq. (5.25) is the moment of inertia of the *total* cross section about the neutral axis.

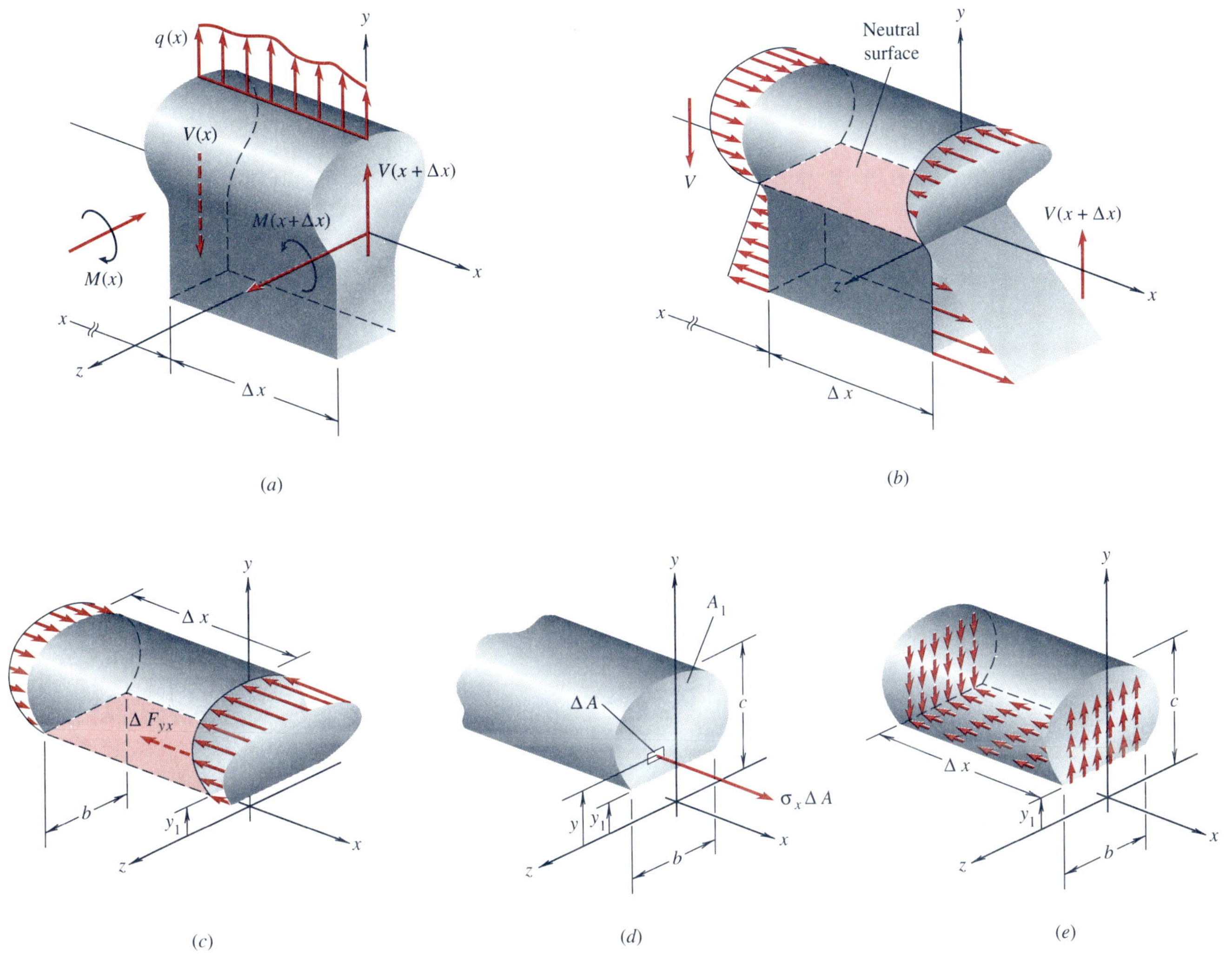

Figure 5.33 Calculation of shear stress τ_{xy} in a symmetric beam from equilibrium of a segment of the beam.

If the force ΔF_{yx} is assumed to arise from an approximately uniform distribution of shear stress τ_{yx} across width b, that is, if

$$\Delta F_{yx} = \tau_{yx}\, b\, \Delta x \tag{5.26}$$

where τ_{yx} is the shear stress acting on the y face in the x direction by analogy with ΔF_{yx}, then

$$\lim_{\Delta x \to 0} \frac{\Delta F_{yx}}{\Delta x} = \tau_{yx} b = \frac{V(x)Q(y_1)}{I_z} \tag{5.27}$$

The value of the uniform shear stress is therefore given by

$$\tau_{yx} = \frac{VQ}{I_z b} \tag{5.28}$$

Equation (5.28) gives the uniform shear stress acting on the horizontal plane $y = y_1$ of width b at a fixed value of x. The shear stress on the perpendicular section (Fig. 5.32) is then given by

$$\tau_{xy} = \tau_{yx} = \frac{VQ}{I_z b} \tag{5.29}$$

where τ_{xy} is the shear stress acting on an x face in the y direction, as shown in Fig. 5.33*e*.

It is important to emphasize that the derivation of the equation for the shear stress distribution τ_{xy} on a section x, Eq. (5.29), used a force balance in the direction of the beam axis. In addition, we have assumed that the shear stresses are uniform across width b at the horizontal plane $y = y_1$. The resultant of the shear stresses on the horizontal plane balances the net force arising from the force resultants of the different normal stresses acting on areas A_1 at each end of the segment Δx.

In the case of a rectangular beam, we can obtain an explicit formula for the distribution of shear stress across the section by using Eq. (5.29) since b is a constant across the section (see Fig. 5.34). The first moment of the crosshatched area A_1 in Fig. 5.34 is

$$Q = \int_{A_1} y\,dA = \int_{-b/2}^{b/2}\int_{y_1}^{h/2} y\,dy\,dz = b\int_{y_1}^{h/2} y\,dy$$
$$= b\left[\frac{y^2}{2}\right]_{y_1}^{h/2} = \frac{b}{2}\left(\frac{h^2}{4} - y_1^2\right) \tag{5.30}$$

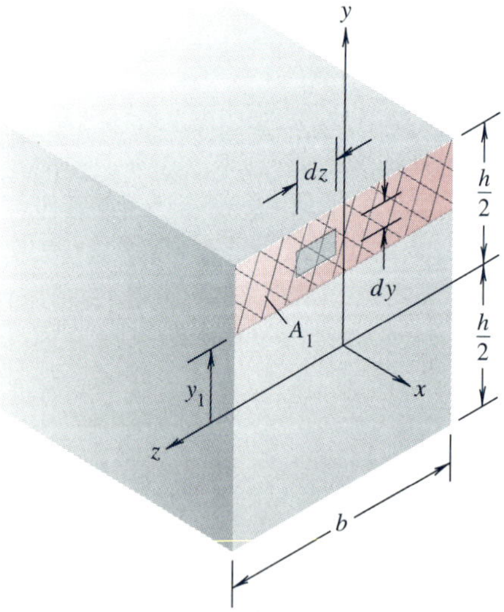

Figure 5.34 Area A_1 for the calculation of the first moment Q.

Therefore, the shear stress distribution, Eq. (5.29), is given by

$$\tau_{xy} = \frac{V}{2I_z}\left(\frac{h^2}{4} - y_1^2\right) \tag{5.31}$$

The shear stress is a maximum at the neutral surface $y_1 = 0$ and falls off parabolically, as shown in Fig. 5.35. The maximum shear stress at $y_1 = 0$ is

$$\tau_{\max} = \frac{Vh^2}{8I_z} = \frac{3V}{2bh} = \frac{3V}{2A} \tag{5.32}$$

where we have used $I_z = bh^3/12$. We see from Eq. (5.32) that the maximum shear stress is 1.5 times the average shear stress V/A;

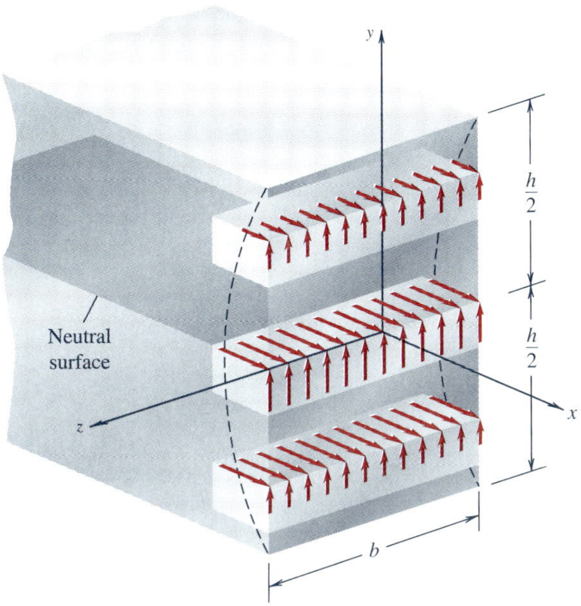

Figure 5.35 Illustration of the parabolic distribution of shear stress τ_{xy} in a rectangular beam.

this result is useful to keep in mind, but it holds *only* for a solid rectangular section. Other cross sections will have different numerical factors relating the maximum shear stress on the section to the average shear stress on the section; e.g., see Eq. (5.34).

For the rectangular section it remains to verify that the resultant of the shear stress distribution is, in fact, equal to the shear force V. The force resultant R of the shear stress distribution on section x (Fig. 5.35) is given by

$$\begin{aligned} R &= \int \tau_{xy}\, dA = \frac{V}{2I_z}\int_{-b/2}^{b/2}\int_{-h/2}^{h/2}\left(\frac{h^2}{4} - y^2\right) dy\, dz \\ &= \frac{Vb}{2I_z}\left[\frac{h^2}{4}y - \frac{y^3}{3}\right]_{-h/2}^{h/2} = \frac{Vbh^3}{12I_z} = V \end{aligned} \qquad (5.33)$$

which verifies the result, i.e., the resultant of the shear stresses is equal to V. This result in fact is true for any *arbitrary cross section* (see Prob. 5.5-4).

EXAMPLE 5.12

A model cantilever beam of rectangular cross section is loaded as shown in Fig. 5.36*a*. Determine the shear stress at points along line AA' on a section 8 in from the built-in end, as shown in Fig. 5.36*b*. The load $P = 1500$ lb, and we neglect the weight of the beam.

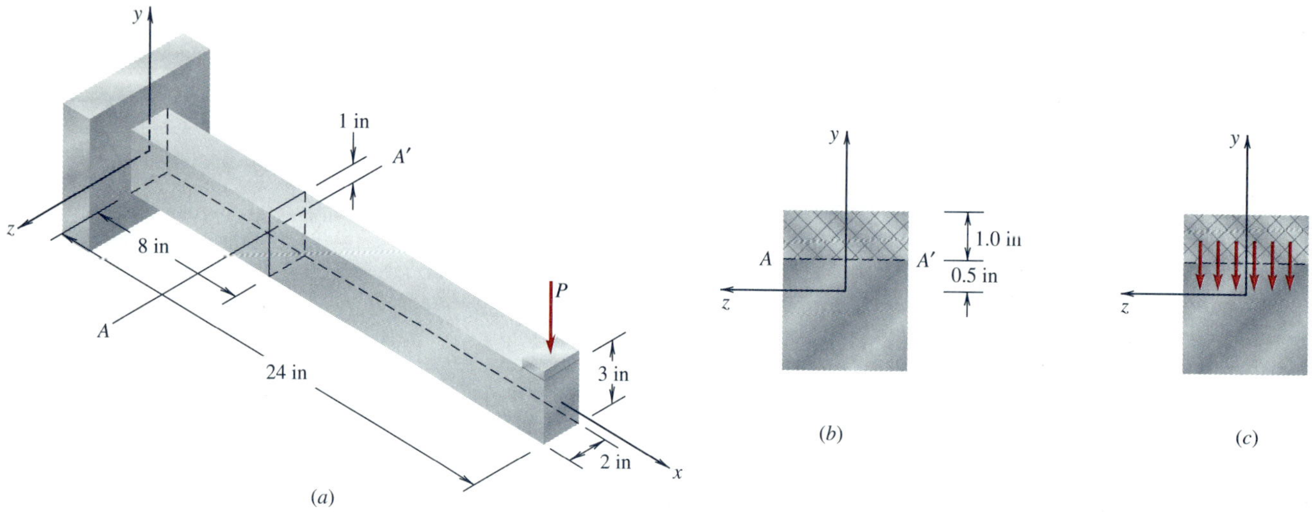

Figure 5.36 Example 5.12

The shear force V is constant along the beam and equal to $-P$. Using Eq. (5.29), we find the shear stress at AA' is

$$\tau_{xy} = \frac{VQ}{I_z b} \qquad (a)$$

where I_z is the moment of inertia of the cross section about the neutral axis; b is the width of the section, equal to 2 in for this case; and Q is the first moment of the area above line AA' (the area A_1 is shown as crosshatched in Fig. 5.36*b*).

The expression for Q is

$$Q = \int_{A_1} y\,dA = \bar{y}_1 A_1 = 1(1 \times 2) = 2\ \text{in}^3 \qquad (b)$$

where $\bar{y}_1$ is the distance to the centroid of area A_1; this result follows from the definition of Q. The moment of inertia is

$$I_z = \tfrac{1}{12} bh^3 = \tfrac{1}{12}(2)(3)^3 = 4.5\ \text{in}^4 \qquad (c)$$

The shear stress follows from Eqs. (*a*) to (*c*) in the form

$$\tau_{xy} = \frac{(-1500)(2)}{(4.5)(2)} = -333\ \text{psi} \qquad (d)$$

which is the required result. The shear stress is shown in Fig. 5.36*c*.

The maximum negative shear stress on this *rectangular section* occurs at the neutral axis, and from Eq. (5.32)

$$\tau_{\max} = \frac{3V}{2A} = \frac{1.5(-1500)}{6} = -375\ \text{psi} \qquad (e)$$

The result in Eq. (*d*) also could have been obtained directly from Eq. (5.31)

$$\tau_{xy} = \frac{V}{2I_z}\left(\frac{h^2}{4} - y_1^2\right) \qquad (f)$$

with $h = 3$ in and $y_1 = 0.5$ in.

EXAMPLE 5.13

A wooden beam of cross section 6 in × 10 in is simply supported, as shown in Fig. 5.37. If the maximum allowable tensile bending stress is 1200 psi and the maximum allowable shear stress is 180 psi, find the maximum allowable load P that can act at the midpoint of the beam. Include the weight of the beam; assume the beam is oak with a specific weight of 45 lb/ft^3.

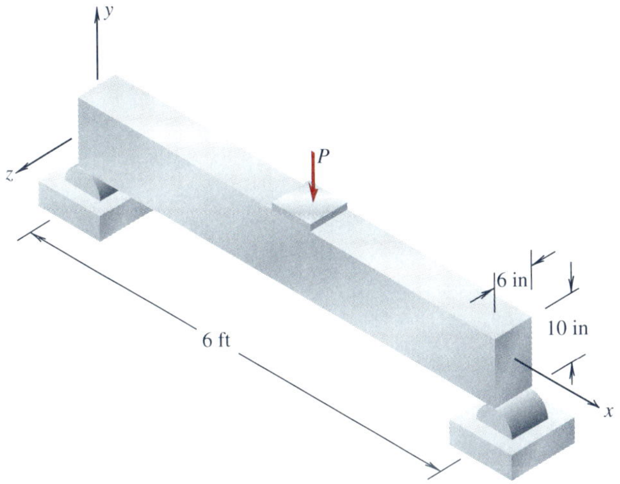

Figure 5.37 Example 5.13

The maximum bending moment due to load P occurs at the center of the beam (Fig. 4.49) and is

$$M = \frac{P(6)(12)}{4} = 18P \quad \text{lb} \cdot \text{in} \tag{a}$$

The maximum bending moment due to the weight of the beam (Fig. 4.49) is

$$M = \frac{wL^2}{8} = \frac{1}{8}\left(\frac{6}{12}\right)\left(\frac{10}{12}\right)(45)(6)^2(12) = 1012.5 \text{ lb} \cdot \text{in} \tag{b}$$

The maximum bending moment at the center of the beam is therefore

$$M_{\max} = 18P + 1012.5 \quad \text{lb} \cdot \text{in} \tag{c}$$

The moment of inertia of the cross section is

$$I_z = \tfrac{1}{12} bh^3 = \tfrac{1}{12}(6 \times 10^3) = 500 \text{ in}^4 \tag{d}$$

The maximum bending stress is found by using

$$\sigma_x = -\frac{My}{I_z} \tag{e}$$

with $M_{\max}$ given by Eq. (c), I_z by Eq. (d), $y_{\max} = -5$ in, and $\sigma_{\max} = 1200$ psi. This leads to the equation

$$1200 = -\frac{(18P + 1012.5)(-5)}{500} \tag{f}$$

or

$$18P + 1012.5 = \frac{500(1200)}{5}$$

and

$$P = 6610 \text{ lb} \tag{g}$$

The maximum shear force occurs at the supports and is given by

$$\begin{aligned} V &= \frac{P}{2} + \frac{wL}{2} = \frac{P}{2} + \frac{45}{2}\left(\frac{6}{12}\right)\left(\frac{10}{12}\right)(6) \\ &= \frac{P}{2} + 56.25 \quad \text{lb} \end{aligned} \tag{h}$$

The maximum shear stress occurs at the neutral axis and for a rectangular beam is given by

$$\tau_{\max} = \frac{3V}{2A} \tag{i}$$

Therefore the maximum allowable shear force in the beam is

$$V_a = \tfrac{2}{3} A \tau_{\max}$$

or

$$0.5P + 56.25 = \tfrac{2}{3}(60 \times 180)$$

and

$$P = 14{,}290 \text{ lb} \tag{j}$$

The maximum allowable load is therefore $P = 6610$ lb, and it is controlled by the maximum bending stress. We note that the total weight of the beam is 112.5 lb and if it had been neglected, the corresponding results would have been 6670 and 14,400 lb, which differ from the above results by less than 1 percent in both cases.

EXAMPLE 5.14

Consider the cantilever beam of homogeneous cross section shown in Fig. 5.38. We wish to determine the ratio of the maximum shear stress to the maximum normal stress in the beam. We neglect the weight of the beam.

The maximum normal stress is given by

$$\sigma_x = -\frac{My}{I_z} \qquad (a)$$

where $M = -PL$, $y = h/2$, and

$$I_z = \frac{1}{12}bh^3 \qquad (b)$$

Therefore

$$\sigma_{\max} = \frac{6PL}{bh^2} = \frac{6P}{A}\frac{L}{h} \qquad (c)$$

Since the beam has a solid rectangular section, the maximum shear stress is given by

$$\tau_{\max} = \frac{3V}{2A} = \frac{3P}{2A} \qquad (d)$$

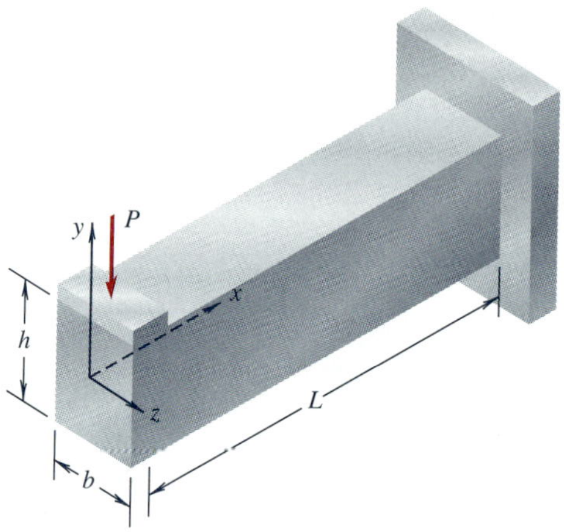

Figure 5.38 Example 5.14

Therefore

$$\frac{\tau_{\max}}{\sigma_{\max}} = \frac{1}{4}\left(\frac{h}{L}\right) \qquad (e)$$

Since for beams (slender members) the ratio h/L is usually less than about $\frac{1}{10}$, we see in this case that the maximum shear stress is usually between 1 and 2 orders of magnitude less than the maximum normal stress.

Thus far in calculating shear stresses for bending problems, we have restricted our attention to beams with rectangular cross sections and have assumed that shear stresses on the cross section act parallel to the y axis. For other cross-sectional geometries such as the circular section shown in Fig. 5.39*a*, we find that shear stresses can be inclined to the y axis. For example, elements located at the boundary of the circle, e.g., at points A and B, must have shear stress components tangentially directed since any radial component would have a companion shear stress on the perpendicular face on the outside surface of the beam, which would contradict the fact that the outside surface is stress-free. By this argument, shear stresses acting on boundary elements around the entire periphery are tangentially directed (Fig. 5.39*a*).

In the derivation of Eq. (5.29) for estimating the shear stress by using axial force equilibrium of a segment defined in Fig. 5.33, the force ΔF_{yx} represented a force resultant in the axial direction acting on a horizontal plane caused by the action of unbalanced bending stresses at the ends of the segment.

If we consider a circular cross section, then for a typical plane defined by $y = y_1$, area A_1 is the part of the circular section shown in Fig. 5.39*b*. As we have shown above, the shear stresses along line *EFG* would not all be directed parallel to the *y* direction even though the overall resultant *V* is so directed. In order to have a simple approximate theory for this class of problems, we again assume [as we did in the discussion leading up to Eq. (5.26)] that τ_{xy} is approximately constant along line *EFG* for the circular section. The average vertical shear stress is found by dividing the shear flow *f* by the width *b* of line *EFG* to obtain

$$\tau_{xy} = \frac{f}{b} = \frac{VQ}{I_z b} \qquad (5.29)$$

Circular sections have the property that the width *b* decreases as the distance y_1 from the neutral axis increases; therefore, for a given value of V/I_z, the factor $1/b$ has the effect of increasing τ_{xy}. However, the first moment *Q* of area A_1 monotonically decreases with increasing y_1. As a consequence, it can be shown (Prob. 5.5-20) that the ratio Q/b takes on its maximum value at $y_1 = 0$. Therefore, to find the maximum shear stress we have at $y_1 = 0$

$$Q = \frac{\pi R^2}{2} \times \frac{4R}{3\pi} = \frac{2}{3}R^3$$

and using

$$I_z = \tfrac{1}{4}\pi R^4 \qquad b = 2R$$

Eq. (5.29) gives

$$\tau_{\max} = \frac{V\left(\frac{2}{3}R^3\right)}{(\pi/4)R^4(2R)} = \frac{4V}{3A} \qquad (5.34)$$

where $A = \pi R^2$ is the area of the section. We will recall that for a solid rectangular section the maximum value of τ_{xy} was found to be $3V/2A$ as compared with the above result $4V/3A$ for the solid circular section.

For a hollow circular section as shown in Fig. 5.40, if we again make use of the equation for the shear stress given by Eq. (5.29), it follows that τ_{xy} takes on its maximum value when $y = 0$ where

$$Q = \tfrac{2}{3}(R_o^3 - R_i^3) \qquad b = 2(R_o - R_i)$$

$$I_z = \frac{\pi}{4}(R_o^4 - R_i^4)$$

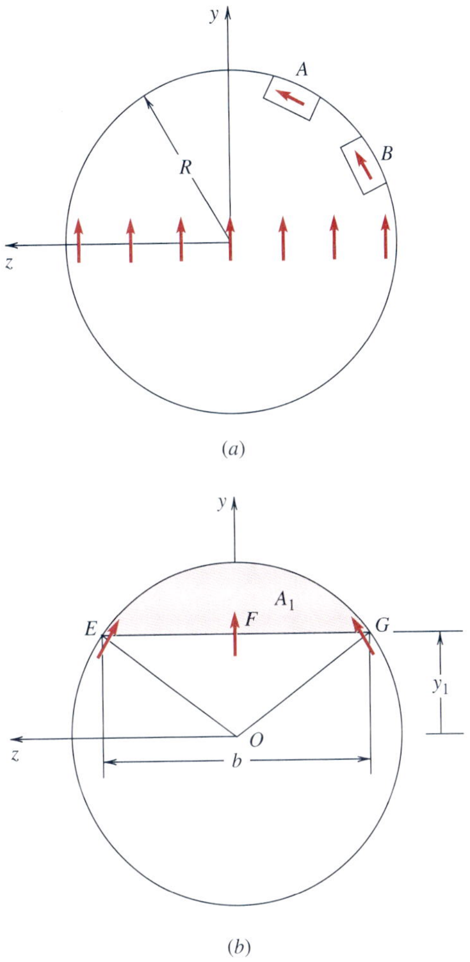

Figure 5.39 Shear stresses on a circular cross section.

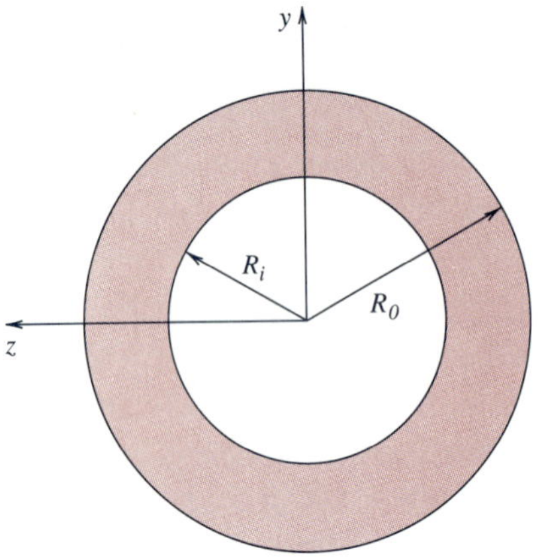

Figure 5.40 Hollow circular section.

so that

$$\tau_{\max} = \frac{VQ}{I_z b} = \frac{\frac{4}{3}V(R_o^3 - R_i^3)}{\pi(R_o^4 - R_i^4)(R_o - R_i)}$$

Since the area of the hollow section is

$$A = \pi(R_o^2 - R_i^2)$$

we have

$$\tau_{\max} = \frac{4V}{3A}\frac{R_o^2 + R_oR_i + R_i^2}{R_o^2 + R_i^2} \tag{5.35}$$

which reduces to the result given in Eq. (5.34) when we set $R_i = 0$ and $R_o = R$.

EXAMPLE 5.15

A hollow pipe and a solid bar of circular cross section of the same material, length, and total weight are loaded as shown in Fig. 5.41. We wish to find the ratios of the maximum normal stresses and maximum shear stresses for the pipe and bar. Let d_s denote the diameter of the solid bar and d_o and d_i denote the outside and inside diameters of the pipe where $d_i = 0.8d_o$. We neglect the weight of the bar and pipe in the calculations of the loads on the bar and pipe.

Since the lengths are the same, the weights will be the same provided that the cross-sectional areas are the same. Therefore

$$\frac{\pi d_s^2}{4} = \frac{\pi}{4}[d_o^2 - (0.8^2)d_o^2] \tag{a}$$

or

$$d_s^2 = 0.36d_o^2$$

The diameter of the solid bar is given therefore by

$$d_s = 0.6d_o \tag{b}$$

The maximum bending moment in both cases occurs at the wall and is given by

$$M_{\max} = PL$$

The maximum bending stress for the pipe is

$$\sigma_{xP} = \frac{PLd_o/2}{(\pi/64)(d_o^4 - 0.8^4d_o^4)}$$

or

$$\sigma_{xP} = \frac{32}{\pi}\frac{PL}{0.590d_o^3} \tag{c}$$

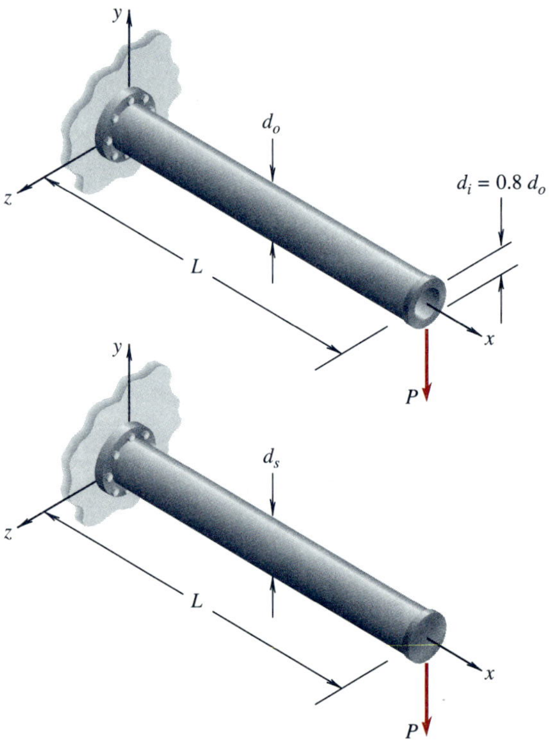

Figure 5.41 Example 5.15

The maximum bending stress for the bar is found by using Eq. (*b*)

$$\sigma_{xB} = \frac{PL\,(0.6d_o/2)}{(\pi/64)\,(0.6)^4 d_o^4}$$

or

$$\sigma_{xB} = \frac{32}{\pi}\frac{PL}{0.216d_o^3} \qquad (d)$$

Therefore, the ratio of the maximum bending stresses becomes

$$\frac{\sigma_{xP}}{\sigma_{xB}} = \frac{(32/\pi)PL/(0.590d_o^3)}{(32/\pi)PL/(0.216d_o^3)} = 0.366 \qquad (e)$$

and we conclude that use of a hollow pipe results in a maximum bending stress which is only 37 percent of that in a solid circular bar of the same weight.

As for the maximum shear stresses, we note first that in each beam V is constant and equal to $-P$. According to Eq. (5.34), the maximum shear stress for the bar with a solid circular cross section is

$$\tau_{\max_B} = \frac{4}{3}\frac{V}{A} = \frac{4}{3}\frac{V}{\pi d_s^2/4} = 1.70\frac{V}{d_s^2} \qquad (f)$$

For the pipe with a hollow section, according to Eq. (5.35),

$$\tau_{\max_P} = \frac{4}{3}\frac{V}{A}\frac{d_o^2 + d_o d_i + d_i^2}{d_o^2 + d_i^2} = \frac{4}{3}\frac{V}{(\pi d_o^4/4)(1-0.8^2)}\frac{d_o^2(1+0.8+0.8^2)}{d_o^2(1+0.8^2)} = 7.02\frac{V}{d_o^2} \qquad (g)$$

Therefore, the ratio of the maximum shear stresses becomes

$$\frac{\tau_{\max_P}}{\tau_{\max_B}} = \frac{7.02V/d_o^2}{1.70V/d_s^2} = 4.13\frac{d_s^2}{d_o^2}$$

or by using Eq. (*b*)

$$\frac{\tau_{\max_P}}{\tau_{\max_B}} = 1.49 \qquad (h)$$

and we see that the hollow pipe has a maximum shear stress which is 49 percent greater than that of a solid bar of the same weight. As we noted before, the shear stresses for slender beams are usually much smaller than the bending stresses for many loadings. In fact, in this case, from Eqs (*c*) and (*g*),

$$\tau_{\max_P} = 0.407\left(\frac{d_o}{L}\right)\sigma_{xP} \qquad (i)$$

The beams considered so far have been of rectangular or circular cross section. However, the equation for the shear stress acting on a section of a beam under transverse loading applies to beams of arbitrary (but symmetric about the y axis) cross sections. If we examine the I beam shown in Fig. 5.42, we can gain further insight into the shear stress distribution in beams.[1] In Fig. 5.42*b* we show a small segment which has been cut from the top flange by a vertical plane through BC. We see that there must be a shear force ΔF_{zx} on the positive z face to maintain equilibrium in the x direction; the subscripts in this case correspond to the force on the z face in the x direction. If we carry out an analysis similar to that which led to Eq. (5.25), we obtain for the shear flow on the positive z face

$$\frac{dF_{zx}}{dx} = -\frac{VQ}{I_z} \qquad (5.36)$$

[1]See, e.g., S. Crandall, N. Dahl, and T. Lardner (eds.), *An Introduction to the Mechanics of Solids*, 2d ed., McGraw-Hill, New York, 1978, p. 438.

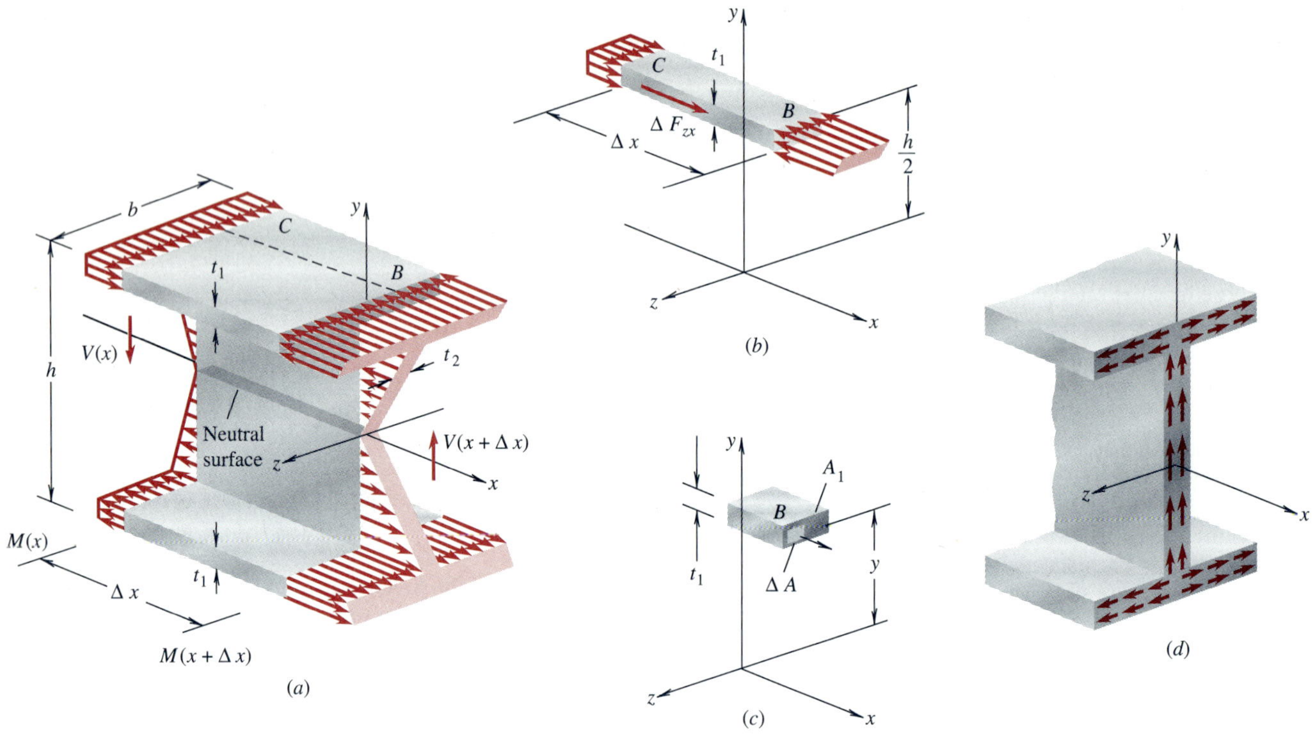

Figure 5.42 Calculation of shear stresses in an I beam.

where Q is the first moment of the shaded area A_1 in Fig. 5.42*c* about the z axis. If we make the assumption that the shear stress is uniform across the thickness t_1 of Fig. 5.42*b*

$$\Delta F_{zx} = \tau_{zx} t_1 \,\Delta x \tag{5.37}$$

we can estimate the shear stress at point B in the flange from Eq. (5.36) and from equality of shear stresses on perpendicular faces to be

$$\tau_{xz} = \tau_{zx} = -\frac{VQ}{t_1 I_z} \tag{5.38}$$

The shear stress in the web can be estimated from Eq. (5.29). In Fig. 5.42*d* we show the shear stress distribution over the cross section of the beam. In each flange, the stress τ_{xz} varies linearly from a maximum at the junction with the web to zero at the edge, while in the web the stress τ_{xy} has a parabolic distribution as in Eq. (5.31). There are also τ_{xy} stresses present in the flanges, but they are small compared with the τ_{xz} stresses illustrated in the sketch. The distribution of stresses at the junction of the web and flange is quite complicated.

5.6 Built-up Beams

In many applications, beams are fabricated or built up from different components, e.g., by gluing rectangular sections together or nailing wooden planks together to create a box beam. Welding or bolting of steel pieces to form a larger cross section is also common. Figure 5.43 shows some typical illustrations of built-up beams. In many situations involving built-up beams, it is necessary to investigate the strength of the joining at the junction between the components. In particular, when a beam with a built-up section is loaded by transverse loads, there is a tendency for the parts of the beam to shear or slide relative to one another along horizontal planes. This sliding action must be restrained by the glue, weld, bolts, or whatever is used to join the pieces. Since the restraining force is a shearing force directed along the axis of the beam, we can use the expression for the shear flow derived earlier, Eq. (5.25). However, we must pay particular attention to the area over which the shear flow acts in order to determine Q and b. We recall that the shear flow f is the shear force per unit length *along* the axis of the beam.

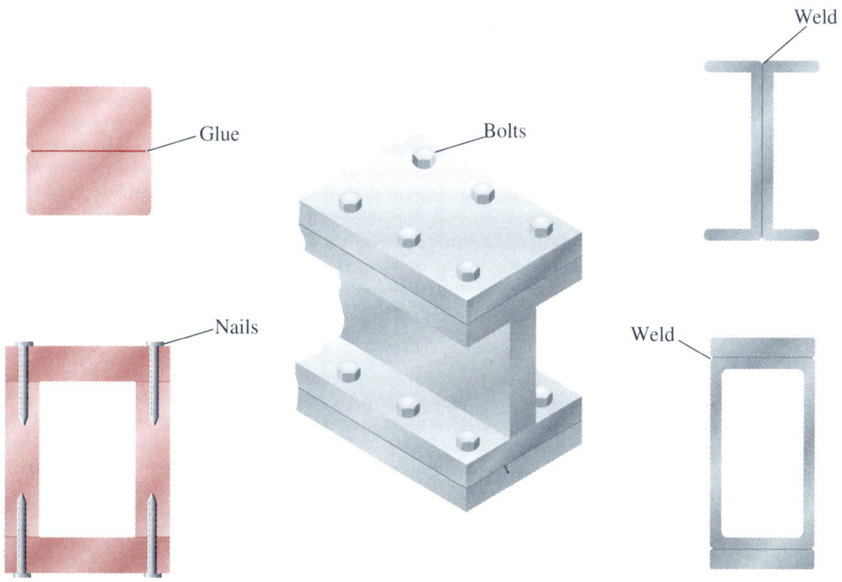

Figure 5.43 Illustrations of built-up beams.

EXAMPLE 5.16

Three 50 mm × 100 mm pine wood members are glued together as shown in Fig. 5.44 and are used as a cantilever beam of length equal to 3.0 m. If the glue strength between the members is $\tau = 100$ kPa, what is the maximum load P that can be applied to the cantilever beam if the glue strength is not to be exceeded? Neglect the weight of the members.

When the beam is loaded, the top member tries to slide across the center member, as shown in Fig. 5.44*c*, and shear stresses are developed in the horizontal glue plane. The magnitude of the shear stress is given by

$$\tau = \frac{VQ}{I_z b} \qquad (a)$$

where for this beam at any section $V = -P$.

The first moment of area A_1 is

$$Q = \int_{A_1} y\,dA = \bar{y}_1 A_1 = 50(50 \times 100) = 2.5 \times 10^5\ \text{mm}^3$$
$$= 2.5 \times 10^{-4}\ \text{m}^3 \qquad (b)$$

The moment of inertia of the total cross section about the neutral axis is

$$I_z = \tfrac{1}{12} bh^3 = \tfrac{1}{12}(100)(150)^3 = 2.8125 \times 10^7\ \text{mm}^4$$
$$= 2.8125 \times 10^{-5}\ \text{m}^4 \qquad (c)$$

and the value of $b = 100$ mm $= 0.1$ m.

Therefore the maximum load P from Eqs. (*a*) to (*c*) is

$$P_{\max} = \frac{\tau I_z b}{Q} = \frac{100 \times 10^3 \times 2.8125 \times 10^{-5} \times 0.1}{2.5 \times 10^{-4}}$$
$$= 1.125\ \text{kN} \qquad (d)$$

At this load the maximum normal stress in the beam is

$$\sigma = \frac{My}{I_z} = \frac{1.125 \times 10^3 \times 3 \times 75 \times 10^{-3}}{2.8125 \times 10^{-5}} = 9.0\ \text{MPa} \qquad (e)$$

and the maximum shear stress at the neutral axis of the rectangular cross section is

$$\tau = 1.5\frac{V}{A} = \frac{1.5 \times 1.125}{1.5 \times 10^{-2}} = 112.5\ \text{kPa} \qquad (f)$$

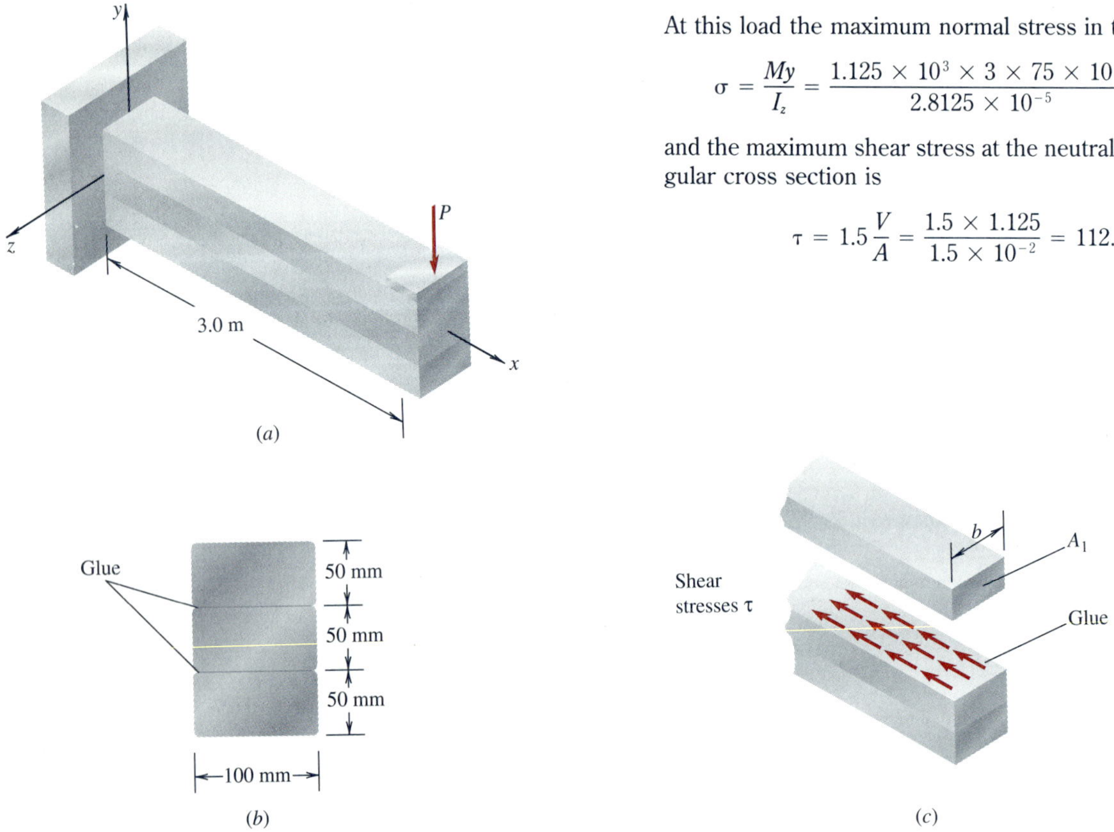

Figure 5.44 Example 5.16

EXAMPLE 5.17

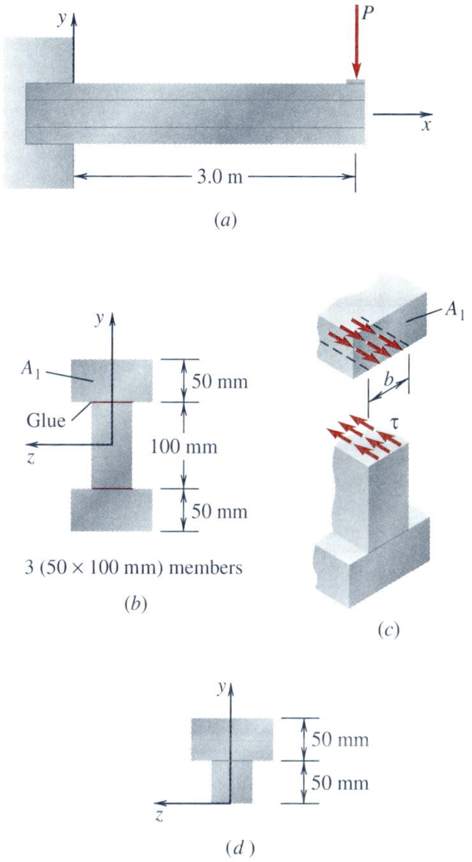

Figure 5.45 Example 5.17

Upon consideration of the results in Example 5.16, it was decided to reconfigure the three members to create an I-beam shape as shown in Fig. 5.45*b*. Also see Fig. 5.44. If the glue strength is still taken as $\tau = 100$ kPa, what is the maximum load P that can be carried by the cantilever beam in the new configuration without exceeding the glue strength? Neglect the weight of the members.

When the beam is loaded, the top member tries to shear across the center member, as shown in Fig. 5.45*c*. The width b corresponds to the width of the area over which the shear stress is acting. The magnitude of the shear stress is given by

$$\tau = \frac{VQ}{I_z b} \qquad (a)$$

where $V = -P$.

For this configuration the first moment of area A_1 is

$$Q = \int_{A_1} y\,dA = \bar{y}_1 A_1 = 75(50 \times 100) = 3.75 \times 10^5 \text{ mm}^3$$
$$= 3.75 \times 10^{-4} \text{ m}^3 \qquad (b)$$

The moment of inertia of the total cross section about the neutral axis is

$$I_z = \tfrac{1}{12}(50)(100)^3 + 2[\tfrac{1}{12}(100)(50)^3 + (75)^2(50 \times 100)]$$
$$= 6.25 \times 10^7 \text{ mm}^4 = 6.25 \times 10^{-5} \text{ m}^4 \qquad (c)$$

The value of $b = 50$ mm $= 5 \times 10^{-2}$ m.

Therefore the maximum load is

$$P_{\max} = \frac{\tau I_z b}{Q} = \frac{100 \times 10^3 \times 6.25 \times 10^{-5} \times 5 \times 10^{-2}}{3.75 \times 10^{-4}}$$
$$= 0.833 \text{ kN} \qquad (d)$$

We note that the maximum allowable load in this case is smaller than that calculated in Example 5.16. The maximum normal stress in the beam at the load $P_{\max}$ is also less

$$\sigma = \frac{My}{I_z} = \frac{0.833 \times 10^3 \times 3 \times 100 \times 10^{-3}}{6.25 \times 10^{-5}}$$
$$= 4.0 \text{ MPa} \qquad (e)$$

The maximum shear stress at the neutral axis of the cross section is given by

$$\tau = \frac{VQ}{I_z b} \qquad (f)$$

The first moment of the area above the neutral surface, $y = 0$, is given by (Fig. 5.45*d*)

$$Q = \int y\,dA = 25(50 \times 50) + 75(50 \times 100)$$

$$= 4.375 \times 10^5 \text{ mm}^3 \qquad (g)$$

$$= 4.375 \times 10^{-4} \text{ m}^3$$

Therefore the maximum shear stress on the section at the neutral surface is given by

$$\tau = \frac{0.833 \times 10^3 \times 4.375 \times 10^{-4}}{6.25 \times 10^{-5} \times 50 \times 10^{-3}} = 117 \text{ kPa} \qquad (h)$$

A summary of the results for the two configurations shown in Figs. 5.44 and 5.45 is given below:

	Maximum load based on τ_g	σ_{max}	τ_{max}
Fig. 5.44	1.125 kN	9.0 MPa	113 kPa
Fig. 5.45	0.833 kN	4.0 MPa	117 kPa

Glue strength $\tau_g = 100$ kPa

We see that the I-shaped configuration of the section in Fig. 5.45 can carry 74 percent of the load for the rectangular section in Fig. 5.44. The maximum bending stress is reduced to 44 percent of the bending stress of the rectangular section.

EXAMPLE 5.18

A built-up aluminum beam is fabricated from two channel sections and two strips with $\frac{1}{4}$-in-diameter bolts with a spacing s between the bolts, as shown in Fig. 5.46*a* and *b*. If each bolt can safely resist a shear force across it of 400 lb, what is the bolt spacing required when the maximum shear force V in the beam arising from the applied transverse load is 10,000 lb?

The centerpiece of the built-up section shown in Fig. 5.46*c* tends to shear between the side pieces because of the unbalance in axial forces arising from the variable bending moment on the beam. Force equilibrium in the axial direction of the centerpiece requires a shear flow of

$$f = \frac{VQ}{I_z} \qquad (a)$$

Recall that the shear flow f is the rate of change of the shear force per unit length along the beam.

The total available shear flow acting on the centerpiece from the action of the bolts on each side of the centerpiece (Fig. 5.46*c* and *d*) is

$$f = \frac{2(\text{allowable shear force in each bolt})}{\text{space between bolts}}$$

$$= \frac{2(400)}{s} \qquad (b)$$

Therefore, the maximum spacing of the bolts follows from balancing the required shear flow with the shear flow available, or, from Eqs. (*a*) and (*b*),

$$\frac{VQ}{I_z} = \frac{800}{s}$$

or

$$s = \frac{800 I_z}{VQ} \qquad (c)$$

The moment of inertia of the total cross section about the neutral axis is

$$I_z = \tfrac{1}{12}[6(10)^3 - 4(8)^3] = 329.3 \text{ in}^4 \qquad (d)$$

The first moment of the area of the centerpiece about the neutral axis is (Fig. 5.46*c*)

$$Q = (1)(2)(4.5) = 9 \text{ in}^3 \qquad (e)$$

Therefore,

$$s = \frac{(800)(329.3)}{(10{,}000)(9)} = 2.93 \text{ in} \qquad (f)$$

This value is the maximum allowable spacing of the bolts.

An alternative way to solve this problem is to first ask, What is the value of the average shear stress on the area of the centerpiece required for axial equilibrium? This shear stress is given by

$$\tau = \frac{VQ}{I_z b} \qquad (g)$$

where b is the total width over which the shear stress acts. In this case, b is twice the thickness of the centerpiece

$$b = 2(1 \text{ in}) = 2 \text{ in} \qquad (h)$$

since the average shear stress acts on both sides of the centerpiece. The required shear stress of Eq. (*g*) is balanced by the

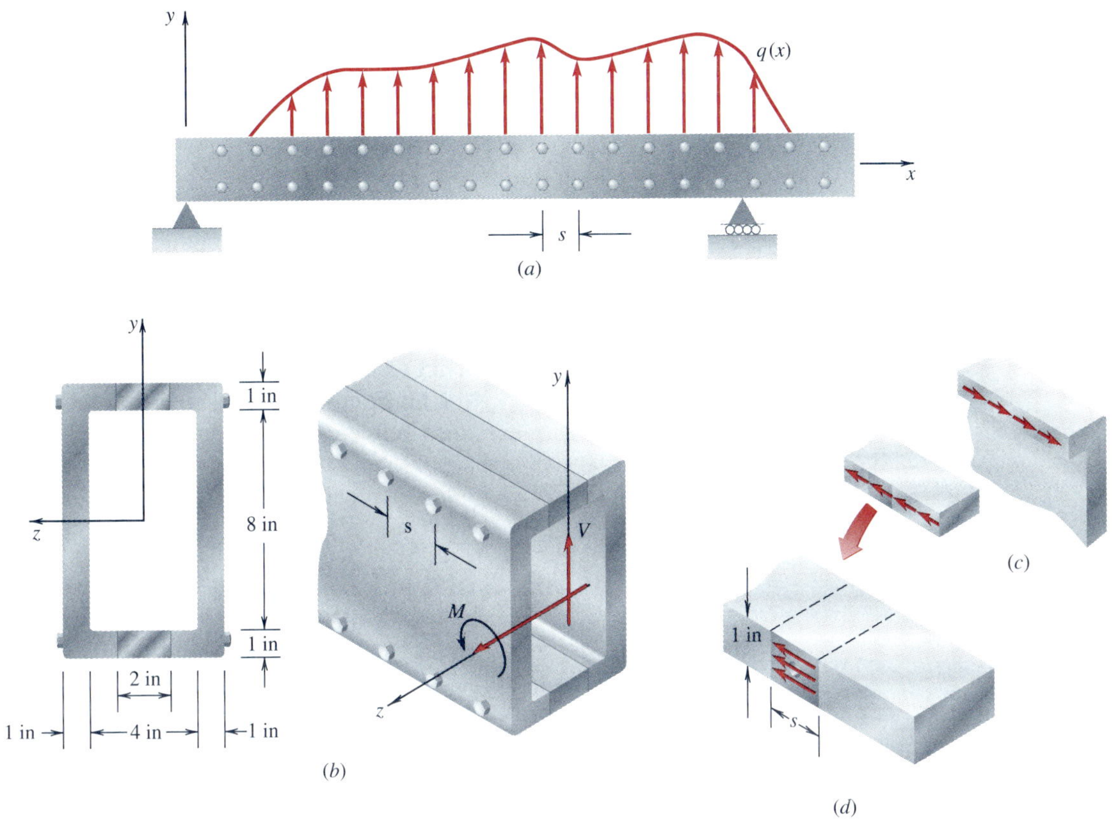

Figure 5.46 Example 5.18

average shear stress arising at each bolt location if we assume that the allowable shear force in the bolt is "spread over" an area given by $(s \cdot 1)$ in^2 around the bolt (Fig. 5.46*d*). Therefore the average shear stress induced by a bolt around the bolt location is

$$\tau_{bolt} = \frac{400}{(s)(1)} \qquad (i)$$

The required shear stress of Eq. (*g*) is balanced by this induced shear stress from the bolt, so we have

$$\frac{VQ}{I_z b} = \frac{400}{(s)(1)}$$

to give

$$s = \frac{400 b I_z}{VQ} \qquad (j)$$

where $b = 2$ in; therefore s is found as before.

EXAMPLE 5.19

A brass beam is fabricated by soldering the box sections to the 6-mm-thick plate as shown in Fig. 5.47*a* and *b*. If the shear stress in the solder is not to exceed 11 MPa, what is the maximum shear force V_{max} that the beam can carry in service?

The solder has to carry the unbalanced bending stress in the axial direction acting on the cross-sectional area of the box shown in Fig. 5.47*c*. If we assume that the shear stress is equal in each solder joint, the shear stress in each joint is given by

$$\tau = \frac{VQ}{I_z 2(3)} \qquad (a)$$

where the width of each solder joint is 3 mm (Fig. 5.47*c*) and

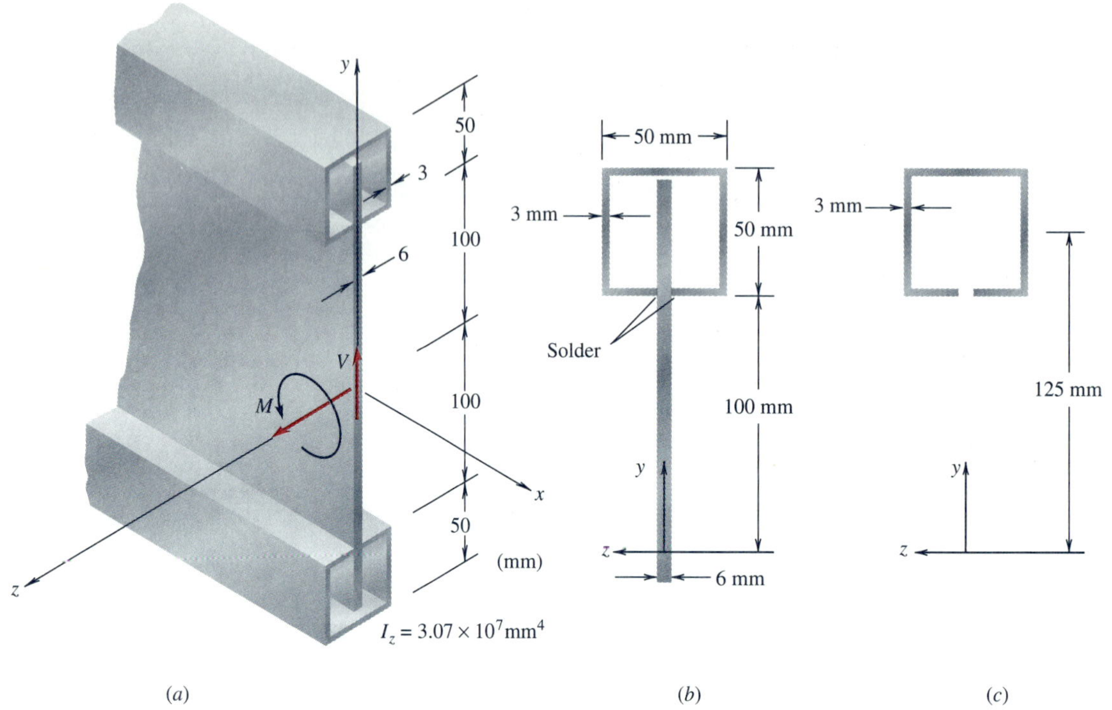

Figure 5.47 Example 5.19

Q is the first moment of the area shown in Fig. 5.47*c* about the neutral axis. If we approximate this area as a complete square hollow section, we have

$$Q = 125[(50)^2 - (44)^2] = 7.05 \times 10^4 \text{ mm}^3 \qquad (b)$$

The maximum shear force V_{max} then follows from Eq. (*a*), by using I_z from Fig. 5.47*a*,

$$V_{max} = \frac{(2)(3\text{ mm})(11\text{ N/mm}^2)(3.07 \times 10^7\text{ mm}^4)}{7.05 \times 10^4\text{ mm}^3} \qquad (c)$$
$$= 2.87 \times 10^4\text{ N} = 28.7\text{ kN}$$

EXAMPLE 5.20

A beam is to be fabricated as a built-up beam from a W12 × 50 wide-flange beam and two C12 × 30 channel sections as shown in Fig. 5.48*a*. The C sections are to be joined to the wide-flange beam by bolts spaced 6 in apart along the beam. If the maximum shear force V in the beam in service is 20 kips, what is the minimum allowable shear force that each bolt must carry? Neglect the weight of the beam.

If we consider the top channel as shearing across the beam (Fig. 5.48*b*), then the shear flow required for equilibrium is

$$f = \frac{VQ}{I_z} \quad \text{lb/in} \qquad (a)$$

where Q is the first moment of the C section about the neutral axis.

The shear flow from the two rows of bolts is

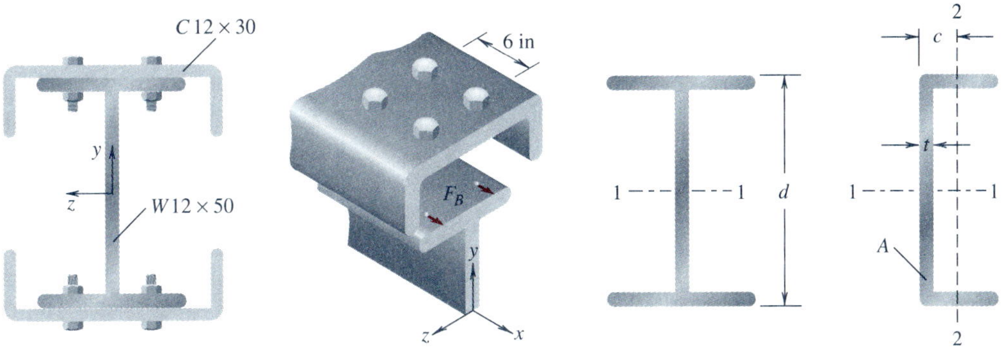

(a) (b) (c) (d)

Figure 5.48 Example 5.20

$$\frac{2F_b}{(6 \text{ in})} \quad \text{lb/in} \tag{b}$$

where F_b is the shear force in pounds in each bolt. Therefore, from Eqs. (a) and (b) we have

$$F_b = 3\frac{VQ}{I_z} \tag{c}$$

where $V = 20$ kips. It remains to calculate Q and I_z. The value of I_z is obtained by using the properties of each section given in App. C. For the W12 × 50 section we have (Fig. 5.48c)

$$I_1 = 394 \text{ in}^4 \qquad d = 12.19 \text{ in} \tag{d}$$

For the C12 × 30 section we have (Fig. 5.48d)

$$\begin{aligned} I_2 &= 5.14 \text{ in}^4 \qquad c = 0.674 \text{ in} \\ A &= 8.82 \text{ in}^2 \qquad t = 0.510 \text{ in} \end{aligned} \tag{e}$$

The moment of inertia of the built-up beam follows from the parallel-axis theorem, by using the above properties,

$$\begin{aligned} I_z &= 394 + 2[5.14 + 8.82(6.095 + 0.510 - 0.674)^2] \\ &= 1025 \text{ in}^4 \end{aligned} \tag{f}$$

The first moment of the C section about the neutral axis is

$$\begin{aligned} Q = \int y\,dA = \bar{y}A &= (6.095 + 0.510 - 0.674)8.82 \\ &= 52.31 \text{ in}^3 \end{aligned} \tag{g}$$

Therefore, from Eq. (c)

$$F_b = \frac{(3)(20{,}000)(52.31)}{1025} = 3062 \text{ lb} \tag{h}$$

is the minimum value of the force which each bolt must carry to support the value of the maximum shear force V.

5.7 Concluding Remarks

In this chapter we have presented methods for the calculation of the normal stress distribution and the shear stress distribution in beams. The equation for the normal stress, Eq. (5.17), is fundamental to the understanding of the load-carrying capacity of members that are exposed to transverse or bending loads. Many preliminary designs of machine components and structures are based on this formula. It is an equation that appears often in engineering design and analysis.

To determine the normal stress due to bending, we need in general to determine the largest bending moment in the beam and the moment of inertia of the cross section about the neutral axis. We noted that each of these steps can be carried out separately, and often computer methods can be used to shorten the work. However, while a computer program may simplify the effort, we must carefully check that the results from a computer solution make sense.

For the beams treated thus far, we have been able to obtain the bending moment distribution in the beam directly from equilibrium equations; i.e., the beam problems were statically determinate. If a beam problem is statically indeterminate, then the bending moment distribution and as a consequence the maximum stress in the beam cannot be found until certain geometric conditions are satisfied along the beam. We will discuss beam deflections in the next chapter.

Equation (5.29) for the shear stress in a uniform beam was used to calculate the shear stress distribution across a rectangular section (Fig. 5.35) and to indicate the shear stress distribution in the web and flange of an I-beam section (Fig. 5.42). We found in Example 5.14, in which a cantilever beam with a rectangular cross section was loaded by an end load, that the maximum shear stress in the beam was an order of magnitude smaller than the maximum normal stress due to bending. This relative magnitude of the stresses holds approximately for homogeneous beams of different cross sections and loading.

The use of the formulas for shear stress. Eq. (5.29), and for shear flow, Eq. (5.25), is important for built-up beams in which sections of a beam may shear relative to one another. Again we emphasize that the shear flow formula arises from considerations of force equilibrium along the axis of the beam.

Finally we note that the arguments leading up to the formulas for homogeneous bars can be applied for the case of symmetric composite beams that have been made by joining two materials of different elastic properties (Prob. 5.7-1) or for reinforced concrete (Prob. 5.7-2).

PROBLEMS

5.1-1 Given a circle of radius R drawn in the xy plane, what are the radius of curvature and the curvature of points on the circle? If radius R is doubled, halved, what are the radius of curvature and the curvature of the new circles?

5.2-1 For the beams shown in Figs. 5.3*a* and 5.4*a*, sketch the shear force diagram.

5.2-2 A steel bar of cross section shown in Fig. P5.2-2 is loaded by constant bending moments M at its ends. Determine the maximum tensile and compressive

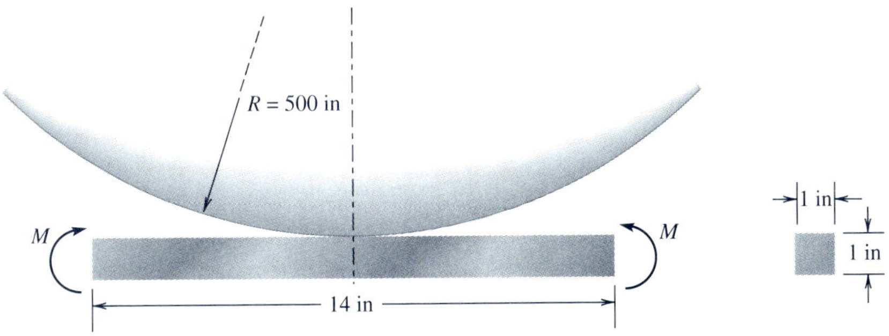

Fig. P5.2-2

strain in the bar if the moments M are such as to bring the bar in contact with a cylinder of radius $R = 500$ in.

5.3-1 Consider Example 5.2, Fig. 5.11. Sketch the normal bending stress distribution across the midsection of the specimen at $x = 19$ mm when $P = 15.55$ N.

5.3-2 Consider Example 5.2, Fig. 5.11. If the values of b and h are doubled while all other geometric lengths remain the same, estimate the value of the load P to cause fracture of the specimen if the fracture stress is approximately 240 MPa. Can we draw any simple conclusions about the effect of changing b and h for a rectangular section on the value of the maximum stress?

5.3-3 Consider Example 5.3, Fig. 5.12. Verify Eqs. (*c*) to (*f*).

5.3-4 Determine the location of the centroid and the moment of inertia about the z axis passing through the centroid for the cross sections shown in Fig. P5.3-4 on the next page. First carry out the calculations by hand and, if appropriate, use the computer program MOMENTS OF INERTIA to check.
(*a*) Fig. P5.3-4*a*
(*b*) Fig. P5.3-4*b*
(*c*) Fig. P5.3-4*c*
(*d*) Fig. P5.3-4*d*
(*e*) Fig. P5.3-4*e*
(*f*) Fig. P5.3-4*f*
(*g*) Fig. P5.3-4*g*
(*h*) Fig. P5.3-4*h*
(*i*) Fig. P5.3-4*i*
(*j*) Fig. P5.3-4*j*

5.4-1 Consider Example 5.4, Fig. 5.15. Sketch the normal bending stress distribution across the section at $x = 9$ ft when $L = 12$ ft. Also sketch the normal bending stress distribution across the section at $x = 3$ ft.

5.4-2 A wooden beam supports two concentrated loads as shown in Fig. P5.4-2. If $P = 5000$ lb, $L = 14$ ft, $a = 4$ ft, $b = 6$ in, and $h = 8$ in, find the maximum compressive and tensile normal stress due to bending in the beam.

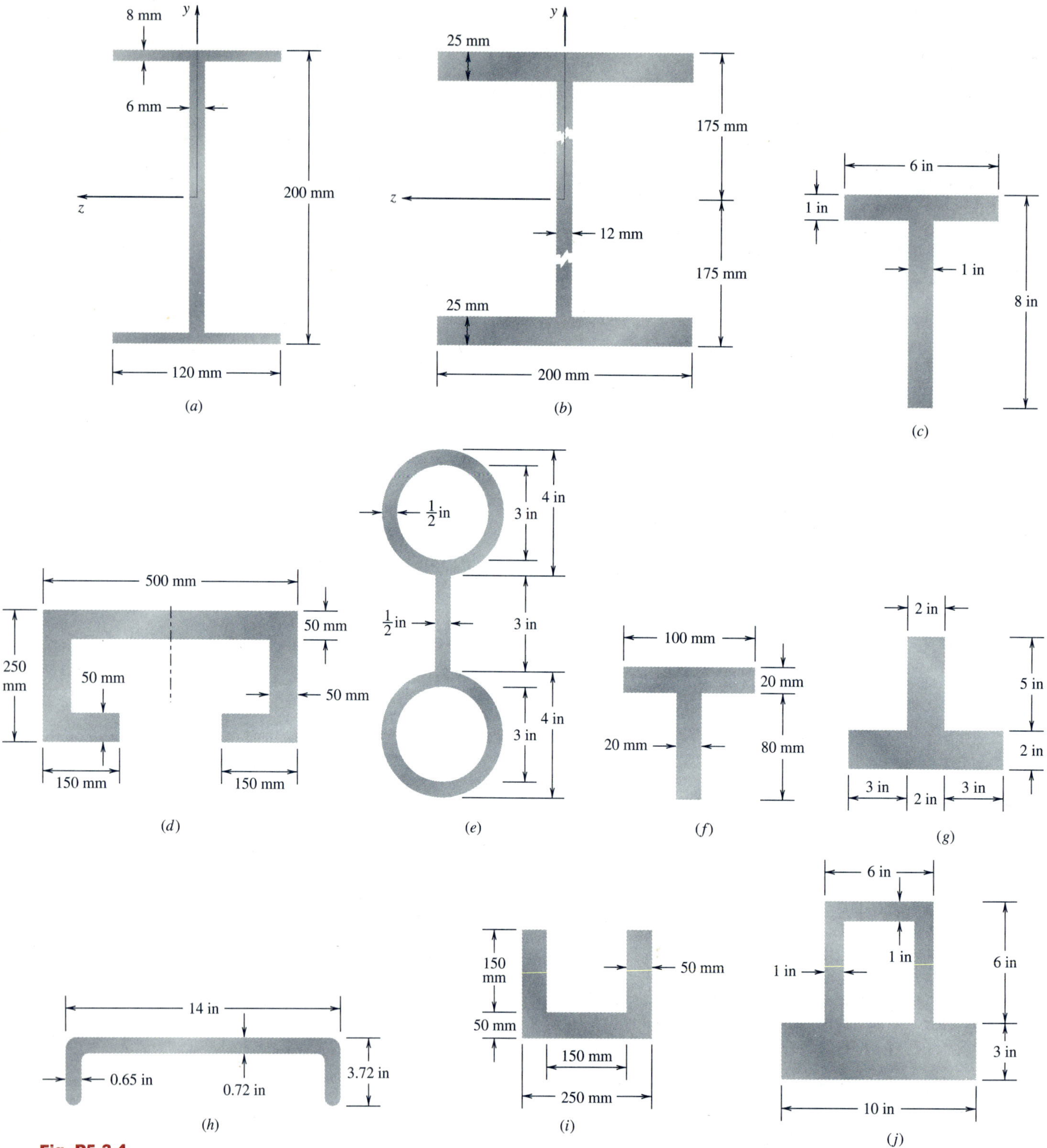

8 mm
y
6 mm
z
200 mm
120 mm
(a)
y
25 mm
175 mm
z
12 mm
175 mm
25 mm
200 mm
(b)
6 in
1 in
1 in
8 in
(c)
500 mm
250 mm
50 mm
50 mm
50 mm
150 mm
150 mm
(d)
$\frac{1}{2}$ in
4 in
3 in
$\frac{1}{2}$ in
3 in
4 in
3 in
(e)
100 mm
20 mm
20 mm
80 mm
(f)
2 in
5 in
2 in
3 in
2 in
3 in
(g)
14 in
0.65 in
0.72 in
3.72 in
(h)
150 mm
50 mm
50 mm
150 mm
250 mm
(i)
6 in
1 in
1 in
6 in
3 in
10 in
(j)

Fig. P5.3-4

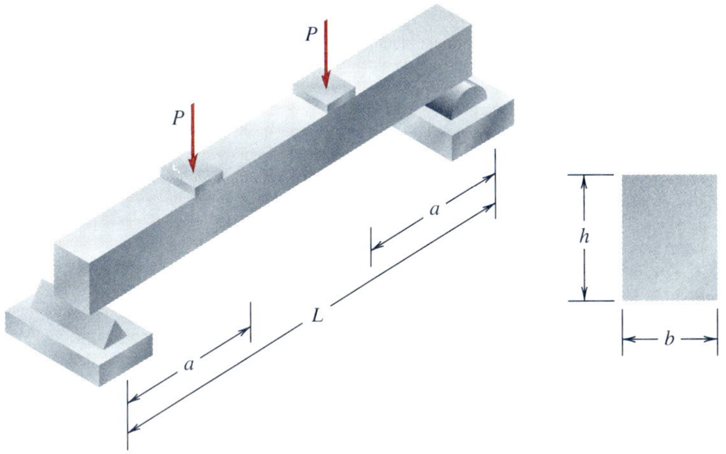

Fig. P5.4-2

5.4-3 In Prob. 5.4-2, the cross section of the beam was 6 in × 8 in. If the net dimensions of a nominal 6-in × 8-in beam are used, namely, 5.5 in × 7.5 in (see App. D), calculate the maximum compressive and tensile normal stress due to bending in the beam. Compare to the values obtained in Prob. 5.4-2.

5.4-4 Consider Fig. P5.4-2. Calculate the maximum bending stresses in the beam if $P = 1750$ lb, $L = 10$ ft, $a = 3$ ft, $b = 5.5$ in, and $h = 9.5$ in.

5.4-5 Consider Example 5.4, Fig. 5.15. Determine the value of L_a including the weight of the beam.

5.4-6 Consider Fig. P5.4-2. If the maximum allowable normal stress due to bending in the wooden beam is 10 MPa, find the maximum allowable span L of the beam. Take $P = 8$ kN, $a = L/3$, $b = 200$ mm, and $h = 300$ mm. Neglect the weight of the beam.

5.4-7 A simply supported wooden beam of rectangular cross section carries a uniform load including the weight of the beam, as shown in Fig. P5.4-7. If $L = 4.0$ m, $w_0 = 5$ kN/m, $b = 150$ mm, and $h = 220$ mm, determine the maximum bending stresses due to bending in the beam.

5.4-8 Consider Fig. P5.4-7. If the applied uniform loading is 40 lb/ft and the maximum allowable normal bending stress in the wooden beam is 800 psi, what is the maximum allowable span for a nominal 2 × 4 piece of lumber? The net dimensions of a 2 × 4 are 1.5 in × 3.5 in, and the weight per foot is approximately 1.3 lb/ft (App. D).

5.4-9 Consider Fig. P5.4-7. If the applied uniform loading is $w_0 = 400$ N/m and the maximum allowable normal bending stress in the wooden beam is 8 MPa, what is the maximum allowable span L when $b = h = 160$ mm?

Fig. P5.4-7

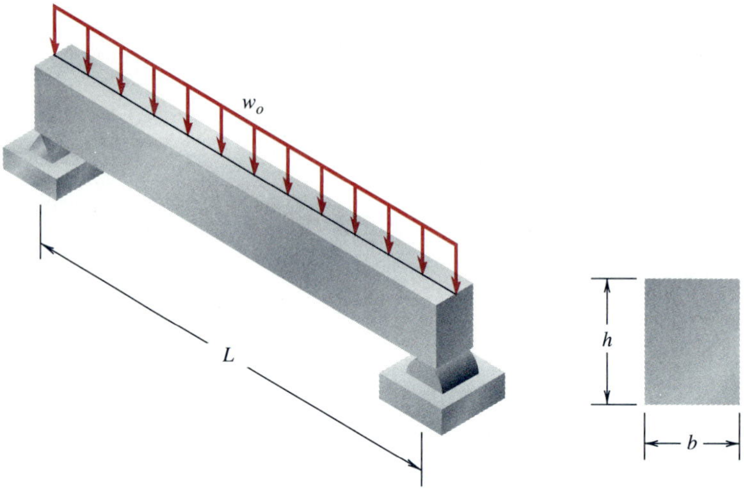

5.4-10 Consider Example 5.5, Fig. 5.16. If the length of the beam is increased to 7.5 m, what is the maximum tensile stress in the beam? All other values of the parameters remain the same. Neglect the weight of the beam.

5.4-11 Consider Example 5.5, Fig. 5.16. The beam shown in Fig. 5.16*b* is not available and is replaced by a beam whose flange thickness is 5.75 mm, and all other values except for the depth of the beam are the same. Find the maximum tensile and compressive stresses in the beam. Neglect the weight of the beam.

5.4-12 Consider the steel cantilever beam shown in Fig. P5.4-12. If $L = 18$ ft and $P = 1200$ lb, find the maximum normal tensile and compressive stresses in the beam. The cross-sectional dimensions are $d = 8.06$ in, $t_w = 0.28$ in, $w = 6.5$ in, and $t_f = 0.46$ in. Neglect the weight of the beam.

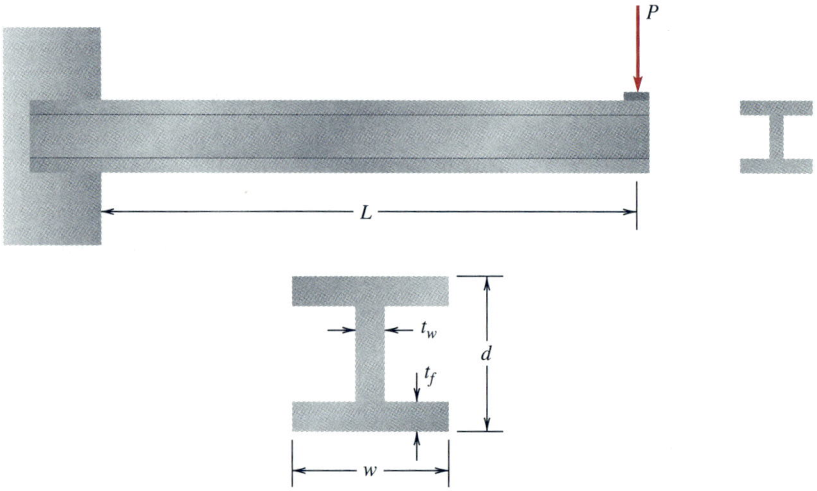

Fig. P5.4-12

5.4-13 Consider Fig. P5.4-12. Calculate the maximum normal bending stresses in the beam if $P = 5$ kN and $L = 5$ m. The cross-sectional dimensions are $d = 200$ mm, $w = 130$ mm, $t_f = 7.75$ mm, and $t_w = 7.75$ mm. Neglect the weight of the beam.

5.4-14 Consider Fig. P5.4-12. The cross-sectional dimensions are $d = 4.00$ in, $t_w = 0.326$ in, $w = 2.796$ in, and $t_f = 0.293$ in; and the load P is 1100 lb. If the maximum allowable stress due to bending is 20 ksi, find the maximum allowable length L. Neglect the weight of the beam.

5.4-15 Consider Example 5.6, Fig. 5.17. Sketch the normal stress distribution across the T section at the midpoint $x = 9$ ft of the beam. What are the net *force* acting on area II and the net *force* acting on area I of Fig. 5.17*c*?

5.4-16 Consider Example 5.6, Fig. 5.17. If the T beam is now turned over so that the stem of the T is on the bottom, determine the maximum tensile and compressive stresses in the beam. If the beam is fabricated from concrete that is weaker in tension than in compression, what configuration of the beam would be better for a parking garage floor application?

5.4-17 A simply supported beam of length L carries a concentrated load P at the midpoint, as shown in Fig. P5.4-17. If $L = 3.0$ m, the maximum allowable tensile stress is 40 MPa, and the maximum allowable compressive stress is 65 MPa, find the maximum allowable load P that can be put on the beam. Neglect the weight of the beam.

5.4-18 A T beam is loaded as shown in Fig. P5.4-18. If the material of the beam has a maximum allowable stress of 5000 psi in tension and 20,000 psi in compression, find the maximum allowable value of P.

5.4-19 A steel box beam supports a constant distributed load of $w = 3000$ lb/ft, as shown in Fig. P5.4-19. Determine the maximum tensile and compressive stresses in the beam. Take $t = \frac{1}{2}$ in, and neglect the weight of the beam.

5.4-20 A steel box beam supports a constant distributed load of $w = 3000$ lb/ft, as shown in Fig. P5.4-19. Determine the maximum tensile and compressive stresses in the beam. Take $t = \frac{3}{4}$ in, and neglect the weight of the beam.

5.4-21 A simply supported beam carries two segments of uniform load, as shown in Fig. P5.4-21. Find the maximum normal bending stresses in the beam. Take $L = 15$ ft, $w_1 = 1$ kip/ft, and $w_2 = 3$ kips/ft; $a = L/2$, $b = L/4$. Neglect the weight of the beam.

5.4-22 A simply supported beam carries two segments of uniform load, as shown in Fig. P5.4-21. Find the maximum normal bending stresses in the beam. Take $L = 20$ ft, $w_1 = 600$ lb/ft, $w_2 = 750$ lb/ft, $a = L/2$, and $b = L/3$. Neglect the weight of the beam.

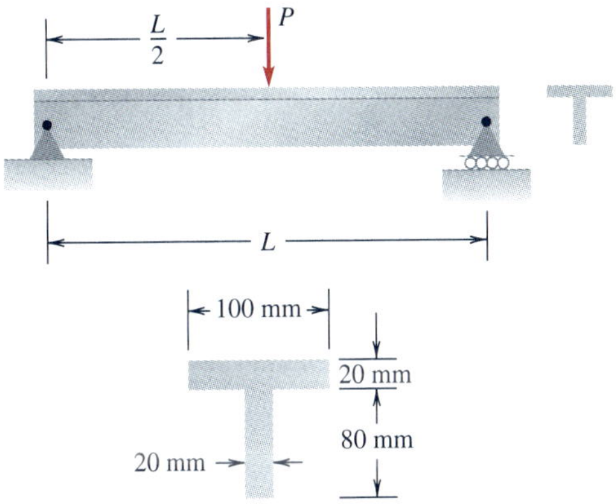

Fig. P5.4-17

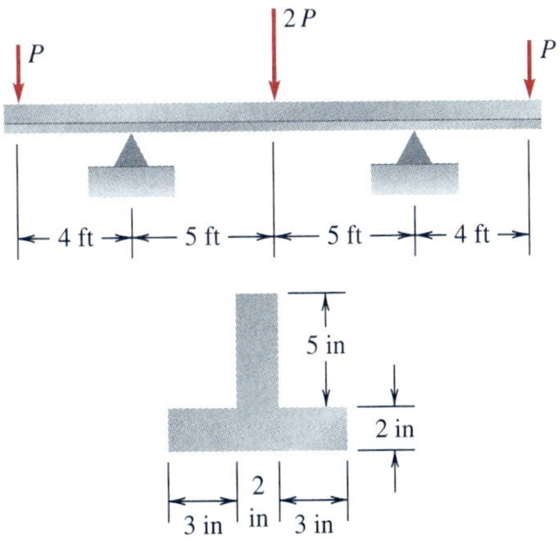

Fig. P5.4-18

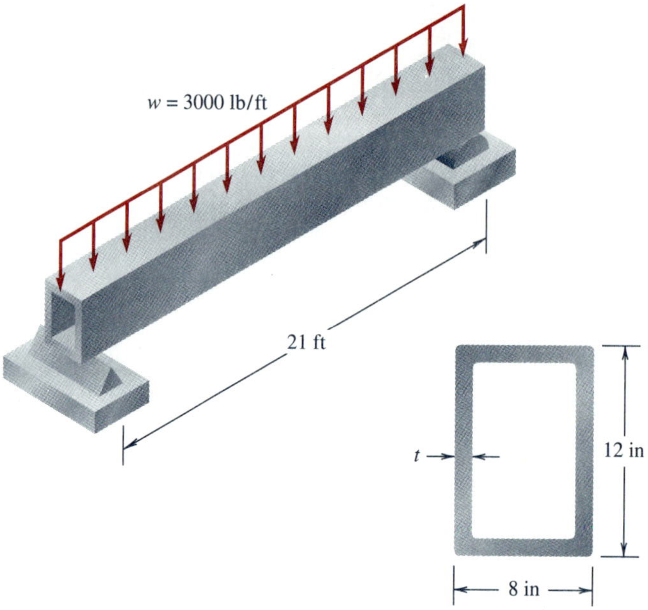

Fig. P5.4-19

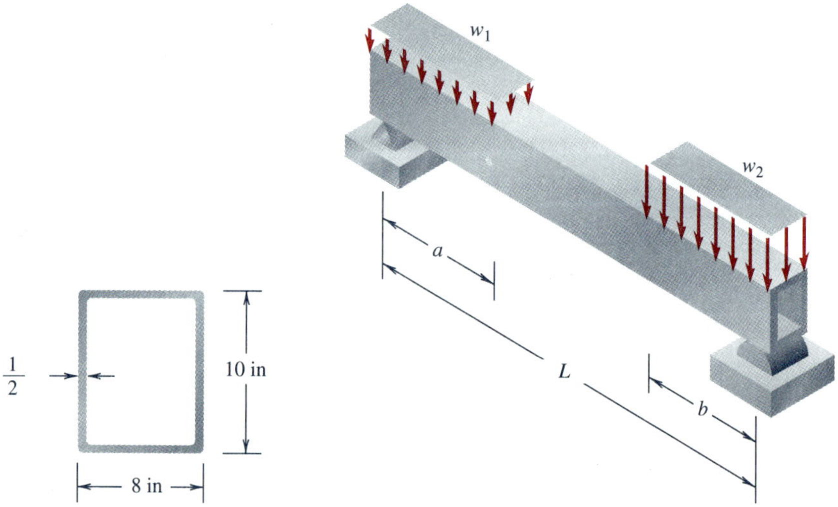

Fig. P5.4-21

5.4-23 A simply supported beam is loaded as shown in Fig. P5.4-23. Find the maximum tensile and compressive stresses in the beam. Neglect the weight of the beam.

5.4-24 Consider Example 5.9, Fig. 5.26. Verify the expression for the moment of inertia given by Eq. (*a*). Note that $I_z = \pi d^4/64$ or $I_z = \frac{1}{2}J$, where J is the polar moment of inertia of the cross section; recall Eq. (3.9) and Fig. 3.11.

5.4-25 Consider Example 5.9, Fig. 5.26. Sketch the distribution of the normal bending stress across section A.

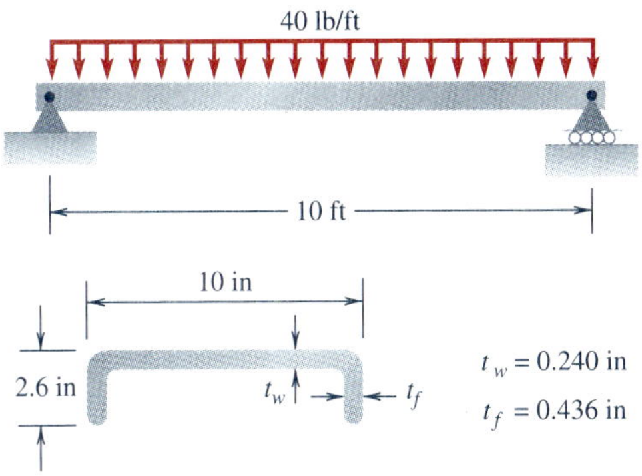

Fig. P5.4-23

5.4-26 A heavy steel member is placed on two supports as shown in Fig. P5.4-26. The member is loaded with a uniform load of 1 kip/ft.

(*a*) Find the distance a such that the maximum bending moment in the beam is as small as possible; see Prob. 4.5-7.

(*b*) If the cross section of the beam is a channel section as shown, find the maximum bending stresses in the beam for the distance a found in part (*a*). Compare the maximum bending stresses to the values of the stresses at $a = 10$ ft.

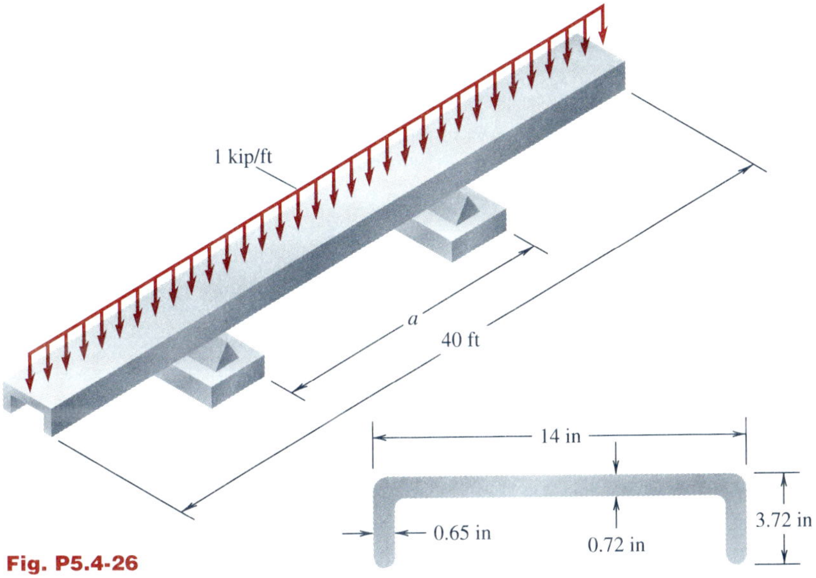

Fig. P5.4-26

5.4-27 A new design of a concrete railroad tie is shown in Fig. P5.4-27. The tie is subjected to two concentrated loads as shown, and the underlying ballast is assumed to give rise to a constant distributed loading q over the length of the tie. Determine the maximum bending stresses in the tie; $P = 45$ kips.

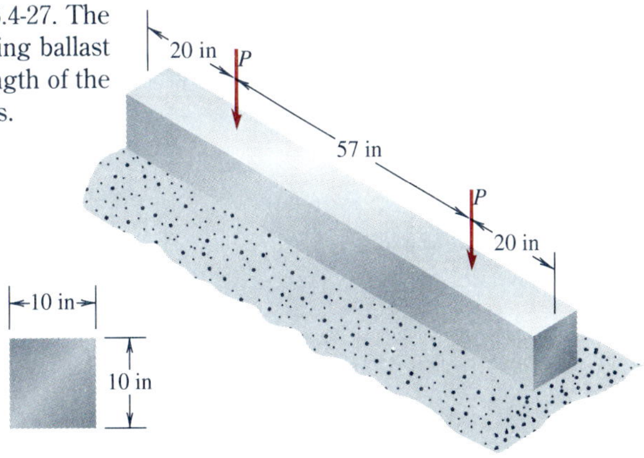

Fig. P5.4-27

5.4-28 A steel beam is loaded as shown in Fig. P5.4-28. Determine the maximum bending stresses in the beam. Neglect the weight of the beam.

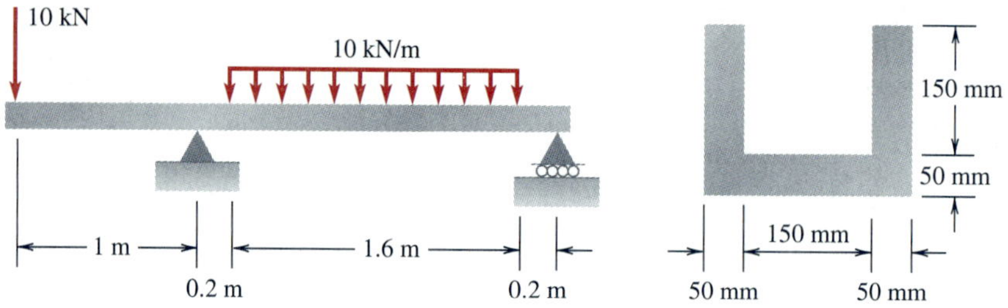

Fig. P5.4-28

5.4-29 A beam is loaded as shown in Fig. P5.4-29. If the maximum allowable normal stress in tension is 20 MPa and the maximum allowable normal stress in compression is 10 MPa, determine the maximum allowable value of w for the beam. Take $L = 3$ m, and neglect the weight of the beam.

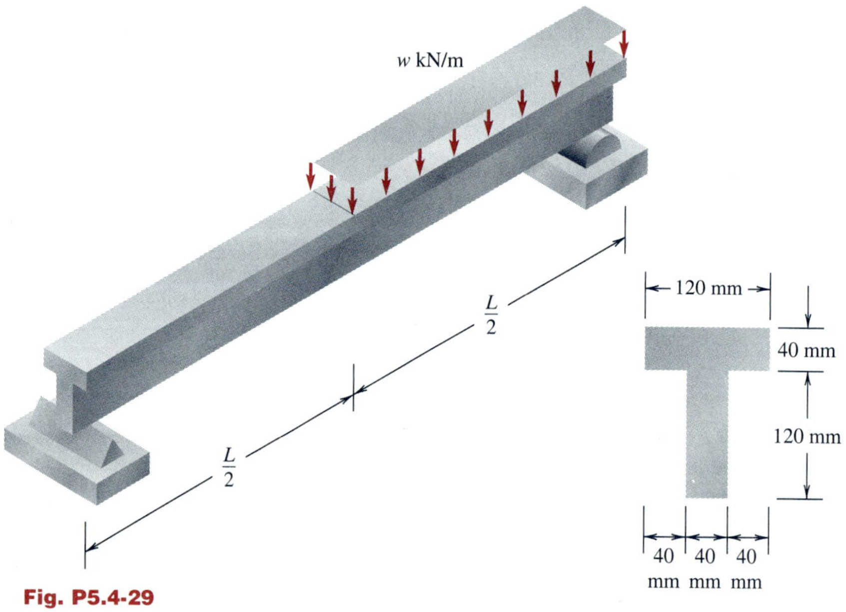

Fig. P5.4-29

5.4-30 A simply supported beam carries the load as shown in Fig. P5.4-30. Find the maximum bending stresses in the beam if the cross section is as shown. Neglect the weight of the beam.

5.4-31 A cantilever beam of length L is loaded as shown in Fig. P5.4-31. If

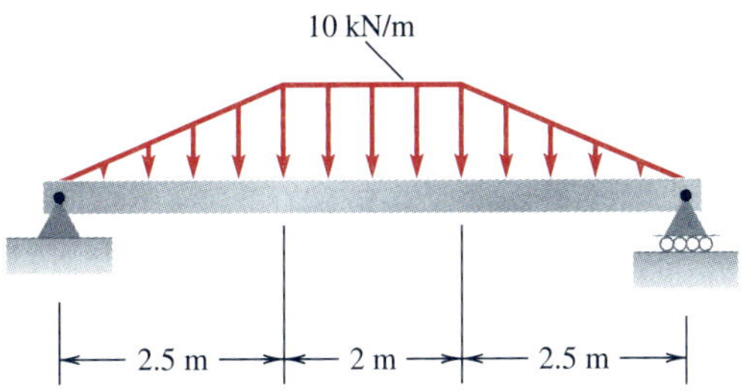

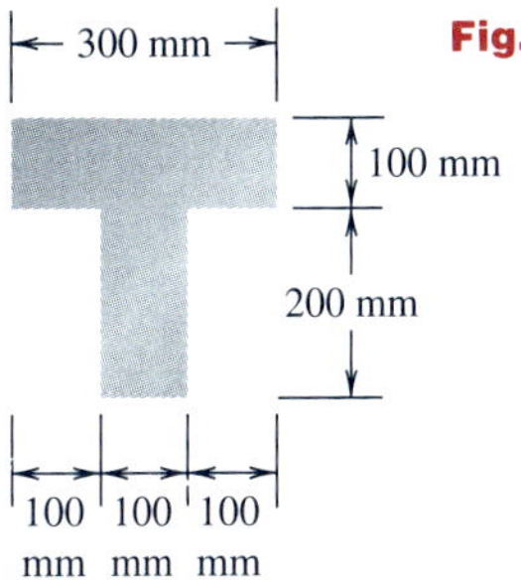

Fig. P5.4-30

L = 8 ft and P = 4000 lb, select a suitable wide-flange beam (App. C) to support the load if the maximum allowable stress in the beam is 18,000 psi. Include the weight of the beam.

5.4-32 A cantilever beam of length L = 6 ft is loaded as shown in Fig. P5.4-31. If P = 3000 lb, select a suitable I beam (S shape, App. C) to support the load if the maximum allowable stress is 15,000 psi. Include the weight of the beam.

5.4-33 A simply supported beam carries a constant distributed load along its length and a concentrated load at its center, as shown in Fig. P5.4-33. Select a suitable I beam (S shape, App. C) to support the load if the maximum allowable normal stress in the beam is 18,000 psi. Include the weight of the beam.

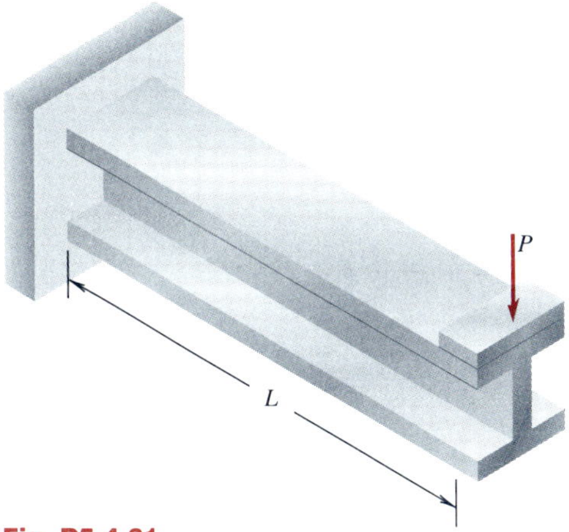

Fig. P5.4-31

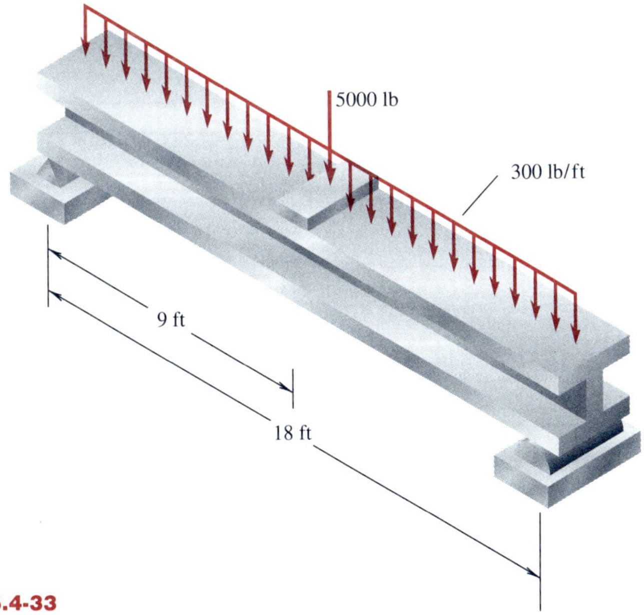

Fig. P5.4-33

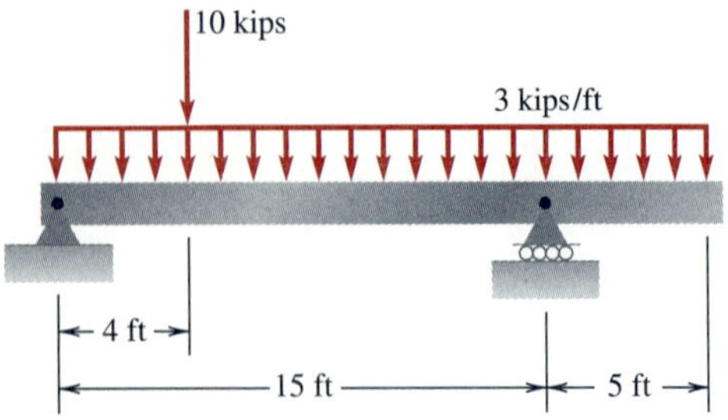

Fig. P5.4-34

5.4-34 A simply supported beam is shown in Fig. P5.4-34. The allowable bending stress in the beam is 20 ksi. Select a wide-flange beam (App. C) that can be used for the loading shown. Include the weight of the beam.

5.4-35 A simply supported beam carries two concentrated loads as shown in Fig. P5.4-35. The allowable stress in the beam is 22 ksi. Select a wide-flange beam (App. C) that can be used; include the weight of the beam.

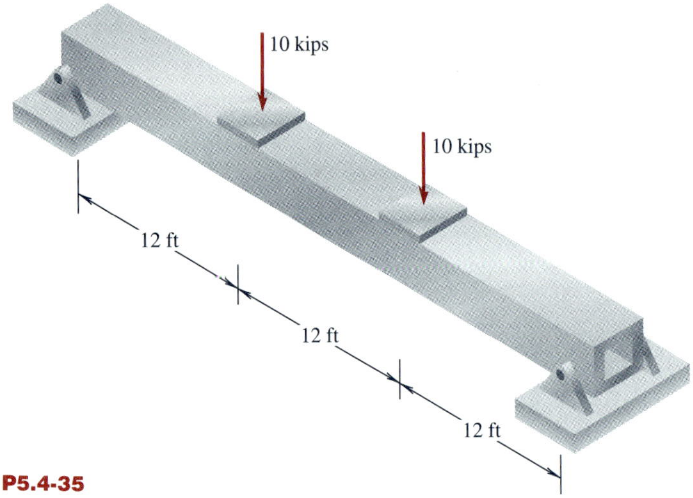

Fig. P5.4-35

5.4-36 Two rolled-steel channels are to be welded back to back so as to form an I-beam shape, as shown in Fig. P5.4-36. The allowable stress in the steel is 20 ksi. Find the channel sections (App. C) that can be used to carry the loads. Take $a = 5$ ft, $w = 3$ kips/ft, $P = 0$, $L = 12$ ft. Include the weight of the beam.

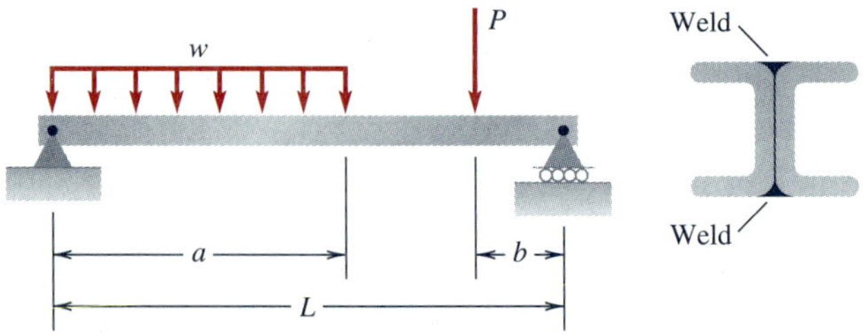

Fig. P5.4-36

5.4-37 Solve Prob. 5.4-36 with $a = L = 10$ ft, $w = 1.5$ kips/ft, $P = 15$ kips, and $b = 4$ ft.

5.4-38 A modern steel chair is fabricated from steel tubing as shown in Fig.

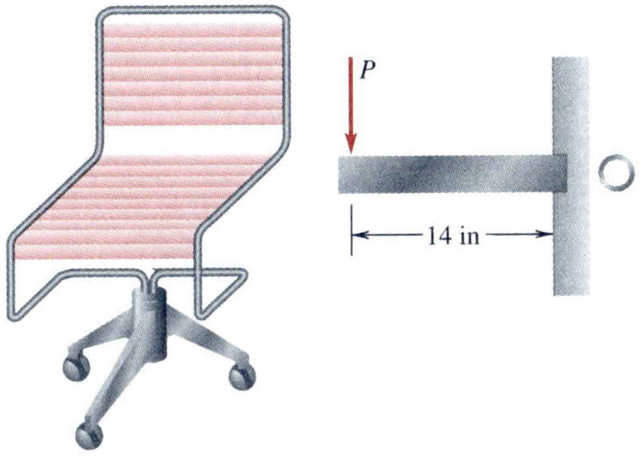

Fig. P5.4-38

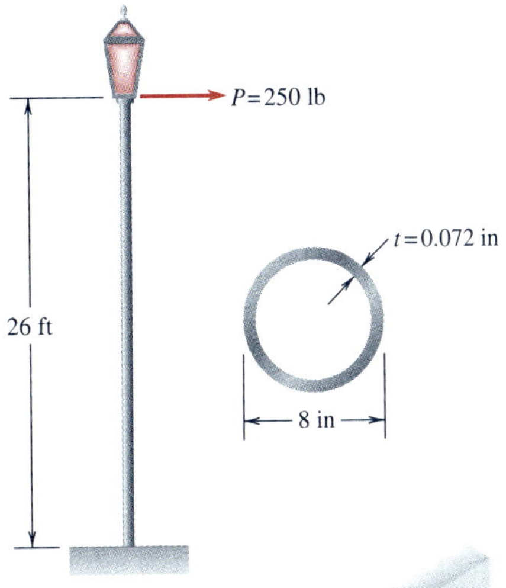

Fig. P5.4-39

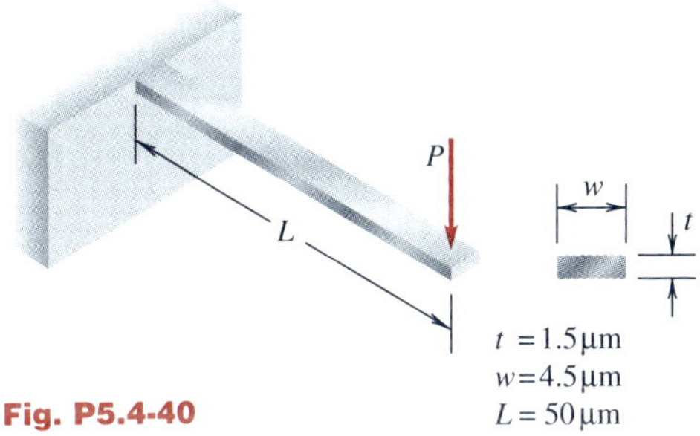

Fig. P5.4-40

P5.4-38. A preliminary design is to select a cross section of extra-strong pipe (App. E) using a model calculation in which a load P = 150 lb is to be carried by a cantilever beam 14 in long, as shown in the figure. Find the required section modulus of the pipe if the allowable stress is 15 ksi and an appropriate cross section.

5.4-39 The design for a lamppost is to be analyzed. The wind pressure on the post is equivalent to a load P of 250 lb acting as shown in Fig. P5.4-39. Find the maximum normal stresses acting in the post when exposed to the wind pressure.

5.4-40 Cantilever SiO_2 microbeams can now be fabricated down to a thickness of order 1.5 μm, as shown in Fig. P5.4-40. (A red blood cell is approximately 7 μm in diameter and 2 μm thick.) If a load of 15 μN is applied to the end of the beam, determine the maximum bending stress in the SiO_2 beam.

5.4-41 A rough sketch of a human femur subjected to a vertical load of 400 N is shown in Fig. P5.4-41. Determine the maximum normal stresses due to bending at section *B-B*. The inner half of the bone consists of "spongy" bone that does not carry appreciable stress.

5.4-42 In a study of the flight characteristics of the extinct species, the cretaceous pterosaur Pteranodon, Fig. P5.4-42, the bending moment distribution of the wing was estimated as shown in the figure. This was one of the largest flying creatures. The cross section of the bone at section *CC* is as shown. Estimate the maximum normal bending stress at section *CC*.

5.4-43 A new theory of Egyptian pyramid building proposes that the large pyramid blocks were lifted onto sledges by the counterweighted wooden lever system shown in Fig. P5.4-43. The sledges were then pulled up the sides of the

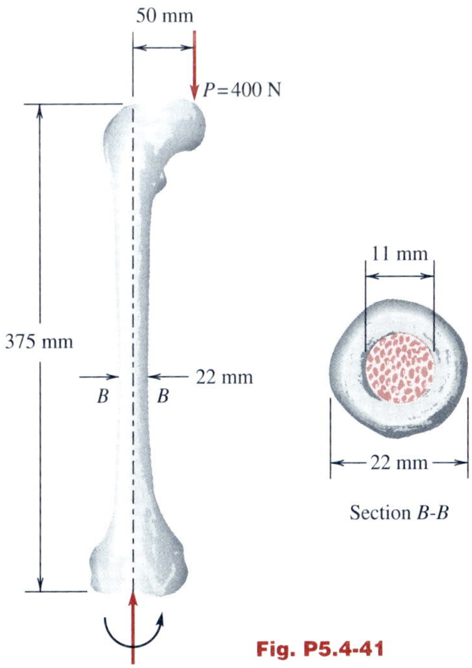

Fig. P5.4-41

pyramid. If the wood in the levers has an ultimate tensile stress of 7500 psi and an ultimate shear stress of 1500 psi, find on the basis of these ultimate stresses the dimensions of the smallest square piece of timber which will support the pyramid blocks as shown.

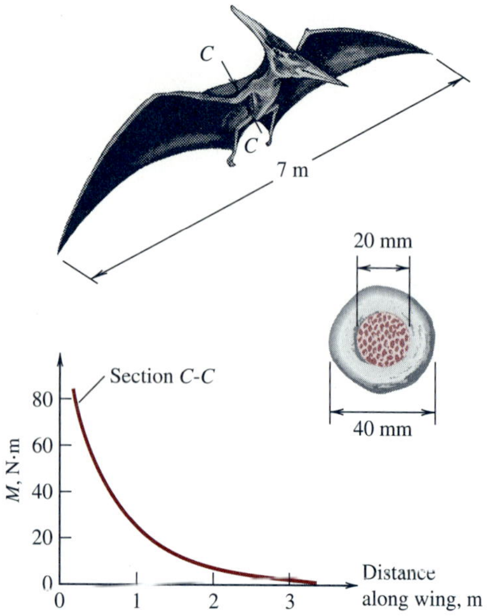

Fig. P5.4-42

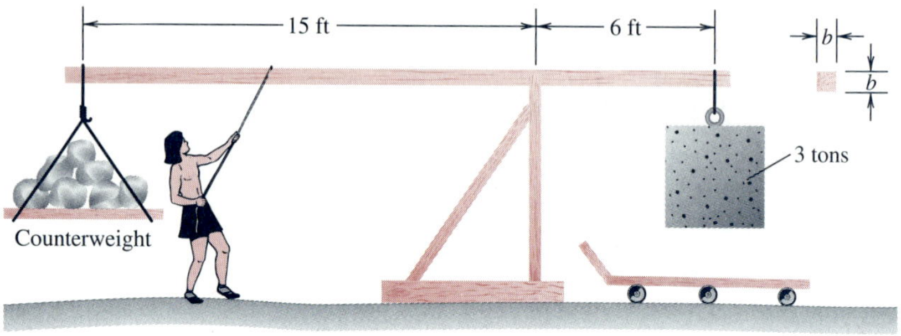

Fig. P5.4-43

5.4-44 A cross section of a cilium (see Prob. 4.5-36) is shown in Fig. P5.4-44. The dark areas are fibrils which are thought to be responsible for the cilium motion. The bending moment at the base of the cilium is estimated to be 5×10^{-7} N · m, and an experimental value of the radius of curvature at the base is 6 μm. Assuming that the bending forces are carried by the fibrils alone, estimate the elastic modulus of the fibrils. The total moment I_z of all the fibril cross-sectional areas is approximately 4×10^{-20} m^4.

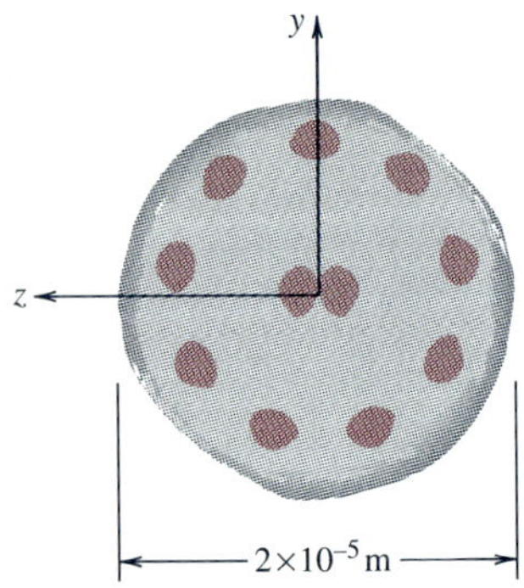

Fig. P5.4-44

5.4-45 A cantilever beam of length $L = 46$ in as shown in Fig. P5.4-45 is loaded by a load $P = 2000$ lb. Find the maximum values of the normal tensile and compressive bending stresses in the beam.

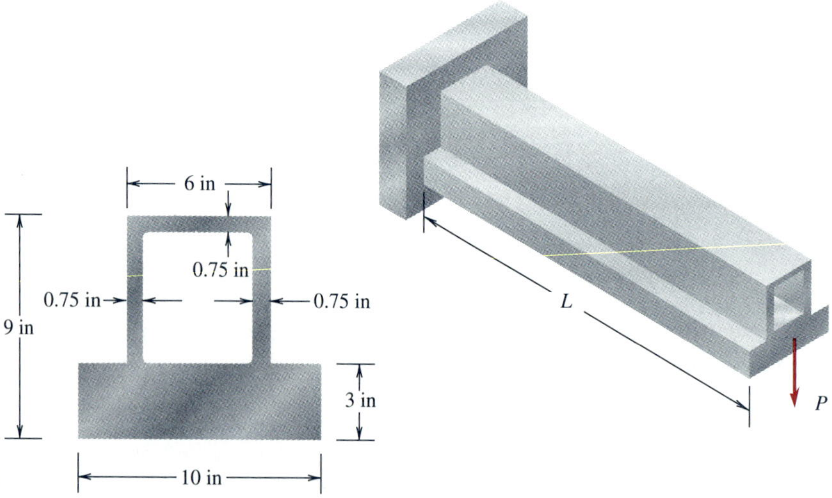

Fig. P5.4-45

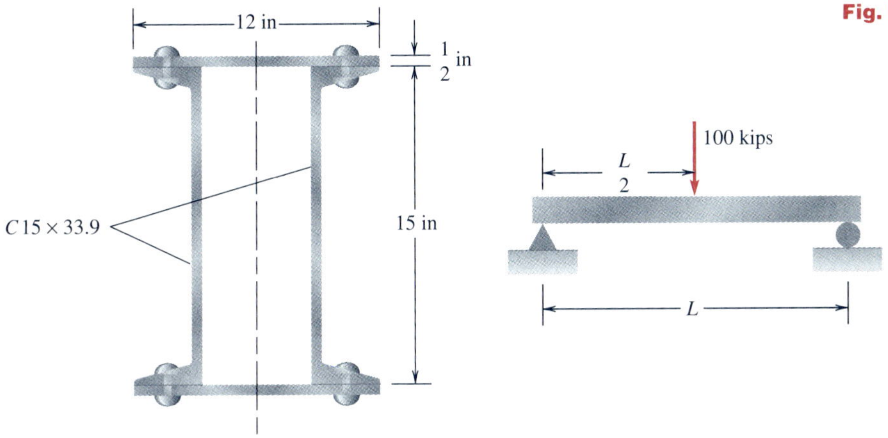

Fig. P5.4-46

5.4-46 A heavy steel beam is fabricated by riveting two C15 × 33.9 sections to two $\frac{1}{2}$-in steel plates as shown in Fig. P5.4-46. The beam is used as a simply supported beam over a span of $L = 12$ ft and subjected at midspan to a load of 100 kips. Find the maximum tensile and compressive stresses in the beam. Neglect the weight of the beam.

5.4-47 If the maximum allowable bending stress in the beam of Prob. 5.4-46 is 19 ksi, what is the longest allowable span L over which the beam can be used?

5.4-48 A large fabricated girder is used to carry a constant distributed loading w_0 over a span L as shown in Fig. P5.4-48. The girder is 64 in deep and is fabricated from a 60-in steel plate $\frac{1}{2}$ in thick, two 24-in × 2-in plates, and four 4-in × 4-in × $\frac{1}{2}$-in angle sections as shown. The section in Fig. P5.4-48 is symmetric about the z axis. Find the maximum tensile and compressive stresses in the girder beam. Neglect the weight of the beam. Take $w_0 = 4000$ lb/ft and $L = 60$ ft.

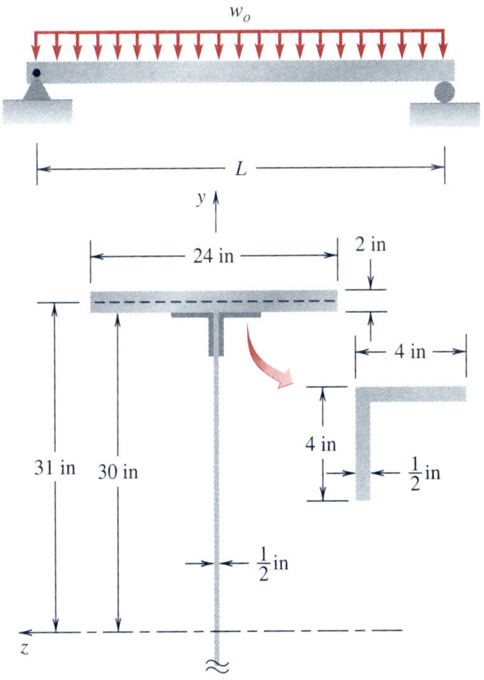

Fig. P5.4-48

5.4-49 If the maximum allowable stress in the beam of Prob. 5.4-48 is 12 ksi, find the maximum allowable loading w_0 if $L = 70$ ft. Neglect the weight of the beam.

5.4-50 Steel belts are used in a belt drive around a pulley of diameter 0.31 m. What is the thickest belt that can be wrapped around the pulley for 180° without exceeding a stress of 250 MPa in the steel belt? How does the maximum stress depend on the thickness? Take $E = 200$ GPa.

5.4-51 A stainless-steel cylindrical tank of radius R, wall thickness t, and length L (Fig. P5.4-51) is used in a dairy to hold milk. Find the maximum bending stresses in the tank. Take $L = 10$ ft, $R = 24$ in, $t = \frac{1}{4}$ in, and the weight density of milk as approximately 70 lb/ft^3. Neglect the weight of the tank itself.

5.4-52 Consider Prob. 5.4-51 but this time with $R = 30$ in. Calculate the new

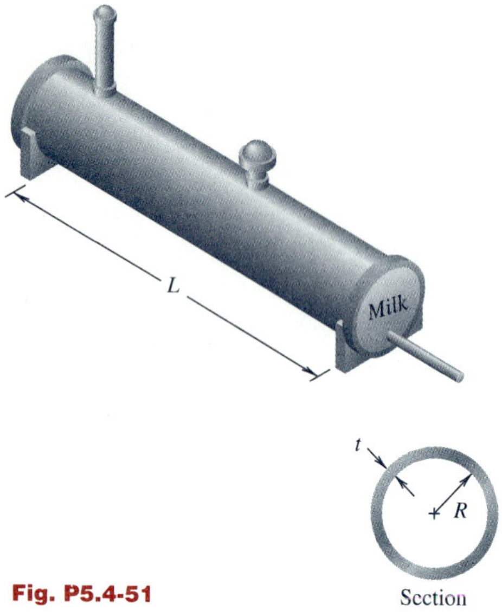

Fig. P5.4-51

maximum bending stresses. Can we draw any conclusions about the effect of the radius on the bending stresses?

5.4-53 A decorative bookshelf is made from plate glass of thickness t as shown in Fig. P5.4-53. The supports are at a distance a from each end located at their optimum positions (see Prob. 4.5-7). Determine the average weight of books per unit length that can be placed along the shelf if the allowable tensile stress of the glass is σ_a. Neglect the weight of the glass, and take $\sigma_a = 7$ MPa, $L = 1$ m, $b = 200$ mm, $t = 6$ mm?

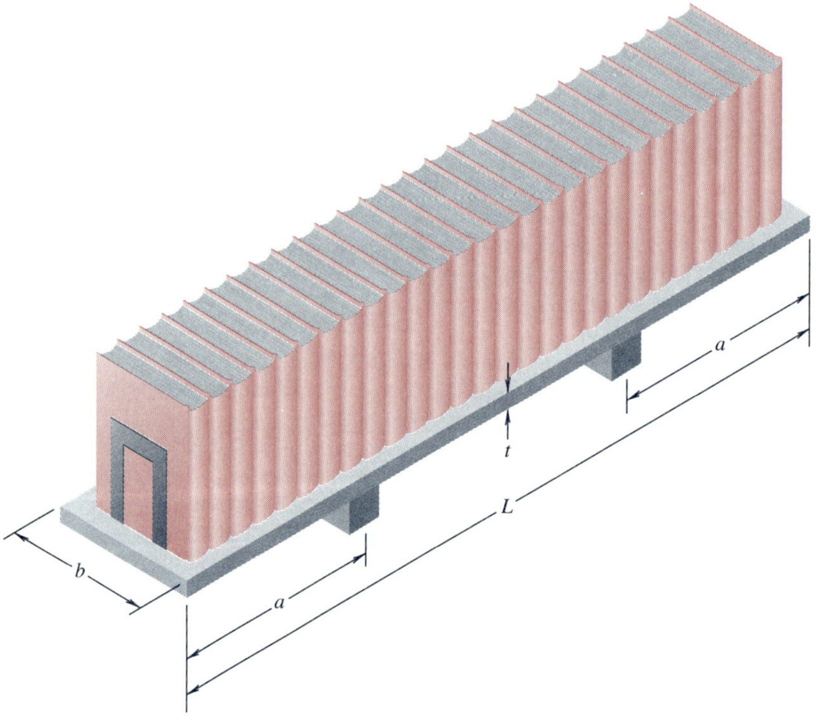

Fig. P5.4-53

5.4-54 Solve Prob. 5.4-53 if $L = 48$ in, $t = \frac{1}{4}$ in, $b = 10$ in, and $\sigma_a = 1200$ psi.

5.4-55 Consider Example 5.10, Fig. 5.27. It is decided to replace the beam found in this example with a rectangular tube beam (Table C.4). Select a tube beam from Table C.4 if the maximum allowable stress is 24 ksi.

5.4-56 Consider Example 5.11, Fig. 5.28. In place of the S6 $\times$ 17.25 beam, use a W beam with an allowable stress of 20,000 psi.

5.4-57 Consider Example 5.8, Fig. 5.21. Determine the size of a W-beam section that can be used to replace the unsymmetric channel section of 50-mm thickness. The maximum allowable stress in the beam is 20 ksi. Note in this problem you will need to change the value of M from SI to English units in order to find the required S value (App. B).

5.4-58 Consider Example 5.7, Fig. 5.19. Find a W-beam section to replace the box beam such that the maximum stress in the beam does not exceed 20 ksi. Include the weight of the beam.

5.4-59 A W8 × 21 section beam is to be used as a cantilever beam as shown in Fig. P5.4-59. What is the maximum allowable length L for the cantilever beam without exceeding the allowable stress of 24 ksi? Include the weight of the beam.

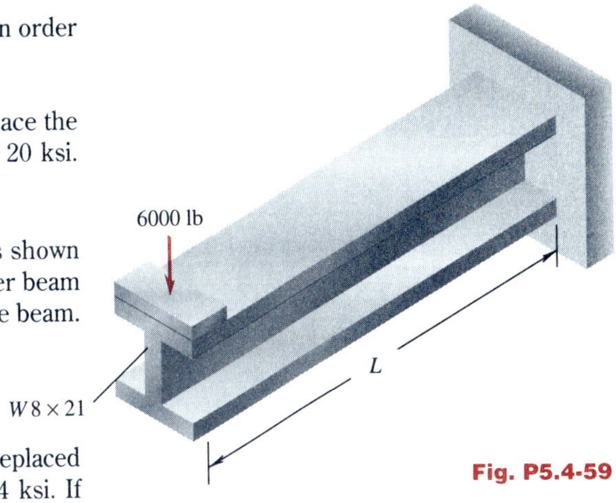

Fig. P5.4-59

5.4-60 The section for the cantilever beam shown in Fig. P5.4-59 is replaced by a square steel tube (App. C) whose maximum allowable stress is 24 ksi. If $L = 8$ ft, find an appropriate steel tube; include the weight of the tube.

5.4-61 A simply supported beam of length L as shown in Fig. P5.4-61 is loaded by a uniform load w_0. If the maximum allowable stress in the beam is 24 ksi, find an appropriate W section beam to carry the load. Include the weight of the beam. Take $L = 14$ ft and $w_0 = 6000$ lb/ft.

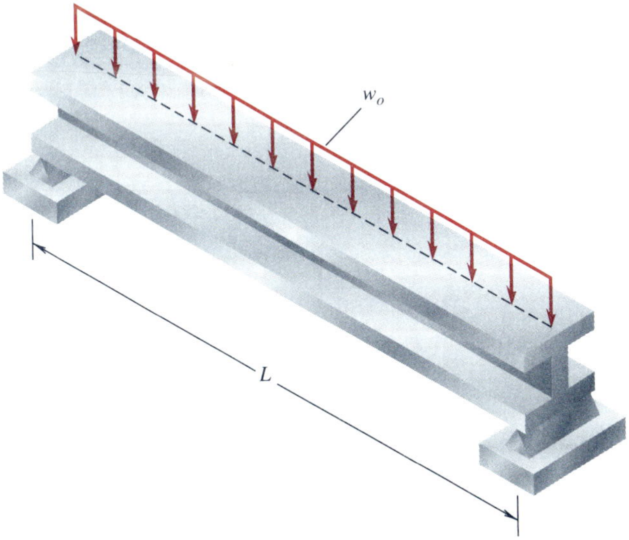

Fig. P5.4-61

5.4-62 A simply supported beam of length L as shown in Fig. P5.4-61 is loaded by a uniform load w_0. If the maximum allowable stress in the beam is 24 ksi, find an appropriate S section beam to carry the load. Include the weight of the beam. Take $L = 10$ ft and $w_0 = 3000$ lb/ft.

5.4-63 Select a rectangular steel tube (Table C.4) to support the loads on the beam shown in Fig. P5.4-63 if the maximum allowable stress in the beam is 20 ksi. Include the weight of the beam.

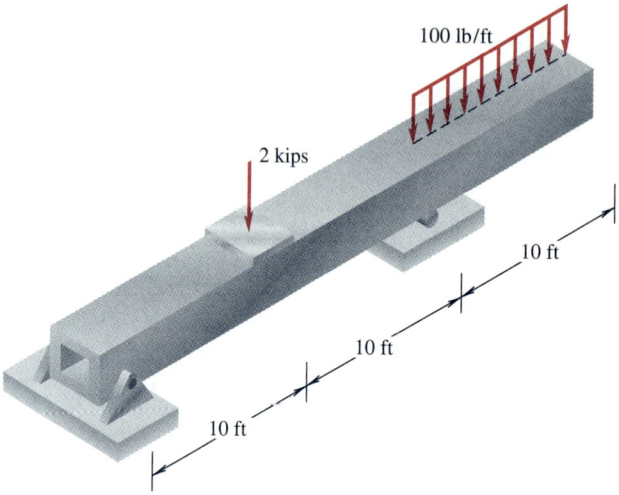

Fig. P5.4-63

5.4-64 A cantilever beam of length $L = 12$ ft is loaded as shown in Fig. P5.4-64. If the allowable stress in the beam is 24 ksi, select an acceptable S beam (Table C.2) to support the loads. Include the weight of the beam.

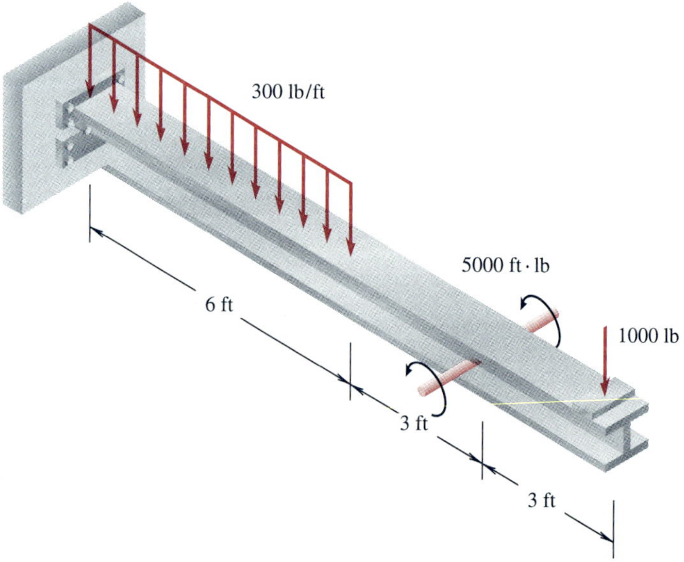

Fig. P5.4-64

5.4-65 A cantilever beam of length $L = 12$ ft is loaded as shown in Fig. P5.4-64. Find the size of two channel sections (Table C.5) that can be used back to back (in a vertical position) to support the loads if the maximum allowable stress is 24 ksi. Include the weights of the channel sections.

5.4-66 A simply supported beam is loaded as shown in Fig. P5.4-66. Determine the maximum allowable value of P if the compressive normal stress due to bending is not to exceed 900 psi.

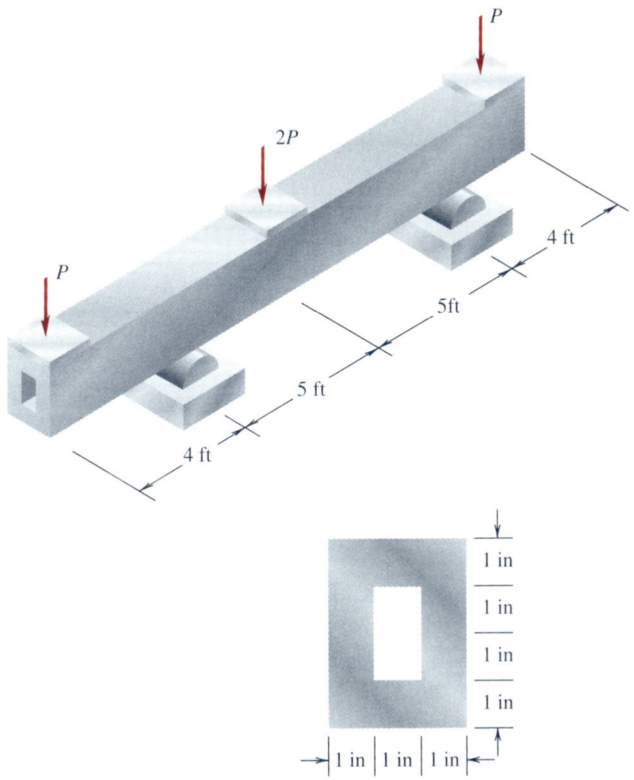

Fig. P5.4-66

5.4-67 A sketch of a portion of a trailer hitch is shown in Fig. P5.4-67. The hitch is loaded in its vertical plane of symmetry and can be treated as a cantilever beam built in at the left end. Determine the maximum normal stresses due to bending at section *A-A*. Neglect the weight of the member.

5.4-68 An I beam as shown in Fig. P5.4-68 is to be used to span 30 ft. What is the maximum value of the constant distributed load w_0 that can be placed on the beam before exceeding a maximum tensile stress due to bending of 34 ksi?

5.4-69 A simply supported beam with a uniform load q_0 over half the beam has the cross section as shown in Fig. P5.4-69. The allowable normal stresses

Fig. P5.4-67

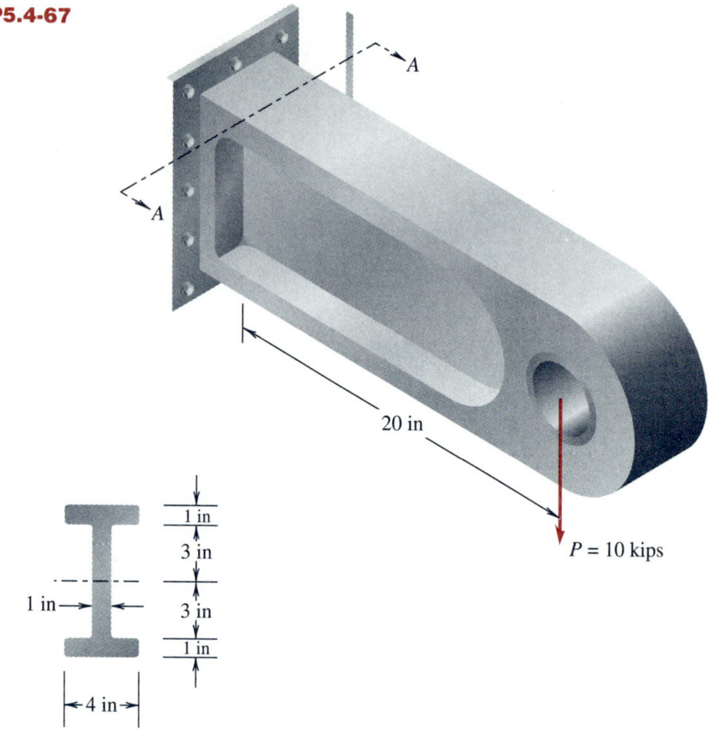

Section A-A

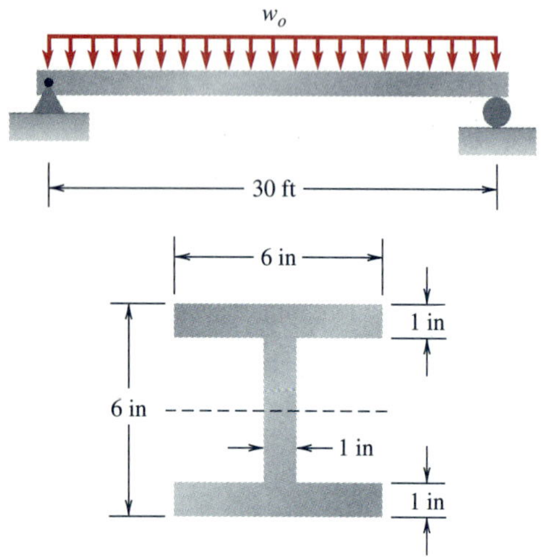

Fig. P5.4-68

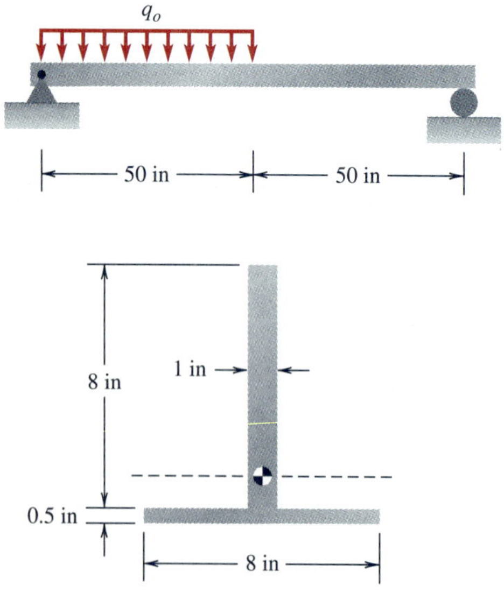

Fig. P5.4-69

on the cross section due to bending are 800 psi in tension and 2500 psi in compression. Calculate the allowable uniform load q_0.

5.4-70 A simply supported beam is loaded as shown in Fig. P5.4-70. Determine the maximum allowable value for P if the compressive normal bending stress is not to exceed 10 MPa.

5.4-71 A beam has a maximum bending moment $M = 10{,}000$ lb · ft and the cross-sectional shape shown in Fig. P5.4-71. Estimate the maximum normal tensile and compressive stresses in the beam.

5.4-72 If a design code for allowable loads on roof decking makes use of the normal stress results for a simply supported beam under a constant distributed loading, calculate the allowable constant loading q_0, lb/ft^2, for a 2-in nominal thickness (1.5-in) of roof decking for the following cases:
(*a*) $L = 8$ and 12 ft, where the allowable stress is $\sigma_a = 1000$ psi
(*b*) $L = 6$ and 8 ft, where the allowable stress is $\sigma_a = 1200$ psi

5.4-73 Consider Example 5.4, Fig. 5.15. The wooden beam of constant cross section is replaced by a beam with a 6-in × 6-in (actual dimensions) section over

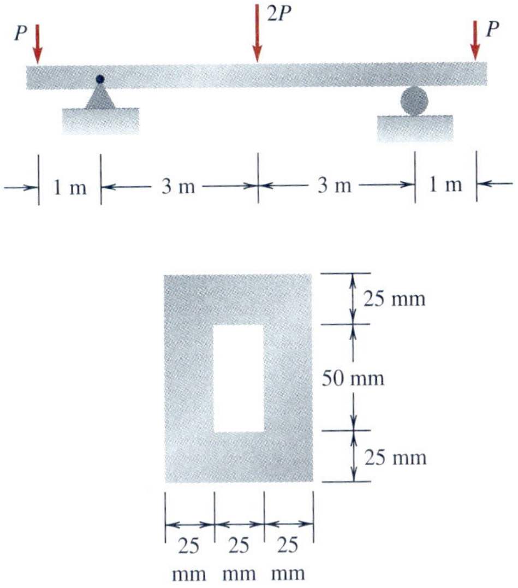

Fig. P5.4-70

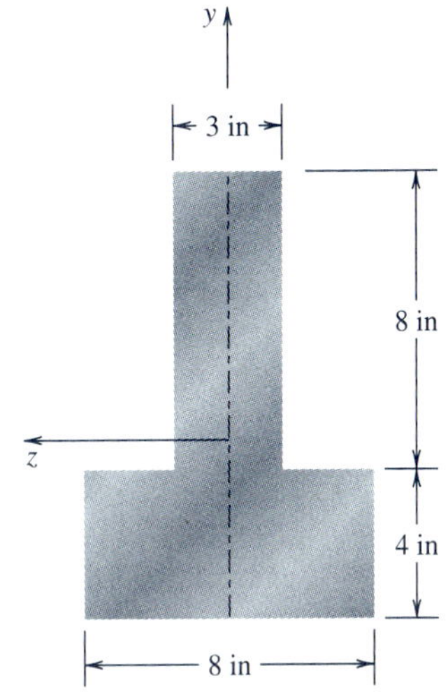

Fig. P5.4-71

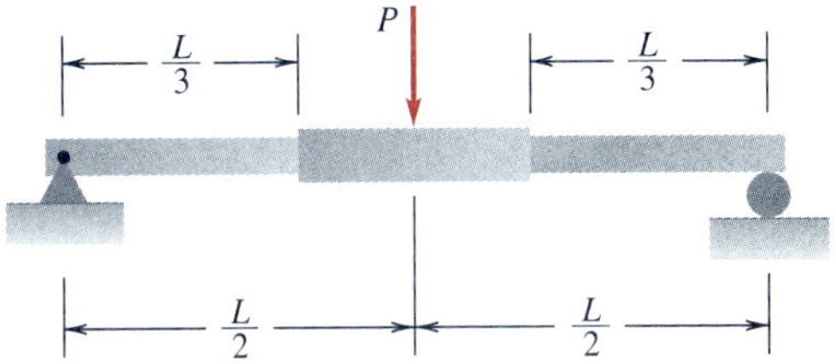

Fig. P5.4-73

the middle portion and 4-in × 4-in (actual dimensions) section over the outer portions as shown in Fig. P5.4-73.

(*a*) Sketch the graph of σ_x at the bottom of the beam as a function of distance from the left support. Take $P = 600$ lb and $L = 9$ ft.

(*b*) What is the maximum value of σ_x in part (*a*), and where does it occur?

(*c*) What is the largest allowable value of L if the allowable value of σ_x is 1200 psi in either tension or compression?

5.5-1 Verify Eq. (5.21) and Eq. (5.25).

5.5-2 Verify that the shear stress distribution across the section of a rectangular beam is parabolic, as given by Eq. (5.31).

5.5-3 Verify that for a beam of a rectangular cross section the resultant of the shear stress distribution equals the shear force V, as outlined in Eq. (5.33).

5.5-4 Show that for an arbitrary symmetric beam the resultant of the shear stress given by

$$\tau_{xy} = \frac{VQ}{I_z b}$$

is equal to the shear force V; that is, show that $\int \tau_{xy}\, dA = V$. Note that $dQ/dy = -by$ with

$$Q(y) = \int_y^c y\, dA.$$

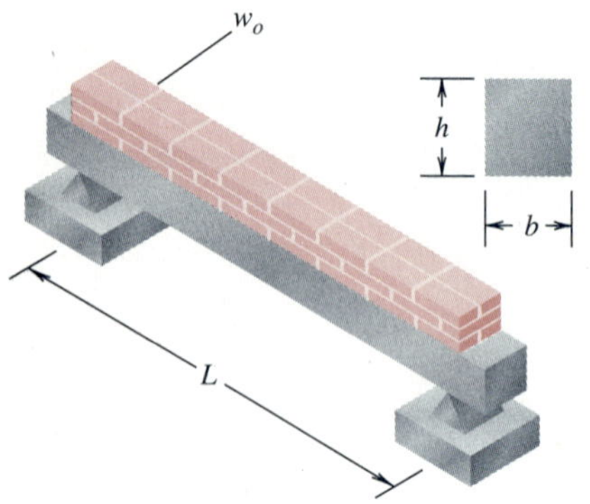

Fig. P5.5-5

5.5-5 A simply supported beam carries a constant distributed loading of intensity w_0 as shown in Fig. P5.5-5. Determine the ratio of the maximum bending stress σ_x to the maximum shear stress τ_{xy} in the beam.

5.5-6 A simply supported beam carries a concentrated load P at midspan as shown in Fig. P5.5-6. The cross section is as shown. Determine the ratio of the maximum bending stress σ_x to the maximum shear stress τ_{xy} in the beam. Neglect the weight of the beam.

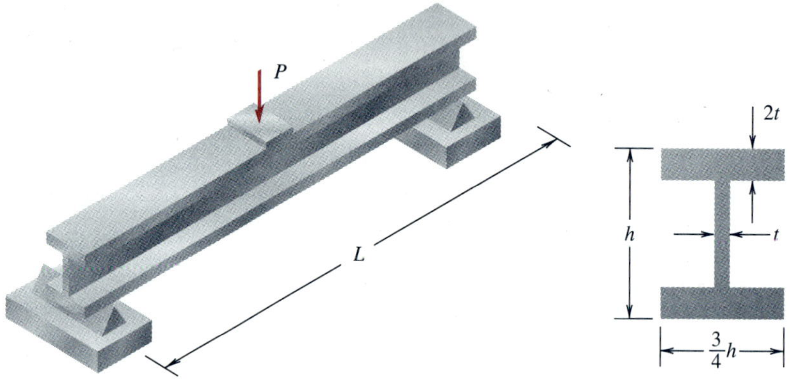

Fig. P5.5-6

5.5-7 Verify Eqs. (5.34) and (5.35).

5.5-8 Consider the I beam in Fig. 5.42 and shown again in Fig. P5.5-8. We wish to obtain an expression for the parabolic shear stress distribution in the web of the beam. Verify that the expression for the first moment of the area above section 1-1 defined by the distance y_1 is given by

$$Q = \frac{bt_1}{2}(h_1 + t_1) + \frac{t_2}{8}(h_1^2 - 4y_1^2) \qquad (a)$$

Therefore, the shear stress τ in the web of the beam is given by

$$\tau = \frac{VQ}{I_z t_2} \qquad (b)$$

Verify that the maximum shear stress at the neutral axis of the beam, $y_1 = 0$, is given by

$$\tau_{max} = \frac{V}{8I_z t_2}[t_2 h_1^2 + 4bt_1(h_1 + t_1)] \qquad (c)$$

An estimate for the maximum shear stress in the web is given by the average shear stress obtained by dividing the shear force by the area of the web

$$\tau_{av} = \frac{V}{t_2 h_1} \qquad (d)$$

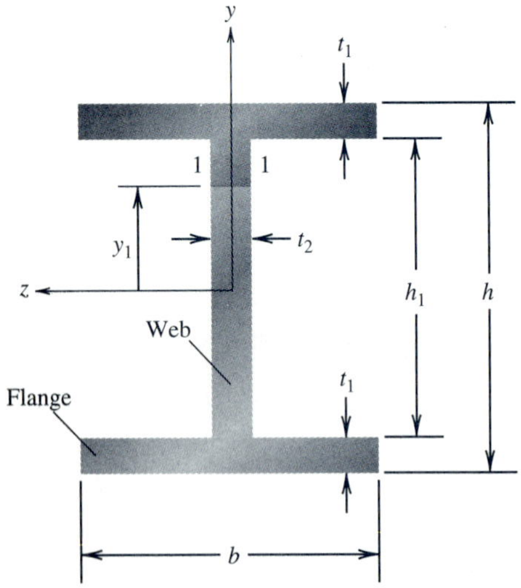

Fig. P5.5-8

5.5-9 A simply supported beam of length L is loaded by a constant distributed loading w as shown in Fig. P5.5-9. The distributed load includes the weight of the beam. Determine the maximum normal stress and the maximum shear stress in the beam. Take $w = 800$ lb/ft, $L = 8$ ft, $b = 8$ in, and $h = 10$ in.

5.5-10 A simply supported beam of length L is loaded by a constant distributed loading w as shown in Fig. P5.5-9. The distributed load includes the weight of the beam. Determine the maximum normal stress and the maximum shear stress in the beam. Take $L = 3$ m, $w = 10$ kN/m, $b = 200$ mm, and $h = 250$ mm.

5.5-11 A simply supported wooden beam of length L is loaded by a concentrated load at midspan as shown in Fig. P5.5-11. If the cross section of the beam is 8 in × 10 in, find the maximum permissible value of load P if the maximum allowable bending stress is $\sigma_x = 1100$ psi and the maximum allowable shear stress is $\tau = 150$ psi. Neglect the weight of the beam; $L = 5$ ft.

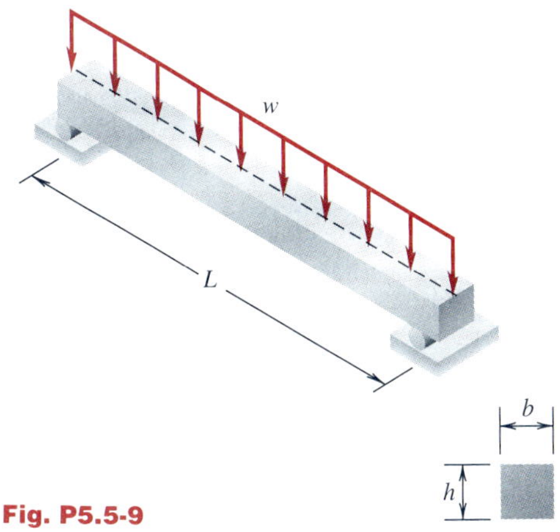

Fig. P5.5-9

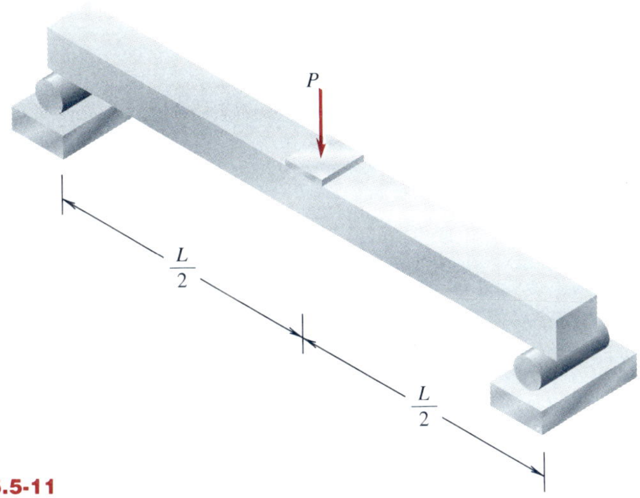

Fig. P5.5-11

5.5-12 Solve Prob. 5.5-11 including the weight of the beam. Take the weight density of the wood making up the beam as 35 lb/ft^3.

5.5-13 Solve Prob. 5.5-11, using the net dimensions (App. D) of the nominal 8-in × 10-in cross section. Neglect the weight of the beam.

5.5-14 Solve Prob. 5.5-11, using the net dimensions (App. D) of the nominal 8-in × 10-in cross section, and include the weight of the beam. Take the weight density of the wood making up the beam as 35 lb/ft^3.

5.5-15 A T beam with cross section as shown in Fig. P5.5-15 transmits a bending moment and a shear force. Determine an expression for the ratio of the maximum bending stress in the stem to that in the flange. Determine an expres-

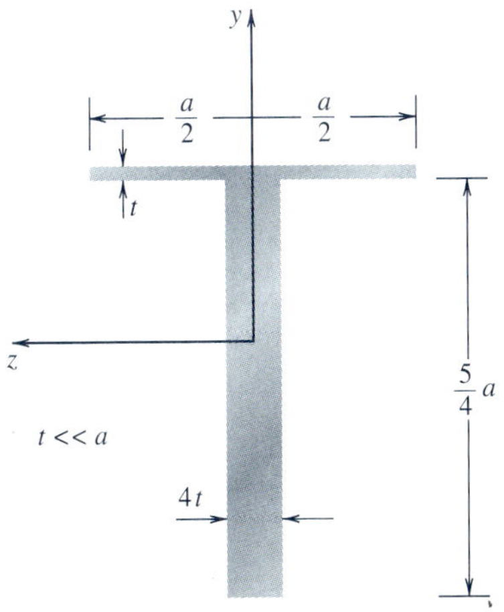

Fig. P5.5-15

sion for the ratio of the maximum shear stress in the stem to the shear stress at the junction of the stem and flange. Finally, obtain an expression for the ratio of the maximum shear stress in the stem to the average value of the shear stress in the stem.

5.5-16 A flooring system as shown in Fig. P5.5-16 carries a design load of 200 lb/ft^2, which includes the deadweight of the flooring. The 2-in × 8-in joists carry this load to the vertical supports over a span of 12 ft. Determine the maximum normal bending stresses in the joists. Neglect the weight of the joists.

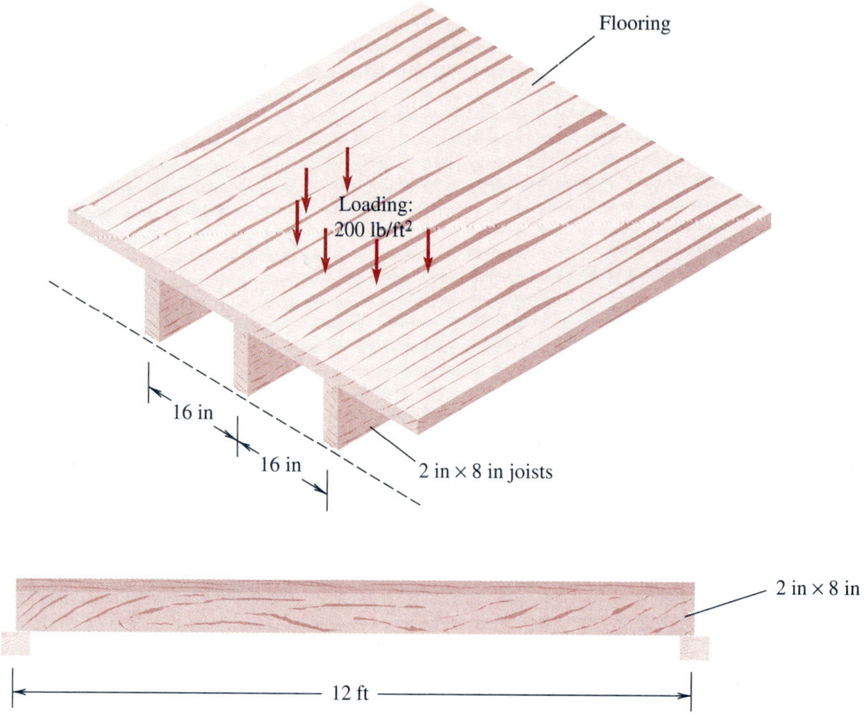

Fig. P5.5-16

5.5-17 Solve Prob. 5.5-16, using the actual dimensions of a nominal 2-in × 8-in beam; see App. D.

5.5-18 A platform shown in Fig. P5.5-18 carries a design load of 60 lb/ft^2 and is supported on steel W14 × 22 beams with a span equal to 20 ft. Determine the maximum normal bending stresses in the beams.

5.5-19 The I-section cantilever beam shown in Fig. P5.5-19 carries a load of 30 kips. Determine an expression for the shear stress distribution in the web of the beam. Sketch the shear stress distribution showing the important values, and show that over 90 percent of the shear load is carried by the web of the beam.

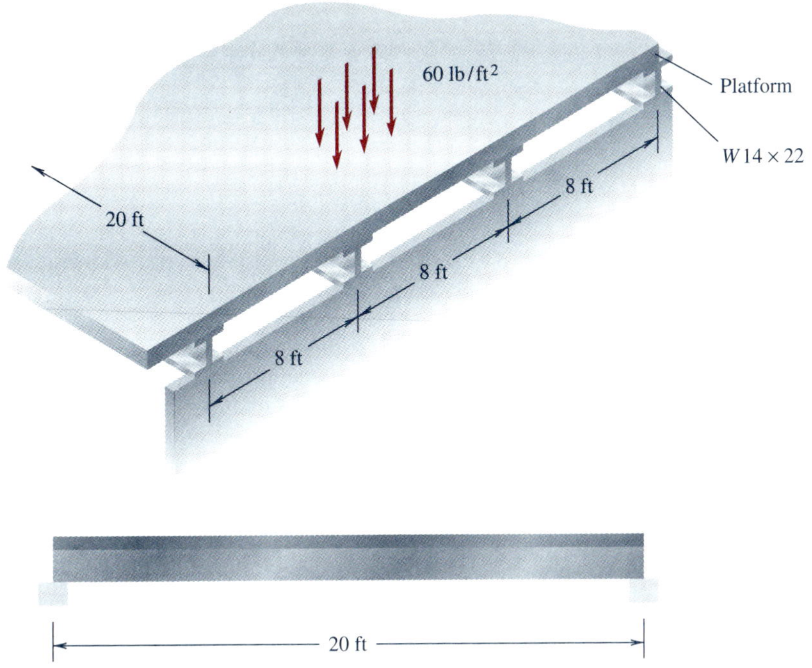

Fig. P5.5-18

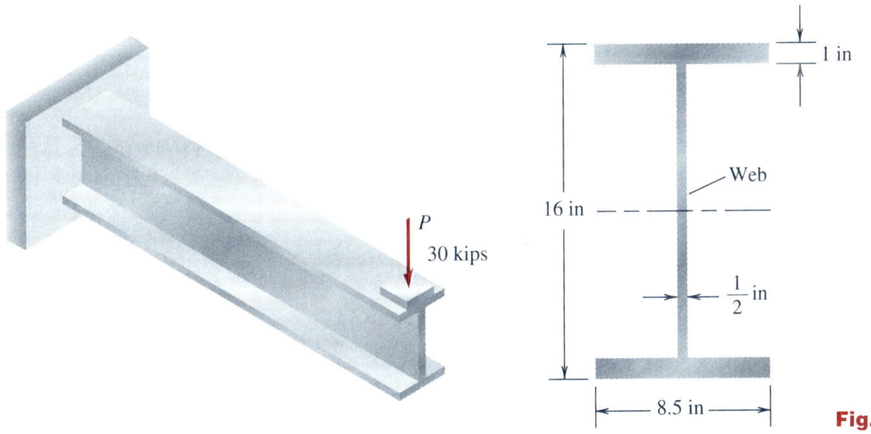

Fig. P5.5-19

5.5-20 Consider a beam of circular cross section as shown in Fig. 5.39*b*. The value of Q for area A_1 is given by

$$Q = \hat{y}_1 A_1 \tag{a}$$

where $\hat{y}_1$ is the distance from the centroid of the total section to the centroid of area A_1. Show that $Q = \frac{2}{3}(b/2)^3$, and so

$$\frac{Q}{b} = \frac{2}{3}\frac{b^2}{8} = \frac{1}{3}(R^2 - y_1^2) \tag{b}$$

Therefore for the case $y_1 = 0$,

$$\frac{Q}{b} = \frac{R^2}{3} \tag{c}$$

and the expression for τ_{max} follows from Eq. (5.34).

5.6-1 Consider Example 5.16, Fig. 5.44. If the glue stress is reduced to 50 kPa, what is the maximum load that can be applied to the beam?

5.6-2 Consider Example 5.17, Fig. 5.45. If the three members to be used to make up the I-beam configuration (Fig. 5.45*b*) are found to have been undersized by 10 percent, i.e., the section is 45 mm × 90 mm, find the maximum allowable load *P* that can be carried by the beam. Also solve for the load *P* if the members instead are oversized by 10 percent, i.e., the section is 55 mm × 110 mm.

5.6-3 Compare the maximum load obtained in Example 5.17 with the loads found in Prob. 5.6-2. When the dimensions of the cross section change by ±10 percent, does the load change by ±10 percent?

5.6-4 Consider Example 5.19, Fig. 5.47. If the thickness of the box sections is decreased from 3 to 2 mm and all other dimensions are held fixed, what is the maximum shear force V_{max} that the beam in service can carry?

5.6-5 Consider Example 5.20, Fig. 5.48. If the C12 × 30 sections are to be replaced by C10 × 20 sections (App. C), what is the minimum allowable shear force that each bolt must carry? All other dimensions of the beam are the same.

5.6-6 Two 2-in × 4-in beams are glued together as shown in Fig. P5.6-6. What is the required glue strength, expressed in terms of *P*, for each of the two configurations shown?

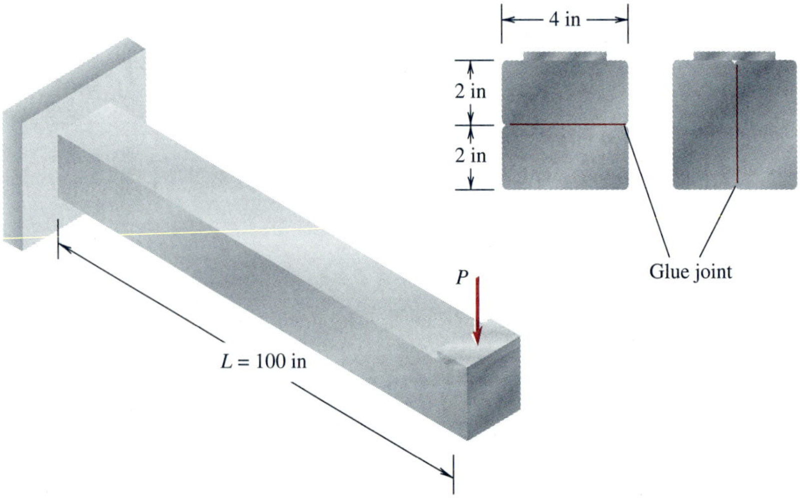

Fig. P5.6-6

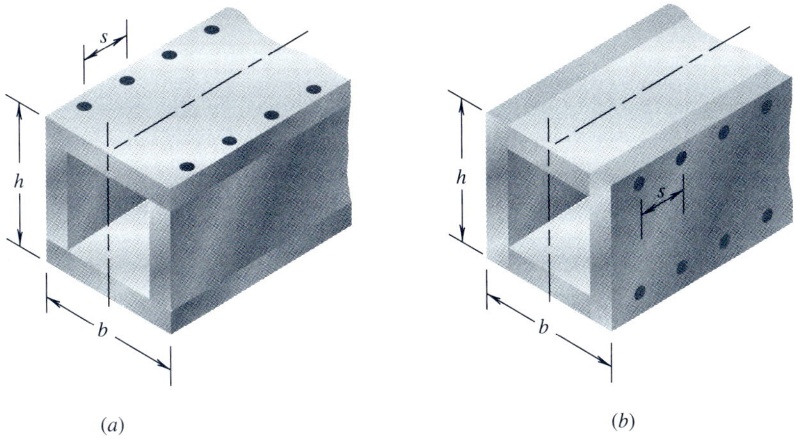

Fig. P5.6-7

5.6-7 Two designs are proposed for building a wooden box beam from four pieces of equal thickness, as shown in Fig. P5.6-7. The dimensions b and h and the spacing s are the same in both designs. If the beam is to carry a loading in the vertical direction, is one design better than the other? Give a quantitative reason to support the conclusion.

5.6-8 A $\frac{1}{4}$-in-thick steel plate is riveted to the top and bottom of an aluminum-alloy I beam whose moment of inertia is 57 in^4, as shown in Fig. P5.6-8. Find the maximum allowable spacing of the rivets along the beam in this composite beam if the maximum shear force in the beam is 6000 lb and if each rivet can safely carry a shear force across it of 450 lb.

5.6-9 A steel girder is fabricated by welding a 1-in × 18-in plate to the top and bottom of a 3/8-in × 68-in web plate as shown in Fig. P5.6-9. If the maximum shear force V acting on the section is 350 kips, find the shear flow required in each weld along the beam.

5.6-10 Solve Prob. 5.6-9 if d now equals 72 in.

5.6-11 A wooden box beam is fabricated from four 1-in × 6-in pieces as shown in Fig. P5.6-11. The pieces are joined together by screws for which the allowable load in shear is 200 lb. Determine the maximum allowable spacing s of the screws if the maximum shear force V on the section is 850 lb.

5.6-12 If the screws in the beam of Prob. 5.6-11 are replaced by gluing the pieces together, estimate the required glue strength to maintain the integrity of the beam.

5.6-13 A wooden box beam is fabricated from four 1-in × 6-in pieces as shown in Fig. P5.6-13. The pieces are joined together by screws for which the allowable load in shear is 200 lb. Determine the maximum allowable spacing s of the screws if the maximum shear force on the section is 850 lb.

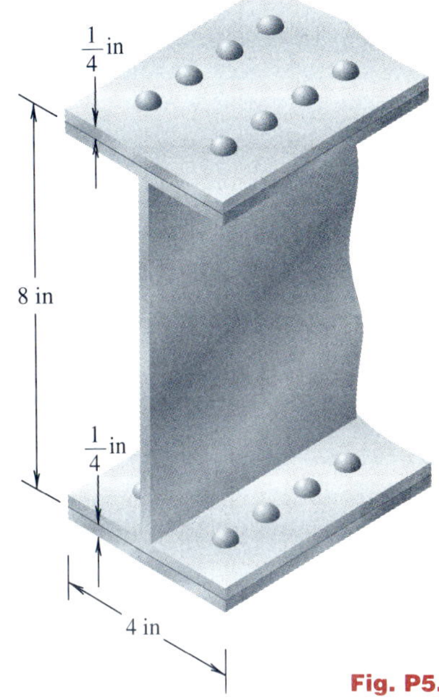

Fig. P5.6-8

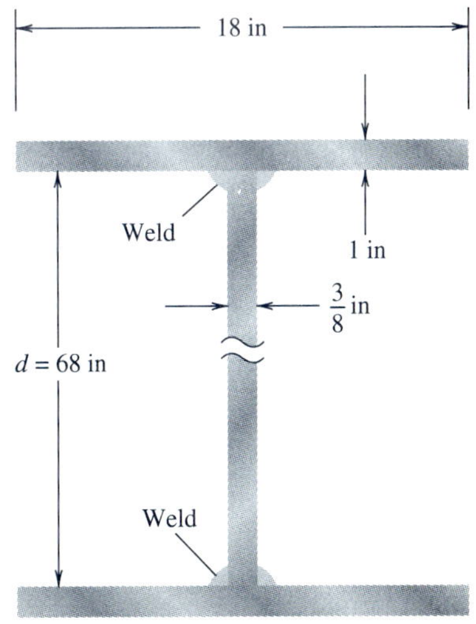

Fig. P5.6-9

Fig. P5.6-11

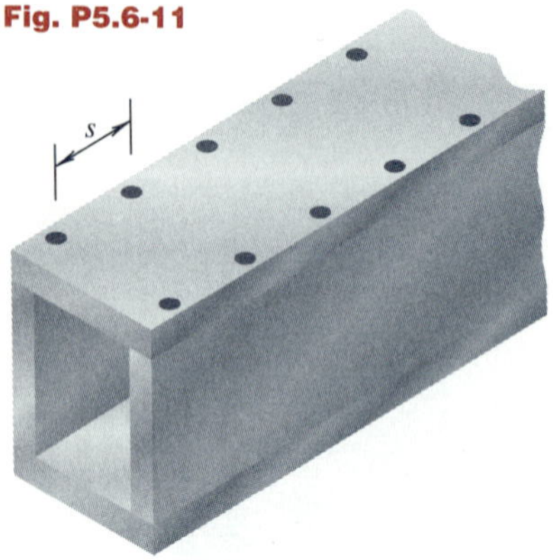

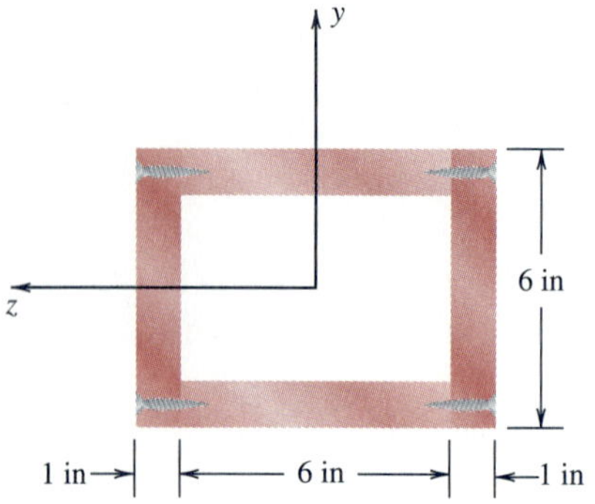

Fig. P5.6-13

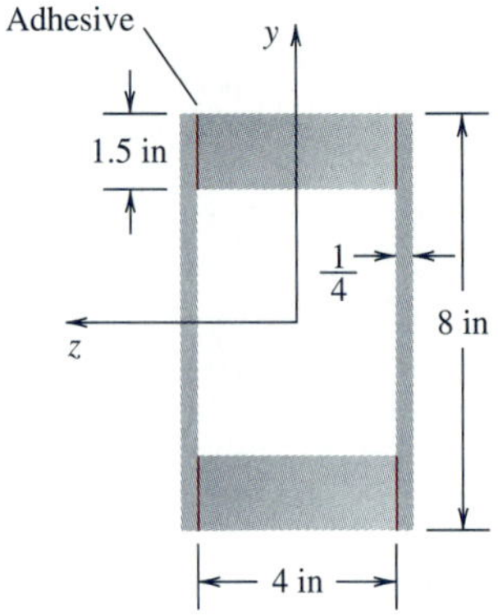

Fig. P5.6-15

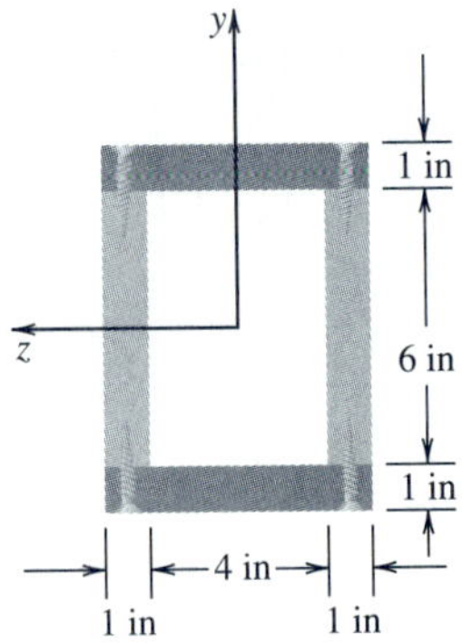

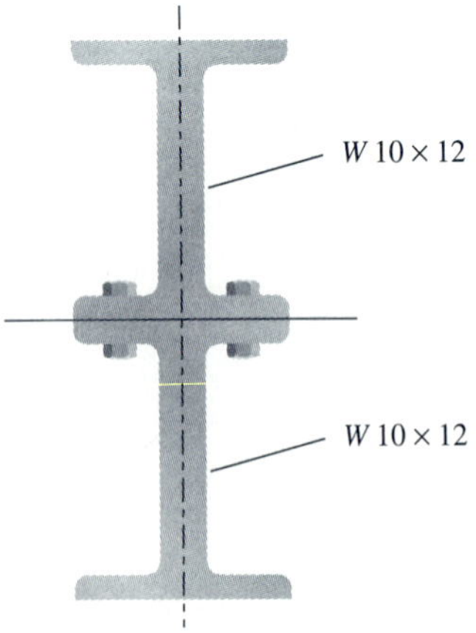

Fig. P5.6-16

5.6-14 If the screws in the beam of Prob. 5.6-13 are replaced by gluing the pieces together, estimate the required glue strength to maintain the integrity of the beam.

5.6-15 A wooden beam is fabricated by gluing together two plywood vertical pieces ($\frac{1}{4}$ in × 8 in) to wooden pieces (1.5 in × 4 in) as shown in Fig. P5.6-15. If the maximum shear force on the section is 200 lb, what is the required glue strength to keep the beam in one piece under load?

5.6-16 A composite beam is fabricated by bolting together two W10 × 12 beam members as shown in Fig. P5.6-16. If the spacing of the bolts can carry a shear flow of 18 kips/in, what is the maximum shear force that the beam can carry?

5.6-17 If the two W10 × 12 members of Prob. 5.6-16 are replaced by two W8 × 18 members, what is the maximum shear force that the beam can carry?

5.6-18 Two different members, as shown in Fig. P5.6-18, are bolted together by $\frac{1}{2}$-in-diameter bolts to form a beam. The bolts have an allowable shear stress of 18 ksi. If the maximum shear force acting on the beam is 25 kips, find the maximum allowable spacing of the bolts along the beam.

5.6-19 Three wooden planks as shown in Fig. P5.6-19 are used to fabricate an I beam with screws. If the maximum allowable shear force in the screws is 3.5 kN and the maximum shear force in the beam is 10 kN, find the required spacing of the screws along the beam.

5.6-20 A box beam is fabricated from four 6-in × 2-in planks as shown in Fig. P5.6-20 by nailing the four planks together. The beam is used as a simply supported beam, and the maximum shear force in the beam is 100 lb. If each nail can support a shear force of 30 lb, determine the maximum allowable spacing s of the nails.

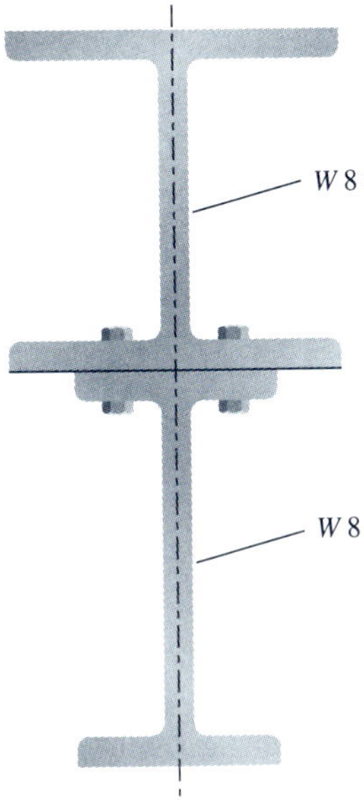

Fig. P5.6-18

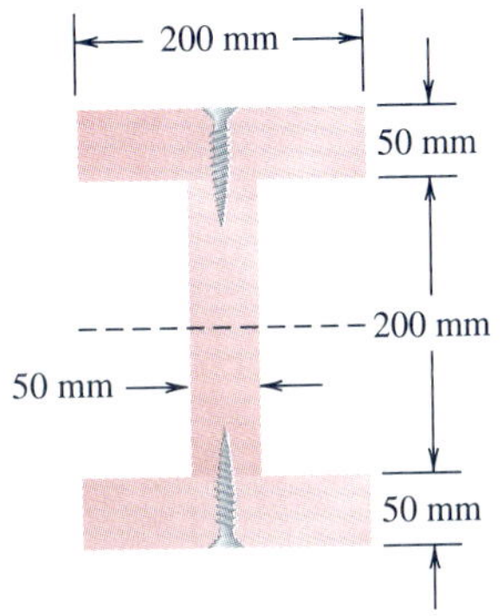

Fig. P5.6-19

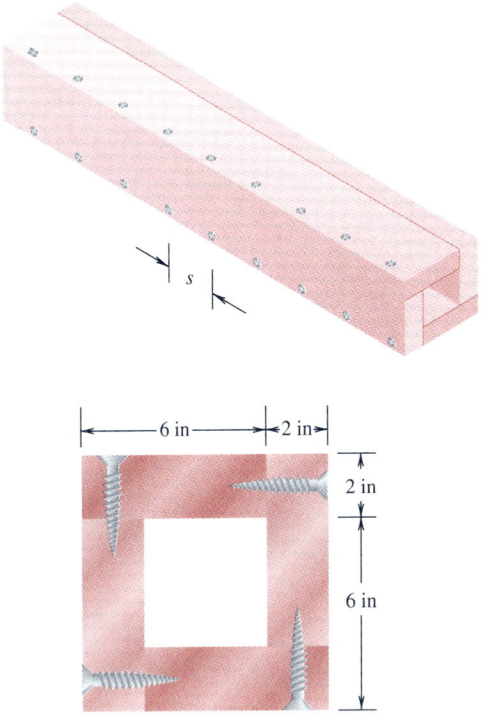

Fig. P5.6-20

5.6-21 A wooden channel beam is fabricated from three pieces of lumber as shown in Fig. P5.6-21. The beam is to carry a maximum shear force of 10 kips, and the allowable shear force in the nails is 1 kip. Find the maximum allowable spacing s of the nails to carry the applied shear force.

5.6-22 A composite beam is fabricated from three pieces of lumber as shown in Fig. P5.6-22. If the beam is used in service to support a maximum shear force of 8000 lb, estimate the maximum average shear force in each of the bolts when the beam is loaded.

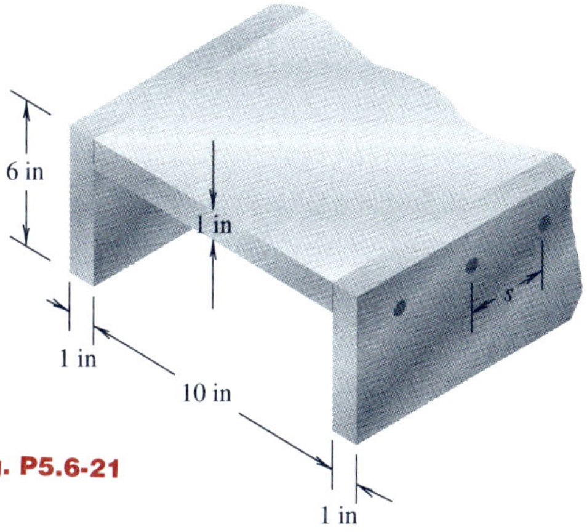

Fig. P5.6-21

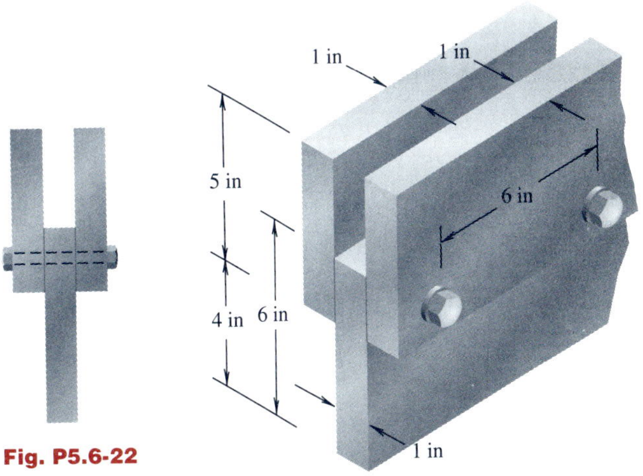

Fig. P5.6-22

5.6-23 A lightweight beam is fabricated of three $\frac{1}{2}$-in $\times$ 1-in sections glued together as shown in Fig. P5.6-23. The simply supported beam carries a concentrated load P at the location shown. If the glue shear strength is 300 psi, what is the maximum load P_m that the beam can carry? What is the corresponding maximum tensile stress in the beam when $P = P_m$?

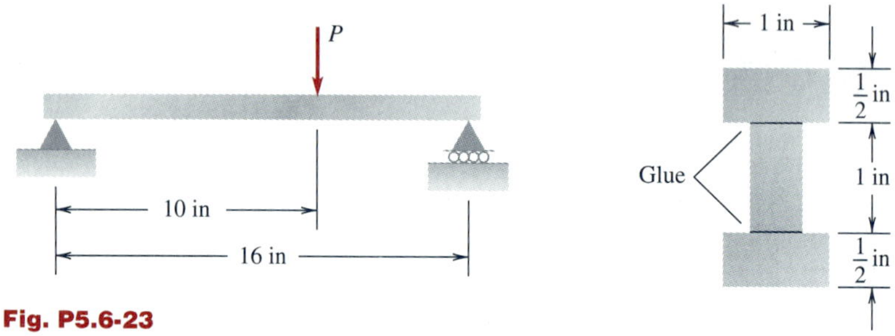

Fig. P5.6-23

5.6-24 A cantilever beam is loaded as shown in Fig. P5.6-24 by the end load P. Determine the maximum load P that can be applied to the beam without exceeding the maximum allowable shear stress in the beam of 560 psi. Neglect the weight of the beam.

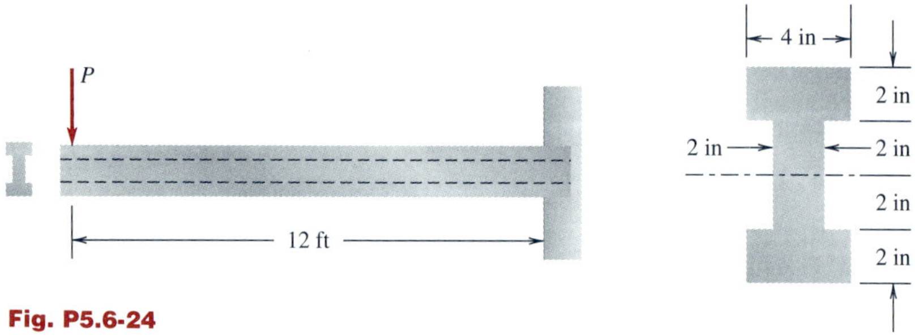

Fig. P5.6-24

5.6-25 A cruciform cross-sectional beam is loaded as shown in Fig. P5.6-25. Find the maximum normal stresses in the beam, the shear stress at the neutral axis of the beam at section A, and the shear stress at A at location a-a on the cross section.

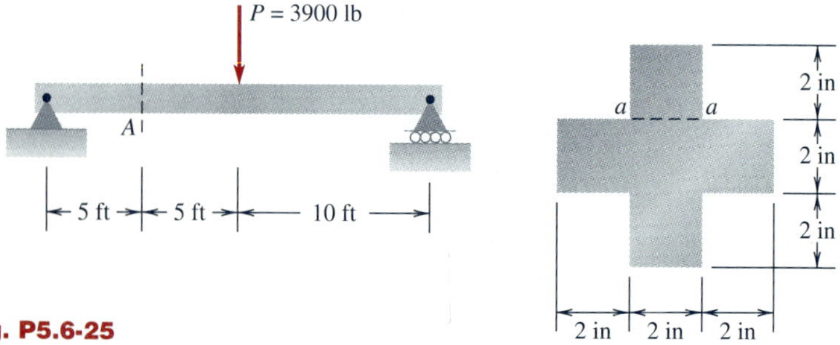

Fig. P5.6-25

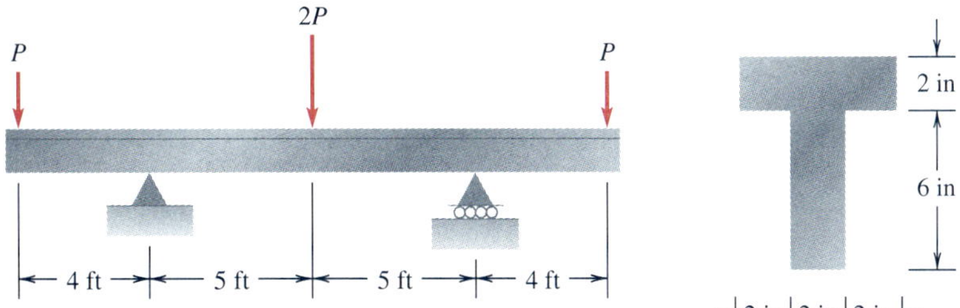

Fig. P5.6-26

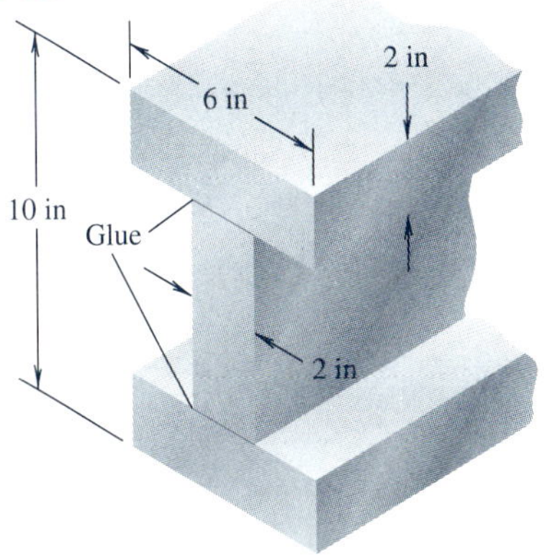

Fig. P5.6-28

5.6-26 A T beam is loaded as shown in Fig. P5.6-26. Determine the maximum allowable value for P if the compressive normal stress due to bending is not to exceed 900 psi and the maximum shearing stress is not to exceed 100 psi.

5.6-27 The cantilever beam of an I cross section as shown in Fig. P5.6-27 carries a load of 2000 lb. Find the shear stress due to bending at a location D as shown at the intersection of the web and flange of the cross section. Also find the maximum shear stress due to bending on the cross section, and compare the maximum shear stress to the value of V/A.

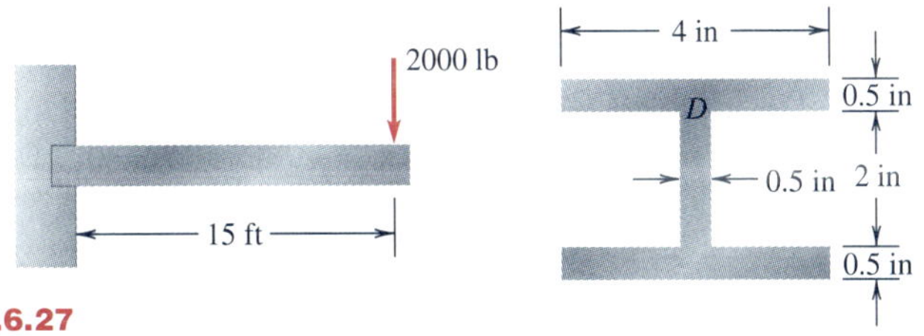

Fig. P5.6.27

Fig. P5.6-29

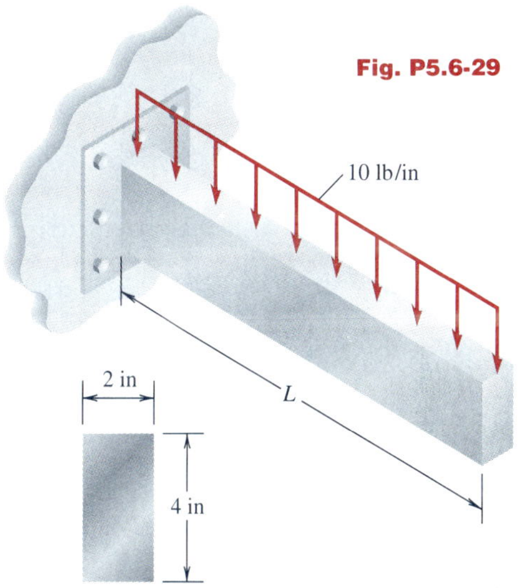

5.6-28 A beam is constructed from three 2-in × 6-in planks glued together as shown in Fig. P5.6-28. If the glue has an allowable shearing strength of 500 psi, what is the value of the maximum allowable transverse shearing force V in the beam?

5.6-29 A cantilever beam carries a uniform load of $q = 10$ lb/in as shown in Fig. P5.6-29. If the allowable stresses in the beam are 500 psi in shear and 1200 psi for normal stress, calculate the maximum allowable length L of the beam.

5.6-30 A T beam is constructed from two 2-in × 8-in wood planks nailed together as shown in Fig. P5.6-30. Each nail has a shear strength of 120 lb, and the nails are spaced 2 in apart. Calculate the maximum allowable shearing force V that can be applied to the beam.

5.6-31 A diver of weight 700 N stands at the end of a diving board, as shown in Fig. P5.6-31. The uniform distributed weight of the diving board is 200 N/m. Find the maximum tensile and compressive stresses in the board due to the weight of the board and the diver.

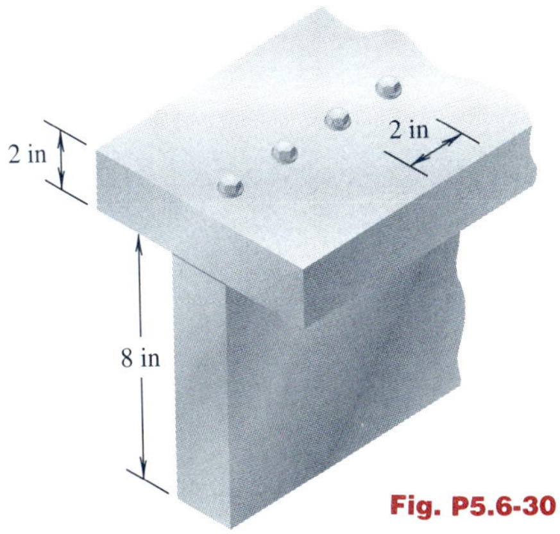

Fig. P5.6-30

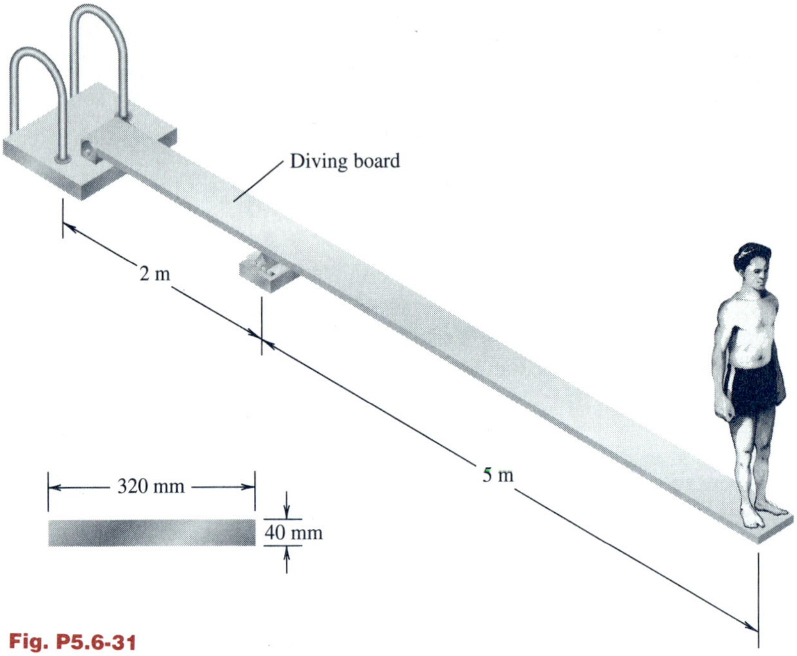

Fig. P5.6-31

5.6-32 The beam shown in Fig. P5.6-32 has the cross section indicated. Draw the shear force and bending moment diagrams for the beam, and then find the maximum bending stresses in the beam and the maximum shear stress due to bending in the beam. Indicate clearly the location of the two maximum stresses. Take $P = 1000$ lb, $w = 200$ lb/ft, and $a = 4$ ft.

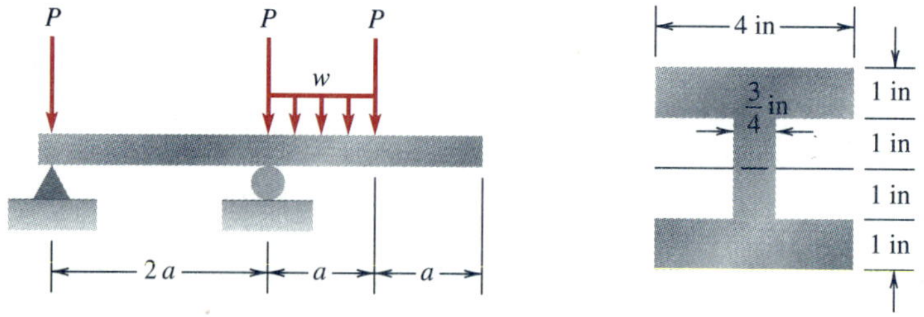

Fig. P5.6-32

5.6-33 A lightweight wooden beam is built up by gluing four 1-in × 6-in planks together as shown in Fig. P5.6-33. If the allowable shear stress in the glue is $\tau_a = 50$ psi, determine the maximum allowable load P so that the glue strength at the joints in the beam is not exceeded if the beam is loaded as shown. Neglect the weight of the beam.

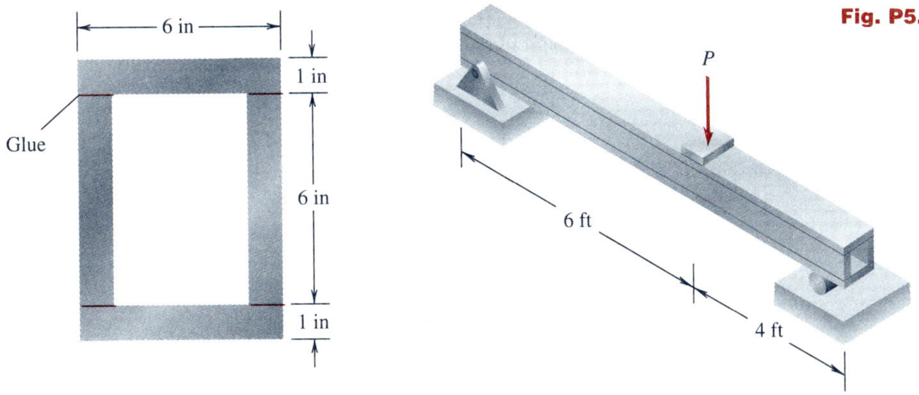

Fig. P5.6-33

5.6-34 An I beam whose cross-sectional area equals 5 in² and whose moment of inertia is 25 in⁴ is strengthened by welding with two continuous welds a cover plate to its upper flange, as shown in Fig. P5.6-34. If the beam is subjected to a maximum shear force of $V = 5000$ lb, calculate the shear force per inch needed for the weld at the top of the I beam.

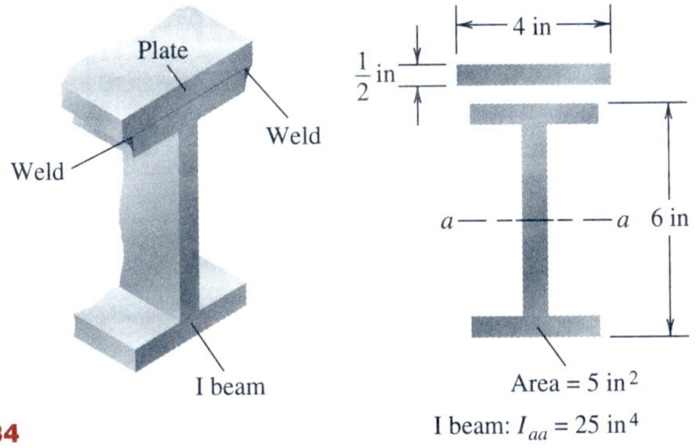

Fig. P5.6-34

5.6-35 A lightweight supporting beam is to be made of two different structural components by gluing the C6 × 8.2 sections (at the top and bottom) to the

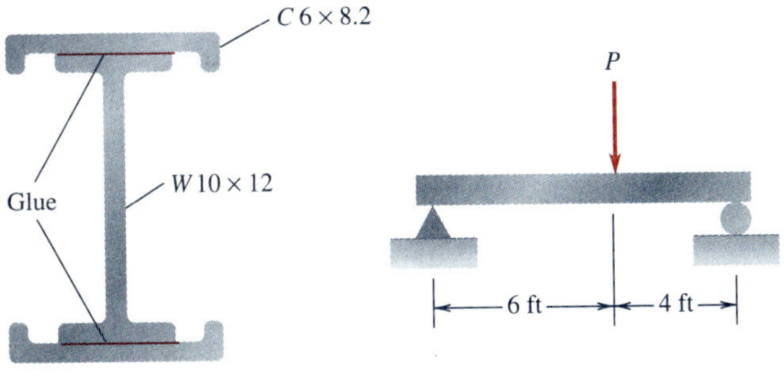

Fig. P5.6-35

I-beam section W10 × 12 as shown in Fig. P5.6-35. The allowable glue strength in shear is $\tau = 50$ psi. Determine the maximum allowable load P on the beam so as not to exceed the allowable shear stress in the glue if the built-up beam is loaded as shown. Neglect the weight of the beam.

5.7-1 Consider the problem of pure bending of a symmetric composite beam which has been made by bonding together two materials of different elastic properties, as shown in Fig. P5.7-1. We can carry out a derivation parallel to that given in Secs. 5.2 to 5.4. Obtain expressions for the deformation and the stresses in the composite beam. Show that the neutral surface [using Eq. (5.11)] is located by the distance y_N, where

$$y_N = \frac{E_1\bar{y}_1A_1 + E_2\bar{y}_2A_2}{E_1A_1 + E_2A_2} \tag{a}$$

and the moment-curvature relation [using Eq. (5.13)] is

$$\frac{d\theta}{ds} = \frac{1}{\rho} = \frac{M_b}{E_1I_{z_1} + E_2I_{z_2}} \tag{b}$$

where I_{z_1} and I_{z_2} are, respectively, the moments of inertia of area A_1 and A_2 about the neutral surface. Finally, show that the bending stress in the beam is given by

$$(\sigma_x)_i = -E_i\frac{M_by}{E_1I_{z_1} + E_2I_{z_2}}$$

where i takes on the value of 1 or 2, depending on the material.

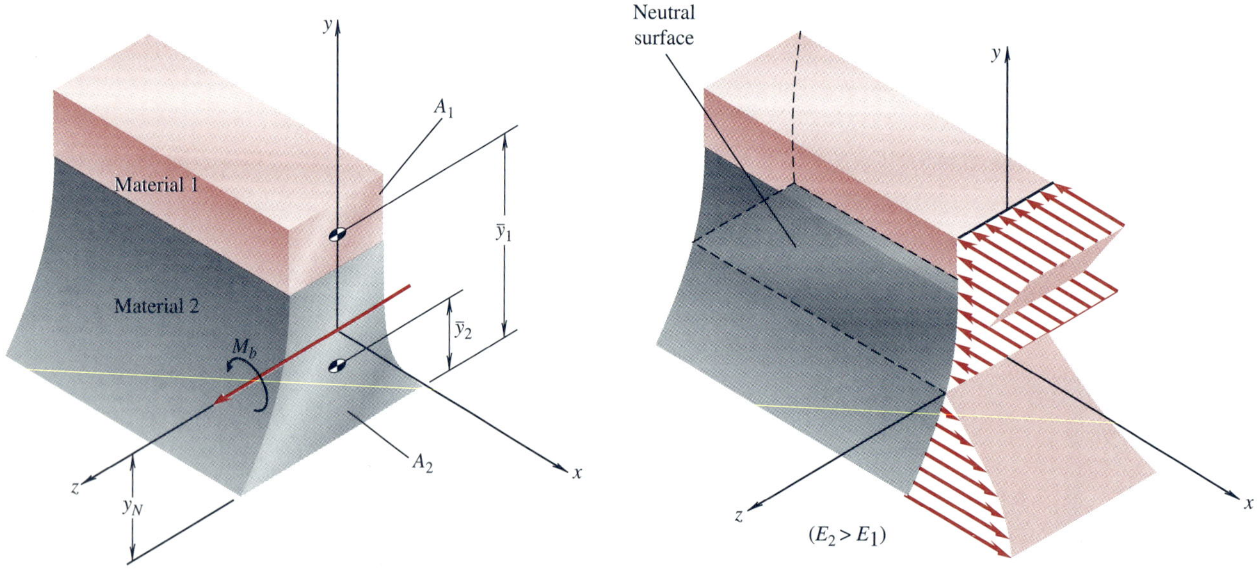

Fig. P5.7-1

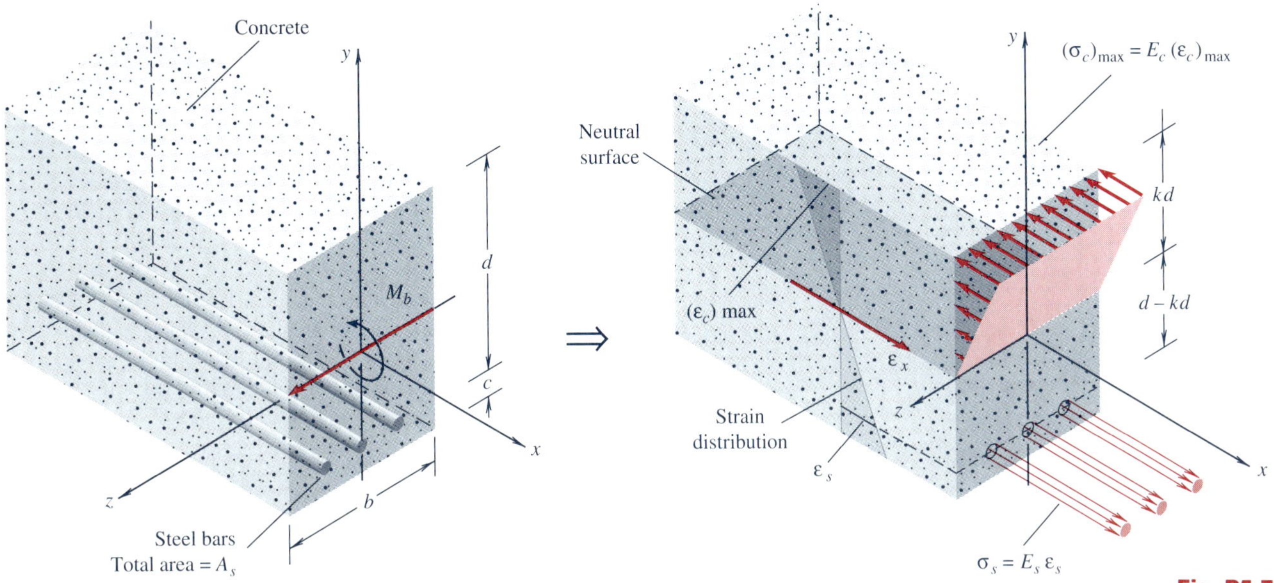

Fig. P5.7-2

5.7-2 Concrete is a brittle material which has good strength in compression but very little strength in tension. Despite its low tensile strength, economic use can be made of concrete in reinforced-concrete construction in which steel bars are embedded in the concrete to provide tensile action. For a reinforced-concrete beam, carry out a development parallel to that given in Secs. 5.2 to 5.4 under the assumptions that no tensile stresses are carried by the concrete and that the tensile stress in the steel is uniform over the bars. Show that the neutral surface is located at a distance kd below the top of the beam, where the factor k is determined by the following quadratic equation (Fig. P5.7-2)

$$E_s(d - kd)A_s - E_c\frac{b(kd)^2}{2} = 0$$

5.7-3 The reinforced-concrete beam shown in the sketch of Fig. P5.7-3 contains five $\frac{3}{4}$-in-diameter steel bars. If the tensile stress in the steel is not to exceed 20,000 psi and the compressive stress in the concrete is not to exceed 1350 psi, what is the maximum bending moment which the beam can transmit? Take $E_c = 1.5 \times 10^6$ psi and $E_s = 30 \times 10^6$ psi.

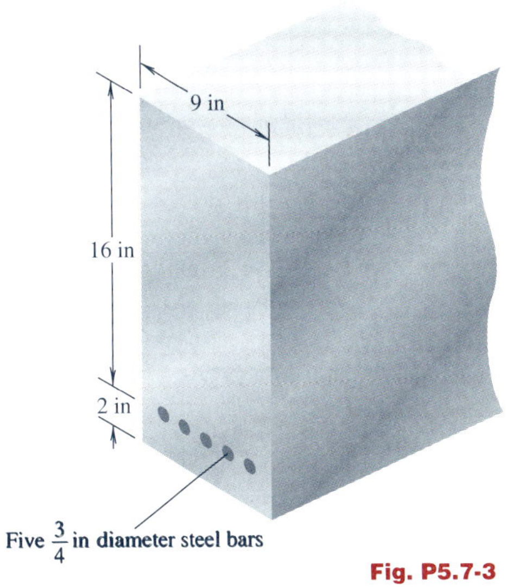

Fig. P5.7-3

5.7-4 Consider the case where a composite beam transmits a shear force in addition to a bending moment, as shown in Fig. P5.7-4. We now wish to use the results of Prob. 5.7-1 and repeat the arguments of Sec. 5.6 to show that the average shear stress at a distance y_0 from the neutral surface is given by

$$\tau_{xy} = \tau_{yx} = \frac{V}{b(E_1 I_{z_1} + E_2 I_{z_2})}\int_{A_0} Ey\,dA$$

where the integral is to be taken over the area shown in part (*b*) of the figure, i.e., over the range from $y = y_0$ to $y = c$.

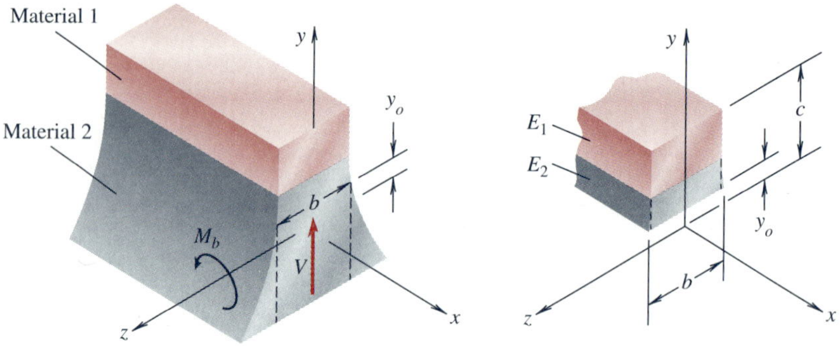

Fig. P5.7-4 (*a*) (*b*)

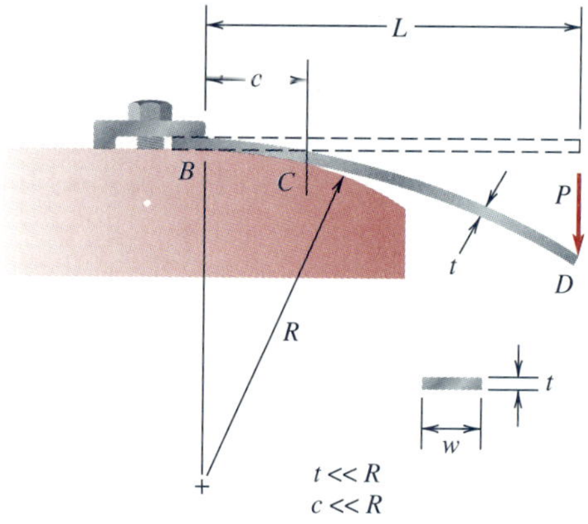

Fig. P5.7-5

5.7-5 An initially straight steel strip of length L is clamped at one end B as shown in Fig. P5.7-5 onto a rigid block of radius R. A load P is applied at the end D, forcing a length c of the strip to come in contact with the rigid block. Determine the value of the contact distance c in terms of the load P and the corresponding dimensions and radius R. Assume that $c \ll R$ and that the weight of the strip can be neglected. Note that the bending moment in the beam is constant in segment BC. A *concentrated* reactive load acting on the strip is required at point C for equilibrium; i.e., a uniform pressure load acting on segment BC will not lead to a constant radius of curvature R in this segment.

5.7-6 A beam of length L, weight w per unit length, and bending modulus EI is placed on a rigid horizontal table such that a short segment CD of length a overhangs the table, as shown in Fig. P5.7-6. We wish to find the reactions at B and C on the beam. (A section of rubber hose can be used conveniently to demonstrate this problem.)

(*a*) First, we observe that the beam has zero curvature in the segment AB so that the bending moment and shear force are zero in this segment. Therefore, the reaction from the table is w.

(*b*) Second, there must be a concentrated load R_B at B to ensure that the moment is positive to lift off the beam from the table at B, Fig. P5.7-6.

(*c*) Finally there is a concentrated reaction R_C at C.

The values of the reactions are

$$R_B = \frac{w(b^2 - a^2)}{2b} \qquad R_C = \frac{w(b + a)^2}{2b}$$

In Prob. 7.9-35 we will find the value of the uplift distance b.

5.7-7 Discuss how the shear force and bending moment program of Chap. 4 could be combined with the data in App. C to select an appropriate beam cross section, given the loading and the allowable normal stress for a statically determinate beam. How would the weight of the beam be included?

5.7-8 A 30-ft portion of two long continuous walkways is shown in Fig. P5.7-8a. The upper walkway is supported by 1.25-in-diameter rods connected to

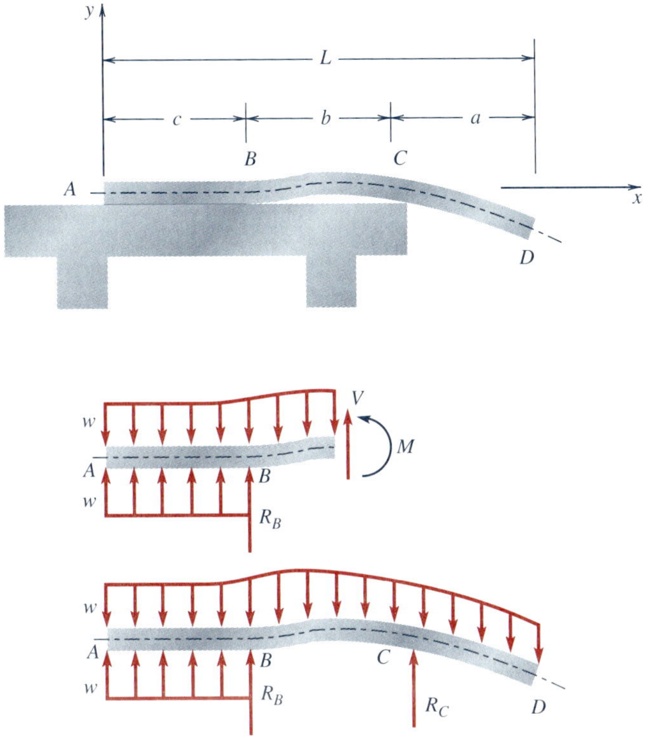

Fig. P5.7-6

the ceiling by washers and nuts attached to the $7 \times 4 \times \frac{1}{2}$ structural tube (see Table C.4), as shown in Fig. P5.7-8*b*. The lower walkway in Fig. P5.7-8*b* hangs on the rods also. The design in the figure was not built, however, and it was replaced by the design in Fig. P5.7-8*c*. Determine the maximum shear force and bending moment and the maximum bending stress in each of the tube beams shown in Fig. P5.7-8*b* and *c*. Comment on the logic of the designs in Fig. P5.7-8*b* and *c*. The loads on the walkway are transmitted to the cross beams as shown.

5.7-9 The material in a beam undergoing pure bending has a stress-strain relation which is elastic–perfectly plastic, as shown in Fig. P5.7-9. For a beam of rectangular section, show that the bending moment M_Y which corresponds to the *onset* of yielding in the beam is given by

$$M_Y = \frac{bh^2Y}{6}$$

where Y is the yield stress of the material, b is the width, and h is the depth of the beam.

5.7-10 For a beam of rectangular section in pure bending, the moment which corresponds to the onset of yielding is M_Y as given in Prob. 5.7-9.

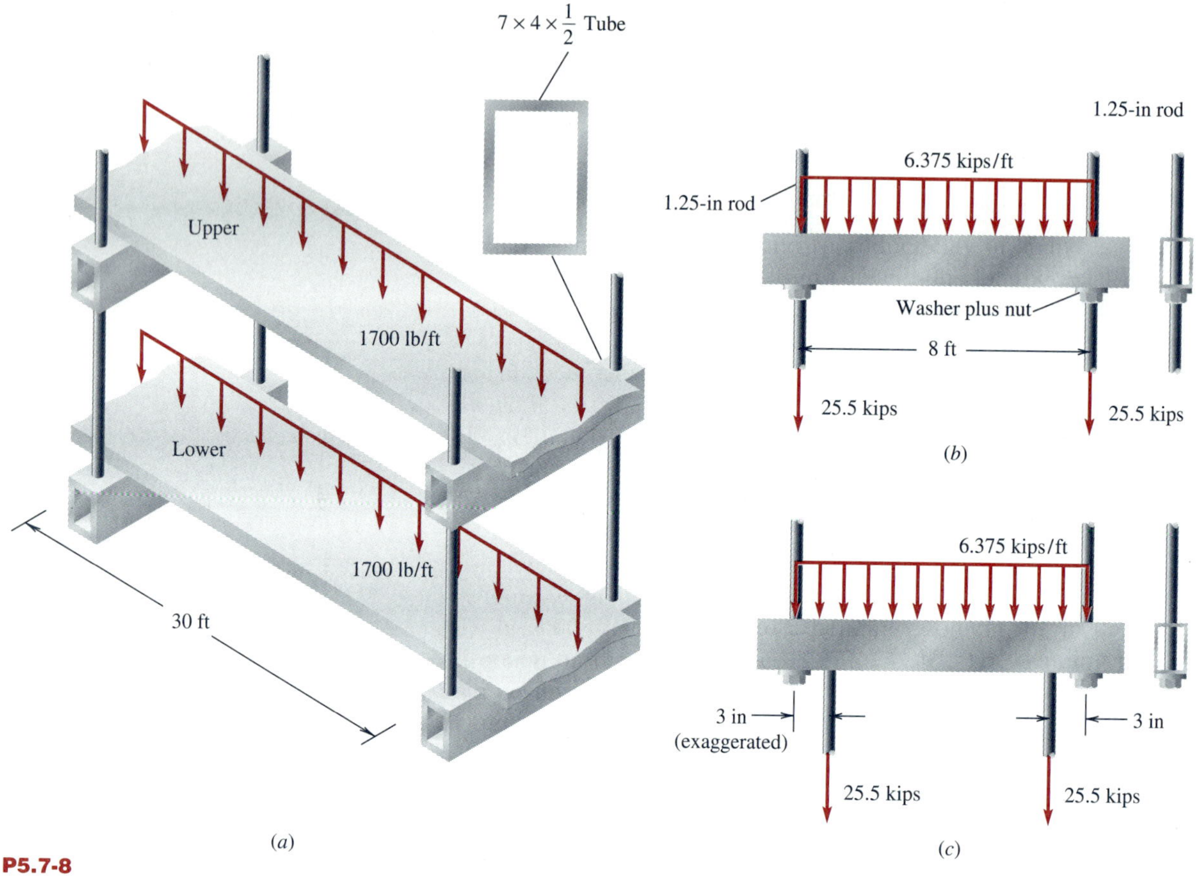

Fig. P5.7-8

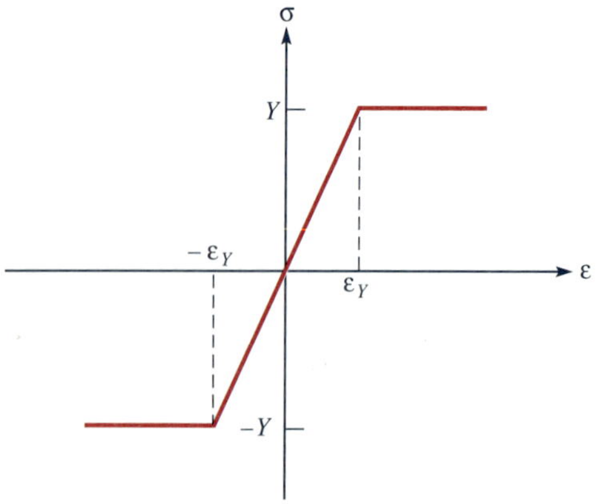

Fig. P5.7-9

(*a*) Show that the corresponding curvature $(1/\rho)_Y$ can be found from Eq. (5.16) with $y = -h/2$ in the form

$$\left(\frac{1}{\rho}\right)_Y = \frac{\epsilon_Y}{h/2}$$

where ϵ_Y is defined in Fig. P5.7-9. For values of $M > M_Y$ the strain-curvature relation

$$\epsilon_x = \frac{-y}{\rho} \qquad (a)$$

remains valid, and the bending stress distribution will be elastic over a central core region and plastic over an outer region, as shown in Fig. P5.7-10*a*, where y_Y defines the outer boundary of the elastic region and

$$\sigma_x = \begin{cases} -Y(y/y_Y) & 0 < y < y_Y \\ -Y & y_Y < y < \dfrac{h}{2} \end{cases} \qquad (b)$$

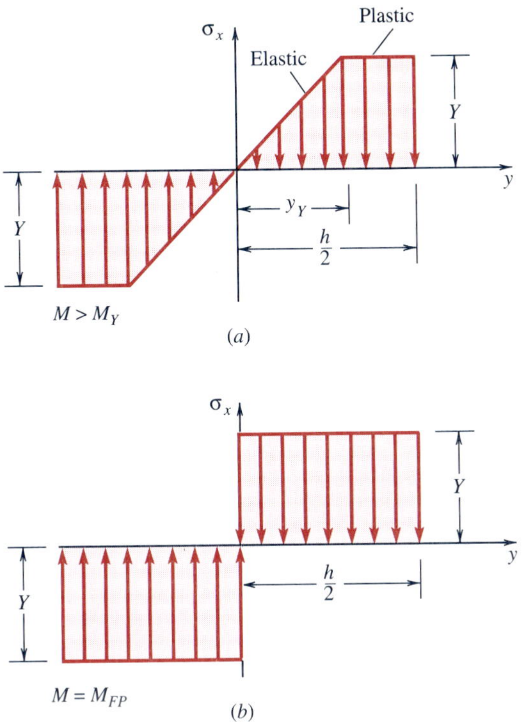

Fig. P5.7-10

(*b*) Show that the bending moment associated with the stress distribution shown in Fig. P5.7-10*a* is given by

$$\begin{aligned} M &= -\int_A \sigma_x y \, dA \\ &= 2\left(-\int_0^{y_Y} \sigma_x y \, b \, dy - \int_{y_Y}^{h/2} \sigma_x y \, b \, dy\right) \end{aligned} \qquad (c)$$

and making use of Eq. (*b*) in Eq. (*c*), show that

$$M = \frac{bh^2}{4} Y\left[1 - \frac{1}{3}\left(\frac{y_Y}{h/2}\right)^2\right] \qquad (d)$$

(*c*) Show that for increasing bending moment, y_Y approaches 0 (see Fig. P5.7-10*b*) and the corresponding fully plastic moment M_{FP} becomes, by using Eq. (*d*),

$$M_{FP} = \frac{bh^2}{4} Y \qquad (e)$$

or from Prob. 5.7-9

$$M_{FP} = \frac{3}{2} M_Y \qquad (f)$$

CHAPTER 6

Deflections of Statically Determinate Beams

6.1 Introduction

When transverse loads are applied to a beam, the beam will bend or deflect. In this chapter we will calculate the deflection of beams caused by transverse loads. While for many structural beams and beam elements in machine parts the amount of deflection is small, there are many situations in which the exact amount of beam deflection must be calculated. For example, machine parts in service often must be designed to close tolerances on deflections in order to avoid interference between moving parts. In structural design it is often a design requirement that the maximum deflection of a beam divided by its length L not exceed $\frac{1}{240}$, or in the case of small house design to avoid cracking plaster in ceilings, the requirement is that this ratio not exceed $\frac{1}{360}$.

We will investigate first the deflection of beams for which the reactions can be found from equilibrium considerations alone; i.e., we will investigate statically determinate beam problems first. In Chap. 7, we will study statically indeterminate beam problems. We will still be guided in solving problems in this chapter and the next chapter by our standard approach of using the three steps of Eq. (2.8). However, for beams we need to obtain the force-deformation relation that constitutes the second step of Eq. (2.8); this we do in the next section.

6.2 Differential Equations for the Deflection of Beams

In Chap. 5 we showed that for an element of a symmetric beam in pure bending as illustrated in Fig. 6.1, the curvature $\kappa = 1/\rho$ of the neutral

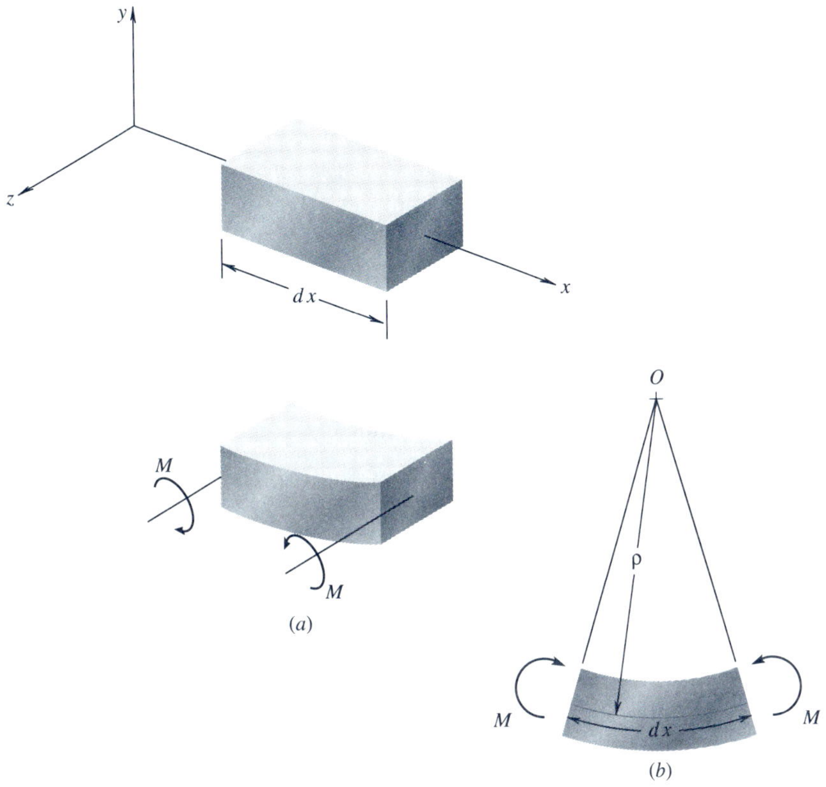

Figure 6.1 (*a*) Deformation of a beam element in pure bending. (*b*) Radius of curvature ρ of neutral axis for an element in pure bending.

axis after bending is related to the applied bending moment M by

$$M = \frac{EI_z}{\rho} \tag{5.13}$$

where E is the modulus of elasticity of the beam, I_z is the moment of inertia of the cross section, and ρ is the radius of curvature of the neutral axis of the bent beam. Equation (5.13) was derived under the assumption that the bending moment is constant along the length of the beam. We rewrite Eq. (5.13) in the form

$$\frac{1}{\rho} = \frac{M}{EI_z} \tag{6.1}$$

to emphasize that a knowledge of the bending moment (on the right-hand side of the equation) leads to the determination of the deformation or curvature of the beam.

In the more general case when the bending moment varies along the beam and shear forces are present, we will assume that the effect of the shear forces on the deformation is negligible. In advanced treatments of the mechanics of solids, it has been shown that for long, slender beams this neglect of the influence of shear forces on beam deformation is a good approximation. Thus according to this assumption, the curvature of the neutral axis at each section along the beam is completely determined by the value of the bending moment M at that section and the product EI_z, as in Eq. (6.1). This assumption is consistent with the engineering theory of stress distributions in beams.

We now need to relate the curvature $1/\rho$ to the deflection of the neutral axis. In Fig. 6.2*a* we show a segment of a loaded symmetric beam whose neutral axis was located along the x axis before loading. The neutral axis after loading is given by the deflection curve $v(x)$, which is positive in the positive y direction. We exaggerate the amount of the beam deflection v in Fig. 6.2; the deflection of the neutral axis is a function of x, that is, $v = v(x)$.

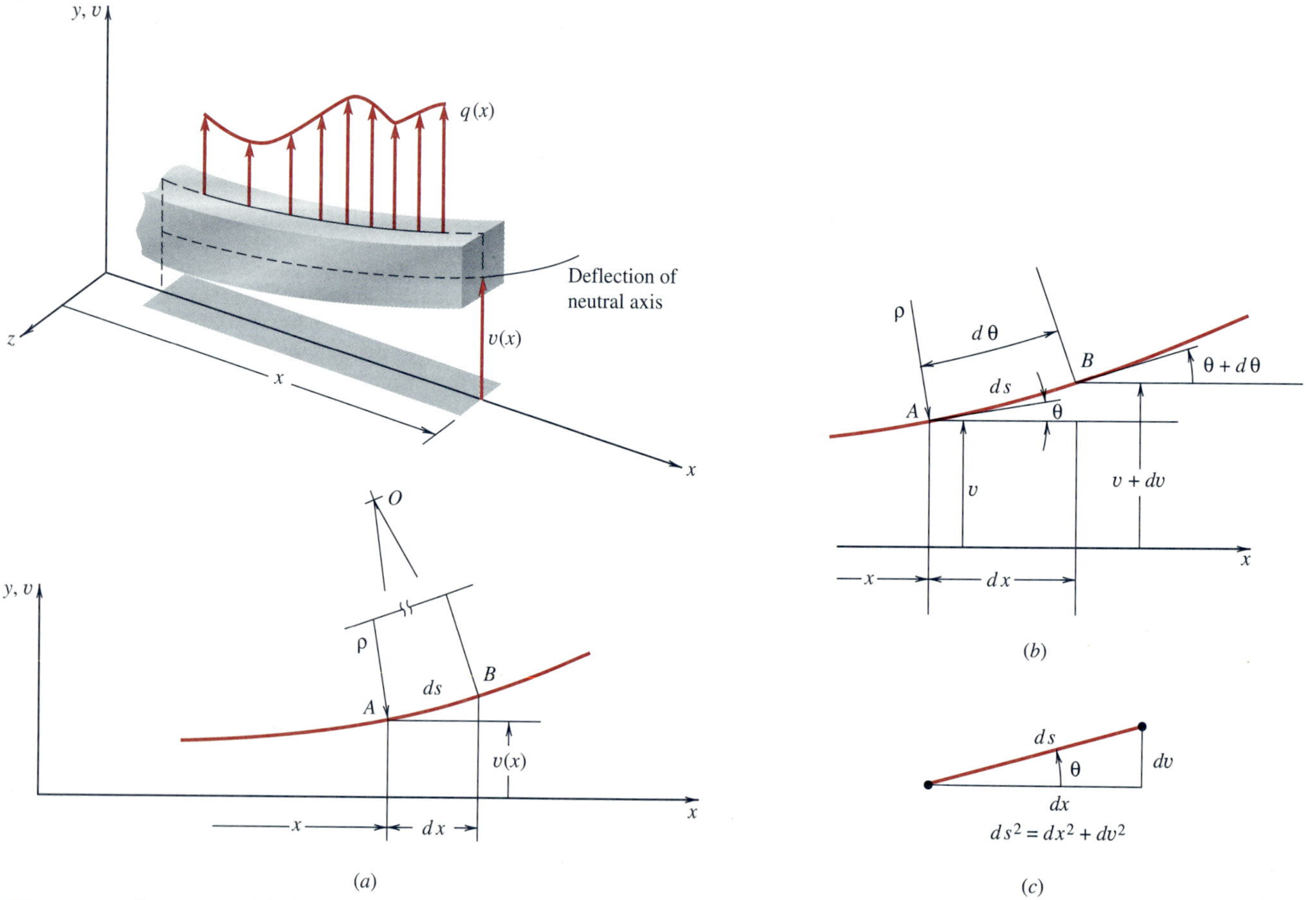

Figure 6.2 Geometry of deformation.

The geometric relations between the deflection v, the slope angle θ, and the curvature $1/\rho$ are shown in Fig. 6.2b and c for a segment ds of the neutral axis of the beam undergoing a deflection v. Two successive normals to the deflection curve at points A and B intersect at the point O. The radius of curvature times the angle $d\theta$ equals the arclength distance ds between A and B, or $ds = \rho d\theta$. Thus the curvature $\kappa = 1/\rho$ can be written

$$\kappa = \frac{1}{\rho} = \frac{d\theta}{ds} \tag{6.2}$$

To relate the curvature κ to the deflection v, we make use of Fig. 6.2c to see that the slope of the deflection curve is given by

$$\frac{dv}{dx} = \tan\theta \tag{6.3}$$

where θ is the angle that the tangent line to the deflection curve makes with the x axis. Since the curvature involves the derivative of θ with respect to s, we differentiate both sides of Eq. (6.3) with respect to s to get

$$\frac{d}{ds}\left(\frac{dv}{dx}\right) = \frac{d^2v}{dx^2}\frac{dx}{ds} = \sec^2\theta\,\frac{d\theta}{ds}$$

and solving for the curvature $d\theta/ds$ gives

$$\frac{d\theta}{ds} = \frac{d^2v}{dx^2}\frac{dx}{ds}\cos^2\theta$$

But using Fig. 6.2c again, we have

$$\cos\theta = \frac{dx}{ds} = \frac{dx}{(dx^2 + dv^2)^{1/2}} = \frac{1}{[1 + (dv/dx)^2]^{1/2}}$$

and we may express the curvature entirely in terms of derivatives of v in the form

$$\frac{1}{\rho} = \frac{d\theta}{ds} = \frac{d^2v/dx^2}{[1 + (dv/dx)^2]^{3/2}} \tag{6.4}$$

Finally, if we combine Eqs. (6.1) and (6.4), we obtain the nonlinear equation

$$\frac{d^2v/dx^2}{[1 + (dv/dx)^2]^{3/2}} = \frac{M}{EI_z} \tag{6.5}$$

relating the derivatives of v to the given M and EI_z.

In most engineering applications of practical interest involving beam deflections, the slope of the deflection curve is small compared to 1. Thus, the term $(dv/dx)^2$ can be neglected compared to 1 in the denominator of the curvature expression in Eq. (6.5), to give a linear relation between the second derivative of the deflection and the bending moment at each section along the beam, i.e.,

$$\frac{d^2v}{dx^2} = \frac{M(x)}{EI_z} \tag{6.6}$$

In the analysis to follow, the moment curvature equation given by Eq. (6.6) will serve as the "force-deformation" relation for beams, the second of the three steps of Eq. (2.8). The product EI_z is often referred to as the ***bending modulus*** of the beam. We often write I for I_z if no confusion can arise.

To obtain an expression for the deflected shape $v(x)$ from Eq. (6.6), we need to first determine from equilibrium considerations an expression for the bending moment along the beam, $M(x)$, and then integrate Eq. (6.6) twice. The first integration gives dv/dx, which is the slope of the deflection curve. Since the slope is small, $\tan\theta \approx \theta = dv/dx$, the first integration gives the slope angle θ in radians of the deflection curve. The second integration gives the deflection curve $v(x)$. The slope dv/dx and deflection $v(x)$ found by integration must satisfy the particular conditions of geometric compatibility for the problem; i.e., all three steps in Eq. (2.8) are to be used to calculate the deflection of a beam.

6.3 Beam Deflections by the Method of Double Integration

For statically determinate beam problems where the bending moment $M(x)$ can be determined completely in terms of the applied loads, we can directly integrate Eq. (6.6) to obtain first the slope dv/dx or the slope angle θ and then the deflection $v(x)$. For each integration, we obtain a constant of integration. The values of the two constants of integration are found from known information about the slope and the deflection. The following examples will illustrate the procedures for this method. Of course, we need to specify a set of coordinate axes and to keep track of positive sign conventions for analyzing beams, as we did in Chaps. 4 and 5.

EXAMPLE 6.1

A wooden beam in a preliminary design of a balcony is modeled as a cantilever beam of rectangular cross section fixed to a rigid wall, as shown in Fig. 6.3*a* and *b*. The equivalent loading on the beam is a concentrated load W as shown. We wish to find the deflection curve of the cantilever beam and the deflection and slope at the free end B. We neglect the weight of the beam and take the bending modulus EI of the beam as constant.

As a first step in finding the deflection curve, we set the coordinate axes through the centroid of the cross section with origin at the fixed end A. To integrate Eq. (6.6), we need to

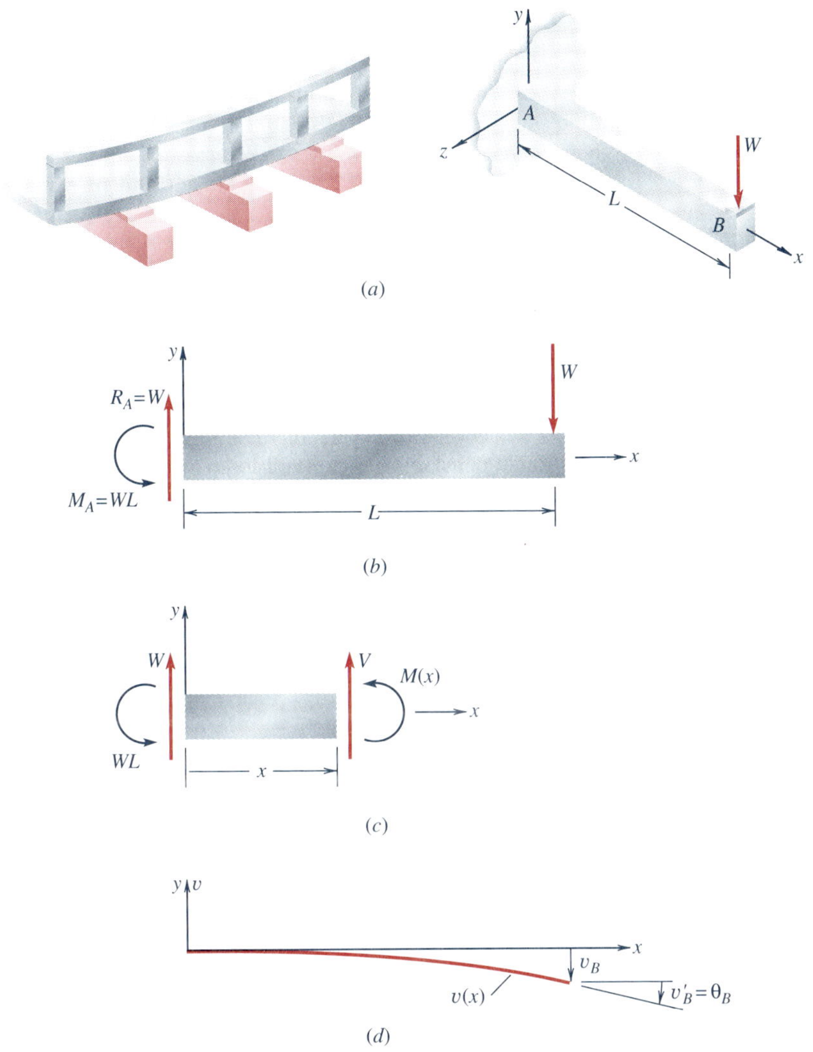

Figure 6.3 Example 6.1

obtain an expression for $M(x)$. To do this, we first compute the reactions acting on the beam at the wall. Force and moment equilibrium of the beam yields the reactions $M_A = WL$ and $R_A = W$, as shown in Fig. 6.3*b*. Moment equilibrium of a segment of the beam of length x as shown in Fig. 6.3*c* gives an expression for the bending moment as a function of x in the form

$$M(x) = Wx - WL \qquad (a)$$

As a check on the expression given in Eq. (*a*), we first set $x = 0$ in the expression to get $M(0) = -WL$, which leads to the correct wall moment reaction in terms of both the sign and the magnitude. At $x = L$, we find $M(L) = 0$ which correctly indicates that there is no applied external moment at the free end of the beam.

Using the moment-curvature relation, Eq. (6.6), in the form $EIv'' = M(x)$ gives

$$EIv'' = Wx - WL \qquad (b)$$

where now we have used primes to indicate derivatives of $v(x)$ with respect to x, for example, $v'' = d^2v/dx^2$. Since EI is a constant, we can integrate Eq. (*b*) twice to obtain

$$EIv' = W\frac{x^2}{2} - WLx + c_1 \qquad (c)$$

$$EIv = W\frac{x^3}{6} - WL\frac{x^2}{2} + c_1x + c_2 \qquad (d)$$

where c_1 and c_2 are constants of integration arising from the two integrations.

Now that we have obtained analytical expressions for the slope dv/dx and the deflection v, we are in a position to make use of information for values of dv/dx and v at particular points along the beam. In the present problem of a cantilever beam, the clamped end at the wall ($x = 0$) does not allow the end section to rotate about the z axis or to undergo any deflection in the y direction, i.e., at $x = 0$

$$v' = 0 \qquad \text{and} \qquad v = 0 \qquad (e)$$

We often refer to the conditions in Eq. (*e*) as "built-in" conditions for a cantilever beam. Making use of Eqs. (*c*) and (*d*), we find in this case

$$c_1 = 0 \qquad c_2 = 0 \qquad (f)$$

The required deflection curve then follows from Eq. (*d*) with c_1 and $c_2 = 0$ in the form

$$v(x) = -\frac{Wx^2}{6EI}(3L - x) \qquad (g)$$

The deflected shape is sketched in Fig. 6.3*d*.

The maximum deflection and the maximum slope occur at the free end of the beam and are found by inserting $x = L$ into Eqs. (*c*) and (*g*) to get

$$v(L) = v_B = -\frac{WL^3}{3EI}$$
$$v'(L) = v'_B = -\frac{WL^2}{2EI} \qquad (h)$$

Since the slope of the deflection curve is assumed to be small compared to 1 in our analysis, the slope angle θ is approximately equal to v'; we show the slope angle as θ_B at the free end B in Fig. 6.3*d*. The minus signs in Eqs. (*h*) indicate that the deflection is in the negative y direction and that the slope angle is measured clockwise from the positive x axis, as shown in Fig. 6.3*d*.

To get an indication of the magnitude of the deflection and slope, let us consider the special case of a wood cantilever beam with

$$L = 9 \text{ ft} = 108 \text{ in} \qquad W = 400 \text{ lb}$$
$$E = 1.5 \times 10^6 \text{ psi} \qquad I = 230.8 \text{ in}^4 \qquad (i)$$

The beam in this case corresponds to a nominal size 4 in × 10 in eastern spruce no. 1 grade timber. The actual cross-sectional dimensions of this "4 × 10" timber are 3.5 in × 9.25 in. With these values we obtain

$$v_B = -\frac{WL^3}{3EI} = -\frac{400(108)^3}{3(1.5 \times 10^6)(230.8)} = -0.485 \text{ in} \qquad (j)$$

$$v'_B = -\frac{WL^2}{2EI} = -\frac{400(108)^2}{2(1.5 \times 10^6)(230.8)} = -0.0067 \text{ rad} = -0.386° \qquad (k)$$

The small value for the maximum slope indicates that the simplification of the curvature expression, Eq. (6.5), which involves neglecting the square of the slope of the beam dv/dx compared to unity is certainly justifiable for the wood beam described by the data in Eq. (*i*).

The deflection value of 0.485 in gives a ratio of the maximum deflection to the total span length L in inches that would not be acceptable in many structural applications. In this case the ratio is $\frac{1}{223}$. For example, for many structural designs the deflection limitation in the design code is $\frac{1}{240}$ of the span; if this were the case for the present problem, then the load would have to be reduced by a factor of $\frac{223}{240} = 0.929$ or to the value of 372 lb, or the span with a load of 400 lb would need to be shortened to 104 in.

EXAMPLE 6.2

A simply supported steel beam is loaded by a constant distributed load q_0 along its length L as shown in Fig. 6.4*a*; also see Example 4.7. We wish to determine the expression for the deflection along the beam and the value of the maximum deflection at the midspan. Assume that the weight of the beam is included in the loading term q_0 and that EI is constant.

The reactions at the ends of the beam are (Fig. 6.4*b*)

$$R_A = R_B = \frac{q_0L}{2} \qquad (a)$$

A direct appeal to moment equilibrium of a segment of length

x gives the moment $M(x)$ at the section x as (Fig. 6.4*c*)

$$M(x) = \frac{q_0 L}{2} x - q_0 \frac{x^2}{2} \qquad (b)$$

The moment-curvature relation, Eq. (6.6), therefore becomes upon use of Eq. (*b*)

$$EIv'' = M(x) = \frac{q_0 L}{2} x - q_0 \frac{x^2}{2} \qquad (c)$$

Upon integration we find the expression for the slope $dv/dx = v'$ to be

$$EIv' = \frac{q_0 L x^2}{4} - q_0 \frac{x^3}{6} + c_1 \qquad (d)$$

where c_1 is the constant of integration. Integration of Eq. (*d*) one more time gives the deflection in the form

$$EIv = \frac{q_0 L x^3}{12} - q_0 \frac{x^4}{24} + c_1 x + c_2 \qquad (e)$$

where c_2 is the second constant of integration.

Thus far we have used equilibrium to obtain the moment expression, Eq. (*b*), and the force-deformation relation of the beam, Eq. (6.6), to obtain the expression given in Eq. (*e*) for the deflection of the beam. It remains to evaluate constants c_1 and c_2 from the geometric conditions restricting the deflection of the beam. In this case, the beam is simply supported at each end so that

$$v(0) = 0 \qquad v(L) = 0 \qquad (f)$$

It follows from Eqs. (*e*) and (*f*) that

$$c_2 = 0 \qquad 0 = \frac{q_0 L^4}{12} - \frac{q_0 L^4}{24} + c_1 L \qquad (g)$$

or

$$c_1 = -\frac{q_0 L^3}{24}$$

Therefore, the deflection expression becomes

$$v(x) = -\frac{q_0 L^4}{24EI}\left[\left(\frac{x}{L}\right)^4 - 2\left(\frac{x}{L}\right)^3 + \frac{x}{L}\right] \qquad (h)$$

By making use of the nondimensional ratio x/L in the terms within the brackets, we see that the coefficient $q_0 L^4/(EI)$ must have the units of deflection. Note that Eq. (*h*) satisfies the conditions $v(0) = 0$ and $v(L) = 0$. If $x = L/2$ is inserted into Eq.

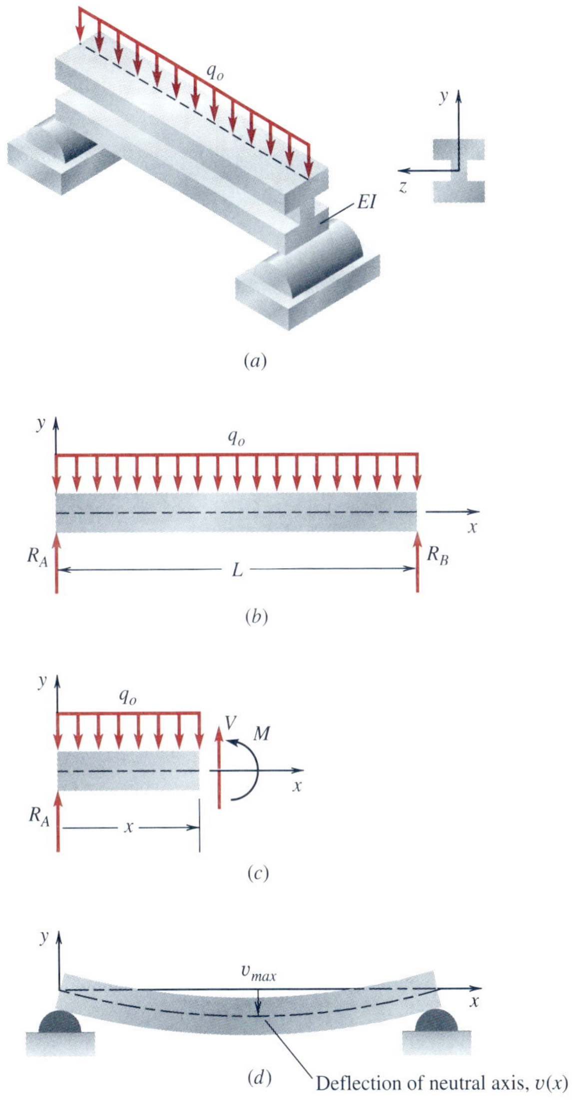

Figure 6.4 Example 6.2

(h), we obtain the maximum deflection

$$v_{\max} = v\left(\frac{L}{2}\right) = -\frac{5}{384}\frac{q_0 L^4}{EI} \qquad (i)$$

The minus sign indicates that the deflection is in the negative y direction (Fig. 6.4d).

For example, a small steel I beam with a depth of 8 in and length $L = 12$ ft, with $I = 48\ \text{in}^4$, and loaded by $q_0 = 1200$ lb/ft gives a maximum deflection of

$$v\left(\frac{L}{2}\right) = -\frac{5(100\ \text{lb/in})(144\ \text{in})^4}{384(30 \times 10^6\ \text{lb/in}^2)(48\ \text{in}^4)} = -0.389\ \text{in} \qquad (j)$$

For this load the ratio of the maximum deflection to the length of the beam is about $\frac{1}{370}$.

EXAMPLE 6.3

A simply supported beam carries a concentrated load P at the position shown in Fig. 6.5a; see Example 4.5. We wish to determine the deflection of the beam, i.e., the deflection curve of the deformed neutral axis. We assume that the weight of the beam can be neglected compared to the value of P and that EI is constant along the length of the beam.

We follow, as in Examples 6.1 and 6.2, the three steps of Eq. (2.8) in that we use first equilibrium considerations to find an expression for the moment $M(x)$ along the beam, then the force-deformation relation of Eq. (6.6) to find $v(x)$, and finally geometric information to evaluate the constants of integration arising from the integration of Eq. (6.6).

From the free-body diagram of the entire beam, we find the values of the reactions at the supports as shown in Fig. 6.5b. We now need to obtain an expression for $M(x)$ which is valid along the entire beam. In Fig. 6.5c we show a segment of the beam of length $x < a$. In Fig. 6.5d we show a segment of the beam of length $x > a$. From moment equilibrium of each segment, we have

$$M(x) = \begin{cases} \dfrac{Pb}{L}x & x < a \\ \dfrac{Pb}{L}x - P(x-a) & x > a \end{cases} \qquad (a)$$

It is more convenient to write a single expression for the moment by appealing directly to the singularity functions of

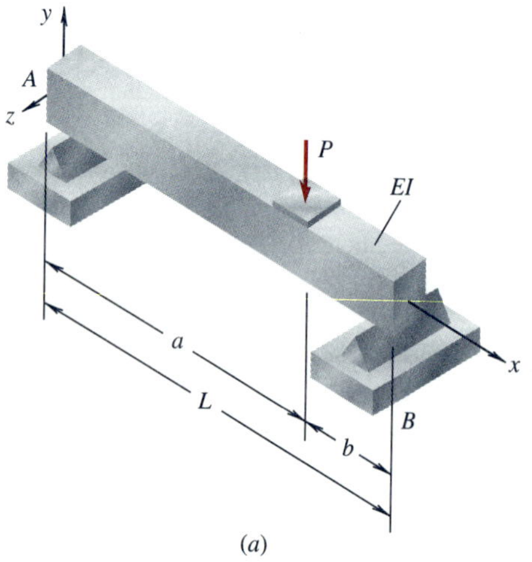

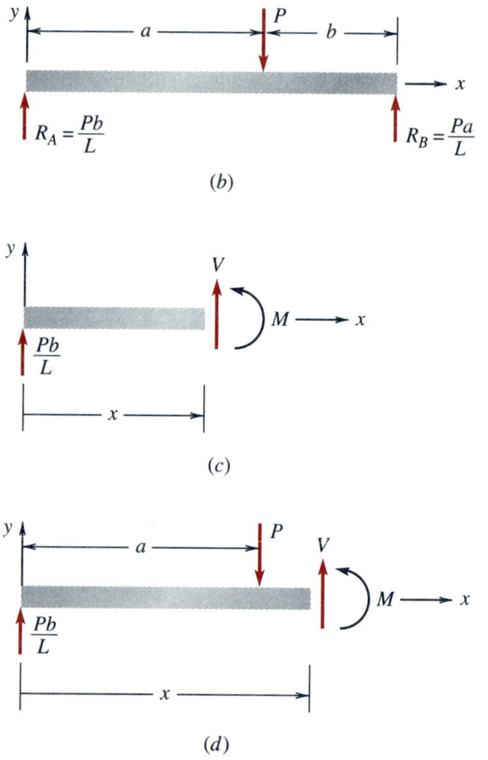

Figure 6.5 Example 6.3

Chap. 4 to obtain

$$M(x) = \frac{Pb}{L}x - P\langle x - a\rangle^1 \qquad (b)$$

where the pointed brackets in Eq. (*b*) correspond to the singularity function $\langle x - a\rangle^1$ where

$$\langle x - a\rangle^1 = \begin{cases} 0 & x \le a \\ (x - a) & x \ge a \end{cases} \qquad (c)$$

Therefore using Eq. (*b*) along with Eq. (6.6), we have

$$EIv'' = M(x) = \frac{Pbx}{L} - P\langle x - a\rangle^1 \qquad (d)$$

Since the bending modulus *EI* is constant along the beam, integration of Eq. (*d*) yields

$$EIv' = \frac{Pbx^2}{2L} - P\frac{\langle x - a\rangle^2}{2} + c_1 \qquad (e)$$

$$EIv = \frac{Pbx^3}{6L} - P\frac{\langle x - a\rangle^3}{6} + c_1x + c_2 \qquad (f)$$

where c_1 and c_2 are the constants of integration.

The geometric boundary conditions for this problem are that there should be zero deflection over the supports at each end, i.e.,

$$v(0) = 0 \qquad v(L) = 0 \qquad (g)$$

These conditions together with Eq. (*f*) give us the following equations for the determination of c_1 and c_2:

$$0 = c_2$$
$$0 = \frac{Pb}{6L}L^3 - \frac{Pb^3}{6} + c_1L \qquad (h)$$

With the values of c_1 and c_2 known, we have for the deflection curve for the neutral axis of the beam

$$v(x) = -\frac{PL^3}{6EI}\left[\frac{bx}{L^2}\left(1 - \frac{b^2}{L^2} - \frac{x^2}{L^2}\right) + \frac{1}{L^3}\langle x - a\rangle^3\right] \qquad (i)$$

The singularity function in Eq. (*i*) does not "turn on" until $x > a$.

The deflection under the load located at $x = a$ is

$$v(a) = -\frac{Pa^2b^2}{3EIL} \qquad (j)$$

The minus sign indicates that the deflection is downward in the negative *y* direction.

As a special case, we consider the case for which $a = b = L/2$, that is, the case of a simply supported beam with an applied concentrated load at the center of the beam, as shown in Fig. 6.6*a*. For this case, we have

$$v(x) = -\frac{PL^3}{6EI}\left[\frac{3x}{8L}\left(1 - \frac{4}{3}\frac{x^2}{L^2}\right) + \frac{1}{L^3}\left\langle x - \frac{L}{2}\right\rangle^3\right] \qquad (k)$$

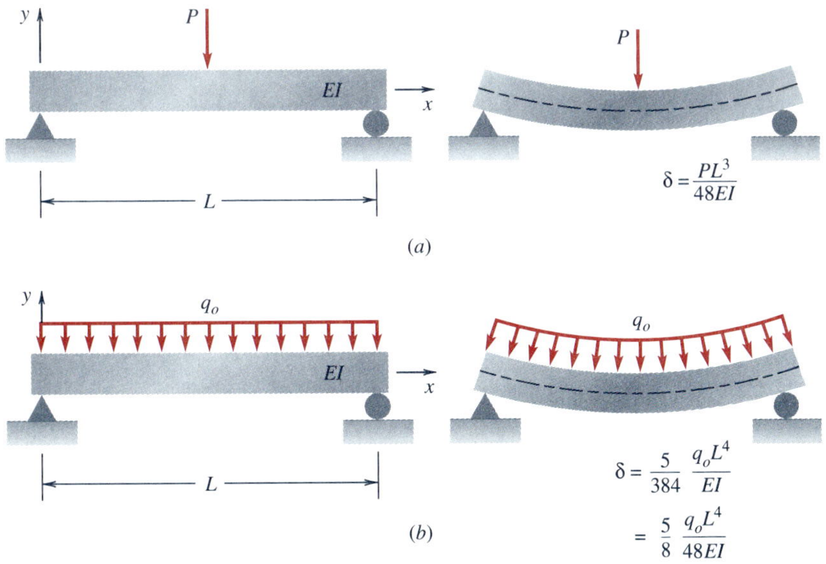

Figure 6.6 Midpoint deflection δ of a simply supported beam with (*a*) a concentrated load *P* and (*b*) a constant distributed loading q_0.

If δ is defined as the downward deflection in the negative y direction, then the maximum deflection at the center is

$$\delta = -v\left(\frac{L}{2}\right) = \frac{PL^3}{48EI} \qquad (l)$$

as shown in Fig. 6.6*a*.

Recall that the maximum deflection for a simply supported beam under a uniform load is, from Example 6.2, Eq. (*i*), and Fig. 6.6*b*,

$$\delta = \frac{5}{384}\frac{q_0L^4}{EI} = \frac{5}{8}\frac{(q_0L)L^3}{48EI} \qquad (m)$$

Keep in mind that the deflected shapes shown in Fig. 6.6 are highly exaggerated. If the resultant of the distributed load q_0L were placed at the center of the beam and Eq. (*l*) used to estimate the deflection with $P = q_0L$, we find in view of Eq. (*m*) that Eq. (*l*) would overestimate the deflection by a factor $\frac{8}{5}$, or by 60 percent

EXAMPLE 6.4

The simply supported beam AB carries a constant distributed load q_0 over the right half of the beam, as shown in Fig. 6.7*a*. We wish (*a*) to find the location and magnitude of the maximum deflection and slope of the deformed neutral axis of the beam and (*b*) if $L = 8$ m, $q_0 = 10$ kN/m, and $E = 200$ GPa, to determine the moment of inertia needed so that the maximum deflection of the beam does not exceed 3 cm. We neglect the weight of the beam.

Force and moment equilibrium for the entire beam shown in Fig. 6.7*a* gives the support reactions

$$R_A = \frac{q_0L}{8} \qquad R_B = \frac{3q_0L}{8} \qquad (a)$$

As in the previous examples, we need to derive an expression for $M(x)$ that is valid over the entire beam and that can be integrated after insertion into the moment-curvature relation, Eq. (6.6). Toward that objective, consider the free-body diagram of a segment of the beam of length x shown in Fig. 6.7*b*. We have the situation such that if x is less than $L/2$, then no distributed load is acting on the free body. But if x is greater than $L/2$, then there is a distributed load acting on a length $x - L/2$. The resultant of this distributed load is given by

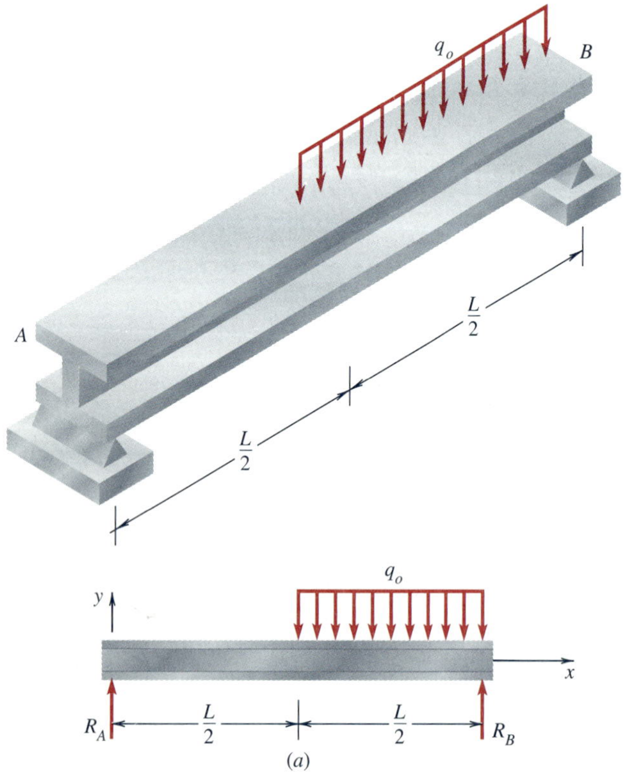

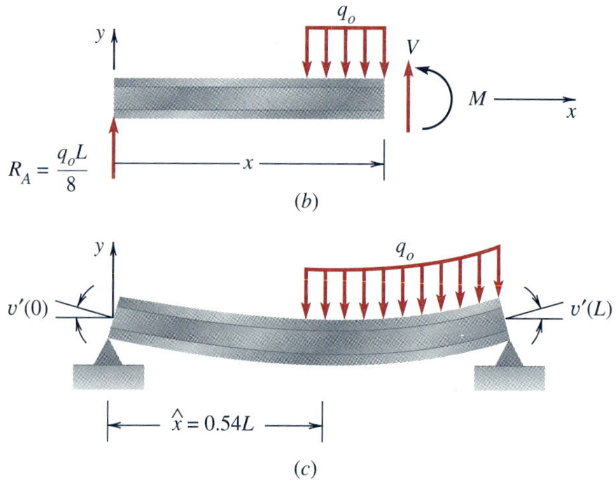

Figure 6.7 Example 6.4

$q_0(x - L/2)$, and it acts through the centroid of the loading diagram. Recalling the definition of the singularity functions discussed in Chap. 4, we see that by writing the resultant in the form

$$q_0\left\langle x - \frac{L}{2}\right\rangle^1$$

we have, according to the definition of the singularity function $\langle x - L/2\rangle^1$, that the resultant will be zero when x is less than or equal to $L/2$ and will be equal to $q_0(x - L/2)$ when x is greater than $L/2$.

By taking moments about the section where the beam was cut to form the free body (Fig. 6.7*b*), we obtain

$$M(x) + q_0\left\langle x - \frac{L}{2}\right\rangle^1 \frac{1}{2}\left\langle x - \frac{L}{2}\right\rangle^1 - R_A x = 0 \qquad (b)$$

where both the resultant and its moment arm are written in terms of the singularity function $\langle x - L/2\rangle^1$ since both are present or not depending upon whether x is greater or less than $L/2$. Thus solving Eq. (*b*) for $M(x)$ gives us the single mathematical expression that we need to find the slope and deflection. Using the moment-curvature relation, Eq. (6.6), we find

$$EIv'' = M(x) = \frac{q_0 L}{8}x - \frac{q_0}{2}\left\langle x - \frac{L}{2}\right\rangle^2 \qquad (c)$$

Integrating twice produces

$$EIv' = \frac{q_0 L}{8}\frac{x^2}{2} - \frac{q_0}{2}\frac{\langle x - L/2\rangle^3}{3} + c_1 \qquad (d)$$

$$EIv = \frac{q_0 L}{8}\frac{x^3}{6} - \frac{q_0}{2}\frac{\langle x - L/2\rangle^4}{12} + c_1 x + c_2 \qquad (e)$$

where c_1 and c_2 are the arbitrary constants of integration.

We can interpret the meaning of these constants by setting $x = 0$ in Eqs. (*d*) and (*e*) to get

$$EIv'(0) = c_1 \qquad EIv(0) = c_2 \qquad (f)$$

Thus we see that for the present case c_1 is proportional to the slope of the deflection curve at the left end of the beam and c_2 is proportional to the deflection at the left end. Since the supports are assumed to be rigid and the deflections are measured relative to the supports, we have

$$v(0) = 0 \qquad c_2 = 0 \qquad (g)$$

However, the slope at the left end is unknown, and we must find some additional geometric condition that can be used to find c_1.

Clearly, the rigid support at the right end of the beam requires that the deflection expression vanish there so that

$$v(L) = 0$$

Therefore, from Eq. (*e*) with $x = L$ we have

$$EIv(L) = 0 = \frac{q_0 L^4}{48} - \frac{q_0}{24}\left(\frac{L}{2}\right)^4 + c_1 L$$

and solving for c_1 results in

$$c_1 = -\tfrac{7}{384}q_0 L^3 \qquad (h)$$

Finally, the expression for the deflection from Eqs. (*e*), (*g*), and (*h*) is

$$v(x) = -\frac{q_0 L^4}{EI}\left[-\frac{1}{48}\left(\frac{x}{L}\right)^3 + \frac{1}{24L^4}\left\langle x - \frac{L}{2}\right\rangle^4 + \frac{7}{384}\frac{x}{L}\right] \qquad (i)$$

To find the location where the deflection has a maximum value, we set dv/dx equal to zero in Eq. (*d*), making use of the value of c_1 given in Eq. (*h*) to obtain

$$0 = \frac{q_0 L}{16}\hat{x}^2 - \frac{q_0}{6}\left(\hat{x} - \frac{L}{2}\right)^3 - \frac{7}{384}q_0 L^3 \qquad (j)$$

where $\hat{x}$ is used to denote the value of x corresponding to the location of the maximum deflection. Also note that we have anticipated that the maximum deflection will take place in the right half of the beam, that is, $\hat{x}/L > 0.5$, and therefore we have changed the pointed brackets to normal parentheses in the second term in Eq. (*j*).

To find the value of $\hat{x}$, we will need to solve Eq. (*j*) numerically. For numerical calculations it is convenient to simplify Eq. (*j*) by multiplying by the factor $384/(q_0 L^3)$ to obtain

$$24\left(\frac{\hat{x}}{L}\right)^2 - 64\left(\frac{\hat{x}}{L} - 0.5\right)^3 - 7 = 0 \qquad (k)$$

where the sought-after value $\hat{x}$ appears as a ratio $\hat{x}/L$; this nondimensional ratio will be convenient in numerical work.

If we use a simple trial-and-error procedure to determine the root or solution $\hat{x}/L$ of Eq. (*k*) or any root-finding routine on a calculator or computer, we find

$$\frac{\hat{x}}{L} = 0.54 \qquad (l)$$

Upon substitution of this value into Eq. (*e*) we find

$$v_{max} = -0.00656\frac{q_0 L^4}{EI} \qquad (m)$$

We have found, therefore, that the maximum deflection occurs near the midpoint of the beam, as expected, and is greater (slightly) than the midpoint deflection of a simply supported beam loaded by a constant uniform load of $q_0/2$ along its entire length; see Example 6.2.

As can be seen from Fig. 6.7*c*, the extreme values of the slope occur at the ends of the beam, $x = 0$ and $x = L$. Using Eq. (*d*), we find that

$$EIv'(0) = c_1 = -\tfrac{7}{384}\, q_0 L^3 \qquad (n)$$

and

$$EIv'(L) = \tfrac{9}{384}\, q_0 L^3 \qquad (o)$$

and therefore the maximum slope occurs at $x = L$. The positive sign for $v'(L)$ indicates a counterclockwise rotation of the right end of the beam. A sketch of the deflected shape is shown in Fig. 6.7*c*.

For the second part of the problem, we need to determine the section moment of inertia so as to meet a deflection design criterion. That is, we are to design a beam of length 8 m with a loading $q_0 = 10$ kN/m and with $E = 200$ GPa, so that the maximum deflection does not exceed 3 cm. Solving Eq. (*m*) for the moment of inertia I and inserting the particular values for this beam, we find

$$\begin{aligned} I &= -0.00656\,\frac{q_0 L^4}{E v_{\max}} \\ &= -\frac{6.56 \times 10^{-3}(10 \times 10^3)(8)^4}{(200 \times 10^9)(-0.03)} \\ &= 4.48 \times 10^{-5}\ \text{m}^4 \end{aligned} \qquad (p)$$

Therefore we conclude that any beam with an I value greater than that given in Eq. (*p*) will give a deflection less than 3 cm.

EXAMPLE 6.5

A beam is subjected to a constant distributed load q_0 as shown in Fig. 6.8*a*. We wish to find the ratio a/L such that the deflection at the tip of each overhang is equal to the deflection at the middle of the beam. The bending modulus of the beam is EI.

Force equilibrium for the whole beam together with symmetry gives the reactions at B and C

$$R_B = R_C = \frac{q_0 L}{2} \qquad (a)$$

In Fig. 6.8*b*, a free body of a segment of the beam of length x is shown. The length x is taken greater than $L - a$ so that both of the reaction loads at B and C are acting on the free body. After replacing the distributed load by the resultant $q_0 x$ acting at the middle of the beam segment, we can sum the moments about the right end of the segment, to obtain

$$M(x) - \frac{q_0 L}{2}\langle x - (L-a)\rangle^1 + q_0\frac{x^2}{2} - \frac{q_0 L}{2}\langle x - a\rangle^1 = 0 \qquad (b)$$

Again using the moment-curvature relation, Eq. (6.6), we have

$$\begin{aligned} EIv'' &= M(x) \\ &= \frac{q_0 L}{2}\langle x - (L-a)\rangle^1 \\ &\quad - \frac{q_0 x^2}{2} + \frac{q_0 L}{2}\langle x - a\rangle^1 \end{aligned} \qquad (c)$$

and after one integration

$$\begin{aligned} EIv' &= \frac{q_0 L}{2}\frac{\langle x - (L-a)\rangle^2}{2} \\ &\quad - \frac{q_0}{2}\frac{x^3}{3} + \frac{q_0 L}{2}\frac{\langle x - a\rangle^2}{2} + c_1 \end{aligned} \qquad (d)$$

From Fig. 6.8*a*, we see that symmetry about the center of the beam gives

$$v' = 0 \qquad \text{at} \qquad x = \frac{L}{2} \qquad (e)$$

In evaluating $v'(L/2)$ by using Eq. (*d*), we note that the first term drops out because for $x = L/2$ the expression inside the pointed brackets is negative. The remaining terms are

$$0 = -\frac{q_0 L^3}{48} + \frac{q_0 L}{4}\left(\frac{L}{2} - a\right)^2 + c_1$$

and we obtain c_1:

$$c_1 = \frac{q_0 L^3}{48} - \frac{q_0 L}{4}\left(\frac{L}{2} - a\right)^2 \qquad (f)$$

Integrating Eq. (*d*) one more time leads to

$$\begin{aligned} EIv(x) &= \frac{q_0 L}{2}\frac{\langle x - (L-a)\rangle^3}{6} \\ &\quad - \frac{q_0 x^4}{2 \cdot 12} + \frac{q_0 L}{2}\frac{\langle x - a\rangle^3}{6} + c_1 x + c_2 \end{aligned} \qquad (g)$$

The constant c_2 can now be found by noting that there can be

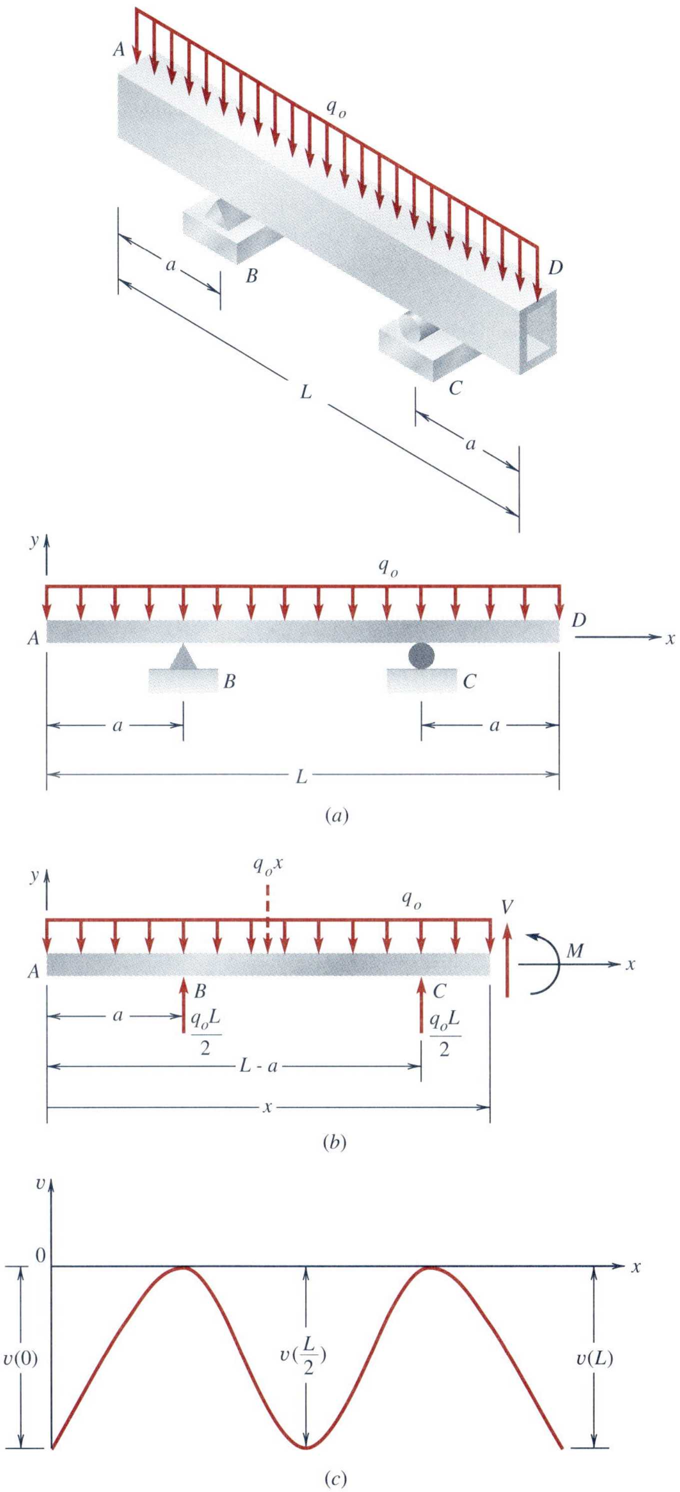

Figure 6.8 Example 6.5

no deflection at the support at B so that

$$v = 0 \quad \text{at} \quad x = a \tag{h}$$

Since the first and third terms in Eq. (g) drop out when $x = a$, it follows that

$$0 = -\frac{q_0 a^4}{24} + c_1 a + c_2$$

so that

$$c_2 = -ac_1 + q_0 \frac{a^4}{24} \tag{i}$$

Now that both arbitrary constants have been found, we can turn our attention to finding the value of the ratio a/L so that the deflection at the tip of either overhang just equals the deflection at the middle of the beam, i.e.,

$$v(0) = v\left(\frac{L}{2}\right) \tag{j}$$

Using the general expression for the deflection from Eq. (g), we have

$$EIv\left(\frac{L}{2}\right) = -\frac{q_0}{384}L^4 + \frac{q_0 L}{12}\left(\frac{L}{2} - a\right)^3 + c_1\frac{L}{2} + c_2 \tag{k}$$

and

$$EIv(0) = c_2 \tag{l}$$

Equating the deflections given in Eqs. (k) and (l) and canceling the c_2 terms lead to

$$-\frac{q_0 L^4}{384} + \frac{q_0 L}{12}\left(\frac{L}{2} - a\right)^3 + c_1\frac{L}{2} = 0 \tag{m}$$

If we cancel common factors upon using the result for c_1 from Eq. (f) and write for the unknown $f = a/L$, we can rewrite Eq. (m) in a form better suited for numerical calculations as

$$32(\tfrac{1}{2} - f)^3 - 48(\tfrac{1}{2} - f)^2 + 3 = 0 \tag{n}$$

By a simple trial-and-error procedure to find f using a hand calculator, we find that

$$f = 0.223$$

or

$$a = 0.223L \tag{o}$$

The deflection curve for the case of an overhang which has length $a = 0.223L$ is sketched in Fig. 6.8c.

To find the value of the deflection at the tip of the overhang now that we have established the overhang length, we return to Eq. (l)

$$EIv(0) = c_2$$

and use Eq. (i) and Eq. (f) to find

$$EIv(0) = -a\left[\frac{q_0L^3}{48} - \frac{q_0L}{4}\left(\frac{L}{2} - a\right)^2\right] + \frac{q_0a^4}{24}$$

With the particular value of $a = 0.223L$, we have

$$v(0) = -2.68 \times 10^{-4}\frac{q_0L^4}{EI} \qquad (p)$$

as the final result.

EXAMPLE 6.6

A machine component modeled as a simply supported beam, shown in Fig. 6.9*a*, carries a load P at midspan between the supports and a "counterweight" Q applied at the end of the overhanging portion of the beam. We wish (*a*) to find the magnitude of Q so that no deflection takes place at B due to the application of loads P and Q and (*b*) to compare the maximum bending stress for the beam with and without the counterweight Q as calculated in part (*a*). We assume that the beam section is symmetric with respect to the neutral axis and has the section modulus S. We neglect the weight of the beam.

We show the idealized beam in Fig. 6.9*b*. If we denote the reactions at A and C by R_A and R_C, respectively, and sum the moments about A, it follows that

$$R_CL - \frac{PL}{2} - Q\frac{3L}{2} = 0$$

or

$$R_C = \frac{P}{2} + \frac{3Q}{2} \qquad (a)$$

Force balance in the y direction gives

$$R_A + R_C = P + Q$$

or

$$R_A = \frac{P}{2} - \frac{Q}{2} \qquad (b)$$

To find an expression for $M(x)$ as a function of position along the beam, we take a free-body segment of the beam, as shown in Fig. 6.9*c*, which is formed by cutting the beam at a section F between C and D so that all forces acting on the beam are included. Taking moments about section F results in

$$M(x) - R_A\langle x\rangle^1 + P\left\langle x - \frac{L}{2}\right\rangle^1 - R_C\langle x - L\rangle^1 = 0 \qquad (c)$$

We are now ready to introduce the moment expression $M(x)$ from Eq. (*c*) into the moment-curvature relation to obtain

$$EIv'' = R_A\langle x\rangle^1 - P\left\langle x - \frac{L}{2}\right\rangle^1 + R_C\langle x - L\rangle^1 \qquad (d)$$

which after two integrations gives the deflection in the form

$$EIv = R_A\frac{\langle x\rangle^3}{6} - P\frac{\langle x - L/2\rangle^3}{6} + R_C\frac{\langle x - L\rangle^3}{6} + c_1x + c_2 \qquad (e)$$

where c_1 and c_2 are the two constants of integration. Substitution of $x = 0$ into Eq. (*e*) shows that c_2 is equal to $EIv(0)$ and is equal to zero since the deflection is zero at support A. To find the other constant c_1, we use the condition that v equals zero at the support C or $v(L) = 0$. Thus for $x = L$ it follows from Eq. (*e*) that

$$0 = \frac{R_AL^3}{6} - \frac{PL^3}{48} + c_1L \qquad (f)$$

or replacing R_A by $(P - Q)/2$, Eq. (*b*), gives

$$c_1 = \frac{QL^2}{12} - \frac{PL^2}{16}$$

To find the particular value of Q which will result in zero deflection at B, we set $v = 0$ on the left side of Eq. (*e*) and evaluate the right side with $x = L/2$ to get

$$0 = \frac{R_AL^3}{48} + \left(\frac{QL^2}{12} - \frac{PL^2}{16}\right)\frac{L}{2}$$

Since R_A is given in terms of P and Q by Eq. (*b*), we can eliminate R_A and find a single equation connecting P and Q, which leads to the rather simple result that

$$Q = \frac{2P}{3} \qquad (g)$$

Using the result $Q = 2P/3$ in Eq. (*e*), we can plot $v(x)$ as shown in Fig. 6.9*d*. Clearly, as seen in Fig. 6.9*d*, for this particular combination of loads, the deflection vanishes at B.

We now turn to the determination of the maximum normal bending stress in the beam. Equation (*c*) together with Eqs. (*a*) and (*b*) gives the explicit dependence of the bending moment M on loads P and Q.

If $Q = 0$, we have from Eqs. (*a*) and (*b*)

$$R_A = R_C = \frac{P}{2}$$

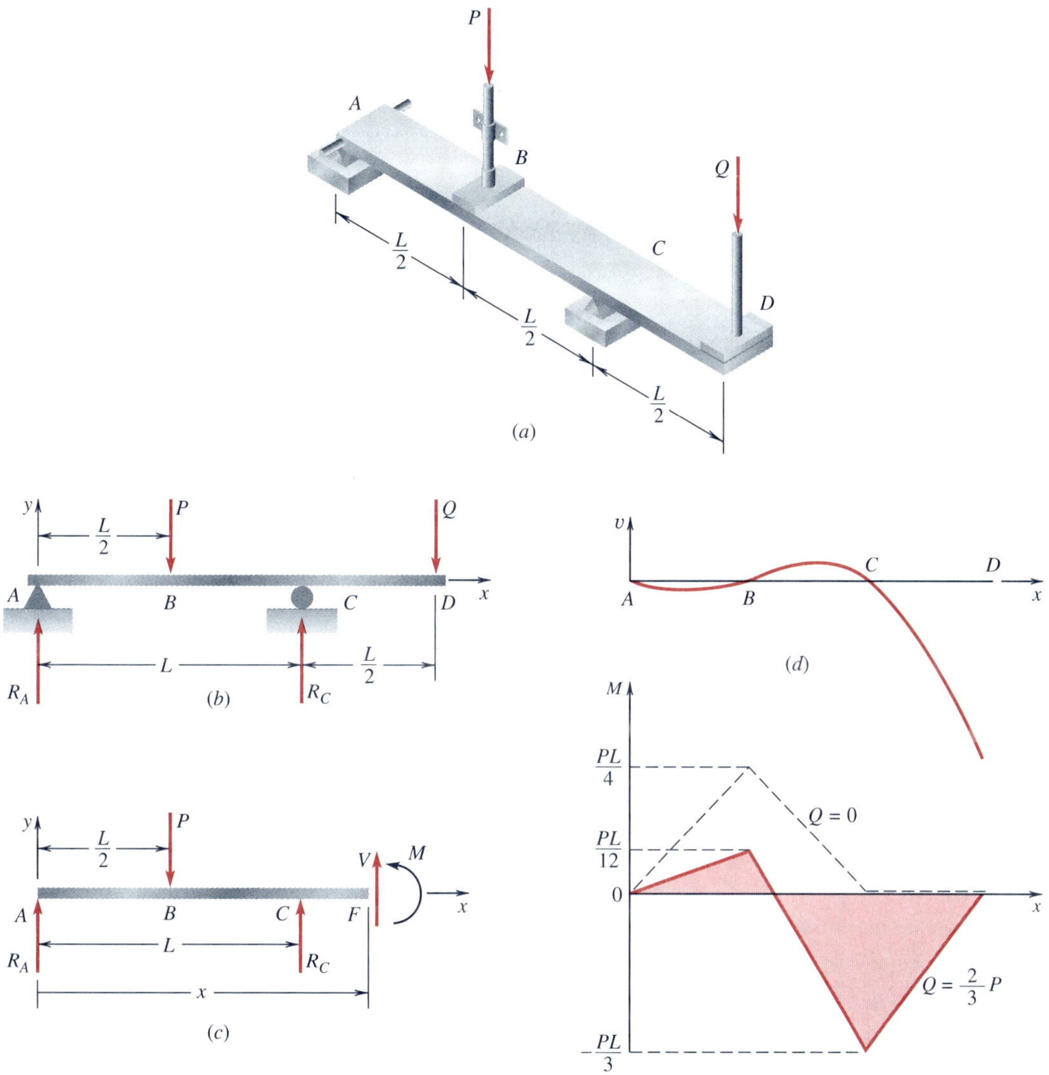

Figure 6.9 Example 6.6

so that

$$M(x) = \frac{P}{2}\langle x\rangle^1 - P\left\langle x - \frac{L}{2}\right\rangle^1 + \frac{P}{2}\langle x - L\rangle^1 \qquad (h)$$

which is plotted as a dashed line in Fig. 6.9*e*. The maximum bending moment for this case is at B

$$M_{\max} = M\left(\frac{L}{2}\right) = \frac{PL}{4} \qquad (i)$$

The maximum bending stress therefore follows in the form

$$\sigma_{\max} = \frac{PL}{4S} \qquad Q = 0 \qquad (j)$$

For the case in which $Q = 2P/3$, from Eqs. (*a*) and (*b*) we have

$$R_A = \frac{P}{2} - \frac{1}{2}\left(\frac{2}{3}P\right) = \frac{P}{6}$$

$$R_C = \frac{P}{2} + \frac{3}{2}\left(\frac{2}{3}P\right) = \frac{3P}{2} \qquad (k)$$

so that according to Eq. (*c*)

$$M(x) = \frac{P}{6}\langle x\rangle^1 - P\left\langle x - \frac{L}{2}\right\rangle^1 + \frac{3P}{2}\langle x - L\rangle^1 \qquad (l)$$

which is plotted as a solid line in Fig. 6.9*e*. From this plot it is easily seen that the moment takes on its largest absolute value at C so that

$$|M_{\text{max}}| = |M(L)| = \frac{PL}{3} \qquad (m)$$

The maximum stress for this case is

$$\sigma_{\text{max}} = \frac{PL}{3S} \qquad Q = \frac{2}{3}P \qquad (n)$$

The suppression of the deflection at B by use of the counterweight $Q = 2P/3$ results in an increase in the maximum bending stress by a factor of $\frac{4}{3}$ (or a 33 percent increase); the stress at B, on the other hand, decreases by a factor of $\frac{1}{3}$.

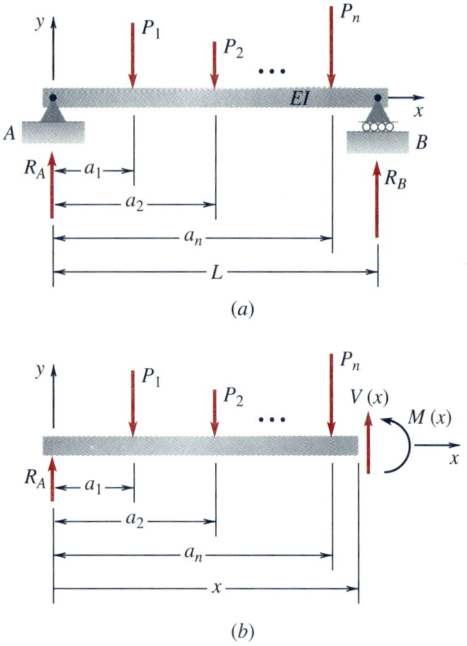

Figure 6.10 (*a*) Simply supported beam of length L with concentrated loads P_i at locations a_i, $i = 1$ to n. (*b*) Segment of beam with $V(x)$ and $M(x)$.

In many of the above examples, we have systematically evaluated the expression for the bending moment along the beam using singularity functions from which the deflection expression $v(x)$ was found by two integrations. The constants of integration were evaluated from information as to how the beam was supported or by using certain symmetry conditions. The advantage of the use of singularity functions in the solution technique should be apparent. A further advantage of the systematic approach as we have shown in Chap. 4 is that often an approach for computer implementation immediately suggests itself. For example, consider as a special case a simply supported beam AB shown in Fig. 6.10*a* on which a number n of concentrated loads are acting. We assume that the weight of the beam is small compared to the loads acting and that the bending modulus EI is constant. We wish to determine the numerical values of the deflection at points along the beam, given the values of the loads, their points of application at a_i, L, and EI.

We first use moment equilibrium about end B to determine reaction R_A; we find (Fig. 6.10*a*)

$$R_A L = P_1(L - a_1) + P_2(L - a_2) + \cdots + P_n(L - a_n) \qquad (6.7)$$

It is convenient to introduce the following summation notation for the right side of Eq. (6.7):

$$R_A L = \sum_{i=1}^{n} P_i(L - a_i) \qquad (6.8)$$

where $\Sigma_{i=1}^{n}$ means "sum with respect to i from $i = 1$ to $i = n$" the terms after the Σ (sigma) symbol. For example,

$$\sum_{i=1}^{3} P_i(L - a_i) = P_1(L - a_1) + P_2(L - a_2) + P_3(L - a_3) \qquad (6.9)$$

The advantage of the summation symbol is that it requires less space for writing the sum of the terms with the concentrated loads.

Returning to Eq. (6.7), we see that once numerical values of P_i, a_i, and L are specified, the value of R_A can be found.

If a cut is made at x to the right of all the concentrated loads (Fig. 6.10*b*), then the moment expression $M(x)$ that is valid for *any* value of x on the beam follows from moment equilibrium and the use of singularity functions

$$M(x) = R_A x - \sum_{i=1}^{n} P_i \langle x - a_i \rangle^1 \tag{6.10}$$

where now the singularity function $\langle x - a_i \rangle^1$ appears in the summation expression. For example, if $x < a_1$,

$$M(x) = R_A x \tag{6.11}$$

As we move across the beam with increasing values of x, contributions from each of the concentrated loads "turn on" in the moment expression, Eq. (6.10).

Upon substitution of the expression for the moment into the moment-curvature relation, Eq. (6.6), we have

$$EIv'' = M(x) = R_A x - \sum_{i=1}^{n} P_i \langle x - a_i \rangle^1 \tag{6.12}$$

Two integrations of Eq. (6.12) give the expression for the deflection of the beam in the form

$$EIv = \frac{R_A x^3}{6} - \sum_{i=1}^{n} P_i \frac{\langle x - a_i \rangle^3}{6} + c_1 x + c_2 \tag{6.13}$$

where c_1 and c_2 are the constants of integration.

The beam is simply supported at $x = 0$ and $x = L$, so that

$$v(0) = 0 \qquad v(L) = 0 \tag{6.14}$$

from which, using the condition at $x = 0$, we have $c_2 = 0$.

To find c_1, we use $v(L) = 0$ to obtain

$$c_1 L = -R_A \frac{L^3}{6} + \sum_{i=1}^{n} P_i \frac{(L - a_i)^3}{6} \tag{6.15}$$

Since the value of R_A is known from Eq. (6.8), the value of c_1 is obtained from Eq. (6.15). With the value of c_1 known and $c_2 = 0$, Eq. (6.13) gives the expression for $v(x)$. To numerically evaluate $v(x)$ along the beam, we divide the beam into a number of points and evaluate v at each point along the beam, with careful attention to the contribution of each of the singularity functions. A brief outline of the steps needed to program this approach is shown in Fig. 6.11. A BASIC program CH6BM is listed in Fig. 6.12; this program, which is on the diskette with this book, can be used to evaluate numerically the value of v along a simply supported beam loaded with concentrated loads, as shown in Fig. 6.10. This specific program uses the notation and equations we have just derived associated with Fig. 6.10; see Probs. 6.3-16 to 6.3-18.

Figure 6.11 Outline of program CH6BM for concentrated loads on a simply supported beam.

```
INPUT:        E, I, L
              n = NUMBER OF LOADS
              Pi, ai (WATCH SIGNS ON Pi; Pi IS POSITIVE AS
                SHOWN IN FIG. 6.10)
              M, INTEGER; NUMBER OF SEGMENTS ALONG THE BEAM
                GIVING RISE TO M + 1 POINTS ON THE BEAM.

CALCULATE:    VALUES OF:
              RA, Eq. (6.8)
              c1, Eq. (6.15)

EVALUATE:     v AT POINTS ALONG BEAM
              FOR s = 0 TO M
              x = sL/M
              v(x), Eq. (6.13)
```

6.4 Beam Deflections by Direct Integration of the Load-Deflection Equation

In the previous section our procedure for the determination of the deflection involved two integrations of the moment-curvature relation, Eq. (6.6), starting with the expression for $M(x)$. In Chap. 4 we found that an analytic expression for the bending moment $M(x)$ could be obtained from two integrations of the loading function $q(x)$ representing the distributed loads, concentrated loads, and concentrated moments including the reaction loads acting on the beam. In this section we combine the procedures of Chap. 4 and Sec. 6.3 to develop a method for finding the deflection function that starts with the loading function $q(x)$. By repeated integration, we are able to obtain expressions for the shear force, bending moment, slope, and deflection that are valid at all sections along the beam.

First, we will derive a relation which connects the fourth derivative of the deflection function to the load function. In Chap. 4, we used force and moment equilibrium arguments on an infinitesimal element of a beam to derive the relations

$$\frac{dV}{dx} + q = 0 \tag{4.4}$$

$$\frac{dM}{dx} + V = 0 \tag{4.5}$$

where V and M are the shear force and bending moment along the beam and q is the loading function.

The moment-curvature relation, Eq. (6.6), can be rewritten in a form to show that the moment can be expressed in terms of the second derivative of the deflection:

$$M = EI\frac{d^2v}{dx^2} = EIv'' \tag{6.16}$$

```
10 PRINT "PROGRAM FOR CONCENTRATED LOADS AS SHOWN IN
   FIGURE 6.10"
20 PRINT "ALL UNITS OF LENGTH MUST BE CONSISTENT: E.G."
30 PRINT "LENGTH IN INCHES, E IN PSI, I IN IN ∧ 4,...."
40 PRINT
50 PROGRAM FOR CONCENTRATED LOADS; P IS POSITIVE AS
   SHOWN IN FIG. 6.10
60 DIM P(40), A(40), T(40), T1(40), T2(40), V(40),
   LP(40)
70 INPUT "LENGTH OF BEAM = "; AL
80 PRINT
90 INPUT  "NUMBER OF LOADS = "; NP
100 PRINT
110 INPUT "MODULUS OF ELASTICITY =  "; E
120 PRINT
130 INPUT "MOMENT OF INERTIA = "; AI
140 PRINT
150 INPUT "THE NUMBER OF INTERVALS ALONG THE BEAM"; N
160 PRINT
170 SUM = 0: SUM1 = 0: SUM2 = 0
180 PRINT "INPUT THE LOCATION AND THE VALUES OF THE LOAD;
    SEE FIGURE 6.10"
190 PRINT
200 FOR I = 1 TO NP
210 PRINT "INPUT LOCATION OF LOAD (";I;") AND VALUE OF
    LOAD (";I;"):"
220 INPUT A(I), P(I)
230 T(I) = P(I)*(AL-A(I))
240 SUM = T(I) + SUM
250 T1(I) = P(I)*(AL-A(I)) ∧ 3/(6*AL)
260 SUM1 = T1(I) + SUM1
270 LP(I) = P(I)
280 SUM2 = LP(I) + SUM2
290 NEXT I
300 RA = SUM/AL
310 PRINT "RA =  "RA
320 RB = SUM2 - RA
330 PRINT  "RB =  "RB
340 C1 = -(RA*(AL*AL)/6) + SUM1
350 SL = AL/N
360 FOR J = 0 TO N
370 X = J*(AL/N)
380 SUM3 = 0
390 FOR I = 1 TO NP
400 IF X < A(I) THEN 410 ELSE 420
410 T2(I) = T2(I): GO TO 430
420 T2(I) = -P(I)*(X-A(I)) ∧ 3/6
430 SUM3 = T2(I) + SUM3
440 NEXT I
450 V(J) = SUM3 + (RA*X ∧ 3/6) + C1*X
460 V(J) = V(J)/(E*AI)
470 PRINT "V("X") =  " V(J)
480 NEXT J
```

Figure 6.12 BASIC program CH6BM.

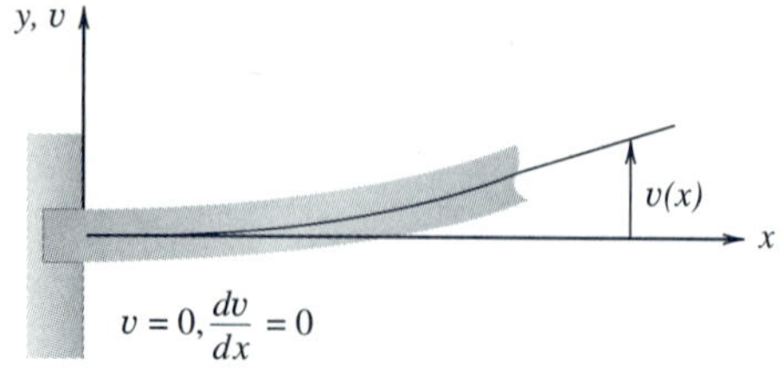

(*a*) Built–in or clamped end

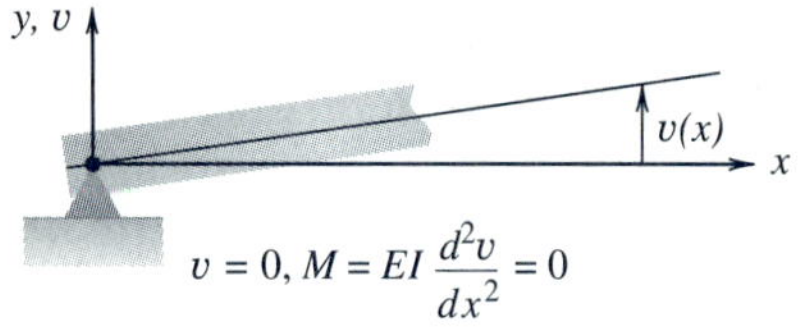

(*b*) Pinned or simply supported end

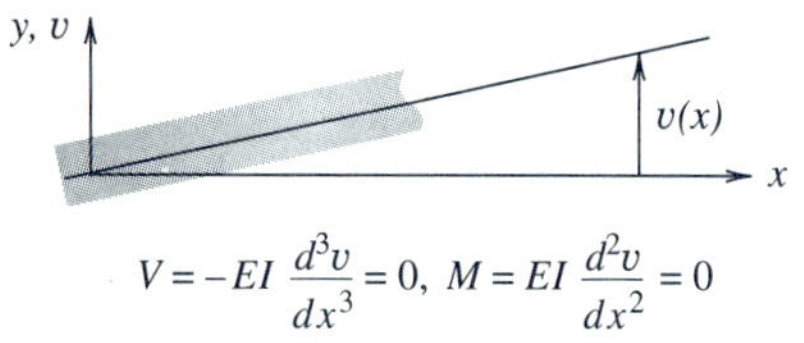

(*c*) Free end

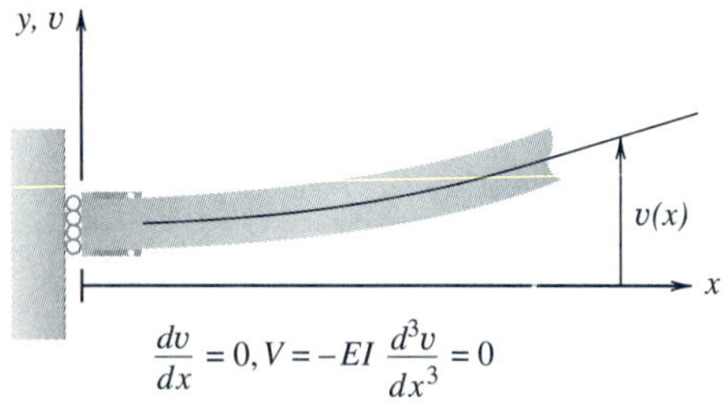

(*d*) End restrained against rotation, but free to deflect

Figure 6.13 Boundary conditions expressed in terms of deflection $v(x)$ and its derivatives; *EI* is constant.

From Eqs. (6.16) and (4.5), we find that the shear force can be expressed in terms of the third derivative of the deflection

$$V = -\frac{dM}{dx} = -\frac{d}{dx}\left(EI\frac{d^2v}{dx^2}\right) \qquad (6.17)$$

When *EI* is constant along the beam, we have

$$V = -EI\frac{d^3v}{dx^3} = -EIv''' \qquad (6.18)$$

Finally from Eqs. (6.17) and (4.4),

$$q = -\frac{dV}{dx} = \frac{d^2}{dx^2}\left(EI\frac{d^2v}{dx^2}\right) \qquad (6.19)$$

When *EI* is constant, Eq. (6.19) takes the form

$$EI\frac{d^4v}{dx^4} = EIv'''' = q(x) \qquad (6.20)$$

Equation (6.20) provides the important relation between the loading $q(x)$ on the beam and the deflection $v(x)$.

Thus, we have reduced the problem of finding the deflection curve of a beam to that of carrying out four integrations of the expression for the loading $q(x)$ in Eq. (6.20).

To determine $v(x)$ from Eq. (6.20), it is first necessary to obtain the expression for $q(x)$. This first step is identical with the procedure used in Chap. 4 in which expressions for V and M were obtained by integration from the $q(x)$ expression. Singularity functions were very useful in setting up the expressions for $q(x)$ in Chap. 4, and in the previous sections of this chapter we used them in writing expressions for the bending moment.

We recall from the examples of Chap. 4 that the two constants of integration arising from the integration of $q(x)$ to $V(x)$ and from the integration of $V(x)$ to $M(x)$, using Eqs. (4.4) and (4.5), were often zero. In general, these two constants of integration will be zero if the loading function $q(x)$ includes all applied loads and all reactive forces and moments. However, *as a general rule,* constants arising from integrations must be evaluated.

When we use the fourth-order equation with all loads included in $q(x)$, Eq. (6.20), and we integrate four times to obtain $v(x)$, the first two constants of integration corresponding to terms in the V and M expression are often zero. The remaining two constants of integration are to be evaluated from the geometric conditions at the supports, as we did in the examples of Sec. 6.3. A summary of four typical boundary or support conditions for constant *EI* is shown in Fig. 6.13. The bending moment and shear force boundary conditions can be expressed in terms of the derivatives of v through Eqs. (6.16) and (6.18); we have, therefore, a means to express all boundary conditions in terms of v and its derivatives.

EXAMPLE 6.7

A simply supported beam of length L is loaded by a uniform load q_0 as shown in Fig. 6.14*a*; see Example 6.2. We wish to use the fourth-order equation, Eq. (6.20), to obtain the expression for the deflection of the beam. Assume that the weight of the beam is included in q_0 and that EI is constant along the beam.

As the first step in the solution procedure, we need to write down an expression for the loading $q(x)$ on the beam. We have adopted the approach in Chap. 4 so that $q(x)$ will include all loads on the beam including all reaction forces. Recall that the advantage of including all reaction forces is that constants of integration arising as we integrate $q(x)$ to obtain $V(x)$ and $M(x)$ often turn out to vanish. For the present beam we have (Fig. 6.14*b*)

$$q(x) = -q_0 + \frac{q_0 L}{2}\langle x\rangle_{-1} + \frac{q_0 L}{2}\langle x - L\rangle_{-1} \qquad (a)$$

where the singularity functions arise due to the concentrated force reactions at the supports. Upon use of Eq. (6.20) we have

$$EIv'''' = -q_0 + \frac{q_0 L}{2}\langle x\rangle_{-1} + \frac{q_0 L}{2}\langle x - L\rangle_{-1} \qquad (b)$$

After integrating Eq. (*b*) once, we have

$$EIv''' = -q_0 x + \frac{q_0 L}{2}\langle x\rangle^0 + \frac{q_0 L}{2}\langle x - L\rangle^0 + a_1 \qquad (c)$$

where a_1 is the constant of integration. Upon integration again we have

$$EIv'' = -q_0\frac{x^2}{2} + \frac{q_0 L}{2}\langle x\rangle^1 + \frac{q_0 L}{2}\langle x - L\rangle^1 + a_1 x + a_2 \qquad (d)$$

The constants a_1 and a_2 are the two constants of integration. These constants are to be evaluated from information on the value of the shear force, Eq. (*c*), or on the bending moment, Eq. (*d*), at specific points along the beam. In other words, we can expect to evaluate the constants from a knowledge of vertical force and moment equilibrium along the beam. When we include all the load terms and support reactions in the function $q(x)$ in Eq. (*a*), we expect further, based on the experience in Sec. 4.5, that these constants will be zero. However, as we noted previously, the general rule is to evaluate all constants of integration since there are loading functions for which they do not turn out to equal zero; see Probs. 4.5-13 and 4.9-1.

To evaluate the constants in the present problem, we note that the bending moment is zero at each of the simple supports at $x = 0$ and $x = L$. Therefore, since $M = EIv''$, we have from Eq. (*d*)

$$M(0) = 0 = EIv''(0) = a_2 \qquad (e)$$

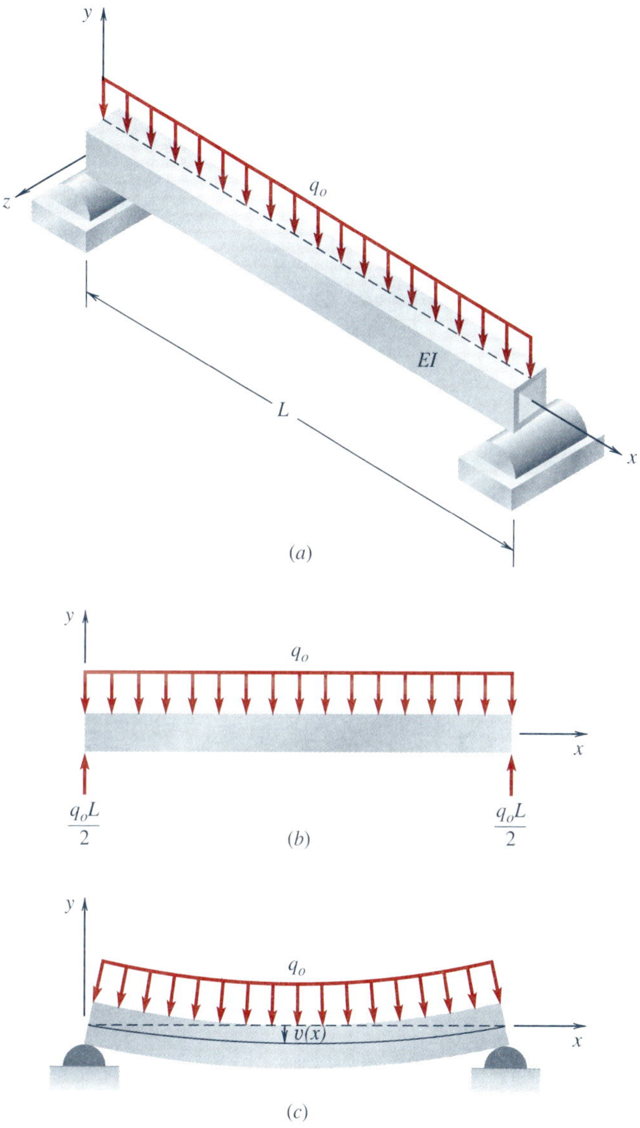

Figure 6.14 Example 6.7

and

$$M(L) = 0 = EIv''(L) = -\frac{q_0 L^2}{2} + \frac{q_0 L^2}{2} + a_1 L \qquad (f)$$

Therefore, constants a_1 and a_2 are zero as expected. Alternatively we could have used information on the value of the shear force, e.g., at $x = 0$, to find a_1. We have from force equilibrium of an infinitesimal element (see, e.g., Fig. 4.19*c*) at the left support and from Eq. (6.18)

$$V(0) = -\frac{q_0 L}{2} = -EIv'''(0) = -\frac{q_0 L}{2} - a_1 \qquad (g)$$

from which a_1 equals zero.

In the remaining examples, we will drop the integration constants arising from the first two integrations of the loading function $q(x)$ with the understanding that we have checked that they are indeed zero.

Equation (*d*) agrees with Eq. (*c*) in Example 6.2 for the moment equation. Integration of Eq. (*d*) twice gives

$$EIv' = -q_0\frac{x^3}{6} + \frac{q_0 L}{4}\langle x\rangle^2 + \frac{q_0 L}{4}\langle x - L\rangle^2 + c_1 \qquad (h)$$

$$EIv = -q_0\frac{x^4}{24} + \frac{q_0 L}{12}\langle x\rangle^3 + \frac{q_0 L}{12}\langle x - L\rangle^3 + c_1 x + c_2 \qquad (i)$$

where c_1 and c_2 are the constants of integration. These constants of integration are to be evaluated from geometric considerations.

The geometric conditions at the supports are $v(0) = v(L) = 0$ from which constants c_1 and c_2 can be found (Fig. 6.14*c*). Evaluating the constants (see Example 6.2), we find

$$v(x) = -\frac{q_0 L^4}{24EI}\left[\left(\frac{x}{L}\right)^4 - 2\left(\frac{x}{L}\right)^3 + \frac{x}{L} - \frac{2}{L^3}\langle x - L\rangle^3\right] \qquad (j)$$

We note that the last term in Eq. (*j*) is zero along the length of the beam $0 < x < L$ and so does not affect the deflection. This term arises from the contribution of the right end reaction force in the expression for $q(x)$. We find that it is more systematic to include all reaction terms in the expression for $q(x)$ even though they may be zero along the length of the beam; this is in keeping with our approach to include all loads in the loading function $q(x)$.

EXAMPLE 6.8

A simply supported beam of length L is loaded by a concentrated load at the midpoint of the beam, as shown in Fig. 6.15. We wish to obtain an expression for the deflection curve of the neutral axis caused by the load P and the maximum deflection of the beam which in this case is under the load. We neglect the weight of the beam and assume that the bending modulus EI is constant along the length of the beam. We will use the fourth-order equation, Eq. (6.20).

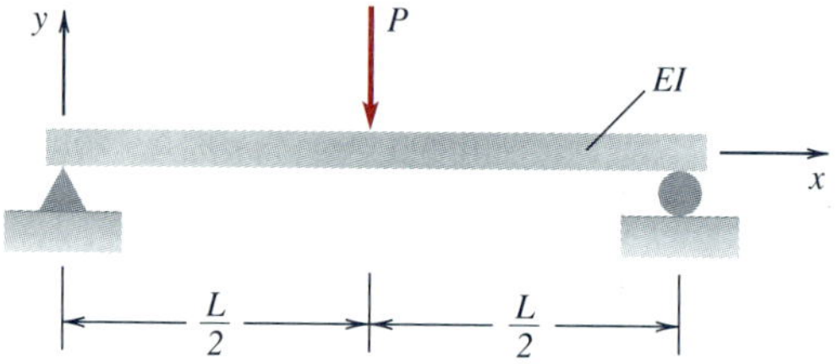

Figure 6.15 Example 6.8

The reaction at each support is $R = P/2$, and the loading $q(x)$ on the beam can be written in the form

$$q(x) = \frac{P}{2}\langle x\rangle_{-1} - P\left\langle x - \frac{L}{2}\right\rangle_{-1} + \frac{P}{2}\langle x - L\rangle_{-1} \qquad (a)$$

The fourth-order equation for the deflection is given by

$$EIv'''' = \frac{P}{2}\langle x\rangle_{-1} - P\left\langle x - \frac{L}{2}\right\rangle_{-1} + \frac{P}{2}\langle x - L\rangle_{-1} \qquad (b)$$

Integration four times gives

$$EIv''' = \frac{P}{2}\langle x\rangle^0 - P\left\langle x - \frac{L}{2}\right\rangle^0 + \frac{P}{2}\langle x - L\rangle^0 \qquad (c)$$

$$EIv'' = \frac{P}{2}\langle x\rangle^1 - P\left\langle x - \frac{L}{2}\right\rangle^1 + \frac{P}{2}\langle x - L\rangle^1 \qquad (d)$$

$$EIv' = \frac{P}{2}\frac{\langle x\rangle^2}{2} - P\frac{\langle x - L/2\rangle^2}{2} + \frac{P}{2}\frac{\langle x - L\rangle^2}{2} + c_1 \qquad (e)$$

$$EIv = \frac{P}{2}\frac{\langle x\rangle^3}{6} - P\frac{\langle x - L/2\rangle^3}{6} + \frac{P}{2}\frac{\langle x - L\rangle^3}{6} + c_1 x + c_2 \qquad (f)$$

where c_1 and c_2 are constants of integration; the first two constants of integration in Eqs. (*c*) and (*d*) are identically zero. The geometric support conditions can be seen from Fig. 6.15 to be $v(0) = v(L) = 0$, from which it follows that $c_2 = 0$ and

$$c_1 L = -\frac{PL^3}{12} + \frac{PL^3}{48} = -\frac{PL^3}{16} \qquad (g)$$

Therefore the expression for the deflection can be written in the form

$$EIv(x) = -\frac{PL^3}{48}\left[3\left(\frac{x}{L}\right) - 4\left(\frac{x}{L}\right)^3 + \frac{8}{L^3}\left\langle x - \frac{L}{2}\right\rangle^3\right] \qquad (h)$$

where we have not included in Eq. (*h*) the $\langle x - L\rangle^3$ term from Eq. (*f*) since it is zero along the beam.

The deflection under the load at $x = L/2$ is

$$v\left(\frac{L}{2}\right) = -\frac{PL^3}{48EI} \qquad (i)$$

The minus sign in Eq. (*i*) indicates that the deflection is downward; also see Fig. 6.6*a*.

EXAMPLE 6.9

A simply supported beam *AB* of length *L* is loaded by a constant distributed load over one-half of its length, as shown in Fig. 6.16*a*. We want to determine the expression for the deflection of the beam due to this loading. In addition, if $q_0 = 50$ lb/in and $L = 96$ in, we wish to find the deflection at $x = L/4$ and $x = L/2$. We assume $EI = 450 \times 10^6$ lb · in^2 and neglect the weight of the beam.

The reactions at each support are shown in Fig. 6.16*b*, and the expression for the loading on the beam is

$$q(x) = \frac{3q_0L}{8}\langle x\rangle_{-1} + \frac{q_0L}{8}\langle x - L\rangle_{-1} - q_0\left(\langle x\rangle^0 - \left\langle x - \frac{L}{2}\right\rangle^0\right) \qquad (a)$$

Remember that for singularity functions with superscripts greater than or equal to zero, once we initiate a load at a point, it will continue over the remaining length of the beam. Thus in Eq. (*a*), since we want the distributed load q_0 to vanish for points $x > L/2$, we introduce an equal but opposite loading function $q_0\langle x - L/2\rangle^0$ which has the effect of turning off the original loading; see also Fig. 4.27. Substitution of $q(x)$ into Eq. (6.20) gives

$$EIv'''' = \frac{3q_0L}{8}\langle x\rangle_{-1} - q_0\left(\langle x\rangle^0 - \left\langle x - \frac{L}{2}\right\rangle^0\right) + \frac{q_0L}{8}\langle x - L\rangle_{-1} \qquad (b)$$

If we integrate four times, we find

$$EIv''' = \frac{3q_0L}{8}\langle x\rangle^0 - q_0\left(\langle x\rangle^1 - \left\langle x - \frac{L}{2}\right\rangle^1\right) + \frac{q_0L}{8}\langle x - L\rangle^0 \qquad (c)$$

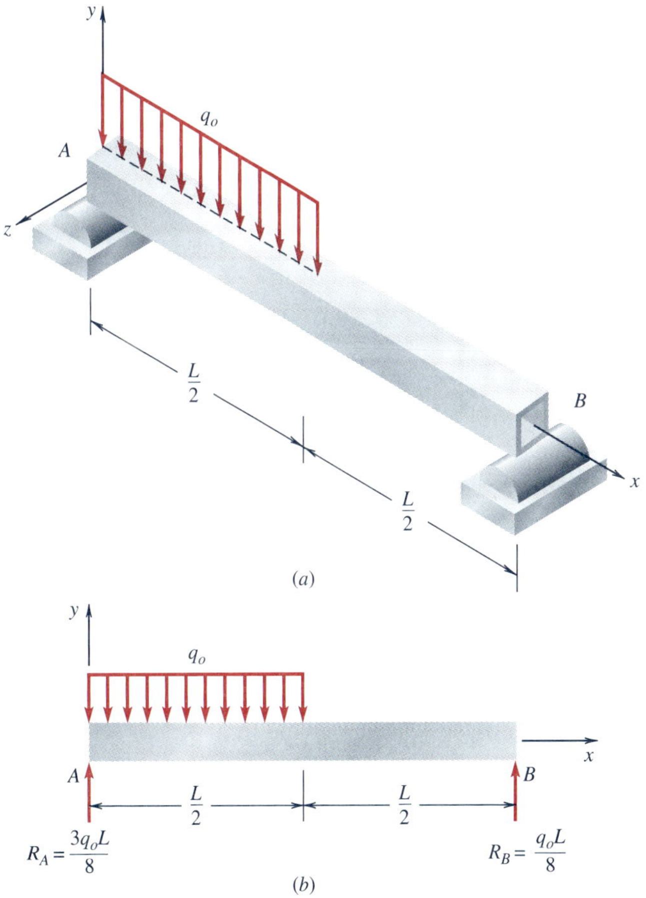

Figure 6.16 Example 6.9

$$EIv'' = \frac{3q_0L}{8}\langle x\rangle^1 - \frac{q_0}{2}\left(\langle x\rangle^2 - \left\langle x - \frac{L}{2}\right\rangle^2\right) + \frac{q_0L}{8}\langle x - L\rangle^1 \qquad (d)$$

$$EIv' = \frac{3q_0L}{8}\frac{\langle x\rangle^2}{2} - \frac{q_0}{6}\left(\langle x\rangle^3 - \left\langle x - \frac{L}{2}\right\rangle^3\right) + \frac{q_0L}{8}\frac{\langle x - L\rangle^2}{2} + c_1 \qquad (e)$$

$$EIv = \frac{3q_0L}{8}\frac{\langle x\rangle^3}{6} - \frac{q_0}{24}\left(\langle x\rangle^4 - \left\langle x - \frac{L}{2}\right\rangle^4\right) + \frac{q_0L}{8}\frac{\langle x - L\rangle^3}{6} + c_1x + c_2 \qquad (f)$$

where c_1 and c_2 are the constants of integration; the first two constants of integration in Eqs. (c) and (d) are identically zero. The geometric support conditions are $v(0) = v(L) = 0$ from which we have $c_2 = 0$ and

$$-c_1L = \frac{q_0L^4}{16} - \frac{q_0L^4}{24}\frac{15}{16} = \frac{9}{384}q_0L^4 \qquad (g)$$

Therefore the expression for the deflection is

$$EIv(x) = \frac{q_0L}{16}x^3 - \frac{q_0}{24}\left(\langle x\rangle^4 - \left\langle x - \frac{L}{2}\right\rangle^4\right) - \frac{9}{384}q_0L^3x \qquad (h)$$

where we have dropped the term with $\langle x - L\rangle^3$ in Eq. (f).

When $x = L/4$ and $x = L/2$, we have

$$v\left(\frac{L}{4}\right) = \frac{q_0L^4}{EI}\left(\frac{1}{16}\cdot\frac{1}{64} - \frac{1}{24}\cdot\frac{1}{256} - \frac{9}{384}\cdot\frac{1}{4}\right) = -\frac{31}{6144}\frac{q_0L^4}{EI} \qquad (i)$$

$$v\left(\frac{L}{2}\right) = \frac{q_0L^4}{EI}\left(\frac{1}{16}\cdot\frac{1}{8} - \frac{1}{24}\cdot\frac{1}{16} - \frac{9}{384}\cdot\frac{1}{2}\right) = -\frac{5}{768}\frac{q_0L^4}{EI} = -\frac{40}{6144}\frac{q_0L^4}{EI} \qquad (j)$$

For the case when $q_0 = 50$ lb/in, $L = 96$ in and $EI = 450 \times 10^6$ lb · in², we have

$$v\left(\frac{L}{4}\right) = -0.0476 \text{ in} \qquad v\left(\frac{L}{2}\right) = -0.0614 \text{ in} \qquad (k)$$

EXAMPLE 6.10

A built-in beam of length L is part of a hydraulic structure and is subjected to a distributed loading that varies linearly with distance from A due to water pressure, as shown in Fig. 6.17a. The loading and the beam can be modeled as a cantilever beam under the triangular loading shown in Fig. 6.17b. The cantilever beam AB is subjected to the distributed loading which varies from q_0 at support A to zero at the free end B. We wish to find the slope and deflection at B. We neglect the weight of the beam and assume EI is constant.

Following the same procedure as used in Chap. 4 to find $q(x)$, we begin by calculating the wall reactions, using overall equilibrium, and then use these reactions in the expression for the load function $q(x)$. Force and moment equilibrium of the free body of the beam in Fig. 6.17c gives the reactions

$$R_A = \frac{q_0L}{2} \qquad M_A = \frac{q_0L^2}{6} \qquad (a)$$

The linearly decreasing load applied to the beam can be represented by $-q_0(1 - x/L)$ so that the complete load expression becomes

$$q(x) = R_A\langle x\rangle_{-1} + M_A\langle x\rangle_{-2} - q_0\left(1 - \frac{x}{L}\right) \qquad (b)$$

The wall reactions are positive loads on the beam consistent with our sign conventions. The load-deflection equation, Eq. (6.20), now takes the form

$$EIv'''' = R_A\langle x\rangle_{-1} + M_A\langle x\rangle_{-2} - q_0\left(1 - \frac{x}{L}\right) \qquad (c)$$

One integration yields

$$EIv''' = R_A\langle x\rangle^0 - M_A\langle x\rangle_{-1} - q_0\left(x - \frac{x^2}{2L}\right) \qquad (d)$$

where attention is paid to the sign arising from the integration of the M_A term. An additional integration gives

$$EIv'' = R_A\langle x\rangle^1 - M_A\langle x\rangle^0 - q_0\left(\frac{x^2}{2} - \frac{x^3}{6L}\right) \qquad (e)$$

The constants of integration for Eqs. (d) and (e) are identically equal to zero. The remaining steps are exactly the same as those used in the previous examples. Integrating Eq. (e) twice results in

$$EIv' = R_A\frac{\langle x\rangle^2}{2} - M_A\langle x\rangle^1 - q_0\left(\frac{x^3}{6} - \frac{x^4}{24L}\right) + c_1 \qquad (f)$$

$$EIv = R_A\frac{\langle x\rangle^3}{6} - M_A\frac{\langle x\rangle^2}{2} - q_0\left(\frac{x^4}{24} - \frac{x^5}{120L}\right) + c_1x + c_2 \qquad (g)$$

where c_1 and c_2 are constants of integration. We must identify two geometric conditions along the beam in order to evaluate these constants. In the present case of a clamped support at the left end of the beam, we have

$$v = v' = 0 \quad \text{at} \quad x = 0 \qquad (h)$$

so that

$$EIv'(0) = 0 = c_1$$
$$EIv(0) = 0 = c_2 \qquad (i)$$

We can complete the calculation and determine the end slope and the end deflection by making use of Eqs. (a) and substituting $x = L$ into Eqs. (f) and (g) to find

$$EIv'(L) = EI\theta_B = \frac{q_0L^3}{4} - \frac{q_0L^3}{6} - \frac{q_0L^3}{8} = -\frac{q_0L^3}{24} \qquad (j)$$

$$EIv(L) = EIv_B = \frac{q_0L^4}{12} - \frac{q_0L^4}{12} - \frac{q_0L^4}{30} = -\frac{q_0L^4}{30} \qquad (k)$$

where θ_B is the slope angle of the deflected shape at end B, as shown in Fig. 6.17d.

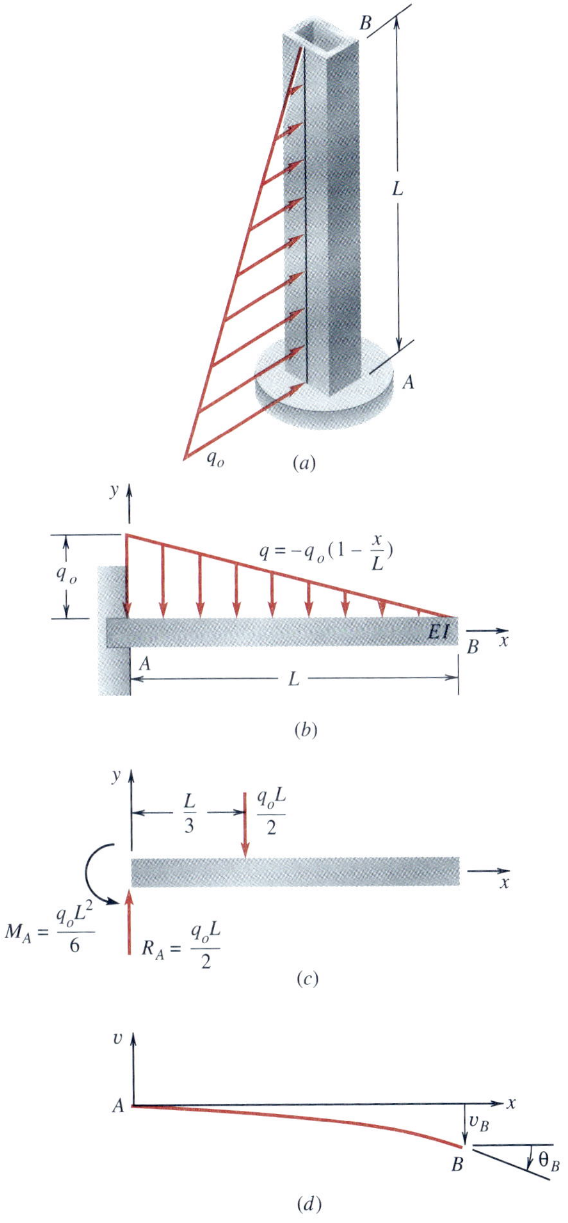

Figure 6.17 Example 6.10

EXAMPLE 6.11

A machine component is to be modeled as a simply supported beam with a concentrated moment M_0 acting at section B, as shown in Fig. 6.18*a*. The beam with the applied moment and the reactions is shown again in Fig. 6.18*b*. We wish to find the slope and deflection at B, the slopes at each end, and, for the special case when $a = \frac{2}{3}L$, the location and magnitude of the maximum absolute value of the deflection. We neglect the weight of the beam and take the bending modulus EI to be constant.

Force and moment equilibrium of the entire beam indicates that the reactions are a pair of equal and opposite forces which produce a countermoment to the applied moment M_0, as shown in Fig. 6.18*b*. Thus the load function $q(x)$ contains two concentrated force terms and one concentrated moment term in the form

$$EIv'''' = \frac{M_0}{L}\langle x\rangle_{-1} + M_0\langle x - a\rangle_{-2} - \frac{M_0}{L}\langle x - L\rangle_{-1} \qquad (a)$$

where the term $\langle x - a\rangle_{-2}$ arises from the concentrated moment. One integration produces

$$EIv''' = \frac{M_0}{L}\langle x\rangle^0 - M_0\langle x - a\rangle_{-1} - \frac{M_0}{L}\langle x - L\rangle^0 \qquad (b)$$

where particular attention is paid to the sign associated with the integration of $\langle x - a\rangle_{-2}$; the constant of integration is equal to zero. An additional integration gives

$$EIv'' = \frac{M_0}{L}\langle x\rangle^1 - M_0\langle x - a\rangle^0 - \frac{M_0}{L}\langle x - L\rangle^1 \qquad (c)$$

where again the constant of integration is equal to zero.

Since $EIv'' = M$ from Eq. (6.6), we see that the first two terms that appear on the right-hand side of Eq. (*c*) can be interpreted as the moment contributions to moment equilibrium of a segment of length $x < L$ of the beam.

The remaining two integrations lead to expressions for the slope and deflection

$$EIv' = \frac{M_0}{L}\frac{\langle x\rangle^2}{2} - M_0\langle x - a\rangle^1 - \frac{M_0}{L}\frac{\langle x - L\rangle^2}{2} + c_1 \qquad (d)$$

$$EIv = \frac{M_0}{L}\frac{\langle x\rangle^3}{6} - M_0\frac{\langle x - a\rangle^2}{2} - \frac{M_0}{L}\frac{\langle x - L\rangle^3}{6} + c_1x + c_2 \qquad (e)$$

The integration constants are found by noting that the supports do not allow deflection in the y direction so that for

$$x = 0: \quad v = 0$$
$$x = L: \quad v = 0 \qquad (f)$$

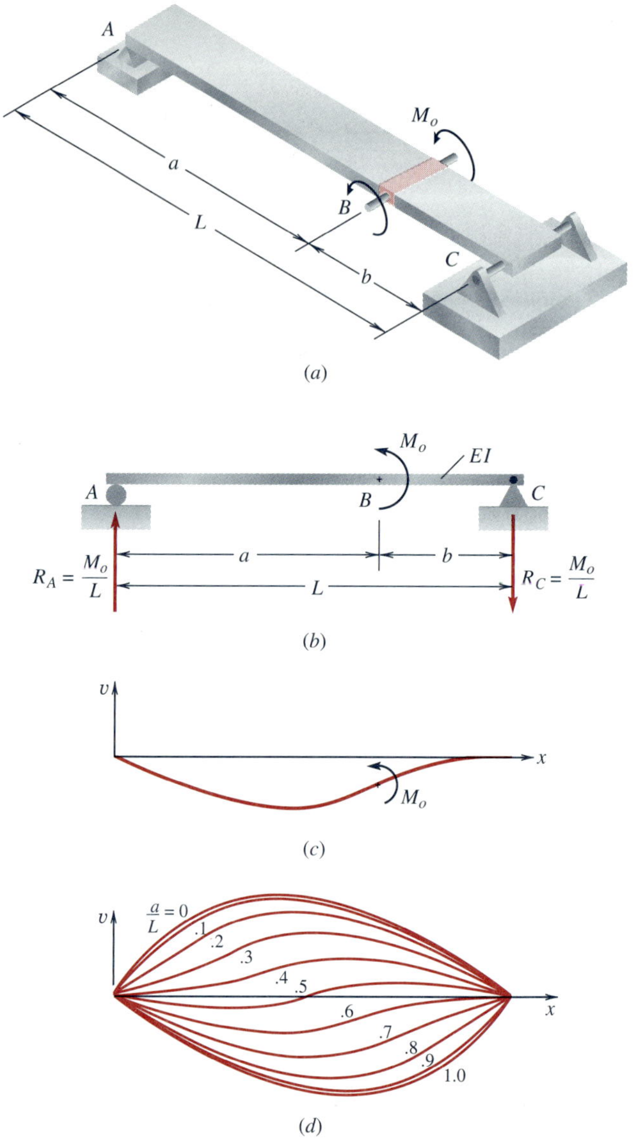

Figure 6.18 Example 6.11

Therefore,

$$c_2 = 0 \qquad (g)$$

and

$$\frac{M_0L^2}{6} - \frac{M_0b^2}{2} + c_1L = 0 \qquad (h)$$

Since Eqs. (*d*) and (*e*) with the value of c_1 given by Eq. (*h*) give values of the slope and deflection for all points along the beam, we can get the specific values at point B by introducing $x = a$ into these equations to obtain

$$v'(a) = \theta_B = \frac{M_0}{6LEI}(3a^2 + 3b^2 - L^2) \qquad (i)$$

and

$$v(a) = v_B = \frac{M_0 a}{6LEI}(a^2 + 3b^2 - L^2) \qquad (j)$$

The values for the end slopes follow by setting $x = 0$ and $x = L$ into Eq. (d) to get

$$v'(0) = \theta_A = -\frac{M_0}{6LEI}(L^2 - 3b^2) \qquad (k)$$

and

$$v'(L) = \theta_C = \frac{M_0}{6LEI}(2L^2 - 6bL + 3b^2) \qquad (l)$$

For the special case for which $a = \frac{2}{3}L$, we have

$$v'(a) = \theta_B = \frac{M_0 L}{6EI}\left(\frac{2}{3}\right) \qquad v'(0) = \theta_A = -\frac{M_0 L}{6EI}\left(\frac{2}{3}\right)$$
$$v(a) = v_B = -\frac{M_0 L^2}{6EI}\left(\frac{4}{27}\right) \qquad v'(L) = \theta_C = \frac{M_0 L}{6EI}\left(\frac{1}{3}\right) \qquad (m)$$

A sketch of the deflected shape for $a/L = \frac{2}{3}$ is shown in Fig. 6.18*c*. Figure 6.18*d* shows the deflected shape of the beam for a/L ranging from 0 to 1 in steps of 0.1; e.g., for $a/L = 0.5$ the shape is antisymmetric.

To determine the maximum value of the deflection when $a = \frac{2}{3}L$, we assume that the location of the maximum value of v will occur in the part of the beam between points A and B. (We could verify that this is a reasonable assumption by applying a moment to the two-thirds point of a ruler or other flexible beam.) Thus if the maximum deflection occurs at $x = \hat{x}$ where $\hat{x} < a = 2L/3$, then we set $v' = 0$ in Eq. (d) in order to find $\hat{x}$:

$$EIv'(\hat{x}) = 0 = \frac{M_0\hat{x}^2}{2L} + \frac{M_0 b^2}{2L} - \frac{M_0 L}{6} \qquad (n)$$

Solving for $\hat{x}$ in Eq. (n) yields

$$\hat{x} = \frac{\sqrt{2}L}{3}$$

Finally, we obtain the value of the maximum deflection by using Eq. (e) with $x = \hat{x}$, $a = 2L/3$, and c_1 and c_2 from Eqs. (g) and (h) to get

$$v(\hat{x}) = -\frac{2\sqrt{2}}{81}\frac{M_0 L^2}{EI} = -\frac{M_0 L^2}{6EI}\left(\frac{4}{27}\right)\sqrt{2} = \sqrt{2}v(a) \qquad (o)$$

EXAMPLE 6.12

A cantilever beam ABC carries a uniform load q_0 acting in opposite directions over each half of the beam, as shown in Fig. 6.19*a*. We wish to find the slope and deflection at the free end; assume that the weight of the beam is negligible compared to the applied load and that EI is constant. We sketch the loading again in Fig. 6.19*b*.

From the overall free-body diagram shown in Fig. 6.19*c*, we find that the wall reaction is a moment M_A

$$M_A = q_0 a^2 \qquad (a)$$

In constructing the load function $q(x)$, we have to start a negative uniform load at $x = 0$ and then stop it at $x = a$. In addition, at $x = a$ we have to start the positive uniform load as shown in Fig. 6.19*b*. Thus $q(x)$ is taken as

$$q(x) = -q_0\langle x\rangle^0 + 2q_0\langle x - a\rangle^0 - M_A\langle x\rangle_{-2} \qquad (b)$$

and the load-deflection equation becomes

$$EIv'''' = q(x) = -q_0\langle x\rangle^0 + 2q_0\langle x - a\rangle^0 - M_A\langle x\rangle_{-2} \qquad (c)$$

Repeated integrations produce

$$EIv''' = -q_0\langle x\rangle^1 + 2q_0\langle x - a\rangle^1 + M_A\langle x\rangle_{-1} \qquad (d)$$

and

$$EIv'' = -q_0\frac{\langle x\rangle^2}{2} + 2q_0\frac{\langle x - a\rangle^2}{2} + M_A\langle x\rangle^0 \qquad (e)$$

where the two constants of integration are equal to zero.

The expression for the slope is

$$EIv' = -q_0\frac{\langle x\rangle^3}{6} + 2q_0\frac{\langle x - a\rangle^3}{6} + M_A\langle x\rangle^1 + c_1 \qquad (f)$$

and we can simplify this expression now by noting that at $x = 0$ the clamped support requires that $EIv'(0) = c_1 = 0$. The final expression for the deflection is

$$EIv = -q_0\frac{\langle x\rangle^4}{24} + 2q_0\frac{\langle x - a\rangle^4}{24} + M_A\frac{\langle x\rangle^2}{2} + c_2 \qquad (g)$$

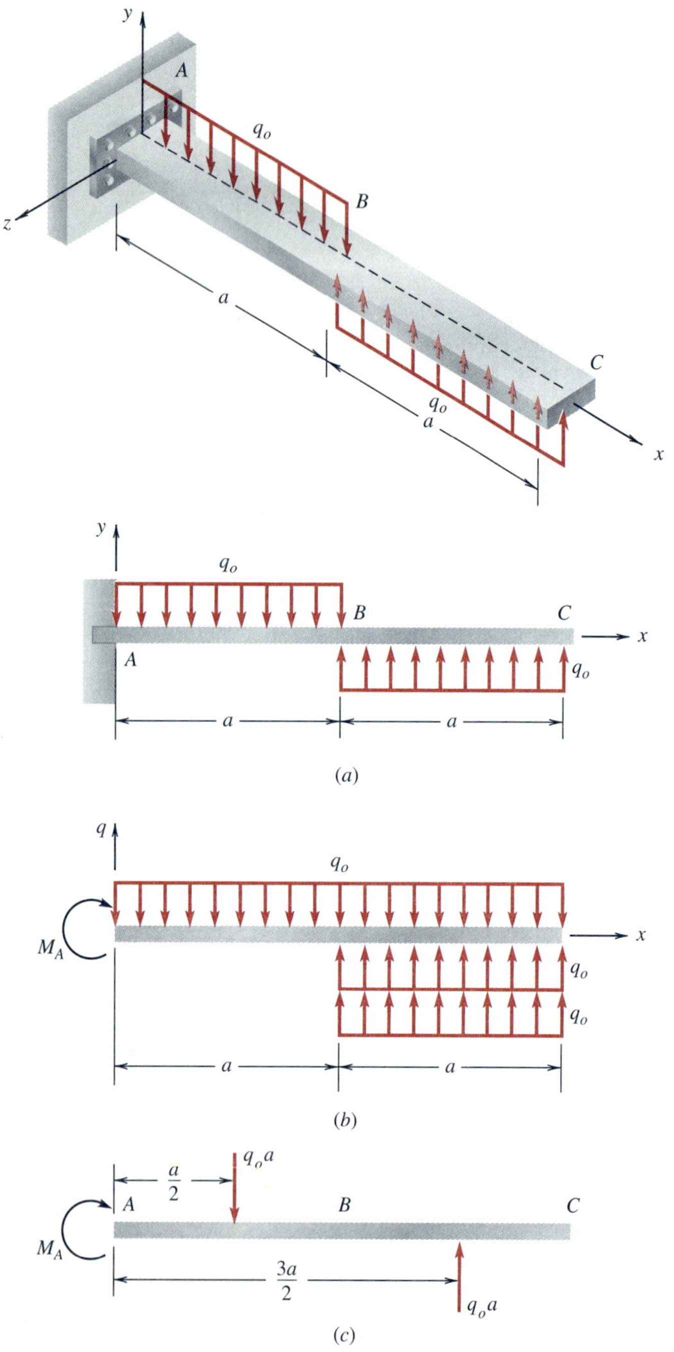

Figure 6.19 Example 6.12

The clamped support at $x = 0$ requires that $EIv(0) = c_2 = 0$.

The slope and deflection at the free end follow by substituting $x = 2a$ into Eqs. (f) and (g) to get the slope

$$EIv'(2a) = -q_0\frac{8a^3}{6} + 2q_0\frac{a^3}{6} + q_0a^2(2a)$$

or

$$\theta_C = v'(2a) = \frac{q_0a^3}{EI} \qquad (h)$$

and the deflection

$$EIv(2a) = -q_0\frac{16a^4}{24} + 2q_0\frac{a^4}{24} + q_0a^2\frac{4a^2}{2}$$

or

$$v_C = v(2a) = \frac{17}{12}\frac{q_0a^4}{EI} \qquad (i)$$

As expected, both the slope and the deflection are positive.

EXAMPLE 6.13

A simply supported steel beam AD with a concentrated load at B and a constant distributed load on the overhang CD is shown in Fig. 6.20*a*. We wish to determine the deflections at $x = 5$, 10, 15, and 30 ft along the beam. The moment of inertia of the beam is $I = 34$ in^4, $E = 30 \times 10^6$ psi, and the weight of the beam can be neglected. See Example 4.17.

The reactions at A and C follow from equilibrium

$$R_A = 7.5 \text{ kips} \qquad R_C = 22.5 \text{ kips} \tag{a}$$

The loading function $q(x)$ for the beam from Fig. 6.20*b* is

$$q(x) = 7.5\langle x\rangle_{-1} - 20\langle x - 10\rangle_{-1} + 22.5\langle x - 20\rangle_{-1} - \langle x - 20\rangle^0 \tag{b}$$

The load-deflection equation for the beam is

$$EIv'''' = 7.5\langle x\rangle_{-1} - 20\langle x - 10\rangle_{-1} + 22.5\langle x - 20\rangle_{-1} - \langle x - 20\rangle^0 \tag{c}$$

Upon integrating twice we have

$$EIv''' = 7.5\langle x\rangle^0 - 20\langle x - 10\rangle^0 + 22.5\langle x - 20\rangle^0 - \langle x - 20\rangle^1 \tag{d}$$

$$EIv'' = 7.5\langle x\rangle^1 - 20\langle x - 10\rangle^1 + 22.5\langle x - 20\rangle^1 - \tfrac{1}{2}\langle x - 20\rangle^2 \tag{e}$$

where the constants of integration are zero.

The moment expression $M = EIv''$ in Eq. (*e*) agrees, of course, with what we would obtain directly from moment equilibrium of a segment of length x, as shown in Fig. 6.20*c*.

Integrating again, we find

$$EIv' = 7.5\frac{\langle x\rangle^2}{2} - 20\frac{\langle x - 10\rangle^2}{2} + 22.5\frac{\langle x - 20\rangle^2}{2} - \frac{1}{6}\langle x - 20\rangle^3 + c_1 \tag{f}$$

$$EIv = \frac{7.5}{6}\langle x\rangle^3 - \frac{20}{6}\langle x - 10\rangle^3 + \frac{22.5}{6}\langle x - 20\rangle^3 - \frac{1}{24}\langle x - 20\rangle^4 + c_1x + c_2 \tag{g}$$

The deflection must be zero at the simple supports at A and C, $v(0) = v(20) = 0$, so that $c_2 = 0$ and

$$c_1 = -\frac{1000}{3} \tag{h}$$

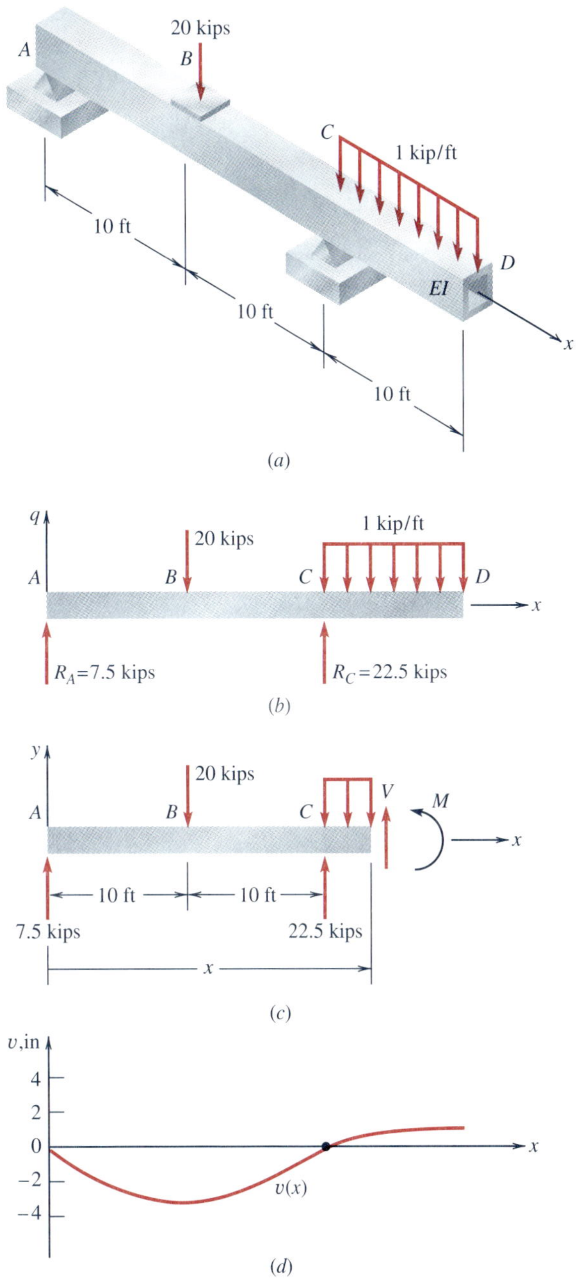

Figure 6.20 Example 6.13

The deflection equation is therefore, from Eqs. (*g*) and (*h*),

$$EIv(x) = \frac{7.5}{6}\langle x\rangle^3 - \frac{20}{6}\langle x - 10\rangle^3 + \frac{22.5}{6}\langle x - 20\rangle^3 - \frac{1}{24}\langle x - 20\rangle^4 - \frac{1000}{3}x \qquad (i)$$

where all distances for x are in feet and loads are in kips. To determine the deflections at points along the beam, we need to evaluate Eq. (*i*) at the different values of x, paying close attention to the units and to the evaluation of the singularity functions.

The deflection at $x = 5$ ft is given by

$$EIv(5) = \frac{7.5}{6}(5)^3 - \frac{1000}{3}(5)$$

or

$$v(5\text{ ft}) = -\frac{1.510 \times 10^3\text{ kips}\cdot\text{ft}^3}{(30 \times 10^6\text{ lb/in}^2)(34\text{ in}^4)(\text{ft}^2/144\text{ in}^2)} = -0.213\text{ ft} = -2.56\text{ in} \qquad (j)$$

In a similar way we find

$$v(10\text{ ft}) = -3.53\text{ in}$$
$$v(15\text{ ft}) = -2.03\text{ in} \qquad (k)$$
$$v(30\text{ ft}) = 0.706\text{ in}$$

We see that the downward deflection under the load at point B is almost exactly 5 times greater in magnitude than the upward deflection at the tip of the overhang. A sketch of the deflected shape is shown in Fig. 6.20*d*.

6.5 Method of Superposition

It is obvious that in using the methods of the previous sections, we could quickly accumulate and tabulate solutions of many standard-type deflection problems. In fact, many tables of solutions for beam deflection problems are available, and the use of these tables can save time and effort when a particular problem of interest happens to be listed in the table. A short table of deflection solutions is given in App. G.

In this section we discuss another important use of beam deflection tables by solving new problems not directly in the tables by combining or adding solutions in the table, according to the principle of superposition. If it is possible to break a given problem down into a combination of simple loading conditions for which the results are given in the table, then the method of superposition allows us to construct the solution for the problem of interest by adding the solutions in the table. We can add the different solutions because all the load-deflection–slope relations that have been derived in this chapter are linear.

This linearity originates in the moment-curvature relationship which made use of the linear relation for stress and strain for the material. Also the moment-curvature equation, Eq. (6.5), was linearized by restricting attention to deflection problems for which the square of the slope in Eq. (6.5) can be neglected compared to 1, thus replacing the nonlinear relation in Eq. (6.5) by the linear relation

$$EI\frac{d^2v}{dx^2} = M \qquad (6.6)$$

The principle of superposition using bending moments follows from the linear equation, Eq. (6.6).

Alternatively when we write the linear fourth-order equation, Eq. (6.20), in terms of the load, we have

$$EIv'''' = q(x) \tag{6.20}$$

For a given beam we see from the linearity of Eq. (6.20) that if $v_1(x)$, $v_2(x), \ldots, v_n(x)$ are the solutions for the deflection corresponding to separate loads $q_1(x), q_2(x), \ldots, q_n(x)$, then the sum of the solutions

$$v(x) = v_1(x) + v_2(x) + \cdots + v_n(x)$$

corresponding to these loads is also a solution corresponding to the loading

$$q(x) = q_1(x) + q_2(x) + \cdots + q_n(x)$$

Examples follow in which some techniques using the method of superposition are given. A real advantage to the use of superposition occurs when *specific* information on deflection or slope at a section is needed in a particular problem.

EXAMPLE 6.14

Consider a cantilever beam of length L subjected to two equal concentrated loads, as shown in Fig. 6.21*a*. One load is applied at the midpoint of the beam, and the other is applied at the tip of the beam as shown. We wish to find the slope and deflection at the tip. We neglect the weight of the beam and assume that EI is constant.

Consulting App. G, we find both the case of a cantilever with a load at the end (case G.1-2 in Table G.1) and of a cantilever with a load located partway along the beam (case G.1-1). In Fig. 6.21*b* and *c*, we have shown the replacement of the original loading in Fig. 6.21*a* by the sum of load systems (1) and (2). The end deflection and slope for the end-loaded cantilever for load system (1) (Fig. 6.21*b*) are given in case G.1-2 as

$$v_1(L) = -\frac{PL^3}{3EI} \tag{a}$$

$$\theta_1 = v_1'(L) = -\frac{PL^2}{2EI} \tag{b}$$

To make use of the results in case G.1-1, we have to specialize the parameters in case G.1-1 so that they apply to load system (2) of Fig. 6.21*c* by taking

$$a = b = \frac{L}{2} \tag{c}$$

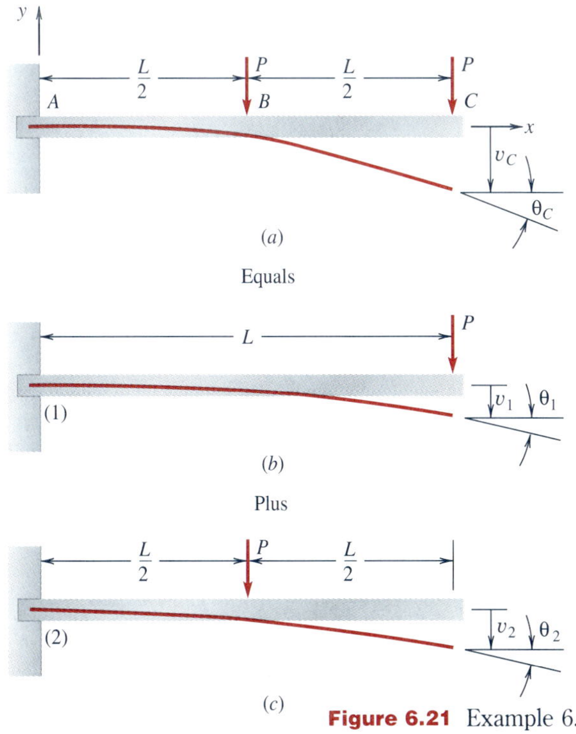

Figure 6.21 Example 6.14

and substituting into the corresponding expressions in case G.1-1 to obtain the deflection at the tip as

$$v_2(L) = -\frac{P(L/2)^2}{6EI}\left(3L - \frac{L}{2}\right) = -\frac{5PL^3}{48EI} \qquad (d)$$

and the slope at the tip

$$\theta_2 = v_2'(L) = -\frac{P(L/2)^2}{2EI} = -\frac{PL^2}{8EI} \qquad (e)$$

By the method of superposition, we now obtain the tip deflection due to both load systems acting on the beam by adding the results given in Eqs. (*a*) and (*d*) to get

$$v_C = v_1(L) + v_2(L) = -\frac{PL^3}{EI}\left(\frac{1}{3} + \frac{5}{48}\right)$$

which reduces to

$$v_C = -\frac{7PL^3}{16EI} \qquad (f)$$

To find the slope at the tip due to both loads, we follow the same general procedure as for the deflection and add the results in Eqs. (*b*) and (*e*), to obtain

$$\theta_C = v_1'(L) + v_2'(L) = -\frac{PL^2}{EI}\left(\frac{1}{2} + \frac{1}{8}\right)$$

or

$$\theta_C = -\frac{5PL^2}{8EI} \qquad (g)$$

In order to compare these results for the case of the two equal loads acting on the beam to the results for a single load acting at the end of the beam, we will compute the ratios of the tip deflections and slopes. First consider the ratio of the tip deflections from Eqs. (*f*) and (*a*)

$$\frac{v_C}{v_1(L)} = \frac{\frac{7}{16}}{\frac{1}{3}} = 1.31$$

That is, there is about a 31 percent increase in the tip deflection when the additional midpoint load is applied to the end-loaded beam. Proceeding in a similar way for the slopes, we compute the ratio of slopes from Eqs. (*g*) and (*b*)

$$\frac{\theta_C}{v_1'(L)} = \frac{\frac{5}{8}}{\frac{1}{2}} = 1.25$$

and find a 25 percent increase in the slope at the tip due to the additional midpoint load.

EXAMPLE 6.15

A cantilever beam shown in Fig. 6.22*a* is loaded by a pair of equal but oppositely directed concentrated loads separated by a distance b near the tip as shown. We wish to use the method of superposition to calculate the deflection at point C. In addition, we would like to compare the tip deflection so obtained for the special case $b = L/10$ with that of a cantilever beam loaded by a concentrated moment Pb at the end, as shown in Fig. 6.22*b*. We neglect the weight of the beam and assume that EI is constant.

This is a good problem for making use of superposition arguments because we are seeking specific information about a deflection at a point. To make use of the principle of superposition, we replace the original pair of loads in Fig. 6.22*a* with the two separate cases shown in Fig. 6.22*c* and *d*. These cases can be found in App. G. From case G.1-1 in Table G.1, we have (Fig. 6.22*c*)

$$v_1 = \frac{Pa^2}{6EI}(3L - a) \qquad (a)$$

and for case G.1-2 (Fig. 6.22*d*)

$$v_2 = -\frac{PL^3}{3EI} \qquad (b)$$

Therefore, the tip deflection for the load in Fig. 6.22*a* is the superposition of the deflections in Eqs. (*a*) and (*b*)

$$v_C = v_1 + v_2 = \frac{P}{6EI}[a^2(3L - a) - 2L^3] \qquad (c)$$

In the special case for which $b = L/10$ and $a = 9L/10$, the deflection of the tip is

$$v_C = \frac{PL^3}{EI}(-0.0498) \qquad (d)$$

The minus sign in Eq. (*d*) indicates that the deflection is downward.

If the pair of loads in Fig. 6.22*a* is replaced by a moment

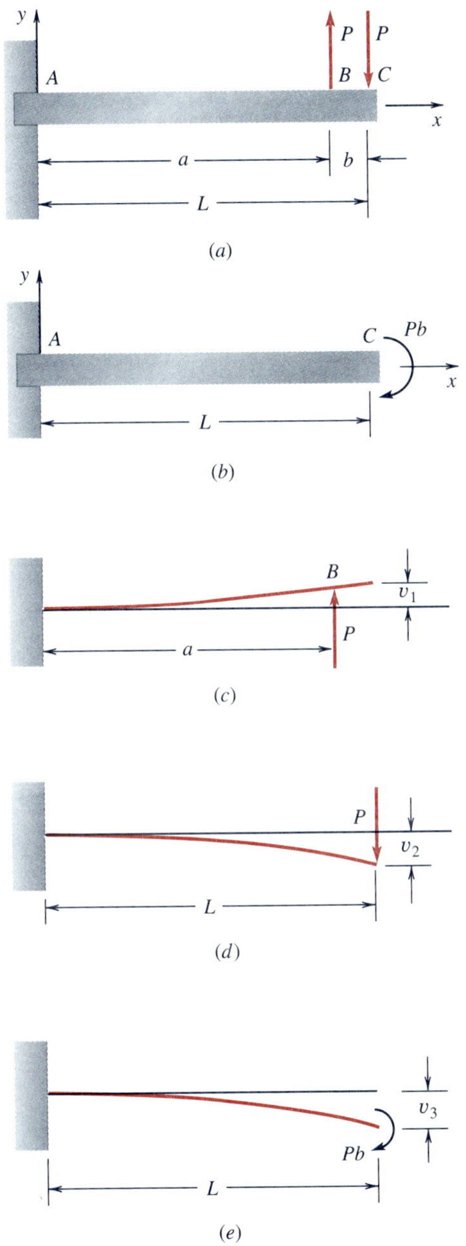

Figure 6.22 Example 6.15

applied to the end of the beam of magnitude Pb, as shown in Fig. 6.22*b*, then the deflection at the tip in this case follows from case G.1-5 in Table G.1 (Fig. 6.22*e*), which gives

$$v_3 = -\frac{PbL^2}{2EI} \tag{e}$$

For $b = L/10$, Eq. (*e*) gives

$$v_C = -\frac{PL^3}{EI}(0.05) \tag{f}$$

If we compare the value of the deflection in Eq. (*f*) to the value from Eq. (*d*), we see that the difference is less than 0.4 percent. In fact, we can increase b to $0.3L$, and the deflections will still agree to about 2.3 percent; however, the deformed shape of the beam will be different near the end.

EXAMPLE 6.16

A cantilever beam is made up of two circular pipes *AB* and *BC*, both with the same moment of inertia *I* but with different elastic moduli E_1 and E_2, as shown in Fig. 6.23*a*. Two loads *P* are applied, one at *B* and the other at *C*. We wish to find the slope and deflection of the tip of the beam at *C*, using the method of superposition. Recall Example 6.14; we neglect the weight of each member.

Since the elastic moduli are different in the two parts *AB* and *BC*, it is only possible to apply the deflection formulas given in App. G to each part separately. Thus we break up the original cantilever beam into two beams as shown in Fig. 6.23*b* and consider each part separately. The internal shear force and bending moment at *B* from part *BC* become external loading at *B* on part *AB*.

Cantilever *AB***:** Cantilever *AB* has an applied force *P*, the shear force *P*, and the bending moment *PL* acting at *B* so that making use of cases G.1-2 and G.1-5 from Table G.1 with $M_0 = PL$, we get

$$v_B = -\frac{(2P)L^3}{3E_1 I} - \frac{(PL)L^2}{2E_1 I} = -\frac{7PL^3}{6E_1 I} \qquad (a)$$

and

$$\theta_B = -\frac{(2P)L^2}{2E_1 I} - \frac{(PL)L}{E_1 I} = -\frac{2PL^2}{E_1 I} \qquad (b)$$

and these values are shown in Fig. 6.23*c*. The deflection v_B causes part *BC* to move down at end *B*, and the rotation θ_B causes part *BC* to rotate about end *B*. Part *BC* is now at an angle θ_B below the horizontal.

Cantilever *BC***:** First, we calculate the deflection and slope of end *C* relative to end *B*, using the formulas for a cantilever beam fixed at *B* and loaded at *C* (case G.1-2), to get

$$v_{C1} = -\frac{PL^3}{3E_2 I} \qquad (c)$$

and

$$\theta_{C1} = -\frac{PL^2}{2E_2 I} \qquad (d)$$

Equations (*c*) and (*d*) give the tip deflection and slope of an initially horizontal cantilever fixed at its left end. However, cantilever *BC* has been rotated by an amount θ_B due to the deflection of part *AB*. Therefore, the total deflection and slope at section *C* are now found by adding the slope and deflection values at section *B*, Eqs. (*a*) and (*b*), to the slope and deflection values for section *C* *relative* to section *B*.

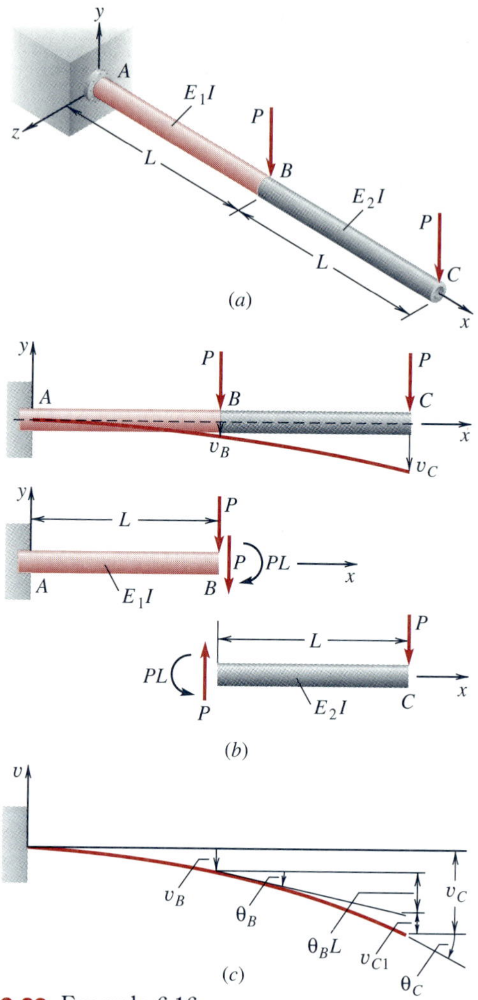

Figure 6.23 Example 6.16

Deflection at *C***:** We see from Fig. 6.23*c* that the total deflection at *C* follows in the form

$$v_C = v_B + \theta_B L + v_{C1}$$

$$= -\frac{7PL^3}{6E_1 I} - \frac{2PL^2(L)}{E_1 I} - \frac{PL^3}{3E_2 I}$$

or

$$v_C = -\frac{PL^3}{I}\left(\frac{19}{6E_1} + \frac{1}{3E_2}\right) \qquad (e)$$

Slope at C: Proceeding in a similar fashion for the slope at C gives

$$\begin{aligned}\theta_C &= \theta_B + \theta_{C1}\\ &= -\frac{2PL^2}{E_1 I} - \frac{PL^2}{2E_2 I}\\ &= -\frac{PL^2}{I}\left(\frac{2}{E_1} + \frac{1}{2E_2}\right)\end{aligned} \quad (f)$$

In the limiting case where $E_1 = E_2 = E$, we have

$$v_C = -\frac{7PL^3}{2EI}$$

and

$$\theta_C = -\frac{5PL^2}{2EI} \quad (g)$$

which are exactly the results obtained in Eqs. (f) and (g) in Example 6.14 provided that L in this example is replaced by $L/2$ to correspond to the beam in Example 6.14.

6.6 Concluding Remarks

In this chapter we have presented three methods for the determination of the deflection of a linearly elastic slender beam of symmetric cross section loaded by transverse loads in the plane of symmetry. In the first method we started from the expression for the bending moment along the beam $M(x)$, and integrated

$$EI\frac{d^2v}{dx^2} = M(x) \quad (6.6)$$

twice to obtain $v(x)$. In the second method we started from the expression for the loading function on the beam $q(x)$ that contained all loads including all reaction loads. We then integrated

$$EI\frac{d^4v}{dx^4} = q(x) \quad (6.20)$$

four times to obtain $v(x)$. In each of these methods we evaluated the constants of integration from information on the support conditions for the beam. Singularity functions allowed us to obtain the expressions for $M(x)$ and $q(x)$ in a systematic fashion. Both methods use the three steps of Eq. (2.8): (1) equilibrium considerations to determine the expression for $M(x)$ or $q(x)$; (2) the use of the appropriate force-deformation relations, Eq. (6.6) or Eq. (6.20), and (3) the use of geometric information to determine the integration constants appearing in the solution. As we have noted often, the three steps of Eq. (2.8) provide the framework for analyzing problems in the mechanics of solids.

The third method made use of superposition and allowed us to add known deflection solutions to obtain a solution to a given problem. Superposition usually works best when a value of a specific deflection or slope at a section is needed.

So far we have treated only statically determinate beams. In the next chapter we will analyze statically indeterminate beams, using techniques similar to those used in this chapter. We will also introduce a beam deflection method that is convenient for computer implementation.

PROBLEMS

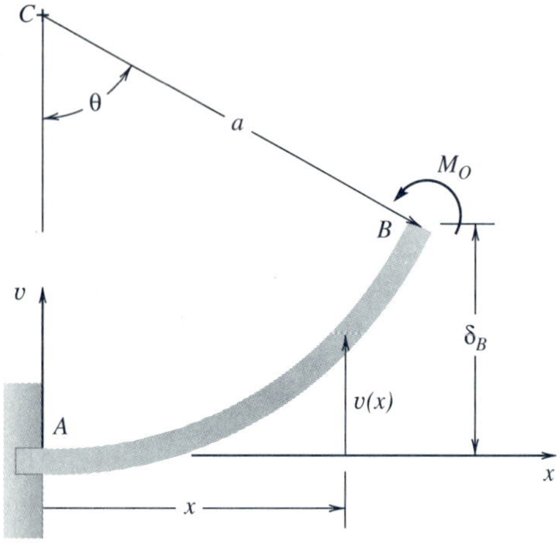

Fig. P6.2-1

6.2-1 A segment of a beam is bent into a circular shape of radius a (Fig. P6.2-1), where

$$(v - a)^2 + x^2 = a^2$$

Show that application of Eq. (6.4) leads to the correct value for the radius of curvature $\rho = a$ at all points along the segment.

6.2-2 The beam shown in Fig. P6.2-1 is loaded by a moment M_0 so as to result in an angle of $\theta = 55°$ for a given bending modulus EI_z.

(*a*) Find the deflection δ_B of the end of the beam due to M_0 by using geometric relations for the circular arc.

(*b*) Use the approximate equation, Eq. (6.6), to calculate δ_B by integrating Eq. (6.6) twice and using the conditions that $v = v' = 0$ at $x = 0$. Also $M(x)$ equals M_0 along the beam.

(*c*) What is the percent error in finding δ_B by using the approximate relation in Eq. (6.6)?

6.2-3 A thin originally straight metal strip AB is wrapped around a semicircular cylinder as shown in Fig. P6.2-3.

(*a*) Derive an expression for the allowable thickness h_a of the strip in terms of the allowable stress due to bending σ_a, the radius R of the cylinder, and the elastic modulus E of the metal.

(*b*) For $E = 200$ GPa and $\sigma_a = 100$ MPa, calculate h_a for $R = 10$, 100, and 1000 mm.

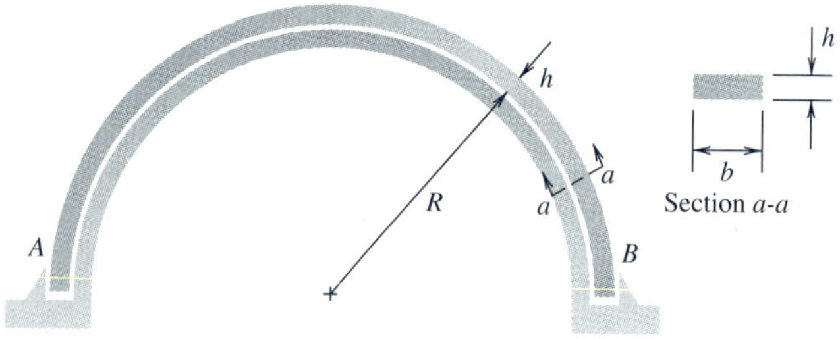

Fig. P6.2-3

6.2-4 A machine part consists of two stiff rigid metal pieces AB and CD separated by a flexible strip BC of bending modulus EI, as shown in Fig. P6.2-4a.

(*a*) Take parts AB and CD to be completely rigid in bending, and calculate the moment M_A applied at section A' which is required to keep AB in a vertical position (see Fig. P6.2-4b).

(*b*) Find the total vertical deflection v_A of section A.

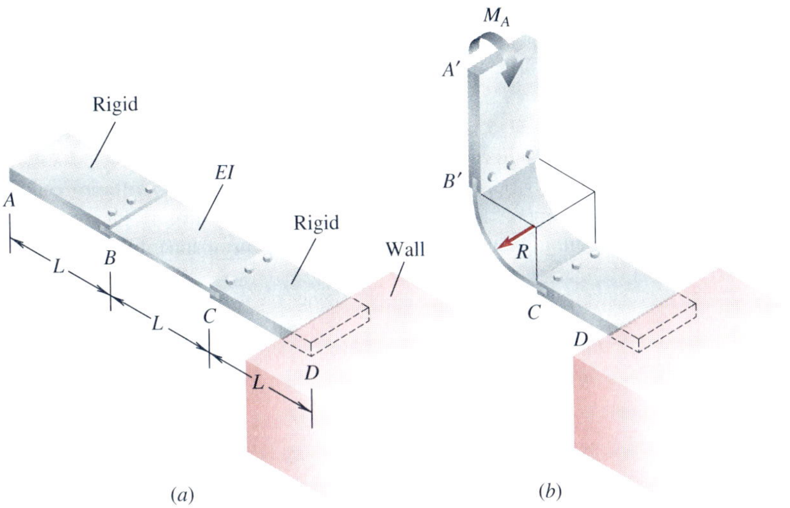

Fig. P6.2-4

6.2-5 Wood strips are formed into arcs of circles in preparation for steam treatment to maintain the shapes, as shown in Fig. P6.2-5.

(*a*) Derive an expression for the maximum stress σ due to bending in terms of the length of the strip L, the angle subtended by the bent strip, the thickness of the strip h, and the elastic modulus E.

(*b*) If the allowable stress in the wood is 1000 psi and $E = 1000$ ksi, calculate the allowable thickness h_a for strips of length 3, 5, and 10 ft for the two cases shown in Fig. P6.2-5*a* and *b*.

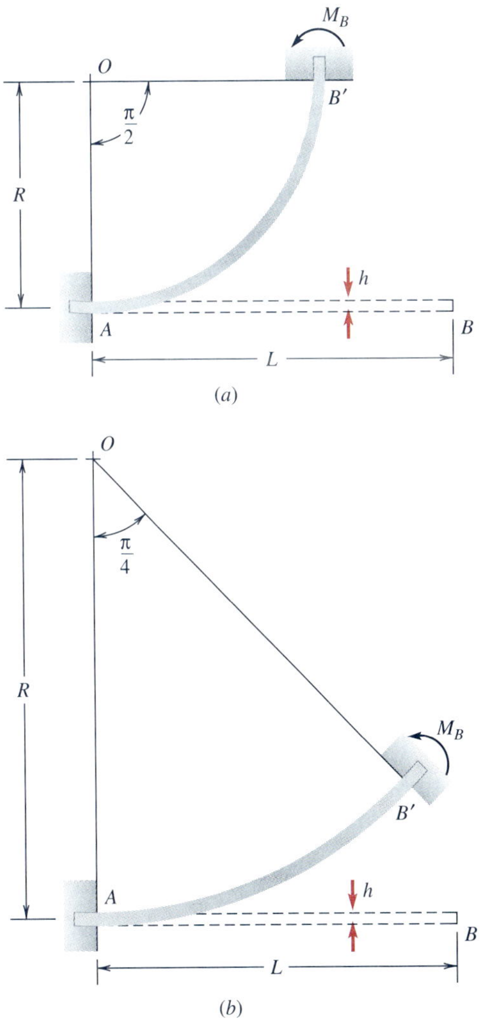

Fig. P6.2-5

Use the method of double integration in solving problems associated with Sec. 6.3. The bending modulus for all beams is EI_z, and the weight of the beam is to be neglected unless otherwise noted.

6.3-1 Consider Example 6.1, Fig. 6.3. Verify that if the load is to stay at 400 lb and the deflection is limited to $L/240$, the span must be reduced to 104 in. See discussion after Eq. (*k*).

6.3-2 Consider Example 6.1, Fig. 6.3. Select an appropriate wide-flange I beam from Table C.1 which could be used to replace the "4 × 10" wood beam used in Example 6.1 such that the tip deflection would be approximately the same or less than the value of 0.485 in. Use $E = 30 \times 10^3$ ksi for the steel beam.

6.3-3 Solve Prob. 6.3-2 but include the weight of the steel beam selected.

6.3-4 Consider Example 6.2, Fig. 6.4. If the I-beam section shown in Fig. 6.4*a* is rotated 90° about the x axis and then the q_0 loading applied, find the new maximum deflection. The moment of inertia about the original horizontal axis is $I_z = 48$ in^4 and for the original vertical axis is $I_y = 3.41$ in^4.

6.3-5 Consider Example 6.2, Fig. 6.4. Determine the expressions for the slopes at the ends of the beam, that is, $v'(0)$ and $v'(L)$. Verify your results with those in App. G, Table G.2, Fig. G.2-3.

6.3-6 Consider Example 6.2, Fig. 6.4. Obtain an expression for the maximum stress in the beam in the form

$$\sigma_{max} = \frac{q_0 L^2}{8} \frac{y_{max}}{I} \qquad y_{max} = \frac{h}{2}$$

where h is the depth of the beam. Relate the maximum stress and maximum deflection δ by deriving the relation

$$\frac{\sigma_{max}}{E} = \frac{48}{5} \frac{y_{max}}{L} \left(\frac{\delta}{L}\right)$$

We note that as the maximum deflection δ increases, so does the maximum stress σ_{max}.

6.3-7 Consider Example 6.3, Fig. 6.5. If $a = b = L/2$, obtain the expressions for the slopes at $x = 0$ and $x = L$. Verify that the results are consistent with those given in Table G.2 (App. G), Fig. G.2-2.

6.3-8 Consider Example 6.3, Fig. 6.5. Obtain the expressions for the slopes at $x = 0$ and $x = L$ for arbitrary values of a and b. Verify that the results agree with those given in Fig. G.2-1 (App. G).

6.3-9 A simply supported steel beam of length L as shown in Fig. P6.3-9 is loaded by a concentrated load P. Find the general expression for the deflection curve and the maximum deflection if $a = 5$ ft, $L = 20$ ft, $E = 30{,}000$ ksi, $I = 70$ in^4, and $P = 2$ kips.

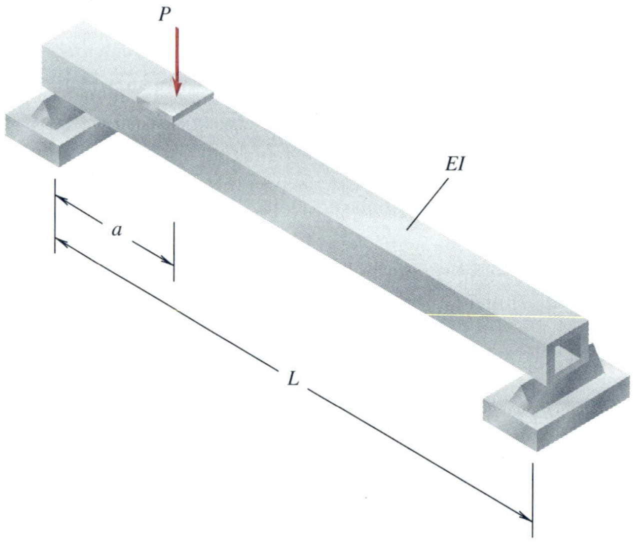

Fig. P6.3-9

6.3-10 Consider Example 6.4, Fig. 6.7. If $L = 4$ m, $q_0 = 5$ kN/m, $I = 2 \times 10^{-5}$ m^4, $E = 200$ GPa, and the depth of the beam is 200 mm, find the values for the maximum deflection and the maximum normal stress due to bending in the beam.

6.3-11 A simply supported beam of length L is loaded by a constant distributed loading over one-half of the span, as shown in Fig. P6.3-11. Determine the expression for the deflection of the beam and the value of the maximum deflection of the beam. Recall the results in Example 6.4.

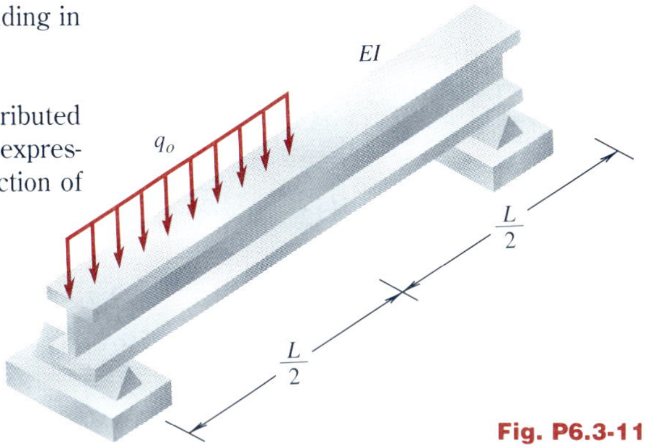

Fig. P6.3-11

6.3-12 A simply supported beam of length L is loaded by a constant distributed loading over a portion b of its length, as shown in Fig. P6.3-12. Determine the expression for the deflection of the beam. Find the value of the maximum deflection of the beam for $a = L/3$. Check the limiting cases of $a = 0$ and $a = L/2$.

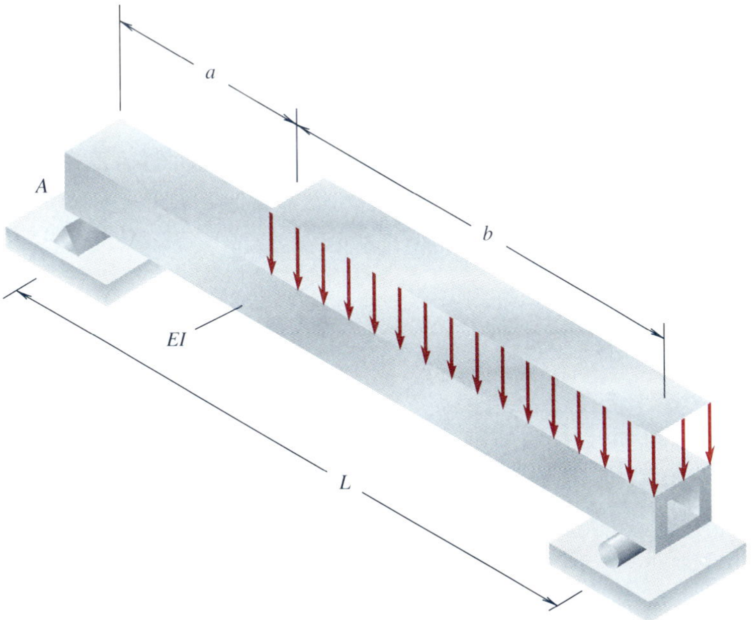

Fig. P6.3-12

6.3-13 Consider Example 6.5, Fig. 6.8. Compare the solution for a/L found in this example to the corresponding result for the location of the supports to give a minimum value for the maximum bending moment, Prob. 4.5-7.

6.3-14 Consider Example 6.5, Fig. 6.8. Verify that the solution given in Eq. (*o*) is correct, i.e., it satisfies Eq. (*n*).

6.3-15 Consider Example 6.6, Fig. 6.9. Obtain an explicit expression for the deflection at point B. Sketch a plot of the deflection as a function of Q with P fixed, and verify that the deflection is zero when $Q = 2P/3$. What is the deflection when $Q = 0$?

6.3-16 Use the program CH6BM on the diskette to solve for the displacement under the load for the simply supported beam shown in Fig. P6.3-16. (To run CH6BM, first go into BASIC and then RUN CH6BM. All sign conventions are consistent with Fig. 6.10.) Take $P = 1000$ lb, $L = 10$ ft, $E = 30{,}000$ ksi, and $I = 10\ \text{in}^4$.

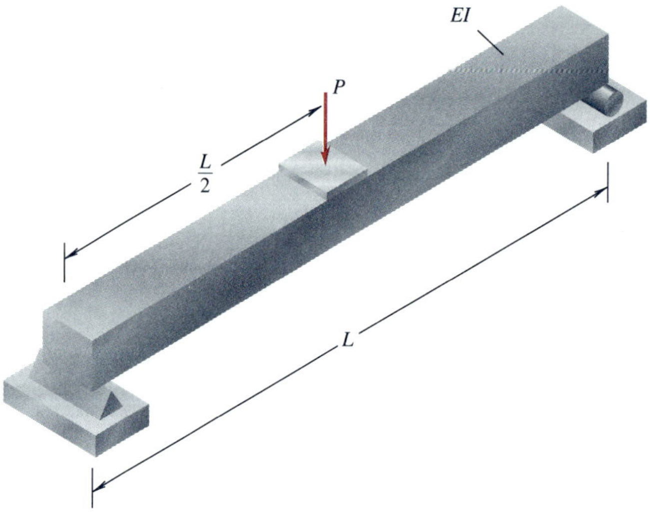

Fig. P6.3-16

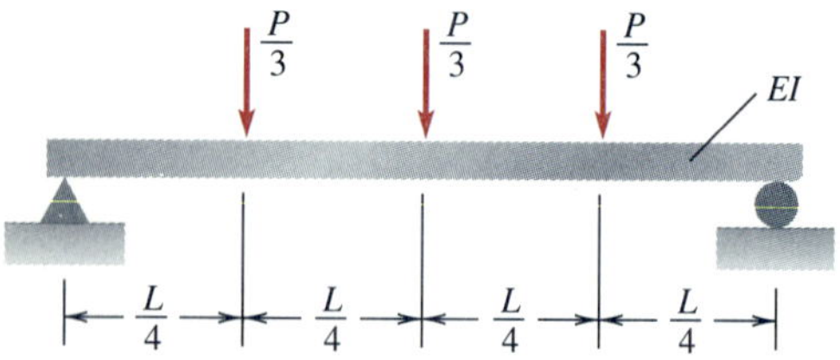

Fig. P6.3-17

6.3-17 Use the program CH6BM on the diskette to determine the midpoint deflection of the simply supported beam shown in Fig. P6.3-17. Take $P = 1000$ lb, $L = 10$ ft, $E = 30{,}000$ ksi, and $I = 10\ \text{in}^4$.

6.3-18 Use the program CH6BM on the diskette to determine the midpoint deflection of the simply supported beam shown in Fig. P6.3-18*a*. Take $P = 1000$ lb, $L = 10$ ft, $E = 30{,}000$ ksi, and $I = 10\ \text{in}^4$.

(*a*) $N = 3$
(*b*) $N = 5$
(*c*) $N = 11$
(*d*) What value of N will give a midpoint deflection within 5 percent of the midpoint deflection for the same beam with a constant distributed loading, as shown in Fig. P6.3-18*b*?

6.3-19 Derive an expression for the maximum deflection of the simply supported beam shown in Fig. P6.3-19 due to a concentrated load P applied at the midpoint. Assume that a is small enough compared to L that the maximum deflection occurs at C. Calculate the ratio of your result to the result given in Eq. (l) in Example 6.3 for the same beam but with $a = 0$.

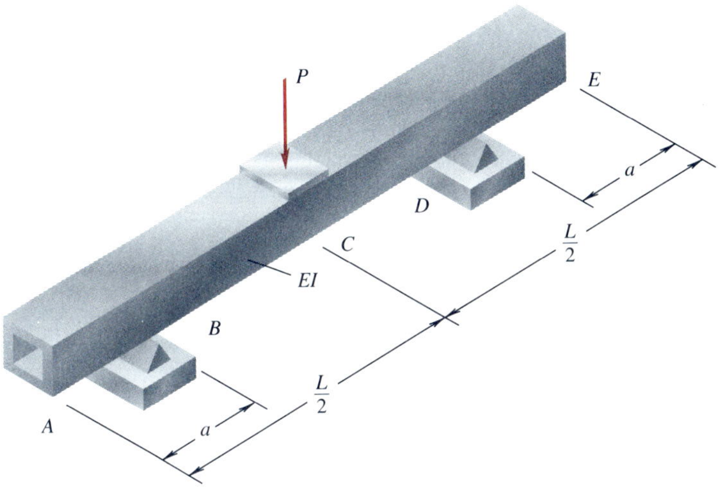

Fig. P6.3-19

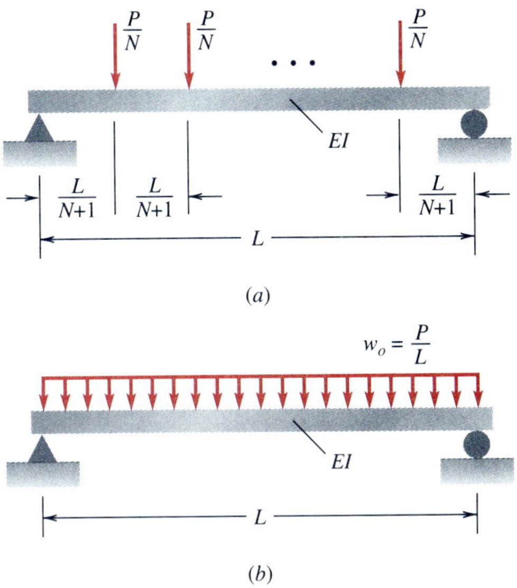

Fig. P6.3-18

6.3-20 A pair of concentrated loads of magnitude P is supported by a beam resting upon a foundation which is assumed to resist the loading of the beam by means of a constant distributed load q_0, as shown in Fig. P6.3-20.

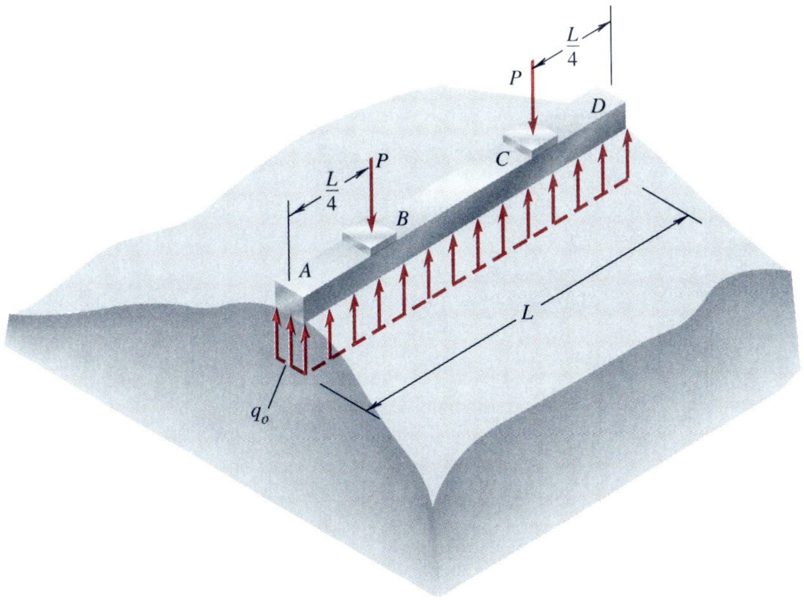

Fig. P6.3-20

(*a*) Use equilibrium to find q_0 in terms of P and L.
(*b*) Find the deflection of the ends of the beam relative to the midpoint of the beam due to the assumed loading. (*Hint:* To establish a reference point, take the deflection equal to zero at the midpoint as well as the slope.)

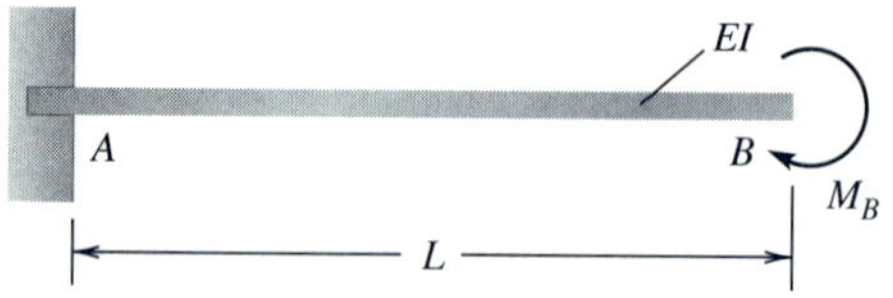

Fig. P6.3-21

6.3-21 A cantilever beam is loaded by a concentrated moment M_B as shown in Fig. P6.3-21. Find the deflection and slope at endpoint B. What are the ratios of the deflection and slope at the midpoint of the beam to the corresponding endpoint values?

6.3-22 A cantilever beam is loaded by a concentrated moment M_B as shown in Fig. P6.3-22. Find the deflection and slope at sections B and C. Draw graphs of the deflection $v(x)$, slope $v'(x)$, and bending moment $M(x)$ over the length of the beam.

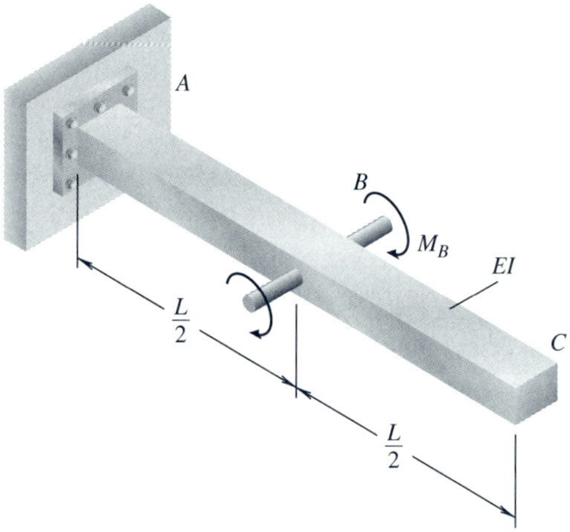

Fig. P6.3-22

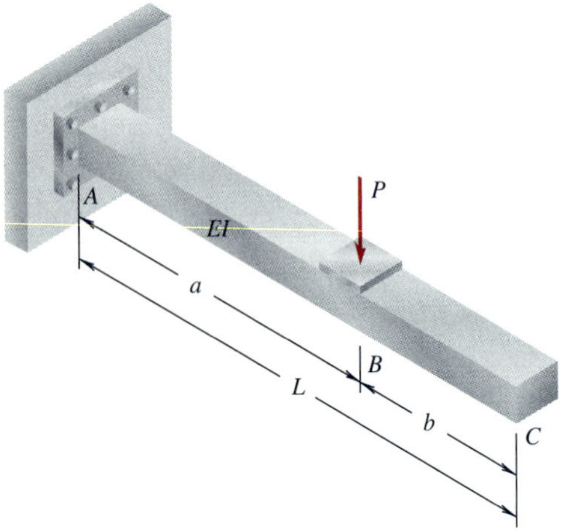

6.3-23 A cantilever beam carries a concentrated load P at B as shown in Fig. P6.3-23. Find the deflection and slope at sections B and C. Draw graphs of the deflection $v(x)$, slope $v'(x)$, and bending moment $M(x)$ over the length of the beam.

6.3-24 A simply supported beam carries a constant distributed load q_0 over a central portion of the beam as shown in Fig. P6.3-24.
(*a*) Find the maximum deflection of the beam due to this loading.
(*b*) Show that as $a \rightarrow L$, that is, $b \rightarrow 0$, the maximum deflection found in part (*a*) approaches the value given by Eq. (*i*) in Example 6.2.

Fig. P6.3-23

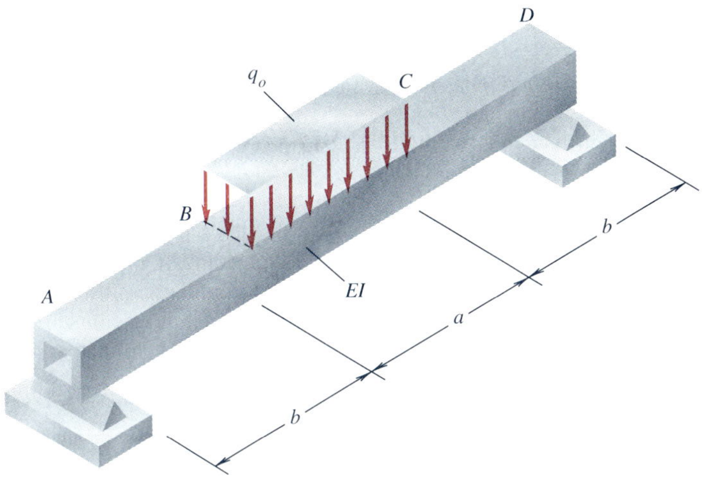

Fig. P6.3-24

6.3-25 A cantilever beam carries a constant distributed load q_0 over the portion BC of the beam as shown in Fig. P6.3-25.

(*a*) Find the slope and deflection at tip C.

(*b*) Show that as b becomes small, the results of part (*a*) approach the corresponding results given in Eq. (*h*), Example 6.1, where $W = q_0 b$.

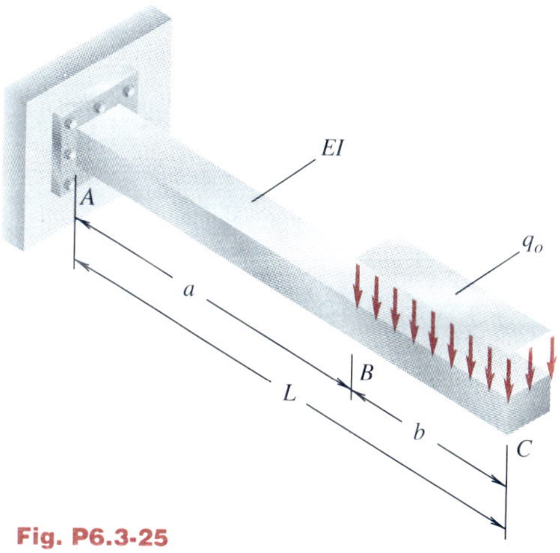

Fig. P6.3-25

6.3-26 A timber construction standard for "an allowable roof load limited by deflection" uses the maximum deflection result for a simply supported beam under a constant distributed loading. For a 2-in-nominal-thickness (actual dimension 1.5 in) roof member, calculate the allowable constant distributed roof loading p lb/ft^2 for

(*a*) $L = 6$ ft, $E = 1000$ ksi, under a standard of a maximum deflection/span $= \frac{1}{180}$;

(*b*) $L = 10$ ft, $E = 1500$ ksi, and the same maximum deflection/span ratio as in part (*a*). Use the results of Example 6.2, and exercise care with units of loading.

6.3-27 For the purposes of deflection analysis, a machine component is modeled as a rigid central beam segment with flexible beam segments on each side, as shown in Fig. P6.3-27. Take advantage of symmetry and the rigidity of segment BD to find the value of the deflection under the load P.

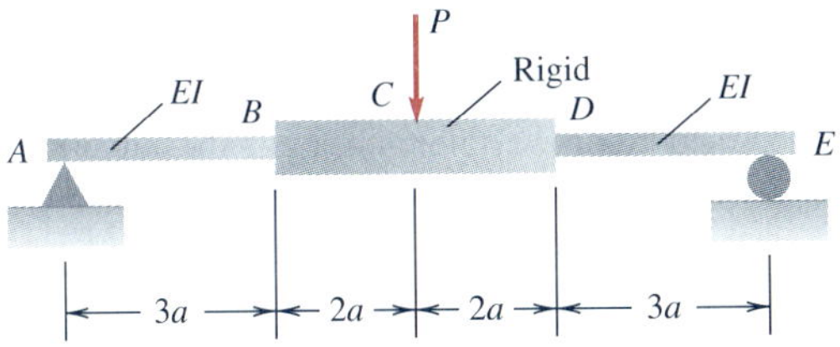

Fig. P6.3-27

6.3-28 A flexible beam *ABC* is pinned at *A* and supported at *B* by a steel circular rod of diameter $\frac{1}{4}$ in, as shown in Fig. P6.3-28. Find the deflection at *C* if *ABC* has a circular section with a diameter of 2.5 in and $P = 300$ lb. (*Hint:* Use equilibrium to find the force in *BD* and then the elongation of *BD*. With the deflection known at *A* and *B*, the deflection curve for *ABC* can be found.) Take $E = 30 \times 10^6$ psi for *BD* and *ABC*.

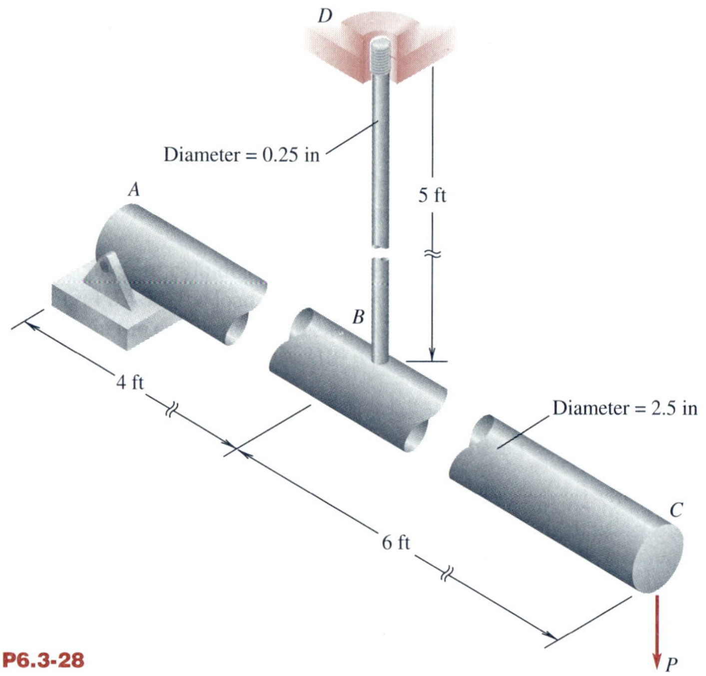

Fig. P6.3-28

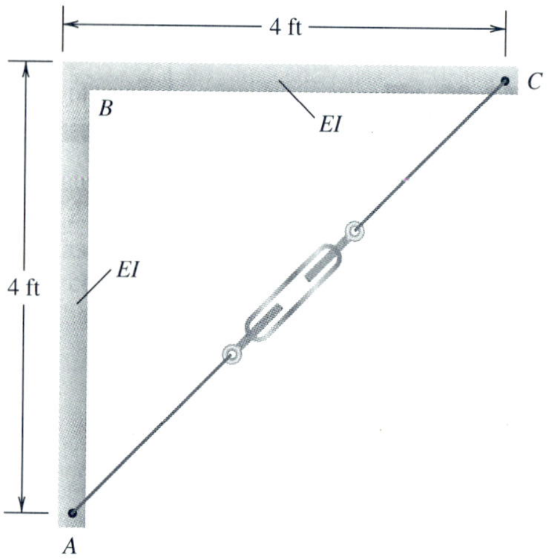

Fig. P6.3-29

6.3-29 Two wood members are joined to form the right-angle piece shown in Fig. P6.3-29. Each member is 2 in × 2 in in cross section and 4 ft in length. Find the maximum allowable shortening of a very stiff wire joining points *A* and *C* if the allowable bending stress in the wood is 1000 psi and $E = 1.5 \times 10^6$ psi. Neglect the effect of axial stresses and axial deformation in the wood pieces.

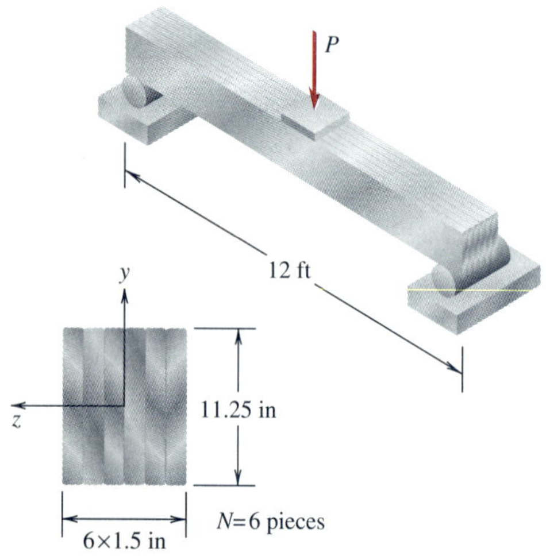

Fig. P6.3-30

6.3-30 A series of wood beams are to be fabricated out of *N* pieces of structural lumber, each having nominal dimensions of 2 in × 12 in. (See App. D.) The cross section for the case $N = 6$ is shown in Fig. P6.3-30. If these beams are to be used in applications where the ends are simply supported, the span is 12 ft, and a concentrated load *P* is applied at midspan, find the allowable load P_a if the allowable maximum normal bending stress is 1200 psi and the allowable maximum deflection is the span length/240, or 0.6 in in this case. Plot P_a versus *N* for $N = 1, 3, 5$, and 9. Take $E = 1.5 \times 10^6$ psi.

6.3-31 Four different cross-sectional configurations for a wood beam made up of four nominal-dimension 2-in × 10-in (see App. D) pieces of structural

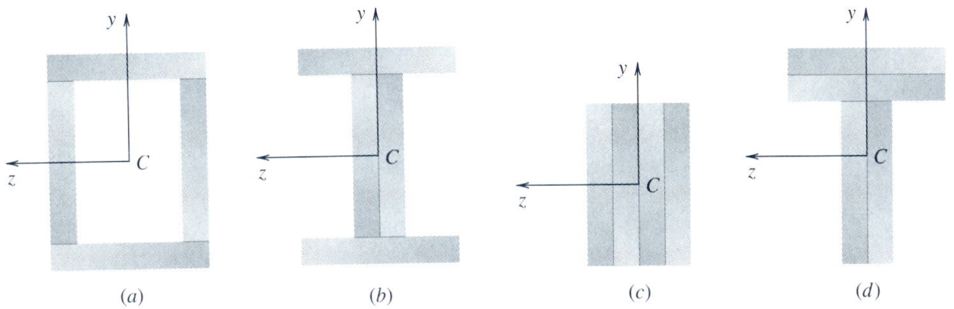

Fig. P6.3-31

lumber are shown in Fig. P6.3-31. If the beam is simply supported at its ends and carries a concentrated load P at midspan and if the length is 10 ft, then calculate the allowable load P_a for all four cases shown in Fig. P6.3-31 when the allowable normal stress due to bending is 1000 psi and the allowable maximum deflection is 0.5 in. Take E = 1500 ksi. Which of the four designs carries the largest load P? Comment on this result.

6.3-32 To estimate the elastic modulus of silicon coatings, a small cantilever beam is etched from the silicon substrate, as shown in Fig. P6.3-32. The cross section of the beam is as shown, and L = 60 μm. If the load P is 1.2 μN, estimate the elastic modulus of the silicon given that the deflection at the tip is 0.5 μm ($\mu = 10^{-6}$).

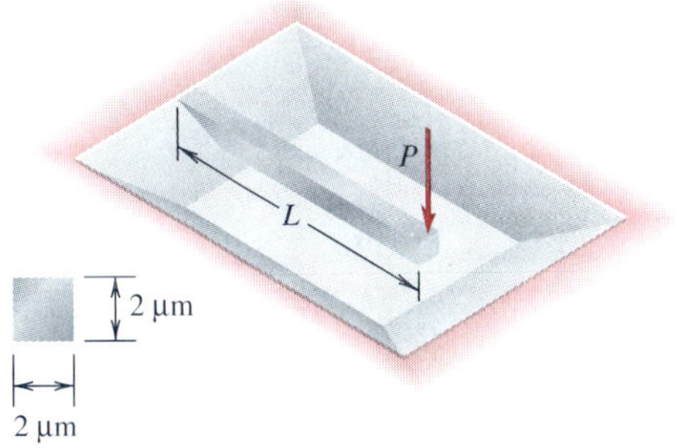

Fig. P6.3-32

6.3-33 Obtain an expression for the deflection of the neutral axis for the beam with bending modulus EI as shown in
(*a*) Fig. P4.2-3*f*
(*b*) Fig. P4.2-5*a*
(*c*) Fig. P4.2-5*d*
(*d*) Fig. P4.4-6*a*
(*e*) Fig. P4.4-6*c*
(*f*) Fig. P4.4-10*b*
(*g*) Fig. P4.4-10*g*
(*h*) Fig. P4.5-5
(*i*) Fig. P4.5-7
(*j*) Fig. P4.5-8
(*k*) Fig. P4.5-22

Use the method of direct integration of the load-deflection equation to solve the problems assigned for Sec. 6.4. The bending modulus for all beams is EI_z, and the weight of the beam is to be neglected unless otherwise noted.

6.4-1 If the bending modulus EI is a function of x, obtain expressions for $V(x)$ and $q(x)$ in terms of the derivatives of v and EI, making use of Eqs. (6.17) and (6.19). Write out your results for the special case where $EI(x) = EI_0\,[1 - x/(2L)]$.

6.4-2 Give arguments for the validity of the expressions for the boundary conditions given in Fig. 6.13.

6.4-3 Consider Example 6.7, Fig. 6.14. Make use of the shear force and moment reactions at the supports to show that the constants a_1 and a_2 in Eqs. (*c*) and (*d*) are zero.

6.4-4 Consider Example 6.9, Fig. 6.16. Take $q_0 = 50$ lb/in, $L = 96$ in, and $EI = 450 \times 10^6$ lb · in². Determine the maximum deflection of the beam, and compare it to a simply supported beam of the same length L with a constant distributed loading of 25 lb/in acting over the entire length.

6.4-5 Consider Example 6.10, Fig. 6.17. If $q_0 = 80$ lb/in, $E = 30{,}000$ ksi, and $L = 10$ ft, find a square steel tube beam (Table C.3) such that the maximum deflection-to-length ratio is $\frac{1}{200}$.

6.4-6 Find the maximum normal bending stress and the maximum shear stress due to bending in the tube selected in Prob. 6.4-5.

6.4-7 Consider Example 6.11, Fig. 6.18. Determine the explicit expression for the deflection, using Eqs. (*e*) and (*h*), and verify the result given in App. G, Table G.2, Fig. G.2-4. Determine the expression for the maximum deflection of the beam, and compare with the result in Fig. G.2-4.

6.4-8 Consider Example 6.11, Fig. 6.18. Verify that the results given in Eqs. (*i*) to (*m*) are correct, and plot the shape of the deflected beam when $a/L = \frac{2}{3}$.

6.4-9 Consider Example 6.11, Fig. 6.18. Find the value of $a/L > \frac{1}{2}$ such that for values of a/L greater than this transition value the deflection is negative over the entire length of the beam.

6.4-10 Consider Example 6.11, Fig. 6.18. For what value a/L is the normal bending stress a maximum, and what is the corresponding maximum deflection of the beam?

6.4-11 Consider Example 6.12, Fig. 6.19. Sketch the deflected shape of the beam.

6.4-12 A cantilever beam of length $2a$ is loaded by two opposing constant distributed loadings q_1 and q_2, as shown in Fig. P6.4-12. Obtain the expression for the deflection of the beam, and investigate the effect of the ratio q_1/q_2 on the value of the deflection at section B.

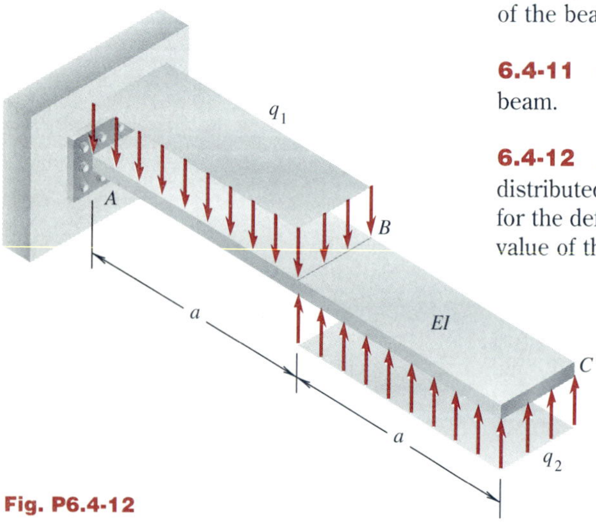

Fig. P6.4-12

6.4-13 Consider Example 6.13, Fig. 6.20. Sketch to scale the shape of the beam deflection curve.

6.4-14 A beam *ABCD* is loaded as shown in Fig. P6.4-14. Find the value of the constant distributed loading q_0 such that the deflection under the 20-kip load is equal to the negative of the deflection at *D*, that is, $v_B = -v_D$.

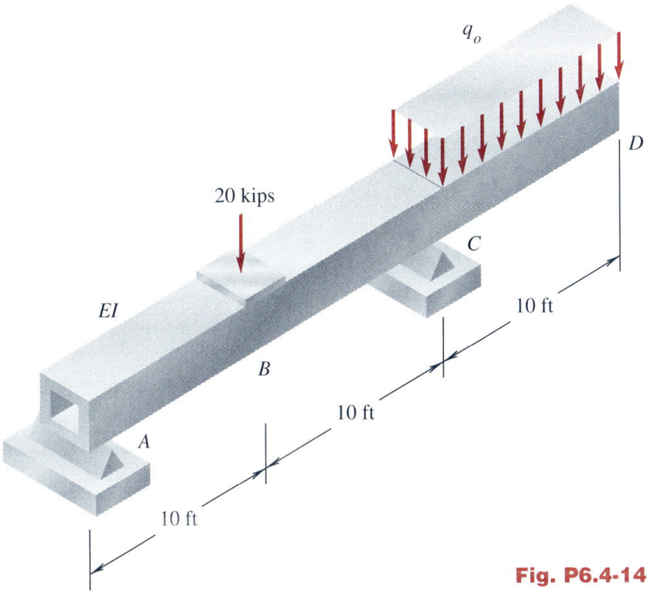

Fig. P6.4-14

6.4-15 A simply supported beam carries the end moment M_A as shown in Fig. P6.4-15. Find the maximum values of the slope and deflection. Verify the results given in Fig. G.2-5 (Table G.2, App. G).

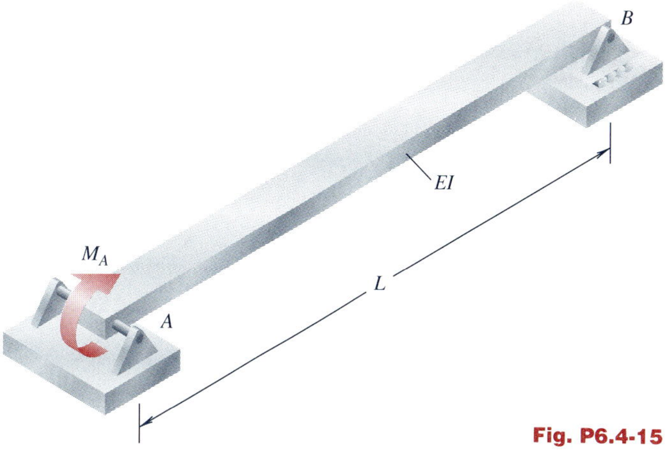

Fig. P6.4-15

6.4-16 A simply supported beam has a vertical member BC fixed to it at point B, as shown in Fig. P6.4-16. A load P is applied at point C as shown. Find the vertical deflection and the horizontal displacement at section B due to load P. Let A be the cross-sectional area of the beam.

Fig. P6.4-16

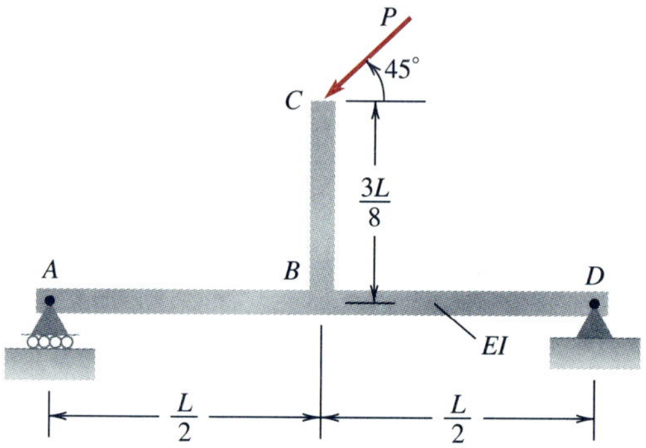

6.4-17 A simply supported beam is subjected to a triangle-shaped distributed loading, as shown in Fig. P6.4-17. Find the expression for the maximum deflection and slope for this beam.

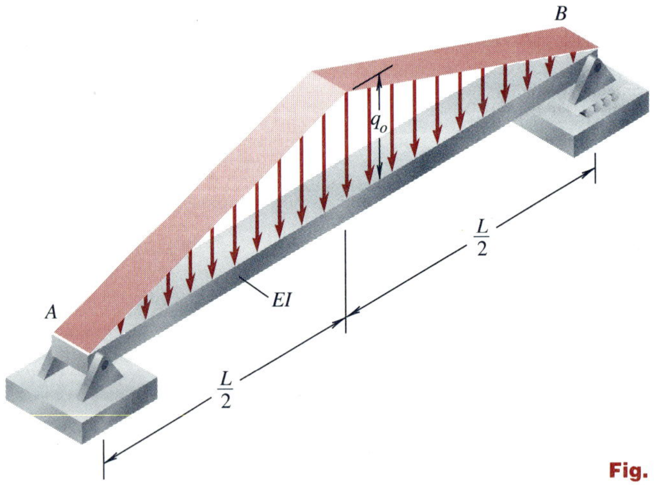

Fig. P6.4-17

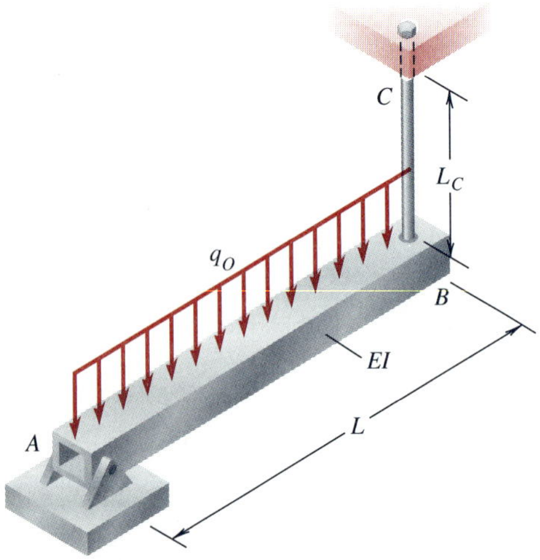

Fig. P6.4-18

6.4-18 A beam is simply supported at A and supported by an elastic cable CB of cross-sectional area A, length L_C, and elastic modulus E, as shown in Fig. P6.4-18. If the beam is subjected to a constant distributed load q_0, find an expression for the deflection curve $v(x)$. In particular, find expressions for the slope and deflection at B.

6.4-19 A beam AB is supported by two steel wires (cross-sectional area A and elastic modulus E) and carries a constant distributed loading q_0, as shown in Fig. P6.4-19. The weight of the beam is included in q_0. Find an expression for the deflection at the midpoint of the beam due to the bending of the beam (bending modulus EI) and the stretching of the wires.

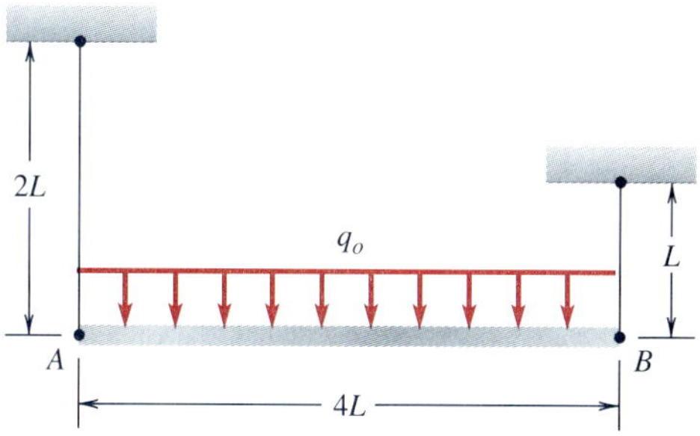

Fig. P6.4-19

6.4-20 Determine the expression for the deflection of the beam shown in Fig. P6.3-20 by direct integration of the load-deflection equation.

6.4-21 Determine the expression for the deflection of the beam shown in Fig. P6.3-21 by direct integration of the load-deflection equation.

6.4-22 Determine the expression for the deflection of the beam shown in Fig. P6.3-22 by direct integration of the load-deflection equation.

6.4-23 Determine the expression for the deflection of the beam shown in Fig. P6.3-23 by direct integration of the load-deflection equation.

6.4-24 Determine the expression for the deflection of the beam shown in Fig. P6.3-24 by direct integration of the load-deflection equation.

6.4-25 Determine the expression for the deflection of the beam shown in Fig. P6.3-25 by direct integration of the load-deflection equation.

6.4-26 In a particular application, long, single pieces of unwelded steel cylinders are to be used as shown in Fig. P6.4-26. In the design process, two different cylinders with outside diameters of 60 in, length 36 ft, and wall thicknesses of 1.5 and 3.5 in are to be compared.

(*a*) Calculate the maximum normal stress and maximum deflection due to the bending of these simply supported cylinders under the loading of self-weight. Take the specific weight for steel as 490 lb/ft^3.

(*b*) If the cylinder contains a fluid with a specific weight of 25 percent that of steel, redo part (*a*) and calculate the stresses and deflections for the two cylinders under consideration. Take $E = 30 \times 10^6$ psi. [*Hint:* The results in part (*b*) can be found by "scaling up" the values found in part (*a*).]

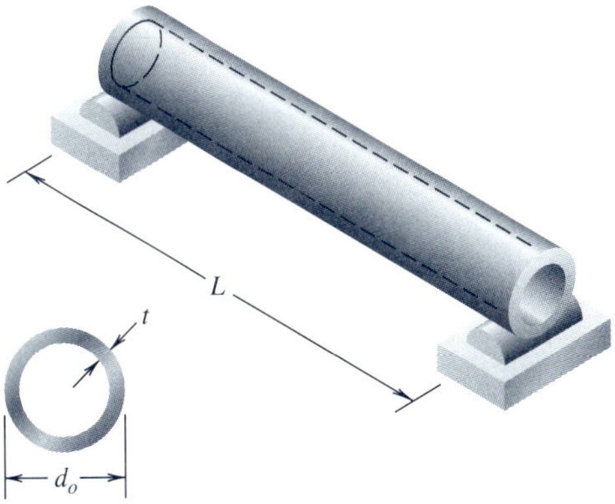

Fig. P6.4-26

6.4-27 A simply supported beam carries a distributed loading $q(x)$ that varies from zero to q_0 and back to zero over the central part of the beam, as shown in Fig. P6.4-27. Find an expression for the maximum deflection of the beam.

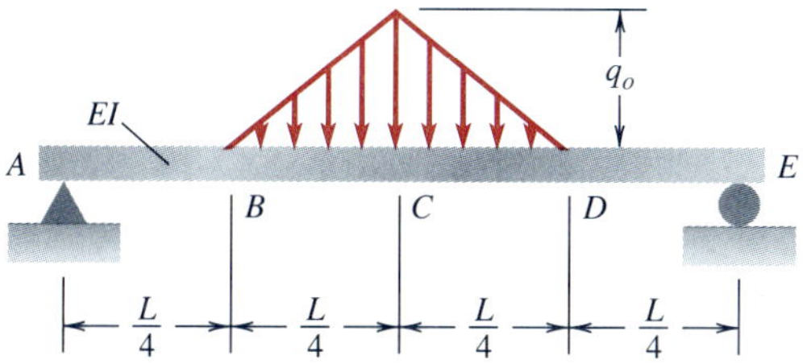

Fig. P6.4-27

6.4-28 A wood diving board is pinned at *A*, supported by a spring at *B*, and carries the weight *W* of a diver at *C*, as shown in Fig. P6.4-28. It is given that $W = 150$ lb, and the spring stiffness $k = 200$ lb/in.

(*a*) Model beam *ABC* as a rigid beam, and calculate the deflection at *C* due to *W*.

(*b*) Repeat part (*a*), but model beam *ABC* as a flexible beam. Take $E = 1500$ ksi, and the actual dimensions of the board are 2 in × 12 in.

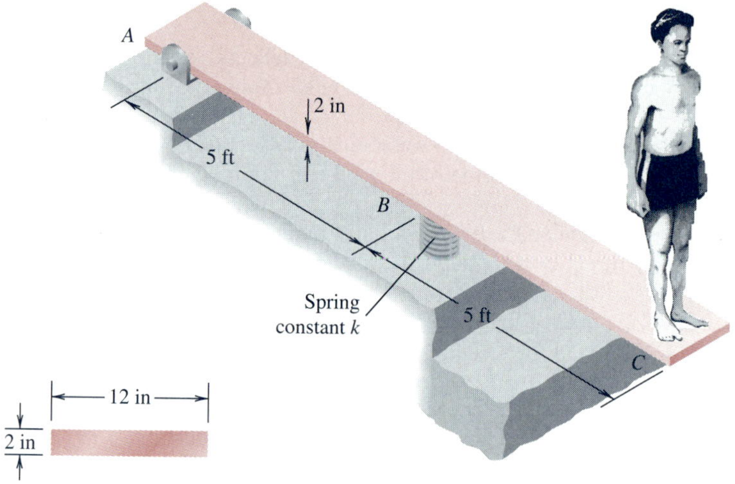

Fig. P6.4-28

6.4-29 A 24-ton utility vehicle is loaded onto a trailer bed, as shown in Fig. P6.4-29. The loads are supported by two W12 × 50 wide-flange I beams as shown. If one-third of the vehicle weight is carried by the wheels at *B* and the rest is carried at *C*, find the location and value of the maximum normal stress due to bending and the maximum deflection for the I beams. Take $E = 30 \times 10^6$ psi.

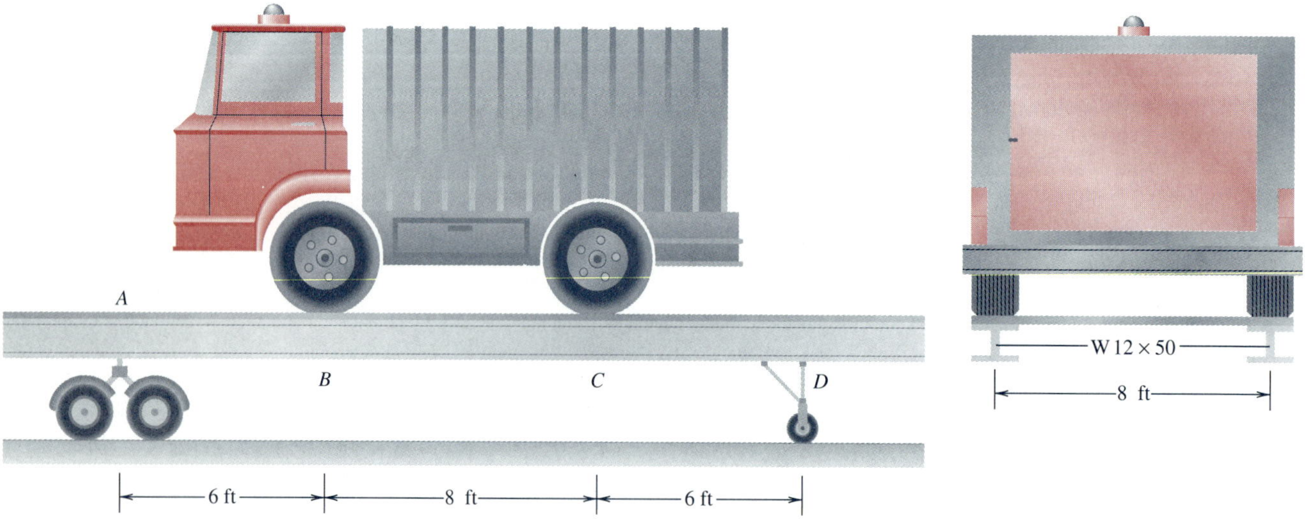

Fig. P6.4-29

6.4-30 Consider a homogeneous tapered cantilever beam as shown in Fig. P6.4-30 with a bending modulus given by $EI = EI_0(1 - x/L)$. Show that the deflection at B due to the load P is 1.5 times the value for the corresponding deflection for a beam with constant $EI = EI_0$.

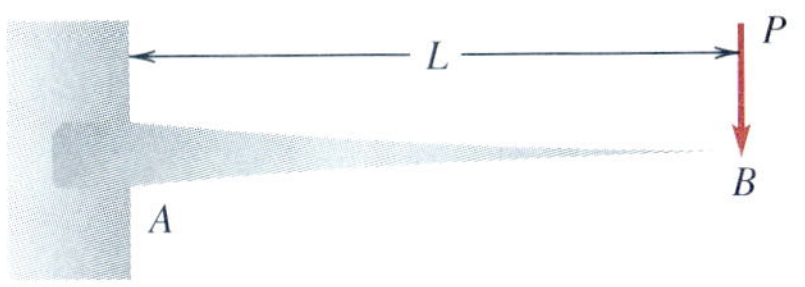

Fig. P6.4-30

6.4-31 A rod ABC of diameter d is to be bent by two stiff wires AD and CD, as shown in Fig. P6.4-31. By using the turnbuckles the wires are both shortened by an amount Δ, where $\Delta = \beta L$. Find an expression for the downward deflection of A relative to point B and the maximum normal stress due to bending in rod ABC in terms of the parameters of the problem. Member BD is rigid.

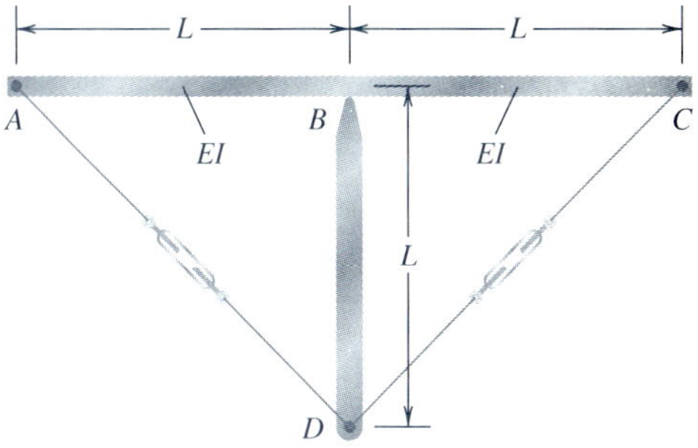

Fig. P6.4-31

6.4-32 A beam is attached to an elastic foundation as shown in Fig. P6.4-32a. The free-body diagram of an element of length Δx is shown in Fig. P6.4-32b, where the reaction of the elastic foundation is taken as an equivalent spring force $-kv(x)\,\Delta x$. The units of k are force/(length)^2.

(a) Follow the steps given in Secs. 4.4 and 4.6 for the derivation of the load-deflection equation, but now with the inclusion of the $-kv(x)$ term in the force equilibrium equation. Show that Eq. (6.20) generalizes in this case to

$$EI\frac{d^4v}{dx^4} + kv = q \tag{a}$$

(b) Show that the solution of Eq. (a) for $q = 0$ is

$$v = e^{px}(c_1 \cos px + c_2 \sin px) + e^{-px}(c_3 \cos px + c_4 \sin px) \tag{b}$$

where $p^4 = k/(4EI)$.

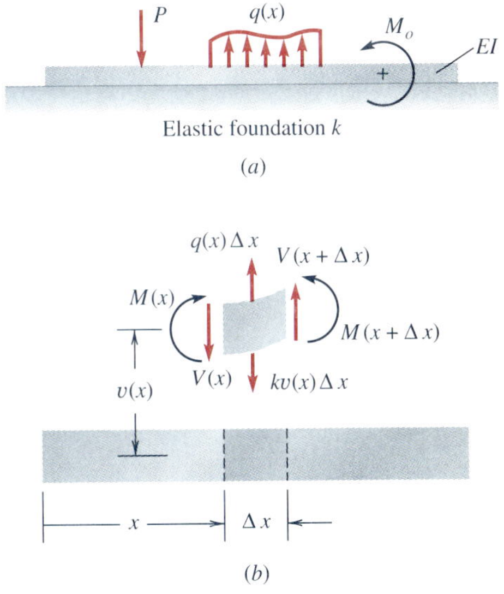

Fig. P6.4-32

6.4-33 A long beam on an elastic foundation is subjected to a concentrated load P acting at section O as shown in Fig. P6.4-33.

(a) Find the deflection v_0 under the load by applying the results of Prob. 6.4-32 to the right half of the beam ($x > 0$) and making use of symmetry conditions and the requirement that v and its derivatives approach zero as $x \to \infty$.

(b) Find the value of the bending moment at $x = 0$.

6.4-34 A long beam on an elastic foundation is subjected to a concentrated moment M_0 at the section O as shown in Fig. P6.4-34. Apply the solution given

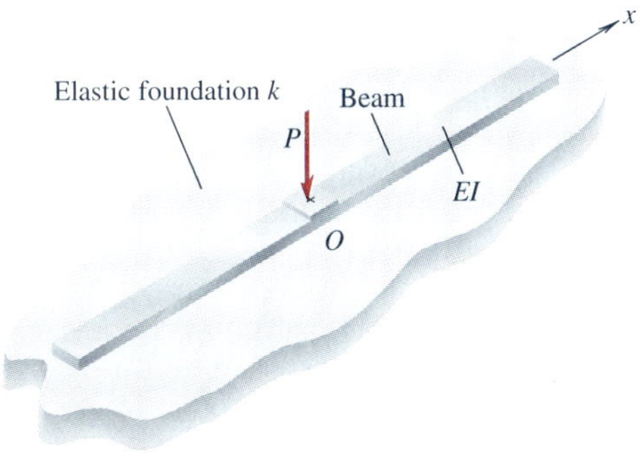

Fig. P6.4-33

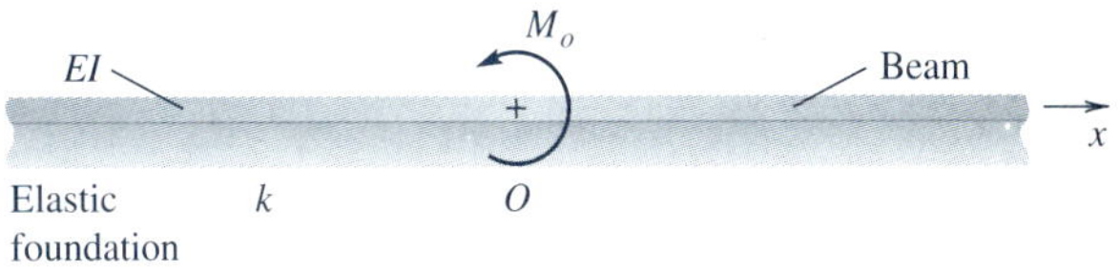

Fig. P6.4-34

in Prob. 6.4-32 to the right half of the beam ($x \geq 0$) to find the slope v_0' and shear force V_0 at point O. (*Hint:* See Prob. 6.4-33 for boundary conditions satisfied as $x \to \infty$, and note that $v = 0$ at $x = 0$.)

6.4-35 A 36-ft ladder is restrained in the x and y directions at its base A, rests against a smooth surface at C, and is inclined at 30° with the vertical, as shown in Fig. P6.4-35*a*.

(*a*) We wish to estimate the horizontal and vertical deflections under the load of a 200-lb person standing in each of positions B and D. The 200 sin 30° = 100-lb component of the weight load bends the ladder and is divided into 50-lb forces acting transverse to the two side beams of the ladder, as shown in Fig. P6.4-35*b*.

(*b*) Calculate the maximum normal stress due to bending in the ladder for the two cases. Neglect the axial deformation in the ladder beams. The side beams of the ladder are nominal-dimension 5-in × 2-in structural tubing of thickness $\frac{1}{4}$ in (Fig. P6.4-35*c*). Take E = 10,000 ksi.

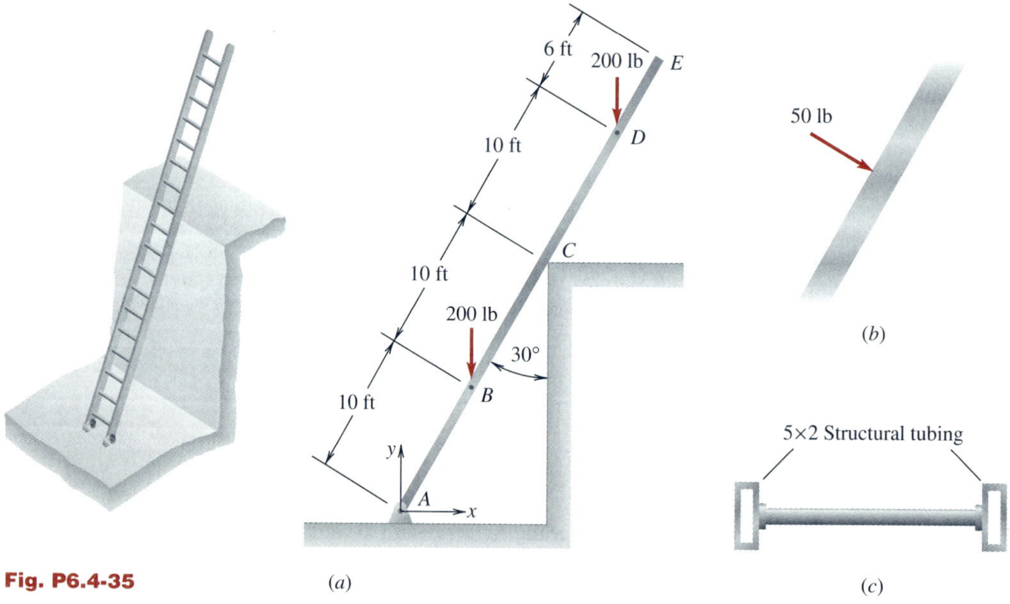

Fig. P6.4-35

6.4-36 Determine the expression for the deflection of the neutral axis for the beam with bending modulus EI as shown in

(*a*) Fig. P6.5-3
(*b*) Fig. P6.5-4
(*c*) Fig. P6.5-18
(*d*) Fig. P6.5-19
(*e*) Fig. P6.5-20
(*f*) Fig. P6.5-21

6.4-37 Obtain the expression for the deflection of the neutral axis for the beam with bending modulus EI as shown in
(*a*) Fig. P4.2-3*a*
(*b*) Fig. P4.2-3*b*
(*c*) Fig. P4.2-3*c*
(*d*) Fig. P4.2-3*d*
(*e*) Fig. P4.2-3*e*
(*f*) Fig. P4.2-3*f*
(*g*) Fig. P4.2-3*g*
(*h*) Fig. P4.4-6*a*
(*i*) Fig. P4.4-6*b*
(*j*) Fig. P4.4-6*c*
(*k*) Fig. P4.4-6*d*

Use the method of superposition to solve the problems associated with Sec. 6.5. The bending modulus for all beams is EI_z, and the weight of the beam is to be neglected unless otherwise noted.

6.5-1 Consider Example 6.15, Fig. 6.22*a*. If $b = 0.3L$, obtain the deflection at the end of the beam and compare it to the value given in Eq. (*e*). Obtain the expression for the deflection $v(x)$ when b is arbitrary, and investigate the shape of the deflection curve as b ranges from 0 to L.

6.5-2 Consider Example 6.16, Fig. 6.23.
(*a*) If we set $E_1I = \infty$, that is, we consider segment AB to be rigid and segment BC to be flexible, what is the value of the deflection at C?
(*b*) If we set $E_2I = \infty$, with E_1I finite, then what is the value of the deflection at C? Can the results found for cases (*a*) and (*b*) be obtained as limiting cases of Eq. (*e*)? Carry out a similar study for the slope at C when segments of the beam are taken as rigid.

6.5-3 A simply supported beam carries a constant distributed load q_0 and an end moment M_B as shown in Fig. P6.5-3. Find the value of the end moment M_B such that the midpoint of the beam will not deflect due to the combined loading.

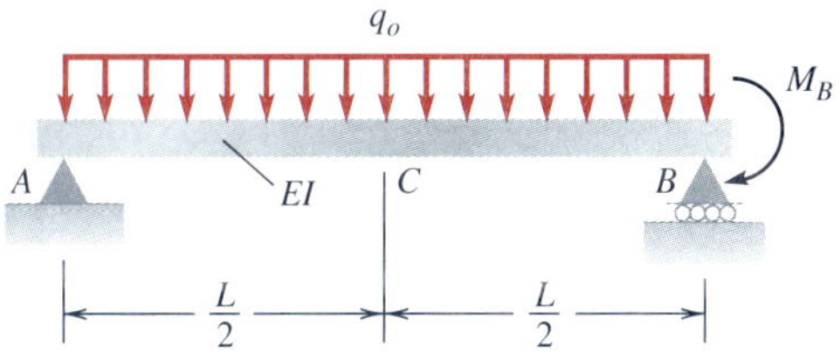

Fig. P6.5-3

6.5-4 A cantilever beam carries two loads, each of magnitude $P/2$, located at the midpoint and endpoint of the beam, as shown in Fig. P6.5-4. Find the slope and deflection at endpoint C. Compare the results for this problem with those found in Example 6.1, where the total load is located at C. Calculate the ratios of the results to those found in Example 6.1.

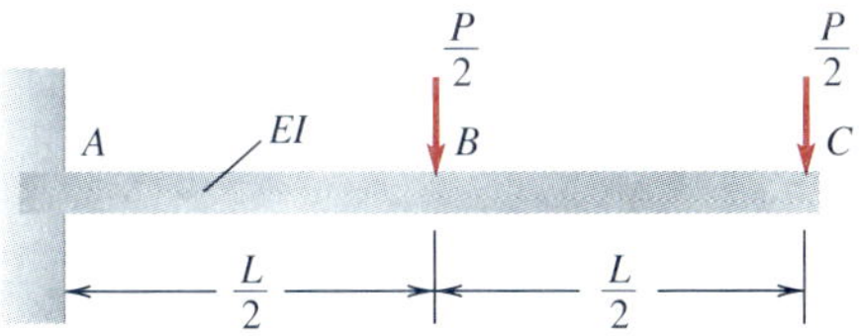

Fig. P6.5-4

6.5-5 Find the vertical deflection at B for beam ABD in Fig. P6.4-16.

6.5-6 The deflection response of a simply supported beam under distributed loading $q(x)$ can be found by superimposing the responses due to infinitesimal concentrated loads $q(x)\,dx$ integrated over the full length of loading, as shown in Fig. P6.5-6. For the case of a beam loaded by a constant distributed load $q(x) = q_0$ over its entire length, show that the result given in Example 6.2 for the maximum deflection of this beam can be obtained by using the result in Table

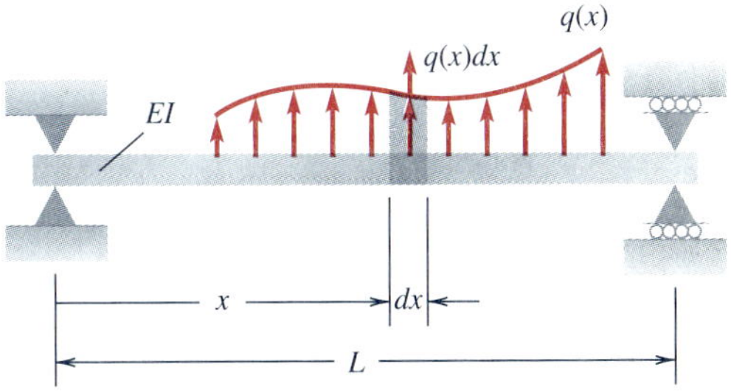

Fig. P6.5-6

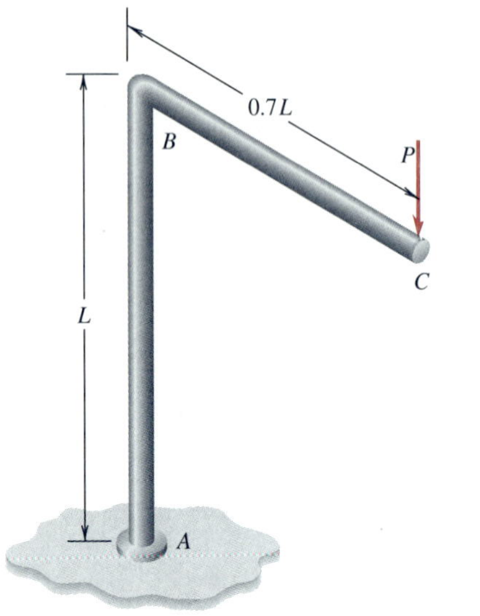

Fig. P6.5-7

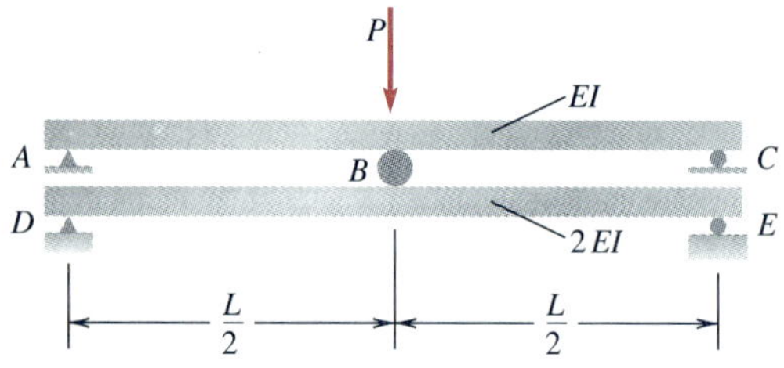

Fig. P6.5-8

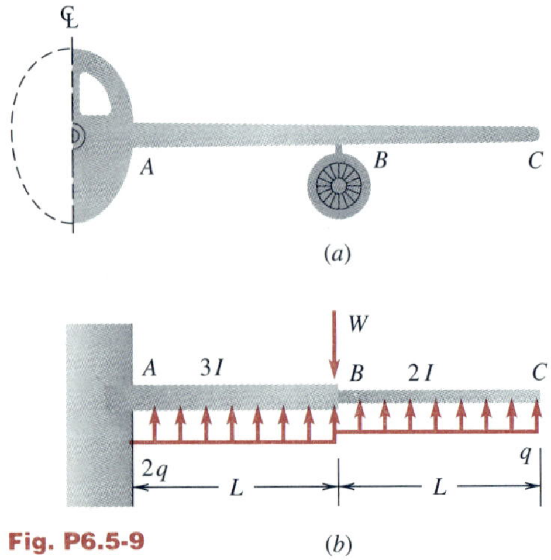

Fig. P6.5-9

G.2 for the deflection of a simply supported beam due to a concentrated load acting at an arbitrary point along the beam (Fig. G.2-1).

6.5-7 A frame ABC carries a vertical load P at C, as shown in Fig. P6.5-7. If steel piping of nominal diameter 4 in (extra strong) is to be used to fabricate this frame, use the results in App. E for section properties of such piping to calculate the horizontal and vertical displacements of section C in terms of load P. Take $L = 8$ ft and $E = 30 \times 10^6$ psi, and neglect the axial elongation in the pipes.

6.5-8 For the beam configuration shown in Fig. P6.5-8, the upper beam carries a concentrated load P at B.

(*a*) If both beams are simply supported at their ends and the lower beam has a moment of inertia twice that of the upper beam, find the deflection under the load. Compare your result with the deflection at B for the case where no contact is made with the lower beam.

(*b*) Compare the value of the maximum normal stress due to bending in each beam with the value of the maximum bending stress for the case of the upper beam carrying the load P alone. Take the depth of each beam as h.

6.5-9 In a preliminary design analysis of the deflection of the wing of a jetplane (Fig. P6.5-9*a*), the cantilever beam model shown in Fig. P6.5-9*b* is to be used, where the distributed forces model the lift forces on the wing and W is the engine weight. If $W = 2qL$, then find the deflections and slopes at sections B and C. Take E as the elastic modulus for AB and BC. (*Hint:* Adapt the solution method of Example 6.16 to this problem.)

6.5-10 A steel strip is to be used in a spring-loaded device. The strip is to be bent and held in shape by three rods, as shown in Fig. P6.5-10. The strip is 1 in wide and $\frac{1}{4}$ in thick, and the rigid rods are $\frac{1}{8}$ in in diameter and are aligned on centers. Find the minimum value for a so that the maximum normal stress due to bending in the strip does not exceed an allowable stress of 20 ksi. Given: $E = 30 \times 10^6$ psi.

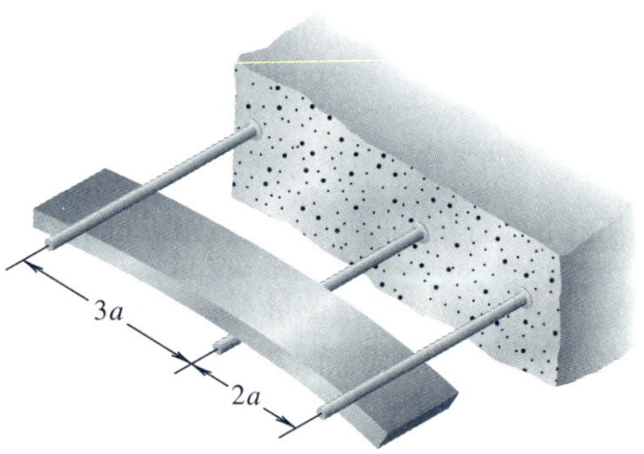

Fig. P6.5-10

6.5-11 A wood board of nominal dimensions 2 in × 8 in (App. D) is to be used as a simply supported beam with the long dimension of the cross section oriented in the vertical direction. If a constant distributed loading of $q_0 = 50$ lb/ft is to be applied, the allowable maximum deflection is limited to $\frac{1}{360}$ of the span, and the maximum normal and shear stresses are 1600 and 100 psi, then find the maximum span L that can be allowed. Take $E = 1500$ ksi.

6.5-12 A cantilever beam is made up of two I beams welded at B and carrying constant distributed loadings, as shown in Fig. P6.5-12. Find the deflection and slope at the free end C. The moments of inertia for the two beam segments differ approximately by a factor of 2. (*Hint:* Adapt the solution method of Example 6.16 to this problem.)

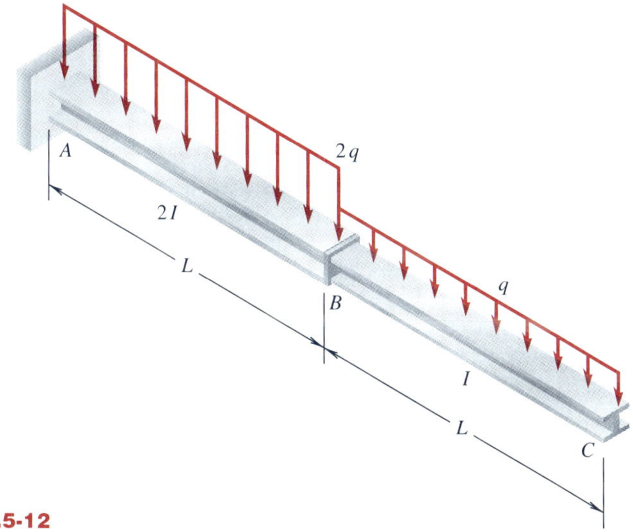

Fig. P6.5-12

6.5-13 A machine of weight W is suspended by a mechanical spring system that consists of four identical flat steel bars clamped to rigid supports at A and D and clamped to the rigid machine at B and C, as shown in Fig. P6.5-13. Assuming that the center of gravity of the machine is midway between points B and C, find the coefficient α in the expression for the downward deflection of the machine

$$v = \alpha \frac{WL^3}{EI}$$

where EI is the bending modulus for the bars. How does α vary with N, where N is the total number of bars?

6.5-14 A cantilever beam is clamped at A and carries a load P at C, as shown in Fig. P6.5-14.

(*a*) If a rigid support B is located a distance $\Delta = L/100$ below the beam at the midpoint B, find the load P_1 which will cause initial contact with the support at B.

Fig. P6.5-13

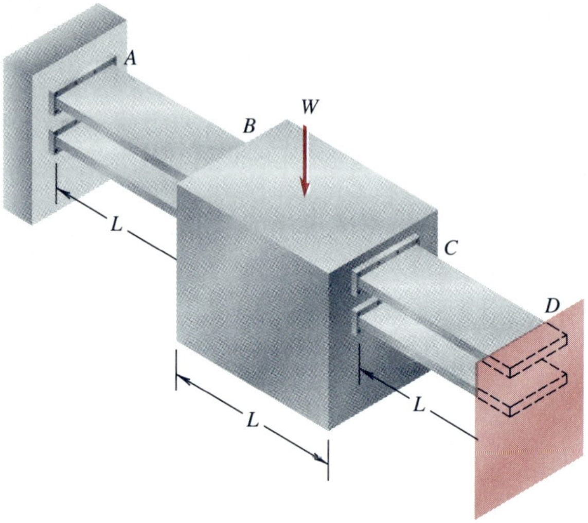

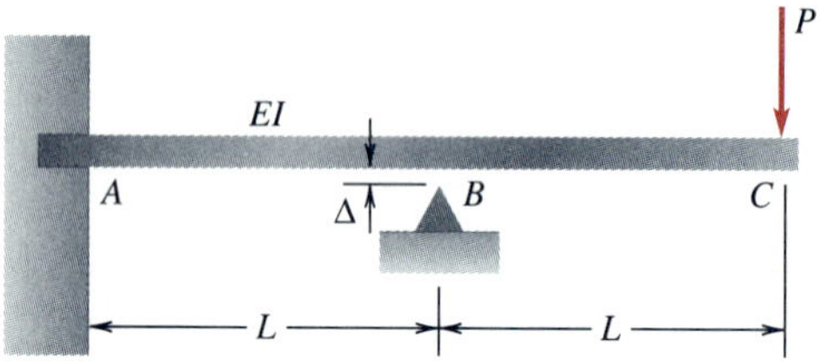

Fig. P6.5-14

(*b*) If the stiffness of the beam at C is defined as the ratio of the applied load to the deflection, then find the stiffness at C for $P < P_1$ and $P > P_1$.

6.5-15 A square frame is deformed by the action of stiff wires and a turnbuckle attached at the midpoints of the top and bottom members of the frame, as shown in Fig. P6.5-15. If the distance between points E and F is reduced by an amount Δ, find the maximum deflection of the side members and the slope of AB at A. Neglect the axial deformation of the members, and assume that the right-angle corners before loading remain right angles after loading. The bending modulus is EI_z for all members of the frame.

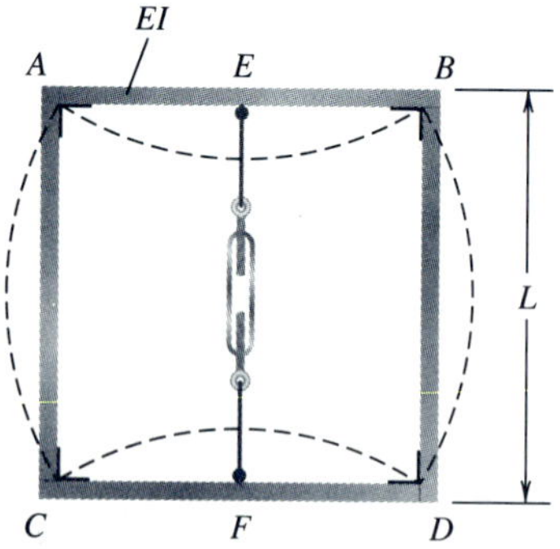

Fig. P6.5-15

6.5-16 Two identical beams are cantilevered from opposite walls and pinned together as shown in Fig. P6.5-16*a*. The two beams carry constant distributed loadings that differ by a factor of 2. The pin transmits only a shear force (Fig. P6.5-16*b*), not a moment.

(*a*) Find the deflection at B.

(*b*) Sketch the shear force and bending moment diagrams for the two beams.

6.5-17 The deflection response of a cantilever beam under distributed loading $q(x)$ can be found by superimposing the response due to infinitesimal concentrated loads $q(x)\,dx$ by integration over the full length of the loading, as shown in Fig. P6.5-17*a*. For the case of a linearly decreasing distributed loading as shown in Fig. P6.5-17*b*, find the slope and deflection at point B, using the method just described, and show that your results agree with those found in Example 6.10.

6.5-18 Use the method described in Prob. 6.5-17 to find the slope and deflection at the free end for the cantilever beam shown in Fig. P6.5-18. Compare your results with those given in Table G.1, Fig. G.1-3, App. G.

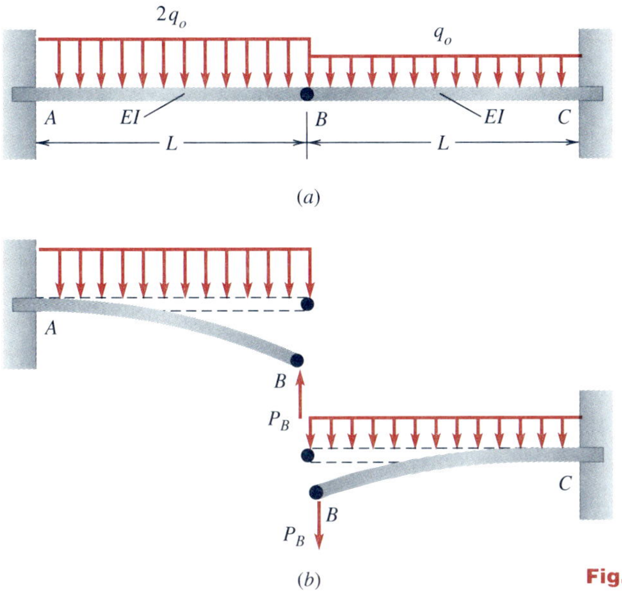

Fig. P6.5-16

6.5-19 A beam is simply supported at B and C and carries a constant distributed loading q_0 as shown in Fig. P6.5-19. For purposes of analysis, replace the overhanging segments of the beam by forces and moments acting on segment BC at B and C. Use the method of superposition to find the slope at B and the center deflection. Find the total deflection and slope at end D by again using the method of superposition.

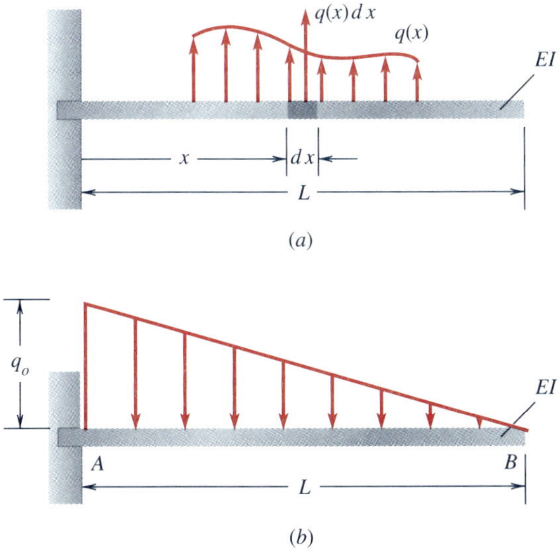

Fig. P6.5-17

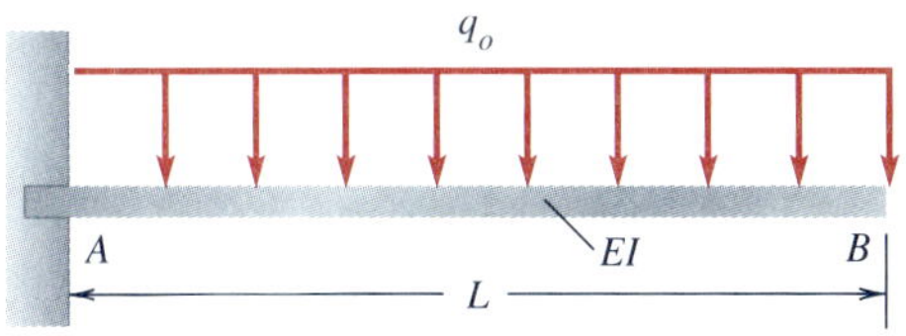

Fig. P6.5-18

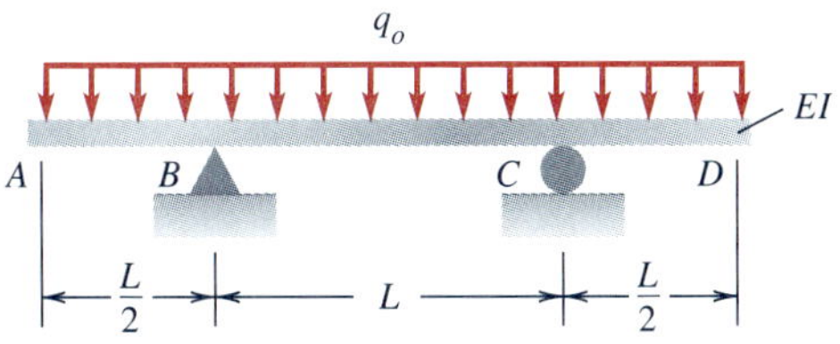

Fig. P6.5-19

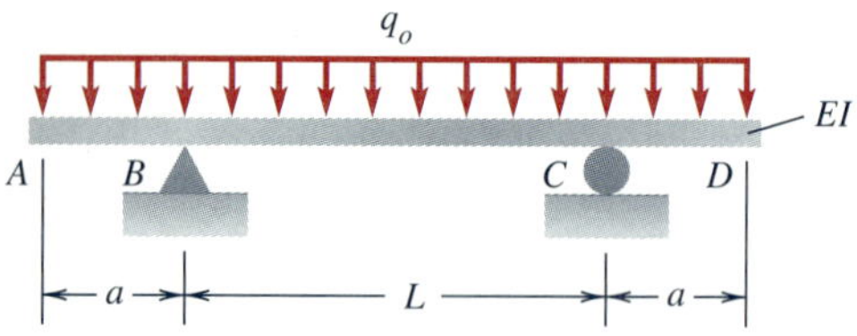

Fig. P6.5-20

6.5-20 A beam is simply supported at B and C and carries a constant distributed loading q_0 as shown in Fig. P6.5-20. For purposes of analysis, replace the overhanging segments of the beam by forces and moments acting on segment BC at B and C. Find the length a of the overhang which results in zero deflection at the midpoint of the beam.

6.5-21 Carry out the solution procedure as presented in Prob. 6.5-20, except the beam is loaded by three concentrated loads P, as shown in Fig. P6.5-21.

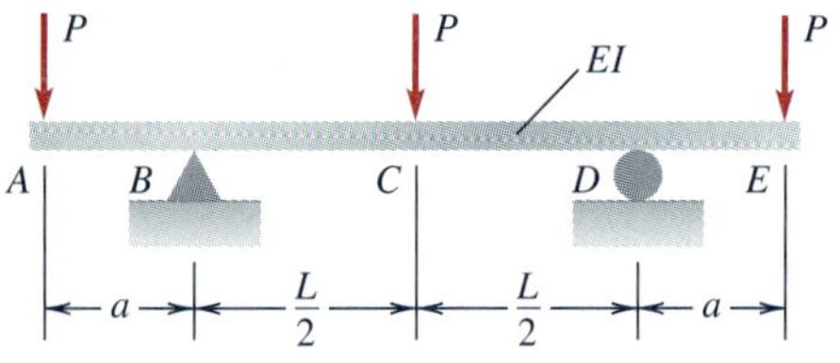

Fig. P6.5-21

6.5-22 A 2-in × 12-in nominal dimension wood plank (App. D) of length 12 ft rests on a table, as shown in Fig. P6.5-22.

(*a*) If the plank is only loaded by its own weight, find the gap that occurs between the middle of the table and the middle of the plank. Take a = 4 ft and E = 1500 ksi and a specific weight of 35 lb/ft^3.

(*b*) Find the length of the overhang a such that the gap between the table and the plank at the midpoint disappears.

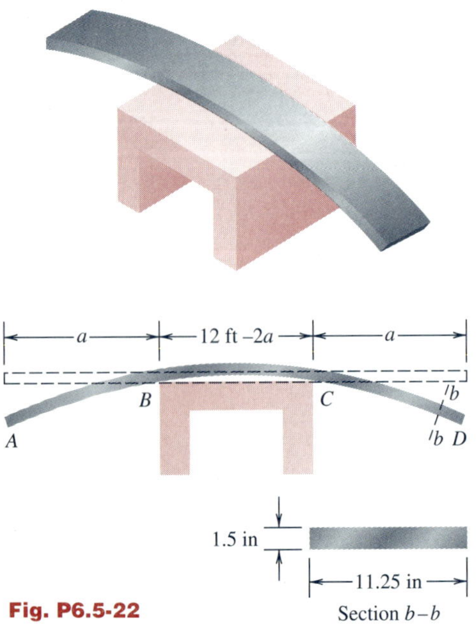

Fig. P6.5-22

6.6-1 Two strips of the same width but in general made of different metals and with different thicknesses are bonded together to form a bimetallic-strip thermostat unit, as shown in Fig. P6.6-1. If the coefficients of thermal expansion α_1 and α_2 are not equal, then the strip will undergo a change in length and curvature upon a change in temperature.

(*a*) If the strain response for the strip is taken in the form

$$\epsilon = \epsilon_0 - \frac{y}{R} \qquad (a)$$

where ϵ_0 is the strain and R is the radius of curvature of the surface $y = 0$, then show that Hooke's law gives

$$\sigma_1 = E_1(\epsilon - \alpha_1 \,\Delta T) \qquad \sigma_2 = E_2(\epsilon - \alpha_2 \,\Delta T) \qquad (b)$$

where E_1 and E_2 are elastic moduli and σ_1 and σ_2 are stresses in materials 1 and 2.

(*b*) Using the axial equilibrium equation

$$\int_A \sigma \, dA = \int_{A_1} \sigma_1 \, dA + \int_{A_2} \sigma_2 \, dA = 0 \qquad (c)$$

show that

$$(\beta\lambda + 1)\epsilon_0 - \frac{1}{2}(\beta\lambda^2 - 1)\frac{h_2}{R} = (\beta\lambda\alpha_1 + \alpha_2)\,\Delta T \qquad (d)$$

where $\lambda = h_1/h_2$ and $\beta = E_1/E_2$.

(*c*) Show that moment equilibrium

$$\int_A \sigma y \, dA = \int_{A_1} \sigma_1 y \, dA + \int_{A_2} \sigma_2 y \, dA = 0 \qquad (e)$$

leads to

$$\frac{1}{2}(\beta\lambda^2 - 1)\epsilon_0 - \frac{1}{3}(\beta\lambda^3 + 1)\frac{h_2}{R} = (\beta\lambda^2\alpha_1 - \alpha_2)\frac{\Delta T}{2} \qquad (f)$$

(*d*) Solve Eqs. (*d*) and (*f*), and show that the curvature $1/R$ can be written as

$$\frac{1}{R} = \frac{6(1+\lambda)^2(\alpha_2 - \alpha_1)\,\Delta T}{h\{3(1+\lambda)^2 + (1+\lambda\beta)[\lambda^2 + 1/(\lambda\beta)]\}} \qquad (g)$$

where $h = h_1 + h_2$.

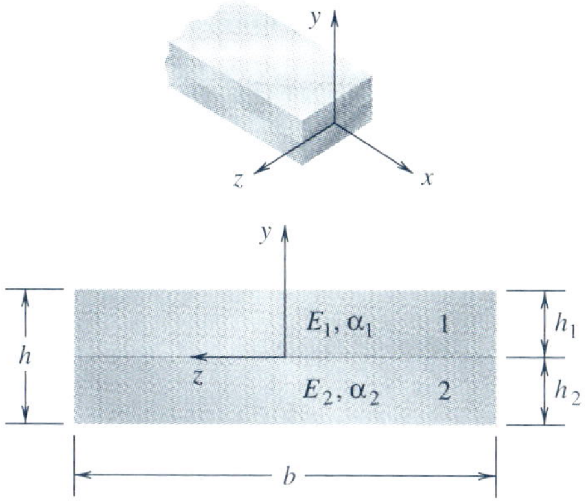

Fig. P6.6-1

6.6-2 A bimetallic thermostat unit (see Fig. P6.6-1) is to be made up of two steel strips with the same elastic modulus $E_1 = E_2 = 200$ GPa, but with different coefficients of thermal expansion $\alpha_1 = 12 \times 10^{-6}/°C$ and $\alpha_2 = 17 \times 10^{-6}/°C$. Several designs are of interest with $h_2 = 3$ mm and $h_1 = 1$, 2, and 3 mm. Use the results of Prob. 6.6-1 to calculate the three values of $1/R$ for each of the two temperature changes $\Delta T = 50$ and 100°C.

6.6-3 For bimetallic strips with both elements of the same thickness $h_1 = h_2 = h/2$ and the same elastic moduli $E_1 = E_2 = E$, use the results of Prob. 6.6-1 to show that

$$\frac{1}{R} = \frac{3}{2}\frac{\Delta T}{h}(\alpha_2 - \alpha_1)$$

6.6-4 For a bimetallic strip as described in Prob. 6.6-3, a change in temperature ΔT results in the strip geometry shown in Fig. P6.6-4.

(*a*) Find an expression for the deflection δ in terms of the radius of curvature R and the length L of the strip before the temperature change.

(*b*) Plot some design curves for δ versus h for $h = 1$, 3, and 5 mm with $\Delta T = 50$°C and $L = 200$ mm. On the same graph plot the curves for $\Delta T = 100$°C and $L = 200$ mm. Take $\alpha_1 = 17 \times 10^{-6}/°C$ and $\alpha_2 = 12 \times 10^{-6}/°C$, and use the equation derived in Prob. 6.6-3.

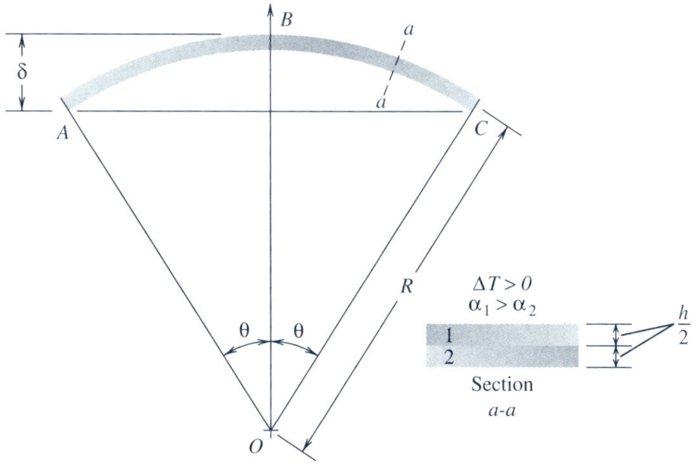

Fig. P6.6-4

6.6-5 When beams undergo deflections in the elastic range of the material behavior such that the slope dv/dx is no longer $\ll 1$, Eq. (6.6) is replaced by

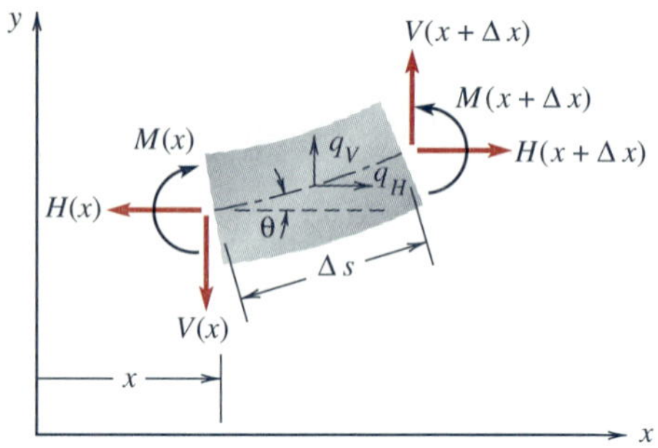

Fig. P6.6-5

$$EI\frac{d\theta}{ds} = M(s) \qquad (a)$$

where θ is the slope angle measured from the x direction and s is the arclength along the neutral axis. To derive a nonlinear elastic beam theory, consider the beam element of length Δs in Fig. P6.6-5 for which equilibrium requires that as $\Delta s \to 0$,

$$\frac{dH}{ds} + q_H(s) = 0 \qquad \frac{dV}{ds} + q_V(s) = 0$$

$$\frac{dM}{ds} + V\cos\theta - H\sin\theta = 0 \qquad (b)$$

where H and V are the horizontal and vertical components of the stress resultants, M is the moment, and q_H and q_V are the horizontal and vertical components of the distributed forces. Show that for $\theta \ll 1$, $ds \approx dx$, $q_V \approx q$, $\cos\theta \approx 1$, $\sin\theta \approx \theta$, Eqs. (*a*) and (*b*) become Eqs. (6.6), (4.4), and (4.5).

6.6-6 A cantilever beam is in a state of pure bending due to a moment M_B applied at the free end, as shown in Fig. P6.6-6.

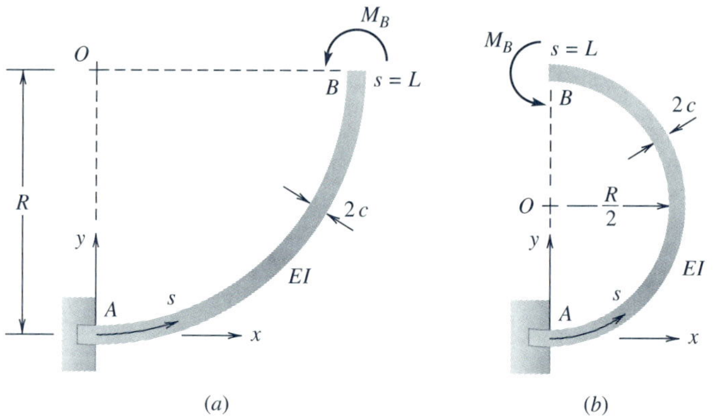

Fig. P6.6-6

(*a*) Integrate Eq. (*a*) of Prob. 6.6-5 for this case, and use the boundary condition at $s = 0$ to show that

$$EI\theta(s) = M_B s$$

where EI is the bending modulus. Find the values of M_B for the beams shown in Fig. P6.6-6*a* and *b*.

(*b*) If the depth of the beam in the plane of bending is $2c$, derive an expression for the ratio c/L in terms of the ratio σ_{max}/E.

(*c*) If $E = 30{,}000$ ksi, the allowable normal stress in bending is 30 ksi, and $L = 100$ in, find the allowable range of values for c so that the beam can be bent into the shapes shown in Fig. P6.6-6*a* and *b*. Do the results seem reasonable, based upon items like sawblades and other thin-metal items which we can bend into quarter- and half-circles without damaging the metal?

6.6-7 For the simply supported beam shown in Fig. P6.6-7*a*, the end loads set up a pure bending state over the central segment *AB* of the beam with $M_0 = Pa$.

(*a*) Using Eq. (*a*) in Prob. 6.6-5, show that

$$EI\theta(s) = EI\theta_0 + M_0 s \qquad (a)$$

where θ_0 is defined in Fig. P6.6-7*b*.

(*b*) Find the value of the moment M_0 required to result in $\theta_0 = \pi/4$. If $2c$ is the depth of the section of the beam, then derive an expression for c/L in terms of σ_{max}/E.

(*c*) If the allowable stress is $\sigma_a = 30$ ksi, $E = 30{,}000$ ksi, and $L = 100$ in, find the allowable values for the depth c for the cases where θ_0 is $\pi/4$ and $\pi/3$.

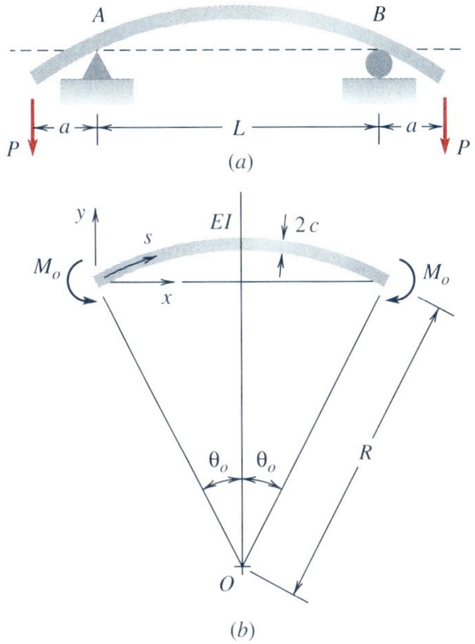

Fig. P6.6-7

6.6-8 The rectangular cross section of a cantilever beam is to be designed so that the maximum normal stress due to bending is constant over the beam for a concentrated load *P* acting at the free end (Fig. P6.6-8).

(*a*) Show that an acceptable design based on beam theory is to take the width in the form $b = b_0(1 - x/L)$ and the depth $h = h_0$ where b_0 and h_0 are constants. Find an expression for σ_{max}.

(*b*) Find the deflection curve $v(x)$ for the beam of part (*a*). Show that it is exactly the same curve as can be found for the corresponding beam of constant bending modulus EI_0 ($I_0 = b_0h_0^3/12$) loaded at its free end by a moment $-PL$.

6.6-9 The rectangular cross section of a cantilever beam is to be designed so that the maximum normal stress due to bending for a constant distributed loading q_0 is constant along the beam (Fig. P6.6-9).

(*a*) Show that an acceptable design is to take the width in the form $b = b_0(1 - x/L)^2$ and the depth $h = h_0$, where b_0 and h_0 are constants. Find an expression for σ_{max}.

(*b*) Find the deflection curve $v(x)$ for the beam of part (*a*). Show that it is exactly

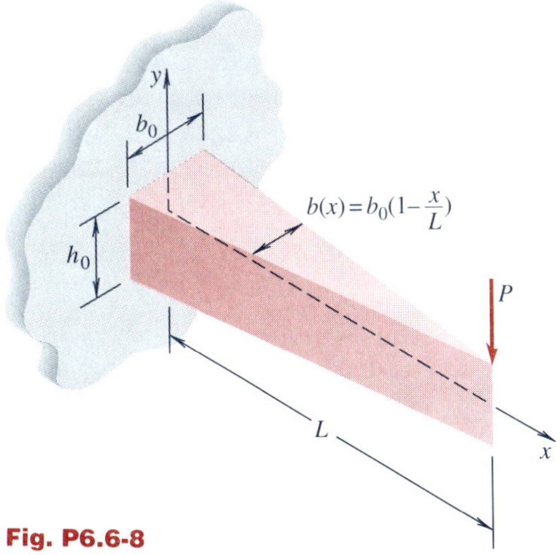

Fig. P6.6-8

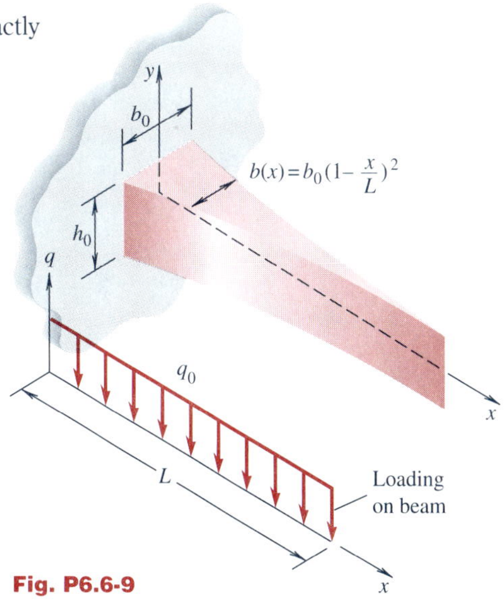

Fig. P6.6-9

the same curve as can be found for the corresponding beam of constant bending modulus EI_0 ($I_0 = b_0 h_0^3/12$) loaded at its free end by a moment $q_0 L^2/2$.

6.6-10 The rectangular cross section of a beam that is simply supported at both ends is to be designed so that the maximum normal stress due to bending remains constant over the length of the beam for a concentrated load acting at the midspan of the beam (Fig. P6.6-10).

(*a*) Show that an acceptable design is to take the width

$$b = \frac{2b_0}{L}\left(\langle x \rangle^1 - 2\left\langle x - \frac{L}{2}\right\rangle^1\right)$$

and the depth $h = h_0$, where b_0 is the width at midspan and h_0 is the constant depth. Find an expression for σ_{max}.

(*b*) Find the deflection curve $v(x)$ for the beam of part (*a*). Show that it is exactly the same curve as can be found for the corresponding beam of constant bending modulus EI_0 ($I_0 = b_0 h_0^3/12$) loaded at each end by a moment $PL/4$.

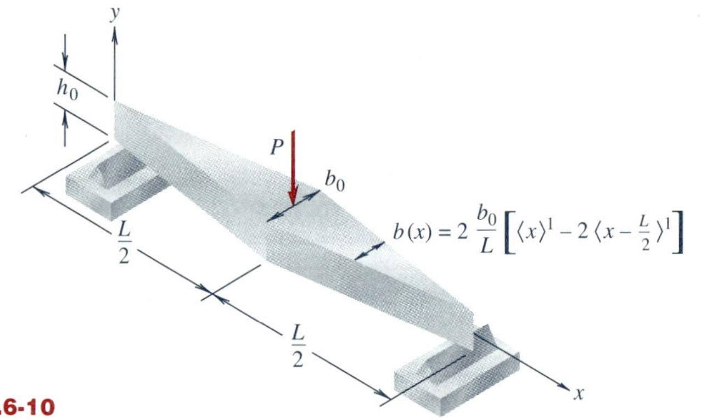

Fig. P6.6-10

6.6-11 The concept behind an atomic force microscope used to scan the atomic structure of surfaces is to fabricate a cantilever beam whose spring constant is smaller than the equivalent spring constant between atoms, which is of the order of 10 N/m. Calculate the equivalent spring constant for the cantilever beam shown in Fig. P6.6-11. It has been claimed that a cantilever spring made from a piece of aluminum foil with the L and w dimensions shown has a spring constant of the order 1 N/m. Is this true? $E = 70$ GPa.

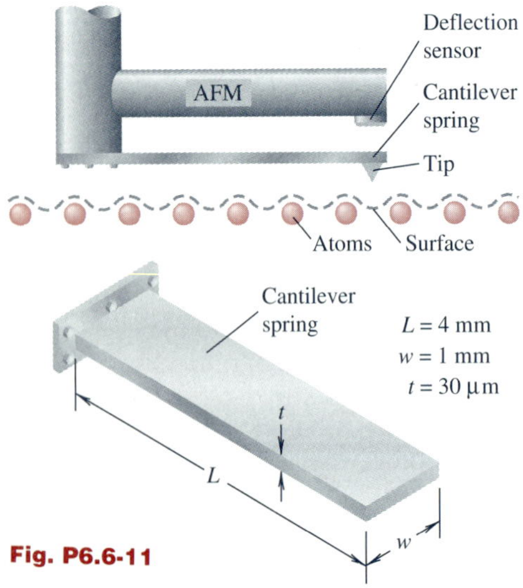

Fig. P6.6-11

6.6-12 A simply supported beam initially at a uniform temperature T_0 is heated so that a linear temperature distribution over its depth is maintained. The upper and lower surfaces are at temperatures T_U and T_L, as shown in Fig. P6.6-12*a*. The thermal strain in the beam can be taken as a uniform axial strain

$$\epsilon_0 = \alpha\left(\frac{T_U + T_L}{2} - T_0\right)$$

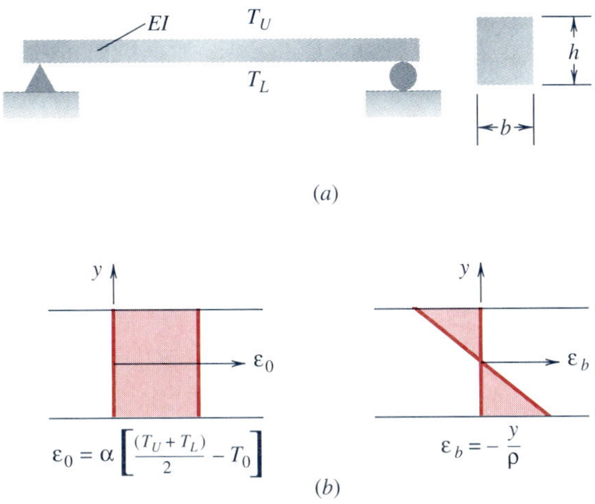

Fig. P6.6-12

plus a bending strain which can be written as $\epsilon_b = -y/\rho$ (Fig. P6.6-12*b*).

(*a*) Show that by equating the thermal strain at the top $\alpha(T_U - T_0)$ to the strain $\epsilon_0 - h/(2\rho)$, we obtain

$$\frac{1}{\rho} = -\frac{\alpha}{h}(T_U - T_L) \qquad (a)$$

(*b*) Show that by integrating the equation

$$\frac{d^2v}{dx^2} = -\frac{\alpha}{h}(T_U - T_L) \qquad (b)$$

and applying the boundary conditions $v = 0$ at $x = 0$ and L, we obtain the following expression for the maximum deflection:

$$v_{\max} = \frac{\alpha L^2}{8h}(T_U - T_L) \qquad (c)$$

6.6-13 The expression for the curvature of a plane curve $v = v(x)$ as given in Eq. (6.4) is the general expression usually obtained in calculus courses. However, its applicability to large (nonlinear) deformation of beams is inappropriate because it assumes that the neutral axis moves only vertically [where $v(x)$ is the

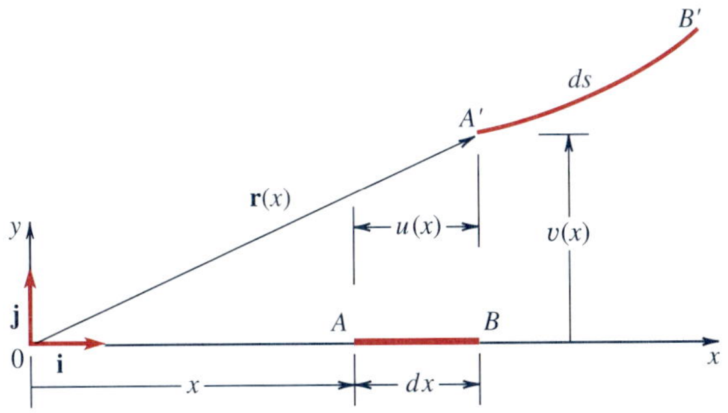

Fig. P6.6-13

vertical deflection of the beam]. We outline a method to obtain an expression for the curvature in terms of the horizontal and vertical deflections of the neutral axis.

Given an element AB of the neutral axis of length dx before loading, Fig. P6.6-13, and let

$$\mathbf{r}(x) = [x + u(x)]\mathbf{i} + v(x)\mathbf{j} \tag{a}$$

describe the location of point A after loading. The unit tangent vector to the displaced curve is

$$\mathbf{t} = \frac{d\mathbf{r}}{ds} = \frac{d\mathbf{r}}{dx}\frac{dx}{ds} = \frac{\mathbf{r}'}{s'} \qquad s'^2 = (1 + u')^2 + v'^2 \tag{b}$$

where ds is the arc length of $A'B'$. The rate of change of the unit tangent vector with respect to arc length gives

$$\frac{d\mathbf{t}}{ds} = \frac{\mathbf{n}}{\rho} \tag{c}$$

where $\mathbf{n}$ is the unit normal vector and $1/\rho$ is the curvature.

(*a*) Show that

$$\frac{\mathbf{n}}{\rho} = \frac{\mathbf{r}''}{s'^2} - \frac{\mathbf{r}'s''}{s'^3} \tag{d}$$

or in terms of $\mathbf{i}$ and $\mathbf{j}$

$$\frac{\mathbf{n}}{\rho} = \mathbf{i}\left[\frac{u''}{s'^2} - \frac{(1 + u')s''}{s'^3}\right] + \mathbf{j}\left[\frac{v''}{s'^2} - \frac{v's''}{s'^3}\right] \tag{e}$$

Therefore

$$\left(\frac{1}{\rho}\right)^2 = \frac{1}{s'^4}(u''^2 + v''^2 - s''^2) \tag{f}$$

(*b*) Show that

$$s'' = \frac{(1 + u')u'' + v'v''}{s'} \tag{g}$$

and thus

$$\frac{1}{\rho} = \frac{u''v' - v''(1 + u')}{[(1 + u')^2 + v'^2]^{3/2}} \tag{h}$$

(*c*) Show that the result in Eq. (6.4) corresponds to dropping terms in u in Eq. (*h*). In this case for large deformations $s' = \sqrt{1 + v'^2}$.

(*d*) Show that for an inextensible neutral axis for which

$$ds = dx \tag{i}$$

we have in general

$$ds^2 = d\mathbf{r} \cdot d\mathbf{r} = [(1 + u')^2 + v'^2]dx^2$$

or

$$\left(\frac{ds}{dx}\right)^2 = 1 = (1 + u')^2 + v'^2 \qquad (j)$$

Therefore show that by using Eq. (j), Eq. (h) becomes

$$\frac{1}{\rho} = u''v' - v''(1 + u')$$

but

$$(1 + u') = \sqrt{1 - v'^2}$$

or finally

$$\frac{1}{\rho} = -\frac{v''}{\sqrt{1 - v'^2}} \qquad (k)$$

CHAPTER 7

Deflections of Statically Indeterminate Beams

7.1 Introduction

In our discussions of beams thus far, we have considered only beam problems that were statically determinate. That is, we found it was possible to calculate the force and moment reactions for the beam from the equations of equilibrium. A knowledge of the reactions then allowed us to calculate the values of the shear force and bending moment along the beam, from which the stresses and the deflections could be found.

In this chapter we extend the previous analysis and consider beam problems for which the equilibrium equations alone are not sufficient for the complete determination of the force and moment reactions. We encountered this same indeterminacy in Chaps. 2 and 3 for problems involving axially loaded bars and circular shafts subjected to twisting moments. In those cases we obtained the additional equations to supplement the equilibrium equations from the force-deformation relations and geometric conditions. This enabled us to find the unknown reactions, using all three steps given in Eq. (2.8). That is, equations using the force-deformation relations and geometric compatibility had to be used together with the equilibrium equations to give the same number of equations as the number of unknowns in the problem. For beam problems, we follow exactly the same general procedure.

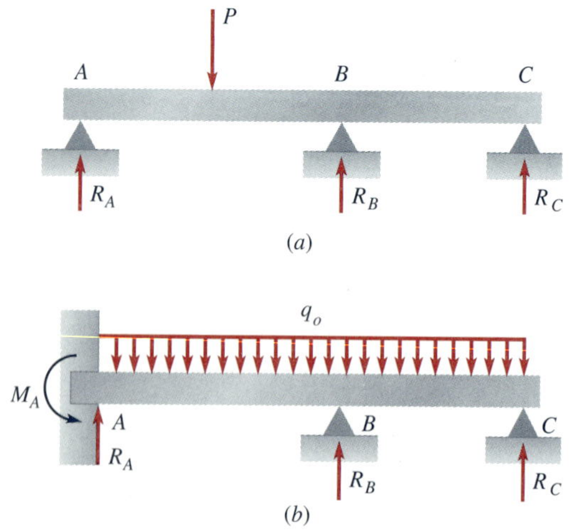

Figure 7.1 Statically indeterminate beams. (*a*) Three unknown reactions. (*b*) Four unknown reactions.

For example, for the statically indeterminate beam of Fig. 7.1*a*, overall force and moment equilibrium will yield only two equations for the three unknown vertical reactions at the supports. However, the requirement that the shape of the beam deflection be compatible with the geometric condition of zero deflection at the three support points *A*, *B*, and *C* gives three additional conditions. We recall that in obtaining the expression for the deflection of a beam, we found in general (in Chap. 6) that two unknown constants of integration arose in the integration process. In this case we have three equations of geometric compatibility arising from the zero deflections at the supports and two equations of equilibrium to determine the three unknown reactions and the two integration constants. In other words, we have enough equations to determine the unknowns.

As a second example of the general procedure, consider the beam shown in Fig. 7.1*b* that has the four unknown reactions as indicated, a reactive force and moment at the wall, and force reactions at points B and C. If we combine these four unknowns with the two integration constants arising from integration to find the beam deflection, we find that four geometric compatibility conditions are needed to augment the two equilibrium equations. At the clamped support A we have both the slope and the deflection equal to zero, and at points B and C we have the deflection equal to zero; therefore we have the four additional conditions.

In this chapter we will solve a number of problems like those shown in Fig. 7.1 and then introduce a procedure that is applicable to both statically indeterminate and determinate beams and which lends itself well to computer implementation. In fact, the computer program will allow us to investigate the shear force and bending moment distributions as well as the deflections of a beam.

7.2 Deflections of Statically Indeterminate Beams

In Chap. 6 the deflection curves for statically determinate beams were found by integrating the moment-curvature equation, Eq. (6.6), twice. The expression for the moment was found by one of two different procedures. One procedure used a free-body diagram of a segment of the beam to derive an expression for the bending moment at an arbitrary section by using moment equilibrium of the segment. The other procedure started with the expression for the distributed loading on the beam $q(x)$ and obtained after two integrations the bending moment $M(x)$. Once $M(x)$ was found by either method, integration of Eq. (6.6) yielded the slope and deflection expressions.

For statically indeterminate problems, we follow the same general procedure used for the statically determinate problems in Chap. 6. The only difference is that whereas for the determinate problems we always found expressions for $M(x)$ that did not involve any unknown reactions (they could be completely determined from the conditions of equilibrium alone), for the indeterminate problems we will find that one or more unknown reactions will appear in the expression for $M(x)$ because the values of all reactions cannot be determined from the equations of equilibrium alone.

The method of superposition was also used in Chap. 6 to solve statically determinate beam problems that could be recognized as a combination of deflection problems previously solved. The method of superposition can be used for the solution of indeterminate problems also and is very fast and convenient, especially when only one or two unknowns are involved.

The following examples illustrate the techniques of solution.

EXAMPLE 7.1

A beam of length L is clamped at both ends A and B and carries a concentrated load P at C, as shown in Fig. 7.2*a*. We wish to find (*a*) the wall reactions, (*b*) the maximum absolute value of the bending moment, and (*c*) the value of the deflection under the load.

We neglect the weight of the beam and take the bending modulus EI as constant. We can see immediately that the beam is statically indeterminate since there are four unknown wall reactions, i.e., the moment and vertical force reactions at both ends A and B. Overall equilibrium of the beam shown in Fig. 7.2*b* gives

$$\begin{aligned} R_A + R_B &= P \\ M_A - Pa - M_B + R_B L &= 0 \end{aligned} \qquad (a)$$

where M_A and M_B are the unknown moment reactions and R_A and R_B are the unknown force reactions at the wall supports.

For this example we will obtain an expression for $M(x)$ by making use of moment equilibrium of a suitable segment of the beam. Consider the free-body diagram shown in Fig. 7.2*c* with M_A and R_A as the unknown reactions at the left end of the beam. The unknown reactions M_B and R_B at the right end can be expressed in terms of the unknowns M_A and R_A from Eqs. (*a*). We will carry M_A and R_A as the two unknown reactions to be determined by geometric compatibility conditions applied to the slope and deflection expressions. Moment equilibrium for the beam segment shown in Fig. 7.2*c* leads to

$$M(x) + P\langle x - a\rangle^1 + M_A - R_A x = 0 \qquad (b)$$

where $M(x)$ is the bending moment at section x. Therefore, making use of Eq. (6.6) gives

$$EIv'' = M(x) = -P\langle x - a\rangle^1 - M_A + R_A x \qquad (c)$$

Integration twice leads to

$$EIv' = -P\frac{\langle x - a\rangle^2}{2} - M_A x + R_A \frac{x^2}{2} + c_1 \qquad (d)$$

and

$$EIv = -P\frac{\langle x - a\rangle^3}{6} - M_A \frac{x^2}{2} + R_A \frac{x^3}{6} + c_1 x + c_2 \qquad (e)$$

where c_1 and c_2 are the two constants of integration.

The two unknown reactions M_A and R_A and the two unknown integration constants c_1 and c_2 can be found provided that we identify four conditions of geometric compatibility. The clamped support of the beam at both ends requires that the beam deflection and the slope be zero at both ends. That is, at $x = 0$

$$v(0) = 0 \qquad \text{and} \qquad v'(0) = 0 \qquad (f)$$

and this gives two conditions, according to Eqs. (*d*) and (*e*), in the form

$$\begin{aligned} EIv'(0) &= 0 - c_1 \\ EIv(0) &= 0 = c_2 \end{aligned} \qquad (g)$$

Therefore c_1 and c_2 are zero. At the other end $x = L$ we have

$$v(L) = 0 \qquad \text{and} \qquad v'(L) = 0 \qquad (h)$$

which becomes, by using Eqs. (*d*) and (*e*),

$$EIv'(L) = 0 = -\frac{Pb^2}{2} - M_A L + R_A \frac{L^2}{2} \qquad (i)$$

$$EIv(L) = 0 = -\frac{Pb^3}{6} - \frac{M_A L^2}{2} + R_A \frac{L^3}{6} \qquad (j)$$

Solving for the unknown reactions gives

$$M_A = \frac{Pb^2}{L^2}(L - b)$$

or

$$M_A = \frac{Pb^2 a}{L^2} \qquad (k)$$

and

$$R_A = \frac{Pb^2}{L^3}(3L - 2b)$$

or

$$R_A = \frac{Pb^2}{L^3}(L + 2a) \qquad (l)$$

The remaining reactions M_B and R_B are now found from Eqs. (*a*), (*k*), and (*l*) to be

$$M_B = \frac{Pa^2b}{L^2} \qquad (m)$$

and

$$R_B = \frac{Pa^2}{L^3}(L + 2b) \qquad (n)$$

The expression for the deflection is given by Eq. (e) with M_A and R_A given in Eqs. (k) and (l).

To determine the maximum absolute value of the bending moment along the beam, we first observe from Eq. (c) that the bending moment diagram consists of two straight-line segments. One segment starts at point A, $x = 0$, with the value $-M_A$ and increases to the value M_C at point C, as shown in Fig. 7.2d. Using Eq. (c), we find

$$M_C = \frac{2Pa^2b^2}{L^3} \qquad (o)$$

The second segment starts at point C with the value M_C and decreases to the value $-M_B$ at point B. The location of the maximum absolute value of the bending moment will depend upon the value of a. From Eqs. (k) and (m) it follows that

$$|M_B| - |M_A| = \frac{Pab}{L^2}(a - b) \qquad (p)$$

so that as long as $a > b$, M_B is larger than M_A, that is, the larger absolute value of the wall moment occurs at the wall nearest the load (point B for the case sketched in Fig. 7.2d). Further, from Eqs. (m) and (o), we have

$$|M_B| - |M_C| = \frac{Pa^2b}{L^3}2\left(\frac{L}{2} - b\right) \qquad (q)$$

Therefore, we conclude that if $b < L/2$, that is, the load P is located to the right of the midpoint as shown in Fig. 7.2a, then the maximum absolute value of the bending moment occurs at the wall nearest to the load and is given by Eq. (m). The bending moment diagram takes the shape shown in Fig. 7.2d.

Finally, to find the deflection under the load at $x = a$, we make use of Eq. (e) with Eqs. (k) and (l) to get

$$v(a) = -\frac{Pa^3b^3}{3EIL^3} \qquad (r)$$

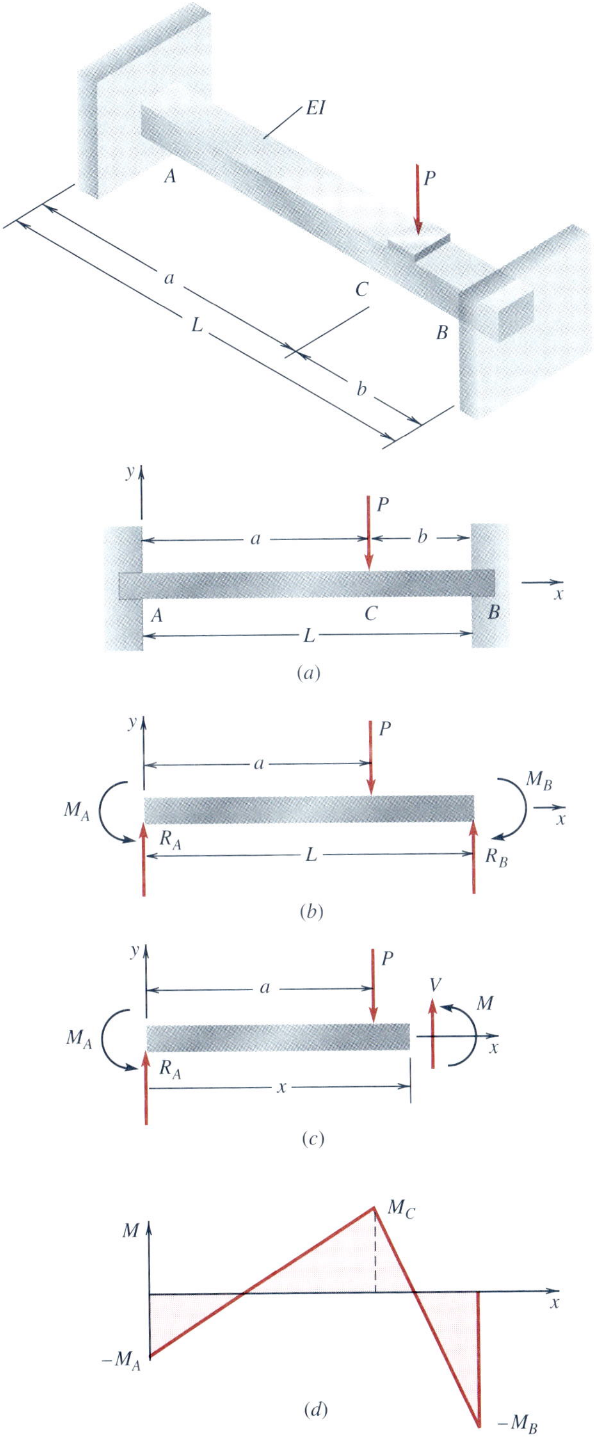

Figure 7.2 Example 7.1

This deflection for a beam clamped at both ends can be compared to the deflection of a beam simply supported at both ends (Example 6.3) for which the deflection under the load is

$$v_{ss}(a) = -\frac{Pa^2b^2}{3EIL} \qquad (s)$$

The ratio of the deflection for the clamped case, Eq. (*r*), to the simply supported case, Eq. (*s*), is

$$\frac{v(a)}{v_{ss}(a)} = \frac{a}{L}\frac{b}{L} \qquad (t)$$

If we consider the special case for which the load is at the midpoint, then $a = b = L/2$, and we find that the ratio is $\frac{1}{4}$. Thus the clamped supports allow one-fourth of the deflection that occurs with simple supports when a beam carries a concentrated load at midspan.

We now consider a second example for which the beam deflections are found by integrating the expression for the bending moment obtained directly from the moment equilibrium equation for a segment of the beam.

EXAMPLE 7.2

Consider a beam of length $2L$ that is simply supported at its midpoint and at both ends and is subjected to a constant distributed loading q_0, as shown in Fig. 7.3*a*. We wish to find the reactions at the supports and to draw the shear force and bending moment diagrams for the beam. We assume that the weight of the beam is included in q_0 and that the bending modulus EI of the beam is constant.

As usual, we will approach the solution by using the three steps of Eq. (2.8). We show the beam with reactions in Fig. 7.3*b*, where symmetry about the section B leads to taking both end reactions the same, namely R_A. Force balance in the y direction leads to

$$2R_A + R_B = 2q_0L \qquad (a)$$

A moment equilibrium equation for the entire beam will not produce an additional independent equation since we have invoked symmetry. Thus we have a statically indeterminate problem. We need to apply a sufficient number of geometric conditions to enable us to calculate the unknown reactions R_A and R_B and the integration constants.

Because the beam is symmetric about its midpoint B, it will be sufficient to study only the left half AB. Moment equilibrium of the beam segment of length x as shown in Fig. 7.3*c* gives

$$M(x) - R_Ax + q_0\frac{x^2}{2} = 0$$

so that Eq. (6.6) becomes

$$EIv'' = M(x) = R_Ax - q_0\frac{x^2}{2} \qquad (b)$$

Repeated integration gives

$$EIv' = R_A\frac{x^2}{2} - q_0\frac{x^3}{6} + c_1 \qquad (c)$$

and

$$EIv = R_A\frac{x^3}{6} - q_0\frac{x^4}{24} + c_1x + c_2 \qquad (d)$$

where c_1 and c_2 are the constants of integration.

The simple supports at A and B require that at

$$x = 0, L: \qquad v = 0 \qquad (e)$$

and the symmetry of the beam about the middle of the beam at B requires a vanishing slope at B, or at

$$x = L: \qquad v' = 0 \qquad (f)$$

In order that $v(0) = 0$ in Eq. (*d*), it is necessary that c_2 be taken equal to zero. The remaining conditions in Eqs. (*e*) and (*f*), when substituted into Eqs. (*c*) and (*d*), lead to

$$0 = R_A\frac{L^2}{2} - q_0\frac{L^3}{6} + c_1 \qquad (g)$$

$$0 = R_A\frac{L^3}{6} - q_0\frac{L^4}{24} + c_1L \qquad (h)$$

These equations are easily solved for R_A and c_1 to obtain

$$R_A = \tfrac{3}{8}q_0L \qquad c_1 = -\tfrac{1}{48}q_0L^3 \qquad (i)$$

Now that R_A has been determined, we can find R_B from the equation for force equilibrium of the entire beam, Eq. (*a*), to get

$$R_B = \frac{5}{4} q_0 L \qquad (j)$$

It is interesting to note that for this beam over the three evenly spaced supports, the middle support carries 62.5 percent of the total load while the outside supports carry only 18.75 percent each.

With all the reactions [Eqs. (i) and (j)] known, it is possible to sketch the shear force and bending moment diagrams as shown in Fig. 7.3d and e.

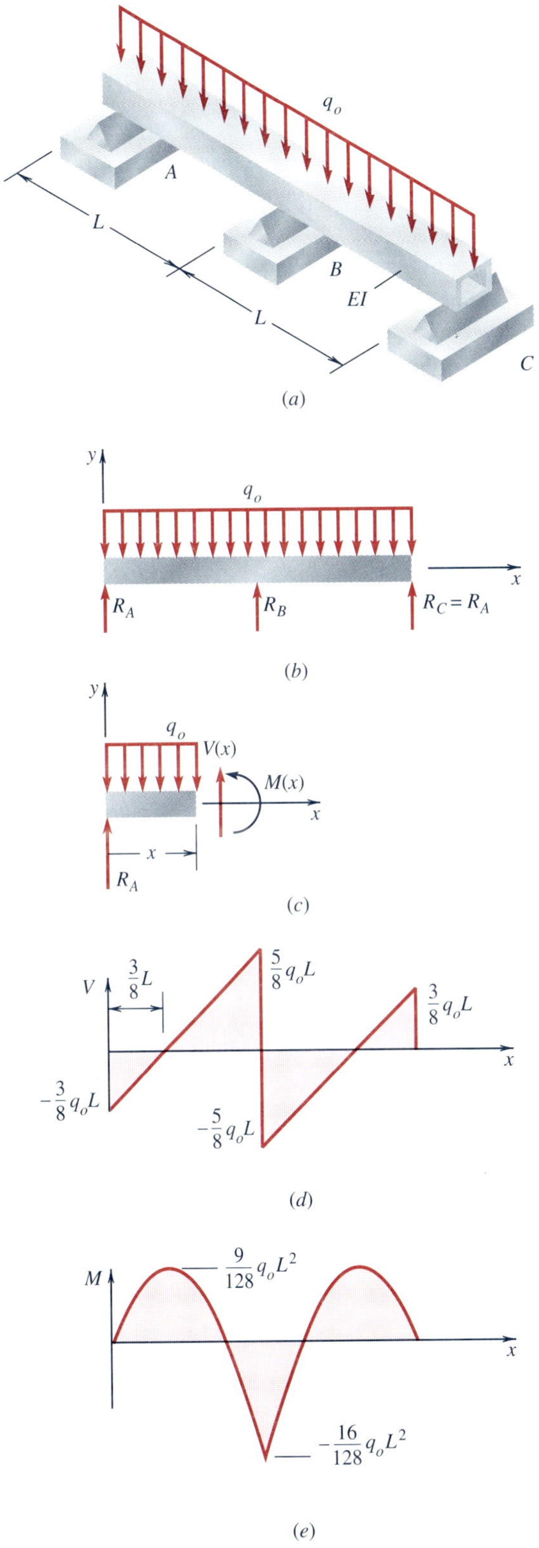

Figure 7.3 Example 7.2

In the next two examples, the expression for $M(x)$ that is to be integrated to obtain the slope and deflection expressions will be derived directly from the expression for the distributed loading $q(x)$ acting on the beam, following procedures given in Sec. 6.4.

EXAMPLE 7.3

The beam of length $2L$ shown in Fig. 7.4*a* is clamped at its left end and simply supported at its midpoint. It carries a constant distributed loading q_0 as shown. We want to find the reactions at A and B, draw the shear force and bending moment diagrams, and find the deflection at C.

We assume that the weight of the beam is included in q_0 and that the bending modulus of the beam is EI.

From the free-body diagram of the entire beam, shown in Fig. 7.4*b*, we see that there are three unknown reactions and only two equations of equilibrium. These equilibrium equations can be written as

$$R_A + R_B - 2q_0L = 0$$
$$M_A + R_BL - 2q_0L^2 = 0 \qquad (a)$$

where $2q_0L$ is the resultant of the distributed loading.

As was the case in previous examples in this section, the additional equations needed to supplement Eqs. (*a*) will come from conditions of geometric compatibility stated in terms of slope and deflection values along the beam. The slope and deflection curves will be derived by starting with an expression for the load $q(x)$ and integrating twice to obtain $M(x)$. The two additional integrations of $M(x)$ will produce the slope and deflection expressions.

In writing the expression for $q(x)$, we will use the three unknown reactions R_A, R_B, and M_A rather than using Eqs. (*a*) at this time to rewrite two of them in terms of a single unknown reaction. Thus referring to Fig. 7.4*b*, we write for the loading on the beam

$$q(x) = R_A\langle x\rangle_{-1} + M_A\langle x\rangle_{-2} - q_0\langle x\rangle^0 + R_B\langle x - L\rangle_{-1} \qquad (b)$$

Upon integrating Eq. (6.20), $EIv'''' = q(x)$, we find

$$EIv''' = R_A\langle x\rangle^0 - M_A\langle x\rangle_{-1} - q_0\langle x\rangle^1 + R_B\langle x - L\rangle^0 \qquad (c)$$

where the constant of integration is equal to zero; integration again produces the bending moment

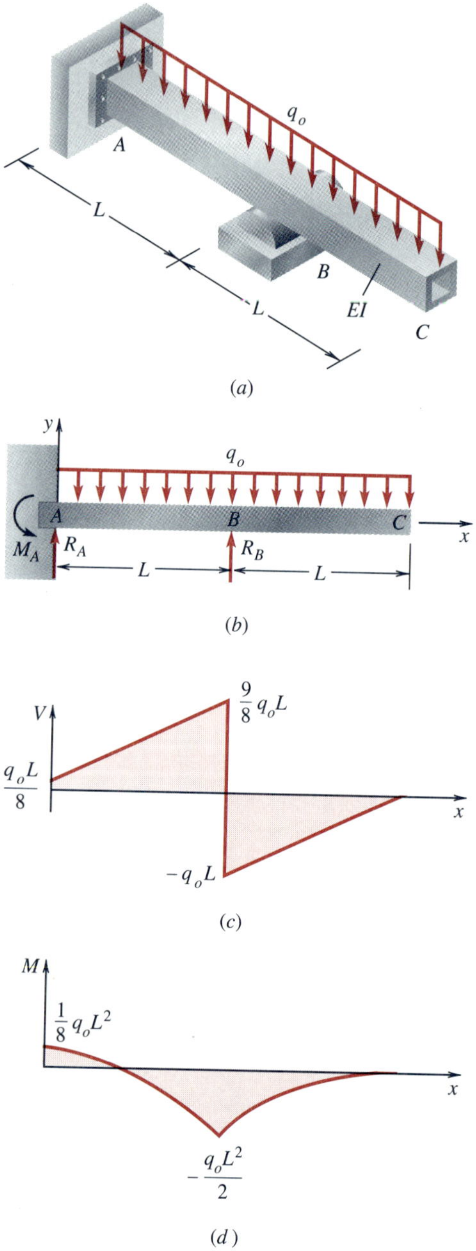

Figure 7.4 Example 7.3

$$EIv'' = M(x) = R_A\langle x\rangle^1 - M_A\langle x\rangle^0 - q_0\frac{\langle x\rangle^2}{2} + R_B\langle x - L\rangle^1 \tag{d}$$

where the constant of integration is equal to zero.

The terms in Eq. (*d*) correspond to the terms that contribute to the value of the moment about the right end of a beam segment of length x. It is useful to verify as an independent check, when you start with the expression for $q(x)$ and integrate to obtain $M(x)$, that the terms in $M(x)$ are as expected. With $M(x)$, we can integrate twice to find the expressions for the slope and the deflection.

Two integrations yield

$$EIv' = R_A\frac{\langle x\rangle^2}{2} - M_A\langle x\rangle^1 - q_0\frac{\langle x\rangle^3}{6} + R_B\frac{\langle x - L\rangle^2}{2} + c_1 \tag{e}$$

$$EIv = R_A\frac{\langle x\rangle^3}{6} - M_A\frac{\langle x\rangle^2}{2} - q_0\frac{\langle x\rangle^4}{24} + R_B\frac{\langle x - L\rangle^3}{6} + c_1x + c_2 \tag{f}$$

Since we have a total of five unknowns R_A, R_B, M_A, c_1, and c_2 and so far only the two equations represented by Eqs. (*a*), we need to identify three additional conditions of geometric compatibility. The clamped support at A, $x = 0$, and the simple support at B, $x = L$, provide the three conditions:

$$v(0) = 0 \qquad v'(0) = 0 \qquad v(L) = 0 \tag{g}$$

The application of the first two conditions in Eqs. (*g*) to Eq. (*e*) and (*f*) requires that the integration constants c_1 and c_2 be zero. The last condition involving $v(L)$ becomes, by using Eq. (*f*),

$$EIv(L) = 0 = R_A\frac{L^3}{6} - M_A\frac{L^2}{2} - q_0\frac{L^4}{24} \tag{h}$$

Now we note that the two equilibrium equations in Eqs. (*a*) and the geometric condition stated in Eq. (*h*) are three equations from which the three unknowns R_A, R_B, and M_A can be found. By multiplying the first equation in Eqs. (*a*) by L and subtracting from the second, it follows that

$$M_A = R_AL \tag{i}$$

which can also be found by taking moments about section B in Fig. 7.4*b*. Using Eqs. (*h*) and (*i*), we get

$$R_A = -q_0\frac{L}{8} \qquad M_A = -q_0\frac{L^2}{8} \tag{j}$$

and from Eq. (*a*),

$$R_B = \tfrac{17}{8}q_0L \tag{k}$$

The shear force and bending moment diagrams can be drawn by using Eqs. (*c*) and (*d*) now that the unknown reactions are all available; recall that $-EIv''' = V$. These are shown in Fig. 7.4*c* and *d*.

For the deflection at the free end, we substitute $x = 2L$ and the values for the reactions given in Eqs. (*j*) and (*k*) into Eq. (*f*) to get

$$v_C = v(2L) = -\frac{11}{48}\frac{q_0L^4}{EI} \tag{l}$$

The minus sign indicates that the deflection is downward. The deflection at $x = L/2$, that is, midway between the wall and the support, is

$$v\left(\frac{L}{2}\right) = \frac{q_0L^4}{96EI} \tag{m}$$

It is of interest to compare the stresses and the deflection response of the beam in Fig. 7.4*a* with the same beam without the support at B. For a cantilever beam of length $2L$, the maximum magnitude of the shear force is $2q_0L$, and the maximum magnitude of the bending moment is $2q_0L^2$. Referring to Fig. 7.4*c*, we see that the ratio of the maximum shear force in the simple cantilever to the propped one is $2q_0L/(9q_0L/8) = 1.78$, while for the maximum bending moment it is $2q_0L^2/(q_0L^2/2) = 4$. Thus the removal of the support at B in Fig. 7.4*a* would increase the maximum shear stress by 78 percent and the maximum normal bending stress by a factor of 4.

Consulting case G.1-3 in Table G.1 in App. G and taking L in the table equal to $2L$, we find that the magnitude of the end deflection for the cantilever beam in Fig. 7.4*a* without the center support at B is $q_0(2L)^4/(8EI)$ or $2q_0L^4/(EI)$. The ratio of this value to the result found in Eq. (*l*) is $[2q_0L^4/(EI)]/[11q_0L^4/(48EI)] = 8.73$. Thus the effect of adding the middle support is to reduce the value of the end deflection by a factor of 8.73 over the value without the support.

EXAMPLE 7.4

A beam of length L is clamped into rigid machine components at A and B, as shown in Fig. 7.5*a*. The component at B moves relative to A in such a way that end B of the beam undergoes an end deflection v_B and a rotation θ_B, as shown in Fig. 7.5*a*. Only end loads are acting on the beam. We wish to find the reactions at both supports and to show that if v_B and θ_B are related by $v_B = \theta_B L/2$, then the beam has no shear force acting at any section along its length and thus is in a state of pure bending. We neglect the weight of the beam and assume that EI is constant. The deflection of the beam arises from the specified deflection and rotation at end B.

Since the beam is clamped at both ends, this problem is statically indeterminate, and we take the unknown reactions as shown in Fig. 7.5*b*. Overall force and moment equilibrium of the beam, as shown in Fig. 7.5*b*, will be satisfied provided that

$$M_B + R_A L - M_A = 0 \tag{a}$$

where we have used the condition that $R_B = -R_A$ from force equilibrium. The unknown reactions are M_A, M_B, and R_A.

An expression for the bending moment $M(x)$ as a function of position along the beam will be derived from $q(x)$ by integration. The load intensity function takes the form

$$\begin{aligned} q(x) &= -M_A\langle x\rangle_{-2} - R_A\langle x\rangle_{-1} \\ &\quad + M_B\langle x - L\rangle_{-2} + R_A\langle x - L\rangle_{-1} \end{aligned} \tag{b}$$

where we have included all reaction forces and moments acting on the beam. Repeated integrations give

$$\begin{aligned} EIv''' = -V(x) &= M_A\langle x\rangle_{-1} - R_A\langle x\rangle^0 \\ &\quad - M_B\langle x - L\rangle_{-1} + R_A\langle x - L\rangle^0 \end{aligned} \tag{c}$$

and

$$\begin{aligned} EIv'' = M(x) &= M_A\langle x\rangle^0 - R_A\langle x\rangle^1 \\ &\quad - M_B\langle x - L\rangle^0 + R_A\langle x - L\rangle^1 \end{aligned} \tag{d}$$

where the constants of integration are equal to zero.

Again we see that the terms in the expression for $M(x)$ derived by integration of the $q(x)$ expression are easily identified with terms that would appear in an expression for $M(x)$ derived from the moment equilibrium of a segment of length x.

Once we have the expression for the moment in Eq. (*d*), we can make use of the moment-curvature relation, Eq. (6.6), in the form

$$\begin{aligned} EIv'' = M(x) &= M_A\langle x\rangle^0 - R_A\langle x\rangle^1 \\ &\quad - M_B\langle x - L\rangle^0 + R_A\langle x - L\rangle^1 \end{aligned} \tag{e}$$

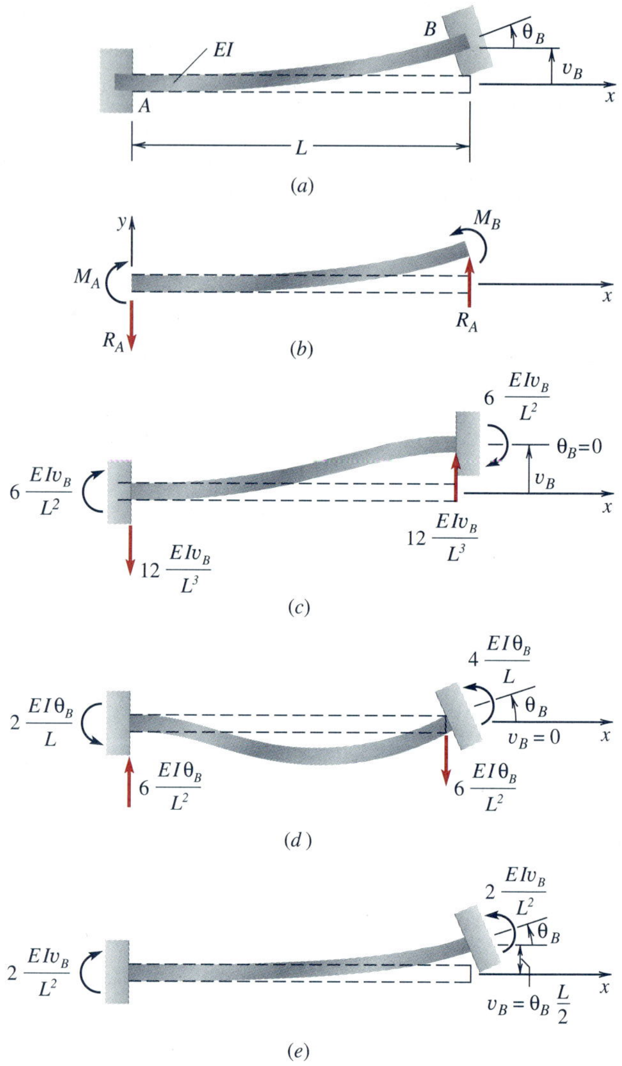

Figure 7.5 Example 7.4

Integrating Eq. (*e*) twice produces

$$\begin{aligned} EIv' &= M_A\langle x\rangle^1 - R_A\frac{\langle x\rangle^2}{2} \\ &\quad - M_B\langle x - L\rangle^1 + R_A\frac{\langle x - L\rangle^2}{2} + c_1 \end{aligned} \tag{f}$$

and

$$EIv = M_A \frac{\langle x \rangle^2}{2} - R_A \frac{\langle x \rangle^3}{6} - M_B \frac{\langle x - L \rangle^2}{2} + R_A \frac{\langle x - L \rangle^3}{6} + c_1 x + c_2 \qquad (g)$$

For the present problem, we have to find the three unknown reactions R_A, M_A, and M_B and the two integration constants c_1 and c_2. From Fig. 7.5*a*, we see that geometric compatibility of the beam at the clamped supports requires that at A

$$x = 0: \qquad v = 0 \quad \text{and} \quad v' = 0 \qquad (h)$$

and at B

$$x = L: \qquad v = v_B \quad \text{and} \quad v' = \theta_B \qquad (i)$$

These four geometric conditions, Eqs. (*h*) and (*i*), together with the equilibrium condition, Eq. (*a*), constitute five equations for the five unknowns in this problem. Using Eqs. (*f*) and (*h*), the conditions at A, $x = 0$, we find

$$EIv'(0) = c_1 = 0 \qquad EIv(0) = c_2 = 0 \qquad (j)$$

The geometric conditions at B will serve to connect the unknown reactions R_A and M_A to the values of the support deflection and slope. Again using Eqs. (*f*) and (*g*) and the conditions in Eqs. (*i*), we find

$$EIv'(L) = EI\theta_B = M_A L - R_A \frac{L^2}{2} \qquad EIv(L) = EIv_B = M_A \frac{L^2}{2} - R_A \frac{L^3}{6} \qquad (k)$$

Solving for the reactions yields

$$M_A = \frac{2EI}{L^2}(3v_B - L\theta_B) \qquad (l)$$

$$R_A = \frac{6EI}{L^3}(2v_B - L\theta_B) \qquad (m)$$

The remaining reaction M_B can be obtained from Eq. (*a*) by using Eqs. (*l*) and (*m*) to get

$$M_B = \frac{2EI}{L^2}(-3v_B + 2L\theta_B) \qquad (n)$$

We are now in a position to consider several special cases of the above problem that are of some interest in practical situations, e.g., in the analysis of support settlement effects in structures. First, consider the case when $\theta_B = 0$; the beam undergoes a relative end deflection without rotation at the supports. From Eqs. (*l*) and (*n*) with $\theta_B = 0$, we get

$$M_A = \frac{6EI}{L^2} v_B = -M_B \qquad (o)$$

and from Eq. (*m*)

$$R_A = 12 \frac{EI}{L^3} v_B \qquad (p)$$

These results are shown in Fig. 7.5*c*. Since both of the support moments are acting in the same direction, the shear forces at the supports are equal and opposite to maintain moment equilibrium.

For another special case we set $v_B = 0$, that is, the right end of the beam undergoes a rotation without any deflection of the endpoint B. With $v_B = 0$, Eqs. (*l*) and (*n*) yield

$$M_A = -2 \frac{EI}{L} \theta_B \qquad (q)$$

$$M_B = -2M_A \qquad (r)$$

$$R_A = -6 \frac{EI}{L^2} \theta_B \qquad (s)$$

The reactions in this case are shown in Fig. 7.5*d*. Again the moment reactions are acting in the same direction so that the shear force reactions must supply the counteracting moment. However, in this case the moment reaction at the end undergoing the rotation is twice as large as the moment at the nonrotating end. Compare this with the previous case where the moments were exactly the same at both ends for a beam undergoing a deflection at one end without any rotation at that end.

Finally, we notice from Eq. (*m*) that if v_B and θ_B are in the relation $2v_B = \theta_B L$, then

$$R_A = 0 \qquad (t)$$

and from Eqs. (*l*) and (*n*)

$$M_A = 2 \frac{EI}{L^2} v_B \qquad (u)$$

$$M_B = 2 \frac{EI}{L^2} v_B = M_A \qquad (v)$$

This case is shown in Fig. 7.5*e*, and we see that the beam is in a state of pure bending, i.e., no shear forces are acting on the beam.

7.3 Method of Superposition

In Sec. 6.5 we introduced the technique of solving beam deflection problems by adding known solutions and using the principle of superposition. All the beam problems treated in that section were statically determinate. In this section we will show how the principle of superposition can be used to advantage to solve statically indeterminate beam problems. Of course, to use the superposition method, we must have available a table of known solutions to beam problems such as those given in App. G.

EXAMPLE 7.5

The beam of length L shown in Fig. 7.6*a* is clamped at A and simply supported at B and is subjected to a constant distributed loading q_0. We wish to find the value of the reaction at support B, to sketch the shear force and bending moment diagrams for the beam, and to find the location and value of the maximum downward deflection of the beam. We assume that the bending modulus EI of the beam is constant over the length.

A free-body diagram of the beam that defines the support reactions is shown in Fig. 7.6*b*. We will use the method of superposition to solve this problem.

First, we notice that the original problem in Fig. 7.6*b* can be replaced by the superposition of the two cantilever beam problems given in Fig. 7.6*c* and *d*. Consulting the table of beam deflection results in App. G, we find that the cantilever beam under uniform loading q_0 acting downward is given by case G.1-3 in Table G.1, and the deflection at the free end in this case is

$$v_{BI} = -\frac{q_0 L^4}{8EI} \qquad (a)$$

Consulting case G.1-2 we find that the deflection at the free end of a cantilever beam under a concentrated load R_B acting at the free end is given by

$$v_{BII} = \frac{R_B L^3}{3EI} \qquad (b)$$

Since no deflection takes place at support B, we require that

$$v_{BI} + v_{BII} = 0 \qquad (c)$$

Using Eqs. (*a*) and (*b*) with Eq. (*c*), we have

$$-\frac{q_0 L^4}{8EI} + \frac{R_B L^3}{3EI} = 0$$

and the reaction at support B follows as

$$R_B = \tfrac{3}{8} q_0 L \qquad (d)$$

Force equilibrium in the y direction (Fig. 7.6*b*) then yields

$$R_A = q_0 L - \tfrac{3}{8} q_0 L = \tfrac{5}{8} q_0 L \qquad (e)$$

and moment equilibrium about section A leads to

$$M_A + R_B L - q_0 \frac{L^2}{2} = 0$$

or using Eq. (*d*), we have

$$M_A = \frac{q_0 L^2}{2} - \frac{3}{8} q_0 L^2 = \frac{1}{8} q_0 L^2 \qquad (f)$$

Now that all of the reactions are known, we can sketch the shear force and bending moment diagrams as given in Fig. 7.6*e* and *f*.

To find the maximum downward deflection for the beam by using superposition, we can construct the deflection curve by a superposition of the deflection curves corresponding to the two cantilever beams in Fig. 7.6*c* and *d* which are given by cases G.1-3 and G.1-2 in App. G. For case G.1-3 (Fig. 7.6*c*), we have

$$v_I = -\frac{q_0 x^2}{24EI}(6L^2 - 4Lx + x^2) \qquad (g)$$

and for case G.1-2 (Fig. 7.6*d*) with $R_B = 3q_0 L/8$, we have

$$v_{II} = \frac{1}{16}\frac{q_0 L x^2}{EI}(3L - x) \qquad (h)$$

where both v_I and v_{II} are positive in the positive y direction.

If we add the deflection curves in Eqs. (*g*) and (*h*), we find

$$v = \frac{q_0 x^2}{48EI}(-3L^2 + 5Lx - 2x^2) \qquad (i)$$

This is the deflection curve that would have been obtained if we had proceeded to solve this problem by the methods given in Sec. 7.2.

To locate the maximum value of the downward deflection, we differentiate v with respect to x and set the derivative equal to zero to get

$$\frac{dv}{dx} = \frac{q_0 \hat{x}}{48EI}(-6L^2 + 15L\hat{x} - 8\hat{x}^2) = 0 \qquad (j)$$

where $\hat{x}$ denotes the x coordinate of points at which the slope vanishes. The solution $\hat{x} = 0$ gives us the obvious result that the slope is zero at the fixed end of the beam. The roots of the quadratic factor in Eq. (j) are

$$\hat{x} = 0.578L \qquad \hat{x} = 1.297L \qquad (k)$$

Since the second root corresponds to a location beyond the right end of the beam, we reject it and take the first root. The deflection $\hat{v}$ corresponding to $\hat{x} = 0.578L$ is found by using Eq. (i) to get

$$\hat{v} = v(0.578L) = -0.260\frac{q_0L^4}{48EI}$$
$$= -0.043\frac{q_0L^4}{8EI} \qquad (l)$$

Equation (l) gives the maximum deflection of the beam.

It is interesting to compare the maximum downward deflection of a cantilever beam without a support at B, as shown in Fig. 7.6c and given by Eq. (a) in the form

$$v_{B1} = -\frac{q_0L^4}{8EI} \qquad (m)$$

with the maximum downward deflection of the cantilever beam with the support at the end as shown in Fig. 7.6a that we have just calculated in Eq. (l). From Eqs. (l) and (m) we see that the support has the effect of reducing the maximum deflection by about 96 percent.

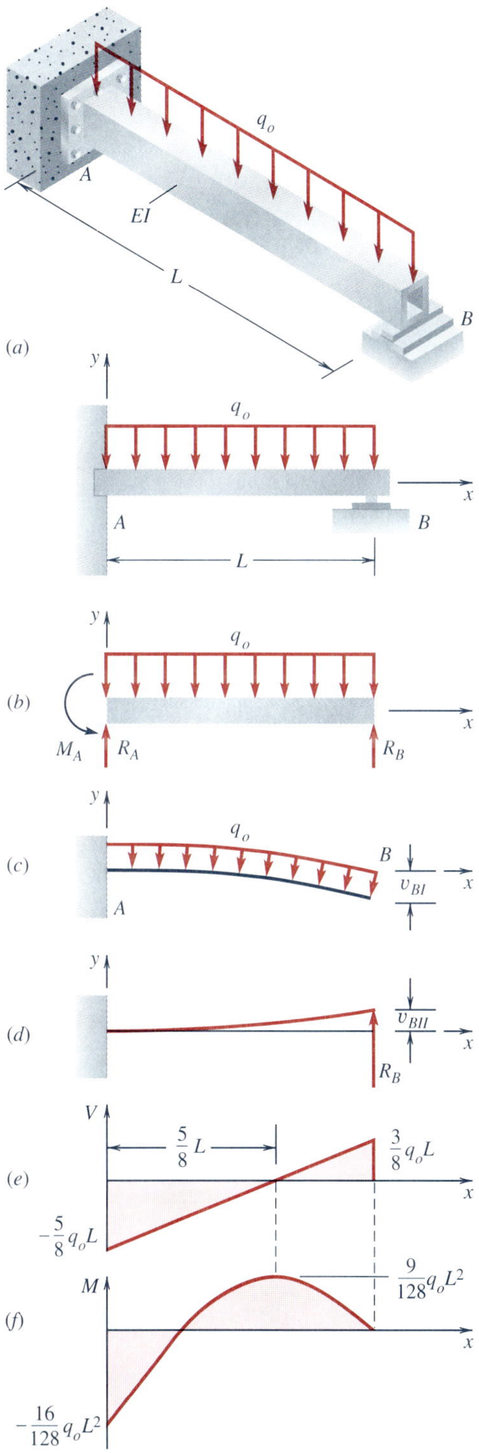

Figure 7.6 Example 7.5

EXAMPLE 7.6

The beam of length L shown in Fig. 7.7*a* is simply supported at both ends and at its midpoint and is subjected to a constant distributed loading q_0. If the bending modulus EI is a constant, we wish to do the following:

(*a*) Find the percentage of the total load q_0L carried by each of the supports,
(*b*) Sketch the shear force and bending moment diagrams, and
(*c*) Determine the maximum deflection of the beam and compare it to the maximum deflection without a central support.

We will use the method of superposition to solve this problem. Note the similarity of this example to Example 7.2 in which the beam has length $2L$.

The beam in Fig. 7.7*a* can be regarded as the superposition of the two beam problems I and II, as shown in Fig. 7.7*b* and *c*. In order that the superposition corresponds to the actual beam in Fig. 7.7*a*, we must enforce the geometric compatibility condition that there should be no deflection at support B. That is, in the present notation

$$v_{BI} + v_{BII} = 0 \tag{a}$$

For case I we can use case G.2-3 of App. G, Table G.2 to get

$$v_{BI} = -\frac{5}{384}\frac{q_0L^4}{EI} \tag{b}$$

and for case II, case G.2-2 gives us

$$v_{BII} = \frac{R_BL^3}{48EI} \tag{c}$$

Substituting the results in Eqs. (*b*) and (*c*) into Eq. (*a*), we find

$$-\frac{5}{384}\frac{q_0L^4}{EI} + \frac{R_BL^3}{48EI} = 0$$

so that

$$R_B = \tfrac{5}{8}q_0L \tag{d}$$

Therefore, we note that the middle support carries 62.5 percent of the total load and that the remaining 37.5 percent is divided equally between the two symmetrically placed supports, as shown in Fig. 7.7*d*. If the beam were to be replaced by two separate beams of length $L/2$, as shown in Fig. 7.7*e*, then the center support would carry $2(q_0L/4)$ or $\frac{4}{8}q_0L$, or less than the value given by Eq. (*d*).

Now that R_B is known, the problem is no longer statically indeterminate and the shear force and bending moment diagrams can be constructed by the methods discussed in Chap. 4. They are given in Fig. 7.8.

We can obtain the deflection curve for the beam of Fig. 7.7*a* by using the deflection curves for cases I and II that are given in App. G. For case I, case G.2-3 in Table G.2, we have

$$v_I = -\frac{q_0x}{24EI}(L^3 - 2Lx^2 + x^3) \tag{e}$$

and for case II, case G.2-2 with $R_B = 5q_0L/8$, we have

$$v_{II} = \frac{5q_0L}{384EI}\left(-4x^3 + 8\left\langle x - \frac{L}{2}\right\rangle^3 + 3L^2x\right) \tag{f}$$

Because of the symmetry of the deflection curve in Fig. 7.7*d* about the center support at B, we need only find the maximum downward deflection for a point between locations A and B. Thus we can drop the bracket term $\langle x - L/2\rangle^3$ in Eq. (*f*) since x is less than $L/2$ for points in segment AB of the beam. Adding expression v_I from Eq. (*e*) to expression v_{II} (without the $\langle x - L/2\rangle^3$ term), we get, for $x < L/2$,

$$v = v_I + v_{II} = \frac{q_0x}{384EI}(-L^3 + 12Lx^2 - 16x^3) \tag{g}$$

The maximum value of the deflection is found by calculating dv/dx and setting the expression equal to zero to get

$$\frac{dv}{dx} = \frac{q_0}{384EI}(-L^3 + 36L\hat{x}^2 - 64\hat{x}^3) = 0 \tag{h}$$

where $\hat{x}$ denotes values of x satisfying Eq. (*h*), i.e., the value of the location between A and B where $v' = 0$. First we observe that symmetry requires that $v' = 0$ at support B. Thus $\hat{x} - L/2$ must be a factor of the cubic polynomial in Eq. (*h*). Carrying out this factorization gives

$$v' = \frac{q_0(\hat{x} - L/2)}{384EI}(2L^2 + 4L\hat{x} - 64\hat{x}^2) = 0 \tag{i}$$

The roots of the quadratic factor are

$$\hat{x} = 0.2108L \qquad \hat{x} = -0.1483L \tag{j}$$

Discarding the negative root, we get the maximum downward deflection by substituting the value $\hat{x} = 0.2108L$ into Eq. (*g*) to find

$$\hat{v} = v(0.2108L) = -\frac{0.130q_0L^4}{384EI} \tag{k}$$

The ratio of the maximum deflection for the beam with the central support given by Eq. (*k*) to the maximum for the same beam but without the center support [see Eq. (*b*)] is

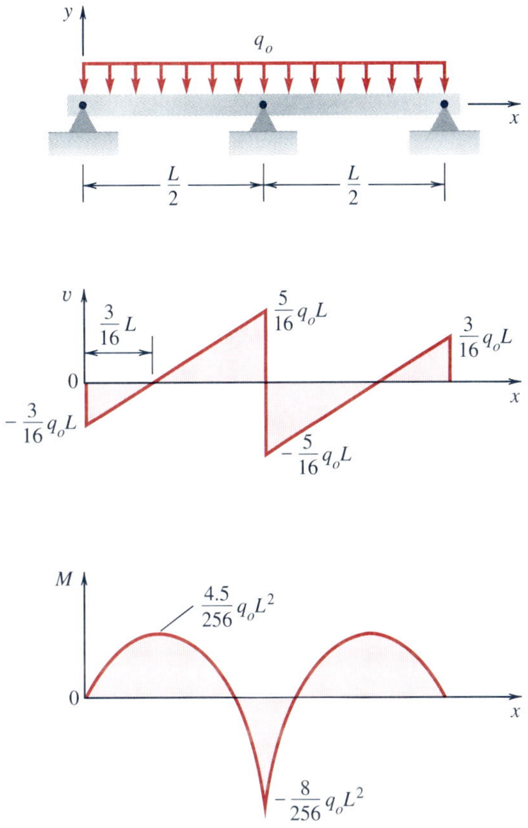

Figure 7.8 Example 7.6: Shear force and bending moment diagrams.

$$\frac{-0.130}{-5} = 0.026 \qquad (l)$$

so that the center support has the effect of reducing the maximum deflection by 97.4 percent.

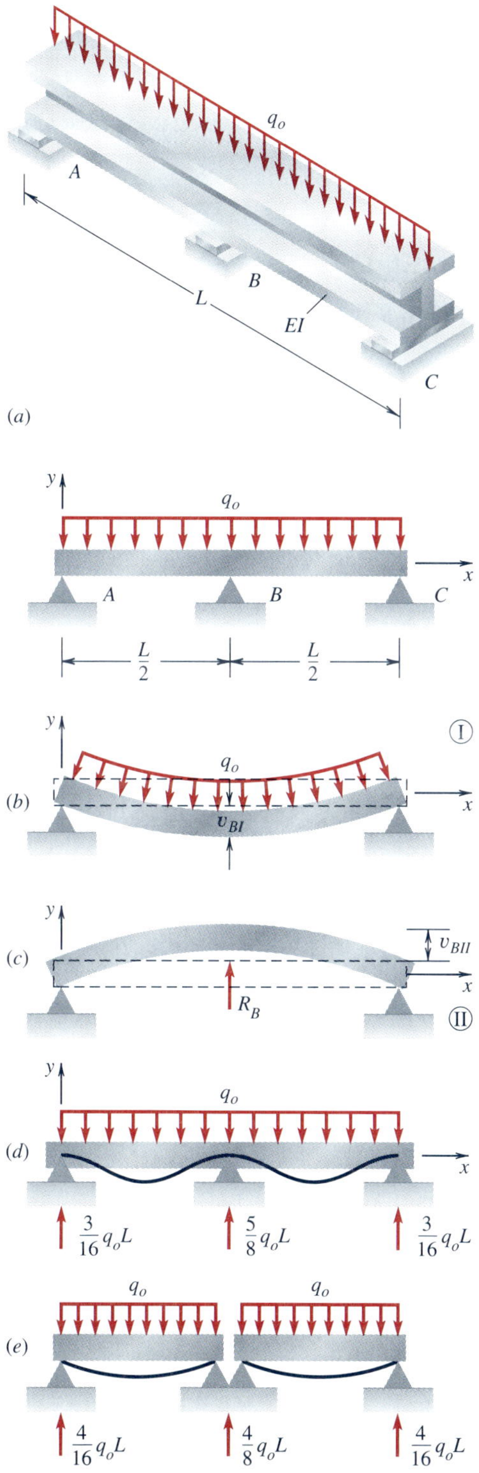

Figure 7.7 Example 7.6

7.4 Displacement Method for Beams

In many of the previous examples we have derived analytical expressions for the deflection of a loaded beam. We were then able to obtain results for the numerical values of the maximum deflection, maximum slope, and the reactions. In engineering practice, often we need to find the deflection at a number of points along a beam or along a machine component modeled as a beam. Furthermore, many problems in structural analysis and machine design involve a large number of interconnected beam members with different loads and supports. The analysis of these problems often reduces to a repetition of fairly simple free-body diagrams with associated equilibrium equations and the use of relations connecting the applied forces and moments to slopes and deflections, in other words, the use of the three steps in Eq. (2.8). It is worthwhile, therefore, to set up a systematic approach and notation which lend themselves to an organized derivation of the equations of equilibrium from which the deflections and slopes of the beam members can be found and which can in turn be used to find the bending moments and shear forces. The coordinate axes, sign conventions, and equations that we plan to use are reviewed in Fig. 7.9. In what follows, it is important to keep these sign conventions clearly in mind.

Consider a beam made up of four distinct segments or elements loaded as shown in Fig. 7.10*a*. Each segment has, in general, a different bending modulus *EI*. We wish to develop a procedure that treats each segment as a unit and then combines the load-response relation of each segment with the applied loads and supports into the load-response relation of the complete beam. To do this, we need to account for segments that have different values of *EI*, segments between concentrated applied loads and moments, and segments with constant distributed loads. We will treat constant distributed loads only in the present formulation; see Probs. 7.9-7 to 7.9-11. The procedure we follow is similar to the displacement method for bars and for circular shafts treated in Chaps. 2 and 3.

Consider a general system of beam segments or elements as shown in Fig. 7.10*b*. In what follows we refer to each segment as a *beam element*. The endpoints of each element are called *nodes*, and we say that each element has a left and right node. The elements may have different cross sections and consist of different materials, but we will restrict attention to beam element configurations where the neutral axis of each beam element can be joined together end to end to form the single neutral axis for the complete beam. The elements may be of different materials, but each element must be homogeneous and must have a constant cross-sectional area over its length.

If we define internal nodes by the junction points between each of the beam elements and add the first and the last nodes to those, then we will have beam elements (1) to $(N - 1)$ defined by node points 1 to

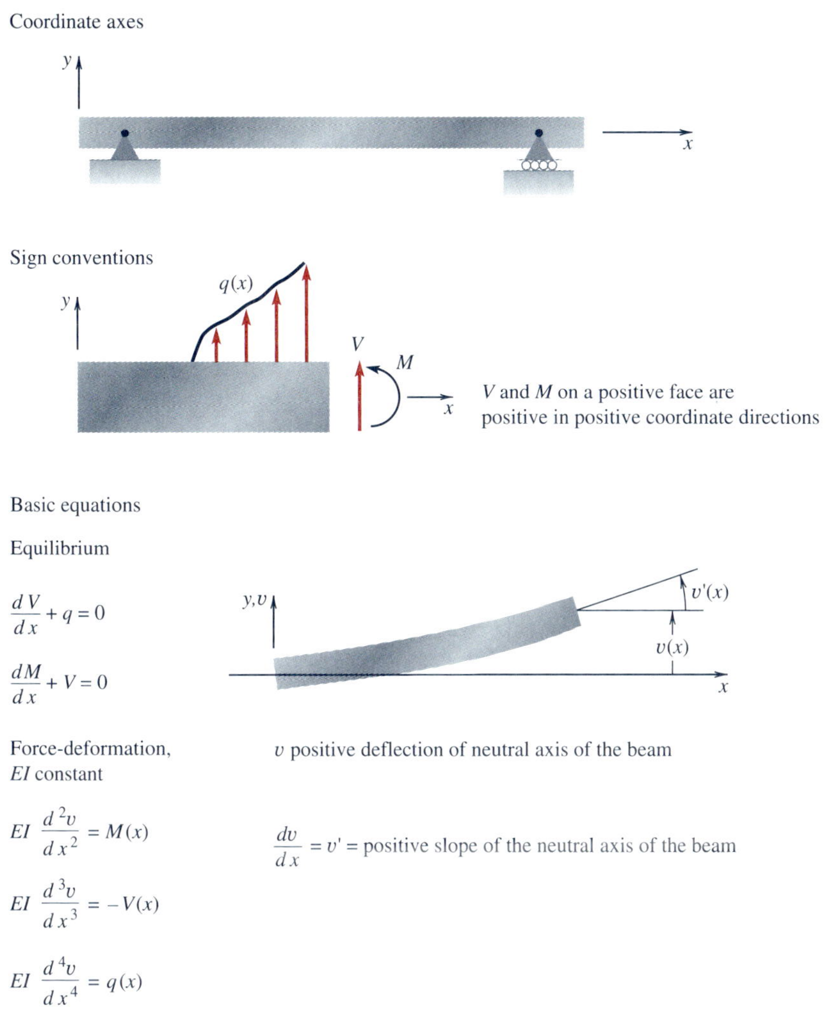

Coordinate axes
y
x
Sign conventions
q(x)
y
V
M
x
V and M on a positive face are positive in positive coordinate directions
Basic equations
Equilibrium
$\frac{dV}{dx} + q = 0$
$\frac{dM}{dx} + V = 0$
y,v
v'(x)
v(x)
x
Force-deformation, EI constant
v positive deflection of neutral axis of the beam
$EI \frac{d^2v}{dx^2} = M(x)$
$\frac{dv}{dx} = v' =$ positive slope of the neutral axis of the beam
$EI \frac{d^3v}{dx^3} = -V(x)$
$EI \frac{d^4v}{dx^4} = q(x)$

Figure 7.9 Basic sign conventions and coordinate axes for beam analysis.

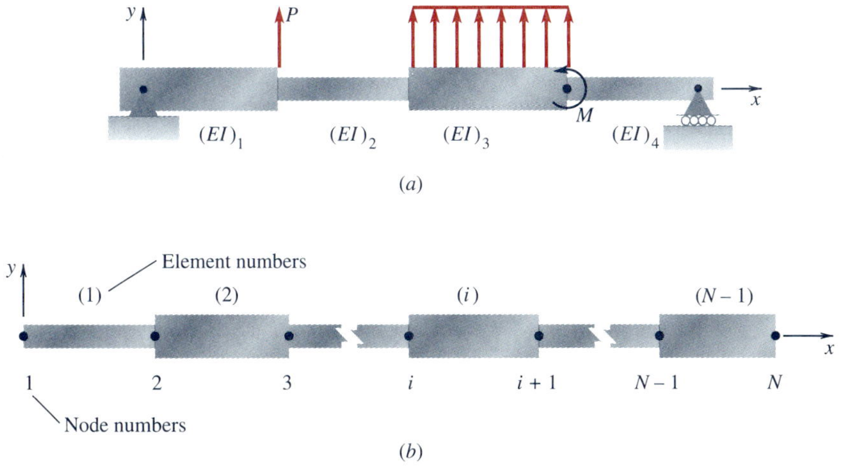

Figure 7.10 Beam made up of beam elements. (*a*) Four elements. (*b*) General system of beam elements.

N, as shown in Fig. 7.10b. For example, beam element (i) has the left end node i and the right end node $i + 1$. We restrict ourselves to beam problems for which the choice of elements and nodes can be made so that the applied loads and supports of the complete beam have the following properties:

1. Applied concentrated loads, moments, and reaction loads from the supports act only at nodes, i.e., at the junction point of two beam elements or at the endpoints. The applied loads are denoted by P_k (lb or N) and M_k (lb · in or N · m), where the subscript k indicates the node number at which the load is applied.
2. A constant distributed transverse load applied to an element is denoted by q (lb/in or N/m).
3. The bending modulus and the length of an element are denoted by EI (lb · in^2 or N · m^2) and L (in or m).

All specified external transverse forces and node deflections are positive in the positive y direction, and specified external moments and node angles are positive in the counterclockwise direction, i.e., are positive in the positive z direction according to the right-hand rule.

In using this approach, it is clearly necessary to introduce nodes at each change in the beam cross-sectional area, at each change in beam material, at each point where a specified force or moment is applied, at each point where a constant distributed load starts and where the load ends, at each support point, and at each point where a deflection or slope is specified (e.g., at a point where the support requires that either the deflection or the slope or both be zero or equal to some specified values).

Our approach to solving problems with a number of beam elements (Fig. 7.10b) under different conditions of loading and support will be to set up an analysis procedure that will generate a system of simultaneous linear algebraic equations for the unknown node deflections and slopes. After solving these linear equations and obtaining the unknown deflections and slopes, we will be able to calculate the internal shear forces and bending moments at the ends of each beam element as well as external reactions at the supports. You will see that the steps in the analysis are a very systematic version of those outlined in Eq. (2.8) and lend themselves to being organized into a computer program. The procedure is similar to the displacement method used for uniaxial loads in Chap. 2 and for twisting moments in Chap. 3.

The approach we will introduce works for both statically determinate and statically indeterminate beam problems. First we need to obtain the force-deformation relation for a beam element.

7.5 Derivation of Equations Relating Element End Shear Forces and Bending Moments to Element End Deflections and Slopes

Consider a free-body diagram (Fig. 7.11*a*) of a typical beam element taken, e.g., from the beam shown in Fig. 7.10*b*. For the purpose of this derivation, we will identify or number the left node and right node of the beam element as 1 and 2, respectively. The end shear force and bending moment at node 1 at the left end of the element are denoted by V_1 and M_1. The shear force and bending moment at node 2 at the right end of the element are denoted by V_2 and M_2. The distributed transverse load per unit length q is equal to a constant value over the element. The node deflections, v_1 at node 1 and v_2 at node 2, and the node slopes v_1' and v_2' are shown in Fig. 7.11*b*. Quantities are positive in the directions as shown in Fig. 7.11.

In the analysis and problems to follow, when we need to refer to a specific single element (i), we will add a superscript (i) to the shear force and bending moment terms shown in Fig. 7.11, for example, $V_1^{(i)}$, $M_2^{(i)}$, etc., so that it will be clear as to which element (in this case the ith element) we are referring. Dropping this superscript notation for a single element will simplify the notation in the derivation to follow.

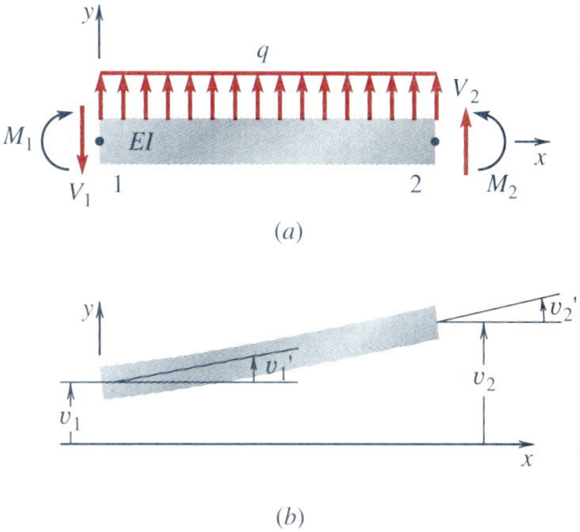

Figure 7.11 (*a*) Positive shear forces and bending moments at the ends of a beam element. (*b*) Positive deflections and slopes at the ends of a beam element.

Force Deformation. The differential relation connecting the deflection $v(x)$ of an element to the constant uniform load q acting on the element was derived in Sec. 6.4, Eq. (6.20), and can be written

$$EI\frac{d^4v}{dx^4} = q \tag{7.1}$$

where x is the coordinate distance (local coordinate system), measured from node 1 at the left end of the element; q is the constant transverse load per unit length; and EI is the bending modulus. Both q and v are positive in the positive y direction. The differential relation Eq. (7.1) is valid along the beam element. Our objective in the analysis to follow is to find four equations which will allow us to express the shear forces and bending moments at the ends of the element in terms of the deflections and slopes at the ends of the element. That is, we wish to find relations between V_1, M_1, V_2, M_2 and v_1, v_1', v_2, v_2'. This will be the force-deformation relation for the element.

Since q is a constant over the beam element, it follows by simple integration of Eq. (7.1) four times that

$$v(x) = a_1 + a_2x + a_3x^2 + a_4x^3 + \frac{qx^4}{24EI} \tag{7.2}$$

where a_1, a_2, a_3, and a_4 are the constants of integration. From the definition of v_1, v_1', v_2, and v_2', we have

$$v(0) = v_1 = a_1 \tag{7.3}$$

$$v'(0) = v_1' = a_2 \tag{7.4}$$

and

$$v(L) = v_2 = v_1 + v_1'L + a_3L^2 + a_4L^3 + \frac{qL^4}{24EI} \tag{7.5}$$

$$v'(L) = v_2' = v_1' + 2a_3L + 3a_4L^2 + \frac{4qL^3}{24EI} \tag{7.6}$$

Solving Eqs. (7.5) and (7.6) for a_3 and a_4 and substituting back into Eq. (7.2), we find

$$\begin{aligned} v(x) = {} & v_1\left(1 - \frac{3x^2}{L^2} + \frac{2x^3}{L^3}\right) + Lv_1'\left(\frac{x}{L} - \frac{2x^2}{L^2} + \frac{x^3}{L^3}\right) \\ & + v_2\left(\frac{3x^2}{L^2} \quad \frac{2x^3}{L^3}\right) + Lv_2'\left(-\frac{x^2}{L^2} + \frac{x^3}{L^3}\right) \\ & + \frac{qL^4}{24EI}\left(\frac{x^2}{L^2} - \frac{2x^3}{L^3} + \frac{x^4}{L^4}\right) \end{aligned} \tag{7.7}$$

Equation (7.7) provides the expression for the transverse deflection of the beam element in terms of the node deflection values v_1 and v_2 and the node slope values v_1' and v_2' for a given loading q and element parameters L and EI.

The next step is to relate the values of the end deflections and slopes to the values of the shear forces and bending moments at the ends of the element. To do this, we calculate the bending moments and shear forces associated with the deflection curve given by Eq. (7.7). Using the basic equation relating the bending moment to the curvature of the deflection curve as derived in Sec. 6.2, Eq. (6.6), we have

$$M(x) = EIv'' \tag{7.8}$$

and using Eq. (7.7), we get

$$\begin{aligned} M(x) = EI\Bigg[& v_1\left(-\frac{6}{L^2} + \frac{12x}{L^3}\right) + Lv_1'\left(-\frac{4}{L^2} + \frac{6x}{L^3}\right) \\ & + v_2\left(\frac{6}{L^2} - \frac{12x}{L^3}\right) + Lv_2'\left(-\frac{2}{L^2} + \frac{6x}{L^3}\right) \\ & + \frac{qL^4}{24EI}\left(\frac{2}{L^2} - \frac{12x}{L^3} + 12\frac{x^2}{L^4}\right)\Bigg] \end{aligned} \tag{7.9}$$

The values of the moment at each end are

$$M(0) = M_1 \qquad M(L) = M_2 \tag{7.10}$$

Using Eq. (7.9) to evaluate $M(0)$ and $M(L)$ and substituting into Eq. (7.10), we find

$$M_1 = \frac{EI}{L^2}(-6v_1 - 4Lv_1' + 6v_2 - 2Lv_2') + \frac{qL^2}{12} \tag{7.11}$$

$$M_2 = \frac{EI}{L^2}(6v_1 + 2Lv_1' - 6v_2 + 4Lv_2') + \frac{qL^2}{12} \tag{7.12}$$

We now wish to relate the values of the shear forces at the ends of the element to the node deflections and slopes. Using Eq. (6.18), we have for a constant EI

$$V(x) = -EIv''' \tag{7.13}$$

and after differentiating $v(x)$ from Eq. (7.7), we obtain

$$V(x) = -\frac{EI}{L^3}\left[12v_1 + 6Lv_1' - 12v_2 + 6Lv_2' + \frac{qL^4}{24EI}\left(-12 + 24\frac{x}{L}\right)\right] \tag{7.14}$$

The values of the shear force at each node are

$$V(0) = V_1 \qquad V(L) = V_2 \tag{7.15}$$

Using Eq. (7.14) to evaluate $V(0)$ and $V(L)$ and substituting into Eq. (7.15), we find

$$V_1 = \frac{EI}{L^3}(-12v_1 - 6Lv_1' + 12v_2 - 6Lv_2') + \frac{qL}{2} \tag{7.16}$$

$$V_2 = \frac{EI}{L^3}(-12v_1 - 6Lv_1' + 12v_2 - 6Lv_2') - \frac{qL}{2} \tag{7.17}$$

Examining Eqs. (7.11) and (7.12) and (7.16) and (7.17), we see that the element end shear forces and moments are completely determined by the end deflections and slopes for a given constant distributed loading q and beam element parameters EI and L. A summary of these beam force-deformation relations is given in Fig. 7.12.

The beam force-deformation relations for a single element shown in Fig. 7.12 provide a building block from which the response of a beam with many different elements and loads can be constructed. Note that once the deflections and slopes at the ends of the element are known, then the deflection expression valid for all points along the element can be found from Eq. (7.7). In addition, once the deflections and slopes at the ends of the element are known, the values of the shear force and bending moment along the element can be found from Eqs. (7.9) and (7.14). In the special case in which $q = 0$, the shear force along the beam element is constant and the bending moment is linear.

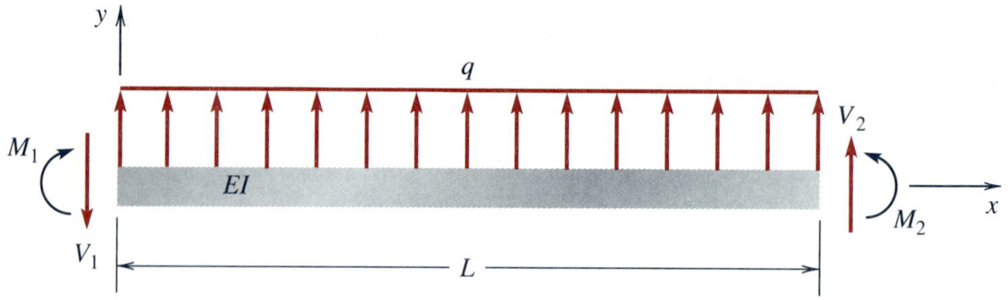

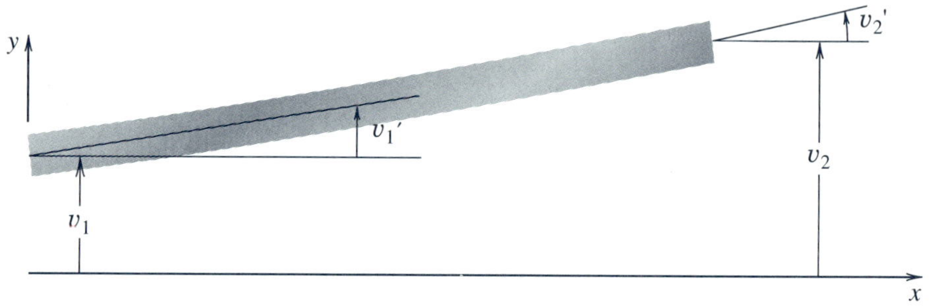

$$M_1 = \frac{EI}{L^2}\{-6v_1 - 4Lv_1' + 6\,v_2 - 2Lv_2'\} + \frac{qL^2}{12}$$

$$V_1 = \frac{EI}{L^3}\{-12v_1 - 6Lv_1' + 12\,v_2 - 6Lv_2'\} + \frac{qL}{2}$$

$$M_2 = \frac{EI}{L^2}\{6v_1 + 2Lv_1' - 6\,v_2 + 4Lv_2'\} + \frac{qL^2}{12}$$

$$V_2 = \frac{EI}{L^3}\{-12v_1 - 6Lv_1' + 12\,v_2 - 6Lv_2'\} - \frac{qL}{2}$$

Figure 7.12 Beam force-deformation relations.

7.6 Application of the Force-Deformation Relations—Single-Element Beam Problems

We will show first how simple single-element beam problems can be easily solved directly by using the force-deformation relations given in Fig. 7.12.

EXAMPLE 7.7

A cantilever beam of length L is fixed at A and carries a concentrated load P at B, as shown in Fig. 7.13*a*. We wish to find the deflection and slope at B, using the basic beam force-deformation relations given in Fig. 7.12. We neglect the weight of the beam.

We treat the beam as a single element with a node at the fixed support A and a node at the free end B, as shown in Fig. 7.13*b*. The coordinate axes are as shown.

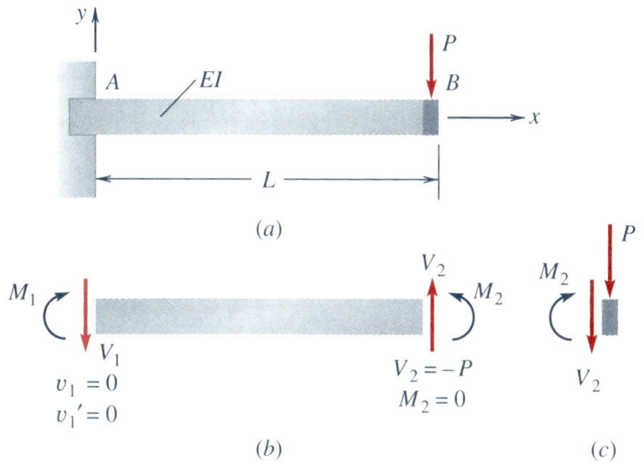

Figure 7.13 Example 7.7

At the fixed support A (node 1), we have the geometric conditions

$$v_1 = 0 \qquad v_1' = 0$$

while at B (node 2), using the equilibrium of the infinitesimal element at node 2 (Fig. 7.13*c*), we have

$$V_2 = -P \qquad M_2 = 0$$

Therefore using the basic force-deformation relations in Fig. 7.12 which contain the known variables V_2 and M_2 with $q = 0$, we find

$$M_2 = 0 = \frac{EI}{L^2}(-6v_2 + 4Lv_2') \tag{a}$$

$$V_2 = -P = \frac{EI}{L^3}(12v_2 - 6Lv_2') \tag{b}$$

The first equation gives

$$v_2 = \tfrac{2}{3}Lv_2' \tag{c}$$

which then can be substituted into Eq. (*b*) to obtain

$$v_2' = -\frac{PL^2}{2EI} \tag{d}$$

and from Eq. (*c*)

$$v_2 = -\frac{PL^3}{3EI} \tag{e}$$

Equations (*d*) and (*e*) give the well-known relations for a cantilever beam with a concentrated load at its end.

EXAMPLE 7.8

Given is the simply supported beam under a constant distributed loading q_0, as shown in Fig. 7.14*a*. We wish to find the reactions, the slope of the deflection curve at each end, and the deflection at the center of the beam. Again we treat this beam as a single beam element with a node at each support.

The element is shown in Fig. 7.14*b* with node slopes v_1' and v_2' and reactions R.

The deflection and moment at each end are zero

$$v_1 = v_2 = 0 \qquad M_1 = M_2 = 0 \tag{a}$$

From the results in Fig. 7.12 we have, with $q = -q_0$,

$$M_1 = 0 = \frac{EI}{L^2}(-4Lv_1' - 2Lv_2') - \frac{q_0L^2}{12} \tag{b}$$

$$M_2 = 0 = \frac{EI}{L^2}(2Lv_1' + 4Lv_2') - \frac{q_0L^2}{12} \tag{c}$$

Solving for v_1' and v_2', we have

$$v_1' = -\frac{q_0L^3}{24EI} \qquad v_2' = \frac{q_0L^3}{24EI} \tag{d}$$

The shear force at each end then follows from the V_1 and V_2 expressions in Fig. 7.12 in the form

$$V_1 = -\frac{q_0L}{2} = -R \qquad V_2 = \frac{q_0L}{2} = R \tag{e}$$

from which both support reactions R are found (Fig. 7.14*c*).

The deflection curve is given by Eq. (7.7) with the values of v_1' and v_2' from Eq. (*d*); we find

$$v(x) = -\frac{q_0L^4}{24EI}\left(\frac{x}{L} - \frac{x^2}{L^2}\right) - \frac{q_0L^4}{24EI}\left(\frac{x^2}{L^2} - 2\frac{x^3}{L^3} + \frac{x^4}{L^4}\right) \tag{f}$$

The deflection at the midpoint of the beam is found by setting $x = L/2$ in Eq. (*f*) to get

$$v\left(\frac{L}{2}\right) = -\frac{5}{384}\frac{q_0L^4}{EI} \tag{g}$$

in agreement with our earlier result, Example 6.2.

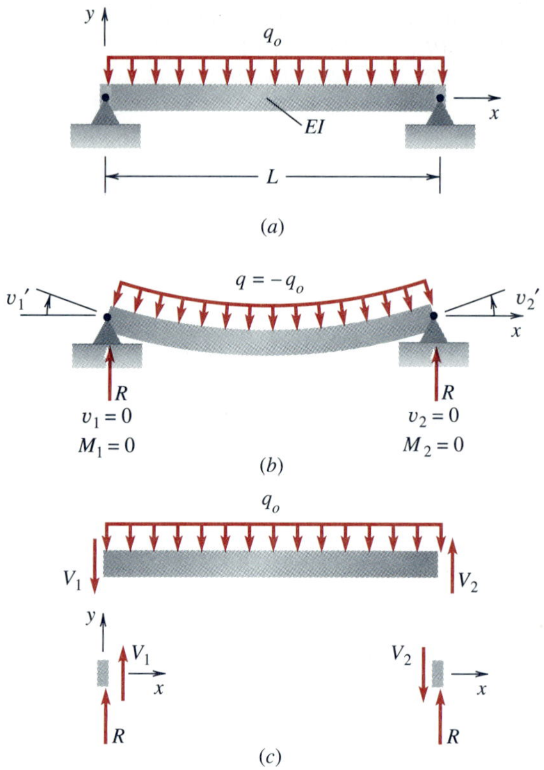

Figure 7.14 Example 7.8

EXAMPLE 7.9

We wish to find the wall reactions and the deflection at the center of a beam of length L with a constant distributed load q_0 and clamped at both ends, as shown in Fig. 7.15*a*.

We show the single beam element in Fig. 7.15*b*.

Since the end slopes and deflections at both ends are zero, Eqs. (7.11) and (7.12) become, with $q = -q_0$,

$$M_1 = -\frac{q_0L^2}{12} \qquad M_2 = -\frac{q_0L^2}{12} \tag{a}$$

We have obtained, therefore, the moment reactions at the walls. Equations (7.16) and (7.17) become

$$V_1 = -\frac{q_0 L}{2} \qquad V_2 = \frac{q_0 L}{2} \tag{b}$$

and the wall reactions take the form shown in Fig. 7.15*c*.

To find the deflection curve, we go back to Eq. (7.7) and introduce the zero values of the end slopes and deflections to get

$$v(x) = -\frac{q_0 L^4}{24EI}\left(\frac{x^2}{L^2} - 2\frac{x^3}{L^3} + \frac{x^4}{L^4}\right) \tag{c}$$

and at the midpoint $x = L/2$ we have

$$v\left(\frac{L}{2}\right) = -\frac{q_0 L^4}{384EI} \tag{d}$$

We note that the center deflection of a simply supported beam under a constant distributed loading, Example 6.2, is 5 times greater than the deflection found in Eq. (*d*) for the same beam but with clamped supports at the ends.

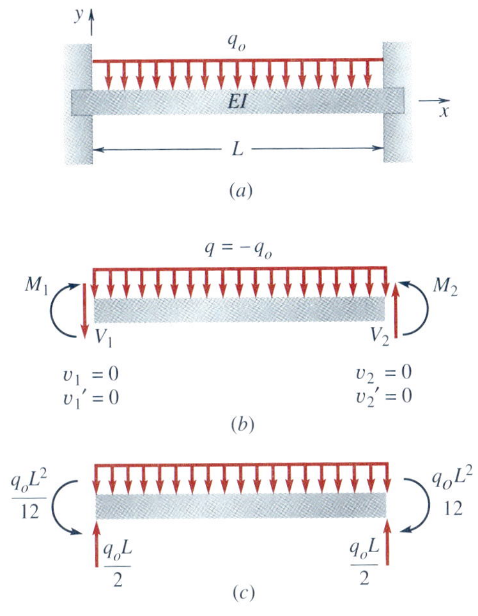

Figure 7.15 Example 7.9

EXAMPLE 7.10

Consider a beam of length L clamped at the left end and simply supported at the right end under a constant distributed loading q_0, as shown in Fig. 7.16*a*. We wish to find the slope and the reaction at the right end of the beam, using the beam force-deformation relations of Fig. 7.12. Note Example 7.5.

For this statically indeterminate beam problem, we have

$$v_1 = 0 \qquad v_1' = 0 \qquad v_2 = 0 \tag{a}$$

and $M_2 = 0$ (Fig. 7.16*b*), since the right end is simply supported with no applied moment.

It follows from the force-deformation relations (Fig. 7.12) that

$$M_2 = 0 = \frac{EI}{L^2}(4Lv_2') + \frac{q_0 L^2}{12} \tag{b}$$

or

$$v_2' = -\frac{q_0 L^3}{48EI} \tag{c}$$

Using the free-body diagram for an infinitesimal element at $x = L$, as shown in Fig. 7.16*b*, we find the reaction at the right end from the value of the shear force V_2 from Fig. 7.12

$$V_2 = \frac{EI}{L^3}(-6Lv_2') - \frac{q_0 L}{2} = -\frac{3}{8}q_0 L$$

Therefore

$$R_2 = -V_2 = \frac{3}{8}q_0 L \tag{d}$$

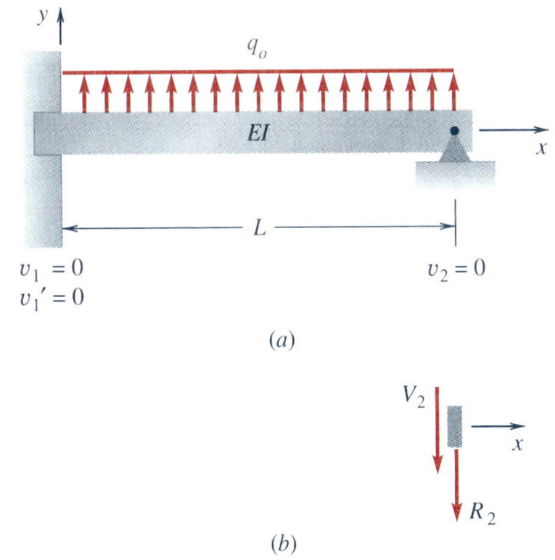

Figure 7.16 Example 7.10

EXAMPLE 7.11

A beam of length L and modulus EI and without applied loads between the endpoints is clamped at each end, as shown in Fig. 7.17*a*. A bending moment and shear force are applied to the right end of the beam so as to cause a specified vertical deflection δ and angle of rotation ϕ. (Think of holding the ends of a flexible beam in each hand and moving one hand up and rotating it relative to the other fixed hand; recall Example 7.4.) We wish to find the wall reactions in terms of the specified values of δ and ϕ; neglect the weight of the beam.

We show the beam as a single element with the prescribed deflection and slope in Fig. 7.17*b*.

In this case we have

$$v_1 = v_1' = 0 \qquad v_2 = \delta \qquad v_2' = \phi \qquad q = 0 \tag{a}$$

It follows immediately from the beam force-deformation relations (Fig. 7.12) that

$$M_2 = \frac{EI}{L^2}(-6\delta + 4L\phi)$$

$$V_2 = \frac{EI}{L^3}(12\delta - 6L\phi) = V_1 \tag{b}$$

$$M_1 = \frac{EI}{L^2}(6\delta - 2L\phi)$$

By using free-body diagrams shown in Fig. 7.17*c*, we can relate the wall reactions to the end shear forces and bending moments

$$M_L = M_1 \qquad M_R = M_2$$
$$R_L = V_1 \qquad R_R = -V_2 \tag{c}$$

Therefore from Eqs. (*b*), the wall reactions are known in terms of the specified values of δ and ϕ.

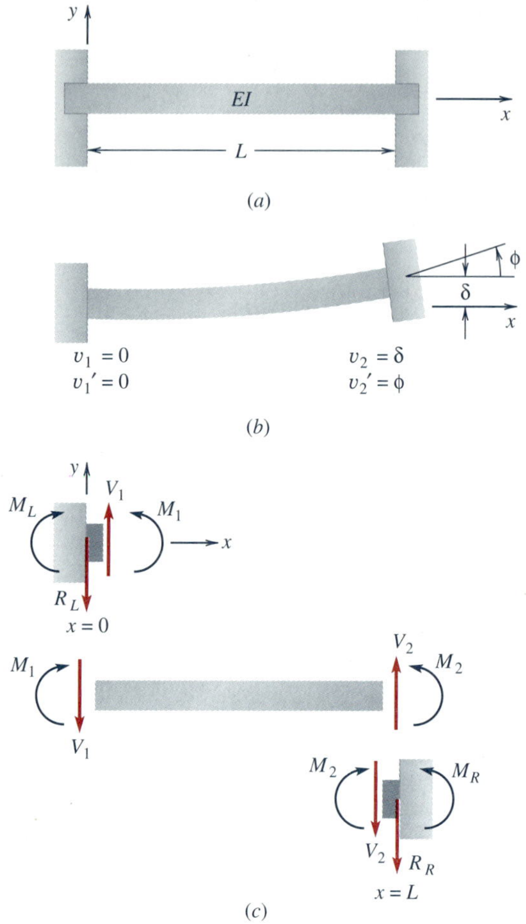

Figure 7.17 Example 7.11

7.7 Application of the Force-Deformation Relations—Two-Element Beam Problems

In Examples 7.7 to 7.11, only a single beam element was used to solve each problem. There was no need to introduce additional elements since there were no changes in the loading or support over the length of the beam. However, as mentioned earlier, if the beam modulus is different in different portions of the beam, or if concentrated loads or moments are applied along the beam, or if there are different segments of uniform loads, then it is necessary to divide the beam into separate elements connected by nodes before the problem can be solved by the displacement method. When a beam is divided into elements, each node has a corresponding node deflection and slope, and each element has a left and right end value of shear force and bending moment. Equilibrium at each node together with specified conditions at the support nodes provides equations for the node deflections and slopes. If the problem leads to only a few linear algebraic equations, then they can be readily solved by hand; otherwise, it is a routine problem to solve such problems by using a computer.

In the next example, we solve a two-element problem; thereafter, we introduce and use a computer program to solve beam problems that have many elements. The following two-element problem shows how the approach can be generalized to a beam with many elements.

EXAMPLE 7.12

We wish to determine the support reactions and the deflection under the concentrated load P applied at the midpoint B of a beam of length L that is clamped at the left end A and simply supported at the right end C, as shown in Fig. 7.18*a*. We neglect the weight of the beam, and we take the bending modulus of the beam to be EI. We note that this is a statically indeterminate problem. Since the applied load P is acting at the midpoint of the beam, it is necessary to introduce two elements, each of length $L/2$, with one node at each support and one node at the midpoint. With this choice of elements, all concentrated forces and reactions occur only at nodes.

In Fig. 7.18*b* we show the two elements numbered (1) and (2) and nodes 1, 2, and 3. For element (1) we designate the end shear forces and bending moments with the superscript "(1)"

$$V_1^{(1)} \quad M_1^{(1)} \quad V_2^{(1)} \quad M_2^{(1)}$$

and similarly for element (2). Remember that for the typical element shown in Fig. 7.12 with a specific *super*script, the *sub*scripts 1 and 2 on the shear force and bending moment refer to the left and right ends of the element. The form of the deflection function in Eq. (7.7) guarantees the interelement continuity for the slope and deflection curves. With two elements we have a total of six node values: v_1, v_1', v_2, v_2', v_3, and v_3', as shown in Fig. 7.18*b*.

Each element has associated with it a left node and a right node with a corresponding deflection and slope; i.e., element (1) has nodes 1 and 2 while element (2) has nodes 2 and 3. Therefore, for element (2), the force-deformation relations of the element will involve deflections v_2 and v_3 and slopes v_2' and v_3'.

From Fig. 7.18*a* we see that at node 1 the beam is clamped, that is, $v_1 = v_1' = 0$, and at node 3 it is simply supported, that

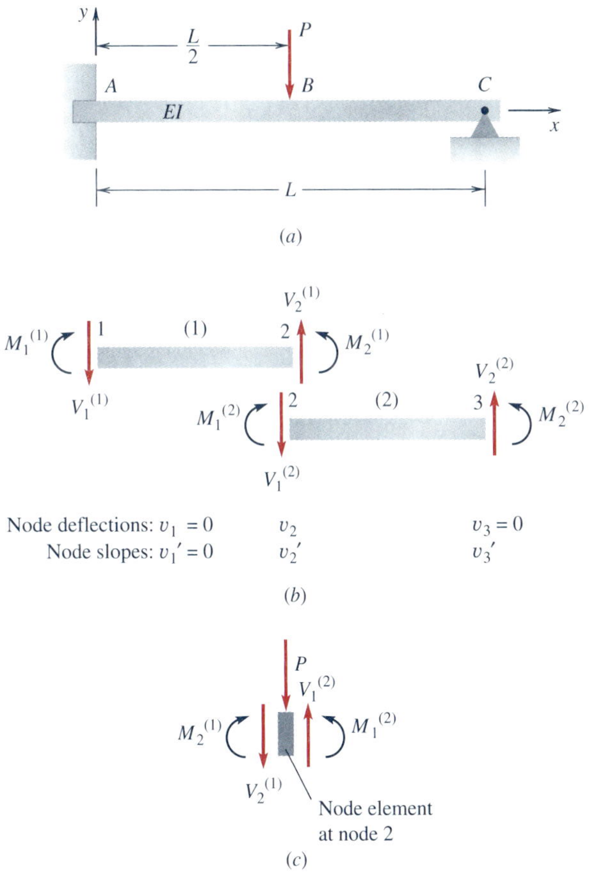

Figure 7.18 Example 7.12

is, $v_3 = 0$. Thus v_2, v_2', and v_3' are the unknown variables for this problem. We need to find three equilibrium equations for the determination of these three unknowns.

In Fig. 7.18*b*, free-body diagrams for beam elements (1) and (2) are shown. To exhibit the equilibrium relations connecting the shear force and bending moment of an element at a node to any applied force or moment acting at that node, we show at node 2 in Fig. 7.18*c* a *node element* of infinitesimal width that represents the node where the applied load is acting on the beam. The shear force and bending moments due to the beam elements on either side of the node element are shown on the free-body diagram of Fig. 7.18*c*.

We now proceed to write node element *force equilibrium equations* corresponding to each *unknown node deflection* and node element *moment equilibrium equations* corresponding to each *unknown node slope.*

Force and moment equilibrium of the node element at node 2 in Fig. 7.18*c* leads to

$$-V_2^{(1)} + V_1^{(2)} - P = 0 \qquad (a)$$

$$M_1^{(2)} - M_2^{(1)} = 0 \qquad (b)$$

In addition, since no applied bending moment exists at node 3, we have from moment equilibrium of a node element at node 3

$$M_2^{(2)} = 0 \qquad (c)$$

Each of Eqs. (*a*) to (*c*) can be written in terms of v_2, v_2', and v_3' by using the force-deformation relations for each element. Considering the set of equilibrium equations in Eqs. (*a*) to (*c*), we see that the equations provide us with the necessary three conditions for determining the three unknown geometric variables v_2, v_2', and v_3'.

The procedure to solve this problem therefore is to use the force-deformation equations (7.11) and (7.12), and (7.16) and (7.17), to express the bending moments and shear forces in the three node equilibrium equations (*a*) to (*c*) in terms of the three unknown geometric variables v_2, v_2', and v_3'. Using Eq. (7.17) for the first element (1), with L as $L/2$, and recalling that $v_1 = v_1' = 0$, we get

$$V_2^{(1)} = \frac{8EI}{L^3}\left(12v_2 - 6\frac{L}{2}v_2'\right) \qquad (d)$$

In a similar fashion, using Eq. (7.16) for the second element in which the left and right ends are now node 2 and node 3, we get

$$V_1^{(2)} = \frac{8EI}{L^3}\left(-12v_2 - 6\frac{L}{2}v_2' - 6\frac{L}{2}v_3'\right) \qquad (e)$$

Substitution of Eqs. (*d*) and (*e*) into Eq. (*a*) gives

$$\frac{8EI}{L^3}(-24v_2 - 3Lv_3') = P \qquad (f)$$

Proceeding in a similar fashion for the moment contribution of elements (1) and (2) to moment equilibrium at node 2, Eq. (*b*), we find

$$M_2^{(1)} = \frac{4EI}{L^2}(-6v_2 + 2Lv_2')$$

$$M_1^{(2)} = \frac{4EI}{L^2}(-6v_2 - 2Lv_2' - Lv_3') \qquad (g)$$

and substitution in Eq. (*b*) yields

$$\frac{4EI}{L^2}(-4Lv_2' - Lv_3') = 0 \qquad (h)$$

Finally, from the vanishing of the moment at node 3, Eq. (*c*), we have

$$\frac{4EI}{L^2}(6v_2 + Lv_2' + 2Lv_3') = 0 \qquad (i)$$

We now have three equations—Eqs. (*f*), (*h*), and (*i*)—for the three unknowns which, when solved, yield

$$v_2 = -\frac{7}{768}\frac{PL^3}{EI} \qquad (j)$$

$$v_2' = -\frac{1}{128}\frac{PL^2}{EI} \qquad (k)$$

$$v_3' = \frac{1}{32}\frac{PL^2}{EI} \qquad (l)$$

A sketch of the deflected shape of the beam showing the values of v_2, v_2', v_3' is found in Fig. 7.19*a*.

Having found the unknown slopes and deflections, we now know all six of the node slopes and deflections. Therefore, using Eqs. (7.11) and (7.12), and Eqs. (7.16) and (7.17), we are now in a position to calculate the end forces and moments for each beam element. If we wish to find the external wall reactions at the clamped end of the beam, we use element (1) and the "left-end" formulas, Eqs. (7.16) and (7.11), with $v_1 = 0$ and $v_1' = 0$ to get

$$V_1^{(1)} = \frac{8EI}{L^3}\left(12v_2 - 6\frac{L}{2}v_2'\right) \qquad (m)$$

and

$$M_1^{(1)} = \frac{4EI}{L^2}\left(6v_2 - 2\frac{L}{2}v_2'\right) \qquad (n)$$

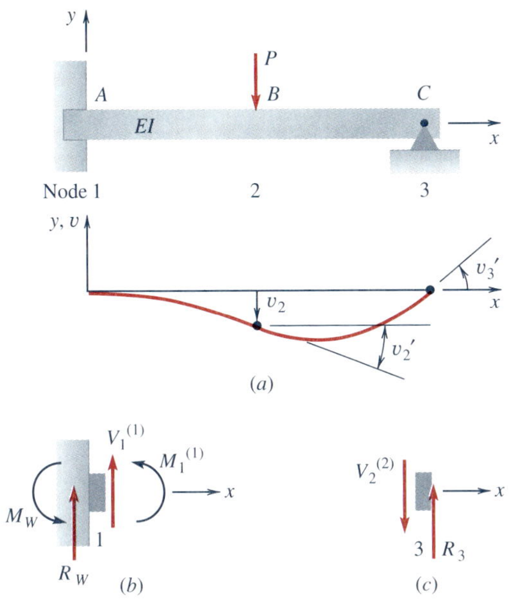

Figure 7.19

If we use the solutions given in Eqs. (j) and (k), we find

$$V_1^{(1)} = -\tfrac{11}{16}P = -R_W \tag{o}$$

$$M_1^{(1)} = -\tfrac{3}{16}PL = -M_W \tag{p}$$

because the relations between the shear force and bending moment at the left end of element (1) and the wall reactive force R_W and the wall reactive moment M_W are as shown in Fig. 7.19b.

To find the reaction at the simple support (node 3), we use the "right-end" formula for element (2), that is, Eq. (7.17) with $v_3 = 0$, to get

$$V_2^{(2)} = \frac{8EI}{L^3}\left(-12v_2 - 6\frac{L}{2}v_2' - 6\frac{L}{2}v_3'\right) \tag{q}$$

Inserting the values of the slopes and deflection from Eqs. (j) to (l), we obtain

$$V_2^{(2)} = \tfrac{5}{16}P$$

or (Fig. 7.19c)

$$R_3 = \tfrac{5}{16}P \tag{r}$$

It is important to note that the method of solution is systematic in employing the beam force-deformation relations in the equilibrium conditions at nodes to obtain equations for the unknown node deflections and slopes. Once these are found, the unknown end shear forces and bending moments are easily calculated and the reactions found.

When a support is at a node between two elements, the value of the reaction at the support is found from the difference, or jump, in the values of the shear force at the node point as one proceeds along the beam across the node point from the element on the left to the element on the right. In addition, once all the node deflections and slopes are found, the deflection, slope, bending moment, and shear force at all points along each element can be obtained. Clearly this approach to the solution of a beam problem can be programmed in a straightforward manner. In the next section we discuss such a program using this approach.

7.8 Use of Computer Program BEAMMECH to Calculate Beam Deflections, Slopes, Shear Forces, Bending Moments, and Maximum Bending Stresses

In the following example, the procedure for using the program BEAMMECH on the diskette with this book to solve beam deflection problems is explained. The displacement method as derived in Sec. 7.5 is used to solve for the beam deflections since it provides a systematic approach to a wide class of beam problems, including statically indeterminate problems, and it can be coded for computer-assisted analysis. We treat beams loaded by concentrated loads and moments and segments of uniform loads.

EXAMPLE 7.13

Consider the steel cantilever beam that is clamped at the left end and simply supported at the right end, as shown in Fig. 7.20*a*. The beam is a wide-flange W8 × 28 for which the depth is 8.06 in and the moment of inertia is 98 in^4. We wish to use the BEAMMECH program to do the following:

(*a*) Plot the deflection curve for this beam and find the deflection values at sections *C* and *D* where the loads are applied. Also find the slope at support *B*.
(*b*) Draw the shear force and bending moment diagrams.
(*c*) Determine the maximum bending stresses in the beam.

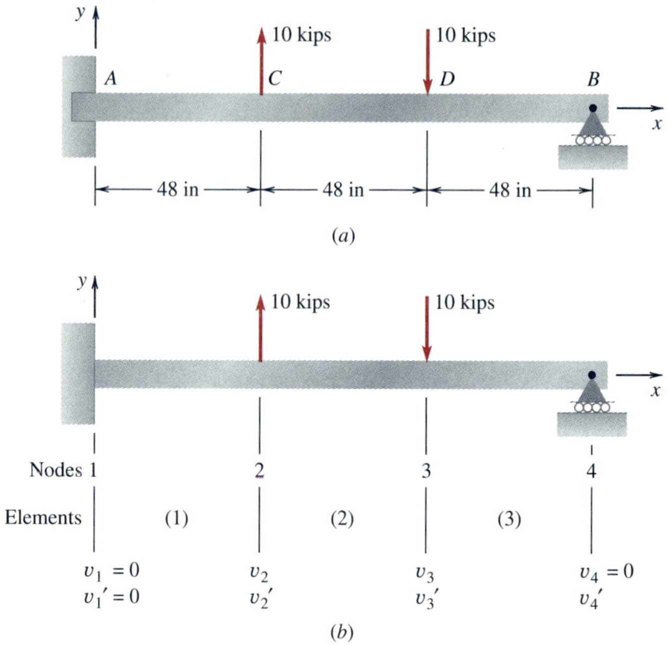

Figure 7.20 Example 7.13

We neglect the weight of the beam.

In this problem we use three elements—(1), (2), and (3)—connected by the four nodes—1, 2, 3, and 4—as shown in Fig. 7.20*b*. The elements are selected so that their nodes correspond to either a support or a section at which a concentrated load is applied. The bending modulus *EI* is constant along the beam. A node deflection is positive in the positive *y* direction, and a node slope angle is positive in the counterclockwise direction or in the positive *z* direction. Since there are four node points, there are a total of four node deflections and four node slopes.

Each element has associated with it a left and right shear force and bending moment. Referring to the beam in Fig. 7.20*a*, we see that the deflection is equal to zero at both ends of the beam, as is the slope at the left end. Therefore, using the notation from Fig. 7.20*b*, we have

$$v_1 = 0 \qquad v_1' = 0 \tag{a}$$

and

$$v_4 = 0 \tag{b}$$

The unknowns in this problem are the remaining five node deflections and slopes:

$$v_2 \qquad v_2' \qquad v_3 \qquad v_3' \qquad v_4' \tag{c}$$

The specified applied forces acting at the nodes are as follows: at node 2 the applied force is +10,000 lb, and at node 3 the applied force is −10,000 lb.

There are no applied moments acting on the beam. The reactive force and moment at node 1 and the reactive force at node 4 are unknown.

The BEAMMECH program is called by selecting from the main MECHMAT menu program 4: DEFLECTIONS OF BEAMS (see Fig. 7.21) to obtain the BENDING OF BEAMS program, (see Fig. 7.22). The data input choice has the same format as that used before in the program BARMECH for axially loaded bars (Sec. 2.7).

If selection 1: Input new data file is chosen, as shown in Fig. 7.22, then in response to the DRIVE USED (A:, B:, or C:) prompt, we should enter the drive letter for the drive on which the input data are to be saved and a file name, as in Fig. 7.22. In the present case we used drive A: and file name EX713.

In preparing a new data file, the first request (Fig. 7.23) is for the value of NM, which is the number of elements used in the model of the beam. We must take enough elements in a problem so that no element has a concentrated load or moment acting at a section between the ends of the element and so that each element has constant cross-sectional properties. An element may have a constant distributed loading along its length. It is clear from Fig. 7.20 that the location of nodes at sections *A*, *C*, *D*, and *B* along the beam satisfies the above requirements. Thus we have NM = 3 for this problem.

In response to the requests for data on element properties (the third line of Fig. 7.23), the information for the element properties is supplied, element by element (Fig. 7.23). The *q* values shown in the last column correspond to the value of the uniform loading (positive in the positive *y* direction) acting on each element. All the *q* values are zero for the present problem. Note the units for *E* are 10^6 psi or MPa, e.g., for steel *E* = 30

```
 *********************************************************

*** *MECHANICS OF MATERIALS* ****

PROGRAM: MECHMAT                mod 5.0

*********************************************************


     MAIN MENU

           1. AXIALLY LOADED MEMBERS
           2. TORSION
           3. V + M DIAGRAMS OF BEAMS
           4. DEFLECTIONS OF BEAMS
           5. MOHR'S CIRCLE
           6. CENTROIDS + MOMENTS OF INERTIA
           7. EXIT

     SELECT NUMBER = ? 4
```

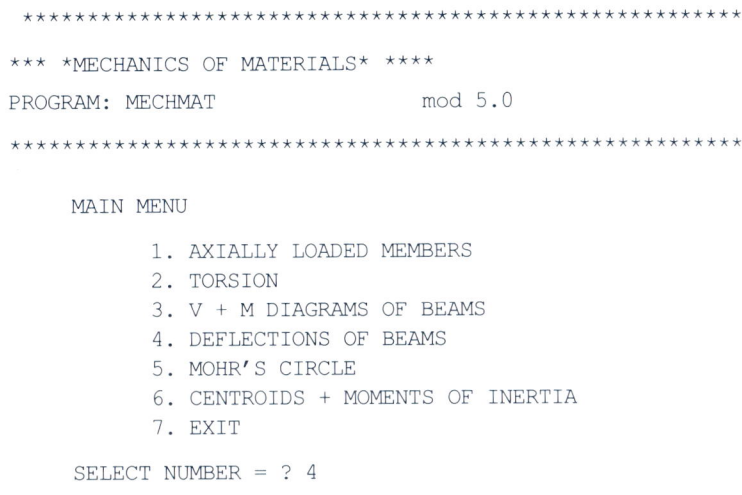

Figure 7.21 MECHMAT menu.

```
*********************************************************

    *** BENDING OF BEAMS ***        PROGRAM: BEAMMECH

*********************************************************


     *** CHOICE OF INPUT ***

           1. INPUT NEW DATA FILE
           2. MODIFY EXISTING DATA FILE
           3. USE EXISTING DATA FILE
       OR  4. RETURN TO MAIN MENU
              SELECT NUMBER? 1


          DRIVE USED (A:, B:, or C:) ? a:
        FILE NAME (OMIT DRIVE LETTER)? ex713
```

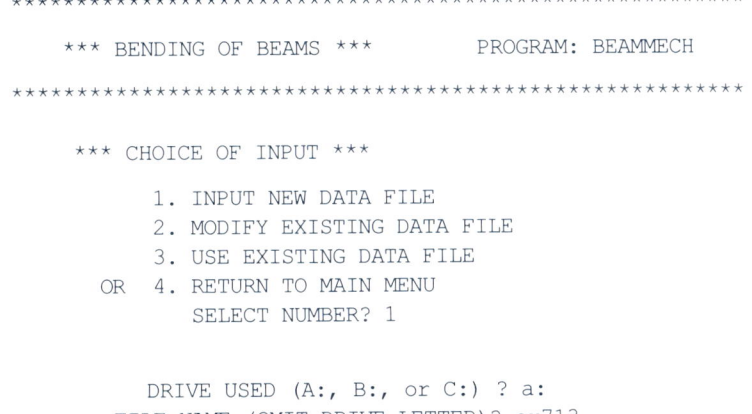

Figure 7.22 Menu for bending of beams: BEAMMECH.

or 200,000. The element properties should be entered in order from element 1 on the left to the last element on the right.

In response to requests for applied forces and moments input and for specified deflections and slopes input, we supply the data shown summarized in Fig. 7.23; care must be taken to follow sign conventions for all input data. The program provides the option of printing a summary of the input data; it is useful to obtain this summary, as shown in Fig. 7.24, so we will be able to perform a check on the correctness of the input data. All units must be consistent.

The computer program follows the same procedure to solve a beam problem as was followed in solving the two-element problem of Example 7.12. The five equilibrium equations needed to determine the five unknown variables v_2, v_2', v_3, v_3', and v_4' are obtained by setting up force and moment equilibrium equations corresponding to each of the unknown deflections and slopes. Referring to the free-body diagrams of the node elements in Fig. 7.25, we find

```
   INPUT THE FOLLOWING DATA:

NUMBER OF ELEMENTS                         NM=? 3

INPUT THE LENGTH, MOMENT OF INERTIA, MODULUS OF
ELASTICITY, AND UNIFORM LOAD:

               L          I            E             q
             (in)      (in**4)    (10**6; psi)    (lb/in)
             (m)       (m**4)        (MPa)         (N/m)

ELEMENT(1)---? 48, 98, 30, 0
ELEMENT(2)---? 48, 98, 30, 0
ELEMENT(3)---? 48, 98, 30, 0

NOTE UNITS OF E, DIRECTION OF q, CONSISTENT UNITS?
  ARE THE ABOVE VALUES CORRECT  Y/N? y

NUMBER OF APPLIED FORCES:  2
NODE NO. , FORCE VALUE: 2,10000
NODE NO. , FORCE VALUE: 3,-10000

NUMBER OF APPLIED MOMENTS:  0

NUMBER OF SPECIFIED DEFLECTIONS:  2
NODE NO. , DEFL. VALUE: 1,0
NODE NO. , DEFL. VALUE: 4,0

NUMBER OF SPECIFIED SLOPES:  1
NODE NO. , SLOPE VALUE: 1,0

   ARE THE ABOVE VALUES CORRECT  Y/N? y
```

Figure 7.23 Example 7.13: Input data format.

At node 2: $$V_1^{(2)} - V_2^{(1)} + 10{,}000 = 0$$

At node 2: $$M_1^{(2)} - M_2^{(1)} = 0$$

At node 3: $$V_1^{(3)} - V_2^{(2)} - 10{,}000 = 0$$

At node 3: $$M_1^{(3)} - M_2^{(2)} = 0$$

At node 4: $$M_2^{(3)} = 0$$

where we use superscripts in parentheses to indicate the element number. The *subscripts* 1 and 2 indicate the left and right end of each element; for example, $M_2^{(2)}$ is the right-end bending moment acting on element number (2).

The computer program substitutes the appropriate $V_k^{(i)}$ and $M_k^{(i)}$ into the equilibrium equations by using Eqs. (7.11), (7.12), (7.16), and (7.17). Thus the equilibrium equations become a set of five linear algebraic equations in terms of the five unknowns. These linear algebraic equations can be solved easily by a standard routine, and the solution can be obtained as given in Fig. 7.26.

Once the simultaneous equations have been solved for the unknown deflections and slopes, the deflection curve for each element is given by Eq. (7.7) by using the appropriate end deflections and slopes for each element. The slope curve follows by taking the x derivative of the deflection curve in Eq. (7.7). The bending moment for each element is given by Eq. (7.9) and the shear force by Eq. (7.14). The results for the values of

```
        ***MECHANICS OF MATERIALS***    DATA FILE: ex713
                         ***BEAM***

   INPUT DATA:

 NUMBER OF ELEMENTS                         NM= 3

  ELEMENT       L          I            E            q
              (in)     (in**4)    (10**6; psi)   (lb/in)
               (m)      (m**4)       (MPa)         (N/m)

     1       48.00     9.80E+01     3.00E+01     0.00E+00
     2       48.00     9.80E+01     3.00E+01     0.00E+00
     3       48.00     9.80E+01     3.00E+01     0.00E+00


   SPECIFIED CONDITIONS:
 *********************************

     NODE NO.      FORCE VALUE

        2            1.00E+04
        3           -1.00E+04

     NODE NO.     MOMENT VALUE

     NODE NO.      DEFL. VALUE

        1           0.00E+00
        4           0.00E+00

     NODE NO.      SLOPE VALUE

        1           0.00E+00
```

Figure 7.24 Example 7.13: Table with summary of input data.

the shear force and bending moment at the left end (L) and right end (R) of each element are given by Fig. 7.27. If a given node is not at a support, then the value of the shear force and the value of the bending moment from adjoining elements at the node will undergo a discontinuity at the node if there are applied loads or moments at the node. If a node is a support node, then the discontinuity in the shear force across the node will give the reaction load at the node.

The BEAMMECH program gives us the choice of different displays and printed output for the quantities of practical interest. For this problem we can obtain:

> A curve of the deflection versus position x along the beam, the maximum and minimum values of the deflection, and their location (Fig. 7.28)
>
> The shear force and bending moment diagrams, shown in Figs. 7.29 and 7.30
>
> The maximum bending stresses in the beam, as shown in Fig. 7.31

The last option of obtaining the maximum normal stresses due to bending is available only for the case of beams with EI constant along the total length of the beam. To obtain the maximum stresses, we need to input the value of the distance from the neutral axis to the top or bottom of the beam; we call these values c_1 and c_2 (Fig. 7.31).

Therefore, we have obtained all the required information for this problem.

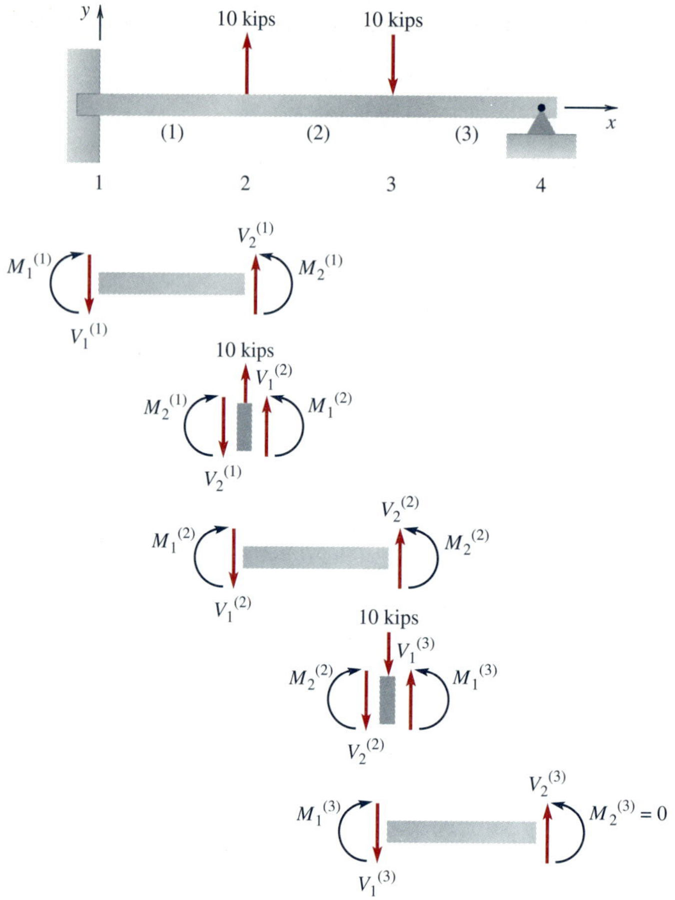

Figure 7.25 Example 7.13: Beam elements and node elements with shear forces and bending moments acting.

SOLUTION:

****

NODE NO.	VALUE OF X	DEFLECTION	SLOPE
1	0.000	0.00E+00	0.00E+00
2	48.000	-2.32E-03	-5.80E-04
3	96.000	-3.95E-02	-1.45E-04
4	144.000	0.00E+00	1.31E-03

TOTAL LENGTH OF THE BEAM = 144.000

Figure 7.26 Example 7.13: Table of solution.

```
*********************************************************
 ELEMENT     SHEAR(L)       SHEAR(R)       MOMENT(L)     MOMENT(R)
*********************************************************
     1        3.70E+03       3.70E+03       5.33E+04     -1.24E+05
     2       -6.30E+03      -6.30E+03      -1.24E+05      1.78E+05
     3        3.70E+03       3.70E+03       1.78E+05      0.00E+00
```

Figure 7.27 Example 7.13: Table of shear forces and bending moments at the left (L) and right (R) ends of the elements.

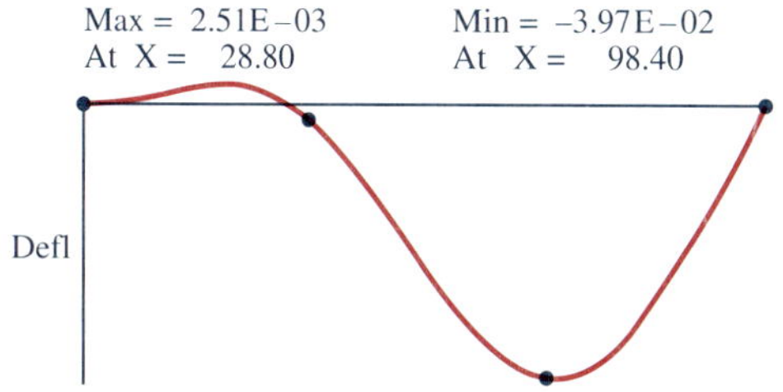

Figure 7.28 Example 7.13: Deflection plot.

```
DO YOU WANT MAX BENDING STRESS [Y/N]: ? y
INPUT C1 (in;m) = ? 4.03
INPUT C2 (in;m) = ? 4.03
C1 = 4.0300E+00 (in;m)
C2 = 4.0300E+00 (in;m)
*********************************************************
MAXIMUM TENSILE STRESS = 7.311E+03
      AT X =     96.00

MAXIMUM COMPRESSIVE STRESS = 7.311E+03
      AT X =     96.00
```

Figure 7.31 Example 7.13: Maximum bending stress.

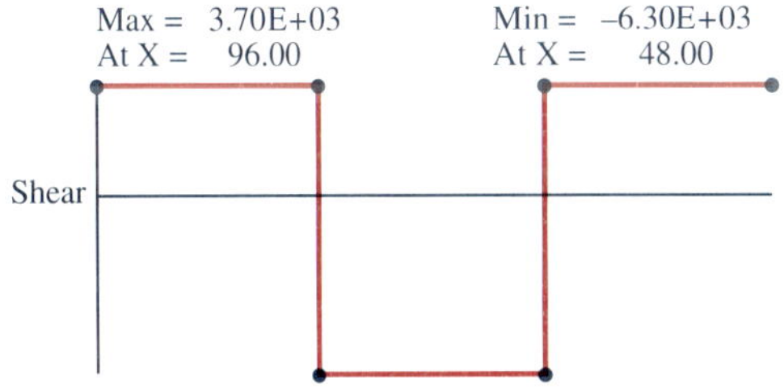

Figure 7.29 Example 7.13: Shear force diagram.

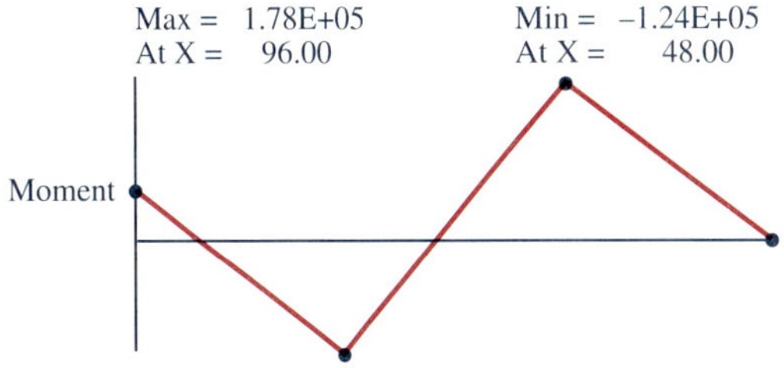

Figure 7.30 Example 7.13: Bending moment diagram.

EXAMPLE 7.14

A steel beam is loaded as shown in Fig. 7.32. We wish to use the BEAMMECH program to determine the maximum deflection of the beam, the deflection curve, and the shear force and bending moment curves. We neglect the weight of the beam and take $E = 200$ GPa and $I = 35 \times 10^6$ mm^4 $= 35 \times 10^{-6}$ m^4.

We divide the beam into three elements with four nodes, as shown in Fig. 7.32, and use the BEAMMECH program to determine the values of the node deflections and slopes.

Figure 7.33 shows the form of the input data. Note that units of I are meters to the fourth power and units of E are in megapascals, so 200 GPa is equal to 200,000 MPa. All units are consistent with newtons and meters.

Figure 7.34 shows the summary of the input data listing the element properties and the specified conditions.

Figure 7.35 shows the values of the deflections in meters and the values of the slope in radians at each of the nodes. A negative deflection is a deflection in the negative y direction.

The plot of the deflection is shown in Fig. 7.36, and the maximum value (largest value of the deflection in the positive y direction) is 1.61 mm at the right end of the beam. The minimum value (largest value in the negative y direction) is -12.1 mm at $x = 3.12$ m. The values of the shear force and the bending moment at the left end and right end of each element are given also in Fig. 7.36. For example, for elements (2) and (3) we have the shear forces and bending moments as shown in Fig. 7.37*a*, where we show the correct direction for each shear force and bending moment on the ends of the element. At node 3, which is a support node and where a reaction force exists, the magnitude and direction of the reaction follow from the jump or discontinuity in the value of the shear force at the node, as shown in Fig. 7.37*a*. The force reaction at node 3 is therefore 46.3 kN.

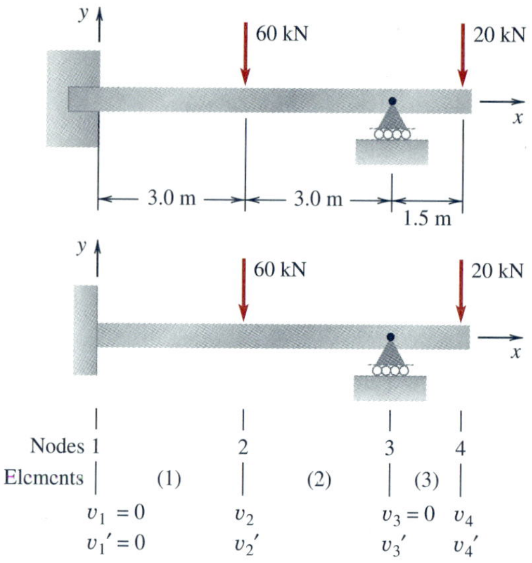

Figure 7.32 Example 7.14

At the wall, we have, as shown in Fig. 7.37*b*,

$$M_W = -M_1^{(1)} = +52.5 \text{ kN} \cdot \text{m}$$

$$R_W = -V_1^{(1)} = +33.8 \text{ kN}$$

The shear force diagram in Fig. 7.38 shows the jumps in the value of the shear force at the supports and at the concentrated loads. The bending moment diagram in Fig. 7.39 shows the values of the bending moment along the beam.

```
   INPUT THE FOLLOWING DATA:

NUMBER OF ELEMENTS                 NM=? 3

INPUT THE LENGTH, MOMENT OF INERTIA, MODULUS OF
ELASTICITY, AND UNIFORM LOAD:

            L        I           E            q
          (in)    (in**4)   (10**6; psi)   (lb/in)
          (m)     (m**4)       (MPa)        (N/m)

ELEMENT(1)---? 3,35e-06, 200000, 0
ELEMENT(2)---? 3,35e-06, 200000, 0
ELEMENT(3)---? 1.5,35e-06, 200000, 0

NOTE UNITS OF E, DIRECTION OF q, CONSISTENT UNITS?
  ARE THE ABOVE VALUES CORRECT  Y/N? y

NUMBER OF APPLIED FORCES:  2
NODE NO. , FORCE VALUE: 2,-60000
NODE NO. , FORCE VALUE: 4,-20000

NUMBER OF APPLIED MOMENTS:  0

NUMBER OF SPECIFIED DEFLECTIONS:  2
NODE NO. , DEFL. VALUE: 1,0
NODE NO. , DEFL. VALUE: 3,0

NUMBER OF SPECIFIED SLOPES:  1
NODE NO. , SLOPE VALUE: 1,0

   ARE THE ABOVE VALUES CORRECT  Y/N? y
```

Figure 7.33 Example 7.14: Input data format.

```
***MECHANICS OF MATERIALS***     DATA FILE: ex714
                     ***BEAM***

INPUT DATA:

NUMBER OF ELEMENTS                     NM= 3

ELEMENT       L           I              E              q
            (in)       (in**4)     (10**6; psi)      (lb/in)
             (m)       (m**4)         (MPa)           (N/m)

  1         3.00      3.50E-05      2.00E+05        0.00E+00
  2         3.00      3.50E-05      2.00E+05        0.00E+00
  3         1.50      3.50E-05      2.00E+05        0.00E+00


SPECIFIED CONDITIONS:
*********************************

  NODE NO.      FORCE VALUE

     2           -6.00E+04
     4           -2.00E+04

  NODE NO.     MOMENT VALUE

  NODE NO.      DEFL. VALUE

     1            0.00E+00
     3            0.00E+00

  NODE NO.      SLOPE VALUE

     1            0.00E+00
```

Figure 7.34 Example 7.14: Table with summary of input data.

```
SOLUTION:

*******************************************************

 NODE NO.   VALUE OF X     DEFLECTION          SLOPE

*******************************************************

    1          0.000        0.00E+00         0.00E+00

    2          3.000       -1.21E-02        -8.04E-04

    3          6.000        0.00E+00         3.21E-03

    4          7.500        1.61E-03         0.00E+00

TOTAL LENGTH OF THE BEAM =    7.500
```

Figure 7.35 Example 7.14: Table of solution.

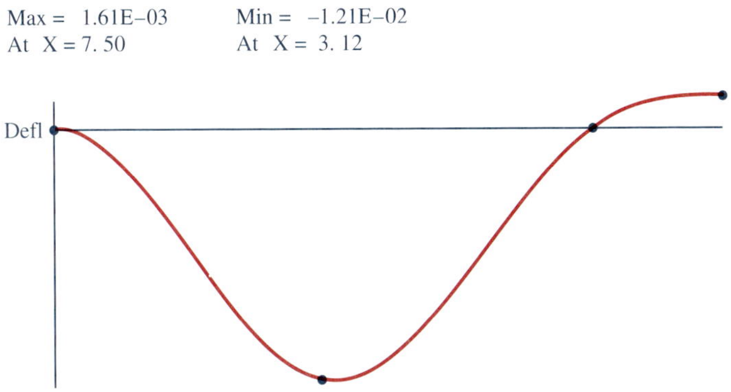

ELEMENT	SHEAR(L)	SHEAR(R)	MOMENT(L)	MOMENT(R)
1	-3.38E+04	-3.38E+04	-5.25E+04	4.88E+04
2	2.63E+04	2.63E+04	4.88E+04	-3.00E+04
3	-2.00E+04	-2.00E+04	-3.00E+04	0.00E+00

Figure 7.36 Example 7.14: Deflection plot and table of element end shear forces and bending moments.

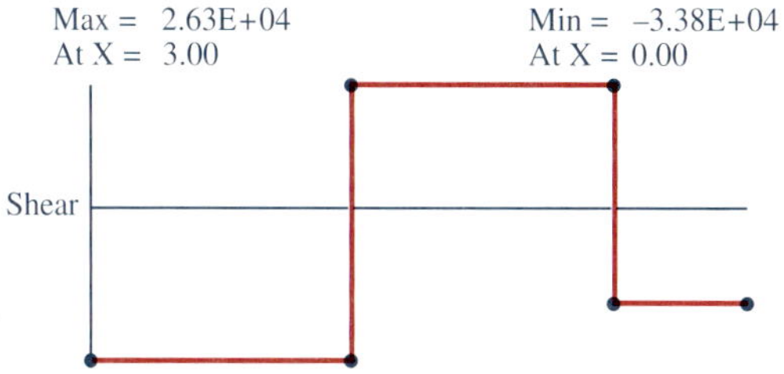

Figure 7.38 Example 7.14: Shear force diagram.

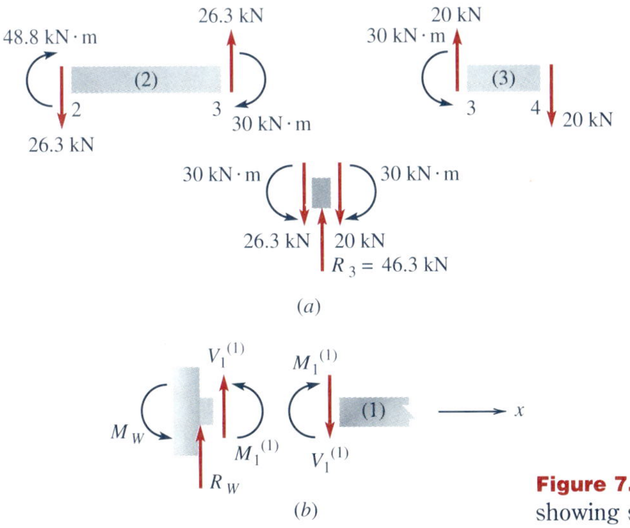

Figure 7.37 Example 7.14: Beam elements showing shear forces and bending moments.

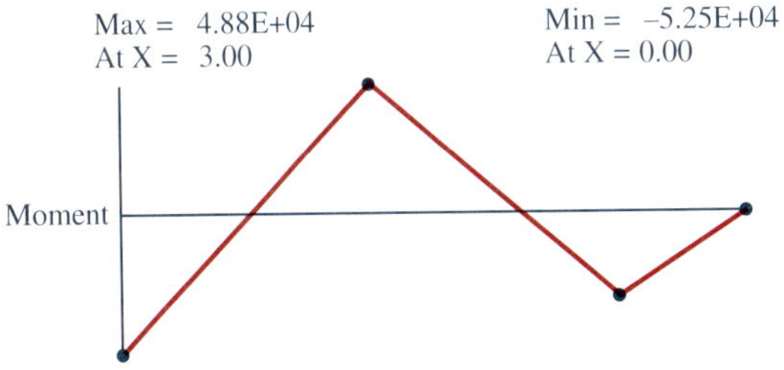

Figure 7.39 Example 7.14: Bending moment diagram.

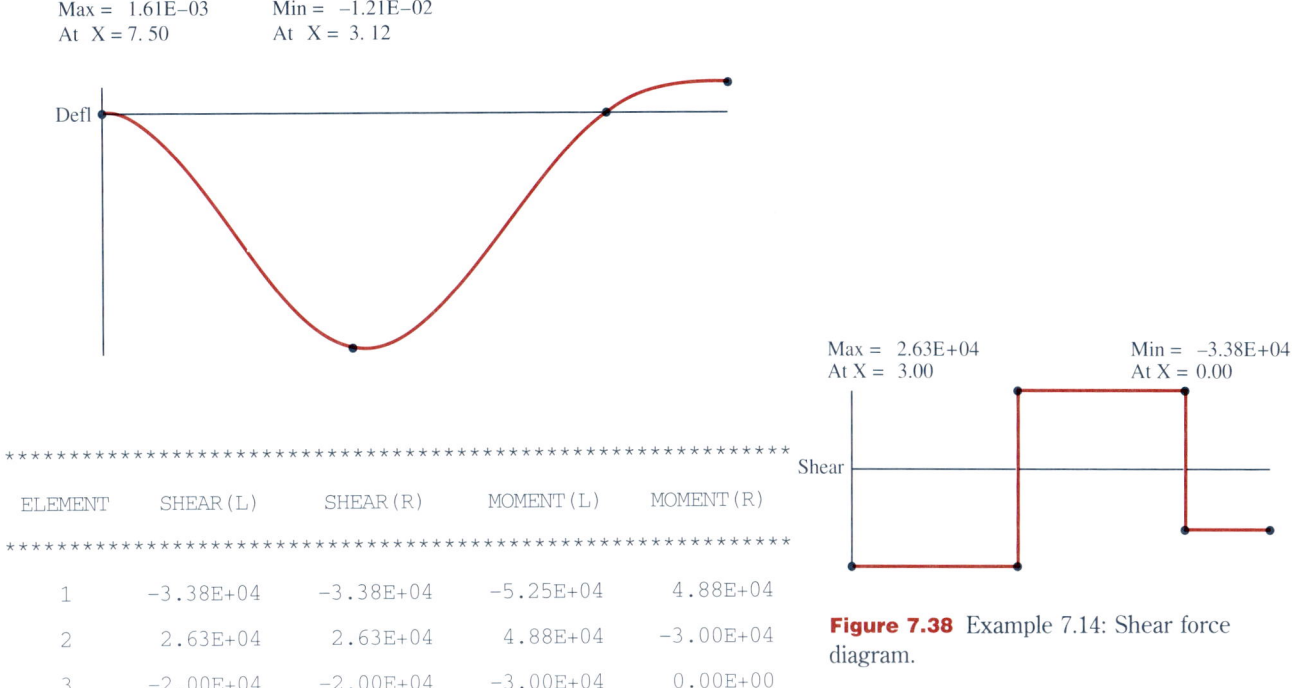

EXAMPLE 7.15

The steel beam shown in Fig. 7.40 is simply supported and loaded with a segment of a uniform load and two concentrated loads, one at each end. In a given application, the load P at the left end can vary from 0 to 3 kips in steps of 1 kip. We wish to determine in each case the shape of the deflection curve and the largest absolute value of the deflection of the beam as P varies from 0 to 3 kips.

The beam is a steel wide-flange beam W12 × 50, $I = 394\ \text{in}^4$. We neglect the weight of the beam, take $E = 30 \times 10^6$ psi, and use the BEAMMECH program.

Our beam model for the BEAMMECH program will make use of four elements with five nodes, as shown in Fig. 7.40. We must first convert all units of length to inches and the uniform load from 1800 lb/ft to 150 lb/in. The prescribed concentrated loads are P lb at node 1 and −8000 lb at node 5. The deflections at node 2 and 4 are zero.

Figure 7.41 shows the input data and the solution for the value $P = 0$. The deflection curve is shown in Fig. 7.42; the largest deflection for this case occurs at $x = 119$ in.

The input data file can be modified for different values of P by using choice 2: Modify existing data file in the menu shown in Fig. 7.22 and following the computer prompts in order to change the value of P in the input file. In this way we can repeat the steps carried out above for the cases $P = 1$, 2, and 3 kips.

The numerically largest values of the deflection along the beam can be found from the deflection curve figures similar to Fig. 7.42, which are displayed for each load case. These are summarized in the table below:

P	$\lvert v\rvert_{max}$	Location from left end
0	2.69×10^{-2} in	119 in
1	2.06×10^{-2} in	120 in
2	1.72×10^{-2} in	240 in
3	2.87×10^{-2} in	0 in

The deflection curves from the four load cases are redrawn on a single graph in Fig. 7.43. The transition in the deflection shapes caused by the increasing downward left-end load P is clearly visible. When $P = 0$, the maximum deflection is due to the sag of the beam between the supports. With increasing load at the left end, there is a transition to a maximum deflection at the right end for $P = 2$ kips, and finally at $P = 3$ kips the maximum occurs at the left end.

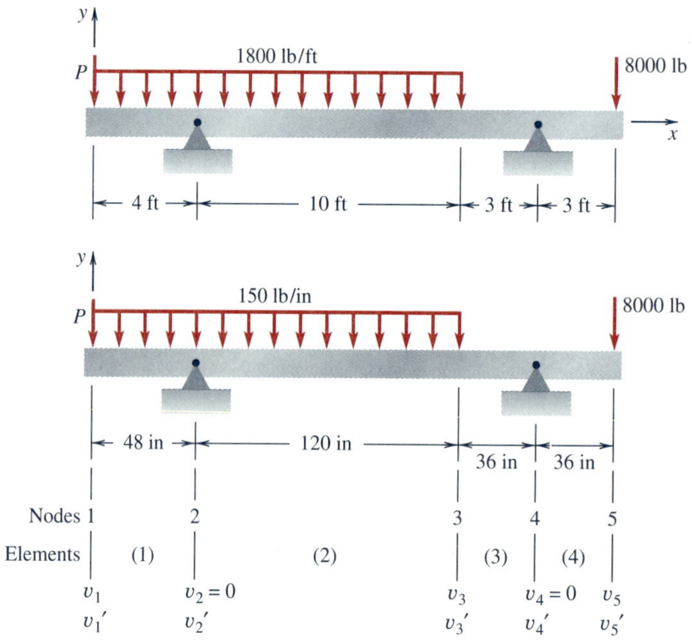

Figure 7.40 Example 7.15

```
        ***MECHANICS OF MATERIALS***    DATA FILE: ex715
                            ***BEAM***

   INPUT DATA:

NUMBER OF ELEMENTS                          NM= 4

  ELEMENT       L          I            E           q
              (in)      (in**4)    (10**6; psi)  (lb/in)
              (m)       (m**4)        (MPa)       (N/m)

     1        48.00    3.94E+02     3.00E+01    -1.50E+02
     2       120.00    3.94E+02     3.00E+01    -1.50E+02
     3        36.00    3.94E+02     3.00E+01     0.00E+00
     4        36.00    3.94E+02     3.00E+01     0.00E+00

    SPECIFIED CONDITIONS:
*********************************

      NODE NO.     FORCE VALUE

         1           0.00E+00
         5          -8.00E+03

      NODE NO.    MOMENT VALUE

      NODE NO.     DEFL. VALUE

         2          0.00E+00
         4          0.00E+00

      NODE NO.     SLOPE VALUE

      SOLUTION:

*******************************************************

   NODE NO.    VALUE OF X      DEFLECTION        SLOPE

*******************************************************

      1           0.000         1.10E-02       -1.72E-04

      2          48.000         0.00E+00       -4.06E-04

      3         168.000        -1.17E-02        4.83E-04

      4         204.000         0.00E+00        2.57E-05

      5         240.000        -9.60E-03       -4.13E-04

TOTAL LENGTH OF THE BEAM =    240.000
```

Figure 7.41 Example 7.15: Table of input data and table of solution.

Max = 1.10E–02 Min = –2.69E–02
At X = 0. 00 At X =118. 80

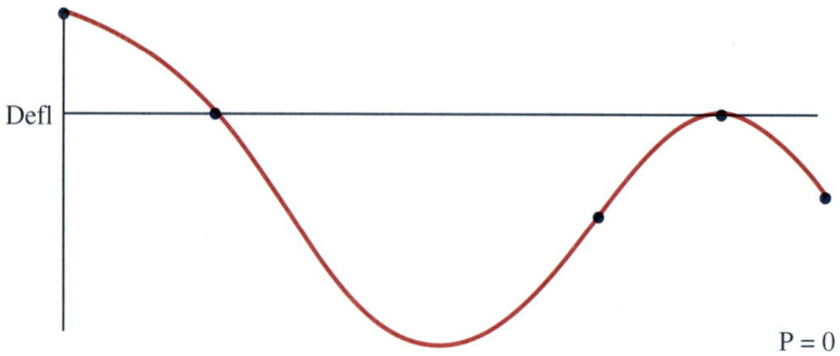


```
************************************************************************
 ELEMENT     SHEAR(L)       SHEAR(R)       MOMENT(L)      MOMENT(R)
************************************************************************
```

ELEMENT	SHEAR(L)	SHEAR(R)	MOMENT(L)	MOMENT(R)
1	-4.88E-04	7.20E+03	-1.56E-02	-1.73E+05
2	-1.03E+04	7.66E+03	-1.73E+05	-1.22E+04
3	7.66E+03	7.66E+03	-1.22E+04	-2.88E+05
4	-8.00E+03	-8.00E+03	-2.88E+05	-6.04E-02

Figure 7.42 Example 7.15: Deflection plot and table of element end shear forces and bending moments for the case $P = 0$.

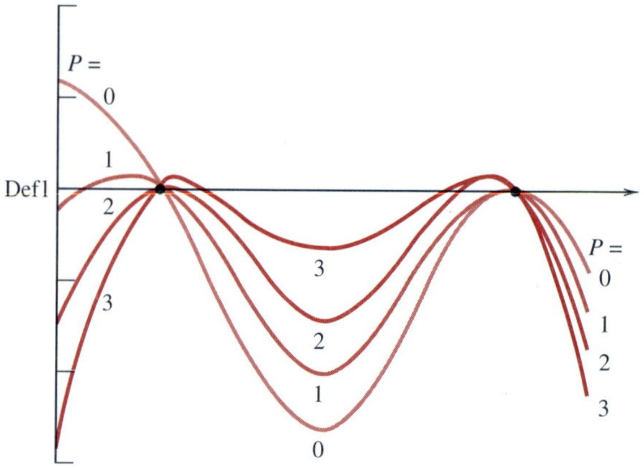

Figure 7.43 Example 7.15: Deflection plots for different values of P.

EXAMPLE 7.16

A long multispan steel beam is shown in Fig. 7.44. The spans are made up of beams of different sizes that have different values for the moments of inertia, as shown in Fig. 7.44. A constant distributed load is applied over the first span, and a 12-kip load is applied at the center of the outer span at section D.

(*a*) Find the deflection curve for the beam, using the BEAMMECH program.

(*b*) If after some years in service, support C is found to have settled (moved downward) because of erosion a distance 0.10 in, we wish to find the new deflection curve and compare it with the one found in part (*a*).

We neglect the weight of the beam; it should be clear from Fig. 7.44 that this beam is statically indeterminate and that the calculation of the deflection curve by hand would be a tedious undertaking.

We divide the beam into four elements and five nodes, as shown in Fig. 7.44. The dimensions should be expressed in inches and the constant distributed loading in pounds per inch. The summary of the input data and the output data containing the solution for no settlement is shown in Fig. 7.45. The deflection curve is shown in Fig. 7.46*a*; the shear force and bending moment diagrams are shown in Fig. 7.46*b* and *c*.

Now if support C (node 3) undergoes a settlement of −0.10 in, we need only make this single change (using option 2 in the data input menu)

NODE NO.	DEFL. VALUE
3	−0.10

We show the input data for part (*b*) in Fig. 7.47. Figure 7.48 gives the numerical and graphical results for the beam deflection and shear force and bending moment diagrams after settlement has occurred.

Comparison between the results for the absolute values of the deflection, shear force, and bending moment for the original condition of the beam and the results for the settled condition shows the following:

	(*a*) No settlement	(*b*) Settlement
|Maximum deflection|	8.83×10^{-2} in	11.8×10^{-2} in
Location	x = 119 in	x = 415 in
|Maximum shear|	72.0 kips	67.5 kips
|Maximum moment|	2.92×10^3 kip·in	2.59×10^3 kip·in

Thus we see that although the settlement of C has increased the value of the maximum deflection and changed the location of the maximum deflection, it has decreased the shear force and bending moment at the wall and therefore reduced the shear and normal stresses there.

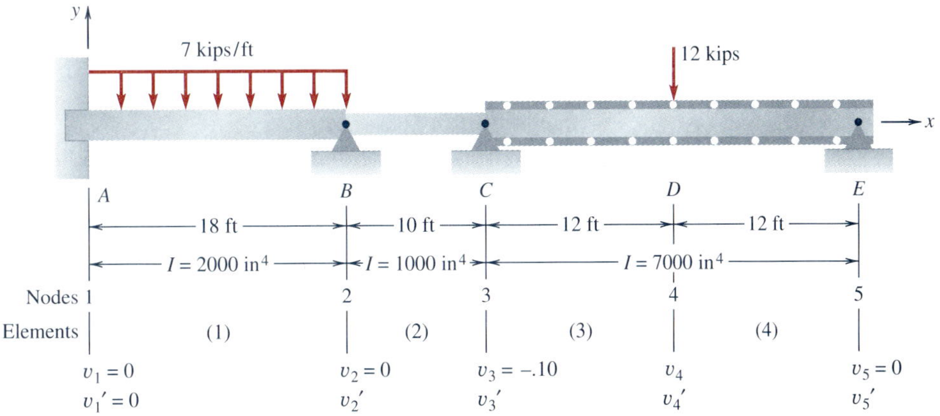

Figure 7.44 Example 7.16

Figure 7.45 Example 7.16: Table of input data and table of solution for no settlement.

```
***MECHANICS OF MATERIALS***     DATA FILE: ex716
                ***BEAM***

INPUT DATA:

NUMBER OF ELEMENTS                    NM= 4

ELEMENT      L         I            E             q
           (in)     (in**4)    (10**6; psi)    (lb/in)
           (m)      (m**4)        (MPa)         (N/m)

   1      216.00   2.00E+03     3.00E+01      -5.83E+02
   2      120.00   1.00E+03     3.00E+01       0.00E+00
   3      144.00   7.00E+03     3.00E+01       0.00E+00
   4      144.00   7.00E+03     3.00E+01       0.00E+00

 SPECIFIED CONDITIONS:
*********************************

 NODE NO.      FORCE VALUE

    4           -1.20E+04

 NODE NO.      MOMENT VALUE

 NODE NO.      DEFL. VALUE

    1           0.00E+00
    2           0.00E+00
    3           0.00E+00
    5           0.00E+00

 NODE NO.      SLOPE VALUE
    1           0.00E+00

 SOLUTION:

*******************************************************

  NODE NO.   VALUE OF X    DEFLECTION        SLOPE

*******************************************************

    1          0.000       0.00E+00        0.00E+00

    2        216.000       0.00E+00        1.17E-03

    3        336.000       0.00E+00       -3.86E-04

    4        480.000      -3.33E-02        1.12E-05

    5        624.000       0.00E+00        3.41E-04

TOTAL LENGTH OF THE BEAM =    624.000
```

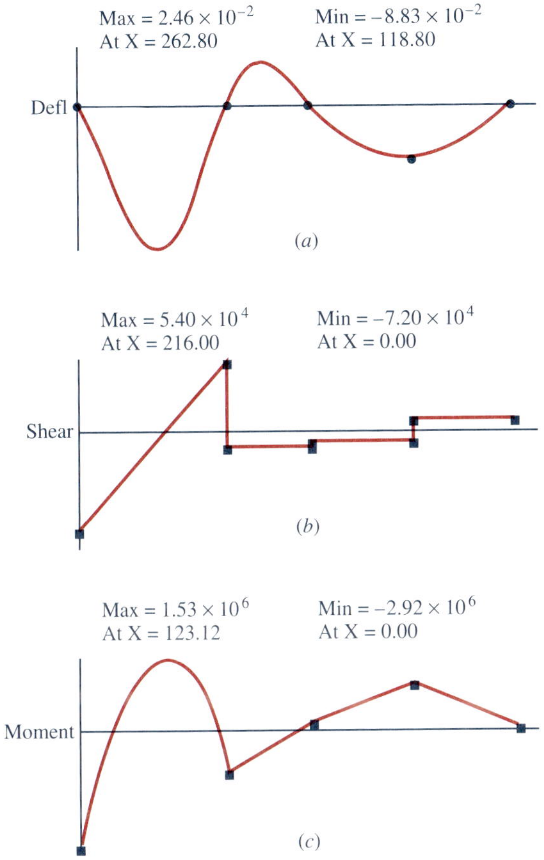

Figure 7.46 Example 7.16: Deflection plot, shear force and bending moment diagrams for no settlement.

MECHANICS OF MATERIALS DATA FILE: ex716

BEAM

INPUT DATA:

NUMBER OF ELEMENTS NM= 4

ELEMENT	L (in) (m)	I (in**4) (m**4)	E (10**6; psi) (MPa)	q (lb/in) (N/m)
1	216.00	2.00E+03	3.00E+01	-5.83E+02
2	120.00	1.00E+03	3.00E+01	0.00E+00
3	144.00	7.00E+03	3.00E+01	0.00E+00
4	144.00	7.00E+03	3.00E+01	0.00E+00

SPECIFIED CONDITIONS:

****

NODE NO.	FORCE VALUE
4	-1.20E+04

NODE NO.	MOMENT VALUE

NODE NO.	DEFL. VALUE
1	0.00E+00
2	0.00E+00
3	-1.00E-01
5	0.00E+00

NODE NO.	SLOPE VALUE
1	0.00E+00

Figure 7.47 Example 7.16: Table of input data for settlement of 0.10 in at node 3.

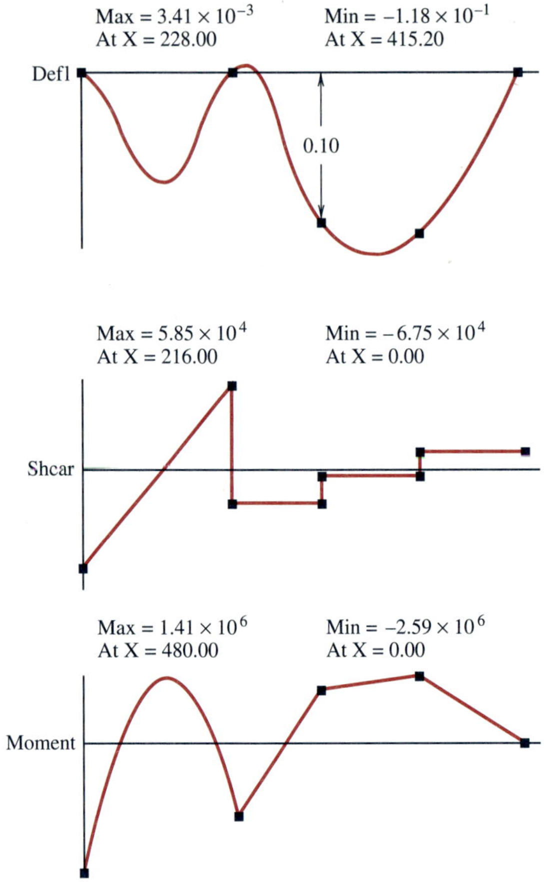

Figure 7.48 Example 7.16: Deflection plot and shear force and bending moment diagrams for settlement of 0.10 in at node 3.

7.9 Concluding Remarks

In this chapter we presented techniques for the calculation of the deflections, slopes, shear forces, and bending moments of statically indeterminate beam problems. The algebraic procedures were the same as those followed for statically determinate beams except that the unknown reactions and the constants of integration were determined simultaneously from the geometric constraints imposed on the deflection curve along with the equilibrium equations.

The use of superposition, as we noted in Chap. 6, provides another method of solution that is useful when specific information for the value of a deflection or the value of a reaction at a point is needed. In this chapter we have seen that the superposition method allows us to calculate indeterminate reactions in a very direct manner for many problems.

We finished the chapter with a presentation of the displacement method for beams. The displacement method provides a technique for analyzing beams that brings together the three steps of solution in a systematic fashion. Force and moment equilibrium equations at nodes, when expressed in terms of node deflections and node slopes, reduce to a system of linear algebraic equations that can be solved easily by computer methods. The program BEAMMECH on the diskette allows us to solve either statically determinate or statically indeterminate problems conveniently and to find the deflection, shear force, and bending moment distributions along the beam. In the case of a beam with a constant bending modulus *EI*, we also obtain the value of the maximum bending stress.

The program has the means to analyze well-defined statically determinate or statically indeterminate beam problems with constant distributed load segments and applied concentrated forces and moments. The input specifications are fairly simple, and the graphical output enables us to understand more clearly the overall deflection of the beam. The program is an introductory version of more advanced structural analysis programs that are available in the engineering profession. The availability of this program does, of course, give one the possibility of going back to the earlier chapters and now using the computer program to solve more easily many of the beam problems encountered there.

PROBLEMS

Solve the problems which are associated with Sec. 7.2 by integrating either the bending moment $M(x)$ or the loading function $q(x)$. Unless otherwise indicated, take the bending modulus as EI_z and neglect the weight of the beam.

7.2-1 A beam is built in at both ends and subjected to a concentrated moment M_B acting at B, as shown in Fig. P7.2-1. Find (*a*) the wall reactions, (*b*) the maximum absolute value of the bending moment, and (*c*) the values of the slope and deflection at section B.

7.2-2 Consider Example 7.2, Fig. 7.3. If the constant distributed load q_0 is now applied over only the left half of the beam in Fig. 7.3, find the reactions at the supports. Draw the shear force and bending moment diagrams for the beam.

7.2-3 Consider Example 7.2, Fig. 7.3. If the center support in Fig. 7.3 settles downward relative to the other supports by an amount Δ_B, then calculate the support reactions in terms of Δ_B.

7.2-4 A propped cantilever beam is subjected to an end moment M_B, as shown

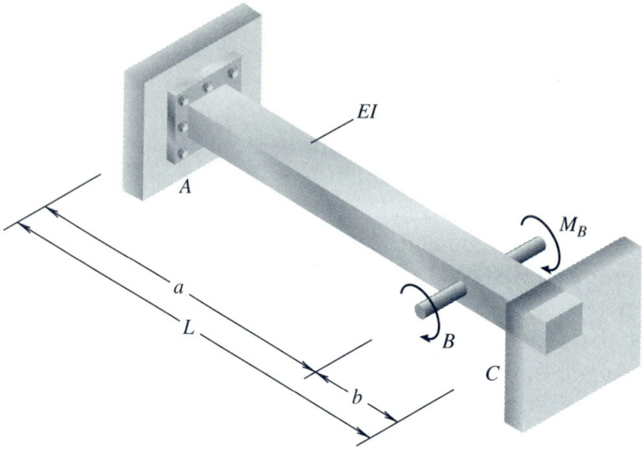

Fig. P7.2-1

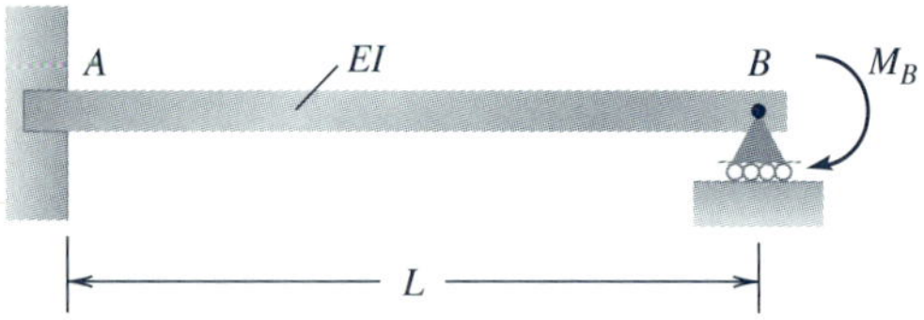

Fig. P7.2-4

in Fig. P7.2-4. (*a*) Find an expression for the deflection $v(x)$, (*b*) find the maximum deflection, and (*c*) draw the shear force and bending moment diagrams for the beam.

7.2-5 A beam is built into rigid walls at both ends, as shown in Fig. P7.2-5. If the wall at the right end B settles downward by an amount Δ_B as shown, find (*a*) an expression for the deflection $v(x)$ and (*b*) the wall reactions. (*c*) If the beam is a W10 × 45 I beam and $E = 30{,}000$ ksi, $L = 20$ ft, and $\Delta_B = 1$ in, then find the maximum normal bending stress in the beam due to the wall settlement.

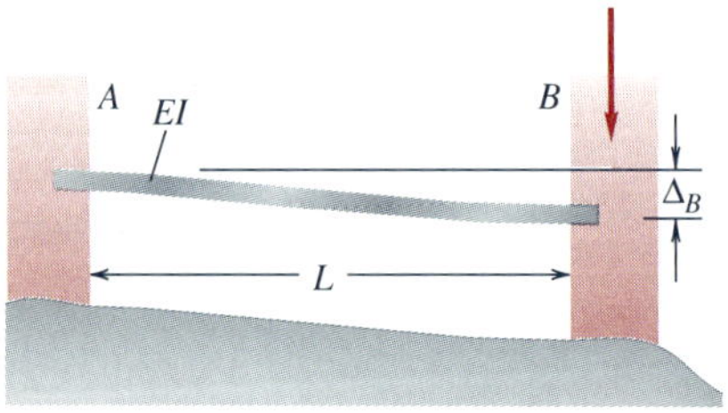

Fig. P7.2-5

7.2-6 A beam is built into the rigid walls at both ends and is subjected to a constant distributed load q_0, as shown in Fig. P7.2-6. Find (*a*) an expression for the deflection $v(x)$ and (*b*) the wall reactions.

7.2-7 A continuous beam is subjected to a linearly increasing distributed loading and is simply supported at both ends and at the midpoint, as shown in Fig. P7.2-7. Find (*a*) the reactions at the supports and (*b*) the shear force and bending moment diagrams for this beam.

7.2-8 A flexible beam ABC is simply supported at one end and is supported by a light rod CD which is pinned to the beam at C and supported at D, as shown in Fig. P7.2-8. If a load P is applied at the midpoint B, show that the deflection under the load is given by

$$v\left(\frac{L}{2}\right) = -\frac{PL^3}{48EI}(1 + 12\lambda)$$

where

$$\lambda = IL_1/(A_1L^3).$$

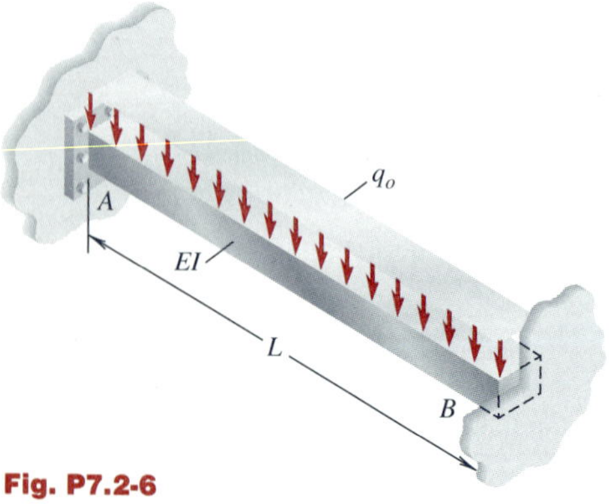

Fig. P7.2-6

Fig. P7.2-7

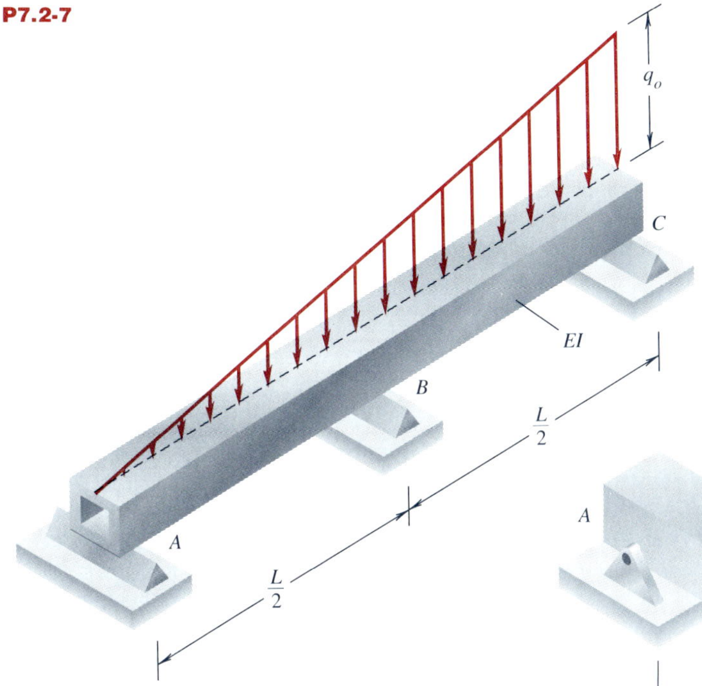

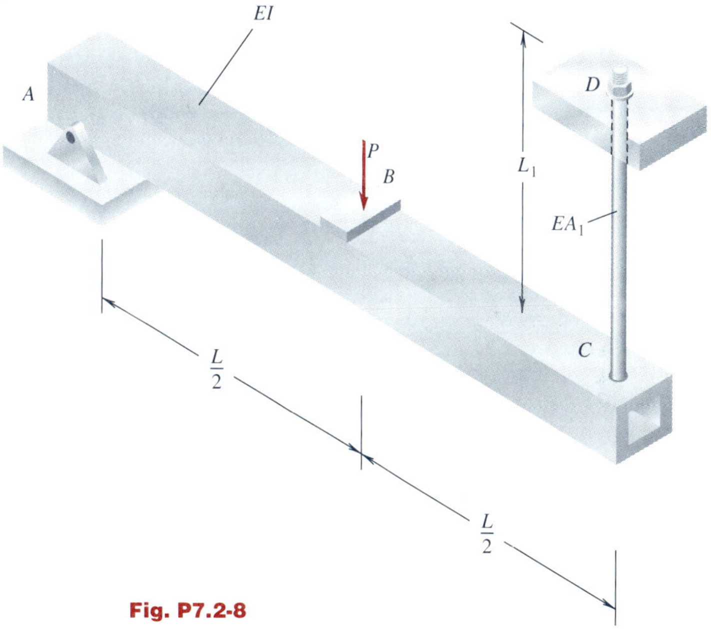

Fig. P7.2-8

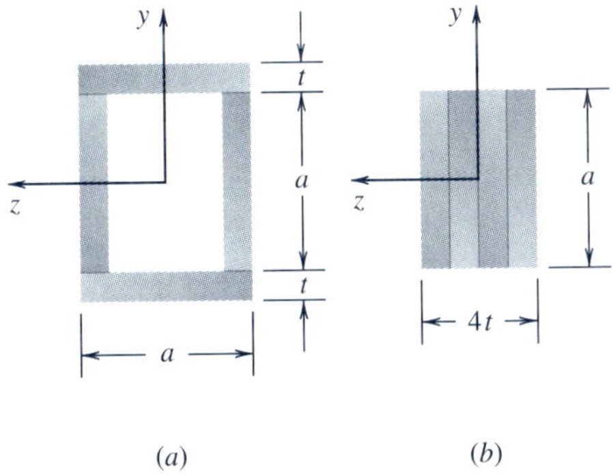

Fig. P7.2-9

7.2-9 Consider Example 7.2, Fig. 7.3. It is proposed to use a wood beam in a given application of the results of Example 7.2. Two beam designs are to be considered, as shown in Fig. P7.2-9. If structural lumber of nominal dimensions 2 in × 12 in is selected (see App. D), find the maximum normal and shear stresses due to bending for each of the proposed designs (*a*) and (*b*). Take $q_0 = 50$ lb/in, and make use of the results of Example 7.2. The planks in each design are glued together, and the length $2L$ is 20 ft.

7.2-10 A cantilever beam ABC supported also at end A is loaded by a concentrated load P, as shown in Fig. P7.2-10. Determine the expression for the deflection of the beam, and sketch the bending moment diagram.

7.2-11 A long uniform rod of length L, weight w per unit length, and bending modulus EI is placed on a rigid horizontal table such that a short segment of length a overhangs the table, as shown in Fig. P7.2-11. (See Prob. 5.7-6.) The reactions on the beam are given in Prob. 5.7-6. Determine the length b of the segment BC that lifts up from the table. The result depends only on length a.

7.2-12 A uniform beam is supported as shown in Fig. P7.2-12. A moment M_0 is applied at the right end as shown. Obtain an expression for the deflection of the beam and the slope angle at C. Compare the slope angle obtained with the slope angle for a beam of length $2L$ without the support at B.

7.2-13 A cantilever beam ABC is supported in part by an elastic spring with spring constant k, as shown in Fig. P7.2-13. When the beam is unloaded, the spring is unstretched. Obtain an expression for the deflection under the load P in terms of EI, L, and the spring constant k.

7.2-14 A beam is supported as shown in Fig. P7.2-14. Obtain an expression for the value of the deflection under the load.

7.2-15 A beam clamped at both ends carries a pair of concentrated loads, as shown in Fig. P7.2-15. Find the deflection and slope at section B.

Fig. P7.2-10

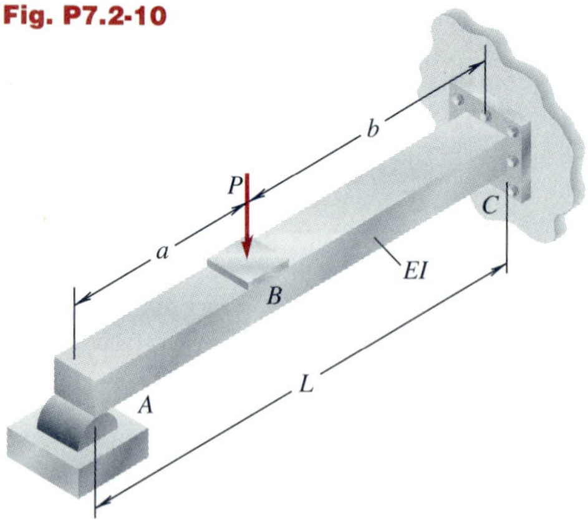

Fig. P7.2-11

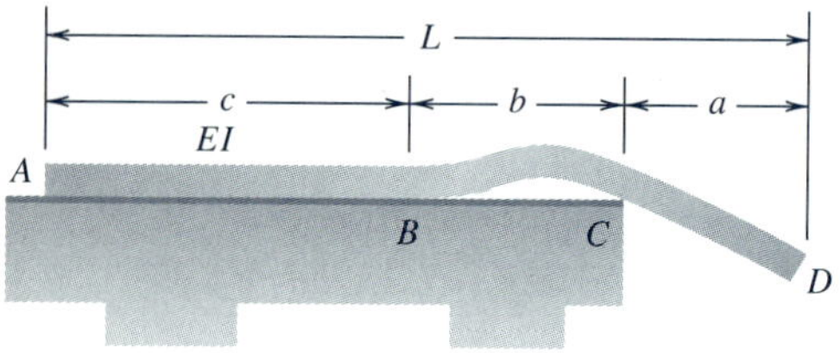

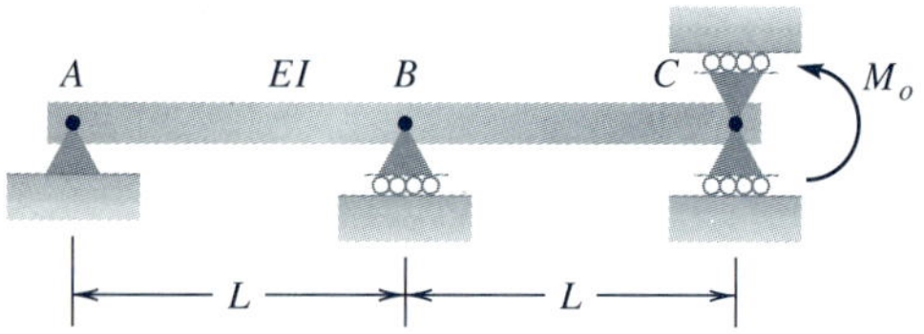

Fig. P7.2-12

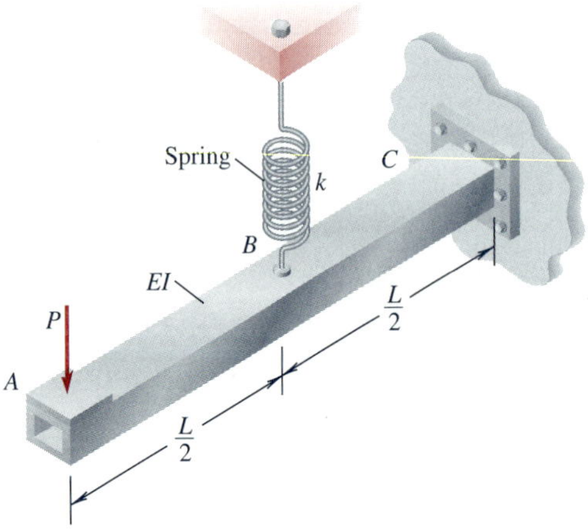

Fig. P7.2-13

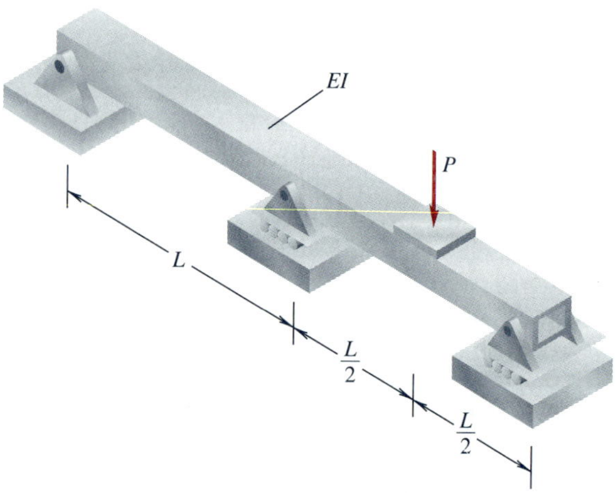

Fig. P7.2-14

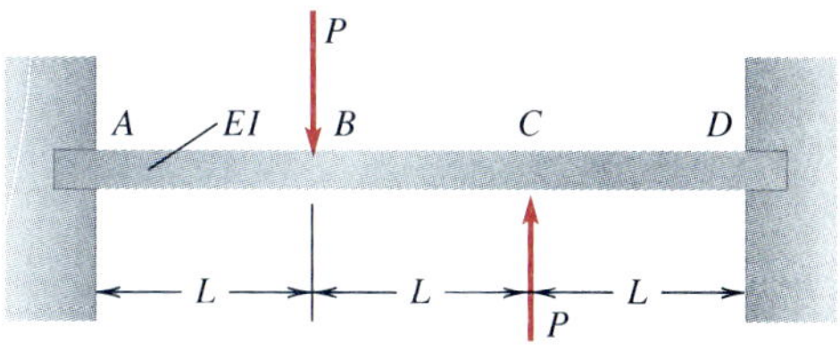

Fig. P7.2-15

7.2-16 A beam clamped at both ends carries a pair of concentrated moments, as shown in Fig. P7.2-16. Find the maximum deflection and maximum normal stress due to bending if the depth of the beam cross section is $L/12.5$.

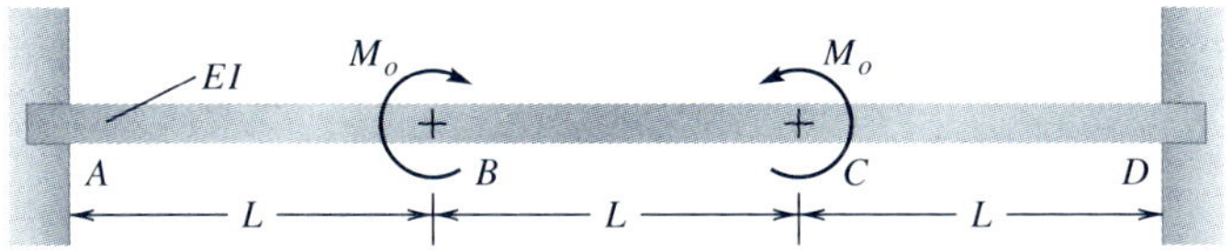

Fig. P7.2-16

7.2-17 A beam which is clamped at one end and is simply supported at the other carries a pair of concentrated loads, as shown in Fig. P7.2-17. Find the deflections at sections B and C and the value of the maximum bending moment and its location.

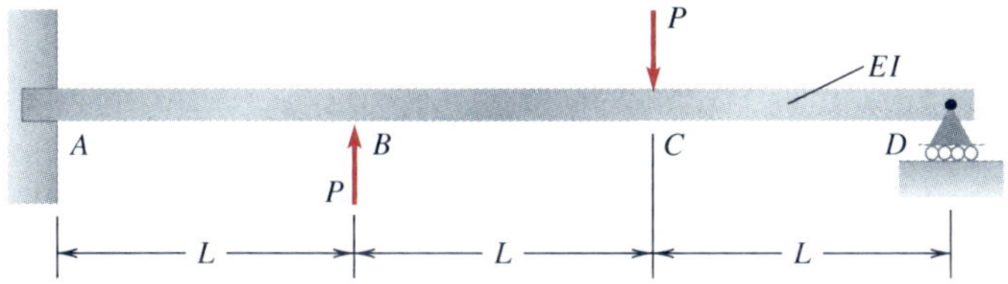

Fig. P7.2-17

7.2-18 A beam under a constant distributed loading rests on three supports, as shown in Fig. P7.2-18. Find the reactions and the maximum bending moment.

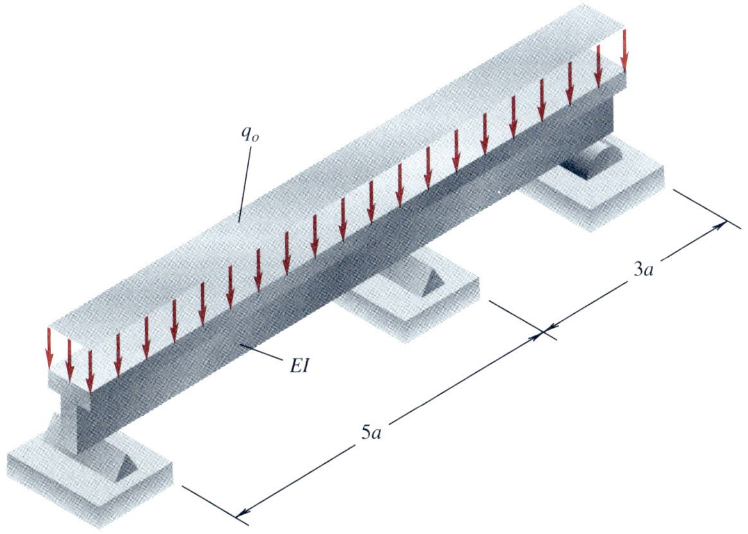

Fig. P7.2-18

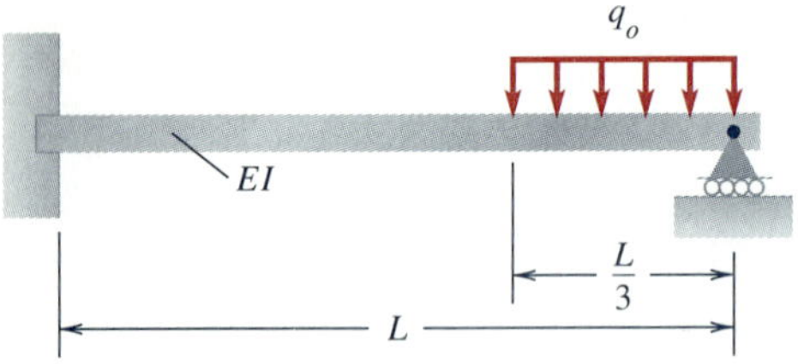

Fig. P7.2-19

7.2-19 A cantilever beam of length L is simply supported at one end and is loaded by a constant distributed loading q_0 over part of the beam, as shown in Fig. P7.2-19. Find the value of the maximum deflection (see Example 7.5).

7.2-20 A supported cantilever beam shown in Fig. P7.2-20 carries a constant distributed loading q_0 and a concentrated load P. Find the deflection under the load. Take $L = 4.0$ m, $a = 3.0$ m, $q_0 = 10$ kN/m, and $P = 18$ kN. The beam is a steel beam with $E = 200$ GPa and $I = 60 \times 10^{-6}$ m^4.

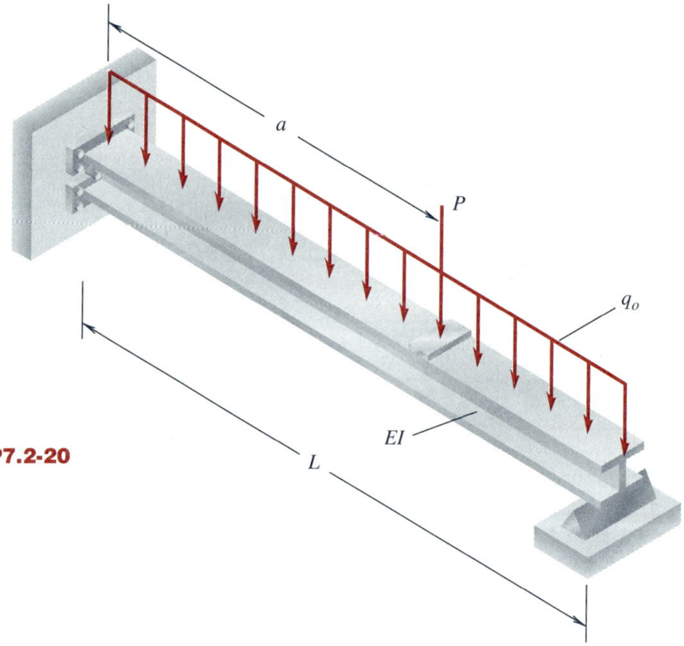

Fig. P7.2-20

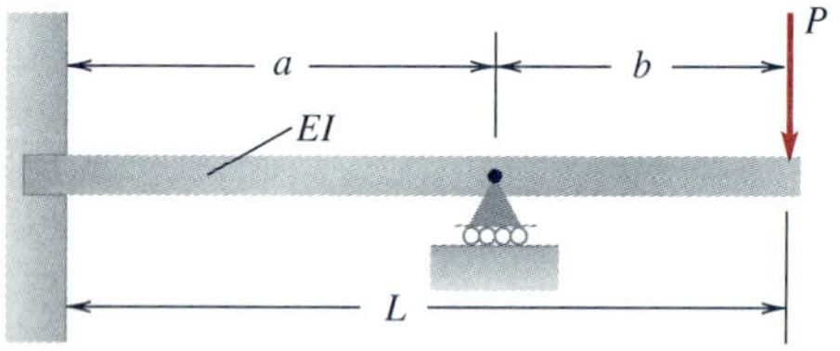

Fig. P7.2-23

7.2-21 Solve Prob. 7.2-20 with $L = 10$ ft, $a = 4$ ft, $q_0 = 500$ lb/ft, $P = 4000$ lb, $E = 30{,}000$ ksi, and $I = 72$ in^4.

7.2-22 Refer to Fig. P7.2-10. If $L = 5$ m, $a = 2$ m, $P = 50$ kN, $E = 205$ GPa, and $I = 80 \times 10^{-6}$ m^4, find the deflection under the load and the value of the reaction at A.

7.2-23 A cantilever beam is supported as shown in Fig. P7.2-23 and carries a concentrated load P at the end. Determine the deflection under the load P if $a = 4$ m, $L = 6$ m, $P = 40$ kN, $I = 100 \times 10^{-6}$ m^4, and $E = 200$ GPa.

7.2-24 Solve Prob. 7.2-23 if the beam is a W-shape beam (App. C) W8 × 35, $a = 10$ ft, $L = 16$ ft, $P = 10$ kips, and $E = 30 \times 10^6$ psi.

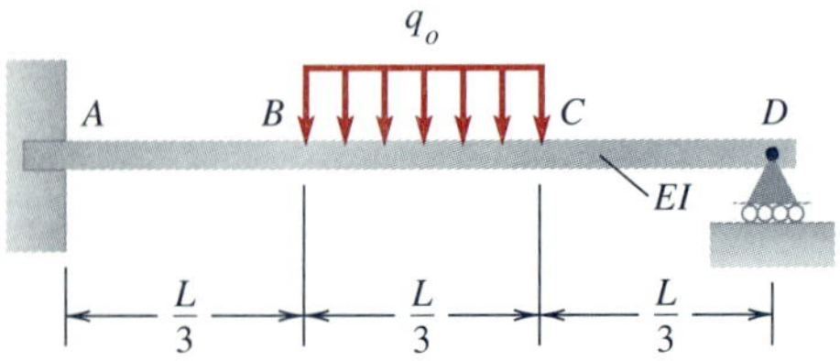

Fig. P7.2-25

7.2-25 A beam of length L carries a constant distributed load over the middle third of the beam, as shown in Fig. P7.2-25. Determine the maximum deflection.

Use the method of superposition to solve the following problems assigned for Sec. 7.3. Unless otherwise indicated, take the bending modulus as EI_z and neglect the weight of the beam.

7.3-1 Two cantilever beams are so configured that they are in contact at their ends, as shown in Fig. P7.3-1. If a constant distributed load q_0 is applied to both beams, find (*a*) the deflection of the contact point B and (*b*) the shear force and bending moment diagrams for each beam.

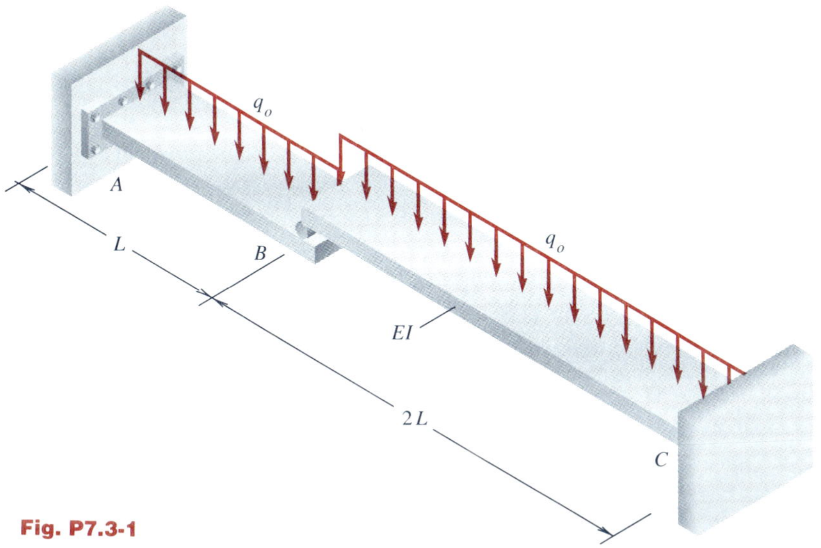

Fig. P7.3-1

7.3-2 Consider Example 7.1, Fig. 7.2. Solve this problem by replacing the original problem by the superposition of three cantilever beam problems: one with the load P acting and a free end at B and the other two with no load P acting but with suitable end loads at B, so as to result in the correct geometric conditions for a built-in support.

7.3-3 Consider Example 7.1, Fig. 7.2. Solve this problem by replacing the original problem by the superposition of a combination of beams that are simply supported at points A and B.

7.3-4 A cantilever beam shown in Fig. P7.3-4 is loaded at C by a concentrated load P. A gap of Δ_B separates the unloaded beam from a support at B. At what load P_1 will the beam just make contact with the support at B? What is the total deflection at C if the load is increased to $2P_1$?

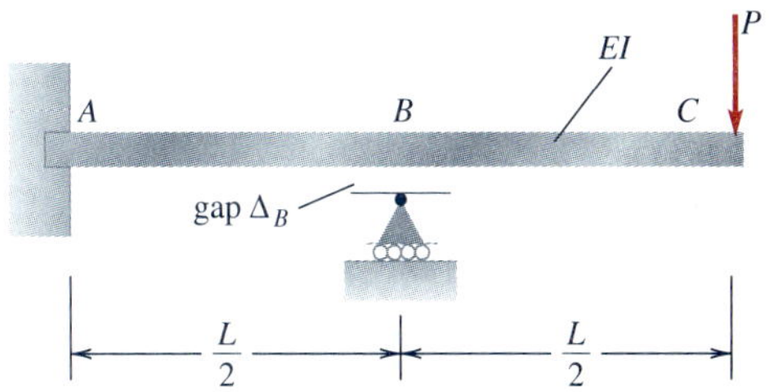

Fig. P7.3-4

7.3-5 Consider Example 7.4, Fig. 7.5. Obtain the results given in Fig. 7.5*c* and *d*.

Fig. P7.3-6

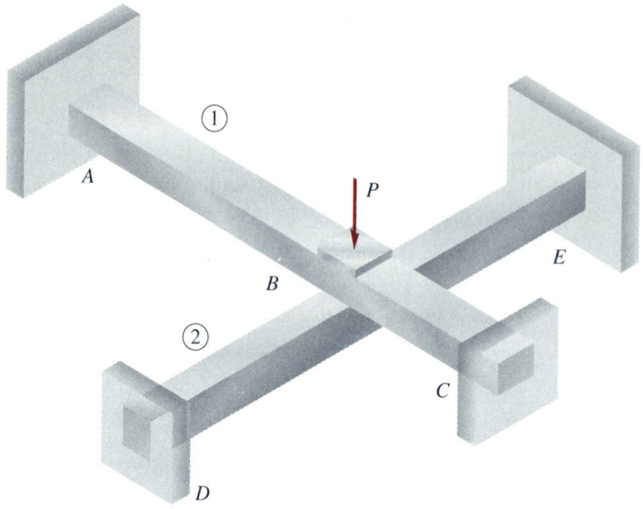

7.3-6 A pair of beams is built into rigid walls, as shown in Fig. P7.3-6. When $P = 0$, the beams are just in contact at B. Find the deflection under the load P at B. For beam 1 the lengths of AB and BC are $5L/8$ and $3L/8$, respectively, and the bending modulus is EI_z. For beam 2 the lengths of DB and BE are both $L/2$, and the bending modulus is $E(2I_z)$.

7.3-7 Solve Prob. 7.2-1 by the method of superposition.

7.3-8 Solve Prob. 7.2-4 by the method of superposition.

7.3-9 Solve Prob. 7.2-5 by the method of superposition.

7.3-10 Solve Prob. 7.2-6 by the method of superposition.

7.3-11 A steel shaft is supported in bearings at A, B, and C and is subjected to transverse concentrated loads, as shown in Fig. P7.3-11. To avoid excessive wear on the bearings, the allowable slopes at the bearings due to the transverse loads must not exceed 0.01° in magnitude. Assume that the bearings can be modeled as simple supports. (*a*) Calculate the slopes at A, B, and C, and find the ratio of the actual slopes to the allowable slope. (*b*) Find the location and magnitude of the maximum bending stress in the shaft. Take $d = 2$ in and $E = 30{,}000$ ksi.

7.3-12 A pipeline crosses a gap between two walls and is supported by a hanger at B so that A, B, and C are in line. To gain clearance at B, the hanger is used to pull the pipe upward by an amount Δ_B. See Fig. P7.3-12.

(*a*) Find an expression for the maximum normal stress due to bending σ_{max} in terms of E, Δ_B, L, and d_o, where d_o is the outside diameter of the pipe.

(*b*) If the pipe has a nominal diameter of 2 in Standard Weight (see App. E), $E = 30{,}000$ ksi, and $L = 20$ ft, find σ_{max} for the values $\Delta_B = 0.5$, 1.0, and 2.0 in. Neglect the weight of the pipe.

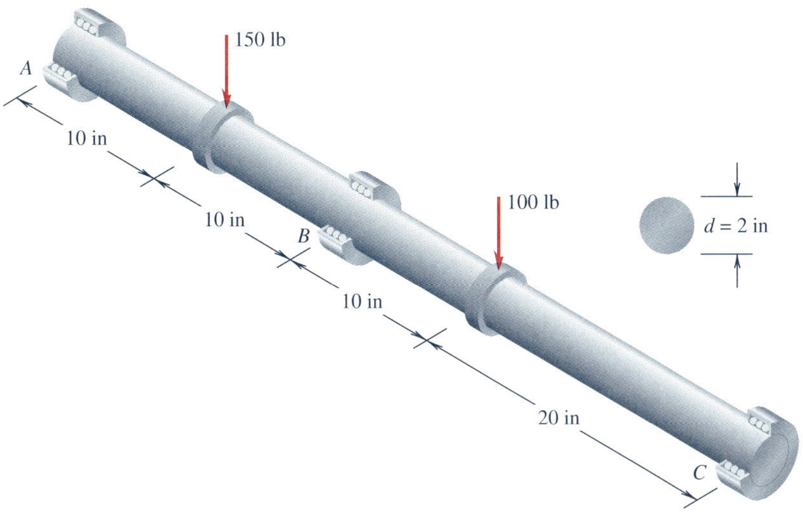

Fig. P7.3-11

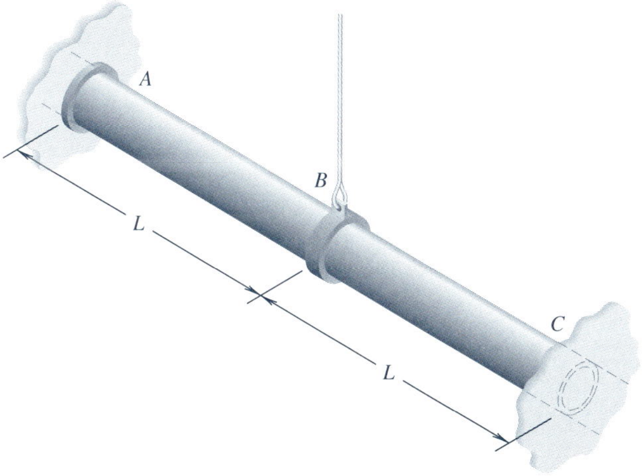

Fig. P7.3-12

7.3-13 In a particular structural application, the stiffness at C, that is, the ratio k of the applied load P to the resulting deflection v_c at C, is varied by locating the cross-beam DE at different distances a from the wall at A. See Fig. P7.3-13.

(*a*) Find an expression for k as a function of L, a, E, and I, where I and $I/2$ are the moments of inertia for AC and DE and E is the elastic modulus for both beams.

(*b*) For the special cases $a/L = \frac{1}{4}, \frac{1}{2}, \frac{3}{4}$, and 1, obtain the corresponding values of the stiffness k.

(*c*) Using the result from Table G.1, Fig. G.1-2 (App. G), that $k = 3EI/L^3$ for $a = 0$, evaluate the ratio R of k for a general value of a to k for $a = 0$, using the results of part (*b*). Plot R versus a on a graph.

Fig. P7.3-13

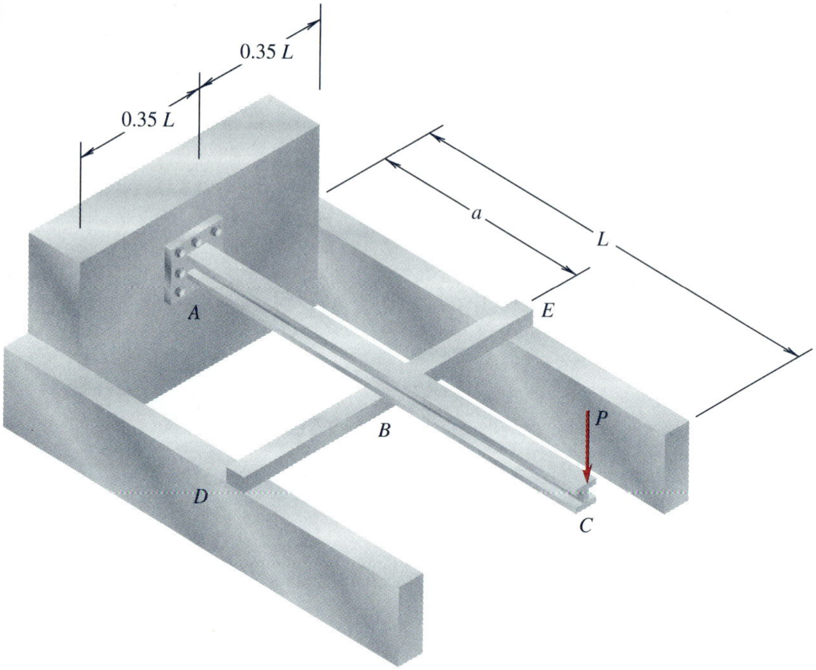

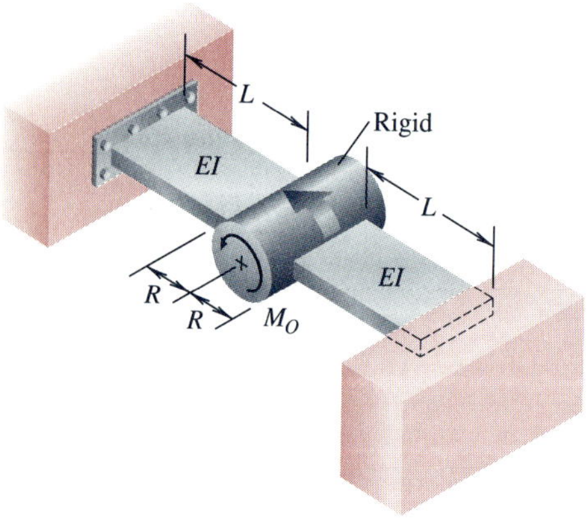

Fig. P7.3-14

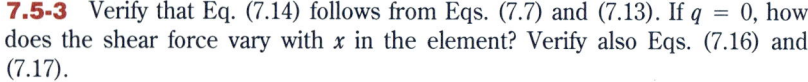

7.3-14 A rigid plug of radius R is supported by two identical elastic beams, as shown in Fig. P7.3-14. Find the torsional stiffness of the system by applying the moment M_0 and dividing the moment by the resulting angle of rotation of the plug.

7.3-15 A small portable bridge is simply supported at both ends and rests on a pontoon support in the middle, as shown in Fig. P7.3-15. If a load P is applied to the bridge at B, find the deflection of the pontoon due to load P. Neglect the weight of the bridge for this analysis, and take the weight per unit volume of the water to be γ and the waterline area of the pontoon to be A.

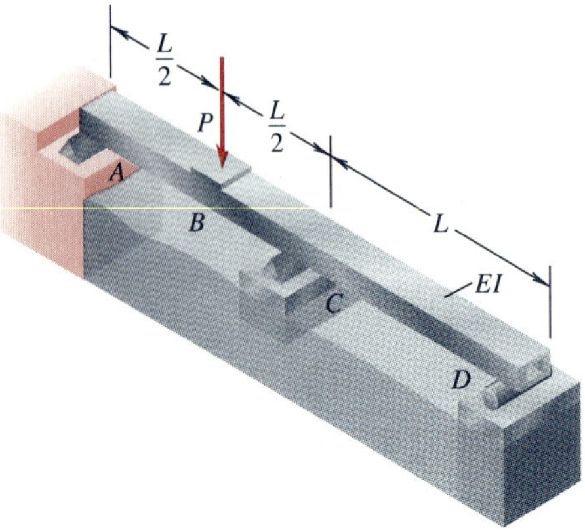

Fig. P7.3-15

7.5-1 Verify that the result of Eq. (7.7) satisfies the four geometric conditions at the ends of the beam as shown in Fig. 7.11 and satisfies the differential equation of equilibrium, Eq. (7.1).

7.5-2 Verify that Eq. (7.9) follows from Eqs. (7.7) and (7.8). If $q = 0$, how does the moment vary with x over the length of the element? Verify also Eqs. (7.11) and (7.12).

7.5-3 Verify that Eq. (7.14) follows from Eqs. (7.7) and (7.13). If $q = 0$, how does the shear force vary with x in the element? Verify also Eqs. (7.16) and (7.17).

Solve the following problems associated with Sec. 7.6 by using the beam force-deformation relations given in Fig. 7.12. Unless otherwise noted, take the bending modulus as EI_z and neglect the weight of the beam.

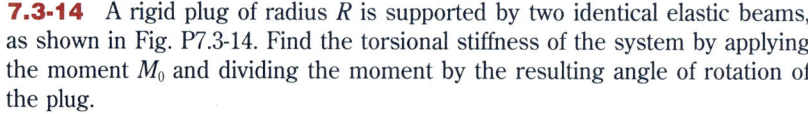

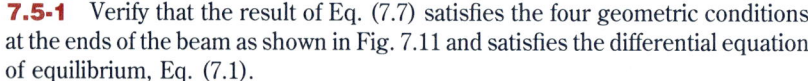

7.6-1 Consider Example 7.7, Fig. 7.13. Use the results in this example and Eqs. (7.7), (7.9), and (7.14) to obtain expressions for the deflection $v(x)$, bending moment $M(x)$, and shear force $V(x)$ of the cantilever beam. Compare your results to those in Fig. G.1-2, App. G.

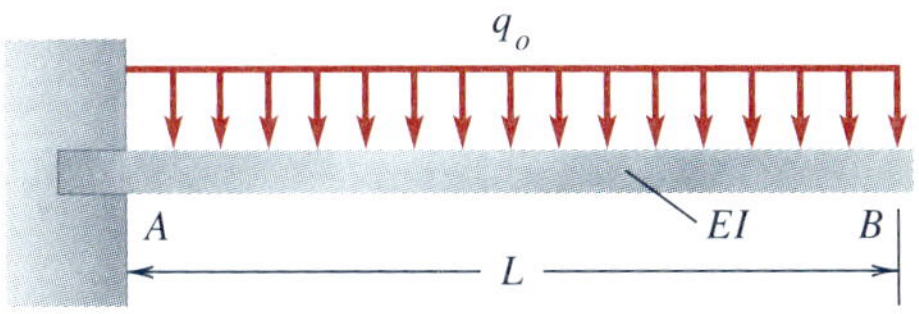

Fig. P7.6-6

7.6-2 Consider Example 7.8, Fig. 7.14. Use the results in this example and Eqs. (7.9) and (7.14) to obtain the expressions for the bending moment $M(x)$ and shear force $V(x)$ for the beam.

7.6-3 Consider Example 7.9, Fig. 7.15. Use the results in this example and Eqs. (7.9) and (7.14) to obtain the expressions for the bending moment $M(x)$ and the shear force $V(x)$ in the beam. Sketch the shear force and bending moment diagrams.

7.6-4 Consider Example 7.10, Fig. 7.16. Use the results in this example and Eqs. (7.9) and (7.14) to obtain the expressions for the bending moment $M(x)$ and shear force $V(x)$ in the beam. Sketch the shear force and bending moment diagrams.

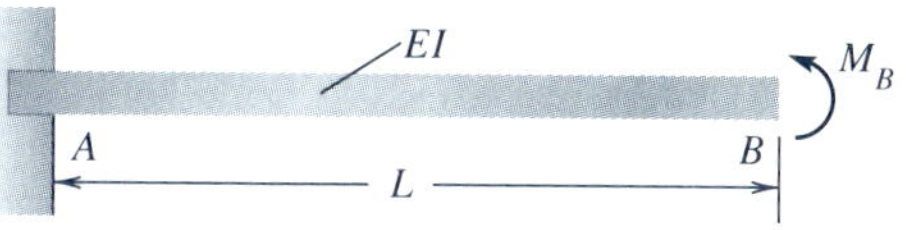

Fig. P7.6-7

7.6-5 Consider Example 7.11, Fig. 7.17. Use the results in this example and Eqs. (7.9) and (7.14) to obtain the expressions for the bending moment $M(x)$ and shear force $V(x)$ in the beam. Consider the special cases of (*a*) $\delta = \frac{2}{3}L\phi$, (*b*) $\delta = \frac{1}{2}L\phi$, and (*c*) $\delta = \frac{1}{3}L\phi$. Sketch the shear force and bending moment diagrams for the three special cases.

7.6-6 A cantilever beam is fixed at A and carries a constant distributed loading q_0, as shown in Fig. P7.6-6. Find the deflection and slope at B. Compare with the results of Fig. G.1-3, App. G.

7.6-7 A cantilever beam is fixed at A and carries a concentrated moment M_B at B, as shown in Fig. P7.6-7. Find the deflection and slope at B. Compare with the results of Fig. G.1-5, App. G.

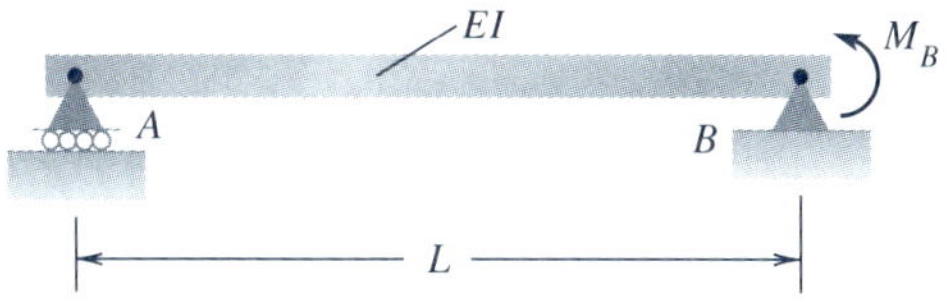

Fig. P7.6-8

7.6-8 A beam is simply supported at both ends A and B and carries a moment M_B at end B, as shown in Fig. P7.6-8. Find (*a*) the slopes at both ends, (*b*) an expression for the deflection $v(x)$ by making use of Eq. (7.7), and (*c*) the maximum deflection and its location along the beam. Compare with the results in Fig. G.2-4, App. G.

7.6-9 A propped cantilever beam is subjected to a moment M_B which is applied to end B, as shown in Fig. P7.6-9. Find (*a*) the slope at B, (*b*) an expression for the deflection $v(x)$ by making use of Eq. (7.7), and (*c*) the maximum deflection and its location along the beam.

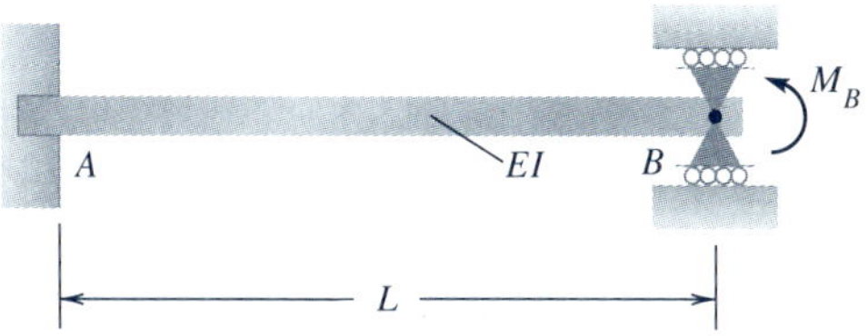

Fig. P7.6-9

Solve the following problems associated with Sec. 7.7 by dividing the beam in each case into two elements and using the methods of Sec. 7.7. Unless otherwise indicated, take the bending modulus as EI_z and neglect the weight of the beam.

7.7-1 A stepped beam of length $2L$ carries a constant distributed load q_0 over its left half, as shown in Fig. P7.7-1. If the bending modulus of segment AB is

Fig. P7.7-1

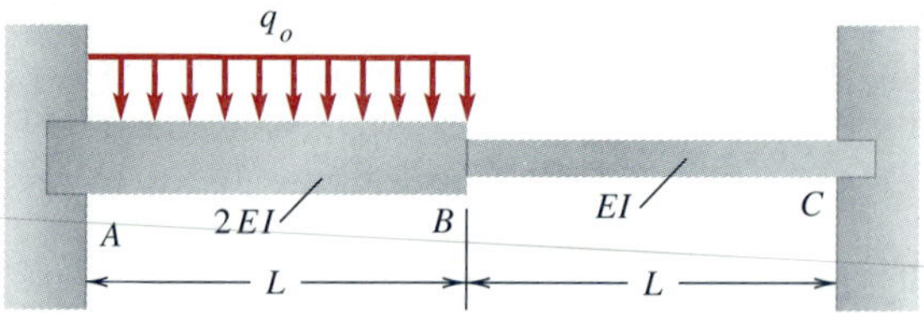

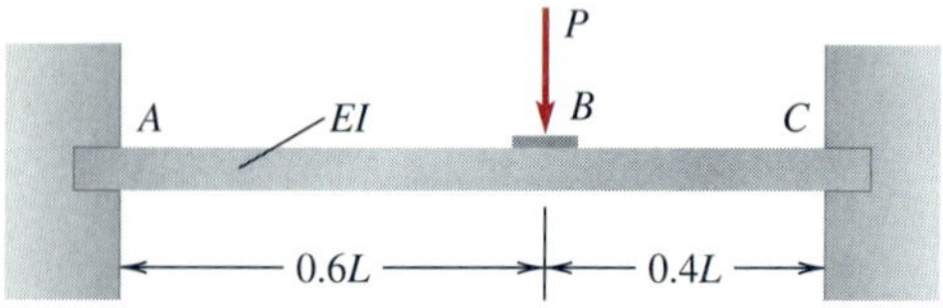

Fig. P7.7-2

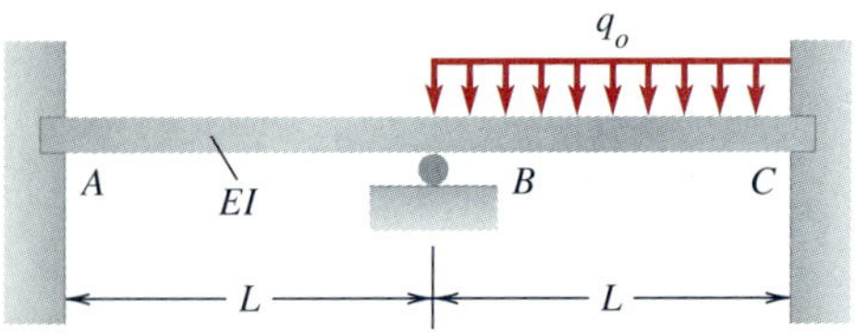

Fig. P7.7-3

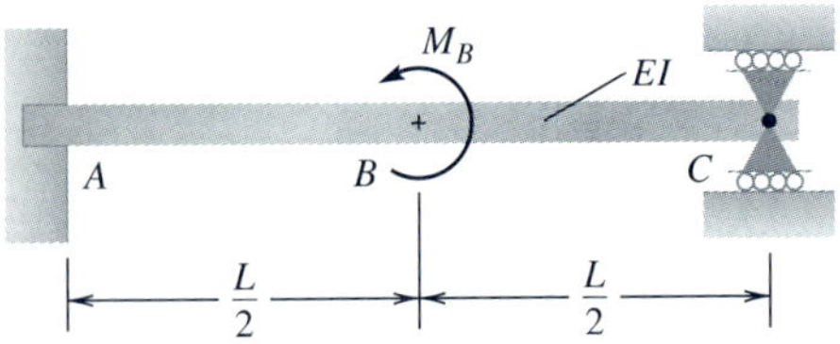

Fig. P7.7-4

2*EI* and for *BC* is *EI*, then use the displacement method to find (*a*) the deflection and slope at section *B* and (*b*) the reactions at *A* and *C*.

7.7-2 A beam is clamped at both ends and carries a concentrated load *P*, as shown in Fig. P7.7-2. Use the displacement method to find the unknown deflection and slope under the load at *B*.

7.7-3 A beam is clamped at both ends and is simply supported at the center, as shown in Fig. P7.7-3. Use the displacement method to derive a single node equilibrium equation for the slope at the center due to a constant distributed loading q_0 over the right half segment of the beam. Find the reactions at both ends and the center.

7.7-4 A propped cantilever beam is subjected to a concentrated moment acting at its midpoint, as shown in Fig. P7.7-4. Use the displacement method to derive a set of three node equilibrium equations for the unknown slopes and deflections of this problem. Solve this system and find the deflection at *B*, the slope at *C*, and the reactions at *A* and *C*.

7.7-5 A beam of length 4*L* extends continuously over five simple supports and is subjected to a constant distributed loading q_0, as shown in Fig. P7.7-5. Take advantage of symmetry about *C*.

(*a*) Solve this problem by using the displacement method, making use of only two elements with two node equilibrium equations for the unknown slopes at sections *A* and *B*.

(*b*) Find the unknown slopes, and calculate the reactions at *A*, *B*, and *C*.

7.7-6 A beam is clamped at both ends and has different bending moduli and constant distributed loadings on its left- and right-hand segments, as shown in Fig. P7.7-6. Use the displacement method to derive two node equilibrium equations for

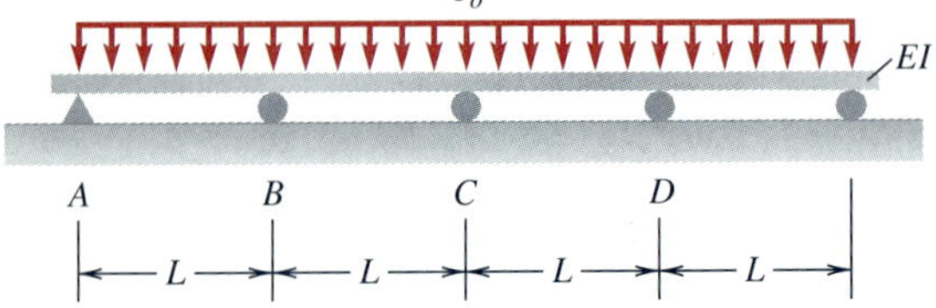

Fig. P7.7-5

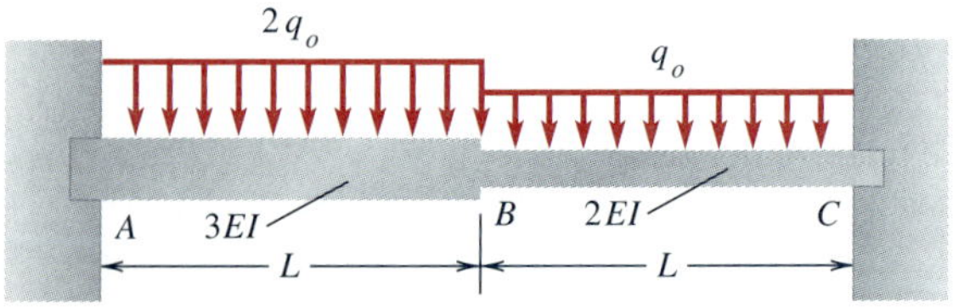

Fig. P7.7-6

the unknown deflection and slope at section B. Solve for these unknowns and find the reactions.

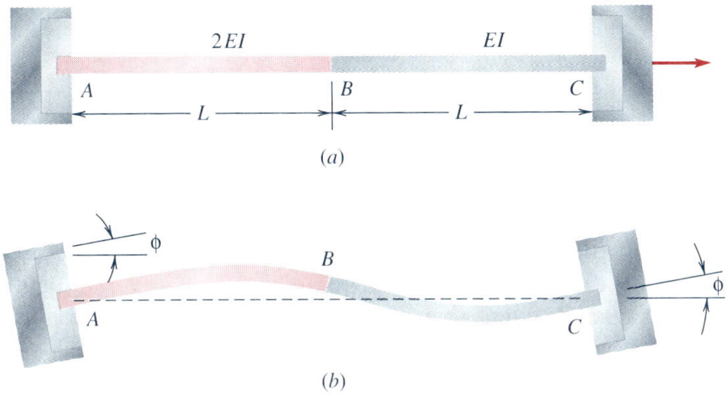

Fig. P7.7-7

7.7-7 A spring component in a machine is made up of different materials in segments AB and BC with elastic moduli $2E$ and E, as shown in Fig. P7.7-7a. In the operation of the machine, the left and right supports rotate by an angle ϕ, as shown in Fig. P7.7-7b.

(a) Use the displacement method to derive two node equilibrium equations for v_B and v'_B and solve for these unknowns.

(b) Find the reactions at the ends.

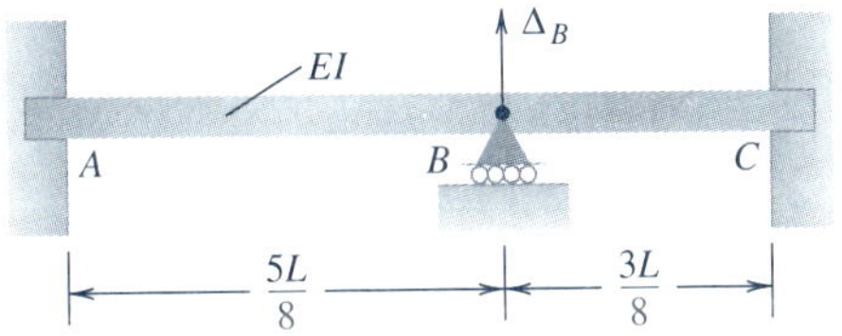

Fig. P7.7-8

7.7-8 A beam is fixed to rigid walls at both ends and is simply supported at B, as shown in Fig. P7.7-8. If the intermediate support at B moves upward by an amount Δ_B relative to the fixed supports at A and C, then use the displacement method to find (a) the slope at B and (b) the reactions at sections A, B, and C.

7.7-9 A component ABC in a machine is clamped at the right end, simply supported at the middle, and restricted at the left end so that vertical deflection is allowed but the slope remains equal to zero under the action of the constant distributed loading q_0. See Fig. P7.7-9.

(a) Use the displacement method to derive two node equilibrium equations for v_A and v'_B.

(b) Solve for these unknowns and find the reactions.

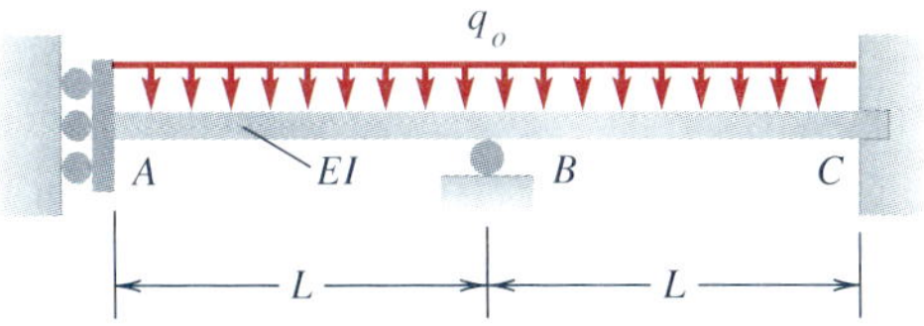

Fig. P7.7-9

7.7-10 Two cantilever beams mounted as shown in Fig. P7.7-10 share the load P. If the bending modulus for beam AB is twice that for beam BC, then use the

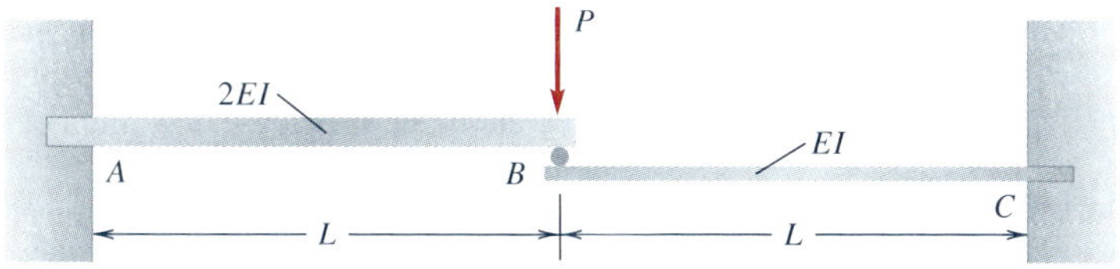

Fig. P7.7-10

displacement method to find (*a*) the deflection under the load, (*b*) the slopes of the two beams at B, and (*c*) the reactions at A and C.

7.7-11 Consider Example 6.14, Fig. 6.21. Use the displacement method to find the slope and deflection at the free end. Note that it is possible to solve this problem in two steps. (1) A free-body of section AB is taken as a separate beam, and the slope and deflection at B are found. (2) Section BC is then taken as a separate beam with specified values for v_B and v'_B as found in part 1. Find values for v_C and v'_C and compare with those given in Example 6.14, Eqs. (*f*) and (*g*).

7.7-12 Consider Example 6.16, Fig. 6.23. Use the procedure outlined in Prob. 7.7-11 to solve this problem. Compare your results with those given in Eqs. (*e*) and (*f*) of Example 6.16.

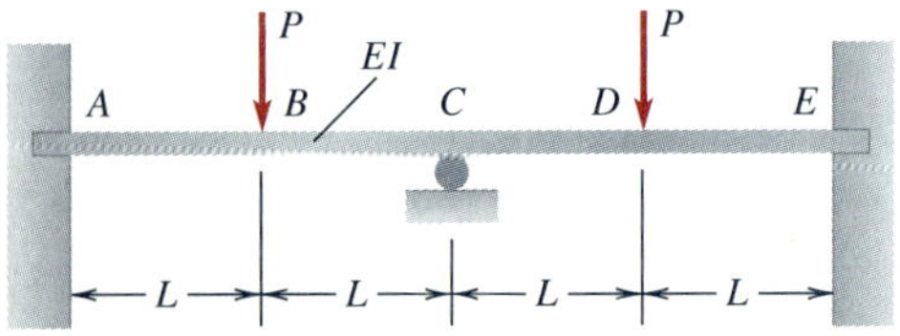

Fig. P7.7-13

7.7-13 A beam is clamped at both ends, is simply supported at the middle, and carries two concentrated loads, as shown in Fig. P7.7-13.
(*a*) Use the displacement method to derive two node equilibrium equations for v_B and v'_B, making use of symmetry about C.
(*b*) Solve for these unknowns and find the reactions at A and C.

7.7-14 Consider Example 7.3, Fig. 7.4. Use the method described in Prob. 7.7-11 to solve this problem. Compare your result with the value given in Eq. (*l*), Example 7.3.

For the following problems associated with Sec. 7.8, use the BEAMMECH computer program on the diskette.

7.8-1 Determine the maximum deflection and the numerically largest values of the bending moment and bending stress for the steel beams shown in the figure. (Neglect the weight of the beam; d is the depth of the beam, take $E = 30 \times 10^6$ psi or 200 GPa.)
(*a*) Fig. P4.2-4*a*; use a 6-in $\times$ 5-in $\times$ $\frac{3}{8}$-in structural tube.
(*b*) Fig. P4.2-4*b*; use a W8 $\times$ 40 shape.
(*c*) Fig. P4.2-4*c*; use $I = 2 \times 10^{-4}$ m^4, $d = 250$ mm.
(*d*) Fig. P4.2-4*d*; use $I = 1.5 \times 10^{-4}$ m^4, $d = 180$ mm.
(*e*) Fig. P4.2-4*e*; use $I = 5.8 \times 10^7$ mm^4, $d = 200$ mm.
(*f*) Fig. P4.2-4*f*; use a W8 $\times$ 18 shape.
(*g*) Fig. P4.2-4*g*; use a W10 $\times$ 33 shape.
(*h*) Fig. P4.2-4*h*; use $I = 9.2 \times 10^{-5}$ m^4, $d = 280$ mm.

7.8-2 Determine the maximum deflection and the numerically largest values of the bending moment and bending stress for the steel beams shown in the figure. (Neglect the weight of the beam; d is the depth of the beam, take $E = 30 \times 10^6$ psi or 200 GPa.)
(*a*) Fig. P4.3-5*a*; use $I = 9.6 \times 10^{-5}$ m^4, $d = 150$ mm.
(*b*) Fig. P4.3-5*b*; use $I = 5 \times 10^{-5}$ m^4, $d = 180$ mm.
(*c*) Fig. P4.3-5*c*; use an S10 $\times$ 25.4 shape.
(*d*) Fig. P4.3-5*d*; use a W12 $\times$ 35 shape.
(*e*) Fig. P4.3-5*e*; use a W8 $\times$ 35 shape.

(*f*) Fig. P4.3-5*f*; use a W8 × 35 shape.
(*g*) Fig. P4.3-5*g*; use $I = 3.1 \times 10^{-5}\ \text{m}^4$, $d = 210$ mm.
(*h*) Fig. P4.3-5*h*; use $I = 9 \times 10^{-4}\ \text{m}^4$, $d = 230$ mm.
(*i*) Fig. P4.3-5*i*; use an S12 × 50 shape.

7.8-3 A continuous W12 × 26 wide-flange beam carries a constant distributed loading and a concentrated load and is supported as shown in Fig. P7.8-3.
(*a*) Find the values of $|v|_{max}$ and $|\sigma|_{max}$ due to bending. Take $E = 30{,}000$ ksi.
(*b*) If the support at B settles by 0.1 in, then find the new values for the quantities calculated in part (*a*).

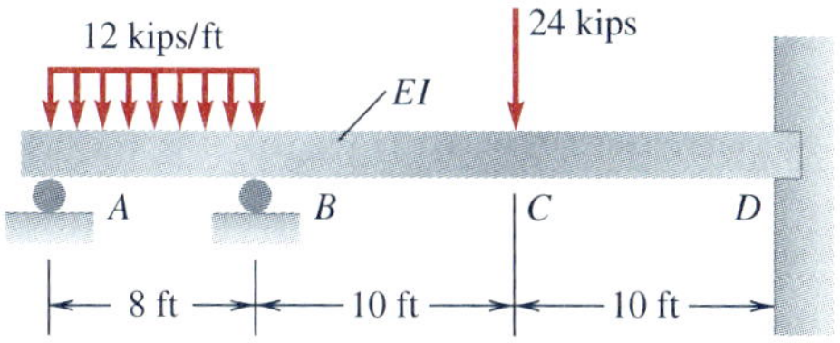

Fig. P7.8-3

7.8-4 A beam of length $2L$ is simply supported at each end and at the middle and is subjected to a constant distributed loading q_0, as shown in Fig. P7.8-4*a*. In Fig. P7.8-4*b* the distributed loading is replaced by concentrated loads at the middle of each span. In each case the maximum deflection and bending moment are of the form

$$|v|_{max} = \beta_1 \frac{q_0 L^4}{EI} \qquad |M|_{max} = \beta_2 q_0 L^2$$

where β_1 and β_2 are numerical values. By giving unit values to q_0, L, E, and I, calculate β_1 and β_2 for each of the two cases shown in Fig. P7.8-4*a* and *b*.

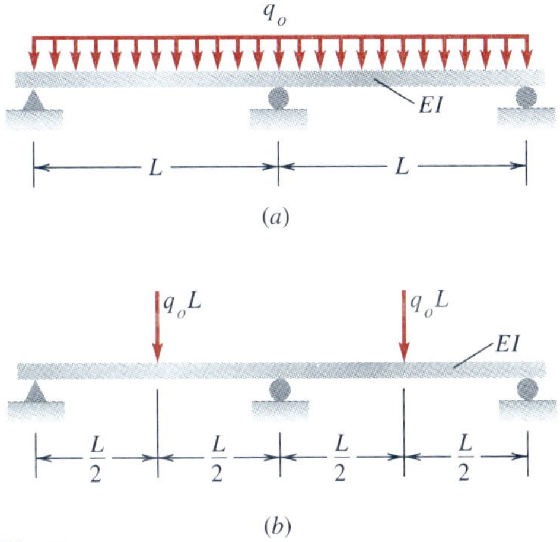

Fig. P7.8-4

7.8-5 Use the BEAMMECH program to verify the results given in Eqs. (*k*) and (*l*) of Example 7.5. Take unit values for q_0, L, E, and I, and locate nodes at $x = 0$, 0.578, and 1. Also verify the results for the reactions.

7.8-6 Use the BEAMMECH program to verify the results given in Eqs. (*d*) and (*k*) of Example 7.6. Take unit values for q_0, L, E, and I.

7.8-7 A stepped beam of length $2L$ carries a constant distributed loading q_0 over its left half, as shown in Fig. P7.7-1. The bending modulus of segment AB is $2EI_z$, and the bending modulus of segment BC is EI_z. We wish to determine the deflection and slope at section B and the reactions at A and C. We note that the deflection and slope at B must be of the form

$$v_B = \lambda_1 \frac{q_0 L^4}{EI_z} \qquad v'_B = \lambda_2 \frac{q_0 L^3}{EI_z}$$

where λ_1 and λ_2 are numerical constants. In a similar way, the reactions are of the form

$$R = \lambda_3 q_0 L \qquad M = \lambda_4 q_0 L^2$$

Select unit values for the parameters q_0, L, E, and I_z, and determine the numerical constants. Neglect the weight of the beam.

7.8-8 A beam is fixed to rigid walls at both ends and simply supported at B as shown in Fig. P7.7-8. If the intermediate support at B moves upward an amount Δ_B relative to the fixed supports at A and C, obtain expressions for the value of the slope at B and the values of the reactions at sections A, B, and C. Note that the slope at B must be of the form

$$v_2' = \lambda_1 \frac{\Delta_B}{L}$$

and the force and moment reactions must be of the form

$$R = \lambda_2 \frac{EI\Delta_B}{L^3} \qquad M = \lambda_3 \frac{EI\Delta_B}{L^2}$$

where λ_1, λ_2, and λ_3 are numerical constants. Neglect the weight of the beam.

7.8-9 Two cantilever beams mounted as shown in Fig. P7.7-10 share the load P. The bending modulus for beam AB is twice that for beam BC. Obtain the deflection under the load and the reactions at A and C. Note that the deflection under an end load F for each beam must be of the form

$$v = \lambda_1 \frac{FL^3}{EI}$$

while the force and moment reactions must be of the form

$$R = \lambda_2 F \qquad M = \lambda_3 FL$$

where the lambdas are numerical constants. Neglect the weight of the beam and use unit values for P, L, E, and I.

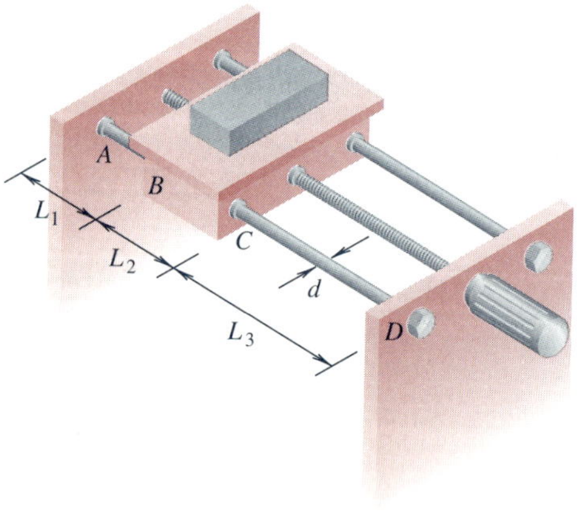

Fig. P7.8-10

7.8-10 A positioning table in a precision instrument moves by sliding on two rods, as shown in Fig. P7.8-10. The rigid table is in contact with the rods at points B and C and it is assumed that one-fourth of the weight W of the table is carried at each of the four contact points. Find (*a*) the slope of the rigid table, (*b*) the deflection of the rod at points B and C, and (*c*) the location and magnitude of the maximum normal stress in the rods due to bending. Take $L_1 = 250$ mm, $L_2 = 250$ mm, $L_3 = 500$ mm, $d = 10$ mm, $W = 100$ N, and $E = 200$ GPa.

7.8-11 A five-span simply supported beam with overhanging ends carries a constant distributed loading q_0, as shown in Fig. P7.8-11. Make use of symmetry and use only four elements. For values of the overhang ratio a/L from 0 to 0.5 in steps of 0.1, make plots of the following quantities as a function of a/L:

(*a*) Maximum absolute value of the deflection $v(x)$ over the left half of the beam

(*b*) Maximum absolute value of the normal stress due to bending over the left half of the beam

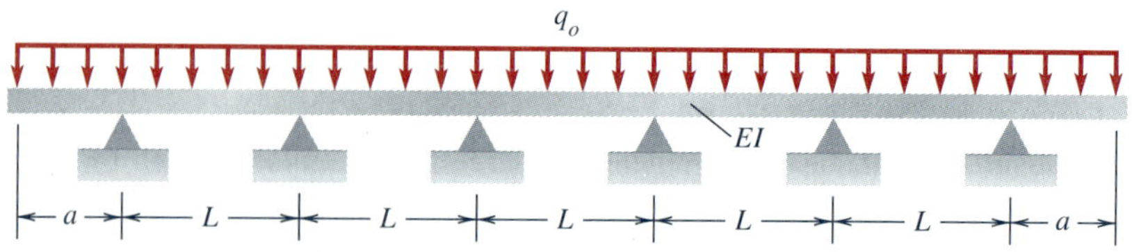

Fig. P7.8-11

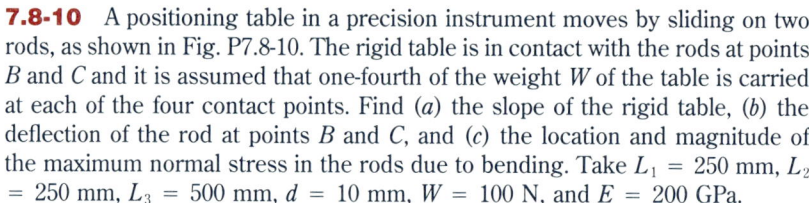

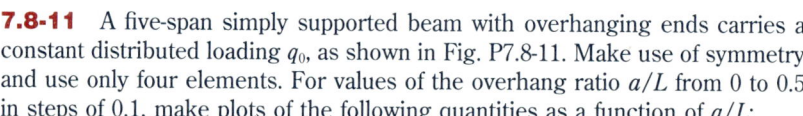

Discuss the results in terms of the advantages or disadvantages of the overhang. Take $E = 30{,}000$ ksi, $q_0 = 1000$ lb/ft, and $L = 20$ ft, and the beam is a W14 × 120 wide-flange I beam.

7.8-12 For the simply supported beam subjected to a constant distributed loading q_0, as shown in Fig. P7.8-12, study the effect of the overhang on the deflection parameters δ_1 and δ_2 as defined in the figure. Define the numerical parameters β_1 and β_2 by $\delta_1 = \beta_1\delta$ and $\delta_2 = \beta_2\delta$, where $\delta = 5q_0L^4/(384EI)$ is the maximum deflection for the case of no overhang, that is, $a/L = 0$. Make computer runs for $a/L = 0$ to 0.5 in steps of 0.1, and plot graphs of β_1 and β_2 as functions of a/L. (*Hint:* for the numerical work normalize δ to the value 1 by suitable choices of the parameters q_0, L, E, and I.)

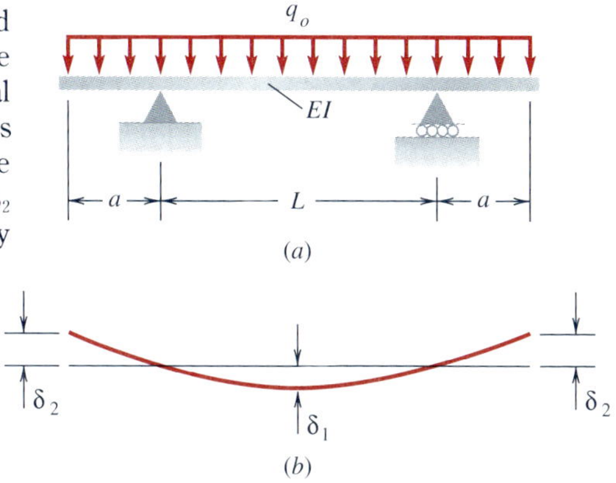

Fig. P7.8-12

7.8-13 A steel cantilever beam ABC consists of two segments AB and BC with moments of inertia of $2I$ and I, respectively, as shown in Fig. P7.8-13. Calculate the coefficient β in the algebraic expression $\beta PL^3/(EI)$ for the deflection at C due to load P. (*Hint:* Take unit values for P, L, E, and I in your computer run.)

7.8-14 A steel cantilever beam has a stepped circular cross section with different moments of inertia in each segment and carries a load P at D, as shown in Fig. P7.8-14. Find expressions for the deflection at sections B, C, and D in the form

$$v_D = -\beta_D \frac{PL^3}{EI} \qquad \text{etc.}$$

Find the coefficients β_B, β_C, and β_D by making a computer run with unit values for P, L, E, and I.

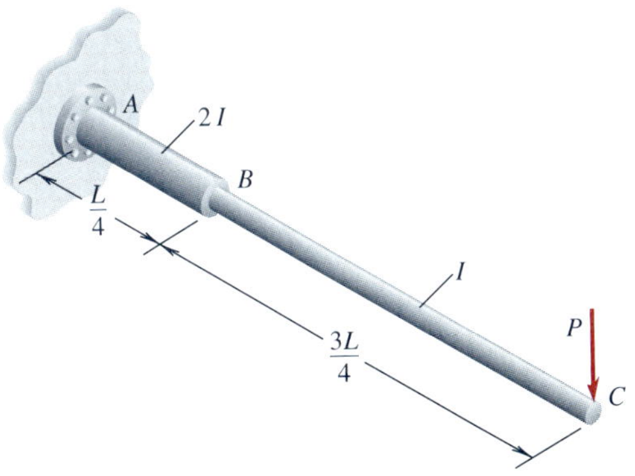

Fig. P7.8-13

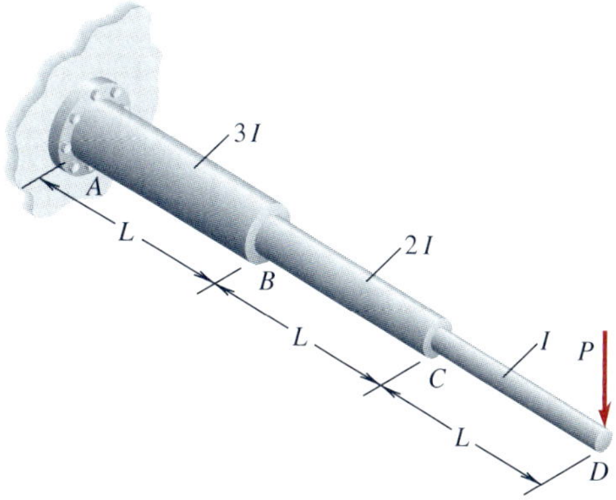

Fig. P7.8-14

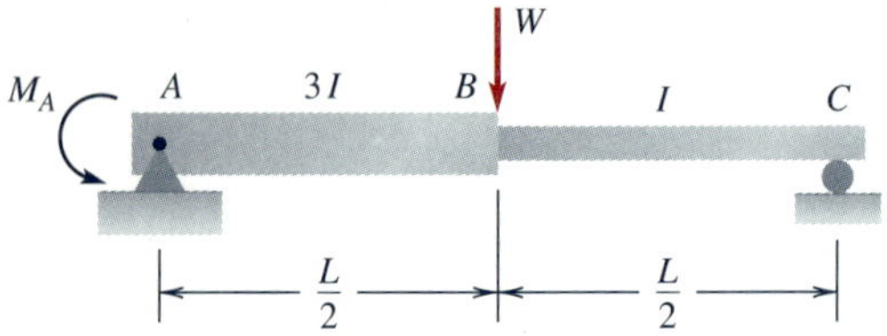

Fig. P7.8-15

7.8-15 A simply supported beam ABC consists of two segments AB and BC with moments of inertia $3I$ and I, respectively, as shown in Fig. P7.8-15. In a particular application, the beam is subjected to a moment M_A applied at end A, and we wish to find the magnitude W of a counterweight to be applied at B so that the net deflection of section B due to the action of both M_A and W will be zero. Find the coefficient α defined by $v_B = \alpha M_A L^2/(EI)$ by using unit values for M_A, L, E, and I. Do the same for β where $\hat{v}_B = -\beta WL^3/(EI)$, and then combine the two deflection results to force the deflection at B to be zero.

7.8-16 A stepped shaft is simply supported at its ends and carries a load P at the midpoint C, as shown in Fig. P7.8-16. If the middle segment has a diameter of 1.5 times the diameter of the outer segments, find (*a*) the deflection under the load at C and (*b*) the maximum normal stress due to bending in segments AB and BC. Find the coefficient α defined by $v_c = -\alpha PL^3/(Ed^4)$ by giving unit values to L, P, E, and d. For the maximum bending stress, find β, defined by $\sigma_{max} = \beta PL/d^3$.

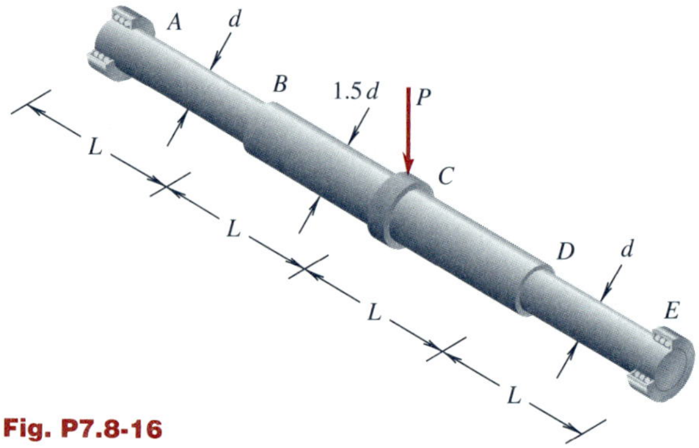

Fig. P7.8-16

7.8-17 Solve Prob. 6.3-27 by using the methods of this section. Use unit values for a, P, I, and E, and take a large value, say 1000, to represent the I value for the rigid middle segment. Compare your result with the answer given for Prob. 6.3-27.

7.8-18 If the load P applied to the beam in Fig. P6.3-27 at C is applied at B instead, then find the deflection at B in the form $v_B = -\alpha Pa^3/(EI)$, where the coefficient α is found by giving unit values to P, a, E, and I. For the rigid member use I equal to 1000.

7.8-19 Solve Prob. 6.5-12, using the methods of this section. In particular, find the coefficients α and β in the expressions $v_c = -\alpha qL^4/(EI)$ and $v_c' = -\beta qL^3/(EI)$ by using unit values for q, L, E, and I.

7.8-20 Use the methods of this section to solve Prob. 6.5-13 by taking only one spring bar on each side and using a large value, say 1000, for the I value for the machine segment and unit values for W, L, E, and I to obtain the coefficient α.

7.8-21 Solve Prob. 7.8-20 again, but this time with the center of gravity of the machine located at one-fourth of the distance from B to C.

7.8-22 Use the methods of this section to solve Prob. 6.3-28. Use equilibrium to find the force in BD and elongation of BD and thus the deflection of point B prior to the computer run.

7.8-23 A beam is simply supported at its ends A and C and at an intermediate section B, as shown in Fig. P7.8-23.

(*a*) If the beam is subjected to loads $4P$ and P at D and E, find the allowable load P_a such that the magnitude of the maximum normal stress due to bending does not exceed 140 MPa.

(*b*) If the support at B settles by an amount 0.06 mm, find P_a.

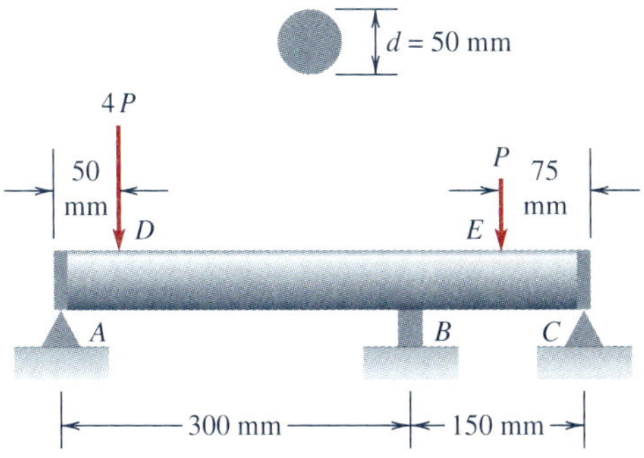

Fig. P7.8-23

7.8-24 A steel beam W12 $\times$ 87 is loaded as shown in Fig. P7.8-24. The load P can vary from 0 to 70 kips. Find the load P such that the deflection under the load is zero. For this load, find the reactions at A and B and the maximum normal stress due to bending in the beam.

7.8-25 A steel beam S15 $\times$ 50 is loaded as shown in Fig. P7.8-24. The load P is 20 kips. Find the deflection and slope at the right end. Obtain the shear force and bending moment diagrams and find the reaction at B.

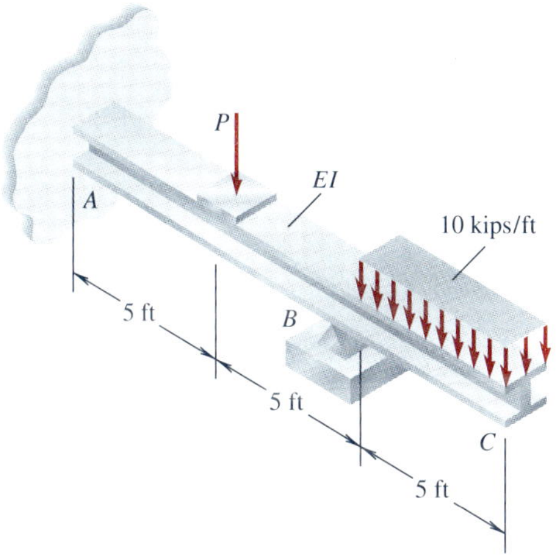

Fig. P7.8-24

7.8-26 A continuous beam is simply supported at both ends and at its midpoint and carries a constant distributed load q_0, as shown in Fig. P7.8-26. It is given that $I_z = 61.2 \times 10^6$ mm^4, $L = 3$ m, $E = 200$ GPa, and $q_0 = 5$ kN/m, and the beam depth is 254 mm.

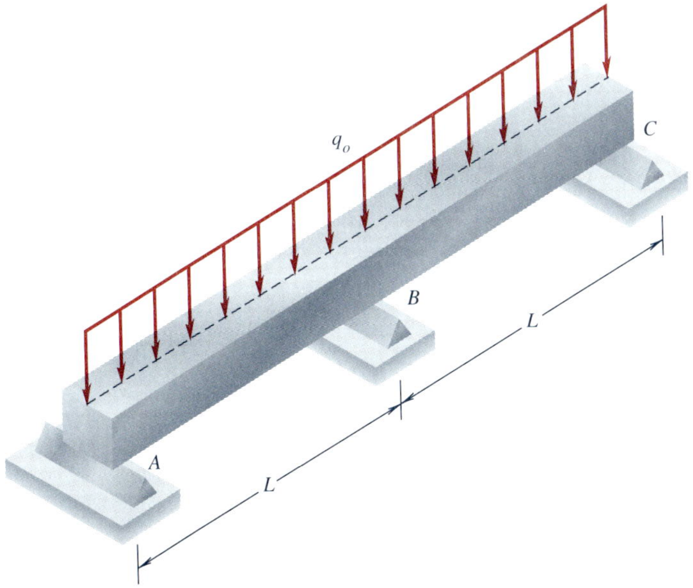

Fig. P7.8-26

(*a*) Find the maximum bending stress and the percentage of the total load $2q_0L$ carried by the center support.

(*b*) If it is known that support B settles (moves downward relative to supports A and C) by 2 mm over each 10-year period, then find the quantities requested in part (*a*) at 10-year intervals until contact between the beam and middle support is lost. How many years will pass until contact is lost at support B?

7.8-27 The cantilever beam shown in Fig. P7.8-27 carries a load P located at a distance kL (k = 1, 2, 3, 4) from the wall support. If the downward deflection at E caused by load P located a distance kL from the wall is denoted by δ_k, then find the ratios

$$R_k = \frac{\delta_k}{\delta_1} \qquad k = 2, 3, 4$$

Take $L = 1$ m, $I_z = 20 \times 10^6$ mm^4, $E = 200$ GPa, and $P = 20$ kN.

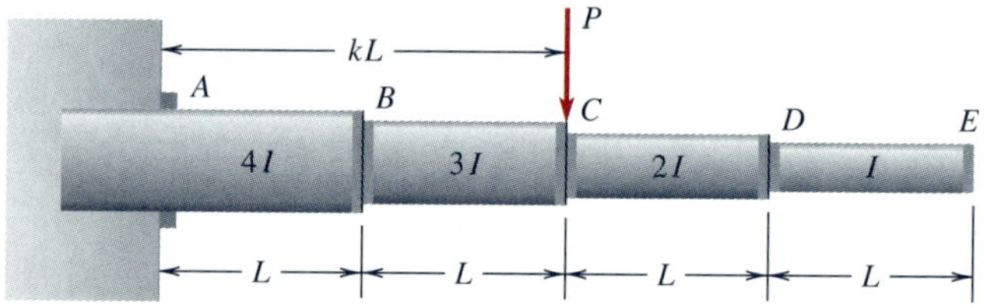

Fig. P7.8-27

7.8-28 Consider the same simply supported beam as in Prob. 7.8-26.

(*a*) Find the maximum load q_0 which can be applied if the allowable bending stress (tension or compression) is 50 MPa and the allowable downward deflection is 5 mm.

(*b*) Repeat part (*a*) except the center support has settled by 3 mm.

In parts (*a*) and (*b*) the stress and deflection restrictions are both to be satisfied for the allowable load q_0.

7.8-29 A continuous beam is simply supported at its ends and at two intermediate points, as shown in Fig. P7.8-29. An S24 × 121 I beam (see Table C.2) with E = 30,000 ksi is to be used.

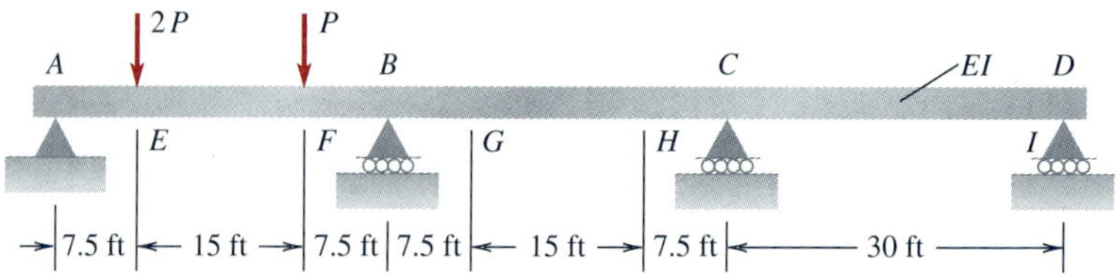

Fig. P7.8-29

(*a*) Find the magnitude and location of the maximum normal stress due to bending and the location and magnitude of the maximum positive and negative values for the deflection due to a load of $P = 40$ kips.
(*b*) Repeat part (*a*) except that the loads at sections E and F are shifted to the right to sections G and H.
(*c*) Compare the results for the locations in parts (*a*) and (*b*), and comment on the differences.

7.8-30 A continuous steel I beam is simply supported at three locations and carries a constant distributed load q_0, as shown in Fig. P7.8-30. It is given that $q_0 = 40$ kN/m, $I_z = 61.2 \times 10^6$ mm^4, and $E = 210$ GPa.
(*a*) Find the length a of the overhang such that the maximum downward deflection is a minimum.
(*b*) Find the length of the overhang such that the maximum of the absolute value of the normal stress due to bending is a minimum. The depth of the beam is 254 mm.

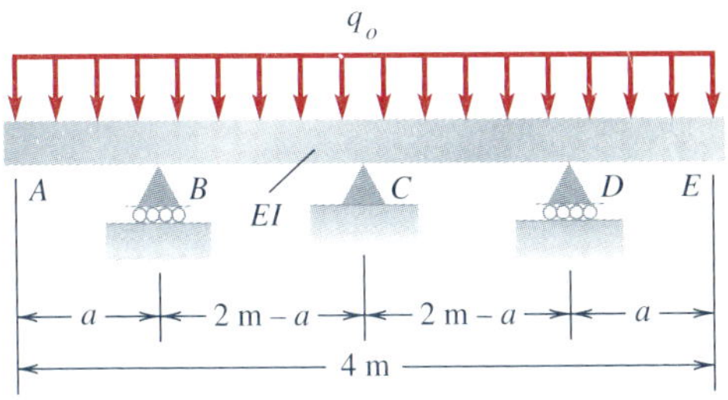

Fig. P7.8-30

7.8-31 A 4-m wood member has a rectangular cross section, as shown in Fig. P7.8-31. If a set of five transverse forces $R_A, \ldots, R_E$ is to be applied to the member so as to force its centerline to take up a shape so that

$$v_A = v_D = 0 \qquad v_B = -10 \text{ mm} \qquad v_C = -5 \text{ mm} \qquad v_E = -15 \text{ mm}$$

find the unknown forces $R_A, \ldots, R_E$ and the value of the maximum bending stress for the cases
(*a*) $b = 60$ mm, $h = 20$ mm
(*b*) $b = 120$ mm, $h = 40$ mm

Explain the stress results. Take $E = 10$ GPa.

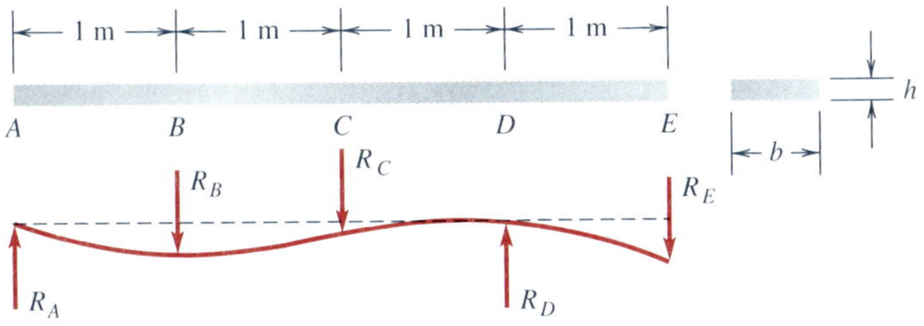

Fig. P7.8-31

Solve the following problems by any of the methods discussed in this chapter, including the use of the BEAMMECH program. Unless otherwise indicated, take the bending modulus as EI_z and neglect the weight of the beam.

7.9-1 A lathe cutting tool is used to cut down the outside diameter of a cylindrical rod, as shown in Fig. P7.9-1*a*. To reduce the deflections of the rod, a live center is used on one end and a chuck support is used at the other end. The rod of length L can be modeled as a cantilever beam of modulus EI with a simple support at end C, as shown in Fig. P7.9-1*b*. The cutting tool is represented by a

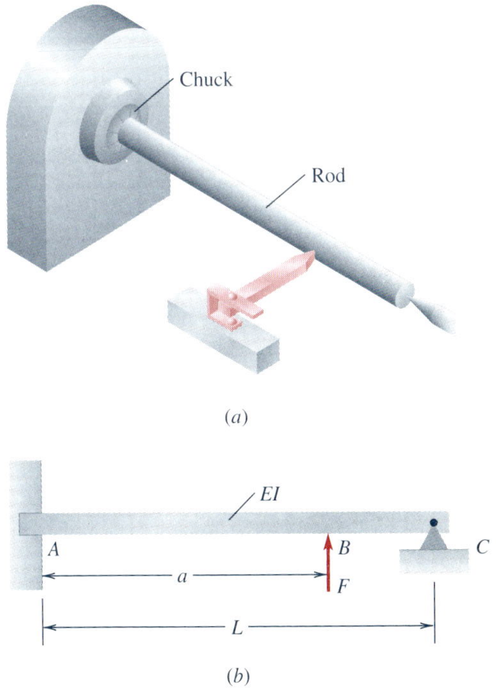

Fig. P7.9-1

concentrated force F, and this concentrated force moves along the rod during the turning operation. Find the location of the force F that will give the largest deflection and find the deflection. [*Hint:* If the BEAMMECH program is used for this problem, use unit values for F, L, E, and I in the computer run to find the coefficient β in the expression $v_{max} = \beta FL^3/(EI)$.]

7.9-2 An 8-in-long rod is turned down in a lathe (see Prob. 7.9-1) from a diameter of 0.25 in to a diameter of 0.20 in. If we assume that the component of the cutting force F acting on the rod is 10 lb, estimate the deflection of the rod under force F for $d = 0.25$ in and $d = 0.20$ in.

7.9-3 A shaft modeled as a beam built in at each end consists of three segments with circular sections and carries a load P at its center, as shown in Fig. P7.9-3. Find the value for the diameter d_1 of the middle segment so that the maximum normal stress due to bending is the same in all three segments of the shaft.

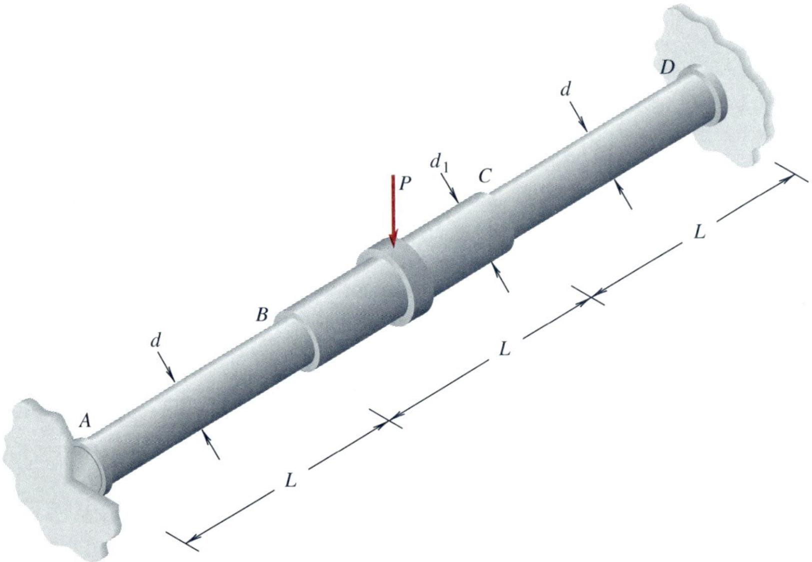

Fig. P7.9-3

7.9-4 It is proposed to lay out a large number of separate simply supported structural beams end to end so that each individual beam $ABCD$ overhangs its supports at B and C so that the end slopes are zero, as shown in Fig. P7.9-4a. Find the ratio a/L so that the beams subjected to self-weight loading will have the zero end slopes, Fig. 7.9-4b.

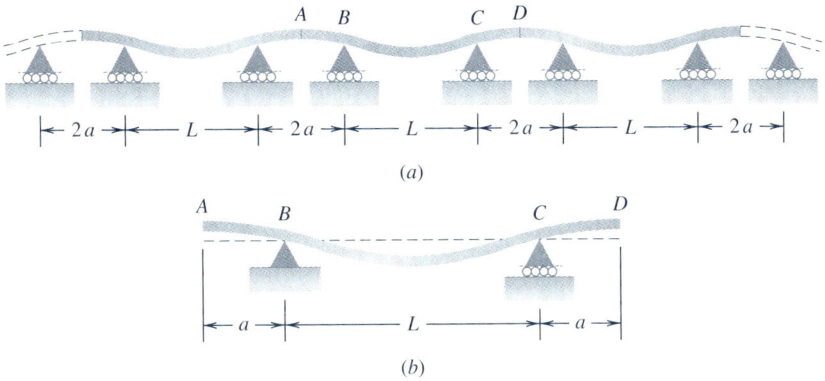

Fig. P7.9-4

7.9-5 A portion of a small elastic flexure system is shown in Fig. P7.9-5. The rigid lower member is designed to move horizontally without changing its angular orientation under a load P. The lower member is supported by two phosphor-bronze strips of the dimensions shown and with a maximum allowable stress of 350 MPa and $E = 100$ GPa. Find the force-deflection relation for the system and the maximum deflection of the system without exceeding the maximum allowable stress.

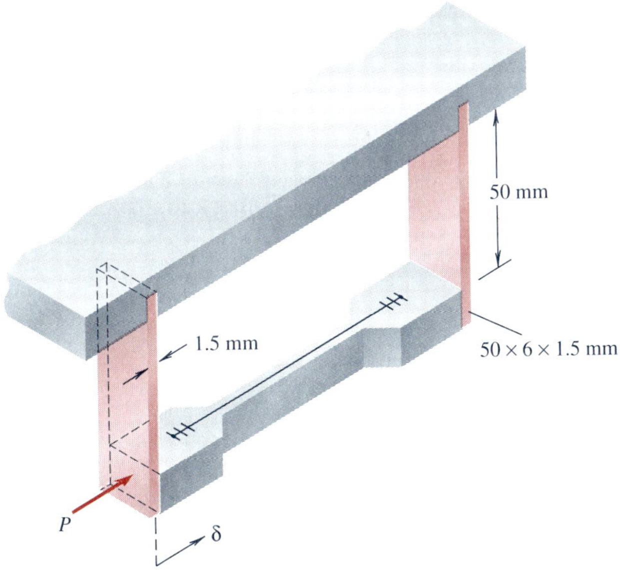

Fig. P7.9-5

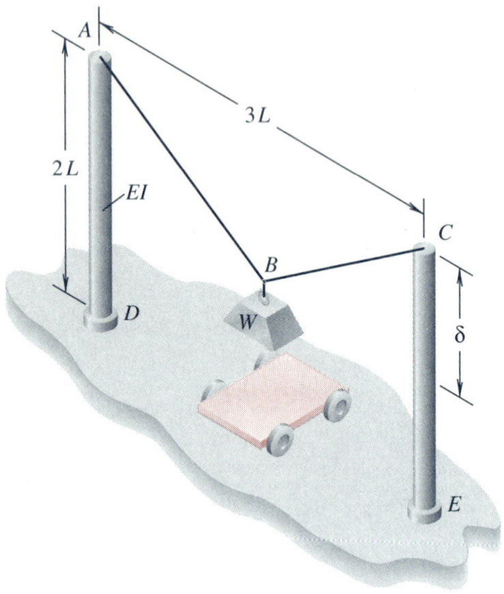

Fig. P7.9-6

7.9-6 A weight W is suspended between two poles of circular cross section, as shown in Fig. P7.9-6. If a certain weight W displaces point B on the cable downward so that $\delta = L$, find an expression for the horizontal deflection of the top ends of the poles and the maximum normal stress due to bending caused by the weight W. Take the bending modulus of the poles as EI.

7.9-7 Consider the linear distributed loading shown acting on the beam element in Fig. P7.9-7. We wish to generalize the results given in Eqs. (7.2) and (7.7) for the case where q is constant over the element to the case where q varies linearly. Starting with the differential relation

$$EI\frac{d^4v}{dx^4} = q(x) = q_1 + \frac{(q_2 - q_1)x}{L} \qquad (a)$$

we integrate four times to obtain

$$v = a_1 + a_2x + a_3x^2 + a_4x^3 + \frac{q_1x^4}{24EI} + \frac{(q_2 - q_1)\,x^5}{120LEI} \qquad (b)$$

Show that two equations for a_3 and a_4 follow in the form

$$v_2 = v_1 + v_1'L + a_3L^2 + a_4L^3 + \frac{q_1L^4}{24EI} + \frac{(q_2 - q_1)\,L^4}{120EI} \qquad (c)$$

$$v_2' = v_1' + 2a_3L + 3a_4L^2 + \frac{q_1L^3}{6EI} + \frac{(q_2 - q_1)\,L^3}{24EI} \qquad (d)$$

Solving for a_3 and a_4 and substituting for a_3 and a_4 into Eq. (b) gives the same result as Eq. (7.7) except that the last term involving q generalizes to

$$\frac{q_1L^4}{120EI}\left(\frac{5x^4}{L^4} + \frac{3x^2}{L^2} - \frac{7x^3}{L^3} - \frac{x^5}{L^5}\right) + \frac{q_2L^4}{120EI}\left(\frac{x^5}{L^5} + \frac{2x^2}{L^2} - \frac{3x^3}{L^3}\right) \qquad (e)$$

7.9-8 Show that Eq. (e) in Prob. 7.9-7 reduces to the last term given in Eq. (7.7) when $q_1 = q_2 = q$.

7.9-9 Show that for the linear distributed loading case, the last term in Eq. (7.9) is replaced by (see Fig. P7.9-7)

$$\frac{q_1L^2}{60EI}\left(\frac{30x^2}{L^2} + 3 - \frac{21x}{L} - \frac{10x^3}{L^3}\right) + \frac{q_2L^2}{60EI}\left(\frac{10x^3}{L^3} + 2 - \frac{9x}{L}\right)$$

Check that for $q_1 = q_2 = q$, the above result reduces to the one given in Eq. (7.9).

7.9-10 Show that for the linear distributed loading case, the last terms in Eq. (7.11) and (7.12) are replaced by

$$(3q_1 + 2q_2)\frac{L^2}{60} \quad \text{and} \quad (2q_1 + 3q_2)\frac{L^2}{60}$$

respectively; see Fig. P7.9-7.

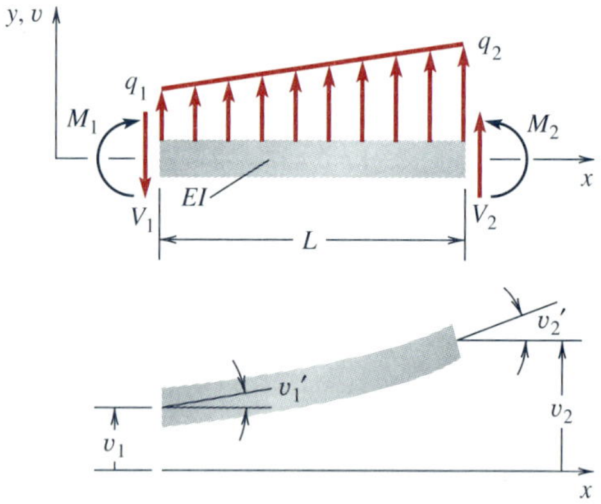

$$M_1 = \frac{EI}{L^2}\{-6v_1 - 4Lv_1' + 6v_2 - 2Lv_2'\} + (3q_1 + 2q_2)\frac{L^2}{60}$$

$$V_1 = \frac{EI}{L^3}\{-12v_1 - 6Lv_1' + 12v_2 - 6Lv_2'\} + (7q_1 + 3q_2)\frac{L}{20}$$

$$M_2 = \frac{EI}{L^2}\{6v_1 + 2Lv_1' - 6v_2 + 4Lv_2'\} + (2q_1 + 3q_2)\frac{L^2}{60}$$

$$V_2 = \frac{EI}{L^3}\{-12v_1 - 6Lv_1' + 12v_2 - 6Lv_2'\} - (3q_1 + 7q_2)\frac{L}{20}$$

Fig. P7.9-7

7.9-11 Show that for the linear distributed loading case, the last terms in Eqs. (7.16) and (7.17) are replaced by

$$(7q_1 + 3q_2)\frac{L}{20} \qquad \text{and} \qquad -(3q_1 + 7q_2)\frac{L}{20}$$

respectively; see Fig. P7.9-7.

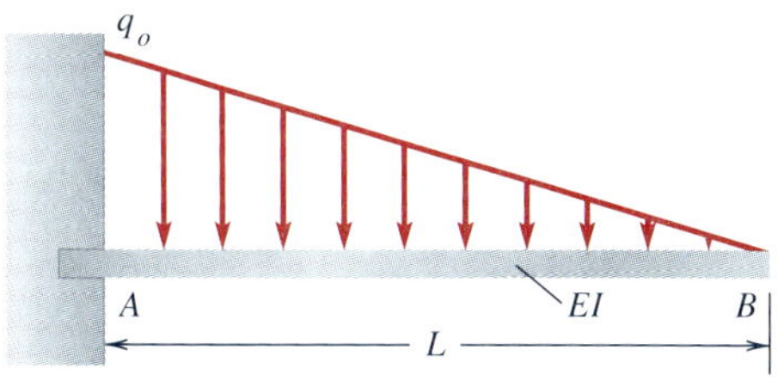

Fig. P7.9-12

7.9-12 A cantilever beam is subjected to a linearly decreasing distributed loading, as shown in Fig. P7.9-12. Use the equations given in Fig. P7.9-7 to find the slope and deflection at B and the reactions at A.

7.9-13 A simply supported beam carries a distributed loading which varies from q_0 to $2q_0$, as shown in Fig. P7.9-13.
(*a*) Use the equations given in Fig. P7.9-7 to find the slopes at the ends of the beam.
(*b*) Use Eq. (7.7) in the general form derived in Prob. 7.9-7 to get an expression for the deflection $v(x)$.

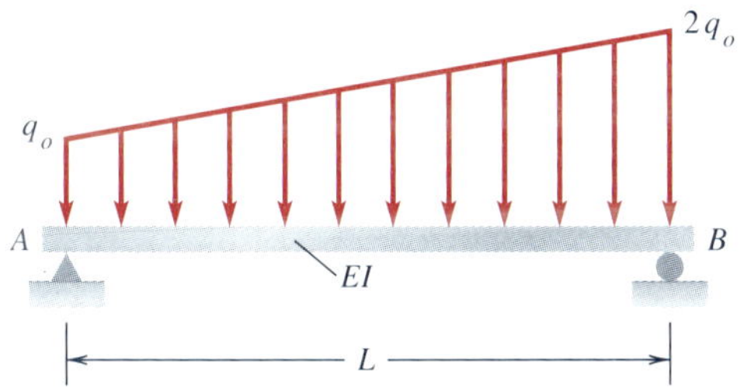

Fig. P7.9-13

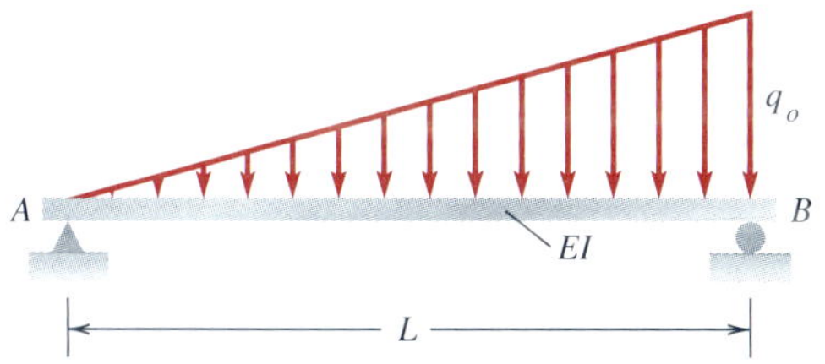

Fig. P7.9-14

7.9-14 A simply supported beam carries a linearly increasing loading, as shown in Fig. P7.9-14.
(*a*) Use the equations given in Fig. P7.9-7 to find the slopes at the ends of the beam.
(*b*) Make use of the results of Prob. 7.9-7 to get an expression for the deflection $v(x)$ in the form

$$v(x) = \frac{q_0 x}{360LEI}(10L^2x^2 - 7L^4 - 3x^4)$$

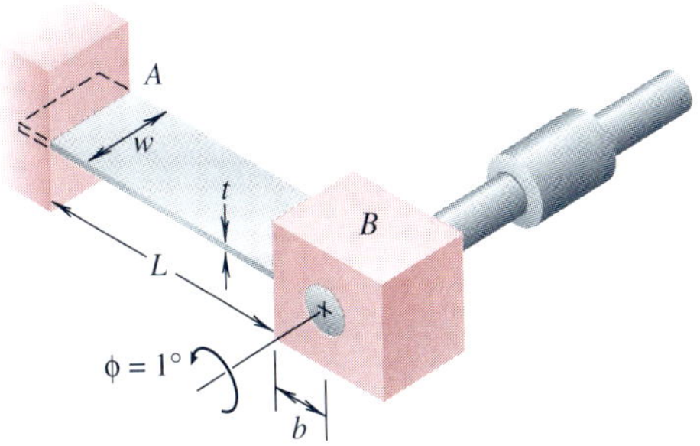

Fig. P7.9-15

7.9-15 A small rotary transducer bends a thin steel beam built in at both ends A and B as shown in Fig P7.9-15. Block B undergoes a rotation of $\phi = 1°$, causing beam AB to bend. Estimate the maximum bending stress in the beam if $t = 0.025$ in, $w = 0.25$ in, $L = 0.8$ in, $b = 0.3$ in, and $E = 30 \times 10^6$ psi.

7.9-16 A precision laser star tracking instrument is placed on a steel beam built in at both ends, as shown in Fig. P7.9-16. Other support fixtures prevent the instrument from being placed at midspan. Take the weight of the instrument W as 4000 lb, L = 14 ft, and the beam is a W10 × 12 shape. Estimate the angle β of the deviation from the vertical in the vertical plane if the instrument is modeled as a concentrated weight acting at $L/4$, as shown in Fig. P7.9-16*a*. An alternative model is to assume that the instrument is mounted on two brackets, each taking one-half of the weight, as shown in Fig. P7.9-16*b*. What is the angle β in this case? Which model of the system is more accurate? Take $E = 30 \times 10^6$ psi.

7.9-17 During processing of silicon crystals, it is necessary to move the crystals on rollers, as shown in Fig. P7.9-17. The crystals are susceptible to fracture if the tensile stresses are too high. Which configuration, (*a*) or (*b*) in Fig. P7.9-17, is better?

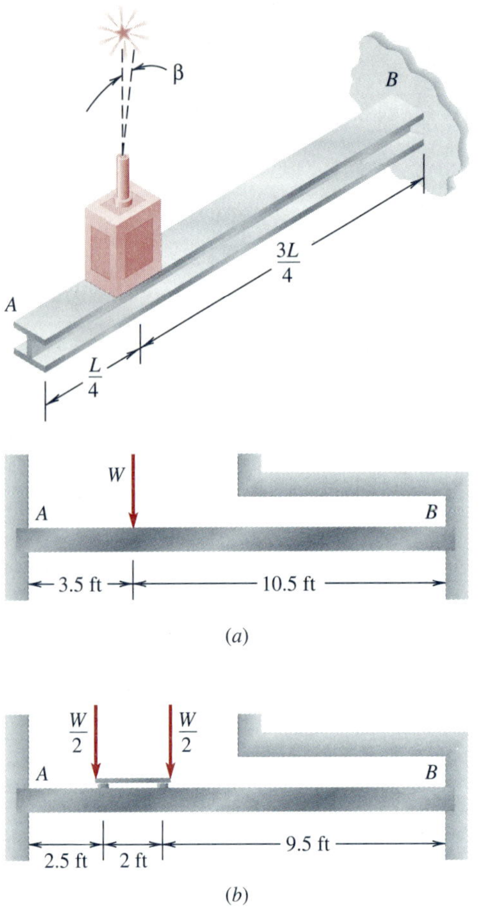

Fig. P7.9-16

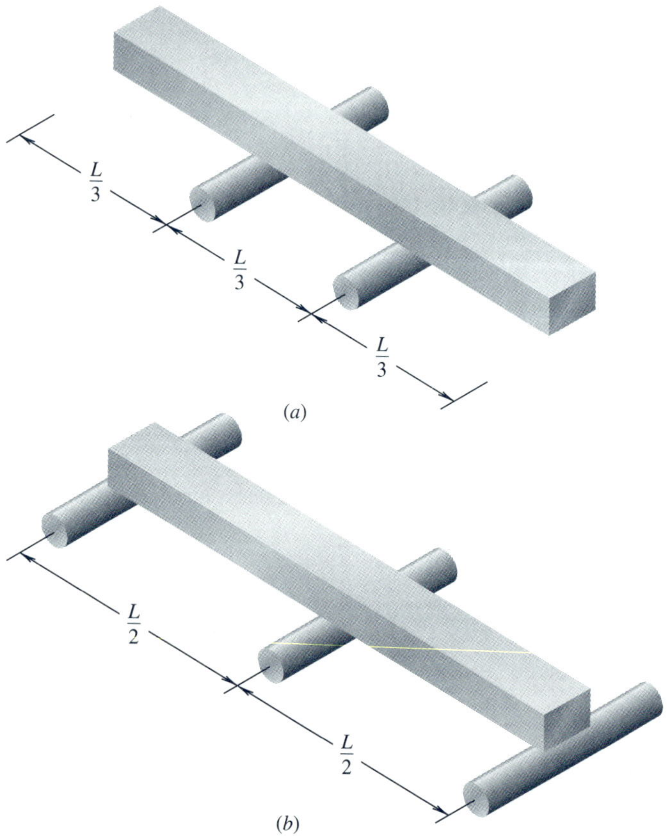

Fig. P7.9-17

7.9-18 An interesting result for linear elastic beams is Maxwell's reciprocal theorem, which states that the deflection at a point $x = x_2$ along a beam due to a load at point x_1 (Fig. P7.9-18) is equal to the deflection at $x = x_1$ when the load is at x_2; that is, $\delta_1(x_2) = \delta_2(x_1)$. Verify the reciprocal theorem for the simply supported beam and cantilever beam shown in cases G.1-1 and G.2-1 of App. G.

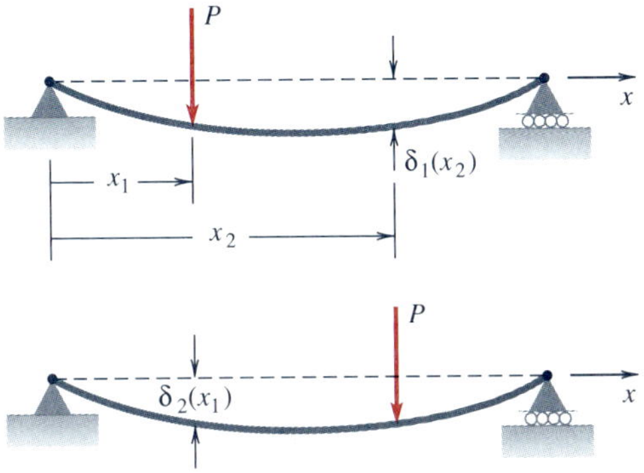

Fig. P7.9-18

7.9-19 A beam of rectangular section is clamped at both ends and initially has a uniform temperature T_0. The beam is heated so that a linear distribution of temperature over the depth is maintained from T_U on the top to T_L on the bottom. See Fig. P7.9-19.

(*a*) Making use of the results given in Prob. 6.6-12, show that the moment-curvature relation becomes

$$EIv'' = M(x) - \frac{\alpha}{h} EI\,(T_U - T_L) \tag{a}$$

where

$$M(x) = M_A + R_A x \tag{b}$$

(*b*) Show that by symmetry $R_A = R_B = 0$ and $M_A = M_B$.

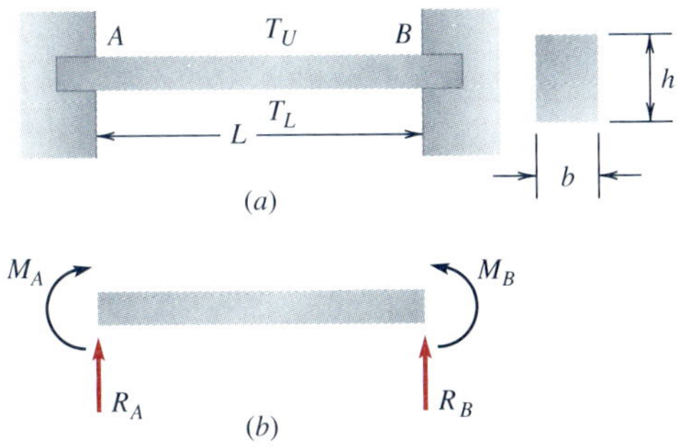

Fig. P7.9-19

(*c*) Integrate Eq. (*a*) and use the conditions that $v = v' = 0$ at $x = 0$ and $x = L$ to show that

$$M_A = \frac{EI\,\alpha\,(T_U - T_L)}{h} \tag{c}$$

(*d*) Show that the beam remains in a straight configuration, that is, v is identically equal to zero, due to the temperature changes.

7.9-20 A frame *ABCD* is clamped at *A* and *D* and carries a horizontal load *P* at *B*, as shown in Fig. P7.9-20. Assume that *BC* is essentially rigid with respect to axial and bending deformation. Find the displacement components of point *B* if we assume that the vertical members are axially rigid, i.e., the vertical members experience bending deformation only.

Fig. P7.9-20

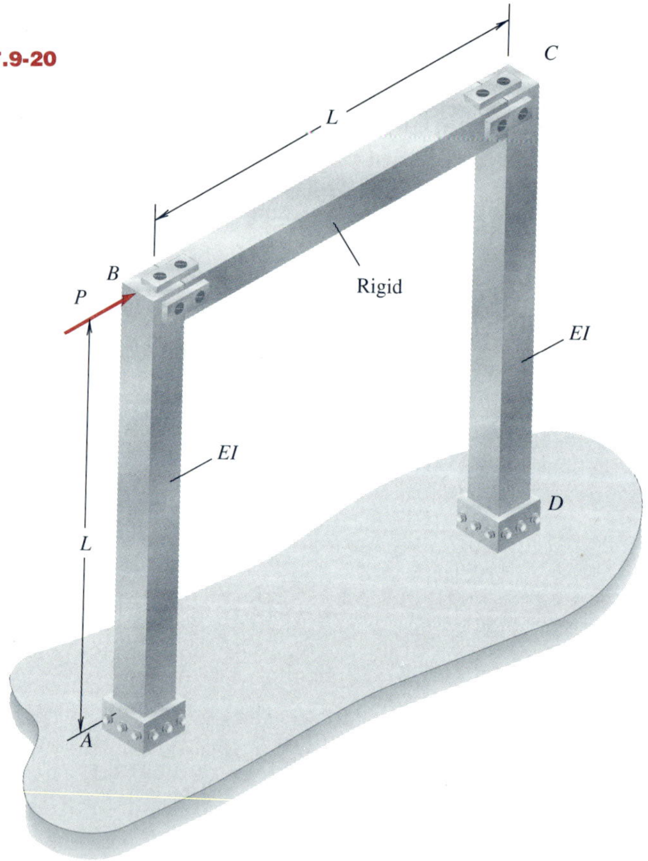

7.9-21 A cylindrical tube of weight w per unit length is used to focus a high-energy beam through two lenses, as shown in Fig. P7.9-21. The tube is clamped at *A* and simply supported at *B*. Find an expression for the deviation angle of the beam for a given *L* and *a*. Assume that the lens at *C* is of weight $wL/5$ and that the bending modulus of the tube is *EI*. Take $a = 0.5L$.

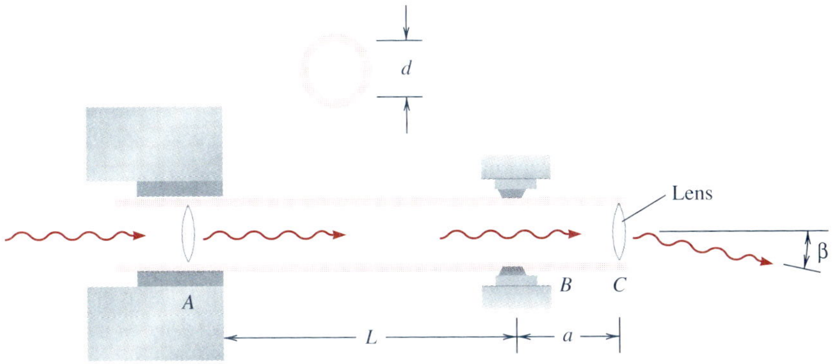

Fig. P7.9-21

7.9-22 A mirror of weight w per unit length is simply supported on a support that is oriented at an angle θ from the horizontal, as shown in Fig. P7.9-22. Find an expression for the maximum slope of the mirror surface relative to the support bed.

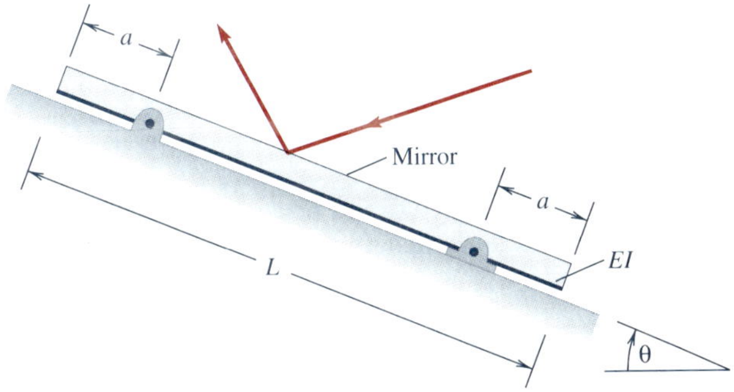

Fig. P7.9-22

7.9-23 Consider Prob. 7.9-22. Find the value of a/L for the location of the supports such that the maximum deflection is as small as possible. For this value of a/L, sketch the slope angle (relative to the support bed) as a function of distance along the mirror.

7.9-24 Two separate cantilever beams of equal bending moduli are built in as shown in Fig. P7.9-24 and are loaded by an end load P. Find the deflection δ under the load P. At B a frictionless roller separates the beams.

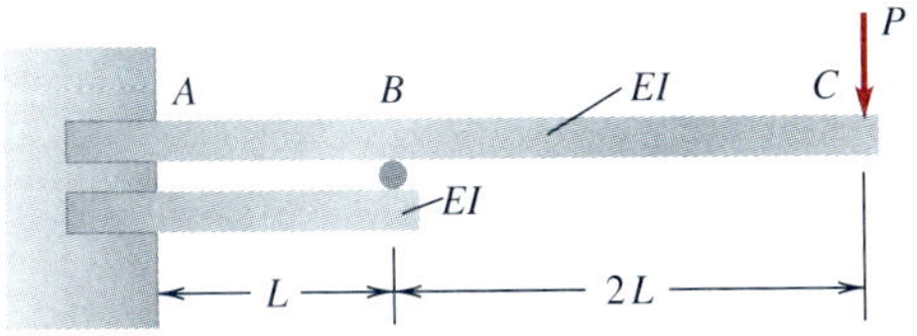

Fig. P7.9-24

7.9-25 A beam shown in Fig. P7.9-25a is supported above a rigid plane by two rollers of small diameter d. Find the value of load P that is sufficient to force the middle of the beam into contact with the surface, as shown in Fig. P7.9-25b.

First note that the moment and shear force in segment CD are zero and that a concentrated reactive force F must act at C and D. Since the bending moments at C and D are zero, $F = 2P/3$. Therefore show that

$$P = \frac{24EId}{5a^3}$$

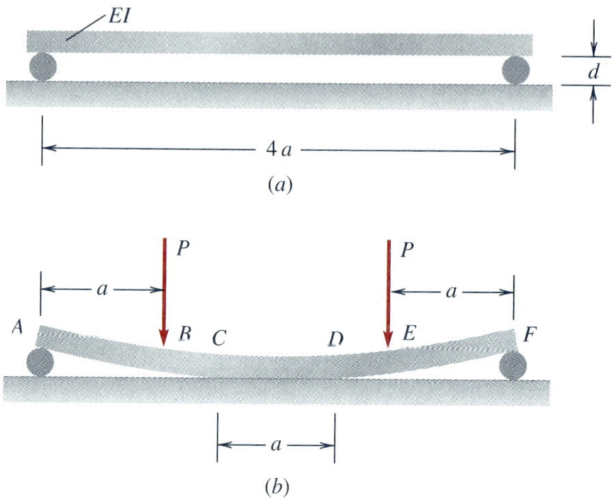

Fig. P7.9-25

7.9-26 A steel shaft 24 mm in diameter passes concentrically through a rigid housing whose inside diameter is 34 mm, as shown in Fig. P7.9-26. The shaft can be considered as simply supported at the bearings. What is the maximum bending moment M_0 which can be applied to the end of the shaft if the clearance between shaft and housing is to be not less than 2.5 mm? Take $E = 200$ GPa.

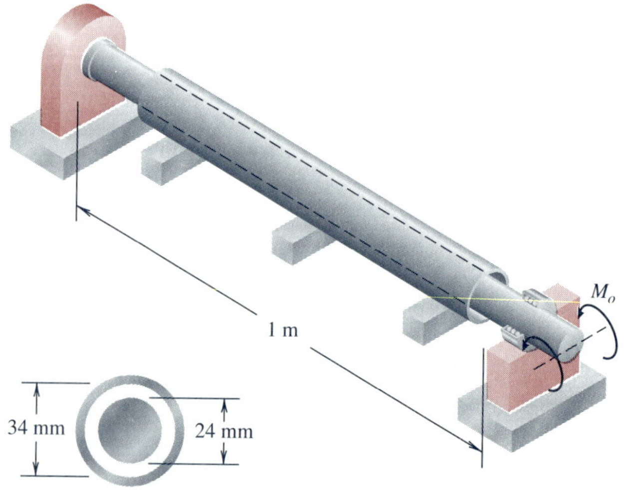

Fig. P7.9-26

7.9-27 A bracket *ABC* made up of two circular rods of diameter $d = 1$ in is built into the wall at *A* and loaded at *C*, as shown in Fig. P7.9-27. Find the horizontal and vertical components of the deflection of point *C* and the rotation angle at *C* due to the two load cases:
(*a*) $P = 100$ lb, $M_c = 0$
(*b*) $P = 0$, $M_c = 1000$ lb · in

Take $E = 30{,}000$ ksi and $L = 3$ ft. Neglect the axial deformation of the rods.

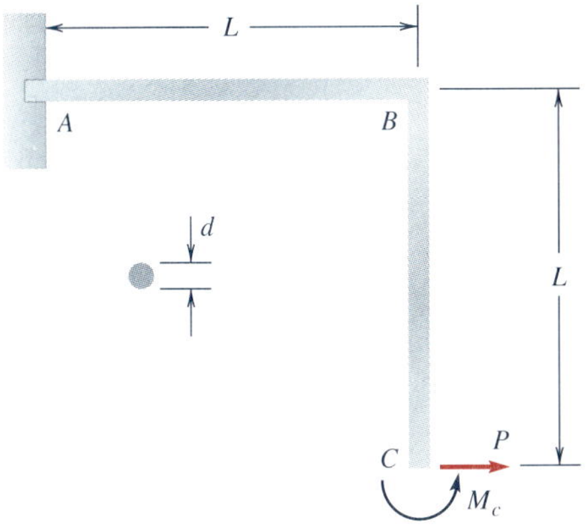

Fig. P7.9-27

7.9-28 A uniform beam of bending modulus *EI* is supported on three identical linear springs with spring constant *k*, as shown in Fig. P7.9-28. Find the forces in the three springs. Comment on the effect of *EI* on the values of the forces.

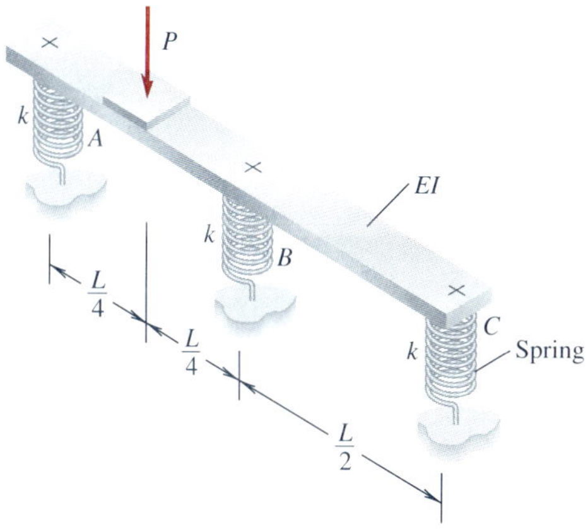

Fig. P7.9-28

7.9-29 A beam is loaded by a constant distributed loading q_0 and is supported as shown in Fig. P7.9-29. Show that the value of the reactive force at *B* can be written in the form

$$R_B = q_0 L\left(3\frac{L}{L_1} + \frac{1}{8}\frac{L_1}{L} - 1\right)$$

Note that Examples 7.5 and 7.3 follow as special cases. For what value of L_1 is R_B exactly equal to $2q_0L$? At what value of L_1 does the moment at the wall equal zero?

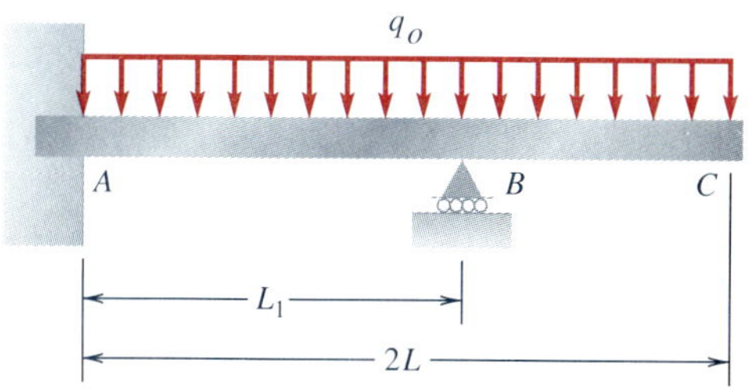

Fig. P7.9-29

7.9-30 Determine the maximum deflection and the numerically largest values of the bending moment and bending stress for the steel beams shown in the figure. (Neglect the weight of the beam and take $E = 30{,}000$ ksi or 200 GPa; *d* is the depth of the beam.)
(*a*) Fig. P4.5-1*a*; use a W8 × 40 shape.
(*b*) Fig. P4.5-1*c*; use a 6-in × 3-in × $\frac{3}{8}$-in structural tube.
(*c*) Fig. P4.5-1*d*; use $I = 1.75 \times 10^{-6}$ m^4, $d = 150$ mm.
(*d*) Fig. P4.5-1*e*; use a W6 × 16 shape.
(*e*) Fig. P4.5-1*f*; use $I = 1.2 \times 10^{-6}$ m^4, $d = 125$ mm.
(*f*) Fig. P4.5-1*g*; use a 6-in × 4-in × $\frac{1}{2}$-in structural tube.
(*g*) Fig. P4.5-1*h*; use $I = 3.7 \times 10^{-5}$ m^4, $d = 180$ mm.
(*h*) Fig. P4.5-1*j*; use a W14 × 82 shape.

7.9-31 Determine the maximum deflection and the numerically largest values of the bending moment and bending stress for the steel beams shown in the figure. (Neglect the weight of the beam and take $E = 30{,}000$ ksi or 200 GPa; *d* is the depth of the beam.)
(*a*) Fig. P4.7-1*a*; use $I = 7 \times 10^5$ mm^4, $d = 180$ mm.
(*b*) Fig. P4.7-1*b*; use a W5 × 19 shape.
(*c*) Fig. P4.7-1*c*; use a W8 × 28 shape.
(*d*) Fig. P4.7-1*e*; use $I = 7.5 \times 10^{-6}$ m^4, $d = 0.2$ m.
(*e*) Fig. P4.7-1*f*; use $I = 26 \times 10^{-6}$ m^4, $d = 240$ mm.
(*f*) Fig. P4.7-1*j*; use an S12 × 31.8 shape.

7.9-32 A very long tube of weight q per unit length is lifted at its center an amount v_0, as shown in Fig. P7.9-32. If the bending modulus of the tube is EI, find an expression for the liftoff distance L, as defined in the figure.

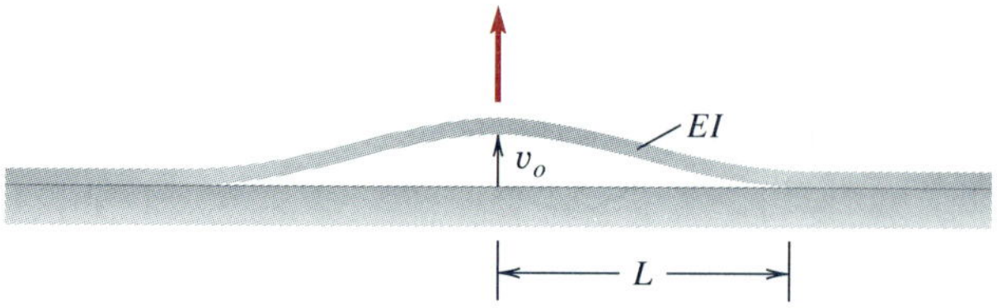

Fig. P7.9-32

7.9-33 Consider the four cases of loading of structural members shown in Fig. P7.9-33: (*a*) Axial force F on a bar, Example 2.7; (*b*) twisting moment T on a circular shaft, Example 3.6; (*c*) a simply supported beam with a concentrated load P, Example 4.1; (*d*) a built-in beam with a concentrated load P, Example 7.1. Determine the reactions at the supports in each case, and comment on the relation of the reactions to the geometry.

7.9-34 Consider Prob. 7.9-33. If the concentrated load P in Fig. P7.9-33*c* and *d* is replaced by a concentrated moment M_0, determine the reactions at each support. Can any conclusions be drawn?

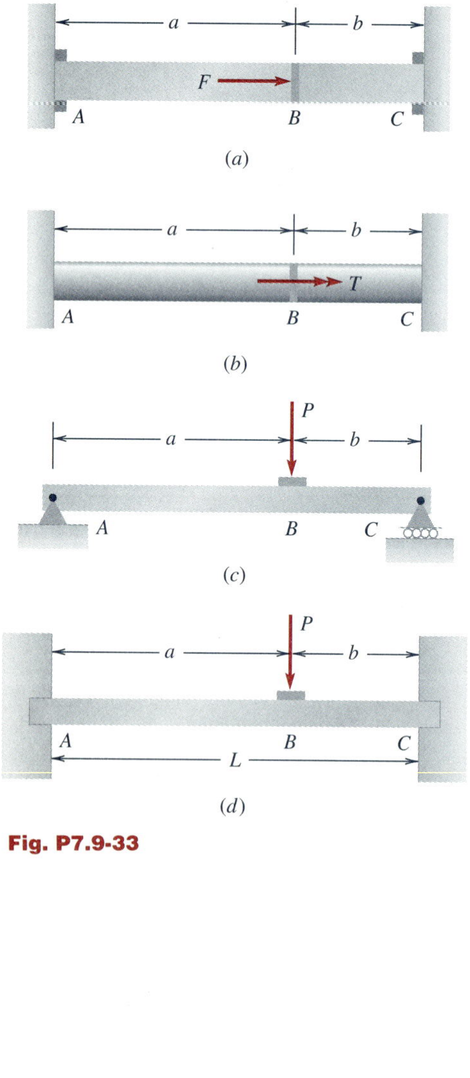

Fig. P7.9-33

CHAPTER 8

Stress and Strain

8.1 Introduction

In our discussions thus far we have analyzed the response of slender members to loads, namely, the axial deformation of bars, the twisting of circular shafts, and the transverse bending of symmetric beams. In each of these cases, the displacement, angle of twist, strain, stress, etc., depended on a single independent variable, and as such these cases are one-dimensional formulations. The results from the analysis of these one-dimensional problems have been found to be sufficiently accurate for most engineering analysis and design purposes.

However, many engineering problems involve two- and three-dimensional stresses and strains, and as a consequence we need to expand our understanding of stress and strain in one dimension to two and three dimensions. When we do this, we are moving into the study of the subject called the *theory of elasticity*. J. E. Gordon, in his book *Structures*,[1] notes that much has been written about the theory of elasticity in the past 150 years by mathematicians and that a great amount of the written material and the accompanying lectures on the material has been obscure, dull, and unhelpful. However, he notes that what is needed for engineering can be understood by any intelligent person who makes the effort. Our intent here is to introduce in a careful way the very basic foundations of the ideas needed for an understanding of stress and strain in two and three dimensions.

We have a pragmatic reason also for wishing to understand stress and strain in two and three dimensions. In previous chapters we have derived formulas for the normal stress and shear stress occurring in a number of structural members under load; e.g., we calculated axial normal

[1]J. E. Gordon, *Structures*, Plenum Press, New York, 1978, p. 28.

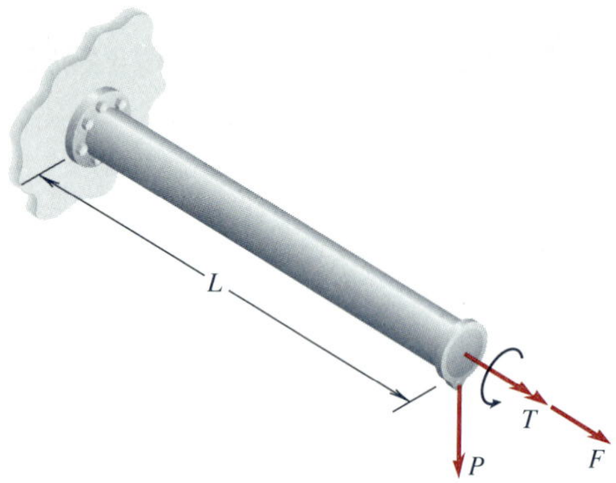

Figure 8.1 An elastic circular shaft with a twisting moment T, axial force F, and transverse load P.

stresses in bars, torsional shear stresses in shafts, and normal and shear stresses in beams due to bending. In many applications the loads acting on a member may be such as to create simultaneously one or more of these basic stress states. For example, Fig. 8.1 shows an elastic circular shaft of length L under the action of a twisting moment T, an axial force F, and a transverse bending load P. In this case we need to know how to combine the stresses arising from each of these loadings to investigate the structural integrity of the shaft. For a general structure, if we wish to determine the combination of applied loads that are safe for a given design, it is necessary to know how to combine the stresses from each of the loadings at critical locations in the structure. To do this leads to a study of the ideas of stress and strain in two and three dimensions.

We divide this chapter into three parts. In Parts A and B, we discuss the ideas of stress and strain in two and three dimensions and learn how to combine different stress states and how to combine different strain states. In Part C we introduce the stress-strain relations in two and three dimensions. We investigate also the conditions for the onset of permanent or plastic deformation in a material. Permanent or plastic deformation often occurs when the material no longer behaves in a linear elastic manner.

In Chap. 9 we will show how to bring many of the concepts introduced in this chapter to bear on the solution of engineering problems involving combined states of loading.

PART A: STRESS

8.2 Stress

When we discussed axial deformation of bars in Chaps. 1 and 2, we introduced the concept of the force intensity or normal stress σ acting on an area normal to the axis of the bar (see Fig. 8.2*a*). However, if we have an arbitrary three-dimensional body under load as shown in Fig. 8.2*b* we need to be careful in introducing force intensity or force per unit area at a point because at each point there are infinitely many areas passing through the point. In addition, we need to ask how the force intensity arises at a given point. Thinking about how the force intensity arises at a point and the need to specify the orientation of the area through the point on which the force intensity acts will lead us to a more general concept of stress.

In Fig. 8.2*b* we show a point O in a body in equilibrium under the

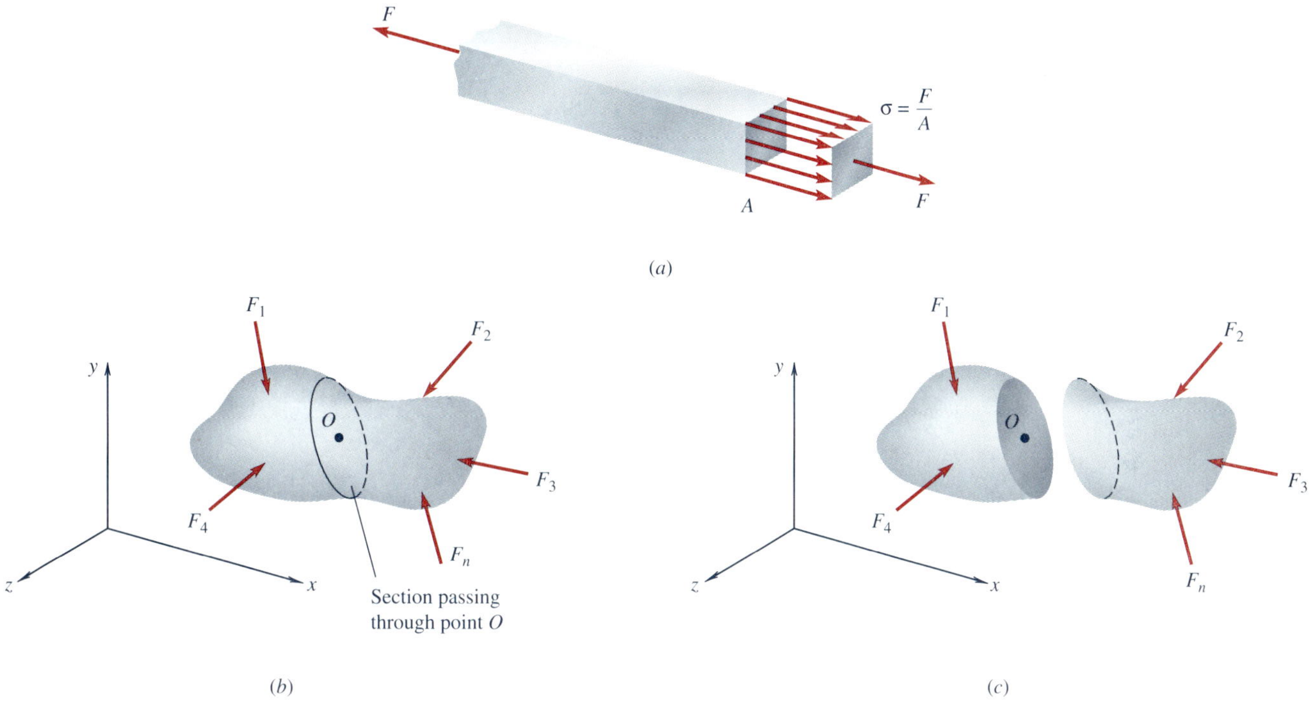

Figure 8.2 (*a*) Normal stress σ acting on the cross section of a bar. (*b*) Continuous body acted on by external forces with section or cut passing through point O. (*c*) Two portions of the continuous body separated at the section through point O.

action of a number of loads. We now pass an imaginary plane section or cut through the point O, cutting the body into two portions, as shown in Fig. 8.2*c*. We have shown the two portions as separated in Fig. 8.2*c* for visual convenience. Each portion of the body interacts with the other portion across this section so that we can expect a distribution of forces acting over the face of the section to ensure equilibrium of each portion. The forces acting on the face of the cut of one portion of the body will be the negative of the forces acting on the face of the other portion. The orientation of the face of one portion at point O will be specified by the outward-pointing unit normal vector $\mathbf{n}$ to the face, as shown in Fig. 8.3*a*. The unit normal vector $\mathbf{n}$ provides the information we need to specify the orientation of the face of the section passing through point O. If we had cut the body with a different section passing through point O, we would have a different vector $\mathbf{n}$ at the point O.

If we now divide the exposed cut face with unit normal $\mathbf{n}$ into a

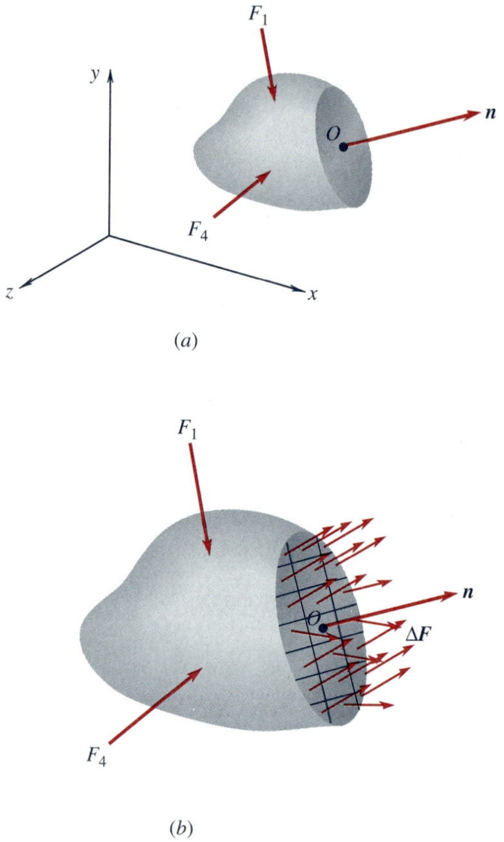

Figure 8.3 (*a*) Portion of body with outward-pointing normal vector **n** acting on a section. (*b*) Internal forces $\Delta\mathbf{F}$ acting on each element of area on the section whose normal is **n**.

number of small elements of area, we will find acting on each element of area ΔA a *force vector* $\Delta\mathbf{F}$. These elemental force vectors arise from the interaction of material across the cut section separating the two portions of the body. The elemental force vectors are not, in general, in the direction of the normal **n** to the elements of area on which they act (Fig. 8.3*b*).

If we are to describe in a quantitative sense the internal force interaction at a point between portions of a body in equilibrium, we need to consider the location of the point at which the interaction occurs, the magnitude and direction of the force vector in the neighborhood of the point, and the orientation of the area on which the force vector acts. Because we wish to treat small elements of area on the section and the corresponding small elements of force vectors acting on the areas, we anticipate that we will need to make use of a limiting process at point O to define the interaction of the material across the section at that point.

We now introduce the fundamental concept of the *stress vector* $\overset{(\mathbf{n})}{\mathbf{T}}$

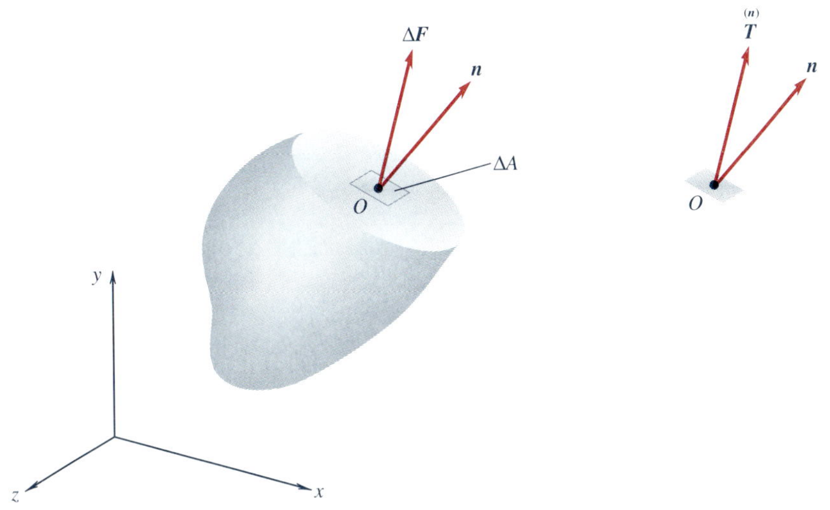

Figure 8.4 Force vector acting on an element of area whose normal is **n** gives rise to stress vector $\overset{(n)}{\mathbf{T}}$.

acting at point O on an element of area whose outward unit normal is **n** passing through O. We define the stress vector as

$$\overset{(n)}{\mathbf{T}} = \lim_{\Delta A \to 0} \frac{\Delta \mathbf{F}}{\Delta A} \tag{8.1}$$

where $\Delta\mathbf{F}$ is the element of force acting on an element of area ΔA containing point O (see Fig. 8.4). We see that $\overset{(n)}{\mathbf{T}}$ is a *force intensity* or *stress* acting on a plane whose normal is **n** at point O. This notation for the stress vector is used to emphasize, first, that $\overset{(n)}{\mathbf{T}}$ is a vector and, second, that it is a vector which acts on a plane passing through point O whose normal is **n** (Fig. 8.4); $\overset{(n)}{\mathbf{T}}$ does *not* act, in general, in the direction of **n**.

Since $\overset{(n)}{\mathbf{T}}$ is a vector, we may write it in terms of its components with respect to the coordinate axes in the form

$$\overset{(n)}{\mathbf{T}} = \overset{(n)}{T}_x\mathbf{i} + \overset{(n)}{T}_y\mathbf{j} + \overset{(n)}{T}_z\mathbf{k} \tag{8.2}$$

where **i, j, k** are the unit vectors parallel to the coordinate axes and $\overset{(n)}{T}_x$, $\overset{(n)}{T}_y$, $\overset{(n)}{T}_z$ are the components of $\overset{(n)}{\mathbf{T}}$ along the coordinate axes (see Fig. 8.5). The components of $\overset{(n)}{\mathbf{T}}$ play an important role in specifying stresses applied to the surface of a body, as we will see later when we discuss two-dimensional bodies.

We must keep in mind four major characteristics of stress: (1) The physical dimensions of stress are force per unit area, (2) the stress vector is defined at a point in a body at an imaginary cut or section separating the material of the body into two parts, (3) the stress vector is a vector

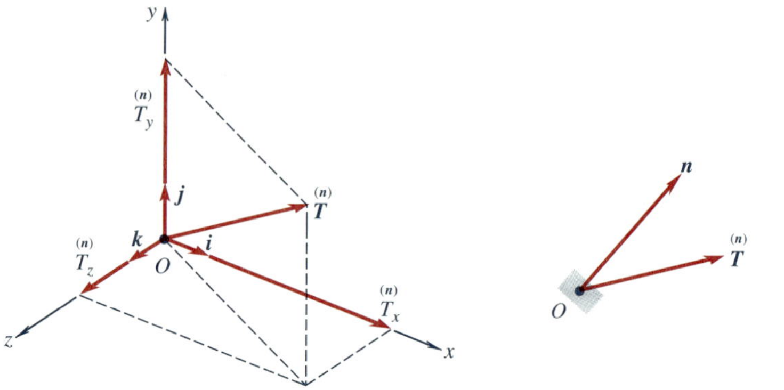

Figure 8.5 Components of stress vector $\overset{(n)}{\mathbf{T}}$.

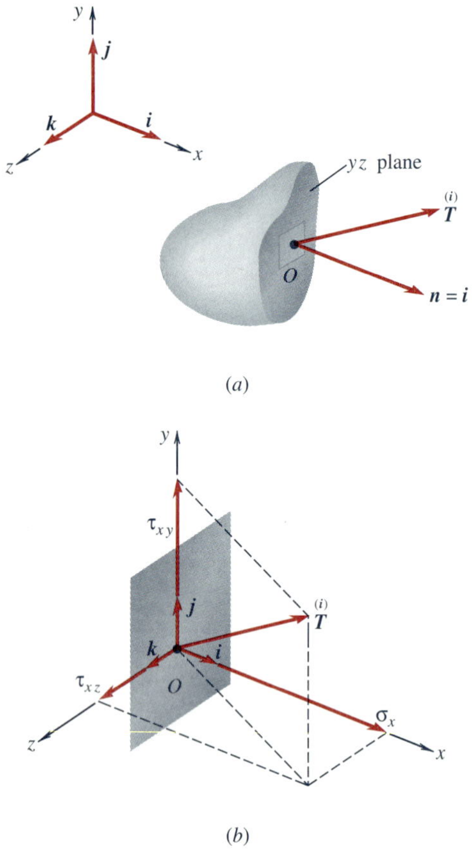

Figure 8.6 (*a*) Stress vector acting on an element of area whose outward-pointing normal is $\mathbf{n} = \mathbf{i}$. (*b*) Components of stress vector $\overset{(i)}{\mathbf{T}}$.

equivalent to the interaction of the material of one portion of the body upon another, and (4) the direction of the stress vector acting on the cut through the point is arbitrary.

In solving problems it is convenient and important, as we have discussed often in previous chapters, to introduce a set of coordinate axes. We have done so in Fig. 8.5 to obtain the components of the stress vector $\overset{(n)}{\mathbf{T}}$ acting at point O on an element of area passing through O with a unit normal vector $\mathbf{n}$. We can also pass through point O planes parallel to the coordinate planes; e.g., in Fig. 8.6 we pass a plane parallel to the yz coordinate plane and obtain a stress vector at point O acting on a plane whose outward-pointing unit normal is in the x direction (Fig. 8.6*a*). We will adopt a convention similar to the one first adopted in Chap. 2: If the outward-pointing unit normal vector to the cut plane is in a positive coordinate direction, then the face exposed by the cut is called a ***positive face.*** Components of a vector acting on a positive face will be positive in the positive coordinate directions. If the outward-pointing unit normal vector on a face is in a negative coordinate direction, then we refer to the face as a ***negative face,*** and the components of a vector acting on a negative face will be positive in the negative coordinate directions.

In Fig. 8.6*a* we show a plane parallel to the yz plane passing through a point O. The outward-pointing unit normal $\mathbf{n}$ to the plane is $\mathbf{i}$. The stress vector acting on the plane is $\overset{(i)}{\mathbf{T}}$, where we have identified the direction of the outward-pointing normal vector in the stress vector quantity. The components of $\overset{(i)}{\mathbf{T}}$ are written in the form (Fig. 8.6*b*).

$$\overset{(i)}{\mathbf{T}} = \sigma_x\mathbf{i} + \tau_{xy}\mathbf{j} + \tau_{xz}\mathbf{k} \tag{8.3}$$

When double subscripts are on a component in Eq. (8.3), the first subscript refers to the coordinate direction of the normal to the area on which the component acts, and the second subscript refers to the coordinate direction

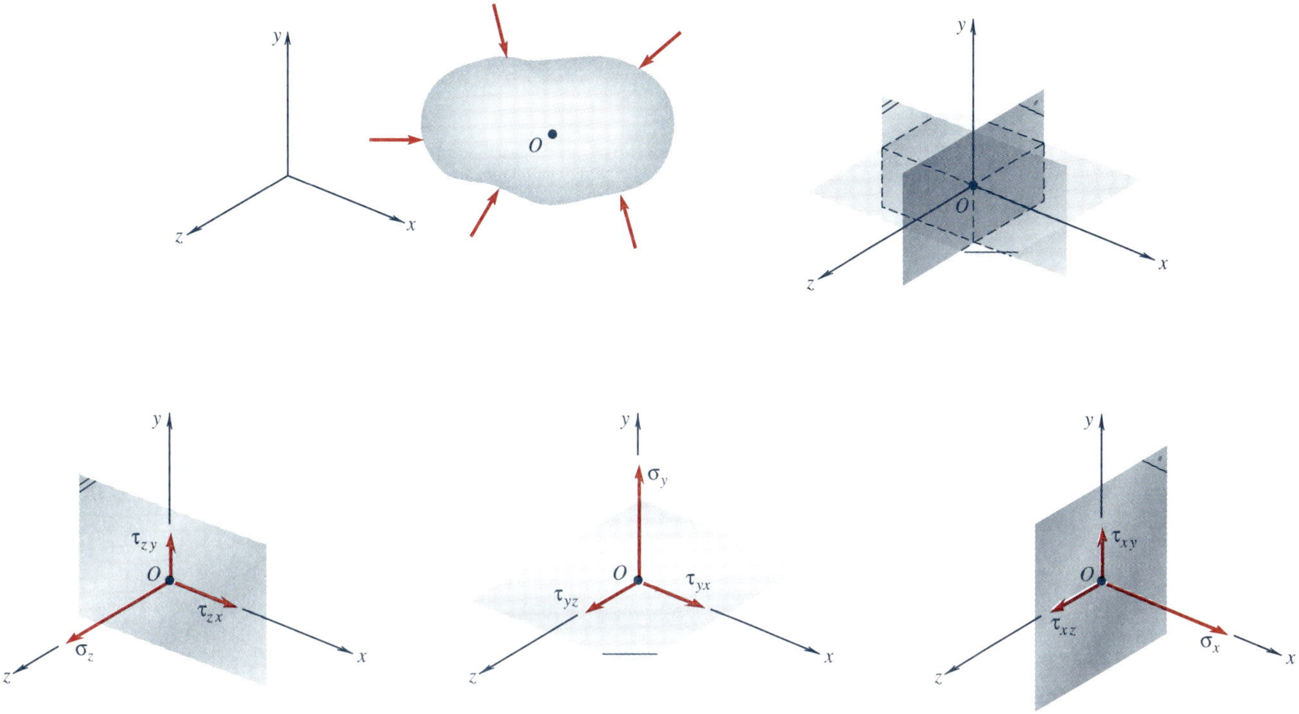

Figure 8.7 Planes parallel to the coordinate planes passing through point O and corresponding stress components.

nate direction in which the component acts; that is, τ_{xz} is the component of the stress vector on a plane whose unit normal is in the x direction acting in the z direction. In Fig. 8.6b we show the positive components of the stress vector $\overset{(i)}{\mathbf{T}}$. The notation for the components follows our earlier usage; σ_x is a stress component *normal* to the surface or a normal stress, while τ_{xy} and τ_{xz} are stress components acting along the surface or *shear* stresses.

If we now pass through point O planes parallel to the xy plane and the xz plane with *outward* normal vectors in the positive coordinate directions, we obtain the components shown in Fig. 8.7. In this way we obtain at a point O the nine components of the three stress vectors acting on planes parallel to the coordinate planes:

$$\begin{matrix} \sigma_x & \tau_{xy} & \tau_{xz} \\ \tau_{yx} & \sigma_y & \tau_{yz} \\ \tau_{zx} & \tau_{zy} & \sigma_z \end{matrix} \tag{8.4}$$

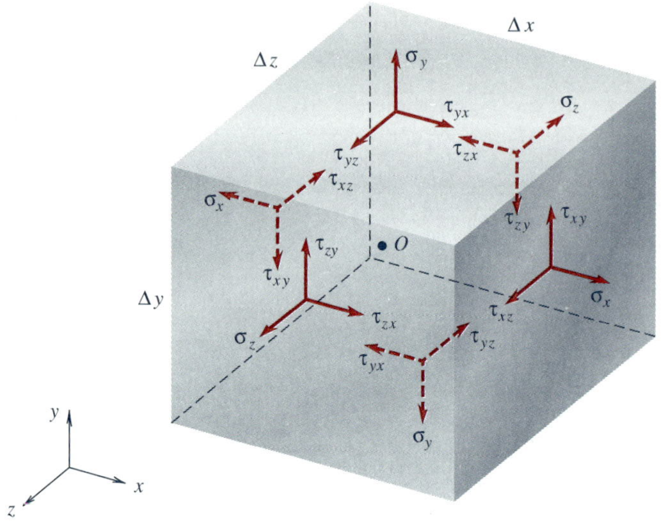

Figure 8.8 Visualization of the general state of stress at a point; stress components at O acting on an element of volume.

We refer to the *state of stress* at a point as given by these nine components; actually, by moment equilibrium arguments we can show that $\tau_{xy} = \tau_{yx}$, $\tau_{yz} = \tau_{zy}$, and $\tau_{xz} = \tau_{zx}$ (recall Sec. 1.3) so that only six of the nine components indicated in Eq. (8.4) are independent.

It is convenient to *visualize* the state of stress at a point in a body by means of a sketch of a small volume element centered at point O. On the faces of the volume element we show the values of the components of the stress vectors acting on the coordinate planes passing through point O (see Fig. 8.8). The totality of the six independent stress components, Eq. (8.4), as shown in Fig. 8.8, gives us what we call the *state of stress* at the point.

The sketch of the small volume element shown in Fig. 8.8 is the means by which we visualize the state of stress at point O. In this way we can see how a small volume of material at point O is experiencing the stress components acting in different directions. For example, if at a point the values of the shear stresses are zero, then the volume element experiences normal stresses only (Fig. 8.9*a*); if, in addition, the normal stresses are equal $\sigma_x = \sigma_y = \sigma_z = \sigma$ (Fig. 8.9*b*), then material in the neighborhood of point O is being pulled equally in the coordinate directions. In this latter case we say that there exists a *hydrostatic tensile state of stress* at point O. If the stress σ is compressive, say, $\sigma = -p$, then the material in the neighborhood of point O is being compressed equally in the coordinate directions, and we say there exists a *hydrostatic compressive state of stress* at point O. The importance of hydrostatic states of stress at a point will arise later when we discuss conditions for the onset of plastic deformation.

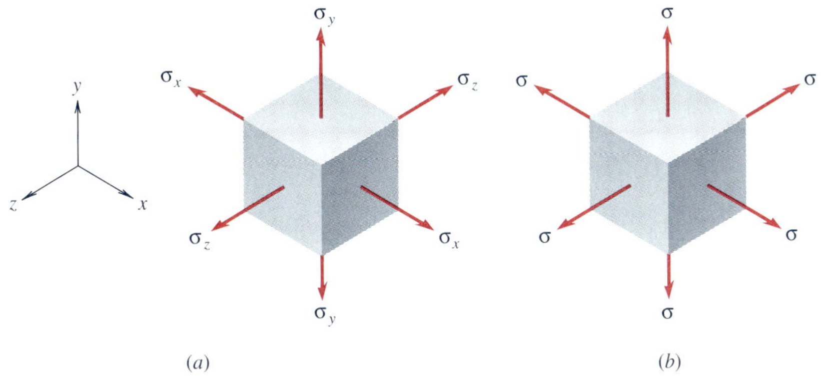

Figure 8.9 (*a*) Normal stresses acting on an element of volume. (*b*) Hydrostatic tensile state of stress.

Even if we knew the values of the stress components acting on the coordinate planes passing through a point, in engineering applications we still need to relate them to the components of a stress vector acting on an arbitrarily oriented element of area, Eq. (8.2), as shown in Fig. 8.5. We will do this in Sec. 8.4 for the case of two-dimensional stress states. Finally, we must remember that, in general, in a body under load, the state of stress will be different at different points in the body. However, the components of the stress vectors acting on the coordinate planes must be related to one another through equations reflecting the requirements of force and moment equilibrium, since each infinitesimal element of the body must be in equilibrium. We derived an equilibrium equation which contained a derivative of the normal stress in the case of uniaxial loading in Sec. 2.8; equilibrium equations for a state of stress in three dimensions can be derived in a similar fashion and will contain derivatives of the stress components; see Probs. 8.2-3 and 8.2-4.

8.3 Plane Stress

We wish to consider a thin sheet of material that is loaded by forces in the plane of the sheet (see Fig. 8.10*a*). If we take the *xy* plane to be the plane of the sheet, then the state of stress at a point *O* in the sheet can be visualized as shown in Fig. 8.10*b*, where for plane stress we assume $\sigma_z = \tau_{xz} = \tau_{yz} = 0$. In this case there are no stresses in the *z* direction. Therefore the state of stress at a point will depend on only the stress components

$$\begin{matrix} \sigma_x & \tau_{xy} \\ \tau_{yx} & \sigma_y \end{matrix} \tag{8.5}$$

where $\tau_{yx} = \tau_{xy}$. Moment equilibrium about the center *O* in the *z* direction

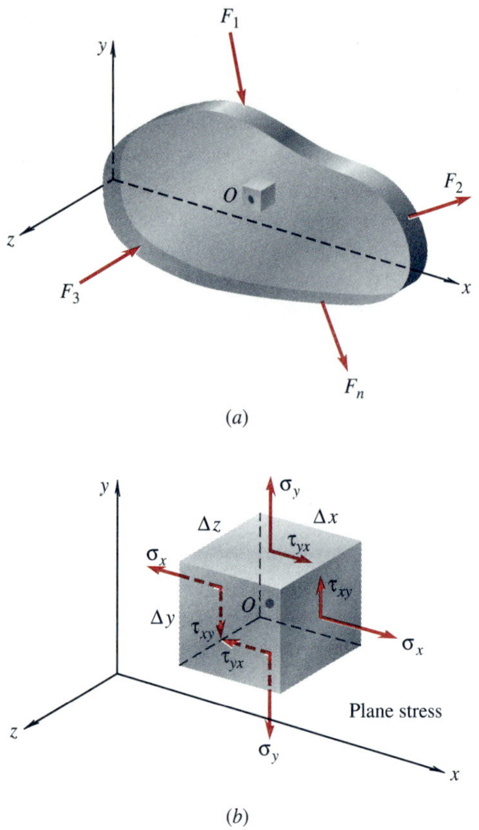

Figure 8.10 (*a*) A thin plate loaded by forces in the plane. (*b*) Stress components which define a state of plane stress in the *xy* plane.

for the element in Fig. 8.10*b* leads to the equality of τ_{yx} and τ_{xy}. If we assume further that for a thin sheet there are no variations in the stress components in Eq. (8.5) in the *z* direction, then the stress components are functions of only *x* and *y*. Under this assumption, the array of stress components in Eq. (8.5) is called ***plane stress*** in the *xy* plane; as we have noted, a state of plane stress at a point in a material is visualized as shown in Fig. 8.10*b*.

We will focus on the case of plane stress in the next three sections.

8.4 Stress Components Associated with Arbitrarily Oriented Faces in Plane Stress

In Sec. 8.2 we introduced the concept of the stress vector acting at a point on a plane with a unit normal vector **n**, Eq. (8.1) and Fig. 8.5. We then introduced stress vectors and their components acting on planes parallel to the coordinate planes, Eq. (8.4) and Fig. 8.7. We now wish to obtain expressions for the components of the stress vector acting at a point *O* on a plane with unit normal **n** in terms of the components of the stress vectors acting on planes parallel to the coordinate planes passing through *O* for the case of plane stress.

In Fig. 8.11*a* we show a body under the conditions of plane stress. The point *O* with stress vector $\overset{(\mathbf{n})}{\mathbf{T}}$ acting on an element of area with unit normal vector **n** is shown. Figure 8.11*b* shows a two-dimensional view. The angle θ gives the orientation of the normal vector to the area. If we know the values of the components σ_x, $\tau_{yx} = \tau_{xy}$, σ_y of the stress vectors acting on coordinate planes passing through *O*, can we find expressions for $\overset{(\mathbf{n})}{T}_x$ and $\overset{(\mathbf{n})}{T}_y$ in terms of these components and the angle θ? The answer is yes, and to do so we need to appeal to equilibrium arguments of an appropriate free body at point *O*. The stress components on the coordinate planes passing through *O* are known, and for convenience we can displace the plane on which we are seeking the components a small distance along the direction of **n.** This gives us a wedge-shaped body, as shown in Fig. 8.11*c*. We will investigate equilibrium of this small wedge made up of faces on which the stress components of interest are acting.

We redraw the wedge in Fig. 8.12*a*. We will consider equilibrium of this wedge in the limit as the size of the wedge shrinks to zero; in so doing we will find the relations between the stress components acting at point *O*. Since we plan to take the limit as the size of the wedge goes to zero, we can treat the stress components acting on each face of the wedge as average values over the face and acting at the midpoint of each face.

Figure 8.12*a* shows the stress components acting on each face of the wedge; ΔA is the area of the face whose unit outward normal vector

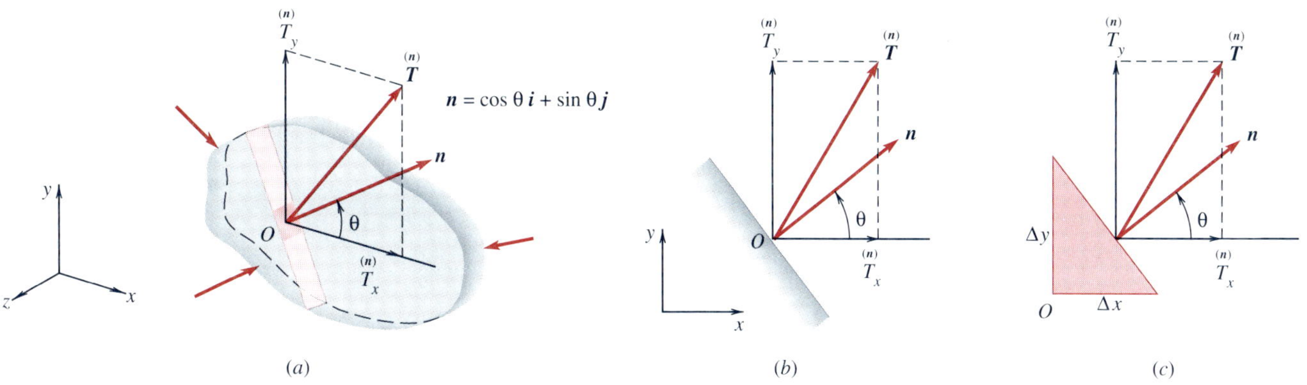

Figure 8.11 (*a*) Stress vector $\overset{(\mathbf{n})}{\mathbf{T}}$ acting at point *O* on a plane whose normal is **n** for a body subjected to plane stress. (*b*) Two-dimensional view of (*a*). (*c*) Wedge-shaped region with $\overset{(\mathbf{n})}{\mathbf{T}}$ acting.

is **n.** We must remember that stress components are force intensities, i.e., force per unit area; to obtain the values of the forces acting on each face of the wedge, we must multiply each stress component by the area of the face on which it acts. Figure 8.12*b* in a two-dimensional view shows the *forces* acting on each face of the wedge. Moment equilibrium of the wedge about point O' in Fig. 8.12*b* gives

$$\tau_{xy}\,\Delta A\cos\theta\,\frac{\Delta x}{2} - \tau_{yx}\,\Delta A\sin\theta\,\frac{\Delta y}{2} = 0 \tag{8.6}$$

In the limit as Δx and Δy approach zero with $\Delta x/\Delta y = \tan\theta$, we find, as expected,

$$\tau_{xy} = \tau_{yx} \tag{8.7}$$

Summation of forces in the *x* and *y* directions gives

$$\begin{aligned} \overset{(\mathbf{n})}{T}_x\,\Delta A &= \sigma_x\,\Delta A\cos\theta + \tau_{yx}\,\Delta A\sin\theta \\ \overset{(\mathbf{n})}{T}_y\,\Delta A &= \tau_{xy}\,\Delta A\cos\theta + \sigma_y\,\Delta A\sin\theta \end{aligned} \tag{8.8}$$

Upon dividing by ΔA, taking the limit as the size of the wedge vanishes, we find

$$\begin{aligned} \overset{(\mathbf{n})}{T}_x &= \sigma_x\cos\theta + \tau_{yx}\sin\theta \\ \overset{(\mathbf{n})}{T}_y &= \tau_{xy}\cos\theta + \sigma_y\sin\theta \end{aligned} \tag{8.9}$$

Equations (8.9) give the components of the stress vector acting on an element of area whose unit normal vector is **n** in terms of the stress

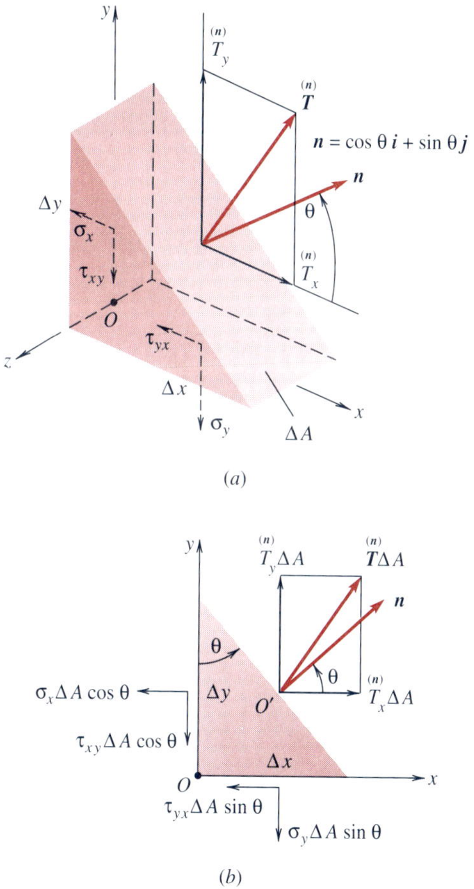

Figure 8.12 (*a*) Stress vector and stress components acting on the faces of a small wedge. (*b*) Two-dimensional view of wedge in (*a*) showing values of forces acting on each face.

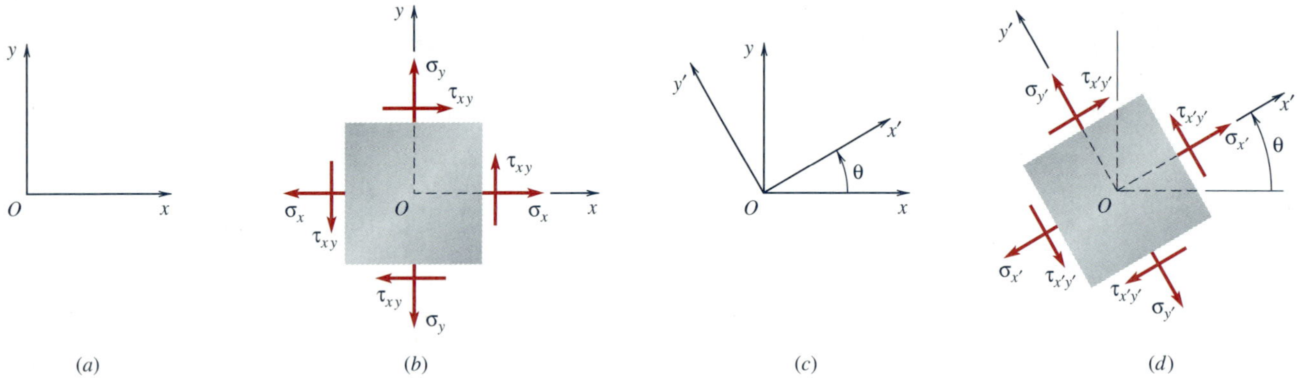

Figure 8.13 Stress components for a set of $x'y'$ axes rotated from the set of xy axes: (*a*) xy axes, (*b*) stress components relative to the xy axes, (*c*) set of $x'y'$ axes rotated from the xy axes by an angle θ, (*d*) stress components relative to the $x'y'$ axes.

components acting on the coordinate planes. The unit normal vector is given by

$$\mathbf{n} = \cos\theta\mathbf{i} + \sin\theta\mathbf{j} \tag{8.10}$$

Equations (8.9) are important results because in formulating stress conditions on the boundary of a body, expressions for the components of the stress vector acting on the boundary in terms of the components on the coordinate planes are required; see Prob. 8.4-2. We have therefore found the required relations, Eqs. (8.9), between the stress components acting on an arbitrary plane defined by angle θ and the stress components acting on the coordinate planes through a point in a body experiencing plane stress.

Another important question to ask about the state of plane stress at a point is, What are the relations between the stress components in one coordinate system and those in a coordinate system rotated through an angle from the first? The interest in this question partially arises from the fact that it may be easier to visualize the state of stress in one set of axes than in another.

In Fig. 8.13*a* we show the original set of xy axes, and in Fig. 8.13*b* we show the state of stress given by the stress components relative to this set of axes. If the axes are rotated through a positive counterclockwise angle θ, we obtain a new set of axes $x'y'$, as shown in Fig. 8.13*c*. The set of stress components relative to this set of $x'y'$ axes is shown in Fig. 8.13*d*. The state of stress at the point is the same; only the means of describing it by different sets of components relative to different sets of axes has changed.

Our objective is to find expressions for the normal stress component $\sigma_{x'}$ and the shear stress component $\tau_{x'y'}$ acting on the face oriented at arbitrary angle θ in terms of the stress components σ_x, σ_y, and τ_{xy} that

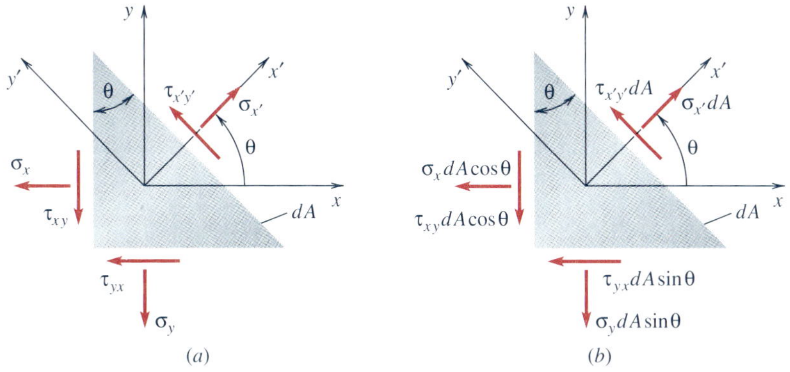

Figure 8.14 (*a*) Stress components acting on the wedge. (*b*) Force components acting on the wedge.

are assumed to be known (Fig. 8.13). Once the expression for $\sigma_{x'}$ is found, if we substitute for the angle θ the value $\theta + \pi/2$, we will obtain the value of $\sigma_{y'}$. In this way we will obtain the components for the state of stress at point O in the $x'y'$ set of axes (Fig. 8.13*d*).

For the derivation of these expressions, we need again a free body which will enable us to write equilibrium equations relating the known stress components σ_x, σ_y, and τ_{xy} to the unknown stress components $\sigma_{x'}$ and $\tau_{x'y'}$ acting on a face oriented at an arbitrary angle θ. Angle θ is measured as positive from the positive x axis in a counterclockwise direction (Fig. 8.13*c*).

A suitable free body is again the wedge-shaped element of Fig. 8.12, now shown in a two-dimensional view as an infinitesimal element in Fig. 8.14*a*. The wedge, as before, is of constant thickness in the z direction, has two edges parallel to the x and y coordinate directions, and has its third edge oriented so that its normal makes an angle of θ with the x direction (Fig. 8.14*a*). To obtain equilibrium equations for this element, we again multiply the stress components by the respective areas upon which they are acting, in order to convert stress components to forces to be used in the equilibrium analysis. Thus if dA is used to designate the area of the inclined face, then $dA \sin\theta$ and $dA \cos\theta$ are the areas of the faces with normals in the negative y and the negative x directions. The forces acting on the faces of the element are shown in Fig. 8.14*b*.

Since we would like to get expressions for $\sigma_{x'}$ and $\tau_{x'y'}$ in terms of σ_x, σ_y, and τ_{xy}, it is convenient to sum force components in the x' and y' directions by writing the forces involving σ_x, σ_y, and τ_{xy} in terms of their components in the x' and y' directions. Summation of forces in the x' direction gives

$$\sigma_{x'}\,dA - \sigma_x\,dA\cos^2\theta - \tau_{xy}\,dA\cos\theta\sin\theta$$
$$- \sigma_y\,dA\sin^2\theta - \tau_{xy}\,dA\sin\theta\cos\theta = 0$$

Summation of forces in the y' direction gives

$$\tau_{x'y'}\,dA + \sigma_x\,dA\cos\theta\sin\theta - \tau_{xy}\,dA\cos^2\theta - \sigma_y\,dA\sin\theta\cos\theta + \tau_{xy}\,dA\sin^2\theta = 0$$

where we have used the relation $\tau_{yx} = \tau_{xy}$, Eq. (8.7). These equations can be rewritten in the form

$$\sigma_{x'} = \sigma_x\cos^2\theta + \sigma_y\sin^2\theta + 2\tau_{xy}\sin\theta\cos\theta \tag{8.11a}$$

$$\tau_{x'y'} = -(\sigma_x - \sigma_y)\sin\theta\cos\theta + \tau_{xy}(\cos^2\theta - \sin^2\theta) \tag{8.11b}$$

Equation (8.11*a*) allows us to calculate the normal stress acting on any face with a normal making an angle of θ with the positive x direction, provided we have the stress components corresponding to the x and y directions. In particular, the normal stress component in the y' direction (as shown in Fig. 8.13*d*) $\sigma_{y'}$ follows by substituting $\theta + \pi/2$ for θ in Eq. (8.11*a*) with $\sin(\theta + \pi/2) = \cos\theta$, $\cos(\theta + \pi/2) = -\sin\theta$ to obtain

$$\sigma_{y'} = \sigma_x\sin^2\theta + \sigma_y\cos^2\theta - 2\tau_{xy}\sin\theta\cos\theta \tag{8.12}$$

The positive directions of the normal stress and the shear stress acting on a face with angle θ are consistent with the positive directions shown in Fig. 8.14.

Addition of $\sigma_{x'}$ and $\sigma_{y'}$ in Eqs. (8.11*a*) and (8.12) yields

$$\sigma_{x'} + \sigma_{y'} = \sigma_x + \sigma_y \tag{8.13}$$

We observe from Eq. (8.13) that the sum of the normal stress components acting on coordinate faces at right angles to each other in a set of axes does not change as we rotate the axes. As a consequence, we say that the sum of the normal stresses is an *invariant* quantity with θ, i.e., a constant independent of the orientation of the axes. We will make use of this result at several points in the discussions to follow.

Equations (8.11) and (8.12) can be rewritten in a simpler form by making use of the trigonometric identities

$$\begin{aligned} 2\sin^2\theta &= 1 - \cos 2\theta \\ 2\cos^2\theta &= 1 + \cos 2\theta \\ 2\sin\theta\cos\theta &= \sin 2\theta \end{aligned} \tag{8.14}$$

which, when substituted into Eqs. (8.11) and (8.12), result in

$$\sigma_{x'} = \frac{\sigma_x + \sigma_y}{2} + \frac{\sigma_x - \sigma_y}{2}\cos 2\theta + \tau_{xy}\sin 2\theta \tag{8.15a}$$

$$\tau_{x'y'} = -\frac{\sigma_x - \sigma_y}{2}\sin 2\theta + \tau_{xy}\cos 2\theta \tag{8.15b}$$

and

$$\sigma_{y'} = \frac{\sigma_x + \sigma_y}{2} - \frac{\sigma_x - \sigma_y}{2}\cos 2\theta - \tau_{xy}\sin 2\theta \tag{8.16}$$

Thus at a point in a body in a state of plane stress, Eqs. (8.15) and (8.16) completely specify the stress components $\sigma_{x'}$, $\sigma_{y'}$, and $\tau_{x'y'}$ acting on faces with normals rotated by angles θ and $\theta + \pi/2$ from the positive x direction, as we show in Fig. 8.13. We will refer to these equations as the *stress transformation equations* for plane stress. Equations (8.15) and (8.16) also can be derived directly from Eqs. (8.9); see Prob. 8.4–4.

Therefore, if we are given the stress components for the state of stress referred to a set of axes at a point, we can find the stress components referred to a new set of axes rotated from the first set by the angle θ, by using Eqs. (8.15) and (8.16); see Fig. 8.13. A number of interesting results arise as we change the value of θ.

EXAMPLE 8.1

A thin plate of steel is subjected to in-plane loads giving rise at a point in the plate to the set of stress components relative to the *xy* axes shown in Fig. 8.15*a*. We wish to find the set of stress components with respect to a set of $x'y'$ axes rotated by a positive angle of 60° from the *xy* axes; i.e., we wish to find the stress components acting on the faces of an element rotated through an angle of $\theta = 60°$, as shown in Fig. 8.15*b*.

Following the sign conventions for the stress components used in the derivation of Eqs. (8.15), we note that for this case

$$\sigma_x = 18{,}000 \text{ psi} \qquad \tau_{xy} = 6000 \text{ psi} \qquad \sigma_y = 8000 \text{ psi} \qquad (a)$$

Therefore inserting these stress components and the angle $\theta = 60°$ into Eq. (8.15*a*) yields

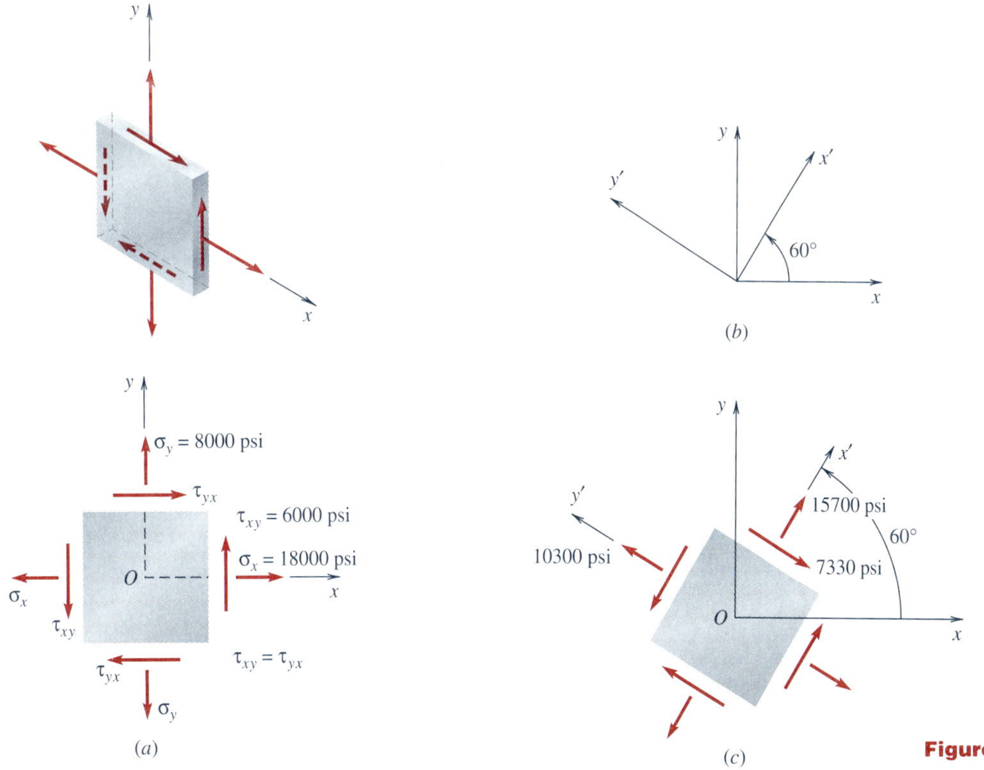

Figure 8.15 Example 8.1

$$\begin{aligned}\sigma_{x'} &= \frac{26{,}000}{2} + \frac{10{,}000}{2}\cos 120 + 6000 \sin 120 \\ &= 15{,}700 \text{ psi}\end{aligned} \tag{b}$$

and similar calculations using Eqs. (8.15*b*) and (8.16) yield

$$\begin{aligned}\tau_{x'y'} &= -\frac{10{,}000}{2}\sin 120 + 6000 \cos 120 \\ &= -\ 7{,}330 \text{ psi}\end{aligned} \tag{c}$$

$$\begin{aligned}\sigma_{y'} &= 13{,}000 - 5000 \cos 120 - 6000 \sin 120 \\ &= 10{,}300 \text{ psi}\end{aligned} \tag{d}$$

As a check on the calculations just carried out, we can compute $\sigma_{x'} + \sigma_{y'} = 26{,}000$ psi and verify that it is indeed equal to $\sigma_x + \sigma_y$, as required by Eq. (8.13).

A sketch of the rotated element with the stress components is given in Fig. 8.15*c*. Figure 8.15*a* and *c* correspond to the same state of stress at point *O* except that the stress components are referred to different sets of axes.

Note that the shear stress which has the value +6000 psi on the *x* face ($\theta = 0$) changes to a negative value of -7330 psi on the x' face with its normal at an angle of $\theta = 60°$ to the positive *x* direction. Since, according to Eq. (8.15*b*), $\tau_{x'y'}$ varies smoothly as a function of the angle θ, there must be an orientation of axes between $\theta = 0°$ and $\theta = 60°$ for which the shear stress vanishes.

In fact, if we plot $\sigma_{x'}$ and $\tau_{x'y'}$ as a function of θ, for $-180° \leq \theta \leq 180°$, we obtain the two curves shown in Fig. 8.16. The curves are periodic with period π; that is, $\sigma_{x'}(\theta \pm \pi) = \sigma_{x'}(\theta)$ and $\tau_{x'y'}(\theta \pm \pi) = \tau_{x'y'}(\theta)$. At $\theta = 60°$, we obtain from the plotted curves the values given by Eqs. (*b*) and (*c*). Since $\sigma_{y'}$ is the value of $\sigma_{x'}$ at $\theta + \pi/2$, Eq. (8.16), we obtain from the graph the value for $\sigma_{y'}$ given in Eq. (*d*). The shear stress $\tau_{x'y'}$ does indeed pass through zero at an angle of approximately $\theta = 25°$ at which angle $\sigma_{x'}$ is a maximum value equal to about 21 ksi. The corresponding value of $\sigma_{y'}$ for $\theta = 25°$ is a minimum equal to 5.0 ksi; $\sigma_{y'} = \sigma_{x'}(25° + 90°)$.

The shear stress $\tau_{x'y'}$ curve in Fig. 8.16 also shows that the shear stress takes on a maximum value and minimum value at an angle of $\pm 45°$ from the angle at which the maximum and minimum values of the normal stress occur.

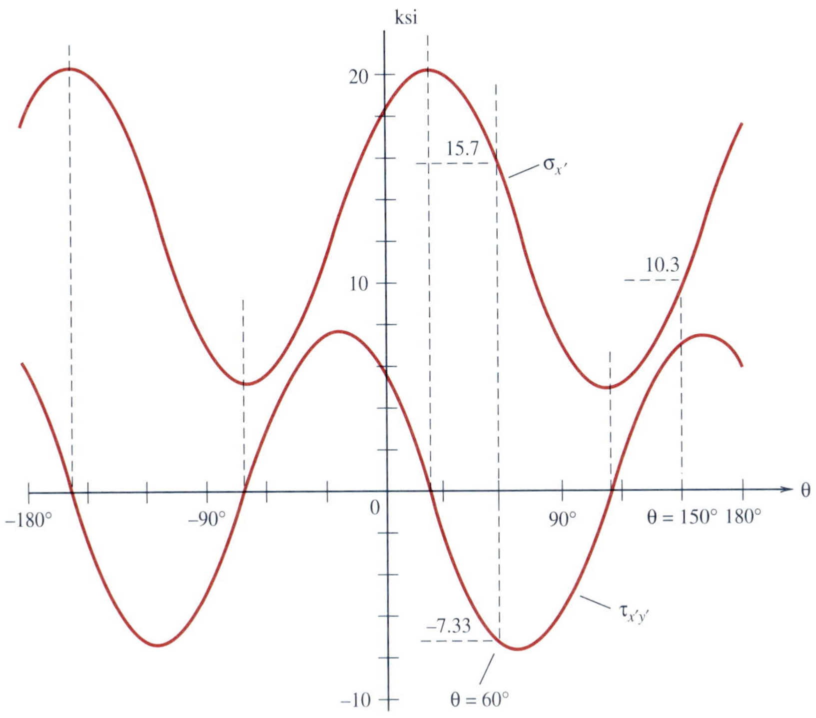

Figure 8.16 Example 8.1: Plot of $\sigma_{x'}$ and $\tau_{x'y'}$ versus θ.

In the next section we will show that in general the faces on which the shear stress vanishes are associated with the maximum and minimum values of the normal stress, and we will find the orientation for the maximum and minimum shear stress.

8.5 Principal Stresses and Maximum Shear Stress

In engineering practice we are often interested in finding the largest and smallest normal stresses and the maximum shear stress acting on an element at a point. The maximum and minimum normal stresses are called the *principal stresses.* One way to find the values of the principal stresses that we used in Example 8.1 is to plot the curve of $\sigma_{x'}$ as a function of θ on a graph and to determine the values of the principal stresses as the maximum and minimum points on the curve. In Example 8.1, the maximum and minimum values of the normal stress were separated on the graph by an angle of 90°. When we plotted on the same graph the stress component $\tau_{x'y'}$ as a function of θ, we found that the principal stresses occurred at those values of θ where the shear stress

$\tau_{x'y'}$ was zero (Fig. 8.16). We will now show that these results hold in general for plane stress at a point.

Principal Stresses

At angles θ for which there is either a maximum or minimum value for $\sigma_{x'}$, the slope of the curve of $\sigma_{x'}$ versus θ is zero. The angles at these values give the orientation of the *principal planes* upon which the principal stresses act. These angles are called the *principal directions*, and they can be found by setting the first derivative of $\sigma_{x'}$ with respect to θ equal to zero. By differentiating Eq. (8.15*a*) we get

$$\frac{d\sigma_{x'}}{d\theta} = -(\sigma_x - \sigma_y) \sin 2\theta_p + 2\tau_{xy} \cos 2\theta_p = 0 \qquad (8.17)$$

where θ_p denotes the angles—the principal directions—satisfying Eq. (8.17). Rewriting Eq. (8.17) produces an equation for θ_p in the form

$$\tan 2\theta_p = \frac{2\tau_{xy}}{\sigma_x - \sigma_y} \qquad (8.18)$$

In the special case when $\sigma_x = \sigma_y$, the angle θ_p follows from Eq. (8.17).

We see immediately from Eqs. (8.17) and (8.15*b*) that when $\theta = \theta_p$, $\sigma_{x'}$ is a maximum or a minimum *and* that the shear stress $\tau_{x'y'}$ acting on that face is zero. The plane corresponding to a principal direction is such that only a normal stress component is acting on it. The principal direction gives the orientation of the element such that *only* a normal stress acts on each of the faces.

To find θ_p, we solve Eq. (8.18) for $2\theta_p$, given σ_x, σ_y, and τ_{xy}. The roots of Eq. (8.18) can be considered as the intersection of the curve of $\tan 2\theta$ plotted as a function of 2θ with the value of the constant $2\tau_{xy}/(\sigma_x - \sigma_y)$ plotted on the same graph as a horizontal line, as shown in Fig. 8.17.

From Fig. 8.17, we see that when $2\tau_{xy}/(\sigma_x - \sigma_y)$ is *positive,* there are an infinite number of roots since $\tan 2\theta$ is periodic of period π and the roots differ from one another by multiples of π or 180°. In calculating the roots of Eq. (8.18) by using a calculator, only one root for $2\theta_p$ is obtained, and it will be in the range of 0° to 90° when $2\tau_{xy}/(\sigma_x - \sigma_y)$ is positive. In Fig. 8.17 we show this positive root as $2\theta_p$ at point A; the next positive root is $2\theta_p + 180°$, shown at A'. From these two roots, the corresponding two values for the principal directions are θ_p in the range of 0° to 45° and $\theta_p + 90°$ in the range of 90° to 135°. The first root, θ_p, corresponds to the normal stress $\sigma_{x'}$ in the x' direction, and the second root, $\theta_p + 90°$, corresponds to the $\sigma_{x'}$ stress in the direction $\theta_p + 90°$ or to the stress $\sigma_{y'}$ in the y' direction. We now need to find the values of the principal stresses given by these directions.

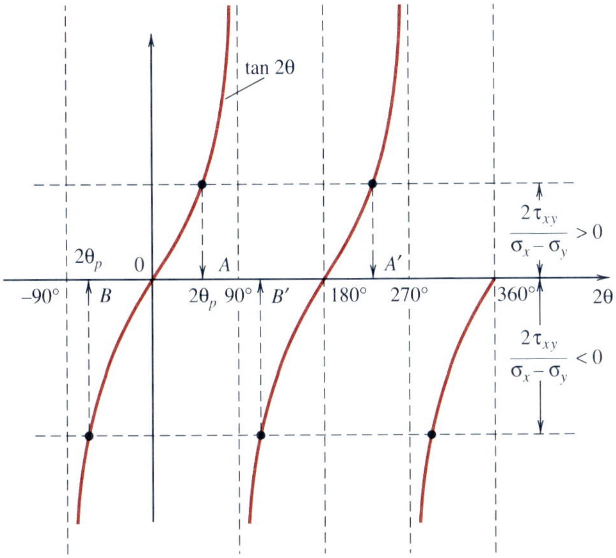

Figure 8.17 Plot to illustrate location of principal directions.

The principal stresses can be found by substituting angles θ_p and $\theta_p + 90°$ into the transformation equation for $\sigma_{x'}$, Eq. (8.15*a*). One stress value will correspond to a maximum value, the other stress value to a minimum value. Upon substitution of θ_p into the stress transformation equation, Eq. (8.15*a*), we have

$$\sigma_{x'}(\theta = \theta_p) = \frac{\sigma_x + \sigma_y}{2} + \frac{\sigma_x - \sigma_y}{2}\cos 2\theta_p + \tau_{xy}\sin 2\theta_p \tag{8.19}$$

It follows from Eq. (8.18) and the trigonometry of Fig. 8.18 with both τ_{xy} and $(\sigma_x - \sigma_y)$ positive that

$$\cos 2\theta_p = \frac{\sigma_x - \sigma_y}{2R} \qquad \sin 2\theta_p = \frac{\tau_{xy}}{R} \tag{8.20}$$

where

$$R = \sqrt{\left(\frac{\sigma_x - \sigma_y}{2}\right)^2 + \tau_{xy}^2} \tag{8.21}$$

Therefore, upon substitution of Eqs. (8.20) into Eq. (8.19), we find

$$\sigma_{x'}(\theta = \theta_p) = \frac{\sigma_x + \sigma_y}{2} + \frac{\sigma_x - \sigma_y}{2}\frac{\sigma_x - \sigma_y}{2R} + \tau_{xy}\frac{\tau_{xy}}{R}$$

or

$$\sigma_{x'}(\theta = \theta_p) = \frac{\sigma_x + \sigma_y}{2} + R \equiv \sigma_1 \tag{8.22}$$

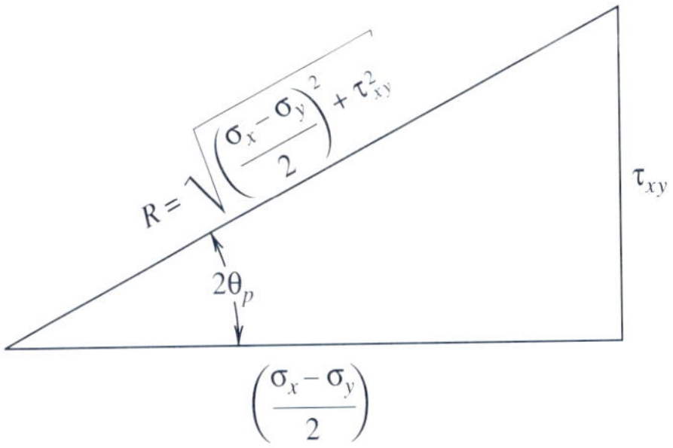

Figure 8.18 Triangle indicating the relation between the angle $2\theta_p$ and the stress components.

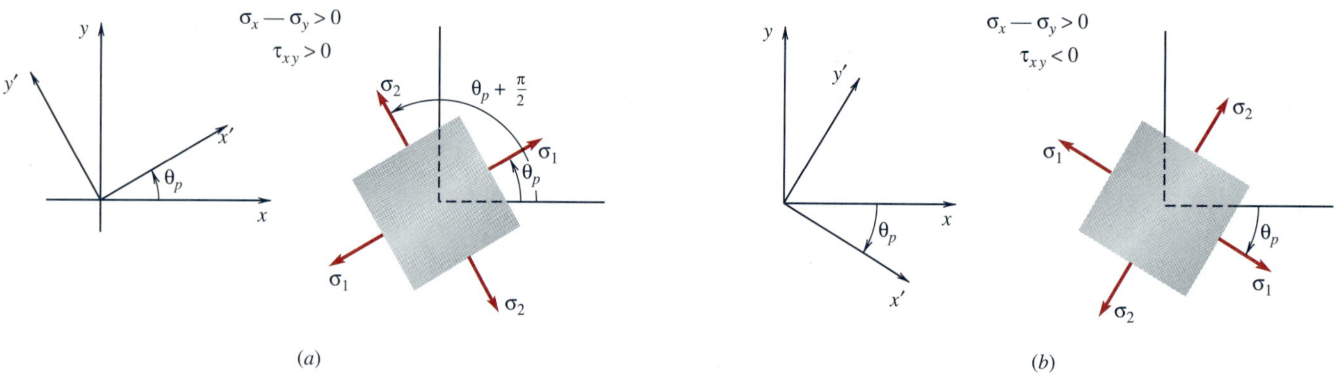

Figure 8.19 (*a*) Principal directions when $\sigma_x - \sigma_y > 0$ and $\tau_{xy} > 0$. (*b*) Principal directions when $\sigma_x - \sigma_y > 0$ and $\tau_{xy} < 0$.

We define σ_1 as the maximum principal stress

$$\sigma_1 = \frac{\sigma_x + \sigma_y}{2} + \sqrt{\left(\frac{\sigma_x - \sigma_y}{2}\right)^2 + \tau_{xy}^2} \tag{8.23}$$

Upon substitution of $\theta_p + 90°$, that is, $\theta_p + \pi/2$, into the expression for $\sigma_{x'}$ given by Eq. (8.19), we obtain

$$\begin{aligned}\sigma_{x'}\left(\theta_p + \frac{\pi}{2}\right) &= \sigma_{y'}(\theta_p) \\ &= \frac{\sigma_x + \sigma_y}{2} - R \equiv \sigma_2\end{aligned} \tag{8.24}$$

We define σ_2 as the minimum principal stress since clearly $\sigma_2 < \sigma_1$

$$\sigma_2 = \frac{\sigma_x + \sigma_y}{2} - \sqrt{\left(\frac{\sigma_x - \sigma_y}{2}\right)^2 + \tau_{xy}^2} \tag{8.25}$$

Figure 8.19*a* shows the orientation of the element corresponding to the principal directions with the principal stresses σ_1 and σ_2 acting on the faces.

We have shown previously, in Eq. (8.13), that the sum of the normal stress components on two perpendicular faces is a constant. It follows, as expected, from the definitions of σ_1 and σ_2 in Eqs. (8.23) and (8.25) that

$$\sigma_1 + \sigma_2 = \sigma_x + \sigma_y \tag{8.26}$$

The above derivation gives σ_1 as the principal stress associated with θ_p and σ_2 as the principal stress associated with $\theta_p + \pi/2$ (Fig. 8.19*a*). However, the argument leading to this result holds *only* if τ_{xy} and

$\sigma_x - \sigma_y$ are positive. If either τ_{xy} or $\sigma_x - \sigma_y$ is negative, or if both are negative, then the signs associated with $\cos 2\theta_p$ and $\sin 2\theta_p$ in Eqs. (8.20) must be accounted for in the expression for the determination of the value of $\sigma_{x'}$ when $\theta = \theta_p$.

For example, if τ_{xy} is negative *and* $\sigma_x - \sigma_y$ is negative, then the evaluation of the expression for $\sigma_{x'}$ when $\theta = \theta_p$ gives $\sigma_{x'}(\theta_p) = \sigma_2$. The evaluation of $\sigma_{x'}$ at $\theta_p + \pi/2$ gives σ_1.

If τ_{xy} is negative and $\sigma_x - \sigma_y$ is positive, then the right-hand side of Eq. (8.18) is negative. If this negative value is now plotted as a horizontal line in Fig. 8.17, this time below the 2θ axis, then the roots are again the intersection points with the curve $\tan 2\theta$. In this case a calculator will give the negative root $2\theta_p$, shown as point B in Fig. 8.17 in the range between $-90°$ and $0°$. A negative value for θ_p means that we rotate the set of axes in a negative or clockwise direction. The second root is $2\theta_p + 180°$, shown as point B' which is positive. In this case the principal stress $\sigma_{x'}(\theta = \theta_p)$ is σ_1, and the value of $\sigma_{x'}$ at $\theta = \theta_p + 90°$ is σ_2. We show this case in Fig. 8.19*b*; again the angle θ_p shown is a negative value. Care needs to be taken in the determination of which of the two principal stresses σ_1 and σ_2 is associated with angle θ_p. Table 8.1 is a summary of the four possible cases that can arise. The results in Table 8.1 can be used to confirm quickly the numerical results; see Prob. 8.5-1.

Maximum Shear Stress

For the particular stress state considered in Example 8.1 and plotted in Fig. 8.16, we found that the maximum and minimum values of the shear

Table 8.1 Principal Stress and Maximum Shear Stress Values for Different Cases of $\sigma_x - \sigma_y$ and τ_{xy}

Case	$\sigma_x - \sigma_y$	τ_{xy}	$\tan 2\theta_p$	θ_p	$\sigma_{x'}(\theta_p)$	$\sigma_{x'}\left(\theta_p + \frac{\pi}{2}\right) = \sigma_{y'}(\theta_p)$
A	+	+	+	$0° < \theta_p < 45°$	σ_1	σ_2
B	−	−	+	$0° < \theta_p < 45°$	σ_2	σ_1
C	+	−	−	$-45° < \theta_p < 0°$	σ_1	σ_2
D	−	+	−	$-45° < \theta_p < 0°$	σ_2	σ_1
	$\sigma_x - \sigma_y$	τ_{xy}	$\tan 2\theta_s$	θ_s	$\tau_{x'y'}(\theta_s)$	$\tau_{x'y'}\left(\theta_s + \frac{\pi}{2}\right)$
A	+	+	−	$-45° < \theta_s < 0°$	R	$-R$
B	−	−	−	$-45° < \theta_s < 0°$	$-R$	R
C	+	−	+	$0° < \theta_s < 45°$	$-R$	R
D	−	+	+	$0° < \theta_s < 45°$	R	$-R$

stress occur on planes that are orthogonal and that the angle for the maximum and minimum values of the shear stress is $\pm 45°$ from the principal directions. We will show now that these same results hold for the general case.

Proceeding in a way similar to that followed to find the location of the maximum and minimum values for the normal stress $\sigma_{x'}$, we differentiate $\tau_{x'y'}$ in Eq. (8.15*b*) with respect to angle θ and equate to zero, to get

$$\frac{d\tau_{x'y'}}{d\theta} = -(\sigma_x - \sigma_y)\cos 2\theta_s - 2\tau_{xy}\sin 2\theta_s = 0 \qquad (8.27)$$

where θ_s denotes the value of angle θ which satisfies Eq. (8.27). It follows that

$$\tan 2\theta_s = -\frac{\sigma_x - \sigma_y}{2\tau_{xy}} \qquad (8.28)$$

Again if we parallel the arguments for principal stress with $\sigma_x - \sigma_y$ and τ_{xy} positive, we find that θ_s will be *negative.* Using the triangle shown in Fig. 8.20*a* with Eq. (8.28), we have

$$\sin 2\theta_s = -\frac{\sigma_x - \sigma_y}{2R} \qquad \cos 2\theta_s = \frac{\tau_{xy}}{R} \qquad (8.29)$$

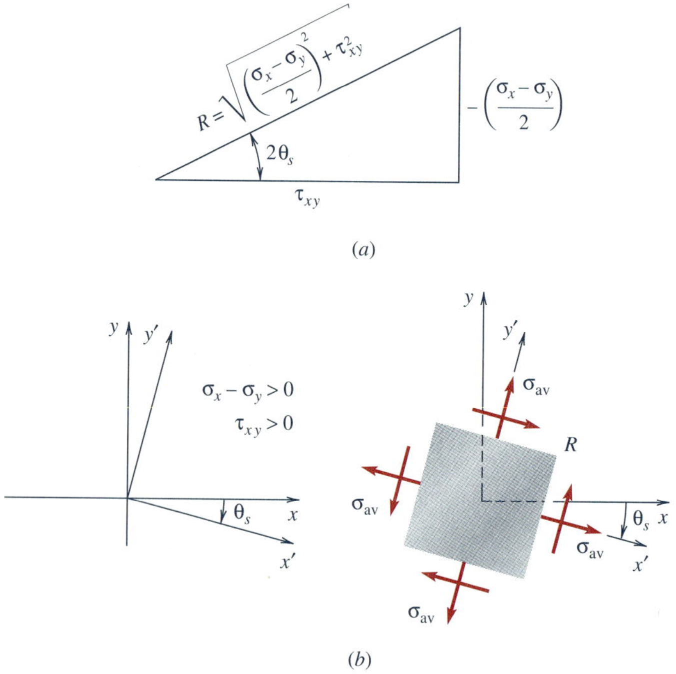

Figure 8.20 (*a*) Triangle indicating the relation between the angle $2\theta_s$ and the stress components. (*b*) Direction of maximum shear stress when $\sigma_x - \sigma_y > 0$ and $\tau_{xy} > 0$.

The corresponding value of the shear stress at θ_s by using Eq. (8.15*b*) is given by

$$\tau_{x'y'}(\theta = \theta_s) = \sqrt{\left(\frac{\sigma_x - \sigma_y}{2}\right)^2 + \tau_{xy}^2} = R \tag{8.30}$$

which is the maximum value. The shear stress $\tau_{x'y'}$ at $\theta = \theta_s + \pi/2$ is equal to $-R$, which is the minimum value.

In addition to the maximum and minimum shear stresses acting on the faces of the element when $\theta = \theta_s$, we have normal stress components from Eqs. (8.15*a*) and (8.16), using Eqs. (8.29):

$$\begin{aligned} \sigma_{x'}(\theta = \theta_s) &= \frac{\sigma_x + \sigma_y}{2} = \sigma_{\text{av}} \\ \sigma_{y'}(\theta = \theta_s) &= \frac{\sigma_x + \sigma_y}{2} = \sigma_{\text{av}} \end{aligned} \tag{8.31}$$

The stress components acting on the faces of an element oriented in the maximum shear stress configuration are shown in Fig. 8.20*b*.

For $\sigma_x - \sigma_y > 0$ and $\tau_{xy} > 0$, angle θ_s will be negative, and the shear stress component $\tau_{x'y'}$ acting on the positive x' face is positive in the positive y' direction. Again we emphasize, as in the case of the discussion for principal stresses, that the above argument giving the value of $\tau_{x'y'}$ at θ_s depends of the signs of $\sigma_x - \sigma_y$ and τ_{xy}. Table 8.1 gives the range for angle θ_s and the corresponding value of $\tau_{x'y'}(\theta = \theta_s)$ which is either the maximum value R or the minimum value $-R$. In many situations we are interested in only the magnitude of the maximum shear stress τ_{max} and in the orientation of the element relative to the xy axes.

The direction corresponding to the maximum shear stress can be explored further by recalling the trigonometric relation

$$\tan\left(2\theta_s \pm \frac{\pi}{2}\right) = \frac{-1}{\tan 2\theta_s} \tag{8.32}$$

If we substitute from Eq. (8.28) and then from Eq. (8.18), we have

$$\tan\left(2\theta_s \pm \frac{\pi}{2}\right) = \frac{2\tau_{xy}}{\sigma_x - \sigma_y} = \tan 2\theta_p \tag{8.33}$$

Therefore we conclude from Eq. (8.33) that

$$\theta_s = \theta_p \pm \frac{\pi}{4} \tag{8.34}$$

That is, the orientation of the element for maximum shear stress is at an angle that is $\pm 45°$ from the principal directions. This result can be used effectively to sketch the orientation of the element for maximum shear.

EXAMPLE 8.2

A state of plane stress at a point is defined by the stress components acting on the coordinate faces, as indicated in Fig. 8.21*a*. We wish (*a*) to find the principal stresses and show them on an element oriented along the principal directions of stress and (*b*) to find the maximum and minimum shear stresses and show them on an element oriented along the directions of maximum and minimum shear stress. We will use the equations we have just derived.

From Fig. 8.21*a* we read off the stress components for this example with attention to the correct algebraic signs according to our sign conventions. In this case

$$\sigma_x = 10{,}500 \text{ psi} \qquad \sigma_y = -5500 \text{ psi}$$
$$\tau_{xy} = -4000 \text{ psi} \qquad (a)$$

In this case $\sigma_x - \sigma_y > 0$ and $\tau_{xy} < 0$, which is case C in Table 8.1.

The directions of the principal stresses can be obtained by using Eq. (8.18) to get

$$\tan 2\theta_p = \frac{2\tau_{xy}}{\sigma_x - \sigma_y} = -\frac{8000}{16{,}000} = -\frac{1}{2} \qquad (b)$$

from which, using a hand calculator for the $\tan^{-1}$ function, we find

$$2\theta_p = -26.57°$$

or

$$\theta_p = -13.28° \qquad (c)$$

The value of R, from Eq. (8.21), is

$$R = \sqrt{\left(\frac{\sigma_x - \sigma_y}{2}\right)^2 + \tau_{xy}^2} = 8944 \text{ psi} \qquad (d)$$

and we have, Eqs. (8.20)

$$\frac{\sigma_x - \sigma_y}{2R} = \frac{10{,}500 - (-5500)}{2R} = 0.8944$$
$$\frac{\tau_{xy}}{R} = -\frac{4000}{R} = -0.4472 \qquad (e)$$

Therefore the principal stress when $\theta = \theta_p$ is, from Eq. (8.19),

$$\sigma_{x'}(\theta = \theta_p) = \frac{\sigma_x + \sigma_y}{2} + \frac{\sigma_x - \sigma_y}{2}\cos 2\theta_p + \tau_{xy}\sin 2\theta_p$$
$$= 11{,}440 \text{ psi} = \sigma_1 \qquad (f)$$

in agreement with case C in Table 8.1. The similar calculation for $\sigma_{x'}(\theta = \theta_p + 90°)$ gives

$$\sigma_{x'}(\theta = \theta_p + 90°) = \sigma_{y'}(\theta_p) = -6440 \text{ psi} = \sigma_2 \qquad (g)$$

The principal stresses are shown in Fig. 8.21*b*. The sum of the principal stresses gives

$$\sigma_1 + \sigma_2 = \sigma_x + \sigma_y = 5000 \text{ psi} \qquad (h)$$

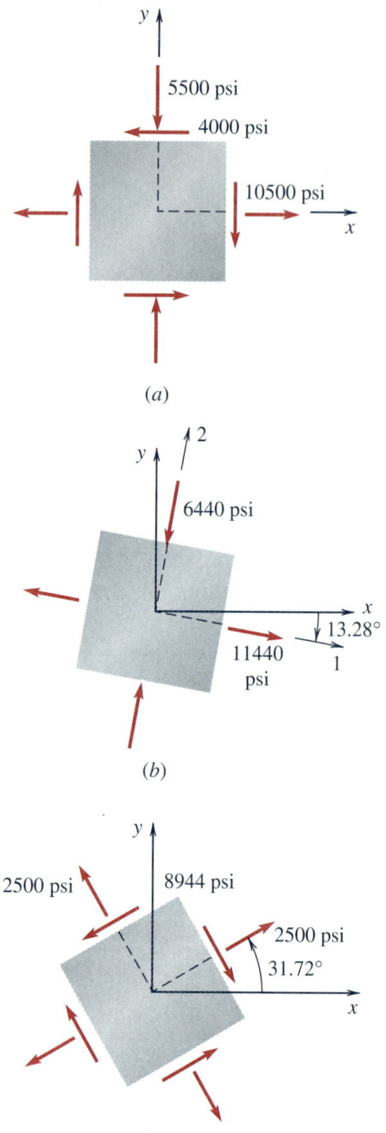

Figure 8.21 Example 8.2

The orientation of the element for maximum shear stress follows from Eq. (8.28)

$$\tan 2\theta_s = -\frac{\sigma_x - \sigma_y}{2\tau_{xy}} = 2 \qquad (i)$$

or

$$2\theta_s = 63.43^\circ \qquad \theta_s = 31.72^\circ \qquad (j)$$

The orientation angle θ_s is positive ($\sigma_x - \sigma_y > 0$, $\tau_{xy} < 0$), and so the shear stress at $\theta = \theta_s$ is

$$\tau_{x'y'}(\theta = \theta_s) = -R = -8944 \text{ psi} \qquad (k)$$

which is a minimum value. The shear stress at $\theta = \theta_s + 90^\circ$ is a maximum equal to $+R$. The normal stresses acting on the faces are given by Eqs. (8.31), and the element oriented to correspond to the directions of maximum and minimum shear stresses is shown in Fig. 8.21*c*.

Finally we note that

$$\theta_s = \theta_p \pm 45^\circ = -13.28^\circ \pm 45^\circ = 31.72^\circ, -58.28^\circ$$

Angle $\theta_s = 31.72^\circ$ gives $\tau_{x'y'} = -R$, and angle $\theta_s = -58.28^\circ$ gives $\tau_{x'y'} = R$, as shown in Fig. 8.21*c*; see case C in Table 8.1.

EXAMPLE 8.3

An element in plane stress is subjected to the stress components shown in Fig. 8.22*a*. We wish (*a*) to find the principal stresses and the maximum and minimum shear stresses and show the results on elements oriented in the directions of the principal stresses and the maximum and minimum shear stresses and (*b*) to find the allowable range for the stress component σ_x so that the magnitude of the maximum shear stress does not exceed 150 MPa. In this case, the stress components σ_y and τ_{xy} are kept equal to the values shown in Fig. 8.22*a* while σ_x varies.

Based upon Fig. 8.22*a*, the stress components have values

$$\sigma_x = -110 \text{ MPa} \qquad \sigma_y = 60 \text{ MPa}$$
$$\tau_{xy} = -40 \text{ MPa} \qquad (a)$$

and using Eqs. (8.21), (8.30), and (8.31), we have

$$R = \sqrt{\left(\frac{\sigma_x - \sigma_y}{2}\right)^2 + \tau_{xy}^2} = 93.94 \text{ MPa}$$
$$\tau_{max} = R = 93.94 \text{ MPa} \qquad (b)$$

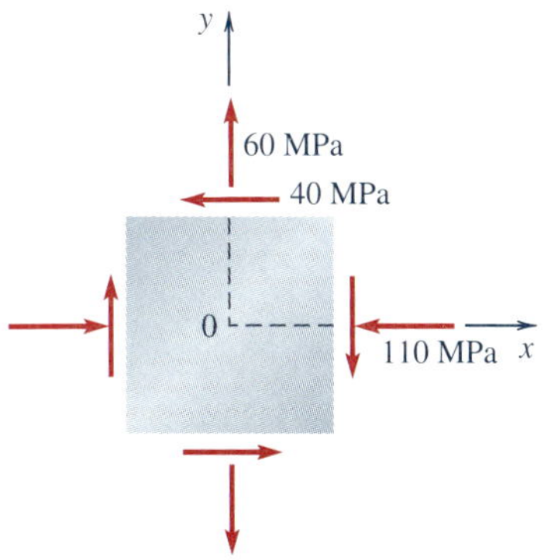

(*a*)

and

$$\sigma_{av} = \frac{-110 + 60}{2} = -25 \text{ MPa} \qquad (c)$$

The magnitudes of the maximum and minimum principal stresses are from Eqs. (8.22) and (8.24),

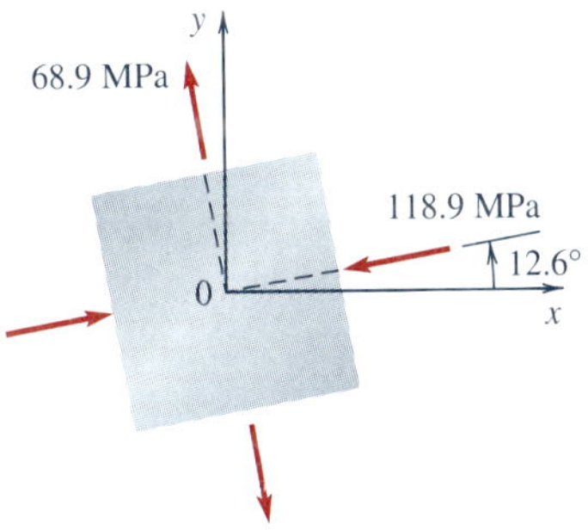

(*b*)

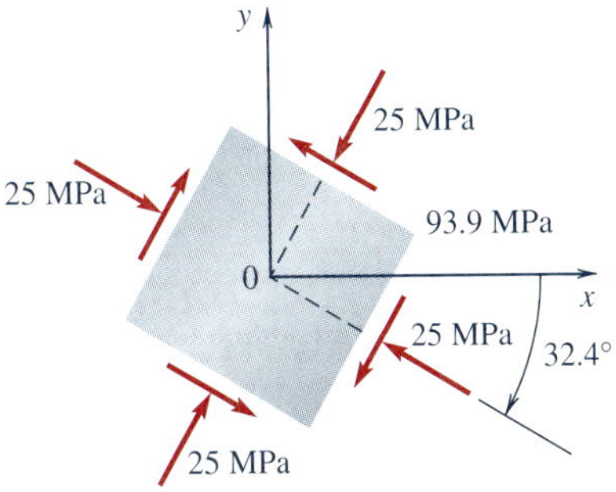

(*c*)

Figure 8.22 Example 8.3

$$\begin{aligned}\sigma_1 &= \sigma_{av} + R \\ &= -25 + 93.9 = 68.9 \text{ MPa}\end{aligned} \tag{d}$$

$$\begin{aligned}\sigma_2 &= \sigma_{av} - R \\ &= -25 - 93.9 = -118.9 \text{ MPa}\end{aligned} \tag{e}$$

According to Eq. (8.18), the angles defining the principal directions satisfy

$$\tan 2\theta_p = \frac{-40}{(-110 - 60)/2} = 0.4706 \tag{f}$$

The solution to Eq. (f) is

$$2\theta_p = 25.20^\circ \qquad \theta_p = 12.60^\circ \tag{g}$$

We have

$$\frac{\sigma_x - \sigma_y}{2R} = \frac{-110 - 60}{2R} = -0.9048 \tag{h}$$

$$\frac{\tau_{xy}}{R} = \frac{-40}{R} = -0.4258 \tag{i}$$

from which we find using Eq. (8.19)

$$\sigma_{x'}(\theta = \theta_p) = \sigma_2 \qquad \sigma_{x'}\left(\theta = \theta_p + \frac{\pi}{2}\right) = \sigma_1 \tag{j}$$

in agreement with case B in Table 8.1. The orientation of the element with the principal stresses acting is shown in Fig. 8.22*b*.

The magnitude of the maximum shear stress is given in Eq. (b); the orientation of the axes of maximum shear stress is given by Eq. (8.28)

$$\tan 2\theta_s = -2.125 \qquad \theta_s = -32.40^\circ \tag{k}$$

Angle θ_s is negative so it is clockwise from the x axis. The shear stress on the face with angle θ_s is, in agreement with Table 8.1, case B,

$$\tau_{x'y'}(\theta = \theta_s) = -R \tag{l}$$

The element with the maximum shear stresses and the average normal stresses is shown in Fig. 8.22*c*.

For the second part of the problem, to find allowable values of σ_x with fixed values of $\sigma_y = 60$ MPa and $\tau_{xy} = -40$ MPa that will lead to a maximum shear stress of 150 MPa, we use the expression for τ_{max}. From Eq. (b) with $\tau_{max} = R = 150$ MPa, we have

$$\tau_{max} = 150 = \sqrt{\left(\frac{\sigma_x - 60}{2}\right)^2 + (-40)^2} \tag{m}$$

Squaring both sides and solving for σ_x results in

$$\left(\frac{\sigma_x - 60}{2}\right)^2 = 20{,}900 \tag{n}$$

with the pair of solutions

$$\sigma_x = -229.1,\ 349.1 \text{ MPa} \tag{o}$$

Thus when σ_x is restricted to the range

$$-229 \le \sigma_x \le 349 \text{ MPa} \tag{p}$$

then τ_{max} will satisfy the condition

$$\tau_{max} \le 150 \text{ MPa} \tag{q}$$

as required.

8.6 Mohr's Circle Representation for Plane Stress

A particular geometric representation for the state of plane stress has been found to be extremely useful in engineering applications. The transformation equations, Eqs. (8.15), can be interpreted as a pair of equations with a single parameter θ that will give a plane curve in a stress coordinate system with the normal stress σ identified with the horizontal axis and the shear stress τ identified with the vertical axis. For each value of the parameter θ, there will be a corresponding point giving a value of $\sigma_{x'}$, $\tau_{x'y'}$ in this stress coordinate system. The value of σ on the horizontal axis is the value of $\sigma_{x'}$, and the value of τ on the vertical axis is the value of $\tau_{x'y'}$. As θ changes, a locus of points for $\sigma_{x'}$ and $\tau_{x'y'}$ or a curve of $\sigma_{x'}$ and $\tau_{x'y'}$ will be traced out. If the parameter θ is eliminated between Eqs. (8.15), then a single equation relating $\sigma_{x'}$ and $\tau_{x'y'}$ will be found in the stress plane, which hopefully can be given a simple geometric interpretation as the curve of all points satisfying Eqs. (8.15).

To carry out the elimination of θ from Eqs. (8.15), we rewrite them in the form

$$\sigma_{x'} - \sigma_{av} = \frac{\sigma_x - \sigma_y}{2}\cos 2\theta + \tau_{xy}\sin 2\theta$$
$$\tau_{x'y'} = -\frac{\sigma_x - \sigma_y}{2}\sin 2\theta + \tau_{xy}\cos 2\theta \qquad (8.35)$$

where $\sigma_{av} = (\sigma_x + \sigma_y)/2$, Eq. (8.31). We then square both sides of Eqs. (8.35) and add the resulting two equations to find

$$(\sigma_{x'} - \sigma_{av})^2 + \tau_{x'y'}^2 = \left(\frac{\sigma_x - \sigma_y}{2}\right)^2 + \tau_{xy}^2 = R^2 \qquad (8.36)$$

This is the equation for a circle with radius R and center at the point $(\sigma_{av}, 0)$ in the σ, τ coordinate system, as shown in Fig. 8.23. We show in Fig. 8.23 the positive σ axis with an arrowhead to the right; for now we do not specify the positive τ axis. We will refer to this circle as the *Mohr's circle representation* for plane stress in honor of the German engineer Otto Mohr who in 1882 emphasized the usefulness of this geometric interpretation. Several important facts about the stress state that were established in the previous section by analytic arguments are immediately clear from the geometry of the circle in Fig. 8.23, namely:

1. The maximum and minimum values for $\sigma_{x'}$ are $\sigma_1 = \sigma_{av} + R$ and $\sigma_2 = \sigma_{av} - R$ with $\tau_{x'y'} = 0$ at these points.
2. The maximum and minimum values of $\tau_{x'y'}$ are $\pm R$ with the corresponding $\sigma_{x'}$ and $\sigma_{y'}$ at these points equal to σ_{av}.

It remains to make a connection between the location of points on

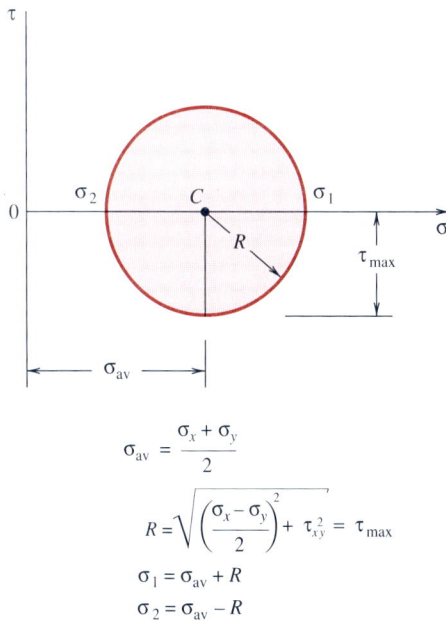

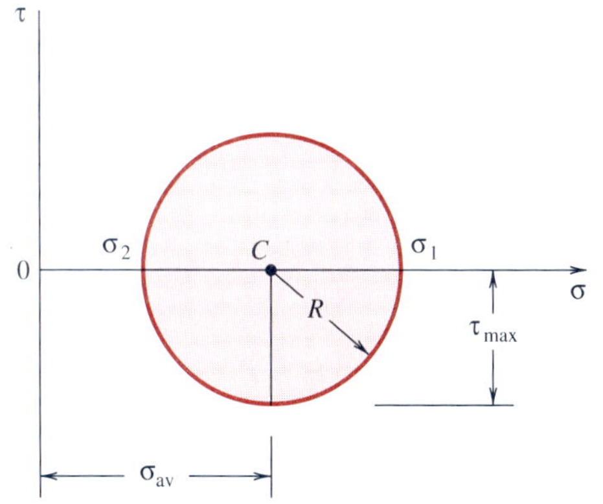

$$\sigma_{av} = \frac{\sigma_x + \sigma_y}{2}$$
$$R = \sqrt{\left(\frac{\sigma_x - \sigma_y}{2}\right)^2 + \tau_{xy}^2} = \tau_{max}$$
$$\sigma_1 = \sigma_{av} + R$$
$$\sigma_2 = \sigma_{av} - R$$

Figure 8.23 Stress components $\sigma_{x'}$ and $\tau_{x'y'}$ lie on a circle of radius R in the σ- τ plane.

the circle to the specific stress components $\sigma_{x'}$ and $\tau_{x'y'}$ on a face at an orientation θ and the stress components σ_x, τ_{xy}, and σ_y at $\theta = 0$.

The transformation equations, Eqs. (8.35), can be recast into a form which helps to clarify their relation to the circle. Using Eqs. (8.20) to replace the terms $(\sigma_x - \sigma_y)/2$ and τ_{xy} in Eqs. (8.35), we find

$$\sigma_{x'} = \sigma_{av} + R\cos 2\theta_p \cos 2\theta + R \sin 2\theta_p \sin 2\theta$$

$$\tau_{x'y'} = -R\cos 2\theta_p \sin 2\theta + R\sin 2\theta_p \cos 2\theta$$

or

$$\sigma_{x'} = \sigma_{av} + R\cos(2\theta_p - 2\theta) \tag{8.37a}$$

$$\tau_{x'y'} = R\sin(2\theta_p - 2\theta) \tag{8.37b}$$

We have already shown that for each angle θ, the point with coordinates $(\sigma_{x'}, \tau_{x'y'})$ lies on Mohr's circle of radius R centered at $(\sigma_{av}, 0)$ for given values of σ_x, τ_{xy}, and σ_y. Equations (8.37) are the parametric representation of this circle in the σ, τ plane.

In the derivation of the transformation equations, Eqs. (8.35), angle θ was measured in a counterclockwise direction from the direction $\theta = 0$ of the positive x axis, (Fig. 8.13*c*). It follows, therefore, that the coordinates of the point on Mohr's circle corresponding to the stress components for the positive x direction can be found by substituting $\theta = 0$ in Eqs. (8.37) to get, by using Eqs. (8.20),

$$\sigma_{x'}(\theta = 0) = \sigma_{av} + R\cos 2\theta_p = \sigma_x \tag{8.38a}$$

$$\tau_{x'y'}(\theta = 0) = R\sin 2\theta_p = \tau_{xy} \tag{8.38b}$$

Since angle θ in the physical plane for the orientation of the rotated axes is taken as positive in the counterclockwise direction, we will set up the location on Mohr's circle of the point $\theta = 0$ in such a way that the parameter 2θ appearing in Eqs. (8.35) for locating the transformed stress components on Mohr's circle will be a rotation also positive in the counterclockwise direction.

To construct the Mohr's circle representation, we use Eqs. (8.37) and (8.38). We start with the stress components at $\theta = 0$, as shown in Fig. 8.24*a*. In Fig. 8.24*b*, we first locate in the σ-τ stress plane the point C with coordinates $((\sigma_x + \sigma_y)/2, 0)$. We then locate the point (σ_x, τ_{xy}) with the convention that if τ_{xy} is positive, we plot its value *below* the horizontal σ axis. This construction is consistent with Eqs. (8.38) and (8.20). This point is marked as x in Fig. 8.24*b*. The convention for plotting the shear stress is opposite to the usual convention for plotting points in an xy coordinate system, but it leads to the same direction of rotation for the angle θ in the physical plane and the angle 2θ in the stress plane for Mohr's circle.

A circle centered at C of radius $Cx = R$ is now drawn to give the circle shown in Fig. 8.24*b*. From the construction, the value of σ_y corresponds

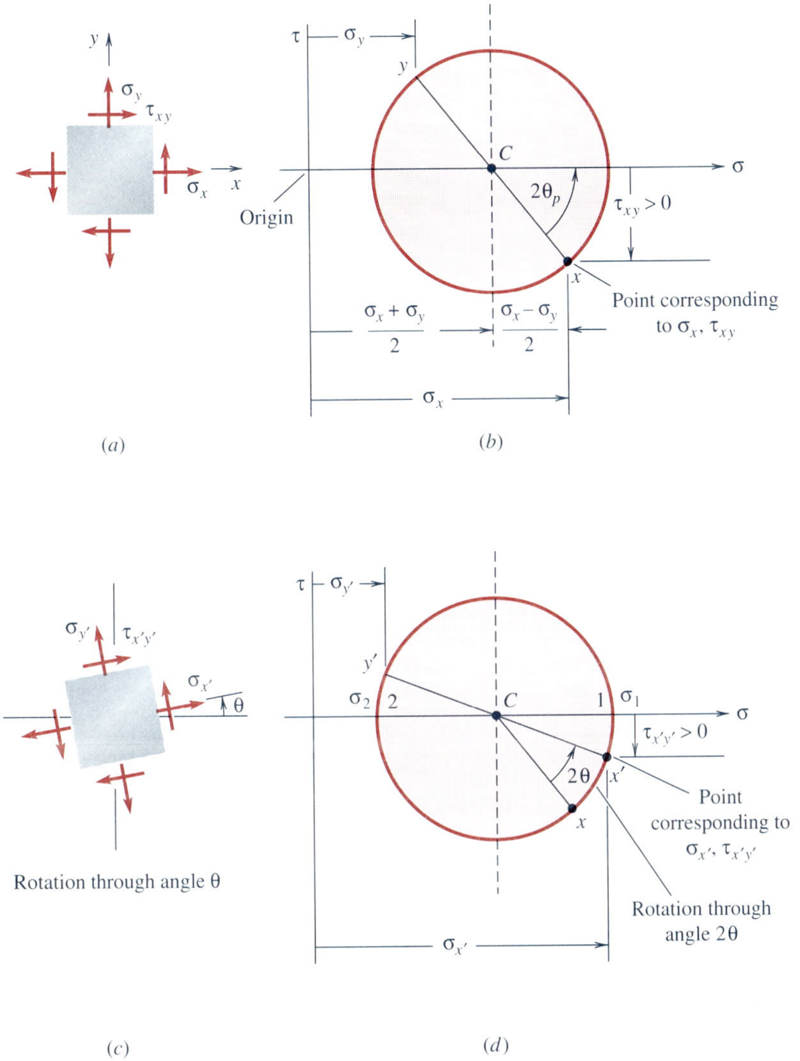

Figure 8.24 Mohr's circle construction for plane stress. A rotation θ of the axes in the physical plane corresponds to a rotation 2θ in the Mohr's circle construction. Positive shear stresses are plotted below the horizontal normal stress axis; negative shear stresses are plotted above the horizontal normal stress axis.

to the value of σ at point *y* on the diameter opposite point *x*. The angle $2\theta_p$ in Fig. 8.24*b* follows from Eq. (8.18).

The point on the circle corresponding to $\sigma_{x'}$ and $\tau_{x'y'}$ for a rotation of axes equal to θ (Fig. 8.24*c*) follows directly from Eqs. (8.37). In Fig. 8.24*d*, the angle $2\theta_p - 2\theta$ in Eqs. (8.37) for locating point x' corresponding to $\sigma_{x'}$ and $\tau_{x'y'}$ is angle $x'C1$. That is, point x' is found by rotation through an angle 2θ from the line Cx to the line Cx' to obtain point x' on the circle. The values of $\sigma_{x'}$ and $\tau_{x'y'}$ follow as the coordinates of point x' consistent with Eqs. (8.37). The value of $\sigma_{y'}$ is the value of σ for point y' at the opposite end of the diameter starting at point x'.

If Mohr's circle is drawn on a piece of graph paper to scale, then all the values of interest can be read off directly from the circle. Alternatively, construction of Mohr's circle can be used as a quick means to establish the formulas for the principal stresses, principal directions, and maximum shear stresses.

To summarize the construction of Mohr's circle in the σ-τ stress plane, we have the following steps (Fig. 8.24):

1. The center of the circle C is located at the point $(\sigma_{av}, 0)$, where $\sigma_{av} = (\sigma_x + \sigma_y)/2$.
2. We locate point x with coordinates σ_x and τ_{xy}. A positive value of τ_{xy} is plotted below the horizontal σ axis. A negative value of τ_{xy} is plotted above the horizontal σ axis.
3. With C as center and Cx as a radius, we draw the circle. The radius of the circle is

$$R = \sqrt{\left(\frac{\sigma_x - \sigma_y}{2}\right)^2 + \tau_{xy}^2}$$

   The value of σ_y is the value of σ for the point at the opposite end of the diameter from point x (Fig. 8.24*b*).
4. We locate the Cx' radius by rotating the Cx radius through the double angle 2θ, as shown in Fig. 8.24*d,* in the *same* direction as the rotation θ of the $x'y'$ axes in Fig. 8.24*c*.
5. With the sign convention for the stress components, we can read off the values of $\sigma_{x'}$ and $\tau_{x'y'}$ as the coordinates of point x' and the value of $\sigma_{y'}$ as the σ coordinate of point y' on the diameter through C and x'.
6. The values of the principal stresses σ_1 and σ_2, the principal directions, and the maximum shear stress are found from the circle.

The following examples illustrate how Mohr's circle for plane stress can be constructed for a specific numerical case.

EXAMPLE 8.4

An element in a state of plane stress has stress components acting on its coordinate faces, as shown in Fig. 8.25*a*. We wish to (*a*) construct Mohr's circle that represents the state of plane stress at this point, (*b*) locate points *a* and *b* on Mohr's circle that give the stress components on planes with normals in the *a* and *b* directions obtained by rotating the *x* and *y* axes by 40° clockwise (Fig. 8.25*b*), and (*c*) show the stress components on an element aligned in the *a* and *b* directions.

In this case $\sigma_{av} = 50$ MPa. We start by plotting in Fig. 8.25*c* point C for the center of the circle and point x for the stress components for the positive x direction. With C as the center and Cx as the radius, we construct the circle. We locate the stress component σ_y on the circle and label the point as y. Since directions x and y in the physical plane differ by 90°, they are 180° apart on Mohr's circle. Thus a line joining x and y is a diameter of Mohr's circle.

The radius of the circle is given by

$$R = \sqrt{\left(\frac{\sigma_x - \sigma_y}{2}\right)^2 + \tau_{xy}^2} = \sqrt{30^2 + 60^2} = 67.1 \text{ MPa} \quad (a)$$

Since the direction a in the xy plane is oriented at 40° clockwise from the positive x direction, it follows that $\theta = -40°$, where the minus sign indicates that a clockwise rotation is in-

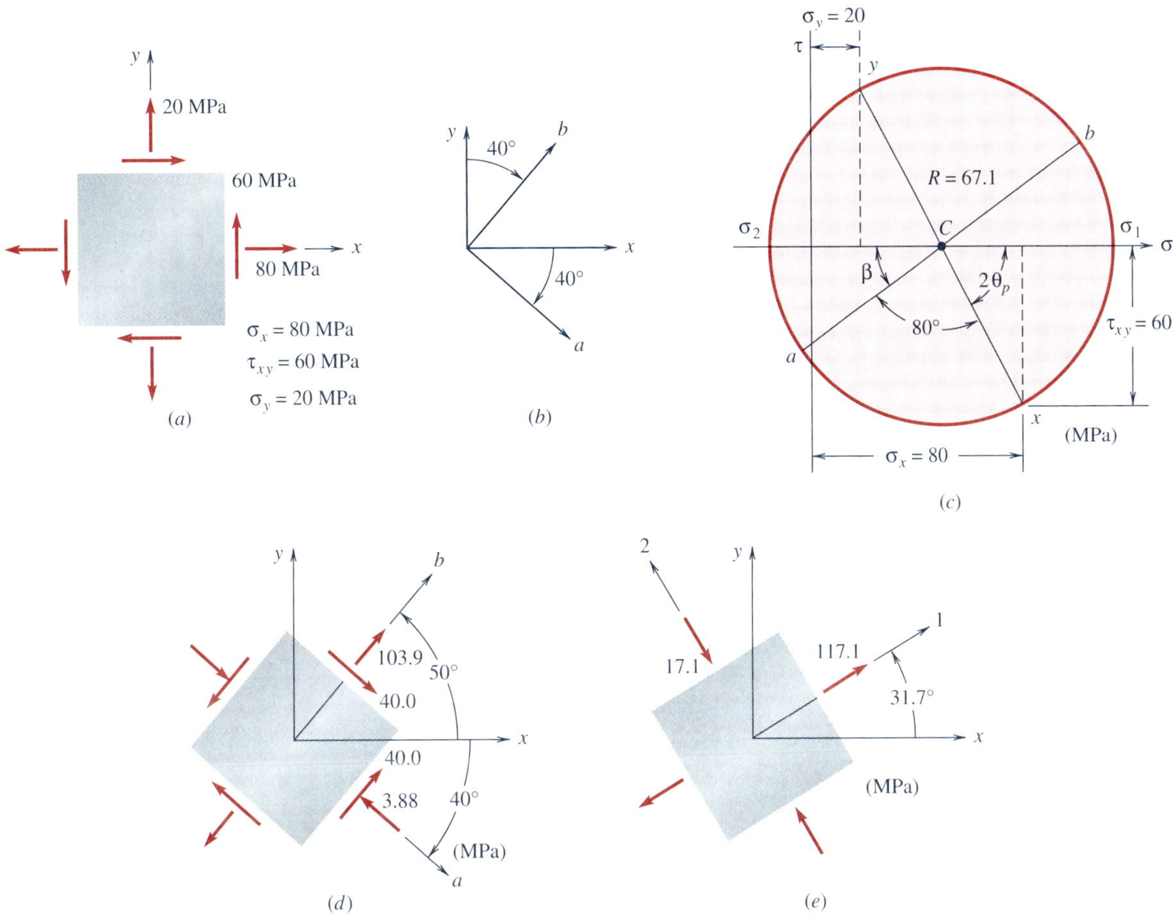

Figure 8.25 Example 8.4

volved. Thus on the stress plane of Mohr's circle, the stress point corresponding to direction a is located at angle $2\theta = -80°$ with respect to the point labeled x. Point a corresponding to the stresses at the angle $-80°$ from point x and point b at $2(90) = 180°$ counterclockwise from point a are both shown on Mohr's circle in Fig. 8.25*c*.

Approximate values for the stress components (σ_a, τ_{ab}) can be estimated graphically from Mohr's circle as constructed in Fig. 8.25*c*. Alternatively, we have from Fig. 8.25*c*

$$\tan 2\theta_p = \frac{60}{30} \qquad 2\theta_p = 63.43°$$

On the circle of Fig. 8.25*c*, the angle $\beta = 36.57°$. Therefore, the stress components at point a are

$$\sigma_a = \sigma_{av} - R\cos\beta = -3.88 \text{ MPa}$$
$$\tau_{ab} = R\sin\beta = 40.0 \text{ MPa} \qquad (b)$$

and the normal stress component at point b is $\sigma_b = 103.9$ MPa. The element with these stresses acting is shown in Fig. 8.25*d*. The principal stresses are $\sigma_1 = 117.1$ MPa, $\sigma_2 = -17.1$ MPa, and $\tau_{max} = 67.1$ MPa. The orientation of the element in the principal directions is shown in Fig. 8.25*e*.

EXAMPLE 8.5

We are given a set of xy axes and the state of stress at a point, as shown in Fig. 8.26*a*. We wish to use Mohr's circle construction to display the components of stress with respect to a set of axes rotated an angle 45° from the original axes (Fig. 8.26*a*) and to determine the principal stresses, principal directions, and the maximum shear stress.

The construction of Mohr's circle is shown in Fig. 8.26*b*, and the orientation of the elements for $\theta = 45°$, $\theta = \theta_p$, and $\theta = \theta_s$ are shown in Fig. 8.26*c*, *d*, and *e*.

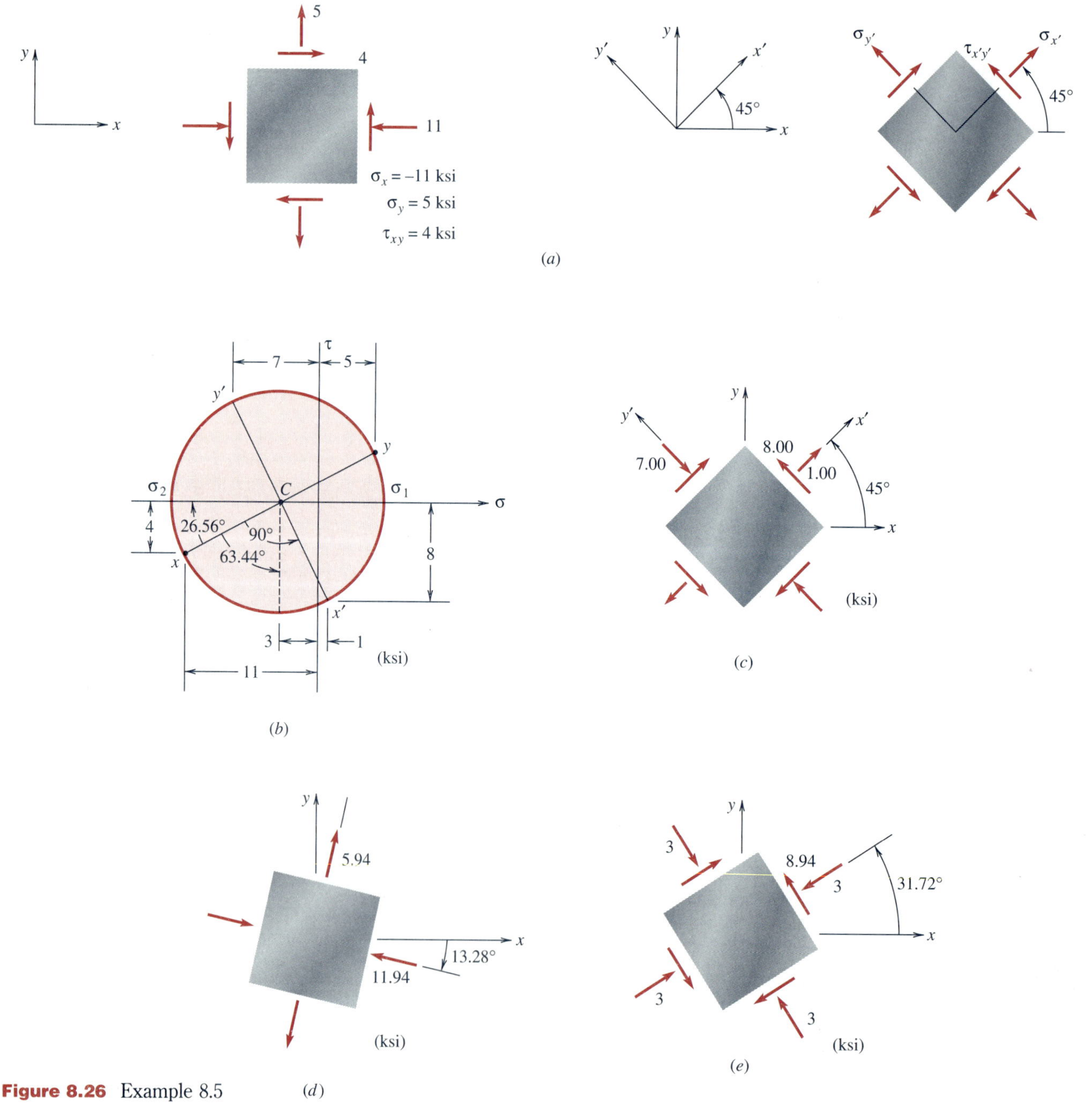

Figure 8.26 Example 8.5

We can set up a computer program to plot Mohr's circle and to print the important stress components. Program 5 in the main menu of MECHMAT (on the diskette with this book) is such a program. Once the stress components for the *xy* axes are given and the angle of rotation θ is specified, the program gives the components of stress at the specified angle θ, the principal stresses, the principal directions, and the maximum shear stress, and it sketches Mohr's circle for the given state of stress.

```
**********************************************************

     *** MOHR'S CIRCLE ***         MOHRP

**********************************************************

         1. PLANE STRESS
         2. PLANE STRAIN
         3. STRAIN ROSETTE MEASUREMENTS
      OR 4. BACK TO MENU

            SELECT NUMBER = ? 1
```

(*a*)

```
     white: SIGMA (X) = -1.10E+01
            SIGMA (Y) =  5.00E+00
             TAU (XY) =  4.00E+00

red:                           blue:
SIGMA ( 45.0) =  1.00E+00      SIGMA1 =  5.94E+00
  TAU ( 45.0) =  8.00E+00      SIGMA2 = -1.19E+01
SIGMA (135.0) = -7.00E+00      THETA1 =  76.72
  TAU  (MAX)  =  8.94E+00      THETA2 = -13.28
```

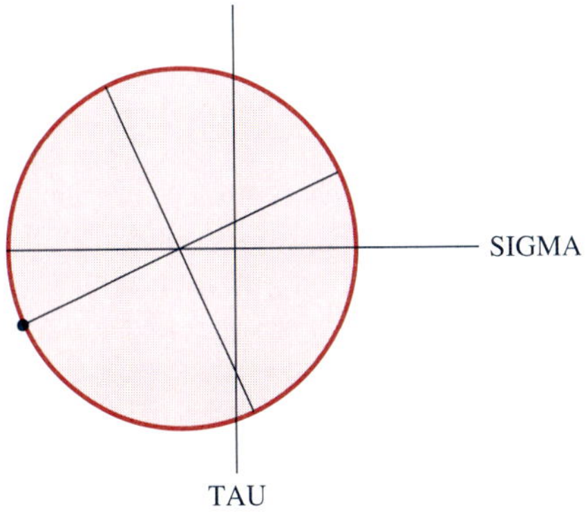

(*b*)

Figure 8.27 (*a*) Mohr's circle menu. (*b*) Solution for Example 8.5.

For example, from the main menu by selecting item 5: Mohr's circle, we obtain the menu for Mohr's circle shown in Fig. 8.27*a*. For plane stress we select item 1. The input to the plane stress program is straightforward. For example, if we input the stress components and θ of Example 8.5 just completed, we obtain the results shown in Fig. 8.27*b*. The "white" values are σ_x, σ_y, and τ_{xy} as the input values; the "red" values are the values at the specified angle θ, namely, $\sigma_{x'}(\theta = 45°)$, $\tau_{x'y'}(\theta = 45°)$, $\sigma_{x'}(\theta = 45° + 90) = \sigma_{y'}$, and τ_{max}. The "blue" values are σ_1 and σ_2; the angle theta1 is the angle to σ_1, and the angle theta2 is the angle to σ_2. The angles in the output are the angles in the physical plane and *not* those in Mohr's circle. The "colors" are those that will appear on a color monitor. The circle as sketched in the program is to approximate scale only.

8.7 Mohr's Circle Representation for a General State of Stress

We now wish to consider the case in which we have a *general* state of stress at a point, as shown in Fig. 8.8. In this situation, in contrast to the case of plane stress in the xy plane, we have stress components in the z direction. Let us turn to finding the stress components associated with a plane whose normal x' lies in the xy plane and makes an angle of θ with the x axis. If we cut out a small wedge as we did for plane stress (Fig. 8.12), we find that the stress components acting on the faces of the wedge are as shown in Fig. 8.28. We note that in addition to the components $\sigma_{x'}$ and $\tau_{x'y'}$ on the $+x'$ face, we have the possibility of another shear stress component $\tau_{x'z}$ acting on the $+x'$ face.

If we repeat the arguments of Sec. 8.4 for force equilibrium of the wedge in the x' and y' directions to obtain expressions for $\sigma_{x'}$ and $\tau_{x'y'}$, we find that Eqs. (8.15) are unchanged. This follows from the fact that for force equilibrium in the x' and y' directions the contributions of components τ_{zx} and τ_{zy} acting on the $+z$ face are exactly balanced by those of components τ_{zx} and τ_{zy} acting on the $-z$ face. The stress σ_z does not enter into the force equilibrium equations in the x' and y' directions.

As a consequence of this analysis of the wedge in Fig. 8.28 for the general state of stress in which σ_z, τ_{zx}, and τ_{zy} are not equal to zero, we conclude that the stress transformation equations, Eqs. (8.15), in the xy plane are still valid and that the Mohr's circle representation of the stress transformation equations still holds.

In general, as shown in Fig. 8.28, when τ_{zx} or τ_{zy} is present, there is a shear stress component $\tau_{x'z}$ acting on the x' inclined face. To see this, we return to equilibrium of the wedge in Fig. 8.28, this time to examine force equilibrium in the z direction. We find from summation of forces

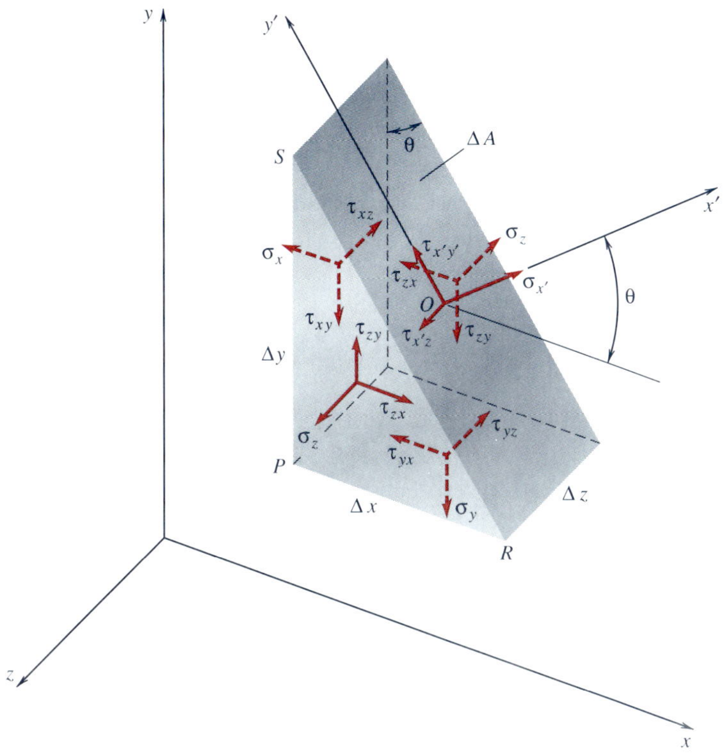

Figure 8.28 Stress components acting on faces of a small wedge cut from a body in a general state of stress.

in the z direction

$$\tau_{x'z}\,\Delta A + \sigma_z \frac{\Delta y\,\Delta x}{2} - \tau_{xz}\,\Delta A \cos\theta - \tau_{yz}\,\Delta A \sin\theta - \sigma_z \frac{\Delta y\,\Delta x}{2} = 0$$

or

$$\tau_{x'z} = \tau_{xz}\cos\theta + \tau_{yz}\sin\theta \tag{8.39}$$

Further, from moment equilibrium for the components of stress at a point (Fig. 8.8) we have

$$\tau_{xz} = \tau_{zx} \qquad \tau_{yz} = \tau_{zy} \tag{8.40}$$

so that Eq. (8.39) can be written in the form

$$\tau_{x'z} = \tau_{zx}\cos\theta + \tau_{zy}\sin\theta \tag{8.41}$$

Therefore with τ_{zx} and τ_{zy} acting on the element there will be no orientation of the face x' of the wedge for which only a normal (principal) stress will be acting on the face of the wedge (in contrast to the situation arising with principal stresses for plane stress in the xy plane when τ_{zx} and τ_{zy} are zero).

If, on the other hand, τ_{zx} and τ_{zy} are both equal to zero, then the stress component $\tau_{x'z}$ acting on the face of the wedge will vanish, Eq. (8.41). Therefore, principal directions and principal stresses in the xy plane can be found.

In Fig. 8.29*a* we show the state of stress of an element relative to a given set of axes for which only a normal stress component σ_z is acting in the z direction. We think of this normal stress σ_z as a principal stress and therefore of the z direction as a principal direction. We will use the symbol σ_3 for this principal stress in the z direction. Since the σ_3 stress component will not affect stress components on faces of a wedge element whose normal is in the xy plane (Fig. 8.28), we can determine the two principal directions 1 and 2 in the xy plane as shown by angle θ_p in Fig. 8.29*b*.

For this special case we now set $\sigma_z = \sigma_3 = 0$; that is, we return to a consideration of plane stress or to what is equivalent to a case when the principal stress in the z direction vanishes in order to investigate stress components referred to axes that do not lie in the xy plane. The element of Fig. 8.29 is redrawn in Fig. 8.30*a*. The stresses σ_1 and σ_2 are the principal stresses acting in the principal directions 1 and 2, and we assume for convenience that $\sigma_1 > \sigma_2$ ($\sigma_3 = 0$). Direction 3 is perpendicular to the 1-2 plane (the xy plane) and thus is parallel to the z axis.

Mohr's circle representation for stress in the xy plane can be drawn immediately from the knowledge of the principal stresses σ_1 and σ_2, as shown in Fig. 8.30*b*.

In addition to the 1-2 plane we could consider stress components in the 1-3 plane. If we "look down" the 2 axis, we have the element shown in Fig. 8.30*c*. The normal stress σ_2 in the 2 direction does not affect the stress components in the 1-3 plane so that the Mohr's circle representation for the 1-3 plane can be drawn from a knowledge this time of the principal stresses σ_1 and σ_3 ($\sigma_3 = 0$), as shown in Fig. 8.30*c*. A similar argument

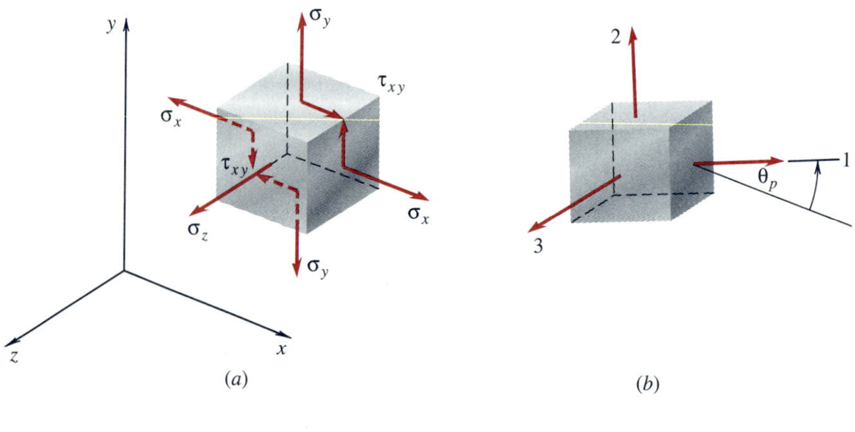

Figure 8.29 (*a*) State of stress with only a normal stress σ_z acting in the z direction. (*b*) Orientation of element in the principal directions 1 and 2.

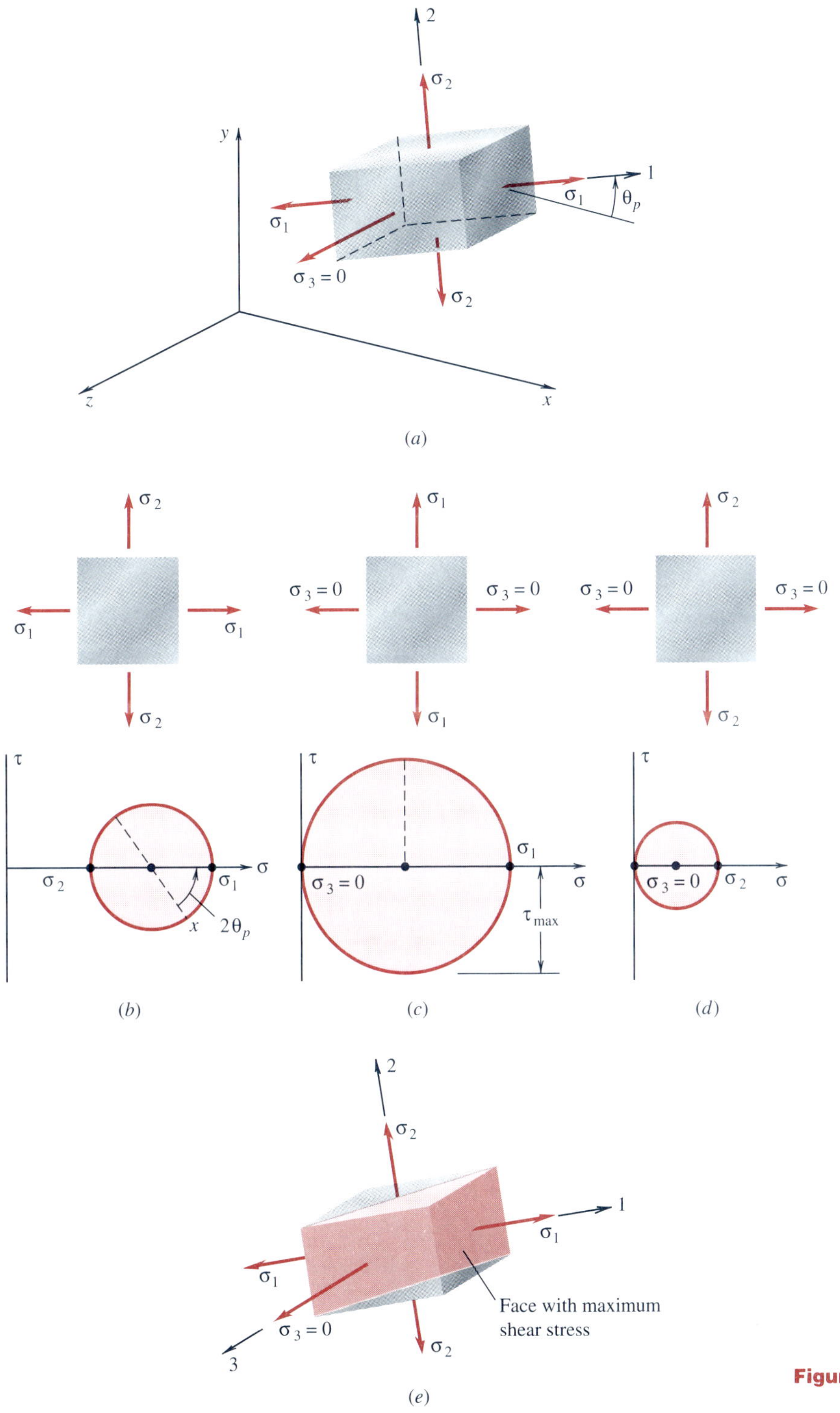

Figure 8.30 Plane stress in the *xy* plane.

Figure 8.31 Plane stress in the *xy* plane.

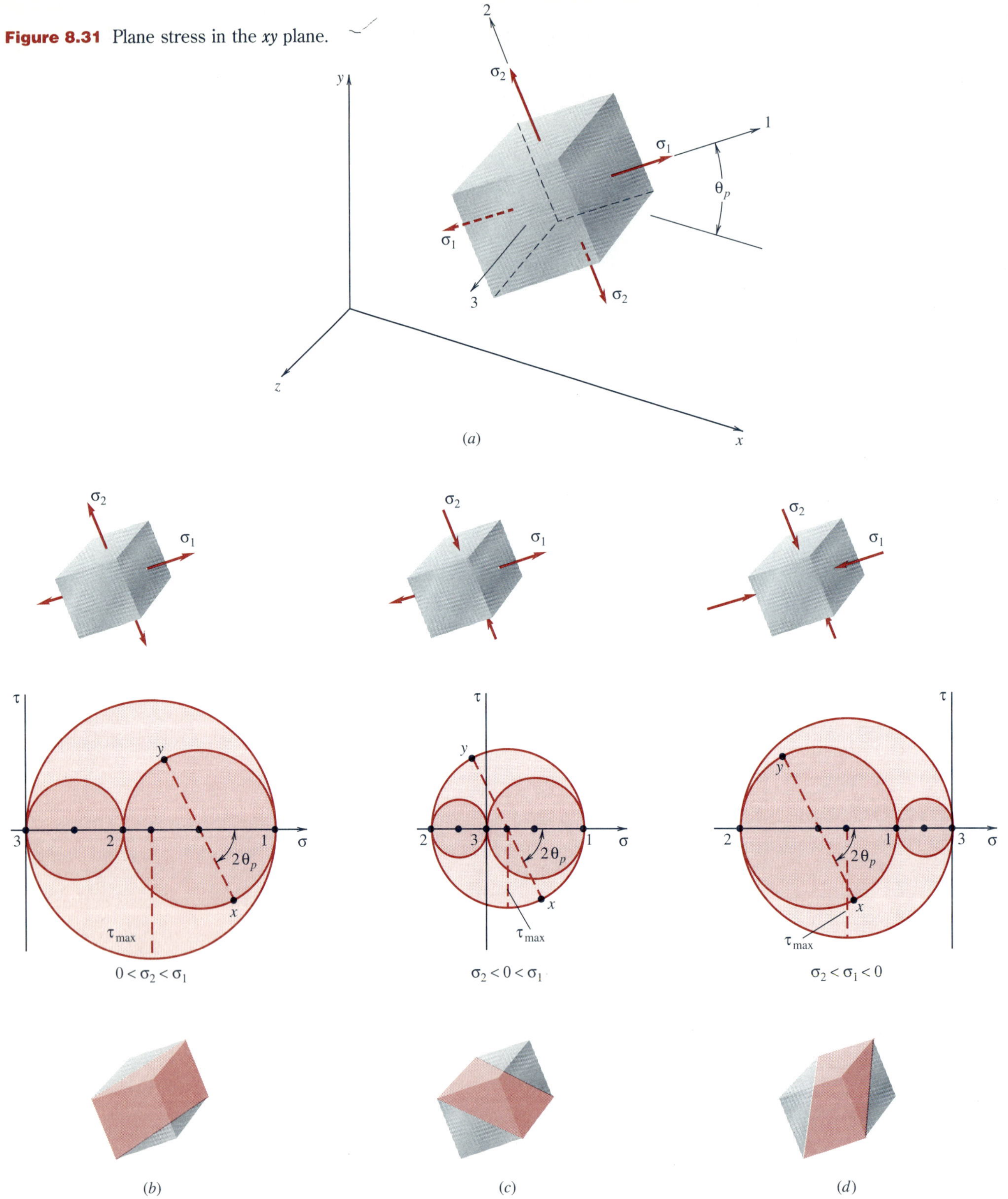

for stress components in the 2-3 plane gives the Mohr's circle representation in Fig. 8.30*d* by using the principal stresses σ_2 and σ_3 ($\sigma_3 = 0$).

We observe from the Mohr's circle representations in Fig. 8.30 the important result that the maximum shear stress $\tau_{\max}$ occurs in the 1-3 plane on a face whose normal is at an angle of $\pm 45°$ to the 1-3 axes; i.e., the maximum shear stress occurs on a face whose normal is inclined at $\pm 45°$ to the *xy plane* of stress, as shown in Fig. 8.30*e*.

When we wish to determine the magnitude of the maximum shear stress at a point experiencing plane stress, we see that we must investigate also the maximum shear stresses associated with faces whose normals lie in planes perpendicular to the plane of stress. The face on which the maximum shear stress will occur will depend on the relative magnitudes and signs of σ_1 σ_2, and σ_3 ($\sigma_3 = 0$). A summary of three possible cases is shown in Fig. 8.31. The three Mohr's circles of Fig. 8.30 are now combined as shown in Fig. 8.31*b*. The faces on which the maximum shear stresses occur are also shown in Fig. 8.31. The value of the maximum shear stress, as we see in Fig. 8.31*b*, *c*, and *d*, is determined by the radius of the largest circle in the Mohr's circle representation, i.e.,

$$\tau_{\max} = \frac{\sigma_{\max} - \sigma_{\min}}{2} \tag{8.42}$$

where $\sigma_{\max}$ is the largest principal stress and $\sigma_{\min}$ is the smallest principal stress. We must exercise care with the signs of the principal stresses. For example, in Fig. 8.31*c* if $\sigma_1 = 10$ MPa and $\sigma_2 = -6$ MPa, then $\tau_{\max} = 8$ MPa.

We now assume that σ_z is not zero (still with τ_{zx} and τ_{zy} equal to zero), Fig. 8.29, and recall that the presence of this normal stress will not affect stresses in the *xy* plane. In this situation we called the σ_z stress a principal stress σ_3 and the *z* direction a principal direction. Therefore we have *three* principal stresses σ_1, σ_2, and $\sigma_3 = \sigma_z$ and *three* principal directions θ_p, $\theta_p + \pi/2$, and the *z* direction.

On the basis of this observation, we anticipate that for a *general* state of stress in which all three stress components σ_z, τ_{xz}, and τ_{yz} act in the *z* direction, there will be *three* mutually perpendicular directions in space for which the stress components acting on the faces of an element so oriented will be normal stresses only. These normal stresses are the principal stresses, and the orientation of the element is given by the principal directions. That is, if the six stress components associated with a set of axes are specified for the state of stress at a point, it is possible to obtain expressions for the values of the principal stresses for the three mutually perpendicular principal directions, as shown in Fig. 8.32*a*. The 1, 2, 3 axes in Fig. 8.32*a* are the principal directions, and the stresses σ_1, σ_2, and σ_3 in those directions are the principal stresses. The Mohr's circle representation relative now to the principal directions is shown in Fig. 8.32*b*. The difference in the Mohr's circle representation from Fig. 8.31 is that now σ_3 is not zero.

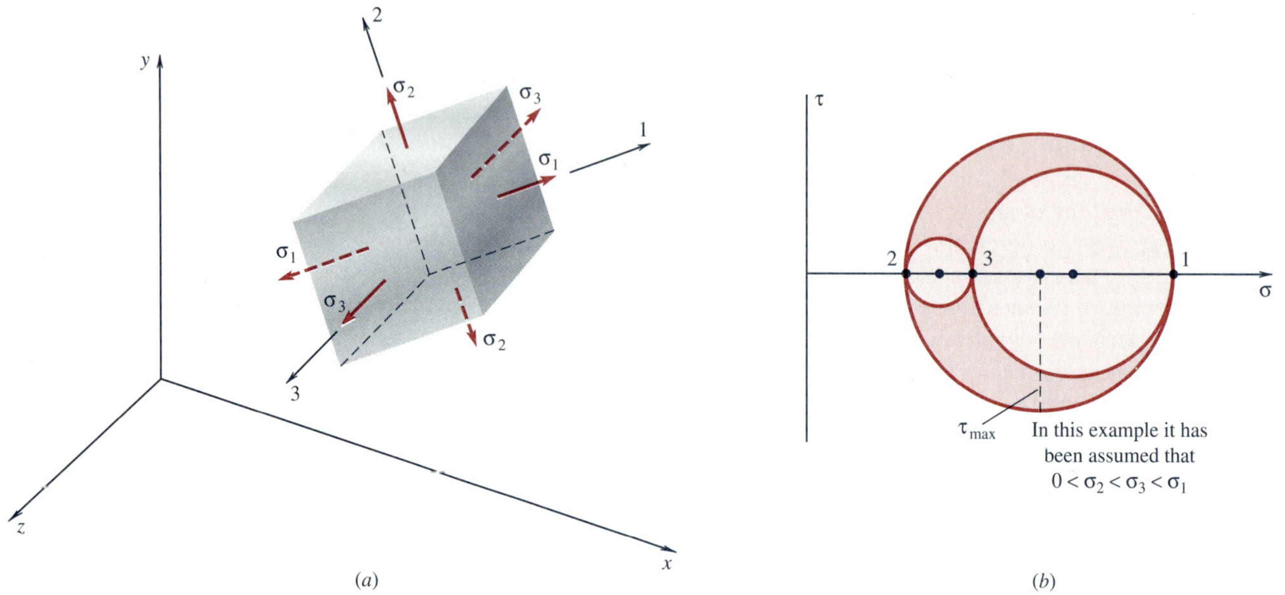

Figure 8.32 (*a*) Three-dimensional state of stress. (*b*) Mohr's circle representation.

It can be shown[2] that the values of the normal and shear stress components on arbitrary faces at a point can be found once the three principal stresses are known. Further, it turns out that the value of the maximum shear stress is again given in Eq. (8.42). Therefore, the value of the maximum shear stress can be determined from a Mohr's circle construction as in Fig. 8.32*b* where in that case $\sigma_{max} = \sigma_1$ and $\sigma_{min} = \sigma_2$ to give, from Eq. (8.42), $\tau_{max} = (\sigma_1 - \sigma_2)/2$.

In many practical situations we shall be dealing with either a case of plane stress or a case in which the direction of one of the principal stress axes is known, e.g., from symmetry. In these cases we will be able to determine the maximum and minimum values of the principal stresses and as a consequence the maximum shear stress.

A review of some simple stress states in the context of the discussion of the general state of stress at a point follows.

---

[2]See S. Crandall, N. Dahl, and T. Lardner (eds.), *An Introduction to the Mechanics of Solids*, 2d ed., McGraw-Hill, New York, 1978, sec. 4.7.

EXAMPLE 8.6

Consider the case of uniaxial tension of a bar, shown in Fig. 8.33*a*. We wish to determine the maximum shear stress in the bar under tension. An element from such a bar is shown in Fig. 8.33*a*.

The stress state is one of plane stress with principal stresses $\sigma_1 = \sigma$ and $\sigma_2 = \sigma_3 = 0$. Mohr's circle construction gives $\tau_{max} = \sigma/2$, as shown in Fig. 8.33*b*. The maximum shear stress acts on a face whose normal makes an angle of 45° to the axis of the bar and lies either in the 1-2 plane or in the 1-3 plane. The plane of the maximum shear stress shown in Fig. 8.33*a* is perpendicular to the 1-2 plane; the normal to this plane lies in the 1-2 plane and is at an angle of 45° to the axis of the bar. In many metals, manifestations of the presence of planes of maximum shear stress occur on the surface of a bar under tension as Lueders' lines. Failure in these materials often takes place when the material reaches a critical value of the shear stress and the material shears apart on the planes of maximum shear stress.

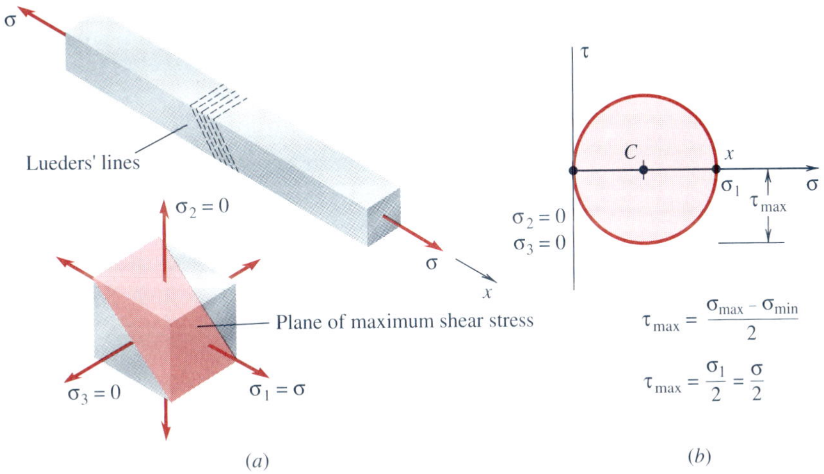

Figure 8.33 Example 8.6

EXAMPLE 8.7

Consider a circular shaft experiencing a twisting moment T at its ends, as shown in Fig. 8.34*a*. The maximum shear stress on an element due to the twisting moment occurs at the outside surface of the shaft, and the element experiences plane stress. The element is experiencing what we call *pure shear*, and the maximum shear stress is τ.

A Mohr's circle representation of the stresses in the xy plane shown in Fig. 8.34*b* indicates that the principal stresses are equal in value in compression and tension and occur on an element at an angle of 45° to the x axis. If, e.g., the material of the shaft is weak in tension, then failure of the material under a large value of the twisting moment may occur by cleavage or separation across the planes of maximum tensile stress. The classical illustration of this type of failure is the twisting of a piece of classroom chalk to give a failure surface, as shown in Fig. 8.34*c*. The cleavage planes are at an angle of approximately 45° to the axis of the chalk where the principal stress is tensile.

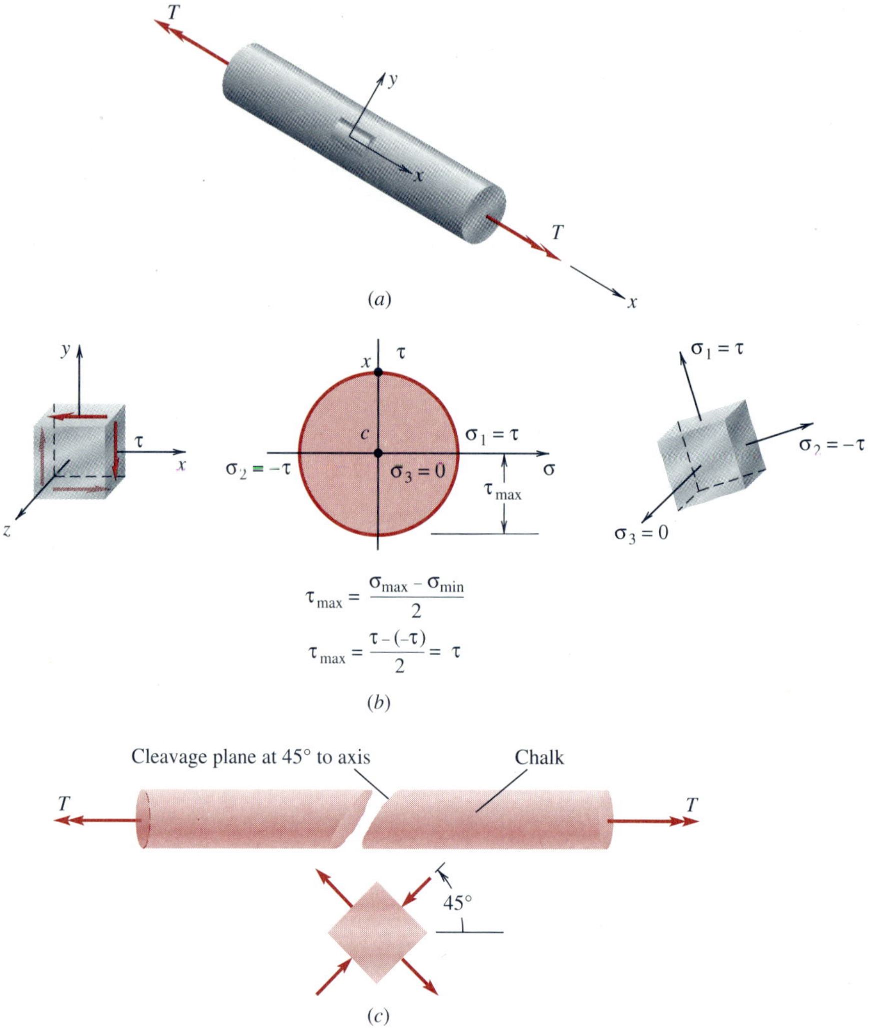

Figure 8.34 Example 8.7

EXAMPLE 8.8

An element at a point in a beam under transverse loading is shown in Fig. 8.35*a*. In general, the element will experience plane stress under a normal stress σ_x and a shear stress τ due to bending. The principal stresses σ_1 and σ_2 are obtained from Mohr's circle construction in Fig. 8.35*b* together with the maximum shear stress τ_{max}. The magnitude and orientation of the principal directions will change with the location of the point selected in the beam. (See Prob. 8.7-12.)

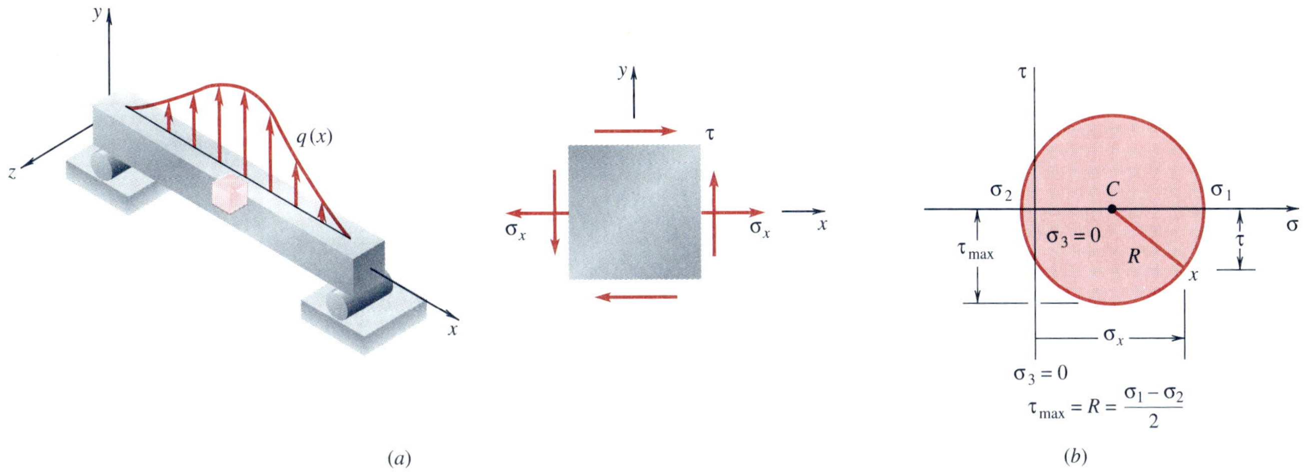

Figure 8.35 Example 8.8

EXAMPLE 8.9

We now wish to consider the case of a biaxial state of plane stress relative to the xy axes, as shown in Fig. 8.36*a*. We will discuss this case in more detail in the next section. In the case of biaxial stress, only normal stresses are acting on the faces of the element; these stresses therefore are the principal stresses in the xy plane. The principal stresses are σ_1 and σ_2, assumed positive with $\sigma_1 > \sigma_2$, and the Mohr's circle representation in Fig. 8.36*b* gives the maximum shear stress as $\sigma_1/2$. The maximum shear stress occurs on a face whose normal is at an angle of 45° to the xy plane, as shown in Fig. 8.36*c*.

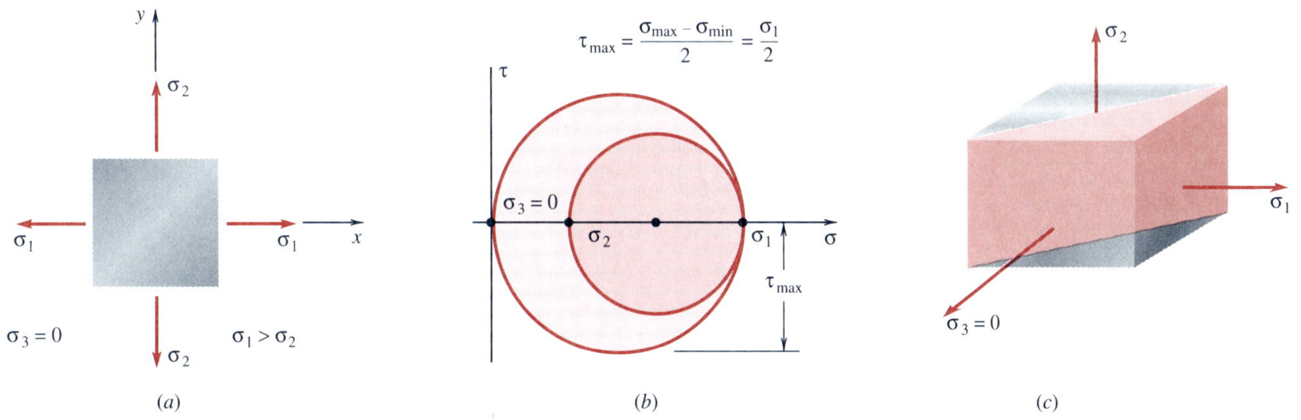

Figure 8.36 Example 8.9

8.8 Stresses in Thin-Walled Pressure Vessels

Pressure vessels are frequently encountered in our technological society. Gas containers, hydraulic cylinders, water tanks, and vacuum chambers are some of the examples that come to mind immediately. The determination of the stresses in pressure vessels is an important step in the design process since failure of a pressure vessel can be often dangerous and sometimes catastrophic. For this reason, numerous design codes exist for pressure and piping applications. In this section we will analyze the stress states in thin-walled pressure vessels; the formulas that we will obtain are the starting point for many design codes. We will find that the state of stress in a pressure vessel is a biaxial state of stress, as discussed in Example 8.9 and Fig. 8.36.

Spherical Pressure Vessels

Consider a thin-walled spherical vessel of inner radius r_i and thickness t with an internal pressure p, as shown in Fig. 8.37*a*. Because of symmetry, the stresses in the wall of the vessel in any two perpendicular directions are equal and can be taken as σ. These stresses, assumed to be constant across the thickness of the vessel, will give rise to forces that balance the forces arising from the internal pressure acting on the inner surface of the vessel. These stresses are sometimes referred to as *membrane*

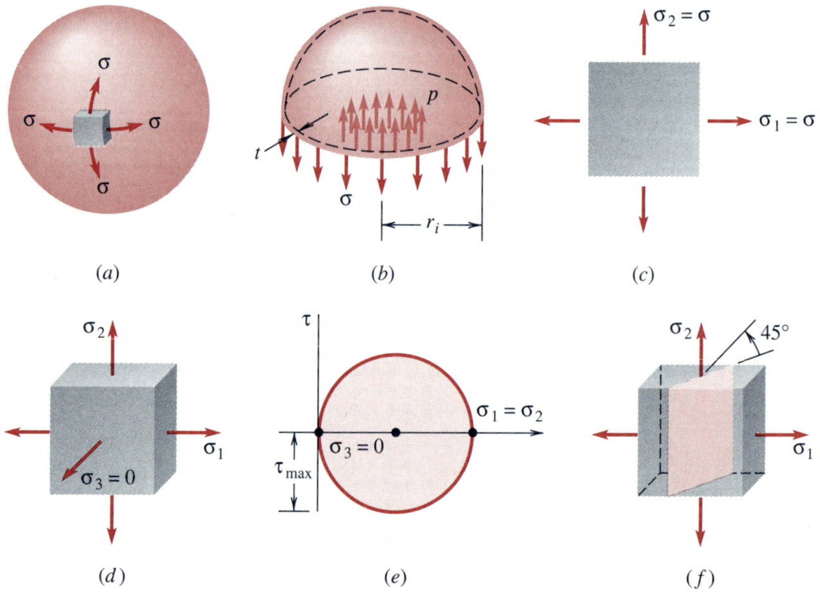

Figure 8.37 State of stress in a thin-walled spherical pressure vessel.

stresses, and we can visualize these stresses as the stresses in the wall of an inflated toy balloon.

To find the value of σ, we take a section of the sphere as shown in Fig. 8.37*b* and consider force equilibrium of the hemispherical portion, which includes the vessel and its contents, as shown. The upward resultant of the distributed pressure forces is the pressure times the total area of the exposed surface in the section, i.e.,

$$p\pi r_i^2 \tag{8.43}$$

where r_i is the inside radius. This upward force due to the pressure is balanced by the downward force arising from the constant stresses σ acting across the thickness of the sphere on the area of the exposed wall

$$\sigma \cdot 2\pi r_i t \tag{8.44}$$

where we have approximated the area of the cross section of the wall by the circumference times the thickness, $2\pi r_i t$, where r_i is the inside radius and t is the thickness of the wall. Since the thickness t is small compared to r_i, this is a good approximation. Equating Eqs. (8.43) and (8.44), we find that

$$\sigma = \frac{pr_i}{2t} \tag{8.45}$$

In view of spherical symmetry, the stresses at any point will be the same in perpendicular directions. Therefore, the state of stress in the sphere is biaxial, as shown in Fig. 8.37*c*, where $\sigma_1 = \sigma_2 = \sigma$ when we neglect the radial stress across the thickness of the wall as compared to the membrane stresses. The radial stress equals p at the inside surface and equals zero at the outside surface and so is of order of magnitude p. However, from Eq. (8.45) we see that the membrane stresses are r_i/t times p, where r_i/t is much larger than 1, and therefore the membrane stresses are much larger than the radial stress.

Therefore, the biaxial state of stress at any point in the wall is given by the three principal stresses

$$\sigma_1 = \frac{pr_i}{2t} \qquad \sigma_2 = \frac{pr_i}{2t} \qquad \sigma_3 \approx 0 \tag{8.46}$$

as shown in Fig. 8.37*d*. The Mohr's circle representation for this state of stress is shown in Fig. 8.37*e*. The maximum shear stress in the wall is then given by, Eq. (8.42)

$$\tau_{\max} = \frac{\sigma_1}{2} = \frac{pr_i}{4t} \tag{8.47}$$

and occurs on a plane at 45° to the plane of the σ_1, σ_2 stress (Fig. 8.37*f*); see also Example 8.9, Fig. 8.36*c*. As we noted, this analysis provides a basis for designing pressurized spherical containers; for thin-walled vessels the inside radius r_i can be replaced by the nominal radius of the vessel.

EXAMPLE 8.10

A spherical pressure vessel shown in Fig. 8.38 is to be designed to carry an industrial gas at an internal pressure of 1.5 MPa. If the material to be used has a yield stress of $\sigma_Y = 250$ MPa, find the allowable value of the thickness t for the pressure vessel if a factor of safety of 3 against yielding is to be used for the design. The inside diameter of the vessel is 10 m.

We assume that the stress state in the vessel is described by the results given in Eqs. (8.46)

$$\sigma_2 = \sigma_1 = \frac{pr}{2t} \tag{a}$$

where r is the inside radius of the vessel.

The allowable stress is found by dividing σ_Y by the factor of safety 3 to get

$$\sigma_a = \frac{\sigma_Y}{3} = \frac{250}{3} = 83.3 \text{ MPa} \tag{b}$$

To find the allowable value of the thickness t_a, we combine Eqs. (*a*) and (*b*)

$$t_a = \frac{pr}{\sigma_a 2}$$

Upon using $r = 5$ m and $p = 1.5$ MPa, we have

$$t_a = \frac{(1.5)(5)}{(250/3)(2)} = 0.045 \text{ m} = 45 \text{ mm} \tag{c}$$

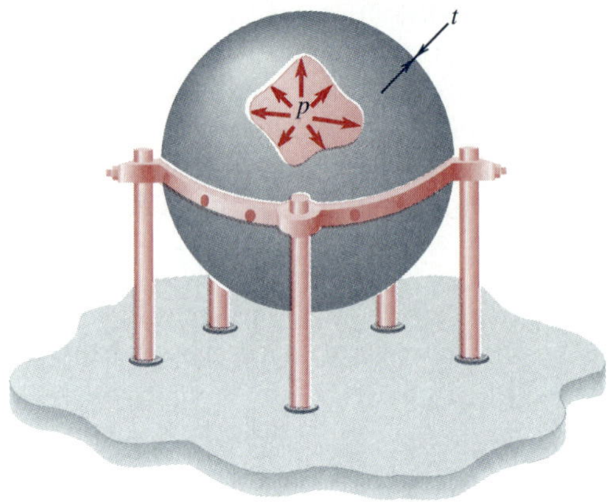

Figure 8.38 Example 8.10

The membrane stress in the pressure vessel when $t = t_a$ is, of course, 83.3 MPa; the vessel is thin-walled with

$$\frac{r}{t} = \frac{5}{45 \times 10^{-3}} = 111 \tag{d}$$

Cylindrical Pressure Vessels

For the case of a spherical pressure vessel, we were able to determine the wall stresses by examining force equilibrium of a hemispherical segment formed by passing a section through the center of the sphere. By symmetry all possible sections give rise to the same equilibrium conditions, and therefore the wall stresses are the same in all directions on the sphere.

In the case of a cylindrical vessel (see Fig. 8.39*a*), sections perpendicular to the axis of the cylinder will cut cylindrical segments from the vessel while sections containing the axis will cut the vessel into two identical half-cylinders. Thus we expect to find different normal stresses acting on the areas exposed by these sections because the geometries of the free-body segments are different. We assume that the normal stresses in the wall are constant across the thickness for thin-walled cylinders.

Suppose we pass a section perpendicular to the axis of the vessel and the contents shown in Fig. 8.39*a* to obtain the segment shown in

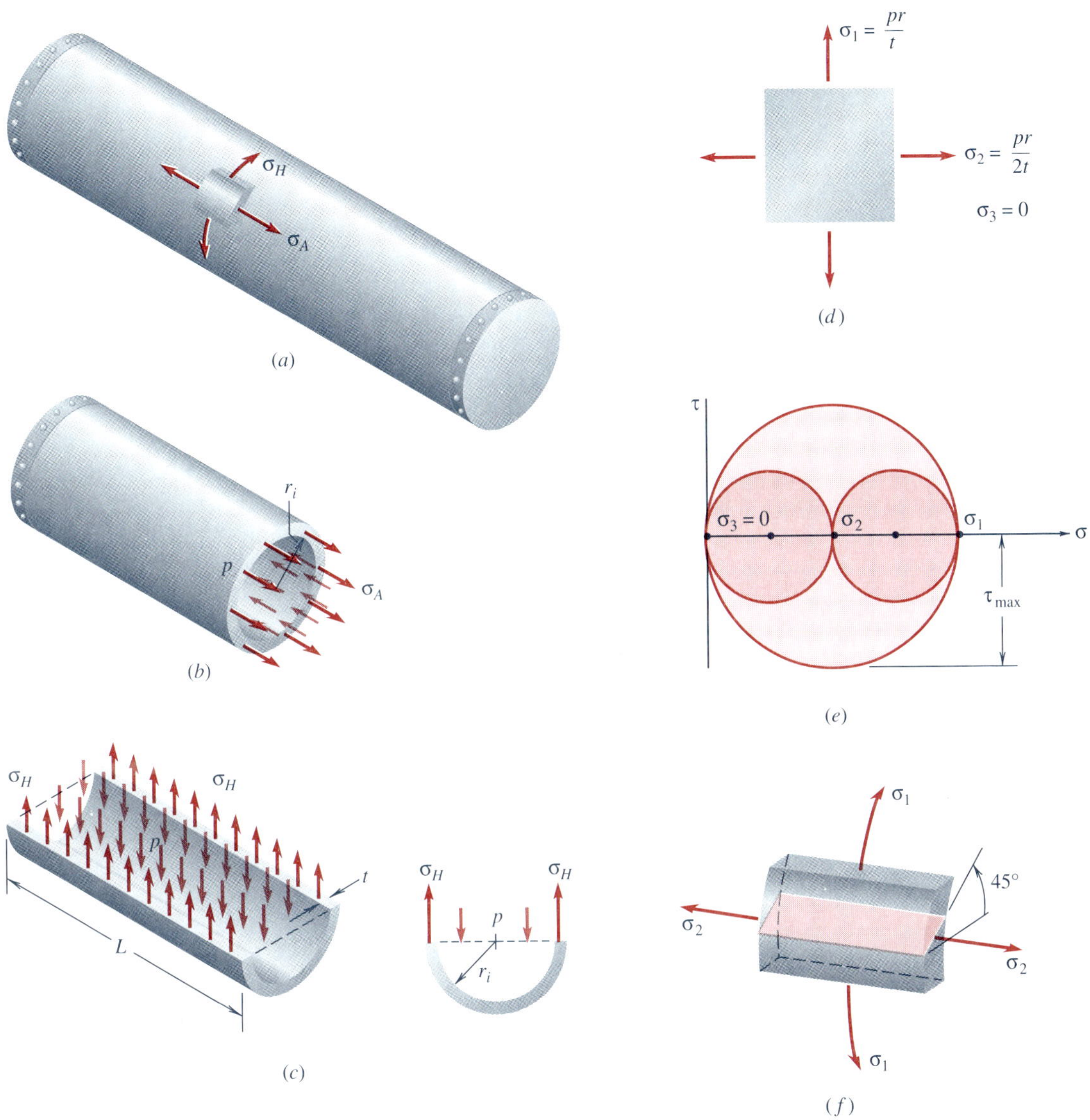

Figure 8.39 State of stress in a thin-walled cylindrical pressure vessel.

Fig. 8.39*b*. (The section should not be taken too near to the end of the vessel where the end cap gives rise to a more complex stress state than is accounted for in the present analysis.) The wall stress in the *axial* direction σ_A is the normal stress acting over the exposed wall thickness. This axial stress is the stress that arises to balance the pressure acting on the contents (Fig. 8.39*b*). By symmetry σ_A does not vary with position around the circumference of the vessel. The total axial force resultant F_A due to the wall stress in Fig. 8.39*b* is found by multiplying σ_A by the wall area over which it acts. That is,

$$F_A = \sigma_A \pi (r_o^2 - r_i^2) \tag{8.48}$$

where r_o and r_i are the outside and inside radii, respectively, of the pressure vessel. For thin-walled vessels we have

$$\begin{aligned} r_o^2 - r_i^2 &= (r_o + r_i)(r_o - r_i) \\ &\approx 2rt \end{aligned}$$

where r is the nominal radius of the vessel and t is the wall thickness. Therefore Eq. (8.48) can be written in the form

$$F_A = \sigma_A(2\pi rt) \tag{8.49}$$

The axial force F_A balances the resultant force of the pressure acting on the exposed contents (Fig. 8.39*b*). This force resultant due to the pressure is given by

$$F_p = p\pi r_i^2 \tag{8.50}$$

Equilibrium requires that

$$F_A = F_p$$

or

$$\sigma_A(2\pi rt) = p\pi r_i^2$$

Therefore, the axial stress is given by

$$\sigma_A = \frac{pr_i^2}{2tr} \tag{8.51}$$

For thin-walled vessels, the inside radius r_i can be approximated by the nominal radius r to obtain

$$\sigma_A = \frac{pr}{2t} \tag{8.52}$$

The axial stress σ_A is sometimes referred to as the *longitudinal stress* in the vessel.

The wall stress acting in the circumferential direction of the cylinder is often called the *hoop stress* σ_H and is similar to the membrane stress in a spherical pressure vessel. We will derive an expression for σ_H based upon the equilibrium of a free-body segment of the vessel obtained by passing a section through the axis and so dividing the vessel into two half-vessels. A segment of length L of a half-vessel is shown in Fig. 8.39*c*; we do not show the stresses or the pressures acting on the ends. We take σ_H as the normal stress over the exposed wall thickness, and we assume that σ_H does not vary with axial position along the vessel except for a region close to the ends, where the influence of the end caps would need to be considered. For the segment shown in Fig. 8.39*c*, the upward force resultant of the σ_H stresses acting on both wall edges is given by

$$F_H = 2\sigma_H L t \tag{8.53}$$

The resultant force $\hat{F}_p$ of the downward pressure in Fig. 8.39*c* is given by the pressure acting on the exposed area $2r_iL$

$$\hat{F}_p = 2r_i L p \tag{8.54}$$

Equilibrium of the vessel segment in Fig. 8.39*c* requires that

$$F_H = \hat{F}_p$$

or

$$2\sigma_H L t = 2r_i L p \tag{8.55}$$

Therefore, the hoop stress in the cylindrical vessel is given by

$$\sigma_H = p\frac{r_i}{t} \tag{8.56}$$

For thin-walled vessels we can replace r_i by the nominal radius r to get

$$\sigma_H = \frac{pr}{t} \tag{8.57}$$

We see upon comparing the results in Eqs. (8.52) and (8.57) that the value of the hoop stress in a cylindrical vessel is twice the value of the axial stress. The hoop stress is sometimes referred to as the *circumferential stress* in the vessel.

The argument we gave for the case of a spherical vessel, leading to the conclusion that the stress in the radial direction across the thickness is 1 to 2 orders of magnitude smaller than the wall stresses in the tangential directions, also is valid for the cylindrical vessel. Thus the stress state for a cylindrical pressure vessel is a biaxial state of stress with the principal stresses

$$\sigma_1 = \sigma_H = \frac{pr}{t}$$

$$\sigma_2 = \sigma_A = \frac{pr}{2t} \tag{8.58}$$

$$\sigma_3 \approx 0$$

as shown in Fig. 8.39*d*.

Mohr's circle for each of the three pairs of principal stresses are shown in Fig. 8.39*e*. The largest shear stress is found on a plane at 45° to the 1-2 plane, as shown in Fig. 8.39*f*, and is given by

$$\tau_{\max} = \frac{pr}{2t} \tag{8.59}$$

We again note that the maximum shear stress in the material occurs in the plane determined by the largest and smallest values of the principal stresses and is given by Eq. (8.42).

The results of this stress analysis for pressurized cylindrical vessels will be applied in the two examples that follow.

EXAMPLE 8.11

A small pressure vessel is fabricated from a segment of a spirally welded pipe with a spriral angle of 50° by welding rigid end plates to each end of the segment, as shown in Fig. 8.40*a*. The inside diameter of the pipe is 250 mm, the wall thickness is 6 mm, and the internal gas pressure is 2 MPa. We wish to find the normal stress perpendicular to the spiral weld and the shear stress parallel to the spiral weld.

We assume that the welded pipe can be treated as a homogeneous cylindrical pressure vessel, and the values of the axial and hoop stresses in the wall of the pipe follow from Eqs. (8.58). Once we find these stresses, we can use Mohr's circle or the stress transformation formulas to find the components of stress in directions perpendicular and parallel to the weld.

For cylindrical pressure vessels, the axial stress away from the ends is given by Eq. (8.52) in the form

$$\begin{aligned}\sigma_A &= \frac{pr}{2t}\\ &= \frac{2(250/2)}{2(6)} = 20.8 \text{ MPa}\end{aligned} \tag{a}$$

The hoop stress given by Eq. (8.57) is twice the axial stress, or

$$\sigma_H = 41.6 \text{ MPa} \tag{b}$$

and these stresses are shown acting on a wall element in Fig. 8.40*b*.

Using these values for σ_A and σ_H, we can sketch Mohr's circle, as shown in Fig. 8.40*c*. In Fig. 8.40*d* we show the stress components acting on an element containing the weld line. To use the Mohr's circle representation for the stress transformation, we need to determine the angle of rotation of the normal to the weld line with respect to the positive axial direction; we see that the angle is −40°. The location of point *a* on Mohr's circle corresponding to the normal to the weld line is located by turning through an angle of 80° clockwise from point *A* corresponding to the axial direction, as shown in Fig. 8.40*c*. The values of the stresses of point *a* on the circle (Fig. 8.40*c*) are

$$\begin{aligned}\sigma_a &= \sigma_{av} - R\cos 80^\circ\\ \tau_a &= -R\sin 80^\circ\end{aligned} \tag{c}$$

where

$$\sigma_{av} = \frac{\sigma_A + \sigma_H}{2} = 31.2 \text{ MPa}$$

$$R = \frac{\sigma_H - \sigma_A}{2} = 10.4 \text{ MPa}$$

We obtain therefore (Fig. 8.40*e*)

$$\begin{aligned}\sigma_a &= 29.4 \text{ MPa}\\ \tau_a &= -10.2 \text{ MPa}\end{aligned} \tag{d}$$

For the integrity of the pressure vessel the weld must be able to carry these stresses.

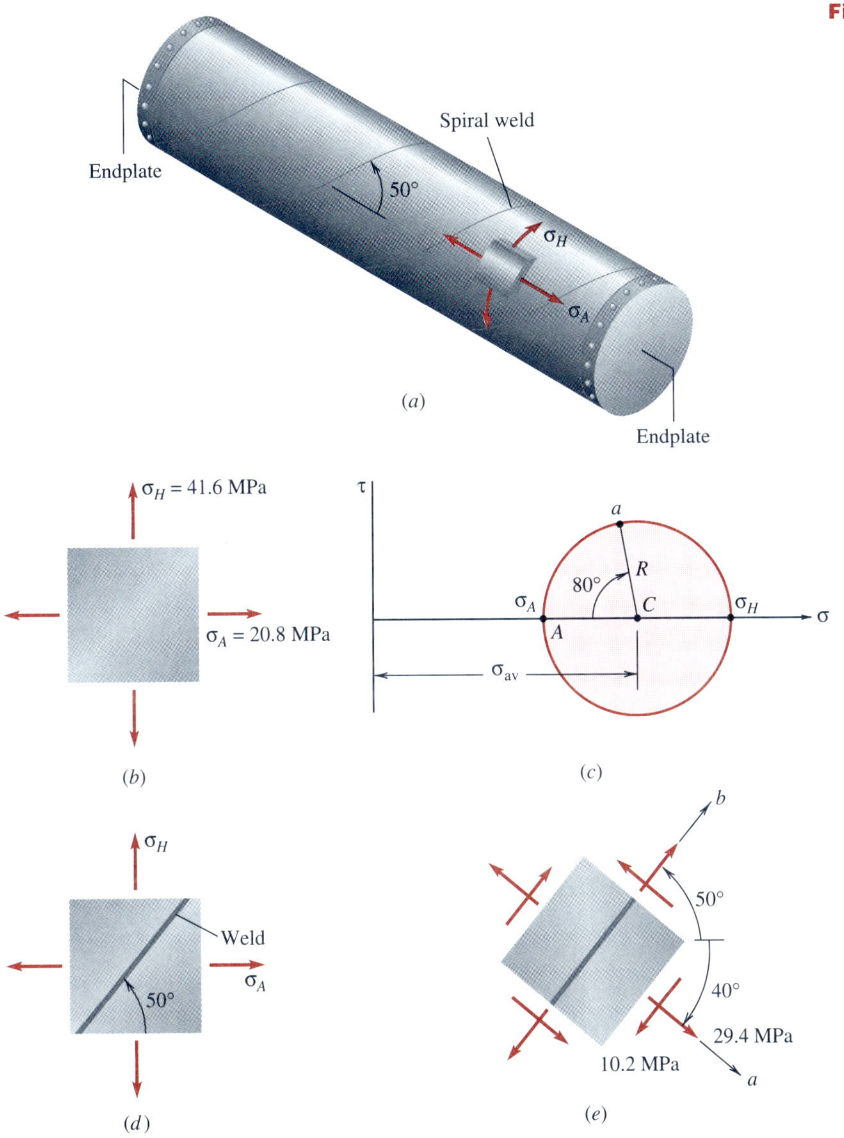

Figure 8.40 Example 8.11

EXAMPLE 8.12

A cylindrical oil-fired pressurized boiler has an outside diameter of 10 ft and a wall thickness of 2 in. If the yield stress of the wall material is equal to 36 ksi, we wish to find (*a*) the maximum tensile stress due to an internal pressure of 290 psi and (*b*) the maximum allowable internal pressure if a factor of safety of 3 against yielding of the material in tension is to be used.

The wall stresses in the boiler can be found from Eqs. (8.52) and (8.57) in the form

$$\sigma_A = \frac{pr}{2t} = \frac{290(60)}{2(2)} = 4350 \text{ psi} \qquad (a)$$

$$\sigma_H = 2\sigma_A = 8700 \text{ psi} \qquad (b)$$

We find therefore that the maximum tensile stress in the wall is the hoop stress of 8700 psi.

To find the maximum allowable internal pressure p_a, we first calculate the allowable stress in tension with the factor of safety

$$\sigma_a = \frac{\sigma_Y}{3} = \frac{36}{3} = 12 \text{ ksi}$$

The maximum allowable pressure p_a follows from

$$\sigma_a = \sigma_H = \frac{p_a r}{t}$$

or

$$p_a = \frac{t}{r}\sigma_a = \frac{2}{60}(12{,}000) = 400 \text{ psi} \qquad (c)$$

Thus we find that the boiler operating pressure of 290 psi is 72.5 percent of the allowable pressure based upon the specified value for the yield stress and a safety factor of 3.

PART B: STRAIN

8.9 Deformation

In Part A we discussed internal force intensities or stresses that arise in a body when it is loaded. These force intensities or stresses occur when the body is deformed or changes shape under load. At a sufficiently fine scale of observation, we find that the molecules in the material of the body under load are displaced from their initial equilibrium positions and as a consequence change their interactive forces with one another. The changes in the interactive forces at a higher scale of observation, or at what we call the continuum level, give rise to the stresses in the body. In this part of the chapter we wish to study the displacements and deformations at points in a body under load.

We introduced the idea of the displacement at a point in Chap. 2 for a uniaxial member under load, and we found that a member under end loads would undergo a deformation or normal strain if there were a relative displacement of the two ends of the member; see Eq. (2.5) and Fig. 2.5. If the displacements at each end of the member were equal, we would conclude that the member either did not move or moved as a rigid body along its axis. The important point here is that relative displacements between points along the axis of the member give rise to deformations. In a similar way, when we discussed twisting of circular shafts, we found that changes in the angle of twist at sections along the shaft gave rise to shear strain deformations or changes in angles between lines that were originally at a right angle to one another (Fig. 3.3). Here, again, it is the difference in the value of the angular displacement at nearby points in the body under load that gives rise to deformation.

If a body translates as a rigid body, e.g., it is picked up and moved a fixed distance without its orientation being changed, then all points in the body experience the same displacement and the body does not undergo deformation; we are neglecting, of course, any changes in gravitational forces acting on the body. In the same way if a body undergoes a rigid body rotation, i.e., it rotates about an axis, then the body will not experience any change in shape or deformation. When a rigid-body motion occurs, all points in the body maintain a fixed distance from one another. Therefore, if deformation of the body is to occur, the displacement at each point in the body should, in general, be different and lead to changes in the distances between points of the body.

To describe deformation in two- and three-dimensional bodies, we need to specify the value of the displacement at each point in the body. Actually, when we solve engineering problems involving deformable bodies, often we are interested in finding the values of the displacement at a number of points in the body. However, because the stresses in the body arise from deformations, we need to consider the stresses and the deformations before we can find the displacements.

8.10 Plane Strain

We now introduce the notion of *plane strain* for a body in which the displacements of all points in the body lie in or are parallel to a single plane, which we take as the xy plane, and depend only on the location of the point in the plane. Figure 8.41 shows a section of the body in the xy plane and a point P whose coordinate location is (x,y). When loads are applied, the body deforms and point P displaces to a new position P'. The *displacement vector* $\mathbf{u}(x,y)$ (see Fig. 8.41) describes the movement of point P from its initial position in the undeformed body to its final position P' in the deformed body. The value of the displacement vector will be different at different points in the body, and we show this by indicating that $\mathbf{u}$ is a function of x and y, that is, $\mathbf{u}(x,y)$. We will find that the deformation of the body can be described from a knowledge of the value of the displacement vector at each point.

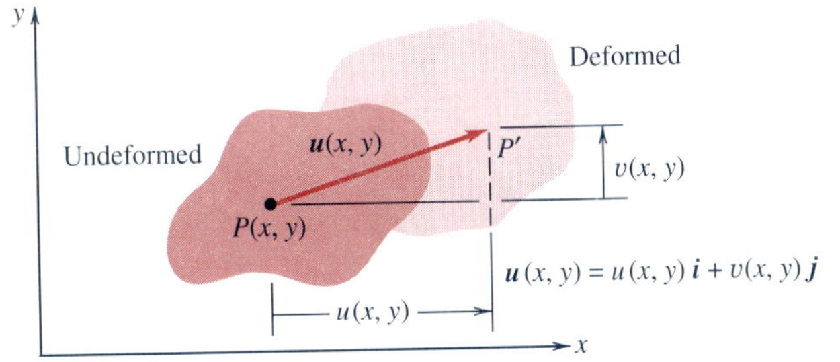

Figure 8.41 Displacement vector $\mathbf{u}(x, y)$ for plane strain.

The components of the displacement vector $\mathbf{u}(x,y)$ along the coordinate axes are defined by

$$\mathbf{u}(x,y) = u(x,y)\mathbf{i} + v(x,y)\mathbf{j} \tag{8.60}$$

where $u(x,y)$ and $v(x,y)$ are the scalar components of the displacement vector along the x and y axes at the point (x,y). In Fig. 8.41, point P moves a distance $u(x,y)$ parallel to the x axis and a distance $v(x,y)$ parallel to the y axis. Once the components are known, the displacement vector is known from Eq. (8.60). We often refer to the collection of displacement vectors in the body as the *vector displacement field* of the body. Since the displacement vectors at each point are different, the body will deform, and we can expect line elements in the body to change their length and angles between intersecting line elements to change. We discuss this deformation and its relation to the displacement field in the next section.

8.11 Relation between Strain and Displacement in Plane Strain

We wish to determine an expression for the normal strain of a line element oriented at an arbitrary angle ϕ to the x axis at a point P in a body that has undergone plane strain. In Fig. 8.42 we show a small line element PQ of length Δs connecting point P to a nearby point Q.

Point P is located at point (x,y), and the nearby point Q is located at point $(x + \Delta x, y + \Delta y)$. From the geometry of Fig. 8.42 we have

$$\cos\phi = \frac{\Delta x}{\Delta s} \qquad \sin\phi = \frac{\Delta y}{\Delta s} \tag{8.61}$$

Angle ϕ can take on any value $0 \le \phi \le 2\pi$.

If every point of the body experiences a nonrigid body displacement given by the vector displacement field $\mathbf{u}(x,y)$, then points P and Q will move to new positions P' and Q', respectively, and the length of PQ will

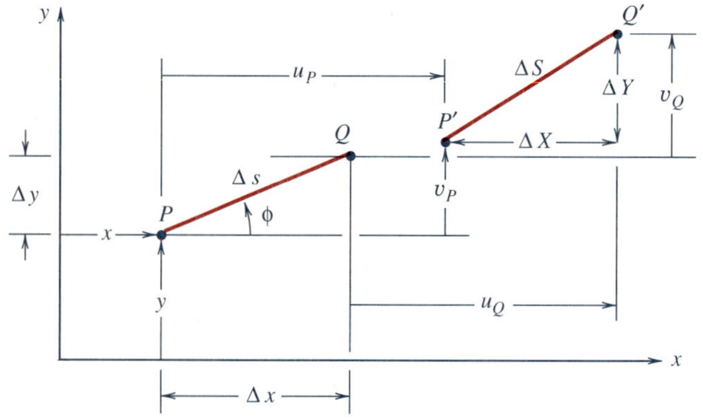

Figure 8.42 Element of length Δs undergoing strain.

change to a new length $P'Q'$ equal to ΔS (Fig. 8.42). The normal strain or change in length per unit length ϵ of the line element PQ is given by the expression

$$\epsilon = \frac{P'Q' - PQ}{PQ} = \frac{\Delta S - \Delta s}{\Delta s} = \frac{\Delta S}{\Delta s} - 1 \tag{8.62}$$

or

$$\Delta S = \Delta s(1 + \epsilon) \tag{8.63}$$

In our analysis we will assume that the normal strain ϵ is small compared to 1 that is, $\epsilon \ll 1$; we often refer to the results we obtain as corresponding to a *small strain theory*. To find the strain, we need to find the value of the length ΔS.

From Fig. 8.42 we have

$$\begin{aligned} (\Delta S)^2 &= (\Delta X)^2 + (\Delta Y)^2 \\ (\Delta s)^2 &= (\Delta x)^2 + (\Delta y)^2 \end{aligned} \tag{8.64}$$

The components of the displacement parallel to the coordinate axes for points P and Q are as shown in Fig. 8.42 so that

$$\begin{aligned} \Delta X &= \Delta x + u_Q - u_P \\ \Delta Y &= \Delta y + v_Q - v_P \end{aligned} \tag{8.65}$$

where

$$\begin{aligned} u_P &= u(x,y) & v_P &= v(x,y) \\ u_Q &= u(x + \Delta x, y + \Delta y) & v_Q &= v(x + \Delta x, y + \Delta y) \end{aligned} \tag{8.66}$$

In the limit as $\Delta x \to 0$ and $\Delta y \to 0$, we can approximate the differences of the displacements in Eqs. (8.65) by a Taylor series approximation, using Eqs. (8.66) to obtain

$$\begin{aligned} \Delta X &= \Delta x + \frac{\partial u}{\partial x}\Delta x + \frac{\partial u}{\partial y}\Delta y = \Delta x\left(1 + \frac{\partial u}{\partial x}\right) + \Delta y\frac{\partial u}{\partial y} \\ \Delta Y &= \Delta y + \frac{\partial v}{\partial x}\Delta x + \frac{\partial v}{\partial y}\Delta y = \Delta x\frac{\partial v}{\partial x} + \Delta y\left(1 + \frac{\partial v}{\partial y}\right) \end{aligned} \tag{8.67}$$

Now to obtain the expression for the strain, we use Eqs. (8.63) and (8.64) to find

$$(\Delta S)^2 = (\Delta s)^2(1 + \epsilon)^2 = (\Delta X)^2 + (\Delta Y)^2 \tag{8.68}$$

Upon substitution from Eqs. (8.67) into Eq. (8.68) and the dropping of the quadratic term in ϵ and the products in which both terms are the derivatives of the displacement components that are assumed to be small, we have

$$(\Delta s)^2(1+\epsilon)^2 \approx (\Delta s)^2(1+2\epsilon) = (\Delta x)^2\left(1+2\frac{\partial u}{\partial x}\right) + 2\,\Delta x\,\Delta y\frac{\partial u}{\partial y} + 2\,\Delta x\,\Delta y\frac{\partial v}{\partial x} + (\Delta y)^2\left(1+2\frac{\partial v}{\partial y}\right) \tag{8.69}$$

Finally, if we use Eq. (8.64)

$$(\Delta s)^2 = (\Delta x)^2 + (\Delta y)^2$$

divide by $(\Delta s)^2$, and use Eqs. (8.61), we find an expression for ϵ in the form

$$\epsilon = \frac{\partial u}{\partial x}\left(\frac{\Delta x}{\Delta s}\right)^2 + \left(\frac{\partial u}{\partial y} + \frac{\partial v}{\partial x}\right)\frac{\Delta x\,\Delta y}{\Delta s\,\Delta s} + \frac{\partial v}{\partial y}\left(\frac{\Delta y}{\Delta s}\right)^2$$

or

$$\epsilon = \frac{\partial u}{\partial x}\cos^2\phi + \left(\frac{\partial u}{\partial y} + \frac{\partial v}{\partial x}\right)\sin\phi\cos\phi + \frac{\partial v}{\partial y}\sin^2\phi \tag{8.70}$$

Equation (8.70) is the expression for the normal strain of line element PQ emanating from point P in the direction ϕ (Fig. 8.42). In general, as angle ϕ changes, the normal strain of line element PQ changes. Equation (8.70) gives the normal strain experienced by an infinitesimal line element connected to point P in terms of the derivatives of the displacement components evaluated at point P and the angle ϕ. The totality of the measures of the normal strain at a point is characterized by the quantities involving the displacement derivatives in Eq. (8.70).

We now define the *state of strain* at a point in terms of these quantities appearing in the expression for ϵ, namely, by the quantities

$$\epsilon_x = \frac{\partial u}{\partial x} \qquad \epsilon_y = \frac{\partial v}{\partial y} \qquad \gamma_{xy} = \frac{\partial u}{\partial y} + \frac{\partial v}{\partial x} \tag{8.71}$$

Equations (8.71) are often referred to as the ***strain components*** or the ***strain-displacement relations.*** Equation (8.70) can then be written in the form

$$\epsilon = \epsilon_x\cos^2\phi + \gamma_{xy}\sin\phi\cos\phi + \epsilon_y\sin^2\phi \tag{8.72}$$

We can interpret geometrically each of the strain components in Eqs. (8.71) by considering special values for ϕ in Eq. (8.72); see Fig. 8.42. In Eq. (8.72), if ϕ is set equal to zero, i.e., line element PQ is oriented parallel to the x axis, then

$$\epsilon = \epsilon_x = \frac{\partial u}{\partial x} \tag{8.73}$$

Therefore ϵ_x is the normal strain of a line element originally parallel to the x axis. We considered this normal strain quantity in Sec. 2.8, Eq.

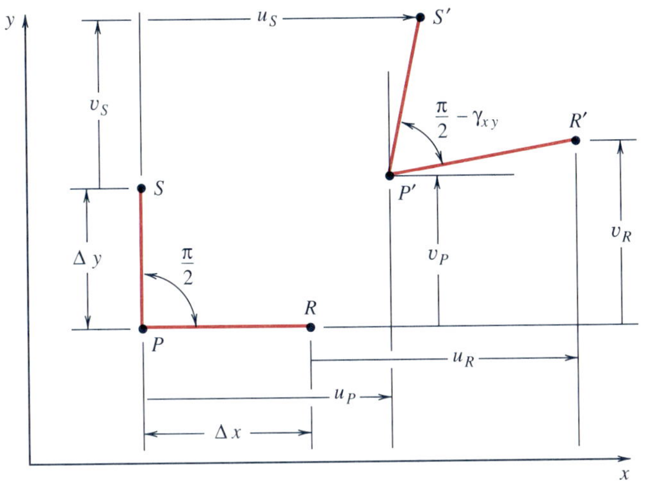

Figure 8.43 Original right angle SPR undergoing shear strain to give angle $S'P'R' = \pi/2 - \gamma_{xy}$.

(2.26). In a similar way, if $\phi = \pi/2$, we see that ϵ_y is the normal strain of a line element originally parallel to the y axis.

The quantity γ_{xy} in Eq. (8.71) is a measure of the shear strain or the change in the right angle of two perpendicular line elements. To see this, we consider a line element PR parallel to the x axis, $\phi = 0$, and a line element PS parallel to the y axis, $\phi = \pi/2$, as shown in Fig. 8.43.

We show the displacement components of points P, R, and S in Fig. 8.43

$$
\begin{aligned}
u_P &= u(x, y) & v_P &= v(x, y) \\
u_R &= u(x + \Delta x, y) & v_R &= v(x + \Delta x, y) \\
u_S &= u(x, y + \Delta y) & v_S &= v(x, y + \Delta y)
\end{aligned}
$$

The vector $P'R'$ can be expressed in the form

$$P'R' = (\Delta x + u_R - u_P)\mathbf{i} + (v_R - v_P)\mathbf{j}$$

which can, in the limit as $\Delta x \to 0$, be written in the form

$$P'R' = \Delta x\left(1 + \frac{\partial u}{\partial x}\right)\mathbf{i} + \Delta x\frac{\partial v}{\partial x}\mathbf{j} \tag{8.74}$$

In a similar way, we have

$$P'S' = \Delta y\frac{\partial u}{\partial y}\mathbf{i} + \Delta y\left(1 + \frac{\partial v}{\partial y}\right)\mathbf{j} \tag{8.75}$$

We exaggerate for clarity the displacements and rotations of the line elements in Fig. 8.43. We now wish to calculate the angle between $P'R'$ and $P'S'$ which we show as $\pi/2 - \gamma_{xy}$ in Fig. 8.43. We recall that the dot product of two unit vectors gives the cosine of the angle between the vectors. Therefore we take the dot product of vectors $P'R'$ and $P'S'$ and divide by the magnitude of each vector

$$\frac{P'R'}{|P'R'|} \cdot \frac{P'S'}{|P'S'|} = \cos\left(\frac{\pi}{2} - \gamma_{xy}\right) = \sin \gamma_{xy} \approx \gamma_{xy} \tag{8.76}$$

where γ_{xy} is assumed small so that $\sin \gamma \approx \gamma$. If we carry out the dot product on the left side of Eq. (8.76), drop products in which both terms are the derivatives of the displacement components, and neglect the normal strains compared to 1 in the product of the expressions

$$|P'R'| = \Delta x\ (1 + \epsilon_x) \qquad |P'S'| = \Delta y\ (1 + \epsilon_y)$$

we find that

$$\gamma_{xy} = \frac{\partial u}{\partial y} + \frac{\partial v}{\partial x} \tag{8.77}$$

Therefore, the strain quantity γ_{xy} is a measure of the change in right angle of two line elements, one parallel to the x axis and the other parallel to the y axis. We call this quantity the ***shear strain*** in the ***xy*** plane.

We show the three quantities ϵ_x, ϵ_y, and γ_{xy} arising from the deformation in Fig. 8.44 where all displacements and rotations of the line elements are highly exaggerated.

We can also visualize the relations for the three strain quantities in terms of the displacement derivatives, Eqs. (8.71), by considering the deformation of a small area element located at point P. In Fig. 8.45 we show an element of area $PRQS$ and the displacements in the x and y directions of each of the P, R, S corner points. The values of the displacements at a point depend on the values of the coordinates of the point. Upon use of Eqs. (8.74) and (8.75), we obtain in addition the components of $P'R'$ and $P'S'$, as shown in Fig. 8.45.

We see in Fig. 8.45 that the area element $PRQS$ is deformed into $P'R'Q'S'$ as a consequence of the normal strains associated with line

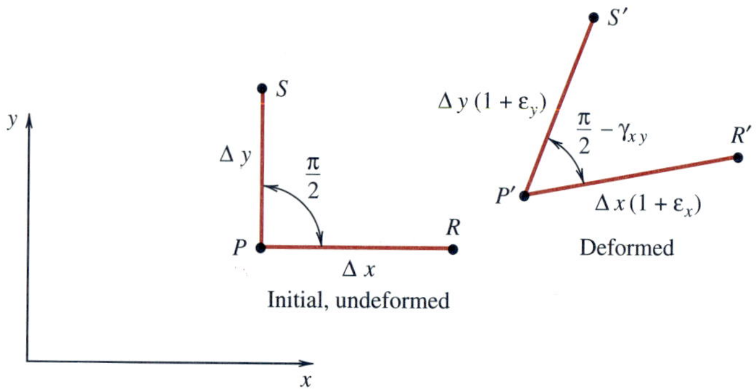

Figure 8.44 Deformation in the neighborhood of point P.

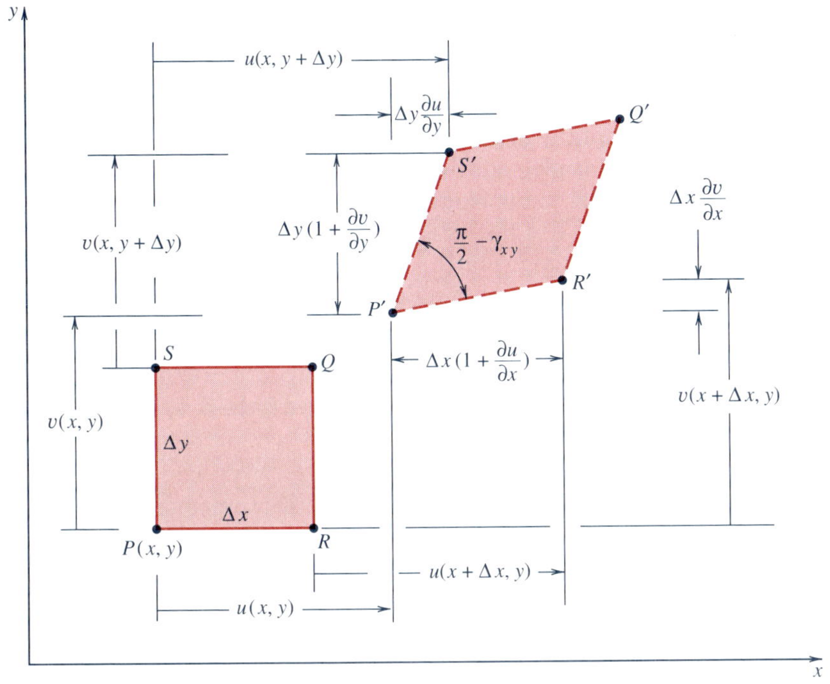

Figure 8.45 Deformation of an element of area at point P.

elements PR and PS and the shear strain γ_{xy}; see also Fig. 8.44. The deformed area of $P'R'Q'S'$ can be approximated by

$$(P'R')(P'S') = \Delta x\,(1 + \epsilon_x)\,\Delta y\,(1 + \epsilon_y)$$

$$= \Delta x\,\Delta y\,(1 + \epsilon_x + \epsilon_y) \tag{8.78}$$

where we neglect the products of the normal strains; see Prob. 8.11-3. The original area of $PRQS$ is given by

$$(PR)(PS) = \Delta x\,\Delta y \tag{8.79}$$

Therefore, the change in area per unit area, or what we call an area strain, can be written in the form

$$\frac{(P'R')(P'S') - (PR)(PS)}{(PR)(PS)} = \epsilon_x + \epsilon_y \tag{8.80}$$

If we know the values of ϵ_x and ϵ_y at a point, we can obtain from Eq. (8.80) an estimate of the change in area per unit area of a small element of area at the point.

In summary, the three strain components at a point

$$\epsilon_x \qquad \epsilon_y \qquad \gamma_{xy} \tag{8.81}$$

define the state of strain at a point in a body undergoing plane strain.

Once these three quantities are known at a point, the normal strain of any line element emanating from the point can be determined by using Eq. (8.72).

Equations (8.71) indicate that the strain components depend linearly on the derivatives of the displacement components. In our derivation of these quantities we neglected all nonlinear terms. In this way we derived a set of linear strain components to be used in a linear theory. As a consequence, we think of the strain components as applicable for *small strains* and *small rotations*. For the case of a rigid-body translation and for small rigid-body rotations, the linear strain components will vanish (Prob. 8.11-4).

8.12 Strain Components Associated with Arbitrary Sets of Axes

If the state of strain at a point P is given by the components ϵ_x, ϵ_y, and γ_{xy} relative to a set of axes, we might ask—in a manner similar to our discussion of the transformation of stress components—how to determine the strain components relative to a set of axes rotated from the original axes. In Fig. 8.46*a* and *b* we show the set of xy axes at a point P with strain components ϵ_x, ϵ_y, and γ_{xy} and the set of axes $x'y'$ rotated from the first by a positive angle θ with strain components $\epsilon_{x'}$, $\epsilon_{y'}$, and $\gamma_{x'y'}$. In Fig. 8.46*c*, we show a line element at point P at an angle ϕ to the positive x axis. This line element has experienced a normal strain

$$\epsilon = \epsilon_x \cos^2 \phi + \gamma_{xy} \sin \phi \cos \phi + \epsilon_y \sin^2 \phi \tag{8.82}$$

expressed in terms of the strain components at point P as a consequence of the deformation. The normal strain of the line element at the point should *not* depend on the set of axes used to determine its value; i.e., the normal strain is a measure of the change in length per unit length, and as a geometric quantity at point P it should not depend on the set of axes used to evaluate it. We say that the normal strain is invariant with respect to the set of axes. Therefore, the expression for the normal strain in terms of the components in the rotated axes (Fig. 8.46*d*) should have the same form as Eq. (8.82), i.e.,

$$\begin{aligned}\epsilon &= \epsilon_{x'} \cos^2(\phi - \theta) + \gamma_{x'y'} \sin(\phi - \theta) \cos(\phi - \theta) + \epsilon_{y'} \sin^2(\phi - \theta)\\ &= \epsilon_{x'} \cos^2 \psi + \gamma_{x'y'} \sin \psi \cos \psi + \epsilon_{y'} \sin^2 \psi \end{aligned} \tag{8.83}$$

where $\epsilon_{x'}$, $\epsilon_{y'}$, and $\gamma_{x'y'}$ are the strain components with respect to the $x'y'$ set of axes and the angle $\psi = \phi - \theta$ measures the angle of the line element from the positive x' axis.

We can now use these observations and Eqs. (8.82) to (8.83) to express the components $\epsilon_{x'}$, $\epsilon_{y'}$, and $\gamma_{x'y'}$ in terms of the components ϵ_x, ϵ_y, and γ_{xy}. To do this, we use first the trigonometric identities

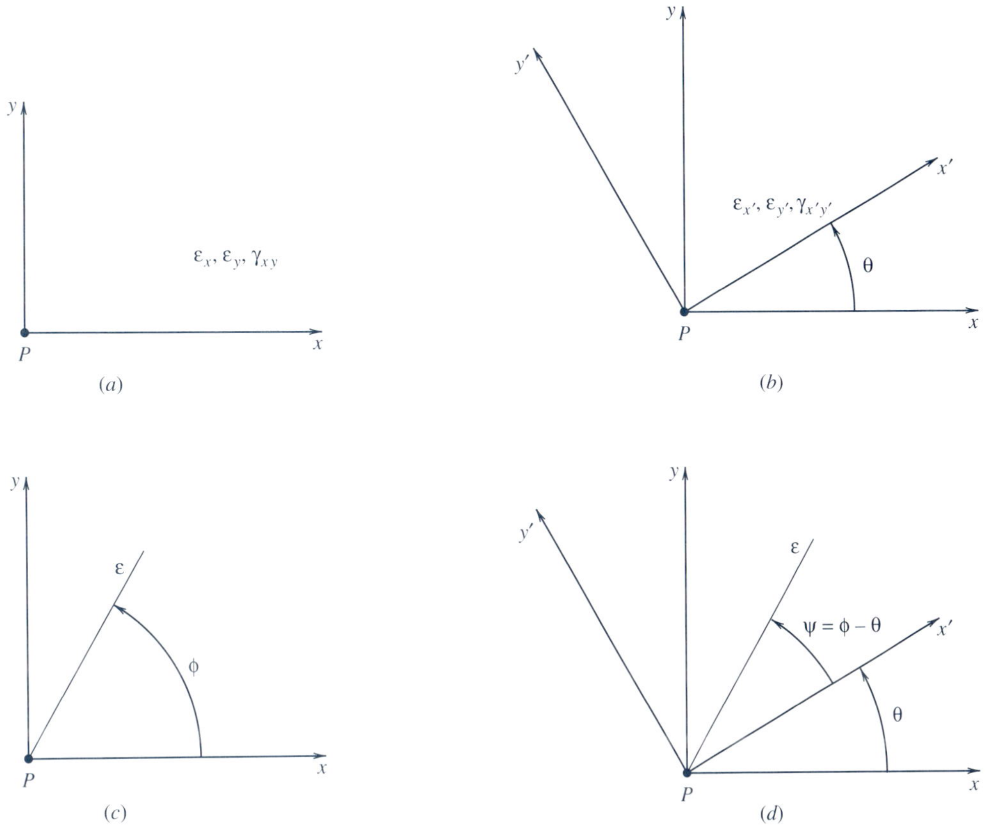

Figure 8.46 Strain components associated with set of *xy* axes and *x′y′* axes.

$$\begin{aligned} \cos\phi &= \cos[(\phi - \theta) + \theta] = \cos\psi\cos\theta - \sin\psi\sin\theta \\ \sin\phi &= \sin[(\phi - \theta) + \theta] = \sin\psi\cos\theta + \cos\psi\sin\theta \end{aligned} \tag{8.84}$$

in Eq. (8.82) and then rewrite the expression so as to take on the form of Eq. (8.83). If we carry out this direct algebra, we find

$$\begin{aligned} \epsilon = {} & (\epsilon_x \cos^2\theta + \epsilon_y \sin^2\theta + \gamma_{xy}\sin\theta\cos\theta)\cos^2\psi \\ & + [-2\epsilon_x \sin\theta\cos\theta + 2\epsilon_y\sin\theta\cos\theta \\ & + \gamma_{xy}(\cos^2\theta - \sin^2\theta)]\sin\psi\cos\psi \\ & + (\epsilon_x\sin^2\theta + \epsilon_y\cos^2\theta - \gamma_{xy}\sin\theta\cos\theta)\sin^2\psi \end{aligned} \tag{8.85}$$

Upon equating Eq. (8.85) to Eq. (8.83), we conclude that the relations for the strain components in the rotated set of axes, $\epsilon_{x'}$, $\epsilon_{y'}$, $\gamma_{x'y'}$, are of the form

$$\begin{aligned} \epsilon_{x'} &= \epsilon_x \cos^2\theta + \epsilon_y \sin^2\theta + \gamma_{xy} \sin\theta\cos\theta \\ \gamma_{x'y'} &= 2(\epsilon_y - \epsilon_x)\sin\theta\cos\theta + \gamma_{xy}(\cos^2\theta - \sin^2\theta) \\ \epsilon_{y'} &= \epsilon_x \sin^2\theta + \epsilon_y \cos^2\theta - \gamma_{xy}\sin\theta\cos\theta \end{aligned} \tag{8.86}$$

If we proceed as in the discussion of the transformation of stress components and we use the trigonometric identities in Eqs. (8.14), we find that the *strain transformation* formulas, Eqs. (8.86), can be written as follows:

$$\begin{aligned} \epsilon_{x'} &= \frac{\epsilon_x + \epsilon_y}{2} + \frac{\epsilon_x - \epsilon_y}{2}\cos 2\theta + \frac{\gamma_{xy}}{2}\sin 2\theta \\ \epsilon_{y'} &= \frac{\epsilon_x + \epsilon_y}{2} - \frac{\epsilon_x - \epsilon_y}{2}\cos 2\theta - \frac{\gamma_{xy}}{2}\sin 2\theta \\ \frac{\gamma_{x'y'}}{2} &= -\frac{\epsilon_x - \epsilon_y}{2}\sin 2\theta + \frac{\gamma_{xy}}{2}\cos 2\theta \end{aligned} \tag{8.87}$$

Equations (8.87) give the relations between the strain components associated with different sets of axes in plane strain. Once the strain components ϵ_x, ϵ_y, γ_{xy} in the original set of axes and the angle θ are specified, the strain components $\epsilon_{x'}$, $\epsilon_{y'}$, $\gamma_{x'y'}$ in the rotated set of axes can be found (Fig. 8.46*b*).

Equations (8.87) are completely analogous to Eqs. (8.15) and (8.16) if we make the substitution of ϵ for σ and $\gamma/2$ for τ. Therefore, the concepts of principal strains, principal directions of strain, maximum shear strain, and the construction of a Mohr's circle representation for plane strain carry over directly from our earlier discussions of stress transformation. We summarize a number of the results from the strain transformation formulas in what follows.

The strain transformation formulas Eqs. (8.87) lead to the idea of principal directions of strain as the orientations of a line element for which the normal strain is a maximum or a minimum. The angles of the principal directions of strain are the roots of

$$\tan 2\theta_p = \frac{\gamma_{xy}}{\epsilon_x - \epsilon_y} \tag{8.88}$$

and the two principal strains are given by

$$\epsilon_{1,2} = \frac{\epsilon_x + \epsilon_y}{2} \pm \sqrt{\left(\frac{\epsilon_x - \epsilon_y}{2}\right)^2 + \left(\frac{\gamma_{xy}}{2}\right)^2} \tag{8.89}$$

where ϵ_1 is associated with the plus sign and ϵ_2 is associated with the minus sign. We note that $\epsilon_1 + \epsilon_2 = \epsilon_x + \epsilon_y$. The sum of the normal strains in two perpendicular directions is an invariant, as we see by adding the expressions for $\epsilon_{x'}$ and $\epsilon_{y'}$ in Eqs. (8.87). This strain invariant is analogous to the stress invariant in plane stress, Eq. (8.13). Thus from Eq. (8.80) it follows that the area strain is an invariant upon rotation of axes.

The value of the maximum shear strain is

$$\frac{\gamma_{\max}}{2} = \sqrt{\left(\frac{\epsilon_x - \epsilon_y}{2}\right)^2 + \left(\frac{\gamma_{xy}}{2}\right)^2} \tag{8.90}$$

and the directions of the maximum and minimum shear strain are given by $\theta_s = \theta_p \pm \pi/4$.

The Mohr's circle representation for plane strain follows the same construction techniques as for plane stress with due regard again for the sign conventions, using $\gamma_{xy}/2$ in place of τ_{xy}, and using ϵ_x and ϵ_y in place of σ_x and σ_y. Figure 8.47 shows the construction to locate the point x and the point x' in parallel with the construction for plane stress in Fig. 8.24.

We present some numerical examples using the construction of Mohr's circle for plane strain in what follows.

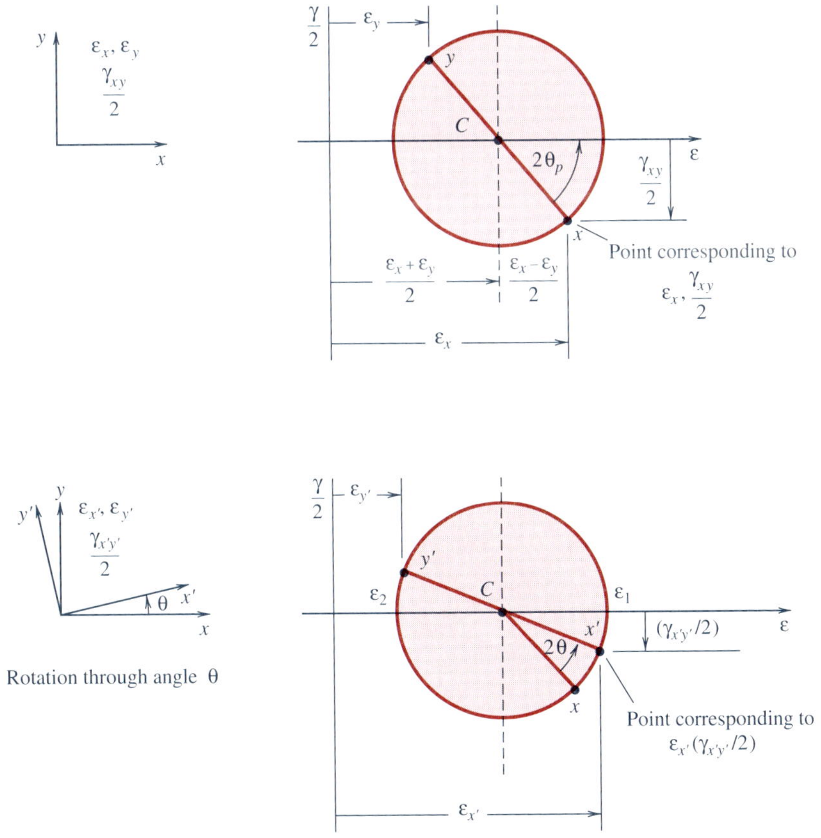

Figure 8.47 Mohr's circle construction for plane strain. A rotation θ of axes in the physical plane corresponds to a rotation 2θ in the Mohr's circle construction.

EXAMPLE 8.13

A sheet of steel is loaded such that the state of strain at a point in the plane of the sheet is given by the strain components related to a set of xy axes, as shown in Fig. 8.48*a*,

$$\epsilon_x = -200 \times 10^{-6}$$
$$\epsilon_y = 1000 \times 10^{-6} \qquad (a)$$
$$\gamma_{xy} = 900 \times 10^{-6}$$

We wish to construct Mohr's circle for plane strain to find the strain components associated with a set of axes $x'y'$ at an angle of 30° clockwise to the xy set of axes (Fig. 8.48*b*), the principal strains, and the maximum shear strain.

Figure 8.48*c* shows Mohr's circle constructed from the strains in Eq. (*a*). The center of the circle in units of 10^{-6} is at $\epsilon = 400$, and point x is located at $(-200, 450)$. Point y is at $(1000, -450)$. The radius R of the circle, Eq. (8.90), is

$$R = \sqrt{600^2 + 450^2} \times 10^{-6} = 750 \times 10^{-6} \qquad (b)$$

and the principal directions of strain satisfy Eq. (8.88)

$$\tan 2\theta_p = \frac{\gamma_{xy}}{\epsilon_x - \epsilon_y} = -\frac{900}{1200}$$

or

$$2\theta_p = -36.87° \qquad \theta_p = -18.43° \qquad (c)$$

Figure 8.48 Example 8.13

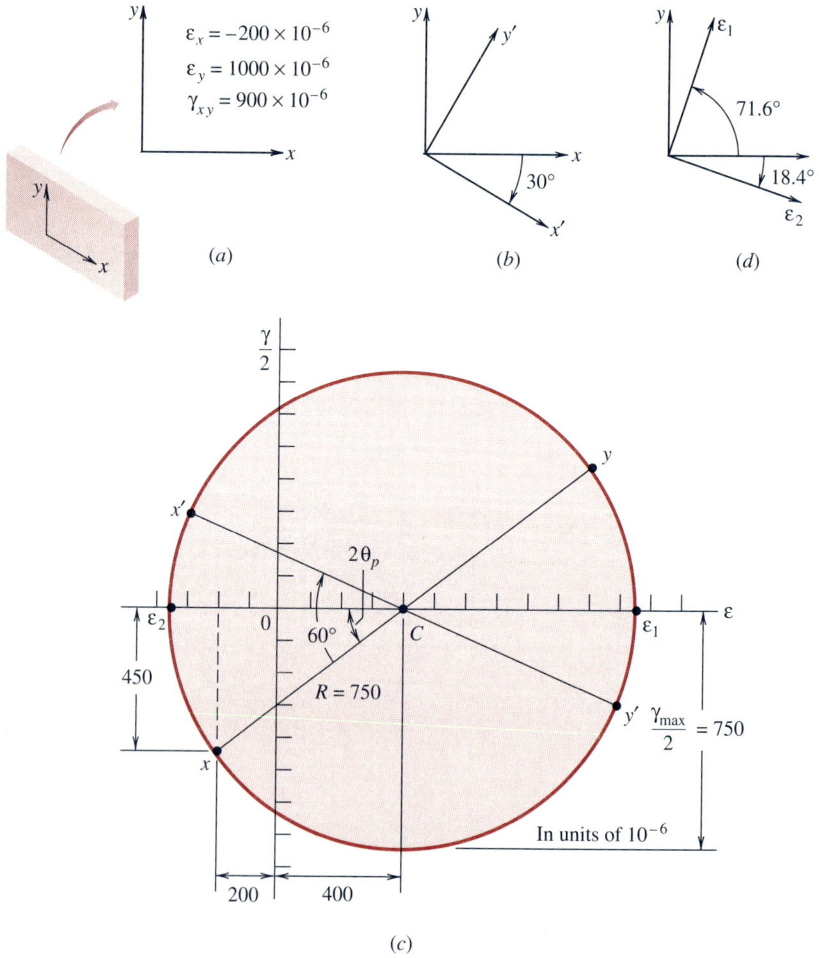

The principal strains follow immediately from Eqs. (8.89) as

$$\epsilon_1 = 400 \times 10^{-6} + R = 1150 \times 10^{-6}$$
$$\epsilon_2 = 400 \times 10^{-6} - R = -350 \times 10^{-6} \qquad (d)$$

The orientation of the axes for the principal directions is shown in Fig. 8.48*d*. The maximum shear strain corresponds to the radius of the circle, and we have

$$\frac{\gamma_{max}}{2} = 750 \times 10^{-6} \qquad \gamma_{max} = 1500 \times 10^{-6} \qquad (e)$$

To obtain the strain components for the $x'y'$ axes of Fig. 8.48*b*, we move in the same direction in Mohr's circle by twice the specified angle, or by 60°, as shown in Fig. 8.48*c*. Point x' lies above the ϵ axis so that we obtain from the circle

$$\epsilon_{x'} = 400 \times 10^{-6} - R\cos(60° - 36.87°) = -290 \times 10^{-6}$$
$$\frac{\gamma_{x'y'}}{2} = -R\sin(60° - 36.87°) = -295 \times 10^{-6} \qquad (f)$$

The value of $\epsilon_{y'}$ follows in the same way as

$$\epsilon_{y'} = 400 \times 10^{-6} + R\cos(60° - 36.87°) = 1090 \times 10^{-6} \qquad (g)$$

In addition, we note that

$$\epsilon_{x'} + \epsilon_{y'} = \epsilon_x + \epsilon_y = 800 \times 10^{-6} \qquad (h)$$

```
*********************************************************

NORMAL STRAIN -- EPSILON (X) = ? -200E-06
NORMAL STRAIN -- EPSILON (Y) = ? 1000E-06
SHEAR STRAIN -- GAMMA (XY) = ? 900E-06
INPUT ANGLE (DEGREE) = ? -30

     white: EPSILON (X)   = -2.00E-04
            EPSILON (Y)   =  1.00E-03
            GAMMA (XY)/2 =  4.50E-04

red:                          blue:
EPSI. (-30.0) = -2.90E-04     EPSI.1 =  1.15E-03
GA. (-30.0)/2 = -2.95E-04     EPSI.2 = -3.50E-04
EPSI. (60.0)  =  1.09E-03     THETA1 =  71.57
GAMMA (MAX)/2 =  7.50E-04     THETA2 = -18.43
```

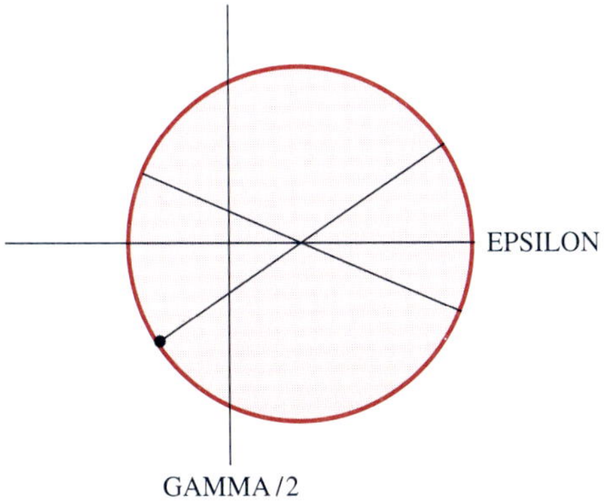

Figure 8.49 Example 8.13: Results from Mohr's circle program.

We can also set up a computer program for plane strain analogous to plane stress to determine principal directions, principal strains, and maximum shear strain given a set of strain components relative to a set of axes; see Sec. 8.6. Program 5: Mohr's circle in the main menu of the MECHMAT program is the one to use; from the Mohr's circle menu we select item 2 for plane strain. The input to the plane strain program is direct, as in the case of plane stress. For illustration, if we enter the strain components and θ of Example 8.13, we obtain the results in Fig. 8.49. Note the shear strains gamma listed in Fig. 8.49 are divided by 2. Angle theta1 is the angle from the positive x direction to the direction of ϵ_1, and angle theta2 is the angle from the positive x direction to the direction of ϵ_2.

8.13 Strain Gage Measurements

A number of techniques exist to measure experimentally the strain on the surface of a machine part or structural component under load. As a consequence, when the operation of a component is critical, or if failures of a component continue to occur in service, analysis by analytical and numerical techniques is often supplemented by experimental efforts to determine the strains and stresses that contribute to failure in the component. If the strains at a point can be measured, then (as we will see later when we discuss stress-strain relations) it will be possible to estimate values of stress components at that point. A knowledge of the strain and stress at a point often suggests new approaches to the design to avoid failure.

The most common method of measuring surface strain is by the use of strain gages bonded to the component under study. The experimental basis for strain gages is the observation that the electrical resistance of a fine wire will change when it is strained, i.e., stretched or shortened. If a fine wire is bonded intimately to the surface of a structural component and the component together with the wire is strained, then a change in the electric resistance of the wire will be detected that is related to the normal strain of the component in the direction of the wire. In this way, normal strains in a given direction can be measured at different points on the component.

The sensing wire can be replaced by foil or a semiconductor element, and the sensing element itself can be mounted in a very thin carrier that permits easy installation at the position at which the strain is to be measured. Figure 8.50*a* shows a sketch of a typical etched foil gage; gages of this sort for measuring normal strain along the axis of the gage come in sizes as small 0.16 in by 0.10 in.

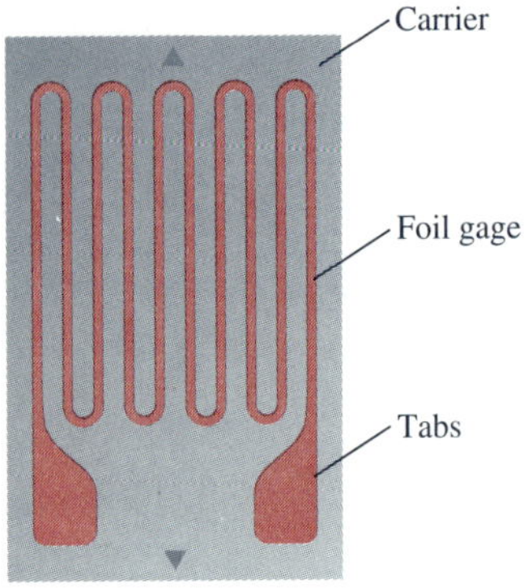

(*a*)

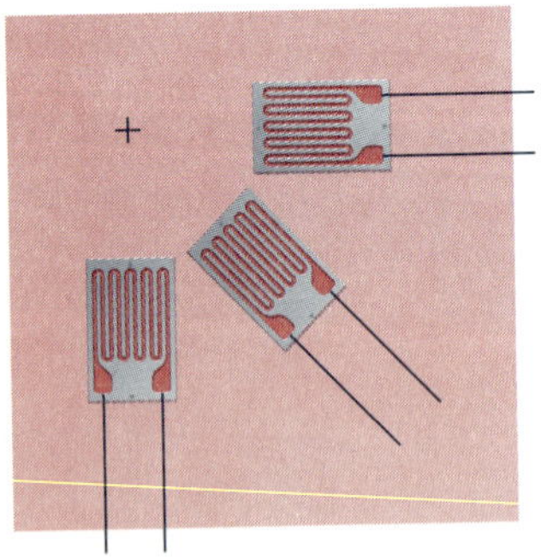

(*b*)

Figure 8.50 (*a*) Strain gage. (*b*) Strain gages mounted in a 45° strain rosette pattern.

The gages are attached to the surface on which the strain is to be measured and are connected into an electric circuit for measuring the change in resistance. Because of their small size and very low mass, gages can be used for static and dynamic applications with frequencies as high as 40 kHz. Instrumentation for specific structures on which gages are mounted can be arranged to provide direct readouts in strain or, e.g., in terms of force or load on the structure. A very common application of the use of strain gages is in load cells in industry.

If we wish to find the values of the principal strains and directions at a point on a surface, we need to know (unless geometry of the loaded component tells us the principal directions) the strain components ϵ_x, ϵ_y, and γ_{xy} with respect to a set of axes on the surface. To find these components, we can use strain gages in a pattern called a *strain rosette* to find normal strains in three directions on the surface. Figure 8.50*b* shows an arrangement of three gages in a pattern called a 45° rosette; each gage will measure the normal strain along its axis.

The general question now is, If we measure the normal strains in three directions defined relative to a set of axes, can we determine the state of strain relative to the set of axes and the principal strains, the principal directions, and the maximum shear strain? In Fig. 8.51 we show the orientation of the directions in which the normal strains are measured relative to the *xy* axes. The normal strain $\epsilon(\theta)$ in a given direction defined by θ is linear in the three strain components ϵ_x, ϵ_y, and γ_{xy}, as given in Eq. (8.72). If we know the normal strain ϵ in three directions, then we can solve for the three strain components. To see this, we assume that the normal strains in the three directions θ_1, θ_2, θ_3 shown in Fig. 8.51 are equal to $\epsilon(\theta_1)$, $\epsilon(\theta_2)$, and $\epsilon(\theta_3)$, respectively. We then have, from Eq. (8.72),

$$\begin{aligned}\epsilon(\theta_1) &= \epsilon_x \cos^2\theta_1 + \gamma_{xy}\sin\theta_1\cos\theta_1 + \epsilon_y\sin^2\theta_1\\ \epsilon(\theta_2) &= \epsilon_x \cos^2\theta_2 + \gamma_{xy}\sin\theta_2\cos\theta_2 + \epsilon_y\sin^2\theta_2\\ \epsilon(\theta_3) &= \epsilon_x \cos^2\theta_3 + \gamma_{xy}\sin\theta_3\cos\theta_3 + \epsilon_y\sin^2\theta_3\end{aligned} \tag{8.91}$$

With $\epsilon(\theta_1)$, $\epsilon(\theta_2)$, $\epsilon(\theta_3)$ and the angles θ_1, θ_2, θ_3 known, we can solve these three equations for ϵ_x, ϵ_y, and γ_{xy}.

The most common configuration available commercially for determining the strain components is $\theta_1 = 0$, $\theta_2 = 45°$, $\theta_3 = 90°$ in a configuration called a 45° strain rosette, shown in Fig. 8.50*b*. In this case, $\epsilon(0) = \epsilon_x$, $\epsilon(45) = (\epsilon_x + \epsilon_y + \gamma_{xy})/2$, $\epsilon(90) = \epsilon_y$; see Prob. 8.13-7. Once the strains ϵ_x, ϵ_y, and γ_{xy} have been found, we can obtain the principal strains, principal directions of strain, and maximum shear strain.

The solution of Eq. (8.91) has been programmed in program 5: Mohr's circle, in the main menu of MECHMAT on the diskette with this book. From the main menu if we select item 5: Mohr's circle, we obtain

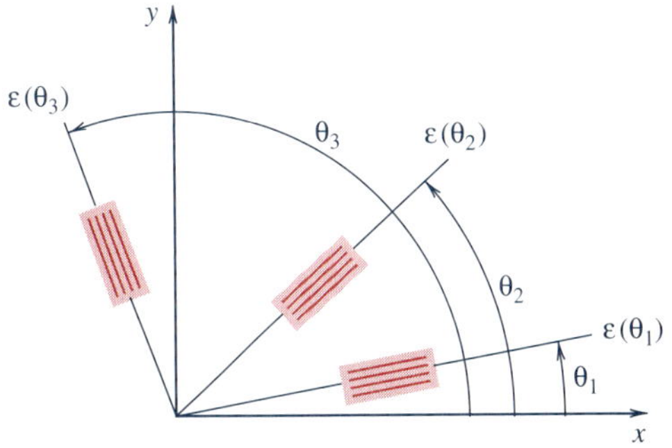

Figure 8.51 General strain rosette pattern.

the menu for Mohr's circle. For strain rosette calculations, we select item 3. The following example demonstrates the program.

EXAMPLE 8.14

A 45° strain rosette, as shown in Fig. 8.52*a*, is bonded to a steel plate, and the normal strains indicated are measured in the direction of the gages. We wish to find the strain components ϵ_x, ϵ_y, and γ_{xy} relative to the xy axes, the principal directions and strains, and the maximum shear strain, using the strain rosette program on the diskette.

From the Mohr's circle menu (Fig. 8.52*b*), we select item 3, strain rosette measurements. The input to the program requires the specification of each of the three direction angles of the gages relative to the x coordinate axis, and the values of the normal strain in each of the three directions. Angles are positive in the counterclockwise direction.

The output from the program is shown in Fig. 8.52*c* and is in a form similar to that of previous outputs. We will return to the determination of the in-plane stresses for this example in Sec. 8.14.

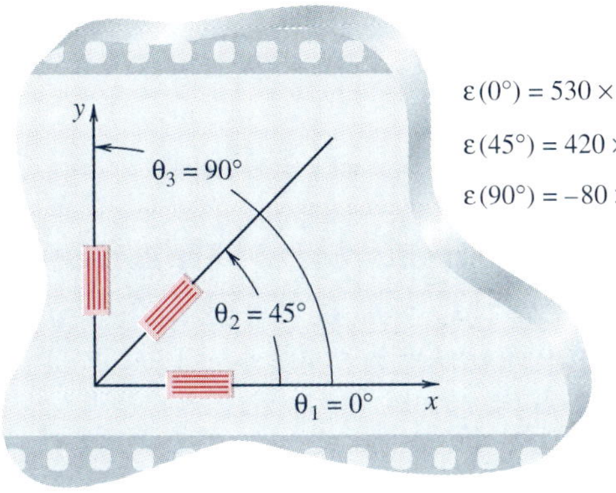

(*a*)

```
*********************************************************

     *** MOHR'S CIRCLE ***        MOHRP

*********************************************************

        1. PLANE STRESS
        2. PLANE STRAIN
        3. STRAIN ROSETTE MEASUREMENTS
     OR 4. BACK TO MENU

           SELECT NUMBER = ? 3
```

(*b*)

```
                STRAIN ROSETTE MEASUREMENTS
          ********************************************

INPUT THE ANGULAR ORIENTATIONS
(IN DEGREES) OF THE THREE STRAIN GAGES: 0,45,90

INPUT THE THREE STRAINS:    530e-06, 420e-06, -80e-06

                            EPSILON(X) =  5.30E-04
                            EPSILON(Y) = -8.00E-05
                             GAMMA(XY) =  3.90E-04

                            EPSILON(1) =  5.87E-04
                            EPSILON(2) = -1.37E-04

                                THETA1 =  16.30
                                THETA2 = 106.30

                          GAMMA(MAX)/2 =  3.62E-04

DO YOU WANT THE CORRESPONDING IN-PLANE STRESSES [Y/N] : n
```

(*c*)

Figure 8.52 Example 8.14

PART C: ELASTIC STRESS-STRAIN

8.14 Elastic Stress-Strain Relations

In Parts A and B we introduced stress and strain for two-dimensional plane stress and plane strain. In those discussions, we did not specify the nature of the material at the point where we investigated stress and strain. The concept of stress follows from arguments about *equilibrium* of an element at a point, and the concept of strain follows from the notions of *geometry* changes associated with relative displacements of points in the body. The concepts of stress and strain at a point apply to any material that can be described in a continuum fashion. The only requirement we introduced thus far is that the strains and rotations be small.

The use of equilibrium and geometry immediately should bring to mind the three steps of Eq. (2.8). What we have been doing for two-dimensional problems thus far is formulating an approach to the solution of two-dimensional load deformation problems in the spirit of the three steps of Eq. (2.8). The third step requires that we introduce a description of the behavior of the material so as to establish at a point a *force-deformation* relation. Recall that in earlier chapters we introduced relations connecting the one-dimensional stress to the one-dimensional strain at a point as a means of describing material behavior. In this section we wish to carry over those ideas of stress-strain relations to two and three dimensions and to construct the relations between stress and strain for linear elastic homogeneous isotropic materials.

In the discussion of the state of stress at a point, we visualized the state of stress in terms of the stress components acting on the faces of a small volume element at a point, as shown in Fig. 8.8. We think of the state of stress at a point as specified by the six independent stress components, given in Eq. (8.4). The three normal stress components σ_x, σ_y, and σ_z tend to expand or compress the small volume of material at the point while the three shear stresses $\tau_{xy} = \tau_{yx}$, $\tau_{xz} = \tau_{zx}$, and $\tau_{yz} = \tau_{zy}$ tend to change the shape of the volume element.

For plane strain in the xy plane, the state of strain at a point was specified by the three strain components ϵ_x, ϵ_y, and γ_{xy}, Eqs. (8.71). The normal strains ϵ_x and ϵ_y cause elongation or shortening of line elements parallel to the coordinate axes at the point while the shear strain γ_{xy} causes changes in right angles of perpendicular line elements parallel to the coordinate axes at the point, or what is equivalent, distortions in the shape of a small area element located at the point (Fig. 8.45). If we go on to three dimensions and consider the state of strain at a point in a three-dimensional body under deformation, the state of strain at a point will be given by the *six* strain components

$$\epsilon_x \qquad \gamma_{xy} \qquad \gamma_{xz} \qquad \epsilon_y \qquad \gamma_{yz} \qquad \epsilon_z \tag{8.92}$$

where ϵ_x, ϵ_y, and ϵ_z are the normal strains of line elements parallel to the coordinate axes and γ_{xy}, γ_{xz}, and γ_{yz} are the shear strains or changes in right angles of perpendicular line elements in planes parallel to the *xy*, *xz*, and *yz* coordinate planes.

Our goal now is to obtain relations between the six components of strain in Eq. (8.92) and the six components of stress in Eq. (8.4) at a point in a material undergoing deformation. These will be the stress-strain relations of the material; the form of the stress-strain relations will depend upon the nature of the material.

In earlier chapters, we considered stress-strain relations for the case of one-dimensional loading. For example, for the analysis of stresses in beam bending, we used Hooke's law for a linear elastic bar in the form

$$\epsilon_x = \frac{\sigma_x}{E} \tag{8.93}$$

where E is the elastic or Young's modulus of the material. Similarly, for the analysis of twisting of elastic circular shafts, we introduced a shear strain–shear stress relation in the form

$$\gamma = \frac{\tau}{G} \tag{8.94}$$

where G is the shear modulus of the material.

In this section we will generalize these one-dimensional relations to obtain relations that connect all six components of stress with all six components of strain for a linear elastic material. By a ***linear elastic*** material we mean one for which the strain is linearly proportional to the stress. Many important engineering materials such as steel, aluminum, and concrete behave in a linear elastic manner over a range of stress values. In addition, we will assume that the material is ***homogeneous***, by which we mean that the physical constants that describe the behavior of the material have the same values at all points in the material. Finally for many materials, the microscopic distribution of atoms and molecules within the material is such that the elastic properties *at a point* will be the same for all possible orientations of axes at that point. We define an *isotropic* elastic material as one whose elastic properties are independent of orientation at a point.

To find the general relations between the six components of stress and the six components of strain at a point, we will consider how the different components of stress cause strain in a linear elastic isotropic material.[3]

Consider at a point in a body undergoing deformation a small element of volume on which there is acting only a normal stress component σ_x, as shown in Fig. 8.53. This normal component of stress will produce a corresponding normal component of strain so that

$$\epsilon_x = \frac{\sigma_x}{E} \tag{8.95}$$

where E is the elastic modulus of the material.

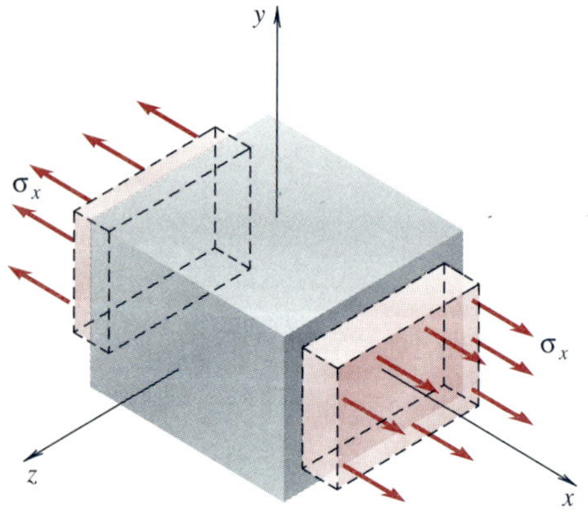

Figure 8.53 Element of volume experiencing normal stress σ_x and transverse contraction.

In addition to the normal component of strain in the x direction, there will be a lateral contraction of the element in the y and z directions, as suggested by the dashed lines in Fig. 8.53. Experimental measurements, as we discussed in Sec 1.5 in the context of tension tests for one-dimensional loading, show that the lateral contraction strain is a fixed fraction of the normal component of strain. This fixed fraction is known as *Poisson's ratio* and often is given the symbol ν (Greek letter nu). For the stress state illustrated in Fig. 8.53, the lateral strains ϵ_y and ϵ_z must be equal because there is no preferred direction in an isotropic material. Therefore, the strains ϵ_y and ϵ_z can be written in the form

$$\epsilon_y = \epsilon_z = -\nu\epsilon_x = -\nu\frac{\sigma_x}{E} \tag{8.96}$$

where ν is Poisson's ratio of the material. The minus sign in Eqs. (8.96) appears because there is contraction or compressive strain in the y and z directions with tensile strain in the x direction, and the value of Poisson's ratio for a material is taken to be a positive number. Values for Poisson's ratio for most isotropic materials range between 0 and $\frac{1}{2}$; typical values for structural steel are about $\frac{1}{3}$.

If normal stresses are now present in the y and z directions, they will give rise to normal strains in the y and z directions and corresponding lateral contraction strains in the perpendicular directions. Further, an appeal to arguments using isotropy of the material leads to the result that normal stresses give rise to only normal strains and shear stresses give rise to only shear strains.

[3]An excellent discussion, some of which we use here, invoking the arguments of symmetry of possible deformations for an isotropic material can be found in Crandall, Dahl, and Lardner, sec. 5.4.

Therefore, if a linear elastic isotropic material has all six components of stress present, the stress-strain relations are

$$\epsilon_x = \frac{1}{E}[\sigma_x - \nu(\sigma_y + \sigma_z)] \qquad \gamma_{xy} = \frac{\tau_{xy}}{G}$$
$$\epsilon_y = \frac{1}{E}[\sigma_y - \nu(\sigma_z + \sigma_x)] \qquad \gamma_{yz} = \frac{\tau_{yz}}{G} \tag{8.97}$$
$$\epsilon_z = \frac{1}{E}[\sigma_z - \nu(\sigma_x + \sigma_y)] \qquad \gamma_{zx} = \frac{\tau_{zx}}{G}$$

where G is the shear modulus of the material.

As may be seen from the above equations for the shear strains, a consequence of isotropy is the fact that the principal axes of strain at a point in a stressed body coincide with the principal axes of stress at that point. This follows because the *xyz* axes are arbitrary, and if the shear stresses relative to the axes are zero, the shear strains are zero. The angular relations in Mohr's circle for stress and in Mohr's circle for strain are therefore identical, and to determine the location of the principal axes corresponding to a given state of stress, one may use either the Mohr's circle for stress or that for strain.

There is one additional result that the requirement of isotropy will provide. A relation among E, G, and ν can be found from the fact that an isotropic material has the same elastic properties in any direction. Consider the state of pure shear in the *xy* plane, shown in Fig. 8.54; recall Example 8.7, Fig. 8.34. From Eqs. (8.97) we have

$$\gamma_{xy} = \frac{\tau}{G} \tag{8.98}$$

It is possible to obtain another expression for γ_{xy} in the manner indicated in Fig. 8.54. To do this, we use the transformation formulas as follows: The shear stress component τ referred to the *xy* axes is equivalent to a principal state of stress with components $\sigma_1 = \tau$ and $\sigma_2 = -\tau$

Figure 8.54 Equivalent states of stress and strain for a state of pure shear.

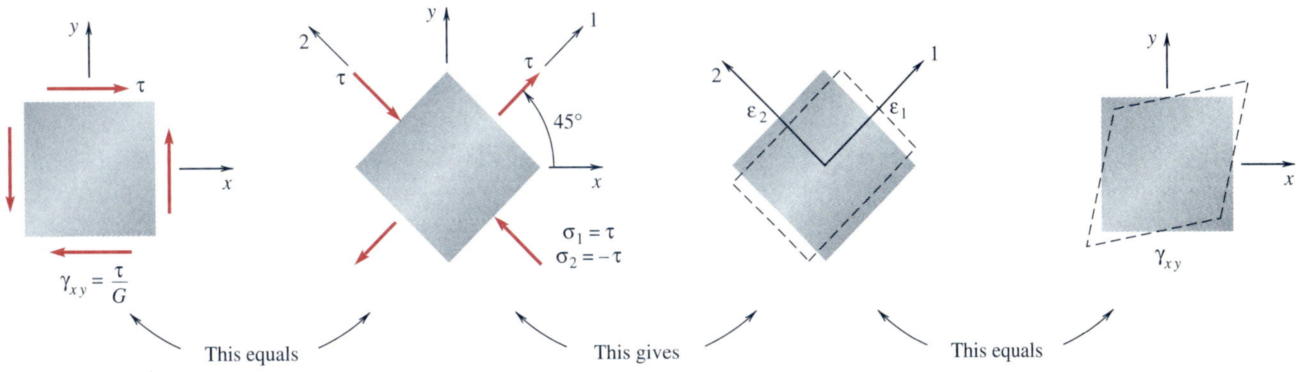

referred to the 1 and 2 coordinate axes. The principal strains referred to the 1 and 2 axes are, from the stress-strain relations, Eqs. (8.97),

$$\epsilon_1 = \frac{\sigma_1}{E} - \nu\frac{\sigma_2}{E} = \frac{\tau(1 + \nu)}{E}$$

$$\epsilon_2 = \frac{\sigma_2}{E} - \nu\frac{\sigma_1}{E} = -\frac{\tau(1 + \nu)}{E}$$

If we now use the strain transformation formulas or Mohr's circle construction, we can obtain an expression for the shearing strain with respect to the xy axes in terms of the principal strains in the form

$$\gamma_{xy} = \epsilon_1 - \epsilon_2 = \frac{2(1 + \nu)}{E}\tau \tag{8.99}$$

Expressions (8.98) and (8.99) for γ_{xy} must be equal, and this identity gives the following relation between the elastic constants:

$$G = \frac{E}{2(1 + \nu)} \tag{8.100}$$

Therefore for an isotropic material we have *two* independent elastic constants. Given values of E and ν for a material, we can find the value of G from Eq. (8.100); e.g., for steel, ν has a value of approximately 0.3 so that G is approximately $E/2.6$. Typical values of G for different materials are given in App. F.

If a situation arises in which a temperature change of the material from a reference temperature T_0 to a temperature T occurs, then, as discussed in Sec. 2.5 for the one-dimensional case, we need to include the effect of the temperature change in the three-dimensional stress-strain relations of Eqs. (8.97). We noted in Sec. 2.5 that the effect of a temperature change on strain appears in two ways: first, by causing a modification in the values of the elastic constants; second, by directly producing a change in the dimensions of the materal, i.e., a strain in the absence of stress, as in Fig. 2.18. The effect of temperature change on the room temperature elastic constants is, in general, small over a temperature range of a hundred degrees Celsius; for this reason we will assume that the values of the elastic constants in the stress-strain relations are equal to their values at room temperature T_0.

The thermal strain of a small element of volume of an isotropic material is either a pure expansion or contraction with no shear strain components referred to any axes. In addition, as in the case of uniaxial deformation, the thermal strain is approximately linear with temperature change. Therefore, the thermal strain at a point in an isotropic material can be written in the form

$$\begin{aligned} \epsilon_x^T = \epsilon_y^T = \epsilon_z^T &= \alpha(T - T_0) \\ \gamma_{xy}^T = \gamma_{yz}^T = \gamma_{zx}^T &= 0 \end{aligned} \tag{8.101}$$

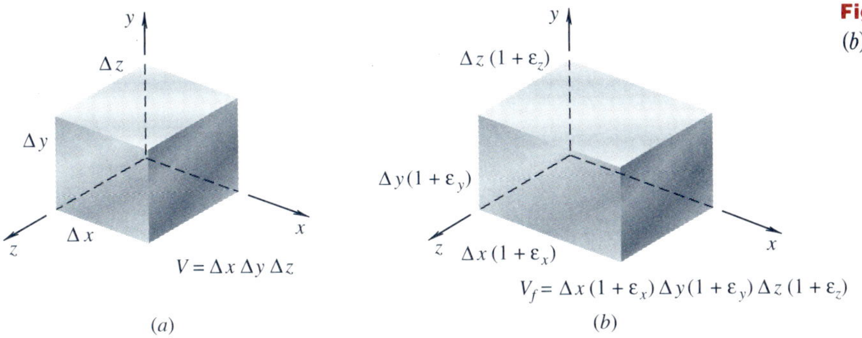

Figure 8.55 (*a*) Initial element of volume. (*b*) Deformed element of volume.

where the superscript T on the strain components is to designate thermal strain. The quantity α in Eqs. (8.101) is the coefficient of thermal expansion with units of 1/temperature. A short table of values of α was given in Sec. 2.5; also see App. F.

Therefore the strain-stress–temperature relations for an elastic isotropic material can be written in the form

$$\begin{aligned}
\epsilon_x &= \frac{1}{E}[\sigma_x - \nu(\sigma_y + \sigma_z)] + \alpha(T - T_0) & \gamma_{xy} &= \frac{\tau_{xy}}{G} \\
\epsilon_y &= \frac{1}{E}[\sigma_y - \nu(\sigma_z + \sigma_x)] + \alpha(T - T_0) & \gamma_{yz} &= \frac{\tau_{yz}}{G} \qquad (8.102) \\
\epsilon_z &= \frac{1}{E}[\sigma_z - \nu(\sigma_x + \sigma_y)] + \alpha(T - T_0) & \gamma_{zx} &= \frac{\tau_{zx}}{G}
\end{aligned}$$

We can gain additional insight into the nature of these stress-strain relations by looking at some special cases.

If we consider a small element with initial volume $V = \Delta x\, \Delta y\, \Delta z$ at a point, as shown in Fig. 8.55*a*, this element will undergo a change in volume arising from strain at the point. The lengths of the sides of the element after deformation are now $\Delta x(1 + \epsilon_x)$, $\Delta y(1 + \epsilon_y)$, and $\Delta z(1 + \epsilon_z)$, and the right angles of the element will change as a consequence of the shear strains (Fig. 8.55*b*). The final value of the volume V_f after deformation can be calculated, upon neglect of the nonlinear terms, as follows (see Problem 8.14-1):

$$V_f = \Delta x(1 + \epsilon_x)\, \Delta y(1 + \epsilon_y)\, \Delta z(1 + \epsilon_z)$$

or

$$V_f = \Delta x\, \Delta y\, \Delta z(1 + \epsilon_x + \epsilon_y + \epsilon_z) \qquad (8.103)$$

Therefore, for small strains the change in volume per unit initial volume or volumetric strain is given by

$$\frac{V_f - V}{V} = \frac{\Delta V}{V} = \epsilon_x + \epsilon_y + \epsilon_z \qquad (8.104)$$

The sum of the normal strains on the right side of Eq. (8.104) is often called the *dilatational strain* since it is a measure of how much the material dilates or changes its volume ΔV at a point under deformation; recall Eq. (8.80). The dilatational strain can be an important consideration in the behavior of a porous material such as soil.

In the absence of a temperature change, we can solve Eqs. (8.102) for the sum of the normal strains in terms of the normal stresses in the form

$$\epsilon_x + \epsilon_y + \epsilon_z = \frac{1 - 2\nu}{E}(\sigma_x + \sigma_y + \sigma_z) \tag{8.105}$$

Equation (8.105) gives an expression for the dilatational strain in terms of the sum of the normal stresses. If at a point in a body the state of stress is one of hydrostatic compression (Fig. 8.9*b*), in which case the normal stresses are

$$\sigma_x = \sigma_y = \sigma_z = -p \tag{8.106}$$

where p is the *hydrostatic pressure* at the point, then the change in volume per unit volume is given by

$$\frac{\Delta V}{V} = \epsilon_x + \epsilon_y + \epsilon_z = -\frac{(1 - 2\nu)3p}{E} = \frac{-p}{B} \tag{8.107}$$

The constant B is defined as the *bulk modulus* of the material, and from Eq. (8.107) we see that it is defined as the ratio of the hydrostatic pressure to the fractional decrease in volume at constant temperature

$$B = \frac{-p}{\Delta V/V} = \frac{E}{3(1 - 2\nu)} \tag{8.108}$$

We note that if $\nu \rightarrow \frac{1}{2}$, the bulk modulus becomes large. The value $\nu = \frac{1}{2}$ corresponds to an incompressible material for which, as we see from Eq. (8.107), the volume change at each point will be identically zero. Natural rubber is an elastic material that is nearly incompressible; in addition, many biological materials behave in a nearly incompressible manner.

EXAMPLE 8.15

An elastic material with a modulus of elasticity E, Poisson's ratio ν, and coefficient of thermal expansion α originally fills a cavity with sides $2a$ and height L in a rigid block, as shown in Fig. 8.56*a*. A rigid cap is placed on top of the elastic material, and a compressive force F is applied to the cap at the same time as the temperature is increased.

We wish to express the relation between the movement of the cap, shown as c in Fig. 8.56*a*, and the force F and temperature increase ΔT. We assume that the sides of the cavity are well lubricated and the effect of friction can be neglected, so that the material does not stick to the walls of the rigid block and cap during deformation. We set up a coordinate system as shown, and we take the movement of the cap c as positive in the positive y direction.

The material in the cavity is linear elastic for which the stress-strain–temperature relations can be written from Eqs. (8.102) in the form

$$E\epsilon_x = \sigma_x - \nu(\sigma_y + \sigma_z) + E\alpha\,\Delta T \tag{a}$$

$$E\epsilon_y = \sigma_y - \nu(\sigma_x + \sigma_z) + E\alpha\,\Delta T \tag{b}$$

$$E\epsilon_z = \sigma_z - \nu(\sigma_x + \sigma_y) + E\alpha\,\Delta T \tag{c}$$

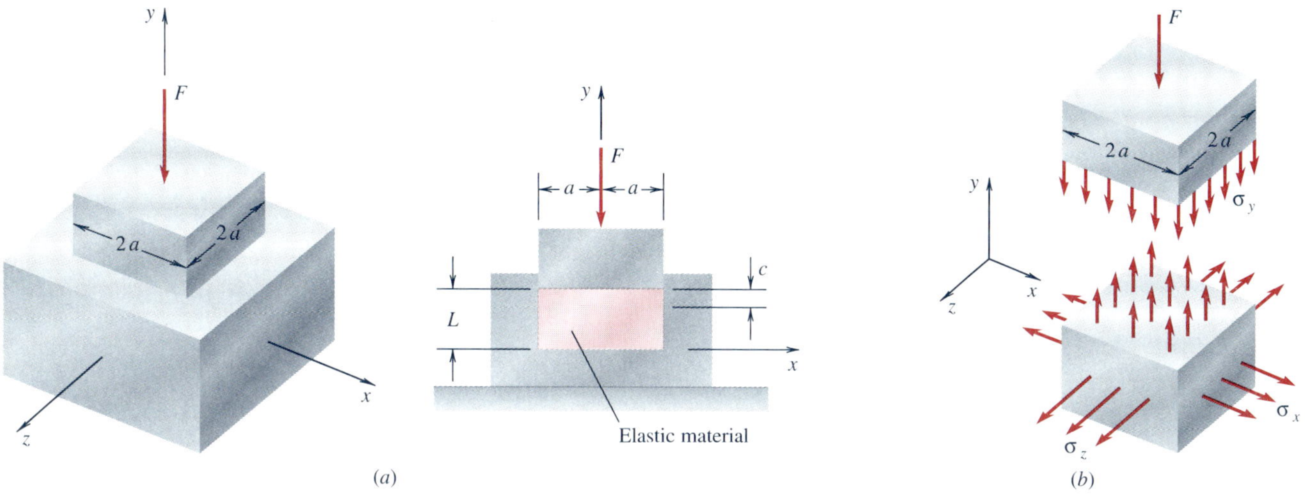

Figure 8.56 Example 8.15

From symmetry, only normal stresses and normal strains occur in the material. In addition, as a consequence of the load and the constraint of the rigid block, only a normal strain in the y direction can take place in the material. Therefore

$$\epsilon_y = \epsilon \qquad \epsilon_x = \epsilon_z = 0 \tag{d}$$

and from Eqs. (a) and (c), $\sigma_x = \sigma_z$. From Eq. (a) we have

$$\sigma_x(1 - \nu) - \nu\sigma_y + E\alpha\,\Delta T = 0$$

or

$$\sigma_x = \frac{\nu}{1 - \nu}\sigma_y - \frac{E\alpha}{1 - \nu}\Delta T \tag{e}$$

Upon substitution of Eq. (e) into Eq. (b) we find

$$E\epsilon = \frac{1 + \nu}{1 - \nu}[(1 - 2\nu)\sigma_y + E\alpha\,\Delta T] \tag{f}$$

The normal strain ϵ in the y direction of the material is given by c/L, where c is the displacement of the cap, assumed positive in the positive y direction. Force equilibrium of the cap in the y direction gives $\sigma_y = -F/(4a^2)$, where F is positive, as shown in Fig. 8.56b, to give a compressive normal stress σ_y in the y direction in the material. Upon solving Eq. (f) for the displacement of the cap, we find

$$\frac{c}{L} = \frac{1}{E}\frac{1 + \nu}{1 - \nu}\left[-(1 - 2\nu)\frac{F}{4a^2} + E\alpha\,\Delta T\right]$$

or

$$\frac{c}{L} = -\frac{1 + \nu}{1 - \nu}\left[\frac{1}{3B}\frac{F}{4a^2} - \alpha\,\Delta T\right] \tag{g}$$

upon using Eq. (8.108) for the definition of the bulk modulus B. Therefore, the displacement-load-temperature change relation is given by Eq. (g). If the temperature change is zero, the load F compresses the material and c is negative. The amount of load required to compress the material a given distance c is sensitive to the value of ν when ν is close to $\frac{1}{2}$.

The force required to keep the cap in its initial position $c = 0$ in the presence of a temperature increase is given by

$$\frac{F}{4a^2} = 3B\alpha\,\Delta T \tag{h}$$

Two-Dimensional Stress-Strain Relations

If we consider the case of plane stress for which $\sigma_z = \tau_{xz} = \tau_{yz} = 0$, the stress-strain relations, Eqs. (8.102), in the absence of a temperature change become

$$\begin{aligned} E\epsilon_x &= \sigma_x - \nu\sigma_y \\ E\epsilon_y &= \sigma_y - \nu\sigma_x \\ G\gamma_{xy} &= \tau_{xy} \end{aligned} \tag{8.109}$$

If we solve for the stresses in terms of the strains, we have

$$\begin{aligned} \sigma_x &= \frac{E}{1-\nu^2}(\epsilon_x + \nu\epsilon_y) \\ \sigma_y &= \frac{E}{1-\nu^2}(\epsilon_y + \nu\epsilon_x) \\ \tau_{xy} &= G\gamma_{xy} = \frac{E}{2(1+\nu)}\gamma_{xy} \end{aligned} \tag{8.110}$$

It follows from Eqs. (8.110) that for plane stress a knowledge of the strain components at a point and the elastic constants of the material will allow us to calculate the values of the stress components at the point. In many situations the strain components can be determined experimentally by using strain gage techniques, as we discussed in Sec. 8.13. Once the strains ϵ_x, ϵ_y, and γ_{xy} are found at a point, we can use Eqs. (8.110) to find the stress components and then determine the maximum shear stress and principal stresses at the point. The computer program for strain rosette measurements discussed in Sec. 8.13 can be used to carry out these calculations; see Example 8.14. We illustrate the use of Eqs. (8.110) and this computer program in the next two examples.

EXAMPLE 8.16

A steel plate used as a structural component in service has a 45° strain gage rosette attached to it at a point, as shown in Fig. 8.57. The angles locating the gages and the measured values of normal strains in the directions of the gages are shown in Fig. 8.57. We wish to find the *normal stress* in the direction of a fixture attached to the component at an angle of 60° to the x axis. To do so, we will use the strain rosette program together with the Mohr's circle program for stress to obtain the required stress component. Take $E = 30 \times 10^6$ psi and $\nu = 0.3$.

The output from the strain rosette program is shown in Fig. 8.58; see Example 8.14 for the calculation of the strains. The components of stress relative to the coordinate axes follow from the output of the program and are

$$\sigma_x = 16{,}700 \text{ psi}$$
$$\sigma_y = 2600 \text{ psi} \qquad (a)$$
$$\tau_{xy} = 4500 \text{ psi}$$

We can now use these stress components to find the normal stress in the 60° direction, $\sigma_{x'}(\theta = 60°)$, from the program for Mohr's circle for plane stress. From the Mohr's circle menu we select item 1, Fig. 8.59*a*, and the values of σ_x, σ_y, and τ_{xy} in Eq. (*a*) as input to the program, to obtain the required normal stress at an angle of 60°, as shown in Fig. 8.59*b*:

$$\sigma_{x'}(\theta = 60°) = 10{,}000 \text{ psi} \qquad (b)$$

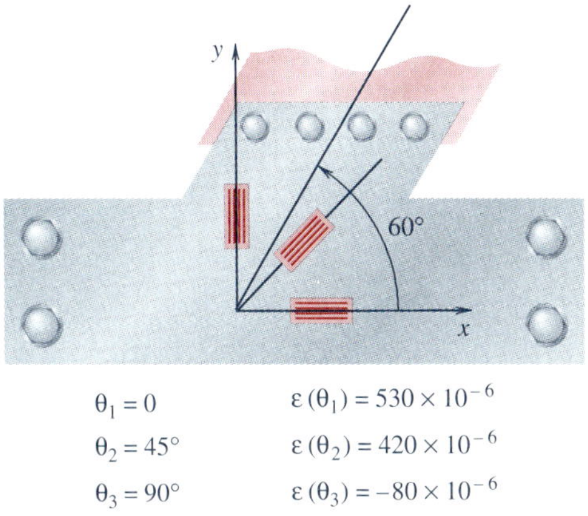

Figure 8.57 Example 8.16

EXAMPLE 8.17

A 60° strain rosette is mounted on a steel plate and oriented relative to a reference system of xy axes at a point, as shown in Fig. 8.60. We wish to find the state of strain associated with the xy axes and the principal strains and directions at this point. Using $E = 200$ GPa and $\nu = 0.3$ for the plate, we wish to find the principal stresses. We will use the strain rosette program. The angles for the strain rosette program are, relative to the reference xy system,

$$\theta_1 = -15° \text{ (negative below axis)}$$
$$\theta_2 = 45° \qquad (a)$$
$$\theta_3 = 105°$$

The output from the program is shown in Fig. 8.61; the principal strains and directions are given. The principal stresses are

$$\sigma_1 = 175 \text{ MPa} \qquad \sigma_2 = 35.1 \text{ MPa} \qquad (b)$$

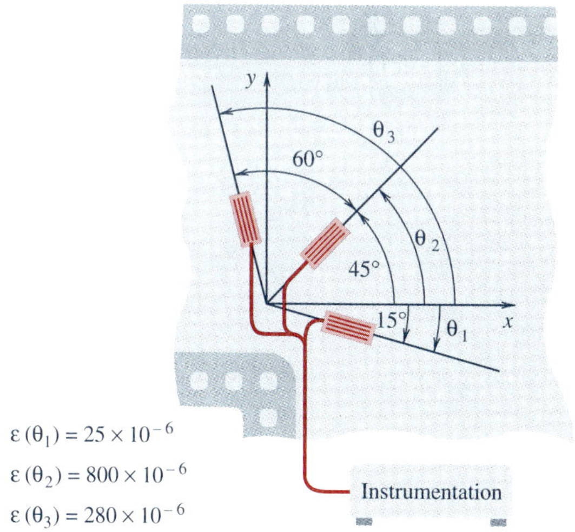

Figure 8.60 Example 8.17

```
                 STRAIN ROSETTE MEASUREMENTS
         *******************************************

INPUT THE ANGULAR ORIENTATIONS
(IN DEGREES) OF THE THREE STRAIN GAGES: 0,45,90

INPUT THE THREE STRAINS:     530e-06, 420e-06, -80e-06

                             EPSILON(X) =  5.30E-04
                             EPSILON(Y) = -8.00E-05
                              GAMMA(XY) =  3.90E-04

                             EPSILON(1) =  5.87E-04
                             EPSILON(2) = -1.37E-04

                                 THETA1 =  16.30
                                 THETA2 = 106.30

                           GAMMA(MAX)/2 =  3.62E-04

DO YOU WANT THE CORRESPONDING IN-PLANE STRESSES [Y/N] : Y


INPUT THE YOUNG'S MODULUS: 30e6

INPUT THE POISSON'S RATIO: .3

                               SIGMA(X) = 1.67E+04
                               SIGMA(Y) = 2.60E+03
                                TAU(XY) = 4.50E+03

                               SIGMA(1) = 1.80E+04
                               SIGMA(2) = 1.29E+03

                                 THETA1 =  16.30
                                 THETA2 = 106.30

                               TAU(MAX) = 8.35E+03
```

Figure 8.58 Example 8.16: Mohr's circle program.

```
**********************************************************

     *** MOHR'S CIRCLE ***        MOHRP

**********************************************************

        1. PLANE STRESS
        2. PLANE STRAIN
        3. STRAIN ROSETTE MEASUREMENTS
     OR 4. BACK TO MENU

      SELECT NUMBER = ? 1
```

(*a*)

```
     white: SIGMA (X) = 1.67E+04
            SIGMA (Y) = 2.60E+03
             TAU (XY) = 4.50E+03

red:                            blue:
SIGMA ( 60.0) =  1.00E+04       SIGMA1 = 1.80E+04
  TAU ( 60.0) = -8.36E+03       SIGMA2 = 1.29E+03
SIGMA (150.0) =  9.28E+03       THETA1 =  16.28
  TAU  (MAX)  =  8.36E+03       THETA2 = 106.28
```

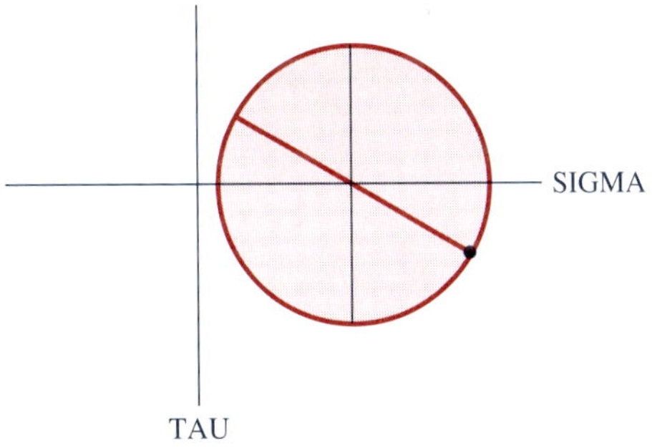

(*b*)

Figure 8.59 Example 8.16: Mohr's circle program.

```
                STRAIN ROSETTE MEASUREMENTS
       *******************************************

INPUT THE ANGULAR ORIENTATIONS
(IN DEGREES) OF THE THREE STRAIN GAGES: -15,45,105

INPUT THE THREE STRAINS:     25e-6, 800e-6, 280e-6

                             EPSILON(X) =  2.21E-04
                             EPSILON(Y) =  5.16E-04
                              GAMMA(XY) =  8.63E-04

                             EPSILON(1) =  8.24E-04
                             EPSILON(2) = -8.77E-05

                                 THETA1 =  54.42
                                 THETA2 = -35.58

                           GAMMA(MAX)/2 =  4.56E-04

DO YOU WANT THE CORRESPONDING IN-PLANE STRESSES [Y/N] : y


INPUT THE YOUNG'S MODULUS: 200e9

INPUT THE POISSON'S RATIO: .3

                               SIGMA(X) = 8.26E+07
                               SIGMA(Y) = 1.28E+08
                                TAU(XY) = 6.64E+07

                               SIGMA(1) = 1.75E+08
                               SIGMA(2) = 3.51E+07

                                 THETA1 =  54.42
                                 THETA2 = -35.58

                               TAU(MAX) = 7.02E+07
```

Figure 8.61 Example 8.17: Mohr's circle program.

8.15 Criteria for Initial Yielding

We now wish to consider what happens at a point experiencing a general state of stress if the material at the point is loaded such that it no longer behaves in a linearly elastic manner. For most metals and many other materials, once the loading response significantly deviates from linearity, *plastic deformation* or plastic flow of the material occurs. We can easily cause plastic deformation, e.g., in a paper clip by bending one side back so as to open the clip into a flat S shape. In such a case the material in the clip has undergone plastic flow and permanent deformation. Plastic flow in general can lead to large deformations under load and to permanent deformations in the material upon removal of load (as with the paper clip); these deformations in turn may lead to an unacceptable response of a component in service.

For this reason it is important to have criteria that indicate when plastic deformation will first occur at a point in a material under a general state of stress. In the case of a uniaxial bar loaded in tension from a zero load, plastic deformation first occurs when the normal stress in the bar reaches the yield stress Y; see the discussion in Sec. 1.5. The value of Y is determined experimentally from the tension stress-strain test. Alternatively, for a bar in tension, we can say that plastic deformation occurs when the maximum shear stress in the bar equals $Y/2$; see Example 8.6. Therefore, the criterion for the onset of plastic deformation or initial yielding in a bar can be stated as yielding occurs either when the normal stress σ_x in the bar equals Y or when the maximum shear stress $\tau_{\max}$ in the bar equals $Y/2$. If the normal stress σ_x is less than Y, the bar will behave in a linear elastic fashion; similarly, if the maximum shear stress $\tau_{\max}$ is less than $Y/2$, the bar will behave in a linear elastic fashion. In a one-dimensional bar these criteria for initial yielding expressed in terms of σ_x and $\tau_{\max}$ are in fact the same; we state them differently in terms of a different stress component.

When a general state of stress occurs at a point in a material and the value of Y from a one-dimensional tension test is known, we would like to know the relation between the six stress components and the value of Y that describes the onset of initial yielding at the point. Often in design, we wish to ensure that no plastic deformation occurs. The behavior of the material *after* plastic deformation occurs is the content of the study of the theory of plasticity.[4]

There is at present no simple theoretical way of deriving a relation

[4]See, e.g., J. Lubliner, *Plasticity Theory*, Macmillan Publishing, New York, 1990; I. H. Shames and F. A. Cozzarelli, *Elastic and Inelastic Stress Analysis*, Prentice-Hall, Englewood Cliffs, N.J., 1992.

between the six components of stress in a general state of stress that will correlate with yielding in a simple test.[5] However, two empirical criteria have been proposed that are fairly simple and at the same time give reasonable agreement with experimental data. Each of these criteria for initial yielding is based on two important observations.

The first observation is that the state of stress at a point can be described completely by the magnitude and orientation of the principal stresses. However, in an isotropic material, the orientation of the principal stresses cannot play a role. Therefore at a point, the state of stress that initiates plastic deformation or initial yielding should depend on only the *magnitude* of the principal stresses. The second observation follows from experiments that have shown that a hydrostatic state of stress at a point (Fig. 8.9*b*) does not influence the onset of initial yielding. Therefore, any criterion for initial yielding should be based not on the absolute magnitude of the principal stresses but rather on the magnitude of the *differences* between the principal stresses so that any hydrostatic state of stress will cancel out.

The first of the two empirical criteria that we will introduce assumes that initial yielding will occur at a point experiencing a three-dimensional state of stress when the root mean square of the differences between the principal stresses reaches the same value that it has when yielding occurs in a tensile test. If Y denotes the stress at which yielding begins in the simple tensile test, the principal stresses are $\sigma_1 = Y$ and $\sigma_2 = \sigma_3 = 0$. Thus in the tensile test, the root mean square of the differences between the principal stresses is

$$\sqrt{\tfrac{1}{3}[(\sigma_1 - \sigma_2)^2 + (\sigma_2 - \sigma_3)^2 + (\sigma_3 - \sigma_1)^2]}$$

$$= \sqrt{\tfrac{1}{3}[(Y - 0)^2 + (0 - 0)^2 + (0 - Y)^2]} = \sqrt{\tfrac{2}{3}}\,Y$$

If we carry the factor $\sqrt{\tfrac{2}{3}}$ over to the left side in this equation for convenience, this criterion can be expressed as follows: For a general state of stress, initial yielding will occur when

$$\sqrt{\tfrac{1}{2}\,[(\sigma_1 - \sigma_2)^2 + (\sigma_2 - \sigma_3)^2 + (\sigma_3 - \sigma_1)^2]} = Y \qquad (8.111)$$

where, again, Y is the value of the stress at which yielding begins in the tensile test. This criterion is known as the *Mises yield criterion.*

In the case of plane stress, $\sigma_3 = 0$, the Mises yield criterion becomes

$$\sqrt{\tfrac{1}{2}[(\sigma_1 - \sigma_2)^2 + \sigma_2^2 + \sigma_1^2]} = Y \qquad (8.112)$$

[5] See additional discussion in Crandall, Dahl, and Lardner, op. cit., sec. 5.11. We follow part of their discussion here.

We can also rewrite Eq. (8.112) in terms of the stress components for plane stress by using Eqs. (8.23) and (8.25), to obtain

$$\sqrt{\tfrac{1}{2}[(\sigma_x - \sigma_y)^2 + \sigma_y^2 + \sigma_x^2] + 3\tau_{xy}^2} = Y \tag{8.113}$$

The second empirical criterion assumes that yielding occurs at a point in the material when the maximum shear stress at the point reaches the value of the maximum shear stress to cause yielding in a tensile test. The maximum shear stress at a point, Eq. (8.42), is one-half the difference between the maximum and minimum principal stresses, and it occurs on a face inclined at 45° to the faces on which the maximum and minimum principal stresses act. In a tensile test the maximum shear stress is $Y/2$, Example 8.6 and Fig. 8.33, so this criterion states that yielding occurs when

$$\tau_{\text{max}} = \frac{\sigma_{\text{max}} - \sigma_{\text{min}}}{2} = \frac{Y}{2} \tag{8.114}$$

This criterion is known as the *Tresca*, or the *maximum shear stress, criterion*. We shall refer to it as the *maximum shear stress yield criterion*.

For plane stress, $\sigma_3 = 0$, we can plot each of the yield criteria, Eqs. (8.112) and (8.114), as a curve in a principal stress coordinate plane in which σ_1 and σ_2 are the horizontal and vertical axes (see Fig. 8.62). The Mises criterion, Eq. (8.112), is represented by an ellipse in this set of axes since

$$\sigma_1^2 - \sigma_1\sigma_2 + \sigma_2^2 = Y^2 \tag{8.115}$$

is the equation of an ellipse.

The maximum shear stress criterion is represented in the same plane by the six-sided polygon shown inscribed inside the Mises ellipse in Fig. 8.62. To verify this, consider the state of stress $\sigma_1 > 0, 0 < \sigma_2 < \sigma_1, \sigma_3 = 0$ for which the maximum shear stress, Eq. (8.114), is $\sigma_1/2$. Therefore in the plane of σ_1 and σ_2 of Fig. 8.62, the maximum shear stress criterion gives $\sigma_1 = Y$, which is the vertical line AB; when $\sigma_2 > \sigma_1$, the maximum shear stress criterion gives $\sigma_2 = Y$, which is the horizontal line BC. When $\sigma_1 > 0$ and $\sigma_2 < 0$, the maximum shear stress is $\tau_{\text{max}} = \frac{1}{2}(\sigma_1 - \sigma_2)$, and yielding occurs when $\tau_{\text{max}} = Y/2$ or $\sigma_1 - \sigma_2 = Y$, which is the line AD in Fig. 8.62. The other three sides of the polygon are found in a similar way. Both the Mises and the maximum shear stress criteria have been found for many engineering materials to give reasonable predictions for the onset of initial yielding when compared to experimental values. In fact, most experimental results for the onset of yielding in metals lie in the regions between the polygon and the ellipse.

In Fig. 8.62, the states of stress given by values of σ_1 and σ_2 inside the polygon or the ellipse are elastic. That is, if the state of stress at a point as determined by its principal stresses lies inside the yield curve,

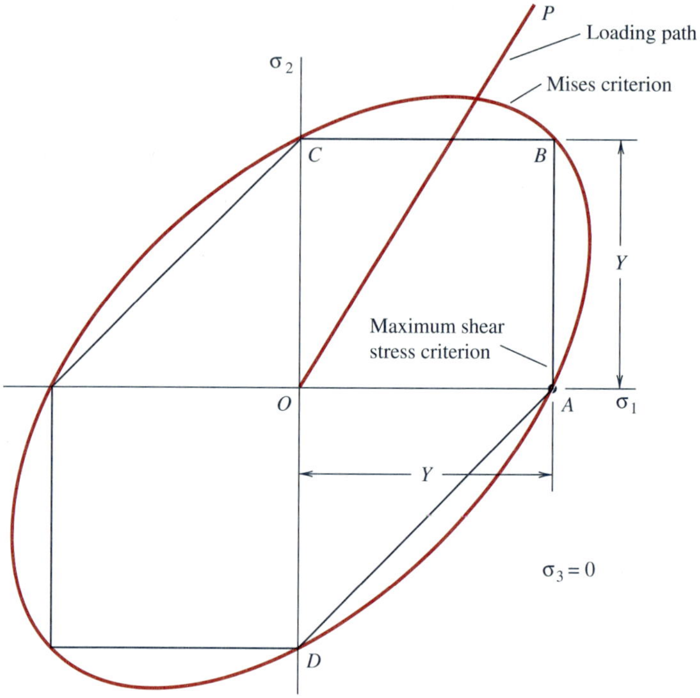

Figure 8.62 Yield surfaces in plane stress.

then the state of stress is elastic. If in plane stress a point experienced a *loading path* shown, e.g., as the line *OP* in Fig. 8.62, we would say that the material at that point behaves elastically until the loading path touched the line *CB* or the Mises ellipse, depending on which yielding criterion we were using. The intersection of the loading path with the curve of the yield criterion gives the values of σ_1 and σ_2 at the onset of initial plastic deformation. Since stresses σ_1 and σ_2 can be expressed in terms of the applied loading, we are able then to determine the loads or combinations of loads acting on the body that give rise to initial yielding at the point.

For our purpose in this text, we will use these criteria to calculate critical loads or load combinations that cause the onset of plastic deformation or initial yielding at a point in a material. The behavior of the material *after* initial yielding is covered in the theory of plasticity. A number of combined states of stress problems will be considered in Chap. 9. However, it is important to realize that the onset of plastic deformation is only one of the many possible "failure" criteria or design criteria for a structural member or machine component. Other failure criteria including fracture and fatigue are often covered in more advanced books on the mechanics of solids.

EXAMPLE 8.18

An alloy yields in uniaxial tension at the stress $Y = 48{,}000$ psi. In service, a component of this material is subjected to the following state of plane stress at a point, $\sigma_3 = 0$:

$$\sigma_x = 20{,}000 \text{ psi} \qquad \sigma_y = -10{,}000 \text{ psi} \qquad \tau_{xy} = 20{,}000 \text{ psi}$$

We wish to know if yielding of the material will occur; we will check for yielding by using the Mises criterion and the maximum shear stress criterion.

To determine if yielding will occur, we need to determine the principal stresses and use Eqs. (8.111) and (8.114).

We first find the principal stresses

$$\sigma_1 = 30{,}000 \text{ psi} \qquad \sigma_2 = -20{,}000 \text{ psi} \qquad \sigma_3 = 0 \qquad (a)$$

According to the Mises criterion, Eq. (8.111), for this state of stress we have

$$\sqrt{\tfrac{1}{2}\left[(\sigma_1 - \sigma_2)^2 + (\sigma_2 - \sigma_3)^2 + (\sigma_3 - \sigma_1)^2\right]} = 43{,}600 \text{ psi} \qquad (b)$$

which is less than Y, so yielding will not occur.

According to the maximum shear stress criterion, we have yielding when

$$\frac{\sigma_{\max} - \sigma_{\min}}{2} = \frac{Y}{2} \qquad (c)$$

or

$$\frac{30{,}000 + 20{,}000}{2} = 25{,}000 \text{ psi} \qquad (d)$$

which is greater than $Y/2$, so yielding will occur according to this criterion.

EXAMPLE 8.19

A 1-in-diameter steel rod, when subjected to an axial force F in the absence of a twisting moment T, Fig. 8.63*a*, yields at an axial stress of 28,000 psi. The force to cause yielding in this case is 22,000 lb. If a twisting moment is applied in addition, the axial force to cause yielding is less than 22,000 lb. We wish to plot the locus of the values of F and T, with F positive,

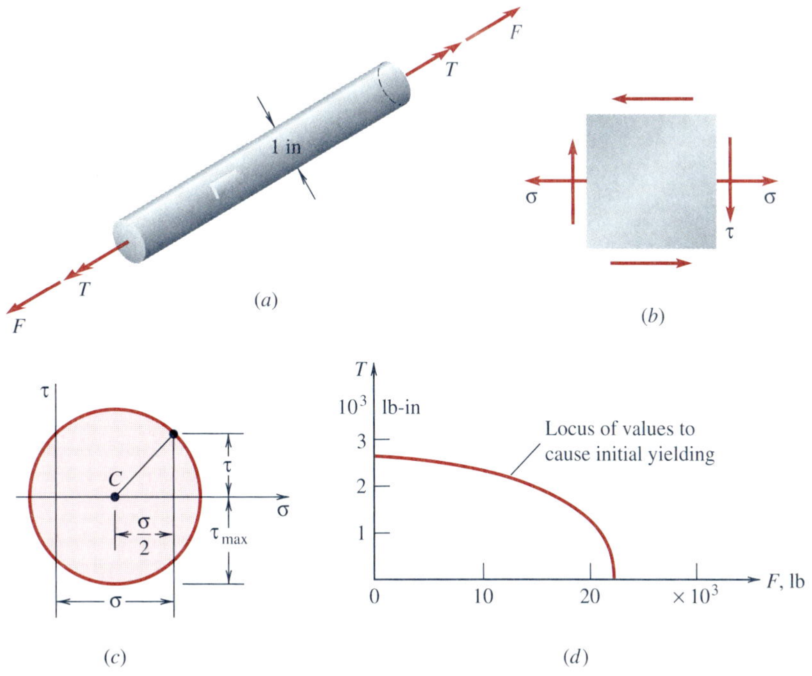

Figure 8.63 Example 8.19

such that initial yielding of the material occurs according to the maximum shear stress criterion.

Yielding of the material occurs according to the maximum shear stress criterion when

$$\tau_{max} = \frac{28{,}000}{2} = 14{,}000 \text{ psi} \qquad (a)$$

If an axial force F and a twisting moment T are applied to the rod, the state of stress at a point on the surface of the rod is as shown in Fig. 8.63*b*. The values of the stress are

$$\sigma = \frac{F}{\pi d^2/4} = \frac{4F}{\pi}$$
$$\tau = \frac{T(d/2)}{\pi d^4/32} = \frac{16T}{\pi} \qquad (b)$$

with $d = 1$ in. The sketch of Mohr's circle for this state of stress is given in Fig. 8.63*c*. The maximum shear stress follows from the radius of the circle

$$\tau_{max} = \sqrt{\left(\frac{\sigma}{2}\right)^2 + \tau^2} \qquad (c)$$

Combining Eqs. (*a*), (*b*), and (*c*), we have

$$\left(\frac{2F}{\pi}\right)^2 + \left(\frac{16T}{\pi}\right)^2 = 1.96 \times 10^8 \qquad (d)$$

Equation (*d*) is the equation for an ellipse if we plot the expression as a function of F and T, as in Fig. 8.63*d*; this equation is the locus of the values of F and T to cause yielding. We see from Eq. (*d*) that the values

$$F = 22{,}000 \text{ lb} \qquad T = 0$$
$$F = 0 \qquad T = 2750 \text{ lb} \cdot \text{in}$$

are the intercepts on the axes. Points on the ellipse will correspond to yielding of the rod, e.g., a load of $F = 15{,}000$ lb and a twisting moment of $T = 2010$ lb · in will cause yielding.

8.16 Concluding Remarks

This chapter provides an introduction to the concepts needed to analyze two- and three-dimensional elastic bodies under load. We have seen that the three-step approach of Eq. (2.8) used so often in the analyses of one-dimensional models of structures and machine components in earlier chapters still provides the guide to organizing our thinking for two- and three-dimensional problems.

The first step, or the use of force and equilibrium ideas, leads to the concepts of internal force intensities or stresses acting on elements of area passing through a point in a deformed body. In Part A, the notion of a stress vector acting on an element of area leads naturally to the idea of stress components acting on coordinate planes and from there to the state of stress at a point expressed in terms of the six independent stress components, as in Fig. 8.8. In the case of bodies experiencing plane stress, we found that by using either the stress transformation formulas or Mohr's circle construction it was possible to obtain stress components with respect to a rotated set of axes. This led naturally to the ideas of principal stresses and principal directions that carry over to three-dimensional bodies. The determination of principal stresses at a point in an elastic body is often an important first step in deciding the integrity under load of the body.

In Part B we turned to the study of geometry changes associated with deformation, and this led to a study of strain in bodies undergoing plane strain. We found that the determination of the normal strain of a line element at a point leads to the idea of strain components at a point. Associated with the strain components are strain transformation formulas and a corresponding Mohr's circle construction. In fact, stress and strain components transform according to the same transformation formulas; a quantity whose components transform in this way upon rotation of axes is called a *tensor.*[6] Principal strains and directions follow as in the case of stress.

Finally in Part C we turned to considerations of the force-deformation response of the material, i.e., the third step of Eq. (2.8). We introduced the stress-strain-temperature change relations for homogeneous isotropic linear elastic materials. We also considered the two-dimensional response of materials and relations between the elastic constants.

If an elastic material is subjected to large values of load, it may yield or plastically deform. The question arises as to how we know if material at a point in a body will yield when subjected to an arbitrary state of stress. We investigated this question briefly by introducing two initial yield criteria.

In the next chapter we will bring together many of the results from previous chapters for the calculation of stress components with the concepts of combining stresses to investigate safe levels of loading before the material yields.

PROBLEMS

8.2-1

(*a*) If the stress vector at a point acting on an element of area whose unit normal is **n** is given by Eq. (8.2), what is the magnitude of the stress vector?

(*b*) If the stress vector at a point acting on an element of area whose unit normal is **i** is given by Eq. (8.3), find an expression for the shear stress acting on the element of area.

(*c*) Write out the vector form of the stress vector, similar to Eq. (8.3), acting on the y and z coordinate planes.

8.2-2

(*a*) Sketch a volume element, showing the state of stress at a point if the state is one of hydrostatic compression with $\sigma_x = \sigma_y = \sigma_z = -100$ MPa.

(*b*) Sketch a volume element, showing the state of stress at a point if $\sigma_x = \sigma_y = \sigma_z = 70$ MPa, $\tau_{xy} = \tau_{yx} = 40$ MPa, and $\tau_{xz} = \tau_{zx} = \tau_{yz} = \tau_{zy} = 0$.

(*c*) Sketch a volume element, showing the state of stress at a point if $\sigma_x = 10$ ksi, $\sigma_y = 15$ ksi, $\sigma_z = -15$ ksi, $\tau_{xy} = \tau_{yx} = 4$ ksi, $\tau_{xz} = \tau_{zx} = 6$ ksi, and $\tau_{yz} = \tau_{zy} = 8$ ksi.

[6] Crandall, Dahl, and Lardner, op. cit., sec. 4.15.

(*d*) If the state of stress at a point is given by $\sigma_x = 100$ MPa, $\sigma_y = 80$ MPa, $\sigma_z = 40$ MPa, $\tau_{xy} = \tau_{yx} = 60$ MPa, and $\tau_{xz} = \tau_{zx} = \tau_{yz} = \tau_{zy} = 0$, sketch the volume element with the stress components acting and decompose the state of stress into the sum of a state of hydrostatic tension and an additional state of stress.

(*e*) Is it possible in general to decompose a state of stress at a point into the sum of a state of hydrostatic tension or compression and an additional state of stress? Is the decomposition unique?

8.2-3 In Sec. 2.8 we derived an equilibrium equation for a one-dimensional state of stress; see Fig. 2.41 and Fig. P8.2-3*a*. Force equilibrium in the *x* direction gives for a small element under a one-dimensional state of stress (Fig. P8.2-3*a*)

$$[\sigma_x(x + \Delta x)\, A - \sigma_x(x)\, A] + q\left(x + \frac{\Delta x}{2}\right) \Delta x\, A = 0$$

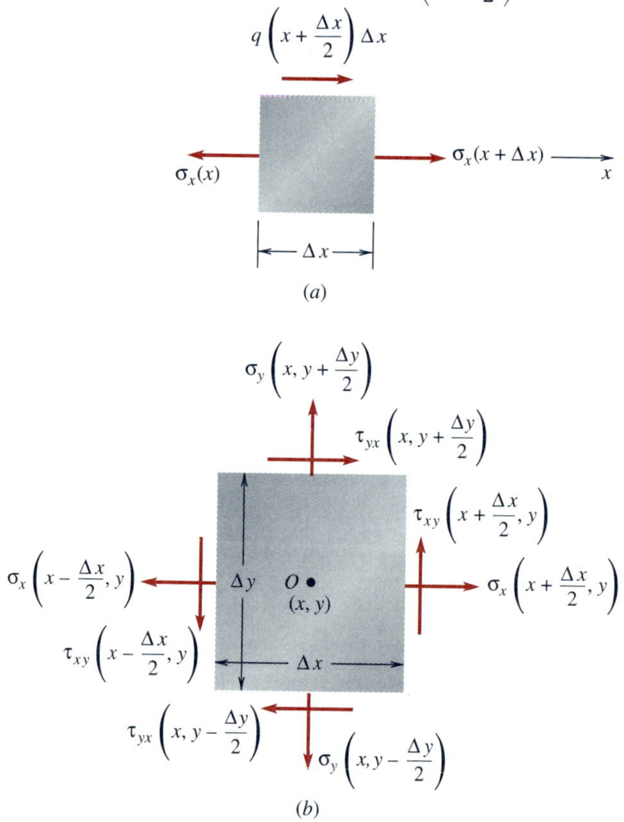

Fig. P8.2-3

where A is the cross-sectional area of the face on which the normal stress σ_x acts and q is the force per unit volume acting on the element of volume. In the limit as the element of volume shrinks to zero,

$$\frac{d\sigma_x}{dx} + q(x) = 0$$

In Fig. P8.2-3*b* we show a small two-dimensional element centered at O with coordinates (x,y) and the components of the stresses acting on coordinate planes displaced distances $\pm\Delta x/2$, $\pm\Delta y/2$ from O. We assume the element has a unit thickness in the direction perpendicular to the page. In the limit show that as the element shrinks to zero, force equilibrium in the x and the y directions gives the two-dimensional equilibrium equations

$$\frac{\partial \sigma_x}{\partial x} + \frac{\partial \tau_{yx}}{\partial y} = 0$$

$$\frac{\partial \tau_{xy}}{\partial x} + \frac{\partial \sigma_y}{\partial y} = 0$$

Remember that to obtain forces from stresses, the stresses must be multiplied by areas.

8.2-4 Show, following the arguments of Prob. 8.2-3, that the three-dimensional equilibrium equations at a point take the form

$$\frac{\partial \sigma_x}{\partial x} + \frac{\partial \tau_{yx}}{\partial y} + \frac{\partial \tau_{zx}}{\partial z} = 0$$

$$\frac{\partial \tau_{xy}}{\partial x} + \frac{\partial \sigma_y}{\partial y} + \frac{\partial \tau_{zy}}{\partial z} = 0$$

$$\frac{\partial \tau_{xz}}{\partial x} + \frac{\partial \tau_{yz}}{\partial y} + \frac{\partial \sigma_z}{\partial z} = 0$$

8.3-1 Show that the three-dimensional equations of equilibrium given in Prob. 8.2-4 reduce to the two-dimensional equations of equilibrium given in Prob. 8.2-3 if the assumptions of plane stress are invoked.

8.3-2 Sketch the stress components acting on an element of volume in plane stress, as in Fig. 8.10*b*, if the stress components are given by
(*a*) $\sigma_x = 10$ ksi, $\sigma_y = 20$ ksi, $\tau_{xy} = 10$ ksi
(*b*) $\sigma_x = \sigma_y = 100$ MPa, $\tau_{xy} = 50$ MPa
(*c*) $\sigma_x = \sigma_y = -70$ MPa, $\tau_{xy} = -30$ MPa
(*d*) $\sigma_x = 20{,}000$ psi, $\sigma_y = -5000$ psi, $\tau_{xy} = 0$
(*e*) $\sigma_x = -5$ ksi, $\sigma_y = 8$ ksi, $\tau_{xy} = 4$ ksi

8.3-3 Decompose the state of plane stress at a point into the sum of a state of stress in which $\sigma_x = \sigma_y$ and $\tau_{xy} = 0$, a state of biaxial stress, and an additional state of stress. Show the volume elements with the states of stress acting for each case:
(*a*) $\sigma_x = 20$ ksi, $\sigma_y = 15$ ksi, $\tau_{xy} = 7$ ksi
(*b*) $\sigma_x = \sigma_y = 130$ MPa, $\tau_{xy} = -30$ MPa
(*c*) $\sigma_x = \sigma_y = -50$ MPa, $\tau_{xy} = 50$ MPa
(*d*) $\sigma_x = 30$ ksi, $\sigma_y = -8$ ksi, $\tau_{xy} = 0$
(*e*) $\sigma_x = -5000$ psi, $\sigma_y = 8000$ psi, $\tau_{xy} = 6000$ psi

8.4-1
(*a*) Verify Eqs. (8.6) to (8.9).

(*b*) Verify that the stress vector can be written in the form

$$\overset{(n)}{\mathbf{T}} = \overset{(n)}{T_x}\mathbf{i} + \overset{(n)}{T_y}\mathbf{j}$$

or

$$\overset{(n)}{\mathbf{T}} = (\sigma_x \cos\theta + \tau_{yx}\sin\theta)\mathbf{i} + (\tau_{xy}\cos\theta + \sigma_y \sin\theta)\mathbf{j}$$

(*c*) Obtain the expression for the magnitude of the normal component of the stress vector acting on the element of area with normal $\mathbf{n} = \cos\theta\,\mathbf{i} + \sin\theta\,\mathbf{j}$

$$T_n = \overset{(n)}{\mathbf{T}} \cdot \mathbf{n}$$

(*d*) Obtain the expression for the shear stress component of the stress vector acting on the element of area with normal **n**

$$T_s = \overset{(n)}{\mathbf{T}} \cdot \mathbf{t}$$

where **t** is the vector lying in the area such that $\mathbf{t} = \mathbf{k} \times \mathbf{n}$.

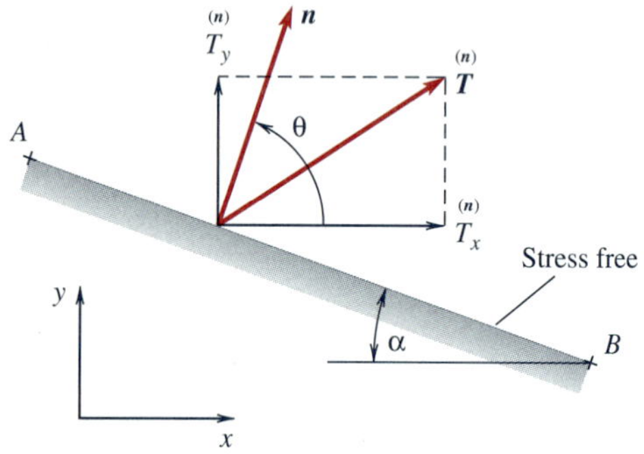

Fig. P8.4-2

8.4-2 If the boundary portion *AB* of a structure under plane stress is stress-free, as shown in Fig. P8.4-2, the stress vector acting on portion *AB* is zero. Express the condition that the components of the stress vector $\overset{(n)}{\mathbf{T}}$ must vanish in terms of the stress components with respect to the coordinate axes and the angle α.

8.4-3
(*a*) Verify Eq. (8.13).
(*b*) Verify Eqs. (8.15) and (8.16).

8.4-4 Show that the $\sigma_{x'}$ and $\tau_{x'y'}$ stress components of Fig. 8.14 can be identified as the normal and shear stress components of $\overset{(n)}{\mathbf{T}}$; see Prob. 8.4-1*c* and *d*. Hence obtain Eqs. (8.11) from which Eqs. (8.12), (8.15), and (8.16) follow.

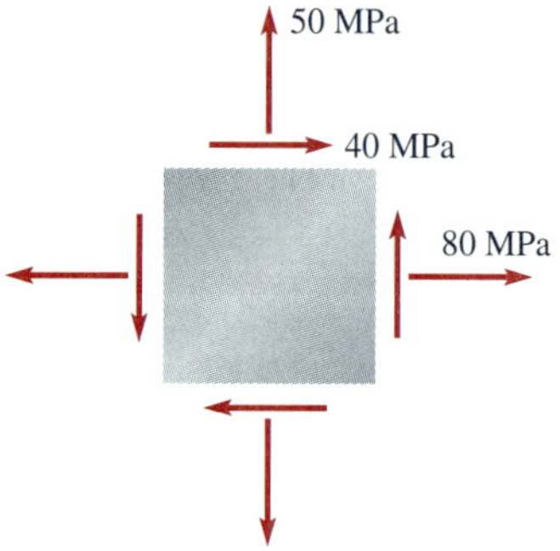

Fig. P8.4-5

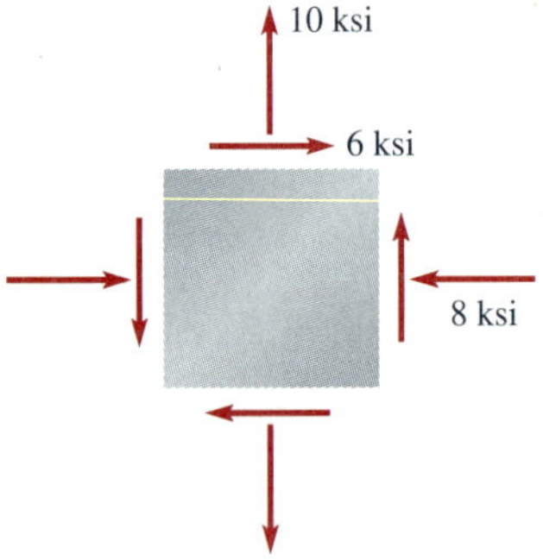

Fig. P8.4-6

In Probs. 8.4-5 to 8.4-9, an element is in a state of plane stress. In each case the stress components σ_x, σ_y, and τ_{xy} are given. Find the normal and shear stress components acting on the faces of an element rotated through an angle of θ from the positive *x* direction and positive in the counterclockwise direction. Show your results on a sketch of the rotated element. See Figs. P8.4-5 to P8.4-9.

Problem	σ_x	σ_y	τ_{xy}	θ
8.4-5	80 MPa	50 MPa	40 MPa	−40°
8.4-6	−8 ksi	10 ksi	6 ksi	70°
8.4-7	−90 MPa	−60 MPa	−75 MPa	−50°
8.4-8	12 ksi	−7 ksi	−8 ksi	20°
8.4-9	−80 MPa	80 MPa	40 MPa	−10°

8.4-10 A thin wood plate *ABCD* is fabricated by gluing two pieces of wood

along surface AD, as shown in Fig. P8.4-10. Stresses $\sigma_x = -400$ psi, $\sigma_y = -200$ psi, and $\tau_{xy} = 0$ are acting in the plate.
(*a*) Find the normal and shear stress components acting on surface AD.
(*b*) Show your result on a sketch of an element that is oriented normal to surface AD.

8.4-11 Consider the wood plate in Prob. 8.4-10. If the magnitude of the shear stress acting on surface AD is limited to 100 psi and σ_y is fixed at the value of -200 psi, find the range of tensile and compressive values that σ_x may have without causing the shear stress along AD to exceed the allowable limit of 100 psi.

8.4-12 The grain in a thin wood plate is inclined 20° with the x direction, as shown in Fig. P8.4-12. Find the shear stress parallel to the grain and the normal-stress perpendicular to the grain.

8.4-13 An element in plane stress is subjected to stresses on its faces as shown in Fig. P8.4-13. Find the normal and shear stress components acting on inclined plane AB.

8.4-14 An element in plane stress is subjected to stresses on its faces as shown in Fig. P8.4-14. Find the normal and shear stress components acting on inclined plane AB.

8.4-15 An element in plane stress is subjected to stresses on its faces as shown in Fig. P8.4-15.
(*a*) Find the orientation of the inclined planes, i.e., the angle that the normal to these planes makes with the x axis, that have the property that $\sigma_{x'} = 0$.
(*b*) What are the values of $\tau_{x'y'}$ on these planes?
(*c*) Show the results on sketches of appropriate elements.

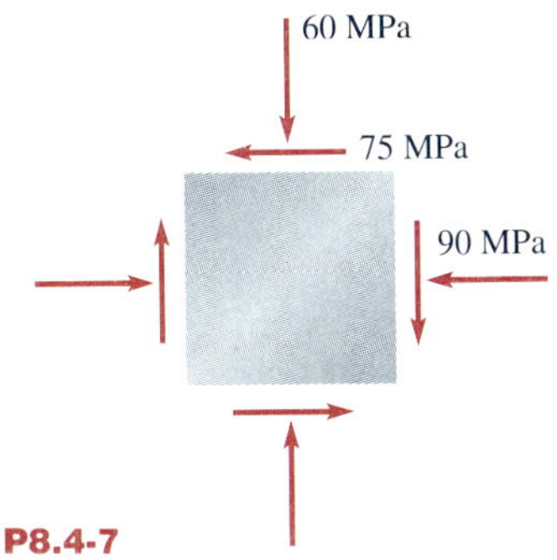

Fig. P8.4-7

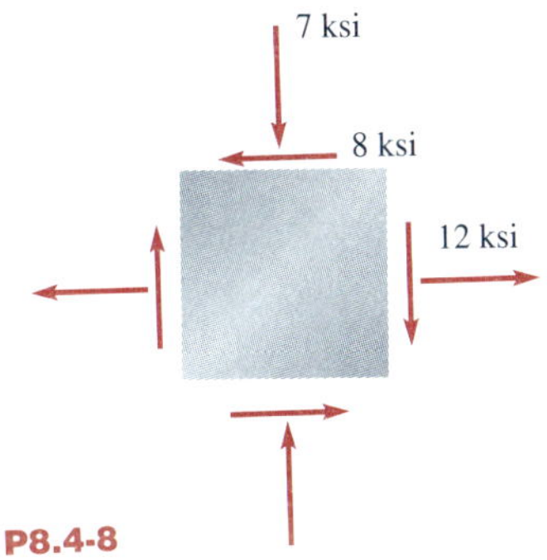

Fig. P8.4-8

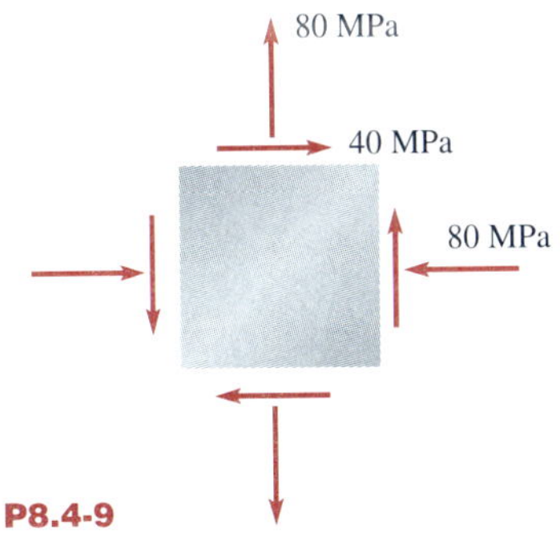

Fig. P8.4-9

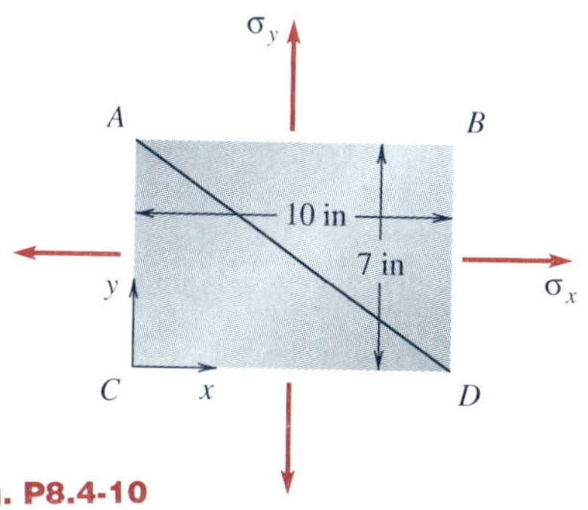

Fig. P8.4-10

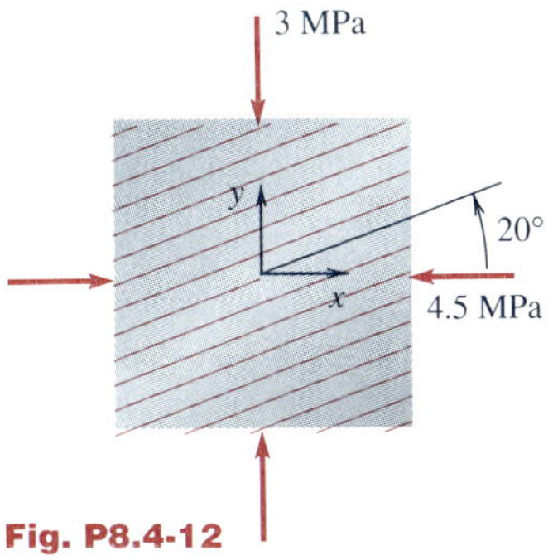

Fig. P8.4-12

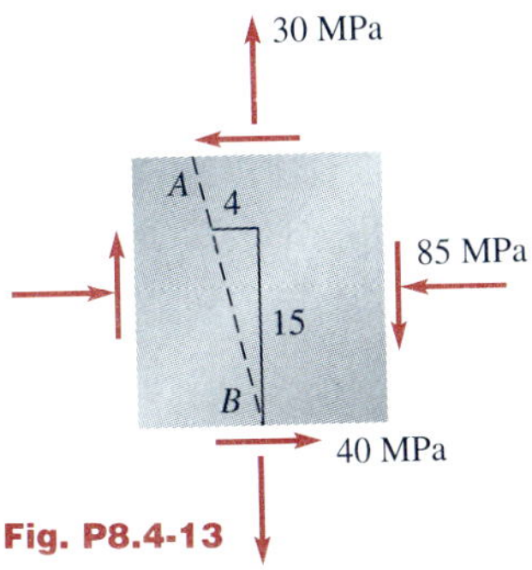

Fig. P8.4-13

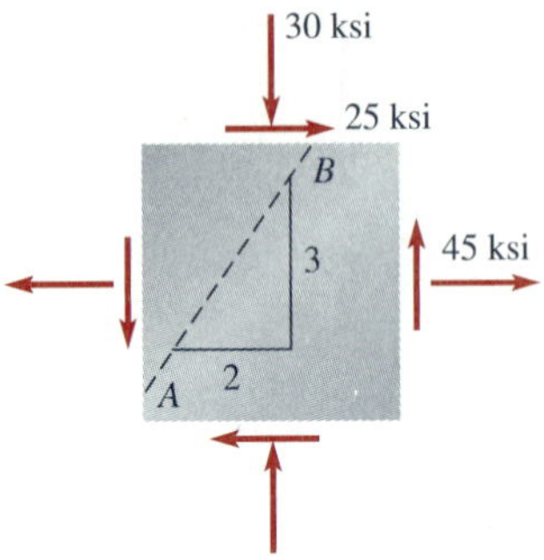

Fig. P8.4-14

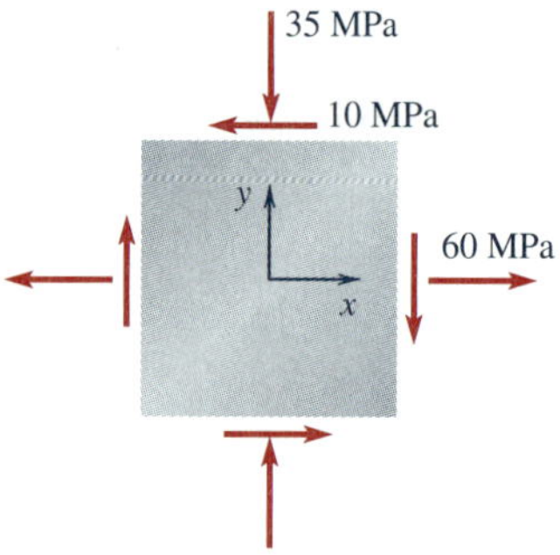

Fig. P8.4-15

8.4-16 Consider the element shown in Fig. P8.4-15.
(*a*) Find the orientation of inclined planes for this element which have zero shear stress.
(*b*) Show the results on a sketch of an appropriate element oriented with respect to the x direction.

8.5-1
(*a*) Verify Eqs. (8.17) and (8.18).
(*b*) Verify Eqs. (8.19) to (8.23).
(*c*) Verify the entries in Table 8.1 for the principal stresses and the maximum shear stresses.
(*d*) Verify Eqs. (8.32) to (8.34).

8.5-2 Consider Example 8.2, Fig. 8.21. If the stress component σ_y in this example is changed to $+5500$ psi, with $\sigma_x = 10{,}500$ psi and $\tau_{xy} = -4000$ psi, find the principal stresses, principal directions, and maximum shear stresses. Show, as in Fig. 8.21*b* and *c*, the orientation of the elements.

8.5-3 Consider Example 8.3, Fig. 8.22. Solve part (*a*) of this example when $\sigma_x = -229$ MPa, showing the orientation of the elements for the principal directions and maximum shear stress.

8.5-4 Consider Example 8.3, Fig. 8.22. Solve part (*a*) of this example when $\sigma_x = 349$ MPa, showing the orientation of the elements for the principal directions and maximum shear stress.

For Probs. 8.5-5 to 8.5-9, a state of plane stress at a point is defined by the stress components indicated in the figure.
(*a*) Find the principal stresses, and show them acting on an element oriented along the principal directions.
(*b*) Find the maximum and minimum shear stresses, and show them on an element oriented along the directions of maximum and minimum shear stress.

8.5-5 See Fig. P8.4-5.

8.5-6 See Fig. P8.4-6.

8.5-7 See Fig. P8.4-7.

8.5-8 See Fig. P8.4-8.

8.5-9 See Fig. P8.4-9.

For Probs. 8.5-10 to 8.5-12, find the principal stresses and the maximum and minimum shear stresses for the stress state shown in the figure. In each case show your results on sketches of elements oriented in the principal directions and in the directions of maximum and minimum shear stress.

8.5-10 A uniaxial stress state $\sigma_x = \sigma_0$, $\sigma_y = \tau_{xy} = 0$ as shown in Fig. P8.5-10.

8.5-11 A pure shear state $\tau_{xy} = \tau_0 > 0$, $\sigma_x = \sigma_y = 0$ as shown in Fig. P8.5-11.

8.5-12 A uniform normal biaxial stress state $\sigma_x = \sigma_y = \sigma_0$, $\tau_{xy} = 0$ as shown in Fig. P8.5-12.

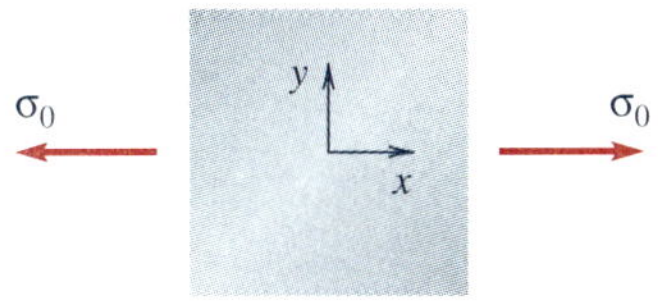

Fig. P8.5-10

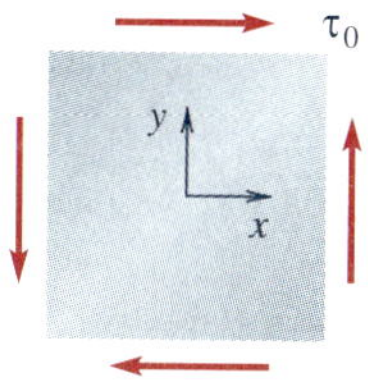

Fig. P8.5-11

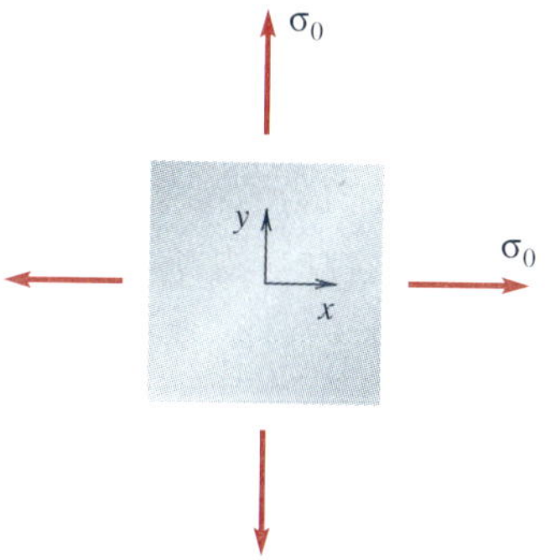

Fig. P8.5-12

For Probs. 8.5-13 to 8.5-15, the elements oriented at an angle with the *x* direction are subjected to the stress components shown in the figures. Find the principal stresses and the maximum and minimum shear stresses for the stress states shown. In each case show the results on sketches of elements oriented in the principal directions and in the directions of maximum and minimum shear.

8.5-13 $\sigma_{x'} = -50$ MPa, $\sigma_{y'} = 40$ MPa, $\tau_{x'y'} = -30$ MPa, Fig. P8.5-13

8.5-14 $\sigma_{x'} = -12$ ksi, $\sigma_{y'} = -20$ ksi, $\tau_{x'y'} = -10$ ksi, Fig. P8.5-14

8.5-15 $\sigma_{x'} = 45$ MPa, $\sigma_{y'} = -30$ MPa, $\tau_{x'y'} = -20$ MPa, Fig. P8.5-15

For Probs. 8.5-16 to 8.5-25, determine the principal stresses, principal directions, maximum shear stress, and orientation of the maximum shear stress. In addition, determine the stress components on a set of axes rotated $\theta = 28°$ from the *x* axis. In each problem show the results on sketches of elements oriented in the principal directions, in the directions of maximum and minimum shear, and at angle θ.

8.5-16 $\sigma_x = 10$ ksi, $\sigma_y = 20$ ksi, $\tau_{xy} = 10$ ksi

8.5-17 $\sigma_x = 100$ MPa, $\sigma_y = 100$ MPa, $\tau_{xy} = 50$ MPa

8.5-18 $\sigma_x = -70$ MPa, $\sigma_y = -70$ MPa, $\tau_{xy} = -30$ MPa

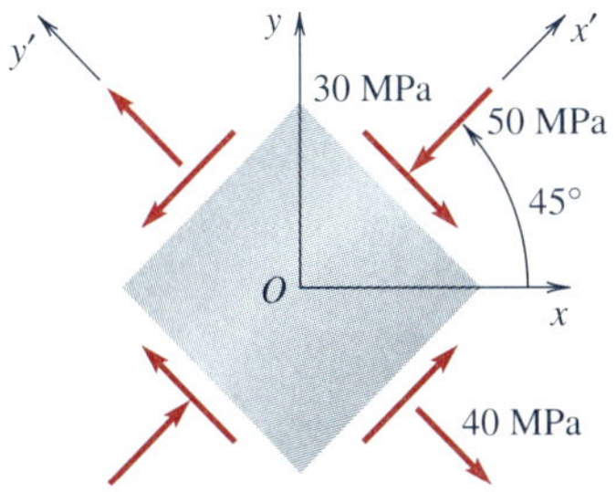

Fig. P8.5-13

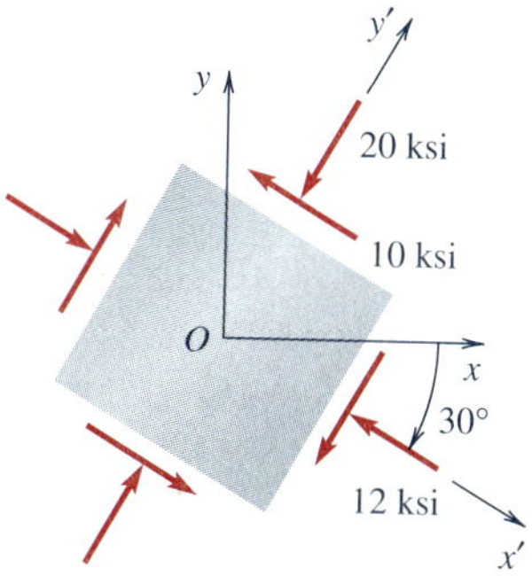

Fig. P8.5-14

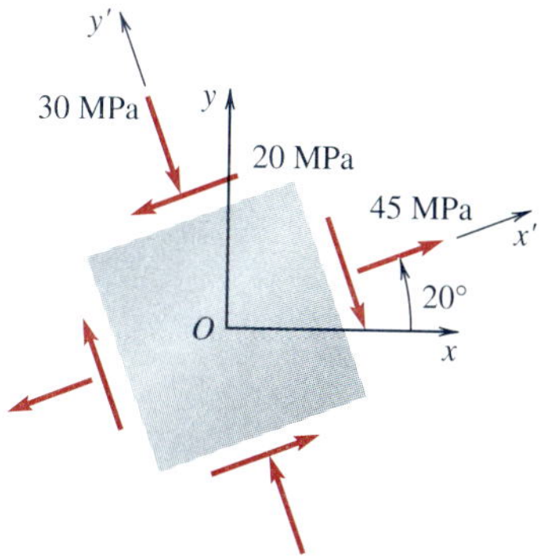

Fig. P8.5-15

8.5-19 $\sigma_x = 20{,}000$ psi, $\sigma_y = -5000$ psi, $\tau_{xy} = 0$

8.5-20 $\sigma_x = -5$ ksi, $\sigma_y = 8$ ksi, $\tau_{xy} = 4$ ksi

8.5-21 $\sigma_x = 20$ ksi, $\sigma_y = 15$ ksi, $\tau_{xy} = 7$ ksi

8.5-22 $\sigma_x = 130$ MPa, $\sigma_y = 130$ MPa, $\tau_{xy} = -30$ MPa

8.5-23 $\sigma_x = -50$ MPa, $\sigma_y = -50$ MPa, $\tau_{xy} = 50$ MPa

8.5-24 $\sigma_x = 30$ ksi, $\sigma_y = -8$ ksi, $\tau_{xy} = 0$

8.5-25 $\sigma_x = -5000$ psi, $\sigma_y = 8000$ psi, $\tau_{xy} = 6000$ psi

8.6-1 Use Mohr's circle construction for plane stress to solve
(*a*) Prob. 8.5-5
(*b*) Prob. 8.5-6
(*c*) Prob. 8.5-7
(*d*) Prob. 8.5-8
(*e*) Prob. 8.5-9
(*f*) Prob. 8.4-13
(*g*) Prob. 8.4-14
(*h*) Prob. 8.4-15
(*i*) Prob. 8.4-16

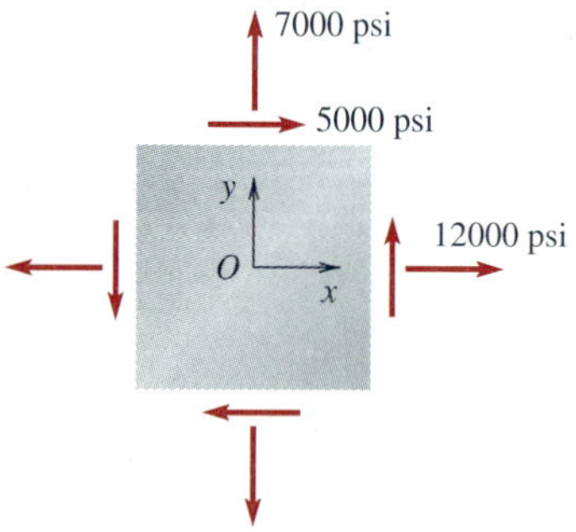

Fig. P8.6-2

8.6-2 For the plane stress element shown in Fig. P8.6-2, (*a*) sketch Mohr's circle for the stress state; (*b*) find the principal stresses, and show your results on a sketch of an element oriented in the principal directions; and (*c*) find the maximum and the minimum shear stresses and show results on a properly oriented sketch.

8.6-3 Find the principal stresses, maximum shear stress, and orientation of the principal axes of stress for the following states of plane stress. Show the orientation of the elements with the corresponding stress components acting in each case.
(*a*) $\sigma_x = 40$ MPa, $\sigma_y = 0$, $\tau_{xy} = 80$ MPa
(*b*) $\sigma_x = 10$ ksi, $\sigma_y = -4$ ksi, $\tau_{xy} = 8$ ksi
(*c*) $\sigma_x = 140$ MPa, $\sigma_y = 20$ MPa, $\tau_{xy} = -60$ MPa
(*d*) $\sigma_x = 10{,}000$ psi, $\sigma_y = 20{,}000$ psi, $\tau_{xy} = -6000$ psi
(*e*) $\sigma_x = 120$ MPa, $\sigma_y = 50$ MPa, $\tau_{xy} = 100$ MPa

8.6-4 Sketch Mohr's circle for stress for each of the special cases of plane stress shown in Fig. P8.6-4. Comment on the nature of Mohr's circle for each case.

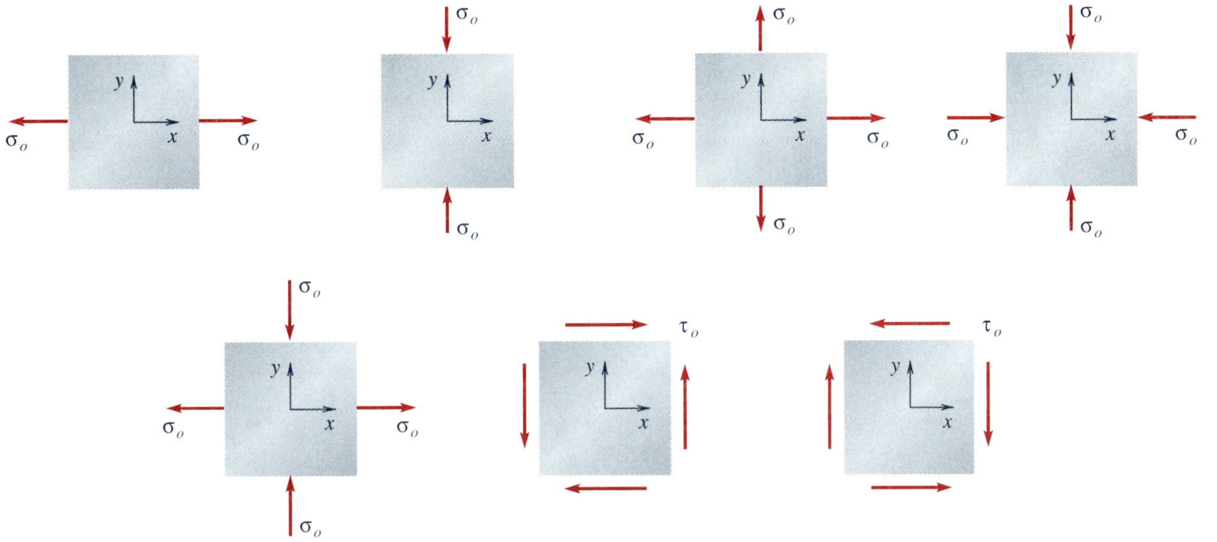

Fig. P8.6-4

8.6-5 If the minimum principal stress at a point is -7 MPa, find σ_x and the principal directions for the case of plane stress shown in Fig. P8.6-5.

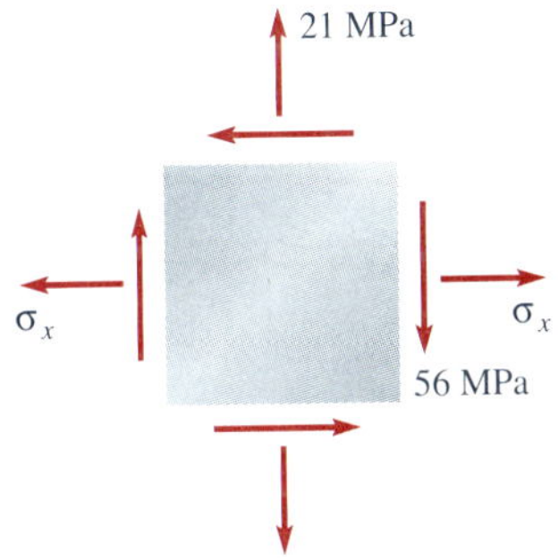

Fig. P8.6-5

8.6-6 The τ_{xy} component of stress at a point is equal to 2 ksi, as shown in Fig. P8.6-6. The principal stresses at the point are $\sigma_1 = 2$ ksi and $\sigma_2 = -6$ ksi. Find the values of the σ_x and σ_y stresses. Are these values unique?

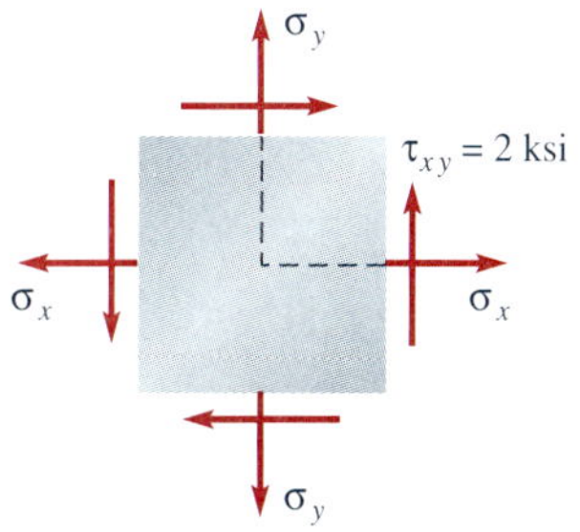

Fig. P8.6-6

8.6-7 The state of stress at a point is given, as shown in Fig. P8.6-7. Determine the principal stresses and principal directions, and show the principal stresses and directions on an appropriate element. Determine the maximum shear stress and the orientation of the axes of maximum shear stress. Show all the stresses acting on the element oriented in the direction of the maximum shear stress.

8.6-8 An element is subjected to different states of plane stress, as shown in Fig. P8.6-8. Construct Mohr's circle for each state of plane stress, and indicate on the circle all important values. In addition, show stress components on an element oriented in the principal directions and the stress on an element oriented in the direction of maximum shear stress.

(*a*) Fig. P8.6-8*a*
(*b*) Fig. P8.6-8*b*
(*c*) Fig. P8.6-8*c*
(*d*) Fig. P8.6-8*d*
(*e*) Fig. P8.6-8*e*
(*f*) Fig. P8.6-8*f*

8.6-9 Construct Mohr's circle for the given state of plane stress, and indicate on the circle all important values. In addition, show stress components on an

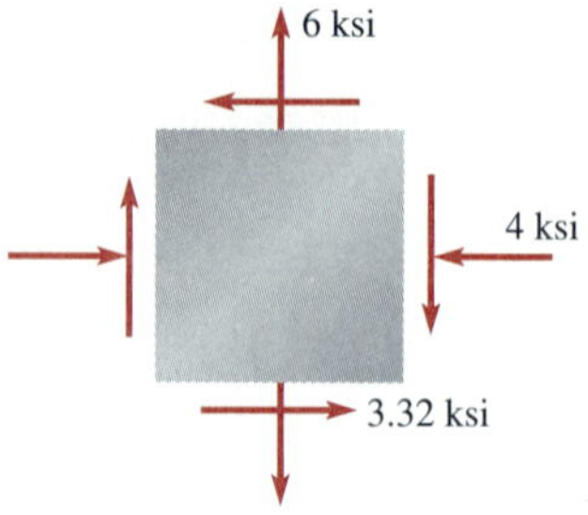

Fig. P8.6-7

element oriented in the principal directions and the stress on an element oriented in the direction of maximum shear stress. Also sketch the element corresponding to angle $\theta = 28°$ from the x axis. The stress state is given in

(*a*) Prob. 8.3-2*a*
(*b*) Prob. 8.3-2*b*
(*c*) Prob. 8.3-2*c*
(*d*) Prob. 8.3-2*d*
(*e*) Prob. 8.3-2*e*
(*f*) Prob. 8.3-3*a*
(*g*) Prob. 8.3-3*b*
(*h*) Prob. 8.3-3*c*
(*i*) Prob. 8.3-3*d*
(*j*) Prob. 8.3-3*e*
(*k*) Prob. 8.5-16
(*l*) Prob. 8.5-17
(*m*) Prob. 8.5-18
(*n*) Prob. 8.5-19
(*o*) Prob. 8.5-20
(*p*) Prob. 8.5-21
(*q*) Prob. 8.5-22
(*r*) Prob. 8.5-23
(*s*) Prob. 8.5-24
(*t*) Prob. 8.5-25

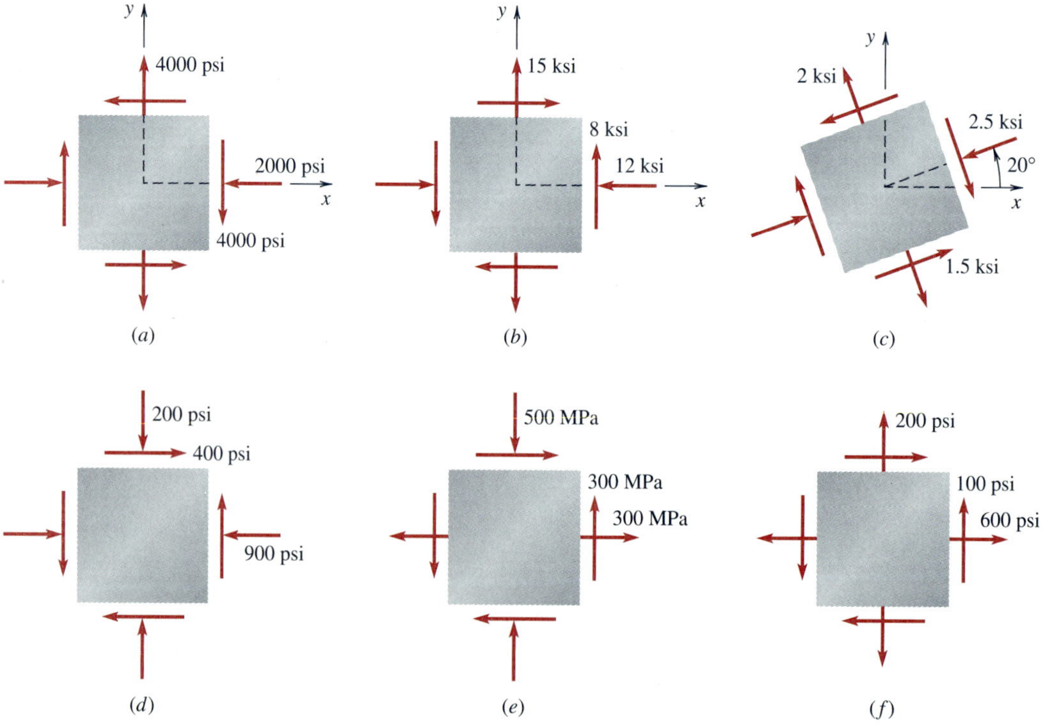

Fig. P8.6-8

8.6-10 Verify Eqs. (8.36) and (8.37).

8.6-11 Construct Mohr's circle for the given state of plane stress, and obtain from the circle the stress components at angle θ from the x axis. The stress state is given in
(*a*) Prob. 8.3-2*a*, $\theta = 20°$
(*b*) Prob. 8.3-2*b*, $\theta = 35°$
(*c*) Prob. 8.3-2*c*, $\theta = -12°$
(*d*) Prob. 8.3-2*d*, $\theta = -60°$
(*e*) Prob. 8.3-2*e*, $\theta = -90°$
(*f*) Prob. 8.3-3*a*, $\theta = 48°$
(*g*) Prob. 8.3-3*b*, $\theta = 35°$
(*h*) Prob. 8.3-3*c*, $\theta = -29°$
(*i*) Prob. 8.3-3*d*, $\theta = 46°$
(*j*) Prob. 8.3-3*e*, $\theta = 180°$

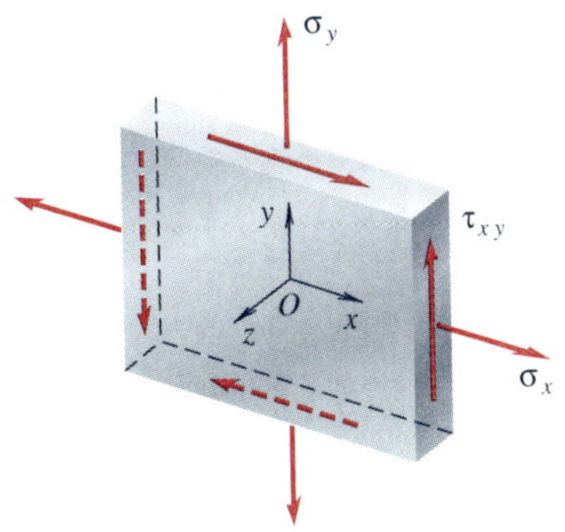

Fig. P8.7-1

For Probs. 8.7-1 to 8.7-4, reference is made to the plane stress element at a point, as shown in Fig. P8.7-1. On a single plot construct Mohr's circles for the states of stress corresponding to the *xy*, *zx*, and *yz* planes. Find the maximum shear stress at this point.

8.7-1 $\sigma_x = 40$ MPa, $\sigma_y = -20$ MPa, and $\tau_{xy} = -30$ MPa

8.7-2 $\sigma_x = -30$ ksi, $\sigma_y = 15$ ksi, and $\tau_{xy} = 20$ ksi

8.7-3 $\sigma_x = -60$ ksi, $\sigma_y = -30$ ksi, and $\tau_{xy} = -30$ ksi

8.7-4 $\sigma_x = 0$ ksi, $\sigma_y = 0$ ksi, and $\tau_{xy} = -20$ ksi

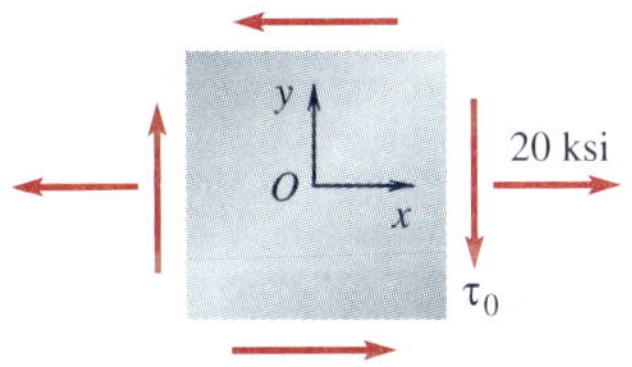

Fig. P8.7-5

8.7-5 An element in a state of plane stress is subjected to stress $\sigma_x = 20$ ksi, as shown in Fig. P8.7-5. It is proposed to add the shear stress τ_0. What is the maximum allowable value of τ_0 if the maximum allowable shear stress at this point (either in plane or out of plane) must not exceed 15 ksi?

8.7-6 The element shown in Fig. P8.7-6 is subjected to the stress state $\sigma_y = -80$ MPa and $\tau_{xy} = -50$ MPa, and it is proposed to subject the element to the additional stress component σ_x.
(*a*) Plot the three Mohr circles corresponding to the in-plane and out-of-plane stresses for the cases $\sigma_x = 0$, 80, and -150 MPa. In each case find the maximum shear stress (either in plane or out of plane).
(*b*) If the maximum allowable shear stress is 100 MPa, then find the range $\sigma_L \leq \sigma_x \leq \sigma_U$ for which the maximum allowable shear stress is not exceeded.

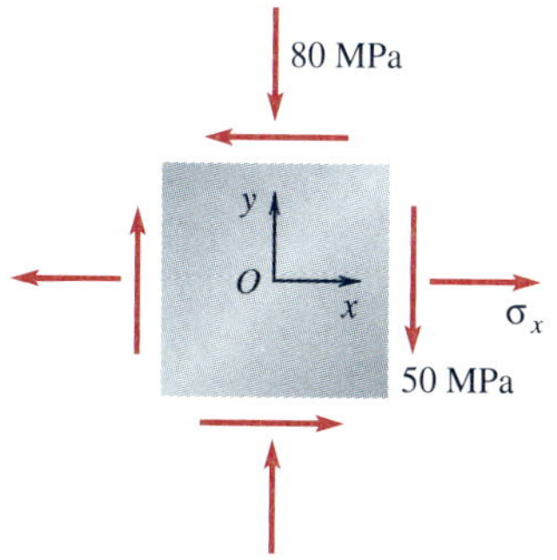

Fig. P8.7-6

8.7-7 Verify Eqs. (8.39) to (8.41).

8.7-8
(*a*) Give an explanation of the statement following Eq. (8.41) that with τ_{zx} and τ_{zy} acting on an element there will be no orientation of the face x' for which only a normal stress will be acting.
(*b*) Verify the result for the maximum shear stress given by Eq. (8.42).

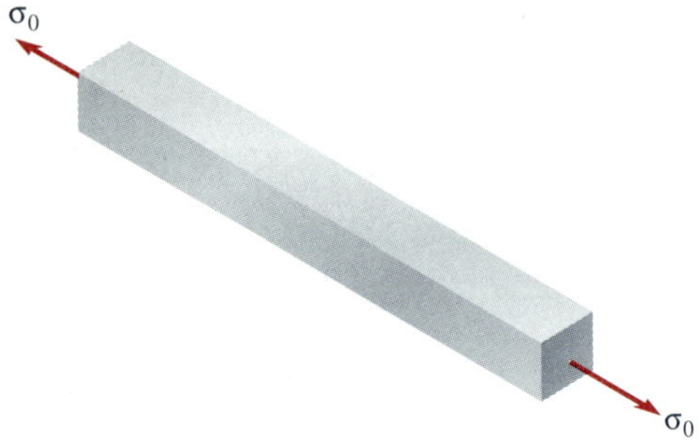

Fig. P8.7-9

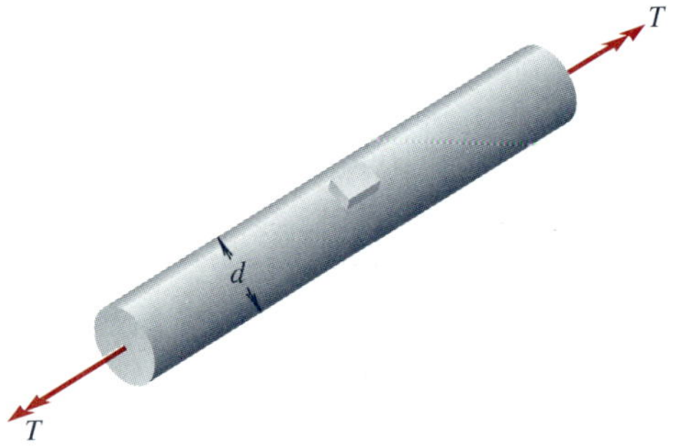

Fig. P8.7-10

8.7-9 A bar shown in Fig. P8.7-9 is loaded in uniaxial tension by a normal stress $\sigma_0 = 5$ ksi. Sketch Mohr's circle for a general state of stress at a point in the bar, and determine the value of the maximum shear stress and the orientation of the face on which the maximum shear stress acts.

8.7-10 A circular shaft of diameter d is under a twisting moment T, as shown in Fig. P8.7-10. Sketch Mohr's circle for a general state of stress at a point on the surface of the shaft. Determine the values of the principal stresses, the principal directions, the maximum shear stress, and the face on which the maximum shear stress acts. What is the value of the maximum normal stress acting at the point? Take $d = 1$ in and $T = 100$ lb · in.

8.7-11 Solve Prob. 8.7-10 with $d = 30$ mm and $T = 150$ N · m.

8.7-12 A simply supported rectangular beam is loaded by a load P at the midspan, as shown in Fig. P8.7-12. If we neglect the weight of the beam, sketch Mohr's circle for the general state of stress at (*a*) point A on the top surface and (*b*) point B on the neutral surface at $x = L/4$. Determine the values of the principal stresses, principal directions, maximum shear stress, and orientation of the face on which the maximum shear stress acts.

8.7-13 Solve Prob. 8.7-12, this time including the weight w (force/length) acting on the beam. The total load on the beam is P and the constant distributed load w over the length of the beam.

8.7-14 Solve Prob. 8.7-12 for a point midway between points A and B at the same section, i.e., at $y = h/4$.

8.7-15 A simply supported rectangular beam is loaded by a load P at the midspan, as shown in Fig. P8.7-12. Sketch Mohr's circle representation for the general state of stress at midspan at a point $y = h/4$. Determine the values of

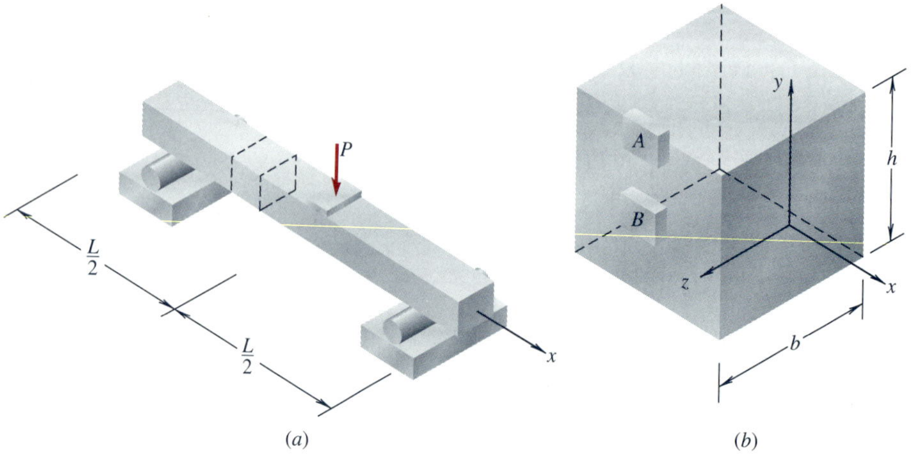

Fig. P8.7-12

the principal stresses, principal directions, maximum shear stress, and orientation of the face on which the maximum shear stress acts. Include the weight of the beam.

8.7-16 A biaxial state of plane stress acting at a point in a body is shown in Fig. P8.7-16. Sketch the Mohr's circle representation for a general state of stress at the point. Determine the value of the maximum shear stress and the orientation of the face on which the maximum shear stress acts.
(*a*) Take $\sigma_x = 1200$ psi, $\sigma_y = 800$ psi
(*b*) Take $\sigma_x = 40$ MPa, $\sigma_y = 40$ MPa
(*c*) Take $\sigma_x = 300$ psi, $\sigma_y = -300$ psi
(*d*) Take $\sigma_x = -100$ psi, $\sigma_y = 100$ psi
(*e*) Take $\sigma_x = 70$ MPa, $\sigma_y = 140$ MPa

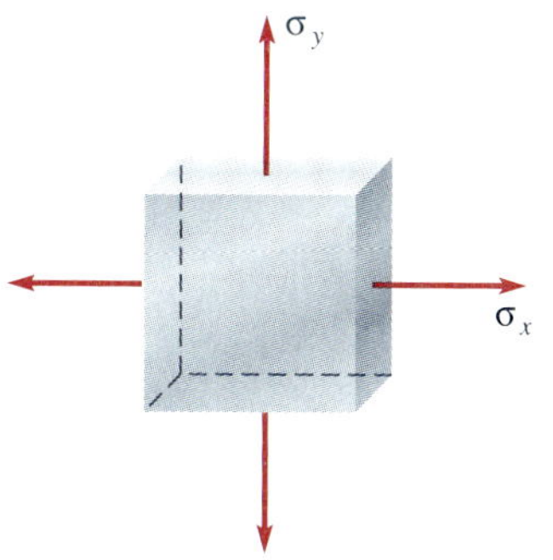

Fig. P8.7-16

8.7-17 Given the element shown in Fig. P8.7-17 in a state of plane stress, determine the principal stresses and the maximum shear stress.

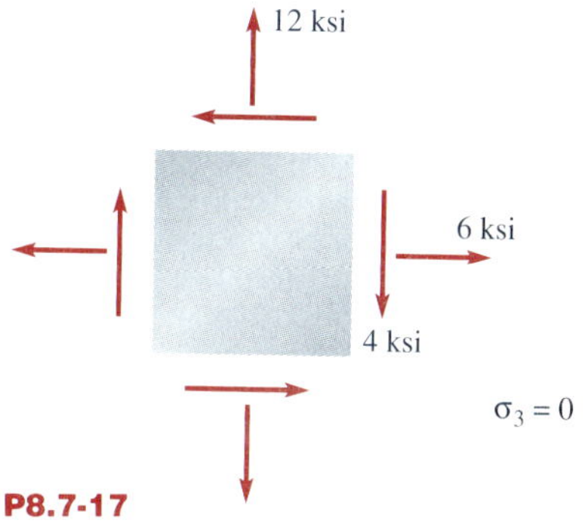

Fig. P8.7-17

8.7-18 A small circular shaft carries an axial load $F = 1200$ N and a twisting moment $T = 0.5$ N · m, as shown in Fig. P8.7-18. Find the maximum shear stress at a point O at the surface of the shaft, and show an element with the maximum shear stress acting. At the same point O show the principal stresses acting on an element oriented in the principal directions.

8.7-19 A section of a circular shaft as shown in Fig. P8.7-19 experiences a twisting moment T and an axial force F. Find the state of stress at a point A on the surface of the shaft. Take $T = 10$ lb · in, $F = 100$ lb, and $d = 0.25$ in.

8.7-20 Obtain the principal stresses and the principal directions for the stress state obtained in Prob. 8.7-19.

8.8-1 Consider Example 8.10, Fig. 8.38. For a vessel of the same material and radius r but with a different thickness $t = 70$ mm, find the allowable internal pressure, using the allowable stress of 100 MPa.

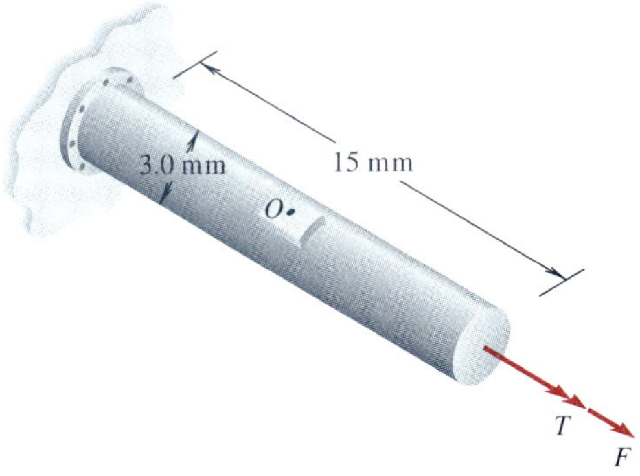

Fig. P8.7-18

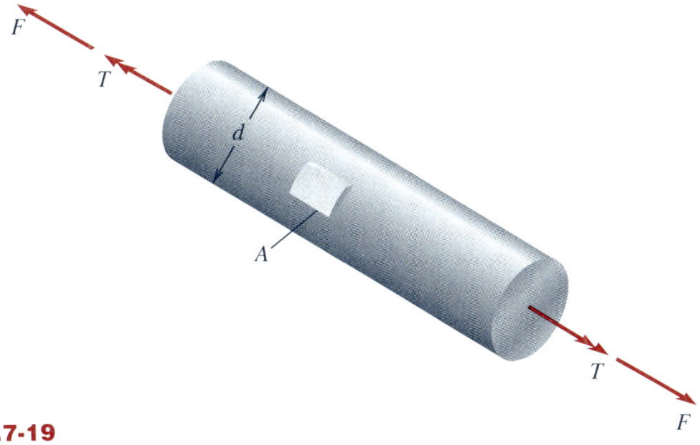

Fig. P8.7-19

8.8-2 Consider Example 8.11, Fig. 8.40*a*. If the spiral angle for the welded pipe is now taken as 60°, find the normal stress perpendicular to the weld and the shearing stress parallel to the weld; compare your results with those in Example 8.11. All other parameters remain the same.

8.8-3 Consider Example 8.12. It is proposed to change the design thickness of the boiler to 1 in. Solve parts (*a*) and (*b*) of Example 8.12, using the new reduced wall thickness. All other parameters remain the same.

8.8-4 A cylindrical boiler with hemispherical ends is made up of two sections welded together at *B*, as shown in Fig. P8.8-4. Find (*a*) the maximum hoop (circumferential) stress in the boiler, (*b*) the maximum tensile stress perpendicular to the weld, and (*c*) the maximum tensile stress in the spherical end sections. Take $d = 5$ m, $t = 50$ mm, and $p = 1.5$ MPa.

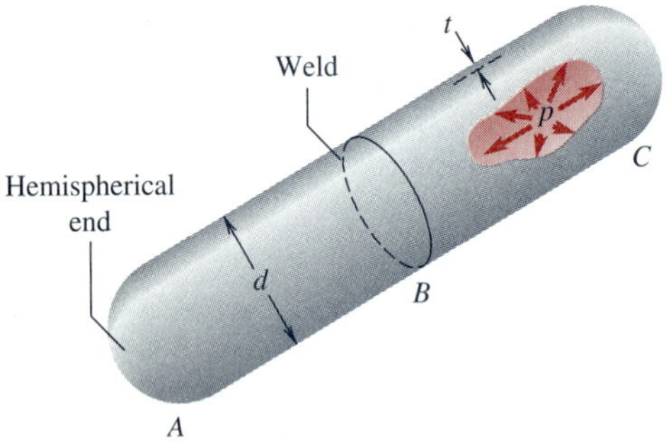

Fig. P8.8-4

8.8-5 A closed thin-walled cylinder of internal radius r and thickness t under an internal pressure p is also subjected to an axial force F, as shown in Fig. P8.8-5. Show that the stresses in the wall are given approximately as

$$\sigma_3 \approx 0 \qquad \sigma_1 = \frac{pr}{2t} + \frac{F}{2\pi rt} \qquad \sigma_2 = \frac{pr}{t}$$

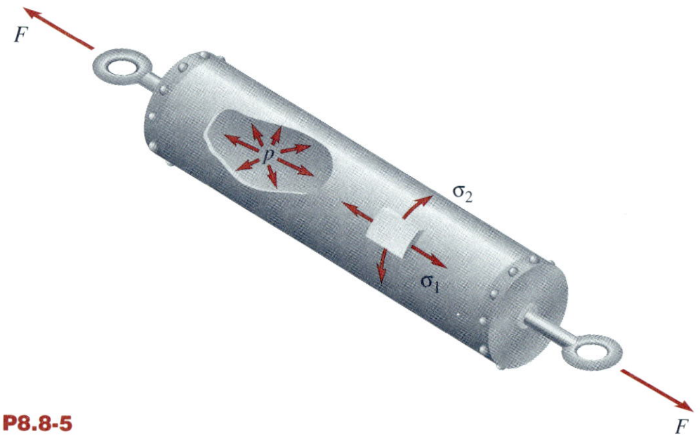

Fig. P8.8-5

8.8-6 Sketch Mohr's circle construction for the state of stress given in Prob. 8.8-5, and determine the value of the maximum shear stress in the wall. Consider the special cases of F greater or less than $\pi r^2 p$.

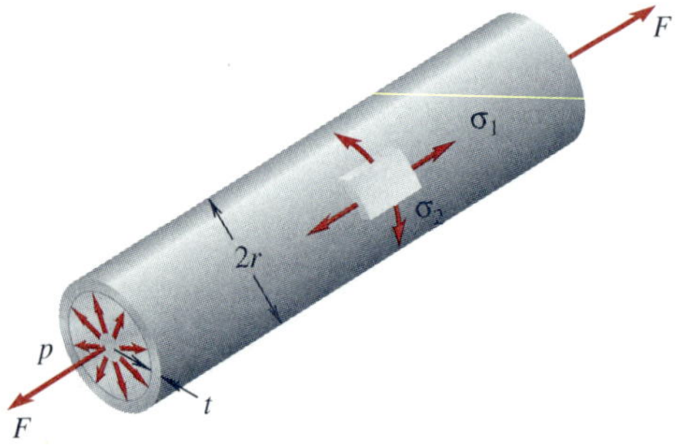

Fig. P8.8-7

8.8-7 A thin-walled open-ended pipe shown in Fig. P8.8-7 is subjected to an internal pressure p and an axial force F. Show that the stresses in the wall of the pipe are given approximately by

$$\sigma_3 \approx 0 \qquad \sigma_1 = \frac{F}{2\pi rt} \qquad \sigma_2 = \frac{pr}{t}$$

8.8-8 If the force F in Fig. P8.8-7 is negative, i.e., the pipe is under compression, find the value of F such that the maximum shear stress in the pipe has the same magnitude as the maximum normal stress component.

8.8-9 A long cylindrical pressure vessel with closed ends is fabricated by rolling a strip of plastic of thickness t and width w into a helix and making a continuous fused joint, as shown in Fig. P8.8-9. What is the maximum allowable width w of the strip if the fused joint is to experience a tensile stress of 80 percent of the maximum tensile stress in the plastic material?

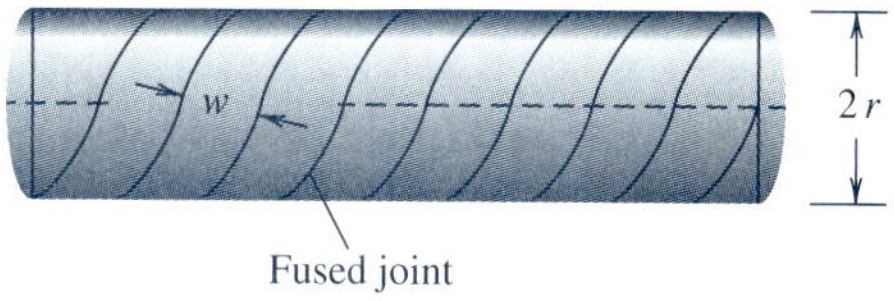

Fig. P8.8-9

8.8-10 A sea urchin egg can be thought of as a thin-walled pressurized sphere. The internal volume of an egg is approximately 35×10^{-5} mm^3, and the wall thickness is $t = 1$ μm. Micropuncture techniques give an internal pressure of 150 mmHg. Estimate the membrane stresses in the wall of the egg.

8.8-11 A closed end lightweight pressure vessel of wall thickness t is constructed by wrapping filaments about the cylinder and using epoxy resin as a binder, as shown in Fig. P8.8-11. Estimate the angle of winding α of the filaments such that the tensile stresses in the filaments are equal when the vessel is subjected to an internal pressure p.

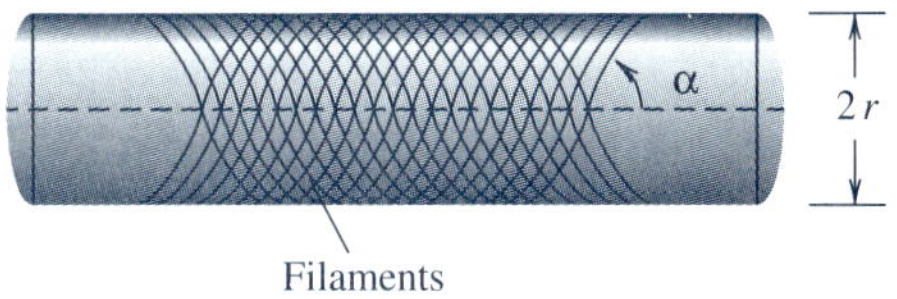

Fig. P8.8-11

8.8-12 An ovarian follicle from which an ovum is released at ovulation can be approximately modeled as a thin-walled spherical membrane under internal pressure, as sketched in Fig. P8.8-12. If the radius of the follicle is approximately 0.60 mm, $t = 0.1$ mm, and the internal pressure is $p = 2.38 \times 10^{-3}$ N/mm^2, estimate the membrane stress in the wall of the follicle.

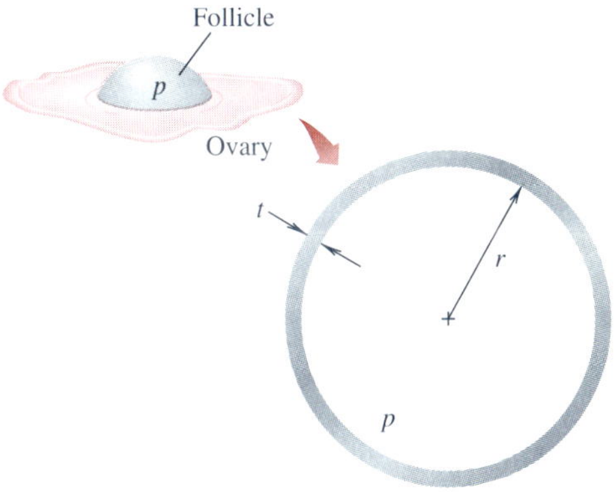

Fig. P8.8-12

8.8-13 A design concept for the oxidation of wet waste material (sludge) is to mix the sludge with oxygen in a sealed reaction chamber at high pressure, as shown in Fig. P8.8-13. The sludge is then converted to relatively clean water in this reaction chamber. If the reactor vessel is 24 in in diameter and 1 in thick, estimate the hoop stress in the vessel if the pressure p is 500 psi.

8.8-14 A typical aluminum-alloy scuba diving tank is shown in Fig. P8.8-14. If the air pressure in the tank is 3000 psi, estimate the hoop and longitudinal stresses in the wall of the tank.

8.8-15 A steel standpipe, shown in Fig. P8.8-15, is fabricated from steel plates and is 8 ft in diameter and 80 ft high. If the thickness of the steel plate is $\frac{3}{8}$ in, what is the value of the hoop stress in the standpipe when it is filled with water?

8.8-16 A cylindrical pressurized steel tank of 1.2-m diameter and wall thickness $t = 20$ mm is subjected to an internal pressure $p = 1800$ kPa and an axial force of $F = 1000$ kN, as shown in Fig. P8.8-16. If the butt weld seams of the cylinder are at an angle of 36° with the axis of the cylinder, find the stress normal to the weld and the maximum shear stress in the steel.

8.8-17 The internal pressure in a combustion test cell during part of a test, as shown in Fig. P8.8-17, is maintained by a force $F = 6$ kN holding the piston

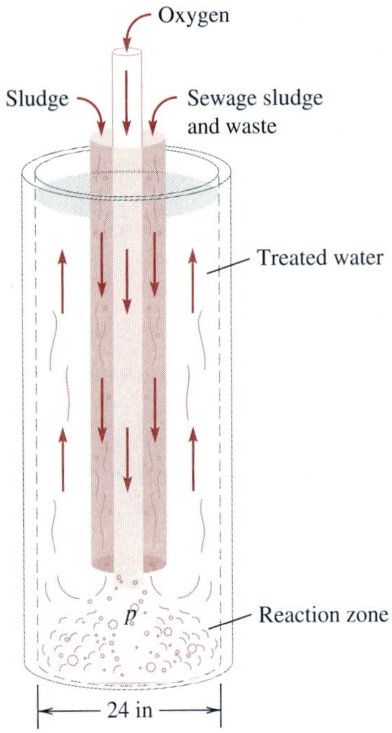

Fig. P8.8-13

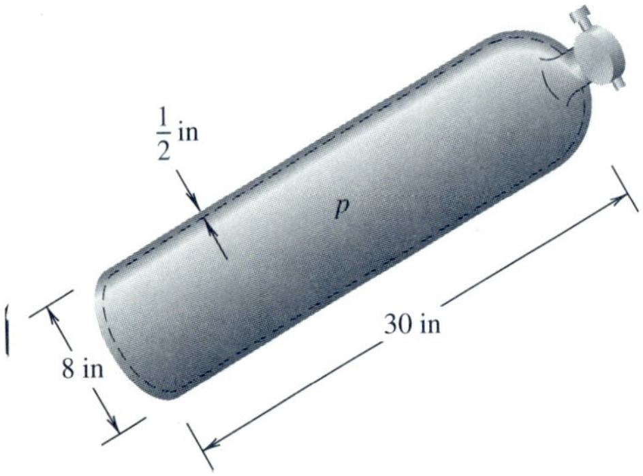

Fig. P8.8-14

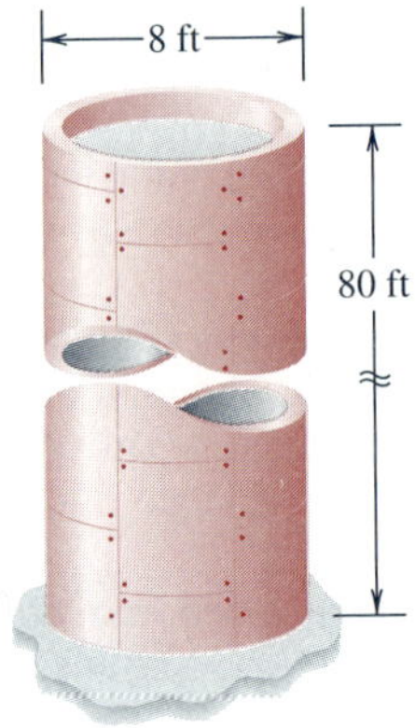

Fig. P8.8-15

Fig. P8.8-16

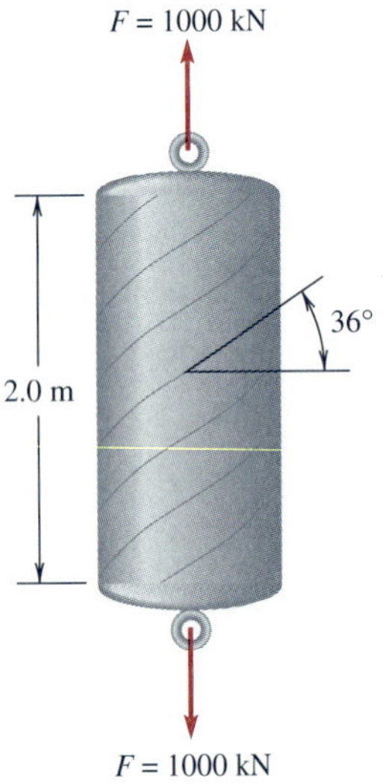

Fig. P8.8-17

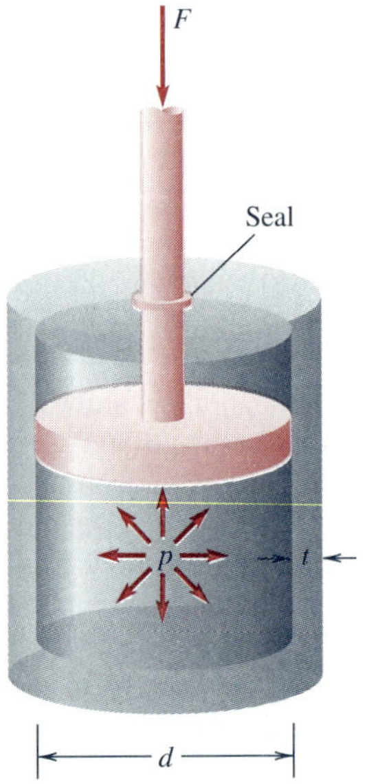

in place. If $d = 100$ mm and $t = 8$ mm, find the hoop stresses in the wall of the cell.

8.8-18 A thin-walled closed-end cylindrical pressure vessel contains a gas at a pressure p and experiences a twisting moment T, as shown in Fig. P8.8-18. Find the value of the maximum shear stress in the material at point A. Take $p = 500$ psi, $t = 0.25$ in, $d = 8.0$ in, and $L = 2.0$ ft.

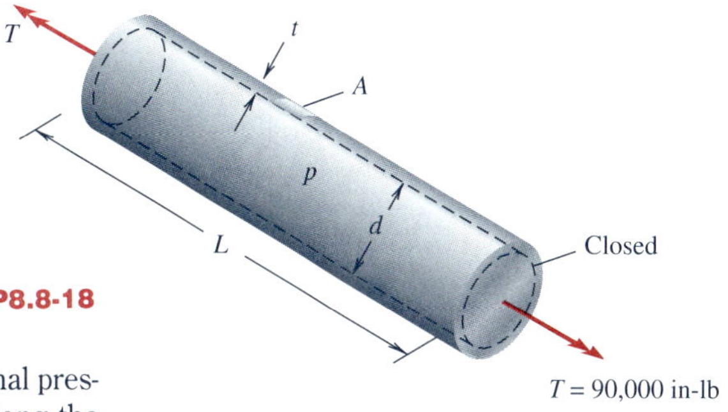

Fig. P8.8-18

8.8-19 A thin-walled closed-end cylindrical pressure vessel with internal pressure p, shown in Fig. P8.8-19, has an additional tensile load F acting along the axis of the vessel. Determine the value of the load F to give a maximum shear stress in the material of $\tau_{max} = 15{,}000$ psi. Take $p = 150$ psi, $d = 20$ in, and $t = 0.1$ in.

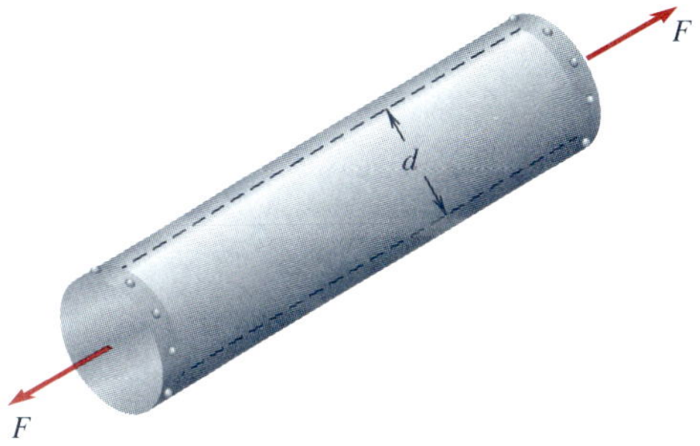

Fig. P8.8-19

8.8-20 Two sections of a cylindrical tube whose outer diameter is 8 in and whose inner diameter is 7.8 in are welded together with a weld angle ψ, as shown in Fig. P8.8-20. The tube is subjected to an internal pressure of 500 psi and a twisting moment $T = 50{,}000$ in · lb. The weld must not be subjected to a normal stress of more than 10,000 psi. Find the smallest positive angle ψ that can be used.

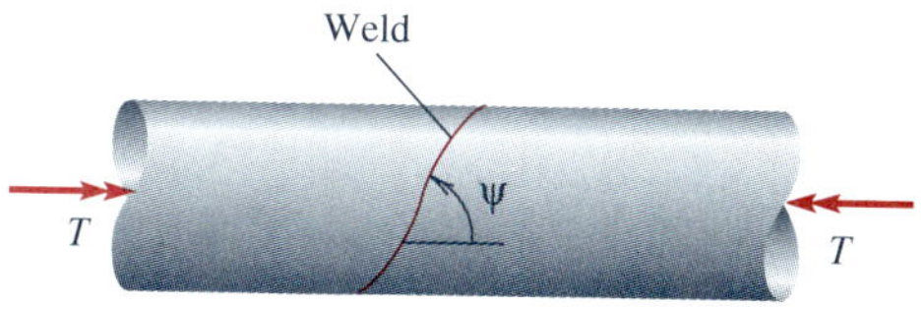

Fig. P8.8-20

8.8-21 A closed cylindrical tank containing compressed air has a wall thickness of $t = 0.25$ in and an inside radius of $r = 10$ in. The stresses in the wall of the tank acting on a rotated element have the values shown in Fig. P8.8-21. Find the air pressure in the tank.

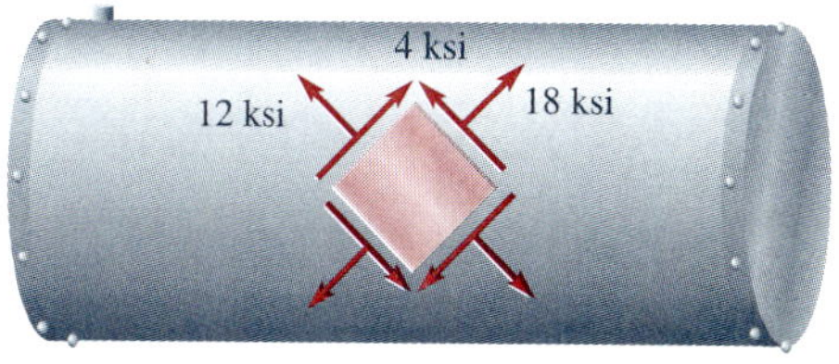

Fig. P8.8-21

8.8-22 A steam pipe is subjected to an internal pressure of 5 MPa, and because of the way it is fabricated, the allowable hoop stress is 70 MPa. Find the minimum allowable thickness for the pipe if the inside diameter is 300 mm.

8.8-23 A water tank is filled to a height of 45 ft. The wall thickness in the lower part of the tank is $\frac{3}{4}$ in. Find the maximum normal stress and maximum shear stress in the wall of the tank due to the water pressure. The inside diameter is 10 ft.

8.8-24 A cylindrical propane tank has an inside diameter of 10 ft and a wall thickness of 1 in. The allowable hoop stress is 15 ksi and the allowable axial stress is 6 ksi. Find the allowable internal pressure.

8.8-25 In the analysis of pressure vessels in Sec. 8.8, we assumed that the principal stress in the radial direction was approximately zero; see Eqs. (8.46) and (8.58). If for a spherical pressure vessel we take σ_3 equal to $-p$, show that the expression for the maximum shear stress in the wall of the vessel can be written in the form [see Eq. (8.47)]

$$\tau_{\max} = \frac{pr}{4t}\left(1 + \frac{2t}{r}\right)$$

What is the effect of the σ_3 stress on $\tau_{\max}$ if $t/r = \frac{1}{20}$ and $\frac{1}{50}$?

8.8-26 Consider Prob. 8.8-25. If for a cylindrical pressure vessel we take σ_3 equal to $-p$, show that the expression for the maximum shear stress in the wall of the vessel can be written in the form [see Eq. (8.59)]

$$\tau_{\max} = \frac{pr}{2t}\left(1 + \frac{t}{r}\right)$$

What is the effect of the σ_3 stress on $\tau_{\max}$ if $t/r = \frac{1}{20}$ and $\frac{1}{50}$?

8.11-1

(*a*) Verify Eqs. (8.67) as following from Eqs. (8.66).
(*b*) Obtain the full nonlinear expression for $(\Delta S)^2$ in Eq. (8.68).
(*c*) Verify Eq. (8.70).

8.11-2 Verify the steps leading up to the expression for shear strain γ_{xy} in Eq. (8.77) as following from the definition given in Eq. (8.76).

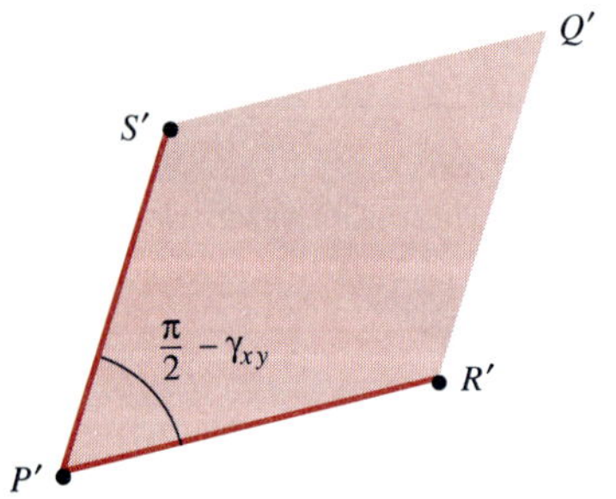

Fig. P8.11-3

8.11-3 The deformed area $P'R'Q'S'$ of Fig. 8.45 and Fig. P8.11-3 can be determined from the magnitude of the vector cross product

$$(P'R') \times (P'S')$$

If vectors $P'R'$ and $P'S'$ are given by Eqs. (8.74) and (8.75), obtain the area of $P'R'Q'S'$ and show that for small strains (neglecting terms that are quadratic in displacement derivatives) the result reduces to Eq. (8.78).

8.11-4 The displacements at a point (x, y) for a rigid-body rotation through an angle β about the z axis are given by

$$u(x,y) = (\cos\beta - 1)x - (\sin\beta)y$$
$$v(x,y) = (\sin\beta)x + (\cos\beta - 1)y$$

Verify that these expressions do indeed represent a rigid-body rotation. Then calculate the strain components in Eqs. (8.71). Do the strain components vanish? What are the values of the strain components when the rotation angle β is small?

8.12-1

(*a*) Verify that Eqs. (8.85) follow from Eqs. (8.82) to (8.84).
(*b*) Show that $\epsilon_{x'} + \epsilon_{y'}$ is an invariant.
(*c*) Show that the principal directions are given by Eq. (8.88).

For each of the following problems, the strain components are given with respect to the xy directions for an element in a state of plane strain, as shown in Fig.

P8.12-2*a*. For a given value of angle θ which orients the rotated element, Fig. 8.12-2*b*, find the strain components with respect to the rotated axes $x'y'$ and sketch the rotated element in its deformed state. ($\mu = 10^{-6}$.)

8.12-2 $\epsilon_x = -800\mu$, $\epsilon_y = 200\mu$, $\gamma_{xy} = -100\mu$, and $\theta = 40°$

8.12-3 $\epsilon_x = 500\mu$, $\epsilon_y = -500\mu$, $\gamma_{xy} = 0$, and $\theta = -45°$

8.12-4 $\epsilon_x = 0$, $\epsilon_y = 0$, $\gamma_{xy} = -500\mu$, and $\theta = 45°$

8.12-5 $\epsilon_x = -1000\mu$, $\epsilon_y = 200\mu$, $\gamma_{xy} = -400\mu$, and $\theta = -80°$

8.12-6 $\epsilon_x = 600\mu$, $\epsilon_y = -1000\mu$, $\gamma_{xy} = 300\mu$, and $\theta = 120°$

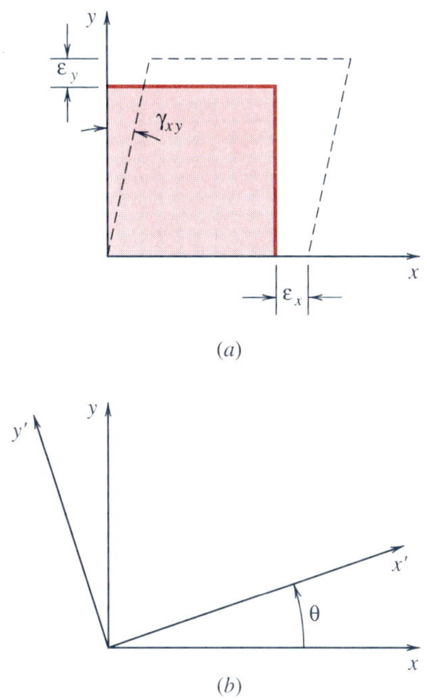

Fig. P8.12-2

8.12-7 Use Mohr's circle construction for plane strain to solve
(*a*) Prob. 8.12-2
(*b*) Prob. 8.12-3
(*c*) Prob. 8.12-4
(*d*) Prob. 8.12-5
(*e*) Prob. 8.12-6

8.12-8 Use Mohr's circle construction for plane strain to obtain the principal directions, principal strains, maximum shear strain, and direction of maximum shear strain for
(*a*) Prob. 8.12-2
(*b*) Prob. 8.12-3
(*c*) Prob. 8.12-4
(*d*) Prob. 8.12-5
(*e*) Prob. 8.12-6

8.12-9 Use the program MOHR'S CIRCLE on the diskette to obtain the principal directions, principal strains, maximum shear strain, and direction of maximum shear strain for
(*a*) Prob. 8.12-2
(*b*) Prob. 8.12-3
(*c*) Prob. 8.12-4
(*d*) Prob. 8.12-5
(*e*) Prob. 8.12-6

8.12-10 At a point in a body in a state of plane strain, the strain components associated with the *xy* axes are

$$\epsilon_x = 900 \times 10^{-6} \qquad \epsilon_y = 150 \times 10^{-6} \qquad \gamma_{xy} = -800 \times 10^{-6}$$

Find the principal directions of strain and the principal strains.

8.12-11 The strain components at a point associated with the *xy* axes are

$$\epsilon_x = -800 \times 10^{-6} \qquad \epsilon_y = -300 \times 10^{-6} \qquad \gamma_{xy} = -600 \times 10^{-6}$$

Determine the principal strains and directions and the strain components associated with an axis rotated 40° relative to the x axis.

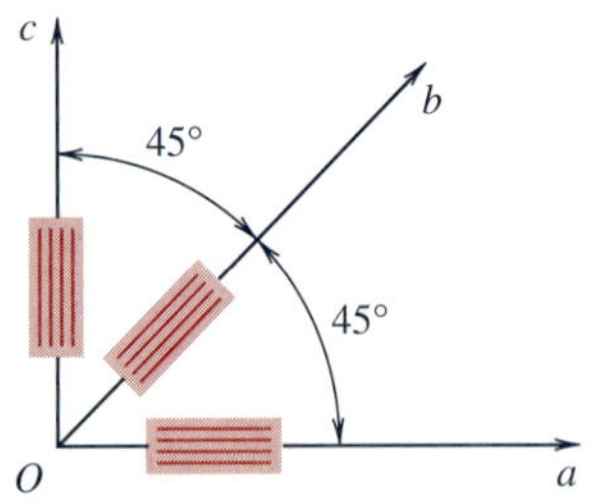

Fig. P8.13-1

8.13-1 For two different load cases, the strain gage rosette shown in Fig. P8.13-1 gives the following readings:
(*a*) $\epsilon_a = -400\mu$, $\epsilon_b = 600\mu$, $\epsilon_c = -900\mu$ ($\mu = 10^{-6}$)
(*b*) $\epsilon_a = -600\mu$, $\epsilon_b = 800\mu$, $\epsilon_c = -1100\mu$ ($\mu = 10^{-6}$)

In each case find the principal strains, maximum shear strain, principal directions, and directions of maximum and minimum shear strain; show their orientation with respect to the directions a and c.

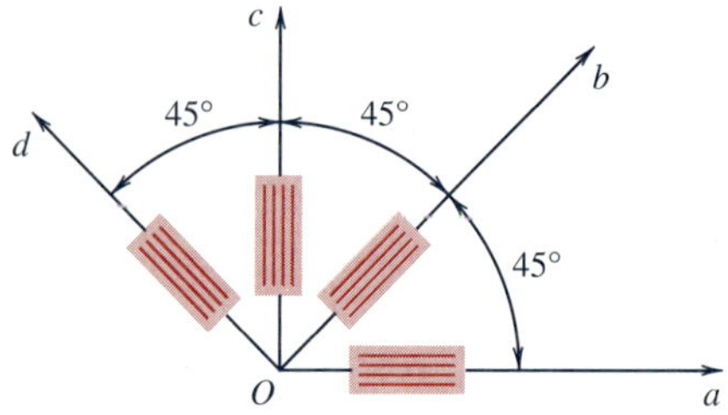

Fig. P8.13-2

8.13-2 As a test of the reliability of the measurements in the directions a, b, and c in Fig. P8.13-1, as given in Prob. 8.13-1, a fourth gage is mounted in the d direction, as shown in Fig. P8.13-2. What should the readings in the direction of the gage d equal for the two cases (*a*) and (*b*) of Prob. 8.13-1 in order to have consistent results?

8.13-3 A 45° strain rosette is mounted on a machine part as shown in Fig. P8.13-3. Find the strain components ϵ_x and ϵ_y for the two sets of readings.
(*a*) $\epsilon_a = 800\mu$, $\epsilon_b = -500\mu$, and $\epsilon_c = 600\mu$ ($\mu = 10^{-6}$)
(*b*) $\epsilon_a = -100\mu$, $\epsilon_b = 600\mu$, and $\epsilon_c = -500\mu$ ($\mu = 10^{-6}$)

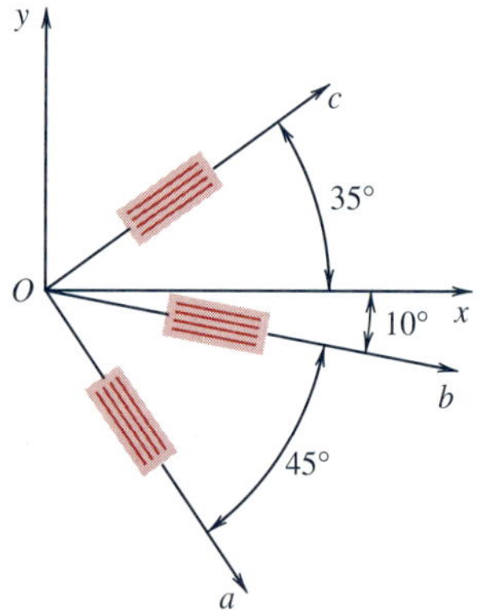

Fig. P8.13-3

8.13-4 The readings from a 45° strain rosette (Fig. P8.13-1) are
(*a*) $\epsilon_a = 100 \times 10^{-6}$ $\epsilon_b = 200 \times 10^{-6}$ $\epsilon_c = 900 \times 10^{-6}$
(*b*) $\epsilon_a = 1200 \times 10^{-6}$ $\epsilon_b = 400 \times 10^{-6}$ $\epsilon_c = 60 \times 10^{-6}$
Find the principal strains and principal directions in the plane of the rosette.

8.13-5 A 45° strain rosette (Fig. P8.13-1) is mounted on the side of a bridge to monitor loadings, as shown in Fig. P8.13-5. If the readings are

$$\epsilon_a = 500 \times 10^{-6} \qquad \epsilon_b = 150 \times 10^{-6} \qquad \epsilon_c = 350 \times 10^{-6}$$

determine the principal strains and principal directions at this point on the bridge.

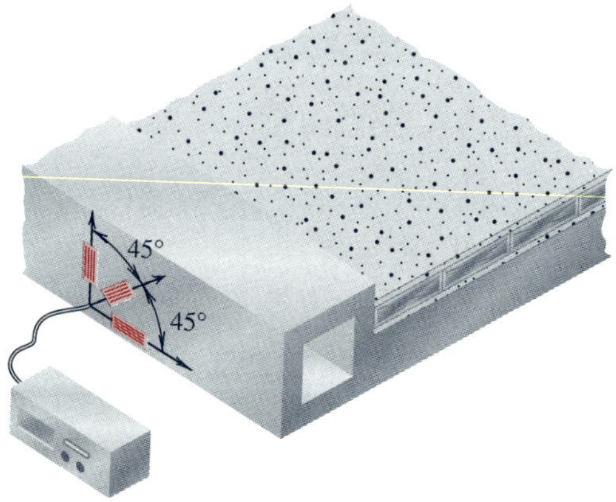

Fig. P8.13-5

8.13-6 A section of a cable lifting system, shown in Fig. P8.13-6, has mounted on its surface a 60° strain rosette. If the strain gage readings are

$$\epsilon_a = -150 \times 10^{-6} \qquad \epsilon_b = 500 \times 10^{-6} \qquad \epsilon_c = 600 \times 10^{-6}$$

find the principal strains and principal directions.

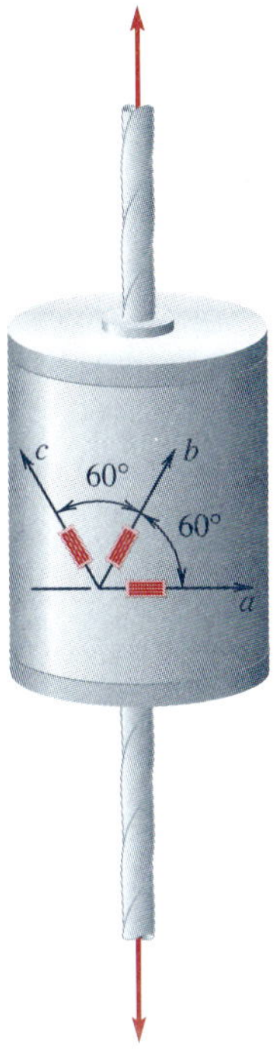

Fig. P8.13-6

8.13-7 Show that for the case of a 45° strain rosette, Eqs. (8.91) reduce to the form

$$\epsilon(0) = \epsilon_x$$
$$\epsilon(45°) = \frac{\epsilon_x + \epsilon_y}{2} + \frac{\gamma_{xy}}{2}$$
$$\epsilon(90°) = \epsilon_y$$

Therefore, the strain components ϵ_x, ϵ_y, and γ_{xy} can be found easily.

8.14-1 Show that the final volume of an infinitesimal element after deformation, Fig. 8.55, can be estimated by the expression given in Eq. (8.103).

8.14-2 For the strain rosette readings given in Prob. 8.13-1*a* and *b*, (*a*) compute the associated ϵ_x, ϵ_y, and γ_{xy} strain components; (*b*) use the stress-strain relations in Eqs. (8.110) to find the associated stress components. Take $E = 30{,}000$ ksi and $\nu = 0.3$.

8.14-3 A hollow cylindrical pipe is cantilevered from a wall and carries load *P* and twisting moment *T* at end *B*, as shown in Fig. P8.14-3. A strain rosette is mounted on the top of the pipe at a section near the wall at *A*. Derive expressions for *P* and *T* in terms of the three strain readings ϵ_z, ϵ_c, and ϵ_x of the strain rosette. The elastic constants are *E* and ν and the inside and outside diameters are d_i and d_o.

8.14-4 The pipe in Prob. 8.14-3 is a 4-in-nominal-diameter (extra-strong) pipe (App. E) of length 10 ft, and the strain gage rosette readings are $\epsilon_x = 500\mu$, $\epsilon_z = -150\mu$, and $\epsilon_c = 425\mu$ ($\mu = 10^{-6}$). If $E = 30{,}000$ ksi and $\nu = 0.3$, calculate values for the loads *P* and *T*, using the expressions derived in Prob. 8.14-3.

8.14-5 A cube of material is placed between lubricated rigid walls in a test fixture, as shown in Fig. P8.14-5. A pair of forces of magnitude *P* which is applied to rigid plates gives rise to a state of uniform compression in the *x* direction.
(*a*) Calculate the stresses σ_y and σ_z and the strains ϵ_x, ϵ_y, and ϵ_z.
(*b*) What additional stresses would be generated by a temperature increase in the material of ΔT? Assume that the elastic constants *E* and ν and the coefficient of thermal expansion α are given.

8.14-6 The following strain data were obtained from strain gages mounted on the surface of a steel member:

$$\epsilon_x = -30 \times 10^{-6} \qquad \epsilon_y = 100 \times 10^{-6} \qquad \gamma_{xy} = 160 \times 10^{-6}$$

Determine the principal stresses, principal directions, maximum shearing stress,

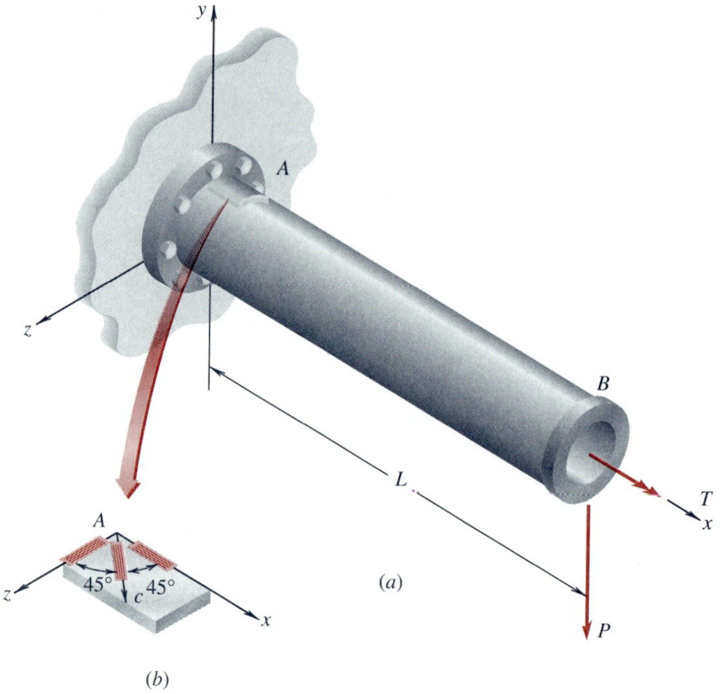

Fig. P8.14-3

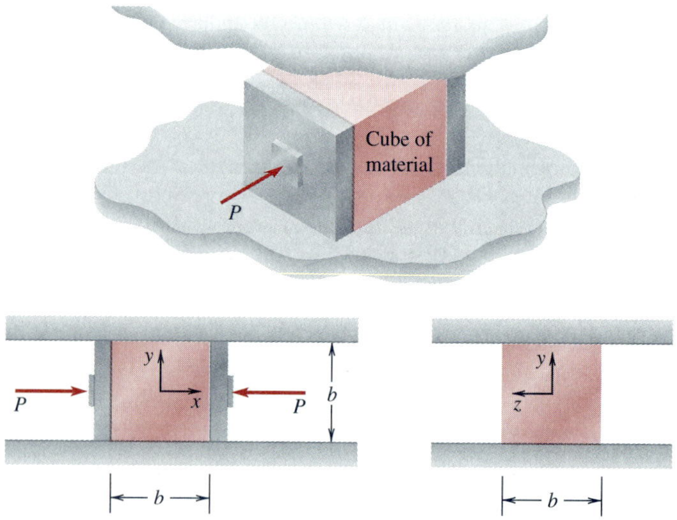

Fig. P8.14-5

and direction of maximum shearing stress. Show the principal stresses and principal directions on an appropriately oriented element. Take $G = 12.5 \times 10^6$ psi and $\nu = 0.3$.

8.14-7 A thin-walled spherical vessel, shown in Fig. P8.14-7, contains a gas under pressure p. The diameter of the sphere is 1 m, and the wall thickness is 2 mm. A strain gage attached to the outer surface of the vessel gives a reading of $\epsilon = 100 \times 10^{-6}$. Estimate the value of the pressure p. Take $E = 200$ GPa and $\nu = 0.3$.

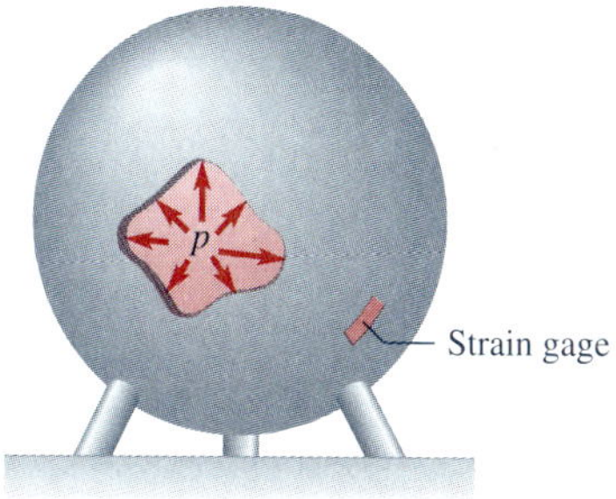

Fig. P8.14-7

8.14-8 A cylindrical pressure vessel, shown in Fig. P8.14-8, was fabricated by welding the edges of a thin steel sheet in a helical weld that makes an angle of 65° with the longitudinal axis of the cylinder. If the nominal radius of the cylinder is 2 in, the sheet thickness is 0.05 in and the internal pressure is 500 psi, find the normal strain perpendicular to the weld. Take $\nu = 0.25$ and $E = 30 \times 10^6$ psi.

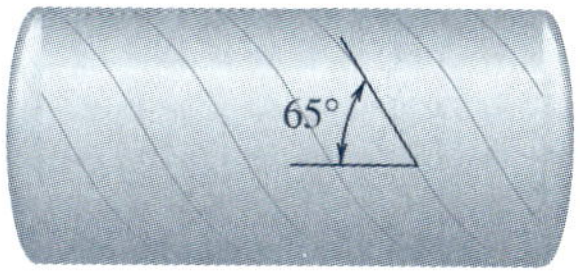

Fig. P8.14-8

8.14-9 A 45° strain gage rosette mounted on an aluminum-alloy panel as shown in Fig. P8.14-9 measures the following strains:

$$\epsilon_1 = 400 \times 10^{-6} \qquad \epsilon_2 = 365 \times 10^{-6} \qquad \epsilon_3 = -200 \times 10^{-6}$$

Find the maximum shear stress in the material at this location. Take $E = 73$ GPa and $\nu = 0.33$.

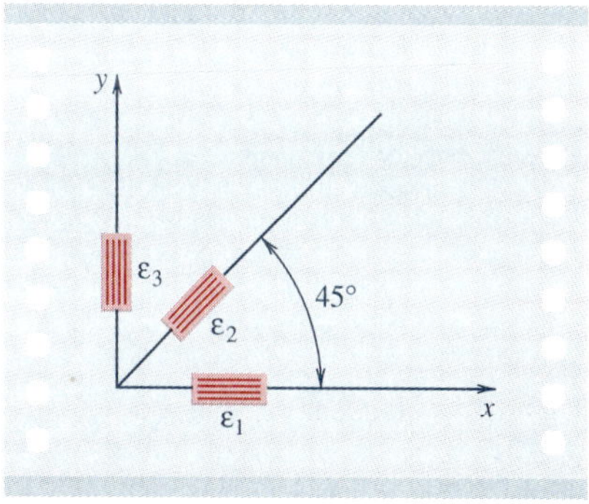

Fig. P8.14-9

8.14-10 A steel torsion bar in a light truck is instrumented with a pair of strain gages, as shown in Fig. P8.14-10. The strain gage at 60° reads a strain of 300×10^{-6} when a twisting moment T is applied to the shaft. If $d = 50$ mm, $L = 0.6$ m, and $G = 70$ GPa, find the value of T.

8.14-11 Find the strain in the second gage located at 90° to the first for the torsion bar in Prob. 8.14-10.

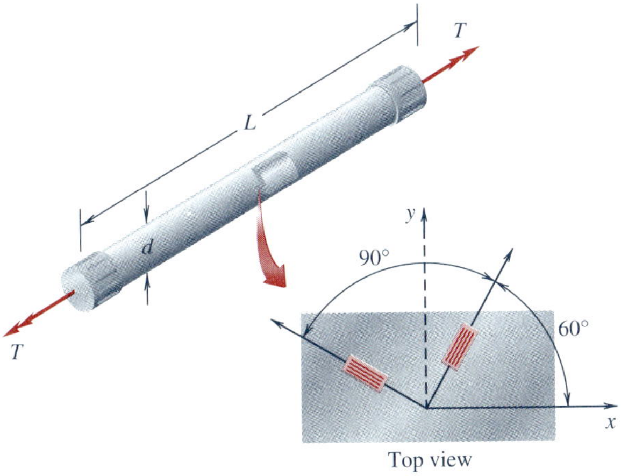

Fig. P8.14-10

8.14-12 The pressure in a low-pressure industrial tank is monitored by a 45° strain rosette mounted on its surface. See Fig. P8.14-12. Unfortunately, during service, gages *a* and *c* have become inoperative. Gage *b* gives a strain reading of 100×10^{-6}. If $d = 1$ m, $t = 18$ mm, $E = 200$ GPa, and $\nu = 0.3$, find the pressure in the tank.

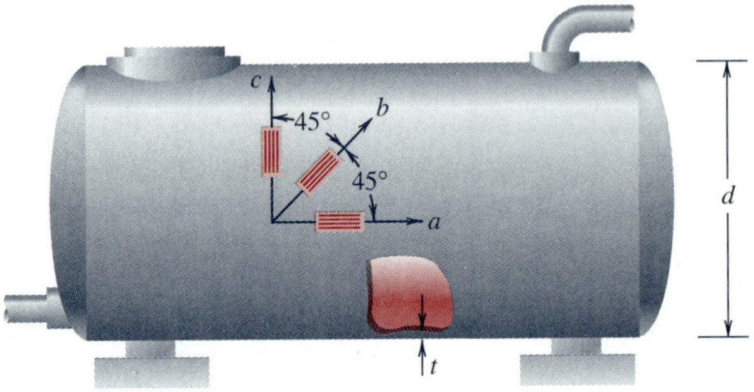

Fig. P8.14-12

8.14-13 At a point in a steel plate in a state of plane stress, the principal stresses are known to be of the form $\sigma_1 = \sigma_0$ and $\sigma_2 = -2\sigma_0$, where σ_0 is unknown. Also strain gage readings $\epsilon_A = 208\mu$ and $\epsilon_B = -442\mu$ ($\mu = 10^{-6}$) are available for a pair of perpendicular directions *A* and *B*, as shown in Fig. P8.14-13, but the angle θ is not known. Find (*a*) the unknown value of σ_0 and (*b*) the maximum shear stress and strain at this point. Take $E = 30{,}000$ ksi and $\nu = 0.3$.

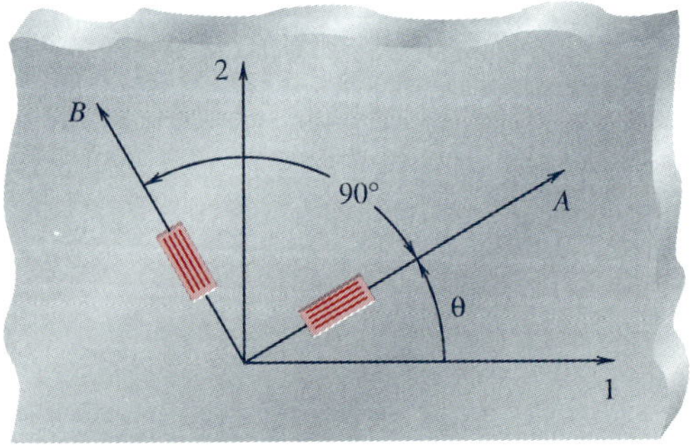

Fig. P8.14-13

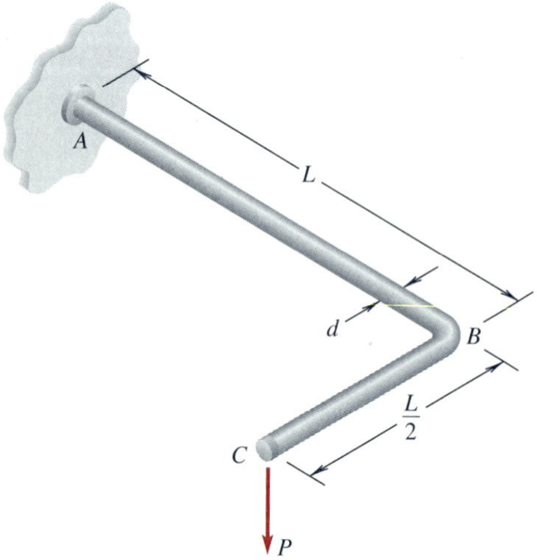

Fig. P8.15-1

8.15-1 An L-shaped rod with solid circular section is clamped in a wall at *A* and carries a load *P* at the free end *C*, as shown in Fig. P8.15-1. If $L = 1$ m and $d = 50$ mm, find the load *P* which corresponds to the onset of yielding in the rod at the top point of the rod at section *A* near the wall, using (*a*) the maximum shear stress criterion and (*b*) the Mises yield criterion.

8.15-2 A 4-in-nominal-diameter (double-extra-strong) pipe (see App. E for section properties) is fixed at end *A*, carries a torque *T* at its midpoint *B*, and carries a load *P* and a torque $2T$ at its endpoint *C*, as shown in Fig. P8.15-2. If $P = 50{,}000$ lb, $L = 36$ in, and the yield stress $Y = 30$ ksi (the stress at which yielding begins in a simple tensile test for this material), find the value of T corresponding to the onset of yielding using the Mises yield criterion.

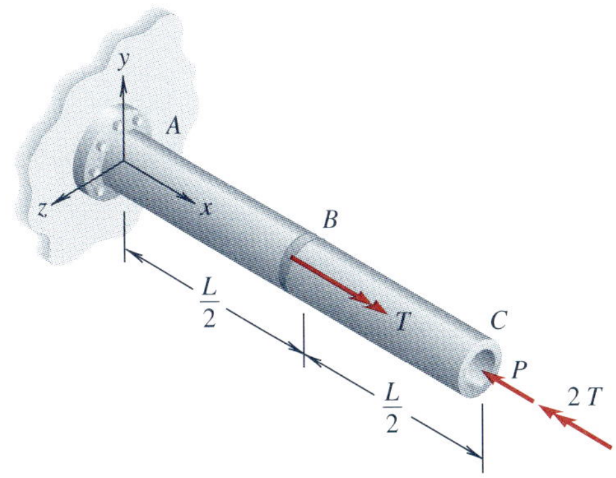

Fig. P8.15-2

8.15-3 A thin-walled closed-end cylinder with internal pressure $p = 1000$ psi is subjected to a twisting moment T, as shown in Fig. P8.15-3. Find the value of the twisting moment which can be applied before yielding of the material occurs. Use the maximum shear stress criterion. Take $Y = 50$ ksi, $r = 10$ in, and $t = 0.25$ in.

8.15-4 A steel plate is clad with a thin layer of soft aluminum on both sides, as shown in Fig. P8.15-4. The temperature of the cladded system is raised by an amount ΔT. Because the cladding is thin and of a lesser stiffness than the steel, the stresses in the plate are negligible. Stresses do arise in the cladding, however, from the mismatch of the coefficients of thermal expansion and the requirement that the axial strains in the plate and cladding be equal.

(*a*) Show that the normal stresses σ_x and σ_y in the cladding are

$$\sigma_x = \sigma_y = \frac{E_c\,(\alpha_p - \alpha_c)}{1 - \nu_c}\,\Delta T$$

(*b*) If $\alpha_c = 12 \times 10^{-6}/°\text{F}$, $\alpha_p = 6 \times 10^{-6}/°\text{F}$, $\nu_c = 0.33$, $E_c = 11 \times 10^6$ psi, and the yield stress of the aluminum is $Y = 5$ ksi, find the change in temperature to cause initial yielding by using the maximum shear stress criterion.

8.15-5 A batch of 2024-T4 aluminum alloy yields in uniaxial tension at a stress $Y = 330$ MPa. If this material is subjected to the state of plane stress given by the components $\sigma_x = 140$ MPa, $\sigma_y = -70$ MPa, and $\tau_{xy} = 140$ MPa, investigate whether it will yield according to the Mises criterion and the maximum shear stress criterion.

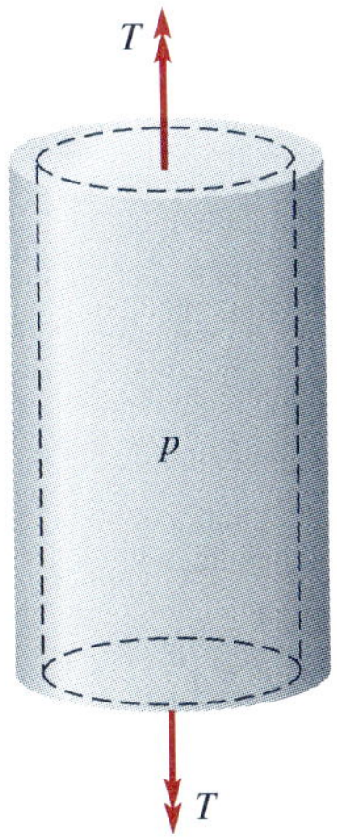

Fig. P8.15-3

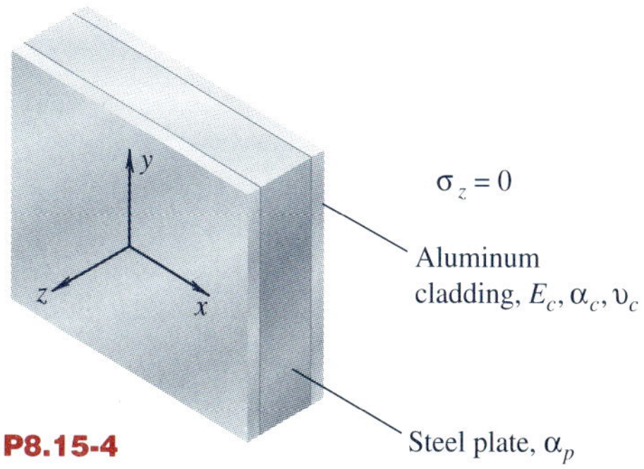

Fig. P8.15-4

8.15-6 A small, thin-walled cylindrical pressure tube is fabricated from a material whose yield strength in tension is 120 MPa. If the outside diameter of the tube is 75 mm and the wall thickness is 2 mm, what internal pressure will cause yielding in the material? Use both the maximum shear stress criterion and the Mises yield criterion to find the pressure.

8.15-7 A machine component in service experiences the state of stress given by $\sigma_x = 80$ MPa, $\sigma_y = 120$ MPa, and $\tau_{xy} = -60$ MPa. If the yield stress in simple tension of the material is 150 MPa, determine whether yielding of the material occurs. Use both the Mises and the maximum shear stress criteria.

8.15-8 A spherical pressure vessel 15 ft in diameter is to contain an industrial gas at a pressure $p = 40$ psi. If the allowable yield strength of the material is 14,000 psi, estimate the required wall thickness of the pressure vessel using the maximum shear stress criterion.

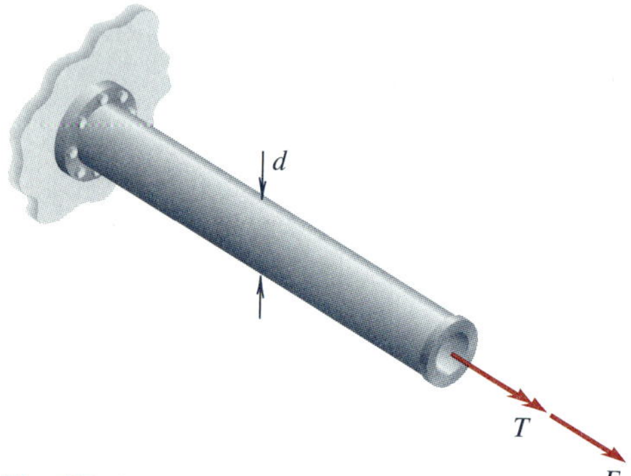

Fig. P8.15-9

8.15-9 A steel bar whose allowable yield strength is 140 MPa is built into a wall and is subjected to an axial force $F = 10$ kN and a twisting moment T, as shown in Fig. P8.15-9. Find the maximum allowable T before the onset of yielding if $d = 20$ mm. Use both the Mises and the maximum shear stress criteria.

8.16-1 A thin-walled spherical membrane of initial radius R_0 and thickness t_0 is under an internal pressure p. We wish to investigate the change in radius of the membrane as the pressure increases, if we assume that the volume of the wall material remains constant as the radius increases. First we note that the stress in the wall is

$$\sigma = \frac{pR}{2t} \tag{a}$$

The thickness at any radius R is given by

$$4\pi R^2 t = 4\pi R_0^2 t_0 = V_w \tag{b}$$

where V_w is the constant volume of the wall. The normal strain in two perpendicular directions in the membrane is given by

$$\epsilon = \frac{R - R_0}{R_0} = \frac{\sigma_1}{E} - \nu\frac{\sigma_2}{E} \tag{c}$$

If $\nu = \frac{1}{2}$ for an incompressible material and $\sigma_1 = \sigma_2 = \sigma$, obtain the pressure-radius relation in the form

$$p = \frac{EV_w (R - R_0)}{\pi R^3 R_0} \tag{d}$$

Sketch the p-R relation and determine p_{max}.

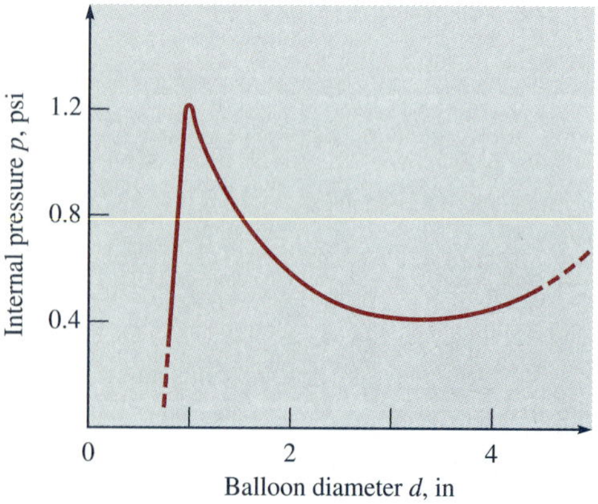

Fig. P8.16-2

8.16-2 If we attempt to blow up a spherical balloon, we find that it becomes easier to continue to inflate once we start the inflation. A typical experimental curve of internal pressure in the balloon versus balloon diameter is shown in Fig. P8.16-2. If the balloon is made of an incompressible linearly elastic material, formulate a simple model of this phenomenon; see Prob. 8.16-1. Obtain a simple analytical expression for the curve of pressure in the balloon versus balloon diameter which is similar to the early portion of the experimental curve in Fig. P8.16-2.

8.16-3 If a material behaves as an elastic–perfectly plastic material (see Fig. 1.19*b*), once the yield stress is reached, the material continues to deform. Two concentric tubes of materials 1 and 2 are connected to a rigid plate and loaded by a load P, as shown in Fig. P8.16-3. We wish to find the load-deflection curve of the system. The deflection of the plate is u.

(*a*) Show that

$$P = \sigma_1 A_1 + \sigma_2 A_2$$

where σ_1 and σ_2 are the stresses in materials 1 and 2.

(*b*) If $\epsilon < \epsilon_{2Y}$, $\sigma_1 = E_1\epsilon$, and $\sigma_2 = E_2\epsilon$, show that

$$P = (E_1A_1 + E_2A_2)\frac{u}{L}$$

(c) If $\epsilon_{2Y} < \epsilon < \epsilon_{1Y}$, what is the load-deflection expression?

(*d*) If $\epsilon_{1Y} < \epsilon$, what is the value of P?

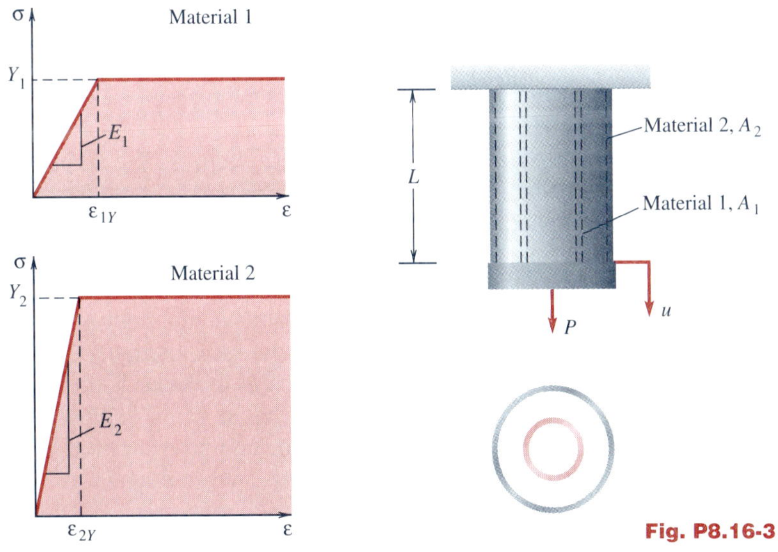

Fig. P8.16-3

8.16-4 A section of a pressurized steel pipe experiences a twisting moment T, as shown in Fig. P8.16-4. If $p = 150$ psi and $T = 140$ kip · in, determine the state of stress at a point A on the outside surface of the pipe. The outside diameter of the pipe is 12 in, and the wall thickness is 0.125 in.

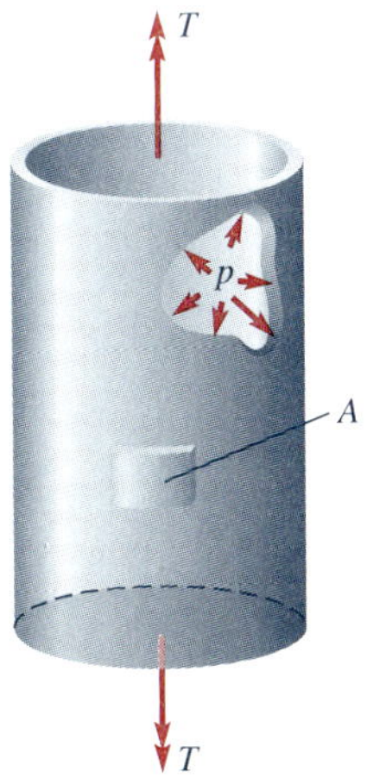

Fig. P8.16-4

CHAPTER 9

Analysis of Combined States of Stress

9.1 Introduction

In Chap. 8 we derived equations which allowed us to combine at a point the states of stress caused by axial, torsional, and transverse bending loads in a slender member. By combining these different stress states, we are able to calculate the principal stresses and the maximum shear stress at a given point. In addition, if the yield stress of the material in simple tension is known, we can use the yield criteria discussed in Sec. 8.15 to investigate whether the level of the applied loads causes the onset of plastic deformation in the member. The steps in this process are shown schematically in Fig. 9.1*a*; we look upon these steps as the steps to be taken to analyze a structural member or machine part to see if it remains elastic under the applied loads. The design criterion that limits the use of the member is the onset of initial plastic deformation at a critical location in the member.

In general, if design criteria for a structural component of a given material are specified, our goal is to investigate the component under the applied loading to see if it meets the design criteria. If the design criteria are exceeded, i.e., we are outside of what some engineers call the design envelope, then the dimensions of the component or the materials used must be changed and the component analyzed again.

In Fig. 9.1*b* we show this general process schematically. The geometry of the structural component is specified by, say, a preliminary design, and we are given the loads and material properties of the component. In addition, we are given some design criteria, e.g., the maximum deflection that can be allowed at any point or the requirement that the stress state at any point not result in plastic deformation. We then analyze the component to see if we can meet the design criteria. If we have, we move on; if we have not, then we must reexamine the component and, if need be, make changes in the geometry or in the material. Usually the loads

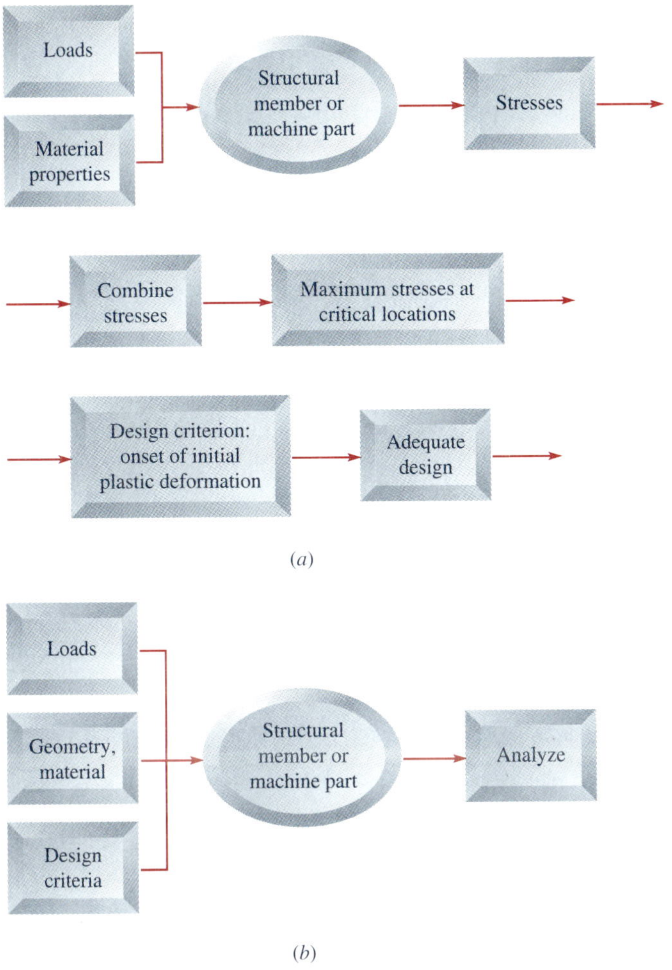

Figure 9.1 Design process for simple structural or mechanical systems.

and design criteria in this process are fixed. We continue this analysis process until we meet the design specifications.

On the basis of what we have learned so far in this book, we can carry out this process for a number of different configurations under design criteria corresponding to, say, a maximum deflection, maximum stresses, or the onset of yielding. In engineering practice, however, this set of criteria would be enlarged to include considerations of fatigue effects, fracture loads, temperature effects, and creep, to name only a few. The care and detail with which a structural component is analyzed will depend on how critical the component is for the overall design. A bolt supporting a jet engine on a commercial airline plane will be analyzed more carefully than the mounting for the catch on the trunk lock of an automobile. It is expected in either case, however, that each part will be engineered to the highest level of care required.

Up to this point, in looking at the schemata in Fig. 9.1, we have emphasized the analysis of a structural component. However, often the most difficult and most creative part of engineering lies in the preliminary design of a component or a structure. There exists a "need," and the engineer is called upon to specify the loads, preliminary shape, materials, design criteria, cost, manufacturability, etc. The list of specifications is long. Simple design problems have presented themselves earlier in this book; e.g., to point out two, find the dimensions of a circular shaft to transmit a specified amount of power and find the dimensions of a wide-flange beam to carry a specified transverse load over a given span. In engineering practice, design problems are more varied and some might say more interesting. However, in mechanical and structural design, much of what we have covered thus far in this book is the basic foundation for preliminary design. We need to remember that in design, careful attention to details is what makes the design successful.

In this chapter we discuss a number of "combined states of stress" problems in which we bring together material from the earlier chapters and treat the problems along the broad outlines sketched in Fig. 9.1.

9.2 Combined Bending and Axial Loads

Frequently slender members in machines and parts of structures are called upon to carry axial loads and bending moments. The normal stresses due to bending and the normal stresses due to the axial loads can be added at points on the cross section of the member.

In the two examples that follow, we indicate how the bending and axial stress states are combined.

EXAMPLE 9.1

The frame of a hacksaw is formed from $\frac{1}{16}$-in-thick steel sheet material to form the cross-sectional shape shown in section *SS* in Fig. 9.2*a*. For a small hacksaw, as shown in Fig. 9.2*a*, the thin blade is connected to the fixed pin at *A* near the handle and to a pin on a bolt at *B*. A wing nut on the bolt at *B*, when turned, tightens the hacksaw blade and puts it into tension. The blade is a two-force member and carries only an axial force.

We see from the free-body diagram in Fig. 9.2*b* that the tension force F in the blade gives rise to a compressive force F and a bending moment $M = Fa$ at any section along the frame. The distance a involved in the bending moment is the distance from the center of the blade to the centroidal axis of the frame. (Recall Example 2.10, Fig. 2.16, in which a nut was tightened on a bolt to put the bolt into tension and the tube on the bolt into compression. In this example, turning the wing nut at *B* puts the blade into tension and the frame into compression *and* bending.) We assume that the compressive force F acts at the centroid of the frame.

If the maximum allowable normal stress in the steel frame is 30 ksi, we wish to find the maximum allowable tension force F in the blade. Further, if the depth d of the channel section of the frame is increased by 20 percent from 0.5 to 0.6 in, we wish to calculate the percentage increase in the allowable blade tension force.

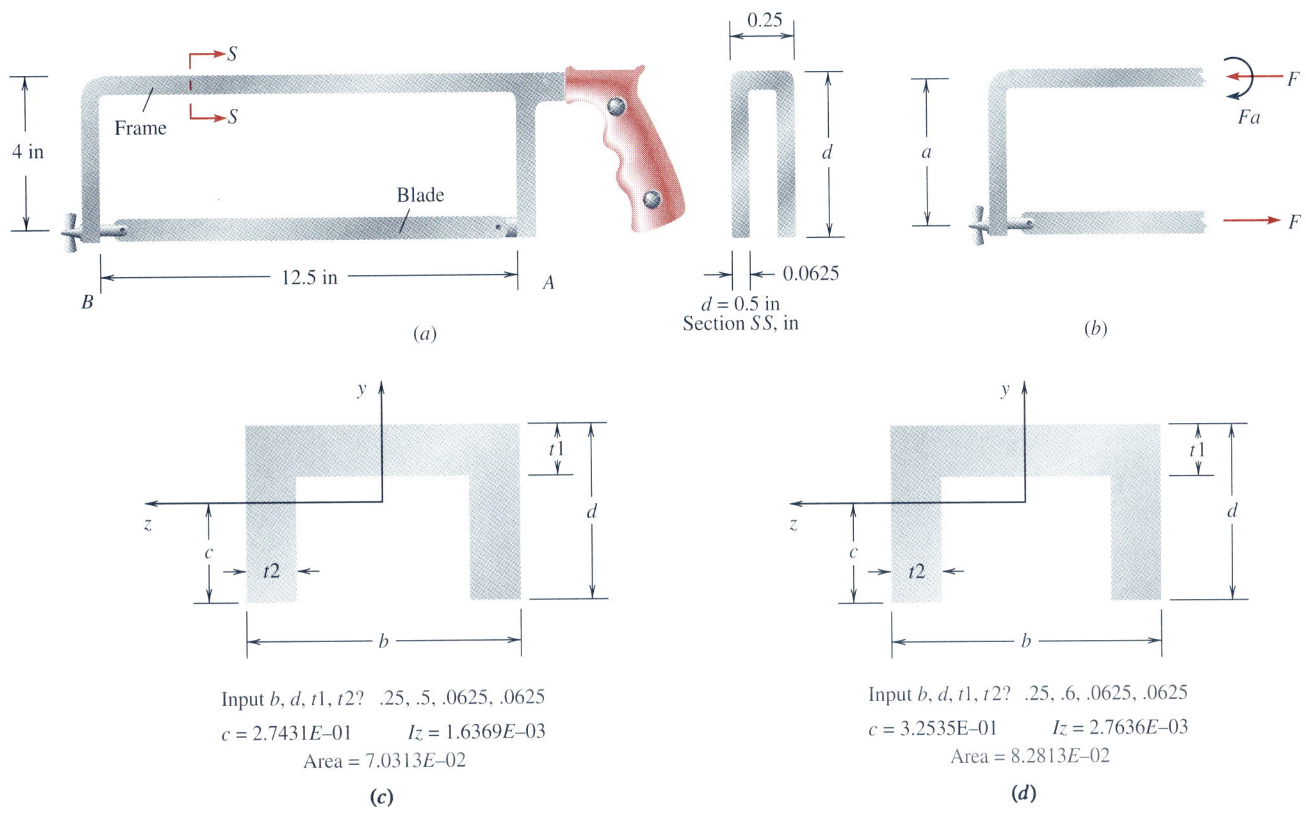

Figure 9.2 Example 9.1

To find the bending stresses in the channel section of the frame, we need to find the centroid of the channel section as well as the area and centroidal moment of inertia; we make use of the CENTROIDS AND MOMENT OF INERTIA option in the MECHMAT program on the diskette. Entering the dimensions from Fig. 9.2*a* into the program, we find the results in the output given in Fig. 9.2*c*.

The moment arm a defined in Fig. 9.2*b* can now be found by using the centroidal distance $c = 0.274$ in and the depth of the channel section $d = 0.5$ in to find

$$\begin{aligned} a &= 4 - (0.5 - 0.274) \\ &= 3.77 \text{ in} \end{aligned} \qquad (a)$$

The expression for the compressive axial stress in the frame is

$$\sigma_A = -\frac{F}{A} = -\frac{F}{7.03 \times 10^{-2}} = -14.2F \qquad (b)$$

where A is the area of the section.

The maximum compressive bending stress, from Eq. (5.17), is

$$\sigma_B = -\frac{My_{\max}}{I_z} \qquad (c)$$

In the present case the maximum compressive bending stress occurs along the bottom edge of the frame and is given by

$$\sigma_{B\max} = -\frac{(-3.77F)(-0.274)}{1.64 \times 10^{-3}} = -630F \qquad (d)$$

The maximum combined normal stress in compression is found by adding σ_A from Eq. (*b*) to $\sigma_{B\max}$ from Eq. (*d*) to obtain

$$\sigma_{\max} = -14.2F - 630F = -644F \qquad (e)$$

We note that the compressive stress due to bending is almost 45 times larger than the stress due to the axial force.

With $\sigma_{\max} = -30$ ksi in Eq. (*e*), it follows that

$$-644F_a = -30{,}000$$

or

$$F_a = 46.6 \text{ lb} \qquad (f)$$

The blade therefore can be tightened to take a force of 47 lb; see Prob. 9.2-5.

If the depth d of the channel section in Fig. 9.2*a* is now increased from 0.5 to 0.6 in, then by using the CENTROIDS AND MOMENT OF INERTIA option of the MECHMAT program, we find (as shown in Fig. 9.2*d*)

$$c = 0.325 \text{ in}$$
$$I_z = 2.76 \times 10^{-3} \text{ in}^4 \qquad (g)$$
$$\text{Area} = 8.28 \times 10^{-2} \text{ in}^2$$

The moment arm for the frame with increased section depth now becomes

$$a = 4 - (0.6 - 0.325) - 3.725 \text{ in} \qquad (h)$$

The compressive axial stress is

$$\sigma_A = -\frac{F}{8.28 \times 10^{-2}} = -12.1F \qquad (i)$$

and the maximum compressive bending stress is

$$\sigma_{B\,\max} = -\frac{(-3.725F)(-0.325)}{2.76 \times 10^{-3}} = -439F \qquad (j)$$

Combining the stresses in Eqs. (*i*) and (*j*) and equating to $\sigma_{\max}$, we find

$$-(439 + 12.1)F_a = -30{,}000 \qquad (k)$$

or

$$F_a = 66.5 \text{ lb} \qquad (l)$$

The 20 percent increase in the depth d of the channel section leads to a 43 percent increase in the allowable tension force in the blade of the hacksaw. This result serves to point out that changes in the depth dimension of the cross section are not linearly related to the allowable force.

EXAMPLE 9.2

A design for a proposed small concrete dam in an irrigation waterway is shown in Fig. 9.3*a*. A part of the design specifications is the requirement that the normal stresses acting on the dam across base AB remain compressive when the height of the water behind the dam is larger than the nominal value of $H = 2$ m shown in Fig. 9.3*a*.

For the preliminary design, the dam can be modeled as a short cantilever beam fixed along base AB and subjected to the water pressure on side BC. The water pressure on the dam gives rise to normal stresses due to bending on the base, and these stresses can be combined with normal stresses due to the weight of the dam to find the total distribution of the normal stresses along base AB. For this part of the preliminary design, we will not consider the shear stresses due to bending on AB, the resultant of which balances the horizontal resultant of the water pressure.

Therefore, we wish to determine first the normal stresses along base AB of the dam when the depth of the water $H = 2$ m, to see if these stresses are compressive. Take the specific weight for concrete as $\gamma_c = 23$ kN/m^3 and for water as $\gamma_w = 9.81$ kN/m^3. The water pressure acting on the dam increases linearly with the depth from the water surface. The peak pressure occurs at the bottom of the dam face at point B and for $H = 2$ m is

$$p_B = 9.81(2) = 19.6 \text{ kN/m}^2 \qquad (a)$$

For a given width b in the direction perpendicular to the plane of the section shown in Fig. 9.3*b*, the load per unit length q (on a cantilever beam of width b) will vary linearly from zero at the waterline to the value

$$q_B = p_B b = 19.6b \qquad \text{kN/m} \qquad (b)$$

at the bottom. This loading is shown in Fig. 9.3*b*. The resultant R of the water pressure acting on the dam is given by the area of the loading diagram

$$R = \tfrac{1}{2} q_B(2) = 19.6b \qquad \text{kN} \qquad (c)$$

The line of action of R passes through the centroid of the triangular loading diagram, Fig. 9.3*b*, so that the moment reaction M_F acting on the dam at base AB is

$$M_F = R(\tfrac{2}{3}) = 13.07b \qquad \text{kN} \cdot \text{m} \qquad (d)$$

An equal and opposite moment acts on the ground supporting the dam.

Therefore, the normal stress distribution due to bending along the base of the dam becomes

$$\sigma_b = -\frac{M_F y}{I_z} = -\frac{13.07by}{\frac{1}{12}b(1.5)^3} = -46.5y \qquad \text{kPa} \qquad (e)$$

where y is the distance from the centroid at base AB. We note that the dimension b drops out of the calculations. The maximum values of σ_b in tension and compression for $y = \mp 0.75$ m are

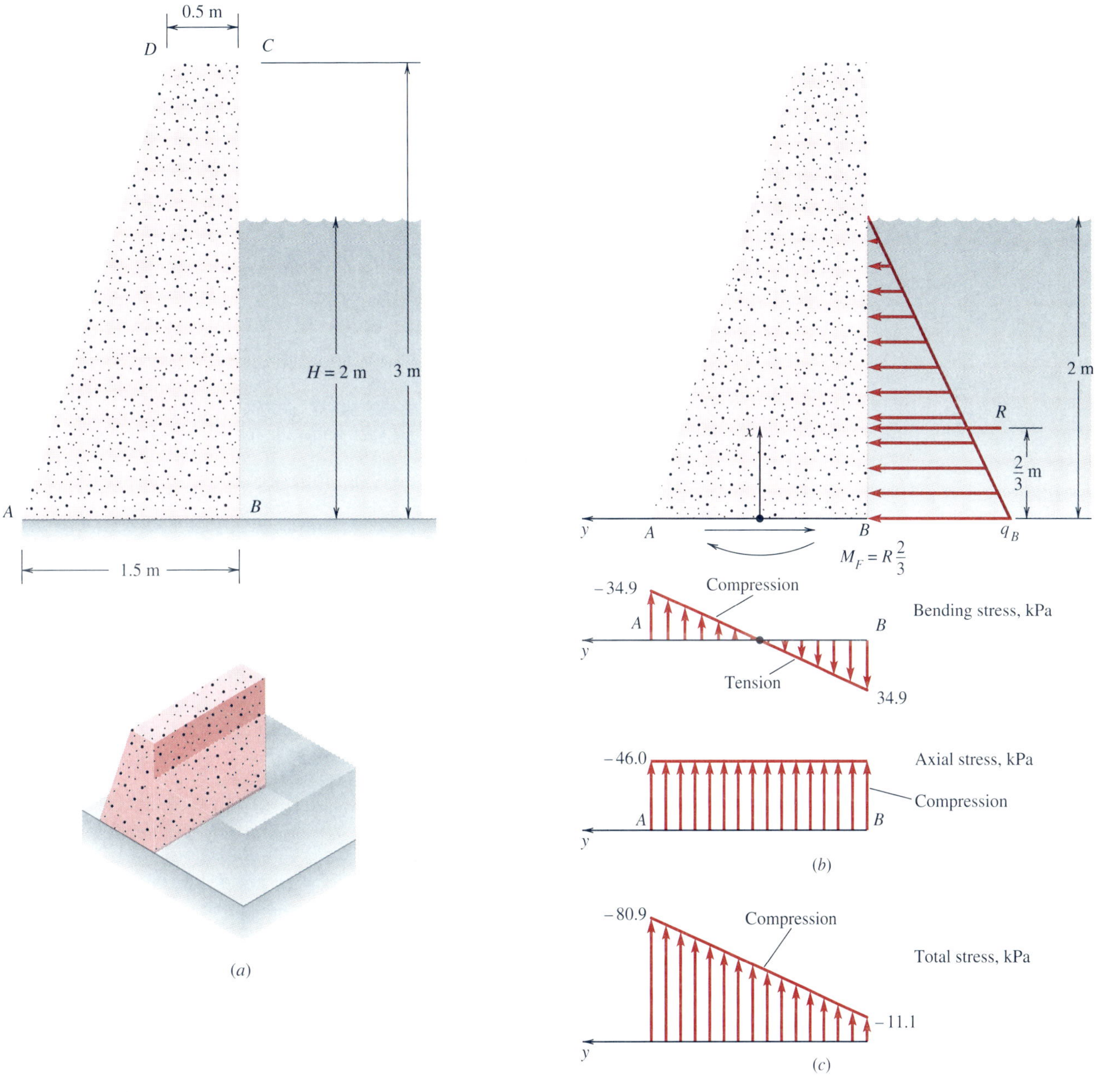

Figure 9.3 Example 9.2

±34.9 kPa and occur at points B and A, as shown in Fig. 9.3b. To these normal stresses due to bending we must add the normal stresses due to the weight of the dam.

The total weight of the dam for a given width b is the area of the cross section multiplied by the width b and the specific weight of 23 kN/m³. From Fig. 9.3a we have

$$A_s = \frac{1}{2}(1.5 + 0.5)(3) = 3 \text{ m}^2 \tag{f}$$

so that the total weight is

$$W_T = 23b(3) = 69b \quad \text{kN} \tag{g}$$

The uniform compressive axial stress along base AB is

$$\sigma_w = -\frac{69b}{1.5b}$$

or

$$\sigma_w = -46 \text{ kPa} \tag{h}$$

Adding the axial and bending stresses along AB shown in Fig. 9.3b results in the stress distribution shown in Fig. 9.3c, and we see that the normal stresses are compressive over the entire length of the base. The minimum value of the compressive normal stress occurs at point B and is equal to 11.1 kPa.

Clearly as the water level increases, the values of the bending stresses will increase until at some level $H = H_a$, the combined stresses of tension and compression along the base give a zero value of the normal stress at point B. Increasing the value of $H > H_a$ will then give tensile stresses along the base. To find the value of H_a, we redo—following Eqs. (a) to (e)—the bending analysis with $H = H_a$.

The pressure at the base is now

$$p_B = 9.81H_a \tag{i}$$

and the loading is

$$q_B = p_B b = 9.81bH_a \tag{j}$$

The resultant becomes

$$R = \frac{1}{2}q_B H_a$$

or

$$R = 4.905bH_a^2 \tag{k}$$

The bending moment reaction at base AB becomes

$$M_F = R\left(\frac{1}{3}\right)H_a$$

or

$$M_F = 1.635bH_a^3 \tag{l}$$

and the normal tensile stress at B due to bending is

$$\sigma_b = \frac{1.635bH_a^3(0.75)}{\frac{1}{12}b(1.5)^3} = 4.36H_a^3 \quad \text{kPa} \tag{m}$$

When $H_a = 2$ m, we find that the value of the stress in Eq. (m) agrees with our previous value of σ_b in Eq. (e) with $y = -0.75$ m.

The axial stress due to self-weight is still given by Eq. (h), and the level H_a can be found by requiring that the sum of the maximum tensile stress due to bending at point B and the compressive stress due to self-weight equal zero. Using Eqs. (h) and (m), we find

$$-46 + 4.36H_a^3 = 0 \tag{n}$$

or

$$H_a^3 = 10.55$$

Therefore,

$$H_a = 2.19 \text{ m} \tag{o}$$

Thus we find that the water level can only be increased from $H = 2$ to 2.19 m or by 19 cm before the normal stress at B changes over from compression to tension. We see that the nominal height of 2 m is approximately 90 percent of the height 2.19 m that causes tension. Therefore, further considerations of the design of the dam may be necessary. Finally we note that the bending stresses are seen from Eq. (m) to be proportional to the third power of the water level H_a. Thus as the water level increases from 1 to 2 m, the bending stresses at B increase by a factor of 8.

9.3 Combined Torsion and Axial Loads

Circular shafts are commonly used to transmit power, and as a consequence the design of circular cylindrical shafts is a common problem in engineering practice. We discussed the relation between the power transmitted by a circular shaft and the maximum shear stress in the shaft due to a twisting moment in Sec. 3.10. We found that once the twisting moment acting on the shaft is known, the shear stress distribution can be determined from Eq. (3.13). In many applications, a shaft experiences an axial load in addition to the twisting moment; this axial load can arise from external axial loads, the weight of components attached to the shaft, or thermal loading arising from a temperature change in service.

We need to be able to analyze the combined state of stress arising from the shear stress due to the twisting moment and the normal stress from the axial load. In particular, if we know the applied twisting moment and the axial load acting on a solid or hollow shaft and the material properties, we should be able to determine the required dimensions of the shaft on the basis of either the maximum-shear-stress criterion or some other criterion for the onset of yielding of the material.

Figure 9.4*a* shows a segment of a hollow cylindrical shaft of outside diameter d_o, inside diameter d_i, and length L loaded by a twisting moment

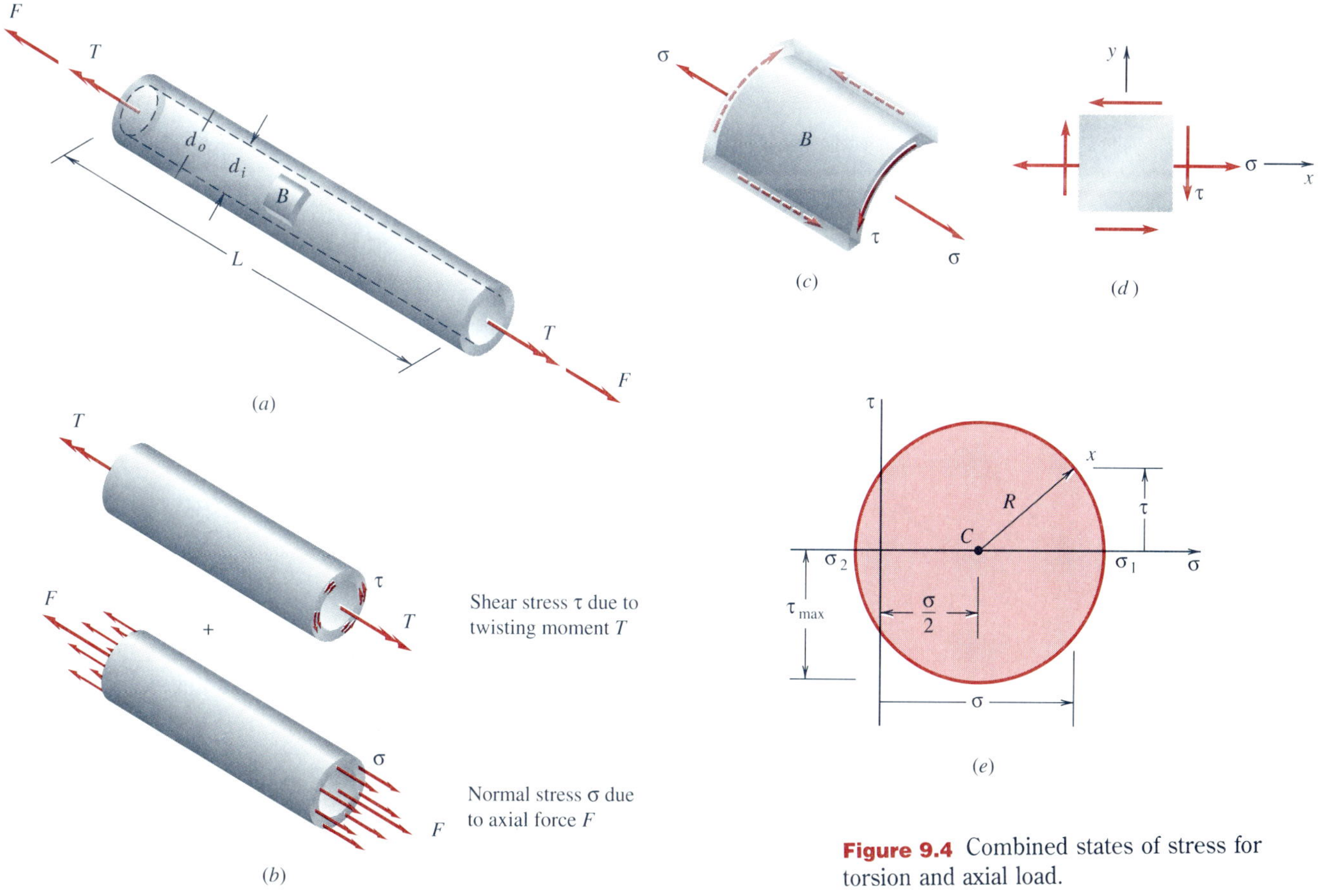

Figure 9.4 Combined states of stress for torsion and axial load.

T and an axial force F. To focus on the combined effects of these two loadings, we proceed to determine the value of the maximum shear stress in the shaft. In discussing this type of problem in general and in analyzing the examples and problems to follow, we must keep in mind that we are carrying out typical preliminary design calculations. We are neglecting, e.g., the effect of stress concentrations, fatigue, and shaft misalignments on the stress state of the shaft; these topics are covered in books on machine design.

The state of stress in the shaft of Fig. 9.4*a* is a combination of the shear stress τ arising from the twisting moment T and the axial normal stress σ arising from the axial load F. We show each of these cases separately in Fig. 9.4*b*. The normal stress σ is constant across the cross section, and the shear stress τ varies linearly with the distance from the center of the shaft and is a maximum at the outside surface. The outside surface is the location of the most critically stressed part of the shaft. Therefore we consider an element of the shaft on the outside surface at B, as shown in Fig. 9.4*a* and *c*.

The maximum shear stress due to the twisting moment T occurs on the outside surface of the shaft and can be determined from Eq. (3.21)

$$\tau = \frac{T(d_o/2)}{J} \tag{9.1}$$

where J is the polar moment of inertia of the cross section, Eq. (3.18),

$$J = \frac{\pi}{32}(d_o^4 - d_i^4) \tag{9.2}$$

The shear stress is shown acting on the element in Fig. 9.4*c*. In addition to the shear stress acting on the element, there is a contribution to the stress state from the axial load F giving rise to the normal stress

$$\sigma = \frac{F}{A} \tag{9.3}$$

where A is the cross-sectional area of the shaft

$$A = \frac{\pi}{4}(d_o^2 - d_i^2) \tag{9.4}$$

The element is shown in a two-dimensional sketch in Fig. 9.4*d*. Corresponding to this element in a state of plane stress, we can draw the Mohr's circle representation shown in Fig. 9.4*e* and obtain the maximum shear stress in the material as

$$\tau_{\max} = R = \sqrt{\left(\frac{\sigma}{2}\right)^2 + \tau^2} \tag{9.5}$$

where σ and τ are given by Eqs. (9.1) and (9.3). If we were given the maximum allowable shear stress τ_{max} as a design specification and values for the loads F and T, we could use Eq. (9.5) to determine the outside diameter d_o of the shaft for a given ratio of d_i/d_o. If, in addition, we were given the maximum allowable angle of twist between the ends of the shaft as a design specification, we would be able to determine the appropriate diameters for a hollow shaft so that both design criteria were satisfied. The following two examples show how we proceed in particular cases.

EXAMPLE 9.3

During the process of drilling holes in a wood piece with an instrumented robotic drilling machine, the $\frac{5}{16}$-in-diameter steel drill bit carries a twisting moment of $T = 10$ lb · in and an axial force $P = 100$ lb, as shown in Fig. 9.5*a*. We wish to estimate the values of the principal stresses and the maximum shear stress due to the combined effects of the axial load and the twisting moment acting on the drill bit. We will model the effect of the drill bit geometry by treating the bit as a $\frac{1}{4}$-in-diameter

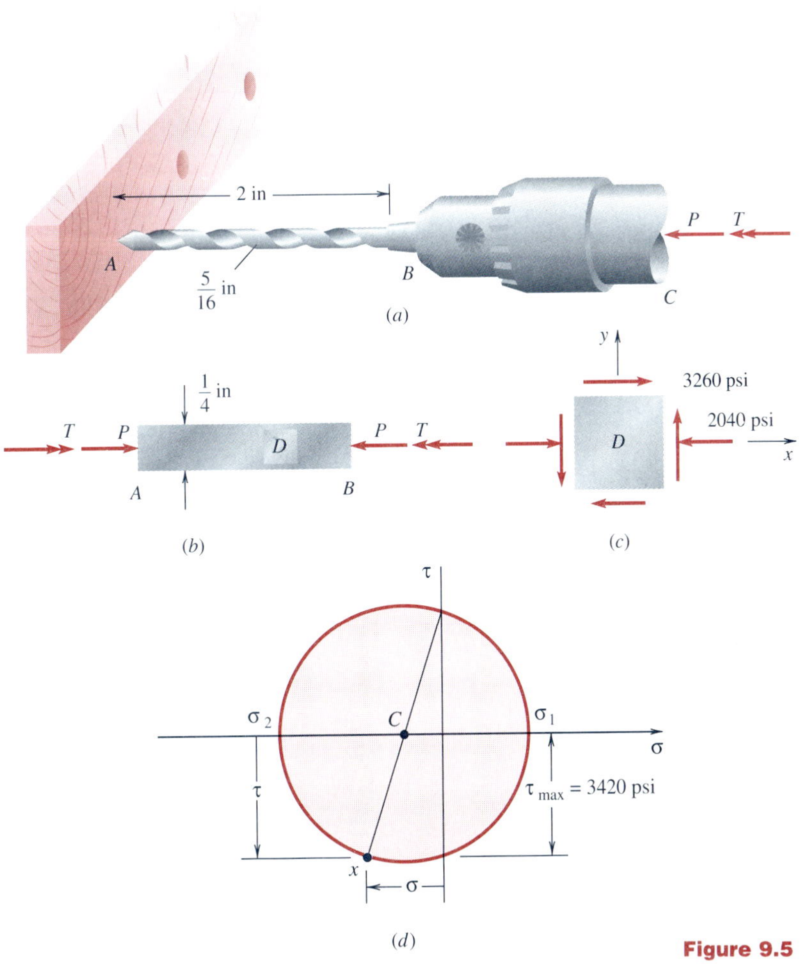

Figure 9.5 Example 9.3

solid shaft. The effective diameter of the shaft is slightly smaller than the maximum diameter of the drill bit in order to account for the helical grooves machined into the bit.

We assume that the axial load P gives rise to a compressive normal stress that is distributed uniformly over the cross section of the shaft and given by

$$\sigma = -\frac{P}{\pi d^2/4} \tag{a}$$

where d is the diameter of the shaft.

The shear stress distribution across a section of the shaft caused by the twisting moment T was derived in Chap. 3, and according to Eq. (3.13) we have

$$\tau = \frac{Tr}{\pi d^4/32} \tag{b}$$

where r is the radial distance from the axis.

As we have noted before, the maximum shear stress in a cylindrical shaft due to a twisting moment T occurs at the outside surface where $r = d/2$, and from Eq. (b) it is given by

$$\tau = \frac{16T}{\pi d^3} \tag{c}$$

A typical surface element along segment AB of the shaft is labeled D, as shown in Fig. 9.5b. Using $d = \frac{1}{4}$ in, $T = 10$ lb · in, and $P = 100$ lb in Eqs. (a) and (c), we find

$$\sigma = -2040 \text{ psi} \tag{d}$$

$$\tau = 3260 \text{ psi} \tag{e}$$

and we show these stresses on a sketch of the element in Fig. 9.5c.

The principal stresses and the maximum shear stress are found by using the Mohr's circle representation for the plane stress element shown in Fig. 9.5c. For this case, we have (Fig. 9.5d)

$$\sigma_{av} = \frac{\sigma}{2} = -\frac{2040}{2} = -1020 \tag{f}$$

$$\tau_{max} = R = \sqrt{\left(\frac{\sigma}{2}\right)^2 + \tau^2} = \sqrt{(-1020)^2 + (3260)^2} = 3420 \text{ psi} \tag{g}$$

Therefore

$$\sigma_1 = \sigma_{av} + R = 2400 \text{ psi} \tag{h}$$

$$\sigma_2 = \sigma_{av} - R = -4440 \text{ psi} \tag{i}$$

EXAMPLE 9.4

The thin-walled closed steel pipe shown in Fig. 9.6a is part of an instrumented pilot plant test facility for synthetic fuel production. The pipe is pressurized with an internal pressure p and is fixed at A and B to rigid supports. During a test, the strain gage readings indicated that the supports had undergone a relative rotation (Fig. 9.6a) of an unknown amount ϕ so as to subject the pipe to an unknown twisting moment T_0. The pipe has a 45° strain rosette on its outside surface at point C, Fig. 9.6b.

We wish to find the unknown twisting moment in the pipe as a consequence of the relative rotation of the supports. The operating pressures in the pipe for the process are expected to be approximately 150 psi, although this value is not certain because of the unknown effect of the relative rotation on the process.

We should be able to deduce the stresses in the pipe from the strain readings, and from the corresponding stress values we should be able to estimate the twisting moment and pressure acting in the pipe. We neglect the effect of temperature (the supports are free to expand in the axial direction).

The readings from the 45° strain rosette oriented as shown in Fig. 9.6b are

$$\epsilon_a = 48 \times 10^{-6}$$

$$\epsilon_b = 343 \times 10^{-6}$$

$$\epsilon_c = 204 \times 10^{-6}$$

We have $E = 30 \times 10^6$ psi, $\nu = 0.3$, $d_o = 12$ in, and $t = \frac{1}{8}$ in.

We have shown in Sec. 8.8 that the internal pressure in a closed thin-walled pressure vessel gives rise to the stresses given by

$$\sigma_x = \frac{pr}{2t} \qquad \sigma_y = \frac{pr}{t} \tag{a}$$

as shown on an element in Fig. 9.6c, where p is the pressure, r is the nominal radius, and t is the wall thickness. Also the unknown twisting moment T_0 gives rise to a shear stress τ_{xy} in the material with

$$\tau_{xy} = \frac{T_0(d_o/2)}{J} \tag{b}$$

where d_o is the outside diameter of the pipe and

$$J = \frac{\pi}{32}(d_o^4 - d_i^4) \tag{c}$$

When we combine these two stress states, we see that the

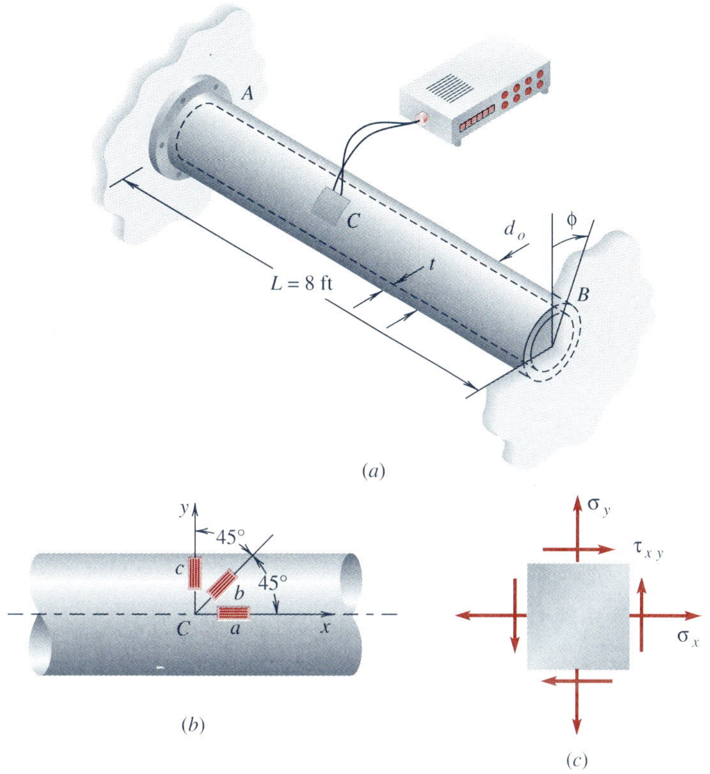

Figure 9.6 Example 9.4

surface element at point C is in the state of plane stress, as shown in Fig. 9.6*c*.

We will first use the three given strain readings to calculate the three associated stresses by means of the stress-strain relations for plane stress. Making use of Eqs. (8.91) with $\theta_1 = 0°$, $\theta_2 = 45°$, and $\theta_3 = 90°$, we find

$$\epsilon(0) = \epsilon_a = \epsilon_x$$

$$\epsilon(45) = \epsilon_b = \frac{\epsilon_x + \epsilon_y}{2} + \frac{\gamma_{xy}}{2} \qquad (d)$$

$$\epsilon(90) = \epsilon_c = \epsilon_y$$

Substituting the given values for ϵ_a, ϵ_b, and ϵ_c into Eqs. (*d*) and solving for ϵ_x, ϵ_y, and γ_{xy}, we get

$$\epsilon_x = 48 \times 10^{-6}$$

$$\epsilon_y = 204 \times 10^{-6} \qquad (e)$$

$$\gamma_{xy} = 434 \times 10^{-6}$$

Equations (8.110) give the stresses for the case of plane stress in terms of the strains in the form

$$\sigma_x = \frac{E}{1 - \nu^2}(\epsilon_x + \nu\epsilon_y)$$

$$\sigma_y = \frac{E}{1 - \nu^2}(\epsilon_y + \nu\epsilon_x) \qquad (f)$$

$$\tau_{xy} = G\gamma_{xy} = \frac{E}{2(1 + \nu)}\gamma_{xy}$$

Using the strains from Eqs. (*e*), we find that Eqs. (*f*) yield

$$\sigma_x = 3600 \text{ psi} \qquad \sigma_y = 7200 \text{ psi} \qquad \tau_{xy} = 5010 \text{ psi} \qquad (g)$$

We could have carried out these calculations directly by using the Mohr's circle strain rosette program on the MECHMAT diskette; the results are shown in Fig. 9.6*d* and agree with Eqs. (*e*) and (*g*). Recall that the maximum shear stresses and strains in the program refer to the in-plane stresses.

For the pipe under consideration $r = d_o/2 = 6$ in, $t = \frac{1}{8}$ in, and either equation in Eqs. (*a*) can be used to find p. For example, we find

$$p = \frac{2t\sigma_x}{r} = \frac{2(\frac{1}{8})(3600)}{6} = 150 \text{ psi} \qquad (h)$$

which is in agreement with the expected value.

To find the unknown moment T_0, we use Eq. (*b*) to get

$$T_0 = \frac{2\tau_{xy}J}{d_o} = \frac{2(5010)}{12}\frac{\pi}{32}[(12)^4 - (11.75)^4]$$

or

$$T_0 = 137 \text{ kip·in} \qquad (i)$$

We should be able now to estimate the angle of rotation ϕ.

The relative angle of rotation between the supports is given by

$$\phi = \frac{T_0 L}{GJ} \qquad (j)$$

where T_0 is given by Eq. (*i*). The value of the shear modulus G follows from Eq. (8.100)

$$G = \frac{E}{2(1+\nu)} = \frac{30 \times 10^6}{2(1+0.3)} = 11.5 \times 10^6 \text{ psi} \qquad (k)$$

Therefore,

$$\phi = \frac{(137 \times 10^3)(96)}{(11.5 \times 10^6)(164.4)} = 6.96 \times 10^{-3} \text{ rad} = 0.399° \qquad (l)$$

The angle of rotation between the ends is small. Finally, the Mohr's circle for the plane stress state corresponding to the stress components given in Eq. (*g*) is sketched in Fig. 9.6*e*. The radius of the circle R is 5320 psi, and the principal stresses are

$$\sigma_1 = \sigma_{av} + R = 5400 + 5320 = 10{,}720 \text{ psi}$$

$$\sigma_2 = \sigma_{av} - R = 5400 - 5320 = 80 \text{ psi} \qquad (m)$$

$$\sigma_3 = 0$$

The magnitude of the maximum shear stress in the material is [Eq. (8.42)]

$$\tau_{max} = \frac{\sigma_{max} - \sigma_{min}}{2} = 5360 \text{ psi} \qquad (n)$$

where we used $\sigma_{max} = \sigma_1$ and $\sigma_{min} = \sigma_3 = 0$.

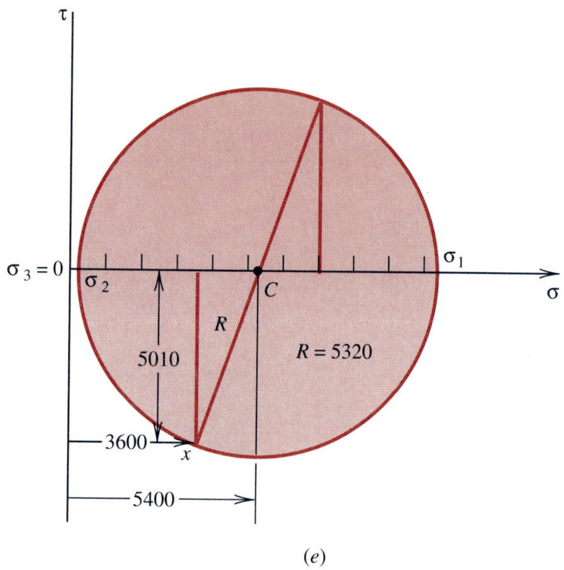

Figure 9.6 *Continued*

```
                    STRAIN ROSETTE MEASUREMENTS
          ********************************************

INPUT THE ANGULAR ORIENTATIONS
(IN DEGREES) OF THE THREE STRAIN GAGES: 0,45,90

INPUT THE THREE STRAINS:     48e-6, 343e-6, 204e-6

                             EPSILON(X) =  4.80E-05
                             EPSILON(Y) =  2.04E-04
                              GAMMA(XY) =  4.34E-04

                             EPSILON(1) =  3.57E-04
                             EPSILON(2) = -1.05E-04

                                 THETA1 =  54.89
                                 THETA2 = -35.11

                           GAMMA(MAX)/2 =  2.31E-04

DO YOU WANT THE CORRESPONDING IN-PLANE STRESSES [Y/N] : y


INPUT THE YOUNG'S MODULUS: 30e6

INPUT THE POISSON'S RATIO: .3

                               SIGMA(X) = 3.60E+03
                               SIGMA(Y) = 7.20E+03
                                TAU(XY) = 5.01E+03

                               SIGMA(1) = 1.07E+04
                               SIGMA(2) = 7.86E+01

                                 THETA1 =  54.89
                                 THETA2 = -35.11

                               TAU(MAX) = 5.32E+03
```

(*d*)

Figure 9.6 *Continued*

9.4 Combined Bending and Torsion of Circular Shafts

To calculate the maximum shear and normal stresses in a circular shaft that is loaded by a twisting moment T and a bending moment M, it is necessary to combine the torsional stress state with the bending stress state. In Fig. 9.7*a*, a segment of a hollow elastic shaft is shown loaded simultaneously by a twisting moment T and a bending moment M. The twisting moment vector $\mathbf{T}$ acts along the x axis while the bending moment vector $\mathbf{M}$ acts along the z axis. The outside and inside diameters of the shaft are d_o and d_i, respectively.

The shear stresses due to the twisting moment which act on the positive x face are shown in Fig. 9.7*b* and are given by Eq. (3.21) in the form

$$\tau = \frac{Tr}{J} \tag{9.6}$$

where J is the polar moment of inertia of the hollow section and is given by

$$J = \frac{\pi}{32}(d_o^4 - d_i^4) \tag{9.7}$$

and r is the radial distance from the axis.

Due to the bending moment M, we have the normal bending stresses acting on the positive x face as shown in Fig. 9.7*b* and given by Eq. (5.17) in the form

$$\sigma = -\frac{My}{I_z} \tag{9.8}$$

where I_z is the moment of inertia of the hollow circular section about the z axis, given by

$$I_z = \frac{\pi}{64}(d_o^4 - d_i^4) \tag{9.9}$$

The polar moment of inertia J and the moment of inertia I_z are related for a circular cross section by

$$J = 2I_z \tag{9.10}$$

The combined state of stress corresponds to the addition of the two separate states of stress shown in Fig. 9.7*b*. The maximum value of the shear stress due to torsion occurs at all points of the outside surface where $r_0 = d_o/2$ in Eq. (9.6). The maximum values for the bending stresses can be seen from Eq. (9.8) to occur along lines on the outer surface where $y = \pm d_o/2$. A typical element along the line $y = d_o/2$ is

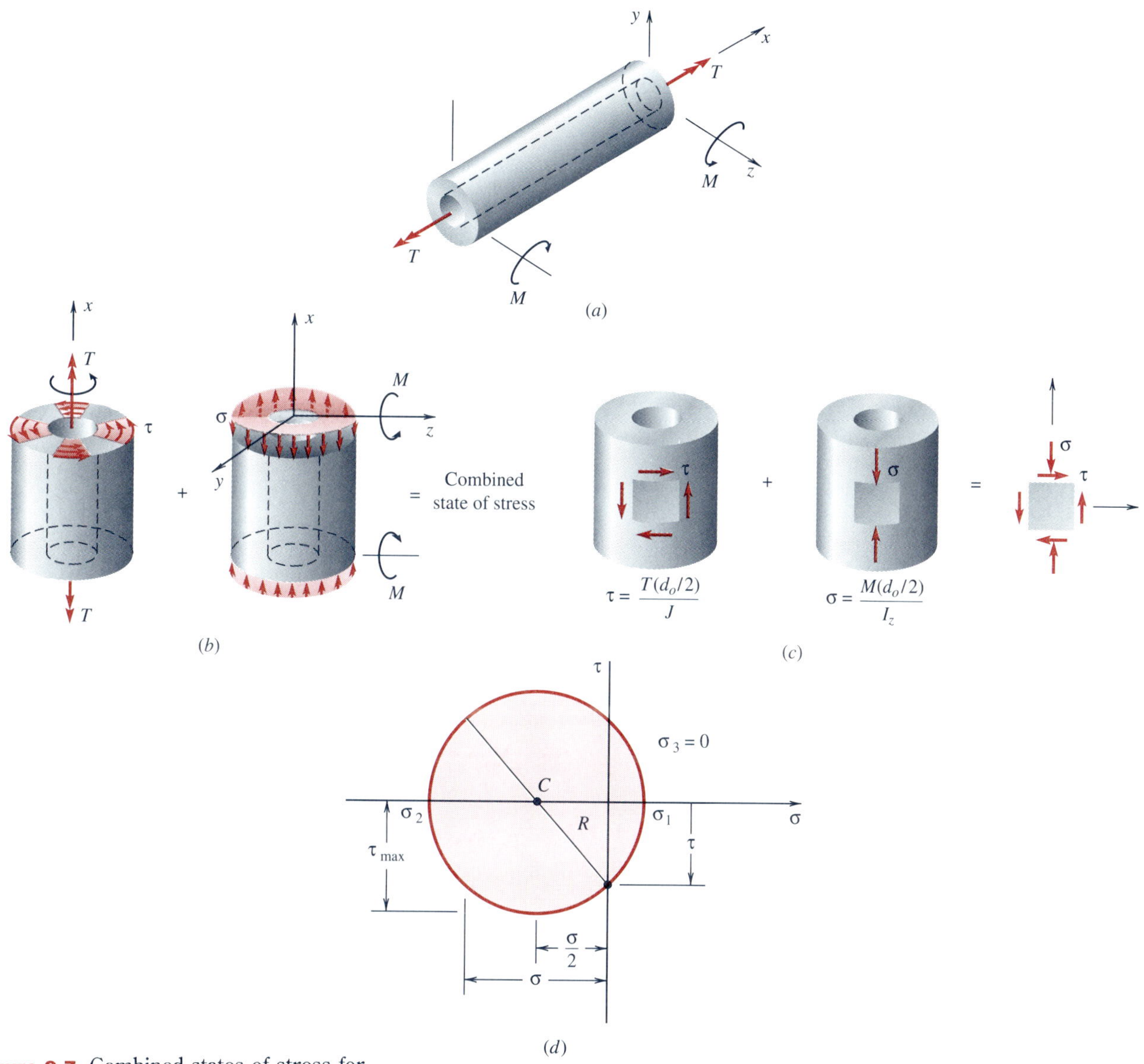

Figure 9.7 Combined states of stress for torsion and bending.

shown in Fig. 9.7*c* with the corresponding stresses acting on the element. On the opposite side of the shaft, the bending stress is tensile.

The surface element may be regarded as a plane stress element, and we may introduce a local coordinate system and show the combined-stress components acting on the plane stress element in Fig. 9.7*c*. With reference to the local axes we obtain the Mohr's circle representation, shown in Fig. 9.7*d*. The radius of the circle is given by

$$R = \tau_{max} = \sqrt{\left(\frac{\sigma}{2}\right)^2 + \tau^2} = \sqrt{\left(\frac{Mr_0}{2I_z}\right)^2 + \left(\frac{Tr_0}{J}\right)^2} \tag{9.11}$$

where $r_0 = d_o/2$, from which the maximum shear stress τ_{max} and the principal stresses σ_1 and σ_2 can be found.

We can also investigate the conditions for the onset of plastic deformations by using the two yield criteria from Sec. 8.15. The maximum shear stress criterion, Eq. (8.114), gives

$$\sqrt{\left(\frac{Mr_0}{2I_z}\right)^2 + \left(\frac{Tr_0}{J}\right)^2} = \frac{Y}{2} \tag{9.12}$$

where Y is the yield stress in tension. The Mises yield criterion, Eq. (8.113), gives

$$\sqrt{\sigma^2 + 3\tau^2} = \sqrt{\left(\frac{Mr_0}{I_z}\right)^2 + 3\left(\frac{Tr_0}{J}\right)^2} = Y \tag{9.13}$$

In general, the loads calculated to give rise to yielding by the two different yield criteria will be different; see Fig. 8.62. However, the differences between the calculated values will be in the range of experimental scatter for the material, and if we are to be conservative, we will select the smaller load value.

We will illustrate cases of combined bending and torsion of circular members in the next two examples.

EXAMPLE 9.5

A small steel bracket shown in Fig. 9.8*a* is to be designed to carry a load P without causing plastic deformation. We wish to determine the load P to cause the onset of plastic deformation as predicted by the Mises and the maximum-shear-stress yield criteria. The yield stress of the steel in simple tension is taken to be $Y = 200$ MPa. We also wish to estimate the deflection at the point of application of the load at the onset of plasticity. We will neglect the weight of the bar and any stress concentration effects that might occur at the wall support.

The first step in the solution is to determine the location on the bracket where the stresses are most likely to be the greatest. A section near A is exposed to a twisting moment, a bending moment, and a shear force, as shown in Fig. 9.8*b*. A section near C experiences a bending moment and a shear force, and a section near B experiences a twisting moment and a shear force. The values of the forces and moments acting on each section are determined from equilibrium of the segments shown in Fig. 9.8*b*. It would appear that the section near A is the most highly loaded and will give rise to the larger stresses. A point on the top surface near section A will experience a normal stress σ due to bending and a shear stress τ due to twisting. The shear stress due to the shear force P on the section is zero on the top (Fig. 9.8*c*).

The normal stress due to bending acting on the element in Fig. 9.8*c* is tensile and is given by

$$\sigma = -\frac{My}{I_z} = \frac{PLa}{I_z} \tag{a}$$

where a is the radius of the bracket. The shear stress due to twisting is given by

$$\tau = \frac{Ta}{J} = \frac{PLa}{J} \tag{b}$$

where $J = 2I_z$ for a circular member.

A sketch of the Mohr's circle representation for the state

Figure 9.8 Example 9.5

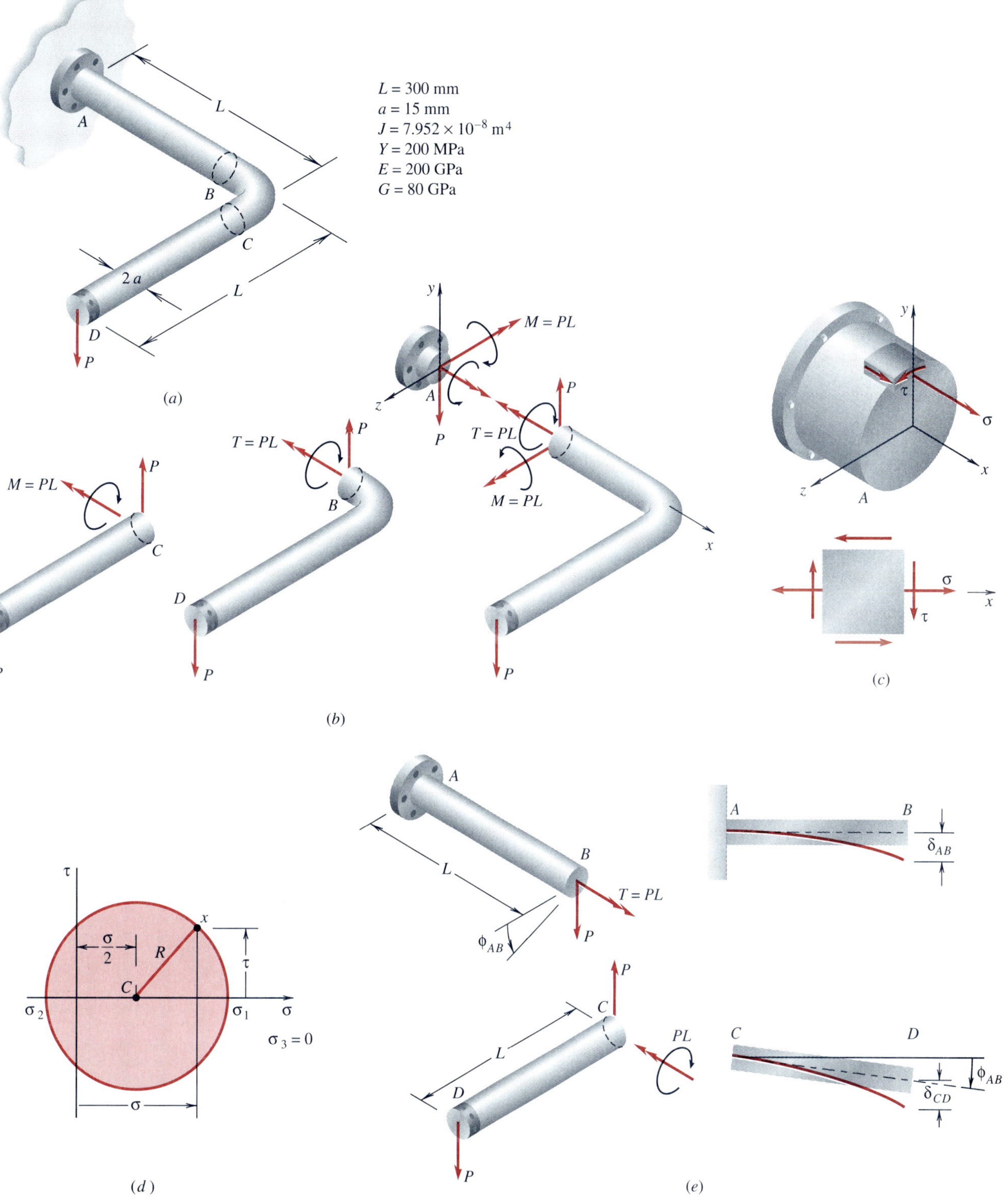

of plane stress in Fig. 9.8*c* is given in Fig. 9.8*d*, from which we have

$$R = \sqrt{\left(\frac{PLa}{2I_z}\right)^2 + \left(\frac{PLa}{J}\right)^2} = \frac{PLa}{J}\sqrt{2} \qquad (c)$$

$$\sigma_1 = \frac{\sigma}{2} + R \qquad \sigma_2 = \frac{\sigma}{2} - R \qquad \sigma_3 = 0 \qquad (d)$$

since $J = 2I_z$ for a circular shaft and

$$J = \frac{\pi}{2}a^4 \qquad (e)$$

To determine the load P to cause the onset of plastic deformation, we use the yield criteria of Sec. 8.15.

The Mises yield criterion, Eq. (8.111), gives

$$\sqrt{\tfrac{1}{2}[(\sigma_1 - \sigma_2)^2 + (\sigma_1 - \sigma_3)^2 + (\sigma_2 - \sigma_3)^2]} = Y \qquad (f)$$

or

$$\frac{P_m La}{J}\sqrt{7} = Y \qquad (g)$$

where P_m denotes the load calculated by the Mises criterion. The maximum shear stress criterion, Eq. (8.114), with Eqs. (*c*) and (*d*) gives

$$\tau_{\max} = \frac{\sigma_1 - \sigma_2}{2} = R = \frac{Y}{2}$$

or

$$\frac{P_s La\sqrt{2}}{J} = \frac{Y}{2} \qquad (h)$$

where P_s denotes the load calculated by the maximum-shear-stress criterion. Therefore, the load to cause yielding is given by either

$$P_m = \frac{J}{La}\frac{Y}{\sqrt{7}} = \frac{\pi a^3}{2L}\frac{Y}{\sqrt{7}} \qquad (i)$$

or

$$P_s = \frac{J}{La}\frac{Y}{\sqrt{8}} = \frac{\pi a^3}{2L}\frac{Y}{\sqrt{8}} \qquad (j)$$

Upon substitution of numerical values into Eqs. (*i*) and (*j*), we find

$$P_m = \frac{\pi(15 \times 10^{-3})^3(200 \times 10^6)}{2(0.300)\sqrt{7}} = 1336 \text{ N} \qquad (k)$$

or

$$P_s = P_m\frac{\sqrt{7}}{\sqrt{8}} = 1250 \text{ N} \qquad (l)$$

The different yield criteria give different estimates of the load P to cause yielding; they differ by about 7 percent. It would be prudent to select the lower value of 1250 N as the design load. It would also be prudent to check the adequacy of the support fixture at the wall to carry this load. It remains to estimate the vertical deflection at the point of application of the load.

The vertical deflection at point D due to load P_s is made up of the three contributions shown in Fig. 9.8*e*. The vertical deflection δ_{AB} arises from the cantilever beam deflection of AB. Segment CD rotates as a rigid body through an angle ϕ_{AB} due to the twisting moment on AB. Finally segment CD deflects as a cantilever beam to give δ_{CD}. The total vertical deflection at D is given by

$$\begin{aligned}\delta_D &= \delta_{AB} + L\phi_{AB} + \delta_{CD} \\ &= \frac{PL^3}{3EI_z} + \frac{L(PL)L}{GJ} + \frac{PL^3}{3EI_z} \qquad (m) \\ &= \frac{PL^3}{GJ}\left(1 + \frac{4}{3}\frac{G}{E}\right) = \frac{YL^2}{\sqrt{8}\,Ga}\left(1 + \frac{4}{3}\frac{G}{E}\right) = 8.13 \text{ mm}\end{aligned}$$

We note that the vertical deflection is about 27 percent of the diameter of the bracket. The total weight of the bracket is approximately 33 N, which is about 2.6 percent of the load $P_s = 1250$ N and can be neglected.

EXAMPLE 9.6

A special highway sign is supported by a pipe with outside and inside diameters of 3.5 and 2.8 in, as shown in Fig. 9.9*a*. The resultant P of the horizontal wind forces acts in the horizontal direction on the sign at the location, as shown. We wish to estimate the maximum shear stresses at points 1, 2, 3, and 4 on the surface of the pipe at the base if the estimated wind force P is 100 lb. The weight of the sign $W = 500$ lb and the weight of the supporting pipes can be neglected.

We would also like to estimate the wind force P that would initiate yielding in the pipe, using the maximum-shear-stress criterion with the allowable yield stress in simple tension $Y = 24$ ksi.

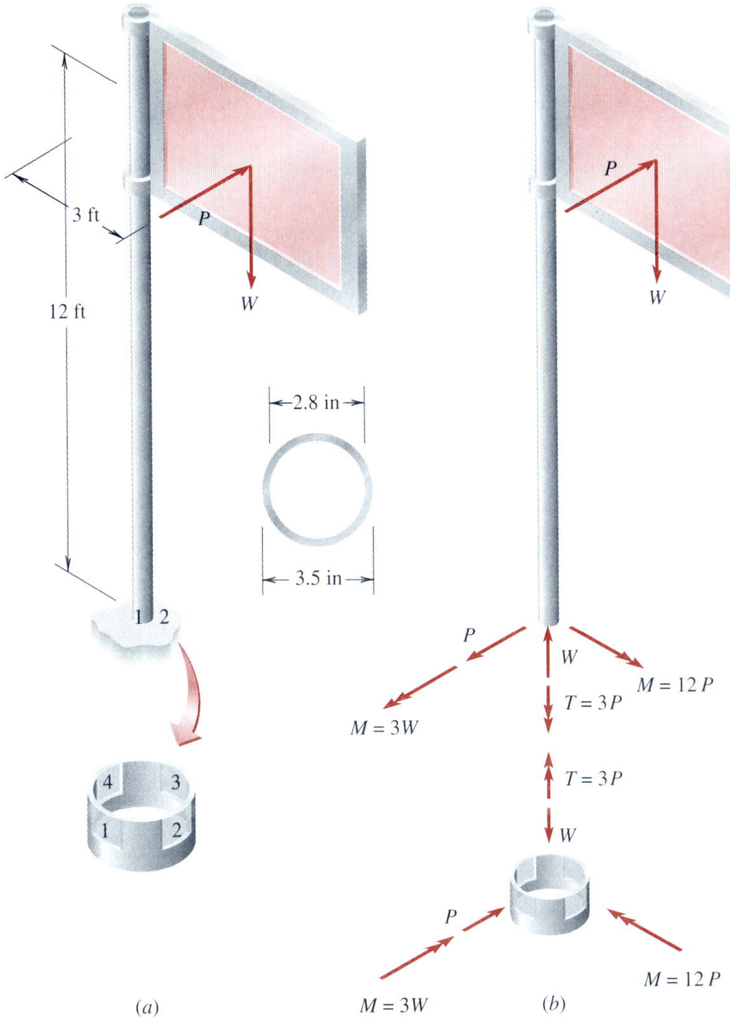

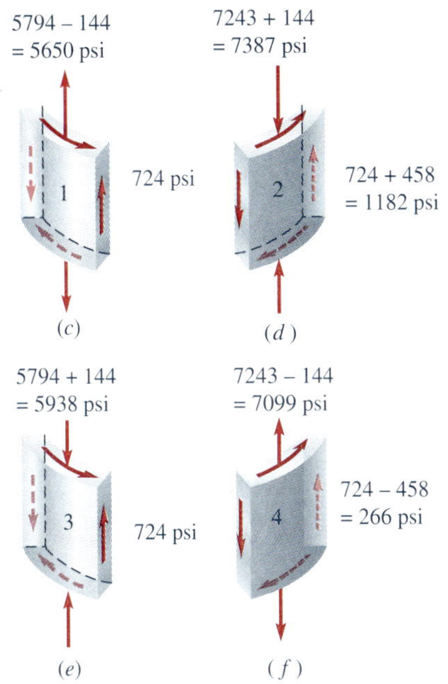

Figure 9.9 Example 9.6

If we neglect the self-weight of the vertical pipe, the wind load P will give resultants at the base of the pipe consisting of a twisting moment $3P$, a shear force P, and a bending moment $12P$, as required by force and moment equilibrium and as shown in Fig. 9.9*b*. The force from the weight W of the sign will give rise to a bending moment of $3W$ and an axial compressive force of W, as shown in Fig. 9.9*b*. We need to pay attention to the units of these force and moment resultants.

A plane stress element on the surface of the pipe at point 1 at the base of the pipe (Fig. 9.9*a*) is shown in Fig. 9.9*c*. The bending moment $12P$ will give rise to a maximum tensile bending stress at point 1 of

$$\sigma_b = \frac{12Pr_0}{I} = \frac{12(12)(100)(3.5/2)}{(\pi/64)(3.5^4 - 2.8^4)} = 5794 \text{ psi} \qquad (a)$$

The axial compressive stress due to $W = 500$ lb is

$$\sigma_a = -\frac{W}{A} = -\frac{500}{(\pi/4)(3.5^2 - 2.8^2)} = -144 \text{ psi} \qquad (b)$$

The twisting moment T acting at the section results in a shearing stress acting on the element at point 1

$$\tau = \frac{3Pr_0}{J} = \frac{3(12)(100)(3.5/2)}{(\pi/32)(3.5^4 - 2.8^4)} = 724 \text{ psi} \qquad (c)$$

The other force and moment acting at the section do not contribute additional stress components at point 1.

The Mohr's circle representation of the stress state for the element at point 1 is given in Fig. 9.9*g*, where

$$R = \sqrt{\left(\frac{5650}{2}\right)^2 + (724)^2} = 2916 \text{ psi} \qquad (d)$$

$$\sigma_{av} = \frac{5650}{2} = 2825 \text{ psi} \qquad (e)$$

and

$$\tau_{max} = R = 2916 \text{ psi} \qquad (f)$$

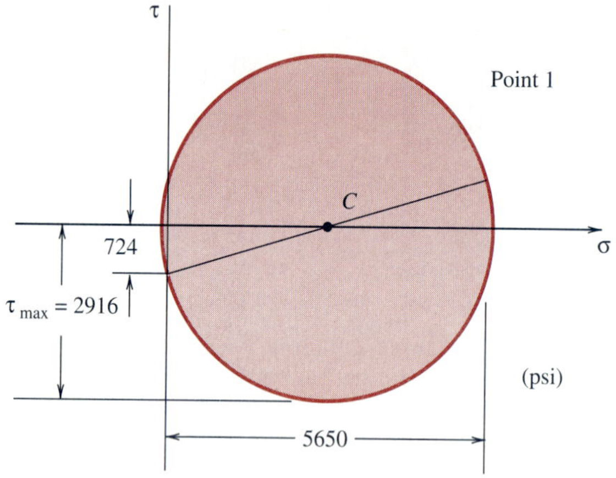

(g)

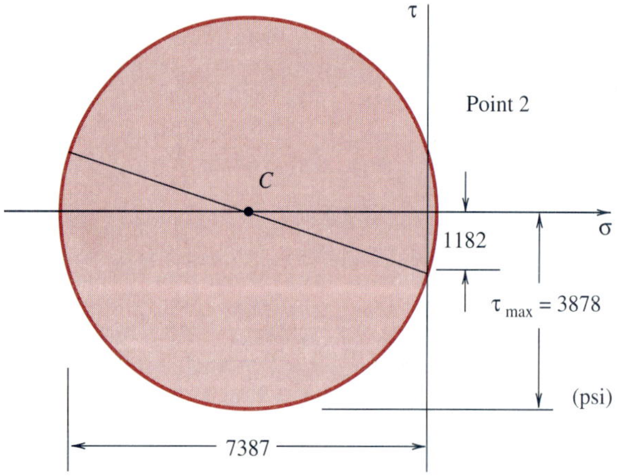

(h)

Figure 9.9 *Continued*

At point 2, at the base, Fig. 9.9*a*, the bending moment $3W$ leads to a compressive stress given by

$$\sigma_b = -\frac{3Wr_0}{I} = \frac{-3(12)(500)(3.5/2)}{(\pi/64)(3.5^4 - 2.8^4)} \qquad (g)$$
$$= -7243 \text{ psi}$$

The axial compressive stress is the same as that calculated for the element at point 1, Eq. (*b*),

$$\sigma_a = -144 \text{ psi}$$

The total compressive stress on the element, Fig. 9.9*d*, is 7387 psi.

Because the shear force P acting on the section gives rise to a shear stress due to bending at point 2, we have with $V = P$

$$\tau = \frac{VQ}{Ib}$$
$$= \frac{100\left[\frac{\pi}{2}(3.5^2)\frac{4(3.5)}{3\pi} - \frac{\pi}{2}(2.8^2)\frac{4(2.8)}{3\pi}\right]}{\left(\frac{\pi}{64}\right)(3.5^4 - 2.8^4)(0.7)} \qquad (h)$$
$$= 458 \text{ psi}$$

Since the shear stress due to the shear force P acts in the same direction as the shear stress due to the twisting moment $3P$ (Fig. 9.9*d*), the total shear stress, by using Eq. (*c*), acting on the element is

$$\tau = 458 + 724 = 1182 \text{ psi} \qquad (i)$$

The Mohr's circle representation for the stress state for the element at point 2 is given in Fig. 9.9*h*. The maximum shear stress is given by

$$\tau_{max} = R = \sqrt{\left(\frac{-7387}{2}\right)^2 + (1182)^2} = 3878 \text{ psi} \qquad (j)$$

Thus we see that the maximum shear stress for the element at point 2, Eq. (*j*), is larger than the maximum shear stress for the element at point 1.

An element located at point 3 is subjected to the same axial stress of −144 psi and the same torsional stress of 724 psi as element 1 (see Fig. 9.9*c*), but the bending stress is now compressive, as shown in Fig. 9.9*e*. The maximum shear stress is given by the radius of Mohr's circle for the stress state shown in Fig. 9.9*e*, that is,

$$\tau_{max} = \sqrt{\left(\frac{5938}{2}\right)^2 + (724)^2} \qquad (k)$$
$$= 3056 \text{ psi}$$

The stress state for an element located at point 4 is shown in Fig. 9.9*f*. Comparison with the stress state for point 2 shown in Fig. 9.9*d* indicates that both the normal and the shear stresses associated with bending are reversed in sign as we move from point 2 to point 4 (see Fig. 9.9*f*). The maximum shear stress at point 4 is

$$\tau_{max} = \sqrt{\left(\frac{7099}{2}\right)^2 + (266)^2} \quad (l)$$
$$= 3559 \text{ psi}$$

The results for the maximum shear stresses at the base of the pipe are given in Table 9.1. The largest stress is associated with point 2 and is mostly due to the bending stress caused by the moment of the weight of the sign.

To find the value of the wind force that will initiate yielding, we first examine the changes in magnitude and location of the maximum shear stress at the base of the pipe as the wind force is increased. We calculate the maximum shear stresses associated with a wind force of $P = 200$ lb (see Prob. 9.4-16 for this calculation), as given in Table 9.1. The doubling of the wind force has approximately doubled the bending and torsional stresses at points 1 and 3 and thus the maximum shear stress at those points. The maximum shear stresses at points 2 and 4 have increased only slightly due to the doubling of the wind force.

It is apparent that as the wind load increases, yielding will be initiated at point 3. An equation for the determination of the wind force P_Y required to initiate yielding at point 3 follows by making use of the maximum-shear-stress criterion to write

$$\tau_{max} = \sqrt{\left(\frac{\sigma}{2}\right)^2 + \tau^2} = \frac{Y}{2} \quad (m)$$

Table 9.1

	$P = 100$ lb	$P = 200$ lb	$P = 399.5$ lb
Point	τ_{max}*	τ_{max}	τ_{max}
1	2916	5902	11,860
2	3878	4385	5,995
3	3056	6043	12,000
4	3559	3589	3,705

*τ_{max} in psi

or

$$\tau_{max} = \sqrt{\left(-\frac{144P_Y r_0}{2I} - \frac{W}{2A}\right)^2 + \left(\frac{36P_Y r_0}{J}\right)^2} = \frac{Y}{2} \quad (n)$$

By squaring both sides of Eq. (n), inserting the values $r_0 = (3.5/2)$ in, $I = 4.349\text{ in}^4$, $J = 2I$, and $W/A = 144$ psi, and solving, we obtain

$$P_Y = 399.5 \text{ lb} \quad (o)$$

The maximum shear stresses at this load are given in Table 9.1; see Prob. 9.4-17.

Thus at wind loads of 100 and 200 lb, the factors of safety against yielding are approximately 4 and 2.

9.5 Other Combined-Stress Examples

In the general case, a given problem may give rise to combinations of several of the basic stress states that we considered in previous chapters. The procedure for the analysis of these more general cases is similar to the one used in previous sections of this chapter. That is, elements associated with high stress levels are identified, and stress contributions due to each of the various loadings, e.g., bending, torsion, axial, pressure, etc., are determined. These stresses are combined, and principal stresses and maximum shear stresses are found by using the stress transformation relations, usually in the form of Mohr's circle.

EXAMPLE 9.7

A large-diameter pressurized steel pipe in a coal gasification plant extends over a number of supports, as shown in Fig. 9.10*a*. We wish to investigate the effect of the weight of the piping on the stresses in the pipe. In particular, is the weight of the pipe an important consideration for estimating the internal pressure that might cause yielding of the pipe? We have $d_o = 1$ m, $t =$

8 mm, $p = 1$ MPa, $L = 20$ m, and $E = 200$ GPa, and the specific weight of the steel is 77 kN/m^3.

If we assume that the pipe is not restrained at the ends, we can neglect the axial stress due to the pressure. The hoop stress in the pipe due to the internal pressure, Eq. (8.57), is

$$\sigma_H = \frac{pr}{t} = \frac{(1)(0.5)}{8 \times 10^{-3}} = 62.5 \text{ MPa} \qquad (a)$$

Thus due to the pressure p inside the pipe, at the outside surface of the pipe we have the plane stress state characterized by the principal stresses

$$\sigma_1 = 62.5 \text{ MPa} \qquad \sigma_2 = 0 \qquad \sigma_3 = 0 \qquad (b)$$

The maximum shear stress τ_{max} without considering the weight of the pipe is

$$\tau_{max} = \frac{\sigma_1 - \sigma_3}{2} = 31.25 \text{ MPa} \qquad (c)$$

The pipe has a zero slope and deflection at each support at least away from the ends of the piping system. Therefore, we can model a segment of length L as a beam clamped at each end and subjected to its own weight q_0, as shown in Fig. 9.10b. The distributed weight loading can be found by multiplying the cross-sectional area

$$A = \frac{\pi}{4}(d_o^2 - d_i^2) = \frac{\pi}{4}(1^2 - 0.984^2) = 0.0249 \text{ m}^2 \qquad (d)$$

by the weight density to obtain

$$q_0 = \gamma A = 77(0.0249) = 1.92 \text{ kN/m} \qquad (e)$$

The moment of inertia of the pipe is

$$I = \frac{\pi}{64}(d_o^4 - d_i^4) = \frac{\pi}{64}(1 - 0.984^4) = 3.067 \times 10^{-3} \text{ m}^4 \qquad (f)$$

We can analyze the model beam segment shown in Fig. 9.10b "by hand," using the single-element equations for a beam from Fig. 7.12 (see Example 7.9, Fig. 7.15) to obtain the value of the maximum bending moment $q_0L^2/12$ at the supports. Alternatively, we can directly use the MECHMAT program (option

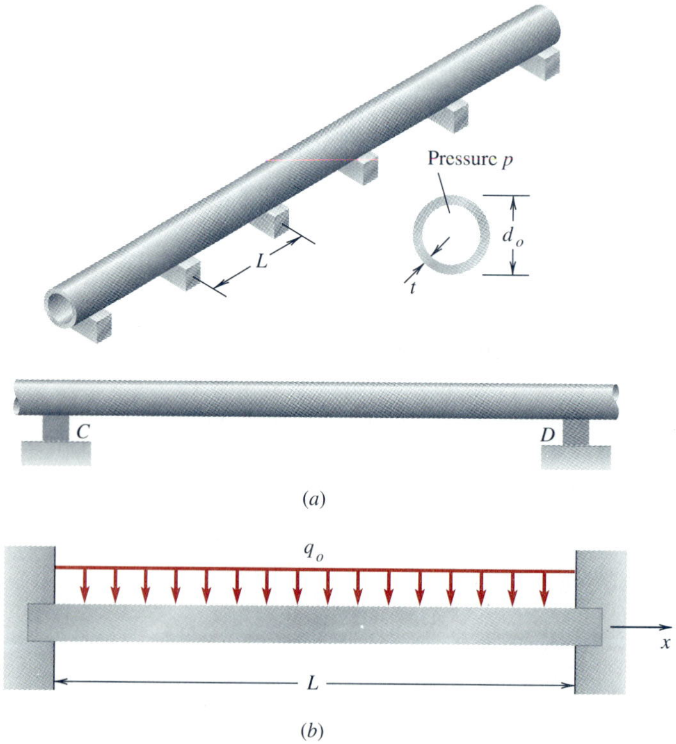

Figure 9.10 Example 9.7

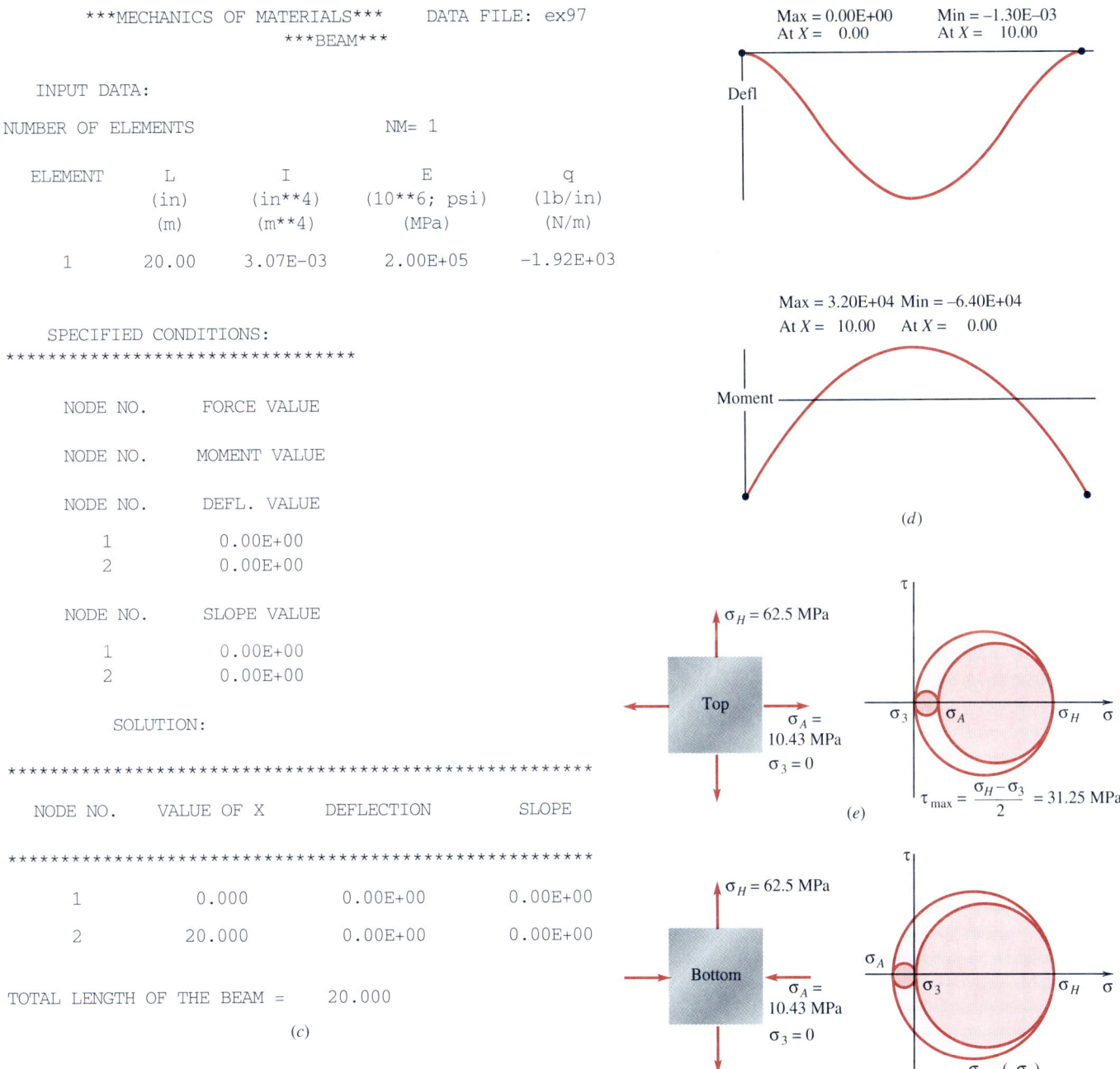

```
     ***MECHANICS OF MATERIALS***     DATA FILE: ex97
                       ***BEAM***

   INPUT DATA:

NUMBER OF ELEMENTS                         NM= 1

  ELEMENT       L          I           E           q
              (in)     (in**4)   (10**6; psi)   (lb/in)
               (m)      (m**4)      (MPa)        (N/m)

     1       20.00    3.07E-03    2.00E+05    -1.92E+03


   SPECIFIED CONDITIONS:
*********************************

     NODE NO.      FORCE VALUE

     NODE NO.     MOMENT VALUE

     NODE NO.      DEFL. VALUE
        1           0.00E+00
        2           0.00E+00

     NODE NO.      SLOPE VALUE
        1           0.00E+00
        2           0.00E+00

          SOLUTION:

*******************************************************
   NODE NO.    VALUE OF X      DEFLECTION        SLOPE

*******************************************************
      1           0.000         0.00E+00       0.00E+00

      2          20.000         0.00E+00       0.00E+00

TOTAL LENGTH OF THE BEAM =     20.000
```

(c)

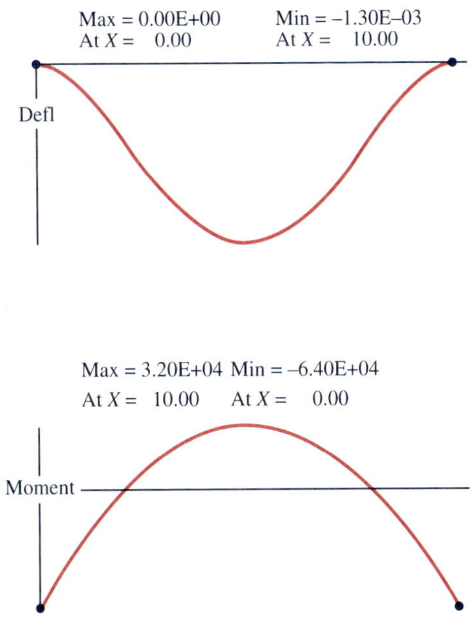

(d)

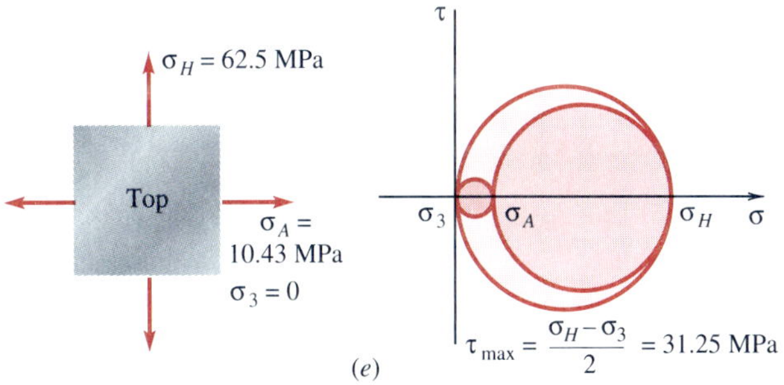

(e)

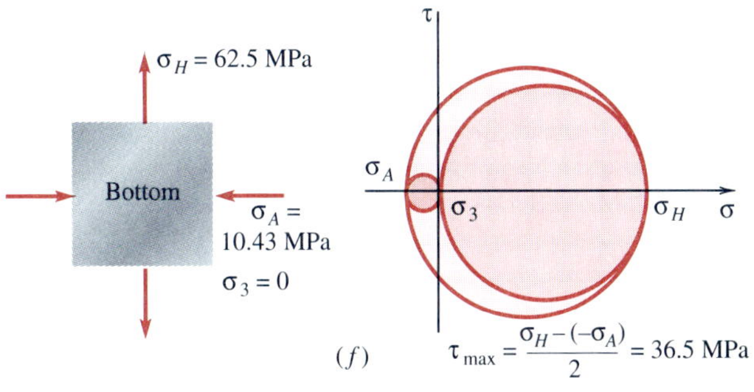

(f)

Figure 9.10 *Continued*

4: Deflections of beams) to obtain the maximum bending stresses and the deflection of the beam. The data supplied to this program for this single-element analysis are summarized in Fig. 9.10*c*. Note the zero slope and deflection values at the end nodes as required by the boundary conditions at both ends. A plot of the deflection curve of the beam is given in Fig. 9.10*d* and reveals that the total sag of the pipe is only 1.3 mm.

A plot of the bending moment in Fig. 9.10*d* shows that the magnitude of M at the supports is nearly twice the value at the midspan. The stress calculation option gives the maximum bending stresses at the top and bottom of the pipe as

$$\sigma_{bT} = 10.43 \text{ MPa} \qquad \sigma_{bC} = -10.43 \text{ MPa} \qquad (g)$$

The maximum tensile and compressive stresses will occur at the top and bottom of the pipe at the supports. The state of stress at a point on the top surface of the pipe at a support is one of biaxial stress, as shown in Fig. 9.10*e*; similarly for the bottom surface of the pipe, as shown in Fig. 9.10*f*. Mohr's circles for these states of stress in Fig. 9.10*e* and *f* show that the maximum shear stress in the material including the effect of the weight is 36.5 MPa, Fig. 9.10*f*, or 5.25 MPa greater than the value obtained by not including the weight, Eq. (*c*). We conclude therefore that the maximum shear stress in the material will be increased by including the effect of the weight of the piping in the analysis. If the yield stress of the material is Y MPa, the maximum internal pressure in the piping to cause yielding (without any consideration of factors of safety) will be reduced by the inclusion of the weight according to

$$p = \frac{t}{r}(Y - 10.43) \qquad \text{MPa} \qquad (h)$$

EXAMPLE 9.8

A model of a portion of a shaft in a dragline excavator consists of a solid shaft AD fixed at both ends (Fig. 9.11*a*). As shown in the figure, transverse loads are applied at sections B and C in addition to the twisting moments applied to the same sections. We wish to find the maximum shear stress in shaft AD caused by the given configuration of loads. Shaft AD is steel with diameter $d = 2$ in, $E = 30 \times 10^6$ psi, $\nu = 0.3$, $L = 72$ in, and $P = 100$ lb. We neglect the weight of the shaft in our analysis.

The shaft can be analyzed as a statically indeterminate beam under two transverse loads and as a statically indeterminate shaft with two applied twisting moments. The shaft is statically indeterminate because we are modeling it as built in at sections A and D. The stress analysis will have three parts.

In view of the relative complexity of this problem, we will approach the solution by using computer programs. First, we calculate the bending stresses, using the program BEAMMECH, then we calculate the torsional stresses, using the TORMECH program. Finally these stresses are combined by using the stress transformation relations in the form of Mohr's circle to obtain the maximum shear stress in the material.

We use the nodes and elements along the shaft shown in Fig. 9.11*a*.

For the bending analysis, we make use of the BEAMMECH program with the moment of inertia for bending given by

$$I_z = \frac{\pi}{64} d^4 = \frac{\pi}{64}(2)^4 = 0.7854 \text{ in}^4 \qquad (a)$$

The data for the elements, specified loads, and geometric conditions are given in Fig. 9.11*b*. Sections A and D are built in. The bending moment diagram is shown in Fig. 9.11*c*. The largest bending stresses occur at section A (the fixed left end of the

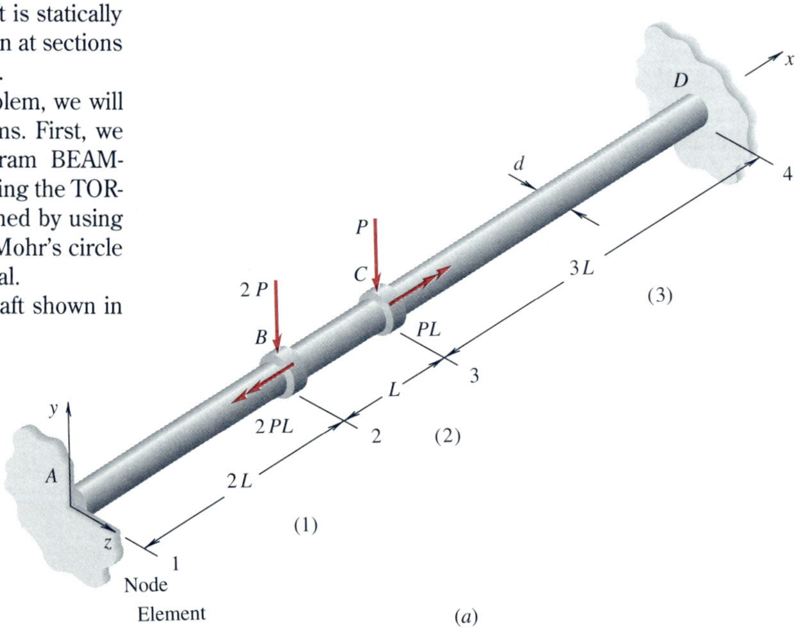

Figure 9.11 Example 9.8

```
        ***MECHANICS OF MATERIALS***    DATA FILE: ex98
                       ***BEAM***

   INPUT DATA:

NUMBER OF ELEMENTS                     NM= 3

   ELEMENT      L          I             E            q
              (in)      (in**4)    (10**6; psi)    (lb/in)
               (m)      (m**4)        (MPa)         (N/m)

      1      144.00    7.85E-01     3.00E+01      0.00E+00
      2       72.00    7.85E-01     3.00E+01      0.00E+00
      3      216.00    7.85E-01     3.00E+01      0.00E+00

    SPECIFIED CONDITIONS:
*********************************

      NODE NO.      FORCE VALUE

         2           -2.00E+02
         3           -1.00E+02

      NODE NO.     MOMENT VALUE

      NODE NO.      DEFL. VALUE

         1            0.00E+00
         4            0.00E+00

      NODE NO.      SLOPE VALUE

         1            0.00E+00
         4            0.00E+00

         SOLUTION:

*******************************************************

  NODE NO.    VALUE OF X      DEFLECTION         SLOPE

*******************************************************

      1          0.000        0.00E+00         0.00E+00

      2        144.000       -3.82E+00        -2.40E-02

      3        216.000       -4.42E+00         7.33E-03

      4        432.000        0.00E+00         0.00E+00

TOTAL LENGTH OF THE BEAM =     432.000
```

(*b*)

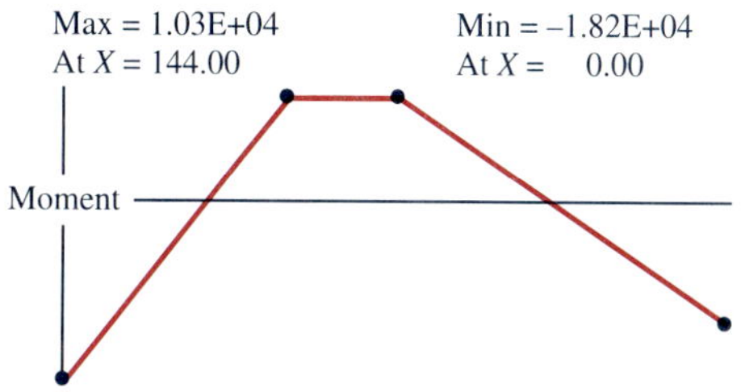


```
Do you want max bending stress [Y/N]: ? y

Input C1 (in;m)  = ? 1.

Input C2 (in;m)  = ? 1.

*****************************************

Maximum tensile stress = 2.317E+04
       At X =        0.00

Maximum compressive stress = 2.317E+04
       At X =        0.00
```

(*c*)

Figure 9.11 *Continued*

```
***MECHANICS OF MATERIALS***    DATA FILE: ex98a
                ***TORSION***

   INPUT DATA:

NUMBER OF ELEMENTS                    NM= 3

  ELEMENT    L (in)     DI (in)      DO (in)     G (10**6; psi)
               (m)        (m)          (m)         (MPa)

      1      144.00     0.0000       2.0000        1.15E+01
      2       72.00     0.0000       2.0000        1.15E+01
      3      216.00     0.0000       2.0000        1.15E+01


    CONDITION
*********************************

        NODE NO.        TORQUE (lb-in;N-m)

            2               -1.44E+04
            3                7.20E+03

        NODE NO.      SPECIFIED ANGLE (rad.)

            1                0.00E+00
            4                0.00E+00

      SOLUTION:

*******************************************************
      NODE         ANGLE (radians)         ANGLE (degrees)
*******************************************************

        1             0.0000E+00              0.0000E+00

        2            -4.7664E-02             -2.7309E+00

        3            -1.4299E-02             -8.1928E-01

        4             0.0000E+00              0.0000E+00

*********************************************************
 ELEMENT    TORQUE     MAX SHEAR STRESS       J            G
*********************************************************

    1    -6.0000E+03    -3.8197E+03    1.5708E+00   1.15E+07

    2     8.4000E+03     5.3476E+03    1.5708E+00   1.15E+07

    3     1.2000E+03     7.6394E+02    1.5708E+00   1.15E+07
```

(*d*)

Figure 9.11 *Continued*

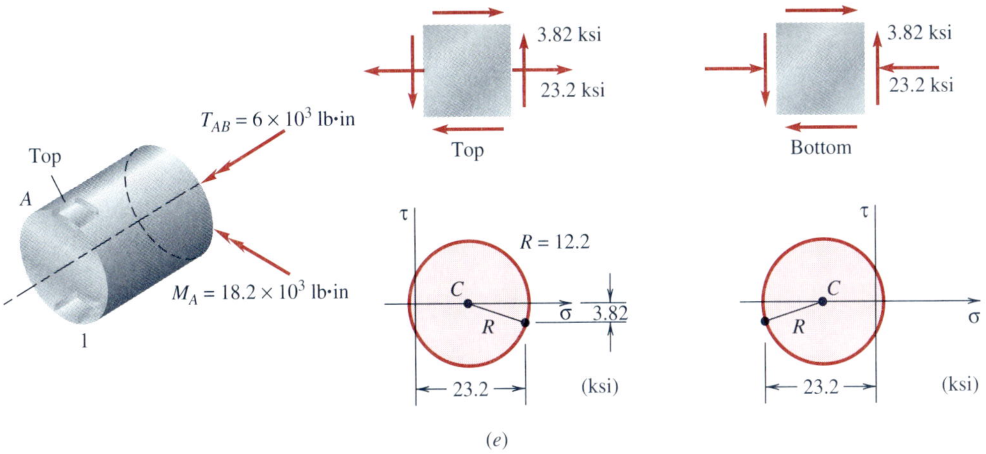

Figure 9.11 *Continued*

shaft at node 1) where, by using the stress calculation option in the program, we get (Fig. 9.11*c*)

$$\sigma_{xA} = \pm 23.2 \text{ ksi} \tag{b}$$

We should also obtain for subsequent combining of bending and torsional stresses the value of the bending stresses at section B, by scaling the values in Eq. (*b*) by the ratio of the moments

$$\sigma_{xB} = \frac{M_B}{M_A}\sigma_{xA} \tag{c}$$

or

$$\sigma_{xB} = \left(\frac{10.3}{-18.2}\right)(\pm 23.2) = \mp 13.1 \text{ ksi} \tag{d}$$

For the torsional analysis of the shaft, the value of the polar moment of inertia is

$$J = 2I = 2(0.7854) = 1.571 \text{ in}^4 \tag{e}$$

and the shear modulus is

$$G = \frac{E}{2(1+\nu)} = \frac{30 \times 10^6}{2(1.3)} = 11.54 \times 10^6 \text{ psi} \tag{f}$$

The element data, specified loads, and geometric conditions needed for use of the TORMECH program are shown in Fig. 9.11*d*. The solutions for the stresses at sections A and B due to the twisting moments are given in Fig. 9.11*d* as

$$\tau_A = -3.82 \text{ ksi} \qquad \tau_B = 5.35 \text{ ksi} \tag{g}$$

The results for the bending and twisting moment distributions along the shaft and the values of the stresses given in Eqs. (*b*), (*d*), and (*g*) suggest that the maximum shear stress in the material will occur at the top and bottom points on the shaft near section A. The states of stress at the top and bottom of the shaft and the corresponding sketches of Mohr's circles are shown in Fig. 9.11*e*. The radius of Mohr's circle in each case is given by Eq. (8.21):

$$R = \sqrt{\left(\frac{\sigma_x - \sigma_y}{2}\right)^2 + \tau_{xy}^2} = \sqrt{\left(\frac{23.2}{2}\right)^2 + (3.82)^2} = 12.2 \text{ ksi} \tag{h}$$

From the sketches of Mohr's circles, we see that the maximum shear stress in the material corresponds to the radius of the circle R, so that the maximum shear stress in the material of the shaft is given by

$$\tau_{\max} = 12.2 \text{ ksi} \tag{i}$$

The maximum shear stress in the material at section B is 8.46 ksi.

9.6 Concluding Remarks

In this chapter we have integrated essentially all the subjects introduced in previous chapters. This has been done by considering problems which involve structures and machine components that are configured and loaded in such a way as to give rise to several of the basic stress states that have been modeled in the previous chapters, i.e., axial, torsional, and bending stress states. In most of the cases, the stressed elements were in a state of plane stress. The equations for the transformation of plane stress derived in Chap. 8 enable us to predict the extreme values for normal and shear stresses for an element and thus provide us with a procedure for designing structural and machine components based upon a specified criterion for initial yielding.

The examples chosen in this chapter ranged from relatively simple problems which combined two basic stress states to problems where several stress states with complicated variations of stress over the structure or machine component were involved. In the latter case it was convenient to make use of one or more of the MECHMAT programs in the analysis.

PROBLEMS

9.2-1 Consider Example 9.1, Fig. 9.2. If the sheet steel used to make the frame is 10 percent thicker than the $\frac{1}{16}$-in nominal thickness, what is the maximum allowable tension in the blade when the depth $d = 0.5$ in? All other parameters remain the same.

9.2-2 Consider Example 9.1, Fig. 9.2. If the sheet steel used to make the frame is 10 percent thinner than the $\frac{1}{16}$-in nominal thickness, what is the maximum allowable tension in the blade when the depth $d = 0.5$ in? All other parameters remain the same.

9.2-3 Solve Prob. 9.2-1 if the depth d is now 0.6 in.

9.2-4 Solve Prob. 9.2-2 if the depth d is now 0.6 in.

9.2-5 The cross-sectional dimensions of a typical hacksaw blade are 1 mm × 13 mm. What is the tensile stress in the blade for the blade force found in Example 9.1 when $d = 0.5$ in?

9.2-6 Consider Example 9.2, Fig. 9.3. Calculate the normal stress distribution on the base when (*a*) $H = 1$ m and (*b*) $H = 3$ m. (*c*) What does a normal tensile stress at the base suggest regarding potential water leakage?

9.2-7 A wooden column AC carries an axial load P at point C and an equipment load F whose line of action is offset by a distance e from the neutral axis of the column, as shown in Fig. P9.2-7. If the combined allowable compressive stress is 1000 psi and $L = 8$ ft, find the minimum dimensions of a square column which can carry these loads. Take $e = L/12$.

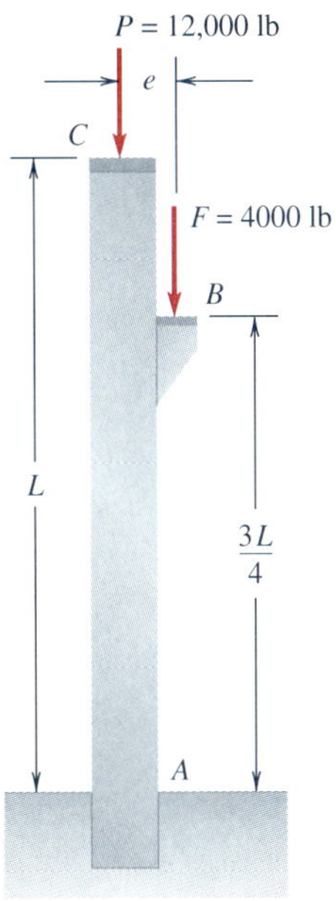

Fig. P9.2-7

9.2-8 A simply supported beam carries a constant distributed load q_0, as shown in Fig. P9.2-8. Derive an expression for the maximum bending stress in the beam due to the load q_0.

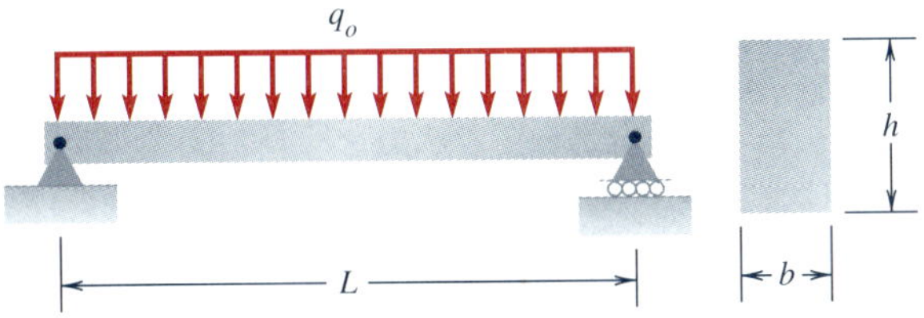

Fig. P9.2-8

9.2-9 A lightweight vertical wood structural member AC of cross-sectional dimensions 300 mm × 300 mm is shown in Fig. P9.2-9. A second structural member BD that is rigid and is 2 mm longer than the distance of 4 m between beam AC and the wall at D is forced into place. Find the maximum bending stresses in member AC caused by the force at B from member BD. Take $E = 14$ GPa for AC.

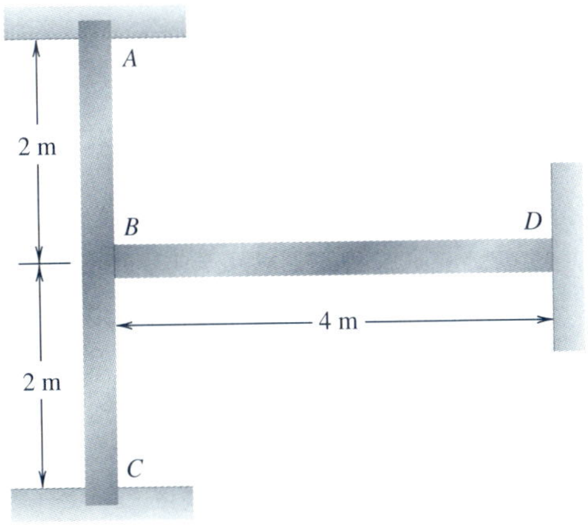

Fig. P9.2-9

9.2-10 A steel rod of circular cross section of radius r is loaded as shown in Fig. P9.2-10. Calculate the maximum shear stress at point A on the top of the section near the built-in wall support. The rod is in the vertical xy plane, and the force P acts in the negative x direction. Take $r = 1$ in, $a = 4$ in, $b = 10$ in, $c = 4$ in, and $P = 1000$ lb.

Fig. P9.2-10

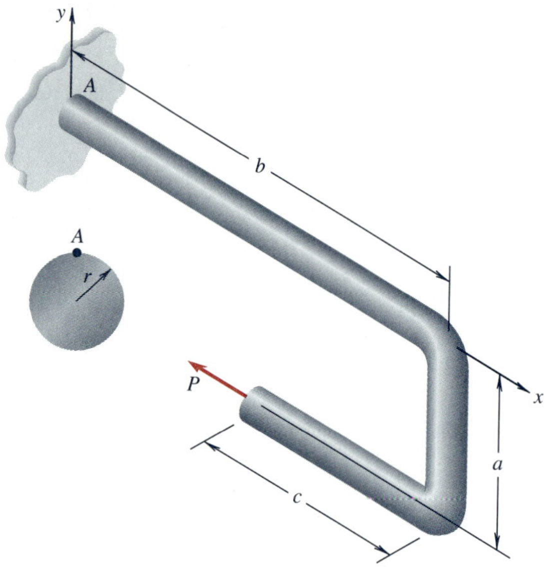

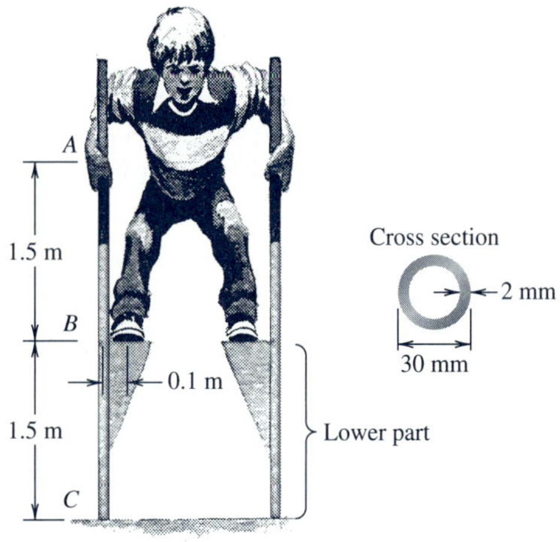

Fig. P9.2-11

9.2-11 A young person of weight equal to 720 N is standing on stilts, as shown in Fig. P9.2-11. The person's weight is equally divided between the two stilts, and the weight of the stilts can be neglected. Estimate the maximum tensile and compressive stresses in the lower part of each stilt.

9.2-12 If the stilts of Prob. 9.2-11 have a maximum allowable compressive stress of 5.6 MPa, what minimum size for a square cross section can be used for the stilts?

9.2-13 A steel W10 × 60 beam during construction is loaded as shown in Fig. P9.2-13. Find the maximum tensile stress at section *A*.

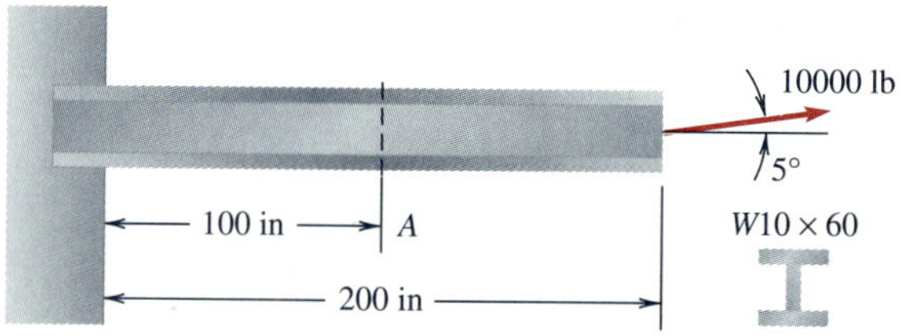

Fig. P9.2-13

9.2-14 An 8000-ft^3-capacity water tower is constructed of a tank 20 × 20 × 20 ft atop a 30-ft-high tube of outside diameter 6 ft and thickness 1.2 in. When the tank is full, it was found that a "strong" wind was blowing normal to one face, gusting up to a wind pressure of 15 lb/ft^2, as shown in Fig. P9.2-14. Assume that the wind pressure is uniformly distributed over only one face of the tank.

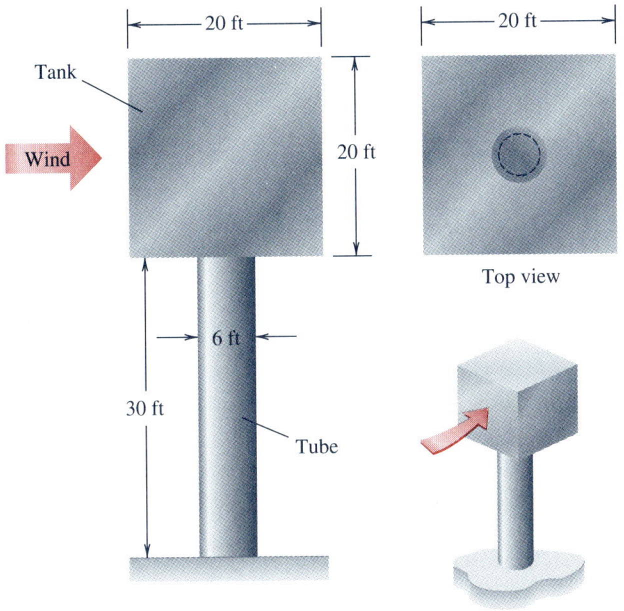

Fig. P9.2-14

(*a*) Determine the location in the tube where the normal stresses are maximum.
(*b*) Find the magnitudes of these stresses.
(*c*) Draw a properly oriented element to show the stresses. Assume that the tube will not buckle and that the weight density of water is 62.4 lb/ft^3. There is no water in the tube.

9.2-15 A single strain gage is mounted at point A in the direction $\theta = 45°$ on a simply supported beam, as shown in Fig. P9.2-15. If a load P is applied as shown, find the strain in the gage as the value of c varies.
(*a*) $c = 30$ mm

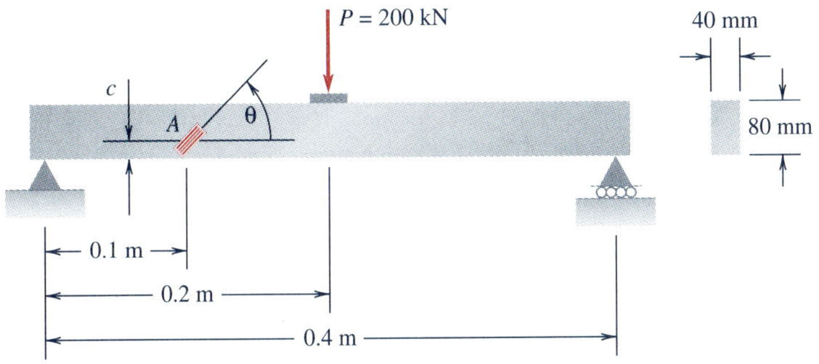

Fig. P9.2-15

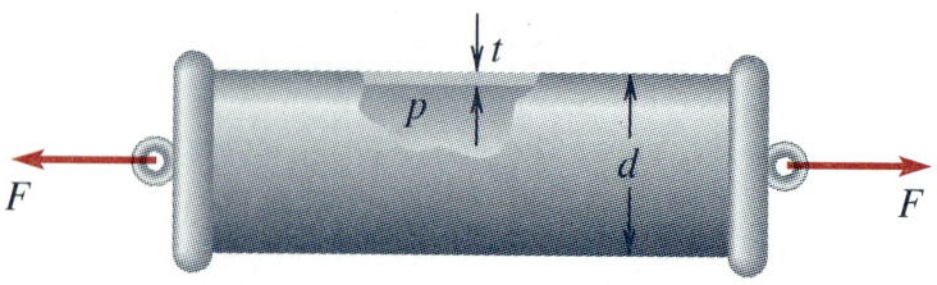

Fig. P9.2-17

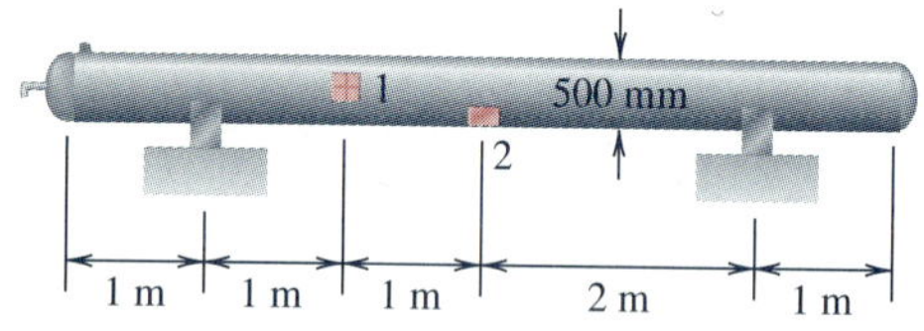

Fig. P9.2-18

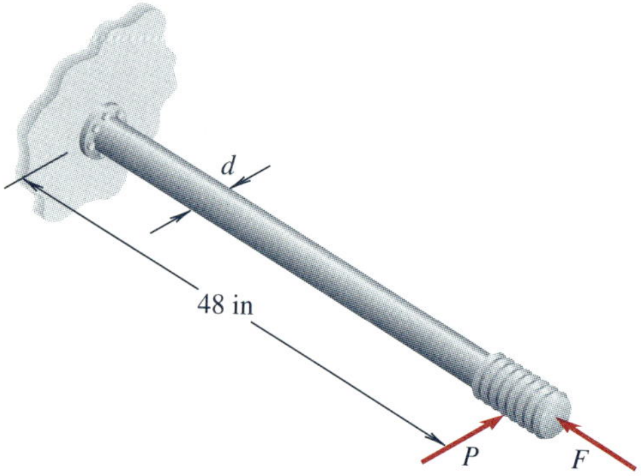

Fig. P9.2-19

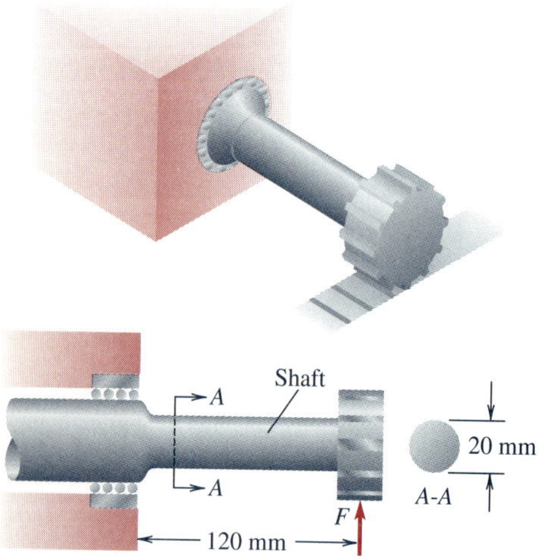

Fig. P9.2-21

(*b*) $c = 40$ mm
(*c*) $c = 50$ mm
Neglect the weight of the beam; take $E = 70$ GPa and $\nu = 0.33$.

9.2-16 Find the principal stresses, principal directions, and maximum shear stress at point A on the simply supported beam shown in Fig. P9.2-15 as the value of c varies.
(*a*) $c = 10$ mm
(*b*) $c = 20$ mm
(*c*) $c = 30$ mm
(*d*) $c = 40$ mm
Neglect the weight of the beam, and take $E = 200$ GPa and $\nu = 0.3$.

9.2-17 A thin-walled closed-end pressure vessel (thin-walled cylinder) has an additional tensile load F acting on the ends, as shown in Fig. P9.2-17. Determine the load F to cause yielding of the material if the yield stress in a simple tension test is $Y = 30$ ksi. Take $p = 150$ psi, $d = 20$ in, and $t = 0.1$ in.

9.2-18 A cylindrical thin-walled tank weighing 1.5 kN/m is supported as shown in Fig. P9.2-18 and contains a gas at a pressure of 1.5 MPa. Find the maximum shear stress in the wall of the tank at positions 1 and 2 along the tank. The thickness of the tank is $t = 10$ mm, and the diameter is 500 mm. Neglect the additional weight of the end caps.

9.2-19 A cylindrical drill rod of diameter d equal to 2 in, as shown in Fig. P9.2-19, is subjected to an axial force F and a transverse force P. Find the maximum compressive stress in the rod. Take $F = 12$ kips and $P = 200$ lb.

9.2-20 A cylindrical drill rod of diameter d is subjected to an axial force F and a transverse force P, as shown in Fig. P9.2-19. The maximum allowable shear stress in the rod is 30 ksi. If $F = 12$ kips and $P = 200$ lb, find the minimum allowable diameter of the shaft so as not to exceed the maximum allowable shear stress.

9.2-21 A structural model for the study of a rotary embossing machine is shown in Fig. P9.2-21. If the maximum allowable normal stress in the shaft is 150 MPa, find the maximum allowable load F acting on the wheel.

9.2-22 A machine fixture is shown in Fig. P9.2-22. A load $P = 1.3$ kN is applied in the direction indicated. Find the maximum normal stress on section A-A indicated. Neglect the weight of the part.

9.2-23 A small punching machine used in the manufacture of computer chips has strain gages mounted as shown in Fig. P9.2-23 in order to monitor the punching load P during operation. The data from the gages during a typical run after the strain readings were used to compute the stresses gave the normal stress at B as 1430 psi tension and at C as 1810 psi compression. Determine the value of the load P for this case.

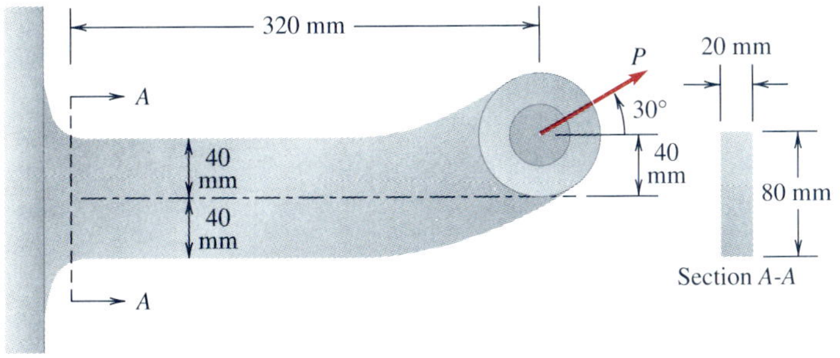

Fig. P9.2-22

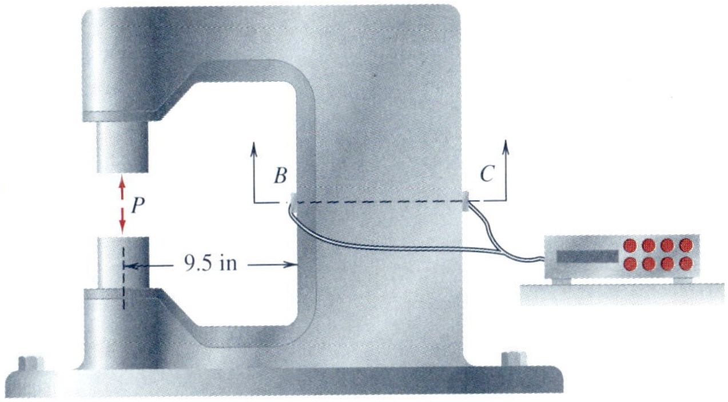

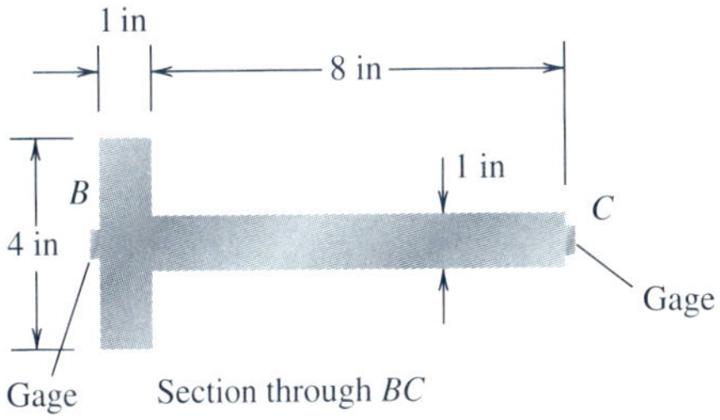

Fig. P9.2-23

9.2-24 A machine bracket shown in Fig. P9.2-24 is loaded by an axial force $F = 300$ kN and a transverse load $P = 200$ kN. Find the maximum normal tensile stress on section *A-A* indicated. Neglect the weight of the part.

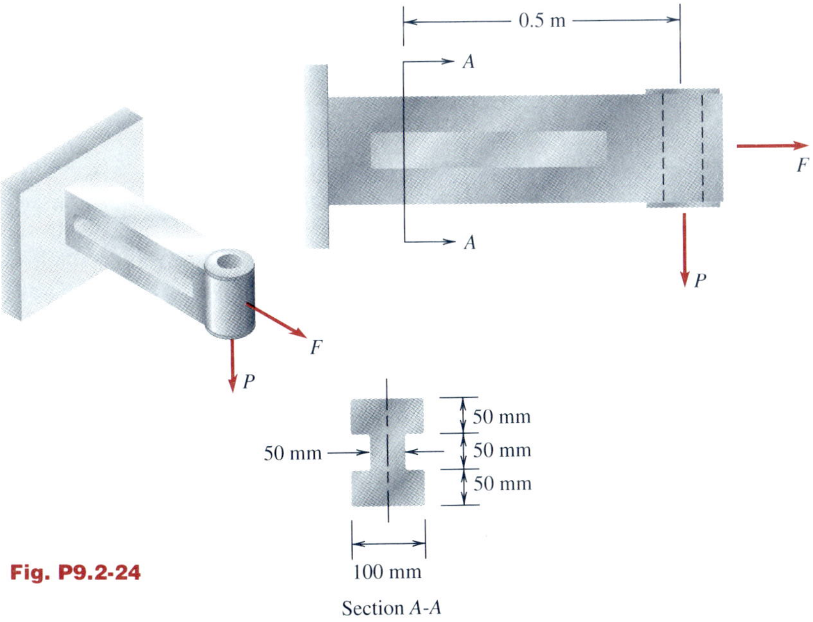

Fig. P9.2-24

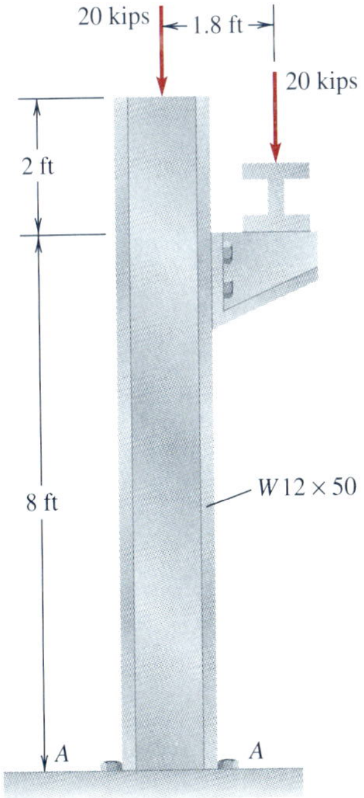

9.2-25 A W12 × 50 column is loaded by two loads as shown in Fig. P9.2-25. Determine the maximum tensile stress at section *A-A* near the base of the column.

Fig. P9.2-25

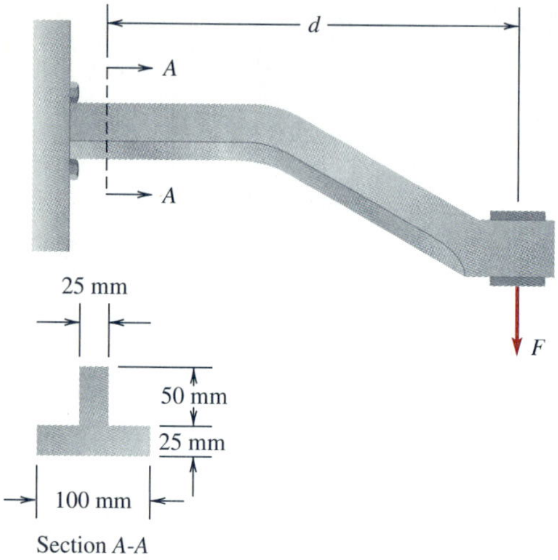

Fig. P9.2-26

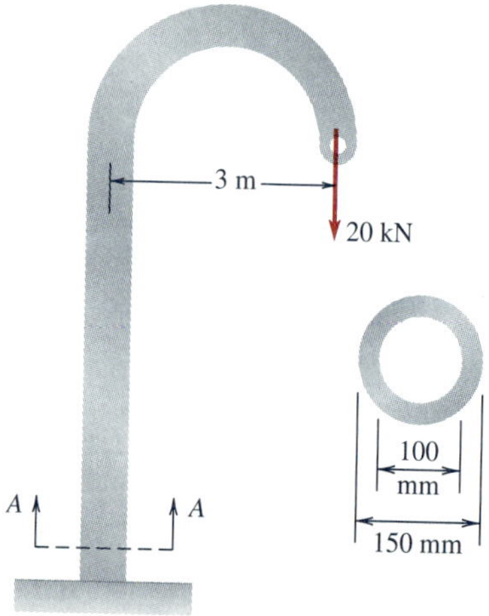

Fig. P9.2-27

9.2-26 A bracket is to be designed to carry a load of $F = 1.25$ kN, as shown in Fig. P9.2-26. If the maximum allowable stress in the bracket is 70 MPa, what is the maximum allowable value of the distance d as shown in the figure?

9.2-27 A bracket carries a load of 20 kN, as shown in Fig. P9.2-27. Determine the maximum shear stress in the material at a section AA near the base. Neglect the weight of the bracket.

9.3-1 Consider Example 9.3, Fig. 9.5. If the loads remain the same as specified in Example 9.3 but the $\frac{1}{4}$-in drill is replaced by an $\frac{1}{8}$-in drill, find the maximum value of the principal stresses and the maximum shear stresses due to the combined loads. Neglect the geometry effect of the grooves.

9.3-2 Given is a circular post of radius r loaded by an axial compressive load P and a twisting moment T, as shown in Fig. P9.3-2.

(*a*) Find the value of the maximum shear stress.
(*b*) Find the values of the principal stresses at point A on the post and the principal directions.
(*c*) Find the shortening of the post.
(*d*) Find the angle of twist of the post relative to the base.

Take $r = 1$ in, $P = 50$ kips, $E = 30{,}000$ ksi, $T = 10$ kip · in, $L = 1$ ft, and $G = 12{,}000$ ksi.

9.3-3 A shaft AB for a machine used for installing telephone poles is shown in Fig. P9.3-3. Shaft AB is subjected to an axial force P and torque T when the machine is operating. Derive expressions for (*a*) the maximum shear stress in the shaft and (*b*) the principal stresses.

9.3-4 A closed section of pressurized pipe is subjected to a twisting moment T, as shown in Fig. P9.3-4. Three strain gages are mounted on the outside surface of the tank with gage a in the x direction and gages b and c at $\pm 120°$ with the x direction. Derive expressions that relate the pressure p and the moment T to the strain readings ϵ_a, ϵ_b, and ϵ_c and the pipe radius r, thickness t, and elastic constants E and ν.

9.3-5 A thin-walled closed-end tube is pressurized with $p = 75$ psi and experiences a twisting moment $T = 4.71 \times 10^5$ lb · in at each end, as shown in Fig. P9.3-5. Find the maximum shear stress in the tube material under these load conditions. Take $a = 10$ in and $t = 0.75$ in.

9.3-6 A cable mechanism for holding back aircraft on an aircraft carrier contains a "holdback bar" which fractures when the aircraft is ready for launch, as shown in Fig. P9.3-6. The bar can be modeled as a circular shaft with an inner hole. During the holdback period, the shaft is subjected to an axial force $F = 63{,}000$ lb and a twisting moment $T = 2000$ lb · in. Determine the maximum shear stress in the shaft due to these combined loads.

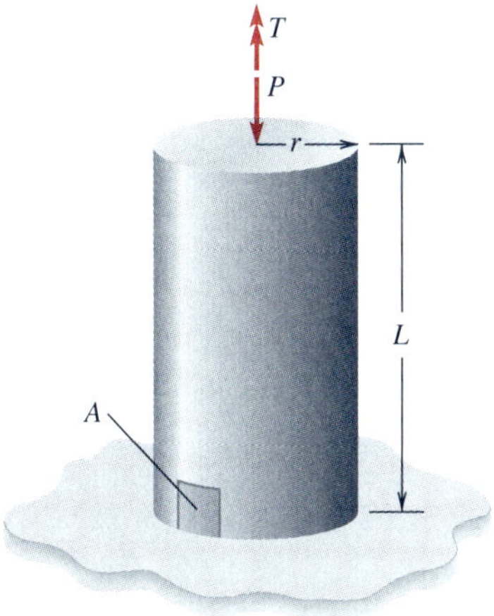

Fig. P9.3-2

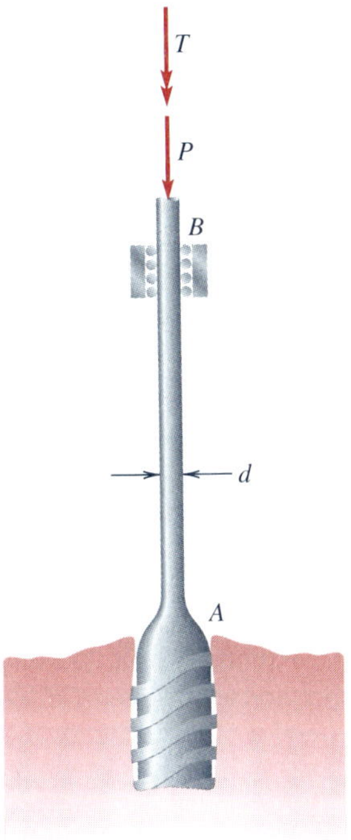

Fig. P9.3-3

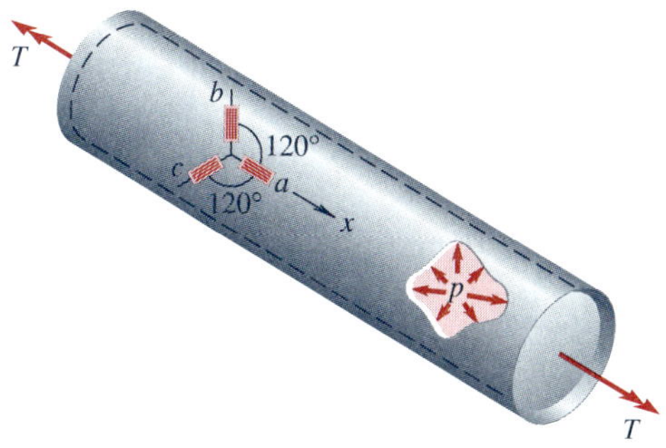

Fig. P9.3-4

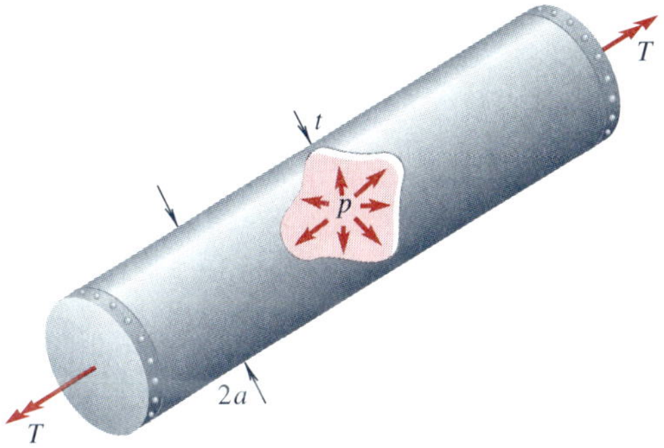

Fig. P9.3-5

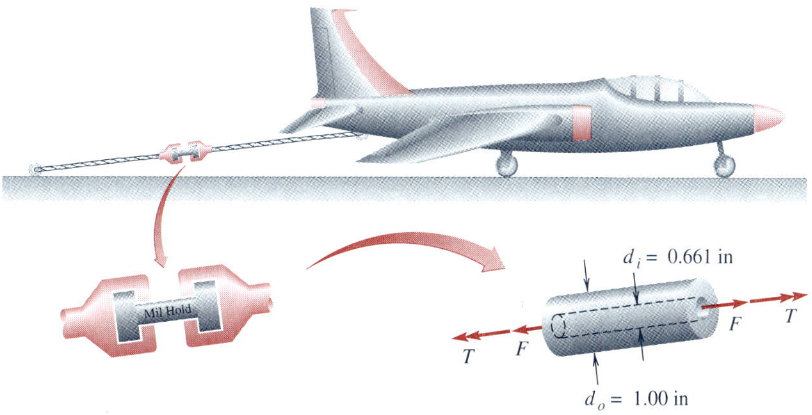

Fig. P9.3-6

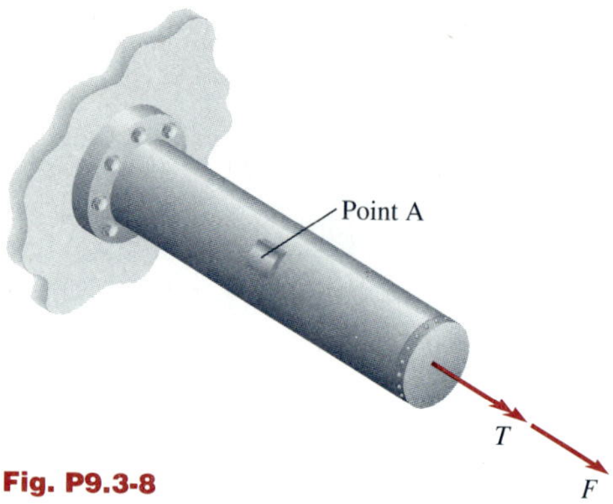

Fig. P9.3-8

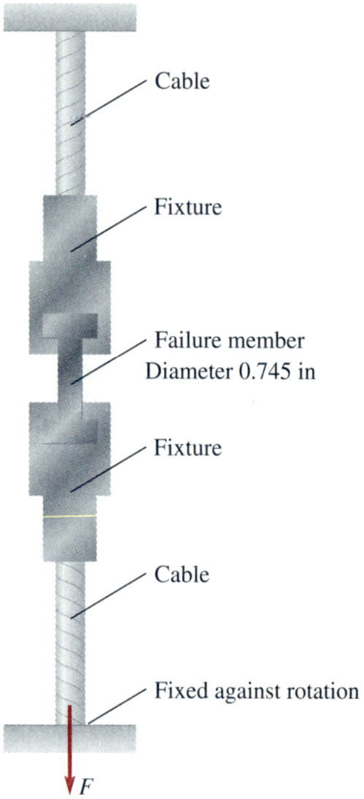

Fig. P9.3-9

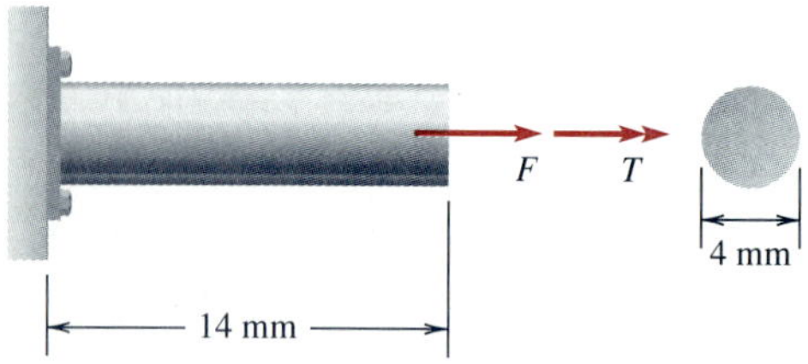

Fig. P9.3-10

9.3-7 A thin-walled closed-end pressure cylinder is under an internal pressure of $p = 500$ psi and a twisting moment $T = 40{,}000$ lb · in, as shown in Fig. P9.3-5. Find the value of the maximum shear stress in the material if $a = 4.0$ in and $t = 0.25$ in.

9.3-8 A solid shaft of circular cross section of 30-mm diameter is loaded by an axial force of $F = 150$ kN and twisted by a twisting moment $T = 265$ N · m, as shown in Fig. P9.3-8. Find the principal stresses and the maximum shear stress for a surface element at point A on the shaft.

9.3-9 A cylindrical member is designed to fail in tension. When the member is loaded through a cable arrangement as shown in Fig. P9.3-9, it is found that untwisting of the cables during loading can put a twisting moment on the member. If the axial force is $F = 20{,}000$ lb and the torque from the cables is 1200 lb · in, find the maximum shear stress in the member and compare the value to the case when no torque is acting.

9.3-10 A small circular shaft is subjected to a twisting moment T and an axial force F, as shown in Fig. P9.3-10. Find the maximum shear stress in the shaft if $F = 1240$ N and $T = 2.4$ N · m.

9.3-11 A portion of a thin-walled cylindrical shell is subjected to a twisting moment T and an axial force F, as shown in Fig. P9.3-11. The shell has a diameter of 16 in and a wall thickness of 0.10 in. Reduction of data from strain gages on the surface of the shell gives the principal stresses $\sigma_1 = 14.8$ ksi and $\sigma_2 = -6.8$ ksi and the principal directions as shown. Determine the value of T and F acting on the shell and the maximum shear stress in the wall.

9.3-12 A wood pole with a blade attached at A is used to clean material out of a horizontal shaft, as shown in Fig. P9.3-12. Each hand of the person using the pole applies a force $P = 20$ lb and a twisting moment $T = 10$ lb · ft to the pole at sections B and C, and the force and twisting moment vectors are directed along the pole in the negative x direction.

(*a*) Sketch a free-body diagram of the wood pole.

(*b*) Calculate the maximum shear stress in the pole, and show the results on an element oriented with respect to the xy axes. Neglect the weight of the pole.

9.3-13 To remove a cover at the bottom of a shaft, a wrench pipe device AB of length 2 m is to be used as shown in Fig. P9.3-13. In a particular application, an axial load $P = 130$ N and a twisting moment $T = 60$ N · m are the resultants of the loads applied by the operator at the top, directed as shown in the figure. Find the maximum shear stress in the hollow pipe AB with outside diameter 33 mm and thickness 3.4 mm. Sketch an element which is subjected to the maximum shear stress, and show its orientation with respect to the axes. Neglect the weight of the pipe.

9.4-1 Consider Example 9.3, Fig. 9.5. In the process of drilling holes, a transverse force Q arises due to misalignment of the drilling machine, as shown in Fig. P9.4-1. If the bending of the drill bit is modeled as a cantilever beam clamped

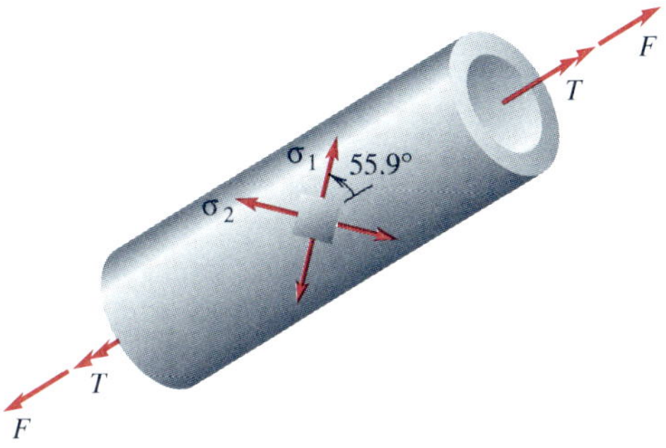

Fig. P9.3-11

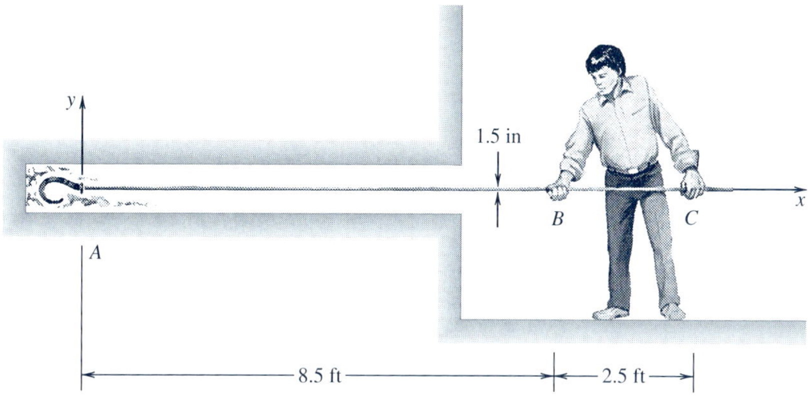

Fig. P9.3-12

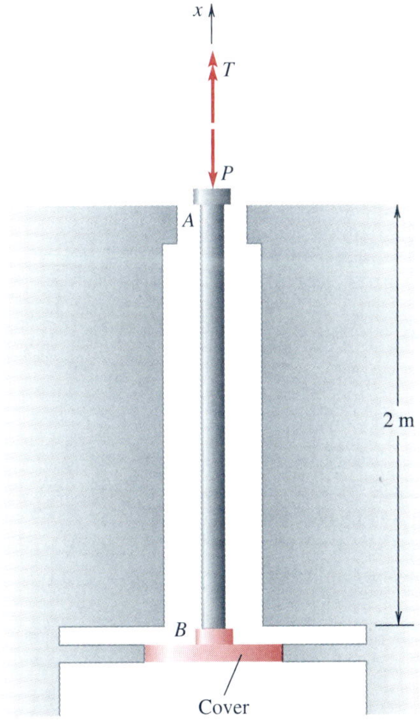

Fig. P9.3-13

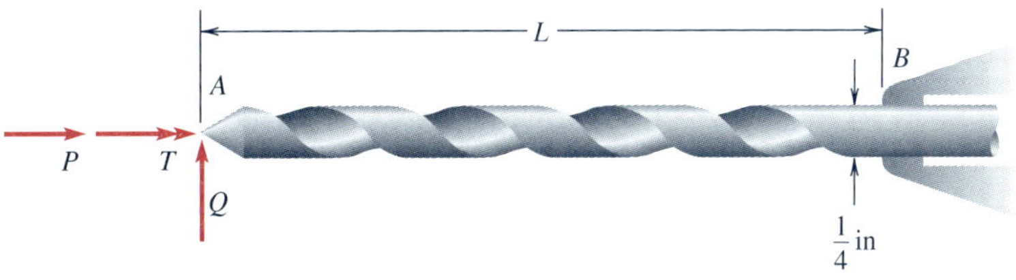

Fig. P9.4-1

at B and subjected to the transverse load Q, find the principal stresses due to the combined effects of P, Q, and T and the maximum shear stress. Take $Q = 10$ lb, $P = 100$ lb, $T = 10$ lb · in, and $L = 2$ in.

9.4-2 Consider Example 9.5, Fig. 9.8. (*a*) A 60° strain gage rosette was attached to the bracket at A on the top and oriented as shown in Fig. P9.4-2*a*. If $P = 1250$ N, what strain readings would be anticipated at A due to the load P? Take $\nu = 0.3$ and all other parameters from Example 9.5 remain the same. (*b*) Repeat (*a*) except that the rosette is located on the side of the bracket oriented as shown in Fig. P9.4-2*b*.

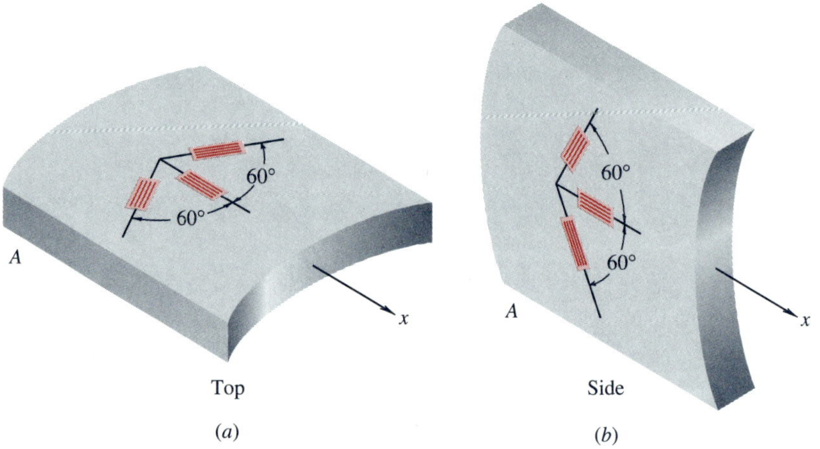

Fig. P9.4-2

9.4-3 Consider Example 9.6, Fig. 9.9. We wish to select one pipe from each of the categories of standard weight and extra-strong in App. E such that the pipe will support the sign in Fig. 9.9 with a factor of safety of 2 against yielding according to the maximum shear stress criterion and with a wind load of $P = 200$ lb. All other parameters remain the same.

9.4-4 A propped cantilever beam of solid circular section of diameter d and length L is subjected to a constant distributed load q_0 and a twisting moment T, as shown in Fig. P9.4-4. Derive an expression for the maximum shear stress in the beam due to the combined loading.

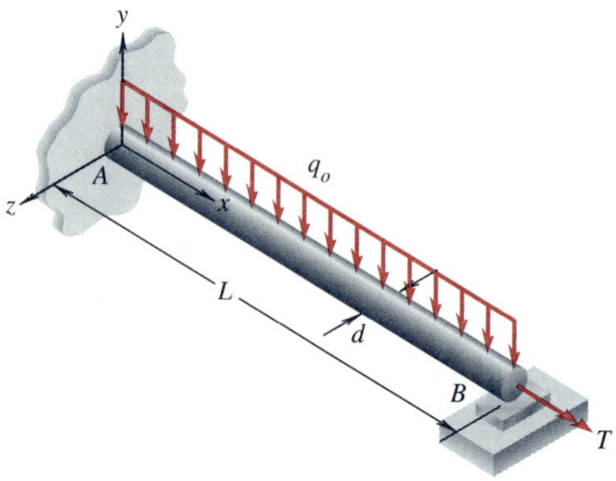

Fig. P9.4-4

9.4-5 In Prob. 9.4-4 if $q_0 = 4$ kN/m, $T = 10$ kN · m, $L = 4$ m, and the allowable value for the maximum shear stress is 100 MPa, find the minimum allowable value for the diameter d.

9.4-6 In Prob. 9.4-4 if $q_0 = 5$ kN/m, $L = 4$ m, $d = 100$ mm, and the allowable shear stress is 100 MPa, find the maximum allowable value for the twisting moment T.

9.4-7 A tubular steel pipe supports a sign, as shown in Fig. P9.4-7. If the maximum wind pressure normal to the sign is 50 lb/ft^2, find the minimum outside diameter of the pipe. The allowable shear stress in the material is 10 ksi, and $d_o = 1.12d_i$. Neglect the weight of the sign itself and the support piping.

9.4-8 A circular shaft is built into a wall and loaded with a twisting moment T and a load P, as shown in Fig. P9.4-8. Determine the maximum shear stress at point A on the shaft. Take $r = 1$ in, $L = 2.5$ ft, $P = 1000$ lb, and $T = 60{,}000$ lb · in.

9.4-9 A steel rod is loaded as shown in Fig. P9.4-9. Calculate the principal stresses, principal directions, and maximum shear stress for an element of material at point C. Neglect the weight of the rod.

9.4-10 A steel rod of circular cross section of radius r is loaded as shown in Fig. P9.4-10. Find the value of the maximum shear stress at point 1 on the section of the rod at the wall. Take $P = 350$ lb, $L = 10$ in, and $r = 1.0$ in. Neglect the weight of the rod.

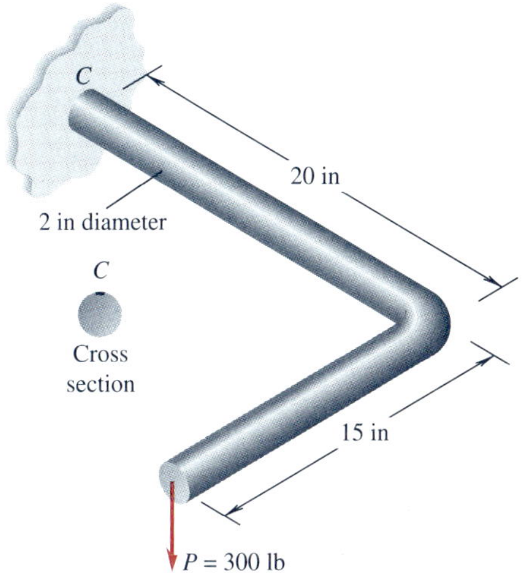

Fig. P9.4-9

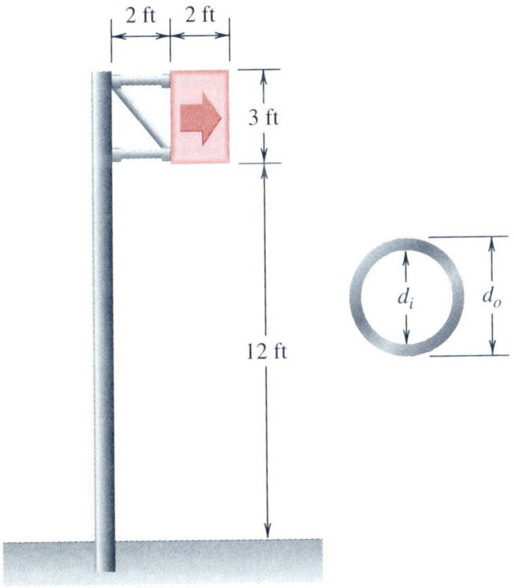

Fig. P9.4-7

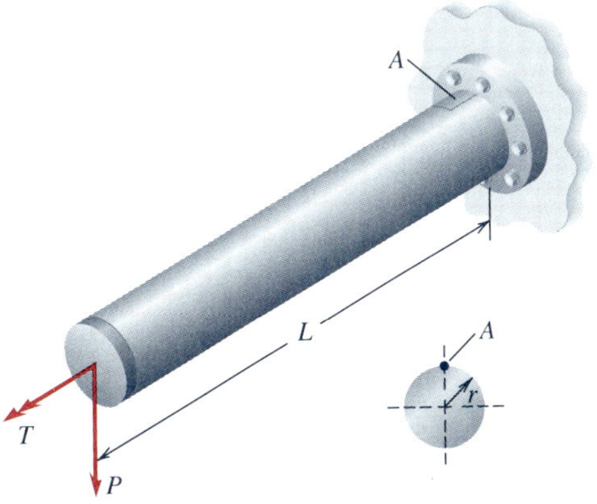

Fig. P9.4-8

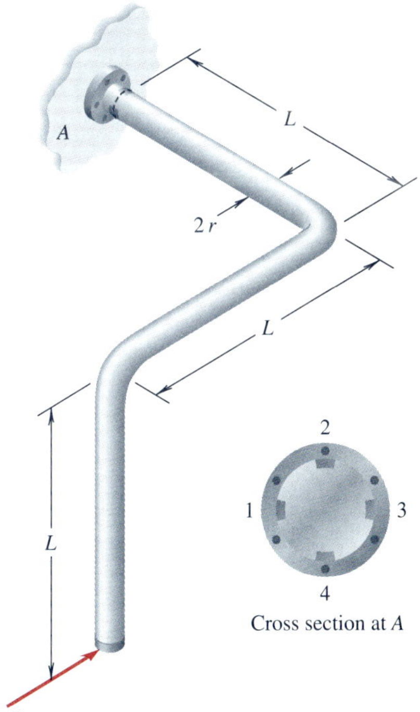

Fig. P9.4-10

9.4-11 A steel rod of circular cross section of radius r is loaded as shown in Fig. P9.4-10. Find the minimum radius of the rod in order to carry the load P = 1.2 kN if L = 0.18 m and the maximum allowable shear stress in the rod is 70 MPa. Neglect the weight of the rod.

9.4-12 A steel rod of circular cross section of radius r is loaded as shown in Fig. P9.4-10. If P = 0.6 kN, L = 0.20 m, and r = 20 mm, find the maximum shear stress at the following points on the section of the rod at the wall:
(*a*) Point 1
(*b*) Point 2
(*c*) Point 3
(*d*) Point 4
Neglect the weight of the rod.

9.4-13 A pipe of length 7.0 m is subjected to a twisting moment T and a transverse load P, as shown in Fig. P9.4-13. Find the maximum shear stress at point A if T = 1 kN · m, P – 1.1 kN, and the section modulus $I/r = 1.6 \times 10^5$ mm^3. Neglect the weight of the pipe.

9.4-14 A portion of a belt-driven shaft is shown in Fig. P9.4-14. Pulleys at sections A and B are driven by belts which exert loads as shown. The connecting shaft is supported by simple support bearings 6 in from the ends. If the shaft has a maximum allowable shear stress of 12 ksi, find the minimum allowable diameter of the shaft. Neglect the weight of all the components.

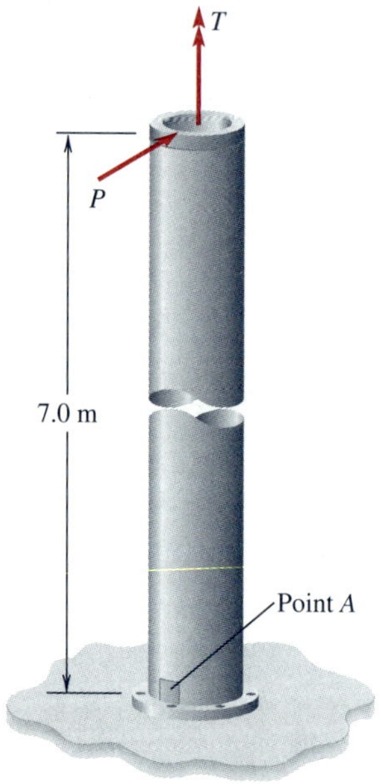

Fig. P9.4-13

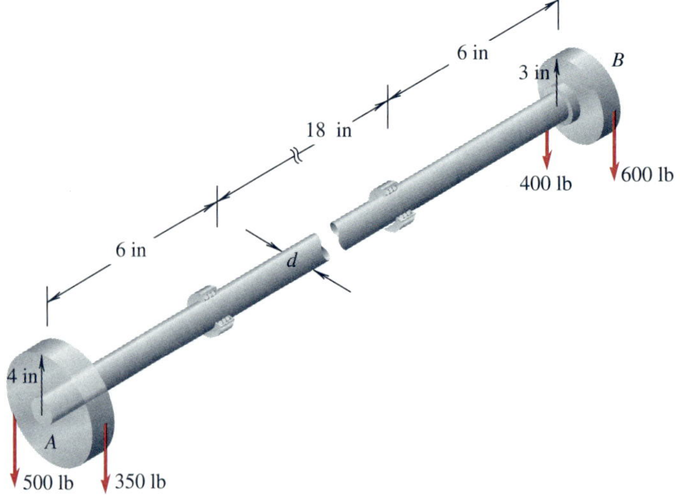

Fig. P9.4-14

9.4-15 A portion of an overhanging shaft system is shown in Fig. P9.4-15. Determine the maximum shear stress at the section 5 in from the hub as shown. Take $T = 250$ ft · lb and $F = 675$ lb.

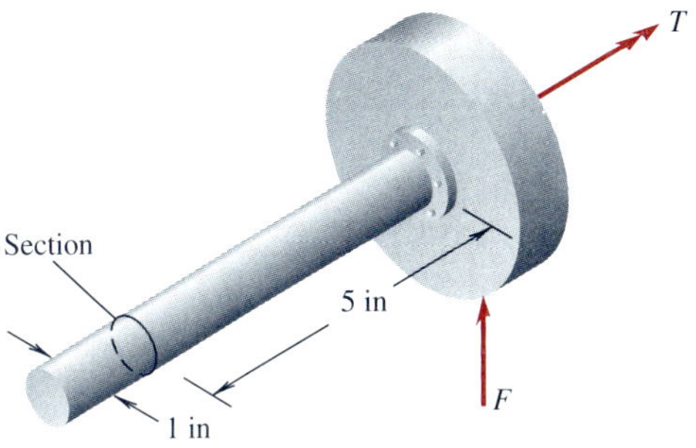

Fig. P9.4-15

9.4-16 Consider Example 9.6, Fig. 9.9. Verify that the values for the maximum shear stress τ_{max} given in Table 9.1 for the case where $P = 200$ lb are correct.

9.4-17 Consider Example 9.6, Fig. 9.9. Derive Eqs. (*n*) and (*o*) and calculate the values of τ_{max} at points 1 to 4, and thus check the values given in Table 9.1 for the case where $P = 399.5$ lb.

9.5-1 Consider Example 9.7, Fig. 9.10. Find the maximum shear stress in the pipe if $t = 12$ mm. All other parameters remain the same.

9.5-2 Consider Example 9.7, Fig. 9.10. Find the maximum span length L such that the maximum shear stress in the pipe material does not exceed 50 MPa. All other parameters remain the same.

9.5-3 An accident involving a piece of hydraulic mining equipment occurred when a large load P pressed against a pressure cylinder, as shown in Fig. P9.5-3. If the cylinder is considered as a closed thin-walled pressure vessel with $p = 1.4$ MPa and the load P is taken as acting on a simply supported beam, estimate the minimum load P to cause failure at point A if the maximum shear stress at failure in the material is estimated to be 50 MPa.

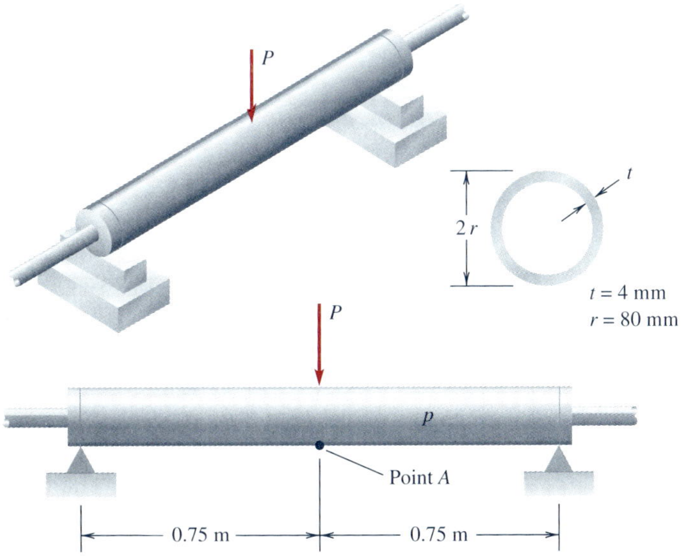

Fig. P9.5-3

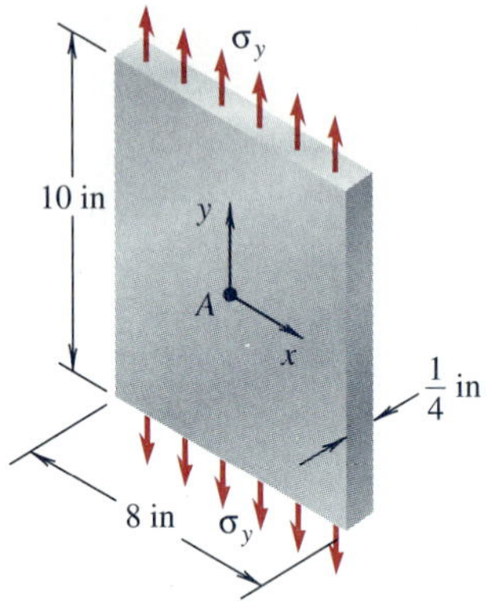

Fig. P9.5-4

9.5-4 A steel plate is initially free of stress at a temperature of 75°F. A stress σ_y is applied to the plate, and the temperature of the plate is changed. Under the new conditions of load and temperature, the measured strains at point A as shown in Fig. P9.5-4 are

$$\epsilon_x = 150 \times 10^{-6} \qquad \epsilon_y = 600 \times 10^{-6}$$

Find the stress σ_y and the new temperature of the plate. Take $E = 30{,}000$ ksi, $\nu = 0.25$, and $\alpha = 6 \times 10^{-6}/°\text{F}$.

9.5-5 As the drill pipe on a floating drilling platform leaves the bottom of the platform, it passes through a guide horn where it bends to a controlled radius of curvature of 150 ft, as shown in Fig. P9.5-5. The pipe at this point also carries an axial tensile load T proportional to the cross-sectional area and length of the pipe below it. It has been determined that in certain circumstances, the normal stress in the pipe will be excessive. Since the radius of curvature of the pipe is fixed at 150 ft, it is suggested that the diameter of the pipe be changed to reduce the stress. If the maximum normal stress in a 10-in-outside-diameter steel pipe with $\frac{1}{2}$-in wall thickness is 100 ksi, what should the diameter of the pipe be changed to in order to reduce the maximum stress to 75 ksi? Assume that the wall thickness of the pipe remains the same. Take $E = 30{,}000$ ksi.

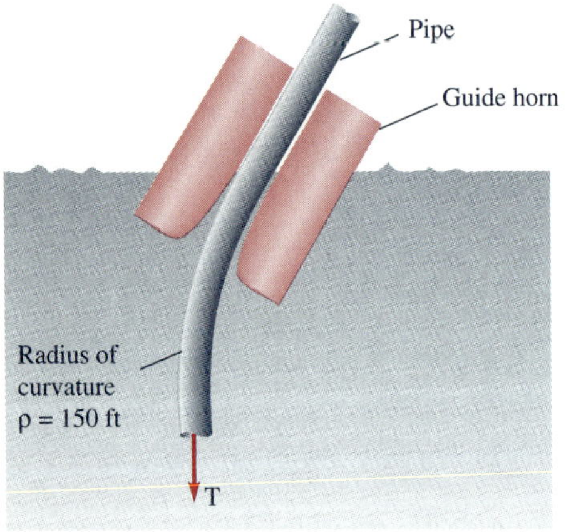

Fig. P9.5-5

9.5-6 A pressurized closed-end cylinder mounted as a cantilever beam is under the action of the transverse load F and the twisting moment T, as shown in Fig. P9.5-6. Find the values of the principal stresses and the maximum shear stress in the material at point A. Neglect the weight of the cylinder and take (*a*) $F = 1500$ lb, $T = 40{,}000$ lb · in, $p = 0$, $t = 0.25$ in, $d = 8.0$ in, and $L = 4.0$ ft and (*b*) the same values as in part (*a*) except $p = 300$ psi.

9.5-7 A steel shaft 4 in in diameter is simply supported in bearings at its ends, as shown in Fig. P9.5-7. Two pulleys, each 24 in in diameter, are keyed to the shaft, and the pulleys carry belts with the belt tensions as shown. Determine the maximum shearing stress at point A. Neglect the weight of the components.

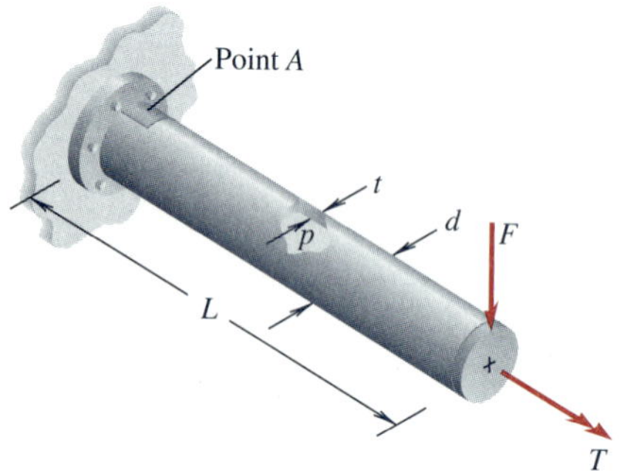

Fig. P9.5-6

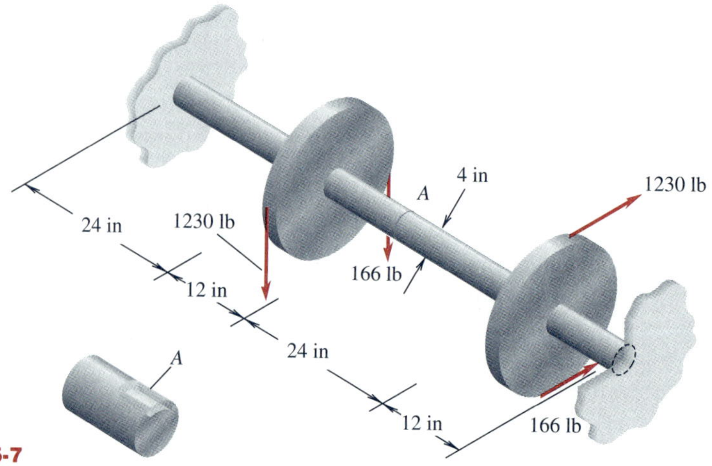

Fig. P9.5-7

9.5-8 The cross-sectional area of a square bar is reduced by one-half over a central portion, as shown in Fig. P9.5-8. Find the maximum tensile and compressive stresses in the reduced section of the bar when a load P is applied. Neglect the weight of the bar and any stress concentration effects.

9.5-9 A model for the study of a structure with a flying buttress is shown in Fig. P9.5-9. The force F due to the arch is 25,000 lb, the weight W of the tower above section AA' is 70,000 lb, the weight density of the masonry column between section AA' and section CC' is 150 lb/ft^3, and cross sections AA', BB', and CC' are 6.5 ft by 4.5 ft.

(a) Find the maximum and minimum normal stresses in section BB'.

(b) At what distance d below section AA' does a section with tensile stress on the cross section first occur?

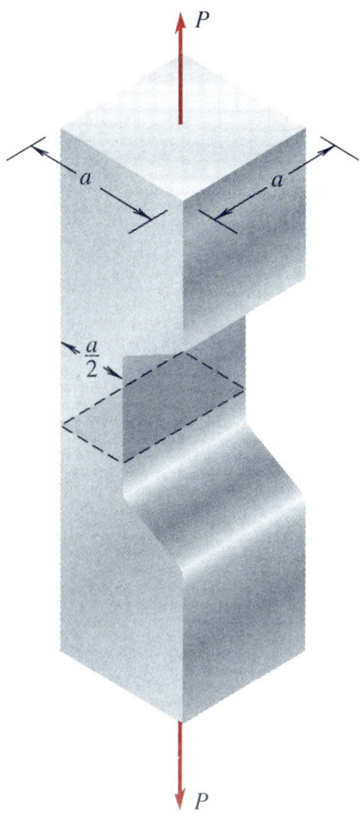

Fig. P9.5-8

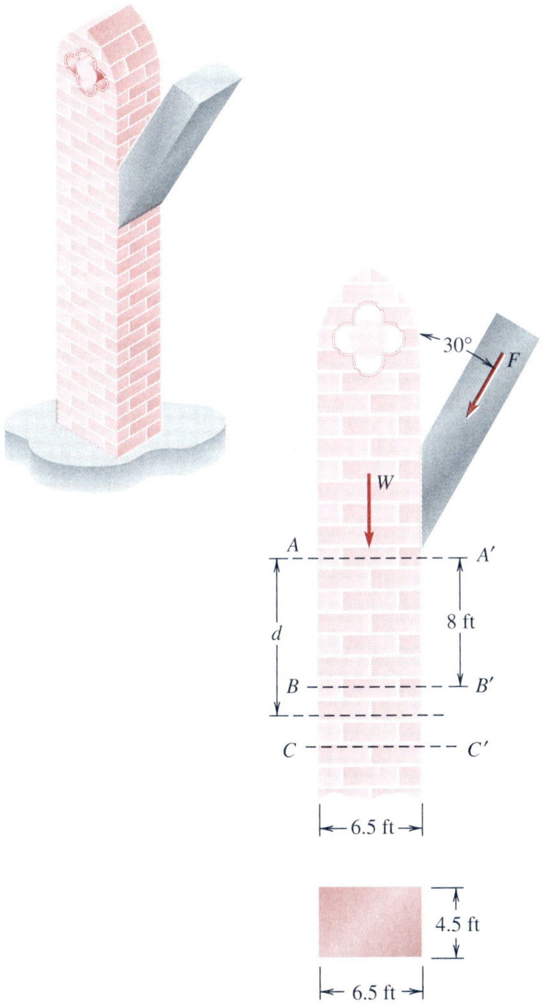

Fig. P9.5-9

9.5-10 A steel bar of diameter d is built into a wall at section A and subjected to a twisting moment T and transverse load P at section B, as shown in Fig. P9.5-10. If the maximum allowable shear stress in the bar must not exceed 40 MPa, find the minimum allowable diameter d_a. Take $E = 200$ GPa.

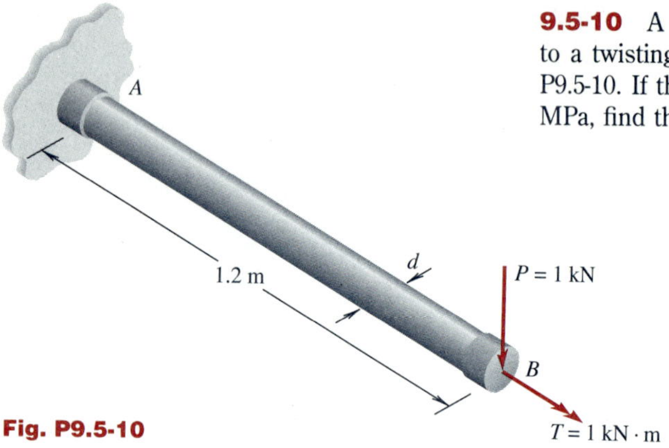

Fig. P9.5-10

9.5-11 A pressurized pipe carries a pressure p, an axial force F, and a twisting moment T, as shown in Fig. P9.5-11. If $p = 150$ psi, $F = 2400$ lb, and $T = 210$ lb · ft, find the maximum shear stress in the pipe.

9.5-12 The maximum allowable shear stress in the pipe of Prob. 9.5-11 is 8 ksi.

(*a*) Find the maximum allowable value of p if $F = 2400$ lb and $T = 210$ lb · ft.
(*b*) Find the maximum allowable value of F if $p = 150$ psi and $T = 210$ lb · ft.
(*c*) Find the maximum allowable value of T if $F = 2400$ lb and $p = 150$ psi.
(*d*) How would we construct a surface in p, F, T space giving the values that cause the maximum allowable shear stress?

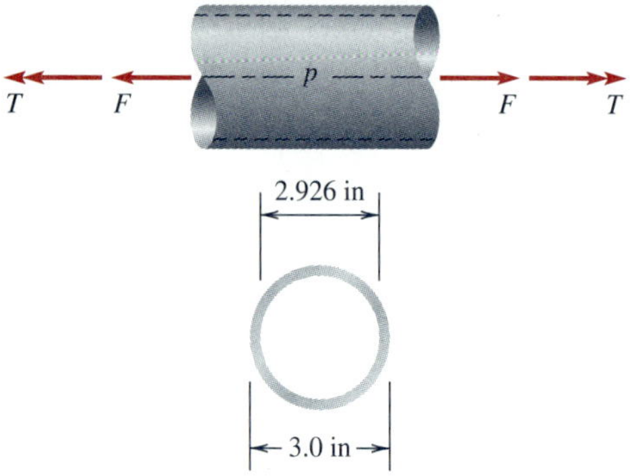

Fig. P9.5-11

9.5-13 A section of piping is to be modeled as shown in Fig. P9.5-13. Find the principal stresses, principal directions, and maximum shear stress at point A if $P = 300$ lb, $a = 10$ in, and $b = 20$ in. In addition, estimate the deflection at the section where the load is applied. Take $E = 30 \times 10^6$ psi and $G = 11.5 \times 10^6$ psi.

9.5-14 A section of a steel shaft of diameter $d = 4$ in is subjected to a twisting moment T and a bending moment M, as shown in Fig. P9.5-14. Strain gages

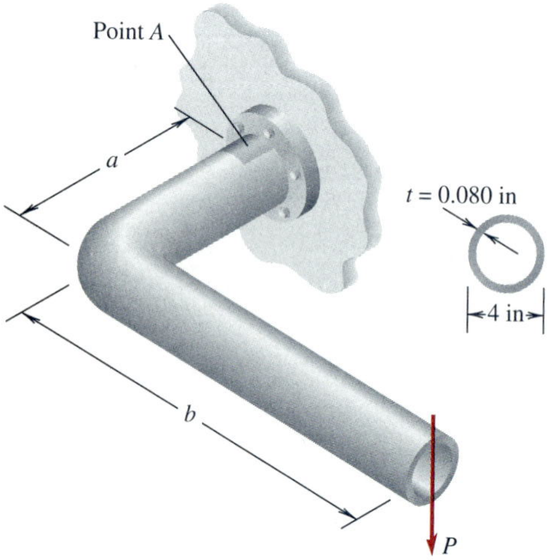

Fig. P9.5-13

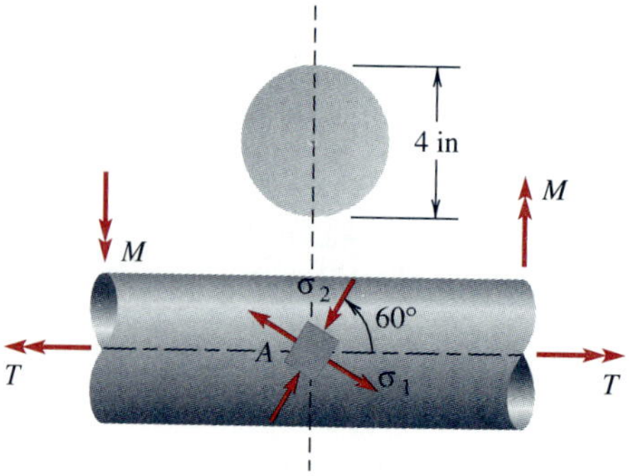

Fig. P9.5-14

placed at the top of the shaft give strain readings which, when converted to stress values, result in the principal stresses and directions as shown, where $\sigma_1 = 1800$ psi and $\sigma_2 = 600$ psi. Find the values of the twisting moment T and bending moment M acting on the shaft.

9.5-15 A closed thin-walled cylindrical pressure vessel with a diameter of 170 mm is subjected to an internal pressure of 2.0 MPa and a twisting moment of 1.2 kN · m, as shown in Fig. P9.3-5. If the maximum allowable stress for the material in tension is 140 MPa, estimate the minimum allowable wall thickness.

9.5-16 A circular rod of diameter d acts as a cantilever beam supporting an axial load of $F = 4000$ lb and a transverse load $P = 200$ lb, as shown in Fig. P9.5-16. If $d = 2$ in and the weight of the rod is neglected, determine the principal stresses at the top of the rod at the wall.

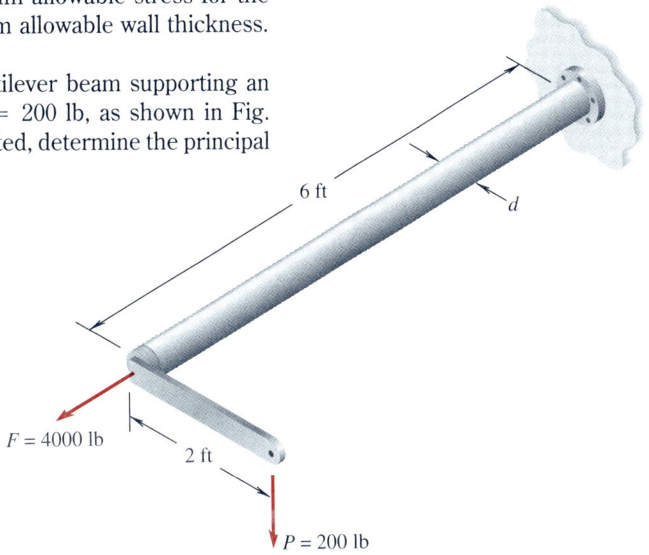

Fig. P9.5-16

9.5-17 If the rod of Prob. 9.5-16 has a maximum allowable shear stress of 15 ksi, find the minimum allowable diameter d for the rod.

9.5-18 A segment of a shaft of diameter 25 mm with a bevel gear, as shown in Fig. P9.5-18, is subjected to the three components of load shown. Find the principal stresses and maximum shear stress at locations A and B on the surface of the shaft.

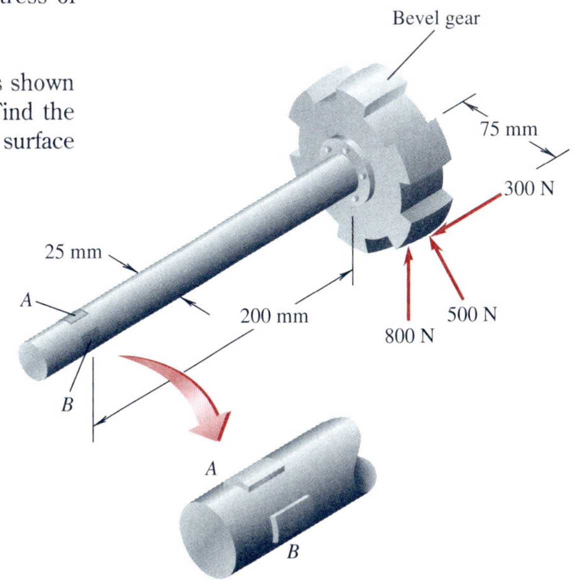

Fig. P9.5-18

9.5-19 A steel shaft 6 ft long is supported at its ends by two bearings, as shown in Fig. P9.5-19, and subjected to a load of 750 lb at the midpoint. If the diameter of the shaft is $d = 3$ in, the angular velocity is $\omega = 150$ rpm, and the maximum allowable shear stress in the shaft is 6 ksi, find the maximum allowable horsepower that the shaft can transmit.

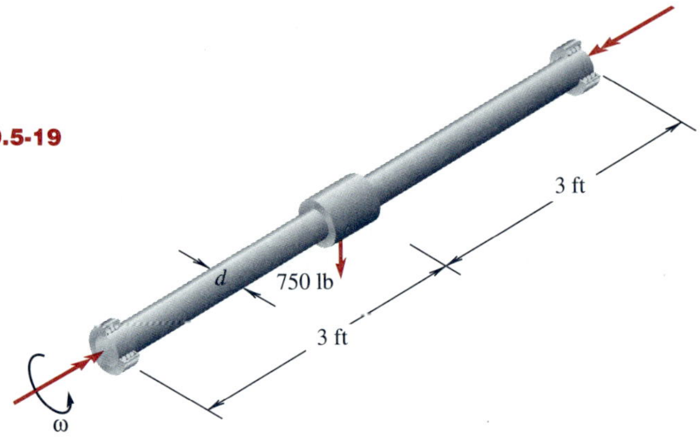

Fig. P9.5-19

9.6-1 A model of a circular steel shaft of diameter d is shown in Fig. P9.6-1. The shaft is supported in bearings at A and D which give rise to force reactions in the y and z directions. Loads normal to the shaft are applied at sections B and C as shown. In addition, a twisting moment T is applied at sections B and C as shown. Find the maximum shear stress in the shaft if $d = 40$ mm, $T = 150\ \text{N} \cdot \text{m}$, and the weight of the shaft is neglected.

(*a*) Show that the reactions at A and D are $Y_A = 0.778$ kN, $Z_A = 0.111$ kN, $Y_D = 0.222$ kN, and $Z_D = 0.389$ kN.

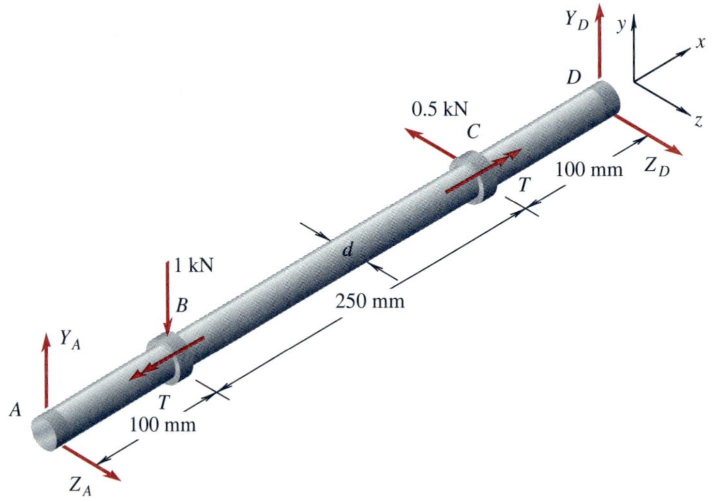

Fig. P9.6-1

(*b*) Sketch the bending moment diagrams in the *xy* plane and the *xz* plane, and determine the value of the *resultant* bending moments at sections *B* and *C*. Thus the maximum normal bending stresses at *B* and *C* can be found.

(*c*) Determine the shear stresses due to the twisting moments acting at sections *B* and *C*.

(*d*) Combine the stresses to obtain the maximum shear stress at sections *B* and *C*.

9.6-2 A circular shaft of diameter 1.75 in supports two pulleys at sections *A* and *C*, as shown in Fig. P9.6-2. The bearings at sections *B* and *D* give rise to concentrated reactive forces only. Determine the maximum shear stress and the principal stresses in the shaft; see Prob. 9.6-1. Neglect the weight of the shaft and the pulleys.

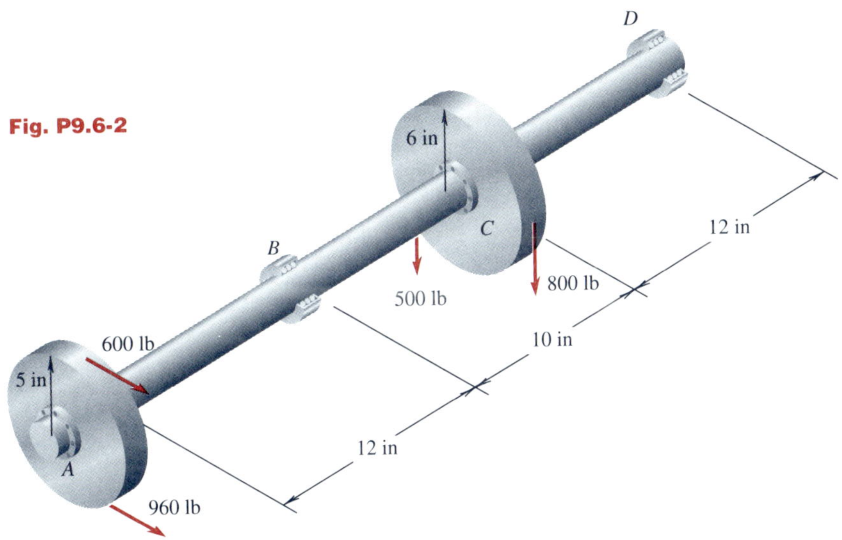

Fig. P9.6-2

9.6-3 A circular rod of diameter *d* is fabricated into the shape of a U to form the structure shown in Fig. P9.6-3. If the value of load *Q* equals zero, find an expression for the value of the vertical load *P* which corresponds to the initiation of yielding at some point in the structure. Use the Mises and maximum-shear-stress criteria to estimate the load *P*. Neglect the weight of the rod and take the yield stress equal to *Y*.

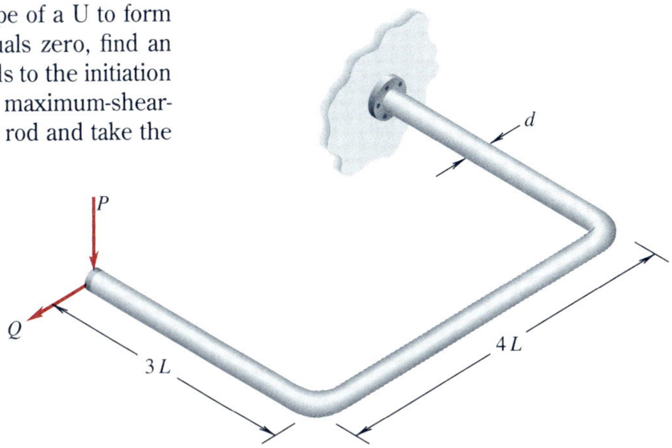

Fig. P9.6-3

9.6-4 Solve Prob. 9.6-3 for the value of load Q to cause initial yielding if the value of load P equals zero.

9.6-5 A 3-in-diameter steel rod of length 240 in is fabricated into the configuration shown in Fig. P9.6-5 and is built into a wall at one end. All angles of the bent rod are right angles. Find the value of the maximum force P that can be applied to the free end in the direction shown without causing initial yielding of the rod. Neglect the weight of the rod. Take $Y = 190$ ksi.

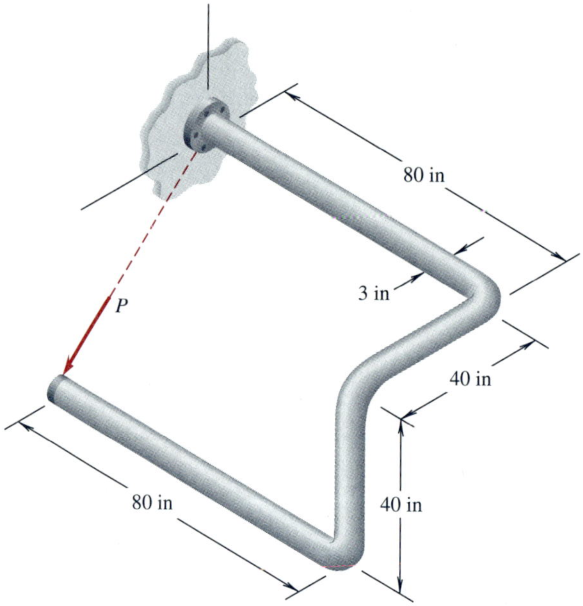

Fig. P9.6-5

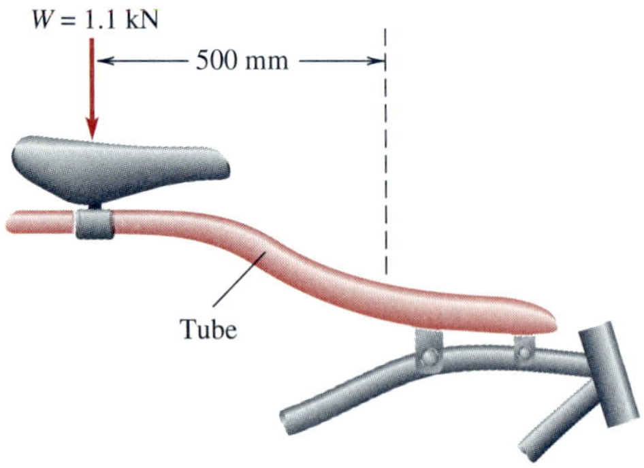

Fig. P9.6-6

9.6-6 A new design of a mountain bike is shown in Fig. P9.6-6. The seat is to be supported by a tube as shown. Select from the pipes listed in App. E the appropriate pipe dimensions that could be used as the tube. Take $E = 240$ GPa and the maximum allowable stress in tension as $\sigma = 300$ MPa.

9.6-7 Part of a suspension system with a torsion bar for the front end of an automobile is shown in Fig. P9.6-7. Failure of the torsion bar occurred in tension across a 45° plane, as shown, and corrosion of the bar lowered the value of the failure stress in tension to about 800 MPa. Estimate the twisting moment on the bar and the angle of twist at the instant of failure. Take $G = 70$ GPa.

9.6-8 Shaft AD, shown in Fig. P9.6-8, has gears attached at sections B and C which carry the tooth forces F_1 and F_2, and the shaft is supported in bearings at A and D, as shown. Find the smallest allowable diameter of the shaft, using the maximum-shear-stress criterion with the allowable stress $\tau_a = 80$ MPa. Take $L_1 = 0.5$ m, $L_2 = 0.4$ m, $L_3 = 0.3$ m, $F_1 = 720$ N, $F_2 = 600$ N, $d_B = 100$ mm, and $d_C = 120$ mm.

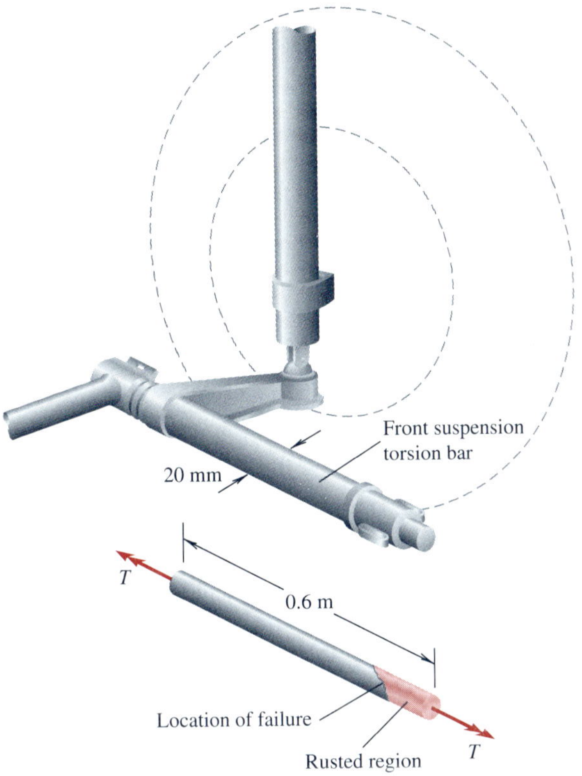

Fig. P9.6-7

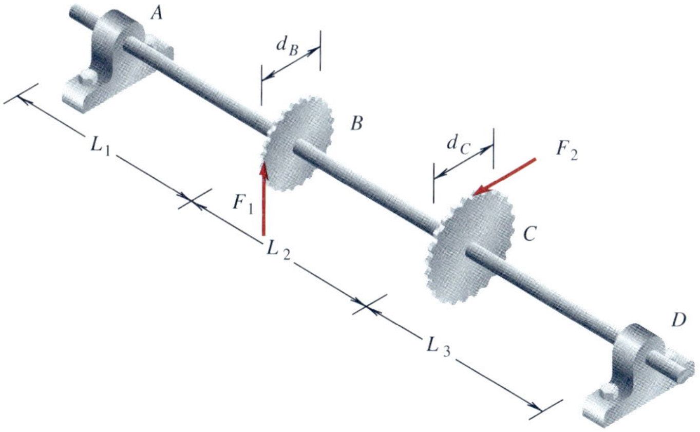

Fig. P9.6-8

9.6-9 Consider Prob. 9.6-8. Solve for smallest allowable diameter of the shaft when $L_1 = 24$ in, $L_2 = 18$ in, $L_3 = 15$ in, $F_1 = 180$ lb, $F_2 = 150$ lb, $\tau_a = 12$ ksi, $d_B = 4$ in, and $d_C = 4.8$ in.

Fig. P9.6-10

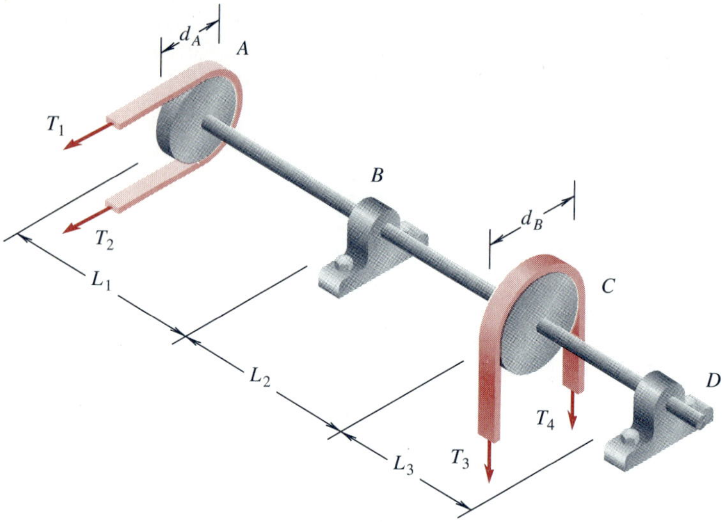

9.6-10 Shaft *AD* has pulleys attached at *A* and *C* and is supported in bearings at *B* and *D*, as shown in Fig. P9.6-10. The pulley tensions are $T_1 = 150$ lb, $T_2 = 100$ lb, $T_3 = 460$ lb, and $T_4 = 500$ lb; the lengths are $L_1 = 20$ in, $L_2 = 40$ in, $L_3 = 40$ in, and the diameters are $d_A = 4$ in and $d_C = 5$ in. Use the maximum-shear-stress criterion to find the minimum allowable diameter of the shaft if the allowable shear stress is $\tau_a = 15$ ksi. (See Prob. 9.6-1.)

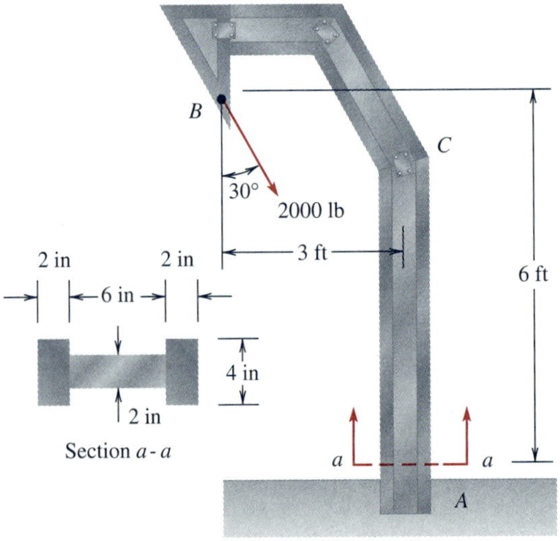

Fig. P9.6-12

9.6-11 Solve for the minimum allowable diameter of the shaft in Prob. 9.6-10, except take the pulley forces as $T_1 = 600$ N, $T_2 = 400$ N, $T_3 = 1840$ N, and $T_4 = 2000$ N; the lengths as $L_1 = 0.5$ m, $L_2 = 1$ m, and $L_3 = 1$ m; and the diameters as $d_A = 100$ mm and $d_C = 125$ mm. Use $\tau_a = 100$ MPa.

9.6-12 A planar stanchion *ACB* is built in at the lower end *A* and carries a 2000-lb load, as shown in Fig. P9.6-12. Find the maximum compressive normal stress and maximum shear stress due to bending acting on a horizontal section *aa* of the vertical segment *A-C* of the stanchion near the end *A*, as shown.

CHAPTER 10

Buckling and Stability

10.1 Introduction

In the examples considered in previous chapters, we have calculated the stresses and displacements of components of structures and machines based upon the use of the three steps: equilibrium of forces, compatibility of displacements, and force-deformation relations. Components were designed and selected based upon strength and rigidity. Common experience indicates that it is necessary to broaden our approach to the analysis of engineering problems by adding stability considerations.

For example, consider a long, slender member, e.g., a wood or metal yardstick, as shown in Fig. 10.1*a*. If an axial compressive load P is applied to a reasonably straight yardstick, we find, for very small values of P, that the yardstick remains straight and vertical in an equilibrium state and in a state of uniform axial compression. However, as P increases, a critical value of P is reached for which the yardstick moves over into a bent configuration or a so-called buckled state (Fig. 10.1*a*). It is important to emphasize that the buckling action can take place entirely within the elastic range of the axial stress state. If the load is removed, the yardstick becomes straight again. For a wood or steel yardstick, the axial stress state when buckling occurs is about 1 percent of the limit of elastic behavior. Therefore, while the stress level is well below the elastic limit, the yardstick loses its load-carrying capacity by buckling or what we call *structural instability.*

To examine this new type of failure, we will ask the following question about a given equilibrium state of a structural system: When the system is slightly disturbed from an equilibrium state, does it tend to return to its original equilibrium state or to move even farther away? This question can be simply illustrated by three equilibrium states involving a small block of weight W resting on frictionless surfaces, as shown in Fig. 10.1*b* to *d*. In all three cases when the block is located at point O, where the

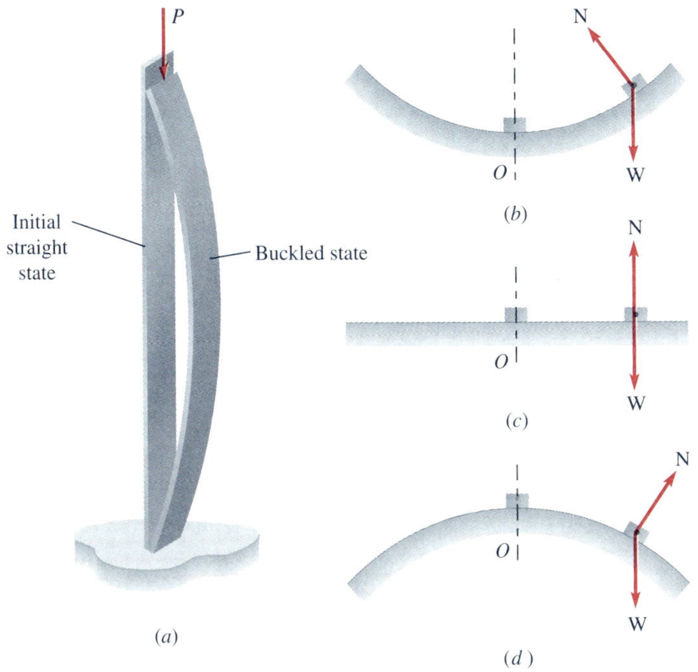

Figure 10.1 (*a*) Instability or buckling of a meter stick and examples of (*b*) stable, (*c*) neutral, and (*d*) unstable equilibrium states.

tangent plane to the surface at O is horizontal, the normal reactive force N and the weight force W are in balance and the block is in force equilibrium.

Suppose the block at position O in Fig. 10.1*b* is given a small disturbance to the right, as shown. Then, because the resultant of the N and W forces acting on the block in the new position would accelerate the block back toward O, the resultant force is a *restoring force.* We refer to this state as a ***stable equilibrium*** state. When the block is disturbed from its equilibrium position at O, it tends to return to its equilibrium position. On the other hand, the block at point O in Fig. 10.1*d* is in an ***unstable equilibrium*** state since the resultant force in the disturbed position is acting so as to accelerate the block away from O; i.e., the resultant force is an *upsetting force.* The case shown in Fig. 10.1*c* is intermediate between the two cases in Fig. 10.1*b* and *d*. Neither a restoring force nor an upsetting force acts on the block in the disturbed position. The block remains in equilibrium in the disturbed position. The block at point O in this case is said to be in a state of ***neutral equilibrium.***

The above discussion suggests a definition for the stability of equilibrium. For a system in equilibrium, if for all possible geometrically admissible small displacements from the equilibrium configuration, restoring forces arise which tend to accelerate the system back toward the original equilibrium configuration, then the system is said to be in a state of stable equilibrium.

Clearly structural components that have been designed to meet criteria for strength but have not been designed to remain stable when exposed to external disturbances may undergo large displacements that can result in sudden and drastic changes in geometry. These changes in geometry can cause catastrophic failure of the part or the structure. Structural failures in practice due to instability tend to be dramatic.

10.2 Examples of Instability

The procedure required for the stability or buckling analysis of equilibrium states often leads to difficult mathematical problems. Therefore, to broaden our understanding of the phenomenon of buckling, we give a qualitative discussion of a few examples of practical interest; the detailed mathematical analysis of these cases is beyond the scope of this text and can be found in advanced books on elastic stability.

Vertical slender members loaded in compression are usually called *columns.* The buckling analysis given in detail in Secs. 10.4 and 10.5 involves single columns supported and loaded at their ends. In many applications, however, columns may be continuous members supported at several points along their length, as shown in Fig. 10.2*a*. The procedure for finding the critical load in a column with intermediate supports not

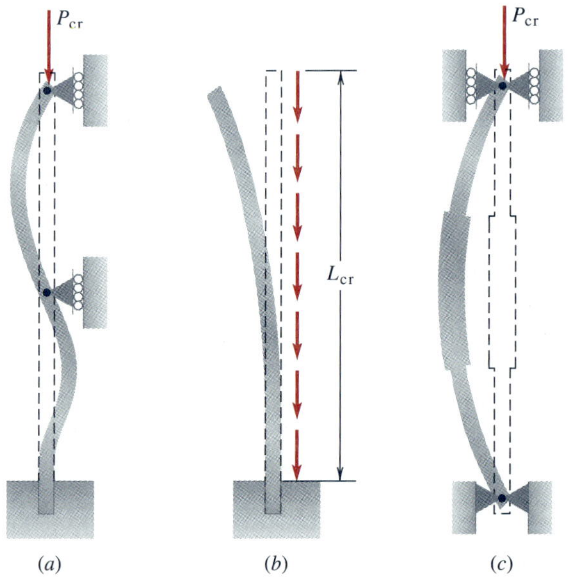

Figure 10.2 Examples of column buckling.

surprisingly leads to more complicated equations to be solved than is the case for the column without the intermediate supports.

In many cases, columns are subjected to distributed loads along their length, and these loads can cause buckling failure. For example, a tall pole under self-weight, for a given cross-sectional area and pole material (Fig. 10.2*b*), can become unstable at a critical value of the pole height L_{cr}.

For some applications a column may have a variable cross section, as shown in Fig. 10.2*c*. The buckling analysis for this case is more complicated than that for the case of constant cross section which we carry out in Sec. 10.4.

The analysis of plane trusses was studied in a course on statics, and values for bar forces in the truss were calculated for given conditions of load and support. For truss members carrying compressive forces, there is the possibility of buckling. For example, the truss shown in Fig. 10.3*a* for certain combinations of loads develops compressive forces in some members, and therefore buckling in those members may occur, as suggested by Fig. 10.3*a*.

A plane frame which consists of two columns built in at their bases and connected at their top ends to a beam is shown in Fig. 10.3*b*. The vertical columns in this case are in compression, and the buckling analysis is complicated by the interaction between the columns and the cross beam. The stability of frame structures is studied in advanced courses on structural stability.

An example of a twist-bend type of buckling for beams is shown in Fig. 10.4*a*. Although the cantilever I beam is loaded in the *y* direction and its primary bending response before buckling is in the *yz* plane, there is a mode of buckling which involves both lateral bending and twisting, as shown in Fig. 10.4*a*. The calculation of the critical buckling load for this case is again complicated.

Thin plates will buckle under end loads, as shown in Fig. 10.4*b*. A plate can be thought of as a set of beamlike strips placed side by side and bonded together. Before we can calculate the buckling load in this case, we need to be familiar with the bending theory of elastic plates.

Finally, a thin-walled cylindrical vessel (e.g., an empty soft drink container) can carry a light axial load by developing uniform axial compressive stresses in the walls (Fig. 10.5*a*). If the load is increased, a buckling load can be reached for which the walls will buckle into a pattern of dimples and finally crumple and lose stiffness (Fig. 10.5*b*).

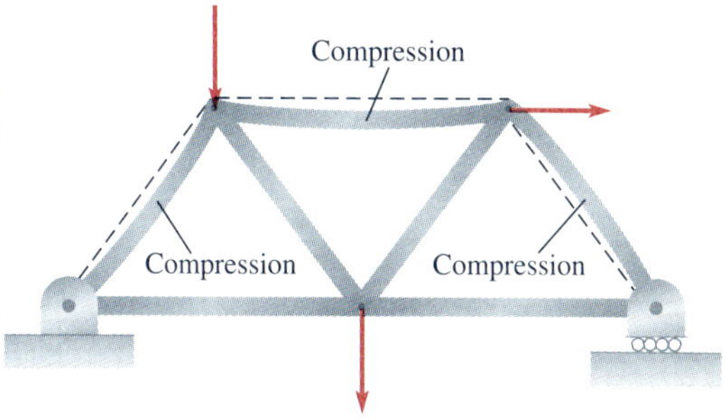

(*a*)

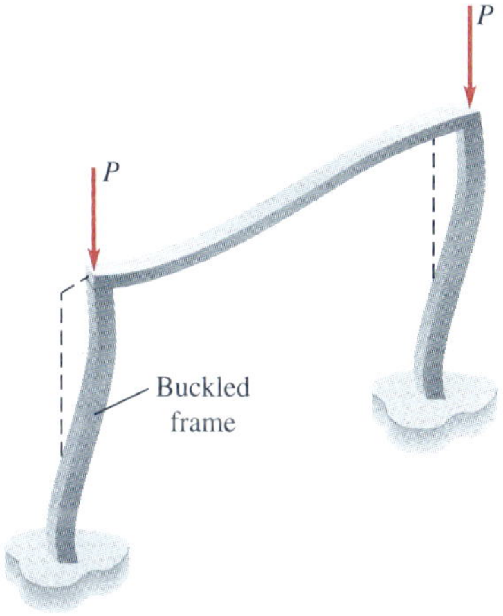

(*b*)

Figure 10.3 Examples of buckling in (*a*) a truss and (*b*) a frame.

10.3 Bar-Spring Models for Stability Analysis

In trying to carry out the stability analysis to predict the loads that might lead to loss of stability of simple structural systems, we will encounter,

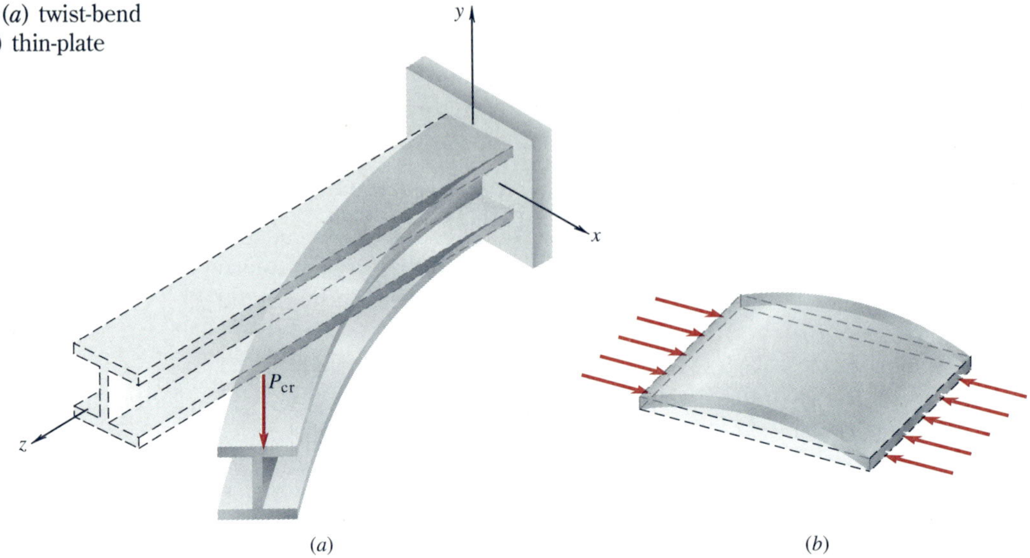

Figure 10.4 Examples of (*a*) twist-bend buckling of a beam and (*b*) thin-plate buckling.

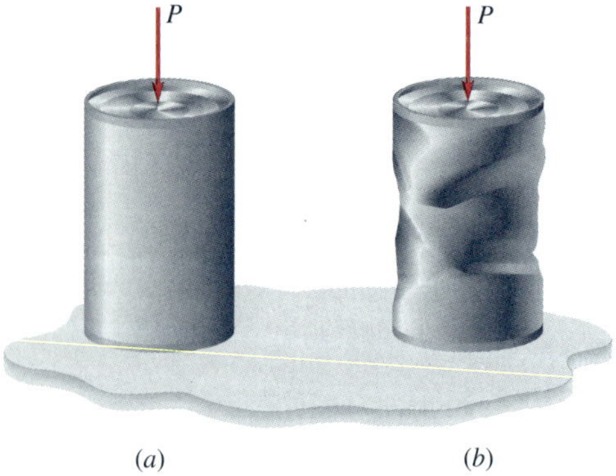

Figure 10.5 Buckling and crumpling of the cylindrical walls of a can subjected to a compressive force *P*.

in general, different mathematical approaches from those used in earlier chapters. For example, the bending deflections associated with the buckled shapes for columns are continuous functions of position along the column, and the equilibrium requirements lead to differential equations which must be solved to find the buckled shapes. This kind of analysis will be taken up in a later section.

In this section we replace some common structural and machine components by rigid-bar-spring models so that the basic concepts of stability analysis can be presented without the mathematical complexities needed if the analysis were done on flexible components. These models also provide insight into the phenomenon of buckling and can be used to provide approximate estimates of buckling loads.

EXAMPLE 10.1

Consider an elastic column that is clamped at the base *B* and carries a vertical load *P* at point *A*, as shown in Fig. 10.6*a*. We wish to estimate the critical value of *P* for which the column ceases to be stable in the unbent shape by making use of a rigid-bar-spring model as a replacement of the original flexible column. We neglect the weight of the column compared to the applied load *P*.

Clearly, for the column of Fig. 10.6*a*, any disturbance which causes a transverse deflection of the load application point *A* will give rise to a restoring moment at the base. A simplified model of the original column that captures the essential mechanical features of the loaded column is shown in Fig. 10.6*b*. The flexible column *AB* is replaced by a rigid bar *AB*. The clamped condition at the base is replaced by a linear "spiral" spring that has the property that it exerts a moment on the bar that opposes rotation of the bar with a magnitude $k\theta$, where θ is the angle of rotation in radians and *k* is the spring constant with the units of pound-inches per radian (newton-meters per radian).

The deflection shapes of the original column of Fig. 10.6*a* are described by continuous curves. The model column of Fig.

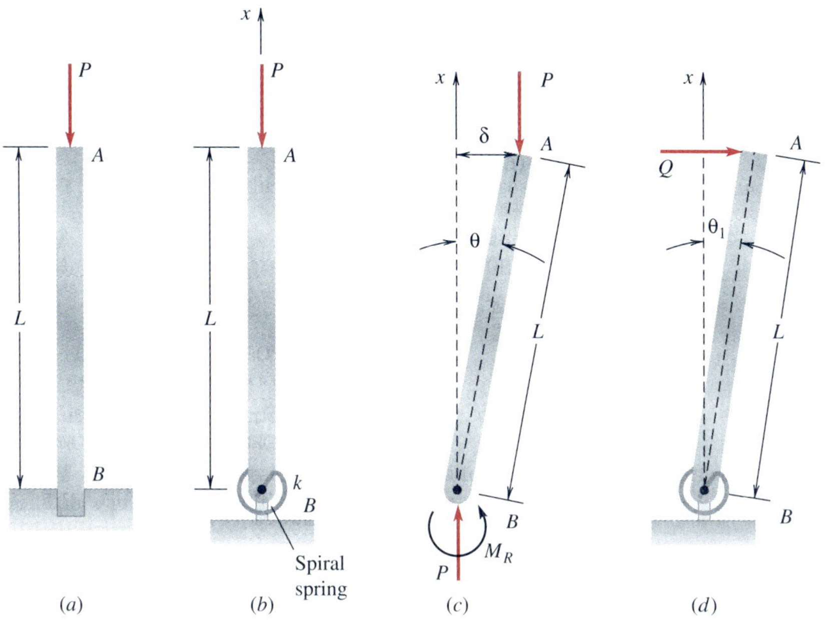

Figure 10.6 Example 10.1

10.6*b* is pinned to the support and attached to the spring at B so that the rigid bar AB can rotate only in the xy plane and the displaced bar only requires a value of θ to completely specify its position.

To carry out a stability analysis of the system from its vertical undeflected position, we give the system a small arbitrary disturbance corresponding to a rotation θ, as shown in Fig. 10.6*c*. We then investigate equilibrium of the system in the disturbed state. The equation for moment equilibrium in the disturbed state about point B is

$$M_R - M_U = 0 \qquad (a)$$

where

$$M_R = k\theta$$
$$M_U = P\delta = PL\sin\theta \approx PL\theta \qquad (b)$$

are the restoring and upsetting moments.

The moment M_R represents the moment arising in the system which attempts to return the disturbed state to the original undisturbed state of $\theta = 0$. The moment M_U is the moment arising from the external load P acting on the disturbed system. In Eqs. (*b*) we will assume that angle θ is small so that $\sin\theta \approx \theta$. Therefore, $M_U = PL\theta$, as shown in Eqs. (*b*). Substituting the relations given by Eqs. (*b*) into Eq. (*a*) with the assumption of small angle θ, we obtain the following expression as the moment equilibrium equation:

$$(k - PL)\theta = 0 \qquad (c)$$

Solutions to Eq. (*c*) for angle θ provide the deflected shape of bar AB (for small angle θ) under the prescribed load P. One solution of Eq. (*c*) is clearly $\theta = 0$, which indicates something that we already know, namely, that the bar will remain in the vertical configuration. As the load P increases from the value 0 and

$$P < \frac{k}{L} \qquad (d)$$

so that $M_U < M_R$, the system is stable because the restoring moment is greater than the upsetting moment, and the system moves back toward the equilibrium state $\theta = 0$.

However, when the value of the load P reaches a value

$$P = \frac{k}{L} \qquad (e)$$

so that

$$M_U = M_R$$

the system can be in equilibrium with $\theta \neq 0$. That is, a solution of Eq. (*c*) is θ arbitrary with the bar in a nonvertical configuration. This is the so-called neutral equilibrium state (Fig. 10.1*c*), and the bar will remain in the disturbed position.

If we continue to increase the load so that

$$P > \frac{k}{L} \qquad (f)$$

then the restoring moment is inadequate to return the bar to its initial state, i.e., the upsetting moment is greater than the restoring moment

$$M_U > M_R$$

and the system is accelerated away from the original state; thus we have an unstable situation. We say that the bar has buckled.

The value of the load P which represents the transition between the stable and unstable states is called the *critical load* or the ***buckling load.*** In the present case, the buckling load is

$$P_{cr} = \frac{k}{L} \qquad (g)$$

Returning to the original elastic column in Fig. 10.6*a*, we realize that the buckling load P_{cr} given by Eq. (g) is of little direct value unless we can relate the spring parameter k for the model to the parameters for the flexible column, namely, E, I, and L. In other words, we need to select the parameters of the model in such a way as to provide a reasonable description of the deflection behavior of the elastic column. One way of doing this is to choose k for the spring constant in the model so that the transverse deflection at the top of the bar due to a transverse load Q at the top of the bar matches the deflection under the same load for the column. From Fig. 10.6*d* we find that moment equilibrium gives

$$k\theta_1 = QL \qquad (h)$$

where Q is the transverse load applied to the top of the bar. The deflection at the top of the bar is given by, for small angles,

$$\delta_1 = L\theta_1 \qquad (i)$$

Therefore we can eliminate θ_1 to obtain

$$\delta_1 = \frac{QL^2}{k} \qquad (j)$$

For a flexible cantilever beam subjected to a transverse load Q at the top, the tip deflection δ_2 is given in App. G in Table G.1, Fig. G.1-2, as

$$\delta_2 = \frac{QL^3}{3EI} \qquad (k)$$

If we set $\delta_1 = \delta_2$ and solve for k, it follows that

$$\frac{QL^2}{k} = \frac{QL^3}{3EI}$$

or

$$k = \frac{3EI}{L} \qquad (l)$$

We take this value of k as the spring constant for the model system. In this way we can estimate the critical or buckling load from Eq. (g) as

$$P_{cr} = \frac{k}{L} = \frac{3EI}{L^2} \qquad (m)$$

It turns out, as we will see in Sec. 10.4, that the exact value of the critical load for the column in Fig. 10.6*a* is

$$P_{cr} = 2.47\frac{EI}{L^2} \qquad (n)$$

We see that the calculations using the model system provide a reasonable estimate of the critical load. Of course, the dependence on EI and L is consistent with what we expect physically and on the basis of dimensional arguments. Also the smallest load to cause buckling of a column will depend on the smallest value of I of the cross section. Finally if, instead of matching the deflections at the top of the bar, we were to match the slopes at the top of the bar, the numerical factor of 3 in Eq. (m) would be 2 (Prob. 10.3-15).

Next we will show how the use of a rigid-bar-spring model can be used to study a different combination of support conditions. In the next example, we model a column with both ends pinned.

EXAMPLE 10.2

An elastic column is pinned at both ends and is subjected to a vertical force P, as shown in Fig. 10.7*a*. We assume that point B is fixed while point A is able to move downward under increasing values of load P. We neglect the weight of the column, assume that no friction is present, and assume that the axis of the smallest value for the moment of inertia I of the cross section

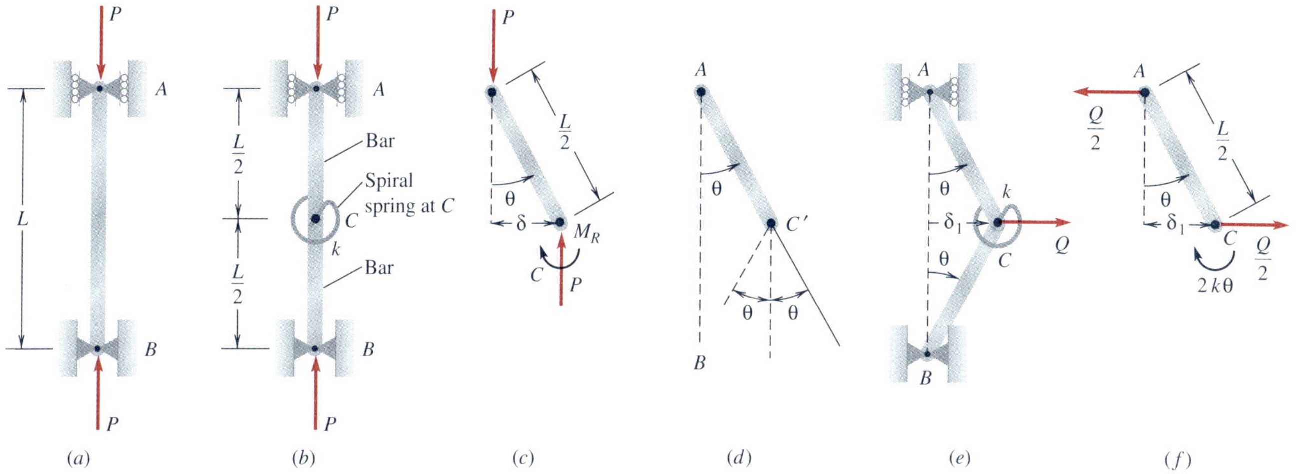

Figure 10.7 Example 10.2

of the column is perpendicular to the plane of the paper. As the load P is increased, the column will remain straight until it reaches a critical or buckling load, at which load transverse deflections will occur in the plane of the paper. We wish to make use of a rigid-bar-spring model of the column to estimate this buckling load.

To allow for the transverse deflection, we introduce a two-bar model, shown in Fig. 10.7*b*, where the two rigid bars are pinned together at point C where a spiral spring with spring constant k resists the *relative* angle change of the attached bars. We assume that all angles are small in the analysis, but we exaggerate all angles in the drawings in the interest of clarity. The bars are pinned at A and B and are free to rotate without friction.

To investigate the stability of the model system, we consider equilibrium of the upper bar after giving point C a small deflection δ, as shown in Fig. 10.7*c*. This deflection gives rise to an upsetting moment caused by load P. The upsetting moment is given by

$$M_U = P\delta = \frac{PL}{2}\sin\theta \approx \frac{PL\theta}{2} \qquad (a)$$

The total relative angle in the spring action at C is 2θ (Fig. 10.7*d*). The restoring moment due to the spring is

$$M_R = k(2\theta) \qquad (b)$$

and moment equilibrium of the upper bar in Fig. 10.7*c* requires that

$$M_R - M_U = 0 \qquad (c)$$

Using Eqs. (*a*) and (*b*), we find for moment equilibrium that

$$\left(2k - \frac{PL}{2}\right)\theta = 0 \qquad (d)$$

If we use the same reasoning as in Example 10.1, we have the stability relations

$$\begin{array}{lll} P < P_{cr} & M_U < M_R & \text{stable} \\ P = P_{cr} & M_U = M_R & \text{neutral} \\ P > P_{cr} & M_U > M_R & \text{unstable} \end{array} \qquad (e)$$

where

$$P_{cr} = \frac{4k}{L} \qquad (f)$$

If $P < P_{cr}$, the column remains straight and no buckling occurs. The separation between no buckling and the buckled configuration is given by the critical load P_{cr} in Eq. (*f*). We see that when the value of $P = P_{cr}$, angle θ in Eq. (*d*) can take on any value, and so the column is in a buckled configuration.

As was also the case in Example 10.1, we would like to make a choice for the spring constant k in the model column shown in Fig. 10.7*b* which reflects the stiffness properties of the flexible column of Fig. 10.7*a*. We wish to choose k so that the transverse deflection at the midpoint of the model column matches that of the flexible column for a transverse load.

A transverse load of magnitude Q is applied to the model column at C and causes a transverse deflection δ_1, as shown in Fig. 10.7*e*. Moment equilibrium of the free-body diagram of bar

AC (Fig. 10.7f) results in

$$2k\theta = \frac{Q}{2}\frac{L}{2}\cos\theta \approx \frac{Q}{2}\frac{L}{2}$$

so that

$$\theta = \frac{QL}{8k} \qquad (g)$$

Writing the deflection δ_1 in terms of θ, using Fig. 10.7f, we find

$$\delta_1 = \frac{\theta L}{2}$$

or from Eq. (g)

$$\delta_1 = \frac{QL^2}{16k} \qquad (h)$$

The deflection of a flexible column (beam) pinned at both ends due to a transverse load Q applied to the midspan of the beam is, according to Table G.2, Fig. G.2-2, App. G

$$\delta_2 = \frac{QL^3}{48EI} \qquad (i)$$

Setting $\delta_1 = \delta_2$, we find

$$\frac{QL^2}{16k} = \frac{QL^3}{48EI}$$

or

$$k = \frac{3EI}{L} \qquad (j)$$

If we select k as given by Eq. (j), the model column and the flexible column will have the same midspan deflection. Comparing the result for the spring constant k given by Eq. (j) for the present example with the result for k in Eq. (l) in Example 10.1 for the beam cantilevered from its base, we find that the procedure which matched the transverse deflections for the model column and the flexible column leads to the same choice for spring constants. Care should be exercised in the use of the units in k.

Now that the spring constant is related to the parameters of the flexible column by Eq. (j), we can insert this value of k into the expression for the critical load, Eq. (f), to get

$$P_{cr} = \frac{4}{L}\frac{3EI}{L} = \frac{12EI}{L^2} \qquad (k)$$

We will return to this predicted critical load in the next section in order to compare it with the exact critical load for the pinned-pinned flexible column. Again we note the dependence of the critical load on the parameters EI and L of the column.

In the analyses carried out in the above examples, we assumed a straight column with a load aligned exactly along the centroidal axis of the column. We can gain additional perspective on the concept of stability if we attempt to account for a geometric imperfection that might occur in a vertically loaded column.

EXAMPLE 10.3

Consider an elastic column clamped at its base and subjected to an axial load P that is applied to the top; the point of application of the load is offset by a distance cL from the centroidal axis of the column, as shown in Fig. 10.8a. The quantity c is nondimensional, and the offset distance is taken in the form of cL for convenience.

We wish to investigate the transverse deflection of the elastic column as we increase the value of the load P. We will use a rigid-bar-spring model, as we did in Examples 10.1 and 10.2, to investigate the stability of this column with the offset load. The flexible column is replaced by the rigid-bar-spring model of Fig. 10.8b, and we investigate the behavior of this model system.

As the load increases from the value zero, the bar will begin to rotate through an angle θ_0 about point A. We use the subscript 0 on the angle to emphasize that this equilibrium state corresponds to a load P; we will investigate the stability of the system by making a small change from this equilibrium angle θ_0. We first obtain the load-angle relation.

The restoring spring at the base balances the moment caused by the load to give an equilibrium angle θ_0 of the bar. If we replace the offset load at the top of the bar by a load P

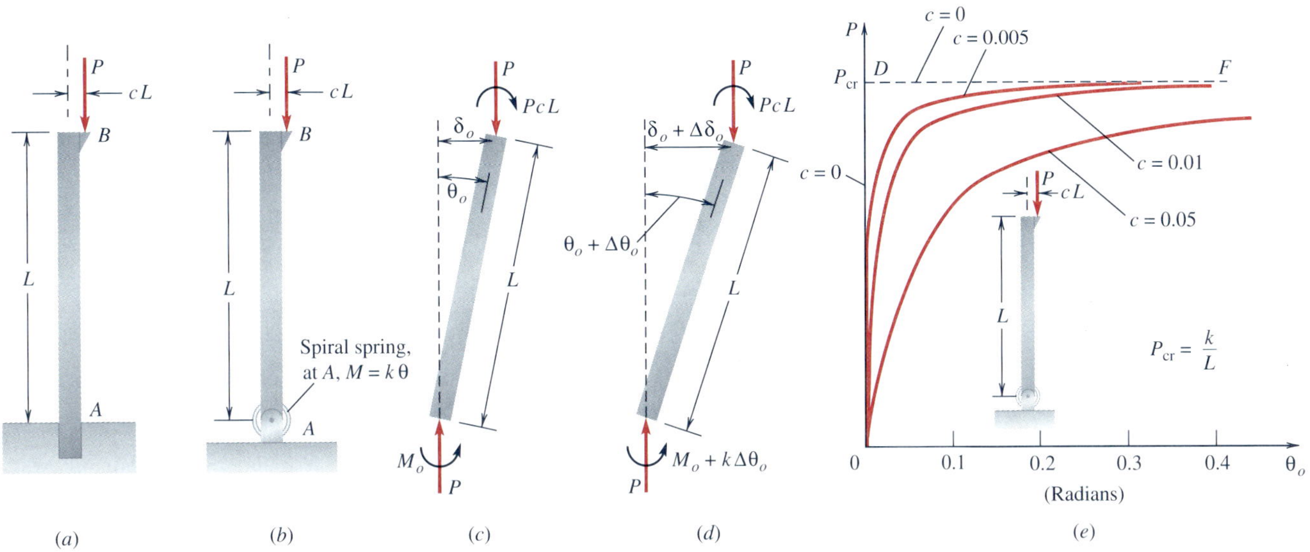

Figure 10.8 Example 10.3

acting along the centroidal axis and a moment PcL at the top and define δ_0 as the deflection at the top, then moment equilibrium of the free body in Fig. 10.8*c* gives

$$P\delta_0 + PcL = M_0 = k\theta_0 \qquad (a)$$

For small values of θ_0 we have

$$\delta_0 = L \sin \theta_0 \approx L\theta_0 \qquad (b)$$

Therefore, it follows from Eqs. (*a*) and (*b*) that

$$(k - PL)\theta_0 = PcL \qquad (c)$$

Solving for P, we obtain the load–angular response relation

$$P = \frac{k}{L}\frac{\theta_0}{c + \theta_0} \qquad (d)$$

The corresponding load–angular response curves are plotted in Fig. 10.8*e* for a range of values of c. These load-response curves of Fig. 10.8*e* predict for very small offset parameter values or load alignment imperfections that as the load increases from zero, a very small angle response occurs. As the load $P = k/L$ is approached from below, a rapid increase of angle occurs, limited by the assumption in this analysis that θ_0 remains small.

These results suggest an alternate definition for the critical load: namely, the value of the load that corresponds to arbitrarily large values for the response angle θ_0 in the presence of an imperfection. That is,

$$P_{cr} = \frac{k}{L} \qquad (e)$$

which is the same critical load that we obtained in Example 10.1 for the bar without the offset load, $c = 0$.

We can therefore rewrite Eq. (*c*) for the response angle θ_0 in terms of the load

$$\theta_0 = \frac{c(P/P_{cr})}{1 - P/P_{cr}} \qquad (f)$$

which clearly shows that as $P \to P_{cr}$, θ_0 becomes unbounded in accordance with the curves shown in Fig. 10.8*e*.

We now turn to the procedure used in Examples 10.1 and 10.2 to investigate stability and apply it to the equilibrium state of the bar given by the angle θ_0 under a load P. That is, we perturb the system from its equilibrium configuration, defined by the angle θ_0. If the perturbation results in restoring moments greater than the upsetting moments, then the bar tends to return to its original configuration, i.e., we have a state of stable equilibrium.

If the inclined bar in Fig. 10.8*c* is given a small disturbance $\Delta\theta_0$ with the load held fixed, as shown in Fig. 10.8*d*, then the upsetting moment is given by

$$M_U = PcL + P(\delta_0 + \Delta\delta_0)$$

or

$$M_U = PcL + PL(\theta_0 + \Delta\theta_0) \qquad (g)$$

We can make use of Eq. (*a*) to rewrite Eq. (*g*) as

$$M_U = M_0 + PL\,\Delta\theta_0 \qquad (h)$$

For the restoring moment acting on the bar of Fig. 10.8*d*, we have, with Eq. (*a*),

$$M_R = k(\theta_0 + \Delta\theta_0) = M_0 + k\,\Delta\theta_0 \qquad (i)$$

Equating the restoring and upsetting moments from Eqs. (*i*) and (*h*), we find

$$M_0 + k\,\Delta\theta_0 = M_0 + PL\,\Delta\theta_0$$

or

$$(k - PL)\,\Delta\theta_0 = 0 \qquad (j)$$

From the same reasoning that was used before for equations in the form of Eq. (*j*), we conclude that neutral stability occurs when

$$P_{cr} = \frac{k}{L} \qquad (k)$$

This stability result for the column with an offset load should be interpreted as follows. For any equilibrium state characterized by an angle of inclination θ_0, if the corresponding load P is less than $P_{cr} = k/L$, then for an arbitrarily small disturbance, the system will generate a restoring moment which will accelerate the system back toward equilibrium; i.e., we have a stable equilibrium state. Each point on the curves shown for different c values in Fig. 10.8*e* corresponds to an equilibrium state for which the stability analysis just carried out is valid. Therefore, for all points on all the curves for which $c > 0$, we can conclude that the equilibrium states are all stable since in all cases $P < P_{cr}$.

The case of the perfectly aligned load with $c = 0$ is also included as a special case of the curves in Fig. 10.8*e*. In this case the initial equilibrium state corresponds to the configuration $\theta_0 = 0$. The perturbation angle $\Delta\theta_0$ from this zero deflection state is equal to θ_0. As a consequence Eq. (*j*) becomes

$$(k - PL)\theta_0 = 0 \qquad (l)$$

and if $0 < P < P_{cr} = k/L$, it follows from Eq. (*l*) that

$$\theta_0 = 0 \qquad (m)$$

and the points that satisfy Eq. (*m*) plot as the vertical line segment from the origin to D in Fig. 10.8*e*. For $P = P_{cr}$, we conclude from Eq. (*l*) that θ_0 is arbitrary, and such points plot as the horizontal line DF in Fig. 10.8*e*.

The load point D on the P axis, P_{cr}, in Fig. 10.8*e* is referred to as a *bifurcation point* on the load-displacement curve since the curve divides into two branches at this point. One branch is the solution $\theta_0 = 0$, and the other branch is line DF with θ_0 arbitrary. It is this possibility for a structure loaded at or near its critical buckling load or bifurcation load to move into another configuration than the one intended in the original design that can sometimes lead to catastrophic results.

10.4 Elastic Stability of Flexible Columns—Some Special Cases

In Sec. 10.3 we calculated the buckling loads for rigid-bar-spring models of flexible columns. The stability analysis was greatly simplified because in every case the deflection response of the bar-spring model could be described by a single angle or the deflection of one point. For a flexible column, however, the deflection curve is continuous and made up of an infinite number of deflection points. In this section the stability analysis for flexible columns is carried out without restrictions on the deflected shape and results in differential equations that must be solved. Flexible columns can be considered as horizontal beams turned in the vertical direction, and the analysis developed for beams carries over to columns. Unless we state otherwise, we will neglect the weight of the column compared to the loads applied. Our goal is to determine the buckling load of a loaded column. This load is also referred to as the *critical* or *bifurcation load*; see Example 10.3.

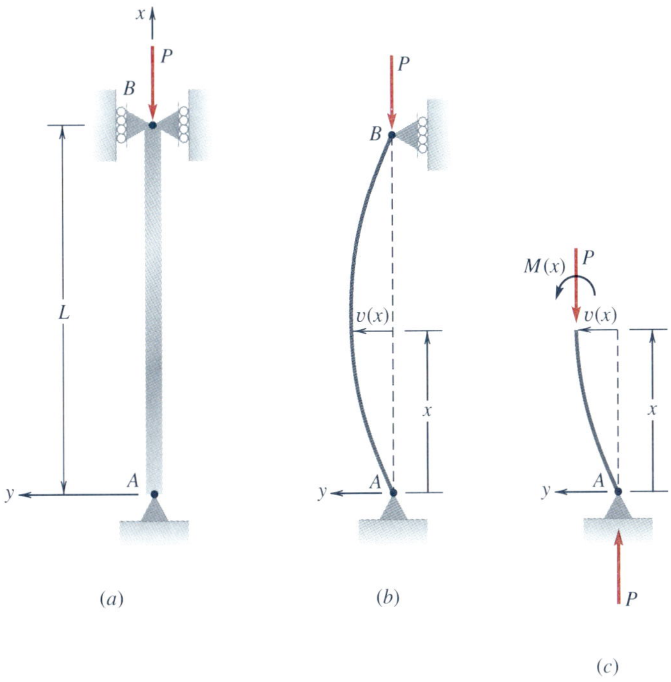

Figure 10.9 A pinned-pinned column subjected to a compressive load P.

Columns with Both Ends Pinned

An elastic column AB of length L, pinned at both ends, is subjected to a downward load P acting at the centroid of the cross section at point B, as shown in Fig. 10.9*a*. We wish to examine the stability of the primary equilibrium state of the column that corresponds to the perfectly straight configuration with a uniform axial compressive stress state. To do this, we investigate the existence of a nearby slightly perturbed state from the perfectly straight configuration. Following the approach outlined in Examples 10.1 to 10.3, we will give the column an arbitrary disturbance $v(x)$ so as not to violate the geometric restraint conditions at the ends; i.e., we must have $v(0) = 0$ and $v(L) = 0$. The minimum load required to maintain the column in a deflected shape in a state of neutral equilibrium will be taken as the buckling load P_{cr}. At this load the upsetting and restoring actions exactly balance. For $P < P_{cr}$, the restoring action exceeds the disturbing action, and the column will move back toward the primary state, i.e., the primary state is a stable equilibrium state.

In Fig. 10.9*b* the column has undergone a small arbitrary deflection, described by a function $v(x)$, from the perfectly straight configuration. The axes and sign conventions are the same as those used for the case of beams in earlier chapters. To find the minimum value of P required to maintain the state of neutral equilibrium shown in Fig. 10.9*b*, it will be necessary to find the shape function for the deflection $v(x)$. When we consider a segment of the column in the deflected shape, as shown in Fig. 10.9*c*, we see from the requirement of moment equilibrium that the bending moment at section x is given by

$$M(x) = -Pv(x) \tag{10.1}$$

where $v(x)$ is the transverse deflection at section x.

If EI_z is the bending modulus of the column, then according to Eq. (6.6),

$$M(x) = EI_z \frac{d^2v}{dx^2} \tag{6.6}$$

Therefore, the equation for moment equilibrium can be written in the form

$$EI_z \frac{d^2v}{dx^2} + Pv = 0 \tag{10.2}$$

Since the unknown function $v(x)$ and its second derivative appear in Eq. (10.2) in a linear form, Eq. (10.2) is referred to as a second-order linear differential equation with constant coefficients. In courses on differential equations, it is shown that all possible solutions of Eq. (10.2) can be found once two linearly independent solutions are found. If each of the two linearly independent solutions is multiplied by a different arbitrary constant and they are added, the resulting expression satisfies the differential equation for arbitrary values of the two constants. Two constants arise because the equation is a second-order linear differential equation.

In the present case with v and d^2v/dx^2 occurring with constant coefficients in Eq. (10.2), we can construct linearly independent solutions by noting that the function $\sin \lambda x$, where λ is any number, has the property that

$$\frac{d^2}{dx^2} \sin \lambda x = -\lambda^2 \sin \lambda x$$

Thus if we try for a solution of Eq. (10.2) in the form

$$v_1(x) = c_1 \sin \lambda x \tag{10.3}$$

where c_1 is an arbitrary constant and λ is unknown as yet and we substitute into Eq. (10.2), then it follows that

$$-EI_z c_1 \lambda^2 \sin \lambda x + Pc_1 \sin \lambda x = 0 \tag{10.4}$$

or

$$(-EI_z \lambda^2 + P)c_1 \sin \lambda x = 0 \tag{10.5}$$

Thus if we define

$$\lambda^2 = \frac{P}{EI_z} \qquad \lambda = \sqrt{\frac{P}{EI_z}} \tag{10.6}$$

then $v_1(x)$ is a solution of Eq. (10.2) for all values of the arbitrary constant c_1.

A similar argument shows that

$$v_2(x) = c_2 \cos \lambda x \tag{10.7}$$

is also a solution of Eq. (10.2) for all c_2.

It is easily verified that the sum of the two independent solutions given by Eqs. (10.3) and (10.7), namely

$$v(x) = c_1 \sin \lambda x + c_2 \cos \lambda x \tag{10.8}$$

is also a solution of Eq. (10.2) for all values of the arbitrary constants c_1 and c_2 with λ given by Eq. (10.6).

Note that the arbitrary disturbance of the equilibrium state, the deflection function $v(x)$ in Eq. (10.8), is now completely specified by the three parameters c_1, c_2, and $\lambda = \sqrt{P/(EI_z)}$.

In addition to satisfying equilibrium requirements, the deflection function in Eq. (10.8) must be compatible with the geometric conditions at the pinned supports, i.e.,

$$\text{at } x = 0, L: \qquad v = 0 \tag{10.9}$$

The requirement that v should take on the value 0 at the point $x = 0$ can be introduced into Eq. (10.8) to obtain

$$c_2 = 0 \tag{10.10}$$

Introducing the condition that $v = 0$ for $x = L$ into Eq. (10.8) reduces to

$$0 = c_1 \sin \lambda L \tag{10.11}$$

One solution of Eq. (10.11) is, of course, $c_1 = 0$, but this solution leads back to $v(x)$ identically zero over the entire column, i.e., no deflection disturbance from the primary state.

The remaining possibility in Eq. (10.11) is to have

$$\sin \lambda L = 0 \tag{10.12}$$

The values of λL which satisfy Eq. (10.12) are the zeros for the sine function

$$\lambda_n L = n\pi \qquad n = 1, 2, \ldots \tag{10.13}$$

where the subscript n on λ indicates that there are an infinity of solutions corresponding to $n = 1, 2, \ldots$. We have discarded $\lambda_0 = 0$ since this value gives $v(x)$ identically zero. Using Eq. (10.6), we have

$$\lambda_n^2 = \frac{P_n}{EI_z} \tag{10.14}$$

or using Eq. (10.13), we can rewrite Eq. (10.14) in the form

$$P_n = n^2\pi^2 \frac{EI_z}{L^2} \tag{10.15}$$

Corresponding to each of the loads P_n there is a neutral equilibrium deflection shape that can be maintained at the load $P = P_n$. To find these shapes or *buckling modes,* we use Eqs. (10.10) and (10.13) to write Eq. (10.8) in the form

$$v(x) = v_n(x) = c_1 \sin\frac{n\pi x}{L} \qquad n = 1, 2, \ldots \tag{10.16}$$

The first three modes of the deflected shape for neutral equilibrium are shown in Fig. 10.10*a* to *c*. We see that as the buckling shapes exhibit more rapid variation of curvature over the length of the column, the axial load for equilibrium increases. For example, the load P_2 required to maintain the buckling mode with zero deflection at the midpoint (Fig.

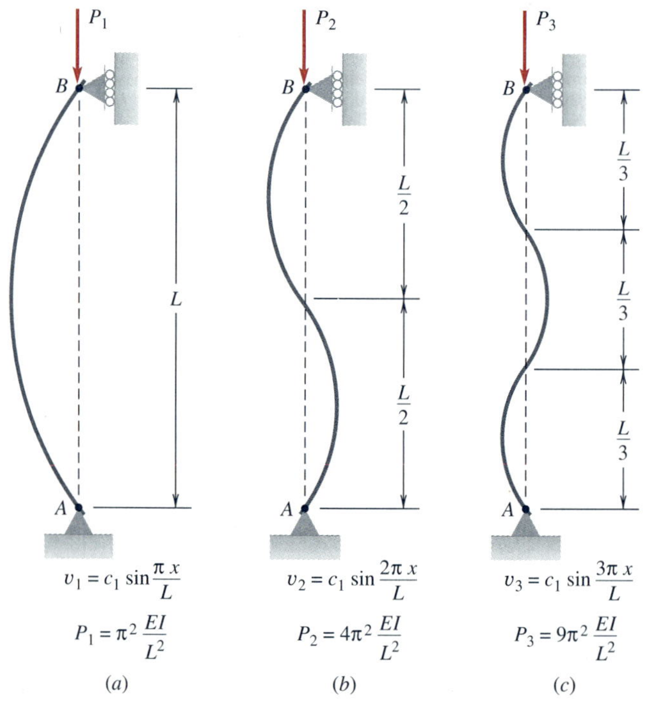

Figure 10.10 Buckled shapes of a pinned-pinned column for different critical loads P.

10.10*b*) is 4 times larger than the load P_1 for the mode with no internal zero deflection value (Fig. 10.10*a*).

The minimum value of these buckling loads is of practical interest for the design of columns, and we refer to it as the *critical buckling load* for a column pinned at both ends and write

$$P_{cr} = P_1 = \pi^2 \frac{EI_z}{L^2} \tag{10.17}$$

This load is often referred to as the *Euler buckling load* of a column in honor of the Swiss scientist-mathematician who first determined this load in 1744.

The critical load P_{cr}, given by Eq. (10.17), separates states of stable equilibrium from states of unstable equilibrium. It is the bifurcation load or the buckling load, as discussed in Example 10.3, that allows transverse deflections to occur. In design and analysis of slender members, a knowledge of the critical load is needed to minimize transverse deflections in members subjected to compressive loads.

It is of some interest to compare the critical buckling load just found for the flexible pinned-pinned column with the load found in Example 10.2, where the same column was modeled by a rigid-bar-spring model. The approximate buckling load was found there to be $12EI_z/L^2$, which is about 22 percent higher than the exact value given in Eq. (10.17).

EXAMPLE 10.4

In a construction application it is proposed to use 2-in × 4-in lumber as columns for the two cases shown in Fig. 10.11. In Fig. 10.11*a*, a single 2 × 4 of length 8 ft is to carry a compressive load of 500 lb. In Fig. 10.11*b*, a pair of 10-ft-long 2 × 4s is nailed together to form a column to carry 2000 lb.

We assume that the columns are to be modeled as pinned-pinned columns. We wish to find the factor of safety compared to the buckling load given in Eq. (10.17) for these two cases. We take the dimensions of the 2 × 4s to be 1.5 in × 3.5 in (see App. D), and $E = 2000$ ksi.

For the single 2 × 4 column we obtain different critical buckling loads from Eq. (10.17) depending upon which value of I_z is used for the cross section. The two values for the moments of inertia about the axes of symmetry are

$$I_{min} = \tfrac{1}{12}(3.5)(1.5)^3 = 0.984 \text{ in}^4 \tag{a}$$

and

$$I_{max} = \tfrac{1}{12}(1.5)(3.5)^3 = 5.36 \text{ in}^4 \tag{b}$$

Since P_{cr} is directly proportional to I, we are interested in the lower critical load when the column is loaded in service. Therefore, using Eq. (10.17), we have

$$P_{cr} = \pi^2 \frac{EI_{min}}{L^2} = \pi^2 \frac{(2 \times 10^6)(0.984)}{(8 \times 12)^2} = 2110 \text{ lb} \tag{c}$$

The factor of safety is therefore

$$f_s = \frac{2110}{500} = 4.2 \tag{d}$$

For the two 2 × 4s nailed together, we have

$$I_{min} = \tfrac{1}{12}(3.5)(3)^3 = 7.88 \text{ in}^4 \tag{e}$$

$$I_{max} = \tfrac{1}{12}(3)(3.5)^3 = 10.7 \text{ in}^4 \tag{f}$$

Again taking the smaller I for calculating P_{cr}, we have

$$P_{cr} = \pi^2 \frac{EI_{min}}{L^2} = \pi^2 \frac{(2 \times 10^6)(7.88)}{(10 \times 12)^2} = 10{,}800 \text{ lb} \tag{g}$$

For this case the factor of safety is

$$f_s = \frac{10{,}800}{2000} = 5.4 \tag{h}$$

In small building construction, both these columns would be adequate (if they met local building codes).

Figure 10.11 Example 10.4

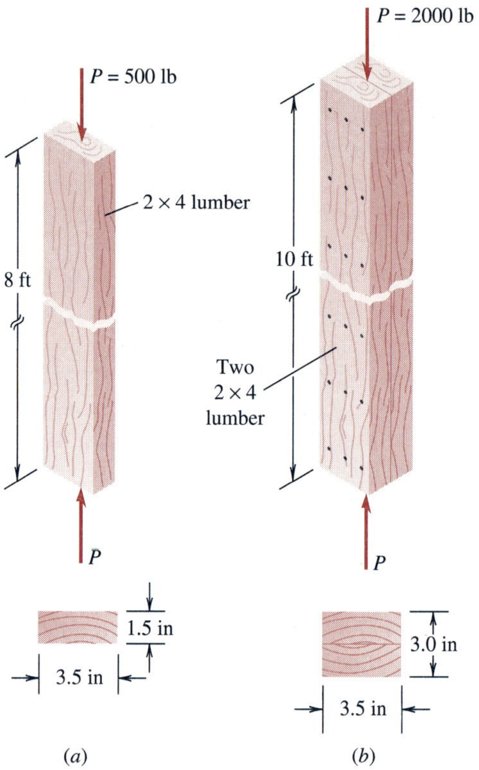

Columns with One End Clamped and the Other End Free

A straight column AB of uniform cross section is clamped at its base A and is loaded at the other end B by a vertical load P applied at the centroid of the section, as shown in Fig. 10.12*a*. We wish to find the critical value of P corresponding to the buckling of this column. Recall the approximate analysis of this column in Example 10.1.

Once again we begin the stability analysis of this column by giving the column a small arbitrary deflection disturbance from the initial, perfectly straight configuration. The deflection $v(x)$ is shown in Fig. 10.12*b*. To relate the load P and the deflection $v(x)$ to the internal bending moment in the column, we set up a free-body segment of the column as shown in Fig. 10.12*c*. Moment equilibrium about section x of the segment gives

$$M(x) = P(\delta - v) \tag{10.18}$$

where v is the value of the deflection at x, that is, $v(x)$, and δ is the value of the deflection at $x = L$. Using Eq. (6.6), we can write Eq. (10.18) in

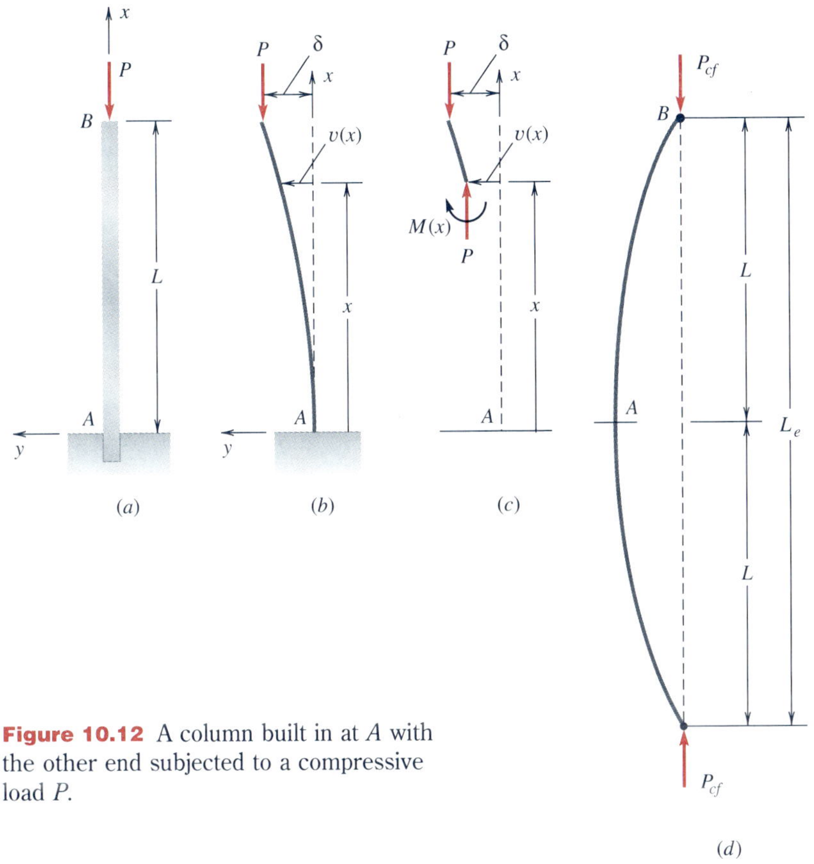

Figure 10.12 A column built in at A with the other end subjected to a compressive load P.

the form

$$EI_z \frac{d^2v}{dx^2} + Pv = P\delta \tag{10.19}$$

Comparison of Eqs. (10.19) and (10.2) shows that they differ only in the term $P\delta$ which appears on the right side of Eq. (10.19). It is known from the theory of linear differential equations that the general solution of Eq. (10.19) consists of the sum of two parts: the homogeneous solution which is found by solving Eq. (10.19) with a zero right side and a particular solution of the equation with the right side retained. The homogeneous solution of Eq. (10.19) is given by Eq. (10.8) and was used in the buckling analysis of the pinned-pinned column. For the particular solution it is seen directly that

$$v_p = \delta \tag{10.20}$$

satisfies Eq. (10.19). Thus combining the homogeneous solution, Eq. (10.8), and the particular solution, Eq. (10.20), we have the general solution of Eq. (10.19)

$$v(x) = c_1 \sin \lambda x + c_2 \cos \lambda x + \delta \tag{10.21}$$

The arbitrary disturbance function $v(x)$ is now completely described by the integration constants c_1 and c_2 and the parameters λ and δ. It remains to choose c_1, c_2, and λ so that the three geometric constraint conditions of support of the column are satisfied, namely,

$$\begin{aligned} x = 0: &\qquad v = \frac{dv}{dx} = 0 \\ x = L: &\qquad v = \delta \end{aligned} \tag{10.22}$$

At $x = 0$, $v = 0$ gives

$$0 = c_1 \cdot 0 + c_2 \cdot 1 + \delta$$

or

$$c_2 = -\delta \tag{10.23}$$

and at $x = 0$, $dv/dx = 0$ gives

$$0 = c_1 \lambda \cdot 1 - \lambda c_2 \cdot 0$$

or

$$c_1 = 0 \tag{10.24}$$

At $x = L$, $v = \delta$ gives

$$\delta = \delta(1 - \cos \lambda L)$$

If δ is not zero, then

$$\cos \lambda L = 0 \tag{10.25}$$

The cosine function takes on zero values at odd multiples of $\pi/2$ so that

$$\lambda_n L = (2n - 1)\frac{\pi}{2} \qquad n = 1, 2, \ldots \tag{10.26}$$

and using Eq. (10.6), we have

$$P_n = (2n - 1)^2 \frac{\pi^2}{4} \frac{EI_z}{L^2} \tag{10.27}$$

Again we take the smallest load as the critical load for the column, i.e.,

$$P_{cr} = P_1 = \frac{\pi^2}{4} \frac{EI_z}{L^2} \tag{10.28}$$

Equation (10.28) gives the Euler buckling load for the column shown in Fig. 10.12*a*.

The buckling mode shape corresponding to P_{cr} can be found from Eq. (10.21) by using the results in Eqs. (10.23), (10.24) and (10.26) to get

$$v(x) = \delta\left[1 - \cos\left(\frac{\pi}{2}\frac{x}{L}\right)\right] \tag{10.29}$$

The stability analysis using a rigid-bar-spring model for this column as presented in Example 10.1 predicted a buckling load of $3EI/L^2$ which turns out to be 22 percent higher than the one given in Eq. (10.28).

Comparison of Eqs. (10.17) and (10.28) shows that the clamped-free column buckles at only one-fourth of the load required to buckle the pinned-pinned column. Another way to interpret this result follows if we rewrite the expression for the critical buckling load for the clamped-free column, denoted by P_{cf} and given in Eq. (10.28) in the form

$$P_{cf} = \pi^2 \frac{EI_z}{L_e^2} \tag{10.30}$$

where L_e denotes an *effective length,* which for this case is given by

$$L_e = 2L \tag{10.31}$$

Equation (10.30) can be interpreted as indicating that a pinned-pinned column of length double that of the original clamped-free column would have the same buckling load. An interpretation of the effective length for the clamped-free column is shown in Fig. 10.12*d.* A pinned-pinned column of length $L_e = 2L$ is shown. One-half of this column, AB, can be viewed as a clamped-free column of length L since $M = 0$ at end B and end A is equivalent to a clamped end with $v' = 0$ because of symmetry. Thus the critical load P_{cr} of Eq. (10.28) can maintain the buckled shape for a clamped-free column of length L or equivalently a pinned-pinned column of length $L_e = 2L$.

10.5 Elastic Stability of Flexible Columns—General Approach

It was possible to determine the critical buckling loads for pinned-pinned and clamped-free columns in Sec. 10.4 because of the simplicity of the support conditions. Simple free-body diagrams were used to derive differential equations from which the buckling loads were found.

In the general case of a column with arbitrary supports, it is still possible to derive a general differential equation for the problem. Bound-

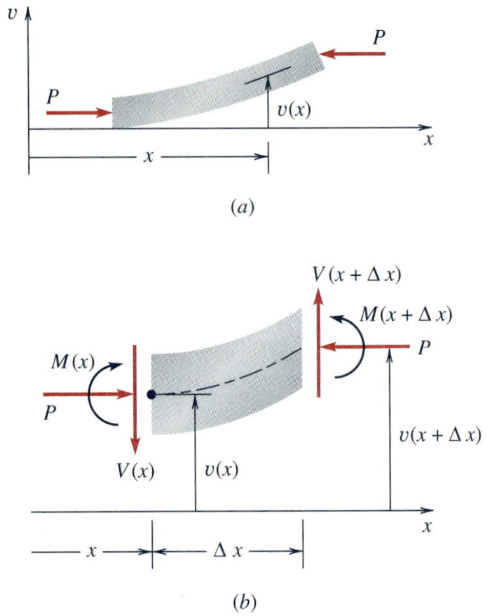

Figure 10.13 (*a*) Beam segment subjected to an axial load *P*, and (*b*) free-body diagram of a small element of the beam.

ary conditions corresponding to a given column can then be used to find the buckling load and associated mode shape.

Consider a segment of a deformed column that carries a constant axial load P, as shown in Fig. 10.13*a*. Suppose we isolate a small element of the column of length Δx, as shown in Fig. 10.13*b*. Force and moment equilibrium for this element in the deformed state under the assumption of small deflections give (similar to the arguments in Sec. 4.4)

$$V(x + \Delta x) - V(x) = 0$$
$$M(x + \Delta x) - M(x) + V(x + \Delta x) \cdot \Delta x + P[v(x + \Delta x) - v(x)] = 0 \tag{10.32}$$

Force equilibrium in the x direction is identically satisfied.

When the size Δx of the element goes to zero, we have, upon dividing by Δx and taking the limit,

$$\frac{dV}{dx} = 0$$
$$\frac{dM}{dx} + V + P\frac{dv}{dx} = 0 \tag{10.33}$$

In the absence of the axial force P, these equations are the same as Eqs. (4.4) and (4.5) with $q = 0$.

The first of Eqs. (10.33) implies that V is constant over the length of the column, and we write upon integration

$$V(x) = V^* \tag{10.34}$$

where the integration constant in this case is written as V^* since the constant has the units of a force. Integrating the second of Eqs. (10.33) once with respect to x, using Eq. (10.34), we find

$$M(x) + V^*x + Pv(x) = M^* \tag{10.35}$$

where the integration constant is written as M^* since it has the units of a moment.

A second-order differential equation for $v(x)$ follows by making use of the moment-curvature relation, Eq. (6.6), to write

$$EI_z\frac{d^2v}{dx^2} = M(x) \tag{6.6}$$

or upon use of Eq. (10.35),

$$EI_z\frac{d^2v}{dx^2} + Pv = -V^*x + M^* \tag{10.36}$$

We have already indicated that the general solution of an equation of the form of Eq. (10.36) can be written as the sum of the homogeneous solution and a particular solution. The homogeneous solution was given in Eq. (10.8) as

$$v_H = c_1 \sin \lambda x + c_2 \cos \lambda x \tag{10.37}$$

where

$$\lambda = \sqrt{\frac{P}{EI_z}} \tag{10.38}$$

It is easily verified by substitution into Eq. (10.36) that

$$v_p = -\frac{V^*}{P}x + \frac{M^*}{P} \tag{10.39}$$

is a particular solution of Eq. (10.36). Therefore, the general solution becomes

$$v(x) = c_1 \sin \lambda x + c_2 \cos \lambda x - \frac{V^*}{P}x + \frac{M^*}{P} \tag{10.40}$$

Since there are four unknown integration constants, namely, c_1, c_2, V^*, and M^*, it will be necessary, for a given column, to specify four boundary conditions for the determination of these four constants. Examples to follow will demonstrate the solution procedure.

EXAMPLE 10.5

An elastic column AB of length L and bending modulus EI_z is clamped at its base A, is pinned at the other end B, and is loaded by the vertical load P, as shown in Fig. 10.14*a* and *b*. Section B is free to move down as the load P is increased. We wish to find the buckling load P_{cr} by making use of the general solution, Eq. (10.40).

Solutions given by Eq. (10.40) for different values of the arbitrary constants represent states of neutral equilibrium since the restoring and upsetting forces are in balance. Thus a load P which satisfies Eq. (10.40) and the geometric and force boundary conditions is a buckling load and corresponds to a state of neutral equilibrium. In general, there are an infinity of buckling loads, and the smallest one is taken as the critical load P_{cr} for the column.

In the present case we note that the conditions at the pinned end $x = L$ are

$$M = 0 \qquad v = 0 \tag{a}$$

which, when they are substituted into Eq. (10.35) with $x = L$, yield

$$V^*L = M^* \tag{b}$$

The clamped condition at $x = 0$ requires that

$$v = 0 \tag{c}$$

$$\frac{dv}{dx} = 0 \tag{d}$$

which, after we use Eqs. (*b*) and (*c*) and $x = 0$ in Eq. (10.40), first results in

$$0 = c_2 + \frac{M^*}{P} = c_2 + \frac{V^*L}{P} \tag{e}$$

If we differentiate Eq. (10.40) and make use of Eq. (*d*), we get

$$0 = c_1\lambda - \frac{V^*}{P} \tag{f}$$

Thus Eqs. (*e*) and (*f*) allow us to write the solution, Eq. (10.40), entirely in terms of V^* in the form

$$v(x) = \frac{V^*}{P}\left(\frac{\sin \lambda x}{\lambda} - L \cos \lambda x - x + L\right) \tag{g}$$

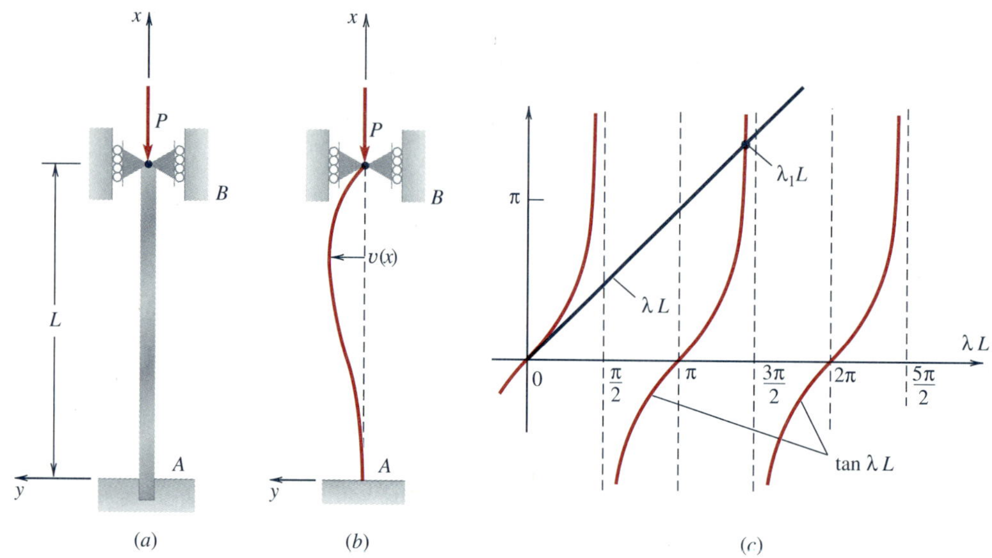

Figure 10.14 Example 10.5

Finally, the condition that at

$$x = L \qquad v = 0 \tag{h}$$

becomes, by using Eq. (*g*),

$$\frac{V^*}{P}\left(\frac{\sin \lambda L}{\lambda} - L\cos \lambda L\right) = 0 \tag{i}$$

One solution of Eq. (*i*) is $V^* = 0$, which corresponds to the original vertical column in a state of uniform compression. The other possible solution is

$$\frac{\sin \lambda L}{\lambda} - L\cos \lambda L = 0 \tag{j}$$

or

$$\tan \lambda L - \lambda L = 0 \tag{k}$$

The solutions or roots of Eq. (*k*) will give the values of the critical load.

Roots of Eq. (*k*), denoted by $\lambda_n L$, can be found by plotting the functions $\tan \lambda L$ and λL on the same graph, as shown in Fig. 10.14*c*, and locating the points of intersection from the graph. Rough estimates of the roots from the graph can be refined by trial-and-error calculations using Eq. (*k*). It is clear from the graph in Fig. 10.14*c* that an infinite number of intersections exist and that for increasing λL they approach odd multiples of $\pi/2$.

For the lowest root if we initially approximate $\lambda_1 L$ as 1.4π, then trial-and-error calculations using Eq. (*k*) yield the more accurate root of Eq. (*k*) as

$$\lambda_1 L = 1.43\pi \tag{l}$$

Making use of Eq. (10.6), we can rewrite Eq. (*l*) to find the buckling load

$$P_{cr} = (1.43\pi)^2 \frac{EI_z}{L^2}$$

or

$$P_{cr} = 2.05\pi^2 \frac{EI_z}{L^2} = 20.2\frac{EI_z}{L^2} \tag{m}$$

The buckled shape corresponding to P_{cr} is found by substituting $\lambda_1 L$ from Eq. (*l*) into the expression for $v(x)$ given by Eq. (*g*)

$$v(x) = c_3\left[\frac{L\sin(1.43\pi x/L)}{1.43\pi} - L\cos\frac{1.43\pi x}{L} - x + L\right] \tag{n}$$

where c_3 is now an arbitrary constant.

We can again ask the question for the present column, What would be the effective length of a pinned-pinned column so that it would buckle at the same value of the load as given by Eq. (*m*)? This effective length L_e can be found by rewriting

Eq. (m) in the form

$$P_{cp} = \pi^2 \frac{EI_z}{L^2/2.05} \qquad (o)$$

where P_{cp} denotes the buckling load of the clamped-pinned column. Comparing P_{cp} to the expression for the buckling load for a pinned-pinned column of length L_e

$$P_{cp} = \pi^2 \frac{EI_z}{L_e^2} \qquad (p)$$

we find that

$$L_e^2 = \frac{L^2}{2.05}$$

or

$$L_e = \frac{L}{1.43} = 0.699L \qquad (q)$$

for the effective length.

EXAMPLE 10.6

An elastic column AB of length L and bending modulus EI_z is clamped at both ends and is loaded by a vertical load P, as shown in Fig. 10.15a and b. The column is free to move downward at point B under load P. We wish to find the buckling load P_{cr} by making use of the general solution given in Eq. (10.40).

The solution follows the same general procedure as that used in Example 10.5 so we only outline the general steps.

First, we note that the condition $v = 0$ at $x = 0$ in, Eq. (10.35), results in

$$M(0) = M^* \qquad (a)$$

Similarly at $x = L$ with $v = 0$, we have from Eq. (10.35)

$$M(L) = -V^*L + M^* \qquad (b)$$

However, from the symmetry of the deflection curve, the bending moments at A and B must be equal, $M(0) = M(L)$, so that from Eqs. (a) and (b) we conclude that $V^* = 0$. Therefore, the general solution, Eq. (10.40), reduces to

$$v(x) = c_1 \sin \lambda x + c_2 \cos \lambda x + \frac{M^*}{P} \qquad (c)$$

The clamped condition at $x = 0$ requires that $v = 0$ and $v' = 0$, which when applied to $v(x)$ from Eq. (c) results in

$$v(0) = 0 = c_2 + \frac{M^*}{P} \qquad (d)$$

$$v'(0) = 0 = c_1\lambda \qquad (e)$$

Equation (c), using the results in Eqs. (d) and (e), therefore takes the form

$$v(x) = \frac{M^*}{P}(1 - \cos \lambda x) \qquad (f)$$

The condition $v = 0$ at $x = L$ in Eq. (f) gives

$$\frac{M^*}{P}(1 - \cos \lambda L) = 0 \qquad (g)$$

The solution $M^* = 0$ of Eq. (g) leads back to the solution that the deflection $v(x)$ is identically zero, which is the undeflected column. We take the other possible choice, which is

$$1 - \cos \lambda L = 0 \qquad (h)$$

Equation (h) becomes an equation for the buckling parameter λ of the form

$$\cos \lambda L = 1 \qquad (i)$$

There is an infinity of values for λL that satisfy Eq. (i), and they may be written as

$$\lambda_n L = 2\pi n \qquad n = 1, 2, \ldots \qquad (j)$$

The remaining boundary condition, $v' = 0$ at $x = L$, by using Eq. (f) is identically satisfied when $\lambda_n L = 2\pi n$.

The smallest value of the buckling loads corresponds to the choice of $n = 1$ in Eq. (j). Since $\lambda_1^2 = P_1/(EI_z)$, it follows that

$$P_{cr} = \lambda_1^2 EI_z = 4\pi^2 \frac{EI_z}{L^2} \qquad (k)$$

and we note that the critical load for the clamped-clamped column is 4 times larger than the critical load for a pinned-pinned column with the same EI_z/L^2 factor

The buckled deflected shape is found by inserting $\lambda_1 L = 2\pi$ into Eq. (f) to get

$$v(x) = c_3 \left[1 - \cos\left(2\pi \frac{x}{L}\right)\right] \qquad (l)$$

where c_3 is now an arbitrary constant.

The effective length for the clamped-clamped column can be calculated by rewriting Eq. (k) as

$$P_{cc} = \pi^2 \frac{EI_z}{L_e^2} \qquad (m)$$

Figure 10.15 Example 10.6

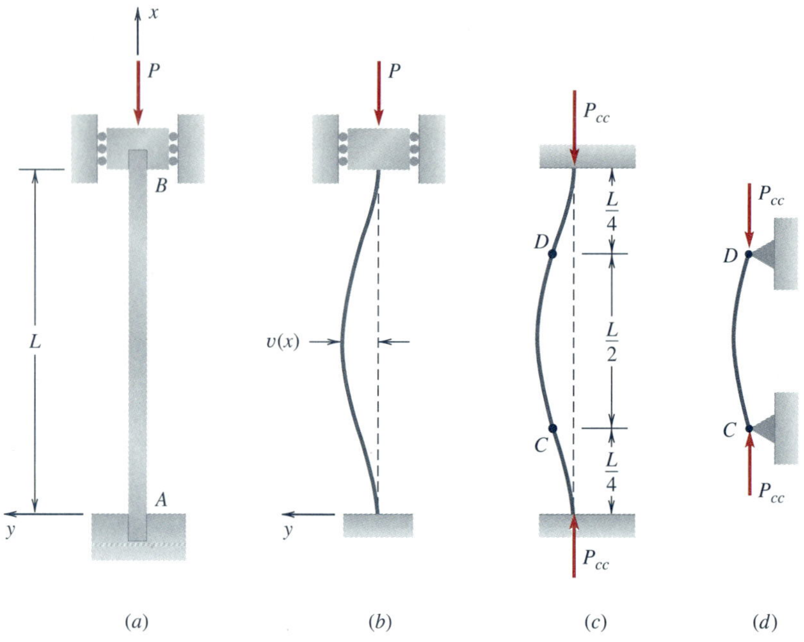

where P_{cc} denotes the critical load for a clamped-clamped column and

$$L_e = \frac{L}{2} \qquad (n)$$

Thus $L_e = L/2$ is the length of a pinned-pinned column which would buckle at the load given by Eq. (k).

An interpretation of the effective length for this column can be seen in Fig. 10.15c. The mode shape $v(x)$ has two inflection points at $x = L/4$ and $x = 3L/4$ (recall that inflection points are those points on a curve where the second derivative or curvature vanishes), as indicated at points C and D in Fig. 10.15c. The middle portion CD of the column has zero curvature at C and D and therefore has zero moments at these points; i.e., the middle portion represents a pinned-pinned column of length $L/2$ loaded by P_{cc}, as shown in Fig. 10.15d.

We have seen from previous examples that the value of the buckling load or critical load of a column is sensitive to the nature of the supports at the ends of the column. We summarize the cases we have investigated in Fig. 10.16. Of course, in practice, the exact nature of the supports may differ from these idealized cases. Many columns have more restraint than a hinged joint but not as much as a clamped joint. Advanced courses in structural design address these matters.

10.6 Columns with Eccentric Loads

In Example 10.3, the stability analysis of a column clamped at its base and subjected to an eccentrically applied vertical load at the top was

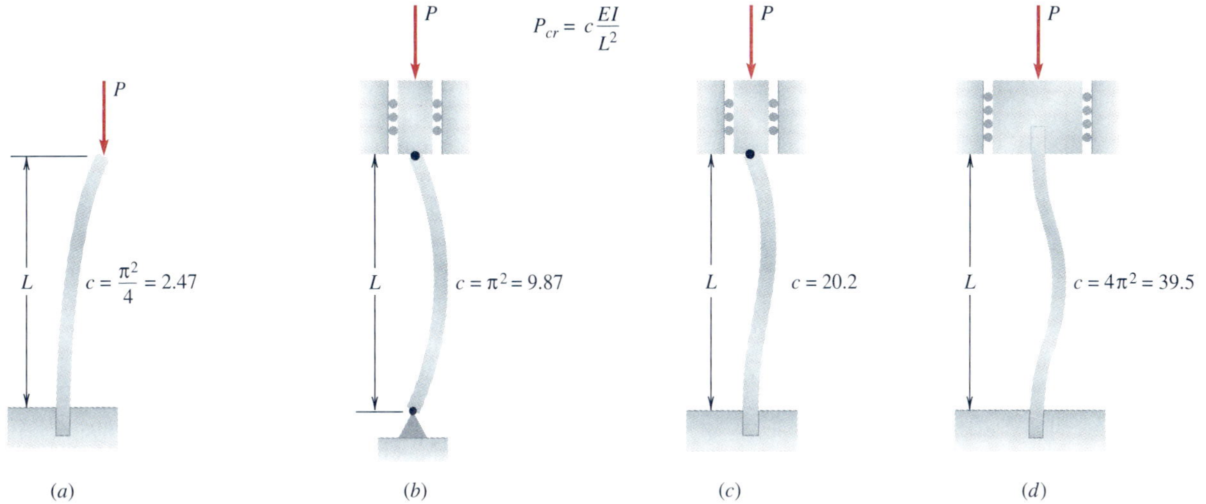

Figure 10.16 Critical loads for (*a*) clamped-free, (*b*) pinned-pinned, (*c*) clamped-pinned, and (*d*) clamped-clamped columns. In each case the constant c shown is to be inserted in the formula $P_{cr} = cEI/L^2$.

carried out by using a bar-spring model. By replacing the flexible column with a bar-spring model, the mathematical complexity was reduced from the problem of solving a differential equation to that of solving a single algebraic equation. In this section we examine the case of eccentric loading of a flexible column.

Consider an elastic column AB of length L and bending modulus EI_z subjected to a load P acting in a direction parallel to the centroidal axis of the column but with a small eccentricity e, as shown in Fig. 10.17*a* and *b*. We wish to find the deflection at the top of the column as a function of load P.

The analysis of this eccentrically loaded column is very similar to that presented in Sec. 10.4 for the same column but with $e = 0$. The deflected shape of the column is shown in Fig. 10.17*b*. Moment equilibrium for the column segment shown in Fig. 10.17*c* gives the bending moment at location x as

$$M(x) = P(\delta + e - v) \tag{10.41}$$

where δ is $v(L)$. Using Eq. (6.6) and Eq. (10.41) we obtain an equation for $v(x)$

$$EI_z \frac{d^2v}{dx^2} + Pv = P(\delta + e) \tag{10.42}$$

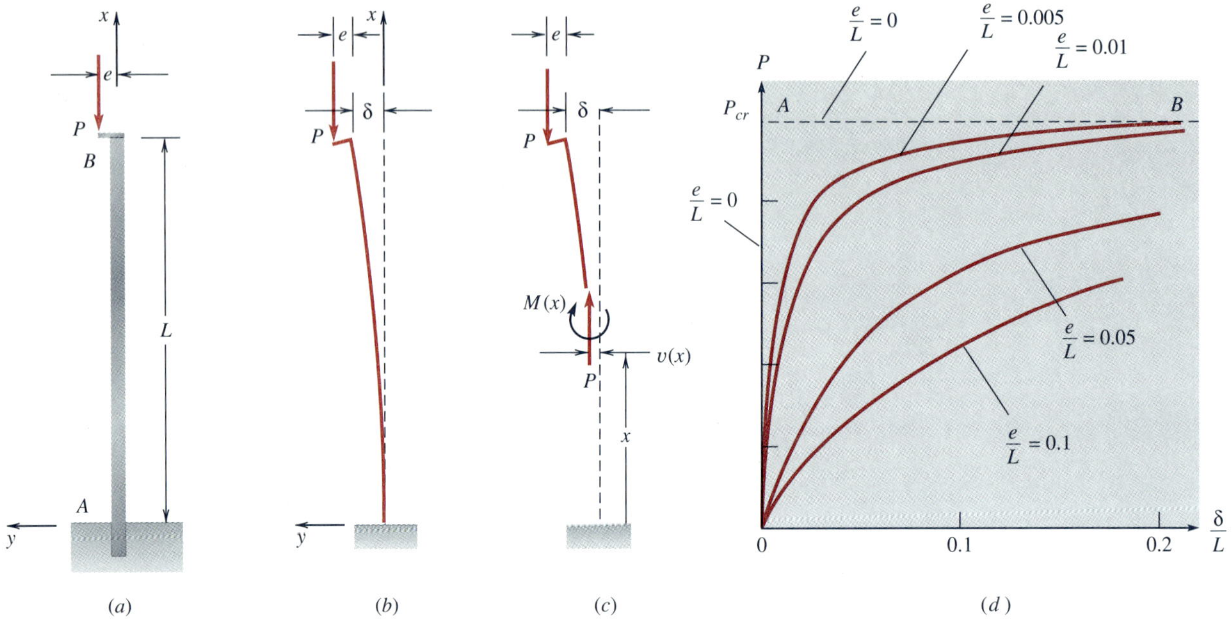

Figure 10.17 Eccentrically loaded column.

We note that for $e = 0$ (no load eccentricity), the differential equation, Eq. (10.42), takes the same form as Eq. (10.19) but with the constant $P(\delta + e)$ on the right side of the equation in place of $P\delta$. Therefore, the particular solution of Eq. (10.42) can be taken as

$$v_p = \delta + e \tag{10.43}$$

and the homogeneous solution is the same as that given in Eq. (10.8). We can write the general solution of Eq. (10.42) as

$$v(x) = c_1 \sin \lambda x + c_2 \cos \lambda x + \delta + e \tag{10.44}$$

where c_1 and c_2 are the two arbitrary integration constants.

The boundary conditions at the base ($x = 0$) are

$$x = 0: \qquad v = 0 \qquad \frac{dv}{dx} = 0 \tag{10.45}$$

Applying these conditions to the general solution, Eq. (10.44), we find

$$v(0) = 0 = c_2 + \delta + e$$

and

$$\frac{dv(0)}{dx} = 0 = \lambda c_1$$

Therefore

$$c_1 = 0 \qquad c_2 = -(\delta + e) \tag{10.46}$$

and Eq. (10.44) can be written as

$$v(x) = (\delta + e)(1 - \cos \lambda x) \tag{10.47}$$

Since δ is the value of the deflection at the top of the column, we have at $x = L$

$$v(L) = \delta = (\delta + e)(1 - \cos \lambda L)$$

or

$$\delta \cos \lambda L = e(1 - \cos \lambda L)$$

and finally

$$\delta = e\,(\sec \lambda L - 1) \tag{10.48}$$

Recalling that $\lambda^2 = P/(EI_z)$, we obtain the expression for the deflection δ as a function of the load P in the form

$$\delta = e\left[\sec\left(\sqrt{\frac{P}{EI_z}}\,L\right) - 1\right] \tag{10.49}$$

For any given value of e, as $\sqrt{P/(EI_z)}\,L \to \pi/2$, the deflection increases without bound. The curves given by Eq. (10.49) for the load P as a function of δ/L for small positive values of e/L are sketched in Fig. 10.17*d*. The case with zero eccentricity ($e = 0$) was studied in Sec. 10.4. In that case, we found that for

$$P \le P_{cr} = \frac{\pi^2}{4}\frac{EI_z}{L^2} \tag{10.50}$$

no transverse deflection occurred so that $\delta = 0$, that is, the solution corresponds to the vertical line from the origin to A in Fig. 10.17*d*. The horizontal line AB corresponds to the load $P = P_{cr}$. The curves for very small eccentricity e are similar to the $e = 0$ case but have a smoothly turning slope for increasing values of δ/L. For the case e greater than zero with P approaching P_{cr}, the transverse deflection increases rapidly.

EXAMPLE 10.7

An elastic column of length L and bending modulus EI_z is pinned at each end and subjected to a vertical load acting parallel to the centroidal axis but applied to the beam cross section at a distance e from the centroidal axis, as shown in Fig. 10.18*a*. We wish to find an expression for the deflection $v(x)$ caused by a given load P; note the nature of the supports at each end.

The deflected shape is shown in Fig. 10.18*b*. When we consider moment equilibrium of the segment of the column shown in Fig. 10.18*c*, we obtain the bending moment at section x

$$M(x) = -P(v + e) \tag{a}$$

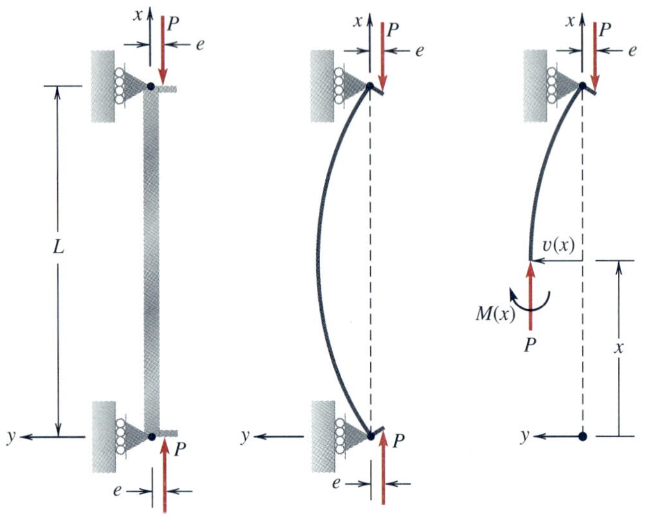

Figure 10.18 Example 10.7

Using Eqs. (6.6) and (*a*), we obtain the equation for $v(x)$

$$EI_z \frac{d^2v}{dx^2} + Pv = -Pe \tag{b}$$

which reduces to Eq. (10.2) for the case of $e = 0$ (no eccentricity of the load). The homogeneous form of Eq. (*b*) (obtained by dropping the $-Pe$ term on the right side of the equation) is the same as Eq. (10.2), so that we can use the homogeneous solution given in Eq. (10.8) and add to it the particular solution

$$v_p = -e \tag{c}$$

The general solution is of the form

$$v(x) = c_1 \sin \lambda x + c_2 \cos \lambda x - e \tag{d}$$

where $\lambda^2 = P/(EI_z)$.

The pinned support at $x = 0$ requires that $v = 0$, or by using Eq. (*d*),

$$v(0) = 0 = c_1 \cdot 0 + c_2 \cdot 1 - e$$

or

$$c_2 = e \tag{e}$$

At the other pinned end ($x = L$), $v = 0$, and by using Eq. (*d*) we find

$$v(L) = 0 = c_1 \sin \lambda L + e(\cos \lambda L - 1)$$

or

$$c_1 = e\frac{1 - \cos \lambda L}{\sin \lambda L} \tag{f}$$

By using the half-angle trigonometric formula for the tangent function, we find that Eq. (*f*) can be rewritten more simply as

$$c_1 = e \tan \frac{\lambda L}{2} \tag{g}$$

so that Eq. (*d*) becomes

$$v(x) = e \tan \frac{\lambda L}{2} \sin \lambda x + e\,(\cos \lambda x - 1) \tag{h}$$

For increasing values of P starting from zero, we can calculate $\lambda L = \sqrt{P/(EI_z)}\, L$ and obtain a deflection shape $v(x)$ from Eq. (*h*). However, if for given values of L and EI_z, P increases until λL approaches π, then the factor $\tan(\lambda L/2)$ in Eq. (*h*) will increase without bound. That is, the deflection will become large.

If we go back to the analysis of the pinned-pinned column without eccentricity of load ($e = 0$) studied in Sec. 10.4, we find that $\lambda L = \pi$ corresponds to the critical buckling load

$$P_{cr} = \pi^2 \frac{EI_z}{L^2} \tag{i}$$

Thus we note from Eq. (*h*) that for any value of the eccentricity e, the deflection shape $v(x)$ will become unbounded as

$$P \rightarrow P_{cr}$$

where P_{cr} given by Eq. (*i*) is the critical buckling load for the same column without eccentricity of loading.

Returning to Eq. (*h*), we see that the maximum deflection δ occurs at the midpoint of the column where $x = L/2$, so that

$$\begin{aligned} \delta &= e \tan \frac{\lambda L}{2} \sin \frac{\lambda L}{2} + e\left(\cos \frac{\lambda L}{2} - 1\right) \\ &= e\left(\sec \frac{\lambda L}{2} - 1\right) \\ &= e\left[\sec\left(\sqrt{\frac{P}{EI_z}}\frac{L}{2}\right) - 1\right] \end{aligned} \tag{j}$$

Comparing the result in Eq. (*j*) for the pinned-pinned column with the eccentrically applied load with the corresponding result obtained for the clamped-free column with the eccentrically applied load, Eq. (10.49), we see that they differ only by a factor of 2 in the argument of the secant function. Therefore, the load deflection curves based upon Eq. (*j*) will have the same general shape as those sketched in Fig. 10.17*d* with P_{cr} now given by Eq. (*i*).

10.7 Secant Formula for the Maximum Stress

In the stability analysis of columns without eccentric loads, we have calculated the critical load associated with a state of neutral equilibrium. It is not possible to relate the amplitude of the transverse deflection to the load when load $P > P_{cr}$ within the framework of our linear small-deflection analysis. Thus the important question of the amount of transverse bending stress in the column for the *postbuckling* load range cannot be examined easily.

However, some indication of the bending stresses caused by buckling can be found by considering the cases of columns with eccentric loading. Bending stresses are present in columns under eccentric loads for all values of P greater than zero, and these stresses intensify as the critical load is approached.

In this section, as a typical example of the bending stress analysis associated with buckling, we examine the bending stresses for a pinned-pinned column loaded by a pair of eccentric loads, as shown in Fig. 10.18*a*. In Eq. (*a*) of Example 10.7, an expression for the bending moment at any point along the column was derived, namely,

$$M(x) = -P(v + e) \tag{10.51}$$

The magnitude of the maximum bending moment at $x = L/2$ with $v = \delta$ is

$$M_{max} = P(\delta + e) \tag{10.52}$$

Using Eq. (*j*) of Example 10.7 and Eq. (10.52), we find

$$M_{max} = Pe \sec\left(\frac{L}{2}\sqrt{\frac{P}{EI_z}}\right) \tag{10.53}$$

As P increases from zero, M_{max} increases monotonically, as shown in Fig. 10.19. For any value of e, M_{max} becomes unbounded as $P \to P_{cr}$, as given by Eq. (10.17). We should note that in all cases of predicted unbounded deflection or stress, the original assumptions in our theory of small slopes, small elastic strains, etc., limit the range of applicability of the results from the analysis.

To find the maximum compressive normal stress in the column under load P, we add the axial stress due to load P to the maximum bending stress to get

$$\sigma_{max} = \frac{P}{A} + \frac{M_{max}\,c}{I_z} \tag{10.54}$$

where c is the distance from the neutral axis of bending to the concave side of the bent column, at which location the compressive bending stress occurs, and A is the cross-sectional area of the section.

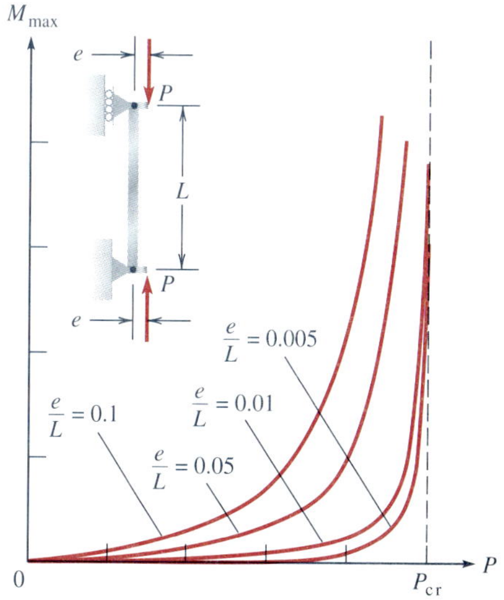

Figure 10.19 Maximum bending moment in an eccentrically loaded column as a function of applied load P.

Using Eq. (10.53), we can rewrite Eq. (10.54) in the form

$$\sigma_{\max} = \frac{P}{A} + \frac{Pec}{I_z}\sec\left(\frac{L}{2}\sqrt{\frac{P}{EI_z}}\right) \tag{10.55}$$

It is helpful for purposes of interpretation to rewrite Eq. (10.55) in terms of nondimensional parameters that characterize the column. First, we define the radius of gyration r of the column cross section by

$$r = \sqrt{\frac{I_z}{A}} \qquad I_z = Ar^2 \tag{10.56}$$

Therefore, the argument appearing in the secant function in Eq. (10.55) can be written as

$$\frac{L}{2}\sqrt{\frac{P}{EI_z}} = \frac{L}{2}\sqrt{\frac{P}{EAr^2}} = \frac{L}{2r}\sqrt{\frac{P}{EA}} \tag{10.57}$$

and Eq. (10.55) takes the form

$$\sigma_{\max} = \frac{P}{A}\left[1 + \frac{ec}{r^2}\sec\left(\frac{L}{2r}\sqrt{\frac{P}{EA}}\right)\right] \tag{10.58}$$

This equation is referred to as the *secant formula* for the maximum stress in an eccentrically loaded column. For no eccentricity of loading, that is, $e = 0$, Eq. (10.58) reduces to the expression for uniform axial stress in a column given by P/A.

Two nondimensional parameters appear in the secant formula, Eq. (10.58). These are the *eccentricity ratio* ER and the *slenderness ratio* SR, defined by

$$\text{ER} = \frac{ec}{r^2} \tag{10.59}$$

$$\text{SR} = \frac{L}{r} \tag{10.60}$$

To get some idea of the magnitude of these two geometric parameters, consider a column that is 10 ft long with a square 6-in × 6-in cross section. For this column

$$r^2 = \frac{I_z}{A} = \frac{1}{12}\frac{(6)(6)^3}{6^2}$$

or

$$r = \sqrt{3} = 1.732 \text{ in} \tag{10.61}$$

For the 6 × 6 cross section, the distance from the neutral axis to the concave side is $c = 3$ in, and we have

$$\text{ER} = \frac{ec}{r^2} = \frac{e(3)}{3} = e \tag{10.62}$$

Thus, if the location of the point of application of the load on the column were to vary from the centroid ($e = 0$) of the section to one edge where $e = 3$ in, the value of ER for this square cross section would lie in the range

$$0 \leq \mathrm{ER} \leq 3 \tag{10.63}$$

The slenderness ratio SR for this column is given by

$$\mathrm{SR} = \frac{L}{r} = \frac{10 \times 12}{\sqrt{3}} = 69.3 \tag{10.64}$$

To see how the secant formula, Eq. (10.58), can be used for the design of columns, we first return to the column with zero eccentricity. For a pinned-pinned column without loading eccentricity ($e = 0$), the critical load associated with buckling is, from Eq. (10.17),

$$P_{cr} = \pi^2 \frac{EI_z}{L^2}$$

The maximum stress in the column is the axial stress [$e = 0$ in Eq. (10.58)], and therefore the stress at buckling is

$$\sigma_{max} = \frac{P_{cr}}{A} = \pi^2 \frac{EI_z}{L^2 A} \tag{10.65}$$

or by using $I_z/A = r^2$,

$$\sigma_{max} = \frac{\pi^2 E}{L^2/r^2} = \frac{\pi^2 E}{(L/r)^2} \tag{10.66}$$

We note in Eq. (10.66) that the maximum stress at the buckling load is inversely proportional to the square of the slenderness ratio. If in Eq. (10.66) the value of E is specified, say, for steel at $E = 200$ GPa, then the maximum stress as a function of L/r can be plotted as shown in Fig. 10.20, as the curve marked Eq. (10.66).

If the yield stress of the same material is $\sigma_Y = 250$ MPa, then there is a value of L/r at which the maximum stress in the column at buckling will equal the yield stress. In Fig. 10.20 this value is indicated as $(L/r)^* = 88.86$. For values of $L/r > (L/r)^*$, the maximum stress at the critical load (with $e = 0$) is less than the yield stress. For example, if $L/r = 150$, the maximum stress in the column at buckling is 87.7 MPa.

On the other hand, if L/r is less than $(L/r)^*$, then the maximum stress at buckling predicted by Eq. (10.66) is greater than the yield stress. Since σ_{max} should not exceed σ_Y, we show a horizontal line $\sigma_{max} = \sigma_Y = 250$ MPa as the upper bound for columns with $L/r < (L/r)^*$.

From Fig. 10.20 the mode of failure of a column is seen to be dependent on the value of L/r. If a column is long, $L/r > (L/r)^*$, it will buckle at the critical buckling load, and the value of the stress at buckling is

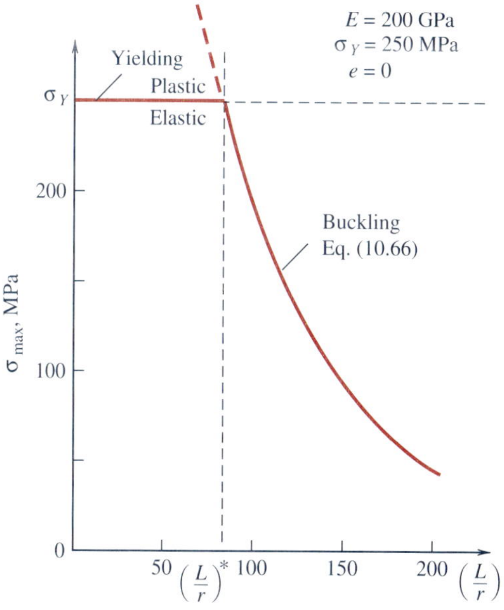

Figure 10.20 Maximum stress in a column at the buckling load as a function of the slenderness ratio L/r.

given by Eq. (10.66). If a column is short, that is, $L/r < (L/r)^*$, then the stress in the column at the critical buckling load will be greater than the yield stress and failure of the column occurs due to yielding and not due to buckling (Fig. 10.20). A plot similar to Fig. 10.20 can be drawn for columns with different values of E and σ_Y.

We now return to the case when a column is eccentrically loaded ($e \neq 0$) and the column begins to bend upon application of load P. When a column is eccentrically loaded, we use Eq. (10.58) to determine the maximum stress σ_{max} in the column as a function of the load P once the values of A, ec/r^2, L/r, and E for the column are specified. As the load P—or what is equivalent, the applied stress P/A—is increased, the maximum stress in the column will increase. The maximum stress, as shown in Fig. 10.21, becomes unbounded as $P/A \to P_{cr}/A$ since the argument of the secant function in Eq. (10.58) approaches $\pi/2$. At a particular value of P/A the maximum stress in the column will equal the yield stress of the material. In this way, we can find the maximum allowable value of P/A that can be applied to an eccentrically loaded column before a maximum stress is reached in the column equal to the yield stress (Fig. 10.21). For example, if $ec/r^2 = 0.1$ for the column in Fig. 10.21 with $E = 200$ GPa, $\sigma_Y = 250$ MPa, and $L/r = 100$, the maximum value of P/A is approximately 157 MPa.

Alternatively, we can use Eq. (10.58) directly to obtain design curves to predict the applied stress P/A to cause yielding in the column. For example, with $\sigma_Y = 250$ MPa and $E = 200$ GPa, we can plot curves for the values of P/A to give a maximum stress equal to the yield stress as

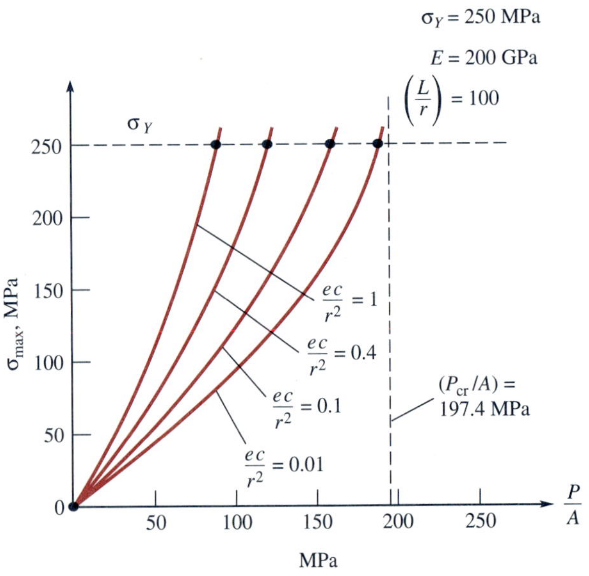

Figure 10.21 Maximum bending stress in an eccentrically loaded column as a function of applied stress P/A. $\sigma_Y = 250$ MPa, $E = 200$ GPa, and $L/r = 100$.

a function of L/r for a fixed value of ec/r^2. These curves are shown in Fig. 10.22. For example, if $ec/r^2 = 0.4$ and $L/r = 100$, we can read from Fig. 10.22 that $P/A \approx 117$ MPa; this value of applied stress P/A gives a maximum bending stress in the column of $\sigma_{max} = \sigma_Y = 250$ MPa; also see Fig. 10.21.

The curves shown in Fig. 10.22 for $e \neq 0$ are bounded on the right by the curve of $(L/2r)\sqrt{P/(EA)} = \pi/2$, or upon solving for P/A, by

$$\frac{P}{A} = \frac{\pi^2 E}{(L/r)^2} \tag{10.67}$$

corresponding to the limiting case of $e = 0$ given by Eq. (10.66). The curves are also bounded above by $P/A = \sigma_Y$.

The curves of Fig. 10.22 can be used for $e \neq 0$ to select either a value of the slenderness ratio L/r for a specified value of applied stress P/A or a value of the applied stress P/A for a specified value L/r with a fixed value of ec/r^2 to give a maximum stress of σ_Y. Similar design curves are used often in advanced courses in structural design.

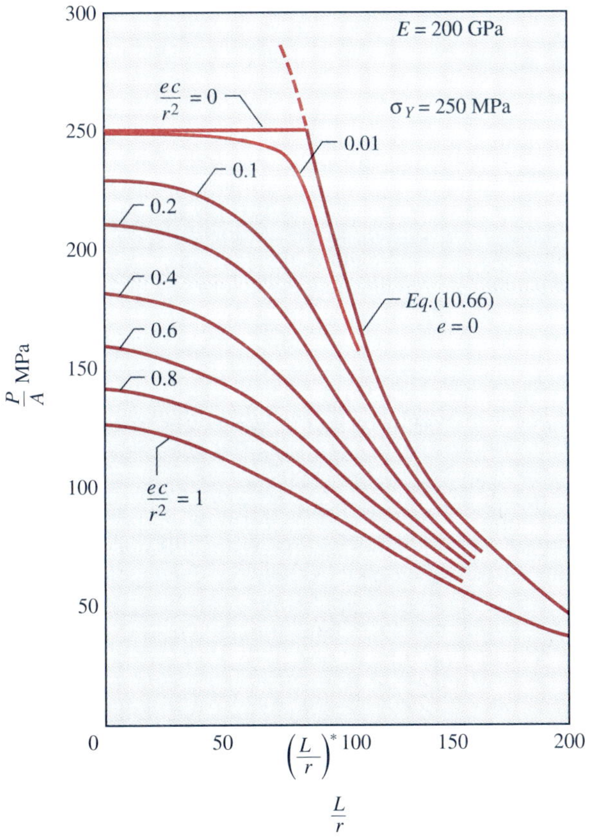

Figure 10.22 Plots of applied stress P/A versus slenderness ratio L/r for different values of ec/r^2 to give a maximum bending stress of $\sigma_{max} = \sigma_Y = 250$ MPa, $E = 200$ GPa.

10.8 Elastic Postbuckling Behavior

In the buckling analysis for columns as carried out in the previous sections, we have restricted ourselves to small deflections and slopes in the bending of the columns. We now wish to examine the behavior of columns after buckling has started. One of the important practical questions to consider is, does the column maintain its load-carrying capacity with increasing load? That is, can the column continue to carry a load after reaching the critical load? In order to begin to answer this question, it is necessary to remove the restriction of small deflections and include geometric nonlinearity. Thus in the moment-curvature relation, Eq. (6.1), the nonlinear form of the curvature, Eq. (6.4), is necessary and this leads to a nonlinear differential equation to be solved in order to analyze the postbuckling behavior of flexible columns.

To avoid undue mathematical complexity however and still learn something about the basic ideas of elastic postbuckling behavior for columns and for simple structures, we again turn to the rigid-bar-spring models of buckling in the following examples.

EXAMPLE 10.8

A pair of rigid bars *AB* and *BC* is pinned together at *B* and pinned to supports at *A* and *C*, as shown in Fig. 10.23*a*. The system is stabilized by attaching a linear spring with stiffness k to point *B*. As point *A* moves down under the application of load *P*, the spring at *B* will stretch or compress. We wish to find the critical load for buckling of the system if only small rotations of the bar are allowed. We then wish to drop the restriction of small rotations and obtain an expression for the load in terms of the rotation angle of the bars.

In Fig. 10.23*b* the bars are shown in a buckled configuration with both bars rotated by angle θ. Since the spring is extended by the amount $L \sin\theta$, there will be a force $kL\sin\theta$ acting at B' as shown. The bars are two-force members, and the compressive force in each bar is $P/\cos\theta$, which results in the force components at A' and C' acting as shown in Fig. 10.23*b*. Force equilibrium of the pin at B' yields

$$2P\tan\theta = kL\sin\theta \tag{a}$$

For the system to pass from its initially straight unbuckled state ($\theta = 0$) to the buckled state shown in Fig. 10.23*b* ($\theta \neq 0$), it must pass through small values of θ. For small values of θ, we have

$$\tan\theta \approx \theta \qquad \sin\theta \approx \theta$$

so that Eq. (*a*) can be written in the form

$$(2P - kL)\theta = 0 \tag{b}$$

To determine the critical buckling load, we conclude from Eq. (*b*) that

$$P_{cr} = \frac{kL}{2} \qquad \theta \neq 0 \tag{c}$$

We now wish to examine the response of this system in terms of its force-displacement characteristics for increasing values of the bar rotation angle θ. Returning to Eq. (*a*) for arbitrary θ and rewriting Eq. (*a*) by dividing both sides of the equation by P_{cr}, we find for the equilibrium states the following relation between the load and the angle of rotation:

$$\left(\frac{P/P_{cr}}{\cos\theta} - 1\right)\sin\theta = 0 \tag{d}$$

Equation (*d*) is satisfied if either of the two factors in Eq. (*d*) is zero. Since $\sin\theta$ is proportional to the lateral displacement $L\sin\theta$, and we are interested in the case of lateral displacement, that is, $\sin\theta \neq 0$, we focus attention on the case where the first factor is zero. That is,

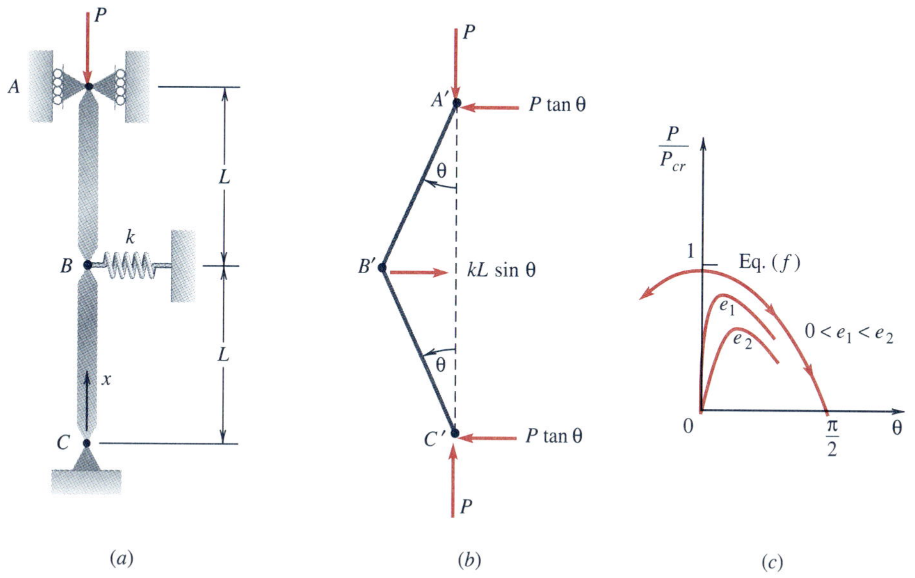

Figure 10.23 Example 10.8

$$\frac{P}{P_{cr}}\frac{1}{\cos\theta} = 1 \qquad (e)$$

where $\cos\theta \neq 0$. Solving Eq. (e) for P, we find

$$\frac{P}{P_{cr}} = \cos\theta \qquad 0 \le \theta < \frac{\pi}{2} \qquad (f)$$

or

$$P = 0 \qquad \theta = \frac{\pi}{2} \qquad (g)$$

A graph of the load–angular displacement curve representing Eq. (f) is given in Fig. 10.23c.

The prebuckling portion of the curve is the segment along the P/P_{cr} axis with $0 < P/P_{cr} < 1$ and $\theta = 0$. At $P/P_{cr} = 1$, there is a bifurcation into two curves, both satisfying equilibrium. One curve corresponds to $\theta = 0$ and $P > P_{cr}$. The other curve, given by Eq. (f), represents the buckling of the system to the right or left since the system is completely symmetric about $\theta = 0$. Examination of the buckling curve in Fig. 10.23c shows that once the system bifurcates, the system loses its load-carrying capacity for any value of $\theta > 0$. For example, for $\theta = 45°$, in order to maintain equilibrium, the load must be reduced to $0.707P_{cr}$ or $0.707kL/2$. Finally at $\theta = 90°$, the bars cannot support a vertical load, and we must reduce P to zero for equilibrium.

If we were to redo this example with an eccentric end load, we would expect, based upon previous results, to find the load-displacement curves for different levels of eccentricity as sketched in Fig. 10.23c and labeled e_1 and e_2. As the parameter $e \to 0$, that is, the eccentricity or imperfection disappears, we obtain the Eq. (f) curve in the limit.

An important observation for any system without an imperfection that shows a postbuckling curve of force displacement similar to the one shown in Fig. 10.23c with $P/P_{cr} < 1$ with $\theta > 0$ is that the maximum load that the system can carry as the load is increasing from zero *in the presence of an imperfection* can be significantly less than P_{cr}. In fact, in more complicated systems, experimental testing shows that even very small imperfections can result in experimental buckling loads that are one-half and sometimes even one-third of the theoretical buckling load calculated for the system without imperfections.

In Example 10.8 we investigated the effect of the nonlinear geometry of deformation on the buckling characteristics of a simple two-bar model in Fig. 10.23. In general, the effect of geometric nonlinearities on the buckling characteristics of real structures is to complicate the analysis. In the next example we wish to discuss again a simplified model whose postbuckling behavior is qualitatively similar to that of many structures, such as columns, plates, and shells. In this example we will follow part of the discussion in Crandall, Dahl, and Lardner, sec. 9.5.[1]

EXAMPLE 10.9

In this example we wish to investigate the stability of the simple model problem shown in Fig. 10.24*a*. This model is similar to the one we discussed in Example 10.3 except that now we assume that the two springs connected to the bar at the top are nonlinear; the pin at B is frictionless. The force in each spring is given by

$$f = kx(1 + \beta x^2/L^2) \qquad (a)$$

where β is a parameter that fixes the nature of the nonlinearity in the spring. When $\beta > 0$, the spring is said to be a *stiffening spring;* when $\beta < 0$, the spring is said to be a *softening spring.* A plot of the nondimensional force $f/(kL)$ versus x/L is shown in Fig. 10.24*c* for $\beta = 10, 0, -10$. We are interested in the behavior of the system as we increase the load from $P = 0$.

If we consider moment equilibrium about point B of the bar for small values of the deflection x at the top (Fig. 10.24*b*), we have

$$Px - 2kLx\left(1 + \frac{\beta x^2}{L^2}\right) = 0 \qquad (b)$$

Therefore, the relations between the load P and the deflection x for possible equilibrium states from Eq. (*b*) are

$$\begin{aligned} x &= 0 \qquad P \neq 0 \\ x &\neq 0 \qquad P = 2kL(1 + \beta x^2/L^2) \end{aligned} \qquad (c)$$

The load-deflection curves corresponding to the results in Eqs. (*c*) are shown in Fig. 10.25. For small values of P, the only stable equilibrium position of the bar is along AB (in Fig. 10.25) where

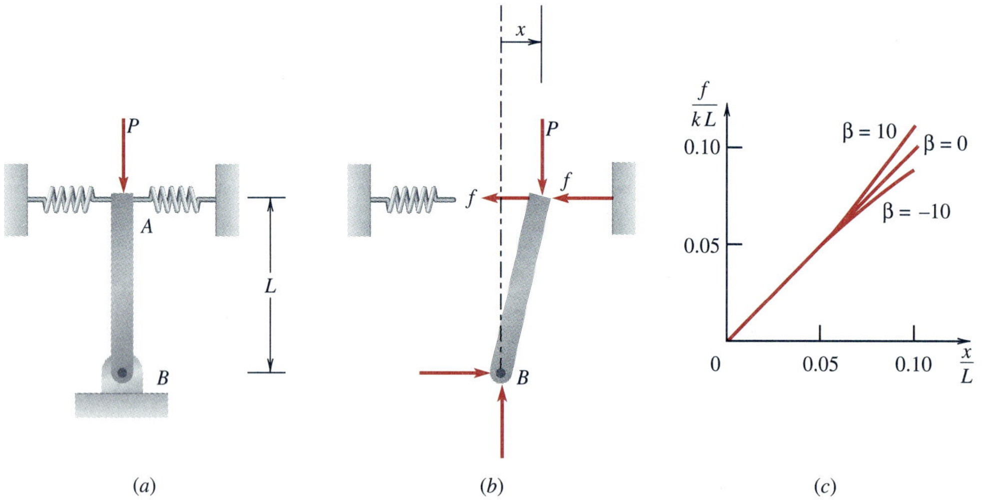

Figure 10.24 Example 10.9

[1]S. H. Crandall, N. C. Dahl, and T. J. Lardner (eds.), *An Introduction to the Mechanics of Solids*, 2d ed., McGraw Hill, New York, 1978.

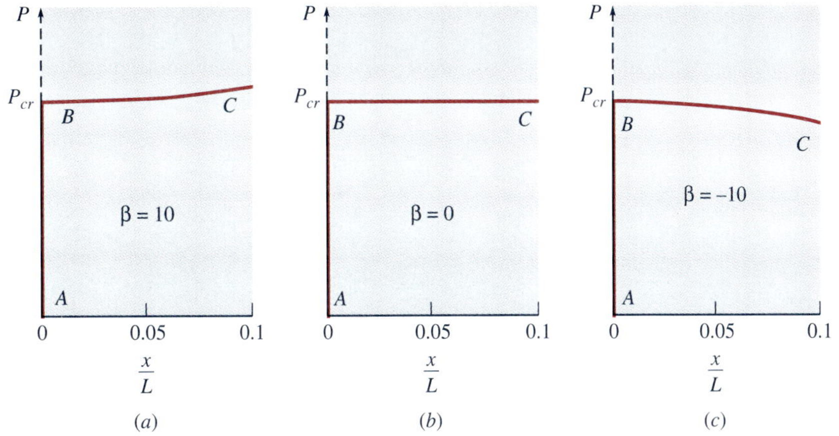

Figure 10.25 Example 10.9: Ideal postbuckling curves.

$x = 0$. When the load reaches the critical value [obtained from Eqs. (*c*) when x is small] $P_{cr} = 2kL$, transverse deflection can occur. Point B is the bifurcation point at $P = P_{cr}$.

If $\beta > 0$, the portion of curve BC represents stable equilibrium positions. If we are in equilibrium at a point on curve BC and disturb the system slightly, a larger value of P is required for equilibrium. As a consequence, unbalanced forces in the new position will cause the bar to move back to its undisturbed equilibrium position. Therefore, the solid line BC in Fig. 10.25*a* represents the locus of stable equilibrium positions. In Fig. 10.25*b*, curve BC represents the locus of neutral equilibrium positions while in Fig. 10.25*c* curve BC represents the locus of unstable equilibrium positions.

If the load is now moved off the centroidal axis of the bar, as shown in Fig. 10.26*a*, by an amount ϵL, we can determine the load-deflection curves for the bar and relate them to the curves of the ideal case when $\epsilon = 0$. The bar in a slightly displaced position is shown in Fig. 10.26*b*, and equilibrium of the free-body diagram in Fig. 10.26*b* gives

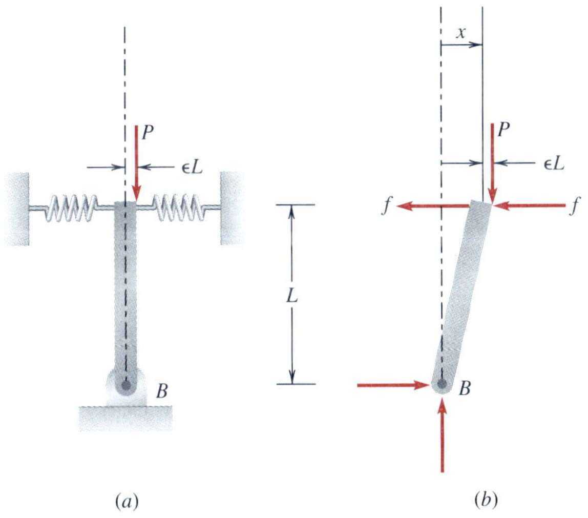

Figure 10.26 Example 10.9: Eccentric load on column.

$$P(x + \epsilon L) = 2kLx\left(1 + \frac{\beta x^2}{L^2}\right) \qquad (d)$$

The relation between P and x for equilibrium given by Eq. (*d*) is plotted for three values of the imperfection parameter ϵ for β equal to 10, 0, and -10 in Fig. 10.27. We see from these plots that for small values of ϵ, the curves approach the postbuckling curves BC with $\epsilon = 0$ in Fig. 10.25.

When $\beta < 0$, the curves of Eq. (*d*) as shown in Fig. 10.27*c* rise to a maximum value of P and decrease. We show this again in Fig. 10.28*a*, where $0MN$ represents a curve from Fig. 10.27*c*. The corresponding equilibrium positions are stable along $0M$ and unstable along MN. In this case there is a maximum load P_{max}. When $P = P_{max}$, a disturbance will cause the bar to move away from the equilibrium position and large deflections will result. Thus when $\beta < 0$, the significant buckling load is not P_{cr} determined from the analysis of the ideal system with $\epsilon =$

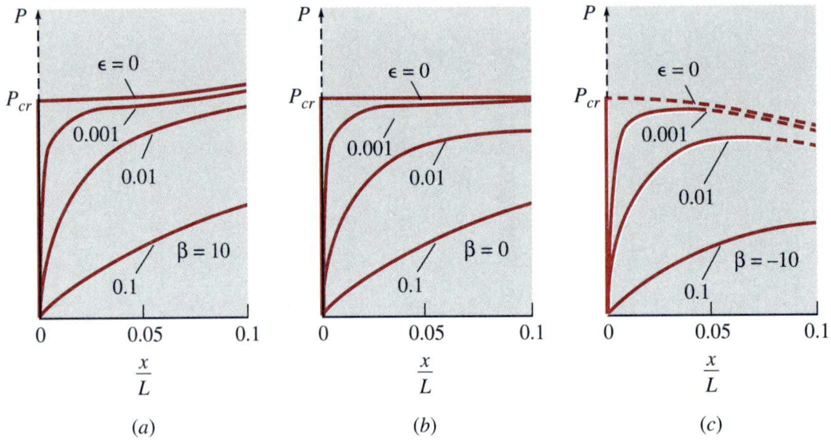

Figure 10.27 Example 10.9: Effect of imperfection ϵ on postbuckling behavior.

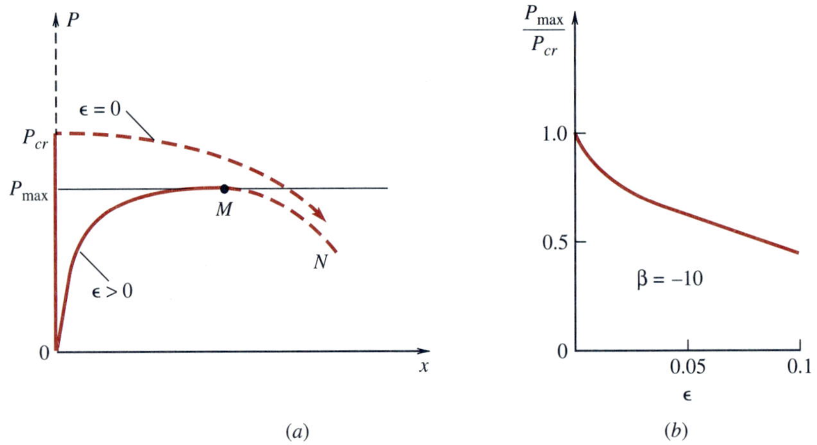

Figure 10.28 Example 10.9: The maximum load for softening nonlinearity (β = −10) depends on the magnitude of the imperfection.

0, but is P_{max} determined from the imperfect system—and it is always less than P_{cr}.

The value of P_{max} depends on the magnitude of the imperfection parameter ϵ, and for this case we can plot P_{max}/P_{cr} as a function of ϵ, as shown in Fig. 10.28*b* (see Probs. 10.8-3 and 10.8-4). We note that P_{max} approaches P_{cr} as ϵ approaches zero but that P_{max} is extremely sensitive to the value of the small imperfection ϵ. Structures whose postbuckling behavior in the ideal case can be modeled by a softening nonlinearity are said to be *imperfection-sensitive*.

As another illustration of the loss of the load-carrying capacity of a structure, we consider the "snap-through" behavior of a structure in the next example.

EXAMPLE 10.10

A pair of identical elastic bars is pinned to supports at A and C and pinned to a frictionless roller block at B that is restrained to move vertically, as shown in Fig. 10.29*a*. A load P is applied vertically to the roller block at B. We neglect friction at all pins and the weight of the bars; in addition, the bars have an elastic modulus E and cross-sectional area A. When the load $P = 0$, the bars make an angle ϕ_0 with the horizontal, as shown in Fig. 10.29*a*, and therefore the initial length of each bar is $L/\cos \phi_0$.

If the angle ϕ_0 is small, we wish to find an expression for P as a function of angle ϕ (Fig. 10.29*b*) that gives the equilibrium position of the bars. We assume that the axial force in the bars is not large enough to cause flexural buckling or yielding in the bars.

A free-body diagram for bar AB' and the pin at B' for an equilibrium configuration described by angle ϕ is shown in Fig. 10.29*b*. Vertical equilibrium gives

$$2F \sin \phi = P \tag{a}$$

where F is the value of the force in the bars.

A second relation for F in terms of ϕ follows by considering the shortening δ in bar AB due to the compressive force F, namely,

$$\delta = \frac{F}{EA} \frac{L}{\cos \phi_0} \tag{b}$$

The geometry of the deformed structure shown in Fig. 10.29*b* requires that

$$AB' \cos \phi = L \tag{c}$$

However, in view of the shortening,

$$AB' = AB - \delta \tag{d}$$

or

$$\frac{L}{\cos \phi} = \frac{L}{\cos \phi_0} - \delta \tag{e}$$

If we solve Eq. (*b*) for F, we have

$$F = \frac{EA}{L} (\cos \phi_0)\, \delta \tag{f}$$

and solving Eq. (*e*) for δ and substituting into Eq. (*f*), we find

$$F = EA\left(1 - \frac{\cos \phi_0}{\cos \phi}\right) \tag{g}$$

We can obtain P as a function of the bar angle ϕ by making use of Eqs. (*a*) and (*g*) to write

$$P = 2EA \sin \phi\left(1 - \frac{\cos \phi_0}{\cos \phi}\right) \tag{h}$$

or

$$P = 2EA \tan \phi(\cos \phi - \cos \phi_0) \tag{i}$$

Since ϕ and ϕ_0 are restricted to small angles, Eq. (*i*) can be written upon use of the series expansions for the trigonometric functions for small angles in the form

$$P = 2EA\left(\phi + \frac{\phi^3}{3} \cdots\right)\left(1 - \frac{\phi^2}{2} \cdots - 1 + \frac{\phi_0^2}{2} \cdots\right) \tag{j}$$

Retaining product terms of the angles up to order 3, we find

$$P = EA\, \phi(\phi_0^2 - \phi^2) \tag{k}$$

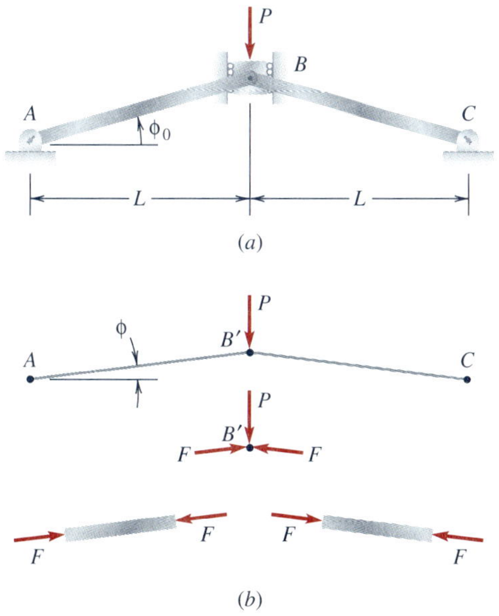

Figure 10.29 Example 10.10

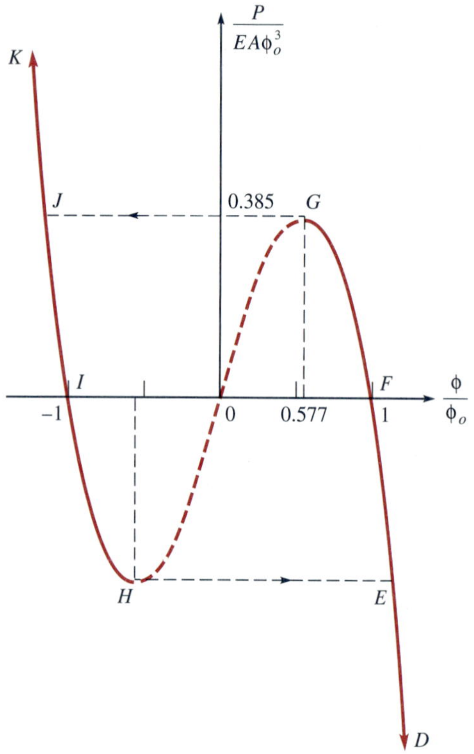

Figure 10.30 Example 10.10

Equation (*k*) is the required load deflection expression for the system.

We see from Eq. (*k*) that P is a cubic function of the bar rotation angle ϕ. The cubic function on the right side of Eq. (*k*) is equal to zero at three angles, namely,

$$\phi = \phi_0 \qquad \phi = 0 \qquad \phi = -\phi_0 \tag{l}$$

These angles correspond to possible equilibrium positions with $P = 0$, that is, no external load. The whole range of equilibrium states for this bar system is shown in Fig. 10.30 in which we plot $P/(EA\phi_0^3)$ versus ϕ/ϕ_0. The points on the curve corresponding to $P = 0$ are F, 0, and I. Point F on the curve corresponds to the state shown in Fig. 10.29*a* with $P = 0$ and $\phi = \phi_0$. At point F, if load P is directed upward, $P < 0$, the equilibrium states are plotted as the locus of points FED on the curve. We note that the system becomes stiffer in the sense that the slope of the curve increases in magnitude with increasing values of the angle.

The more interesting case, however, occurs for downward values of the load, $P > 0$. As the load is increased from point F, angle ϕ decreases over the range FG until it reaches the value at G where $dP/d\phi = 0$ at $\phi = \phi_0/\sqrt{3}$. At this point the system has lost its load-carrying capacity.

The states corresponding to the segment $G0H$ of the curve in Fig. 10.30 are in equilibrium, but are not stable, as can be seen, e.g., from the special case at point 0 where $\phi = 0$ and $P = 0$. In this case we have a pair of compressed bars in line with no external load acting; this is definitely an unstable state.

The equilibrium state at point I in Fig. 10.30 corresponds to an unloaded system in a configuration geometrically similar to the initial state at F but located below the supports at angle $-\phi_0$. From this state at point I, if the downward loading is continued over the segment IJK, the system becomes stiffer.

If an experiment were conducted starting at $P = 0$ and $\phi = \phi_0$ (at point F in Fig. 10.30) in which the downward load P were increased slowly, when the system reached the state at point G ($P_G = 0.385EA\phi_0^3$) and P increased to a value slightly above P_G, no "local" equilibrium state would be possible. Therefore we expect that the bars would accelerate downward or "jump" and vibrate about an equilibrium state just above point J on the curve in Fig. 10.30. This instability phenomenon is called *snap-through buckling*. Additional downward loading would result in stable states along segment JK in Fig. 10.30. If now the downward load were reduced through point I, the snap-through behavior would be repeated for upward loading ($P < 0$) at point H, and a "jump" to point E on the FD portion of the curve would occur.

The loss of load-carrying capacity in the snap-through buckling case of Example 10.10 takes a different form from those studied in other examples in this chapter. For example, for the straight column without eccentricity of loading considered in Sec. 10.4, the basic equilibrium state prior to buckling was a state of uniform axial compression. In the stability analysis for each load P, nearby equilibrium states involving bending were checked for possible neutral equilibrium states. Not until a certain load level was reached at P_{cr} could one of these neutral states be sustained. For values of $P < P_{cr}$, the restoring action dominated the upsetting action, and the column returned to the primary compressive state. The search for the nearby states was restricted to small-deflection shapes since we were interested only in the initiation of buckling as a departure from the straight vertical compressed column. Thus the differential equation governing the buckling analysis was linear, and a complete mathematical solution was comparatively simple. At $P = P_{cr}$, we found that more than one state of equilibrium was possible: the straight column without bending and the buckled shape involving bending. This phenomenon was called *bifurcation buckling*.

If we restrict the angular displacement in the snap-through problem of Example 10.10 to small departures from the initial state at point F in Fig. 10.30, no snap-through buckling is predicted. In general, it is only when the changing geometry of a structure is allowed to enter into the analysis that particular kinds of snap-through buckling mechanisms are revealed.

10.9 Concluding Remarks

In this chapter we have discussed the stability of simple structures under compressive loads. We have found that in general for structures without imperfections there is a critical load below which the structure is stable and above which the structure is unstable if exposed to small disturbances from an equilibrium state. The critical load corresponds to a state of neutral equilibrium at which large deflections may occur. Large deflections in most structures are synonymous with failure of the structure.

In the case of elastic columns under compressive loads, we found that the critical or buckling load of the column depended on the nature of the supports at the ends of the column (Fig. 10.16).

In practice, we must be aware that members under compressive forces may buckle, and for this reason we consider buckling and instability as a potential mode of failure of a structure. This is especially true of lightweight structures such as thin-walled plates and shells.

If a structure is loaded by an offset eccentric load, we found from investigations of simple model structures that if the postbuckling character of the perfect structure was imperfection-sensitive, then the load to cause buckling could be less than the critical buckling load. For this reason, it is important to have an analysis of the postbuckling characteristics of a structure. There are many interesting questions regarding buckling that are important in engineering practice and continue to be studied.

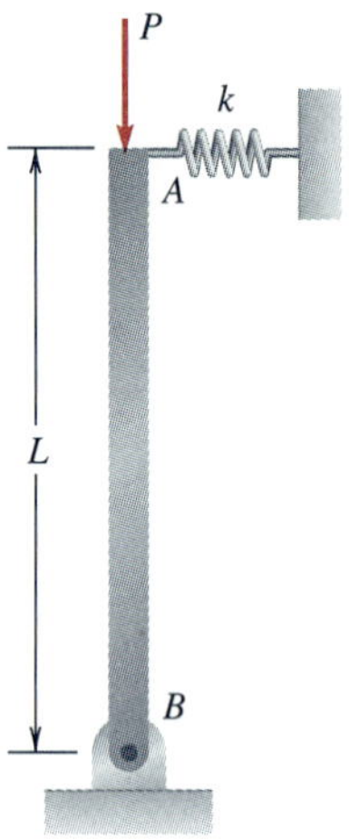

Fig. P10.3-1

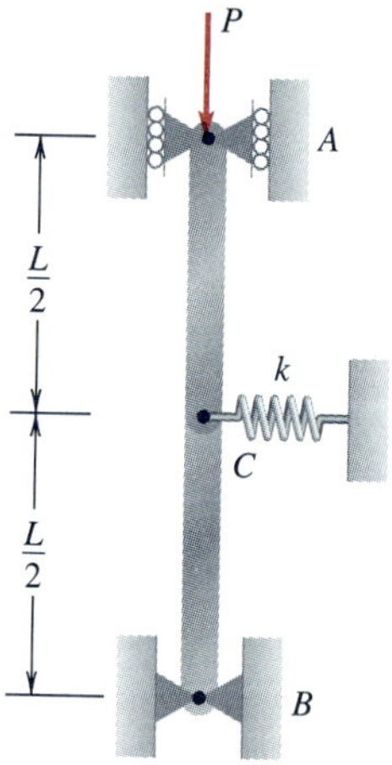

Fig. P10.3-2

PROBLEMS

10.3-1 A rigid bar in a vertical position is pinned at its base at B and is stabilized by a linear spring of stiffness k at the top A, as shown in Fig. P10.3-1. If a vertical load P is applied at A, find the critical value of P for which the bar ceases to be stable.

10.3-2 A pair of rigid bars AC and CB is pinned at A and B, pinned together at C, and stabilized at C by a linear spring, as shown in Fig. P10.3-2. If an axial load P is applied at A, find the critical value of P for which the bar configuration ceases to be stable.

10.3-3 For the rigid-bar-spring model of Prob. 10.3-2:

(*a*) Find the value of the spring constant k_e such that the bar system will have the same midspan stiffness as a simply supported flexible beam of length L. (The stiffness k is defined by $F = k\delta$, where F is the midspan transverse force and δ is the midspan transverse deflection.)

(*b*) Combine the result from Prob. 10.3-2 with the result from part (*a*) to obtain the result

$$P_{cr} = \frac{12EI}{L^2}$$

as an estimate for the critical load for a flexible column.

10.3-4 Consider Example 10.1, Fig. 10.6. To allow more flexibility in the rigid-bar-spring model for the cantilever beam of Fig. 10.6*a*, a two-bar model, stabilized at points B and C by spiral springs, is taken as shown in Fig. P10.3-4.

(*a*) Find the critical value of P for which the system of bars ceases to be stable in the vertical configuration.

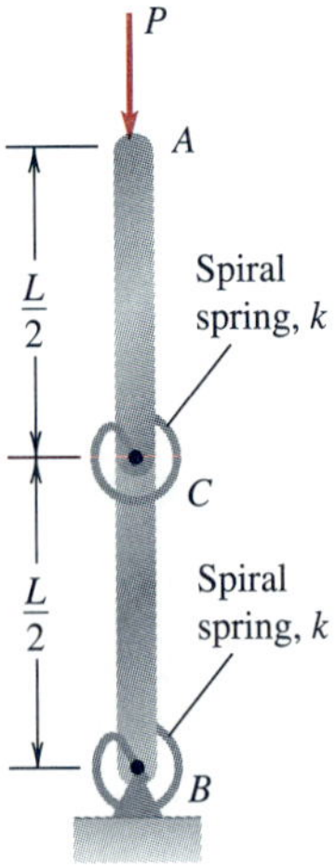

Fig. P10.3-4

(*b*) Compare the result with the result in Eq. (*g*) in Example 10.1. The spiral springs at *B* and *C* have spring constants *k*.

10.3-5 For the bar system shown in Fig. P10.3-4 find the stiffness of the system at *A*, that is, *k*, where $F = k\delta$ and *F* is a horizontal force applied at *A* and δ is the horizontal (small) deflection due to *F* with $P = 0$. Find the value of the spring constant k_e such that the bar system in Fig. P10.3-4 will have the same stiffness at *A* as the corresponding flexible cantilever beam shown in Fig. 10.6*a*.

10.3-6 Combine the results of Probs. 10.3-4 and 10.3-5 to calculate an estimate for the critical load for the flexible cantilever column based on the model of Fig. P10.3-4. Compare your result with the result given in Eq. (*m*) of Example 10.1 for a model with only one rigid bar.

10.3-7 Bar *AB* is stabilized in the vertical position by a system of linear springs, as shown in Fig. P10.3-7. If a pair of vertical loads of magnitude *P* is applied so as to remain vertical, then find the critical value of *P* for which the system ceases to be stable in the vertical position. Consider only small deflections.

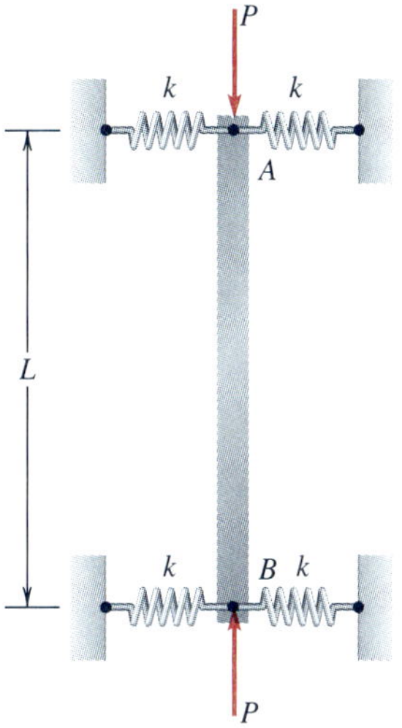

Fig. P10.3-7

10.3-8 A rigid bar *ABC* is stabilized at its midpoint by a linear spiral spring, as shown in Fig. P10.3-8. If a pair of forces of magnitude *P* is applied at the ends and remains vertical, then find the critical value of *P* for which the system ceases to be stable.

10.3-9 A rigid bar *BD* carries a vertical load *P* and is stabilized by an elastic steel shaft *ABC*, as shown in Fig. P10.3-9. Derive an expression for the critical load *P* in terms of the diameter *d* of the shaft. Take $G = 80$ GPa. Compare P_{cr} for the diameter choices (*a*) 10 mm and (*b*) 20 mm. The elastic shaft is built in at sections *A* and *C* and is supported by a frictionless bearing at *B*.

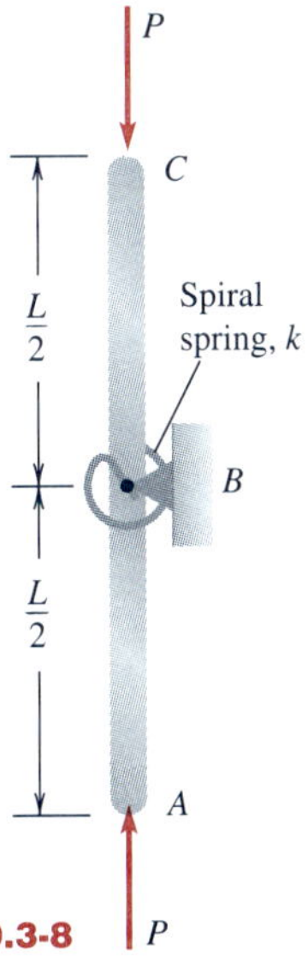

Fig. P10.3-8

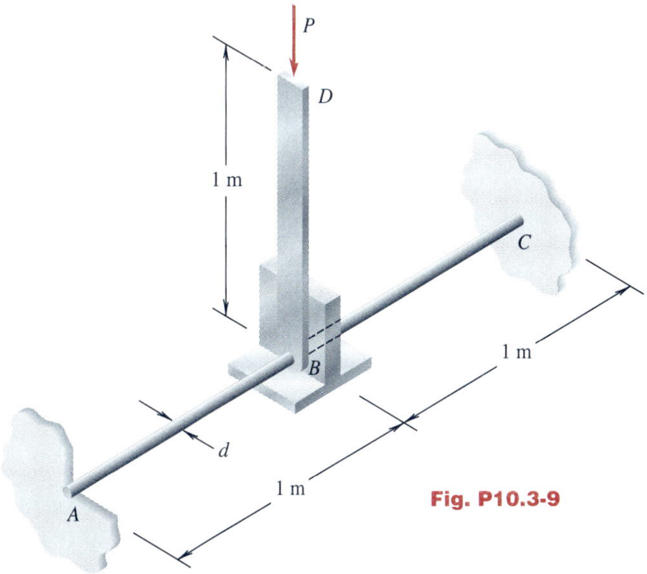

Fig. P10.3-9

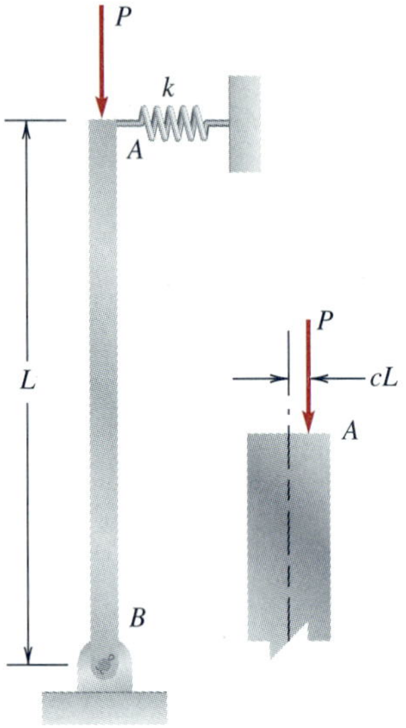

Fig. P10.3-11

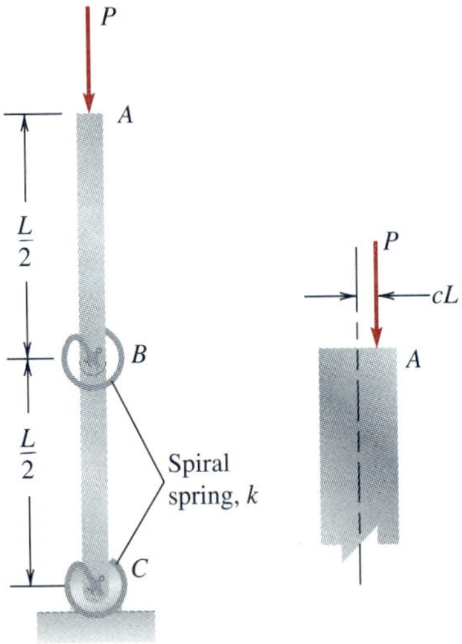

Fig. P10.3-12

10.3-10 For the bar shaft system in Prob. 10.3-9, what shaft diameter should be chosen if a given design requires that $P_{cr} = 500$ N?

10.3-11 A rigid-bar-spring model for a column carrying an axial load P which is offset by a distance cL from the centroid of the section is shown in Fig. P10.3-11. Carry out an analysis similar to the one presented in Example 10.3, and find results that correspond to Eqs. (*d*) and (*e*).

10.3-12 A two-bar-spring model for a column carrying an axial load P which is offset from the centroid of the section by a distance cL is shown in Fig. P10.3-12. Find an expression for the horizontal deflection of A as a function of P, c, L, and k.

10.3-13 A rigid bar pinned at end A carries a vertical load P at end B and is embedded in an elastic foundation, as shown in Fig. P10.3-13*a*. For small displacements from the vertical, the foundation exerts a force $dF = kx\, ds$ on a segment ds of the bar which has displaced an amount x, as shown in Fig. P10.3-13*b*. Find the critical load P_{cr} for the bar in terms of k and L.

10.3-14 In a model of a machine, two identical bars AB and CD are pinned at their bases at A and C and carry identical vertical loads P at the top ends, as shown in Fig. P10.3-14. The bars are stabilized by linear springs as shown.

(*a*) Give small angular disturbances θ_1 and θ_2 (clockwise) to bars AB and CD, and show that the moment equilibrium equations about points A and C for the bars take the form

$$(2kL - P)\delta_1 - kL\delta_2 = 0$$

$$-kL\,\delta_1 + (2kL - P)\delta_2 = 0$$

where $\delta_1 = L\theta_1$ and $\delta_2 = L\theta_2$.

(*b*) Solve for δ_2 in the first equation and substitute into the second to obtain

$$(3kL - P)(kL - P)\frac{\delta_1}{kL} = 0$$

and conclude that the critical values of the load P are $P_{cr_1} = kL$ and $P_{cr_2} = 3kL$.

(*c*) Return to the equations in part (*a*), and interpret the neutral equilibrium states associated with the critical loads found in part (*b*).

10.3-15 Consider Example 10.1. Instead of matching the deflection value at the top of the bar due to a transverse load Q to the corresponding deflection of a flexible column in order to calculate the equivalent stiffness k, match the slopes at the top for the two cases. Show that for this choice $k = 2EI/L$ and $P_{cr} = 2EI/L^2$.

10.4-1 Find the critical axial load for a wooden yardstick with a 1-in $\times$ $\frac{1}{8}$-in cross section. Take $E = 1800$ ksi, and assume that both ends are pinned and that the column can buckle in any direction.

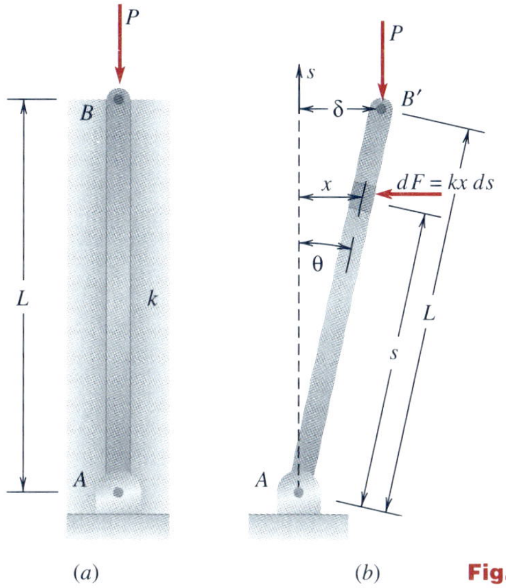

Fig. P10.3-13

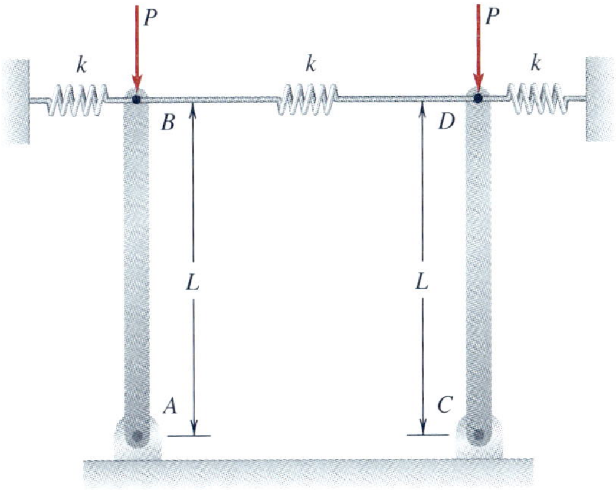

Fig. P10.3-14

10.4-2 Find the critical axial load for a hollow metal circular tube pinned at both ends with an inside and an outside diameter of 6 and 8 mm and a length of 915 mm. Take $E = 100$ GPa.

10.4-3 A steel rod of solid circular cross section has a diameter of 13 mm, a length of 1.02 m, and an elastic modulus of 200 GPa. Find the critical axial load if it is pinned at both ends.

10.4-4

(*a*) Use the table in App. E for common piping sections to calculate the critical buckling loads for pinned-pinned pipes (standard weight) with nominal diameters of 1, 3, 6, and 12 in in terms of the length L and the elastic modulus E.

(*b*) For each pipe calculate the diameter of a pipe of solid section which has the same cross-sectional area as the corresponding hollow pipe.

(*c*) For each pipe calculate the ratio of the critical load for the hollow pipe to the critical load for the solid pipe.

10.4-5 An aluminum rod of solid circular section and elastic modulus of 72 GPa is to have the same buckling load as a steel rod of solid circular section with diameter 15 mm and elastic modulus of 200 GPa.

(*a*) Find the diameter of the aluminum rod.

(*b*) Compare the weights of the two rods for a length of 1 m. Both rods are pinned at their ends and have the same length (see Table F.3).

10.4-6 An 8-ft pinned-pinned column is to be composed of four 2-in × 6-in structural lumber members (see App. D).

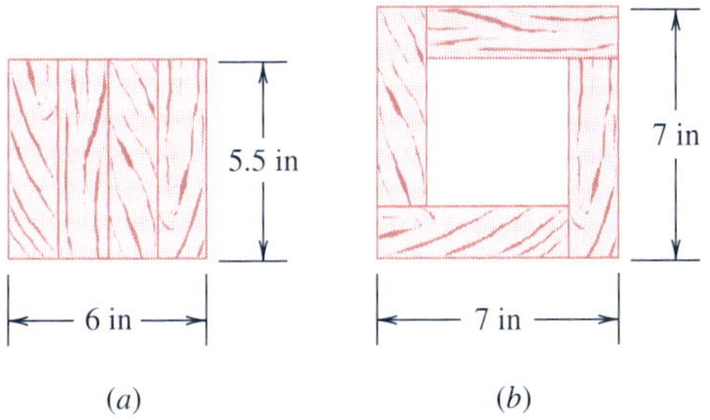

Fig. P10.4-6

(*a*) For the two cross-sectional configurations shown in Fig. P10.4-6*a* and *b*, find the critical loads in each case. Take $E = 1800$ ksi.
(*b*) By what percentage does the larger load exceed the smaller?

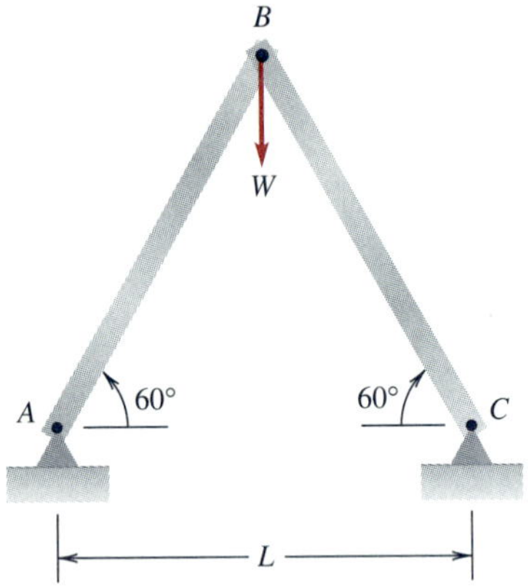

Fig. P10.4-7

10.4-7 A weight W is supported by a truss ABC, as shown in Fig. P10.4-7. If members AB and BC are 6-in-nominal-diameter steel pipes (standard weight), then find the allowable weight W_a if the factor of safety against buckling is to be 2.5. Take $L = 15$ ft and $E = 30{,}000$ ksi. Assume that motion perpendicular to plane ABC is restricted.

10.4-8 Consider Example 10.4, Fig. 10.11. If the factor of safety against buckling is taken as 3, find the allowable length of the columns in cases (*a*) and (*b*). All other parameters remain the same.

10.4-9 Consider Example 10.4, Fig. 10.11. Solve for the two cases again, but use 2-in × 6-in lumber (see App. D) instead of 2 × 4s. All other parameters remain the same.

10.4-10 A 10-ft length of a steel I beam, S6 × 17.25, is to be used as a column. If the boundary conditions are assumed to be pinned-pinned, find the buckling load for the column. Take $E = 30{,}000$ ksi.

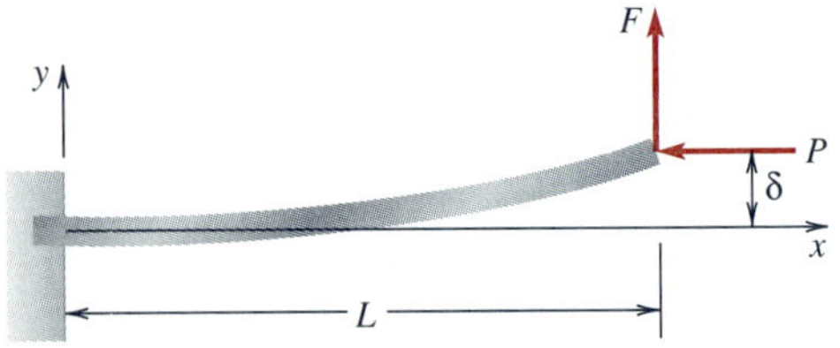

Fig. P10.4-11

10.4-11 If the clamped-free column shown in Fig. 10.12*a* and analyzed in Sec. 10.4 is loaded by the small transverse load F as shown in Fig. P10.4-11, obtain an expression for the tip deflection δ by means of an analysis similar to that presented in Sec. 10.4. Plot δ versus P for a fixed value of F.

10.4-12 A variation on the problem of the stability of a clamped-free column under axial load, as studied in Sec. 10.4, Fig. 10.12*a*, is shown in Fig. P10.4-12 where an end moment M_0 has been added. Use the methods of Sec. 10.4 to find an expression for the tip deflection δ.

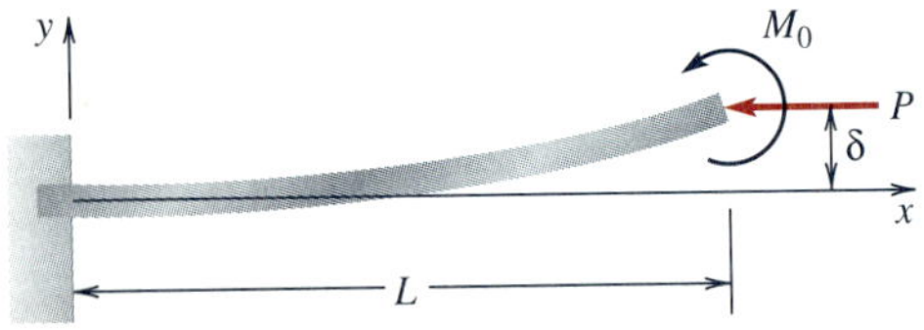

Fig. P10.4-12

10.4-13 The simply supported beam shown in Fig. P10.4-13*a* is subjected to an axial load P and the transverse load F at midspan.
(*a*) Using symmetry and equilibrium of the beam segment shown in Fig. P10.4-13*b*, show that

$$EIv'' + Pv = -\frac{Fx}{2} \qquad (a)$$

where the relation $M = EIv''$ has been used.
(*b*) Show that the general solution of Eq. (*a*) is

$$v(x) = -\frac{Fx}{2P} + c_1 \cos \lambda x + c_2 \sin \lambda x \qquad (b)$$

where $\lambda = \sqrt{P/(EI)}$. Apply the conditions $x = 0$, $v = 0$ and $x = L/2$, $v' = 0$ to get the deflection expression

$$v(x) = \frac{F}{2P}\left[\frac{\sin \lambda x}{\lambda \cos (\lambda L/2)} - x\right] \qquad (c)$$

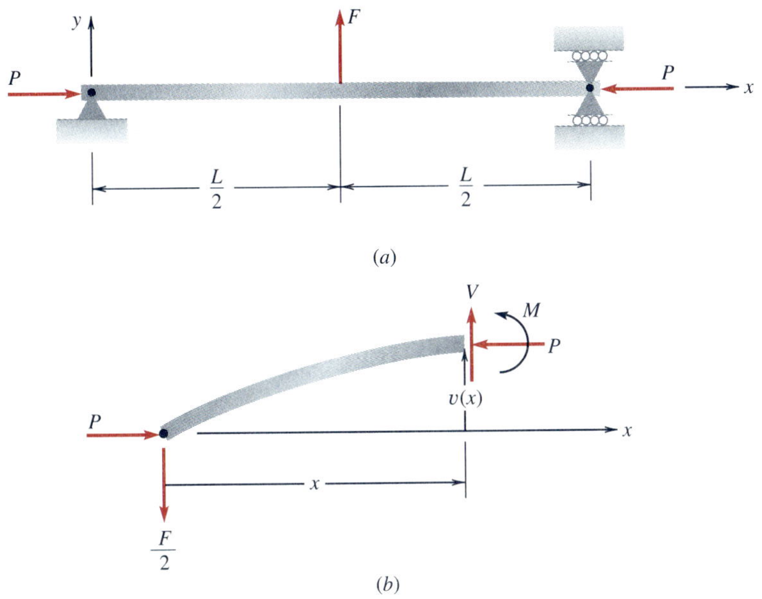

Fig. P10.4-13

(*c*) Show that the maximum deflection is given by

$$v\left(\frac{L}{2}\right) = v_{\max} = \frac{F}{2P}\left[\frac{\tan(\lambda L/2)}{\lambda} - \frac{L}{2}\right] \qquad (d)$$

10.4-14 For the axial load $P = \pi^2 EI/(4L^2) = P_{cr}/4$ or $\lambda L = \pi/2$, show that Eq. (*d*) of Prob. 10.4-13 predicts that the maximum deflection due to the load F at the center of a simply supported beam will increase by a factor of 1.33 due to the axial load P.

10.4-15 Show that as the axial load $P = \lambda^2 EI$ goes to zero, the expression for $v_{\max}$ given by Eq. (*d*) of Prob. 10.4-13 approaches

$$v_{\max} = \frac{FL^3}{48EI}$$

which is the result given in App. G, Table G.2, Fig. G.2-2 for a simply supported beam with a load F at midspan. (*Hint*: use

$$\tan\frac{\lambda L}{2} \approx \frac{\lambda L}{2} + \frac{1}{3}\frac{\lambda^3 L^3}{8}$$

which is valid for small λL.)

10.4-16 The simply supported beam shown in Fig. P10.4-16*a* is subjected to an axial load P and a concentrated moment M_0 applied at midspan.
(*a*) Show that equilibrium of the beam segments shown in Fig. P10.4-16*b* and *c*

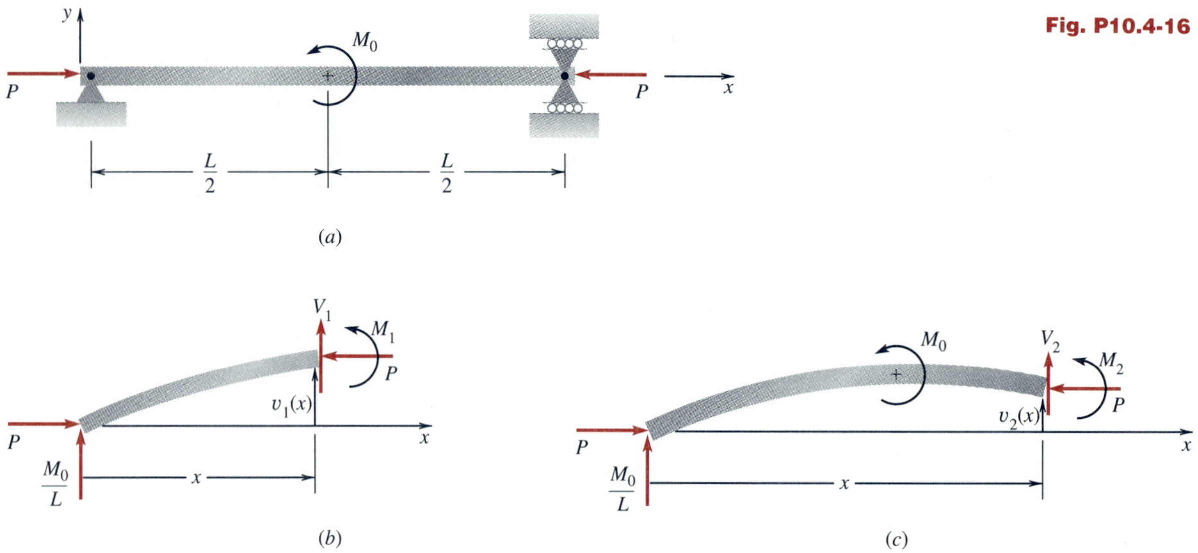

Fig. P10.4-16

require that

$$EIv_1'' + Pv_1 = M_0 \frac{x}{L} \qquad 0 \le x \le L/2 \tag{a}$$

and

$$EIv_2'' + Pv_2 = M_0\left(\frac{x}{L} - 1\right) \qquad L/2 \le x \le L \tag{b}$$

where the relation $M = EIv''$ has been used.

(*b*) Show that the general solutions of Eqs. (*a*) and (*b*) are

$$v_1(x) = \frac{M_0}{P}\frac{x}{L} + c_1 \cos \lambda x + c_2 \sin \lambda x \tag{c}$$

and

$$v_2(x) = \frac{M_0}{P}\left(\frac{x}{L} - 1\right) + c_3 \cos \lambda x + c_4 \sin \lambda x \tag{d}$$

where $\lambda = \sqrt{P/EI}$. Apply the conditions that at $x = 0$: $v_1 = 0$; $x = L$: $v_2 = 0$; $x = L/2$: $v_1 = v_2$, $v_1' = v_2'$ to obtain four equations for the unknown constants c_1 to c_4. Solve these equations to get

$$c_3 = \frac{M_0}{P} \cos\left(\frac{\lambda L}{2}\right) \tag{e}$$

$$c_4 = -\frac{M_0}{P} \cot \lambda L \cos\left(\frac{\lambda L}{2}\right) \tag{f}$$

$$c_2 = -\frac{M_0}{P}\left[\cot \lambda L \cos\left(\frac{\lambda L}{2}\right) + \sin\left(\frac{\lambda L}{2}\right)\right] \qquad (g)$$

(*c*) Show that the slope $v_2'(L)$ takes the form

$$v_2'(L) = \frac{M_0}{PL}\left(1 - \lambda L \frac{\cos(\lambda L/2)}{\sin \lambda L}\right) \qquad (h)$$

10.4-17 For the axial load $P = \pi^2 EI/(4L^2) = P_{cr}/4$ or $\lambda L = \pi/2$, show that Eq. (*h*) of Prob. 10.4-16 predicts that the slope at the right end of a simply supported beam due to a moment M_0 at midspan will increase by a factor of 1.08 due to the axial load P.

10.4-18 Show that as the axial load $P = \lambda^2 EI$ goes to zero, the expression for $v'(L)$ in Eq. (*h*) of Prob. 10.4-16 approaches $v'(L) = -M_0 L/(24EI)$, which is the value given in App. G, Table G.2, Fig. G.2-4, with $a = b = L/2$.

10.4-19 The proposed cross sections for two alternate designs for a column are shown in Fig. P10.4-19*a* and *b*. Find the critical buckling load in terms of E and L for each of the columns proposed for the case of both ends pinned. Calculate the ratio of the larger to the smaller of the two buckling loads.

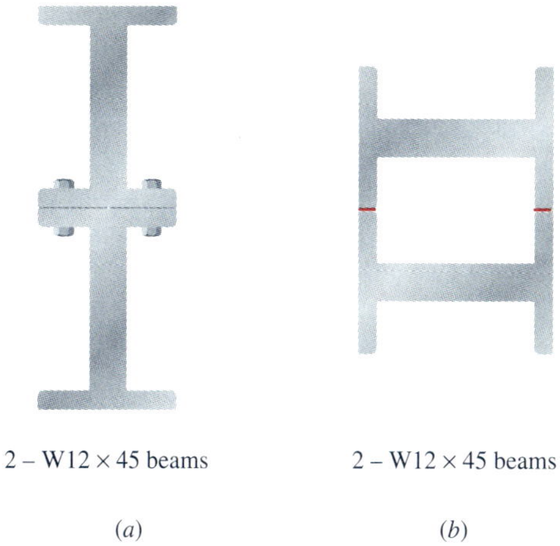

Fig. P10.4-19

10.4-20 A truss carries a 4-kN load at joint D and is supported at joints A and E as shown in Fig. P10.4-20. If all the members are identical circular bars with $d = 30$ mm, $L = 3$ m, and $E = 200$ GPa, find the factor of safety against buckling for all members in compression.

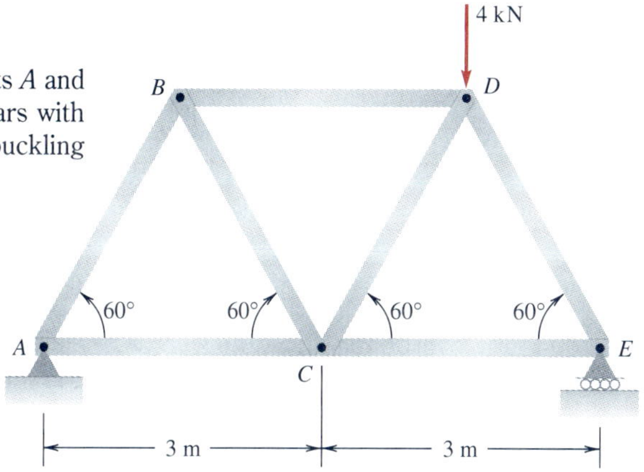

Fig. P10.4-20

10.4-21 For the truss shown in Fig. P10.4-20, if the vertical load acting at D is replaced by a horizontal load at D of magnitude 5 kN and acting to the right, find the factor of safety against buckling for all members in compression. Use the member properties given in Prob. 10.4-20.

10.4-22 A truss carries a 3-kN load at joint C and is supported at joints A and D as shown in Fig. P10.4-22. If all members are circular bars of diameter 40 mm and $E = 200$ GPa, find the factor of safety against buckling for all members in compression.

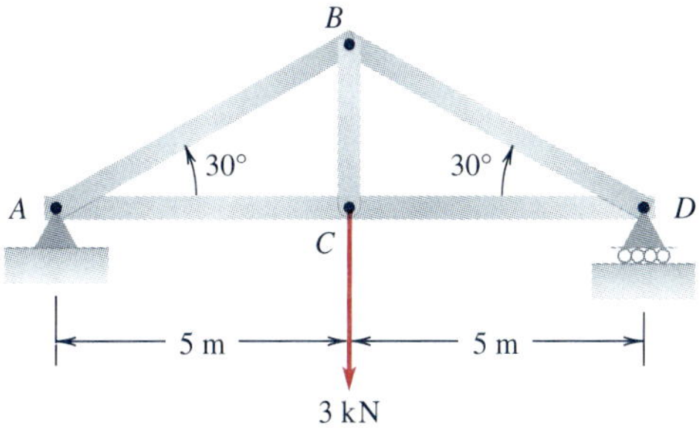

Fig. P10.4-22

10.4-23 For the truss shown in Fig. P10.4-22, if the load at joint C is removed and replaced by a load of 5 kN acting at joint B and directed to the right, find the factor of safety against buckling for all members in compression. Use the member properties given in Prob. 10.4-22.

10.5-1 Use the general approach presented in Sec. 10.5 to find the critical buckling load for a pinned-pinned column. In particular, apply four boundary conditions at $x = 0$ and $x = L$ to Eq. (10.40) to get the critical load.

10.5-2 Use the general approach presented in Sec. 10.5 to find the critical buckling load P_{cr} for a column with one end clamped and the other end free with an axial load P applied. In particular, apply four boundary conditions at $x = 0$ and $x = L$, making use of Eq. (10.40) to get P_{cr}.

10.5-3 A clamped-pinned column is fabricated by bolting together two L = 12-ft-long C10 × 30 channels as shown in Fig. P10.5-3. Find the critical buckling load. Take E = 30,000 ksi.

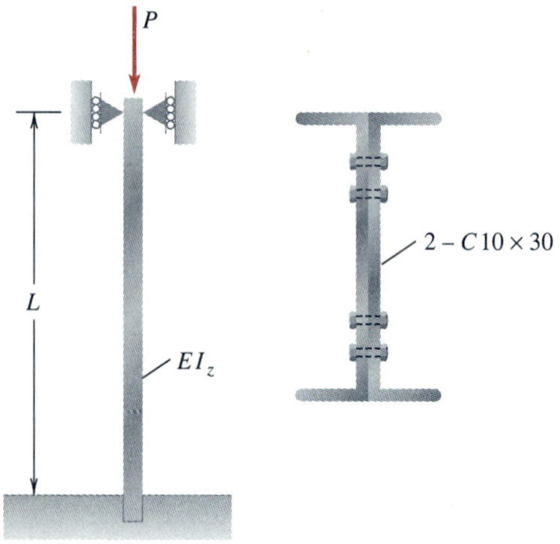

Fig. P10.5-3

10.5-4 Two wood 2 × 4s are nailed together to form a column with a 3-in × 3.5-in cross section. If a factor of safety of 3 is used, E = 1500 ksi, and P = 2000 lb, find the allowable length for the four cases: (*a*) pinned-pinned, (*b*) clamped-free, (*c*) clamped-pinned, and (*d*) clamped-clamped.

10.5-5 Calculate the buckling loads for pinned-pinned wood columns of nominal cross-sectional dimensions (in inches) of 2 × 4, 3 × 6, 4 × 8, and 6 × 12, all of length 10 ft. Take E = 1500 ksi.

10.5-6 In the case of a beam continuously attached to an elastic foundation, the beam element in Fig. 10.13*b* is subjected to the additional downward force of $-kv\Delta x$, where k is the distributed foundation stiffness.

(*a*) Show that Eqs. (10.33) become

$$\frac{dV}{dx} - kv = 0 \qquad (a)$$

$$\frac{dM}{dx} + V + P\frac{dv}{dx} = 0 \qquad (b)$$

(*b*) Using the moment-curvature relation, Eq. (6.6), show that by differentiating Eq. (*b*) and using Eq. (*a*) we get, for constant EI

$$EI\frac{d^4v}{dx^4} + P\frac{d^2v}{dx^2} + kv = 0 \qquad (c)$$

10.5-7 A pinned-pinned column is attached to an elastic foundation and is subjected to end loads P, as shown in Fig. P10.5-7.

(*a*) Show that if a solution of Eq. (*c*) in Prob. 10.5-6 is attempted in the form

$$v = A \sin \lambda x \qquad (a)$$

then to obtain a deflection shape other than that corresponding to a straight beam, λ must satisfy the equation

$$EI\lambda^4 - P\lambda^2 + k = 0 \qquad (b)$$

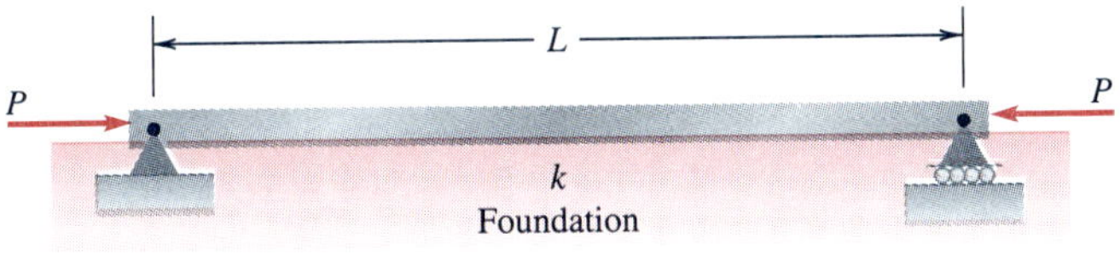

Fig. P10.5-7

(*b*) Show that the pinned-end boundary conditions ($x = 0, L$; $v = 0$, $v'' = 0$) will be satisfied provided that

$$\sin \lambda L = 0 \tag{c}$$

and therefore the roots λ_n of Eq. (*c*) are given by

$$\lambda_n L = n\pi \qquad n = 1, 2, \ldots \tag{d}$$

(*c*) Show that the critical buckling loads P_n can be found by substituting $\lambda_n = n\pi/L$ into Eq. (*b*) and solving for P_n to get

$$P_n = \frac{EIn^2\pi^2}{L^2} + \frac{kL^2}{n^2\pi^2} \tag{e}$$

(*Note:* If k is taken as zero (no foundation) in Eq. (*e*), we return to the earlier result for a pinned-pinned column which is not attached to an elastic foundation.)

10.5-8 Two copper pipes and one steel pipe are built into rigid walls which do not move with temperature changes, as shown in Fig. P10.5-8. The following data are given for the pipes:

	Type	d_o, mm	d_i, mm	L, m	α, /°C	E, GPa
AB	Copper	33	25	3.1	9.7×10^{-6}	110
CD	Steel	25	19	5.0	6.5×10^{-6}	200
EF	Copper	66	50	5.0	9.7×10^{-6}	110

If the pipes are free of stress at room temperature, find the increase in temperature that will cause buckling for each pipe.

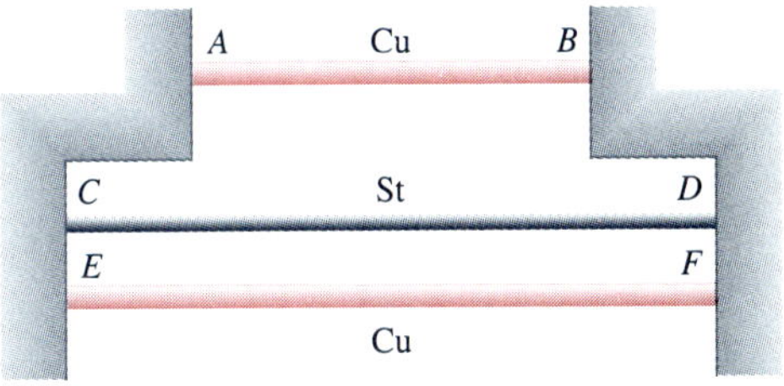

Fig. P10.5-8

10.5-9 Show that the increase in temperature ΔT required to buckle the clamped-clamped beam shown in Fig. P10.5-9 is given by

$$\Delta T = \frac{4\pi^2}{\alpha}\left(\frac{r}{L}\right)^2$$

where α is the coefficient of thermal expansion and r is defined by Eq. (10.56).

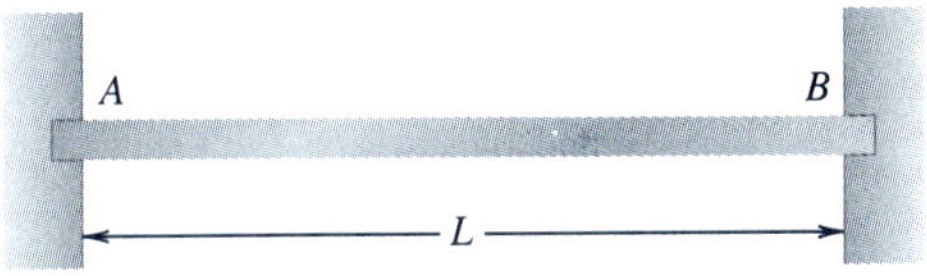

Fig. P10.5-9

10.5-10 A W10 × 77 wide-flange steel beam is clamped at end A and pinned at B as shown in Fig. P10.5-10. Find the increase in temperature ΔT required to buckle this beam if L = 30 ft and $\alpha = 6.5 \times 10^{-6}$/°F. Take E = 30,000 ksi.

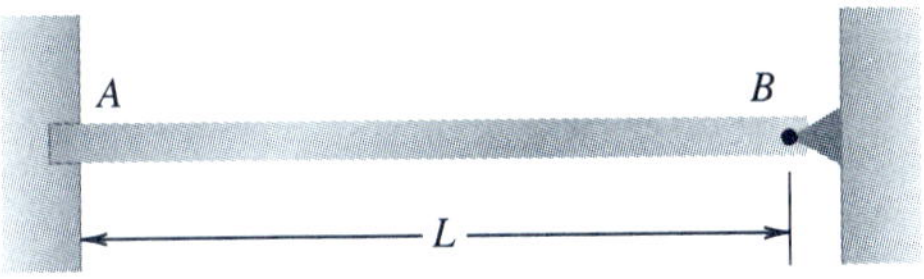

Fig. P10.5-10

10.5-11 The proposed cross sections for two alternate designs for a pinned-clamped column are shown in Fig. P10.5-11. Find the critical buckling loads for

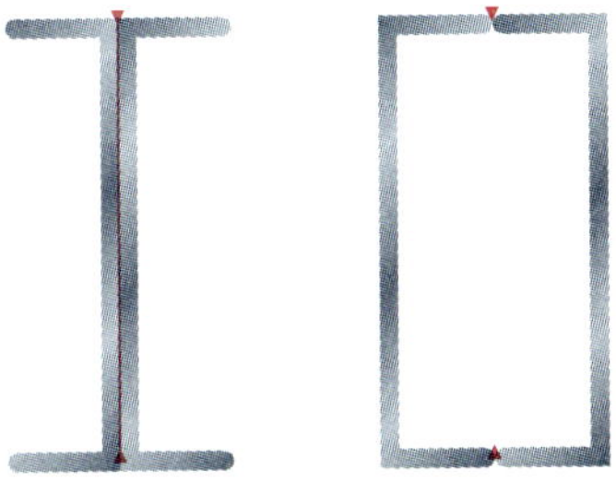

2 – C12 × 25   2 – C12 × 25

Fig. P10.5-11

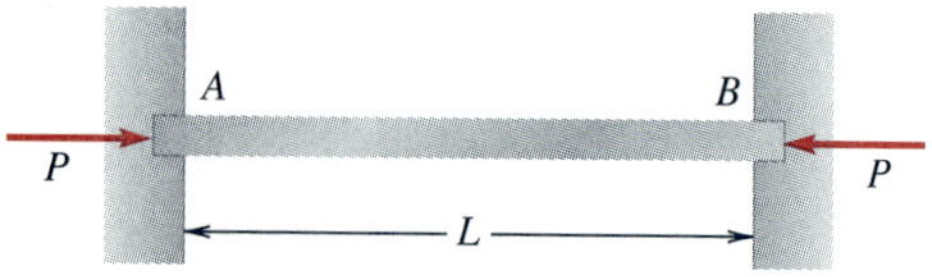

Fig. P10.5-13

each of the proposed columns. Calculate the ratio of the larger to the smaller of the two buckling loads.

10.5-12 A clamped-clamped column consists of a 12-ft length of 7-in × 4-in × $\frac{1}{2}$-in structural tubing. Find the buckling load P_{cr} for an axially loaded column. Take E = 30,000 ksi.

10.5-13 Estimate the buckling load of a small aircraft oil cooler circular tube as shown in Fig. P10.5-13. The length is L = 4 in, the outer diameter is $\frac{1}{8}$ in, and the wall thickness is 0.01 in. Take E = 10,000 ksi.

10.5-14 Estimate the buckling load of a needle modeled as a clamped-pinned beam, as shown in Fig. P10.5-14. The length L = 1 in, diameter d = 0.015 in, wall thickness t = 0.001 in, and E = 30,000 ksi.

10.5-15 A rigid bar $ABCD$ is pinned at A, loaded at D by a vertical load P, and supported by two vertical columns pinned at B and C; the bottom ends of the columns are attached at E and F, as shown in Fig. P10.5-15. Find the ratio L_1/L_2 so that columns BE and CF will both buckle at the same buckling load P_{cr}. Both columns have the same circular cross section and are made of the same material.

10.5-16 A homogeneous rigid machine component of total weight W is supported by three columns of circular cross section, as shown in Fig. P10.5-16. (*a*) Find the weight W_1 which will initiate buckling for at least one of the columns. (*b*) Find the weight W_2 which will cause buckling in all three columns.

Assume that a buckled column continues to carry its buckling load independent of the end displacements. Take d = 25 mm, L = 1 m, and E = 200 GPa; assume that the bars are pinned at A, C, D, and E and are clamped at B and F.

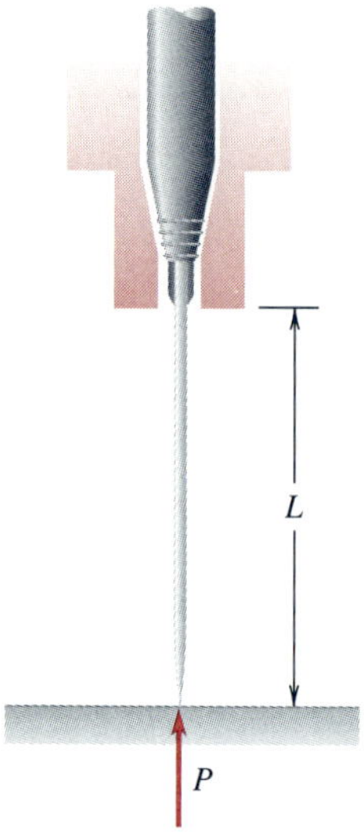

Fig. P10.5-14

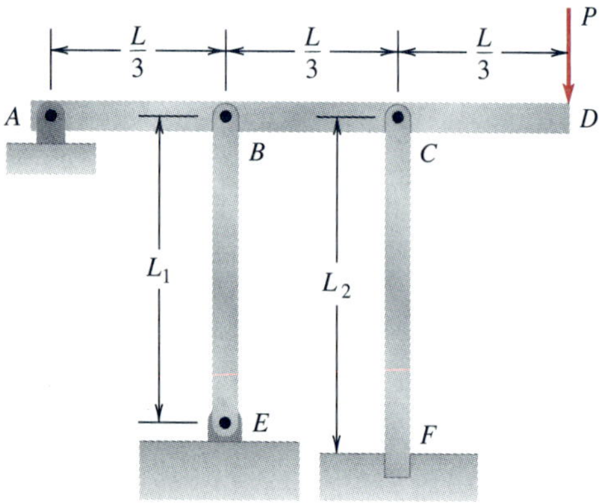

Fig. P10.5-15

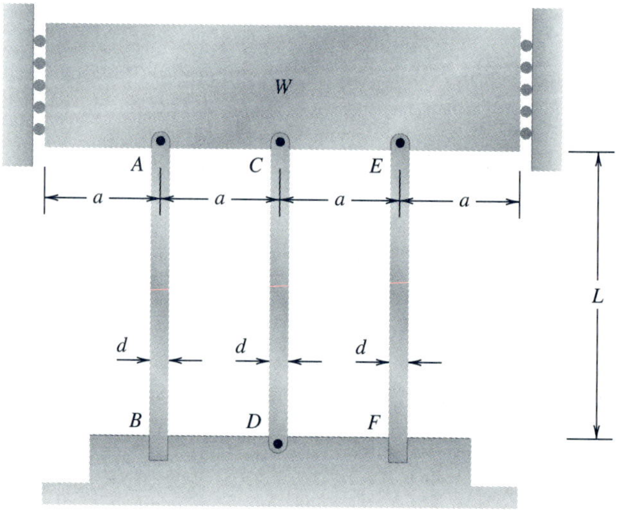

Fig. P10.5-16

10.6-1 A column is clamped at A and is subjected to an eccentric axial load P at the free end B, as shown in Fig. P10.6-1. Make use of two boundary conditions at each end of this column together with Eq. (10.40) to find an expression for the tip deflection δ.

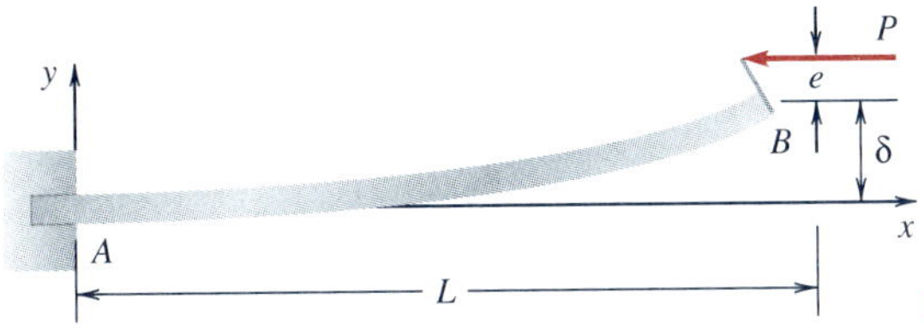

Fig. P10.6-1

10.6-2 A steel bar of 25-mm × 50-mm section is clamped at one end and is subjected to an axial force P at the free end which is applied at the midpoint B of the long side and outside edge of the cross section, as shown in Fig. P10.6-2. Use Eq. (10.49) to plot a graph of P versus δ. In particular, plot values of δ for P ranging from 0 up to 95 percent of the value of P_{cr} for the column with the load at the centroid. Take $E = 200$ GPa.

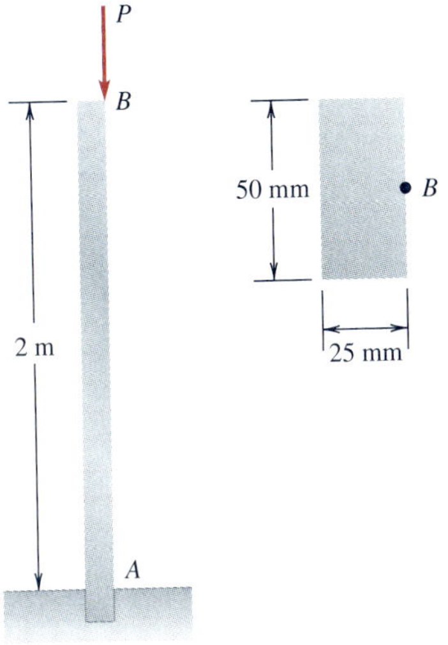

Fig. P10.6-2

10.6-3 A steel column of section shape S10 × 35 is clamped at base A and eccentrically loaded at the free end at point B, as shown in Fig. P10.6-3.

(*a*) Use Eq. (10.49) to calculate the deflection δ in the y direction at the top of the column for a load $P = 0.5P_{cr}$ $[P_{cr} = \pi^2 EI/(4L^2)]$. Take $E = 30{,}000$ ksi and $L = 20$ ft.

(*b*) Find the maximum normal stress in the column due to combined bending and axial stresses.

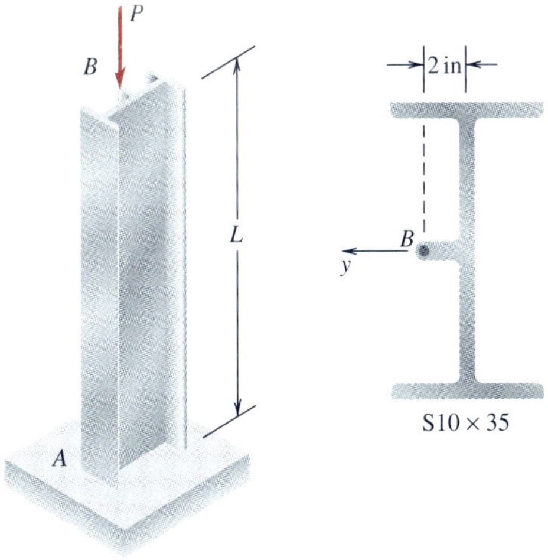

Fig. P10.6-3

10.6-4 A simply supported column is eccentrically loaded at its ends as shown in Fig. 10.18, Example 10.7. The column carries an axial load of $P = 0.5P_{cr}$ $(P_{cr} = \pi^2 EI/L^2)$, $e = 2$ in, $L = 24$ ft, and the column is fabricated by bolting 4-in × $\frac{1}{4}$-in steel plates to the top and bottom of a W10 × 15 steel I beam, as shown in Fig. P10.6-4.

(*a*) Find the deflection δ in the y direction at the midspan of the column, making use of Eq. (j) in Example 10.7.

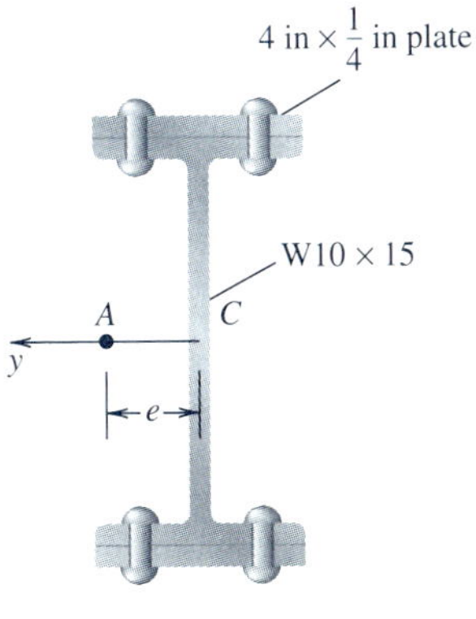

Fig. P10.6-4

(*b*) Find the maximum normal stress in the column due to the combined bending and axial stresses. Use Eq. (*a*) in Example 10.7 to calculate $M_{max} = -P(v_{max} + e)$. Take $E = 30{,}000$ ksi.

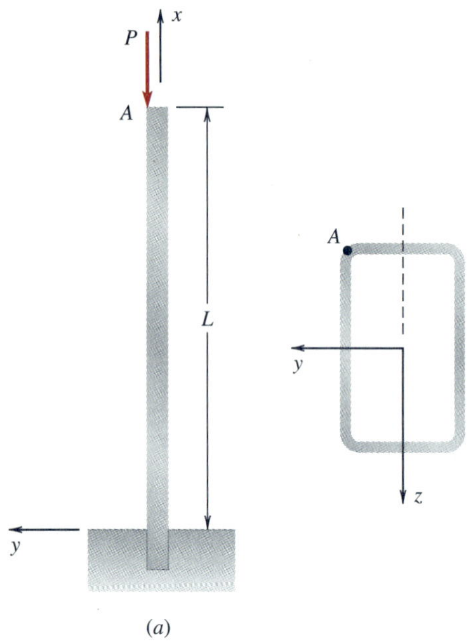

Fig. P10.6-5

10.6-5 An eccentrically loaded column as shown in Fig. P10.6-5 consists of a 20-ft length of rectangular structural tubing of nominal size 7 in × 4 in × $\frac{1}{2}$ in. If the load P is applied at point A at the corner of the section, find the deflections in the y and z directions if $P = 0.5P_{cr}$, where P_{cr} is the buckling load for the same column with the load applied at the centroid of the section $[P_{cr} = \pi^2EI/(4L^2)]$. Take $E = 30{,}000$ ksi.

10.6-6 Two poles are separated by a distance of one-half of the diameter of the poles, as shown in Fig. P10.6-6. If two loads of magnitude P are applied at points A and B as shown, find the value of f such that $P = fP_{cr}$ $[P_{cr} = \pi^2EI/(4L^2)]$ will cause points A and B to just touch.

10.7-1
(*a*) For the steel bar described in Prob. 10.6-2, if the load P is applied at the centroid, find the stress in the bar when $P = P_{cr}$.
(*b*) If a load of $P = 7$ kN is applied eccentrically as described in Prob. 10.6-2, then find the maximum stress in the bar.

10.7-2 A 10-ft section of 6-in × 6-in structural tubing with wall thickness 0.5 in (see App. C, Table C.3) is clamped at its base and free at the top end.
(*a*) If a centroidal load P is to be applied at the top end with a factor of safety of 2.5 against buckling, then find the allowable value of the load P_a.
(*b*) If the load P_a is applied at point D of the cross section as shown in Fig. P10.7-2, find the deflection of the top of the column and the maximum normal stress in the column. Take $E = 30{,}000$ ksi and $e = 2$ in.

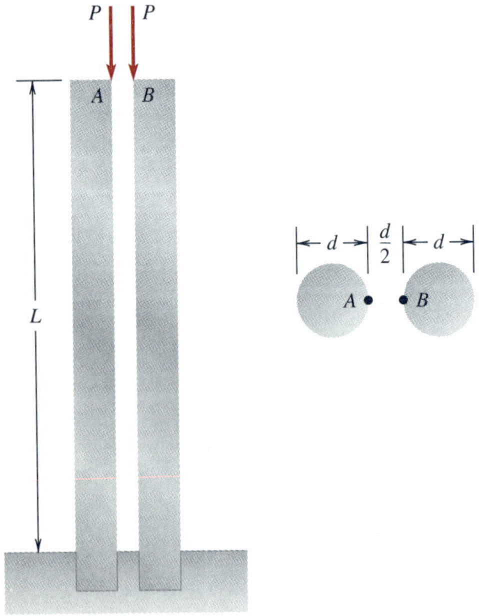

Fig. P10.6-6

10.7-3 A 10-ft section of 6-in-diameter (standard weight) pipe (see App. E) is used as a column and is pinned at both ends as shown in Fig. P10.7-3. If the yield stress for the pipe material is $Y = 32$ ksi and a factor of safety of 2 against yielding is to be used, find the allowable load P for two cases of loading: (*a*) $e = 0.4$ in and (*b*) $e = 0.6$ in. Take $E = 30{,}000$ ksi.

10.7-4 Find the slenderness ratio of a nominal size 6-in × 10-in timber beam which is 12 ft long.

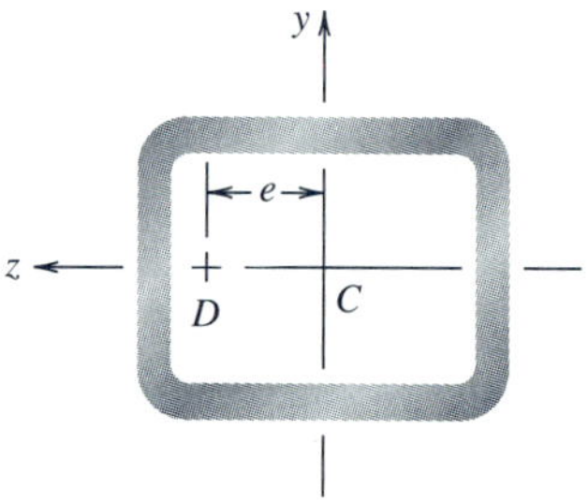

Fig. P10.7-2

10.7-5 Find the slenderness ratio for a 10-ft length of structural tubing of nominal size 6 in × 4 in × $\frac{1}{2}$ in.

10.7-6 Compare the slenderness ratios for two columns both of length 12 ft: (*a*) S12 × 50 (S shape) and (*b*) W12 × 50 (W shape).

10.7-7 For the column with an eccentric axial load as shown in Fig. 10.17*a*, the deflection δ of the top end was found to be

$$\delta = e\left(\sec\left[\sqrt{\frac{P}{EI_z}}L\right] - 1\right) \qquad (10.49)$$

(*a*) Carry out a derivation of an expression for the maximum compressive normal stress in a form analogous to the one given in Eq. (10.55).

(*b*) Show that a secant formula for the column which is of the form of Eq. (10.58) for the pinned-pinned column is

$$\sigma_{\max} = \frac{P}{A}\left[1 + \frac{ec}{r^2}\sec\left(\frac{L}{r}\sqrt{\frac{P}{EA}}\right)\right] \qquad (a)$$

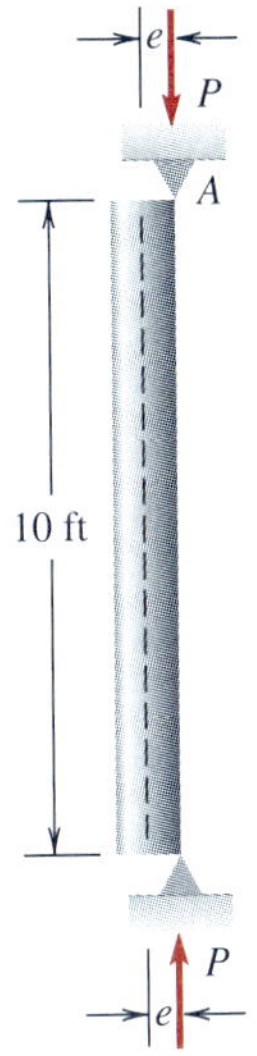

Fig. P10.7-3

10.7-8 A wood pole is firmly embedded in a foundation at its base A and carries a concentrated load P at its top end applied at a point B midway between the center and the outside edge of its circular section, as shown in Fig. P10.7-8. Use Eq. (*a*) in Prob. 10.7-7 to plot a curve of $\sigma_{\max}$ versus P/A for the two cases with $L = 10$ and 20 ft, $d = 12$ in, and $E = 1500$ ksi. From the curves estimate the values of P/A corresponding to allowable values of $\sigma_{\max}$ of 1000 and 1200 psi. The plots will be similar to those given in Fig. 10.21.

10.7-9 An 8-ft length of structural tubing of square section is to be used as a built-in column and will be subjected to a 10,000-lb axial load with an eccentricity

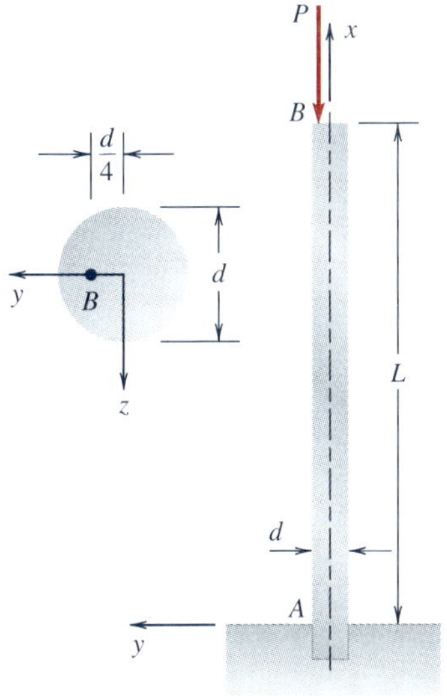

Fig. P10.7-8

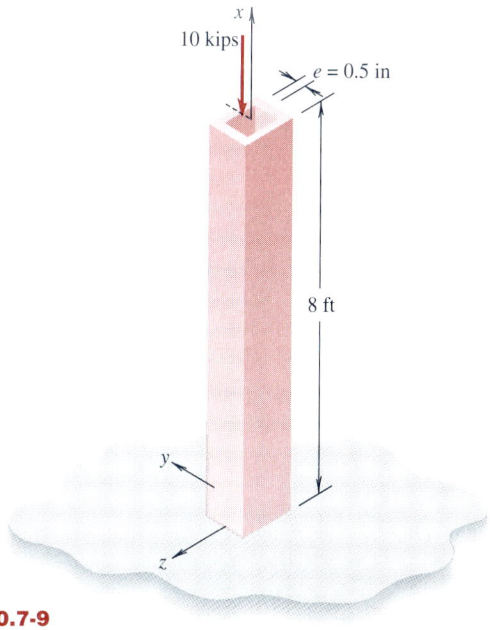

Fig. P10.7-9

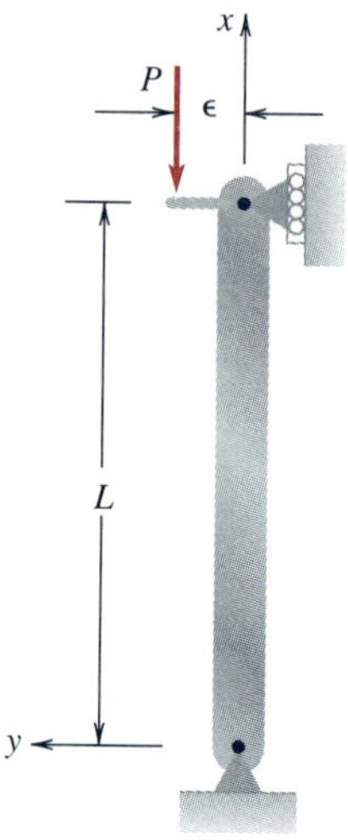

Fig. P10.7-10

$e = 0.5$ in, as shown in Fig. P10.7-9. If the allowable maximum compressive normal stress is 22 ksi, find the lightest allowable tube from Table C.3 which can be used if $E = 30{,}000$ ksi.

10.7-10 An elastic column supported as shown in Fig. P10.7-10 is loaded by an eccentric load P. The supports in this case will give rise to horizontal reactions. Show that the moment at any section at a distance x from the base is $M(x) = -Pv + P(\epsilon x/L)$ and that the deflection is given by

$$v(x) = \epsilon\left(\frac{x}{L} - \frac{\sin \lambda x}{\sin \lambda L}\right)$$

Obtain the expression for the maximum deflection and bending moment in the column for $P = (\frac{1}{4})P_{cr} = \pi^2 EI/4L^2$.

10.8-1 A rigid bar AB is pinned at its lower end at B, is supported by a linear stabilizing spring k, and carries a vertical load P at the top at point A, as shown in Fig. P10.8-1. The right end of the spring can move down.

(*a*) If the spring remains horizontal in all positions, find an expression for the load P in terms of k, L, and the rotation angle θ of the bar.

(*b*) Plot P versus θ for $0° \leq \theta \leq 90°$.

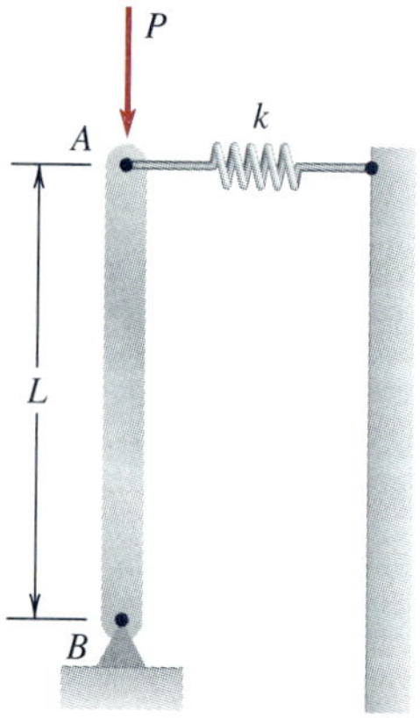

Fig. P10.8-1

10.8-2 Consider the rigid-bar-spring model of Fig. 10.6*b*. In Example 10.1 this model was analyzed for the case of θ restricted to small angles.

(*a*) For $P > P_{cr} = k/L$, show that moment equilibrium about B requires that $P = k\theta/(L \sin \theta)$.

(*b*) Sketch a graph of P versus θ (in radians) for $0 \leq \theta \leq \pi$. Discuss the mechanical reason for the large increase in P required for moment equilibrium as θ increases.

10.8-3 For the bar stabilized by nonlinear springs as shown in Fig. 10.24*a*, Example 10.9, consider the case where $\beta < 0$.

(*a*) Show that the maximum point M in Fig. 10.28*a* for any value of the imperfection parameter ϵ/L lies on the curve

$$1 - \frac{P_{max}}{P_{cr}} = -3\beta\frac{x^2}{L^2}$$

(*b*) Verify that the equation for the curve shown in Fig. 10.28*b* is

$$\left(1 - \frac{P_{max}}{P_{cr}}\right)^{3/2} = \frac{3\sqrt{-3\beta}}{2}\frac{\epsilon}{L}\frac{P_{max}}{P_{cr}}$$

10.8-4 For the bar stabilized by nonlinear springs as shown in Fig. 10.24*a*, Example 10.9, consider the new spring relation

$$f = kx\left(1 + \alpha\frac{x}{L}\right)$$

in place of Eq. (*a*) of Example 10.9.

(*a*) Sketch the postbuckling curves for $\alpha > 0$ and $\alpha < 0$.

(*b*) Work Prob. 10.8-3 again for the case $\alpha < 0$ and show that for $x > 0$, the maximum point corresponding to M in Fig. 10.28*a* lies on the curve

$$1 - \frac{P_{max}}{P_{cr}} = -2\alpha \frac{x}{L}$$

and the equation for the curve corresponding to the one in Fig. 10.28*b* is

$$\left(1 - \frac{P_{max}}{P_{cr}}\right)^2 = -4\alpha \frac{\epsilon}{L} \frac{P_{max}}{P_{cr}}$$

10.8-5 For the control device shown in Fig. P10.8-5*a*, two rigid bars *AB* and *BC* are pinned at their ends and are forced to move in horizontal and vertical tracks as shown. Linear springs at *A* and *C* resist the end movements.

(*a*) Show that vertical equilibrium at *B* gives, Fig. P10.8-5*c* and *d*

$$2F \sin\theta = P \qquad (a)$$

where *F* is the force in either bar. Show that equilibrium at *A* requires that

$$F \cos\theta = k\delta \qquad (b)$$

where δ is the horizontal displacement of *A*.

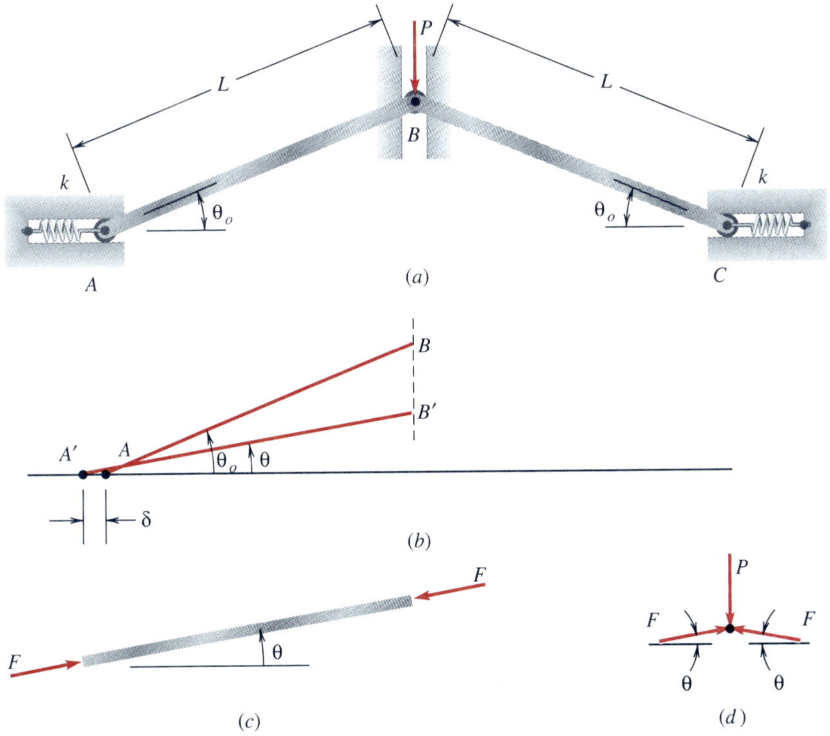

Fig. P10.8-5

(*b*) Show that the relation between δ and θ is, Fig. P10.8-5*b*

$$L(\cos\theta - \cos\theta_0) = \delta \tag{c}$$

and that Eqs. (*a*) through (*c*) can be combined to get

$$P = 2kL\tan\theta(\cos\theta - \cos\theta_0) \tag{d}$$

(*c*) For $\theta_0 = \pi/6$, $k = 100$ lb/in, and $L = 10$ in, make a sketch of P versus θ for $0 \leq \theta \leq \pi/6$ and calculate the value of P which will cause the device to "snap through," i.e., the maximum value of P over the range $0 \leq \theta \leq \pi/6$.

10.8-6 A control device consists of two encased linear springs *AB* and *BC* which are attached to a slide block at *B*. See Fig. P10.8-6. A vertical load *P* is applied to *B*.

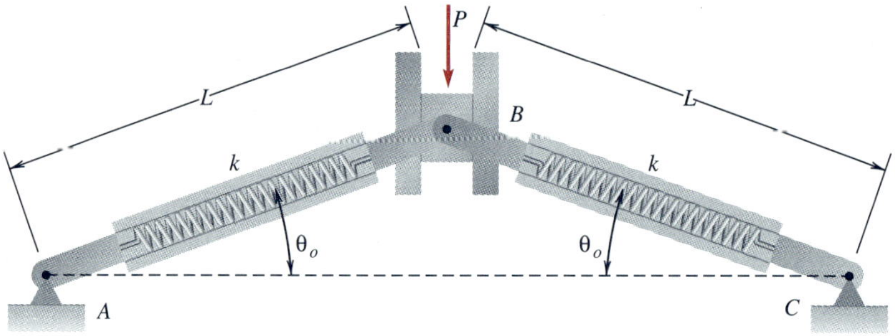

Fig. P10.8-6

(*a*) Derive an expression for P in terms of k, L, the initial angle θ_0 for no load, and the angle θ which the links *AB* and *BC* make with the horizontal for a given P value. Assume that θ_0 and θ are limited to the range from 0° to 60°. (*Hint*: Follow the general approach suggested in Prob. 10.8-5.)
(*b*) If $\theta_0 = \pi/6$, $k = 100$ lb/in, and $L = 10$ in, sketch P versus θ for $0 \leq \theta \leq \pi/8$.

10.9-1 A staple designed for use in an industrial staple gun is U-shaped with two 9-mm legs and with cross-sectional dimensions of 1.4 mm × 0.5 mm, as shown in Fig. P10.9-1.
(*a*) Find the buckling load for each leg if the end where the force is applied is

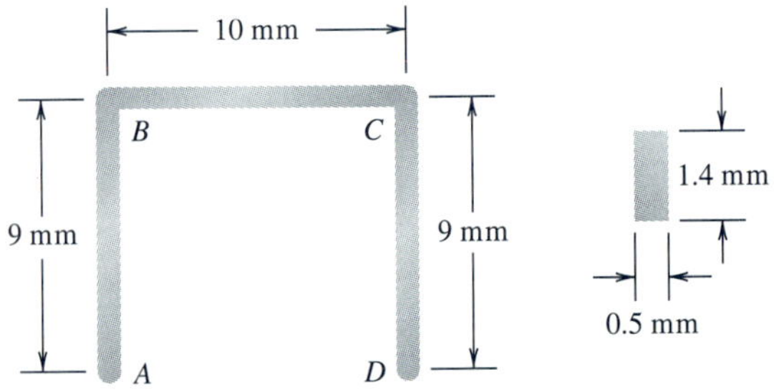

Fig. P10.9-1

assumed to be clamped and the other end is assumed to be pinned. Take $E = 200$ GPa.

(*b*) If a staple gun is assumed to apply the equivalent of a constant static distributed loading along portion *BC* of the staple with a resultant force of 500 N, find the factor of safety against buckling for the staple legs.

10.9-2 One type of hypodermic tubing used in medical applications is made of stainless steel with outside and inside diameters of 0.018 and 0.012 in. For a particular hypodermic needle the tubing is 1.5 in long. Find the buckling load for the needle if the end applied to the skin is assumed to be pinned and the other end is clamped. Take $E = 30{,}000$ ksi.

APPENDIX A

Centroids and Moments of Inertia for Plane Areas

A.1 First Moments and Centroids of Plane Areas

For a given plane area A (Fig. A.1), if x and y are the coordinates of an element of area dA, then we define the first moment of A with respect to the x axis by

$$Q_x = \int_A y\, dA \tag{A.1}$$

and in a similar fashion the first moment of A with respect to the y axis is defined by

$$Q_y = \int_A x\, dA \tag{A.2}$$

The units of Q_x and Q_y are area × length, or cubic inches, meters, etc.

The location of the *centroid of area A* is defined as the location of the point C with coordinates x_C and y_C with the property that

$$A x_C = \int_A x\, dA \tag{A.3}$$

and

$$A y_C = \int_A y\, dA \tag{A.4}$$

where A is the total area. Using Eqs. (A.1) and (A.2) in Eqs. (A.3) and (A.4), we can write

$$x_C = \frac{Q_y}{A} = \frac{\int_A x\, dA}{A} \tag{A.5}$$

$$y_C = \frac{Q_x}{A} = \frac{\int_A y\, dA}{A} \tag{A.6}$$

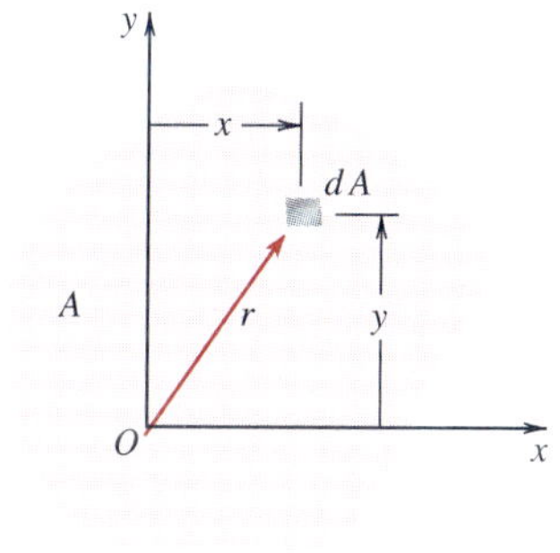

Figure A.1 A plane area A with element of area dA at point x, y.

In the analysis of the bending of beams, we have restricted our attention to the bending of symmetric beams, i.e., to beams whose cross section is symmetric with respect to the plane of transverse loading of the beam. Therefore, if the y axis is taken along the axis of symmetry, then the contribution $x\,dA$ to the total first moment of each area element dA with $x > 0$ is canceled by the corresponding $x\,dA$ for $x < 0$, and therefore $Q_y = 0$. It follows from Eq. (A.5) that $x_C = 0$. This is a general result: *If a plane area has an axis of symmetry, then the centroid lies on that axis.*

In Table A.1 the centroidal coordinates are given for a number of commonly occurring plane sections.

In practice, plane areas often consist of a collection of subareas that are of simple geometric form, e.g., Fig. A.2. In the next section we present a procedure for the calculation of first moments and centroids for such composite areas.

A.2 First Moments and Centroids of Composite Areas

Suppose a given plane area can be decomposed into subareas such that the area and centroid of each subarea are known (Fig. A.2). We can find the first moment Q_x of the total area with respect to the x axis by making use of the property of plane areas that

$$Q_x = \int_A y\,dA = \int_{A_1} y\,dA + \int_{A_2} y\,dA + \cdots \tag{A.7}$$

Now by using Eq. (A.6) for each subarea and inserting into Eq. (A.7), we obtain

$$Q_x = A_1 y_{C_1} + A_2 y_{C_2} + \cdots \tag{A.8}$$

In general, for the case of n subareas we have

$$Q_x = \sum_{k=1}^{n} A_k y_{C_k} \tag{A.9}$$

where A_k and y_{C_k} are the areas and centroids of each individual subarea. A similar argument for Q_y leads to

$$Q_y = \sum_{k=1}^{n} A_k x_{C_k} \tag{A.10}$$

To obtain the coordinates of the centroid C for the total area, we use Eqs. (A.5) and (A.6) to get

$$x_C = \frac{Q_y}{A} = \frac{\sum_{1}^{n} A_k x_{C_k}}{A} \tag{A.11}$$

Table A.1

	AREA	MOMENT OF INERTIA
1. 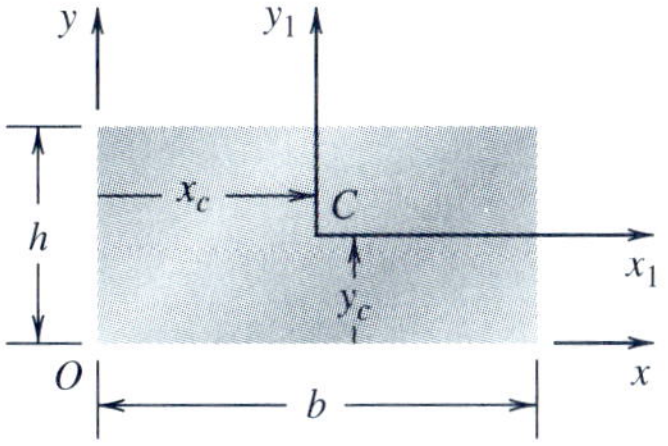	$A = bh$ $x_c = \frac{b}{2}$ $y_c = \frac{h}{2}$	$I_x = \frac{bh^3}{3}$ $I_y = \frac{hb^3}{3}$ $I_{xc} = \frac{bh^3}{12}$ $I_{yc} = \frac{hb^3}{12}$
2. 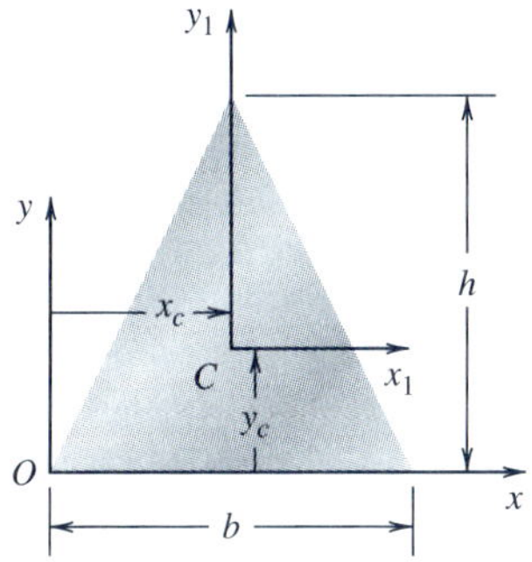	$A = \frac{bh}{2}$ $x_c = \frac{b}{2}$ $y_c = \frac{h}{3}$	$I_{xc} = \frac{bh^3}{36}$ $I_{yc} = \frac{hb^3}{48}$
3. 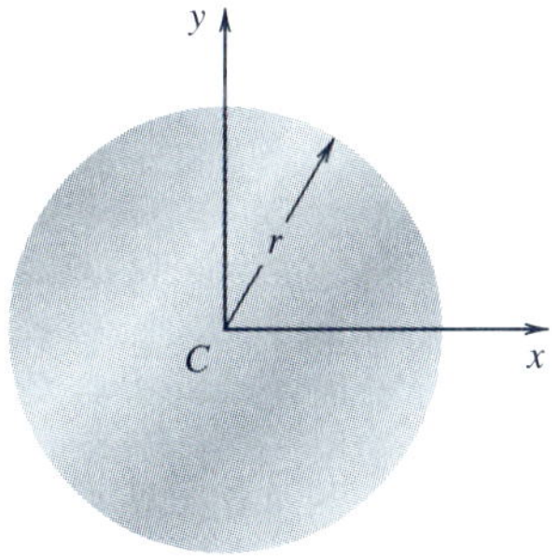	$A = \pi r^2$ $A = \frac{\pi d^2}{4}$ $d = 2r$	$I_{xc} = I_{yc} = \frac{\pi r^4}{4}$ $= \frac{\pi d^4}{64}$ $J_c = I_{xc} + I_{yc}$ $= \frac{\pi r^4}{2} = \frac{\pi d^4}{32}$
4. 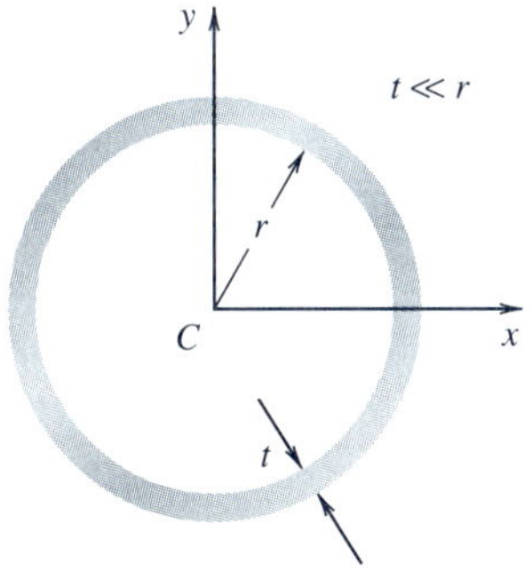	$A = 2\pi rt$ $A = \pi dt$ $d = 2r$	$I_{xc} = I_{yc} = \pi r^3 t$ $= \frac{\pi d^3 t}{8}$ $J_c = 2\pi r^3 t$ $= \frac{\pi d^3 t}{4}$

Table A.1 *Continued*

	AREA	MOMENT OF INERTIA
5. 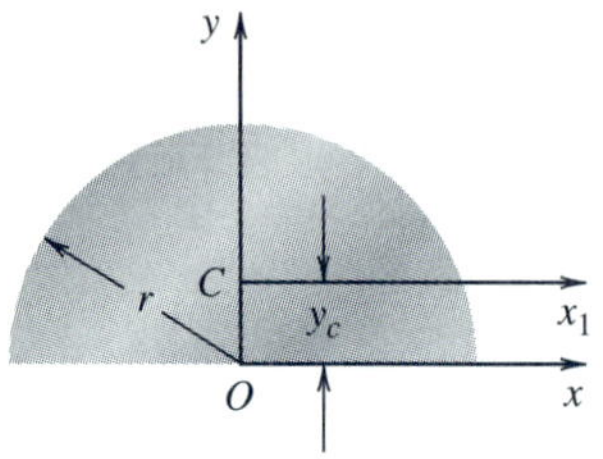	$A = \dfrac{\pi r^2}{2}$ $y_c = \dfrac{4r}{3\pi}$	$I_x = \dfrac{\pi r^4}{8}$ $I_{xc} = \dfrac{(9\pi^2 - 64)\, r^4}{72\pi}$ $I_y = \dfrac{\pi r^4}{8}$
6. 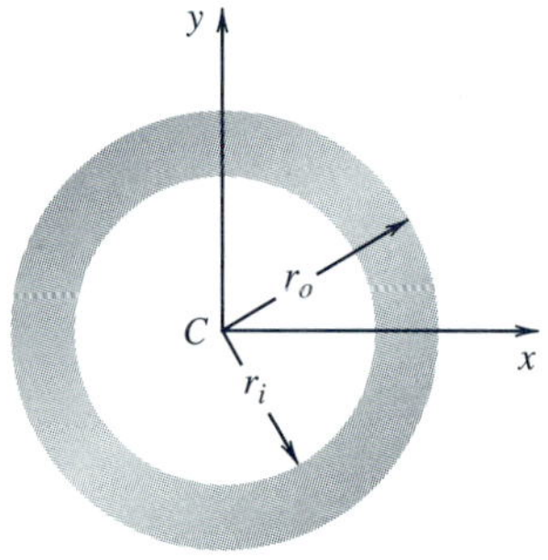	$A = \pi\,(r_o^2 - r_i^2)$ $= \dfrac{\pi}{4}\,(d_o^2 - d_i^2)$	$I_{xc} = \dfrac{\pi}{4}\,(r_o^4 - r_i^4)$ $= \dfrac{\pi}{64}(d_o^4 - d_i^4)$ $J_c - I_{xc} + I_{yc}$ $= \dfrac{\pi}{2}\,(r_o^4 - r_i^4)$ $= \dfrac{\pi}{32}(d_o^4 - d_i^4)$
7. 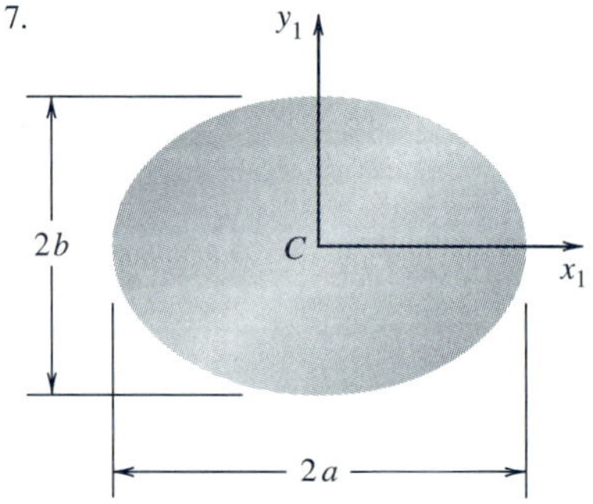	$A = \pi ab$	$I_{xc} = \dfrac{\pi ab^3}{4}$ $I_{yc} = \dfrac{\pi a^3 b}{4}$
8. 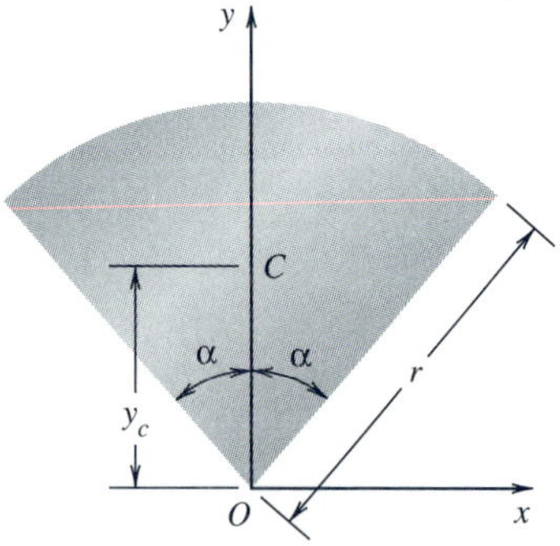	$A = \alpha r^2$ $y_c = \dfrac{2r \sin\alpha}{3\alpha}$	$I_x = \dfrac{r^4}{4}\,(\alpha + \sin\alpha \cos\alpha)$ $I_y = \dfrac{r^4}{4}\,(\alpha - \sin\alpha \cos\alpha)$ $J_o = \dfrac{\alpha r^4}{2}$

and

$$y_C = \frac{Q_x}{A} = \frac{\sum_1^n A_k y_{C_k}}{A} \tag{A.12}$$

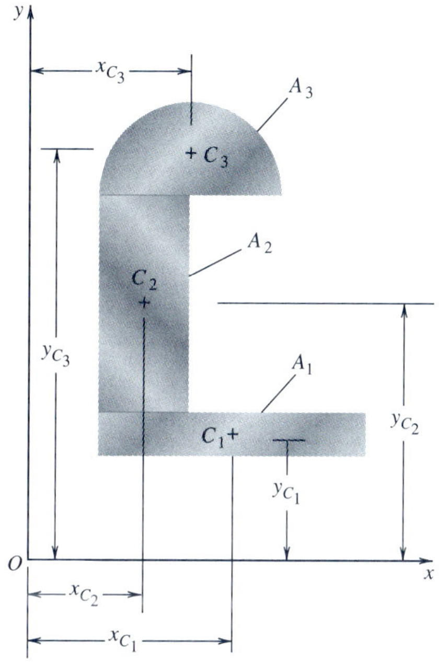

Figure A.2 A plane area with three subareas. The centroid coordinates for each subarea A_i are given by x_{C_i}, y_{C_i}, $i = 1, 2, 3$.

EXAMPLE A.1

Suppose that we wish to find the location of the centroid C for the I section shown in Fig. A.3. We subdivide the area of the I section into three areas as shown. The reference coordinate system x,y is introduced. Symmetry of the section about the y axis means that C lies on that axis and therefore $x_C = 0$.

The centroids of the three rectangular areas are located at the intersection of their lines of symmetry and are given in Fig. A.3. Thus y_C can be found by using Eq. (A.12) to get

$$y_C = \frac{(1 \times 6)(6.5) + (1 \times 5)(3.5) + (1 \times 4)(0.5)}{1 \times 6 + 1 \times 5 + 1 \times 4}$$

$$= \frac{58.5}{15} = 3.9 \text{ in}$$

The centroid C with coordinates $x_C = 0$ and $y_C = 3.9$ in is shown in Fig. A.3.

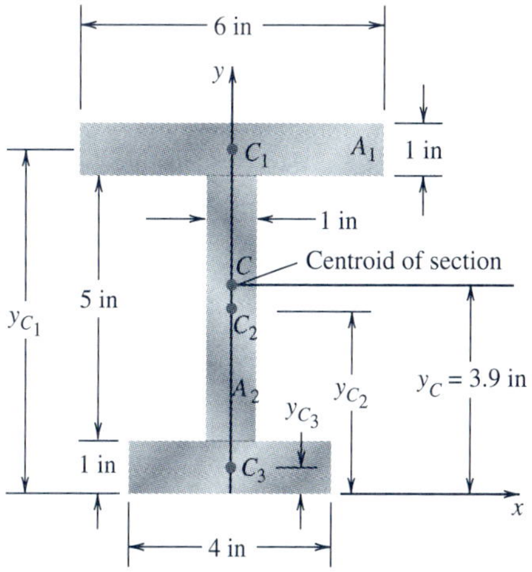

Figure A.3 Example A.1

EXAMPLE A.2

Find the location of the centroid of the I section given in Fig. A.3 by making use of the computer program MECHMAT on the diskette supplied with this book. The main menu of MECHMAT contains option 6: Centroids and moments of inertia (see Fig. A.4*a*) which can be used to calculate the centroids of some commonly occurring composite areas (see the menu in Fig. A.4*b*). For example, if we would like to use this program to find the centroid for the I section shown in Fig. A.3, we can select item 4 in the menu and obtain the drawing shown in Fig. A.5. If the appropriate six dimensions are provided after the computer prompt, then the program prints out the area and distance to the centroid C from the ***bottom of the section*** (Fig. A.5). The results for the area and centroid agree with the results of the calculation carried out in Example A.1. Further discussion of the program can be found in Example 5.3.

```
**********************************************************

*** *MECHANICS OF MATERIALS* ***

PROGRAM: MECHMAT                 mod 5.0

**********************************************************


     MAIN MENU

           1. AXIALLY LOADED MEMBERS
           2. TORSION
           3. V + M DIAGRAMS OF BEAMS
           4. DEFLECTIONS OF BEAMS
           5. MOHR'S CIRCLE
           6. CENTROIDS + MOMENTS OF INERTIA
           7. EXIT

     SELECT NUMBER = ?
```

(*a*)

```
**********************************************************

    *MOMENTS OF INERTIA*

**********************************************************

      1. CHANNEL SECTION
      2. T BEAM
      3. I BEAM
      4. UNSYMMETRIC I BEAM
      5. BACK TO MENU

   SELECT NUMBER = ?


                      **************************************
                      * TO OBTAIN PRINT-OUT OF RESULTS     *
                      * ON NEXT SCREEN, USE PRINT SCREEN   *
                      * AFTER NEXT SCREEN.                 *
                      * AFTER RESULTS, PRESS ANY KEY TO    *
                      * RETURN TO MENU                     *
                      **************************************
```

(*b*)

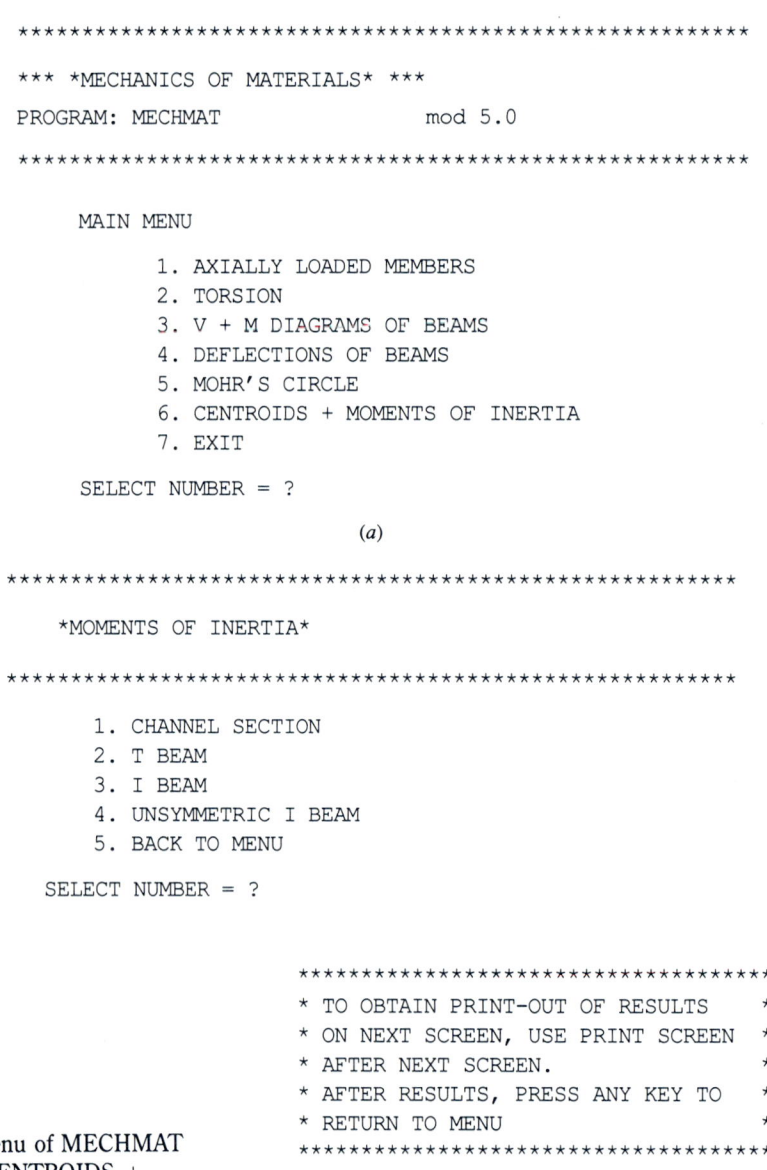

Figure A.4 (*a*) Main menu of MECHMAT program, (*b*) menu for CENTROIDS + MOMENTS OF INERTIA program.

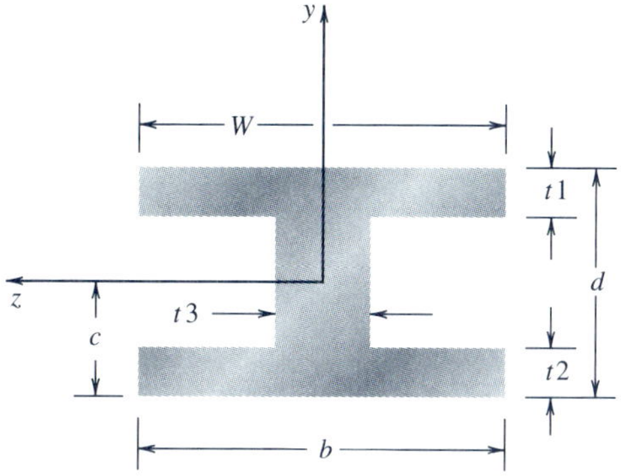

Input W, b, d, $t1$, $t2$, $t3$? 6, 4, 7, 1, 1, 1

c = 3.9000E+00 Iz = 9.8850E+01

Area = 1.5000E+01

Figure A.5 Example A.3

A.3 Second Moments or Moments of Inertia of Plane Areas

In calculating the resultant bending moment associated with the assumed stress distribution over the cross section of a beam, we encounter integrals of the form

$$I_x = \int_A y^2 \, dA \tag{A.13}$$

where y and dA are defined in Fig. A.1 for a given cross-sectional area A of the beam. We refer to I_x as the second moment or moment of inertia of the area about the x axis. The label *moment of inertia* originates in the subject area of dynamics where the integral defined in Eq. (A.13) also occurs. We usually refer to I_x as the moment of inertia of area A with respect to the x axis. In a similar way we have

$$I_y = \int_A x^2 \, dA \tag{A.14}$$

and refer to I_y as the moment of inertia of area A with respect to the y axis.

In the study of torsion of circular shafts, we encounter the so-called polar moment of inertia of an area with respect to a point O. For the area shown in Fig. A.1, we define the polar moment J by the integral

$$J = \int_A r^2 \, dA \tag{A.15}$$

where r is the distance from the point O to the element of area dA (Fig.

A.1). A connection between moments of inertia I_x and I_y and the polar moment J can be derived by using the geometric relation

$$r^2 = x^2 + y^2$$

to write

$$\begin{aligned} J &= \int_A r^2\, dA = \int_A (x^2 + y^2)\, dA \\ &= I_y + I_x \end{aligned} \tag{A.16}$$

In Chap. 10 in our discussion of the bending stresses in long columns, we encounter the ratio of the moment of inertia to the area of a section. These ratios may be written in the form

$$r_x^2 = I_x/A \tag{A.17}$$

$$r_y^2 = I_y/A \tag{A.18}$$

$$r_0^2 = J_0/A \tag{A.19}$$

where r_x, r_y, and r_0 are called the ***radii of gyration*** and J_0 is the polar moment of inertia about the origin O. The radii of gyration can be found from Eqs. (A.17) to (A.19) in the form

$$r_x = \sqrt{I_x/A} \tag{A.20}$$

$$r_y = \sqrt{I_y/A} \tag{A.21}$$

$$r_0 = \sqrt{J_0/A} \tag{A.22}$$

A.4 Parallel-Axis Theorem

For many calculations of moments of inertia for composite areas, it is convenient to make use of a relation between the moment of inertia of a plane area about a centroidal axis and the moment about an arbitrary axis parallel to the centroidal axis.

Consider the plane area A shown in Fig. A.6. Let x_1 be a centroidal axis and x be an axis parallel to x_1 with d_x as the distance separating the two axes, as shown in Fig. A.6. The moment of inertia of A about the x axis is defined by

$$I_x = \int_A y^2\, dA \tag{A.23}$$

From Fig. A.6 we see that the y distance in Eq. (A.23) may be written in the form

$$y = d_x + y_1 \tag{A.24}$$

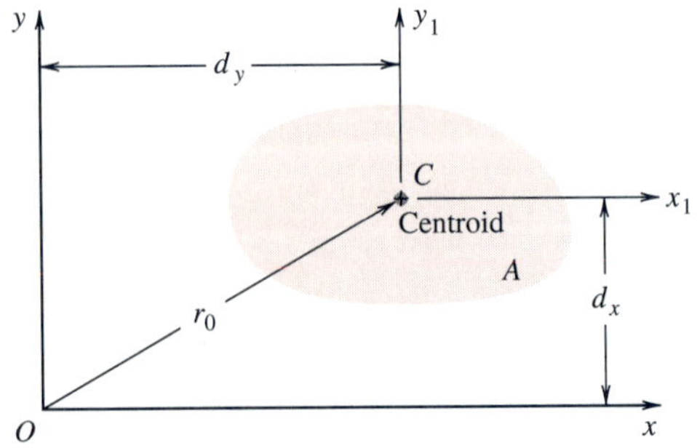

Figure A.6 Plane area with axes x_1y_1 through the centroid C and a set of parallel axes xy separated by d_x and d_y.

and when it is substituted into Eq. (A.23), we get

$$I_x = \int_A (d_x + y_1)^2\, dA$$
$$= A\, d_x^2 + 2d_x \int_A y_1\, dA + \int_A y_1^2\, dA \qquad \text{(A.25)}$$

But since C is the centroid, we have

$$\int_A y_1\, dA = 0 \qquad \text{(A.26)}$$

and Eq. (A.25) reduces to

$$I_x = A\, d_x^2 + I_{xC} \qquad \text{(A.27)}$$

where I_{xC} denotes the moment of inertia with respect to an axis parallel to the x axis and passing through the centroid.

Thus according to Eq. (A.27), the moment of inertia I_x of a plane area A with respect to an arbitrary axis x is equal to the moment of inertia of the area with respect to a centroidal axis parallel to the x axis plus the product of the area and the square of the distance d_x between the parallel axes. The expression in Eq. (A.27) is often referred to as the *parallel-axis theorem.* Additional discussion of the parallel-axis theorem can be found in Example 5.3. It is useful in problems for which we already know the centroidal moments of inertia for one or more areas and wish to make use of these known moments to find the moments for a composite area, e.g., Example A.3.

A similar argument leads to the relation

$$I_y = A\, d_y^2 + I_{yC} \qquad \text{(A.28)}$$

and making use of Eq. (A.16), we write

$$I_x + I_y = A(d_x^2 + d_y^2) + I_{xC} + I_{yC}$$

or

$$J_0 = Ar_0^2 + J_C \qquad \text{(A.29)}$$

Equation (A.29) states that the polar moment of inertia with respect to the point O can be found by adding the product of the area and the square of the distance between point O and the centroid C to the polar moment of the area about the centroid C.

EXAMPLE A.3

For the I section shown in Fig. A.3 use the parallel-axis theorem to find the moment of inertia about an axis x_C which is parallel to the x axis and passes through the centroid of the section.

In Examples A.1 and A.2, the y coordinate of the centroid was found to be $y_C = 3.9$ in. We make use of the known y centroidal coordinates shown in Fig. A.3 for each of the three areas involved, and we use the parallel-axis theorem to calculate the contribution of each subarea to the total moment of inertia

I_{xC}. That is,

$$I_{xC} = \sum_{k=1}^{3} (I_{xC_k} + A\, d_{x_k}^2)$$
$$= \tfrac{1}{12}(6)(1)^3 + (6 \times 1)(6.5 - 3.9)^2$$
$$+ \tfrac{1}{12}(1)(5)^3 + (5 \times 1)(3.5 - 3.9)^2$$
$$+ \tfrac{1}{12}(4)(1)^3 + (4 \times 1)(0.5 - 3.9)^2 = 98.85 \text{ in}^4$$

which agrees with the value given in Fig. A.5. Further discussion of the parallel-axis theorem and its application to moments of inertia of composite sections can be found in Example 5.3.

APPENDIX B

Conversion Factors Useful for Mechanics of Solids

Length
1 mm = 10^{-3} m
1 in = 0.0254 m
1 ft = 0.3048 m
1 m = 39.37 in
1 m = 3.28 ft

Loading
1 N = 0.225 lb
1 lb = 4.45 N
1 kN = 225 lb
1 kip = 1000 lb
1 lb/in = 175.2 N/m
1 lb/ft = 14.6 N/m
1 kip/ft = 83.33 lb/in
1 kip/ft = 14.6 kN/m
1 kN/m = 68.53 lb/ft
1 kN/m = 5.71 lb/in
1 lb · in = 0.113 N · m
1 lb · ft = 1.356 N · m
1 kip · ft = 1.356 kN · m
1 N · m = 0.7376 lb · ft
1 N · m = 8.85 lb · in
1 kN · m = 0.7376 kip · ft
1 kN · m = 8.85 kip · in

Stress/pressure
1 Pa = 1 N/m^2
1 psi = 6.895 kPa
1 ksi = 6.895 MPa
1 psf = 47.88 Pa
1 MPa = 145 psi
1 kPa = 20.88 psf
1 GPa = 1.45×10^5 psi

Moment of inertia
1 in^4 = 0.416×10^{-6} m^4
1 mm^4 = 2.40×10^{-6} in^4

APPENDIX C

Properties of Selected Structural-Steel Shapes

The following tables for properties of cross sections have been adopted with permission from the *Manual of Steel Construction, Allowable Stress Design,* 9th ed., American Institute of Steel Construction, Inc.

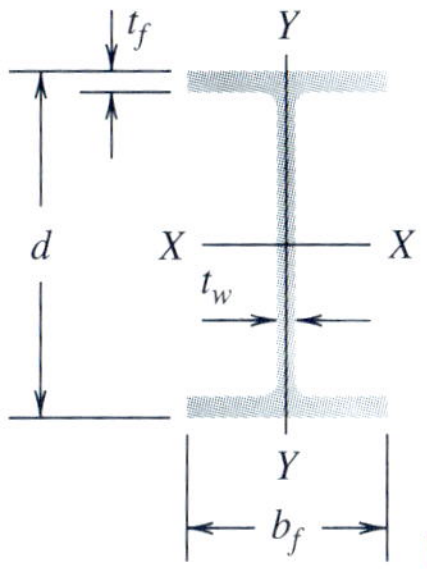

Table C.1 W Shapes: Dimensions

	Area	Depth		Web			Flange				Elastic Properties					
				Thickness		$\frac{t_w}{2}$	Width		Thickness		Axis XX			Axis YY		
	A	d		t_w			b_f		t_f		I	S	r	I	S	r
Designation	in^2	in		in		in	in		in		in^4	in^3	in	in^4	in^3	in
W16 × 31	9.12	15.88	$15\frac{7}{8}$	0.275	$\frac{1}{4}$	$\frac{1}{8}$	5.525	$5\frac{1}{2}$	0.440	$\frac{7}{16}$	375	47.2	6.41	12.4	4.49	1.17
× 26	7.68	15.69	$15\frac{3}{4}$	0.250	$\frac{1}{4}$	$\frac{1}{8}$	5.500	$5\frac{1}{2}$	0.345	$\frac{3}{8}$	301	38.4	6.26	9.59	3.49	1.12
W14 × 730	215.0	22.42	$22\frac{3}{8}$	3.070	$3\frac{1}{16}$	$1\frac{9}{16}$	17.890	$17\frac{7}{8}$	4.910	$4\frac{15}{16}$	14,300	1,280	8.17	4,720	527	4.69
× 665	196.0	21.64	$21\frac{5}{8}$	2.830	$2\frac{13}{16}$	$1\frac{7}{16}$	17.650	$17\frac{5}{8}$	4.520	$4\frac{1}{2}$	12,400	1,150	7.98	4,170	472	4.62
× 605	178.0	20.92	$20\frac{7}{8}$	2.595	$2\frac{5}{8}$	$1\frac{5}{16}$	17.415	$17\frac{3}{8}$	4.160	$4\frac{3}{16}$	10,800	1,040	7.80	3,680	423	4.55
× 550	162.0	20.24	$20\frac{1}{4}$	2.380	$2\frac{3}{8}$	$1\frac{3}{16}$	17.200	$17\frac{1}{4}$	3.820	$3\frac{13}{16}$	9,430	931	7.63	3,250	378	4.49
× 500	147.0	19.60	$19\frac{5}{8}$	2.190	$2\frac{3}{16}$	$1\frac{1}{8}$	17.010	17	3.500	$3\frac{1}{2}$	8,210	838	7.48	2,880	339	4.43
× 455	134.0	19.02	19	2.015	2	1	16.835	$16\frac{7}{8}$	3.210	$3\frac{3}{16}$	7,190	756	7.33	2,560	304	4.38
W14 × 426	125.0	18.67	$18\frac{5}{8}$	1.875	$1\frac{7}{8}$	$\frac{15}{16}$	16.695	$16\frac{3}{4}$	3.035	$3\frac{1}{16}$	6,600	707	7.26	2,360	283	4.34
× 398	117.0	18.29	$18\frac{1}{4}$	1.770	$1\frac{3}{4}$	$\frac{7}{8}$	16.590	$16\frac{5}{8}$	2.845	$2\frac{7}{8}$	6,000	656	7.16	2,170	262	4.31
× 370	109.0	17.92	$17\frac{7}{8}$	1.655	$1\frac{5}{8}$	$\frac{13}{16}$	16.475	$16\frac{1}{2}$	2.660	$2\frac{11}{16}$	5,440	607	7.07	1,990	241	4.27
× 342	101.0	17.54	$17\frac{1}{2}$	1.540	$1\frac{9}{16}$	$\frac{13}{16}$	16.360	$16\frac{3}{8}$	2.470	$2\frac{1}{2}$	4,900	559	6.98	1,810	221	4.24
× 311	91.4	17.12	$17\frac{1}{8}$	1.410	$1\frac{7}{16}$	$\frac{3}{4}$	16.230	$16\frac{1}{4}$	2.260	$2\frac{1}{4}$	4,330	506	6.88	1,610	199	4.20
× 283	83.3	16.74	$16\frac{3}{4}$	1.290	$1\frac{5}{16}$	$\frac{11}{16}$	16.110	$16\frac{1}{8}$	2.070	$2\frac{1}{16}$	3,840	459	6.79	1,440	179	4.17
× 257	75.6	16.38	$16\frac{3}{8}$	1.175	$1\frac{3}{16}$	$\frac{5}{8}$	15.995	16	1.890	$1\frac{7}{8}$	3,400	415	6.71	1,290	161	4.13
× 233	68.5	16.04	16	1.070	$1\frac{1}{16}$	$\frac{9}{16}$	15.890	$15\frac{7}{8}$	1.720	$1\frac{3}{4}$	3,010	375	6.63	1,150	145	4.10
× 211	62.0	15.72	$15\frac{3}{4}$	0.980	1	$\frac{1}{2}$	15.800	$15\frac{3}{4}$	1.560	$1\frac{9}{16}$	2,660	338	6.55	1,030	130	4.07
× 193	56.8	15.48	$15\frac{1}{2}$	0.890	$\frac{7}{8}$	$\frac{7}{16}$	15.710	$15\frac{3}{4}$	1.440	$1\frac{7}{16}$	2,400	310	6.50	931	119	4.05
× 176	51.8	15.22	$15\frac{1}{4}$	0.830	$\frac{13}{16}$	$\frac{7}{16}$	15.650	$15\frac{5}{8}$	1.310	$1\frac{5}{16}$	2,140	281	6.43	838	107	4.02
× 159	46.7	14.98	15	0.745	$\frac{3}{4}$	$\frac{3}{8}$	15.565	$15\frac{5}{8}$	1.190	$1\frac{3}{16}$	1,900	254	6.38	748	96.2	4.00
× 145	42.7	14.78	$14\frac{3}{4}$	0.680	$\frac{11}{16}$	$\frac{3}{8}$	15.500	$15\frac{1}{2}$	1.090	$1\frac{1}{16}$	1,710	232	6.33	677	87.3	3.98
W14 × 132	38.8	14.66	$14\frac{5}{8}$	0.645	$\frac{5}{8}$	$\frac{5}{16}$	14.725	$14\frac{3}{4}$	1.030	1	1,530	209	6.28	548	74.5	3.76
× 120	35.3	14.48	$14\frac{1}{2}$	0.590	$\frac{9}{16}$	$\frac{5}{16}$	14.670	$14\frac{5}{8}$	0.940	$\frac{15}{16}$	1,380	190	6.24	495	67.5	3.74
× 109	32.0	14.32	$14\frac{3}{8}$	0.525	$\frac{1}{2}$	$\frac{1}{4}$	14.605	$14\frac{5}{8}$	0.860	$\frac{7}{8}$	1,240	173	6.22	447	61.2	3.73
× 99	29.1	14.16	$14\frac{1}{8}$	0.485	$\frac{1}{2}$	$\frac{1}{4}$	14.565	$14\frac{5}{8}$	0.780	$\frac{3}{4}$	1,110	157	6.17	402	55.2	3.71

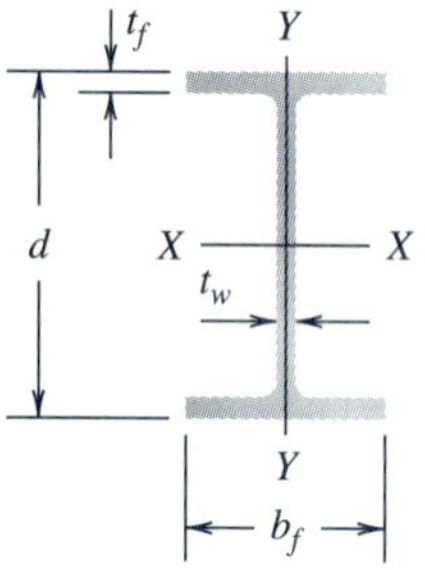

Table C.1 *Continued*

Designation	Area A	Depth d		Web Thickness t_w		Web $\frac{t_w}{2}$	Flange Width b_f		Flange Thickness t_f		Axis XX I	Axis XX S	Axis XX r	Axis YY I	Axis YY S	Axis YY r
	in^2	in		in		in	in		in		in^4	in^3	in	in^4	in^3	in
× 90	26.5	14.02	14	0.440	$\frac{7}{16}$	$\frac{1}{4}$	14.520	$14\frac{1}{2}$	0.710	$\frac{11}{16}$	999	143	6.14	362	49.9	3.70
W14 × 82	24.1	14.31	$14\frac{1}{4}$	0.510	$\frac{1}{2}$	$\frac{1}{4}$	10.130	$10\frac{1}{8}$	0.855	$\frac{7}{8}$	882	123	6.05	148	29.3	2.48
× 74	21.8	14.17	$14\frac{1}{8}$	0.450	$\frac{7}{16}$	$\frac{1}{4}$	10.070	$10\frac{1}{8}$	0.785	$\frac{13}{16}$	796	112	6.04	134	26.6	2.48
× 68	20.0	14.04	14	0.415	$\frac{7}{16}$	$\frac{1}{4}$	10.035	10	0.720	$\frac{3}{4}$	723	103	6.01	121	24.2	2.46
× 61	17.9	13.89	$13\frac{7}{8}$	0.375	$\frac{3}{8}$	$\frac{3}{16}$	9.995	10	0.645	$\frac{5}{8}$	640	92.2	5.98	107	21.5	2.45
W14 × 53	15.6	13.92	$13\frac{7}{8}$	0.370	$\frac{3}{8}$	$\frac{3}{16}$	8.060	8	0.660	$\frac{11}{16}$	541	77.8	5.89	57.7	14.3	1.92
× 48	14.1	13.79	$13\frac{3}{4}$	0.340	$\frac{5}{16}$	$\frac{3}{16}$	8.030	8	0.595	$\frac{5}{8}$	485	70.3	5.85	51.4	12.8	1.91
× 43	12.6	13.66	$13\frac{5}{8}$	0.305	$\frac{5}{16}$	$\frac{3}{16}$	7.995	8	0.530	$\frac{1}{2}$	428	62.7	5.82	45.2	11.3	1.89
W14 × 38	11.2	14.10	$14\frac{1}{8}$	0.310	$\frac{5}{16}$	$\frac{3}{16}$	6.770	$6\frac{3}{4}$	0.515	$\frac{1}{2}$	385	54.6	5.87	26.7	7.88	1.55
× 34	10.0	13.98	14	0.285	$\frac{5}{16}$	$\frac{3}{16}$	6.745	$6\frac{3}{4}$	0.455	$\frac{7}{16}$	340	48.6	5.83	23.3	6.91	1.53
× 30	8.85	13.84	$13\frac{7}{8}$	0.270	$\frac{1}{4}$	$\frac{1}{8}$	6.730	$6\frac{3}{4}$	0.385	$\frac{3}{8}$	291	42.0	5.73	19.6	5.82	1.49
W14 × 26	7.69	13.91	$13\frac{7}{8}$	0.255	$\frac{1}{4}$	$\frac{1}{8}$	5.025	5	0.420	$\frac{7}{16}$	245	35.3	5.65	8.91	3.54	1.08
× 22	6.49	13.74	$13\frac{3}{4}$	0.230	$\frac{1}{4}$	$\frac{1}{8}$	5.000	5	0.335	$\frac{5}{16}$	199	29.0	5.54	7.00	2.80	1.04
W12 × 336*	98.8	16.82	$16\frac{7}{8}$	1.775	$1\frac{3}{4}$	$\frac{7}{8}$	13.385	$13\frac{3}{8}$	2.955	$2\frac{15}{16}$	4,060	483	6.41	1,190	177	3.47
× 305*	89.6	16.32	$16\frac{3}{8}$	1.625	$1\frac{5}{8}$	$\frac{13}{16}$	13.235	$13\frac{1}{4}$	2.705	$2\frac{11}{16}$	3,550	435	6.29	1,050	159	3.42
× 279*	81.9	15.85	$15\frac{7}{8}$	1.530	$1\frac{1}{2}$	$\frac{3}{4}$	13.140	$13\frac{1}{8}$	2.470	$2\frac{1}{2}$	3,110	393	6.16	937	143	3.36
× 252*	74.1	15.41	$15\frac{3}{8}$	1.395	$1\frac{3}{8}$	$\frac{11}{16}$	13.005	13	2.250	$2\frac{1}{4}$	2,720	353	6.06	828	127	3.34
× 230*	67.7	15.05	15	1.285	$1\frac{5}{16}$	$\frac{11}{16}$	12.895	$12\frac{7}{8}$	2.070	$2\frac{1}{16}$	2,420	321	5.97	742	115	3.31
× 210*	61.8	14.71	$14\frac{3}{4}$	1.180	$1\frac{3}{16}$	$\frac{5}{8}$	12.790	$12\frac{3}{4}$	1.900	$1\frac{7}{8}$	2,140	292	5.89	664	104	3.28
× 190	55.8	14.38	$14\frac{3}{8}$	1.060	$1\frac{1}{16}$	$\frac{9}{16}$	12.670	$12\frac{5}{8}$	1.735	$1\frac{3}{4}$	1,890	263	5.82	589	93.0	3.25
× 170	50.0	14.03	14	0.960	$\frac{15}{16}$	$\frac{1}{2}$	12.570	$12\frac{5}{8}$	1.560	$1\frac{9}{16}$	1,650	235	5.74	517	82.3	3.22
× 152	44.7	13.71	$13\frac{3}{4}$	0.870	$\frac{7}{8}$	$\frac{7}{16}$	12.480	$12\frac{1}{2}$	1.400	$1\frac{3}{8}$	1,430	209	5.66	454	72.8	3.19
× 136	39.9	13.41	$13\frac{3}{8}$	0.790	$\frac{13}{16}$	$\frac{7}{16}$	12.400	$12\frac{3}{8}$	1.250	$1\frac{1}{4}$	1,240	186	5.58	398	64.2	3.16
× 120	35.3	13.12	$13\frac{1}{8}$	0.710	$\frac{11}{16}$	$\frac{3}{8}$	12.320	$12\frac{3}{8}$	1.105	$1\frac{1}{8}$	1,070	163	5.51	345	56.0	3.13
× 106	31.2	12.89	$12\frac{7}{8}$	0.610	$\frac{5}{8}$	$\frac{5}{16}$	12.220	$12\frac{1}{4}$	0.990	1	933	145	5.47	301	49.3	3.11

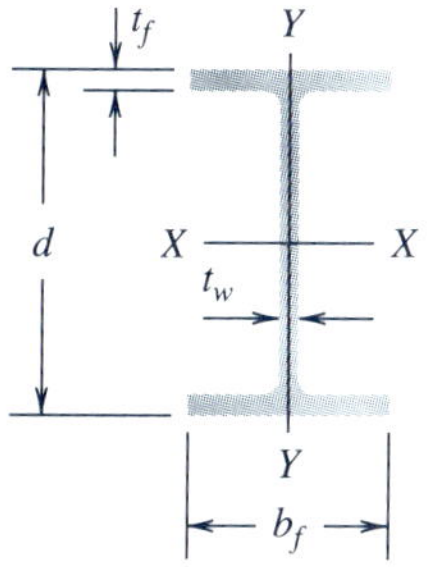

Table C.1 *Continued*

	Area	Depth		Web			Flange				Elastic Properties					
											Axis *XX*			Axis *YY*		
	A	d		Thickness t_w		$\frac{t_w}{2}$	Width b_f		Thickness t_f		I	S	r	I	S	r
Designation	in^2	in		in		in	in		in		in^4	in^3	in	in^4	in^3	in
× 96	28.2	12.71	$12\frac{3}{4}$	0.550	$\frac{9}{16}$	$\frac{5}{16}$	12.160	$12\frac{1}{8}$	0.900	$\frac{7}{8}$	833	131	5.44	270	44.4	3.09
× 87	25.6	12.53	$12\frac{1}{2}$	0.515	$\frac{1}{2}$	$\frac{1}{4}$	12.125	$12\frac{1}{8}$	0.810	$\frac{13}{16}$	740	118	5.38	241	39.7	3.07
× 79	23.2	12.38	$12\frac{3}{8}$	0.470	$\frac{1}{2}$	$\frac{1}{4}$	12.080	$12\frac{1}{8}$	0.735	$\frac{3}{4}$	662	107	5.34	216	35.8	3.05
× 72	21.1	12.25	$12\frac{1}{4}$	0.430	$\frac{7}{16}$	$\frac{1}{4}$	12.040	12	0.670	$\frac{11}{16}$	597	97.4	5.31	195	32.4	3.04
× 65	19.1	12.12	$12\frac{1}{8}$	0.390	$\frac{3}{8}$	$\frac{3}{16}$	12.000	12	0.605	$\frac{5}{8}$	533	87.9	5.28	174	29.1	3.02
W12 × 58	17.0	12.19	$12\frac{1}{4}$	0.360	$\frac{3}{8}$	$\frac{3}{16}$	10.010	10	0.640	$\frac{5}{8}$	475	78.0	5.28	107	21.4	2.51
× 53	15.6	12.06	12	0.345	$\frac{3}{8}$	$\frac{3}{16}$	9.995	10	0.575	$\frac{9}{16}$	425	70.6	5.23	95.8	19.2	2.48
W12 × 50	14.7	12.19	$12\frac{1}{4}$	0.370	$\frac{3}{8}$	$\frac{3}{16}$	8.080	$8\frac{1}{8}$	0.640	$\frac{5}{8}$	394	64.7	5.18	56.3	13.9	1.96
× 45	13.2	12.06	12	0.335	$\frac{5}{16}$	$\frac{3}{16}$	8.045	8	0.575	$\frac{9}{16}$	350	58.1	5.15	50.0	12.4	1.94
× 40	11.8	11.94	12	0.295	$\frac{5}{16}$	$\frac{3}{16}$	8.005	8	0.515	$\frac{1}{2}$	310	51.9	5.13	44.1	11.0	1.93
W12 × 35	10.3	12.50	$12\frac{1}{2}$	0.300	$\frac{5}{16}$	$\frac{3}{16}$	6.560	$6\frac{1}{2}$	0.520	$\frac{1}{2}$	285	45.6	5.25	24.5	7.47	1.54
× 30	8.79	12.34	$12\frac{3}{8}$	0.260	$\frac{1}{4}$	$\frac{1}{8}$	6.520	$6\frac{1}{2}$	0.440	$\frac{7}{16}$	238	38.6	5.21	20.3	6.24	1.52
× 26	7.65	12.22	$12\frac{1}{4}$	0.230	$\frac{1}{4}$	$\frac{1}{8}$	6.490	$6\frac{1}{2}$	0.380	$\frac{3}{8}$	204	33.4	5.17	17.3	5.34	1.51
W12 × 22	6.48	12.31	$12\frac{1}{4}$	0.260	$\frac{1}{4}$	$\frac{1}{8}$	4.030	4	0.425	$\frac{7}{16}$	156	25.4	4.91	4.66	2.31	0.847
× 19	5.57	12.16	$12\frac{1}{8}$	0.235	$\frac{1}{4}$	$\frac{1}{8}$	4.005	4	0.350	$\frac{3}{8}$	130	21.3	4.82	3.76	1.88	0.822
× 16	4.71	11.99	12	0.220	$\frac{1}{4}$	$\frac{1}{8}$	3.990	4	0.265	$\frac{1}{4}$	103	17.1	4.67	2.82	1.41	0.773
× 14	4.16	11.91	$11\frac{7}{8}$	0.200	$\frac{3}{16}$	$\frac{1}{8}$	3.970	4	0.225	$\frac{1}{4}$	88.6	14.9	4.62	2.36	1.19	0.753
W10 × 112	32.9	11.36	$11\frac{3}{8}$	0.755	$\frac{3}{4}$	$\frac{3}{8}$	10.415	$10\frac{3}{8}$	1.250	$1\frac{1}{4}$	716	126	4.66	236	45.3	2.68
× 100	29.4	11.10	$11\frac{1}{8}$	0.680	$\frac{11}{16}$	$\frac{3}{8}$	10.340	$10\frac{3}{8}$	1.120	$1\frac{1}{8}$	623	112	4.60	207	40.0	2.65
× 88	25.9	10.84	$10\frac{7}{8}$	0.605	$\frac{5}{8}$	$\frac{5}{16}$	10.265	$10\frac{1}{4}$	0.990	1	534	98.5	4.54	179	34.8	2.63
× 77	22.6	10.60	$10\frac{5}{8}$	0.530	$\frac{1}{2}$	$\frac{1}{4}$	10.190	$10\frac{1}{4}$	0.870	$\frac{7}{8}$	455	85.9	4.49	154	30.1	2.60
× 68	20.0	10.40	$10\frac{3}{8}$	0.470	$\frac{1}{2}$	$\frac{1}{4}$	10.130	$10\frac{1}{8}$	0.770	$\frac{3}{4}$	394	75.7	4.44	134	26.4	2.59
× 60	17.6	10.22	$10\frac{1}{4}$	0.420	$\frac{7}{16}$	$\frac{1}{4}$	10.080	$10\frac{1}{8}$	0.680	$\frac{11}{16}$	341	66.7	4.39	116	23.0	2.57
× 54	15.8	10.09	$10\frac{1}{8}$	0.370	$\frac{3}{8}$	$\frac{3}{16}$	10.030	10	0.615	$\frac{5}{8}$	303	60.0	4.37	103	20.6	2.56
× 49	14.4	9.98	10	0.340	$\frac{5}{16}$	$\frac{3}{16}$	10.000	10	0.560	$\frac{9}{16}$	272	54.6	4.35	93.4	18.7	2.54

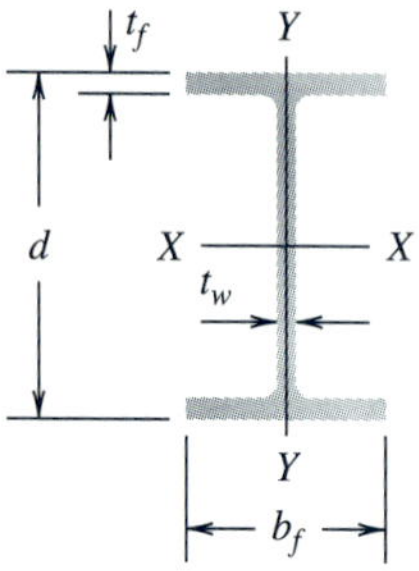

Table C.1 *Continued*

	Area	Depth		Web			Flange				Elastic Properties					
											Axis *XX*			Axis *YY*		
	A	d		Thickness t_w		$\frac{t_w}{2}$	Width b_f		Thickness t_f		I	S	r	I	S	r
Designation	in^2	in		in		in	in		in		in^4	in^3	in	in^4	in^3	in
W10 × 45	13.3	10.10	$10\frac{1}{8}$	0.350	$\frac{3}{8}$	$\frac{3}{16}$	8.020	8	0.620	$\frac{5}{8}$	248	49.1	4.32	53.4	13.3	2.01
× 39	11.5	9.92	$9\frac{7}{8}$	0.315	$\frac{5}{16}$	$\frac{3}{16}$	7.985	8	0.530	$\frac{1}{2}$	209	42.1	4.27	45.0	11.3	1.98
× 33	9.71	9.73	$9\frac{3}{4}$	0.290	$\frac{5}{16}$	$\frac{3}{16}$	7.960	8	0.435	$\frac{7}{16}$	170	35.0	4.19	36.6	9.20	1.94
W10 × 30	8.84	10.47	$10\frac{1}{2}$	0.300	$\frac{5}{16}$	$\frac{3}{16}$	5.810	$5\frac{3}{4}$	0.510	$\frac{1}{2}$	170	32.4	4.38	16.7	5.75	1.37
× 26	7.61	10.33	$10\frac{3}{8}$	0.260	$\frac{1}{4}$	$\frac{1}{8}$	5.770	$5\frac{3}{4}$	0.440	$\frac{7}{16}$	144	27.9	4.35	14.1	4.89	1.36
× 22	6.49	10.17	$10\frac{1}{8}$	0.240	$\frac{1}{4}$	$\frac{1}{8}$	5.750	$5\frac{3}{4}$	0.360	$\frac{3}{8}$	118	23.2	4.27	11.4	3.97	1.33
W10 × 19	5.62	10.24	$10\frac{1}{4}$	0.250	$\frac{1}{4}$	$\frac{1}{8}$	4.020	4	0.395	$\frac{3}{8}$	96.3	18.8	4.14	4.29	2.14	0.874
× 17	4.99	10.11	$10\frac{1}{8}$	0.240	$\frac{1}{4}$	$\frac{1}{8}$	4.010	4	0.330	$\frac{5}{16}$	81.9	16.2	4.05	3.56	1.78	0.844
× 15	4.41	9.99	10	0.230	$\frac{1}{4}$	$\frac{1}{8}$	4.000	4	0.270	$\frac{1}{4}$	68.9	13.8	3.95	2.89	1.45	0.810
× 12	3.54	9.87	$9\frac{7}{8}$	0.190	$\frac{3}{16}$	$\frac{1}{8}$	3.960	4	0.210	$\frac{3}{16}$	53.8	10.9	3.90	2.18	1.10	0.785
W8 × 67	19.7	9.00	9	0.570	$\frac{9}{16}$	$\frac{5}{16}$	8.280	$8\frac{1}{4}$	0.935	$\frac{15}{16}$	272	60.4	3.72	88.6	21.4	2.12
× 58	17.1	8.75	$8\frac{3}{4}$	0.510	$\frac{1}{2}$	$\frac{1}{4}$	8.220	$8\frac{1}{4}$	0.810	$\frac{13}{16}$	228	52.0	3.65	75.1	18.3	2.10
× 48	14.1	8.50	$8\frac{1}{2}$	0.400	$\frac{3}{8}$	$\frac{3}{16}$	8.110	$8\frac{1}{8}$	0.685	$\frac{11}{16}$	184	43.3	3.61	60.9	15.0	2.08
× 40	11.7	8.25	$8\frac{1}{4}$	0.360	$\frac{3}{8}$	$\frac{3}{16}$	8.070	$8\frac{1}{8}$	0.560	$\frac{9}{16}$	146	35.5	3.53	49.1	12.2	2.04
× 35	10.3	8.12	$8\frac{1}{8}$	0.310	$\frac{5}{16}$	$\frac{3}{16}$	8.020	8	0.495	$\frac{1}{2}$	127	31.2	3.51	42.6	10.6	2.03
× 31	9.13	8.00	8	0.285	$\frac{5}{16}$	$\frac{3}{16}$	7.995	8	0.435	$\frac{7}{16}$	110	27.5	3.47	37.1	9.27	2.02
W8 × 28	8.25	8.06	8	0.285	$\frac{5}{16}$	$\frac{3}{16}$	6.535	$6\frac{1}{2}$	0.465	$\frac{7}{16}$	98.0	24.3	3.45	21.7	6.63	1.62
× 24	7.08	7.93	$7\frac{7}{8}$	0.245	$\frac{1}{4}$	$\frac{1}{8}$	6.495	$6\frac{1}{2}$	0.400	$\frac{3}{8}$	82.8	20.9	3.42	18.3	5.63	1.61
W8 × 21	6.16	8.28	$8\frac{1}{4}$	0.250	$\frac{1}{4}$	$\frac{1}{8}$	5.270	$5\frac{1}{4}$	0.400	$\frac{3}{8}$	75.3	18.2	3.49	9.77	3.71	1.26
× 18	5.26	8.14	$8\frac{1}{8}$	0.230	$\frac{1}{4}$	$\frac{1}{8}$	5.250	$5\frac{1}{4}$	0.330	$\frac{5}{16}$	61.9	15.2	3.43	7.97	3.04	1.23
W8 × 15	4.44	8.11	$8\frac{1}{8}$	0.245	$\frac{1}{4}$	$\frac{1}{8}$	4.015	4	0.315	$\frac{5}{16}$	48.0	11.8	3.29	3.41	1.70	0.876
× 13	3.84	7.99	8	0.230	$\frac{1}{4}$	$\frac{1}{8}$	4.000	4	0.255	$\frac{1}{4}$	39.6	9.91	3.21	2.73	1.37	0.843
× 10	2.96	7.89	$7\frac{7}{8}$	0.170	$\frac{3}{16}$	$\frac{1}{8}$	3.940	4	0.205	$\frac{3}{16}$	30.8	7.81	3.22	2.09	1.06	0.841

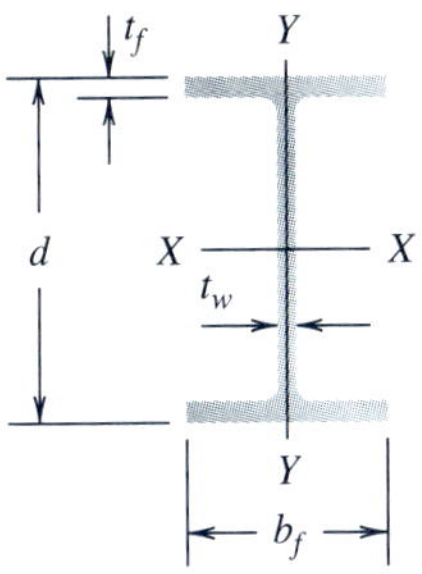

Table C.1 *Continued*

Designation	Area A	Depth d		Web Thickness t_w		Web $\frac{t_w}{2}$	Flange Width b_f		Flange Thickness t_f		Axis XX I	Axis XX S	Axis XX r	Axis YY I	Axis YY S	Axis YY r
	in²	in		in		in	in		in		in⁴	in³	in	in⁴	in³	in
W6 × 25	7.34	6.38	$6\frac{3}{8}$	0.320	$\frac{5}{16}$	$\frac{3}{16}$	6.080	$6\frac{1}{8}$	0.455	$\frac{7}{16}$	53.4	16.7	2.70	17.1	5.61	1.52
× 20	5.87	6.20	$6\frac{1}{4}$	0.260	$\frac{1}{4}$	$\frac{1}{8}$	6.020	6	0.365	$\frac{3}{8}$	41.4	13.4	2.66	13.3	4.41	1.50
× 15	4.43	5.99	6	0.230	$\frac{1}{4}$	$\frac{1}{8}$	5.990	6	0.260	$\frac{1}{4}$	29.1	9.72	2.56	9.32	3.11	1.46
W6 × 16	4.74	6.28	$6\frac{1}{4}$	0.260	$\frac{1}{4}$	$\frac{1}{8}$	4.030	4	0.405	$\frac{3}{8}$	32.1	10.2	2.60	4.43	2.20	0.966
× 12	3.55	6.03	6	0.230	$\frac{1}{4}$	$\frac{1}{8}$	4.000	4	0.280	$\frac{1}{4}$	22.1	7.31	2.49	2.99	1.50	0.918
× 9	2.68	5.90	$5\frac{7}{8}$	0.170	$\frac{3}{16}$	$\frac{1}{8}$	3.940	4	0.215	$\frac{3}{16}$	16.4	5.56	2.47	2.19	1.11	0.905
W5 × 19	5.54	5.15	$5\frac{1}{8}$	0.270	$\frac{1}{4}$	$\frac{1}{8}$	5.030	5	0.430	$\frac{7}{16}$	26.2	10.2	2.17	9.13	3.63	1.28
× 16	4.68	5.01	5	0.240	$\frac{1}{4}$	$\frac{1}{8}$	5.000	5	0.360	$\frac{3}{8}$	21.3	8.51	2.13	7.51	3.00	1.27
W4 × 13	3.83	4.16	$4\frac{1}{8}$	0.280	$\frac{1}{4}$	$\frac{1}{8}$	4.060	4	0.345	$\frac{3}{8}$	11.3	5.46	1.72	3.86	1.90	1.00

*Shapes not available from domestic producers.

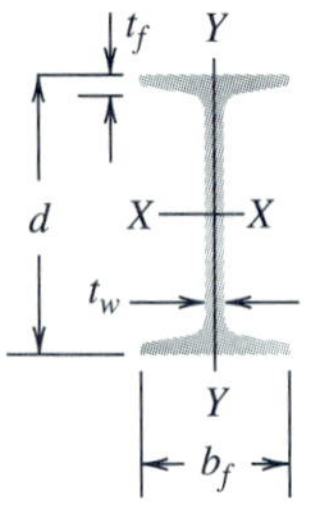

Table C.2 S Shapes: Dimensions

Designation	Area A	Depth d		Web Thickness t_w		Web $\frac{t_w}{2}$	Flange Width b_f		Flange Thickness t_f		Elastic Properties Axis XX I	Axis XX S	Axis XX r	Axis YY I	Axis YY S	Axis YY r
	in²	in		in		in	in		in		in⁴	in³	in	in⁴	in³	in
S24 × 121	35.6	24.50	$24\frac{1}{2}$	0.800	$\frac{13}{16}$	$\frac{7}{16}$	8.050	8	1.090	$1\frac{1}{16}$	3,160	258	9.43	83.3	20.7	1.53
× 106	31.2	24.50	$24\frac{1}{2}$	0.620	$\frac{5}{8}$	$\frac{5}{16}$	7.870	$7\frac{7}{8}$	1.090	$1\frac{1}{16}$	2,940	240	9.71	77.1	19.6	1.57
S24 × 100	29.3	24.00	24	0.745	$\frac{3}{4}$	$\frac{3}{8}$	7.245	$7\frac{1}{4}$	0.870	$\frac{7}{8}$	2,390	199	9.02	47.7	13.2	1.27
× 90	26.5	24.00	24	0.625	$\frac{5}{8}$	$\frac{5}{16}$	7.125	$7\frac{1}{8}$	0.870	$\frac{7}{8}$	2,250	187	9.21	44.9	12.6	1.30
× 80	23.5	24.00	24	0.500	$\frac{1}{2}$	$\frac{1}{4}$	7.000	7	0.870	$\frac{7}{8}$	2,100	175	9.47	42.2	12.1	1.34
S20 × 96	28.2	20.30	$20\frac{1}{4}$	0.800	$\frac{13}{16}$	$\frac{7}{16}$	7.200	$7\frac{1}{4}$	0.920	$\frac{15}{16}$	1,670	165	7.71	50.2	13.9	1.33
× 86	25.3	20.30	$20\frac{1}{4}$	0.660	$\frac{11}{16}$	$\frac{3}{8}$	7.060	7	0.920	$\frac{15}{16}$	1,580	155	7.89	46.8	13.3	1.36
S20 × 75	22.0	20.00	20	0.635	$\frac{5}{8}$	$\frac{5}{16}$	6.385	$6\frac{3}{8}$	0.795	$\frac{13}{16}$	1,280	128	7.62	29.8	9.32	1.16
× 66	19.4	20.00	20	0.505	$\frac{1}{2}$	$\frac{1}{4}$	6.255	$6\frac{1}{4}$	0.795	$\frac{13}{16}$	1,190	119	7.83	27.7	8.85	1.19
S18 × 70	20.6	18.00	18	0.711	$\frac{11}{16}$	$\frac{3}{8}$	6.251	$6\frac{1}{4}$	0.691	$\frac{11}{16}$	926	103	6.71	24.1	7.72	1.08
× 54.7	16.1	18.00	18	0.461	$\frac{7}{16}$	$\frac{1}{4}$	6.001	6	0.691	$\frac{11}{16}$	804	89.4	7.07	20.8	6.94	1.14
S15 × 50	14.7	15.00	15	0.550	$\frac{9}{16}$	$\frac{5}{16}$	5.640	$5\frac{5}{8}$	0.622	$\frac{5}{8}$	486	64.8	5.75	15.7	5.57	1.03
× 42.9	12.6	15.00	15	0.411	$\frac{7}{16}$	$\frac{1}{4}$	5.501	$5\frac{1}{2}$	0.622	$\frac{5}{8}$	447	59.6	5.95	14.4	5.23	1.07
S12 × 50	14.7	12.00	12	0.687	$\frac{11}{16}$	$\frac{3}{8}$	5.477	$5\frac{1}{2}$	0.659	$\frac{11}{16}$	305	50.8	4.55	15.7	5.74	1.03
× 40.8	12.0	12.00	12	0.462	$\frac{7}{16}$	$\frac{1}{4}$	5.252	$5\frac{1}{4}$	0.659	$\frac{11}{16}$	272	45.4	4.77	13.6	5.16	1.06
S12 × 35	10.3	12.00	12	0.428	$\frac{7}{16}$	$\frac{1}{4}$	5.078	$5\frac{1}{8}$	0.544	$\frac{9}{16}$	229	38.2	4.72	9.87	3.89	0.980
× 31.8	9.35	12.00	12	0.350	$\frac{3}{8}$	$\frac{3}{16}$	5.000	5	0.544	$\frac{9}{16}$	218	36.4	4.83	9.36	3.74	1.00
S10 × 35	10.3	10.00	10	0.594	$\frac{5}{8}$	$\frac{5}{16}$	4.944	5	0.491	$\frac{1}{2}$	147	29.4	3.78	8.36	3.38	0.901
× 25.4	7.46	10.00	10	0.311	$\frac{5}{16}$	$\frac{3}{16}$	4.661	$4\frac{5}{8}$	0.491	$\frac{1}{2}$	124	24.7	4.07	6.79	2.91	0.954
S8 × 23	6.77	8.00	8	0.441	$\frac{7}{16}$	$\frac{1}{4}$	4.171	$4\frac{1}{8}$	0.426	$\frac{7}{16}$	64.9	16.2	3.10	4.31	2.07	0.798
× 18.4	5.41	8.00	8	0.271	$\frac{1}{4}$	$\frac{1}{8}$	4.001	4	0.426	$\frac{7}{16}$	57.6	14.4	3.26	3.73	1.86	0.831
S7 × 20	5.88	7.00	7	0.450	$\frac{7}{16}$	$\frac{1}{4}$	3.860	$3\frac{7}{8}$	0.392	$\frac{3}{8}$	42.4	12.1	2.69	3.17	1.64	0.734
× 15.3	4.50	7.00	7	0.252	$\frac{1}{4}$	$\frac{1}{8}$	3.662	$3\frac{5}{8}$	0.392	$\frac{3}{8}$	36.7	10.5	2.86	2.64	1.44	0.766

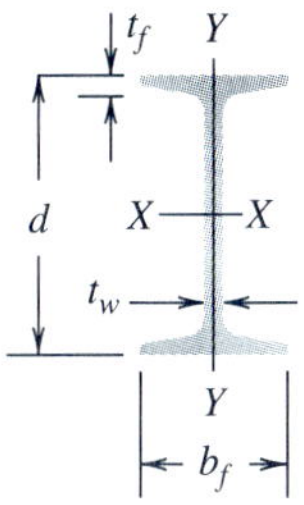

Table C.2 *Continued*

				Web			Flange				Elastic Properties					
											Axis *XX*			Axis *YY*		
	Area A	Depth d		Thickness t_w		$\frac{t_w}{2}$	Width b_f		Thickness t_f		I	S	r	I	S	r
Designation	in²	in		in		in	in		in		in⁴	in³	in	in⁴	in³	in
S6 × 17.25	5.07	6.00	6	0.465	$\frac{7}{16}$	$\frac{1}{4}$	3.565	$3\frac{5}{8}$	0.359	$\frac{3}{8}$	26.3	8.77	2.28	2.31	1.30	0.675
× 12.5	3.67	6.00	6	0.232	$\frac{1}{4}$	$\frac{1}{8}$	3.332	$3\frac{3}{8}$	0.359	$\frac{3}{8}$	22.1	7.37	2.45	1.82	1.09	0.705
S5 × 14.75	4.34	5.00	5	0.494	$\frac{1}{2}$	$\frac{1}{4}$	3.284	$3\frac{1}{4}$	0.326	$\frac{5}{16}$	15.2	6.09	1.87	1.67	1.01	0.620
× 10	2.94	5.00	5	0.214	$\frac{3}{16}$	$\frac{1}{8}$	3.004	3	0.326	$\frac{5}{16}$	12.3	4.92	2.05	1.22	0.809	0.643
S4 × 9.5	2.79	4.00	4	0.326	$\frac{5}{16}$	$\frac{3}{16}$	2.796	$2\frac{3}{4}$	0.293	$\frac{5}{16}$	6.79	3.39	1.56	0.903	0.646	0.569
× 7.7	2.26	4.00	4	0.193	$\frac{3}{16}$	$\frac{1}{8}$	2.663	$2\frac{5}{8}$	0.293	$\frac{5}{16}$	6.08	3.04	1.64	0.764	0.574	0.581
S3 × 7.5	2.21	3.00	3	0.349	$\frac{3}{8}$	$\frac{3}{16}$	2.509	$2\frac{1}{2}$	0.260	$\frac{1}{4}$	2.93	1.95	1.15	0.586	0.468	0.516
× 5.7	1.67	3.00	3	0.170	$\frac{3}{16}$	$\frac{1}{8}$	2.330	$2\frac{3}{8}$	0.260	$\frac{1}{4}$	2.52	1.68	1.23	0.455	0.390	0.522

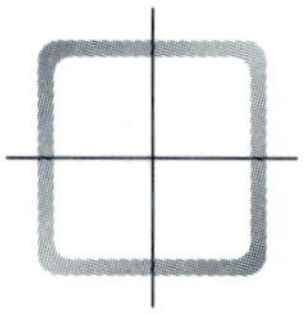

Table C.3 Square Structural Tubing: Dimensions and Properties

Dimensions				Properties*					
Nominal† Size	Wall Thickness		Weight per Foot	Area	*I*	*S*	*r*	*J*	*Z*
in	in		lb	in^2	in^4	in^3	in	in^4	in^3
8 × 8	0.6250	$\frac{5}{8}$	59.32	17.4	153	38.3	2.96	258.	47.2
	0.5625	$\frac{9}{16}$	54.17	15.9	143	35.7	3.00	238	43.6
	0.5000	$\frac{1}{2}$	48.85	14.4	131	32.9	3.03	217	39.7
	0.3750	$\frac{3}{8}$	37.69	11.1	106	26.4	3.09	170	31.3
	0.3125	$\frac{5}{16}$	31.84	9.36	90.9	22.7	3.12	145	26.7
	0.2500	$\frac{1}{4}$	25.82	7.59	75.1	18.8	3.15	118	21.9
	0.1875	$\frac{3}{16}$	19.63	5.77	58.2	14.6	3.18	90.6	16.8
7 × 7	0.5625	$\frac{9}{16}$	46.51	13.7	91.4	26.1	2.59	154	32.3
	0.5000	$\frac{1}{2}$	42.05	12.4	84.6	24.2	2.62	141	29.6
	0.3750	$\frac{3}{8}$	32.58	9.58	68.7	19.6	2.68	112	23.5
	0.3125	$\frac{5}{16}$	27.59	8.11	59.5	17.0	2.71	95.6	20.1
	0.2500	$\frac{1}{4}$	22.42	6.59	49.4	14.1	2.74	78.3	16.5
	0.1875	$\frac{3}{16}$	17.08	5.02	38.5	11.0	2.77	60.2	12.7
6 × 6	0.5625	$\frac{9}{16}$	38.86	11.4	54.1	18.0	2.18	92.9	22.7
	0.5000	$\frac{1}{2}$	35.24	10.4	50.5	16.8	2.21	85.6	20.9
	0.3750	$\frac{3}{8}$	27.48	8.08	41.6	13.9	2.27	68.5	16.8
	0.3125	$\frac{5}{16}$	23.34	6.86	36.3	12.1	2.30	58.9	14.4
	0.2500	$\frac{1}{4}$	19.02	5.59	30.3	10.1	2.33	48.5	11.9
	0.1875	$\frac{3}{16}$	14.53	4.27	23.8	7.93	2.36	37.5	9.24
5 × 5	0.5000	$\frac{1}{2}$	28.43	8.36	27.0	10.8	1.80	46.8	13.7
	0.3750	$\frac{3}{8}$	22.37	6.58	22.8	9.11	1.86	38.2	11.2
	0.3125	$\frac{5}{16}$	19.08	5.61	20.1	8.02	1.89	33.1	9.70
	0.2500	$\frac{1}{4}$	15.62	4.59	16.9	6.78	1.92	27.4	8.07
	0.1875	$\frac{3}{16}$	11.97	3.52	13.4	5.36	1.95	21.3	6.29
4.5 × 4.5	0.2500	$\frac{1}{4}$	13.91	4.09	12.1	5.36	1.72	19.7	6.43
	0.1875	$\frac{3}{16}$	10.70	3.14	9.60	4.27	1.75	15.4	5.03

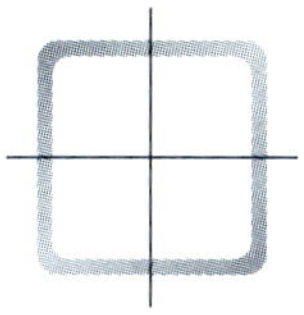

Table C.3 *Continued*

Dimensions				Properties*					
Nominal† Size	Wall Thickness		Weight per Foot	Area	I	S	r	J	Z
in	in		lb	in^2	in^4	in^3	in	in^4	in^3
4 × 4	0.5000	$\frac{1}{2}$	21.63	6.36	12.3	6.13	1.39	21.8	8.02
	0.3750	$\frac{3}{8}$	17.27	5.08	10.7	5.35	1.45	18.4	6.72
	0.3125	$\frac{5}{16}$	14.83	4.36	9.58	4.79	1.48	16.1	5.90
	0.2500	$\frac{1}{4}$	12.21	3.59	8.22	4.11	1.51	13.5	4.97
	0.1875	$\frac{3}{16}$	9.42	2.77	6.59	3.30	1.54	10.6	3.91
3.5 × 3.5	0.3125	$\frac{5}{16}$	12.70	3.73	6.09	3.48	1.28	10.4	4.35
	0.2500	$\frac{1}{4}$	10.51	3.09	5.29	3.02	1.31	8.82	3.70
	0.1875	$\frac{3}{16}$	8.15	2.39	4.29	2.45	1.34	6.99	2.93
3 × 3	0.3125	$\frac{5}{16}$	10.58	3.11	3.58	2.39	1.07	6.22	3.04
	0.2500	$\frac{1}{4}$	8.81	2.59	3.16	2.10	1.10	5.35	2.61
	0.1875	$\frac{3}{16}$	6.87	2.02	2.60	1.73	1.13	4.28	2.10
2.5 × 2.5	0.2500	$\frac{1}{4}$	7.11	2.09	1.69	1.35	0.899	2.92	1.71
	0.1875	$\frac{3}{16}$	5.59	1.64	1.42	1.14	0.930	2.38	1.40
2 × 2	0.2500	$\frac{1}{4}$	5.41	1.59	0.766	0.766	0.694	1.36	1.00
	0.1875	$\frac{3}{16}$	4.32	1.27	0.668	0.668	0.726	1.15	0.840

*Properties are based upon a nominal outside corner radius equal to two times the wall thickness.
†Outside dimensions across flat sides.

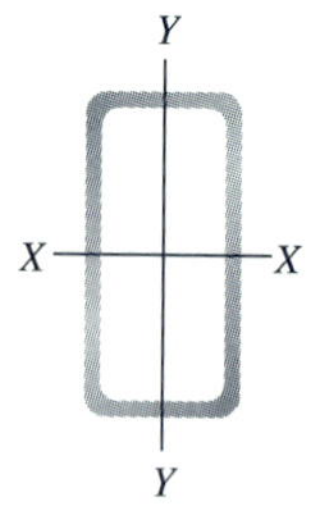

Table C.4 Rectangular Structural Tubing: Dimensions and Properties

Dimensions				Properties*									
					XX Axis				*YY* Axis				
Nominal† Size	Wall Thickness		Weight per Foot	Area	I_x	S_x	Z_x	r_x	I_y	S_y	Z_y	r_y	J
in	in		lb	in^2	in^4	in^3	in^3	in	in^4	in^3	in^3	in	in^4
7 × 4	0.5000	$\frac{1}{2}$	31.84	9.36	52.9	15.1	19.8	2.38	21.5	10.8	13.3	1.52	53.0
	0.3750	$\frac{3}{8}$	24.93	7.33	44.0	12.6	16.0	2.45	18.1	9.06	10.8	1.57	43.3
	0.3125	$\frac{5}{16}$	21.21	6.23	38.5	11.0	13.8	2.49	16.0	7.98	9.36	1.60	37.5
	0.2500	$\frac{1}{4}$	17.32	5.09	32.3	9.23	11.5	2.52	13.5	6.75	7.78	1.63	31.2
	0.1875	$\frac{3}{16}$	13.25	3.89	25.4	7.26	8.91	2.55	10.7	5.34	6.06	1.66	24.2
7 × 3	0.5000	$\frac{1}{2}$	28.43	8.36	42.3	12.1	16.6	2.25	10.5	6.99	8.84	1.12	29.8
	0.3750	$\frac{3}{8}$	22.37	6.58	35.7	10.2	13.5	2.33	9.08	6.05	7.32	1.18	25.1
	0.3125	$\frac{5}{16}$	19.08	5.61	31.5	9.00	11.8	2.37	8.11	5.41	6.40	1.20	22.0
	0.2500	$\frac{1}{4}$	15.62	4.59	26.6	7.61	9.79	2.41	6.95	4.63	5.36	1.23	18.5
	0.1875	$\frac{3}{16}$	11.97	3.52	21.1	6.02	7.63	2.45	5.57	3.71	4.21	1.26	14.6
7 × 2	0.2500	$\frac{1}{4}$	13.91	4.09	20.9	5.98	8.10	2.26	2.69	2.69	3.19	0.812	8.36
	0.1875	$\frac{3}{16}$	10.70	3.14	16.7	4.77	6.36	2.31	2.21	2.21	2.54	0.839	6.74
6 × 5	0.5000	$\frac{1}{2}$	31.84	9.36	42.9	14.3	18.1	2.14	32.1	12.8	16.0	1.85	62.9
	0.3750	$\frac{3}{8}$	24.93	7.33	35.6	11.9	14.7	2.21	26.8	10.7	12.9	1.91	50.9
	0.3125	$\frac{5}{16}$	21.21	6.23	31.2	10.4	12.7	2.24	23.5	9.40	11.2	1.94	43.9
	0.2500	$\frac{1}{4}$	17.32	5.09	26.2	8.74	10.5	2.27	19.8	7.91	9.26	1.97	36.3
	0.1875	$\frac{3}{16}$	13.25	3.89	20.6	6.87	8.15	2.30	15.6	6.23	7.20	2.00	28.1
6 × 4	0.5000	$\frac{1}{2}$	28.43	8.36	35.3	11.8	15.4	2.06	18.4	9.21	11.5	1.48	42.1
	0.3750	$\frac{3}{8}$	22.37	6.58	29.7	9.90	12.5	2.13	15.6	7.82	9.44	1.54	34.6
	0.3125	$\frac{5}{16}$	19.08	5.61	26.2	8.72	10.9	2.16	13.8	6.92	8.21	1.57	30.1
	0.2500	$\frac{1}{4}$	15.62	4.59	22.1	7.36	9.06	2.19	11.7	5.87	6.84	1.60	25.0
	0.1875	$\frac{3}{16}$	11.97	3.52	17.4	5.81	7.06	2.23	9.32	4.66	5.34	1.63	19.5
6 × 3	0.3750	$\frac{3}{8}$	19.82	5.83	23.8	7.92	10.4	2.02	7.78	5.19	6.34	1.16	20.3
	0.3125	$\frac{5}{16}$	19.96	4.98	21.1	7.03	9.11	2.06	6.98	4.65	5.56	1.18	17.9
	0.2500	$\frac{1}{4}$	13.91	4.09	17.9	5.98	7.62	2.09	6.00	4.00	4.67	1.21	15.1
	0.1875	$\frac{3}{16}$	10.70	3.14	14.3	4.76	5.97	2.13	4.83	3.22	3.68	1.24	11.9
6 × 2	0.3750	$\frac{3}{8}$	17.27	5.08	17.8	5.94	8.33	1.87	2.84	2.84	3.61	0.748	8.72

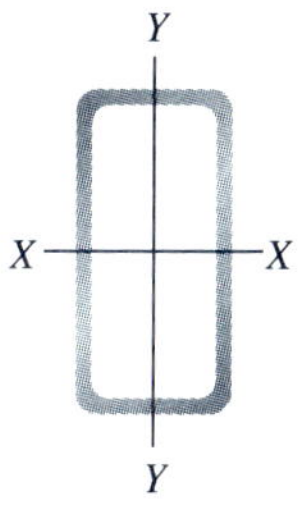

Table C.4 *Continued*

Nominal† Size (in)	Wall Thickness (in)		Weight per Foot (lb)	Area (in^2)	I_x (in^4)	S_x (in^3)	Z_x (in^3)	r_x (in)	I_y (in^4)	S_y (in^3)	Z_y (in^3)	r_y (in)	J (in^4)
	Dimensions			Properties*	*XX* Axis				*YY* Axis				
	0.3125	5/16	14.83	4.36	16.0	5.34	7.33	1.92	2.62	2.62	3.22	0.775	7.94
	0.2500	1/4	12.21	3.59	13.8	4.60	6.18	1.96	2.31	2.31	2.75	0.802	6.88
	0.1875	3/16	9.42	2.77	11.1	3.70	4.88	2.00	1.90	1.90	2.20	0.829	5.56
5 × 4	0.3750	3/8	19.82	5.83	18.7	7.50	9.44	1.79	13.2	6.58	8.08	1.50	26.3
	0.3125	5/16	16.96	4.98	16.6	6.65	8.24	1.83	11.7	5.85	7.05	1.53	22.9
	0.2500	1/4	13.91	4.09	14.1	5.65	6.89	1.86	9.98	4.99	5.90	1.56	19.1
	0.1875	3/16	10.70	3.14	11.2	4.49	5.39	1.89	7.96	3.98	4.63	1.59	14.9
5 × 3	0.5000	1/2	21.63	6.36	16.9	6.75	9.20	1.63	7.33	4.88	6.35	1.07	18.2
	0.3750	3/8	17.27	5.08	14.7	5.89	7.71	1.70	6.48	4.32	5.35	1.13	15.6
	0.3125	5/16	14.83	4.36	13.2	5.27	6.77	1.74	5.85	3.90	4.72	1.16	13.8
	0.2500	1/4	12.21	3.59	11.3	4.52	5.70	1.77	5.05	3.37	3.99	1.19	11.7
	0.1875	3/16	9.42	2.77	9.06	3.62	4.49	1.81	4.08	2.72	3.15	1.21	9.21
5 × 2	0.3125	5/16	12.70	3.73	9.74	3.90	5.31	1.62	2.16	2.16	2.70	0.762	6.24
	0.2500	1/4	10.51	3.09	8.48	3.39	4.51	1.66	1.92	1.92	2.32	0.789	5.43
	0.1875	3/16	8.15	2.39	6.89	2.75	3.59	1.70	1.60	1.60	1.86	0.816	4.40
4 × 3	0.3125	5/16	12.70	3.73	7.45	3.72	4.75	1.41	4.71	3.14	3.88	1.12	9.89
	0.2500	1/4	10.51	3.09	6.45	3.23	4.03	1.45	4.10	2.74	3.30	1.15	8.41
	0.1875	3/16	8.15	2.39	5.23	2.62	3.20	1.48	3.34	2.23	2.62	1.18	6.67
4 × 2	0.3125	5/16	10.58	3.11	5.32	2.66	3.60	1.31	1.71	1.71	2.17	0.743	4.58
	0.2500	1/4	8.81	2.59	4.69	2.35	3.09	1.35	1.54	1.54	1.88	0.770	4.01
	0.1875	3/16	6.87	2.02	3.87	1.93	2.48	1.38	1.29	1.29	1.52	0.798	3.26
3.5 × 2.5	0.2500	1/4	8.81	2.59	3.97	2.27	2.88	1.24	2.33	1.86	2.28	0.948	4.99
	0.1875	3/16	6.87	2.02	3.26	1.86	2.31	1.27	1.93	1.54	1.83	0.977	4.02
3 × 2	0.2500	1/4	7.11	2.09	2.21	1.47	1.92	1.03	1.15	1.15	1.44	0.742	2.63
	0.1875	3/16	5.59	1.64	1.86	1.24	1.57	1.06	0.977	0.977	1.18	0.771	2.16

*Properties are based upon a nominal outside corner radius equal to two times the wall thickness.
†Outside dimensions across flat sides.

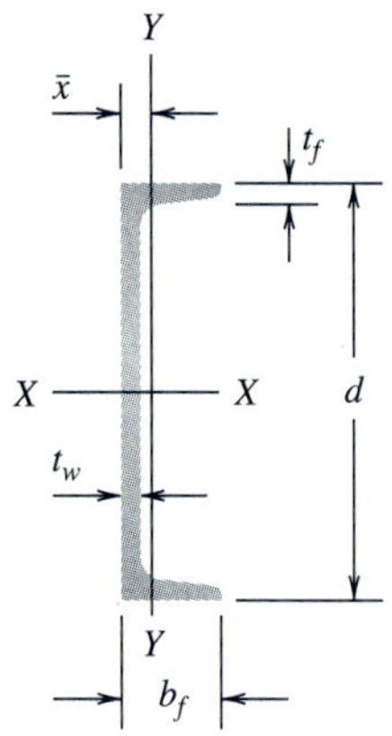

Table C.5 American Standard Channels: Dimensions

			Web			Flange				Axis XX			Axis YY			
	Area A	Depth d	Thickness t_w		$\frac{t_w}{2}$	Width b_f		Average Thickness t_f		I	S	r	I	S	r	$\bar{x}$
Designation	in^2	in	in		in	in		in		in^4	in^3	in	in^4	in^3	in	in
C15 × 50	14.7	15.00	0.716	$\frac{11}{16}$	$\frac{3}{8}$	3.716	$3\frac{3}{4}$	0.650	$\frac{5}{8}$	404	53.8	5.24	11.0	3.78	0.867	0.798
× 40	11.8	15.00	0.520	$\frac{1}{2}$	$\frac{1}{4}$	3.520	$3\frac{1}{2}$	0.650	$\frac{5}{8}$	349	46.5	5.44	9.23	3.37	0.886	0.777
× 33.9	9.96	15.00	0.400	$\frac{3}{8}$	$\frac{3}{16}$	3.400	$3\frac{3}{8}$	0.650	$\frac{5}{8}$	315	42.0	5.62	8.13	3.11	0.904	0.787
C12 × 30	8.82	12.00	0.510	$\frac{1}{2}$	$\frac{1}{4}$	3.170	$3\frac{1}{8}$	0.501	$\frac{1}{2}$	162	27.0	4.29	5.14	2.06	0.763	0.674
× 25	7.35	12.00	0.387	$\frac{3}{8}$	$\frac{3}{16}$	3.047	3	0.501	$\frac{1}{2}$	144	24.1	4.43	4.47	1.88	0.780	0.674
× 20.7	6.09	12.00	0.282	$\frac{5}{16}$	$\frac{1}{8}$	2.942	3	0.501	$\frac{1}{2}$	129	21.5	4.61	3.88	1.73	0.799	0.698
C10 × 30	8.82	10.00	0.673	$\frac{11}{16}$	$\frac{5}{16}$	3.033	3	0.436	$\frac{7}{16}$	103	20.7	3.42	3.94	1.65	0.669	0.649
× 25	7.35	10.00	0.526	$\frac{1}{2}$	$\frac{1}{4}$	2.886	$2\frac{7}{8}$	0.436	$\frac{7}{16}$	91.2	18.2	3.52	3.36	1.48	0.676	0.617
× 20	5.88	10.00	0.379	$\frac{3}{8}$	$\frac{3}{16}$	2.739	$2\frac{3}{4}$	0.436	$\frac{7}{16}$	78.9	15.8	3.66	2.81	1.32	0.692	0.606
× 15.3	4.49	10.00	0.240	$\frac{1}{4}$	$\frac{1}{8}$	2.600	$2\frac{5}{8}$	0.436	$\frac{7}{16}$	67.4	13.5	3.87	2.28	1.16	0.713	0.634
C9 × 20	5.88	9.00	0.448	$\frac{7}{16}$	$\frac{1}{4}$	2.648	$2\frac{5}{8}$	0.413	$\frac{7}{16}$	60.9	13.5	3.22	2.42	1.17	0.642	0.583
× 15	4.41	9.00	0.285	$\frac{5}{16}$	$\frac{1}{8}$	2.485	$2\frac{1}{2}$	0.413	$\frac{7}{16}$	51.0	11.3	3.40	1.93	1.01	0.661	0.586
× 13.4	3.94	9.00	0.233	$\frac{1}{4}$	$\frac{1}{8}$	2.433	$2\frac{3}{8}$	0.413	$\frac{7}{16}$	47.9	10.6	3.48	1.76	0.962	0.669	0.601
C8 × 18.75	5.51	8.00	0.487	$\frac{1}{2}$	$\frac{1}{4}$	2.527	$2\frac{1}{2}$	0.390	$\frac{3}{8}$	44.0	11.0	2.82	1.98	1.01	0.599	0.565
× 13.75	4.04	8.00	0.303	$\frac{5}{16}$	$\frac{1}{8}$	2.343	$2\frac{3}{8}$	0.390	$\frac{3}{8}$	36.1	9.03	2.99	1.53	0.854	0.615	0.553
× 11.5	3.38	8.00	0.220	$\frac{1}{4}$	$\frac{1}{8}$	2.260	$2\frac{1}{4}$	0.390	$\frac{3}{8}$	32.6	8.14	3.11	1.32	0.781	0.625	0.571
C7 × 14.75	4.33	7.00	0.419	$\frac{7}{16}$	$\frac{3}{16}$	2.299	$2\frac{1}{4}$	0.366	$\frac{3}{8}$	27.2	7.78	2.51	1.38	0.779	0.564	0.532
× 12.25	3.60	7.00	0.314	$\frac{5}{16}$	$\frac{3}{16}$	2.194	$2\frac{1}{4}$	0.366	$\frac{3}{8}$	24.2	6.93	2.60	1.17	0.703	0.571	0.525
× 9.8	2.87	7.00	0.210	$\frac{3}{16}$	$\frac{1}{8}$	2.090	$2\frac{1}{8}$	0.366	$\frac{3}{8}$	21.3	6.08	2.72	0.968	0.625	0.581	0.540

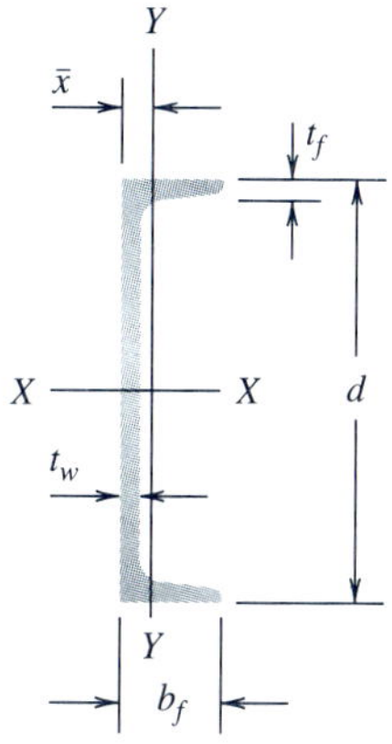

Table C.5 *Continued*

	Area	Depth	Web			Flange				Axis XX			Axis YY			
	Area A	Depth d	Thickness t_w		$\frac{t_w}{2}$	Width b_f		Average Thickness t_f		I	S	r	I	S	r	$\bar{x}$
Designation	in^2	in	in		in	in		in		in^4	in^3	in	in^4	in^3	in	in
C6 × 13	3.83	6.00	0.437	$\frac{7}{16}$	$\frac{3}{16}$	2.157	$2\frac{1}{8}$	0.343	$\frac{5}{16}$	17.4	5.80	2.13	1.05	0.642	0.525	0.514
× 10.5	3.09	6.00	0.314	$\frac{5}{16}$	$\frac{3}{16}$	2.034	2	0.343	$\frac{5}{16}$	15.2	5.06	2.22	0.866	0.564	0.529	0.499
× 8.2	2.40	6.00	0.200	$\frac{3}{16}$	$\frac{1}{8}$	1.920	$1\frac{7}{8}$	0.343	$\frac{5}{16}$	13.1	4.38	2.34	0.693	0.492	0.537	0.511
C5 × 9	2.64	5.00	0.325	$\frac{5}{16}$	$\frac{3}{16}$	1.885	$1\frac{7}{8}$	0.320	$\frac{5}{16}$	8.90	3.56	1.83	0.632	0.450	0.489	0.478
× 6.7	1.97	5.00	0.190	$\frac{3}{16}$	$\frac{1}{8}$	1.750	$1\frac{3}{4}$	0.320	$\frac{5}{16}$	7.49	3.00	1.95	0.479	0.378	0.493	0.484
C4 × 7.25	2.13	4.00	0.321	$\frac{5}{16}$	$\frac{3}{16}$	1.721	$1\frac{3}{4}$	0.296	$\frac{5}{16}$	4.59	2.29	1.47	0.433	0.343	0.450	0.459
× 5.4	1.59	4.00	0.184	$\frac{3}{16}$	$\frac{1}{16}$	1.584	$1\frac{5}{8}$	0.296	$\frac{5}{16}$	3.85	1.93	1.56	0.319	0.283	0.449	0.457
C3 × 6	1.76	3.00	0.356	$\frac{3}{8}$	$\frac{3}{16}$	1.596	$1\frac{5}{8}$	0.273	$\frac{1}{4}$	2.07	1.38	1.08	0.305	0.268	0.416	0.455
× 5	1.47	3.00	0.258	$\frac{1}{4}$	$\frac{1}{8}$	1.498	$1\frac{1}{2}$	0.273	$\frac{1}{4}$	1.85	1.24	1.12	0.247	0.233	0.410	0.438
× 4.1	1.21	3.00	0.170	$\frac{3}{16}$	$\frac{1}{16}$	1.410	$1\frac{3}{8}$	0.273	$\frac{1}{4}$	1.66	1.10	1.17	0.197	0.202	0.404	0.436

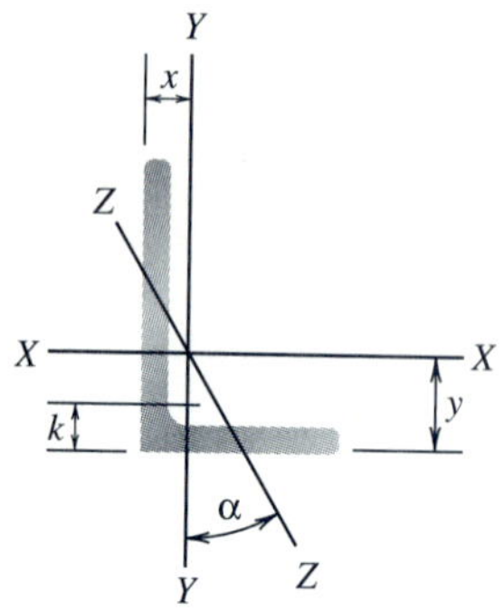

Table C.6 Angles: Equal and Unequal Legs and Properties for Designing

Size and Thickness	k	Weight per Foot	Area	Axis XX				Axis YY				Axis ZZ	
				I	S	r	y	I	S	r	x	r	
in	in	lb	in^2	in^4	in^3	in	in	in^4	in^3	in	in	in	Tan α
L5 × $3\frac{1}{2}$ × $\frac{3}{4}$	$1\frac{1}{4}$	19.8	5.81	13.9	4.28	1.55	1.75	5.55	2.22	0.977	0.996	0.748	0.464
$\frac{5}{8}$	$1\frac{1}{8}$	16.8	4.92	12.0	3.65	1.56	1.70	4.83	1.90	0.991	0.951	0.751	0.472
$\frac{1}{2}$	1	13.6	4.00	9.99	2.99	1.58	1.66	4.05	1.56	1.01	0.906	0.755	0.479
$\frac{7}{16}$	$\frac{15}{16}$	12.0	3.53	8.90	2.64	1.59	1.63	3.63	1.39	1.01	0.883	0.758	0.482
$\frac{3}{8}$	$\frac{7}{8}$	10.4	3.05	7.78	2.29	1.60	1.61	3.18	1.21	1.02	0.861	0.762	0.486
$\frac{5}{16}$	$\frac{13}{16}$	8.7	2.56	6.60	1.94	1.61	1.59	2.72	1.02	1.03	0.838	0.766	0.489
$\frac{1}{4}$	$\frac{3}{4}$	7.0	2.06	5.39	1.57	1.62	1.56	2.23	0.830	1.04	0.814	0.770	0.492
L5 × 3 × $\frac{5}{8}$	1	15.7	4.61	11.4	3.55	1.57	1.80	3.06	1.39	0.815	0.796	0.644	0.349
$\frac{1}{2}$	1	12.8	3.75	9.45	2.91	1.59	1.75	2.58	1.15	0.829	0.750	0.648	0.357
$\frac{7}{16}$	$\frac{15}{16}$	11.3	3.31	8.43	2.58	1.60	1.73	2.32	1.02	0.837	0.727	0.651	0.361
$\frac{3}{8}$	$\frac{7}{8}$	9.8	2.86	7.37	2.24	1.61	1.70	2.04	0.888	0.845	0.704	0.654	0.364
$\frac{5}{16}$	$\frac{13}{16}$	8.2	2.40	6.26	1.89	1.61	1.68	1.75	0.753	0.853	0.681	0.658	0.368
$\frac{1}{4}$	$\frac{3}{4}$	6.6	1.94	5.11	1.53	1.62	1.66	1.44	0.614	0.861	0.657	0.663	0.371
L4 × 4 × $\frac{3}{4}$	$1\frac{1}{8}$	18.5	5.44	7.67	2.81	1.19	1.27	7.67	2.81	1.19	1.27	0.778	1.000
$\frac{5}{8}$	1	15.7	4.61	6.66	2.40	1.20	1.23	6.66	2.40	1.20	1.23	0.779	1.000
$\frac{1}{2}$	$\frac{7}{8}$	12.8	3.75	5.56	1.97	1.22	1.18	5.56	1.97	1.22	1.18	0.782	1.000
$\frac{7}{16}$	$\frac{13}{16}$	11.3	3.31	4.97	1.75	1.23	1.16	4.97	1.75	1.23	1.16	0.785	1.000
$\frac{3}{8}$	$\frac{3}{4}$	9.8	2.86	4.36	1.52	1.23	1.14	4.36	1.52	1.23	1.14	0.788	1.000
$\frac{5}{16}$	$\frac{11}{16}$	8.2	2.40	3.71	1.29	1.24	1.12	3.71	1.29	1.24	1.12	0.791	1.000
$\frac{1}{4}$	$\frac{5}{8}$	6.6	1.94	3.04	1.05	1.25	1.09	3.04	1.05	1.25	1.09	0.795	1.000
L4 × $3\frac{1}{2}$ × $\frac{1}{2}$	$\frac{15}{16}$	11.9	3.50	5.32	1.94	1.23	1.25	3.79	1.52	1.04	1.00	0.722	0.750
$\frac{7}{16}$	$\frac{7}{8}$	10.6	3.09	4.76	1.72	1.24	1.23	3.40	1.35	1.05	0.978	0.724	0.753
$\frac{3}{8}$	$\frac{13}{16}$	9.1	2.67	4.18	1.49	1.25	1.21	2.95	1.17	1.06	0.955	0.727	0.755
$\frac{5}{16}$	$\frac{3}{4}$	7.7	2.25	3.56	1.26	1.26	1.18	2.55	0.994	1.07	0.932	0.730	0.757
$\frac{1}{4}$	$\frac{11}{16}$	6.2	1.81	2.91	1.03	1.27	1.16	2.09	0.808	1.07	0.909	0.734	0.759

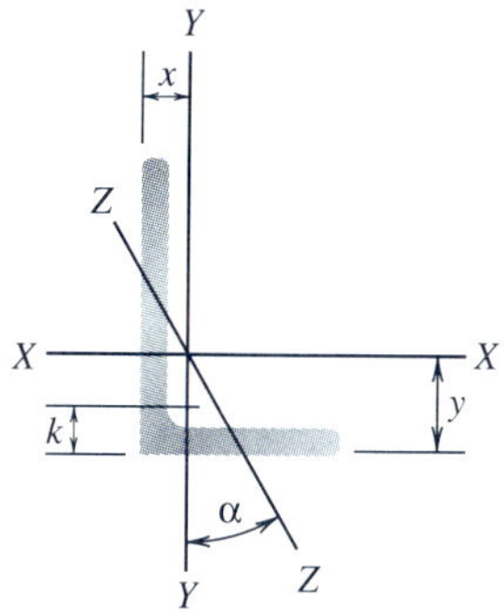

Table C.6 *Continued*

Size and Thickness		k	Weight per Foot	Area	Axis XX				Axis YY				Axis ZZ	
					I	S	r	y	I	S	r	x	r	Tan α
in		in	lb	in^2	in^4	in^3	in	in	in^4	in^3	in	in	in	
L4 × 3 ×	1/2	15/16	11.1	3.25	5.05	1.89	1.25	1.33	2.42	1.12	0.864	0.827	0.639	0.543
	7/16	7/8	9.8	2.87	4.52	1.68	1.25	1.30	2.18	0.992	0.871	0.804	0.641	0.547
	3/8	13/16	8.5	2.48	3.96	1.46	1.26	1.28	1.92	0.866	0.879	0.782	0.644	0.551
	5/16	3/4	7.2	2.09	3.38	1.23	1.27	1.26	1.65	0.734	0.887	0.759	0.647	0.554
	1/4	11/16	5.8	1.69	2.77	1.00	1.28	1.24	1.36	0.599	0.896	0.736	0.651	0.558
L3½ × 3½ ×	1/2	7/8	11.1	3.25	3.64	1.49	1.06	1.06	3.64	1.49	1.06	1.06	0.683	1.000
	7/16	13/16	9.8	2.87	3.26	1.32	1.07	1.04	3.26	1.32	1.07	1.04	0.684	1.000
	3/8	3/4	8.5	2.48	2.87	1.15	1.07	1.01	2.87	1.15	1.07	1.01	0.687	1.000
	5/16	11/16	7.2	2.09	2.45	0.976	1.08	0.990	2.45	0.976	1.08	0.990	0.690	1.000
	1/4	5/8	5.8	1.69	2.01	0.794	1.09	0.968	2.01	0.794	1.09	0.968	0.694	1.000
L3½ × 3 ×	1/2	15/16	10.2	3.00	3.45	1.45	1.07	1.13	2.33	1.10	0.881	0.875	0.621	0.714
	7/16	7/8	9.1	2.65	3.10	1.29	1.08	1.10	2.09	0.975	0.889	0.853	0.622	0.718
	3/8	13/16	7.9	2.30	2.72	1.13	1.09	1.08	1.85	0.851	0.897	0.830	0.625	0.721
	5/16	3/4	6.6	1.93	2.33	0.954	1.10	1.06	1.58	0.722	0.905	0.808	0.627	0.724
	1/4	11/16	5.4	1.56	1.91	0.776	1.11	1.04	1.30	0.589	0.914	0.785	0.631	0.727
L3½ × 2½ ×	1/2	15/16	9.4	2.75	3.24	1.41	1.09	1.20	1.36	0.760	0.704	0.705	0.534	0.486
	7/16	7/8	8.3	2.43	2.91	1.26	1.09	1.18	1.23	0.677	0.711	0.682	0.535	0.491
	3/8	13/16	7.2	2.11	2.56	1.09	1.10	1.16	1.09	0.592	0.719	0.660	0.537	0.496
	5/16	3/4	6.1	1.78	2.19	0.927	1.11	1.14	0.939	0.504	0.727	0.637	0.540	0.501
	1/4	11/16	4.9	1.44	1.80	0.755	1.12	1.11	0.777	0.412	0.735	0.614	0.544	0.506
L3 × 3 ×	1/2	13/16	9.4	2.75	2.22	1.07	0.898	0.932	2.22	1.07	0.898	0.932	0.584	1.000
	7/16	3/4	8.3	2.43	1.99	0.964	0.905	0.910	1.99	0.954	0.905	0.910	0.585	1.000
	3/8	11/16	7.2	2.11	1.76	0.833	0.913	0.888	1.76	0.833	0.913	0.888	0.587	1.000
	5/16	5/8	6.1	1.78	1.51	0.707	0.922	0.865	1.51	0.707	0.922	0.865	0.589	1.000
	1/4	5/16	4.9	1.44	1.24	0.577	0.930	0.842	1.24	0.577	0.930	0.842	0.592	1.000
	3/16	1/2	3.71	1.09	0.962	0.441	0.939	0.820	0.962	0.441	0.939	0.820	0.596	1.000

Section Properties of Sawn Lumber and Timber

Nominal Size $b \times h$	Standard Dressed Size $b \times h$	Area of Section $A = bh$	Moment of Inertia $I = \frac{bh^3}{12}$	Section Modulus $S = \frac{bh^2}{6}$	Weight per Linear Foot (Specific Weight = 40 lb/ft^3)
in	in	in^2	in^4	in^3	lb
2 × 4	1.5 × 3.5	5.250	5.359	3.063	1.5
2 × 6	1.5 × 5.5	8.250	20.80	7.563	2.3
2 × 8	1.5 × 7.25	10.88	47.63	13.14	3.0
2 × 10	1.5 × 9.25	13.88	98.93	21.39	3.9
2 × 12	1.5 × 11.25	16.88	178.0	31.64	4.7
3 × 4	2.5 × 3.5	8.750	8.932	5.104	2.4
3 × 6	2.5 × 5.5	13.75	34.66	12.60	3.8
3 × 8	2.5 × 7.25	18.12	79.39	21.90	5.0
3 × 10	2.5 × 9.25	23.12	164.9	35.65	6.4
3 × 12	2.5 × 11.25	28.12	296.6	52.73	7.8
4 × 4	3.5 × 3.5	12.25	12.50	7.146	3.4
4 × 6	3.5 × 5.5	19.25	48.53	17.65	5.3
4 × 8	3.5 × 7.25	25.38	111.1	30.66	7.0
4 × 10	3.5 × 9.25	32.38	230.8	49.91	9.0
4 × 12	3.5 × 11.25	39.38	415.3	73.83	10.9
6 × 6	5.5 × 5.5	30.25	76.26	27.73	8.4
6 × 8	5.5 × 7.5	41.25	193.4	51.56	11.5
6 × 10	5.5 × 9.5	52.25	393.0	82.73	14.5
6 × 12	5.5 × 11.5	63.25	697.1	121.2	17.6
8 × 8	7.5 × 7.5	56.25	263.7	70.31	15.6
8 × 10	7.5 × 9.5	71.25	535.9	112.8	19.8
8 × 12	7.5 × 11.5	86.25	950.5	165.3	24.0
10 × 10	9.5 × 9.5	90.25	678.8	142.9	25.1
10 × 12	9.5 × 11.5	109.2	1204	209.4	30.3
12 × 12	11.5 × 11.5	132.2	1458	253.5	36.7

Section Properties of Common Piping Sections

Pipe: Dimensions and Properties

Dimensions				Weight per Foot Plain Ends	Properties				Schedule No.
Nominal Diameter in	Outside Diameter in	Inside Diameter in	Wall Thickness in	lb	A in^2	I in^4	S in^3	r in	
Standard Weight									
$\frac{1}{2}$	0.840	0.622	0.109	0.85	0.250	0.017	0.041	0.261	40
$\frac{3}{4}$	1.050	0.824	0.113	1.13	0.333	0.037	0.071	0.334	40
1	1.315	1.049	0.133	1.68	0.494	0.087	0.133	0.421	40
$1\frac{1}{4}$	1.660	1.380	0.140	2.27	0.669	0.195	0.235	0.540	40
$1\frac{1}{2}$	1.900	1.610	0.145	2.72	0.799	0.310	0.326	0.623	40
2	2.375	2.067	0.154	3.65	1.07	0.666	0.561	0.787	40
$2\frac{1}{2}$	2.875	2.469	0.203	5.79	1.70	1.53	1.06	0.947	40
3	3.500	3.068	0.216	7.58	2.23	3.02	1.72	1.16	40
$3\frac{1}{2}$	4.000	3.548	0.226	9.11	2.68	4.79	2.39	1.34	40
4	4.500	4.026	0.237	10.79	3.17	7.23	3.21	1.51	40
5	5.563	5.047	0.258	14.62	4.30	15.2	5.45	1.88	40
6	6.625	6.065	0.280	18.97	5.58	28.1	8.50	2.25	40
8	8.625	7.981	0.322	28.55	8.40	72.5	16.8	2.94	40
10	10.750	10.020	0.365	40.48	11.9	161	29.9	3.67	40
12	12.750	12.000	0.375	49.56	14.6	279	43.8	4.38	—
Extra-Strong									
$\frac{1}{2}$	0.840	0.546	0.147	1.09	0.320	0.020	0.048	0.250	80
$\frac{3}{4}$	1.050	0.742	0.154	1.47	0.433	0.045	0.085	0.321	80
1	1.315	0.957	0.179	2.17	0.639	0.106	0.161	0.407	80
$1\frac{1}{4}$	1.660	1.278	0.191	3.00	0.881	0.242	0.291	0.524	80

Pipe: Dimensions and Properties (*Continued*)

Dimensions				Weight per Foot Plain Ends lb	Properties				Schedule No.
Nominal Diameter in	Outside Diameter in	Inside Diameter in	Wall Thickness in		A in^2	I in^4	S in^3	r in	
$1\frac{1}{2}$	1.900	1.500	0.200	3.63	1.07	0.391	0.412	0.605	80
2	2.375	1.939	0.218	5.02	1.48	0.868	0.731	0.766	80
$2\frac{1}{2}$	2.875	2.323	0.276	7.66	2.25	1.92	1.34	0.924	80
3	3.500	2.900	0.300	10.25	3.02	3.89	2.23	1.14	80
$3\frac{1}{2}$	4.000	3.364	0.318	12.50	3.68	6.28	3.14	1.31	80
4	4.500	3.826	0.337	14.98	4.41	9.61	4.27	1.48	80
5	5.563	4.813	0.375	20.78	6.11	20.7	7.43	1.84	80
6	6.625	5.761	0.432	28.57	8.40	40.5	12.2	2.19	80
8	8.625	7.625	0.500	43.39	12.8	106	24.5	2.88	80
10	10.750	9.750	0.500	54.74	16.1	212	39.4	3.63	80
12	12.750	11.750	0.500	65.42	19.2	362	56.7	4.33	—
Double Extra-Strong									
2	2.375	1.503	0.436	9.03	2.66	1.31	1.10	0.703	—
$2\frac{1}{2}$	2.875	1.771	0.552	13.69	4.03	2.87	2.00	0.844	—
3	3.500	2.300	0.600	18.58	5.47	5.99	3.42	1.05	—
4	4.500	3.152	0.674	27.54	8.10	15.3	6.79	1.37	—
5	5.563	4.063	0.750	38.55	11.3	33.6	12.1	1.72	—
6	6.625	4.897	0.864	53.16	15.6	66.3	20.0	2.06	—
8	8.625	6.875	0.875	72.42	21.3	162	37.6	2.76	—

The listed sections are available in conformance with ASTM Specification A53 Grade B or A501. Other sections are made to these specifications. Consult with pipe manufacturers or distributors for availability. Adopted from the *Manual of Steel Construction*, 9th ed.

Typical Mechanical Properties of Selected Materials

Tables F.1 through F.3 provide an abbreviated tabulation of mechanical properties of many common engineering materials. However, the values are approximate, and care must be exercised in using these numbers for design and analysis. A number of different handbooks and references exist and depending on the applications should be consulted for more extensive listings of material and mechanical properties.

Table F.1 Selected Properties: Modulus of Elasticity E, Poisson's Ratio ν, and Coefficient of Thermal Expansion α

	E			α	
Materials	10^6 psi	GPa	ν	$10^{-6}/°F$	$10^{-6}/°C$
Aluminum	10	70	0.33	13	23
Aluminum alloys	10–11.5	70–79	0.33	13	23
2014-T6	10.6	73	0.33	13	23
6061-T6	10.0	70	0.33	13	23
7075-T6	10.4	72	0.33	13	23
Brass	14–16	96–110	0.34	10.6–11.8	19.1–21.2
Brick*	1.5–3.5	10–24		3–4	5–7
Bronze	14–17	96–120	0.34	9.9–11.6	18–21
Cast iron	12–25	83–170	0.2–0.3	5.5–6.6	9.9–12
Concrete*	3.6–4.5	25–30		4–8	7–14
Copper	16–18	110–120	0.33–0.36	9.2–9.8	16.6–17.6
Glass	7–12	48–83	0.20–0.27	3–6	5–11
Granite	6–10	41–69	0.20–0.30	3–5	5–9
Magnesium alloy	6.5	45	0.35	14.5–16.0	26.1–28.8
Nickel	30	210	0.31	7.2	13
Steel					
Structural	29	200	0.33	6.5	12
High-strength alloy	29	200	0.33	8.0	14
Stainless	29	200	0.33	9.6	17
Titanium alloys	15.5	110	0.33	4.5–5.5	8–10
Tungsten	50–55	340–380	0.2	2.4	4.3
Wood (bending)					
Ash	1.5–1.6	10–11	—	—	—
Douglas fir	1.6–1.9	11–13	—	—	—
Oak	1.6–1.8	11–12	—	—	—
Southern pine	1.6–2.0	11–14	—	—	—
Wrought iron	28	190	0.3	6.5	12

*Denotes compression.
Shear modulus $G = E/[2(1 + \nu)] \approx \frac{3}{8}E$ with $\nu = 0.33$.

Table F.2 Yield Stress, Ultimate Stress, and Percent Elongation

Materials	Yield Stress		Ultimate Stress		Percent Elongation (2-in Gage Length)
	ksi	MPa	ksi	MPa	
Aluminum	3	20	10	70	60
Aluminum alloys	5–70	35–500	15–80	100–550	1–45
2014-T6	60	410	70	480	13
6061-T6	40	270	45	310	17
7075-T6	70	480	80	550	11
Brass	10–80	70–550	30–90	200–620	4–60
Brick*			1–10	7–70	
Bronze	12–100	82–690	30–120	200–830	5–60
Cast iron†	17–42	120–290	10–70	69–480	0–1
Concrete*			1.5–10	10–70	
Copper	8–48	55–330	33–55	230–380	10
Glass			5–150	30–1000	
Granite*			10–40	70–280	
Magnesium alloy	12–40	80–280	20–50	140–340	2–50
Nickel	20–90	140–620	45–110	310–760	2–50
Steel					
Structural	30–100	200–700	50–120	340–830	10–40
High-strength alloy	50–150	340–1000	80–180	550–1200	5–25
Stainless	40–100	280–700	60–150	400–1000	5–40
Titanium alloys	110–130	760–900	130–140	900–970	10
Tungsten			200–600	1400–4000	0–4
Wood (bending)					
Ash	6–10	40–70	8–14	50–100	
Douglas fir	5–8	30–50	8–12	50–80	
Oak	6–9	40–60	8–14	50–100	
Southern pine	6–9	40–60	8–14	50–100	
Wrought iron	30	210	50	340	35

*Denotes compression
†Denotes tension.

Table F.3 Specific Weight

Materials	Specific Weight lb/ft^3	Specific Weight kN/m^3
Aluminum	169	26.6
Aluminum alloys	160–180	26–28
2014-T6	175	28
6061-T6	170	26
7075-T6	175	28
Brass	520–540	82–85
Brick	110–140	17–22
Bronze	510–550	80–86
Cast iron	435–460	68–72
Concrete	145–150	23–24
Copper	550	87
Glass	150–180	24–28
Granite	160	26
Magnesium alloy	110–114	17–18
Nickel	550	87
Steel	490	77
Titanium alloys	280	44
Tungsten	1200	190
Wood (dry)	35–45	5.5–7.1
Wrought iron	470	74

Figure F.1 Bar chart of data for Young's modulus *E*; GFRPs and CFRPs are glass fiber and carbon fiber reinforced polymers. (*Courtesy Pergamon Press, from Engineering Materials, vol. 1, M. F. Ashby and D. R. H. Jones, 1980.*)

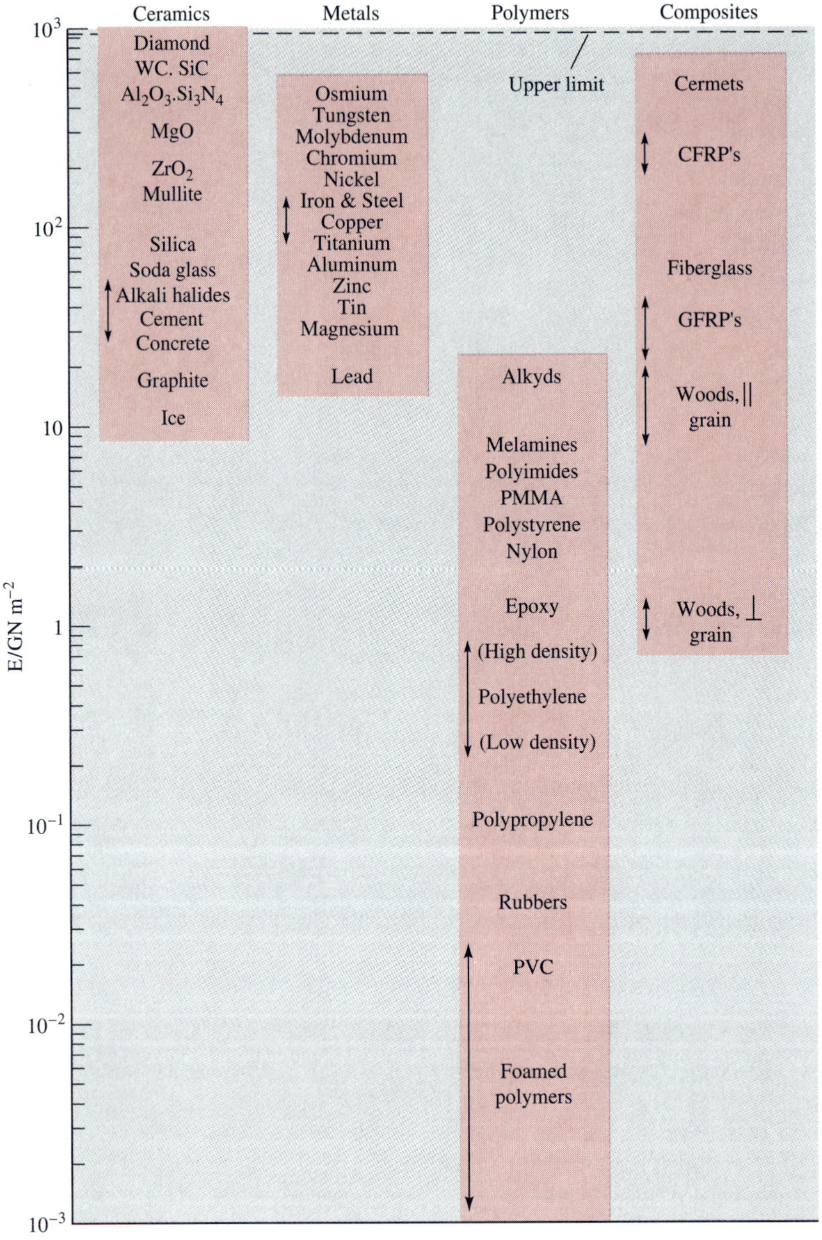

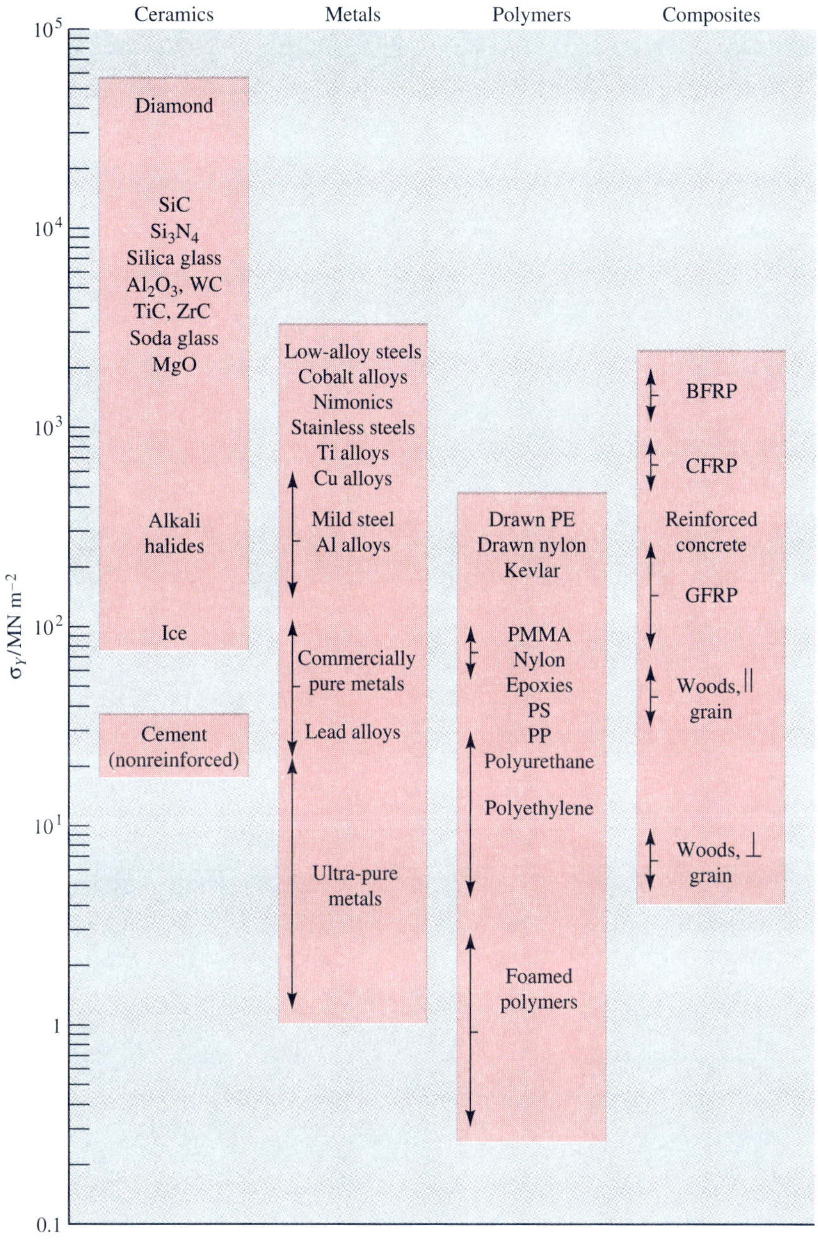

Figure F.2 Bar chart of data for yield stress (see Fig. F.1).

APPENDIX G

Deflections and Slopes of Beams

Table G.1 Cantilever Beams of Length L*

1.

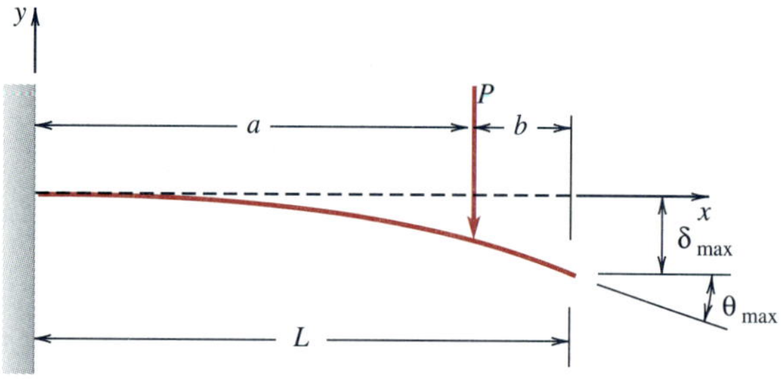

Figure G.1-1

$$\delta = \frac{P}{6EI}(\langle x - a\rangle^3 - x^3 + 3x^2a)$$

$$\delta_{max} = \frac{Pa^2(3L - a)}{6EI}$$

$$\theta_{max} = \frac{Pa^2}{2EI}$$

2.

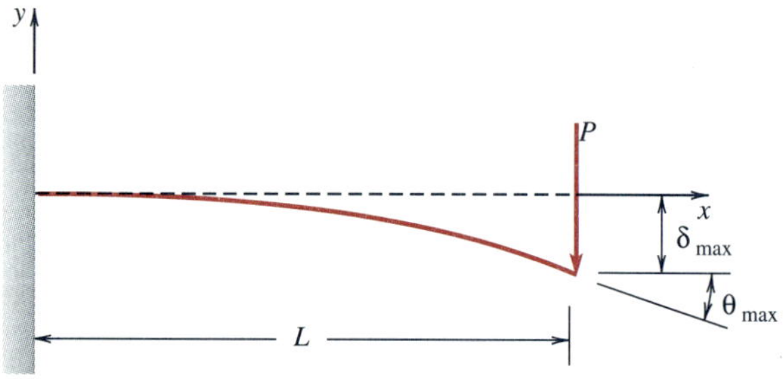

Figure G.1-2

$$\delta = \frac{Px^2}{6EI}(3L - x)$$

$$\delta_{max} = \frac{PL^3}{3EI}$$

$$\theta_{max} = \frac{PL^2}{2EI}$$

3.

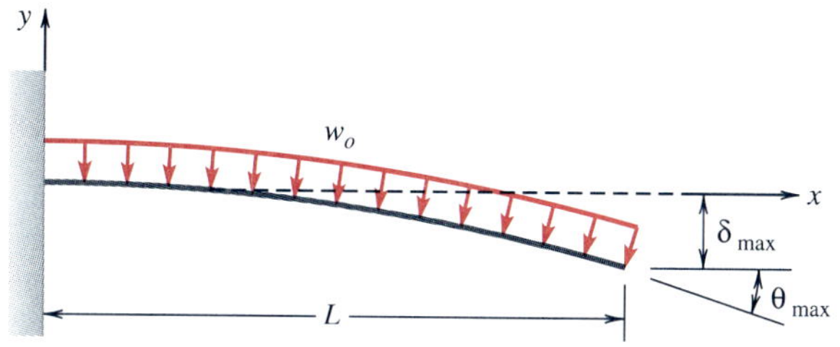

Figure G.1-3

$$\delta = \frac{w_0 x^2}{24EI}(x^2 + 6L^2 - 4Lx)$$

$$\delta_{max} = \frac{w_0 L^4}{8EI}$$

$$\theta_{max} = \frac{w_0 L^3}{6EI}$$

4.

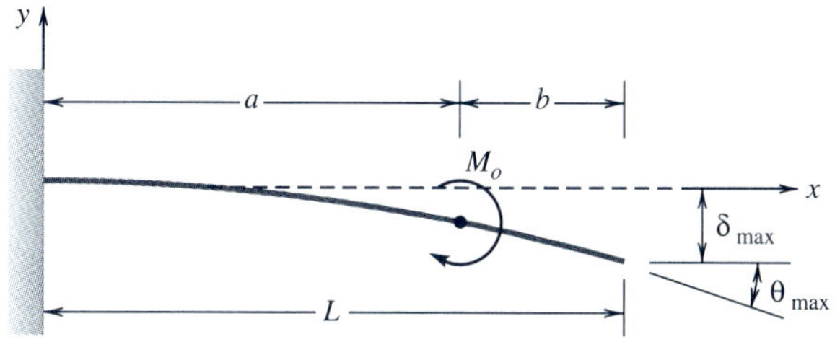

Figure G.1-4

$$\delta = \frac{M_0}{2EI}(x^2 - \langle x - a\rangle^2)$$

$$\delta_{max} = \frac{M_0 a(2L - a)}{2EI}$$

$$\theta_{max} = \frac{M_0 a}{EI}$$

5.

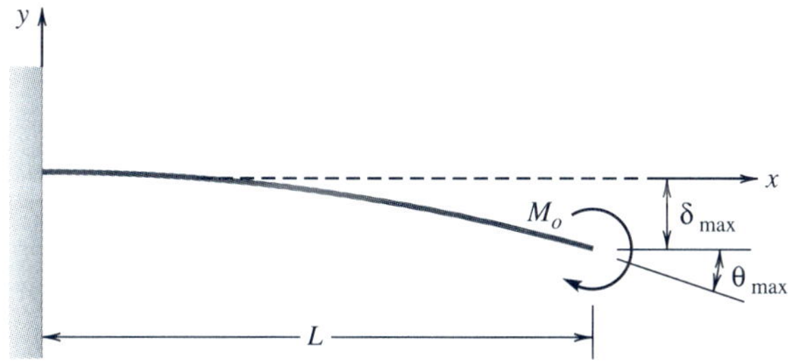

Figure G.1-5

$$\delta = \frac{M_0 x^2}{2EI}$$

$$\delta_{max} = \frac{M_0 L^2}{2EI}$$

$$\theta_{max} = \frac{M_0 L}{EI}$$

*δ is positive downward; maximum deflections and slopes are as shown.

Table G.2 Simply Supported Beams of Length L*

1.

$$\delta = \frac{Pb}{6LEI}\left[\frac{L}{b}\langle x - a\rangle^3 - x^3 + (L^2 - b^2)x\right]$$

$$\delta_{max} = \frac{Pb(L^2 - b^2)^{3/2}}{9\sqrt{3}\,LEI} \quad \text{at} \quad x = \sqrt{\frac{L^2 - b^2}{3}} \quad \text{with} \quad a \geq b$$

$$\theta_1 = \frac{Pab(2L - a)}{6LEI}$$

$$\theta_2 = \frac{Pab(2L - b)}{6LEI}$$

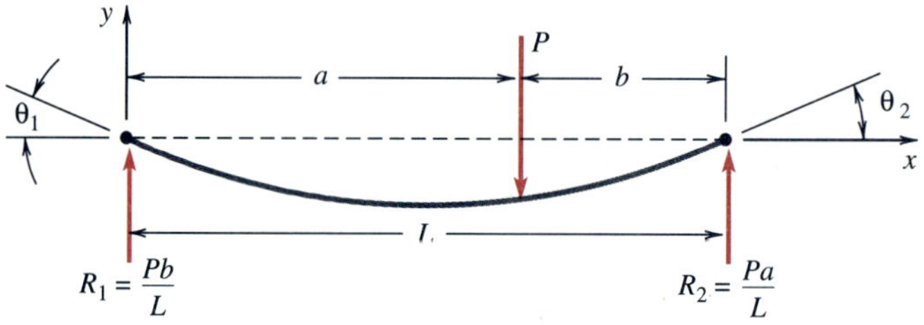

Figure G.2-1

2.

$$\delta = \frac{P}{48EI}\left(8\left\langle x - \frac{L}{2}\right\rangle^3 - 4x^3 + 3L^2x\right)$$

$$\delta_{max} = \frac{PL^3}{48EI}$$

$$\theta_1 = \theta_2 = \frac{PL^2}{16EI}$$

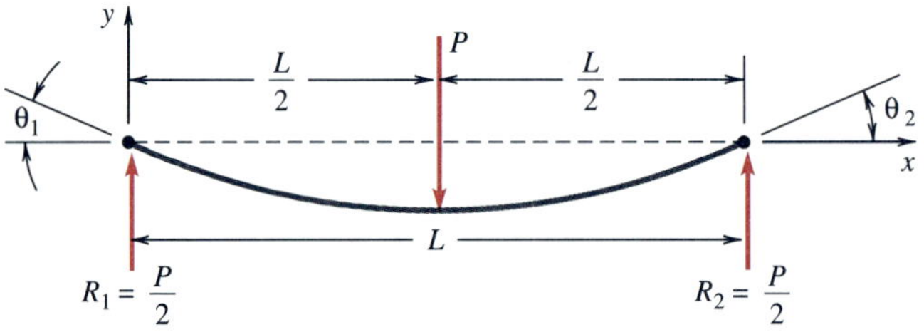

Figure G.2-2

3.

$$\delta = \frac{w_0x}{24EI}(L^3 - 2Lx^2 + x^3)$$

$$\delta_{max} = \frac{5w_0L^4}{384EI}$$

$$\theta_1 = \theta_2 = \frac{w_0L^3}{24EI}$$

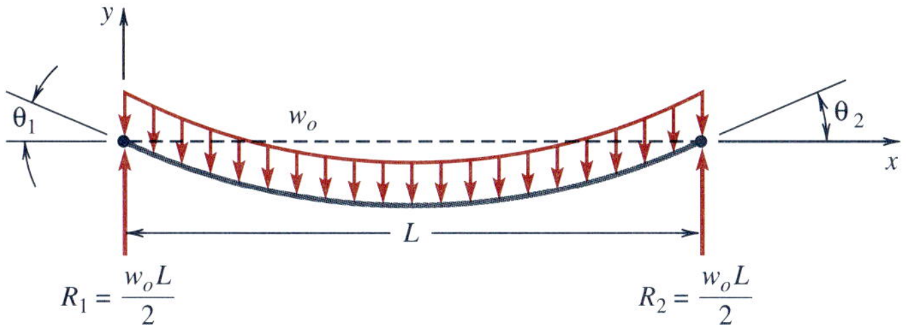

Figure G.2-3

4.

$$\delta = \frac{M_0}{6LEI}[3L\langle x - a\rangle^2 + x(L^2 - 3b^2) - x^3]$$

$$\delta_{max} = \frac{M_0(L^2 - 3b^2)^{3/2}}{9\sqrt{3}\,LEI} \quad \text{at} \quad x = \sqrt{\frac{L^2 - 3b^2}{3}} \quad \text{with} \quad a \geq b$$

$$\theta_1 = \frac{M_0}{6LEI}(L^2 - 3b^2)$$

$$\theta_2 = \frac{M_0}{6LEI}(2L^2 - 6bL + 3b^2)$$

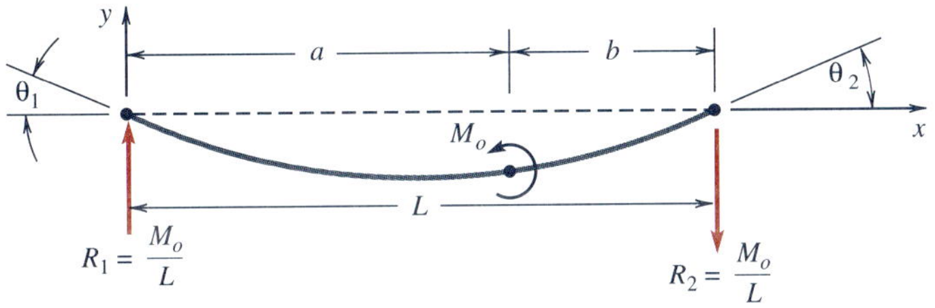

Figure G.2-4

5.

$$\delta = \frac{M_0 Lx}{6EI}\left(1 - \frac{x^2}{L^2}\right)$$

$$\delta_{max} = \frac{M_0 L^2}{9\sqrt{3}\,EI} \quad \text{at} \quad x = \frac{L}{\sqrt{3}}$$

$$\theta_1 = \frac{M_0 L}{6EI}$$

$$\theta_2 = \frac{M_0 L}{3EI}$$

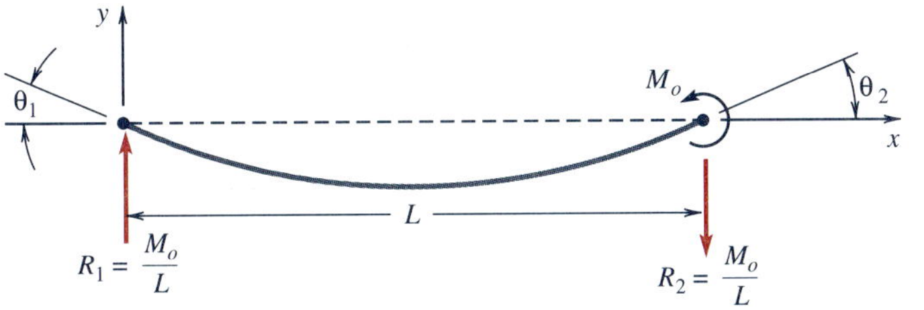

Figure G.2-5

*δ is positive downward; deflections and slopes are as shown.

APPENDIX H

Instructions for Running the Programs Given on the Diskette

The programs on the MECHMAT disk are supplied as both executable program files and then BASIC source code files.

The system requirements are an IBM personal computer or an IBM-compatible computer with a color graphics adapter. If you wish to print hard copies of screen graphics on a printer, you will need the GRAPHICS program supplied with MS DOS or PC DOS.

All programs start from the MECHMAT menu, or they can be run individually.

To reach the MECHMAT menu on typical computer systems with a hard disk drive, proceed as follows:

1. Do *not* insert the MECHMAT diskette in drive A.
2. Start up the system.
3. Once the system is up, insert MECHMAT diskette in drive A.
4. Change to drive A. Type **A:**
5. Type **GRAPHICS** and press enter. (This command loads a program into memory that allows MS DOS to print a hard copy of the screen if it contains graphics. The use of some printers requires a different command, e.g., if a Hewlett-Packard Deskjet printer is used, then type **GRAPHICS DESKJET**.)
6. Type **MECHMAT** and press enter.
7. MECHMAT menu appears.

The menus for each program are given in App. I.

On some keyboards it is only necessary to hit the "print screen" key for a hard copy; it is not necessary to hit "shift-print screen."

If you wish to see the BASIC programs in order to modify them or to run them with a BASIC interpreter, the BASIC source code files for the MECHMAT programs have been archived into a file named SOURCE.EXE. To retrieve these files, first copy SOURCE.EXE either to the hard disk or to a floppy disk with at least 200,000 bytes of free space. With the new disk, then type **SOURCE** and press enter to extract the nine BASIC files. These BASIC files can be loaded and run with most modern BASIC interpreters (QBASIC or QUICKBASIC, for example) or modified and recompiled if desired.

APPENDIX I

Summary of Computer Programs on MECHMAT Diskette, Menus for Each Program

The 11 files on the MECHMAT diskette are as follows:

BRUN45
SOURCE
MECHMAT
BARMECH
BEAMMECH
MOHRP
BEAM
CANB
TORMECH
INERTIA
CH6BM

The class of problems that can be solved by each program and reference to the corresponding chapter are as follows:

MECHMAT: main program	
BARMECH: axially loaded members	Chap. 2
TORMECH: torsion of circular shafts	Chap. 3
BEAM, CANB: shear force and bending moment diagrams	Chap. 4
INERTIA: centroids and moments of inertia	Chap. 5
BEAMMECH: deflections of beams	Chap. 7
MOHRP: Mohr's circle	Chap. 8
CH6BM: beam deflections (Fig. 6.10)	Chap. 6

Figure I.1 sketches typical problems for which the computer programs can be used. The menus for each of the programs follow.

```
**********************************************************************
*** *MECHANICS OF MATERIALS* **** PROGRAM : MECHMAT            mod 5.0
**********************************************************************
   MAIN MENU
         1. AXIALLY LOADED MEMBERS
         2. TORSION
         3. V + M DIAGRAMS OF BEAMS
         4. DEFLECTIONS OF BEAMS
         5. MOHR'S CIRCLE
         6. CENTROIDS + MOMENTS OF INERTIA
         7. EXIT
   SELECT NUMBER =?
```

```
***************************************************
        ***AXIALLY LOADED MEMBERS***
              ***BARMECH***
***************************************************
   *** DATA INPUT CHOICE ***
     1. INPUT NEW DATA FILE
     2. MODIFY EXISTING DATA FILE
     3. USE EXISTING DATA FILE
OR   4. RETURN TO MAIN MENU
        SELECT NUMBER?
```

```
***************************************************
    *** TORSION OF BARS ***
        *** TORMECH ***
***************************************************

    *** DATA INPUT CHOICE ***
     1. INPUT NEW DATA FILE
     2. MODIFY EXISTING DATA FILE
     3. USE EXISTING DATA FILE
OR   4. RETURN TO MAIN MENU
        SELECT NUMBER?
```

```
**************************************************
  BENDING OF BEAMS                     PROGRAM

          1. BEAM                        BEAM
          2. CANT BEAM                   CANB
  OR      3. RETURN TO MAIN MENU

  SELECT NUMBER =?

**************************************************
    *MOMENTS OF INERTIA*
**************************************************

      1. CHANNEL SECTION
      2. T BEAM
      3. I BEAM
      4. UNSYMMETRIC I BEAM
      5. BACK TO MENU
    SELECT NUMBER =?
                   ******************************
                   * TO OBTAIN PRINTOUT OF RESULTS *
                   * ON NEXT SCREEN, USE PRINT SCREEN*
                   * AFTER NEXT SCREEN.             *
                   * AFTER RESULTS, PRESS ANY KEY TO *
                   * RETURN TO MENU                 *
                   ******************************


**************************************************
    *** BENDING OF BEAMS ***      PROGRAM : BEAMMECH
**************************************************
    *** DATA INPUT CHOICE ***

      1. INPUT NEW DATA FILE
      2. MODIFY EXISTING DATA FILE
      3. USE EXISTING DATA FILE
OR    4. RETURN TO MAIN MENU
         SELECT NUMBER?
```

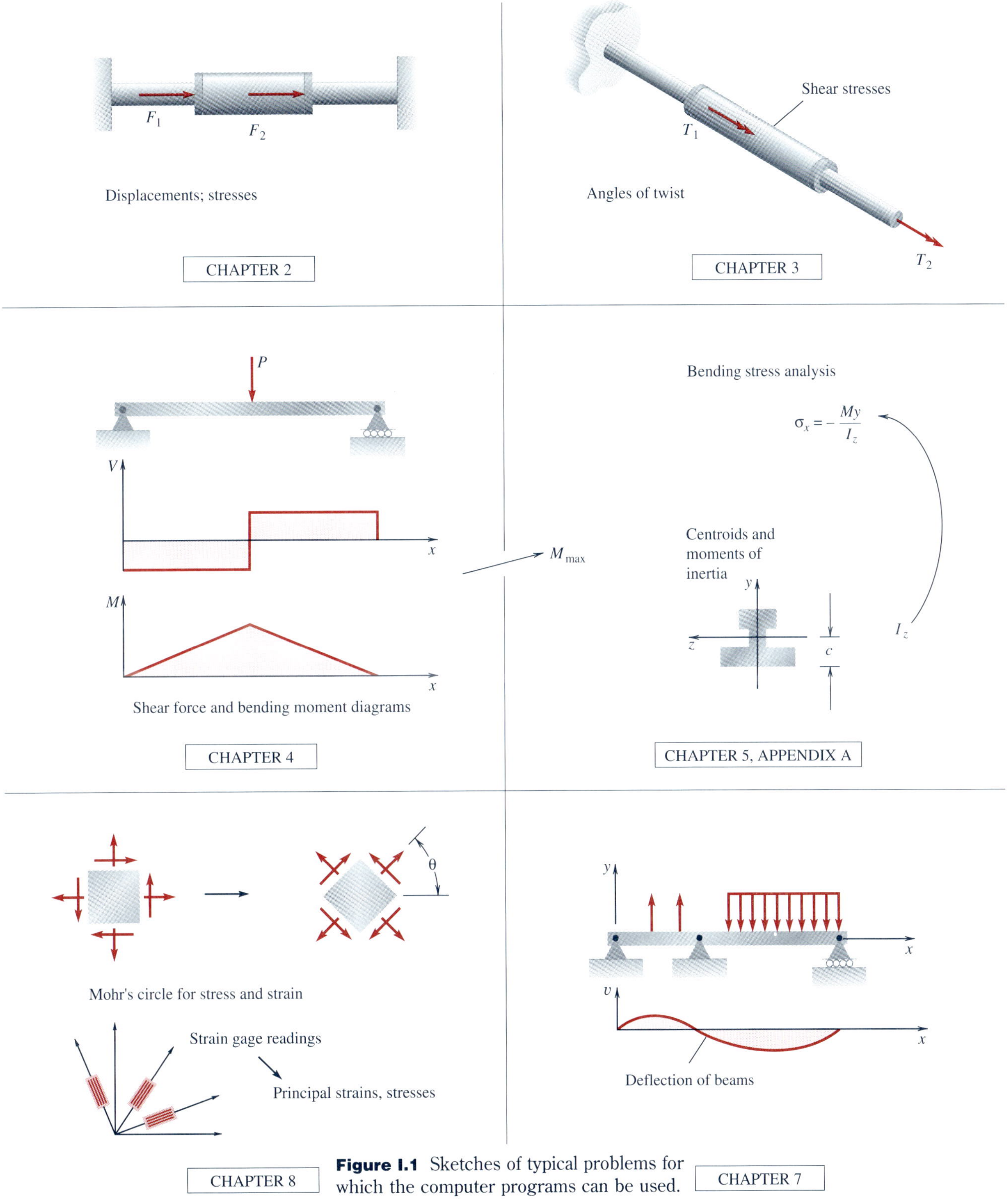

Figure I.1 Sketches of typical problems for which the computer programs can be used.

```
**********************************************
   *** MOHR'S CIRCLE ***               MOHRP
**********************************************

   1. PLANE STRESS
   2. PLANE STRAIN
   3. STRAIN ROSETTE MEASUREMENTS
OR 4. BACK TO MENU
   SELECT NUMBER?
```

APPENDIX J

Instructions for Running Specific Programs on the Diskette with Reference to Text Examples

J.1 Introduction

The following comments provide a *quick reminder* on how to run specific programs on the diskette. All programs start from the main menu of MECHMAT; see App. H and I. Alternatively, programs can be run individually using whatever version of BASIC is available.

J.2 Axially Loaded Members, BARMECH Program

See Examples 2.17, 2.18, and 2.19.

Note that the units of E in the input are 10^6 psi or MPa; that is, 30×10^6 psi is 30, 200 GPa is 200,000. All applied forces and displacements are defined relative to the x coordinate system, whose origin is at the left end of the member.

J.3 Torsion of Circular Shafts, TORMECH Program

See Examples 3.8, 3.9, and 3.10.

The same general comments apply as for the case of axially loaded members.

J.4 Shear Force and Bending Moment Diagrams, BEAM or CANB Program

See Examples 4.22, 4.23, and 4.24.

All quantities are defined relative to the xy coordinate axes: The x

axis is positive along the beam, with the origin at the left end of the beam; the y axis is positive in the ***upward*** vertical direction. The program treats only statically determinate beams and loadings in the form of concentrated forces, concentrated moments, and linear distributed loads, as shown in Fig. 4.36. Input for the distributed loads must be such that segments of the distributed loads do ***not*** overlap, and input for each of the distributed loads must start at the left end of the load segment. Linear distributed loads that change sign should be separated into positive and negative load segments.

For a cantilever beam, the built-in end at the wall must be at the left end. We need to redraw the beam configuration if it is not; care should be taken with directions of concentrated moments on the beam.

The number of points N to select along the beam is usually 20 times the length. However, if units are such that N is less than 100, we should rescale the length. If an error message should appear, rescale the length of the beam to give a larger number, e.g., if $L = 1$ foot, and this message appears, use $L = 12$ in. Values of N equal to about 500 are usually adequate. The maximum value permitted in the program is 700. The maximum value of N can be changed in the dimension statements.

Again all quantities are positive in the positive coordinate directions.

J.5 Centroids and Moments of Inertia, INERTIA Program

See Example 5.3 and App. A.

This program will calculate the location of the centroid and the value of the moment of inertia about the z axis for a channel section, a T beam, an I beam, or an unsymmetric I beam. To obtain results as printed output, use the Print Screen key.

J.6 Deflection of Beams, BEAMMECH Program

See Examples 7.13, 7.14, 7.15, and 7.16.

The sign convention is again such that all node quantities are positive in the positive coordinate directions. The beams treated can be statically indeterminate. Loading on the beam can be concentrated forces, concentrated moments, and segments of uniform loads. Units of E are 10^6 psi or MPa. Specified values of slopes are in *radians*. All units between lengths, areas, and uniform distributed loading must be consistent.

The uniform load, if any, is specified in the input data for each beam element; positive values are positive in the positive y direction. Reactions are found from the differences in the values of the shear forces and the moments at the ends of the adjoining elements.

J.7 Mohr's Circle for Stress and Strain, MOHRP Program

See Examples 8.5, 8.13, 8.14, 8.16, and 8.17.

This program will sketch Mohr's circle for stress and strain. In addition, it can be used to analyze strain rosette measurements. The circle as sketched in the output is only approximately to scale because of the plotting routine.

Solutions to Selected Problems

Chapter 1

1.2-1 $\epsilon_T = 1.59 \times 10^{-4}$

1.2-2 $\sigma_{st} = 205$ MPa, $\sigma_{al} = 69$ MPa

1.2-3 $\sigma = 0.26$ MPa

1.2-4 $\sigma_{AB} = 1990$ psi, $\sigma_{BC} = 791$ psi

1.2-5 $\sigma_{AB} = 1.80$ MPa,
$\sigma_{BC} = 7.20$ MPa

1.2-6 $\sigma = 2550$ psi

1.2-7 $\sigma_{AB} = 4W/(\pi d^2 \sin\theta)$

1.2-8 $\sigma = 0.8W/A$

1.2-9 $\delta = \sqrt{\dfrac{PL^2}{AE}}$

1.2-10 $\sigma_{AB} = 84.7$ MPa,
$\sigma_{BC} = -42.5$ MPa
(compression)

1.3-1 $\tau = 12.7$ ksi

1.3-2 $P = 118 \times 10^3$ lb

1.3-3 $T = 6185$ lb · in

1.3-4 $\sigma_{DB} = 26{,}680$ psi,
$\delta_A = 5.32 \times 10^{-2}$ in

1.3-5 $F = 8.6$ lb

1.3-6 $F = 5740$ lb

1.3-7 $F = 1964$ lb

1.3-8 $\tau = 27.3$ MPa

1.3-9 $\tau = 255$ MPa

1.3-10 $P = 1104$ lb

1.3-11 $\tau = 115$ MPa

1.3-12 $p = 7$ kPa

1.3-13 $\tau_L = 1132$ psi, $\tau_R = 679$ psi

1.3-14 $P = 7854$ lb

1.4-1 $n = 11$ (use 12) bolts

1.4-2 $d = 15.9$ mm

1.4-3 $P = 2830$ lb

1.4-4 $d = 0.74$ in

1.4-5 $\delta = 3.5$ mm

1.4-6 $\tau = 260$ psi

1.4-7 $d_i = 1.60$ in, $d_o = 2.39$ in

1.4-8 $\sigma = 5990$ psi

1.4-9 $P_{all} = 10{,}000$ lb

1.4-10 $P = 131$ lb

1.4-11 $P = 47.1$ kN

1.4-12 $P = 1993$ lb

1.4-13 $P = 1993$ lb

1.4-14 $P = 53.0 \times 10^3$ lb

1.4-15 $d = 14.7$ mm

1.4-16 $d = 15.6$ mm

1.4-17 $0 \le x \le 21.42$ in

1.4-18 $P = 7.54 \times 10^5$ lb

1.4-19 $l = 5.26$ km paper
$l = 7.08$ km linerboard

1.4-20 $p = 223$ psi, linearly with L_1

1.4-21 $A_{14} = 0.33$ in^2, $A_{57} = 0.267$ in^2

1.4-22 $d = 1.35$ in

1.4-23 $d = 30.6$ mm

1.4-24 $\sigma = 107{,}000$ psi; 7.5 mi

1.4-25 $A_1 = P/\sigma_a$, $A_2 = 0.8P/\sigma_a$

Chapter 2

2.3-1 $u_A = 3.60$ mm, $u_B = 2.70$ mm

2.3-2 $u_A = 4.95$ mm, $u_B = 2.70$ mm

2.3-3 $F_{AB} = -900$ kN, $u_A = 9.90$ mm, $u_B = 5.40$ mm

2.3-4 $u_A = 0.204$ in, $u_B = 0.144$ in, $\sigma_{AB} = -12.5$ ksi

2.3-5 $u_A = u_B + \dfrac{P_A L_1}{(EA)_{AB}}$, $u_B = \dfrac{(P_A + P_B)L_2}{(EA)_{BC}}$

2.3-6 $\sigma_{AB} = \sigma_{BC} = -1200$ psi, $u_B = -0.48 \times 10^{-3}$ in

2.3-7 $F_{AB} = 500$ lb, $u_B = 0.2 \times 10^{-3}$ in, $u_C = 0$

2.3-8 $u_B = 1.2 \times 10^{-4}$ in, $u_C = u_D = -1.6 \times 10^{-4}$ in

2.3-9 $\sigma_{AB} = 52$ MPa, $u_B = 0.078$ mm, $u_C = 0.111$ mm

2.3-10 $P_B = -22$ kN

2.3-11 $u_B = 3.58 \times 10^{-3}$ in, $u_C = 7.88 \times 10^{-3}$ in

2.3-12 $u_C = 1.56$ mm

2.3-13 $\sigma_{AB} = -400$ psi, $u_B = -80 \times 10^{-6}$ in

2.3-14 (*a*) $F_{AB} = 110$ kN, $F_{BC} = 130$ kN, $u_B = 0.0289$ mm
(*b*) $F_{AB} = F_{BC} = 60$ kN, $u_B = 0.0158$ mm
(*c*) $F_{AB} = F_{BC} = 0$, $u_D = u_E = -0.0497$ mm
(*d*) $F_{AB} = F_{BC} = 50$ kN, $u_B = 0.0132$ mm
(*e*) $F_{AB} = F_{BC} = 100$ kN, $u_B = 0.0263$ mm

2.3-15 $\sigma_1 = 18.5$ MPa, $\sigma_2 = 55.6$ MPa, $u_P = 0.351$ mm

2.3-16 $x = 1.64$ mm

2.3-17 $u_P = 3.88$ mm

2.3-18 $d_{BC} = 16$ mm

2.3-19 $A_{\text{min}} = \frac{1}{4}$ in^2, $u_B = 0.048$ in, $u_C = 0.0736$ in

2.3-20 $d_{AB} = 23$ mm, $d_{BC} = 12.6$ mm, $d_{CD} = 36.7$ mm

2.3-21 (*a*) $F_{AB} = 350$ kN, $F_{BC} = 100$ kN, $u_B = 0.7$ mm
(*b*) $F_{AB} = 40$ kips, $F_{BC} = -10$ kips, $u_B = 0.0133$ in

2.3-22 $\sigma_A = 100$ MPa, $\sigma_B = 44.4$ MPa, $u_{\text{top}} = 0.785$ mm

2.3-24 $\delta_V = 0.0182$ in, $\theta = 0.589°$

2.3-25 $\sigma_{AB} = 5$ ksi, $\sigma_{BC} = 4$ ksi, $u_B = 0.006$ in, $u_C = 0.0092$ in

2.3-26 $x = 37.5$ in, $\sigma_{AD} = 8.33$ ksi

2.4-1 Slope of load displacement is 3.47×10^9 with concrete and 2.10×10^9 without.

2.4-2 $P = 50.25 \times 10^6 A_f \dfrac{u}{L}$ (comp.), $P = 30 \times 10^6 (A_f + A_m) \dfrac{u}{L}$ (steel)

2.4-3 $u = 1.00 \times 10^{-4}$ in, $\sigma_C = 22$ psi, $\sigma_S = 335$ psi

2.4-4 $u = \dfrac{PL}{(E_1 A_1 + E_2 A_2)}$
$P = 2(E_1 A_1 + E_2 A_2) \dfrac{u}{L}$

2.4-5 $u = 6.67 \times 10^{-4}$ in, $\sigma_1 = 200$ psi, $\sigma_2 = 2000$ psi

2.4-6 $P < P_1$, $u = \dfrac{PL}{E_1 A_1}$

where $P_1 = \dfrac{E_1A_1\Delta}{L}$

$P > P_1$, $\dfrac{u - \Delta}{L} = \dfrac{P - P_1}{E_1A_1 + E_2A_2}$

2.4-7 $u_D = \dfrac{2PL_1}{E\pi d^2}$, $u_C = \dfrac{PL_1}{2E\pi d^2}$

2.4-8 $u_D = \dfrac{4PL_1(1 - x/L)}{E\pi d^2}$

$u_C = \dfrac{x}{L} \cdot \dfrac{PL_1}{E\pi d^2}$

2.4-9 $u_1 = 0.280$ in, $u_2 = 0.440$ in, $u_A = 0.493$ in
$\theta = 0.0107$ rad

2.4-10 $u_1 = 0.856$ in,
$u_2 = 0.488$ in, $u_A = 0.660$ in
$\theta = -0.02453$ rad

2.4-11 $u_1 = 0.048(30 - x) - 1.24$,
$\theta = 0.00293(30 - x) - 0.048$

2.4-12 $u_B = 0.0549$ mm,
$R_A = 18.75$ kN,
$\sigma_{AB} = 18.75$ MPa

2.4-13 $u_B = 0.121$ mm, $\sigma_{AB} =$ 82.7 MPa, $\sigma_{BC} = -18.9$ MPa

2.4-15 $u_B = \dfrac{PL_1L_2}{EA(L_1 + L_2)}$,

$R_A = \dfrac{PL_2}{L_1 + L_2}$

2.4-16 $u_B = 2.04 \times 10^{-3}$ in,
$\sigma_{AB} = 4.07$ ksi,
$\sigma_{BC} = -4.07$ ksi

2.4-17 $u_B = 0.444$ mm,
$\sigma_{AB} = 44.5$ MPa,
$\sigma_{BC} = -89$ MPa

2.4-18 $F_1 = \dfrac{2EA\Delta}{L}$,

$F_2 = 0$, $R_A = F_1$, $R_B = 0$

$F_1 = F_2 = EA\dfrac{\Delta}{L}$,

$R_A = R_B = EA\dfrac{\Delta}{L}$

2.4-19 $P_C = EA\dfrac{s}{L}$, $\Delta R_A = \dfrac{\Delta PL_2}{L}$,

$\Delta R_C = \dfrac{-\Delta PL_1}{L}$

2.4-20 $u_E = 1.65 \times 10^{-4}$ in,
$\theta = (0.94 \times 10^{-3})°$

2.4-21 $u_E = 3.82 \times 10^{-4}$ in,
$\theta = (3.82 \times 10^{-5})°$

2.4-22 $\sigma_{AB} = -1150$ psi,
$u_B = -1.38 \times 10^{-3}$ in,
$u_C = -0.831 \times 10^{-4}$ in

2.4-23 $F_a = 720$ lb

2.4-24 $F_B = 15.6$ kips (T),
$F_T = 15.6$ kips (C)

2.4-25 0.265 times one turn

2.4-26 $\sigma_B = 32.5$ ksi, $\sigma_T = -11.5$ ksi

2.4-27 $\sigma_B = 129$ MPa,
$\sigma_T = 42.9$ MPa

2.4-28 σ_T increases

2.4-29 $u_B = 0.863$ mm

2.4-30 $\phi = 0°$ (max),
$\phi = 54.7°$ (min)

2.4-31 u_A (down) $= 0.03194$ in

2.4-32 $u = 5.27 \times 10^{-3}$ in,
$\sigma_C = 479$ psi, $\sigma_S = 7185$ psi

2.4-33 $u = 4.58 \times 10^{-3}$ in,
$\sigma_C = 416$ psi, $\sigma_S = 6245$ psi

2.4-34 $P_{max} = 114.2$ kips

2.4-35 $b = 3.23$ ft

2.4-38 $b = 3.39$ ft

2.4-39 $u_A = \dfrac{3 - 4\lambda}{5}\dfrac{P}{k}$,

$\theta = \dfrac{3\lambda - 1}{5}\dfrac{P}{ka}$

2.4-42 (*a*) $\sigma_1 = \sigma_2 = -19.23$ MPa,
$u = 0.048$ mm
(*b*) $\sigma_1 = -38.46$ MPa,
$\sigma_2 = 0$, $u = 0.096$ mm

2.4-43 $P = 595$ lb

2.4-44 $c = \frac{1}{3}$ m

2.4-45 $c = \frac{1}{3}$ m

2.4-46 $F_B = 1.08$ lb, $F_s = 0.54$ lb

2.4-47 $\sigma_{BE} = 12.2$ MPa,
$\sigma_{CF} = 24.4$ MPa,
$u_D = 0.274$ mm

2.4-48 $F_B = 240$ lb, $F_C = 120$ lb,
$u_A = 0.144$ in

2.4-49 $v_c = \dfrac{8\sqrt{2}}{3}\dfrac{PL}{EA}$

2.4-50 $P_a = 560$ kips

2.4-51 $\sigma_s = 11.4$ ksi, $\sigma_A = 6.33$ ksi,
$\delta = 0.0456$ in

2.4-52 $F_s = -0.685P$

2.4-53 $u_D = 0.009$ in

2.5-1 $\Delta L_c - -12.6$ mm to $\Delta L_c =$ 11.2 mm

2.5-2 Mirror rotates $(6.67 \times 10^{-3})°$ counterclockwise.

2.5-3 $\Delta = 0.374$ in at 50°F, $\Delta =$ 0.608 in

2.5-4 $\Delta_G = 0.132$ in, $\sigma = -330$ psi

2.5-5 $\sigma = 25.1$ MPa

2.5-6 $\sigma = \dfrac{F}{A} - E\alpha\,\Delta T$

2.5-7 $F = -471$ kN

2.5-8 $F = -12.9$ kips

2.5-9 $\Delta T = 26.5°$F to close gap,
$\sigma = -13.9$ ksi

2.5-10 $u_B = 0.0168$ mm,
$R_A = 21.2$ kN (to right)

2.5-11 (*a*) $v = \dfrac{Wa}{2EA}$,

(*b*) $\Delta T = -\dfrac{W}{3\alpha_T EA}$

2.5-12 (*a*) $\Delta T = 41.7°$C,
(*b*) $\Delta T_a = 75°$C

2.5-13 (*a*) $\Delta T_1 = 63.5°$F

2.5-14 $\tau_{av} = \dfrac{F}{2A_p}$ where

$F = \Delta T(\alpha_B - \alpha_A)\dfrac{E_C E_B A_C A_B}{E_B A_B + E_C A_C}$

2.5-15 $\sigma_{AB} = -13$ ksi,
$\sigma_{BC} = -26$ ksi,
$u_B = 0.00867$ in

2.5-16 $P = EA\left(3\dfrac{u}{L} + 3\alpha\,\Delta T\right)$

2.5-17 $u_A = -0.014$ in, $F_1 = 20$ kips (C), $F_2 = 40$ kips (T)

2.5-18 (*a*) $u = 0.0676$ mm,
(*b*) $\Delta T = -17.8°$C

2.5-19 $u = \dfrac{PL}{2EA} + \dfrac{3}{2}\alpha_1\,\Delta T \quad \Delta T < 0$

2.5-20 (*a*) $\Delta T = -8.33°$F,
(*b*) $\Delta T = 30.9°$F

2.5-22 Brass: −16.5 ksi; copper: −16.2 ksi

2.7-1 $\sigma_{AB} = 25$ ksi, $\sigma_{BC} = 5$ ksi,
$\sigma_{CD} = 15$ ksi,
$u_{max} = -0.0497$ in

2.7-2 $\sigma_{AB} - -13$ ksi,
$\sigma_{BC} = -26$ ksi,
$u_B = 8.67 \times 10^{-3}$ in

2.7-3 (*a*) 39.0°C,
(*b*) $\sigma_{AB} = -135$ MPa,
B': 0.354 mm to left

2.7-4 $u_B = 2.72 \times 10^{-2}$ in,
$u_D = 9.05 \times 10^{-2}$ in

2.7-5 $P = 1.65$ kips, $u_B = -5.04 \times 10^{-3}$ in, $\sigma_{AB} = -2100$ psi

2.7-6 $P_{max} = 60.8$ kips,
$P_{max} = 1.02$ kips

2.7-7 (*a*) $\sigma_{AB} = -41.9$ MPa,
$\sigma_{BC} = -107.5$ MPa,
(*b*) $\Delta T_{max} = 65.0°$C

2.7-8 (*a*) $d = 0.564$ in,
(*b*) $\Delta T = -51.2°$F

2.7-9 $P = 31.5$ kips, $\Delta T = 34.5°$F

2.7-10 (*a*) $P = 10.6$ kN,
(*b*) $\sigma_A = -33.6$ MPa,
$\sigma_B = -134.6$ MPa,
(*c*) $u_B = 0.0192$ mm (down),
(*d*) $\Delta T = -45.87°$F,
(*e*) $\Delta T = -22.9°$C

2.8-2 19.2 ft

2.8-3 $u(L) = 1.09 \times 10^{-3}$ in, $\sigma_{max} = 136$ psi

2.8-4 $u\left(\frac{L}{4}\right) = -5.85 \times 10^{-3}$ in, $u\left(\frac{L}{2}\right) = -7.80 \times 10^{-3}$ in
$\sigma = -9750$ psi
Uniform ΔT: $u(x) = 0$ for all x, $\sigma = -19{,}500$ psi

2.8-5 $u_B = \alpha \Delta T_B \frac{L}{2}$

2.8-6 $-\frac{PL}{EA} \ln 2$

2.8-7 $u_C = \alpha\, \Delta T\, L[(\ln 2)(\ln 0.75) + 0.5]$
$\sigma_{max} = \alpha\, \Delta T\, E \ln 2$

2.8-8 $\sigma_{max} = 0.832 \frac{P}{wt}$
$u_C = 0.335 \frac{PL}{Etw}$

2.9-3 $F_T = 1.08$ kN, $u_R = 0.18$ mm

2.9-4 $F_{T_B} = 5.22 \times 10^{-4}$ N, $F_{T_S} = 20.2 \times 10^{-4}$ N, $u_{R_B} = 8.31 \times 10^{-11}$ in

2.9-5 $u_{RA} = 1.5 \times 10^{-3}$ in

2.9-7 $F = 1320$ N

2.9-8 $F_1 = 6000$ lb, $F_2 = 8000$ lb

2.9-9 $P_1 = 66.7$ kips, $P_2 = 91.2$ kips

Chapter 3

3.4-1 $k = 9.41 \times 10^4$ lb · in; $k = 1.50 \times 10^6$ lb · in

3.5-2 13.8 percent difference

3.5-3 $\tau_{maxH} = 0.745\tau_{maxS}$; $\phi_H = 2\phi_S/3$

3.6-1 $T_1 = 1.72$ kN · m, $T_2 = 8T_1$

3.6-2 $T = 1140$ N · m

3.6-3 $\phi = 9.40 \times 10^{-3}$ rad

3.6-4 $\phi = 5 \times 10^{-2}$ rad

3.6-5 $d_o = 146.8$ mm, $d_i = 124.0$ mm

3.6-6 (*a*) $d_i = 1.77$ in, (*b*) $d_i = 1.39$ in

3.6-7 $\tau_{max} = 84.5$ MPa

3.6-8 $\tau_{max} = 19.8$ MPa

3.6-9 $d = 327$ mm

3.6-10 $\phi = 4.17 \times 10^{-2}$ rad

3.6-11 (*a*) $s = -1$, (*b*) $s = -3$

3.6-12 $k = 9.41 \times 10^4$ lb · in

$$T = \frac{\phi}{\left(\frac{L}{GJ}\right)_{CD} + \left(\frac{L}{GJ}\right)_{AB}}$$

In CD: $\tau_{max} = 5.09T$
In AB: $\tau_{max} = 0.407T$

3.6-13 $\phi_B = -1.27 \times 10^{-2}$ rad, $\phi_C = -9.42 \times 10^{-2}$ rad, $\tau_{max} = 13.58$ ksi

3.6-14 $T_B = 2.5$ kip · in, $\tau_{max} = 13.6$ ksi

3.6-15 $T_B = 10.5$ kip · in

3.6-16 $\phi_B = 0.406°$, $\phi_C = 1.05°$, $\tau_{AB} = 20$ MPa, $\tau_{BC} = 26.5$ MPa

3.6-17 $\phi_B = 1.50 \times 10^{-3}$ rad, $\phi_C = -9.80 \times 10^{-3}$, $\tau_{AB} = 4.22$ MPa, $\tau_{BC} = -26.5$ MPa

3.6-18 $T = 1484$ lb · in

3.6-19 $\tau_{AB} = 102$ MPa, $\tau_{BC} = 76.4$ MPa, $\phi_C = -18.2°$

3.6-20 $\phi_B = 0.873°$, $\phi_C = 1.75°$, $\tau_{BC} = 10.4$ MPa

3.6-21 $d = 12.02$ mm

3.6-22 $d = 11.93$ mm

3.6-23 $\phi_C = 2.07°$

3.6-24 $\phi_C = T\left[\left(\frac{L}{GJ}\right)_{BC} + \left(\frac{r_U}{r_L}\right)^2\left(\frac{L}{GJ}\right)_{AB}\right]$

3.6-25 $d_o = 225.3$ mm,
$d_i = 112.6$ mm; 25 percent

3.6-26 $\phi = 1.7 \times 10^{-2}$ rad,
$\tau = 32$ MPa,
$d = 399$ mm

3.6-27 $\phi = 8.692 \times 10^{-2}$ rad,
$\tau_{max} = 6.79$ ksi

3.6-28 $T = 7.43 \times 10^5$ lb · in

3.6-29 $\phi = 1.92 \times 10^{-2}$ rad

3.6-30 $T = 1581$ lb · in

3.6-31 $\phi_A = \frac{T_0 L}{\pi}\left[\frac{32}{G_1 d^4} + \frac{1}{G_2(2a^3t)}\right]$

3.6-32 $\phi_A = -0.78°$, $\phi_B = -5.56°$
$\tau_{AB} = 26.1$ ksi, $\tau_{BC} = 16.2$ ksi

3.6-33 $k = 497$ lb/in, $\tau_{max} = 27.2$ ksi

3.6-34 $k = 921$ lb/in, $\tau_{max} = 17.1$ ksi

3.6-35 $T_A = T_B$, $\phi_A = 1.17°$

3.6-36 $\tau = 23.6$ MPa, $\phi = 0.84°$

3.6-37 $\tau_1 = 60.4$ ksi, $\tau_2 = 15.1$ ksi,
$\phi = 2.88°$

3.6-38 $\phi = 1.5°$

3.6-39 $\phi_B = \frac{TL}{G}\frac{32}{\pi d^4}\frac{31}{30}$, $\tau_{max} =$
$\frac{16T}{\pi d^3}\frac{16}{15}$

3.6-40 $T_2 = -3354$ lb · in,
$T_1 = 5318$ lb · in, $\phi = 3.18°$

3.6-41 $T = 2000$ lb · in

3.6-42 $\phi_A = \frac{326T_0L}{G\pi d^4}$, $\tau_{CB} = \frac{6T_0}{\pi d^3}$
$\tau_{BA} = \frac{80T_0}{\pi d^3}$

3.6-43 $T = -414$ lb · in

3.7-1 $T = 109$ lb · ft

3.7-2 $d = 0.97$ in

3.7-3 $\tau_{max} = 1173$ psi

3.7-4 $\tau_{max} = 286$ psi

3.7-5 $d = 1.5$ in

3.7-8 $T = 67{,}570$ lb · in

3.7-9 $T = 13.51 \times 10^4$ lb · in,
$\phi_C = 3.2 \times 10^{-2}$ rad

3.7-10 $T_1 = 46.7$ N · m,
$\tau_{max} = 29.7$ MPa, $\phi = 0.567°$

3.7-11 $T_1 = 46.7$ N · m,
$\tau_{max} = 8.80$ MPa, $\phi = 0.112°$

3.7-12 $T = 25.9$ N · m

3.7-13 $T = 158.4$ lb · in

3.7-14 $T = 39.2$ N · m

3.7-15 $T = 33.4$ N · m

3.7-16 $T = 2290$ lb

3.7-17 (*a*) $d = 0.75$ in,
(*b*) $T = 1074$ lb · in,

3.7-18 $T_1 = 322$ N · m,
$T_3 = 122$ N · m,
$\phi_2 = 1.57°$, $\phi_3 = -0.594°$

3.7-19 $\phi = 0.413°$, $\tau_{br} = 3.24$ MPa,
$\tau_{st} = 12.6$ MPa

3.7-20 $\phi = 0.02°$, $\tau_{max} = 209$ psi

3.7-21 $\phi = 0.016°$, $\tau_{max} = 166$ psi

3.7-22 $\phi_3 = 0.0544°$,
$\tau_{max} = 1.58$ MPa

3.7-23 $\phi_3 = 0.0592°$,
$T_1 = T_2 = 80.3$ N · m,
$T_3 = -619.7$ N · m

3.7-24 (*a*) $\tau_{max} = 1220$ psi,
$\phi_B = 0.267°$,
$\phi_C = 0.110°$
(*b*) $\tau_{max} = 1080$ psi,
$\phi_B = 1.92 \times 10^{-3}$ rad,
$\phi_C = 5.76 \times 10^{-3}$ rad
(*c*) $\tau_{max} = 1900$ psi,

$\phi_B = 7.24 \times 10^{-3}$ rad,
$\phi_C = 9.66 \times 10^{-3}$ rad

3.7-25 $k = 3.14\ \mathrm{N \cdot m}$

3.7-26 $\tau_{max} = 63.7$ MPa

3.7-27 $\phi_A =$

$$T\left[\left(\frac{L_1}{GJ}\right) + \frac{1}{\left(\frac{GJ}{L_2}\right)_{tube} + \left(\frac{GJ}{L_3}\right)_2}\right]$$

3.7-28 $\Delta\phi = 3.01 \times 10^{-4}$ rad,
$\tau_s = 0.68$ MPa, $\tau_a = 1.14$ MPa

3.7-29 $T_a = -17{,}000\ \mathrm{lb \cdot in}$,
$T_b = -5000\ \mathrm{lb \cdot in}$,
$T_C = 19{,}000\ \mathrm{lb \cdot in}$

3.7-30 $\tau_{max} = 12.1$ ksi

3.7-31 $d = 27$ mm

3.7-32 $\phi_A = \dfrac{Ta}{16GJ}, J = \dfrac{\pi d^4}{2}$

3.7-33 $\tau_{AC} = 1280$ psi, $\tau_{CB} = 0$

3.8-2 $x = 41.25$ in

3.9-4 $\phi_C = -7.385 \times 10^{-2}$ rad,
$\tau_{max} = 13.58$ ksi

3.9-5 $T = 1407\ \mathrm{lb \cdot in}$

3.9-6 $\tau_{max} = 21.88$ ksi,
$T = 3.17 \times 10^3\ \mathrm{lb \cdot in}$

3.9-7 (*a*) $T = 3700\ \mathrm{lb \cdot in}$,
(*b*) $d = 1.17$ in

3.9-8 (*a*) $d_{AB} = 76$ mm,
$d_{BC} = 98$ mm,
(*b*) $\phi_{AB} = 1.9 \times 10^{-2}$ rad,
$\phi_{CB} = 6.0 \times 10^{-2}$ rad

3.9-9 $\tau = 18.6$ MPa

3.9-10 $\tau_{max} = 68.1$ MPa,
$T = 361\ \mathrm{N \cdot m}$, $\phi_B = 0.62°$

3.9-11 (*a*) $495\ \mathrm{N \cdot m}$,
(*b*) $-660\ \mathrm{N \cdot m} < T_C < 0$,
(*c*) $742\ \mathrm{N \cdot m}$

3.9-12 (*a*) $d = 0.0298$ m,
(*b*) $T = 476\ \mathrm{N \cdot m}$

3.9-13 (*a*) $\tau_{max} = 382$ MPa
(*b*) $\phi_B = -0.12°$, $\phi_D = 21.8°$

3.10-1 $P = 2490$ hp,
$\phi = 4.8 \times 10^{-2}$ rad

3.10-2 $\tau_{max} = 27.4$ ksi

3.10-3 $d = 10.4$ in

3.10-4 $\tau_{max} = 4790$ psi,
$\phi = 3.06 \times 10^{-2}$ rad

3.10-5 $d_i = 8.85$ in, $d_o = 14.2$ in

3.10-6 $d = 3.77$ in,
$\phi = 4.1 \times 10^{-2}$ rad

3.10-7 $W_S = 1.55W_H$

3.10-8 $d_o = 11.4$ in

3.10-9 $d_o = 11.55$ in

3.11-2 $\phi_B = 3.18 \times 10^{-2}$ rad

3.11-4 $L = [\pi r^3/(2q_0)]\tau_{max}$,
$\phi = \tau_{max} L/(G2r)$

3.11-5 $\phi = 62.4°$

3.12-1 $\dfrac{d\phi}{dx} = \dfrac{49T^2k}{16\pi^2 r^7}$, $\tau_{max} = \dfrac{7T}{4\pi r^3}$

3.12-4 $T = \dfrac{4}{3}T_Y\left(\dfrac{31}{32}\right)$

$\tau_{r\,max} = 0.354\tau_Y$,
$\tau_{r\,min} = 0.292\tau_Y$

Chapter 4

4.2-2 $V(4) = -50$ kN, $V(14) = 0$;
$M(4) = 200\ \mathrm{kN \cdot m}$,
$M(14) = 300\ \mathrm{kN \cdot m}$

4.2-3 (*a*) $V = -P, M = -\frac{2}{3}PL$
(*b*) $V = 0, M = 0$

(c) $V = \frac{P}{2}, M = \frac{PL}{8}$
(d) $V = 0, M = \frac{PL}{4}$
(e) $V = -P, M = P(L - c)$
(f) $V = 0, M = -Pa$
(g) $V = -4P, M = -\frac{5}{6}PL$
(h) $V = \frac{-P\sqrt{2}}{2}$,
$M = \frac{-PL\sqrt{2}}{2}$,
$N = \frac{P\sqrt{2}}{2}$

4.2-4 (a) $V = 0.9$ kips, $M = 19.5$ kip · ft
(b) $V = 8.5$ kips, $M = 51$ kip · ft
(c) $V = 102.9$ kN, $M = 34.3$ kN · m
(d) $V = -10$ kN, $M = -20$ kN · m
(e) $V = 0, M = 180$ kN · m
(f) $V = -5$ kips, $M = -15$ kip · ft
(g) $V = -4$ kips, $M = 12$ kip · ft
(h) $V = 8.57$ kN, $M = 42.86$ kN · m

4.2-5 (a) $V = W, M = WL/8$
(b) $V = 2.4$ kN, $M = 2.4$ kN · m
(c) $V = 333$ lb, $M = 667$ lb · ft
(d) $V = 0, M = 0$

4.2-6 $V_D = -400$ lb, $M_D = 900$ lb · ft, $V_C = 307$ lb, $M_C = 2707$ lb · ft

4.2-7 $V = -Q \cos\theta, N = Q \sin\theta$, $M = -Qa(1 - \sin\theta)$

4.2-8 $V = P \sin\theta$, $M = P(a - b) \sin\theta - Pd \cos\theta$, $N = P \cos\theta$

4.2-9 $V = -F \cos(\phi - \theta)$, $N = F \sin(\phi - \theta)$, $M = -Fa(1 - \cos\theta) \sin\phi - Fa \cos\theta \cos\phi$

4.2-10 (c) $V = -P \cos(\phi - \theta)$, $N = P \sin(\phi - \theta)$, $M = aP \sin\phi \sin\theta - aP \cos\phi\,(1 + \cos\theta)$

4.3-1 $a = L/2$

4.3-3 $V = -\frac{w_0}{2L}(L - x)^2$, $M = -\frac{w_0}{6L}(L - x)^3$

4.3-6 $M_{max} = 12w_0L^2/128$

4.3-7 $M_{max} = 5.4$ kN · m

4.3-8 $R = \frac{a - b}{2a}P$, $M = Ra(1 - \cos\theta)$, $N = R \cos\theta, V = -R \sin\theta$, arbitrary b; $0 \le \theta < \theta_b$

4.4-3 $V_{max} = \frac{w_0L}{2}$, $M_{max} = \frac{w_0kL^2}{8}$

4.4-4 $M_{max} = 5.06$ kN · m

4.4-5 $M_{max} = 3.42 \times 10^4$ lb · in

4.4-6 (e) $M_{max} = 30$ N · m

4.4-7 $R_1 = 84.375$ lb, $R_2 = 165.625$ lb, $M_{max} = 274$ lb · ft

4.4-9 (a) $M_{max} = 11.25 \times 10^4$ lb · ft
(b) $M_{max} = 96.6 \times 10^3$ lb · ft
(c) $M_{max} = 81.45 \times 10^3$ lb · ft
(d) $M_{max} = 57.0$ kN · m
(e) $M_{max} = 22.0$ kN · m, $V_{max} = 11.5$ kN

4.4-14 $M_{max} = W\left(a - \frac{s}{2}\right)$

4.4-15 $V_{max} = 3200$ lb, $M_{max} = 9650$ lb · ft

4.5-1 (a) $q(x) = R_A\langle x\rangle_{-1} - 1.5(\langle x\rangle^0 - \langle x - 12\rangle^0) - 5\langle x - 16\rangle_{-1} + R_B\langle x - 20\rangle_{-1}$

(b) $q(x) = M_0\langle x\rangle_{-2} + R_A\langle x - a\rangle_{-1} + R_B\langle x - (a + b)\rangle_{-1} - P\langle x - (a + b + c)\rangle_{-1}$

(c) $q(x) = R_A\langle x\rangle_{-1} - 200(\langle x\rangle^0 - \langle x - 5\rangle^0) + R_B\langle x - 10\rangle_{-1} - 1000\langle x - 16\rangle_{-1}$

(d) $q(x) = R_A\langle x\rangle_{-1} - 8\langle x - 1\rangle_{-1} + R_B\langle x - 3\rangle_{-1} - 3\langle x - 4\rangle_{-2}$

(e) $q(x) = R_A\langle x\rangle_{-1} - 300(\langle x\rangle^0 - \langle x - 5\rangle^0) + 15{,}000\langle x - 10\rangle_{-2} - 300\langle x - 15\rangle^0 + R_B\langle x - 20\rangle_{-1}$

(f) $q(x) = -5\langle x\rangle^0 + R_A\langle x - 1\rangle_{-1} + R_B\langle x - 3\rangle_{-1}$

(g) $q(x) = R_w\langle x\rangle_{-1} - M_w\langle x\rangle_{-2} - \langle x - 4\rangle_{-1} + 10\langle x - 7\rangle_{-2}$

(h) $q(x) = -2\langle x\rangle_{-1} - 2\langle x - 1.6\rangle^0 + R_w\langle x - 2.6\rangle_{-1} - M_w\langle x - 2.6\rangle_{-2}$

(i) $q(x) = -7.5\langle x\rangle_{-1} + q_0(\langle x - 14\rangle^0 - \langle x - 18\rangle^0) - 6\langle x - 36\rangle_{-1}$

(j) $q(x) = -800\langle x\rangle^0 + R_A\langle x\rangle_{-1} - 20\langle x - 30\rangle_{-1} + R_B\langle x - 45\rangle_{-1} - 10\langle x - 60\rangle_{-1}$

4.5-2 (a) $q(x) = M_w\langle x\rangle_{-2} + R_w\langle x\rangle_{-1} - P\langle x - L\rangle_{-1}$

(b) $q(x) = M_w\langle x\rangle_{-2} + R_w\langle x\rangle_{-1} - P\left\langle x - \frac{L}{2}\right\rangle_{-1}$

(c) $q(x) = \frac{P}{2}\langle x\rangle_{-1} - P\left\langle x - \frac{L}{2}\right\rangle_{-1} + \frac{P}{2}\langle x - L\rangle_{-1}$

(d) $q(x) = P\langle x\rangle_{-1} - P\left\langle x - \frac{L}{4}\right\rangle_{-1} - P\left\langle x - \frac{3L}{4}\right\rangle_{-1} + P\langle x - L\rangle_{-1}$

(e) $q(x) = P\langle x\rangle_{-1} - Q\langle x - (L - b)\rangle_{-1} + R_w\langle x - L\rangle_{-1} - M_w\langle x - L\rangle_{-2}$

(f) $q(x) = -P\langle x\rangle_{-1} + P\langle x - a\rangle_{-1} + P\langle x - (L - a)\rangle_{-1} - P\langle x - L\rangle_{-1}$

(g) $q(x) = M_w\langle x\rangle_{-2} + R_w\langle x\rangle_{-1} - 3P\left\langle x - \frac{L}{3}\right\rangle_{-1} - 2P\left\langle x - \frac{2L}{3}\right\rangle_{-1} - P\langle x - L\rangle_{-1}$

4.5-3 (a) $q(x) = 7.1\langle x\rangle_{-1} - 8\langle x - 3\rangle_{-1} - 5\langle x - 7\rangle_{-1} + 5.9\langle x - 10\rangle_{-1}$

(b) $q(x) = 6.5\langle x\rangle_{-1} - 5\langle x - 6\rangle_{-1} - 10\langle x - 12\rangle_{-1} + 8.5\langle x - 20\rangle_{-1}$

(c) $q(x) = R_A\langle x\rangle_{-1} - 150\langle x - 4\rangle_{-1} + R_B\langle x - 7\rangle_{-1} - 40\langle x - 10\rangle_{-1}$

(d) $q(x) = R_A\langle x\rangle_{-1} - 40\langle x - 3\rangle_{-1} + R_B\langle x - 7\rangle_{-1} + 20\langle x - 7\rangle_{-1} - 10\langle x - 10\rangle_{-1}$

(e) $q(x) = -30\langle x\rangle_{-1} + 30\langle x - 3\rangle_{-1} + 30\langle x - 12\rangle_{-1} - 30\langle x - 15\rangle_{-1}$

(f) $q(x) = M_w\langle x\rangle_{-2} + R\langle x\rangle_{-1} - 3\langle x - 8\rangle_{-1} - 5\langle x - 14\rangle_{-1}$

(g) $q(x) = 4\langle x\rangle_{-1} - 20\langle x - 4.5\rangle_{-1} + R\langle x - 8.4\rangle_{-1} - M_w\langle x - 8.4\rangle_{-2}$

(h) $q(x) = R_A\langle x - 3\rangle_{-1} -$

$60\langle x-4\rangle_{-1} -$
$15\langle x-7\rangle_{-1} -$
$45\langle x-7\rangle_{-2} +$
$R_B\langle x-10\rangle_{-1}$

4.5-7 $a/L = 0.586$

4.5-8 $R_A = w_0(L-a)$,
$M_A = (w_0/2)(L^2 - a^2)$,
$M(x) = -M_A\langle x\rangle^0 + R_A\langle x\rangle^1 - (w_0/2)\langle x-a\rangle^2$

4.5-9 $V_{max} = 130$ lb,
$M_{max} = 408$ lb · in

4.5-10 (*b*) $M(x) = M_0\langle x\rangle^0 - M_0\langle x - L\rangle^0$

4.5-11 $V_{min} = -20$ kN,
$M_{min} = -20$ kN · m

4.5-12 $M_{max} = 39.6$ kN · m

4.5-13 $M(x) = -\dfrac{w_0 L}{\pi}\left(x - \dfrac{L}{\pi}\sin\dfrac{\pi x}{L}\right) + \dfrac{w_0 L}{\pi}\langle x-a\rangle^1 + \dfrac{w_0 L}{\pi}\langle x-(L-a)\rangle^1$, $a = L/\pi$

4.5-16 $x = 3.6$ ft, 13.9 ft

4.5-18 $V_{min} = -4.40$ kN,
$M_{max} = 8.80$ kN · m

4.5-19 $q(x) = 100\langle x\rangle_{-1} - 100\langle x-1\rangle_{-1} - 100\langle x-4\rangle_{-1} + 100\langle x-5\rangle_{-1}$

4.5-20 $a = 14.65$ ft, $M = 1390$ lb · ft

4.5-21 $F(L-a) =$ constant

4.5-22 $M_{max} = (\sqrt{3}/27)\, w_0 L^2$,
$\dfrac{M_{max}}{M(2L/3)} = 1.04$

4.5-23 $V_{max} = 9$ kips,
$M_{max} = 120$ kip · ft

4.5-24 $V_{max} = 12$ kips,
$M_{max} = 180$ kip · ft

4.5-25 $q = 2P/L$

4.5-26 $M_w = 6.31 \times 10^5$ lb · in,
$R_w = 6120$ lb

4.5-27 $M_{max} = 115$ kip · ft

4.5-28 $V_{max} = 120$ kips,
$V_{min} = -120$ kips,
$M_{max} = 400$ kip · ft,
$M_{min} = -2600$ kip · ft

4.5-29 $M_{max} = 30.2$ kN · m

4.5-30 $M_{max} = 10.4$ kip · ft

4.5-31 $V(x) = \dfrac{2w_0}{3}\langle x - 0.75\rangle^2 - \dfrac{w_0}{16}$
$M(x) = -\frac{2}{9} w_0\langle x-0.75\rangle^3 + \dfrac{w_0 x}{16}$
$M_{max} = 263$ N · m

4.5-32 $M_{max} = 5010$ lb · in

4.5-33 (*a*) $M_{max} = 42$ kN · m
(*b*) $M_{max} = 53.3$ kN · m
(*c*) $M_{max} = 35.6$ kN · m

4.5-35 $R_D = 0.678P$, $R_A = 0.736P$

4.5-36 (*a*) $M = 66 \times 10^{-15}$ N · m
(*b*) $M = \omega\eta \int_0^L x^2 f(x)\, dx$

4.7-4 $M_{max} = 51.6$ N · m

4.8-1 (*a*) $M_z = -Qa$, (*b*) $M_x = -QL$

4.8-2 $V_{max} = 1000$ lb,
$M_{max} = 2000$ lb · in

4.8-3 $\mathbf{F} = P\,\mathbf{k}$, $\mathbf{M}_c = Pa \sin\theta\, \mathbf{e}_r + Pa(1 + \cos\theta)\mathbf{e}_\theta$

4.8-5 $\mathbf{F}_{AB} = \dfrac{0.584}{6.423}(-5\mathbf{i} + 3.5\mathbf{j} + 2\mathbf{k})$,
$\mathbf{F}_{DC} = \dfrac{1.201}{3.775}(-2.5\mathbf{i} - 2\mathbf{j} + 2\mathbf{k})$,
$|M|_{max} = 0.81$ kN · m

4.8-6 $|M|_{max} = 308$ lb · in

4.8-7 $F = 100$ lb,
$|M|_{max} = 1400$ lb · in

4.9-1 $V(x) = R_w - M_w\langle x\rangle_{-1} - q_0 \dfrac{2L}{\pi}\sin\dfrac{\pi x}{2L}$
$M(x) = -R_w x + M_w + q_0\left(\dfrac{2L}{\pi}\right)^2\left(1 - \cos\dfrac{\pi x}{2L}\right)$

$R_w = q_0 \dfrac{2L}{\pi};$

$M_w = q_0\left(\dfrac{2L}{\pi}\right)^2\left(\dfrac{\pi}{2} - 1\right)$

4.9-2 $R_A = R_C = \frac{1}{2}P$

4.9-3 $V(x) = -R_A\langle x\rangle^0 + \dfrac{q_0 L}{2\pi}\cos\dfrac{2\pi x}{L} - \dfrac{q_0 L}{2\pi}$

$M(x) = R_A\langle x\rangle^1 - q_0\left(\dfrac{L}{2\pi}\right)^2 \sin\dfrac{2\pi x}{L} - \dfrac{q_0 L}{2\pi}x$

$R_A = -R_B = -q_0\dfrac{L}{2\pi}$

4.9-4 $R_A = -P,\ R_C = 2P$

4.9-5 $|V|_{\max} = 13$ kN,

4.9-6 $N_G = -\dfrac{P}{5},\ V_G = -\dfrac{P}{10},$

$M_G = \dfrac{PL}{10},\ N_H = -\dfrac{P}{10},$

$V_H = \dfrac{P}{5},\ M_H = \dfrac{PL}{10}$

$N_I = -\dfrac{9P}{10},\ V_I = -\dfrac{P}{5},\ M_I = \dfrac{PL}{5}$

4.9-7 $X_A = -23.2$ kN,
$Y_A = 86.5$ kN,
$X_C = -63.4$ kN,
$Y_C = 63.4$ kN

4.9-8 $X_B = 2.5$ kips,
$X_C = -3.5$ kips,
$M_{\max} = 6.0$ kip · ft,
$M_{\min} = -2.0$ kip · ft

Chapter 5

5.1-1 $\rho = R,\ \kappa = \dfrac{1}{R}$

5.2-2 $\epsilon = \pm 9.99 \times 10^{-4}$

5.3-2 $P = 124\ N$

5.3-4 (a) $I_z = 2.08 \times 10^7$ mm^4
(b) $I_z = 2.92 \times 10^8$ mm^4
(c) $I_z = 80.78$ in^4
(d) $I_z = 4.18 \times 10^8$ mm^4
(e) $I_z = 136.9$ in^4
(f) $I_z = 3.14 \times 10^6$ mm^4
(g) $I_z = 101.5$ in^4
(h) $I_z = 13.1$ in^4
(i) $I_z = 9.89 \times 10^7$ mm^4
(j) $I_z = 351.6$ in^4

5.4-2 $\sigma_x = 3750$ psi

5.4-3 $\sigma_x = 4654$ psi

5.4-4 $\sigma_x = \pm 761$ psi

5.4-5 $L_a = 11.5$ ft

5.4-6 $L = 11.25$ m

5.4-7 $\sigma_x = 8.26$ MPa

5.4-8 $L = 75.5$ in

5.4-9 $L = 10.4$ m

5.4-10 $\sigma_x = 242$ MPa

5.4-11 $\sigma_{\max} = 249$ MPa

5.4-12 $\sigma_x = 11$ ksi

5.4-13 $\sigma_x = \pm 110$ MPa

5.4-14 $L = 61.1$ in

5.4-15 $F = 20{,}490$ lb

5.4-16 $\sigma_x = 2220$ psi,
$\sigma_x = -4400$ psi

5.4-17 $P_{\max} = 2.47$ kN

5.4-18 $P_{\max} = 2270$ lb

5.4-19 $\sigma_{\max} = \pm 31{,}700$ psi

5.4-20 $\sigma_{\max} = \pm 22{,}630$ psi

5.4-21 $\sigma_{\max} = \pm 4970$ psi

5.4-22 $\sigma = \pm 5.89$ ksi

5.4-23 $\sigma_{xt} = 3970$ psi, $\sigma_{xc} = -1520$ psi

5.4-26 (*a*) $a = 23.44$ ft, (*b*) $\sigma_t = 27.7$ ksi, $\sigma_c = -89.4$ ksi

5.4-27 $\sigma_x = \pm 1150$ psi

5.4-28 $\sigma_{xt} = 12.1$ MPa, $\sigma_{xc} = -8.09$ MPa

5.4-29 $w = 5.725$ kN/m

5.4-30 $\sigma_{xt} = 15.4$ MPa, $\sigma_{xc} = -26.7$ MPa

5.4-31 W8 × 28

5.4-32 S8 × 23

5.4-33 S10 × 35

5.4-34 W8 × 67

5.4-35 W10 × 68

5.4-36 C7 × 14.75

5.4-37 C10 × 25

5.4-38 1-in nominal diameter

5.4-39 $\sigma_x = \pm 21.55$ ksi

5.4-40 $\sigma_{max} = 444$ MPa

5.4-41 $\sigma_x = 19.0$ MPa, $\sigma_x = -21.8$ MPa

5.4-42 $\sigma_{max} = 13.6$ MPa

5.4-43 $b = 7.02$ in

5.4-44 $E = 75$ MPa

5.4-45 $\sigma_x = 1800$ psi, $\sigma_x = -912$ psi; $I_z = 305$ in^4

5.4-46 $\sigma_x = \pm 21.32$ ksi

5.4-47 $L = 10.7$ ft

5.4-48 $\sigma = \pm 6060$ psi

5.4-49 $w_0 = 5820$ lb/ft

5.4-50 $t = 0.388$ mm

5.4-51 $\sigma = \pm 292$ psi

5.4-52 $\sigma = \pm 292$ psi; independent of R

5.4-53 $w = 390$ N/m

5.4-54 $w = 30$ lb/ft

5.4-55 $6 \times 5 \times \frac{1}{2}$ tube

5.4-56 W5 × 16

5.4-57 W10 × 77

5.4-58 W4 × 13

5.4-59 $L = 72.7$ in

5.4-60 $8 \times 8 \times \frac{3}{8}$

5.4-61 W10 × 68

5.4-62 S10 × 25.4

5.4-63 $5 \times 3 \times \frac{5}{16}$

5.4-64 S6 × 12.5

5.4-65 C5 × 9

5.4-66 $P = 144$ lb

5.4-67 $\sigma = \pm 6860$ psi

5.4-68 $w_0 = 683$ lb/ft

5.4-69 $q_0 = 33.6$ lb/in

5.4-70 $P = 599$ N

5.4-71 $\sigma_t = 826$ psi, $\sigma_c = -1340$ psi

5.4-72 (*a*) $q_0 = 46.9$ lb/ft^2, $L = 8$ ft; $q_0 = 20.8$ lb/ft^2, $L = 12$ ft (*b*) $q_0 = 100$ lb/ft^2, $L = 6$ ft; $q_0 = 56.2$ lb/ft^2, $L = 8$ ft

5.4-73 (*b*) $\sigma_x = 1012$ psi at $x = 3$ ft (*c*) $L = 10.67$ ft

5.5-5 $(\sigma_x/\tau)_{max} = L/h$

5.5-6 $\dfrac{\sigma_x}{\tau_{xy}} = \dfrac{2}{7}\dfrac{L}{h}$

5.5-9 $\tau_{max} = 60$ psi, $\sigma_{max} = 576$ psi

5.5-10 $\tau_{max} = 0.45$ MPa, $\sigma_{x\,max} = 5.4$ MPa

5.5-11 $P = 9780$ lb

5.5-12 $P = 9780$ lb

5.5-13 $P = 8270$ lb

5.5-14 $P = 8270$ lb

5.5-15 $c = 0.73a$,
$\sigma_s/\sigma_f = 1.40$, $\tau_s/\tau_f = 1.02$

5.5-16 $\sigma = \pm 2700$ psi

5.5-17 $\sigma = \pm 4380$ psi

5.5-18 $\sigma = \pm 9930$ psi

5.5-19 τ at flange $= 3570$ psi,
$\tau_{max} = 4254$ psi

5.6-1 $P = 562.5$ N

5.6-2 $P = 0.325$ kN, $P = 1.0$ kN

5.6-4 $V_{max} = 23.2$ kN

5.6-5 $F_b = 2570$ lb

5.6-6 $\tau = 9.375 \times 10^{-2} P$; no glue required in second configuration

5.6-7 (*b*) is better

5.6-8 $s = 3.31$ in

5.6-9 $f = 2060$ lb/in for each weld

5.6-10 $f = 1930$ lb/in for each weld

5.6-11 $s = 4.12$ in

5.6-12 $\tau = 48.5$ psi

5.6-13 $s = 3.51$ in

5.6-14 $\tau = 57$ psi

5.6-15 $\tau = 8.67$ psi

5.6-16 $V_{max} = 288$ kips

5.6-17 $V_{max} = 250$ kips

5.6-18 $s = 4.30$ in

5.6-19 $s = 98$ mm

5.6-20 $s = 2.0$ in

5.6-21 $s = 1.0$ in

5.6-22 $F_B = 3530$ lb

5.6-23 $P = 400$ lb, $\sigma_{max} = 3600$ psi

5.6-24 $P = 6.4$ kips

5.6-25 $\sigma = \pm 18.16$ ksi, $\tau = 75.6$ psi,
$\tau = 151$ psi

5.6-26 $P = 510$ lb

5.6-27 $\tau_D = 1500$ psi, $\tau_{max} = 1650$ psi

5.6-28 $V_{max} = 8920$ lb

5.6-29 $L = 35.8$ in

5.6-30 $V_{max} = 436$ lb

5.6-31 $\sigma_{max} = \pm 35.1$ MPa

5.6-32 $\sigma_{max} = \pm 7010$ psi,
$\tau_{max} = 798$ psi, at right support

5.6-33 $P = 1460$ lb

5.6-34 $f = 289$ lb/in, each weld

5.6-35 $P = 4690$ lb

5.7-3 $k = 0.475$,
$M_{max} = 5.92 \times 10^5$ lb · in

5.7-8 Case (*b*) $\sigma_{max} = 40.5$ ksi,
case (*c*) $\sigma_{max} = 45.6$ ksi

Chapter 6

6.2-2 (*a*) $0.426 \dfrac{M_0 a^2}{EI}$

(*b*) $0.461 \dfrac{M_0 a^2}{EI}$

(*c*) 8.2%

6.2-3 (*a*) $h_a = \dfrac{2R\sigma_a}{E}$

(*b*) $h_a = 0.01, 0.1, 1$ mm

6.2-4 (*a*) $M_A = \dfrac{EI}{R}$ (*b*) $v_a =$

$\frac{4}{\pi^2}\left(1 + \frac{\pi}{2}\right)\frac{M_A L^2}{EI}$

6.2-5 (*a*) $\sigma = \frac{Eh\theta}{2L}$ (*b*) $h_a = 0.046$, 0.076, 0.153 in

6.3-2 W6 × 9

6.3-3 W6 × 9

6.3-4 −5.48 in

6.3-9 $v(x) = \frac{P}{6EI}\left[\frac{b}{L}x^3 - \langle x - a\rangle^3 + \frac{b}{L}(b^2 - L^2)\,x\right]$

$v_{max} = -0.192$ in

6.3-10 $v_{max} = 2.10 \times 10^{-3}$ m, $\sigma_{max} = 28.1$ MPa

6.3-11 $EIv(x) = \frac{q_0 L x^3}{16} - \frac{q_0}{24}x^4 + \frac{q_0}{24}\left\langle x - \frac{L}{2}\right\rangle^4 - \frac{3}{128}q_0 L^3 x$

$v_{max} = 6.56\,\frac{q_0 L^4}{EI} \times 10^{-3}$ at $x = 0.46L$

6.3-12 $EIv(x) = \frac{q_0 b^2}{2L}x^3 - \frac{q_0}{24}\langle x - a\rangle^4 + \frac{q_0 b^2}{24L}(b^2 - 2L^2)x$

$v_{max} = -9.82\,\frac{q_0 L^4}{EI} \times 10^{-3}$ at $x = 0.52L$

6.3-13 $a/L = 0.207$, $a/L = 0.223$

6.3-15 $v_B = -0.0208\,\frac{PL^3}{EI} + 0.0313\,\frac{QL^3}{EI}$

$Q = 0$: $v_B = -0.0208\,\frac{PL^3}{EI}$

6.3-16 −0.12 in

6.3-17 −0.0949 in

6.3-18 (*a*) −0.0949 in
(*b*) −0.0880 in
(*c*) −0.0814 in

6.3-19 $EIv(x) = -\frac{P}{6}\left\langle x - \frac{L}{2}\right\rangle^3 + \frac{P}{12}\langle x - a\rangle^3 + \frac{P}{12}\langle x - (L - a)\rangle^3 - \frac{P}{4}\left(\frac{L}{2} - a\right)^2 (x - a)$,

$\frac{v(L/2)}{v(L/2)_{a=0}} = \left(1 - \frac{2a}{L}\right)^3$,

$v\left(\frac{L}{2}\right) = -\frac{P}{48}(L - 2a)^3$

6.3-20 (*a*) $q_0 = \frac{2P}{L}$

(*b*) $2.60 \times 10^{-3}\,\frac{PL}{EI}$

6.3-21 $v_B = -\frac{M_b L^2}{2EI}$, $v'_B = -\frac{M_b L}{EI}$,

$\frac{v_M}{v_B} = \frac{1}{4}$, $\frac{v'_M}{v'_B} = \frac{1}{2}$

6.3-22 $v_B = -\frac{M_B L^2}{8EI}$, $v'_B = -\frac{M_B L}{2EI}$,

$v_C = -\frac{3M_B L^2}{8EI}$, $v'_C = -\frac{M_B L}{2EI}$

6.3-23 $v_B = -\frac{Pa^3}{3EI}$, $v'_B = -\frac{Pa^2}{2EI}$,

$v_C = -\frac{Pa^2(3L - a)}{6EI}$,

$v'_C = -\frac{Pa^2}{2EI}$

6.3-24 $EIv_{max} = \frac{q_0 a L^3}{96} - \frac{q_0}{24}\left(\frac{L}{2} - b\right)^4 + \left[\frac{q_0}{6}\left(\frac{L}{2} - b\right)^3 - q_0\frac{aL^2}{16}\right]\frac{L}{2}$

6.3-25 $EIv'_C = q_0\frac{bL^2}{2} - q_0 b\left(a + \frac{b}{2}\right)L - q_0\frac{b^3}{6}$

$EIv_C = q_0\frac{bL^3}{6} - q_0 b\left(a + \frac{b}{2}\right)\frac{L^2}{2} - \frac{q_0 b^4}{24}$

6.3-26 (*a*) $p = 46.3$ lb/ft^2
(*b*) $p = 15$ lb/ft^2

6.3-27 $v_C = -4.5 \dfrac{Pa^3}{EI}$

6.3-28 $v_C = -1.16$ in

6.3-29 $\Delta_{AC} = -0.724$ in

6.3-30 $P_\sigma = 586b$, $N = 1$, $P = 879$ lb, $N = 9$, $P = 7911$ lb

6.3-31 Case (*a*) or (*b*) gives $P_{\max} = 10.9$ kips

6.3-32 $E = 130$ GPa

6.3-33 (*a*) $EIv = -\dfrac{Px^3}{6} + \dfrac{P}{6}\langle x - a\rangle^3 + c_1 x + c_2,\ 0 \le x \le \dfrac{L}{2}$

where $c_1 = \dfrac{Pa}{2}(L - a)$,

$c_2 = -c_1 a + \dfrac{Pa^3}{6}$

(*b*) $EIv = W\dfrac{x^3}{6} - \dfrac{W}{6}\left\langle x - \dfrac{L}{4}\right\rangle^3 - \dfrac{3}{32}WL^2x$

(*c*) $EIv = \dfrac{Pa^3}{24}\left(-4\dfrac{x^3}{a^3} + 8\dfrac{\langle x - a\rangle^3}{a^3} - \dfrac{4}{a^3}\langle x - 2a\rangle^3 + 23\dfrac{x}{a} - 19\right)$

(*d*) $EIv = \dfrac{M_0L^2}{6}\left[-\dfrac{x^3}{L^3} + 3\dfrac{\langle x - a\rangle^2}{L^2} + \dfrac{x}{L}\left(1 - 3\dfrac{b^2}{L^2}\right)\right]$

(*e*) $EIv = \dfrac{M_0}{2}(x^2 - \langle x - a\rangle^2)$

(*f*) $EIv = \dfrac{w_0L^4}{24}\left[2\left(1 - \dfrac{a^2}{L^2}\right)\dfrac{x}{L}\left(\dfrac{x^2}{L^2} - 1\right) + \dfrac{x}{L} - \dfrac{x^4}{L^4} + 2\left(1 - \dfrac{a^2}{L^2}\right)\dfrac{\langle x - L\rangle^3}{L^3}\right]$

(*g*) $EIv = \dfrac{W}{6}\left[2x^3 - 3Lx^2 - \left\langle x - \dfrac{L}{3}\right\rangle^3 - \left\langle x - \dfrac{2L}{3}\right\rangle^3\right]$

(*h*) $EIv = \dfrac{w_0}{24}\left[2(b - a)(2L - b - a)\left(\dfrac{x^3}{L} - Lx\right) - \langle x - a\rangle^4 + (L - a)^4\dfrac{x}{L} + \langle x - b\rangle^4 - (L - b)^4\dfrac{x}{L}\right]$

(*i*) $EIv = \dfrac{w_0}{48}\left(4L\langle x - b\rangle^3 - 2x^4\right) + c_1x + c_2$, $c_1 = \dfrac{w_0}{48}\left[L^3 - 12L\left(\dfrac{L}{2} - b\right)^2\right]$

$c_2 = \dfrac{w_0}{48}\left[2b^4 - bL^3 + 12\,Lb\left(\dfrac{L}{2} - b\right)^2\right]$

where $b = \dfrac{L - a}{2}$ and $0 \le x \le \dfrac{L}{2}$

(*j*) $EIv = \dfrac{w_0}{24}\left[4bx^3 - 12b\left(L - \dfrac{b}{2}\right)x^2 - \langle x - a\rangle^4\right]$

(*k*) $EIv = \dfrac{w_0}{360}\left(-3\dfrac{x^5}{L} + 10Lx^3 - 7xL^3\right)$

6.4-1 $V(x) = -\dfrac{d}{dx}\left(EI\dfrac{d^2v}{dx^2}\right)$,

$q = \dfrac{d^2}{dx^2}\left(EI\dfrac{d^2v}{dx^2}\right)$,

$V = -EI_0\left[\left(1 - \dfrac{x}{2L}\right)\dfrac{d^3v}{dx^3} - \dfrac{1}{2L}\dfrac{d^2v}{dx^2}\right]$

6.4-4 $v_{max} = 0.0619$ in, $v_{max} = 0.0614$ in

6.4-5 6 in × 6 in × $\frac{5}{16}$ in

6.4-6 $\sigma_{max} = 15.9$ ksi, $\tau_{max} = 16.1$ ksi

6.4-9 $a/L = 0.577$

6.4-10 $\dfrac{a}{L} = 0$ or 1, $v_{max} = \dfrac{M_0L^2}{9\sqrt{3}\,EI}$

6.4-12 $EIv = -M_A\dfrac{x^2}{2} + R_A\dfrac{x^3}{6} - q_1\dfrac{x^4}{24} + q_2\dfrac{\langle x-a\rangle^4}{24}$, $R_A = (q_1 - q_2)a$, $M_A = \dfrac{a^2}{2}(q_1 - 3q_2)$

6.4-14 $q_0 = 42.7$ lb/in

6.4-16 $v_B = -\dfrac{1}{48}\dfrac{PL^3}{\sqrt{2}\,EI}$, $u_B = -\dfrac{1}{2\sqrt{2}}\dfrac{PL}{EA}$

6.4-17 $v_{max} = -\dfrac{q_0L^4}{120EI}$, $v'_{max} = -\dfrac{5}{192}\dfrac{q_0L^3}{EI}$

6.4-18 $EIv = \dfrac{q_0}{24}\left[2Lx^3 - x^4 - L^3x - 12\dfrac{(L_cI)}{A}x\right]$

$v_B = -\dfrac{q_0}{2}\dfrac{LL_c}{EA}$,

$v'_B = \dfrac{q_0L^3}{24EI}\left(1 - 12\dfrac{L_cI}{L^3A}\right)$

6.4-19 $v_{mid} = -\dfrac{q_0}{E}\left(\dfrac{10}{3}\dfrac{L^4}{I} + 3\dfrac{L^2}{A}\right)$

6.4-20 $EIv = \dfrac{q_0x^4}{24} - \dfrac{q_0L}{12}\left\langle x - \dfrac{L}{4}\right\rangle^3 - \dfrac{q_0L^3x}{192} + \dfrac{q_0L^4}{768}$, $0 \le x \le \dfrac{L}{2}$

6.4-21 $v(x) = -\dfrac{M_Bx^2}{2EI}$

6.4-22 $v(x) = -\dfrac{M_B}{2EI}\left(x^2 - \left\langle x - \dfrac{L}{2}\right\rangle^2\right)$

6.4-23 $v(x) = -\dfrac{P}{6EI}\left(\langle x - a\rangle^3 - x^3 + 3x^2a\right)$

6.4-24 $v(x) = \dfrac{q_0}{EI}\left[\dfrac{ax^3}{12} - \dfrac{\langle x-b\rangle^4}{24} + \dfrac{a}{48}(a^2 - 3L^2)\,x\right]$, $0 \le x \le \dfrac{L}{2}$

6.4-25 $v(x) = \dfrac{q_0}{EI}\left[\dfrac{bx^3}{6} - b\left(a + \dfrac{b}{2}\right)\dfrac{x^2}{2} - \dfrac{\langle x-a\rangle^4}{24}\right]$

6.4-26

	Without Fluid		With Fluid	
t, in	σ_{max}, psi	v_{max}, in	σ_{max}, psi	v_{max}, in
1.5	464	0.0100	1540	0.0331
3.5	496	0.0107	936	0.0202

6.4-27 $v_{max} = -\dfrac{151}{30{,}720}\dfrac{q_0L^4}{EI}$

6.4-28 (*a*) $v_C = -3$ in, (*b*) $v_C = -4.8$ in

6.4-29 $\sigma_{max} = 15.2$ ksi at 168 in from *A*, $v_{max} = 0.464$ in at 124 in from *A*

6.4-30 $v_B = -\dfrac{PL^3}{2EI_0}$

6.4-31 $\delta = \sqrt{2}\Delta$, $\sigma_{max} = \dfrac{3\beta E}{\sqrt{2}}\dfrac{d}{L}$

6.4-33 $v_0 = -\dfrac{P}{8EIp^3}$, $M_0 = \dfrac{P}{4p}$

6.4-34 $v'(0) = \dfrac{M_0}{4pIE}$, $V(0) = \dfrac{-pM_0}{2}$

6.4-35 (*a*) 0.147 (right), 0.085 (down) (*b*) 885 psi

6.4-36 (*a*) $EIv(x) = \left(\frac{q_0 L}{2} - \frac{M_0}{L}\right)\left(\frac{x^3}{6} - \frac{L^2 x}{6}\right) + \frac{q_0}{24}(L^3 x - x^4)$

(*b*) $EIv(x) = \frac{Px^3}{6} - \frac{3PLx^2}{8} - \frac{P}{12}\left\langle x - \frac{L}{2}\right\rangle^3$

(*c*) $EIv(x) = q_0\frac{Lx^3}{3} - q_0\frac{L^2x^2}{4} - q_0\frac{x^4}{24}$

(*d*) $EIv(x) = -q_0\frac{x^4}{24} + q_0\frac{L}{6}\left\langle x - \frac{L}{2}\right\rangle^3 + q_0\frac{L^3 x}{24} - \frac{7}{384}q_0 L^4,\ 0 \le x \le L$

(*e*) $EIv(x) = -q_0\frac{x^4}{24} + q_0\left(a + \frac{L}{2}\right)\langle x - a\rangle^3 + c_1 x + c_2,\ 0 \le x \le a + \frac{L}{2}$

$c_1 = \frac{q_0}{24}\left[4\left(a + \frac{L}{2}\right)^3 - 3\left(a + \frac{L}{2}\right)L^2\right]$

$c_2 = \frac{q_0 a}{24}\left\{a^3 - \left(a + \frac{L}{2}\right)\left[4\left(a + \frac{L}{2}\right)^2 - 3L^2\right]\right\}$

(*f*) $EIv(x) = -\frac{Px^3}{6} + \frac{P}{4}\langle x - a\rangle^3 + c_1 x + c_2,\ 0 \le x \le a + \frac{L}{2}$

$c_1 = \frac{P}{16}\left[8\left(a + \frac{L}{2}\right)^2 - 3L^2\right],$

$c_2 = \frac{Pa^3}{6} - c_1 a$

6.4-37 (*a*) $v(x) = -\frac{Px^2}{6EI}(3L - x)$

(*b*) $v(x) = -\frac{P}{6EI}\left(\left\langle x - \frac{L}{2}\right\rangle^3 - x^3 + 3\frac{x^2 L}{2}\right)$

(*c*) $v(x) = -\frac{P}{48EI}\left(8\left\langle x - \frac{L}{2}\right\rangle^3 - 4x^3 + 3L^2 x\right)$

(*d*) $v(x) = \frac{P}{EI}\left(\frac{x^3}{6} - \frac{\langle x - L/4\rangle^3}{6} - \frac{3}{32}L^2 x\right),$

$0 \le x \le L/2$

(*e*) $v(x) = \frac{1}{EI}\left[\frac{Px^3}{6} - \frac{Q}{6}\langle x - a\rangle^3 + \left(\frac{Qb^2}{2} - \frac{PL^2}{2}\right)x + c_2\right],\ c_2 = \frac{P}{6}b^3 + \frac{P}{3}L^3 - \frac{Qb^2 L}{2},\ a = L - b$

(*f*) $v(x) = \frac{P}{EI}\left(-\frac{x^3}{6} + \frac{\langle x - a\rangle^3}{6} + c_1 x + c_2\right),$

$0 \le x \le \frac{L}{2}$

$c_1 = \frac{PL^2}{8} - \frac{P}{2}\left(\frac{L}{2} - a\right)^2,$

$c_2 = -ac_1 + \frac{Pa^3}{6}$

(*g*) $v(x) = \frac{P}{6EI}\left(6x^3 - 10Lx^2 - 3\left\langle x - \frac{L}{3}\right\rangle^3 - 2\left\langle x - \frac{2L}{3}\right\rangle^3\right)$

(*h*) $v(x) = \frac{M_0}{6LEI}\left[3L\langle x - a\rangle^2 + x(L^2 - 3b^2) - x^3\right]$

(*i*) $v(x) = \frac{1}{EI}\left[R_1\frac{x^3}{6} - \frac{P}{6}\langle x - a\rangle^3 - \frac{M_0}{2}\langle x - (a + b)\rangle^2 + c_1 x\right]$

$R_1 = P\left(1 - \frac{a}{L}\right) + \frac{M_0}{L}, c_1 =$
$-\frac{R_1 L^2}{6} + \frac{P}{6L}(L - a)^3 + \frac{M_0 c^2}{2L}$

(*j*) $v(x) = \frac{M_0}{2EI}(x^2 - \langle x - a\rangle^2)$

(*k*) $v(x) =$
$\frac{M_0}{2EI}\left[x^2 - \langle x - (a + b)^2\right] +$
$\frac{P}{6EI}(x^3 - \langle x - a\rangle^3 - 3ax^2)$

6.5-1 $v(x) = \frac{P}{6EI}$
$(\langle x - a\rangle^3 - 3Lx^2 + 3x^2 a),$
$v_C = -0.1455\frac{PL^3}{EI}$

6.5-2 (*a*) $v_C = -\frac{PL^3}{3E_2 I}$
(*b*) $v_C = -\frac{19}{6}\frac{PL^3}{E_1 I}$

6.5-3 $M_0 = \frac{5}{24}q_0 L^2$

6.5-4 $v_C = -\frac{7}{32}\frac{PL^3}{EI}$
$v'_C = -\frac{5}{16}\frac{PL^2}{EI}, \frac{5}{8}$

6.5-5 $v_B = -\frac{1}{48\sqrt{2}}\frac{PL^3}{EI}$

6.5-7 $u_C = 0.350\frac{PL^3}{EI}$
$v_C = -0.604\frac{PL^3}{EI}$

6.5-8 (*a*) $v_B = -\frac{PL^3}{144EI}$
(*b*) σ_{max} (upper beam alone) = $3 \times \sigma_{max}$ (both)

6.5-9 $v_B = 0.0556\frac{qL^4}{EI}$
$v_C = 0.229\frac{qL^4}{EI}$
$v'_B = 0.111\frac{qL^3}{EI}$
$v'_C = 0.194\frac{qL^3}{EI}$

6.5-10 $a = 3.423$ in

6.5-11 $L = 154$ in or 12.8 ft

6.5-12 $v_C = -\frac{29}{24}\frac{qL^4}{EI}, v'_C = -\frac{5}{6}\frac{qL^3}{EI}$

6.5-13 $\alpha = \frac{1}{48}, \alpha = \frac{1}{12}\frac{1}{2M};$
M is number of pairs of bars

6.5-14 (*a*) $P_1 = \frac{6EI\Delta}{5L^3}$
(*b*) $P < P_1, P_C = \frac{3}{8}\frac{EI}{L^3}$

6.5-15 $v_{max} = \frac{3\Delta}{10}, \theta_A = \frac{6}{5}\frac{\Delta}{L}$

6.5-16 (*a*) $v_B = -\frac{3}{16}\frac{q_0 L^4}{EI}$

6.5-19 $v'_B = \frac{1}{48}\frac{qL^3}{EI}$
$v_M = \frac{qL^4}{384EI}, v_D = -\frac{7}{384}\frac{qL^4}{EI}$

6.5-20 $a = \sqrt{\frac{5}{24}}L = 0.456L$

6.5-21 $a = L/6$

6.5-22 (*a*) $v_M = 0.0189$ in
(*b*) $a = 34.36$ in

6.6-2 $h_1 = 3$ mm,
$\Delta T = 50°C$,
$\frac{1}{R} = 0.141\ m^{-1}$
$h_1 = 3$ mm,
$\Delta T = 100°C$,
$\frac{1}{R} = 0.125\ m^{-1}$

6.6-4 $\delta = R\left(1 - \cos\frac{L}{2R}\right)$ with $R = \frac{2h}{3\Delta T(\alpha_2 - \alpha_1)}$

6.6-6 (*a*) $M_B = \frac{EI\pi}{2L}$ and $\frac{EI\pi}{L}$
(*b*) $\frac{c}{L} = \frac{2\sigma}{\pi E}$ and $\frac{\sigma}{\pi E}$
(*c*) $c = 0.0637$ and 0.0318 in

6.6-7 (*b*) $M_0 = -\frac{\pi}{2}\frac{EI}{L}, \frac{c}{L} = \frac{2}{\pi}\frac{\sigma_{max}}{E}$

(*c*) $c = 0.0637$ in for $\theta_0 = \dfrac{\pi}{4}$

6.6-8 (*a*) $\sigma_{\max} = \dfrac{6PL}{bh_0^2}$

(*b*) $v(x) = -\dfrac{PLx^2}{2EI_0}$

6.6-9 (*a*) $\sigma_{\max} = \dfrac{3q_0L^2}{b_0h_0^2}$

(*b*) $v(x) = -\dfrac{q_0L^2x^2}{4EI_0}$

6.6-10 (*a*) $\sigma_{\max} = \dfrac{3PL}{b_0h_0^2}$

(*b*) $v(x) = \dfrac{PLx}{8EI_0}(x - L)$

6.6-11 $k = 7.38$ N/m

Chapter 7

7.2-1 (*a*) $M_A = \dfrac{-M_Bb}{L^2}(2L - 3b)$,

$M_C = \dfrac{-M_Bb}{L^2}\left(3b - 4L + \dfrac{L^2}{b}\right)$

$R_A = -6M_B\dfrac{ba}{L^3}$, $R_C = -R_A$

(*b*) $|M_{\max}|$ at $x = a$

(*c*) $EIv(a) = \dfrac{M_Bba^2}{2L^2}\left(2L - 3b - \dfrac{2a^2}{L}\right)$

7.2-2 $R_A = \frac{7}{16}q_0L$, $R_B = \frac{5}{8}q_0L$,

$R_C = -\frac{1}{16}q_0L$

7.2-3 $R_A = R_C = \frac{3}{8}q_0L + 3\Delta_B\dfrac{EI}{L^3}$

$R_B = \frac{5}{4}q_0L - 6\Delta_B\dfrac{EI}{L^3}$

7.2-4 $v(x) = \dfrac{M_Bx^2}{4EI}\left(1 - \dfrac{x}{L}\right)$

$v_{\max} = \dfrac{M_BL^2}{27EI}$

7.2-5 (*a*) $v(x) = -3\Delta_B\dfrac{x^2}{L^2} + 2\Delta_B\dfrac{x^3}{L^3}$

(*b*) $M_A = -6\dfrac{EI}{L^2}\Delta_B$,

$R_A = \dfrac{-12EI\Delta_B}{L^3}$

7.2-6 (*a*) $v(x) = \dfrac{q_0L^4}{24EI}\dfrac{x^2}{L^2}\left(2\dfrac{x}{L} - 1 - \dfrac{x^2}{L^2}\right)$

(*b*) $M_A = \dfrac{q_0L^2}{12}$, $R_A = \dfrac{q_0L}{2}$

7.2-7 (*a*) $R_A = \dfrac{q_0L}{96}$

$R_B = \dfrac{30}{96}q_0L$, $R_C = \dfrac{17}{96}q_0L$

7.2-9 (*a*) $\sigma_{\max} = 370$ psi,
$\tau_{\max} = 112$ psi
(*b*) $\sigma_{\max} = 711$ psi,
$\tau_{\max} = 83$ psi

7.2-10 $EIv(x) = \dfrac{Pb^2}{2}(3L - b)\left(\dfrac{x^3}{6} - \dfrac{L^2x}{2}\right) + P\left(\dfrac{b^2x}{2} - \dfrac{\langle x - a\rangle^3}{6}\right)$

7.2-11 $b = \sqrt{2}\,a$

7.2-12 $v(x) = \dfrac{M_0}{24LEI}(-x^3 + 6\langle x - L\rangle^3 + L^2x)$

7.2-13 $v_{\text{end}} = \dfrac{-PL^3}{3EI}\left(1 - \dfrac{25}{48}\dfrac{1}{1 + 16/\beta}\right)$, $\beta = \dfrac{kL^3}{EI}$

7.2-14 $v_p = \dfrac{-23}{1536}\dfrac{PL^3}{EI}$

7.2-15 $v_B = \frac{-5}{162}\frac{PL^3}{EI}$, $v'_B = \frac{1}{54}\frac{PL^2}{EI}$

7.2-16 $v_{\max} = \frac{-1}{4}\frac{M_0L^2}{EI}$, $\sigma_{\max} = \frac{2}{75}\frac{M_0L}{I}$

7.2-17 $v_B = \frac{-1}{162}\frac{PL^3}{EI}$

$v_C = \frac{-17}{162}\frac{PL^3}{EI}$

7.2-18 $R_A = 2.03\,q_0a$,
$R_B = 5.27\,q_0a$,
$R_C = 0.708\,q_0a$
$M_{\max} = -2.38q_0a^2$ at $x = 5a$

7.2-19 $v_{\max} = -1.80 \times 10^{-3}\frac{q_0L^4}{EI}$

7.2-20 -1.62 mm

7.2-21 -0.0397 in

7.2-22 -3.73 mm, 21.6 kN

7.2-23 -1.33×10^{-2} in

7.2-24 -0.735 in

7.2-25 $v_{\max} = -2.91 \times 10^{-3}\frac{q_0L^4}{EI}$

7.3-1 $\frac{-1}{3}\frac{q_0L^4}{EI}$

7.3-4 $P_1 = \frac{48EI}{5L^3}\Delta_B$,

$\delta_c = \frac{39}{10}\Delta_B$

7.3-6 $\frac{PL^3}{612EI}$

7.3-11 (*a*) $v'_A = -8.02 \times 10^{-5}$
$|\sigma|_{\max} = 711$ psi at $x = 20$ in

7.3-12 (*a*) $\sigma_{\max} = 3E\frac{\Delta}{L}\frac{d_0}{L}$

7.3-13 (*a*) $k = \frac{-P}{v_C}$ where $v_C =$
$\frac{PL^3}{3EI}\left[-1 + \frac{2a^4(3L - a)^2}{8a^3 + 0.343L^3}\right]$

7.3-14 $\frac{8EI}{L^3}(3R^2 + 3RL + L^2)$

7.3-15 $\frac{-11PL^3}{96EI\,[1 + \gamma AL^3/(6EI)]}$

7.6-2 $M(x) = \frac{q_0L^2}{2}\frac{x}{L}\left(1 - \frac{x}{L}\right)$,

$V(x) = \frac{q_0L}{2}\left(2\frac{x}{L} - 1\right)$

7.6-3 $M(x) = \frac{-q_0L^2}{12}$
$\left(1 - 6\frac{x}{L} + 6\frac{x^2}{L^2}\right)$,

$V(x) = \frac{q_0L}{2}\left(2\frac{x}{L} - 1\right)$

7.6-4 $M(x) = \frac{q_0L^2}{8}\left(1 - 5\frac{x}{L} + 4\frac{x^2}{L^2}\right)$

$V(x) = q_0L\left(\frac{5}{8} - \frac{x}{L}\right)$

7.6-5 $M(x) = \frac{EI}{L^2}\left[\delta\left(6 - 12\frac{x}{L}\right) + L\phi\left(-2 + 6\frac{x}{L}\right)\right]$

7.6-9 (*a*) $v'_B = \frac{M_BL}{4EI}$

(*b*) $v(x) = \frac{M_BL^2}{4EI}\frac{x^2}{L^2}\left(-1 + \frac{x}{L}\right)$

7.7-1 (*a*) $v_B = \frac{-1}{72}\frac{q_0L^4}{EI}$, $v'_B = 0$

(*b*) $R_A = \frac{5}{6}q_0L$, $M_A = -\frac{q_0L^2}{4}$

7.7-2 $v_B = -0.00461\frac{PL^3}{EI}$

$v'_B = 0.00576\frac{PL^2}{EI}$

7.7-3 $v'_B = -\frac{1}{96}\frac{q_0L^4}{EI}$
$R_A = \frac{1}{16}q_0L$, $M_A = \frac{1}{48}q_0L^2$

7.7-4 $v_B = 0.00781\frac{M_BL^2}{EI}$

$v'_C = -0.0625\frac{M_BL}{EI}$

$R_A = 1.13\frac{M_B}{L}$

7.7-5 $v'_A = \frac{-1}{42}\frac{q_0L^3}{EI}$

$v'_B = \frac{1}{168}\frac{q_0L^3}{EI}$

$R_A = \frac{33}{84} q_0 L$

$R_B = 8 \frac{q_0 L}{7}$

7.7-6 $v_B = \frac{-179}{6984} \frac{q_0 L^4}{EI}$

$v'_B = \frac{-1}{291} \frac{q_0 L^3}{EI}$

$R_A = \frac{361}{194} q_0 L$

7.7-7 $v_B = \frac{L\phi}{11}$

$v'_B = \frac{-10}{22} \phi$

$M_A = \frac{-5.09 EI\phi}{L}$

7.7-8 $v'_B = \frac{-8}{5} \frac{\Delta_B}{L}$

$R_A = -73.7 \frac{EI}{L^3} \Delta_B$

$M_A = 20.5\, EI \frac{\Delta_B}{L^2}$

7.7-9 $v_A = \frac{1}{15} \frac{q_0 L^4}{EI}$

$v'_B = \frac{1}{20} \frac{q_0 L^3}{EI}$

$M_A = \frac{13}{60} q_0 L^2$

7.7-10 (*a*) $v_B = \frac{-1}{9} \frac{PL^3}{EI}$

(*b*) $v'_B = \frac{-1}{6} \frac{PL^2}{EI}$

(beam *AB*)

7.7-13 $v_B = \frac{-1}{24} \frac{PL^3}{EI}$, $v'_B = 0$,

$R_A = \frac{P}{2}$, $M_A = \frac{-PL}{4}$

7.8-1 (*a*) 0.347 in,
2.56×10^5 lb · in, 21.5 ksi
(*b*) 0.881 in,
7.20×10^5 lb · in,
20.3 ksi
(*c*) 65.4 mm,
3.26×10^5 N · m, 204 MPa
(*d*) 6.30 mm,
5.57×10^4 N · m,
33.4 MPa
(*e*) 128 mm,
9.00×10^4 N · m, 155 MPa
(*f*) 5.27 in,
1.13×10^6 lb · in,
74.2 ksi
(*g*) 0.0981 in,
5.33×10^5 lb · in, 15.2 ksi
(*h*) 21.3 mm,
7.07×10^4 N · m,
107.6 MPa

7.8-2 (*a*) 26.4 mm,
7.50×10^4 N · m,
58.6 MPa
(*b*) 37.1 mm,
5.91×10^4 N · m,
106.3 MPa
(*c*) 0.234 in,
4.80×10^5 lb · in, 19.4 ksi
(*d*) 0.386 in,
1.2×10^6 lb · in, 26.3 ksi
(*e*) 1.15 in,
8.91×10^5 lb · in, 28.5 ksi
(*f*) 0.693 in,
6.24×10^5 lb · in, 20.0 ksi
(*g*) 30.0 mm,
4.22×10^4 N · m,
143 MPa
(*h*) 1.96 mm,
1.00×10^5 N · m,
12.8 MPa
(*i*) 0.931 in,
1.30×10^6 lb · in,
25.5 ksi

7.8-3 (*a*) 0.239 in, 26.1 ksi
(*b*) 0.303 in, 24.1 ksi

7.8-4 Distributed case:
$\beta_1 = 0.00542$, $\beta_2 = 0.125$
Point-load case:
$\beta_1 = 0.00932$, $\beta_2 = 0.188$

7.8-7 $v_B = \frac{-1}{72} \frac{q_0 L^4}{EI}$

$v'_B = 0$, $R_A = \frac{5}{6} q_0 L$

$M_A = \frac{-q_0 L^4}{4}$

7.8-8 $v_B' = \dfrac{-8}{5}\dfrac{\Delta_B}{L}$

$R_A = -73.7\dfrac{EI}{L^3}\Delta_B$

$M_A = 20.5\dfrac{EI}{L^2}\Delta_B$

7.8-9 $v_B = \dfrac{-1}{9}\dfrac{PL^3}{EI}$

7.8-10 (*a*) -0.00308 rad
(*b*) -1.22 mm
(*c*) -1.99 mm

7.8-11 $\dfrac{a}{L} = 0.5$:
$|v|_{max} = 0.0924$ in,
$|\sigma|_{max} = 3.15$ ksi

7.8-12 $\dfrac{a}{L} = 0.2$:
$\beta_1 = 0.808$, $\beta_2 = -0.471$

7.8-13 $\beta = -0.237$

7.8-14 $\beta_D = 3.61$

7.8-15 $W = 2.496\dfrac{M_A}{L}$

7.8-16 $\alpha = 8.09$, $\beta = 10.2$

7.8-17 $4.52\dfrac{Pa^3}{EI}$

7.8-18 $\alpha = 5.23$

7.8-19 $\alpha = 1.21$, $\beta = 0.833$

7.8-20 $\alpha = 0.0209$

7.8-21 $v_B = -0.0233\dfrac{WL^3}{EI}$,

$v_{cg} = -0.0221\dfrac{WL^3}{EI}$

7.8-22 -1.16 in

7.8-23 (*a*) 11.14 kN
(*b*) 8.80 kN

7.8-24 $P = 42$ kips

7.8-25 $v_C = -0.241$ in,
$v_C' = -4.53 \times 10^{-3}$

7.8-26 (*a*) $\sigma_{max} = 11.7$ MPa
(*b*) 10 yr, $\sigma_{max} = 14.4$ MPa,
34.5 yr

7.8-27 $R_2 = 3.88$,
$R_3 = 8.52$,
$R_4 = 14.63$

7.8-28 (*a*) $q_0 = 21.4$ kN/m
(*b*) $q_0 = 30.3$ kN/m

7.8-29 (*a*) $\sigma_{max} = 21.8$ ksi at
$x = 90$ in,
$v_{max} = 0.209$ in at
$x = 497$ in
(*b*) $\sigma_{max} = 14.5$ ksi at
$x = 450$ in,
$v_{max} = 0.222$ in at
$x = 211$ in

7.8-30 (*a*) $a = 0.569$ m
(*b*) $a = 0.575$ m

7.8-31 (*a*) 3.05 MP
(*b*) 6.11 MPa

7.9-1 $x = 0.5858L$,
$\beta = 0.00981$

7.9-2 0.00874 in
for diameter of 0.25 in

7.9-3 $d_1 = d$

7.9-4 $a = 0.367L$

7.9-5 $P = k\delta$, $k = 32.4$ kN/m,
$\delta_{max} = 1.94$ mm

7.9-6 $u_A = \dfrac{4}{3}\dfrac{WL^3}{EI}\cot\theta$,

$\sigma = \dfrac{16}{\pi}\dfrac{WL}{d^3}\cot\theta$

7.9-12 $v_B = \dfrac{-q_0L^4}{30EI}$

$v_B' = \dfrac{-q_0L^3}{24EI}$

$M_A = -\dfrac{q_0L^2}{6}$

7.9-13 $v_A' = \dfrac{-11}{180}\dfrac{q_0L^3}{EI}$

$v_B' = \dfrac{23}{360}\dfrac{q_0L^3}{EI}$

7.9-14 (*a*) $v_A' = -\dfrac{7q_0L^3}{360EI}$

$v_B' = \dfrac{q_0L^3}{45EI}$

7.9-15 $\sigma_{max} = 51.1$ ksi

7.9-16 $\beta = -6.15 \times 10^{-4}$ rad for single-load case, $\beta = -5.88 \times 10^{-4}$ rad for double-load case

7.9-17 Configuration (*b*) is best.

7.9-20 $\delta_H = \dfrac{PL^3}{24EI}$, $\delta_V = 0$

7.9-21 $\beta = -0.0813 \dfrac{wL^3}{EI}$

7.9-22 (Origin at center of mirror) If v'_{max} at $x = \hat{x}$ and $\hat{x} > L/2 - a$, then $\hat{x} = \dfrac{L}{2}$.

$$EIv'_{max} = \frac{-\hat{w}L^3}{48} + \frac{\hat{w}La^2}{4} + \frac{\hat{w}L^2}{4}\left(\frac{L}{4} - a\right)$$

If $\hat{x} < \dfrac{L}{2} - a$,

$$EIv'_{max} = \frac{\hat{w}L^3}{3}\left(\frac{1}{4} - \frac{a}{L}\right)^{3/2}$$

7.9-23 $a/L = 0.223$

7.9-24 $\delta = \dfrac{19}{6}\dfrac{PL^3}{EI}$

7.9-26 $M_0 = 127$ N · m

7.9-27 (*a*) $u_C = 4.22$ in, $u_C = 1.58$ in, $v_C = 0.132$ rad (counterclockwise) (*b*) $u_C = 1.32$ in, $v_C = 0.440$ in, $\theta_C = 0.0489$ rad (counterclockwise)

7.9-28 $F_A = (7 + 26N)\dfrac{P}{\beta}$

$F_B = (-2 + 44N)\dfrac{P}{\beta}$

$F_C = (3 - 6N)\dfrac{P}{\beta}$

$N = \dfrac{kL^3}{384EI}$

$\beta = 8\,(1 + 8N)$

7.9-30 (*a*) $v_{max} = -0.997$ in, $M_{max} = 7.40 \times 10^5$ lb · in, $\sigma_{max} = 20.9$ ksi (*b*) $v_{max} = 0.412$ in, $M_{max} = 72$ kip · in, $\sigma_{max} = 9.08$ ksi (*c*) $v_{max} = -6.56$ mm, $M_{max} = 4.33$ kN · m, $\sigma_{max} = 186$ MPa (*d*) $v_{max} = 0.375$ in, $M_{max} = 1.35 \times 10^5$ lb · in, $\sigma_{max} = 13.1$ ksi (*e*) $v_{max} = -2.60$ mm, $M_{max} = -2.5$ kN · m, $\sigma_{max} = 130$ MPa (*f*) $v_{max} = 0.326$ in, $M_{max} = 1.20 \times 10^5$ lb · in, $\sigma_{max} = 10.2$ ksi (*g*) $v_{max} = 1.69$ mm, $M_{max} = 6.20 \times 10^3$ N· m, $\sigma_{max} = 15.1$ MPa (*h*) $v_{max} = -2.92$ in, $M_{max} = 2.88 \times 10^6$ lb · in, $\sigma_{max} = 23.4$ ksi

7.9-31 (*a*) $v_{max} = 16.7$ mm, $M_{max} = -1.69$ kN · m, $\sigma_{max} = 217$ MPa (*b*) $v_{max} = 0.241$ in, $M_{max} = 63.0$ kip · in, $\sigma_{max} = 6.18$ ksi (*c*) $v_{max} = 0.504$ in, $M_{max} = 291$ kip · in, $\sigma_{max} = 12.2$ ksi (*d*) $v_{max} = 11.9$ mm, $M_{max} = -12$ kN · m, $\sigma_{max} = 160$ MPa (*e*) $v_{max} = 18.6$ mm, $M_{max} = 20.8$ kN · m, $\sigma_{max} = 96.2$ MPa (*f*) $v_{max} = 0.273$ in, $M_{max} = 367$ kip · in, $\sigma_{max} = 10.1$ ksi

7.9-32 $L^4 = 72\dfrac{EIv_0}{q}$

7.9-33 (*a*), (*c*) $R_A = \dfrac{Pb}{L}$, $R_B = \dfrac{Pa}{L}$,

(*b*) $T_A = \dfrac{Tb}{L}$, $T_B = \dfrac{Ta}{L}$

(*d*) $R_A = \dfrac{Pb^2}{L^3}(L + 2a)$,

$M_A = \dfrac{Pab^2}{L^2}$

7.9-34 Case (c): $R_A = \frac{M}{L}$, $R_B = -R_A$

Case (d): $R_A = \frac{6Mab}{L^3}$, $R_B = -R_A$; $M_A = \frac{Mb}{L^2}(L - 3a)$, $M_B = \frac{Ma}{L^2}(3b - L)$

Chapter 8

8.2-1 (a) $\sqrt{\overset{(n)}{T}{}_x^2 + \overset{(n)}{T}{}_y^2 + \overset{(n)}{T}{}_z^2}$

(b) $\sqrt{\tau_{xy}^2 + \tau_{xz}^2}$

(c) $\overset{(j)}{\mathbf{T}} = \tau_{yx}\mathbf{i} + \sigma_y\mathbf{j} + \tau_{yz}\mathbf{k}$

8.3-3 (a) (1) $\sigma_x = \sigma_y = 17.5$ ksi, $\tau_{xy} = 0$
(2) $\sigma_x = 2.5$ ksi, $\sigma_y = -2.5$ ksi, $\tau_{xy} = 7$ ksi
(b) (1) $\sigma_x = \sigma_y = 130$ MPa, $\tau_{xy} = 0$
(2) $\sigma_x = \sigma_y = 0$, $\tau_{xy} = -30$ MPa
(c) (1) $\sigma_x = \sigma_y = -50$ MPa, $\tau_{xy} = 0$
(2) $\sigma_x = \sigma_y = 0$, $\tau_{xy} = 50$ MPa
(d) (1) $\sigma_x = \sigma_y = 11$ ksi, $\tau_{xy} = 0$
(2) $\sigma_x = 19$ ksi, $\sigma_y = -19$ ksi, $\tau_{xy} = 0$
(e) (1) $\sigma_x = \sigma_y = 1500$ psi, $\tau_{xy} = 0$
(2) $\sigma_x = -6500$ psi, $\sigma_y = 6500$ psi, $\tau_{xy} = 6000$ psi

8.4-1 (c) $T_n = \sigma_x \cos^2\theta + \sigma_y \sin^2\theta + 2\tau_{xy} \sin\theta \cos\theta$
(d) $T_s = (\sigma_y - \sigma_x) \sin\theta \cos\theta + \tau_{xy}(\cos^2\theta - \sin^2\theta)$

8.4-2 $\overset{(n)}{T}{}_x = \sigma_x \sin\alpha + \tau_{yx} \cos\alpha$
$\overset{(n)}{T}{}_y = \tau_{xy} \sin\alpha + \sigma_y \cos\alpha$

8.4-5 $\sigma_{x'} = 28.2$ MPa, $\tau_{x'y'} = 21.7$ MPa

8.4-6 $\sigma_{x'} = 11.8$ ksi, $\tau_{x'y'} = 1.19$ ksi

8.4-7 $\sigma_{x'} = 1.47$ MPa, $\tau_{x'y'} = -1.75$ MPa

8.4-8 $\sigma_{x'} = 4.64$ ksi, $\tau_{x'y'} = -12.2$ ksi

8.4-9 $\sigma_{x'} = -88.9$ MPa, $\tau_{x'y'} = 10.2$ MPa

8.4-10 $\sigma_{x'} = -266$ psi, $\tau_{x'y'} = 94.0$ psi

8.4-11 $-413 \le \sigma_x < 12.8$ (psi)

8.4-12 y' axis is perpendicular to grain, $\sigma_{y'} = -3.18$ MPa, $\tau_{x'y'} = -0.482$ MPa

8.4-13 $\sigma_{x'} = -97.3$ MPa, $\tau_{x'y'} = -6.06$ MPa

8.4-14 $\sigma_{x'} = -1.15$ ksi, $\tau_{x'y'} = 44.2$ ksi

8.4-15 (a) $\theta_1 = -58.4°$, $\theta_2 = 46.5°$
(b) $\theta = \theta_1$: $\tau_{x'y'} = 46.9$ MPa, $\theta = \theta_2$: $\tau_{x'y'} = -46.9$ MPa

8.4-16 (a) $\theta_1 = -5.94°$, $\theta_2 = 84.06°$

8.5-2 $\sigma_1 = 12.7$ ksi, $\sigma_2 = 3.28$ ksi, $\theta_p = -29.0°$

8.5-3 $\sigma_1 = 65.4$ MPa, $\sigma_2 = -234$ MPa, $\theta_p = 7.74°$

8.5-4 $\sigma_1 = 354$ MPa, $\sigma_2 = 54.6$ MPa, $\theta_p = -7.74°$

8.5-5 (a) $\sigma_1 = 108$ MPa, $\sigma_2 = 22.3$ MPa, $\tau_{max} = 42.7$ MPa

8.5-6 $\sigma_1 = 11.8$ MPa,

$\sigma_2 = -9.82$ MPa, $\tau_{max} = 10.82$ MPa

8.5-7 $\sigma_1 = 1.49$ MPa, $\sigma_2 = -152$ MPa, $\tau_{max} = 76.5$ MPa

8.5-8 $\sigma_1 = 14.9$ ksi, $\sigma_2 = -9.92$ ksi, $\tau_{max} = 12.4$ ksi

8.5-9 $\sigma_1 = 89.4$ MPa, $\sigma_2 = -89.4$ MPa, $\tau_{max} = 89.4$ MPa

8.5-10 $\sigma_1 = \sigma_0$, $\sigma_2 = 0$, $\tau_{max} = \dfrac{\sigma_0}{2}$

8.5-11 $\sigma_1 = \tau_0$, $\sigma_2 = -\tau_0$, $\tau_{max} = \tau_0$

8.5-12 $\sigma_1 = \sigma_2 = \sigma_0$, $\tau_{max} = 0$

8.5-13 $\sigma_1 = 49.1$ MPa, $\sigma_2 = -59.1$ MPa, $\tau_{max} = 54.1$ MPa

8.5-14 $\sigma_1 = -5.23$ ksi, $\sigma_2 = -26.8$ ksi, $\tau_{max} = 10.8$ ksi

8.5-15 $\sigma_1 = 50$ MPa, $\sigma_2 = -35$ MPa, $\tau_{max} = 42.5$ MPa

8.5-16 $\sigma_1 = 26.2$ ksi, $\sigma_2 = 3.82$ ksi, $\theta_p = 31.7°$ $\tau_{max} = 11.2$ ksi, $\sigma_{x'} = 20.5$ ksi, $\sigma_{y'} = 9.51$ ksi

8.5-17 $\sigma_1 = 150$ MPa, $\sigma_2 = 50$ MPa, $\theta_p = 45°$ $\tau_{max} = 50$ MPa, $\sigma_{x'} = 141$ MPa, $\sigma_{y'} = 58.5$ MPa

8.5-18 $\sigma_1 = -40$ MPa, $\sigma_2 = -100$ MPa, $\theta_p = -45°$ $\tau_{max} = 30$ MPa, $\sigma_{x'} = -94.9$ MPa, $\sigma_{y'} = -45.1$ MPa

8.5-19 $\sigma_1 = 20$ ksi, $\sigma_2 = -5$ ksi, $\theta_p = 0°$ $\tau_{max} = 12.5$ ksi, $\sigma_{x'} = 14.5$ ksi, $\sigma_{y'} = 0.51$ ksi

8.5-20 $\sigma_1 = 9.13$ ksi, $\sigma_2 = -6.13$ ksi, $\theta_p = -15.8°$ $\tau_{max} = 7.63$ ksi, $\sigma_{x'} = 1.18$ ksi, $\sigma_{y'} = 1.82$ ksi

8.5-21 $\sigma_1 = 24.9$ ksi, $\sigma_2 = 10.1$ ksi, $\theta_p = 35.17°$ $\tau_{max} = 7.43$ ksi, $\sigma_{x'} = 24.7$ ksi, $\sigma_{y'} = 10.3$ ksi

8.5-22 $\sigma_1 = 160$ MPa, $\sigma_2 = 100$ MPa, $\theta_p = -45°$ $\tau_{max} = 30$ MPa, $\sigma_{x'} = 105$ MPa, $\sigma_{y'} = 155$ MPa

8.5-23 $\sigma_1 = 0$, $\sigma_2 = -100$ MPa, $\theta_p = 45°$ $\tau_{max} = 50$ MPa, $\sigma_{x'} = -8.55$ MPa, $\sigma_{y'} = -91.5$ MPa

8.5-24 $\sigma_1 = 30$ ksi, $\sigma_2 = -8$ ksi, $\theta_p = 0$ $\tau_{max} = 19$ ksi, $\sigma_{x'} = 21.6$ ksi, $\sigma_{y'} = 0.375$ ksi

8.5-25 $\sigma_1 = 10.3$ ksi, $\sigma_2 = -7.35$ ksi, $\theta_p = -21.35°$ $\tau_{max} = 8.85$ ksi, $\sigma_{x'} = 2.84$ ksi, $\sigma_{y'} = 0.161$ ksi

8.6-2 (*b*) $\sigma_1 = 15.1$ ksi, $\sigma_2 = 3.91$ ksi (*c*) $\tau_{max} = 5.59$ ksi

8.6-3 (*a*) $\sigma_1 = 102$ MPa, $\sigma_2 = -62.5$ MPa, $\tau_{max} = 82.5$ MPa, $\theta_p = 38.0°$ (*b*) $\sigma_1 = 13.6$ ksi, $\sigma_2 = -7.63$ ksi, $\tau_{max} = 10.6$ ksi, $\theta_p = 24.4°$ (*c*) $\sigma_1 = 165$ MPa, $\sigma_2 = -4.85$ MPa, $\tau_{max} = 84.9$ MPa, $\theta_p = -22.5°$ (*d*) $\sigma_1 = 22.8$ ksi,

$\sigma_2 = 7.19$ ksi,
$\tau_{max} = 7.81$ ksi, $\theta_p = 25.1°$
(*e*) $\sigma_1 = 191$ MPa,
$\sigma_2 = -21$ MPa,
$\tau_{max} = 106$ MPa,
$\theta_p = 35.35°$

8.6-5 $\sigma_x = 105$ MPa, $\theta_p = -26.6°$

8.6-6 $\sigma_x = -5.46$ ksi or 1.46 ksi

8.6-7 $\sigma_1 = 7$ ksi, $\sigma_2 = -5$ ksi,
$\theta_p = 16.8°$, $\tau_{max} = 6$ ksi

8.6-8 (*a*) $\sigma_1 = 6$ ksi, $\sigma_2 = -4$ ksi,
$\theta_p = 26.6°$
(*b*) $\sigma_1 = 17.2$ ksi,
$\sigma_2 = -14.2$ ksi,
$\theta_p = -15.3°$
(*c*) $\sigma_1 = 2.45$ ksi,
$\sigma_2 = -2.95$ ksi,
$\theta_p = 16.8°$
(*d*) $\sigma_1 = -18.5$ psi,
$\sigma_2 = -1080$ psi,
$\theta_p = -24.4°$
(*e*) $\sigma_1 = 400$ MPa,
$\sigma_2 = -600$ MPa,
$\theta_p = 18.4°$
(*f*) $\sigma_1 = 624$ psi,
$\sigma_2 = 176$ psi, $\theta_p = 13.3°$

8.6-9 (*a*) $\sigma_1 = 26.2$ ksi,
$\sigma_2 = 3.82$ ksi,
$\theta_p = -31.7°$
$\theta = 28°$: $\sigma_{x'} = 20.5$ ksi,
$\tau_{x'y'} = 9.74$ ksi
(*b*) $\sigma_1 = 150$ MPa,
$\sigma_2 = 50$ MPa, $\theta_p = 45°$
$\theta = 28°$: $\sigma_{x'} = 141$ MPa,
$\tau_{x'y'} = 28.0$ MPa
(*c*) $\sigma_1 = -40$ MPa,
$\sigma_2 = -100$ MPa,
$\theta_p = -45°$; $\theta = 28°$:
$\sigma_{x'} = -94.9$ MPa,
$\tau_{x'y'} = -16.8$ MPa
(*d*) $\sigma_1 = 20$ MPa,
$\sigma_2 = -5$ MPa,
$\theta_p = 0°$
$\theta = 28°$: $\sigma_{x'} = 14.5$ MPa,
$\tau_{x'y'} = -10.4$ MPa
(*e*) $\sigma_1 = 9.13$ ksi,
$\sigma_2 = -6.13$ ksi,
$\theta_p = 74.2°$
$\theta = 28°$: $\sigma_{x'} = 1.18$ ksi,
$\tau_{x'y'} = 7.63$ ksi
(*f*) $\sigma_1 = 24.9$ ksi,
$\sigma_2 = 10.1$ ksi,
$\theta_p = 35.2°$
$\theta = 28°$: $\sigma_{x'} = 24.7$ ksi,
$\tau_{x'y'} = 1.84$ ksi
(*g*) $\sigma_1 = 160$ MPa,
$\sigma_2 = 100$ MPa,
$\theta_p = -45°$
$\theta = 28°$: $\sigma_{x'} = 105$ MPa,
$\tau_{x'y'} = -16.8$ MPa
(*h*) $\sigma_1 = 0$,
$\sigma_2 = -100$ MPa,
$\theta_p = 45°$
$\theta = 28°$:
$\sigma_{x'} = -8.55$ MPa,
$\tau_{x'y'} = 28.0$ MPa
(*i*) $\sigma_1 = 30$ ksi,
$\sigma_2 = -8$ ksi, $\theta_p = 0°$
$\theta = 28°$: $\sigma_{x'} = 21.6$ ksi,
$\tau_{x'y'} = -15.8$ ksi
(*j*) $\sigma_1 = 10.3$ ksi,
$\sigma_2 = -7.35$ ksi,
$\theta_p = 68.65°$
$\theta = 28°$: $\sigma_{x'} = 2.84$ ksi,
$\tau_{x'y'} = 8.74$ ksi
(*k*) $\sigma_1 = 26.2$ ksi,
$\sigma_2 = 3.82$ ksi,
$\theta_p = -31.2°$
$\theta = 28°$: $\sigma_{x'} = 20.5$ ksi,
$\tau_{x'y'} = 9.74$ ksi
(*l*) $\sigma_1 = 150$ MPa,
$\sigma_2 = 50$ MPa,
$\theta_p = 45°$
$\theta = 28°$: $\sigma_{x'} = 141$ MPa,
$\tau_{x'y'} = 28.0$ MPa
(*m*) $\sigma_1 = -40$ MPa,
$\sigma_2 = -100$ MPa,
$\theta_p = 45°$; $\theta = 28°$:
$\sigma_{x'} = -94.9$ MPa,
$\tau_{x'y'} = -16.8$ MPa
(*n*) $\sigma_1 = 20$ ksi,
$\sigma_2 = -5$ ksi, $\theta_p = 0°$
$\theta = 28°$: $\sigma_{x'} = 14.5$ ksi,
$\tau_{x'y'} = -10.4$ ksi
(*o*) $\sigma_1 = 9.13$ ksi,

$\sigma_2 = -6.13$ ksi,
$\theta_p = -15.8°$
$\theta = 28°$: $\sigma_{x'} = 1.18$ ksi,
$\tau_{x'y'} = 7.63$ ksi
(*p*) $\sigma_1 = 24.9$ ksi,
$\sigma_2 = 10.1$ ksi,
$\theta_p = 35.2°$
$\theta = 28°$: $\sigma_{x'} = 24.7$ ksi,
$\tau_{x'y'} = 1.84$ ksi
(*q*) $\sigma_1 = 160$ MPa,
$\sigma_2 = 100$ MPa,
$\theta_p = -45°$
$\theta = 28°$: $\sigma_{x'} = 105$ MPa,
$\tau_{x'y'} = -16.8$ MPa
(*r*) $\sigma_1 = 0$,
$\sigma_2 = -100$ MPa,
$\theta_p = 45°$; $\theta = 28°$:
$\sigma_{x'} = -8.55$ MPa,
$\tau_{x'y'} = 28.0$ MPa
(*s*) $\sigma_1 = 30$ ksi,
$\sigma_2 = -8$ ksi, $\theta_p = 0°$
$\theta = 28°$: $\sigma_{x'} = 21.6$ ksi,
$\tau_{x'y'} = -15.8$ ksi
(*t*) $\sigma_1 = 10.3$ ksi,
$\sigma_2 = -7.35$ ksi,
$\theta_p = -21.35°$
$\theta = 28°$: $\sigma_{x'} = 2.84$ ksi,
$\tau_{x'y'} = 8.74$ ksi

8.6-11 (*a*) $\sigma_{x'} = 17.6$ ksi,
$\tau_{x'y'} = 10.9$ ksi
(*b*) $\sigma_{x'} = 147$ MPa,
$\tau_{x'y'} = 17.1$ MPa
(*c*) $\sigma_{x'} = -57.8$ MPa,
$\tau_{x'y'} = -27.4$ MPa
(*d*) $\sigma_{x'} = 1.25$ ksi,
$\tau_{x'y'} = 10.8$ ksi
(*e*) $\sigma_{x'} = 8$ ksi, $\tau_{x'y'} = -4$ ksi
(*f*) $\sigma_{x'} = 24.2$ ksi,
$\tau_{x'y'} = -3.22$ ksi
(*g*) $\sigma_{x'} = 102$ MPa,
$\tau_{x'y'} = -10.3$ MPa
(*h*) $\sigma_{x'} = -92.4$ MPa,
$\tau_{x'y'} = 26.5$ MPa
(*i*) $\sigma_{x'} = 10.3$ ksi,
$\tau_{x'y'} = -19.0$ ksi
(*j*) $\sigma_{x'} = -5$ ksi,
$\tau_{x'y'} = 6$ ksi

8.7-1 $\sigma_1 = 52.4$ MPa,
$\sigma_2 = -32.4$ MPa,
$\sigma_3 = 0$, $\tau_{max} = 42.4$ MPa

8.7-2 $\sigma_1 = 22.6$ ksi,
$\sigma_2 = -37.6$ ksi, $\sigma_3 = 0$,
$\tau_{max} = 30.1$ ksi

8.7-3 $\sigma_1 = -11.5$ MPa,
$\sigma_2 = -78.5$ MPa, $\sigma_3 = 0$,
$\tau_{max} = 39.25$ MPa

8.7-4 $\sigma_1 = 20$ ksi, $\sigma_2 = -20$ ksi,
$\sigma_3 = 0$, $\tau_{max} = 20$ ksi

8.7-5 $\tau_0 = \pm 11.2$ ksi

8.7-6 (*a*) $\sigma_x = 0$: $\tau_{max} = 64.0$ MPa
$\sigma_x = 80$ MPa:
$\tau_{max} = 94.3$ MPa
$\sigma_x = -150$ MPa:
$\tau_{max} = 88$ MPa
(*b*) $-257.5 < \sigma_x < 93.2$ (MPa)

8.7-9 $\tau_{max} = 2500$ psi at $\theta = -45°$

8.7-10 $\sigma_1 = 509$ psi,
$\sigma_2 = -509$ psi, $\theta_p = -45°$
$\tau_{max} = 509$ psi

8.7-11 $\sigma_1 = 28.3$ MPa,
$\sigma_2 = -28.3$ MPa, $\theta_p = -45°$
$\tau_{max} = 28.3$ MPa

8.7-12 (*a*) $\sigma_2 = -\dfrac{3}{4}\dfrac{PL}{bh^2}$, $\sigma_1 = 0$,
$\tau_{max} = -\dfrac{\sigma_2}{2}$
(*b*) $\sigma_1 = \dfrac{3}{4}\dfrac{P}{bh}$, $\sigma_2 = -\dfrac{3}{4}\dfrac{P}{bh}$,
$\tau_{max} = \dfrac{3}{4}\dfrac{P}{bh}$

8.7-13 (*a*) $\sigma_2 = -\dfrac{L}{16bh^2}(12P + 9w_0L)$,
$\sigma_1 = 0$, $\tau_{max} = \dfrac{|\sigma_2|}{2}$
(*b*) $\tau_{max} = \dfrac{3}{2bh}\left(\dfrac{P}{2} + \dfrac{w_0L}{4}\right)$,
$\sigma_1 = \tau_{max}$, $\sigma_2 = -\tau_{max}$

8.7-14 $\sigma_1 = \sigma_{av} + \tau_{max}$, $\sigma_2 = \sigma_{av} - \tau_{max}$, $\sigma_{av} = -\dfrac{3}{16}\dfrac{PL}{bh^2}$

$\tau_{max} = \frac{3}{16}\frac{P}{bh}\sqrt{9 + \frac{L^2}{h^2}}$

8.7-15 $\sigma_1 = \sigma_{av} + \tau_{max}$, $\sigma_2 = \sigma_{av} - \tau_{max}$, $\sigma_{av} = -\frac{3}{8}\frac{PL}{bh^2}$

$\tau_{max} = \frac{3}{16}\frac{P}{bh}\sqrt{\frac{4L^2}{h^2} + 9}$

8.7-16 (*a*) $\tau_{max} = 600$ psi, face inclined 45° to *xy* plane
(*b*) $\tau_{max} = 20$ MPa, face inclined 45° to *xy* plane
(*c*) $\tau_{max} = 300$ psi, face inclined 45° to *xz* plane
(*d*) $\tau_{max} = 100$ psi, face inclined 45° to *xz* plane
(*e*) $\tau_{max} = 70$ MPa, face inclined 45° to *xy* plane

8.7-17 $\sigma_1 = 14$ ksi, $\sigma_2 = 4$ ksi, $\tau_{max} = 7$ ksi

8.7-18 $\tau_{max} = 127$ MPa

8.7-19 $\sigma_x = 2040$ psi, $\sigma_y = 0$, $\tau_{xy} = -3260$ psi

8.7-20 $\sigma_1 = 4440$ psi, $\sigma_2 = -2400$ psi, $\theta_p = -36.3°$

8.8-1 $p_a = 2.8$ MPa

8.8-2 $\sigma_{x'} = 26$ MPa, $\tau_{x'y'} = -9.01$ MPa

8.8-3 (*a*) $\sigma_A = 17{,}400$ psi
(*b*) $p_a = 200$ psi

8.8-4 (*a*) $\sigma_H = 75$ MPa
(*b*) $\sigma_A = 37.5$ MPa
(*c*) $\sigma = 37.5$ MPa

8.8-6 If $F > \pi p r^2$, $\tau_{max} = \frac{pr}{4t} + \frac{F}{4\pi rt}$

If $F < \pi p r^2$, $\tau_{max} = \frac{pr}{2t}$

8.8-8 $F = -2\pi p r^2$

8.8-9 $\frac{w}{r} \leq 4.86$

8.8-10 $\sigma = 0.437$ MPa

8.8-11 $\alpha = 5.5°$

8.8-12 $\sigma = 7.14$ kPa

8.8-13 $\sigma_H = 6000$ psi

8.8-14 $\sigma_H = 24$ ksi, $\sigma_A = 12$ ksi

8.8-15 $\sigma_H = 4.44$ ksi

8.8-16 45 MPa normal to weld, $\tau_{max} = 27$ MPa

8.8-17 $\sigma_H = 4.78$ MPa

8.8-18 $\tau_{max} = 4.41$ ksi

8.8-19 $F = -47.1$ kips

8.8-20 $\psi = 44.0°$

8.8-21 $p = 1000$ psi

8.8-22 $t_{min} = 10.7$ mm

8.8-23 $\sigma_{max} = 1560$ psi, $\tau_{max} = 780$ psi

8.8-24 $p_a = 200$ psi

8.11-4 $\epsilon_x = \epsilon_y = \cos\beta - 1$, $\gamma_{xy} = 0$

8.12-2 $\epsilon_{x'} = -436\mu$, $\epsilon_{y'} = -164\mu$, $\gamma_{x'y'} = 968\mu$, $\mu = 10^{-6}$

8.12-3 $\epsilon_{x'} = \epsilon_{y'} = 0$, $\gamma_{x'y'} = 1000\mu$, $\mu = 10^{-6}$

8.12-4 $\epsilon_{x'} = -250\mu$, $\epsilon_{y'} = 250\mu$, $\gamma_{x'y'} = 0$, $\mu = 10^{-6}$

8.12-5 $\epsilon_{x'} = 232\mu$, $\epsilon_{y'} = -1030\mu$, $\gamma_{x'y'} = -34.6\mu$, $\mu = 10^{-6}$

8.12-6 $\epsilon_{x'} = -730\mu$, $\epsilon_{y'} = 330\mu$, $\gamma_{x'y'} = 1236\mu$, $\mu = 10^{-6}$

8.12-8 (*a*) $\epsilon_1 = 202\mu$, $\epsilon_2 = -802\mu$, $\theta_p = 2.86°$, $\mu = 10^{-6}$
(*b*) $\epsilon_1 = 500\mu$, $\epsilon_2 = -500\mu$, $\theta_p = 0°$, $\mu = 10^{-6}$
(*c*) $\epsilon_1 = 250\mu$, $\epsilon_2 = -250\mu$, $\theta_p = -45°$, $\mu = 10^{-6}$

(*d*) $\epsilon_1 = 232\mu$, $\epsilon_2 = -1030\mu$, $\theta_p = 4.73°$, $\mu = 10^{-6}$
(*e*) $\epsilon_1 = 614\mu$, $\epsilon_2 = -1010\mu$, $\theta_p = 5.31°$, $\mu = 10^{-6}$

8.12-10 $\epsilon_1 = 1070\mu$, $\epsilon_2 = -23.3\mu$, $\theta_p = -23.3°$, $\mu = 10^{-6}$

8.12-11 $\epsilon_1 = -159\mu$, $\epsilon_2 = -941\mu$, $\theta_p = 25.1°$, $\mu = 10^{-6}$, $\epsilon_{x'} = -889\mu$, $\epsilon_{y'} = -211\mu$, $\gamma_{x'y'} = 388\mu$, $\mu = 10^{-6}$

8.13-1 (*a*) $\epsilon_1 = 625\mu$, $\epsilon_2 = -1920\mu$, $\theta_p = 39.35°$, $\mu = 10^{-6}$
(*b*) $\epsilon_1 = 820\mu$, $\epsilon_2 = -2520\mu$, $\theta_p = 40.7°$, $\mu = 10^{-6}$

8.13-2 (*a*) $\epsilon_d = -1900\mu$, $\mu = 10^{-6}$
(*b*) $\epsilon_d = -2500\mu$, $\mu = 10^{-6}$

8.13-3 (*a*) $\epsilon_x = -462\mu$, $\epsilon_y = 1860\mu$, $\mu = 10^{-6}$
(*b*) $\epsilon_x = 477\mu$, $\epsilon_y = -1080\mu$, $\mu = 10^{-6}$

8.13-4 (*a*) $\epsilon_1 = 1000\mu$, $\epsilon_2 = 0$, $\theta_p = 18.43°$, $\mu = 10^{-6}$
(*b*) $\epsilon_1 = 1240\mu$, $\epsilon_2 = 15.3\mu$, $\theta_p = -10.99°$, $\mu = 10^{-6}$

8.13-5 $\epsilon_1 = 710\mu$, $\epsilon_2 = 140\mu$, $\theta_p = -37.4°$, $\mu = 10^{-6}$

8.13-6 $\epsilon_1 = 787\mu$, $\epsilon_2 = -154\mu$, $\theta_p = 93.5°$, $\mu = 10^{-6}$

8.14-2 (*a*) $\gamma_{xy} = 2500\mu$, $\mu = 10^{-6}$, $\sigma_x = -22.1$ ksi, $\sigma_y = -33.6$ ksi, $\tau_{xy} = 28.8$ ksi
(*b*) $\gamma_{xy} = 3300\mu$, $\mu = 10^{-6}$, $\sigma_x = -30.7$ ksi, $\sigma_y = -42.2$ ksi, $\tau_{xy} = 38.1$ ksi

8.14-3 $P = 2\dfrac{EI}{Ld_o}\epsilon_x$,
$T = \dfrac{2GJ}{d_o}(2\epsilon_c - \epsilon_x - \epsilon_z)$,
$J = \dfrac{\pi}{32}(d_o^4 - d_i^4)$, $I = \dfrac{J}{2}$

8.14-4 $P = 534$ lb, $T = 49.3$ kip · in

8.14-5 (*a*) $\sigma_x = -\dfrac{P}{b^2}$, $\sigma_y = \dfrac{-\nu P}{b^2}$, $\sigma_z = 0$
(*b*) $\sigma_y = -\nu\dfrac{P}{b^2} - \alpha E\,\Delta T$

8.14-6 $\sigma_1 = 4.20$ ksi, $\sigma_2 = -0.954$ ksi, $\theta_p = -25.45°$

8.14-7 $p = 229$ kPa

8.14-8 276×10^{-6}

8.14-9 58.4 MPa

8.14-10 -1190 N · m

8.14-11 -300×10^{-6}

8.14-12 1.37 MPa

8.14-13 (*a*) $\sigma_0 = 10$ ksi
(*b*) $\tau_{max} = 15$ ksi, $\gamma_{max} = 1300 \times 10^{-6}$

8.15-1 (*a*) $1.10 \times 10^{-3}\,Y$
(*b*) $1.12 \times 10^{-3}\,Y$

8.15-2 115 kip · in

8.15-3 8320 kip · in

8.15-4 ±50.8°F

8.15-5 No (Mises criterion), yes (maximum-shear criterion)

8.15-6 7.39 MPa (Mises), 6.40 MPa (maximum shear)

8.15-7 No (Mises), yes (maximum shear)

8.15-8 0.129 in

8.15-9 124 N · m (Mises), 97.9 N · m (maximum shear)

8.16-1 $p_{max} = \dfrac{4}{27}\dfrac{EV}{\pi R_0^3}$

8.16-2 $p = \frac{8Et_o}{D_0}\left[\left(\frac{D_0}{D}\right)^2 - \left(\frac{D_0}{D}\right)^3\right]$

8.16-4 $\sigma_1 = 17.1$ ksi, $\sigma_2 = 4.55$ ksi

Chapter 9

9.2-1 $F = 50.3$ lb

9.2-2 $F = 42.3$ lb

9.2-3 $F = 72.3$ lb

9.2-4 $F = 60.6$ lb

9.2-5 $\sigma = 15.95$ MPa $= 2310$ psi

9.2-6 (*a*) $\sigma_B = -41.6$ MPa,
$\sigma_A = -50.4$ MPa
(*b*) $\sigma_B - 71.7$ MPa,
$\sigma_A = -163.7$ MPa

9.2-7 Width $= 6.67$ in

9.2-8 $\sigma = \frac{3}{4}\frac{q_0L^4}{bh^2}$

9.2-9 $\sigma = \pm 0.39$ MPa

9.2-10 $\tau_{max} = 2.39$ ksi

9.2-11 $\sigma_C = -33.3$ MPa,
$\sigma_t = 29.2$ MPa

9.2-12 $b = 34.5$ mm

9.2-13 $\sigma = 1870$ psi

9.2-14 $\sigma_{max} = 2490$ psi, compression

9.2-15 (*a*) $\epsilon_x = 1.25 \times 10^{-3}$
(*b*) $\epsilon_x = 0.89 \times 10^{-3}$
(*c*) $\epsilon_x = 0.42 \times 10^{-3}$

9.2-16

	σ_1	σ_2	θ	τ
(*a*)	178	−2.36	6.56°	90.2
(*b*)	127	−9.70	15.5°	68.3
(*c*)	82.1	−23.5	28.1°	52.8
(*d*)	46.9	−46.9	45.0°	46.9
	MPa	MPa		MPa

9.2-17 $F = 141.3 \times 10^3$ lb

9.2-18 $\tau_{max} = 18.75$ MPa

9.2-19 $\sigma_C = 16$ ksi, compressive

9.2-20 $d = 1.25$ in

9.2-21 $F = 982$ N

9.2-22 $\sigma = 8.34$ MPa, tensile

9.2-23 $P = 2685$ lb

9.2-24 $\sigma_{max} = 296$ MPa

9.2-25 $\sigma_{max} = 9.4$ ksi

9.2-26 $d = 1.74$ m

9.2-27 $\tau_{max} = 114$ MPa

9.3-1 $\sigma_1 = 22.3$ ksi, $\sigma_2 = -30.4$ ksi,
$\tau = 26.3$ ksi

9.3-2 (*a*) $\tau_{max} = 10.2$ ksi
(*b*) $\sigma_1 = 2230$ psi,
$\sigma_2 = -18.1$ ksi, $\theta = 19.3°$
(*c*) $\delta = 6.37 \times 10^{-3}$ in
(*d*) $\phi = 6.37 \times 10^{-2}$ rad

9.3-3 $\tau_{max} = R$
$R = \sqrt{[P/(2A)]^2 + [16T/(\pi d^3)]^2}$
$\sigma_{1,2} = -\frac{P}{2A} \pm R$

9.3-5 $\tau_{max} = 510$ psi

9.3-6 $\tau_{max} = 72.4$ ksi

9.3-7 $\tau_{max} = 4278$ psi

9.3-8 $\sigma_1 = 223$ MPa, $\sigma_2 = -11$ MPa,
$\tau_{max} = 117$ MPa

9.3-9 $\tau_{max} = 2.72 \times 10^4$ psi,
$\tau_{max} = 2.29 \times 10^4$ psi when $T = 0$

9.3-10 $\tau_{max} = 197$ MPa

9.3-11 $F = 40.2$ kips, $T = 403$ kip · in, $\tau_{max} = 10.8$ ksi

9.3-12 $\tau_{max} = 362.3$ psi

9.3-13 $\tau_{max} = 14.1$ MPa

9.4-1 $\sigma_1 = 15.75$ ksi, $\sigma_2 = -675$ psi, $\tau_{max} = 8214$ psi

9.4-2 (*a*) $\epsilon(0) = 7.08 \times 10^{-4}$, $\epsilon(60°) = 4.01 \times 10^{-4}$, $\epsilon(-60°) = -3.65 \times 10^{-4}$
(*b*) $\epsilon(0) = 0$, $\epsilon(\pm 60°) = \mp 3.96 \times 10^{-4}$

9.4-3 From App. E, standard weight: nominal diameter 4.0 in, extra-strong: nominal diameter 3.5 in

9.4-4 $\tau_{max} = \sqrt{\left(\frac{2q_0 L^2}{\pi d^3}\right)^2 + \left(\frac{16T}{\pi d^3}\right)^2}$

9.4-5 $d = 86.7$ mm

9.4-6 $T = 16.9$ kN · m

9.4-7 $d_o = 4.11$ in

9.4-8 $\tau_{max} = 42.7$ ksi

9.4-9 $\tau_{max} = 4770$ psi, $\sigma_1 = 8590$ psi, $\sigma_2 = -952$ psi

9.4-10 $\tau_{max} = 3150$ psi

9.4-11 $d = 28.1$ mm

9.4-12 (*a*) $\tau_{max} = 13.5$ MPa
(*b*) 8.9 MPa
(*c*) $\tau_{max} = 13.5$ MPa
(*d*) 10.2 MPa

9.4-13 $\tau_{max} = 24.26$ MPa

9.4-14 $d = 1.37$ in

9.4-15 $\tau_{max} = 23$ ksi

9.5-1 $\tau_{max} = 26.1$ MPa

9.5-2 $L = 37.9$ m

9.5-3 P = 16.7 kN

9.5-4 $\sigma_y = 10.8$ ksi, $T = 115$°F

9.5-5 $d = 7$ in

9.5-6 (*a*) $\sigma_1 = 6140$ psi, $\sigma_2 = -410$ psi, $\tau_{max} = 3276$ psi
(*b*) $\sigma_1 = 8770$ psi, $\sigma_2 = 4160$ psi, $\tau_{max} = 4385$ psi

9.5-7 $\tau_{max} = 2850$ psi

9.5-8 $\sigma_t = \frac{8P}{a^2}$, $\sigma_c = -\frac{4P}{a^2}$

9.5-9 (*a*) $\sigma_{min} = -51.9$ psi, $\sigma_{max} = -8.1$ psi
(*b*) $d = 12.8$ ft

9.5-10 $d = 5.84$ mm

9.5-11 $\tau_{max} = 5774$ psi

9.5-12 (*a*) $p = 325$ psi
(*b*) $F = 4650$ lb
(*c*) $T = 395$ lb·ft

9.5-13 $\sigma_1 = 4830$ psi, $\sigma_2 = -1840$ psi, $\tau_{max} = 3340$ psi, $\theta_p = 31.7°$, $\delta = 4.09 \times 10^{-2}$ in

9.5-14 $T = 13.06 \times 10^3$ lb·in, $M = 3770$ lb·in

9.5-15 $t = 1.27$ mm

9.5-16 $\sigma_1 = 20.1 \times 10^3$ psi, $\sigma_2 = -466$ psi

9.5-17 $d = 1.70$ in

9.5-18 At A: $\sigma_1 = -110$ MPa, $\sigma_2 = 5.4$ MPa, $\tau_{max} = 57.9$ MPa
At B: $\sigma_1 = 56.7$ MPa, $\sigma_2 = -6.7$ MPa, $\tau_{max} = 31.7$ MPa

9.5-19 $H = 40$ hp

9.6-1 At B: $\tau_{max} = 13.5$ MPa; at C: $\tau_{max} = 12.5$ MPa

9.6-2 At C: $\tau_{max} = 17.9$ ksi, $\sigma_1 = 35.7$ ksi, $\sigma_2 = -0.1$ ksi

9.6-3 $P = 0.40 \frac{IY}{Ld}$

9.6-4 $Q = \frac{2}{3}\frac{IY}{Ld}\frac{1}{1 + \frac{1}{24}\frac{d}{L}}$

9.6-5 $P = 4750$ lb

9.6-6 Standard 1-in pipe

9.6-7 $T = 400\ \text{N} \cdot \text{m}$, $\phi = 12.5°$

9.6-8 $d = 24.3$ mm

9.6-9 $d = 1.05$ in

9.6-10 $d = 1.87$ in

9.6-11 $d = 46.2$ mm

9.6-12 $\sigma_C = 94.3$ psi, $\tau_{max} = 68.9$ psi

Chapter 10

10.3-1 $P_{cr} = kL$

10.3-2 $P_{cr} = \dfrac{kL}{4}$

10.3-3 (*a*) $k_e = 48\dfrac{EI}{L^3}$

10.3-4 (*a*) $P_{cr} = \dfrac{k}{L}(3 - \sqrt{5})$

10.3-5 $k_e = \dfrac{15}{4}\dfrac{EI}{L}$

10.3-6 $P_{cr} = \dfrac{15}{4}(3 - \sqrt{5})\dfrac{EI}{L^2} = 2.47\dfrac{EI}{L^2}$

10.3-7 $P_{cr} = kL$

10.3-8 $P_{cr} = \dfrac{k}{L}$

10.3-9 $P_{cr} = 157$ N

10.3-10 $d = 13.4$ mm

10.3-11 $P = kL\dfrac{\theta_0}{c + \theta_0}$

10.3-12 $\delta = \dfrac{PLc(6k - PL)}{4k^2 - 6PLk + P^2L^2}$

10.3-13 $P_{cr} = \dfrac{kL^2}{3}$

10.4-1 2.23 lb

10.4-2 162 N

10.4-3 2.66 kN

10.4-4 (*a*) For 1-in pipe: $P_{cr} = 0.859E/L^2$; for 3-in pipe: $P_{cr} = 29.8E/L^2$

10.4-5 (*a*) $d_A = 19.4$ mm

10.4-6 (*a*) $P_{cr,\,a} = 160$ kips, $P_{cr,\,b} = 345$ kips
(*b*) 215%

10.4-7 $W_a = 178$ kips

10.4-8 (*a*) 9.48 ft

10.4-9 (*a*) $f_s = 6.62$

10.4-10 47.5 kips

10.4-11 $\delta = \dfrac{FL}{P}\dfrac{\sin\lambda L - \lambda L\cos\lambda L}{\lambda L\cos\lambda L}$

10.4-12 $\delta = \dfrac{M_0}{\lambda^2 EI}\dfrac{1 - \cos\lambda L}{\cos\lambda L}$

10.4-19 $P_{cr,\,a} = 100\pi^2\dfrac{E}{L^2}$

$P_{cr,\,b} = 527\pi^2\dfrac{E}{L^2}$, 5.27

10.4-20 Members *AB*, *BD*, *CD*; $f_s = 7.6$, *DE*; $f_s = 2.5$

10.4-21 Members *DE* and *BC*: $f_s = 3.5$

10.4-22 Members *AB* and *BD*: $f_s = 2.5$

10.4-23 *BD*: $f_s = 2.6$

10.5-3 447 kips

10.5-4 (*a*) 20.1 ft (*b*) 10.1 ft
(*c*) 28.7 ft (*d*) 40.2 ft

10.5-5 1010 lb, 7360 lb

10.5-8 *AB*: 45.3°C

10.5-10 163°F

10.5-11 5.87

10.5-12 1230 kips

10.5-13 148.5 lb

10.5-14 0.654 lb

10.5-15 $L_1 = 0.978L_2$

10.5-16 (*a*) $W_1 = 114$ kN,
(*b*) $W_2 = 193$ kN

10.6-1 $\delta = e(\sec \lambda L - 1)$

10.6-3 (*a*) 2.5 in (*b*) −15.0 ksi

10.6-4 (*a*) 2.5 in (*b*) −41.8 ksi

10.6-5 $\delta_z = 2.5$ in, $\delta_y = 1.11$ in

10.6-6 $f = \left(\frac{2}{\pi}\cos^{-1}\frac{2}{3}\right)^2 = 0.287$

10.7-1 (*a*) −6.42 MPa
(*b*) −172 MPa

10.7-2 (*a*) $P_a = 104$ kips
(*b*) $\delta = 1.66$ in,
$\sigma_{max} = -32.6$ ksi

10.7-3 (*a*) 70 kips (*b*) 64 kips

10.7-4 90.6

10.7-5 81.1

10.7-6 (*a*) 140 (*b*) 73.8

10.7-7 (*a*) $\sigma_{max} = \frac{P}{A} + \frac{Pec}{I_z}\sec\left(L\sqrt{\frac{P}{EI_z}}\right)$

10.7-9 2.5 in × 2.5 in × $\frac{1}{4}$ in

10.8-1 $P_{cr} = kL \cos \theta$

10.8-5 (*c*) $P_{max} = 55.2$ lb

10.8-6 $P = 2kL \sin \theta \left(1 - \frac{\cos \theta_0}{\cos \theta}\right)$

10.9-1 (*a*) 728 N (*b*) 2.9

10.9-2 1.11 lb

Index

Boldface numbers refer to the problems at the end of each chapter and are not page references.

Index

Boldface numbers refer to the problems at the end of each chapter and are not page references.